Spectroscopic Properties of Inorganic and Organometallic Compounds

Volume 25

A Specialist Periodical Report

Spectroscopic Properties of Inorganic and Organometallic Compounds

Volume 25

A Review of the Recent Literature Published up to Late 1991

Senior Reporter
G. Davidson, *Department of Chemistry, University of Nottingham*

Reporters
J. H. Carpenter, *University of Newcastle upon Tyne*
S. J. Clark, *City University, London*
K. B. Dillon, *University of Durham*
J. D. Donaldson, *Brunel University, Uxbridge*
S. M. Grimes, *Brunel University, Uxbridge*
B. E. Mann, *University of Sheffield*
D. W. H. Rankin, *University of Edinburgh*
H. E. Robertson, *University of Edinburgh*

ROYAL SOCIETY OF CHEMISTRY

ISBN 0-85186-243-8
ISSN 0584-8555

Published by The Royal Society of Chemistry,
Thomas Graham House, Science Park, Cambridge CB4 4WF

Printed by Bookcraft (Bath) Ltd.

Preface

It gives me great pleasure to introduce the 25th volume of "Spectroscopic Properties of Inorganic and Organometallic Compounds". Over the past quarter of a century this series has become an essential database and reference tool for spectroscopists and others working in important areas of modern chemical research. Long may it continue!

The structure of this year's volume continues as before, and I am grateful to my fellow Reporters for their excellent work, and for producing a much more standardised format for the various chapters.

George Davidson

July, 1992

Contents

Conversion Factors

1 kJ mol^{-1}

2.3901×10^{-1} kcal mol^{-1}
1.0364×10^{-2} eV $atom^{-1}$
8.3593×10 cm^{-1}
2.5061×10^{6} MHz

1 kcal mol^{-1}

4.1840 kJ mol^{-1}
4.3364×10^{-2} eV $atom^{-1}$
3.4976×10^{2} cm^{-1}
1.0486×10^{7} MHz

1 cm^{-1}

1.1963×10^{-2} kJ mol^{-1}
2.8592×10^{-3} kcal mol^{-1}
1.2399×10^{-4} eV atom
2.9979×10^{4} MHz

1 MHz

3.9903×10^{-7} kJ mol^{-1}
9.5370×10^{-8} kcal mol^{-1}
4.1357×10^{-9} eV $atom^{-1}$
3.3356×10^{-5} cm^{-1}

1 eV $atom^{-1}$

9.6485×10 kJ mol^{-1}
2.3060×10 kcal mol^{-1}
8.0655×10^{3} cm^{-1}
2.4180×10^{8} MHz

Moessbauer spectra: E_v (^{57}Fe) = 14.413 keV

1 mm s^{-1}

4.639×10^{-6} kJ mol^{-1}
1.109×10^{-6} kcal mol^{-1}
4.808×10^{-8} eV $atom^{-1}$
3.878×10^{-4} cm^{-1}
1.162×10 MHz

For other Moessbauer nuclides, multiply the above conversion factors by E_v(KeV)/14.413.

1
Nuclear Magnetic Resonance Spectroscopy

BY B. E. MANN

1 Introduction

Following the criteria established in earlier volumes, only books and reviews directly relevant to this chapter are included, and the reader who requires a complete list is referred to the Specialist Periodical Reports 'Nuclear Magnetic Resonance',[1] where a complete list of books and reviews is given. Reviews which are of direct relevance to a section of this Report are included in the beginning of that section rather than here. Papers where only ^{1}H, ^{2}H, ^{19}F, and/or ^{31}P n.m.r. spectroscopy is used are only included when they make a non-routine contribution, but complete coverage of relevant papers is still attempted where nuclei other than these are involved. In view of the greater restrictions on space, and the ever growing number of publications, many more papers in marginal areas have been omitted. This is especially the case in the sections on solid-state n.m.r. spectroscopy, silicon and phosphorus.

One book has been published which is relevant to this review:- 'Multinuclear Magnetic Resonance in Liquids and Solids - Chemical Applications', by P. Granger and R.K. Harris.[2]

Several relevant reviews have been published, including 'N.m.r. properties of nuclei',[3] 'Structural determination of inorganic compounds',[4] 'Para hydrogen-induced polarization: a new spin on reactions with molecular hydrogen',[5] 'Solid state and solution n.m.r. of nonclassical transition-metal polyhydrides',[6] 'Metal n.m.r. of organometallic (*d*-block) systems',[7] 'Two-dimensional n.m.r. and its application in coordination compounds',[8] 'Chemistry in a spin: multinuclear magnetic resonance',[9] 'Macrocyclic complexes of lanthanide ions',[10] 'Structure, reactivity, spectra, and redox properties of cobalt(III) hexaamines', which contains ^{59}Co n.m.r. spectroscopy,[11] 'Diphosphazanes as ligands. Symbiosis of phosphorus chemistry and organometallic chemistry', which contains ^{1}H, ^{13}C, and ^{31}P n.m.r. spectra,[12] 'N.m.r. under high gas pressure',[13] 'Nuclear magnetic resonance at high

1 'Nuclear Magnetic Resonance', ed. G.A. Webb (Specialist Periodical Reports), The Royal Society of Chemistry, London, 1991, Vol. 20; 1992, Vol. 21.
2 P. Granger and R.K. Harris, 'Multinuclear Magnetic Resonance in Liquids and Solids - Chemical Applications', NATO ASI Series, Ser. C: Mathematical and Physical Sciences, Vol. 322, Kluwer, Dordrecht, Netherlands, 1990.
3 K. Hallenga and A.I. Popov, *Pract. Spectrosc.*, 1991, **11**, 1 (*Chem. Abstr.*, 1991, **114**, 176 641).
4 J.W. Akitt, *Pract. Spectrosc.*, 1991, **11**, 427 (*Chem. Abstr.*, 1991, **114**, 133 140).
5 R. Eisenberg, *Acc. Chem. Res.*, 1991, **24**, 110.
6 K.W. Zilm and J.M. Millar, *Adv. Magn. Opt. Reson.*, 1990, **15**, 163 (*Chem. Abstr.*, 1991, **115**, 83 729).
7 D. Rehder, *Coord. Chem. Rev.*, 1991, **110**, 161.
8 X. Mao and X. You, *Huaxue Tongbao*, 1990, 16 (*Chem. Abstr.*, 1991, **114**, 34 456).
9 G. Van Koten, *NATO ASI Ser., Ser. C*, 1990, **322**, 1 (*Chem. Abstr.*, 1991, **114**, 112 009).
10 G.Y. Adachi and Y. Hirashima, *Cation Binding Macrocycles*, 1990, 701. Ed. Y. Inoue and G.W. Gokel (*Chem. Abstr.*, 1991, **114**, 177 017).
11 P. Hendry and A. Ludi, *Adv. Inorg. Chem.*, 1990, **35**, 117 (*Chem. Abstr.*, 1991, **114**, 177 007).
12 M.S. Balakrishna, T.K. Prakasha, and S.S. Krishnamurthy, *Adv. Organomet., Proc. Indo-Sov. Symp. "Organomet. Chem.", 1st 1988*, (Pub. 1989), 205 (*Chem. Abstr.*, 1991, **115**, 125 452).
13 I.T. Horváth and J.M. Millar, *Chem. Rev.*, 1991, **91**, 1339.

temperature',[14] 'Alkali metal n.m.r. studies of synthetic and natural ionophore complexes',[15] 'Behaviour of ion and water in lyotropic liquid crystals. N.m.r. study',[16] 'Analytical chemical applications of high resolution solid state n.m.r. spectroscopy',[17] and 'Nuclear magnetic resonance spectroscopy', which contains the application of pulsed ^{1}H, ^{13}C, and ^{29}Si n.m.r. spectroscopy to the analysis of organosilicon compounds and polymers.[18]

A number of papers have been published which are too broadly based to fit into a later section and are included here. The n.m.r. line shapes of double and triple quantum coherences have been calculated for spin $\frac{3}{2}$ nuclei over a wide range of reorientation times.[19] A low temperature probe (1.2 K) for ^{1}H, ^{2}H, ^{3}He, ^{11}B, ^{13}C, ^{19}F, ^{27}Al, ^{29}Si, ^{115}Sn, ^{117}Sn, ^{119}Sn, and ^{133}Cs has been reported.[20] The scope and limitations of the stepped frequency experiment have been investigated using ^{2}H and ^{27}Al n.m.r. spectroscopy.[21] Qualitative models for ^{1}H and ^{13}C chemical shifts of interstitial atoms in clusters have been presented.[22] A correlation of the ^{13}C and ^{15}N chemical shifts with compression has been demonstrated for interstitial carbide and nitride in transition metal clusters.[23] Periodicities in the correlation of ^{11}B, ^{13}C, ^{15}N, and ^{17}O nuclear magnetic shielding and the correlation of interstitial atoms in metal clusters have been reported.[24] ^{13}C T_1 values and chemical shifts have been determined for 19 selected organometallic half-sandwich and sandwich complexes of Cr, Mn, Fe, and Co. Large variations in T_1 were found.[25] New coordination modes of octafluorocyclo-octatetraene bound to Rh, Co, Ni, Pd, Pt, and Mn have been reported. The n.m.r. spectra were reported.[26] Results of relaxation in protein solutions and in silica particle suspensions show that the old and simple notion of bound water must be redefined to be of any help for interpreting relaxation results.[27] ^{13}C n.m.r. data have also been reported for (salicylidene-2-aminoacetophenone)-2-thenoylhydrazone complexes of VO, Mn, Co, Ni, Cu, and Zn,[28] 2-acetylfuran-2-thenoylhydrazone complexes of VO, Mn, Co, Ni, Cu, Zn, Cd, Hg,[29] saccharin complexes of Mg, Zn, Cd, Hg, and Pb,[30] $R^1C(SH)=NNCR^2R^3$ complexes of Co, Ni, Pd, Cu, Zn, Cd, and Hg,[31] and some thiocyanato complexes of Co, Ni, Pd,

14 J.F. Stebbins, *Chem. Rev.*, 1991, **91**, 1353.

15 C. Detellier, H.P. Graves, and K.P. Briere, *Isot. Phys. Biomed. Sci.*, 1991, **2**, 159 (*Chem. Abstr.*, 1991, **115**, 293 389).

16 M. Iida, *Hyomen*, 1991, **29**, 524 (*Chem. Abstr.*, 1991, **115**, 244 273).

17 C.A. Fyfe, *NATO ASI Ser., Ser. C*, 1990, **322**, 241 (*Chem. Abstr.*, 1991, **115**, 40 211).

18 R.B. Taylor, B. Parbhoo, and D.M. Fillmore, *Chem. Anal. (N.Y.)*, 1991, **112**, 347 (*Chem. Abstr.*, 1991, **114**, 248 099).

19 U. Eliav and G. Navron, *J. Chem. Phys.*, 1991, **95**, 7114.

20 P.L. Kuhns, S.H. Lee, C. Coretsopoulos, P.C. Hammel, O. Gonen, and J.S. Waugh, *Rev. Sci. Instrum.*, 1991, **62**, 2159 (*Chem. Abstr.*, 1991, **115**, 148 750).

21 M.A. Kennedy, R.L. Vold, and R.R. Vold, *J. Magn. Reson.*, 1991, **92**, 320.

22 M.J. Duer and D.J. Wales, *Polyhedron*, 1991, **10**, 1749.

23 J. Mason, *J. Am. Chem. Soc.*, 1991, **113**, 24.

24 J. Mason, *J. Am. Chem. Soc.*, 1991, **113**, 6056.

25 H. Schumann, *J. Organomet. Chem.*, 1991, **421**, C7.

26 R.P. Hughes, *Report*, 1989, **AFOSR-TR-89-1819**; Order No. AD-A216415, 12 pp. Avail. NTIS. From *Gov. Rep. Announce. Index (U.S.)*, 1990, **90**, Abstr. No. 016,630 (*Chem. Abstr.*; 1991, **114**, 82 076).

27 P. Gilles, *Phys. Mag.*, 1990, **12**, 295 (*Chem. Abstr.*, 1991, **115**, 148 716).

28 B. Singh and A.K. Srivastav, *Proc. - Indian Acad. Sci., Chem. Sci.*, 1990, **102**, 503 (*Chem. Abstr.*, 1991, **114**, 93 989).

29 B. Singh and A.K. Srivastav, *Synth. React. Inorg. Metal-Org. Chem.*, 1991, **21**, 513.

30 E. Kleinpeter, D. Stroehl, G. Jovanovski, and B. Soptrajanov, *J. Mol. Struct.*, 1991, **246**, 185.

31 K.N. Zelenin, L.A. Khorseeva, V.V. Alekseev, M.T. Toshev, and Kh.B. Dustov, *Zh. Obshch. Khim.*, 1990, **60**, 2348 (*Chem. Abstr.*, 1991, **115**, 148 906).

Cu, Zn, Cd, and Hg.[32]

2 Stereochemistry

This section is subdivided into eleven parts which contain n.m.r. information about Groups IA and IIA and transition-metal complexes presented by Groups according to the Periodic Table. Within each Group, classification is by ligand type.

Complexes of Groups IA and IIA.—Reviews entitled 'A review of ^{23}Na nuclear magnetic resonance spectroscopy for the *in vitro* study of cellular sodium metabolism'[33] and 'Nuclear magnetic resonance spectroscopy of chlorophyll'[34] have appeared.

The IGLO method has been applied to calculate the ^{13}C chemical shifts of numerous organolithium compounds.[35] Homonuclear J(^{6}Li-^{6}Li) has been measured directly for *E*-2-Li-1-Ph-1-*o*-lithiophenylpent-1-ene for the first time using INADEQUATE.[36] Theoretical calculations on allyl alkali metals have been used to discuss ^{1}J(^{13}C-^{1}H).[37] J(^{13}C-^{13}C) has been measured on 1:1 adducts of Bu^{t}Li and butadiene.[38] ^{6}Li and ^{13}C n.m.r. spectroscopy has been used to examine 2-Li-$C_6H_4PPh_2$ in solution and it was concluded that it is a monomer.[39] The low frequency shift of ^{6}Li in (1) has been attributed to ring currents. The ^{13}C n.m.r. spectrum was also reported.[40] The ^{7}Li n.m.r. chemical shift of $[Li(dme)_2]^+[1,2,4,5\text{-}(Me_3Si)_4\text{-benzenide}]^-$ has revealed a large high frequency shift indicating anti-aromaticity. The ^{13}C and ^{29}Si n.m.r. spectra were also reported.[41] The ^{7}Li n.m.r. spectrum of $[Ph_3SnLi\{NMe(CH_2CH_2NMe_2)_2\}]$ shows the first recorded example of ^{1}J(^{119}Sn-^{7}Li) and ^{1}J(^{117}Sn-^{7}Li).[42] The ^{23}Na n.m.r. spectrum of $[(C_5Me_5)Na]$ has a line width of 420 Hz which was attributed to the quadrupole relaxation due to the electric field gradient. The ^{13}C n.m.r. spectrum was also reported.[43] T_1 and T_2 of ^{23}Na in Na^- have been determined.[44] N.m.r. data have also been reported for $[(1,4\text{-}Me_2\text{-}1,4\text{-diazacyclohexane})Li(\mu\text{-}H)AlBu^t\{CH(SiMe_3)_2\}_2]$, (^{13}C),[45] $[(tmeda)LiCH(SiMe_2)_2CH_2]$, (^{13}C, ^{29}Si),[46] $[(tmeda)LiCH(SiMe_3)Ph]$, (^{13}C),[47] $[Li(tmeda)]_2[(Me_3SiCHCH)_2]$,

32 M.T.H. Tarafder and K. Fatema, *J. Bangladesh Chem. Soc.*, 1988, **1**, 77 (*Chem. Abstr.*, 1991, **114**, 155 856).
33 H. Nissen, J.P. Jacobsen, and M. Hoerder, *Scand. J. Clin. Lab. Invest.*, 1990, **50**, 497 (*Chem. Abstr.*, 1991, **114**, 224 684).
34 R.J. Abraham and A.E. Rowan, *Chlorophylls*, 1991, 797. Ed. H. Scheer (*Chem. Abstr.*, 1991, **115**, 275 680).
35 M. Bühl, N.J.R. Van Eikema Hommes, P.v.R. Schleyer, U. Fleischer, and W. Kutzelnigg, *J. Am. Chem. Soc.*, 1991, **113**, 2459.
36 O. Eppers, H. Günther, K.-D. Klein, and A. Maercker, *Mag. Res. Chem.*, 1991, **29**, 1065.
37 N.J.R.E. Hommes, M. Bühl, P.v.R. Schleyer, and Y.-D. Wu, *J. Organomet. Chem.*, 1991, **409**, 307.
38 S. Bywater, D.J. Worsfold, and P.E. Black, *Makromol. Chem., Macromol. Symp.*, 1991, **47**, 203 (*Chem. Abstr.*, 1991, **115**, 280 644).
39 S. Harder, L. Brandsma, J.A. Kanters, A. Duisenberg, and J.H. van Lenthe, *J. Organomet. Chem.*, 1991, **420**, 143.
40 W. Bauer, G.A. O'Doherty, P.v.R. Schleyer, and L.A. Paquette, *J. Am. Chem. Soc.*, 1991, **113**, 7093.
41 A. Sekiguchi, K. Ebata, C. Kabuto, and H. Sakurai, *J. Am. Chem. Soc.*, 1991, **113**, 7081; A. Sekiguchi, K. Ebata, C. Kabuto, and H. Sakurai, *J. Am. Chem. Soc.*, 1991, **113**, 1464.
42 D. Reed, D. Stalke, and D.S. Wright, *Angew. Chem., Int. Ed. Engl.*, 1991, **30**, 1459.
43 G. Rabe, H.W. Roesky, D. Stalke, F. Pauer, and G.M. Sheldrick, *J. Organomet. Chem.*, 1991, **403**, 11.
44 P.P. Edwards, A.S. Ellaboudy, D.M. Holton, and N.C. Pyper, *Mol. Phys.*, 1990, **69**, 209.
45 W. Uhl and J.E.O. Schnepf, *Z. Anorg. Allg. Chem.*, 1991, **595**, 225.
46 J.L. Robison, W.M. Davis, and D. Seyferth, *Organometallics*, 1991, **10**, 3385.
47 W. Zarges, M. Marsch, K. Harms, W. Koch, G. Frenking, and G. Boche, *Chem. Ber.*, 1991, **124**, 543.

(^{7}Li, including ^{1}H-^{7}Li HOESY, ^{13}C),[48] [LiC(SiMe$_3$)$_3$], (^{13}C),[49] [(dioxan)Li(Me$_3$SiCH$_2$)$_4$Ga], (^{13}C),[50] [Li{PhCH$_2$P(O)(OEt)$_2$}], (^{13}C),[51] (2), (^{13}C),[52] [(2-Ph-C$_3$H$_4$)M], (M = Li, K; ^{13}C),[53] [(Bu^{t}NCHCHCHCH$_2$)Li], (^{13}C),[54] [(Ph$_2$CNCHCHCH$_2$)Li], (^{13}C),[55] [2,6-(Bu^{t}O)$_2$C$_6$H$_3$Li]$_3$, (^{13}C),[56] [C$_{60}$Li$_4$], [C$_{70}$Li$_4$], (^{7}Li, ^{13}C),[57] [(η-C$_6$H$_6$)KOSiMe$_2$Ph]$_4$, (^{29}Si),[58] K$^-$[K(18-crown-6)]$^+$, (^{39}K),[59] [(η^8-C$_8$H$_8$)Er(μ-η^8-C$_8$H$_8$)K(μ-η^8-C$_8$H$_8$)Er(μ-η^8-C$_8$H$_8$)K(THF)$_4$], (^{13}C),[60] and [Cs(*p*-Bu^t-calix[4]arene-H)(MeCN)], (^{133}Cs).[61]

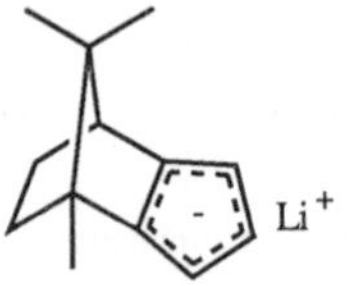

(1)

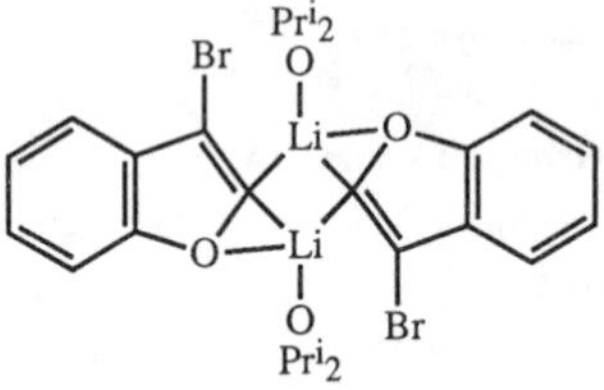

(2)

^{6}Li and ^{15}N n.m.r. studies with [^{6}Li]-Li(2,2,6,6-tetramethylpiperidide) and [^{6}Li,^{15}N]-Li(2,2,6,6-tetramethylpiperidide) have supported an earlier suggestion that Li(2,2,6,6-tetramethylpiperidide) exists as a dimer-monomer mixture in THF.[62] ^{6}Li and ^{15}N n.m.r. spectroscopic studies on [^{6}Li,^{15}N]-LiNPr$^i_2$ have shown that [^{6}Li]-lithium pinacolate forms a 1:1 cyclic mixed aggregate with LiNPr$^i_2$.[63] Two dimensional ^{6}Li-^{1}H heteronuclear Overhauser enhancement spectroscopy has been applied to Li$^+$ complexes of lasolocid A.[64] A ^{23}Na triple quantum dynamic shift has been observed for the Na$^+$ complex of 4,7,13,16,21-pentaoxa-1,10-diazabicyclo-[8.8.5]-tricosane.[65] The influence of cysteamine on the entropy of fluctuations in DNA thermal *trans*-conformation has been studied by ^{23}Na n.m.r. spectroscopy.[66] CsBPh$_4$ dissolved in pyridine has been studied by ^{133}Cs-^{1}H HOESY.[67] N.m.r. data have also been reported for [Bu$^t_2$NLi], (^{6}Li, ^{13}C),[68] [Li{N(CH$_2$CH$_2$NMeCH$_2$-

48 L.D. Field, M.G. Gardiner, C.H.L. Kennard, B.A. Messerle, and C.L. Raston, *Organometallics*, 1991, **10**, 3167.
49 W. Hiller, M. Layh, and W. Uhl, *Angew. Chem., Int. Ed. Engl.*, 1991, **30**, 324.
50 W. Uhl, K.-W. Klinkhammer, M. Layb, and W. Massa, *Chem. Ber.*, 1991, **124**, 279.
51 W. Zarges, M. Marsch, K. Harms, F. Haller, G. Frenking, and G. Boche, *Chem. Ber.*, 1991, **124**, 861.
52 F. Bosold, P. Zulauf, M. Marsch, K. Harms, J. Lohrenz. and G. Boche, *Angew. Chem., Int. Ed. Engl.*, 1991, **30**, 1454.
53 O. Desponds and M. Schlosser, *J. Organomet. Chem.*, 1991, **409**, 93.
54 G. Wolf and E.-U. Würthwein, *Chem. Ber.*, 1991, **124**, 889.
55 G. Wolf and E.-U. Würthwein, *Chem. Ber.*, 1991, **124**, 655.
56 S. Harder, J. Boersma, L. Brandsma, J.A. Kanters, A.J.M. Duisenberg, and J. H. van Lenthe, *Organometallics*, 1991, **10**, 1623.
57 J.W. Bausch, G.K.S. Prakash, G.A. Olah, D.S. Tse, D.C. Lorents, Y.K. Bae, and R. Malhotra, *J. Am. Chem. Soc.*, 1991, **113**, 3205.
58 G.R. Fuentes, P.S. Coan, W.E. Streib, and K.G. Caulton, *Polyhedron*, 1991, **10**, 2371.
59 Z. Jedliński, A. Misiolek, A. Jankowski, and H. Janeczek, *J. Chem. Soc., Chem. Commun.*, 1991, 1513.
60 J. Xia, Z. Jin, and W. Chen, *J. Chem. Soc., Chem. Commun.*, 1991, 1214.
61 J.M. Harrowfield, M.I. Ogden, M.R. Richmond, and A.H. White, *J. Chem. Soc., Chem. Commun.*, 1991, 1159.
62 P.L. Hall, J.H. Gilchrist, A.T. Harrison, D.J. Fuller, and D.B. Collum, *J. Am. Chem. Soc.*, 1991, **113**, 9575.
63 A.S. Galiano-Roth, Y.J. Kim, J.H. Gilchrist, A.T. Harrison, D.J. Fuller, and D.B. Collum, *J. Am. Chem. Soc.*, 1991, **113**, 5053.
64 M. Tokles, F.J. Swiecinski, and P.L. Rinaldi, *J. Magn. Reson.*, 1991, **91**, 222.
65 U. Eliav, H. Shinar, and G. Navon, *J. Magn. Reson.*, 1991, **94**, 439.
66 G. Mallet, J. Lematre, and D. Vasilescu, *Physiol. Chem. Phys. Med. N.M.R.*, 1991, **23**, 51 (*Chem. Abstr.*, 1991, **115**, 227 251).
67 W. Bauer, *Magn. Reson. Chem.*, 1991, **29**, 494.
68 P.L. Hall, J.H. Gilchrist, and D.B. Collum, *J. Am. Chem. Soc.*, 1991, **113**, 9571.

$CH_2)_3N\}]^+$, (^{13}C),[69] [M{1,4,7-tris(pyrazol-1-ylmethyl)-1,4,7-triazacyclononane}]$^+$, (M = Li, Na; ^{7}Li, ^{13}C, ^{23}Na),[70] [LiN(SiMe$_3$)(2,6-Pr$^i_2$C$_6$H$_3$)]$_2$, (^{7}Li),[71] [PhSi{N(2,4,6-Bu$^t_3$C$_6$H$_2$)}$_2$Li(solvent)], (^{13}C, ^{29}Si),[72] [(THF)$_2$Li{N(=C=C[BH$_3$]PPh$_3$)}$_2$Li(THF)$_2$], (^{7}Li, ^{11}B, ^{13}C),[73] (3), (^{7}Li),[74] [M{N(SiMe$_3$)}$_2$SPh]$_2$, {M = Li, (THF)Cs; ^{7}Li, ^{13}C, ^{29}Si, ^{133}Cs},[75] (4), (^{7}Li, ^{13}C),[76] [{MeN-(CH$_2$CH$_2$NMe$_2$)$_2$}LiNBu^{t}CPh=Ni(η^2-C$_2$H$_4$)$_2$], (^{13}C),[77] and [(THF)$_2$LiN(SiBu$^t_2$F)SiBu$^t_2$Me], (^{7}Li, ^{13}C, ^{29}Si).[78]

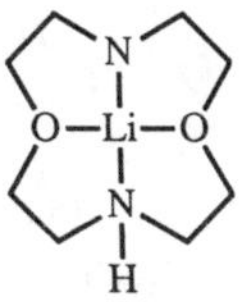

(3)

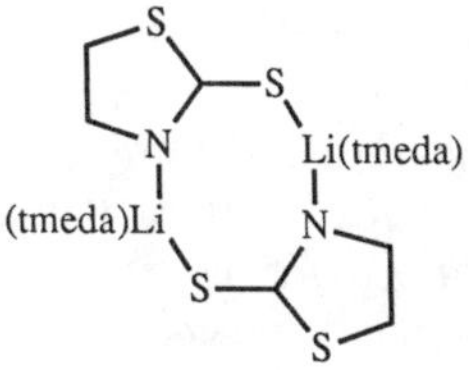

(4)

The pharmocokinetics of Li$^+$ uptake in the brain and muscle have been measured by ^{7}Li n.m.r. spectroscopy.[79] Multiple quantum filtered ^{23}Na n.m.r. spectroscopy has been explored in model systems.[80] Triple quantum ^{23}Na imaging has been reported.[81] Soy sauce has been investigated by ^{13}C and ^{23}Na n.m.r. spectroscopy.[82] ^{23}Na n.m.r. spectroscopy has been used to measure intracellular Na$^+$ in hearts,[83] red blood cells,[84] kidney,[85] liver,[86] and muscle.[87] The effect of head-group structure and counterion condensation on phase equilibria in anionic phospholipid-water systems has been studied by ^{2}H, ^{23}Na, and ^{31}P n.m.r. spectroscopy.[88] The Burg maximum entropy

69 A. Bencini, A. Bianchi, S. Chimichi, M. Ciampolini, P. Dapporto, E. Garcia-España, M. Micheloni, N. Nardi, and P. Paoli, *Inorg. Chem.*, 1991, **30**, 3687.

70 M. Di Vaira, B. Cosimelli, F. Mani, and P. Stoppioni, *J. Chem. Soc., Dalton Trans.*, 1991, 331.

71 D.K. Kennepohl, S. Brooker, G.M. Sheldrick, and H.W. Roesky, *Chem. Ber.*, 1991, **124**, 2223.

72 G.E. Underiner, R.P. Tan, D.R. Powell, and R. West, *J. Am. Chem. Soc.*, 1991, **113**, 8437.

73 H.J. Bestmann, T. Röder, M. Bremer, and D. Löw, *Chem. Ber.*, 1991, **124**, 199.

74 D. Barr, D.J. Berrisford, L. Méndez, A.M.Z. Slawin, R. Snaith, J.F. Stoddart, D.J. Williams, and D.S. Wright, *Angew. Chem., Int. Ed. Engl.*, 1991, **30**, 82.

75 P. Pauer and D. Stalke, *J. Organomet. Chem.*, 1991, **418**, 127.

76 D.R. Armstrong, R.E. Mulvey, D. Barr, R.W. Porter, P.R. Raithby, T.R.E. Simpson, R. Snaith, D.S. Wright, K. Gregory, and P. Mikulcik, *J. Chem. Soc., Dalton Trans.*, 1991, 765.

77 B. Gabor, C. Krüger, B. Marczinke, R. Mynott, and G. Wilke, *Angew. Chem., Int. Ed. Engl.*, 1991, **30**, 1666.

78 S. Walter, U. Klingebiel, and D. Schmidt-Bäse, *J. Organomet. Chem.*, 1991, **412**, 319.

79 R.A. Komoroski, J.E.O. Newton, E. Walker, D. Cardwell, N.R. Jagannathan, S. Ramaprasad, and J. Sprigg, *Magn. Reson. Med.*, 1990, **15**, 347 (*Chem. Abstr.*, 1991, **114**, 135 465).

80 G.S. Payne and P. Styles, *J. Magn. Reson.*, 1991, **95**, 253.

81 S. Wimperis and B. Wood, *J. Magn. Reson.*, 1991, **95**, 428.

82 K. Matsushita, *Nippon Shoyu Kenkyusho Zasshi*, 1990, **16**, 251 (*Chem. Abstr.*, 1991, **115**, 7117).

83 M.M. Pike, M. Kitakaze, and E. Marban, *Am. J. Physiol.*, 1990, **259**, H1767 (*Chem. Abstr.*, 1991, **114**, 79 507); T. Konishi, I. Mori, C.S. Apstei, and J.S. Ingwall, *Yakuri to Chiryo*, 1990, **18**, S3451 (*Chem. Abstr.*, 1991, **114**, 156 916); S.E. Anderson, E. Murphy, C. Steenbergen, R.E. London, and P.M. Cala, *Am. J. Physiol.*, 1990, **259**, C940 (*Chem. Abstr.*, 1991, **114**, 79 500); C.J.A. Van Echteld, J.H. Kirkels, M.H.J. Eijgelshoven, P. Van der Meer, and T.J.C. Ruigrok, *J. Mol. Cell. Cardiol.*, 1991, **23**, 297 (*Chem. Abstr.*, 1991, **114**, 245 324).

84 H. Shinar and G. Navon, *Biophys. J.*, 1991, **59**, 203 (*Chem. Abstr.*, 1991, **114**, 117 961).

85 M. Maeda, Y. Seo, M. Murakami, S. Kuki, H. Watari, S. Iwasaki, and H. Uchida, *Magn. Reson. Med.*, 1990, **16**, 361 (*Chem. Abstr.*, 1991, **114**, 202 575).

86 M. Brauer, R.A. Towner, and D.L. Foxall, *Magn. Reson. Imaging*, 1990, **8**, 459 (*Chem. Abstr.*, 1991, **115**, 86 849).

87 R.J. Buist, R. Deslauriers, J.K. Saunders, and G.W. Mainwood, *Can. J. Physiol.* Pharmacol., 1991, **69**, 1663 (*Chem. Abstr.*, 1991, **115**, 275 000).

88 G. Lindblom, L. Rilfors, J.B. Hauksson, I. Brentel, M. Sjoelund, and B. Bergenstahl, *Biochemistry*, 1991, **30**,

method has been applied to estimate ^{39}K n.m.r. spectra of mung bean root tips.[89] The complexation of K^+ by tetraglyme and 18-crown-6 has been investigated by ^{39}K n.m.r. spectroscopy.[90] N.m.r. data have also been reported for $[LiOCMe_2Ph]_6$, (^7Li, ^{13}C),[91] $[As(RCO)_2Li(THF)_2]$, (^{13}C),[92] $[PhOLi.hmpa]_4$, (^7Li, ^{31}P),[93] $[LiO\text{-}2,6\text{-}Me_2\text{-}C_6H_3]_2$, ($^6Li$, ^7Li, ^{13}C),[94] $[Li(\mu\text{-}OSiPh_3)(\eta^2\text{-}DME)]_2$, ($^{13}C$),[95] $[M(OBu^t)]_4$, (M = K, Rb, Cs; ^{13}C),[96] and $[RCO_2Cs]$, (^{133}Cs).[97]

The purity of Grignard reagents has been assayed by ^{13}C n.m.r. spectroscopy.[98] ^{14}N, ^{15}N, and ^{17}O n.m.r. spectroscopy has been used to study the mode of binding of some amino acids to Ca^{2+}.[99] N.m.r. data have also been reported for (5), (^{13}C),[100] $[(C_8H_{17})_2Mg]$, (^{13}C),[101] $[CH_2{=}CHMgBr(THF)]$, (^{13}C),[102] $[(Bu^tC_5H_4)_2Mg]$, (^{13}C),[103] $[(2\text{-}MeOC_6H_4)_2M]$, (M = Mg, Hg; ^{13}C),[104] $[2\text{-}MeOCH_2CH_2OCH_2C_6H_4M]$, (M = MgBr, HgBr, $SnMe_3$; ^{13}C),[105] $[(1,2\text{-}C_6H_4)Mg(THF)]_4$, ($^{13}C$),[106] $[Ca\{CH(SiMe_3)_2\}_2(THF)_3]$, ($^{13}C$),[107] $[Ca(Sn\{CH(SiMe_3)_2\}_2)_2]$, ($^{13}C$, ^{29}Si, ^{119}Sn),[108] $[M\{N(SiMe_3)_2\}_2]$, (^{13}C, ^{29}Si),[109] (6), (M = Mg, Zn; ^{13}C),[110] Ca^{2+}, Sr^{2+}, and Ba^{2+} complexes of $H_2NCH_2CH_2OCH_2CH_2OCH_2CH_2NH_2$, ($^{13}C$),[111] $[Ba\{Bu^tC(O)CH\text{-}C(O)Bu^t\}_2(NH_3)_2]_2$, ($^{13}C$),[112] $[Ba\{N(SiMe_3)_2\}_2]$, (^{13}C),[113] $[Ba\{N(SiMe_3)_2\}_2(THF)_2]$, ($^{13}C$),[114] $[Ba(2,6\text{-}Bu^t_2C_6H_3O)_2(HOCH_2CH_2NMe_2)_4].2C_7H_8$, ($^{13}C$),[115] $[Ba\{N(C_2H_4O)(C_2H_4OH)_2\}_2]$, ($^{13}C$),[116] $[Be(2\text{-}OC_6H_4CO_2)(H_2O)_2]$, ($^9Be$),[117] [Mg(adenylate kinase)(ATP)], (^{25}Mg),[118] Ca^{+2}

10938 (*Chem. Abstr.*, 1991, **115**, 226 545).
89 T. Uchiyama and H. Minamitani, *J. Magn. Reson.*, 1991, **94**, 449.
90 J. Grobelny, M. Sokól, and Z.J. Jedliński, *Magn. Reson. Chem.*, 1991, **29**, 679.
91 M.H. Chisholm, S.R. Drake, A.A. Naiini, and W.E. Streib, *Polyhedron*, 1991, **10**, 805.
92 G. Becker, W. Becker, M. Schmidt, W. Schwarz, and M. Westerhausen, *Z. Anorg. Allg. Chem.*, 1991, **605**, 7.
93 P.R. Raithby, D. Reed, R. Snaith, and D.S. Wright, *Angew. Chem., Int. Ed. Engl.*, 1991, **30**, 1011.
94 L.M. Jackman and E.F. Rakiewicz, *J. Am. Chem. Soc.*, 1991, **113**, 1202.
95 M.J. McGeary, K. Folting, W.E. Streib, J.C. Huffman, and K.G. Caulton, *Polyhedron*, 1991, **10**, 2699.
96 M.H. Chisholm, S.R. Drake, A.A. Naiini, and W.E. Streib, *Polyhedron*, 1991, **10**, 337.
97 T.A. Mirnaya, G.G. Yaremchuk, and V.V. Trachevskii, *Zh. Neorg. Khim.*, 1991, **36**, 1269 (*Chem. Abstr.*, 1991, **115**, 83 943.
98 K.V. Baker, J.M. Brown, N. Hughes, A.J. Skarnulis, and A. Sexton, *J. Org. Chem.*, 1991, **56**, 698.
99 M. Maeda, K. Okada, and K. Ito, *J. Inorg. Biochem.*, 1991, **41**, 143.
100 N. Kuhn, M. Schulten, R. Boese, and D. Bläser, *J. Organomet. Chem.*, 1991, **421**, 1.
101 B. Bogdanović, P. Bons, M. Schwickardi, and K. Seevogel, *Chem. Ber.*, 1991, **124**, 1041.
102 R. Han and G. Parkin, *Organometallics*, 1991, **10**, 1010.
103 K.H. Thiele and V. Lorenz, *Z. Anorg. Allg. Chem.*, 1990, **591**, 195.
104 P.R. Markies, G. Schat, A. Villena, O.S. Akkerman, F. Bickelhaupt, W.J.J. Smeets, and A.L. Spek, *J. Organomet. Chem.*, 1991, **411**, 291.
105 P.R. Markies, G. Schat, S. Griffioen, A. Villena, O.S. Akkerman, F. Bickelhaupt, W.J.J. Smeets, and A.L. Spek, *Organometallics*, 1991, **10**, 1531.
106 M.A.G.M. Tinga, O.S. Akkerman, F. Bickelhaupt, E. Horn, and A.L. Spek, *J. Am. Chem. Soc.*, 1991, **113**, 3604.
107 F.G.N. Cloke, P.B. Hitchcock, M.F. Lappert, G.A. Lawless, and B. Royo, *J. Chem. Soc., Chem. Commun.*, 1991, 724.
108 M. Westerhausen and T. Hildenbrand, *J. Organomet. Chem.*, 1991, **411**, 1.
109 M. Westerhausen, *Inorg. Chem.*, 1991, **30**, 96.
110 L.M. Engelhardt, P.C. Junk, W.C. Patalinghug, R.E. Sue, C.L. Raston, B.W. Skelton, and A.H. White, *J. Chem. Soc., Chem. Commun.*, 1991, 930.
111 M.L. Tul'chinskii, L.Kh. Minacheva, G.G. Sadikov, V.A. Trofimov, V.G. Sakharova, A.Yu. Tsivadze, and M.A. Porai-Koshits, *Koord. Khim.*, 1990, **16**, 1299 (*Chem. Abstr.*, 1991, **114**, 113 947).
112 W.S. Rees, jun., M.W. Carris, and W. Hesse, *Inorg. Chem.*, 1991, **30**, 4479.
113 J.M. Boncella, C.J. Coston, and J.K. Cammack, *Polyhedron*, 1991, **10**, 769.
114 B.A. Vaartstra, J.C. Huffman, W.E. Streib, and K.G. Caulton, *Inorg. Chem.*, 1991, **30**, 121.
115 K.G. Caulton, M.H. Chisholm, S.R. Drake, and K. Folting, *Inorg. Chem.*, 1991, **30**, 1500.
116 O. Poncelet, L.G. Hubert-Pfalzgraf, L. Toupet, and J.C. Daran, *Polyhedron*, 1991, **10**, 2045.
117 H. Schmidbaur, O. Kumberger, and J. Riede, *Inorg. Chem.*, 1991, **30**, 3101.
118 H. Yan and M.D. Tsai, *Biochemistry*, 1991, **30**, 5539 (*Chem. Abstr.*, 1991, **114**, 243 407).

complex of $EtC\{CH_2O(CH_2)_nCONHCHBu^tCONEt_2\}_3$, ($^{13}C$),[119] Ca^{2+} complexes of 1-Ph-3-Me-4-acylpyrazolone-5, (^{13}C),[120] $[Ba\{O(CH_2CH_2O)_nCH_3\}_2]$, ($n$ = 2 or 3; ^{13}C),[121] $[Ba_4Ti_{13}(\mu_3\text{-}O)_{12}(\mu_5\text{-}O)_6(\mu_1\text{-}\eta^1\text{-}OCH_2CH_2OCH_3)_{12}(\mu_1,\mu_3\text{-}\eta^2\text{-}OCH_2CH_2OCH_3)_{12}]$, ($^{13}C$, ^{17}O),[122] $[KBa_2(OSiPh_3)_5\text{-}(CH_3OCH_2CH_2OCH_3)_2]$, ($^{29}Si$),[123] $[Ba_4(\mu_4\text{-}O)\{\mu_2\text{-}OC_6H_2(CH_2NMe_2)_3\text{-}2,4,6\}_6]$, ($^{13}C$),[124] [Ba(18-crown-6)(hfac)$_2$], (^{13}C),[125] and [Ba(2,5,8,11,14-pentaoxaheptadecane)(acac)$_2$], (^{13}C).[126]

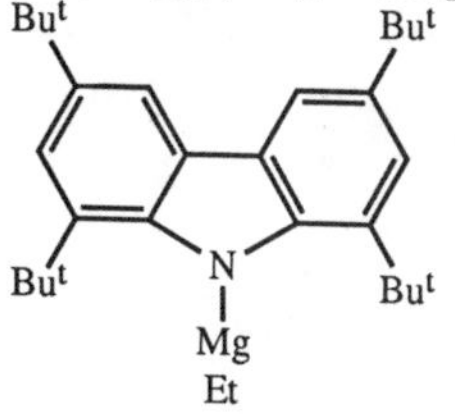

(5)

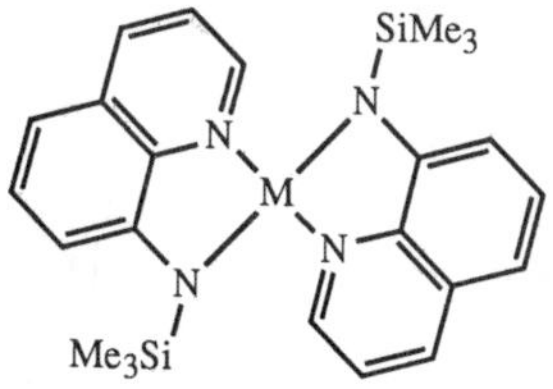

(6)

Complexes of Groups IIIA, the Lanthanides, and Actinides.—A review entitled 'Solution chemistry of dioxouranium(VI) ion' which contains n.m.r. spectroscopy has appeared.[127]

Magnetic moments have been measured for ^{91m}Y, ^{95}Zr, and ^{97}Nb.[128] N.m.r. data have also been reported for $[\{Y(\eta^5\text{-}C_5Me_5)OC_6H_3Bu^t_2\}(\mu\text{-}H)]_2$, $[Y(\eta^5\text{-}C_5Me_5)\{CH(SiMe_3)_2\}(OC_6H_3Bu^t_2)_2]$, ($^{13}C$, ^{29}Si, ^{89}Y),[129] $[(\eta^5\text{-}C_5Me_5)_2Sm(\mu\text{-}H)(\mu\text{-}CH_2C_5H_4\text{-}\eta^5)Sm(\eta^5\text{-}C_5Me_5)]$, ($^{13}C$),[130] $[(MeCN)_4\text{-}Yb\{(\mu\text{-}H)_3BH\}_2]$, ($^{11}B$),[131] [(octaethylporphyrin)Y(μ-Me)$_2$AlMe$_2$], (^{13}C),[132] $[(\eta^5\text{-}C_5Me_5)_2Sc\text{-}C{\equiv}CH]$, ($^{13}C$),[133] $[(\eta^5\text{-}C_5Me_5)_2La(\mu\text{-}\eta^1,\eta^3\text{-}C_4H_6)La(\eta^5\text{-}C_5Me_5)_2(THF)]$, ($^{13}C$),[134] $[(\eta^5\text{-}C_5Me_5)_2\text{-}MCH(SiMe_3)_2]$, (M = Y, La, Ce; ^{13}C),[135] $[(\eta^5\text{-}C_5Me_5)_3Sm(C_8H_8)]$, ($^{13}C$),[136] $[Yb\{Sn(CH_2\text{-}CMe_3)_3\}_2(THF)_2]$, ($^{13}C$, ^{119}Sn, ^{171}Yb),[137] and $[(\eta^5\text{:}\eta^5\text{-}C_{13}H_{20})Yb(THF)_2]$, ($^{13}C$).[138]

^1H and ^{13}C n.m.r. spectroscopy has been used to indicate that the nitrogen atom of cryptand 221 has greater affinity to lanthanide(III) and $[UO_2]^{2+}$ ions than does the oxygen atom and the planarities of the

119 I. Dayan, J. Libman, A. Shanzer, C.E. Felder, and S. Lifson, *J. Am. Chem. Soc.*, 1991, **113**, 3431.
120 E.C. Okafor and B.A. Ozoukwu, *Synth. React. Inorg. Metal-Org. Chem.*, 1991, **21**, 1375.
121 W.S. Rees, jun. and D.A. Moreno, *J. Chem. Soc., Chem. Commun.*, 1991, 1759.
122 J.F. Campion, D.A. Payne, H.K. Chae, J.K. Maurin, and S.R. Wilson, *Inorg. Chem.*, 1991, **30**, 3244.
123 P.S. Coan, W.E. Streib, and K.G. Caulton, *Inorg. Chem.*, 1991, **30**, 5019.
124 K.F. Tesh and T.P. Hanusa, *J. Chem. Soc., Chem. Commun.*, 1991, 879.
125 J.A.T. Norman and G.P. Pez, *J. Chem. Soc., Chem. Commun.*, 1991, 971.
126 R. Gardiner, D.W. Brown, P.S. Kirlin, and A.L. Rheingold, *Chem. Mater.*, 1991, **3**, 1053 (*Chem. Abstr.*, 1991, **115**, 246 647).
127 U.S. Jung, *Hwahak Kwa Kongop Ui Chinbo*, 1991, **31**, 80 (*Chem. Abstr.*, 1991, **115**, 169 064).
128 I. Berkes, M. De Jesus, B. Hlimi, M. Massaq, E.H. Sayouty, and K. Heyde, *Phys. Rev. C: Nucl. Phys.*, 1991, **44**, 104 (*Chem. Abstr.*, 1991, **115**, 100 730).
129 C.J. Schaverien, J.H.G. Frijns, H.J. Heeres, J.R. Van den Hende, J.H. Teuben, and A.L. Spek, *J. Chem. Soc., Chem. Commun.*, 1991, 642.
130 W.J. Evans, T.A. Ulibarri, and J.W. Ziller, *Organometallics*, 1991, **10**, 134.
131 J.P. White, tert., H. Deng, and S.G. Shore, *Inorg. Chem.*, 1991, **30**, 2337.
132 C.J. Schaverein, *J. Chem. Soc., Chem. Commun.*, 1991, 458.
133 M. St. Clair, W.P. Schaefer, and J.E. Bercaw, *Organometallics*, 1991, **10**, 525.
134 A. Scholz, A. Smola, J. Scholz, J. Loebel, H. Schumann, and K.-H. Theile, *Angew. Chem., Int. Ed. Engl.*, 1991, **30**, 435.
135 H.J. Heeres and J.H. Teuben, *Organometallics*, 1991, **10**, 1980.
136 W.J. Evans, S.L. Gonzales, and J.W. Ziller, *J. Am. Chem. Soc.*, 1991, **113**, 7423.
137 F.G.N. Cloke, C.I. Dalby, P.B. Hitchcock, H. Karamallakis, and G.A. Lawless, *J. Chem. Soc., Chem. Commun.*, 1991, 779.
138 W.-q. Weng, K. Kunze, A.M. Arif, and R.D. Ernst, *Organometallics*, 1991, **10**, 3643.

ring are lost by complexation.[139] N.m.r. data have also been reported for [MR{N($SiMe_2CH_2$-PMe_2)$_2$}$_2$], (M = Y, La, Lu; ^{13}C),[140] Y complexes of N-benzoylglycine hydrazide and acetone (N-benzoyl)glycyl hydrazone, (^{13}C),[141] lanthanide complexes of 2-hydroxy-2-p-bromophenylmaleimide, (^{13}C),[142] [Ln(acac)$_2$(salicylaldehydeate)], (^{13}C),[143] lanthanide complexes of (7), (^{13}C),[144] [{Ph_2P-[N($SiMe_3$)]$_2$}$_2$Yb(THF)$_2$], (^{171}Yb),[145] [(octaethylporphyrin)Y(μ-OMe)$_2$AlMe_2], (^{13}C, ^{17}O),[146]

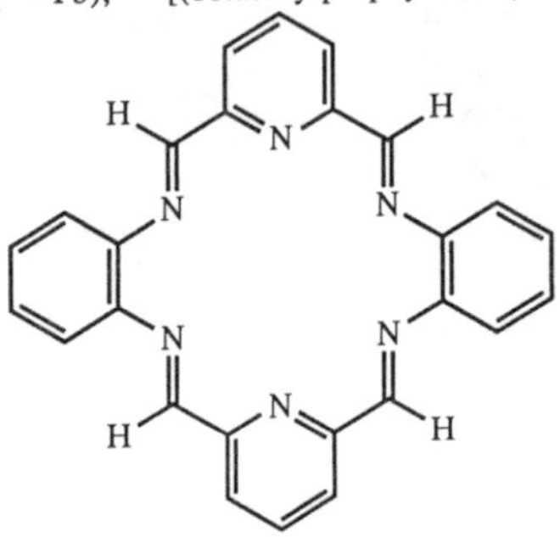

(7)

[(Bu^tO$SiMe_2$NBu^t)$_2$Yb(THF)$_2$], (^{171}Yb),[147] [Y_5(μ_5-O)(μ_3-OPr^i)$_4$(μ_2-OPr^i)$_4$(OPr^i)$_5$], (^{13}C, ^{89}Y),[148] [Y_2($OSiPh_3$)$_6$], (^{13}C, ^{29}Si),[149] [La_3(OBu^t)$_9$(Bu^tOH)$_2$], (^{13}C),[150] [La(O_2CCH_3)$_2$(OH)], (^{13}C),[151] [M(ascorbate)$_n$], (M = Al, La; n = 3; M = Pb; n = 2; ^{13}C),[152] [Cl_4La_2{Zr_2(OPr^i)$_9$}$_2$], (^{13}C),[153] [La($OSiPh_3$)$_2$(μ-$OSiPh_3$)]$_2$, (^{13}C, ^{29}Si),[154] La^{3+} and UO_2^{2+} complexes of macrocyclic Schiff bases, (^{13}C),[155] lanthanide complexes of polyhydoxycarboxylates, (^{13}C),[156] and 3-acetyl-4-hydroxy-6-methyl-2H-pyran-2-one, (^{13}C),[157] mixed acetato-dehydroacetato complexes of lanthanides, (^{13}C),[158]

139 O.J. Jung, C.N. Choi, and H.J. Jung, *Bull. Korean Chem. Soc.*, 1991, **12**, 130 (*Chem. Abstr.*, 1991, **115**, 173 304).
140 M.D. Fryzuk, T.S. Haddad, and S.J. Rettig, *Organometallics*, 1991, **10**, 1026.
141 T.R. Rao and G. Singh, *Asian J. Chem.*, 1990, **2**, 111 (*Chem. Abstr.*, 1991, **114**, 239 236).
142 D. Yang, Y. Zhu, Y. Liu, and S. Kong, *Gaodeng Xuexiao Huaxue Xuebao*, 1990, **11**, 710 (*Chem. Abstr.*, 1991, **114**, 93 996).
143 S.K. Ramalingam and S.A. Samath, *Transition Met. Chem. (London)*, 1990, **15**, 374 (*Chem. Abstr.*, 1991, **114**, 34 846).
144 F. Benetollo, G. Bombieri, K.K. Fonda, A. Polo, J.R. Quagliano, and L.M. Vallarino, *Inorg. Chem.*, 1991, **30**, 1345.
145 A. Recknagel, A. Steiner, M. Noltemeyer, S. Brooker, D. Stalke, and F.T. Edelmann, *J. Organomet. Chem.*, 1991, **414**, 327.
146 C.J. Schaverien and A.G. Orpen, *Inorg. Chem.*, 1991, **30**, 4968.
147 A. Recknagel, A. Steiner, S. Brooker, D. Stalke, and F.T. Edelmann, *J. Organomet. Chem.*, 1991, **415**, 315.
148 C.J. Page, S.K. Sur, M.C. Lonergan, and G.K. Prashar, *Magn. Reson. Chem.*, 1991, **29**, 1191.
149 P.S. Coan, M.J. McGeary, E.B. Lobkovsky, and K.G. Caulton, *Inorg. Chem.*, 1991, **30**, 3570.
150 D.C. Bradley, H. Chudzynska, M.B. Hursthouse, and M. Motevalli, *Polyhedron*, 1991, **10**, 1049.
151 M. Inoue, H. Kominami, H. Otsu, and T. Inui, *Nippon Kagaku Kaishi*, 1991, 1254 (*Chem. Abstr.*, 1991, **115**, 293 472).
152 H.A. Tajmir-Riahi, *J. Inorg. Biochem.*, 1991, **44**, 39.
153 U.M. Tripathi, A. Singh, and R.C. Mehrotra, *Polyhedron*, 1991, **10**, 949.
154 W.J. Evans, R.E. Golden, and J.W. Ziller, *Inorg. Chem.*, 1991, **30**, 4963.
155 A. Aguiari, E. Bullita, U. Casellato, P. Guerriero, S. Tamburini, and P.A. Vigato, *J. Solid State Inorg. Chem.*, 1991, **28**, 299.
156 J. Van Westrenen, J.A. Peters, H. van Bekkum, E.N. Rizkalla, and G.R. Choppin, *Inorg. Chim. Acta*, 1991, **181**, 233 (*Chem. Abstr.*, 1991, **115**, 113 806).
157 S. Sitran, D. Fregona, and G. Faraglia, *J. Coord. Chem.*, 1990, **22**, 229.
158 S. Sitran, D. Fregona, and G. Faraglia, *J. Coord. Chem.*, 1991, **24**, 127.

[La(S_2CNEt_2){OP(NMe_2)$_3$}$_5$]$^{2+}$, (^{13}C),[159] [$Ce_4O(OPr^i)_{13}(Pr^iOH)$], (^{13}C),[160] [$UO_2$(4-$X_3$FC(O)3-Me-1-Ph-pyrazol-5-onate)$_2$], (^{13}C),[161] and [Yb(SC_6H_2-2,4,6-Pr^i_3)$_2$($MeOCH_2CH_2OMe$)], (^{13}C).[162]

Complexes of Groups IVA—N.O.e. and ^{13}C INADEQUATE experiments have provided unambiguous assignments of the α and β proton and ^{13}C nuclei of the cyclopentadienyl group in [(η^5-C_5H_4Me)$PtMe_3$] and [(η^5-C_5H_4Me)$_2$Zr(CD_3)$_2$].[163] The relative sign of $^2J(^2H-^1H)$ in [(CH_2D)$TiCl_3$] has been found to be negative.[164] The ^{13}C n.m.r. spectrum of (8) shows an agostic interaction with the β-vinylic hydrogen with $^1J(^{13}C-^1H)$ reduced to 123 Hz.[165] N.m.r. data have also been reported for [{CH_2-2-(4-Me-6-Bu^t-C_6H_2O)$_2$}Ti{(μ-H)$_3$BH}$_2$], (^{11}B),[166] [(η^5-C_5H_5)$_2$ZrH($SiPh_3$)(PMe_3)], (^{13}C, ^{29}Si),[167] [{1,2-(η^5-1-indenyl)$_2$ethane}MH(μ-H)]$_2$, (M = Zr, Hf; ^{13}C),[168] [(η^5-C_5H_5)Zr-(BH_4)$_3$], (^{11}B),[169] [(η^5-$Me_3SiC_5H_4$)$_2$HfCl(BH_4)], (^{11}B, ^{13}C),[170] [(η^5-C_5Me_4H)$_2$$TiMe_2$], (^{13}C),[171] (9), (^{13}C),[172] (10), (^{13}C),[173] [(η^5-C_5H_5)$_2$Zr(Me)(μ-Me)B(C_6F_5)$_3$], (^{11}B, ^{13}C),[174] [Me_2Si(η^5-C_5H_4)$_2$Zr(μ-CH_3)(μ-C_2H_4)$AlMe_2$], (^{13}C, ^{27}Al),[175] (11), (^{13}C),[176] *anti*-[(η^5:η^5-$C_{10}H_8$){(η^5-C_5H_5)-Zr(η^1-CH_2CHCH_2)(η^3-CH_2CHCH_2)}$_2$], (^{13}C),[177] [(η^5-C_5H_5)$_2$ZrCH$_2$CH$_2$CH(C_8H_{17})CH$_2$],

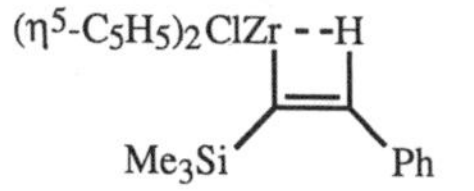

(8)

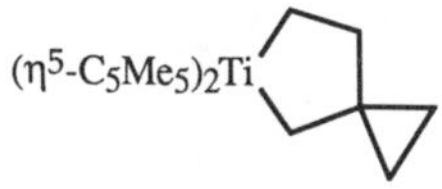

(9)

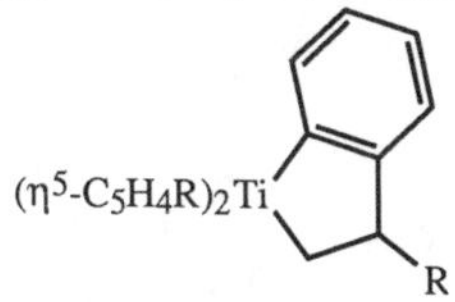

(10)

(^{13}C),[178] [(η^5-C_5H_5)$_2$Zr{C(CMe=CH_2)=CC≡CCMe=CH_2}(C≡CCMe=CH_2)], (^{13}C),[179] [(η^5-C_5H_5)$_2$ZrOCH$_2$(η^5-C_5H_5)$_2$ZrOCH$_2$C(=NBu^t)S], (^{13}C),[180] (12), (^{13}C),[181] [(η^5-C_5H_4Me)$_2$Zr{Sn-

159 V.V. Skopenko, A.F. Savost'yanova, V.V. Trachevskii, A.D. Gorbalyuk, and T.A. Sukhan, *Koord. Khim.*, 1991, **17**, 128 (*Chem. Abstr.*, 1991, **114**, 172 514).
160 K. Yunlu, P.S. Gradeff, N. Edelstein, W. Kot, G. Shalimoff, W.E. Streib, B.A. Vaartstra, and K.G. Caulton, *Inorg. Chem.*, 1991, **30**, 2317.
161 B.A. Uzoukwu, *Indian. J. Chem., Sect. A: Inorg., Bio-inorg., Phys., Theor. Anal. Chem.*, 1991, **30A**, 372 (*Chem. Abstr.*, 1991, **115**, 40 680).
162 E. Gretz, W.M. Vetter, H.A. Stecker, and A. Sen, *Isr. J. Chem.*, 1990, **30**, 327.
163 R.A. Newmark, L.D. Boardman, and A.R. Siedle, *Inorg. Chem.*, 1991, **30**, 853.
164 M.L.H. Green and A.K. Hughes, *J. Chem. Soc., Chem. Commun.*, 1991, 1231.
165 G. Erker and R. Zwettler, *J. Organomet. Chem.*, 1991, **409**, 179.
166 F. Corazza, C. Floriani, A. Chiesi-Villa, and C. Guastini, *Inorg. Chem.*, 1991, **30**, 145.
167 K.A. Kreutzer, R.A. Fisher, W.M. Davis, E. Spaltenstein, and S.L. Buchwald, *Organometallics*, 1991, **10**, 4031.
168 R.B. Grossman, R.A. Doyle, and S.L. Buchwald, *Organometallics*, 1991, **10**, 1501.
169 A.G. Császár, L. Hedberg, K. Hedberg, R.C. Burns, A.T. Wen, and M.J. McGlinchey, *Inorg. Chem.*, 1991, **30**, 1371.
170 J.B. Hoke and E.W. Stern, *J. Organomet. Chem.*, 1991, **412**, 77.
171 K. Mach, V. Varga, V. Hanuš, and P. Sedmera, *J. Organomet. Chem.*, 1991, **415**, 87.
172 K. Mashima, N. Sakai, and H. Takaya, *Bull. Chem. Soc. Jpn.*, 1991, **64**, 2475.
173 G. Erker, U. Korek, R. Petrenz, and A.L. Rheingold, *J. Organomet. Chem.*, 1991, **421**, 215.
174 X. Yang, C.L. Stern, and T.J. Marks, *J. Am. Chem. Soc.*, 1991, **113**, 3623.
175 A.R. Siedle, R.A. Newmark, J.N. Schroepfer, and P.A. Lyon, *Organometallics*, 1991, **10**, 400.
176 M. Akita, H. Yasuda, H. Yamamoto, and A. Nakamura, *Polyhedron*, 1991, **10**, 1.
177 Y. Wielstra, R. Duchateau, S. Gambarotto, C. Bensimon, and E. Gabe, *J. Organomet. Chem.*, 1991, **418**, 183.
178 T. Takahashi, T. Seki, Y. Nitto, M. Saburi, C.J. Rousset, and E.-i. Negishi, *J. Am. Chem. Soc.*, 1991, **113**, 6266.
179 K. Takagi, C.J. Rousset, and E.-i. Negishi, *J. Am. Chem. Soc.*, 1991, **113**, 1440.
180 G. Erker, M. Mena, C. Krüger, and R. Noe, *J. Organomet. Chem.*, 1991, **402**, 67.
181 G. Erker, M. Albrecht, C. Krüger, and S. Werner, *Organometallics*, 1991, **10**, 3791.

$[CH(SiMe_3)_2]_2\}_2]$, (^{13}C),[182] $[(Ph_3Sn)_4M(CO)_4]^{2-}$, (M = Zr, Hf; ^{13}C, ^{119}Sn),[183] $[(\eta^5\text{-}C_5Me_5)_2ThMe]^+[B(C_6F_5)_4]^-$, ($^{11}B$, ^{13}C),[184] $[(\eta^5\text{-}C_5H_5)Ti(CH_2Ph)(4\text{-}MeOC_6H_4NCMe{=}CMe\text{-}NC_6H_4OMe\text{-}4)]$, ($^{13}C$),[185] $[Ti(OC_6H_3Ph_2\text{-}2,6)_2(C_4Et_4C_6H_4NNC_6H_4)_2]$, ($^{13}C$),[186] $[(\eta^5\text{-}C_5H_5)_2\text{-}\overline{ZrNC_5H_4\text{-}2\text{-}CH_2CH(CH_2SiMe_3)C}H_2]$, ($^{13}C$),[187] $[(\eta^5\text{-}C_5H_5)_2\overline{ZrNC_5H_3\text{-}2\text{-}Me\text{-}6\text{-}CH_2CH_2C}H\text{-}2\text{-}C_5H_4N]^+$, ($^{13}C$),[188] $[(\mu\text{-}\eta^5{:}\eta^5\text{-}C_{10}H_8)\{(\mu\text{-}\eta^1{:}\eta^2\text{-}CN)Zr(\eta^5\text{-}C_5Me_5)\}_2]$, ($^{13}C$),[189] $[(\eta^5\text{-}C_5H_5)_2\text{-}\overline{ZrCH(CH{=}CH_2)CH_2B\{NBu^t(SiMe_3)\}{=}N}Bu^t]$, ($^{13}C$),[190] $[(\eta^5\text{-}C_5H_5)_2Zr\{P(SiMe_3)_3\}Me]$, ($^{13}C$, ^{29}Si),[191] $[(\eta^5\text{-}C_5H_5)_2\overline{ZrCH_2CPh{=}C}Ph(PMe_3)]$, ($^{13}C$),[192] $[(\eta^5\text{-}C_5H_5)_2\overline{ZrCH_2C}Me_2(PMe_3)]$, ($^{13}C$),[193] $[(\eta^5\text{-}C_5H_5)_2Ti(OR)Me]$, ($^{13}C$),[194] $[(\eta^5\text{-}C_5Me_5)_2Ti(CH_3)OC(CH_3){=}CH_2]$, ($^{13}C$),[195] $[(\eta^5\text{-}C_5Me_5)_2Ti(OMe)Me]$, ($^{13}C$),[196] $[(2,6\text{-}Ph_2C_6H_3O)_2\overline{TiCH_2CHR^1CHR^2C}H_2]$, ($^{13}C$),[197] $[(\eta^5\text{-}C_5Me_5)_2\overline{TiCH_2CH_2CHRO}]$, ($^{13}C$),[198] $[(\eta^5\text{-}C_5Me_5)_2\overline{TiCH_2CH_2C\{{=}M(CO)_5\}O}]$, (M = Cr, Mo, W; ^{13}C),[199] $[(\eta^5\text{-}C_5H_5)_2Ti(\mu\text{-}CH_2)(\mu\text{-}2\text{-}MeOC_6H_4)Rh(\eta^4\text{-}1,5\text{-}C_8H_{12})]$, ($^{13}C$),[200] $[\{SiMe_2(\eta^5\text{-}C_5H_4)_2\}Ti(CH_2SiMe_3)(THF)]^+$, ($^{13}C$),[201] (13), ($^{13}C$),[202] $[(\eta^5\text{-}C_5Me_5)_2\overline{ZrCH_2CH_2CH_2O}]$,

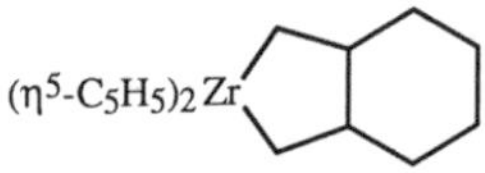

(11)

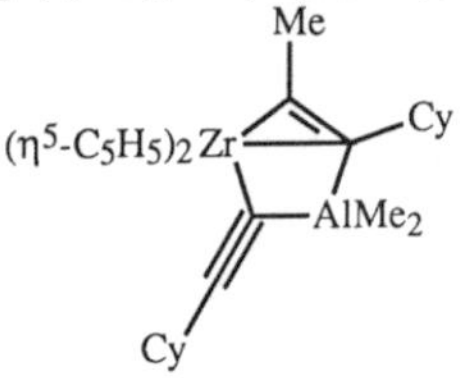

(12)

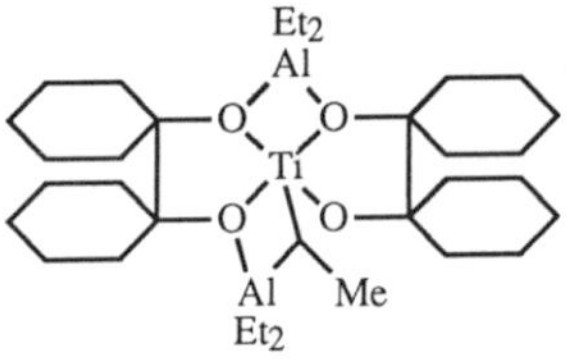

(13)

(^{13}C),[203] (14), $\{ML_n$ = $Mo(CO)_5$, $W(CO)_5$, $(\eta^5\text{-}C_5H_5)Co(CO)$, $(\eta^5\text{-}C_5H_5)Rh(CO)$, $(\eta^5\text{-}$

182 R.M. Whittal, G. Ferguson, J.F. Gallagher, and W.E. Piers, *J. Am. Chem. Soc.*, 1991, **113**, 9867.
183 J.E. Ellis, K.M. Chi, A. DiMaio, S.R. Frerichs, J.R. Stenzel, A.L. Rheingold, and B.S. Haggerty, *Angew. Chem., Int. Ed. Engl.*, 1991, **30**, 194.
184 X. Yang, C.L. Stern, and T.J. Marks, *Organometallics*, 1991, **10**, 840.
185 J. Scholz, A. Dietrich, H. Schumann, and K.-H. Thiele, *Chem. Ber.*, 1991, **124**, 1035.
186 J.E. Hill, P.E. Fanwick, and I.P. Rothwell, *Inorg. Chem.*, 1991, **30**, 1143.
187 A.S. Guram, R.F. Jordan, and D.F. Taylor, *J. Am. Chem. Soc.*, 1991, **113**, 1833.
188 A.S. Guram and R.F. Jordan, *Organometallics*, 1991, **10**, 3470.
189 W.A. Herrmann, B. Menjón, and E. Herdtweck, *Organometallics*, 1991, **10**, 2134.
190 H. Braunschweig, I. Manners, and P. Paetzold, *Z. Naturforsch., B*, 1990, **45**, 1453 (*Chem. Abstr.*, 1991, **114**, 43 015).
191 E. Hey-Hawkins, M.F. Lappert, J.L. Atwood, and S.G. Bott, *J. Chem. Soc., Dalton Trans.*, 1991, 939.
192 P. Binger, P. Müller, A.T. Herrmann, P. Philipps, B. Gabor, F. Langhauser, and C. Krüger, *Chem. Ber.*, 1991, **124**, 2165.
193 D.R. Swanson and E.-i. Negishi, *Organometallics*, 1991, **10**, 825.
194 R. Schobert, *J. Organomet. Chem.*, 1991, **405**, 201.
195 C.P. Gibson and D.S. Bem, *J. Organomet. Chem.*, 1991, **414**, 23.
196 G.A. Luinstra and J.H. Teuben, *J. Organomet. Chem.*, 1991, **420**, 337.
197 J.E. Hill, P.E. Fanwick, and I.P. Rothwell, *Organometallics*, 1991, **10**, 15; J.E. Hill, G.J. Balaich, P.E. Fanwick, and I.P. Rothwell, *Organometallics*, 1991, **10**, 3428.
198 K. Mashima, H. Haraguchi, A. Ohyoshi, N. Sakai, and H. Takaya, *Organometallics*, 1991, **10**, 2731.
199 K. Mashima, K. Jyodoi, A. Ohyoshi, and H. Takaya, *Bull. Chem. Soc. Jpn.*, 1991, **64**, 2065.
200 J.W. Park, L.M. Henling, W.P. Schaefer, and R.H. Grubbs, *Organometallics*, 1991, **10**, 171.
201 D.M. Amorose, R.A. Lee, and J.L. Petersen, *Organometallics*, 1991, **10**, 2191.
202 C. Krüger, R. Mynott, C. Siedenbiedel, L. Stehling, and G. Wilke, *Angew. Chem., Int. Ed. Engl.*, 1991, **30**, 1668.
203 K. Mashima, M. Yamakawa, and H. Takaya, *J. Chem. Soc., Dalton Trans.*, 1991, 2851.

$C_5H_5)V(CO)_3$, $(C_6F_5)_2Pt(CO)$; $^{13}C\}$,[204] $[(\eta^5\text{-}C_5H_5)_2ZrPh(THF)]^+$, ($^{13}C$),[205] $[(\eta^5\text{-}C_5H_5)_2(\eta^2\text{-}R_2CHC{=}S)ZrCl]$, ($^{13}C$),[206] $[(\eta^5\text{-}C_5H_5)_2M(CH_2SiMe_2SiMe_3)Cl]$, ($^{13}C$, ^{29}Si),[207] $[\{Me_2Si(\eta^5\text{-}C_5H_4)_2\}TiRCl]$, ($^{13}C$),[208] $[(\eta^5\text{-}C_5Me_5)\{\eta^5{:}\eta^1\text{-}C_5H_4CH_2C({=}NC_6H_3Me_2\text{-}2,6)\}TiCl]$, ($^{13}C$),[209] $[(\eta^4\text{-}Bu^tCH{=}CHCH{=}CHBu^t)_2Ti]$, ($^{13}C$),[210] $[(\eta^5\text{-}C_5H_5)_2Ti(S_4N_4)]^+$, ($^{14}N$),[211] $[(\eta^5\text{-}C_5H_5)(OC)_2M^1CO_2M^2(Cl)(\eta^5\text{-}C_5H_5)_2]$, ($^{13}C$),[212] $[(\eta^5\text{-}C_5H_5)_2Ti{=}NCBu^t{=}CH_2PMe_3]$, ($^{13}C$),[213] $[(\eta^5\text{-}C_5H_5)_2TiSe_nS_{5-n}]$, ($^{77}Se$),[214] $[(\eta^5\text{-}C_5H_5)_4Ti_4(\mu_2\text{-}Se)_3(\mu_3\text{-}Se)_3]$, ($^{77}Se$),[215] $[(\eta^5\text{-}C_5H_4SiMe_3)(\eta^8\text{-}C_8H_7SiMe_3)Ti]$, ($^{13}C$),[216] $[\{\eta^5\text{-}C_5H_3(SiMe_3)_2\}MCl_3]$, (M = Ti, Zr, Hf; ^{13}C),[217] $[\{\eta^5\text{-}C_5H_2(SiMe_3)_3\}TiCl_2(OMe)]$, ($^{13}C$),[218] (15), ($^{13}C$),[219] $[(\eta^5\text{-}C_5H_4CR{=}CRC_5H_4\text{-}\eta^5)TiCl_2]$, ($^{13}C$),[220]

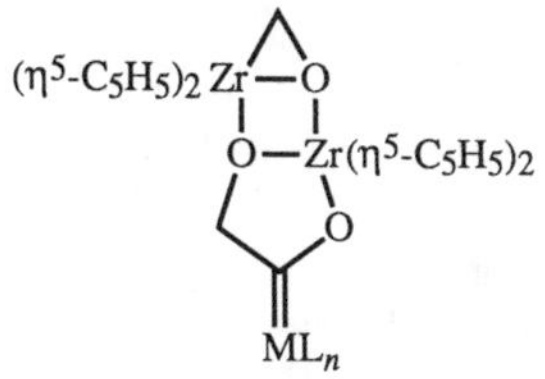

(14)

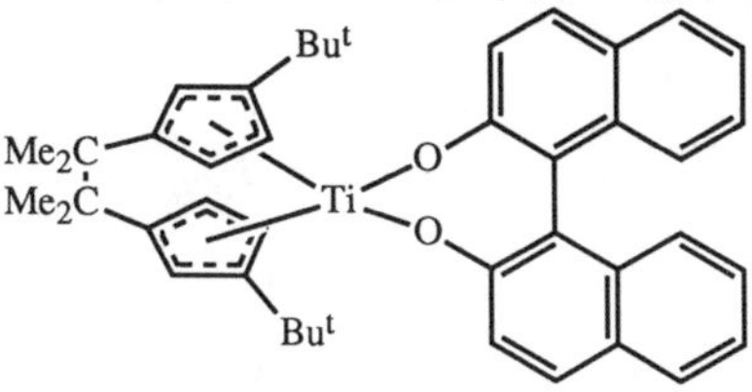

(15)

$[(\eta^5\text{-}1,2,4\text{-}Me_3C_5H_2)MCl_2]$, (M = Ti, Zr, Hf; ^{13}C),[221] $[Ti\{(OCHPh)_2C_5H_4(CH_2)\}Cl_2]$, ($^{13}C$),[222] (16), (M = Ti, Zr, Hf; ^{13}C),[223] $[(\eta^5\text{-}C_5H_4SiMe_2CH_2CH_2SiMe_2C_5H_4\text{-}\eta^5)MCl_2]$, (M = Ti, Zr, Hf; ^{13}C),[224] $[\{2,2'\text{-}(\eta^5\text{-}C_5H_4CH_2)_2\text{-}1,1\text{-binaphthyl}\}MCl_2]$, (M = Ti, Zr; ^{13}C),[225] $[\{2,2'\text{-}(\eta^5\text{-indenyl-}2\text{-}CH_2)_2\text{-}1,1\text{-binaphthyl}\}MCl_2]$, (M = Ti, Zr; ^{13}C),[226] (17), (M = Ti, Zr; ^{13}C),[227] $[(\eta^7\text{-}C_7H_7)(\eta^5\text{-}C_7H_9)ZrPMe_2CH_2CH_2PMe_2Zr(\eta^7\text{-}C_7H_7)(\eta^5\text{-}C_7H_9)]$, ($^{13}C$),[228] $[(\eta^5\text{-}C_5H_5)_2Zr(N{=}CMe_2)(N{\equiv}C\text{-}Me)]^+$, ($^{13}C$),[229] $[(\eta^5\text{-}C_5H_5)_2Zr(NCO)(NPh_2)]$, ($^{13}C$),[230] $[(\eta^5\text{-}C_5H_5)_2ZrOCH_2SiMe_2NBu^t]$, ($^{13}C$,

204 G. Erker, M. Mena, U. Hoffmann, B. Menjón, and J.L. Petersen, *Organometallics*, 1991, **10**, 291.
205 S.L. Borkowsky, R.F. Jordan, and G.D. Hinch, *Organometallics*, 1991, **10**, 1268.
206 W. Ando, T. Ohtaki, T. Suzuki, and Y. Kabe, *J. Am. Chem. Soc.*, 1991, **113**, 7782.
207 S. Sharma, R.N. Kapoor, F. Cervantes-Lee, and K.H. Pannell, *Polyhedron*, 1991, **10**, 1177.
208 R. Gomez, T. Cuenca, P. Royo, and E. Hovestreydt, *Organometallics*, 1991, **10**, 2516.
209 R. Fandos, A. Meetsma, and J.H. Teuben, *Organometallics*, 1991, **10**, 2665.
210 F.G.N. Cloke and A. McCamley, *J. Chem. Soc., Chem. Commun.*, 1991, 1470.
211 P.K. Gowik, T.M. Klapoetke, and S. Cameron, *J. Chem. Soc., Dalton Trans.*, 1991, 1433.
212 J.C. Vites, B.D. Steffey, M.E. Giuseppetti-Dery, and A.R. Cutler, *Organometallics*, 1991, **10**, 2827.
213 K.M. Doxsee, J.B. Farahi, and H. Hope, *J. Am. Chem. Soc.*, 1991, **113**, 8889.
214 P. Pekonen, Y. Hiltunen, R.S. Laitinen, and J. Valkonen, *Inorg. Chem.*, 1991, **30**, 1874.
215 F. Bottomley and R.W. Day, *Organometallics*, 1991, **10**, 2560.
216 M.D. Rausch, M. Ogasa, R.D. Rogers, and A.N. Rollins, *Organometallics*, 1991, **10**, 2084.
217 C.H. Winter, X.-X. Zhou, D.A. Dobbs, and M.J. Heeg, *Organometallics*, 1991, **10**, 210.
218 J. Okuda and E. Herdtweck, *Inorg. Chem.*, 1991, **30**, 1516.
219 M.S. Erickson, F.R. Fronczek, and M.L. McLaughlin, *J. Organomet. Chem.*, 1991, **415**, 75.
220 P. Burger and H.H. Brintzinger, *J. Organomet. Chem.*, 1991, **407**, 207.
221 P.G. Gassman and C.H. Winter, *Organometallics*, 1991, **10**, 1592.
222 R. Fandos, J.H. Teuben, G. Helgesson, and S. Jagner, *Organometallics*, 1991, **10**, 1637.
223 J.A. Bandy, M.L.H. Green, I.M. Gardiner, and K. Prout, *J. Chem. Soc., Dalton Trans.*, 1991, 2207.
224 H. Lang and D. Seyferth, *Organometallics*, 1991, **10**, 347.
225 S.L. Colletti and R.L. Halterman, *Organometallics*, 1991, **10**, 3438.
226 M.J. Burk, S.L. Colletti, and R.L. Halterman, *Organometallics*, 1991, **10**, 2998.
227 Z. Chen, K. Eriks, and R.L. Halterman, *Organometallics*, 1991, **10**, 3449.
228 J.C. Green, M.L.H. Green and N.M. Walker, *J. Chem. Soc., Dalton Trans.*, 1991, 173.
229 Y.W. Alelyunas, R.F. Jordan, S.F. Echols, S.L. Borkowsky, and P.K. Bradley, *Organometallics*, 1991, **10**, 1406.
230 P.J. Walsh, M.J. Carney, and R.G. Bergman, *J. Am. Chem. Soc.*, 1991, **113**, 6343.

^{29}Si),[231] [$(\eta^5-C_5H_5)_2Zr\{N{=}CHP(NPr^i{}_2)_2\}Cl$], ($^{13}C$),[232] (18), ($^{13}C$),[233] [$(\eta^5-C_5H_5)_2ZrOCH_2C({=}NR)]_2$, ($^{13}C$),[234] [$(OC)_5CrPPh_2C({=}CH_2)OZr(\eta^5-C_5H_5)_2Cl$], ($^{13}C$),[235] [$(\eta^5-C_5H_5)Ru(CO)_2(CH_2O)Zr(\eta^5-C_5H_5)_2Cl$], ($^{13}C$),[236] [$(\eta^5-C_5H_5)_2Zr(OBu^t)(THF)]^+$, ($^{13}C$),[237] [$\{(\eta^5-C_5H_5)_2Zr(\mu-SCH_2CH_2S)_2Zr(\eta^5-C_5H_5)_2\}Ag]^+$, ($^{13}C$),[238] [$\{(\eta^5-C_5H_5)_2ZrCl\}_2SC(O)C(O)S$], ($^{13}C$),[239] [$Ph_2NC\{CH_2Cr(CO)_5\}OZr(\eta^5-C_5H_5)_2Cl$], ($^{13}C$),[240] [$(\eta^5-C_5H_4CH_2CH{=}CH_2)_2MCl_2$], (M = Zr, Hf;

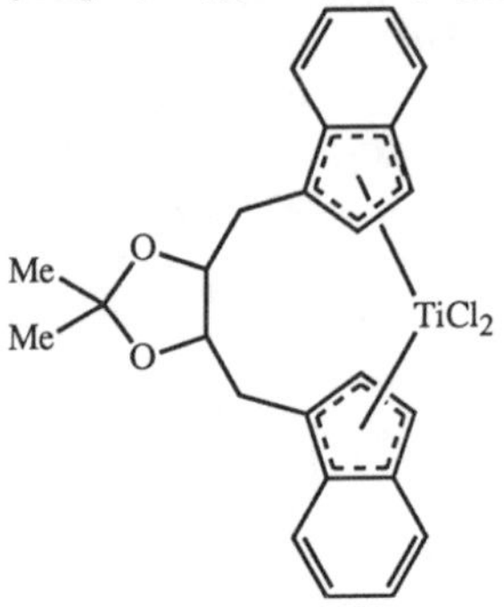
(16)

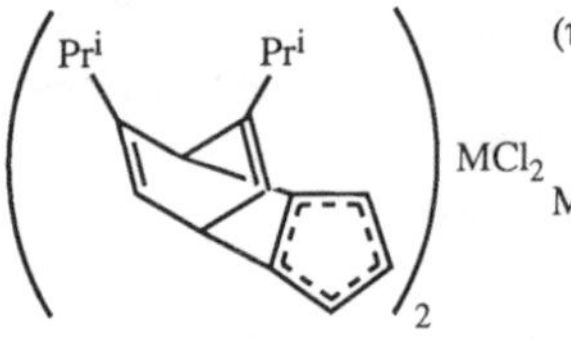
(17)

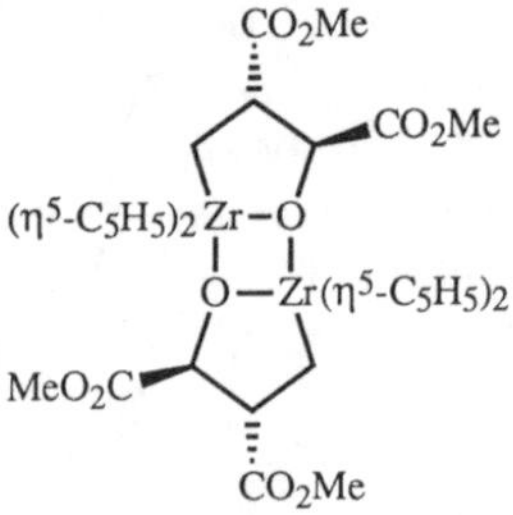
(18)

^{13}C),[241] [$(\eta^5-C_5H_4CHMeCy)_2ZrCl_2$], ($^{13}C$),[242] [$\{(\eta^5-C_5H_3R)_2C_2H_4\}ZrCl_2$], ($^{13}C$),[243] [$\{CHBu^t(\eta^5-C_5H_4)(\eta^5$-fluorenyl$)\}ZrCl_2$], ($^{13}C$),[244] and [$(\eta^6-C_6H_5Me)Zr(PMe_3)_2Cl_2$], ($^{13}C$).[245]

Ab initio calculations of ^{47}Ti and ^{49}Ti shieldings for $Ti(OH)_4$, $[Ti(OH_2)_6]^{2+}$, $[Ti(OH)_4O]^{2-}$, $[Ti(OH)_3O]^-$, TiF_4, $[TiF_5]^-$, and $TiCl_4$ have been reported.[246] N.m.r. data have also been reported for [$(Me_3SiNCH_2CH_2NSiMe_3)_2M$], (M = Ti, Si; ^{13}C, ^{29}Si),[247] [(7,16-dihydro-6,8,15,17-Me$_4$-dibenzo-[*b,i*]-[1,4,8,11]-tetraazacyclotetradecine)$TiO_2CRe(\eta^5-C_5Me_5)(NO)(CO)$], ($^{13}C$),[248] [$(Bu^t{}_3SiNH)_3TiX$], ($^{13}C$),[249] [$(\{(Me_3Si)_2N\}(tmeda)ClTi)_2N_2$], ($^{13}C$),[250] [$(tmeda)Ti(NPh)Cl_2$], ($^{13}C$),[251] [{TiCl-

231 L.J. Procopio, P.J. Carroll, and D.H. Berry, *J. Am. Chem. Soc.*, 1991, **113**, 1870.
232 F. Boutonnet, N. Dufour, T. Straw, A. Igau, and J.-P. Majoral, *Organometallics*, 1991, **10**, 3939.
233 G. Erker, S. Dehnicke, M. Rump, C. Krûger, S. Werner, and M. Nolte, *Angew. Chem., Int. Ed. Engl.*, 1991, **30**, 1349.
234 G. Erker, M. Mena, C. Krüger, and R. Noe, *Organometallics*, 1991, **10**, 1201.
235 P. Veya, C. Floriani, A. Chiesi-Villa, and C. Guastini, *Organometallics*, 1991, **10**, 2991.
236 B.D. Steffey, J.C. Vites, and A.R. Cutler, *Organometallics*, 1991, **10**, 3432.
237 S. Collins, B.E. Koene, R. Ramachandran, and N.J. Taylor, *Organometallics*, 1991, **10**, 2092.
238 D.W. Stephan, *Organometallics*, 1991, **10**, 2037.
239 C.A. Hester, M. Draganjac, and A.W. Cordes, *Inorg. Chim. Acta*, 1991, **184**, 137.
240 P. Veya, C. Floriani, A. Chiesi-Villa, and C. Guastini, *J. Chem. Soc., Chem. Commun.*, 1991, 1166.
241 G. Erker and R. Aul, *Chem. Ber.*, 1991, **124**, 1301.
242 G. Erker, R. Nolte, R. Aul, S. Wilker, C. Krüger, and R. Noe, *J. Am. Chem. Soc.*, 1991, **113**, 7594.
243 S. Collins, W.J. Gauthier, D.A. Holden, B.A. Kuntz, N.J. Taylor, and D.G. Ward, *Organometallics*, 1991, **10**, 2061.
244 G.S. Herrmann, H.G. Alt, and M.D. Rausch, *J. Organomet. Chem.*, 1991, **401**, C5.
245 G.M. Diamond, M.L.H. Green, and N.M. Walker, *J. Organomet. Chem.*, 1991, **413**, C1.
246 J.A. Tossell, *J. Magn. Reson.*, 1991, **94**, 301.
247 W.A. Herrmann, M. Denk, R.W. Albach, J. Behm, and E. Herdtweck, *Chem. Ber.*, 1991, **124**, 683.
248 C.E. Housmekerides, R.S. Pilato, G.L. Geoffroy, and A.L. Rheingold, *J. Chem. Soc., Chem. Commun.*, 1991, 563.
249 C.C. Cummins, C.P. Schaller, G.D. Van Duyne, P.T. Wolczanski, A.W.E. Chan, and R. Hoffmann, *J. Am. Chem. Soc.*, 1991, **113**, 2985.
250 R. Duchateau, S. Gambarotta, N. Beydoun, and C. Bensimon, *J. Am. Chem. Soc.*, 1991, **113**, 8986.
251 R. Duchateau, A.J. Williams, S. Gambarotta, and M.Y. Chiang, *Inorg. Chem.*, 1991, **30**, 4863.

(Schiff base)}$_2$O], (^{13}C),[252] [Ti{tris[3-(2,3-dihydroxyphenoxy)propyl]amine}], (^{13}C),[253] [Me$_2$Et-COTi(OCH$_2$CH$_2$)$_3$N], (^{13}C),[254] [{(Bu$^t_3$SiNH)$_3$Zr}$_2$(μ-OCH$_2$)], (^{13}C),[255] [Zr(salen)$_2$], (^{13}C),[256] [Ti(OC$_6$HMe$_4$-2,3,5,6)$_4$], (^{13}C),[257] [M(O$_2$CNR$_2$)$_4$], (M = Ti, Zr, Hf; ^{13}C),[258] [Ti(OBu)(O$_2$CR)$_3$], (^{13}C),[259] [Ti$_7$O$_4$(OEt)$_{20}$], (^{17}O),[260] [{(η^5-C$_5$H$_4$)(CH$_2$)$_3$O}TiCl$_2$], (^{13}C),[261] [TiCl$_2$(OCH$_2$CH$_2$X)$_2$-(HOCH$_2$CH$_2$X)]$_2$, (X = Cl, Br, I; ^{13}C),[262] [{TiCl$_2$(O$_2$CBu^t)(MeCO$_2$Et)}$_2$O], (^{13}C, ^{17}O),[263] [Ti$_2$-(OR)$_8$MX$_2$]$_2$, (M = Mg, Zn; X = Cl, Br, I; ^{13}C),[264] [BaZr$_2$(OPr^i)$_{10}$]$_2$, (^{13}C),[265] [Zr(O$_2$C-CHR^{1}R^2)$_n$(OPr^i)$_{4-n}$], (^{13}C),[266] and [(Pr^{i}O)$_3$Zr(*O*,*O*-alkylene dithiophosphate)], (^{13}C).[267]

Complexes of V, Nb, and Ta.—A review entitled 'Vanadium-51 n.m.r.' has appeared.[268]

The ^{1}H n.m.r. spectrum of [(η^5-C$_5$H$_5$)$_2$MH$_3$], M = Nb, Ta, Mo$^+$, W$^+$, has been examined as a function of temperature and is generally AB$_2$ with *J*(A-B) = 7 to 10 Hz. When M = Nb, *J*(A-B) is temperature dependent, decreasing with decreasing temperature, becoming zero at 243 K, and reappearing at 173 K. For M = Mo$^+$, *J*(A-B) = 1000 Hz at 203 K and decreases to 450 Hz at 153 K.[269] N.m.r. data have also been reported for [M(crown)][(η^5-C$_5$H$_5$)$_2$NbH$_{2-n}$(SnMe$_3$)$_n$], (M = Li, Na, K; ^{13}C),[270] [(η^5-C$_5$H$_4$SiMe$_3$)$_2$Nb(μ-H){Au(PPh$_3$)}$_2$]$^+$, (^{13}C),[271] [Ta(OC$_6$H$_3$Pr$^i_2$-2,6)$_2$-(H)$_3$(PMe$_2$Ph)$_2$], (^{13}C),[272] [(η^5-C$_5$H$_5$)$_2$Ta(H)$_2$(SiH$_3$)], (^{13}C, ^{29}Si),[273] [(η^5-C$_5$Me$_5$)Ta(=CHPMe$_2$)-(H)$_2$(PMe$_3$)], (^{13}C),[274] [(η^5-C$_5$Me$_5$)$_2$TaH(=C=CHMe)], (^{13}C),[275] [TaH(HCOCCOH)(dmpe)$_2$Cl]$^+$, (^{13}C),[276] [(η^5-C$_5$Me$_5$)Ta(dmpe)(H)(η^2-CH$_2$PMe$_3$)], (^{13}C),[277] [(η^5-C$_5$H$_5$)$_2$M(CH$_2$Me)(alkyne)], (M = Nb, Ta; ^{13}C),[278] [(η^5-C$_5$H$_4$Me)$_2$Nb(=O)CH$_2$SiMe$_3$], (^{13}C),[279] [M(COSiPr$^i_3$)(CO)(dmpe)$_2$], (M =

252 C. Sasaki, K. Nakajima, M. Kojima, and J. Fujita, *Bull. Chem. Soc. Jpn.*, 1991, **64**, 1318.
253 F.E. Hahn, S. Rupprecht, and K.H. Moock, *J. Chem. Soc., Chem. Commun.*, 1991, 224.
254 A.A. Naiini, W.M.P.B. Menge, and J.G. Verkade, *Inorg. Chem.*, 1991, **30**, 5009.
255 C.C. Cummins, G.D. Van Duyne, C.P. Schaller, and P.T. Wolczanski, *Organometallics*, 1991, **10**, 164.
256 P.K. Mishra, V. Chakravortty, and K.C. Dash, *Transition Met. Chem. (London)*, 1991, **16**, 73.
257 R.T. Toth and D.W. Stephan, *Can. J. Chem.*, 1991, **69**, 172.
258 F. Calderazzo, S. Ianelli, G. Pampaloni, G. Pelizzi, and M. Sperrle, *J. Chem. Soc., Dalton Trans.*, 1991, 693.
259 N. Yoshino, Y. Shiraishi, and H. Hirai, *Bull. Chem. Soc. Jpn.*, 1991, **64**, 1648.
260 V.W. Day, T.A. Eberspacher, W.G. Klemperer, C.W. Park, and F.S. Rosenberg, *J. Am. Chem. Soc.*, 1991, **113**, 8190.
261 R. Fandos, A. Meetsma, and J.H. Teuben, *Organometallics*, 1991, **10**, 59.
262 C.H. Winter, P.H. Sheridan, and M.J. Heeg, *Inorg. Chem.*, 1991, **30**, 1962.
263 N.W. Alcock, D.A. Brown, T.F. Illson, S.M. Roe, and M.G.H. Wallbridge, *J. Chem. Soc., Dalton Trans.*, 1991, 873.
264 L. Abis, G. Bacchilega, S. Spera, U. Zucchini, and T. Dall'Occo, *Makromol. Chem.*, 1991, **192**, 981 (*Chem. Abstr.*, 1991, **114**, 247 865).
265 B.A. Vaarstra, J.C. Huffman, W.E. Streib, and K.G. Caulton, *Inorg. Chem.*, 1991, **30**, 3068.
266 A.K. Saxena, S. Saxena, and A.K. Rai, *Indian J. Chem., Sect. A*, 1990, **29**, 1027 (*Chem. Abstr.*, 1991, **114**, 113 931).
267 J.S. Yadav, R.K. Mehrotra, and G.K. Srivastava, *Indian J. Chem., Sect. A*, 1990, **29**, 1212 (*Chem. Abstr.*, 1991, **114**, 74 145).
268 O.W. Howarth, *Prog. Nucl. Magn. Reson. Spectrosc.*, 1990, **22**, 453.
269 D.M. Heinekey, *J. Am. Chem. Soc.*, 1991, **113**, 6074.
270 M.L.H. Green, A.K. Hughes, and P. Mountford, *J. Chem. Soc., Dalton Trans.*, 1991, 1699.
271 A. Antiñolo, F.A. Jalón, A. Otero, M. Fajardo, B. Chaudret, F. Lahoz, and J.A. López, *J. Chem. Soc., Dalton Trans.*, 1991, 1861.
272 B.C. Ankianiec, P.E. Fanwick, and I.P. Rothwell, *J. Am. Chem. Soc.*, 1991, **113**, 4710.
273 Q. Jiang, P.J. Carroll, and D.H. Berry, *Organometallics*, 1991, **10**, 3648.
274 V.C. Gibson, T.P. Kee, S.T. Carter, R.D. Sanner, and W. Clegg, *J. Organomet. Chem.*, 1991, **418**, 197.
275 V.C. Gibson, G. Parkin, and J.E. Bercaw, *Organometallics*, 1991, **10**, 220.
276 R.N. Vrtis, S.G. Bott, R.L. Rardin, and S.J. Lippard, *Organometallics*, 1991, **10**, 1364.
277 H.M. Anstice, H.H. Fielding, V.C. Gibson, C.E. Housecroft, and T.P. Kee, *Organometallics*, 1991, **10**, 2183.
278 H. Yasuda, H. Yamamoto, T. Arai, A. Nakamura, J. Chen, Y. Kai, and N. Kasai, *Organometallics*, 1991, **10**,

Nb, Ta; ^{13}C),[280] [(η^5-C_5H_5)$_2$Nb(SnMe$_3$)(CH$_2$=CHPh)], (^{2}H, ^{13}C, ^{117}Sn, ^{119}Sn),[281] [Ta(OC$_6$H$_2$-Bu^t-4-Me-CMe$_2$CH$_2$)Cl$_2$(CH$_2$Bu^t)], (^{13}C),[282] [(η^5-C_5Me_5)$_2$MeTaCH$_2$CH$_2$O], (^{13}C),[283] [(2,6-Ph$_2$C$_6$H$_3$O)$_2$Ta(CH$_2$C$_6$H$_4$X-4)$_3$], (^{13}C),[284] and (19), (^{13}C).[285]

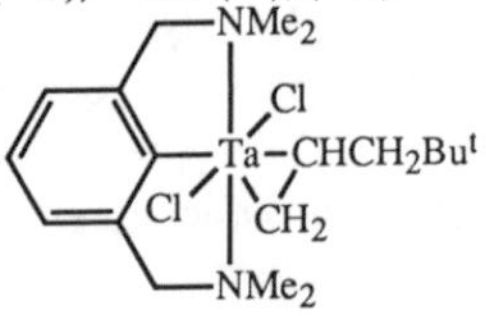

(19)

^{1}H and ^{2}H n.m.r. spectroscopy has provided strong evidence for a stereoselective *exo* addition of D$^+$ to [(η^5-C_5H_5)V(η^8-C_8H_8)]$^-$ and a subsequent metal assisted hydrogen migration.[286] N.m.r. data have also been reported for [V(Me$_3$SiOC≡COSiMe$_3$)(dmpe)$_2$Br], (^{13}C),[287] [Nb(η^5-C_5H_4SiMe$_3$)$_2$-Cl{η^2-(*C,N*)-EtPhHCCNPh}]$^+$, (^{13}C),[288] [HB(3,5-Me$_2$C$_3$HN$_2$)$_3$NbCl$_2$(PhC≡CMe)], (^{13}C),[289] [(η^5-C_5H_5)Nb(μ-Cl){(4-MeC$_6$H$_4$C≡CC$_6$H$_4$Me-4)}], (^{13}C),[290] [{(Bu$^t_3$SiO)$_3$Ta}$_2${μ-η^2(1,2);-η^2(4,5)-C$_6$H$_6$}], (^{13}C),[291] [(2,6-Pr$^i_2$C$_6$H$_3$O)$_3$Ta(Me$_3$SiC≡CSiMe$_3$)], (^{13}C),[292] [(η^5-C_5H_4Me)$_2$-Nb(η^3-C_3H_5)], (^{13}C, ^{93}Nb),[293] [{(η^5-C_5Me_5)V}$_4$(μ-O)$_6$], (^{13}C, ^{51}V),[294] [(η^5-C_5Me_5)$_2$V$_2$Te$_3$O], (^{51}V),[295] [Nb(η^6-C_6H_5Me)(η^5-C_7H_9)(PMe$_3$)], (^{13}C),[296] [{(η^5-1,3-Bu$^t_2$C$_5$H$_3$)Nb}$_2${As$_8$Cr(CO)$_5$}], (^{13}C),[297] [(η^5-C_5H_5)Nb(CO)$_3$(ES$_4$H$_8$)], (E = S, Se, Te; ^{13}C),[298] [(η^6-C_6Et_6)Ta(OC$_6$H$_3$Pr$^i_2$-2,6)$_2$I], (^{13}C),[299] and [RV(CO)$_4$(dppe)], (^{51}V).[300]

^{51}V n.m.r. spectroscopy has been used to study the formation of vanadate complexes with peptides.[301] The interaction of vanadate with bromo/iodoperoxidase has been investigated by ^{51}V n.m.r. spectroscopy.[302] Heteronuclear shift correlation for ^{17}O and ^{51}V has been applied to

4058.
279 P.-f. Fu, M.A. Khan, and K.M. Nicholas, *Organometallics*, 1991, **10**, 382.
280 R.N. Vrtis, S. Liu, Ch.P. Rao, S.G. Bott, and S.J. Lippard, *Organometallics*, 1991, **10**, 275.
281 M.L.H. Green, A.K. Hughes, and P. Mountford, *J. Chem. Soc., Dalton Trans.*, 1991, 1407.
282 A.-S. Baley, Y. Chauvin, D. Commereuc, and P.B. Hitchcock, *New J. Chem.*, 1991, **15**, 609.
283 L.L. Whinnery, jun., L.M. Henling, and J.E. Bercaw, *J. Am. Chem. Soc.*, 1991, **113**, 7575.
284 R.W. Chesnut, G.G. Jacob, J.S. Yu, P.E. Fanwick, and I.P. Rothwell, *Organometallics*, 1991, **10**, 321.
285 H.C.L. Abbenhuis, N. Feiken, H.F. Haarman, D.M. Grove, E. Horn, H. Kooijman, A.L. Spek, and G. van Koten, *Angew. Chem., Int. Ed. Engl.*, 1991, **30**, 996.
286 B. Bachmann and J. Heck, *Organometallics*, 1991, **10**, 1373.
287 J.D. Protasiewicz and S.J. Lippard, *J. Am. Chem. Soc.*, 1991, **113**, 6564.
288 A. Antiñolo, M. Fajardo, C. Lopez-Mardomingo, P. Martin-Villa, A. Otero, M.M. Kubicki, Y. Mourad, and Y. Mugnier, *Organometallics*, 1991, **10**, 3435.
289 M. Etienne, P.S. White, and J.L. Templeton, *Organometallics*, 1991, **10**, 3801.
290 D. Kwon, J. Real, M.D. Curtis, A. Rheingold, and B.S. Haggerty, *Organometallics*, 1991, **10**, 143.
291 K.J. Covert, D.R. Neithamer, M.C. Zonnevylle, R.E. La Point, C.P. Schaller, and P.T. Wolczanski, *Inorg. Chem.*, 1991, **30**, 2494.
292 J.R. Strickler, P.A. Wexler, and D.E. Wigley, *Organometallics*, 1991, **10**, 118.
293 L.K. Cheatham, J.J. Graham, A.W. Apblett, and A.R. Barron, *Polyhedron*, 1991, **10**, 1075.
294 F. Bottomley, C.P. Magill, and B. Zhao, *Organometallics*, 1991, **10**, 1946.
295 M. Herberhold and M. Schrepfermann, *J. Organomet. Chem.*, 1991, **419**, 85.
296 M.L.H. Green, D. O'Hare, P. Mountford, and J.G. Watkin, *J. Chem. Soc., Dalton Trans.*, 1991, 1705.
297 O.J. Scherer, R. Winter, G. Heckmann, and G. Wolmershäuser, *Angew. Chem., Int. Ed. Engl.*, 1991, **30**, 850.
298 J.W. Freeman and F. Basolo, *Organometallics*, 1991, **10**, 256.
299 P.A. Wexler, D.E. Wigley, J.B. Koerner, and T.A. Albright, *Organometallics*, 1991, **10**, 2319.
300 D. Rehder, F. Suessmilch, W. Priebsch, and M. Fornalczyk, *J. Organomet. Chem.*, 1991, **411**, 357.
301 J.S. Jaswal and A.S. Tracey, *Can. J. Chem.*, 1991, **69**, 1600.
302 D. Rehder, H. Holst, W. Priebsch, and H. Vilter, *J. Inorg. Biochem.*, 1991, **41**, 171.

[$(OC)_5Cr=C(OMe)CH_2CH_2CH_2Re(CO)_5$], ($^{13}C$),[355] [$(OC)_5Cr=C(OMe)CH=CHR$], ($^{13}C$),[356] [$(OC)_5Cr=OCHMeCH_2C=CHPr^n$], ($^{13}C$),[357] [$(OC)_5Cr=C(OEt)CH=C(NMe_2)CMe_2OEt$], ($^{13}C$),[358] [$(OC)_5Cr=OCR^1R^2CHR^3CH_2$], ($^{13}C$),[359] [$(OC)_5M=C(OMe)CH=CHCH=CHCH=CHMe$], (M = Cr, W; ^{13}C),[360] [$(OC)_5Cr=C(OMe)Ar$], (^{13}C),[361] [$(OC)_5Cr=C(SR)Ar$], (^{13}C),[362] [$(OC)_5MSn(\mu\text{-}OBu^t)_3InM(CO)_5$], (M = Cr, Mo; ^{119}Sn),[363] [$(\eta^5\text{-}C_5H_{5\text{-}n}Et_n)Mo(CO)_3Me$], ($^{13}C$),[364] [$(\eta^5\text{-}C_5H_4I)M(CO)_3Me$], (M = Mo, W; ^{13}C),[365] [1,2-{$Me(OC)_3M(\eta^5\text{-}C_5H_4SiMe_2O)$}$_2C_6H_4$], (M = Mo, W; ^{13}C, ^{29}Si),[366] [$(\eta^5;\eta^5\text{-}Bu^t{}_4C_{10}H_4)Mo_2(CO)_6Me_2$], ($^{13}C$),[367] [$(\eta^5\text{-}C_5H_5)_2Mo(CH_2Bu^t)_2$], ($^{13}C$),[368] [$(\eta^5\text{-}C_5H_5)Mo(CO)_3(CH_2CH=NPr^i)$], ($^{13}C$),[369] [$Mo(C_6H_2Me_3\text{-}2,4,6)_2(CH_2PBu^n{}_3)(O)_2$], ($^{13}C$),[370] [(4,4'-$Me_2$-bipy)$_2Mo(O)_2(CH_2C_6H_4Me\text{-}4)_2$], ($^{13}C$),[371] [(bipy)$Mo(O)_2(CH_2C_6H_4Me\text{-}2)_2$], ($^{13}C$),[372] (20), ($^{13}C$),[373] [$(\eta^5\text{-}C_5H_5)Mo(CO)_2(PPh_3)CHMeOMe$], ($^{13}C$),[374] [$(\eta^5\text{-}C_5Me_5)$-$M(CO)_n(C_6H_3\text{-}2\text{-}R\text{-}6\text{-}N=NC_6H_4R\text{-}4)$], {$M(CO)_n$ = $Mo(CO)_2$, Fe(CO); ^{13}C},[375] (21), (^{13}C),[376] [$Mo(CHBu^t)(NC_6H_3Pr^i{}_2\text{-}2,6)${$OC(CF_3)_2$}$_2(PMe_3)$], ($^{13}C$),[377] [Mo(CCHR){$P(OMe)_3$}$_2(\eta^5\text{-}C_5H_5)$]$^-$, ($^{13}C$),[378] [$(\eta^5\text{-}C_5H_5)M(=CCH_2CH_2CH_2NMe)${C(O)Ph}$(CO)_2$], (M = Mo, W; ^{13}C),[379] [$(OC)_5Mo=C(OMe)C_6H_4Me\text{-}4$], ($^{13}C$),[380] [$(\eta^5\text{-}C_5H_5)M${=C(OEt)R}$(CO)_2$], (M = Mo, W; ^{13}C),[381] [$(\eta^5\text{-}C_5H_5)Mo(CO)_3(=CF_2)$], ($^{13}C$),[382] [{$HB(3,5\text{-}Me_2C_3N_2)_3$}$M(\equiv CH)(CO)_2$], (M = Mo, W; ^{13}C),[383] [$(\eta^5\text{-}C_5H_5)(OC)${$(PhO)_3P$}$M\equiv CPh$], (M = Mo, W; ^{13}C),[384] [{HB(3,5-

[355] J. Breimair, T. Weidmann, B. Wagner, and W. Beck, *Chem. Ber.*, 1991, **124**, 2431.
[356] A. Wienand and H.-U. Reißig, *Chem. Ber.*, 1991, **124**, 957.
[357] L. Lattuada, E. Licandro, S. Maiorana, and A. Papagni, *J. Chem. Soc., Chem. Commun.*, 1991, 437.
[358] F. Stein, M. Duetsch, R. Lackmann, M. Noltemeyer, and A. de Meijere, *Angew. Chem., Int. Ed. Engl.*, 1991, **30**, 1658.
[359] L. Lattuada, E. Licandro, S. Maiorana, H. Molinari, and A. Papagni, *Organometallics*, 1991, **10**, 807.
[360] H. Le Bozec, C. Cosset, and P.H. Dixneuf, *J. Chem. Soc., Chem. Commun.*, 1991, 881.
[361] M.E. Bos, W.D. Wulff, R.A. Miller, S. Chamberlin, and T.A. Brandvold, *J. Am. Chem. Soc.*, 1991, **113**, 9293.
[362] K.H. Dötz and V. Leue, *J. Organomet. Chem.*, 1991, **407**, 337.
[363] M. Veith and K. Kunze, *Angew. Chem., Int. Ed. Engl.*, 1991, **30**, 95.
[364] D. Stein and H. Sitzmann, *J. Organomet. Chem.*, 1991, **402**, 249.
[365] C. Lo Sterzo, *J. Organomet. Chem.*, 1991, **408**, 253.
[366] H. Plenio, *Chem. Ber.*, 1991, **124**, 2185.
[367] P. Jutzi, J. Schnittger, B. Neumann, and H.-G. Stammler, *J. Organomet. Chem.*, 1991, **410**, C13.
[368] J.M. Atkinson and P.B. Brindley, *J. Organomet. Chem.*, 1991, **411**, 139.
[369] G.-M. Yang, G.-H. Lee, S.-M. Peng, and R.-S. Liu, *Organometallics*, 1991, **10**, 1305.
[370] R. Lai, S. Mabille, A. Croux, and S. Le Bot, *Polyhedron*, 1991, **10**, 463.
[371] W.M. Vetter and A. Sen, *Organometallics*, 1991, **10**, 244.
[372] H. Arzoumanian, H. Krentzien, and H. Teruel, *J. Chem. Soc., Chem. Commun.*, 1991, 55.
[373] G.C. Bazan, J.H. Oskam, H.-N. Cho, L.Y. Park, and R.R. Schrock, *J. Am. Chem. Soc.*, 1991, **113**, 6899.
[374] H. Adams, N.A. Bailey, J.T. Gauntlett, I.M. Harkin, M.J. Winter, and S. Woodward, *J. Chem. Soc., Dalton Trans.*, 1991, 1117.
[375] H. Kisch and D. Garn, *J. Organomet. Chem.*, 1991, **409**, 347.
[376] H.G. Alt, J.S. Han, and H.E. Maisel, *J. Organomet. Chem.*, 1991, **409**, 197.
[377] R.R. Schrock, W.E. Crowe, G.C. Bazan, M.D. Mare, M.B. O'Regan, and M.H. Schofield, *Organometallics*, 1991, **10**, 1832.
[378] R.G. Beevor, M.J. Freeman, M. Green, C.E. Morton, and A.G. Orpen, *J. Chem. Soc., Dalton Trans.*, 1991, 3021.
[379] H. Adams, N.A. Bailey, C.E. Tattershall, and M.J. Winter, *J. Chem. Soc., Chem. Commun.*, 1991, 912.
[380] K.H. Dötz and H. Larbig, *J. Organomet. Chem.*, 1991, **405**, C38.
[381] H. Adams, N.A. Bailey, G.W. Bentley, G. Hough, M.J. Winter, and S. Woodward, *J. Chem. Soc., Dalton Trans.*, 1991, 749.
[382] J.D. Koola and D.M. Roddick, *Organometallics*, 1991, **10**, 591.
[383] G.M. Jamison, A.E. Bruce, P.S. White, and J.L. Templeton, *J. Am. Chem. Soc.*, 1991, **113**, 5057.
[384] J.D. Carter, K.B. Kingsbury, A. Wilde, T.K. Schoch, C.J. Leep, E.K. Pham, and L. McElwee-White, *J. Am. Chem. Soc.*, 1991, **113**, 2947.

$Me_2C_3N_2)_3\}Mo(CO)_2\{\equiv CFe(\eta^5\text{-}C_5H_5)(CNBu^t)_2\}]$, ($^{13}C$),[385] *trans*-[$Mo(CNH_2)Cl(dppe)_2$], ($^{13}C$),[386] [$Mo\{\equiv CN(Bu^t)(SiMe_2Bu^t)\}(CNBu^t)_5]^+$, ($^{13}C$),[387] [$(\eta^5\text{-}C_5H_5)Mo(CO)_3InCl(THF)$], ($^{13}C$),[388] [$(\eta^5\text{-}C_5Me_5)WMe_3(NLi)$], ($^6Li$, ^7Li, ^{13}C, ^{15}N),[389] [$(\eta^5\text{-}C_5H_5)Co\{P(O)(OEt)_2\}_3$-$W(CO)_3Me$], ($^{13}C$),[390] [$WMe(acac)(CO)_2(PMe_3)_2$], ($^{13}C$),[391] [$(\eta^5\text{-}C_5H_5)W(CO)_2Me(NCMe)_2]^{2+}$, ($^{13}C$),[392] [$(\eta^5\text{-}C_5H_4Bu^t)(\eta^3\text{-}C_3H_5)_2(\eta^1\text{-}C_3H_5)W$], ($^{13}C$),[393] [$(\eta^5\text{-}C_5H_5)W(CO)_3CH_2$-$C\{COFe(CO)_2CO\}CCH{=}CH_2$], ($^{13}C$),[394] [$\{(\eta^5\text{-}C_5H_5)W(CO)_2P(OMe)_3\}_2CH_2$], ($^{13}C$),[395] [$RW(CO)_5]^-$, ($^{13}C$),[396] [$(\eta^5\text{-}C_5H_5)_2W(CH{=}CPh_2)(O_2CCHPh_2)$], ($^{13}C$),[397] [$(\eta^5\text{-}C_5H_5)W(MePhP$-$CH_2CH_2PMePh)(CR{=}C{=}O)$], ($^{13}C$),[398] [$(\eta^5\text{-}C_5H_5)W\{C({=}NMe)CMe{=}NEt\}(CO)_3$], ($^{13}C$),[399] [$(\eta^5$-$C_5H_5)W(CO)_2\{CH_2C(O)CH{=}CPh\text{-}\eta^2\}Ni(\eta^5\text{-}C_5Me_5)$], ($^{13}C$),[400] [$\{HB(3,5\text{-}Me_2C_3HN_2)_3\}W({=}O)$-$({=}CHBu^t)Cl$], ($^{13}C$),[401] [$(OC)_5W\{C(NPh{=}CHPh){=}CPh_2\}$], ($^{13}C$),[402] (22), ($^{13}C$),[403]

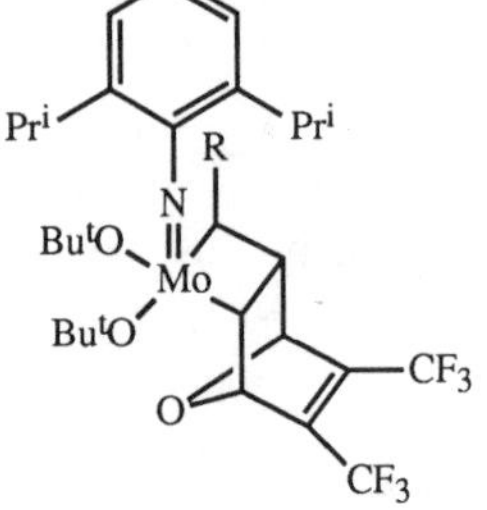

(20)

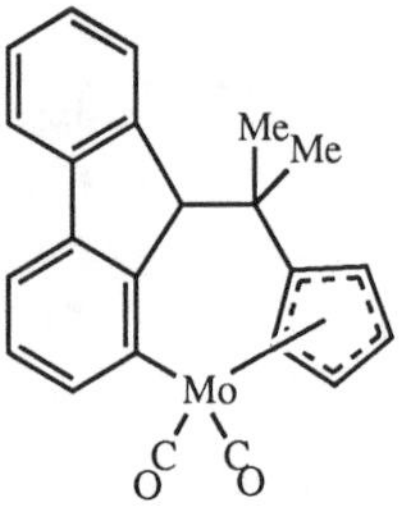

(21)

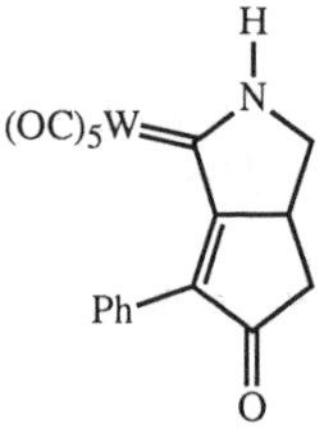

(22)

[$(OC)_5W{=}C(NEt_2)CMe{=}CHCPh{=}CHPh$], ($^{13}C$),[404] [$(OC)_5W\{C(OMe)CH_2CH_2CH{=}C(OMe)$-$SiMe_3\}$], ($^{13}C$),[405] [$(OC)_5W{=}CR^1OR^2$], ($^{13}C$),[406] [$\{HB(3,5\text{-}Me_2C_3HN_2)_3\}W(CO)_2(\equiv CR)$],

385 M. Etienne, P.S. White, and J.L. Templeton, *J. Am. Chem. Soc.*, 1991, **113**, 2324.
386 A. Hills, D.L. Hughes, C.J. Macdonald, M.Y. Mohammed, and C.J. Pickett, *J. Chem. Soc., Dalton Trans.*, 1991, 121.
387 E.M. Carnahan and S.J. Lippard, *J. Chem. Soc., Dalton Trans.*, 1991, 699.
388 L.M. Clarkson, W. Clegg, D.C.R. Hockless, N.C. Norman, L.J. Farrugia, S.G. Bott, and J.L. Atwood, *J. Chem. Soc., Dalton Trans.*, 1991, 2241.
389 T.E. Glassman, M.G. Vale, and R.R. Schrock, *Organometallics*, 1991, **10**, 4046.
390 W. Kläui and H. Hamers, *Z. Anorg. Allg. Chem.*, 1991, **595**, 151.
391 E. Carmona, L. Contreras, E. Gutiérrez-Puebla, A. Monge, and L.J. Sánchez, *Organometallics*, 1991, **10**, 71.
392 V. Skagestad and M. Tilset, *Organometallics*, 1991, **10**, 2110.
393 M.C. Azevedo, T.H. Brock, P.W. Jolly, A. Rufińska, and G. Schroth, *Polyhedron*, 1991, **10**, 459.
394 M.-H. Cheng, G.-H. Lee, S.-M. Peng, and R.-S. Liu, *Organometallics*, 1991, **10**, 3600.
395 M.-C. Chen, Y.-J. Tsai, C.-T. Chen, Y.-C. Lin, T.-W. Tseng, G.-H. Lee, and Y. Wang, *Organometallics*, 1991, **10**, 378.
396 R. Aumann, H. Heinen, M. Dartmann, and B. Krebs, *Chem. Ber.*, 1991, **124**, 2343.
397 G.E. Herberich and K. Linn, *J. Organomet. Chem.*, 1991, **418**, 409.
398 J.C. Jeffery, D.W.I. Sams, and K.D.V. Weerasuria, *J. Organomet. Chem.*, 1991, **418**, 395.
399 A.C. Filippou, C. Völkl, and P. Kiprof, *J. Organomet. Chem.*, 1991, **415**, 375.
400 M.J. Chetcuti, P.E. Fanwick, and B.E. Grant, *Organometallics*, 1991, **10**, 3003.
401 L.L. Blosch, K. Abboud, and J.M. Boncella, *J. Am. Chem. Soc.*, 1991, **113**, 7066.
402 H. Fischer, A. Schlageter, W. Bidell, and A. Früh, *Organometallics*, 1991, **10**, 389.
403 F. Camps, J.M. Moretó, S. Ricart, and J.M. Viñas, *Angew. Chem., Int. Ed. Engl.*, 1991, **30**, 1470.
404 H. Fischer, J. Hofmann, and E. Mauz, *Angew. Chem., Int. Ed. Engl.*, 1991, **30**, 998.
405 D.W. Macomber, P. Madhukar, and R.D. Rogers, *Organometallics*, 1991, **10**, 2121.
406 J.W. Herndon, G. Chatterjee, P.P. Patel, J.J. Matasi, S.U. Turner, J.J. Harp, and M.D. Reid, *J. Am. Chem. Soc.*, 1991, **113**, 7808.

(^{13}C),[407] [(Bu^{t}O)$_3$W≡C(CH$_2$)$_n$C≡W(OBu^t)$_3$], (^{13}C),[408] [(η^5-C$_5$H$_5$)W(CO)L(≡CNEt$_2$)], (^{13}C),[409] [(η^5-C$_5$Me$_5$)W(≡CNEt$_2$)(CNEt)$_2$], (^{13}C),[410] [(η^5-C$_5$Me$_5$)WL(CO)(≡CNEt$_2$)], (^{13}C),[411] and [{HB(3,5-Me$_2$C$_3$HN$_2$)$_3$}W(CO)(≡CPMe$_2$Ph)]$^+$, (^{13}C).[412]

T_1 and n.O.e. values have been determined for [(η^4-norbornadiene)Cr(CO)$_4$] and the chemical shift anisotropy of the carbonyls calculated.[413] ^{19}F shifts in [(η^6-arene)$_2$Cr], where the arene contains CF$_3$ substituents, have been discussed in terms of back-donation.[414] N.m.r. data have also been reported for [W$_2$(OSiBu^{t}Me$_2$)(μ-C$_2$H$_2$)(NC$_5$H$_5$)], (^{13}C),[415] [(η^5-C$_5$H$_5$)WOs$_3$(CO)$_{11}$(μ$_3$-CC$_6$H$_4$Me-4)], (^{13}C),[416] [W$_2$Ru$_2$(CO)$_9${CC(C$_6$H$_4$F-4)CC(C$_6$H$_4$F-4)}(η^5-C$_5$H$_5$)$_2$], (^{13}C),[417] [WOs$_3$(η^5-C$_5$H$_5$)(CO)$_8$(μ$_3$-CPh){C(C$_6$H$_4$Me-4)C(C$_6$H$_4$Me-4)CC(C$_6$H$_4$Me-4)C(C$_6$H$_4$Me-4)}], (^{13}C),[418] [(η^4-norbornadiene)Cr(CO)$_3$(η^2-C$_2$H$_4$)], (^{13}C),[419] (23), (M = Cr, W; ^{13}C),[420] [(η^2-PhC≡CPh)$_2$-Cr(PMe$_3$)$_2$], (^{13}C),[421] [(η^5-C$_5$H$_5$)(η^2-C$_2$H$_4$)Mo{(MeO)$_2$PCH$_2$CH$_2$P(OMe)$_2$}]$^+$, (^{13}C),[422] [Mo(NAr){OCMe(CF$_3$)$_2$}$_2$(^{13}C$_4$H$_8$)], (^{13}C),[423] [(η^5-C$_5$H$_5$)M(SR)(CO)(η^2-PhC≡CPh)], (M = Mo, W; ^{13}C),[424] [(η^5-C$_5$H$_5$)$_2$Mo(η^2-Me$_3$SiC≡CSiMe$_3$)], (^{13}C),[425] [(η^5-C$_5$Me$_5$)Mo(η^2-EtC=NEt)-(CO)$_2$], (^{13}C),[426] [(η^5-C$_5$Me$_5$)Mo(η^2-PhHC=NHEt)(CO)$_2$], (M = Mo, W; ^{13}C),[427] [(η^5-C$_5$H$_5$)$_2$Mo$_2$(μ-η^2-MeCHO)(CO)$_4$], (^{13}C),[428] [(η^5-C$_5$H$_5$)Mo(NO)(η^4-CH$_2$CH=CHCH$_2$COMe$_2$)], (^{13}C),[429] *syn*-[(η^2-CS$_3$)Mo(O)(μ-S)$_2$Mo(O)(η^2-S$_4$]$^{2-}$, (^{13}C),[430] [(η^5-C$_5$H$_5$)Mo{η^2-MeSC(Ph)-PMe$_3$}(CO)(PMe$_3$)], (^{13}C),[431] [Mo$_2$(μ-4-RC$_6$H$_4$C≡CH)(μ-O$_2$CMe)(en)$_4$]$^{3+}$, (^{13}C),[432] [(OC)$_3$Fe(μ-η^2,η^3-RC=C=CH$_2$)M(η^5-C$_5$H$_5$)(CO)$_2$], (M = Mo, W; ^{13}C),[433] [Co$_2$Mo$_2$(η^5-C$_5$H$_5$)$_2$(μ$_3$-S)$_2$(μ$_4$-

407 J.C. Jeffery, F.G.A. Stone, and G.K. Williams, *Polyhedron*, 1991, **10**, 215; P.K. Byers and F.G.A. Stone, *J. Chem. Soc., Dalton Trans.*, 1991, 93.
408 M.H. Chisholm, K. Folting, J.C. Huffman, and E.A. Lucas, *Organometallics*, 1991, **10**, 535.
409 A.C. Filippou, W. Grünleitner, and E.O. Fischer, *J. Organomet. Chem.*, 1991, **401**, C37.
410 A.C. Filippou, W. Grünleitner, C. Völkl, and P. Kiprof, *Angew. Chem., Int. Ed. Engl.*, 1991, **30**, 1167.
411 A.C. Filippou and W. Grünleitner, *J. Organomet. Chem.*, 1991, **407**, 61.
412 G.M. Jamison, P.S. White, and J.L. Templeton, *Organometallics*, 1991, **10**, 1954.
413 A. Gryff-Keller, H. Krawczyk, and P. Szczeciński, *J. Organomet. Chem.*, 1991, **402**, 77.
414 H. Ssekaalo, J.J. Lagowski, and C.M. Seymour, *Inorg. Chim. Acta*, 1991, **183**, 91.
415 M.H. Chisholm, C.M. Cook, J.C. Huffman, and W.E. Streib, *J. Chem. Soc., Dalton Trans.*, 1991, 929.
416 J.T. Park and J.R. Shapley, *Bull. Korean Chem. Soc.*, 1990, **11**, 531 (*Chem. Abstr.*, 1991, **114**, 185 716).
417 D.-K. Hwang, P.-J. Lin, Y. Chi, S.-M. Peng, and G.-H. Lee, *J. Chem. Soc., Dalton Trans.*, 1991, 2161.
418 Y. Chi, H.-F. Hsu, S.-M. Peng, and G.-H. Lee, *J. Chem. Soc., Chem. Commun.*, 1991, 1019.
419 D. Chmielewski, F.-W. Grevels, J. Jacke, and K. Schaffer, *Angew. Chem., Int. Ed. Engl.*, 1991, **30**, 1343.
420 H. Fischer and J. Hofmann, *Chem. Ber.*, 1991, **124**, 981.
421 P.W. Jolly and U. Zakrzewski, *Polyhedron*, 1991, **10**, 1427.
422 S.E. Kegley, D.T. Bergstrom, L.S. Crocker, E.P. Weiss, W.G. Berndt, and A.L. Rheingold, *Organometallics*, 1991, **10**, 567.
423 J. Robbins, G.C. Bazan, J.S. Murdzek, M.B. O'Regan, and R.R. Schrock, *Organometallics*, 1991, **10**, 2902.
424 J.L. Davidson and F. Sence, *J. Organomet. Chem.*, 1991, **409**, 219.
425 G.E. Herberich, U. Englert, and W. Faßbender, *J. Organomet. Chem.*, 1991, **420**, 303.
426 A.C. Filippou, W. Grünleitner, E.O. Fischer, W. Imhof, and G. Huttner, *J. Organomet. Chem.*, 1991, **413**, 165.
427 H. Adams, N.A. Bailey, M.J. Winter, and S. Woodward, *J. Organomet. Chem.*, 1991, **410**, C21.
428 H. Adams, N.A. Bailey, J.T. Gauntlett, M.J. Winter, and S. Woodward, *J. Chem. Soc., Dalton Trans.*, 1991, 2217.
429 N.J. Christensen, P. Legzdins, J. Trotter, and V.C. Yee, *Organometallics*, 1991, **10**, 4021.
430 D. Coucouvanis, A. Toupadakis, J.D. Lane, S.M. Koo, C.G. Kim, and A. Hadjikyriacou, *J. Am. Chem. Soc.*, 1991, **113**, 5271.
431 F.R. Kreißl and C.M. Stegmair, *Chem. Ber.*, 1991, **124**, 2747.
432 M.C. Kerby, B.W. Eichhorn, L. Doviken, and K.P.C. Vollhardt, *Inorg. Chem.*, 1991, **30**, 156.
433 G.H. Young, M.V. Raphael, A. Wojcicki, M. Calligaris, G. Nardin, and N. Bresciani-Pahor, *Organometallics*, 1991, **10**, 1934.

$CF_3C_2Me)(CO)_4]$, (^{13}C),[434] $[(\eta^2\text{-}Ph_2PCH_2CH_2CH_2CH_2CH{=}CH_2)W(CO)_4(HNC_5H_{10})]$, ($^{13}C$),[435] $[(\eta^5\text{-}C_5H_5)W\{1,2\text{-}H_2PC_6H_4PMeC(OMe){=}CHC_6H_4OMe\text{-}4\}(CO)]$, ($^{13}C$),[436] $[(\eta^5\text{-}C_5Me_5)W(\eta^2\text{-}HC{\equiv}CBu^t)_2(CO)]$, ($^{13}C$),[437] $[WF_5(\eta^2\text{-}PhC{\equiv}CC{\equiv}CSiMe_3)]^-$, ($^{13}C$),[438] $[WF_5(\eta^2\text{-}PhC{\equiv}CH)]$, ($^{13}C$),[439] $[W\{\eta^2\text{-}C(O)R\}Cl(CO)(PMe_3)_3]$, ($^{13}C$),[440] $[(\eta^5\text{-}C_5H_5)W(\eta^2\text{-}S{=}CRPMe_3)(CO)_2]$, ($^{13}C$),[441] $[(\eta^5\text{-}C_5H_5)W(\eta^2\text{-}MeSCPh)(CO)(PMe_3)]^+$, ($^{13}C$),[442] $[(OC)_5W\{\eta^2\text{-}PhHC{=}TeW(CO)_5\}]$, ($^{13}C$),[443] $[(\eta^5\text{-}C_5H_4Pr^i)_2W_2Cl_3(\mu\text{-}Cl)(\mu\text{-}EtC{\equiv}CEt)(PMe_3)]$, ($^{13}C$),[444] $[(\eta^5\text{-}C_5H_4Me)CoW(\eta^2\text{-}PhC{\equiv}CH)_2(CO)_4]$, ($^{13}C$),[445] $[W_2Cl_4(\mu\text{-}Cl)_2(\mu\text{-}R^1C{\equiv}CR^2)(THF)_2]$, ($^{13}C$),[446] $[(\eta^3\text{-}C_3H_5)Mo\text{-}(S_2CNEt_2)(CO)_2]$, ($^{13}C$),[447] $[Mo_2(\mu_2\text{-}\eta^3\text{-allyl})_2(acac)_2]$, ($^{13}C$),[448] $[(\eta^5\text{-}C_5H_5)Mo(\eta^3\text{-}1\text{-}C_3H_4\text{-}CO_2Me)(CO)_2]$, ($^{13}C$),[449] $[(\eta^3\text{-}C_7H_7)Mo(CO)(CNBu^t)_4]^+$, ($^{13}C$),[450] $[(\eta^5\text{-}C_5H_5)Mo(\eta^3\text{-cyclopentenonyl})(CO)_2]$, ($^{13}C$),[451] $[(\eta^5\text{-indenyl})(\eta^3\text{-}CH_2CHCHCHO)Mo(CO)_2]$, ($^{13}C$),[452] (24), ($^{13}C$),[453] $[(\eta^5\text{-}C_5H_5)Mo(\eta^3\text{-}H_2CCHNBu^t)(CO)_2]$, ($^{13}C$),[454] $[Mo\{\eta^3\text{-}CH_2C(CONR^1R^2)C{=}CH_2\}\text{-}X(CO)_2(bipy)]$, ($^{13}C$),[455] $[(\eta^5\text{-}C_5H_5)Mo(CO)_2\{Ph_2PCH{=}C(Ph)CO\}]$, ($^{13}C$),[456] $[(\eta^5\text{-}C_5H_5)Ni\{\mu\text{-}\eta^1,\eta^3\text{-}C(CMe_2)CH_2\}M(CO)_2(\eta^5\text{-}C_5H_5)]$, (M = Mo, W; ^{13}C),[457] $[W_2(OBu^t)_6(MeC_3H_3)]$, (^{13}C),[458] $[(\eta^5\text{-}C_5H_5)W\{\eta^3\text{-}PhCH(OEt)\}(CO)_2]$, ($^{13}C$),[459] $[WOs_3(\eta^5\text{-}C_5H_5)(CO)_{10}(CMeCMeCCPh)]$, ($^{13}C$),[460] (25), ($^{13}C$),[461] $[Mo(\eta^4\text{-hexa-1,5-diene})(CO)_4]$, ($^{13}C$),[462] $[(\eta^5\text{-}C_5Me_5)Mo(\eta^4\text{-penta-1,3-diene})(NO)]$, ($^{13}C$),[463] $[(\eta^5\text{-}C_5H_5)(\eta^4\text{-3,4-}Me_2\text{-hexa-2,4-diene})MoBr\{P(OMe)_3\}]$, ($^{13}C$),[464] $[\{\eta^4\text{-}$

434 F. Robin, R. Rumin, F.Y. Pétillon, K. Foley, and K.W. Muir, *J. Organomet. Chem.*, 1991, **418**, C33.
435 I.-H. Wang, P.H. Wermer, C.B. Dobson, and G.R. Dobson, *Inorg. Chim. Acta*, 1991, **183**, 31.
436 J.E. Denison, J.C. Jeffery, S. Harvey, and K.D.V. Weerasuria, *J. Chem. Soc., Dalton Trans.*, 1991, 2677.
437 A.K. McMullen, J.P. Selegue, and J.-G. Wang, *Organometallics*, 1991, **10**, 3421.
438 A. Werth, K. Dehnicke, D. Fenske, and G. Baum, *Z. Anorg. Allg. Chem.*, 1991, **591**, 125.
439 P. Neumann, K. Dehnicke, D. Fenske, and G. Baum, *Z. Naturforsch., B*, 1991, **46**, 999 (*Chem. Abstr.*, 1991, **115**, 280 195).
440 E. Carmona, L. Contreras, M.L. Poveda, L.J. Sánchez, J.L. Atwood, and R.D. Rogers, *Organometallics*, 1991, **10**, 61.
441 F.R. Kreissl, N. Ullrich, A. Wirsing, and U. Thewalt, *Organometallics*, 1991, **10**, 3275.
442 W. Schütt, N. Ullrich, and F.R. Kreißl, *J. Organomet. Chem.*, 1991, **408**, C5.
443 H. Fischer, A. Früh, and C. Troll, *J. Organomet. Chem.*, 1991, **415**, 211.
444 Q. Feng, M. Ferrer, M.L.H. Green, P.C. McGowan, P. Mountford, and V.S.B. Mtetwa, *J. Chem. Soc., Chem. Commun.*, 1991, 552.
445 M.J. Chetcuti, P.E. Fanwick, and J.C. Gordon, *Inorg. Chem.*, 1991, **30**, 4710.
446 S.G. Bott, D.L. Clark, M.L.H. Green, and P. Mountford, *J. Chem. Soc., Dalton Trans.*, 1991, 471.
447 K.-B. Shiu, K.-H. Yih, S.-L. Wang, and F.-L. Liao, *J. Organomet. Chem.*, 1991, **420**, 359.
448 R.J. Blau, M.S. Goetz, and R.J. Tsay, *Polyhedron*, 1991, **10**, 605.
449 W.-J. Vong, S.-M. Peng, S.-H. Lin, W.-J. Lin, and R.-S. Liu, *J. Am. Chem. Soc.*, 1991, **113**, 573.
450 J.R. Hinchliffe, A. Ricalton, and M.W. Whiteley, *Polyhedron*, 1991, **10**, 267.
451 L.S. Liebeskind and A. Bombrun, *J. Am. Chem. Soc.*, 1991, **113**, 8736.
452 S.A. Benyunes, A. Binelli, M. Green, and M.J. Grimshire, *J. Chem. Soc., Dalton Trans.*, 1991, 895.
453 G.-M. Yang, G.-H. Lee, S.-M. Peng, and R.-S. Liu, *J. Chem. Soc., Chem. Commun.*, 1991, 478.
454 E. Carmona, P.J. Daff, A. Monge, P. Palma, M.L. Poveda, and C. Ruiz, *J. Chem. Soc., Chem. Commun.*, 1991, 1503.
455 B.J. Brisdon, R.J. Deeth, A.G.W. Hodson, C.M. Kemp, M.F. Mahon, and K.C. Molloy, *Organometallics*, 1991, **10**, 1107.
456 H. Adams, N.A. Bailey, A.N. Day, M.J. Morris, and M.M. Harrison, *J. Organomet. Chem.*, 1991, **407**, 247.
457 M.J. Chetcuti, P.E. Fanwick, S.R. McDonald, and N.N. Rath, *Organometallics*, 1991, **10**, 1551.
458 S.T. Chacon, M.H. Chisholm, K. Folting, J.C. Huffman, and M.J. Hampden-Smith, *Organometallics*, 1991, **10**, 3722.
459 H. Adams, N.A. Bailey, M.J. Winter, and S. Woodward, *J. Organomet. Chem.*, 1991, **418**, C39.
460 Y. Chi, H.-Y. Hsu, S.-M. Peng, and G.-H. Lee, *J. Chem. Soc., Chem. Commun.*, 1991, 1023.
461 R. Aumann and B. Trentmann, *Chem. Ber.*, 1991, **124**, 2335.
462 R.J. Blau and U. Siriwardane, *Organometallics*, 1991, **10**, 1627.
463 N.J. Christensen, P. Legzdins, F.W.B. Einstein, and R.H. Jones, *Organometallics*, 1991, **10**, 3070.
464 M. Green, M.F. Mahon, K.C. Molloy, C.B. Nation, and C.M. Woolhouse, *J. Chem. Soc., Chem. Commun.*,

R-(+)-pulegone}$_2$M(CO)$_2$], (M = Mo, W; ^{13}C),[465] (26), (^{11}B, ^{13}C),[466] [(η^5-C_5H_5)(η^4-2-MeO_2C-penta1,3-diene)W(CO)$_2$], (^{13}C),[467] [$W_2(OCH_2Bu^t)_6(NC_5H_5)(C_4H_6)$], (^{13}C),[468] [($\eta^5$-$C_5H_5$)($\eta^5$-$C_5H_7$)Cr(CO)], (^{13}C),[469] [(η^5-C_5H_7)Mo(CO)$_2$L$_2$]$^+$, (^{13}C),[470] [{(η^5-C_5H_5)Cr(CO)$_2$}$_2$As$_2$],

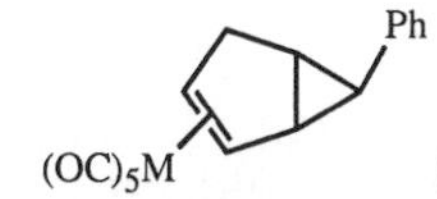
(23)

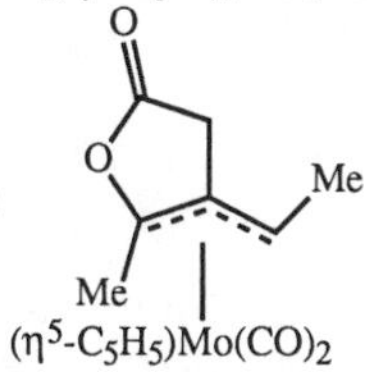
(24)

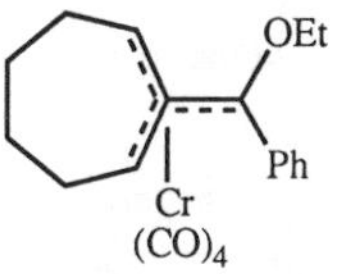
(25)

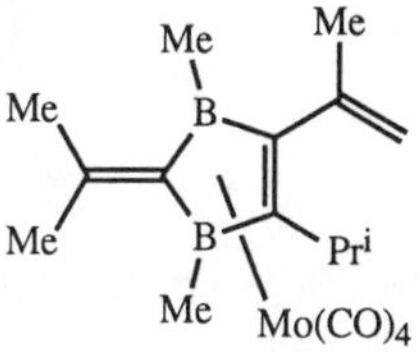
(26)

(^{13}C),[471] [(η^5-C_5H_5)Cr(CO)$_2$(NO)], (^{13}C),[472] [(OC)$_2$(NO)Cr{η^5-C_5H_4C(O)-η^5-C_5H_4}Fe(η^5-C_5H_5)], (^{13}C),[473] [(OC)$_2$(ON)Cr(η^5-$C_5H_4CH_2C_5H_4$-η^5)Fe{η^5-C_5H_4C(O)-η^5-C_5H_4}Cr(CO)$_2$-(NO)], (^{13}C),[474] [(η^5-C_5H_5)Mo(CO)$_3$W(CO)$_3$(η^5-C_5H_5)], (^{13}C),[475] [(η^5-C_5H_5)-Mo(CO)$_3$PCH(SiMe$_3$)C(SiMe$_3$)$_2$], (^{13}C),[476] [(η^5-C_5H_5)Mo(CO)$_2$L$_2$]$^+$, (^{13}C, ^{95}Mo),[477] [(η^5-C_5H_5)Mo(η^2-RPPPh$_2$)(CO)$_2$], (^{13}C),[478] [(η^5-C_5H_5)Mo(CO)$_2${μ-P(C_6H_4Me-4)$_2$}Ir(C_6H_4Me-4)-{P(C_6H_4Me-4)$_3$}(CO)$_2$], (^{13}C),[479] [Pt_2Mo_2(η^5-C_5H_4Me)$_2$(CO)$_6$(PCy$_3$)$_2$], (^{195}Pt),[480] [(η^5-C_5H_5)$_2$M(CO)$_4$(μ-η^2-Bi$_2$)], (M = Mo, W; ^{13}C),[481] [(η^5-1,3-Bu$^t_2$$C_5H_3$)Mo$_2(CO)_6$], (^{13}C),[482] [($\eta^5$-$C_5H_5$)M(CO)Cl$_2$]$_2$, (M = Mo, W; ^{13}C),[483] (27), (^{13}C),[484] [(η^5-C_5H_5)M(=NBu^t)$_2$Cl], (M = Mo, W; ^{13}C),[485] [(η^5-C_5Me_5)Mo(O)$_2$Cl], (^{13}C),[486] [{(η^5-C_5Me_5)Mo}$_2$(S_2CH_2)(μ-S)(μ-SCH$_2$Cl)]$^+$, (^{13}C),[487] [{(η^5-C_5H_5)Mo}$_2$(S_2CH_2){μ-SC(=CH$_2$)CH$_2$SC(CH$_3$)(=CH$_2$)}]$^+$, (^{13}C),[488] (28),

1991, 1587.
465 T. Schmidt, C. Krüger, and P. Betz, *J. Organomet. Chem.*, 1991, **402**, 97.
466 M. Enders, A. Krämer, H. Pritzkow, and W. Siebert, *Angew. Chem., Int. Ed. Engl.*, 1991, **30**, 84.
467 M.-H. Cheng, Y.-H. Ho, G.-H. Lee, S.-M. Peng, and R.-S. Liu, *J. Chem. Soc., Chem. Commun.*, 1991, 697.
468 M.H. Chisholm, J.C. Huffman, E.A. Lucas, and E.B. Lubkovsky, *Organometallics*, 1991, **10**, 3424.
469 J.W. Freeman, N.C. Hallinan, A.M. Arif, R.W. Gedridge, R.D. Ernst, and F. Basolo, *J. Am. Chem. Soc.*, 1991, **113**, 6509.
470 J.R. Hinchliffe and M.W. Whiteley, *J. Organomet. Chem.*, 1991, **402**, C50.
471 L.Y. Goh, R.C.S. Wong, W.-H. Yip, and T.C.W. Mak, *Organometallics*, 1991, **10**, 875.
472 Y.-P. Wang, Y.-H. Yang, S.-L. Wang, and F.-L. Liao, *J. Organomet. Chem.*, 1991, **419**, 325.
473 Y.-P. Wang and J.-M. Hwu, *Tung-hai Hsueh Pao*, 1990, **31**, 493 (*Chem. Abstr.*, 1991, **114**, 102 318).
474 Y.-P. Wang, J.-M. Hwu, S.-L. Wang, and Y.-J. Wu, *J. Organomet. Chem.*, 1991, **414**, 33.
475 F. Calderazzo, A. Juris, R. Poli, and F. Ungari, *Inorg. Chem.*, 1991, **30**, 1274.
476 H. Brombach, E. Niecke, and M. Nieger, *Organometallics*, 1991, **10**, 3949.
477 H. Schumann, J.H. Enemark, M.J. Labarre, M. Bruck, and P. Wexler, *Polyhedron*, 1991, **10**, 665.
478 E. Lindner and M. Heckmann, *Chem. Ber.*, 1991, **124**, 1715.
479 J.M. McFarland, M.R. Churchill, R.F. See, C.H. Lake, and J.D. Atwood, *Organometallics*, 1991, **10**, 3530.
480 P. Braunstein, C. De Meric de Bellefon, S.E. Bouaoud, D. Grandjean, J.F. Halet, and S.Y. Saillard, *J. Am. Chem. Soc.*, 1991, **113**, 5282.
481 W. Clegg, N.A. Compton, R.J. Errington, G.A. Fisher, N.C. Norman, and T.B. Marder, *J. Chem. Soc., Dalton Trans.*, 1991, 2887.
482 M. Scheer, K. Schuster, K. Schenzel, E. Herrmann, and P.G. Jones, *Z. Anorg. Allg. Chem.*, 1991, **600**, 109.
483 N.H. Dryden, P. Legzdins, R.J. Batchelor, and F.W.B. Einstein, *Organometallics*, 1991, **10**, 2077.
484 L.Y. Kuo, M.G. Kanatzidis, M. Sabat, A.L. Tipton, and T.J. Marks, *J. Am. Chem. Soc.*, 1991, **113**, 9027.
485 J. Sundermeyer, *Chem. Ber.*, 1991, **124**, 1977.
486 M.K. Trost and R.G. Bergman, *Organometallics*, 1991, **10**, 1172.
487 M.M. Farmer, R.C. Haltiwanger, F. Kvietok, and M.R. DuBois, *Organometallics*, 1991, **10**, 4066.
488 J. Birnbaum, R.C. Haltiwanger, P. Bernatis, C. Teachout, K. Parker, and M.R. Dubois, *Organometallics*, 1991, **10**, 1779.

(^{13}C),[489] [(η^5-C_5H_5)W(CNEt)$_2$(CO)I], (^{13}C),[490] [(η^5-C_5Me_5)W(=O)$_2$(OCPh$_3$)], (^{13}C),[491] (29),

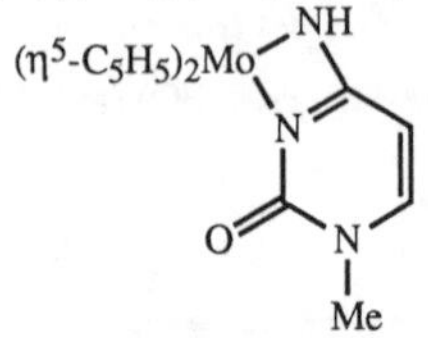

(27)

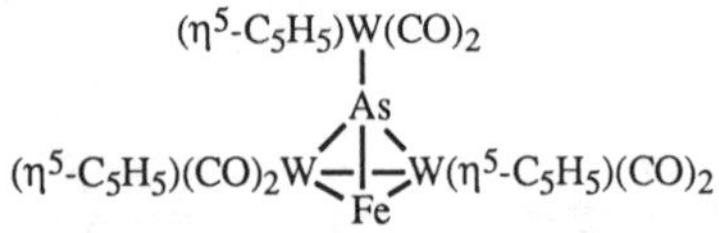

(28)

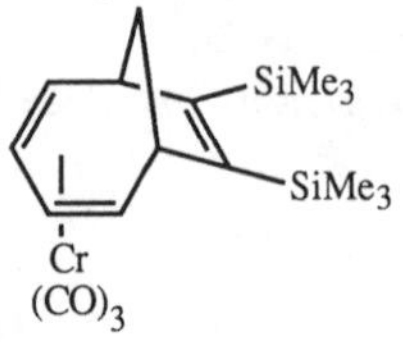

(29)

(^{13}C),[492] [(η^6-C_6H_5R)Cr(CO)$_3$], (^{13}C),[493] [(OC)$_3$Cr(η^6-C_6H_5CH=CHCH=CHC$_6H_5$-η^6)Cr(CO)$_3$], (^{13}C),[494] [(OC)$_3$Cr(η^6-C_6H_5PdL$_2$Cl)], (^{13}C),[495] [$N_3P_3X_5${(η^6-C_6H_5)Cr(CO)$_3$}], (^{13}C),[496] [(η^6-2-FC$_6H_4$CHO)Cr(CO)$_3$], (^{13}C),[497] Cr(CO)$_3$ complexes of tripticene, (^{13}C),[498] (30), (^{13}C),[499] [(η^6-cyclobutabenzene)Cr(CO)$_3$], (^{13}C),[500] [(OC)$_3$Cr(η^6-indenyl-η^5)Rh(CO)$_2$], (^{13}C),[501] [(η^6-1-MeO-naphthalene)Cr(CO)$_3$], (^{13}C),[502] [(η^6-naphthalene)Cr(CO)$_2${PPhOCH(CO$_2$Et)CH(CO$_2$Et)O}],

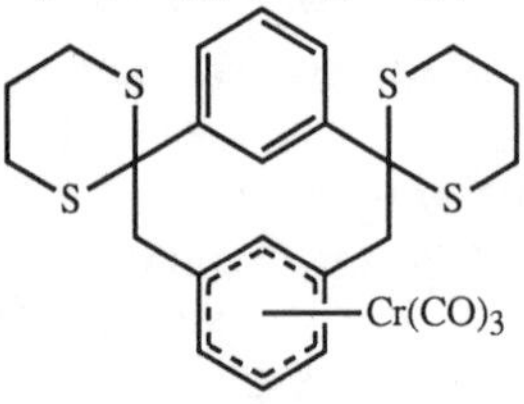

(30)

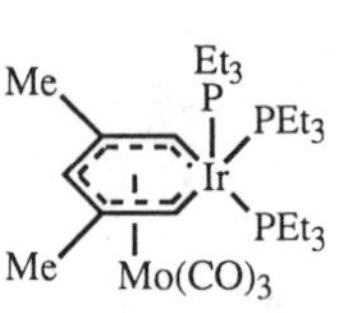

(31)

(^{13}C),[503] [(η^6-1,3,5-Me$_3C_6H_3$)Cr(CO)$_2$(CNCF$_3$)], (^{13}C),[504] [{(OC)$_3$Cr(η^6-2,3-Me$_2$-4-Pr$^i_3$Si-C$_6H_2$)}$_2$], (^{13}C),[505] [(η^6-arene)Cr(CO)$_3$], (^{13}C),[506] [(η^5-C_5H_4COMe)Fe{η^5-C_5H_4C(O)CH$_2$-η^6-

489 C. Caballero, D. Lehne, B. Nuber, and M.L. Ziegler, *Chem. Ber.*, 1991, **124**, 1327.
490 A.C. Filippou, W. Grüleitner, and E.O. Fischer, *J. Organomet. Chem.*, 1991, **411**, C21.
491 M.S. Rau, C.M. Kretz, A. Mercando, G.L. Geoffroy, and A.L. Rheingold, *J. Am. Chem. Soc.*, 1991, **113**, 7420.
492 I. Fischler, F.-W. Grevels, J. Leitich, and S. Özkar, *Chem. Ber.*, 1991, **124**, 2857.
493 H. Rudler, A. Parlier, R. Goumont, J.C. Daran, and J. Vaissermann, *J. Chem. Soc., Chem. Commun.*, 1991, 1075; I.T. Badejo, H. Choi, C.M. Hockensmith, R. Karaman, A.A. Pinkerton, and J.L. Fry, *J. Org. Chem.*, 1991, **56**, 4688.
494 R.D. Rieke, K.P. Daruwala, and M.W. Forkner, *Organometallics*, 1991, **10**, 2946.
495 V. Dufaud, J. Thivolle-Cazat, J.-M. Basset, R. Mathieu, J. Jaud, and J. Waissermann, *Organometallics*, 1991, **10**, 4005.
496 H.R. Allcock, A.A. Dembek, J.L. Bennett, I. Manners, and M. Parvez, *Organometallics*, 1991, **10**, 1865.
497 A. Solladié-Cavallo and M. Bencheqroun, *J. Organomet. Chem.*, 1991, **403**, 159.
498 A. Ceccon, A. Gambaro, F. Manoli, A. Venzo, D. Kuck, T.E. Bitterwolf, P. Ganis, and G. Valle, *J. Chem. Soc., Perkin Trans. 2*, 1991, 233.
499 F. Vögtle, J. Schulz, and M. Nieger, *Chem. Ber.*, 1991, **124**, 1415.
500 H.G. Wey, P. Betz, and H. Butenschon, *Chem. Ber.*, 1991, **124**, 465.
501 A. Ceccon, A. Gambaro, S. Santi, and A. Venzo, *J. Mol. Catal.*, 1991, **69**, L1.
502 E.P. Kündig, M. Inage, and G. Bernardinelli, *Organometallics*, 1991, **10**, 2921.
503 P.C. Nirchio and D.J. Wink, *Organometallics*, 1991, **10**, 336.
504 D. Lentz and R. Marschall, *Z. Anorg. Allg. Chem.*, 1991, **593**, 181.
505 F. Rose-Munch, O. Bellot, L. Mignon, A. Sermra, F. Robert, and Y. Jeannin, *J. Organomet. Chem.*, 1991, **402**, 1.
506 M. Persson, U. Hacksell, and I. Csöregh, *J. Chem. Soc., Perkin Trans. 1*, 1991, 1453; A. Solladié-Cavallo and M. Bencheqroun, *J. Organomet. Chem.*, 1991, **406**, C15; J. Schulz, M. Nieger, and F. Vögtle, *Chem. Ber.*, 1991, **124**, 2797; G. Ganesh and A. Sarkar, *Tetrahedron Lett.*, 1991, **32**, 1085; F. Rose-Munch, K. Aniss, E. Rose, and J. Vaisserman, *J. Organomet. Chem.*, 1991, **415**, 223.

$C_7H_7\}Cr(CO)_3]$, (^{13}C),[507] (31), (^{13}C),[508] and $[(\eta^5\text{-}C_5H_5)M^1(CO)_2M^2(CO)_2(\eta^7\text{-}C_7H_7)]$, ($M^1$ = Fe, Ru; M^2 = Mo, W; ^{13}C).[509]

The ^{31}P n.m.r. spectrum of $[Cr(CO)_3(dppm)(Ph_2PCH_2PPh_2Ag)]^+$ shows $^1J(^{109}Ag\text{-}^{31}P)$.[510] The ^{13}C and ^{31}P n.m.r. spectra of $[Mo(CO)_4(Ph_2PCH{=}CMeCHDPPh_2)]$ show secondary isotope effects.[511] The ^{17}O T_1 of $[W(CO)_6]$ has been measured and used to determine the reorientational diffusion coefficient as a function of temperature.[512] The ^{15}N chemical shift of N_α in $[(OC)_5W{=}NNMe_2]$ is 420.1 which is very deshielded. At low temperature, the ^1H n.m.r. spectrum shows inequivalent methyl groups. The ^{13}C n.m.r. spectrum was also recorded.[513] N.m.r. data have also been reported for $[(OC)_5Cr(CNR)]$, (^{13}C),[514] $[MeC\{CH_2NCM(CO)_5\}_3]$, (M = Cr, W; ^{13}C),[515] $[MMn(CO)_9L]^-$, (M = Cr, W; ^{13}C),[516] $[(OC)_5CrCNCH_2PR^1{}_2R^2]^+$, ($^{13}C$),[517] $[HP\{PPh_2Cr(CO)_5\}_2]$, (^{13}C),[518] (32), (^{13}C),[519] $[(OC)_4M(Ph_2PCH_2)_2C{=}CH_2]$, (M = Cr, Mo, W; ^{13}C, ^{77}Se, ^{95}Mo, ^{183}W),[520] (33), (M = Cr, Mo; ^{13}C),[521] $[\{(Bu^tPhPNMe)_2C{=}O\}M(CO)_4]$, (M = Cr, Mo, W; ^{13}C),[522] $[M(CO)_4(Pr^i{}_2NPO)_4]$, (M = Cr, Mo, W; ^{13}C),[523] (34), (^{13}C),[524] [(1,4,8,11-tetra-aza-cyclotetradeane)$Cr(CO)_3$], (^{13}C),[525] (35), (^{13}C),[526] *cis*-$[Mo(CO)_4\{Ph_2P(CH_2CH_2O)_nCH_2\text{-}CH_2PPh_2\}]$, ($^{13}C$),[527] [{biphenyl-2,2'-$(CH_2CH_2PPh_2)_2\}Mo(CO)_4]$, ($^{13}C$),[528] [{*cis*-1,5-$(Ph_2PCH_2)_2$-3-oxacyclo[3.3.0]octane}$M(CO)_4$], (M = Cr, Mo, W; ^{13}C),[529] $[M(\mu\text{-}I)(CO)_3\{S_2CN(CH_2Ph)_2\}]_2$, ($^{13}C$),[530] *fac*-[M(2-picolinecarboxylate)$(CO)_3(PR_3)]^-$, (^{13}C),[531] [{1,2-$(2\text{-}SC_6H_4S)_2$ethane}$Mo(CO)_3$], (^{13}C),[532] $[MI_2(CO)_3(S_2CPCy_3)]$, (M = Mo, W; ^{13}C),[533] $[ClM^1(\mu\text{-}SO_2)(\mu\text{-}CO)(\mu\text{-}dppm)_2M^2(CO)_2]$, ($M^1$ = Rh, Ir; M^2 = Mo, W; ^{13}C),[534] $[M(SC_6H_2Me_3\text{-}$

507 J. Breimair, M. Wieser, B. Wagner, K. Polborn, and W. Beck, *J. Organomet. Chem.*, 1991, **421**, 55.
508 J.R. Bleeke, Y.-F. Xie, L. Bass, and M.Y. Chiang, *J. Am. Chem. Soc.*, 1991, **113**, 4703.
509 R.L. Beddoes, E.S. Davies, M. Helliwell, and M.W. Whiteley, *J. Organomet. Chem.*, 1991, **421**, 285.
510 R.N. Bagchi, A.M. Bond, R. Colton, I. Creece, K. McGregor, and T. Whyte, *Organometallics*, 1991, **10**, 2611.
511 B.L. Shaw and J.D. Vessey, *J. Chem. Soc., Dalton Trans.*, 1991, 3303.
512 S.P. Wang and M. Schwartz, *J. Mol. Liq.*, 1990, **47**, 121 (*Chem. Abstr.*, 1991, **114**, 129 698).
513 H.F. Sleiman, B.A. Arndtsen, and L. McElwee-White, *Organometallics*, 1991, **10**, 541.
514 W.P. Fehlhammer, S. Ahn, and G. Beck, *J. Organomet. Chem.*, 1991, **411**, 181; R.F. Johnston, *Report*, 1989, **Order No. AD-A212875**, 3 pp. Avail. NTIS. From *Gov. Rep. Announce. Index (U.S.)*, 1990, **90**, Abstr. No. 005,427 (*Chem. Abstr.*, 1991, **115**, 148 887).
515 F.E. Hahn and M. Tamm, *J. Organomet. Chem.*, 1991, **410**, C9.
516 Y.K. Park, S.J. Kim, J.H. Kim, I.S. Han, C.H. Lee, and H.S. Choi, *J. Organomet. Chem.*, 1991, **408**, 193.
517 S. Ahn, G. Beck, and W.F. Fehlhammer, *J. Organomet. Chem.*, 1991, **418**, 365.
518 M. Scheer, S. Gremler, E. Herrmann, and P.G. Jones, *J. Organomet. Chem.*, 1991, **414**, 337.
519 K.H. Dötz and A. Rau, *J. Organomet. Chem.*, 1991, **418**, 219.
520 J.L. Bookham and W. McFarlane, *Polyhedron*, 1991, **10**, 2381.
521 L. Weber, U. Sonnenberg, H.-G. Stammler, and B. Neumann, *Z. Anorg. Allg. Chem.*, 1991, **605**, 87.
522 R. Vogt, P.G. Jones, A. Kolbe, and R. Schmutzler, *Angew. Chem., Int. Ed. Engl.*, 1991, **30**, 2705.
523 E.H. Wong, X. Sun, E.J. Gabe, F.L. Lee, and J.-P. Charland, *Organometallics*, 1991, **10**, 3010.
524 P. Le Floch, D. Carmichael, L. Ricard, and F. Mathey, *J. Am. Chem. Soc.*, 1991, **113**, 667.
525 J.-J. Yaouanc, N. LeBris, G. Le Gall, J.-C. Clément, H. Handel, and H. des Abbayes, *J. Chem. Soc., Chem. Commun.*, 1991, 206.
526 D.V. Khasnis, M. Lattman, and U. Siriwardane, *Organometallics*, 1991, **10**, 1326.
527 A. Varshney and G.M. Gray, *Inorg. Chem.*, 1991, **30**, 1748.
528 W.A. Herrmann, C.W. Kohlpaintner, E. Herdtweck, and P. Kiprof, *Inorg. Chem.*, 1991, **30**, 4271.
529 M.R. Mason, Y. Su, R.A. Jacobson, and J.G. Verkade, *Organometallics*, 1991, **10**, 2335.
530 P.K. Baker, S.G. Fraser, and D. ap Kendrick, *J. Chem. Soc., Dalton Trans.*, 1991, 131.
531 A.E. Sanchez-Pelaez and M.F. Perpinan, *J. Organomet. Chem.*, 1991, **405**, 101.
532 D. Sellmann, F. Grasser, and M. Moll, *J. Organomet. Chem.*, 1991, **415**, 367.
533 P.K. Baker and D. ap Kendrick, *J. Organomet. Chem.*, 1991, **411**, 215.
534 O. Heyke, W. Hiller, and I.-P. Lorenz, *Chem. Ber.*, 1991, **124**, 2217.

2,4,6)$_2$(^{13}CO)$_2$(PMe$_2$Ph)$_2$], (M = Mo, W; ^{13}C),[535] [(η^4-norbornadiene)Rh(μ-CO)(μ-Cl)(μ-dppm)-M(CO)$_2$Cl], (M = Mo, W; ^{13}C),[536] [(3,5-Bu^{t}C$_6$H$_2$-1,2-S$_2$)$_2$Mo(CO)(PPh$_3$)], (^{13}C),[537] [(CH$_2$=C=CHPH$_2$)W(CO)$_5$], (^{13}C),[538] [W(CO)$_5${P(C$_6$H$_4$SO$_3$-3)$_3$}]$^{3-}$, (^{13}C),[539] [(CH$_2$CH$_2$P-NCy$_2$)W(CO)$_5$], (^{13}C),[540] [{CH$_2$CH$_2$CH$_2$C(Me)=P}W(CO)$_5$], (^{13}C),[541] [(OC)$_5$WPHPhPr^i], (^{13}C),[542] [(4-Me-phosphabenzene-2-PPh$_2$)W(CO)$_5$], (^{13}C),[543] [(CH$_2$CPh=CPhPPh)W(CO)$_5$], (^{13}C),[544] [(OC)$_5$W(η^4-PhPCHCHCHEt)Fe(CO)$_3$], (^{13}C),[545] [(OC)$_5$WP=CMeCH$_2$P(CMe=CH$_2$)-W(CO)$_5$], (^{13}C),[546] [(OC)$_5$WPHCH$_2$P(OH){W(CO)$_5$}CPh=CPh], (^{13}C),[547] [W(CO)$_4${*cyclo*-[PW(CO)$_5$]$_4$}], (^{13}C),[548] [MeCS$_2$W(CO)$_4$]$^-$, (^{13}C),[549] [(η^5-P$_5$)W(CO)$_3$]$^-$, (^{13}C),[550] and [{HB(3,5-Me$_2$C$_3$HN$_2$)$_3$}W(O)(CO)X], (^{13}C).[551]

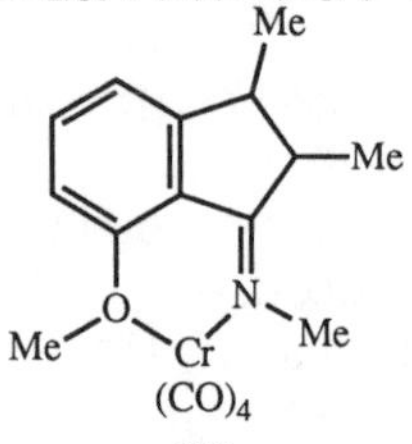
(32)

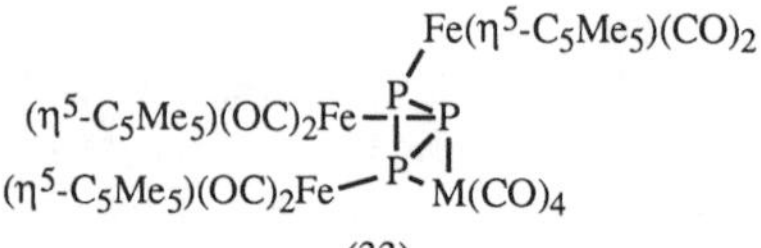
(33)

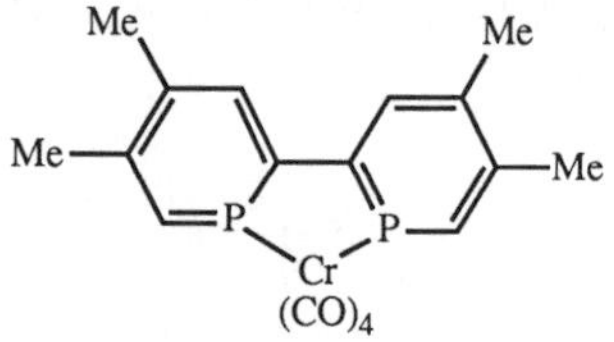
(34)

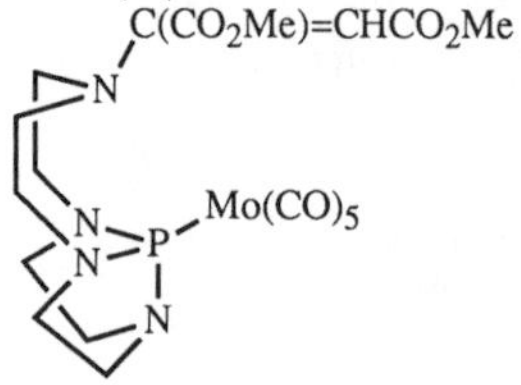
(35)

Ab initio m.o. studies have shown that ^{17}O, ^{33}S, and ^{95}Mo chemical shieldings in [MoO$_n$S$_{4-n}$]$^{2-}$, [Mo(CO)$_6$], and [MoNCl$_4$]$^-$ are dominated by the paramagnetic contribution.[552] All ^{183}W n.m.r. signals and connectivities have been unambiguously assigned using two dimensional COSY and

535 T.E. Burrow, A.J. Lough, R.H. Morris, A. Hills, D.L. Hughes, J.D. Lane, and R.L. Richards, *J. Chem. Soc., Dalton Trans.*, 1991, 2519.
536 M. Cano, J.V. Heras, P. Ovejero, E. Pinilla, and A. Monge, *J. Organomet. Chem.*, 1991, **410**, 101.
537 D. Sellmann, F. Grasser, F. Knoch, and M. Moll, *Angew. Chem., Int. Ed. Engl.*, 1991, **30**, 1311.
538 J.-C. Guillemin, P. Savignac, and J.-M. Denis, *Inorg. Chem.*, 1991, **30**, 2170.
539 D.J. Darensbourg, C.J. Bischoff, and J.H. Reibenspies, *Inorg. Chem.*, 1991, **30**, 1144.
540 F. Mercier, L. Ricard, F. Mathey, and M. Regitz, *J. Chem. Soc., Chem. Commun.*, 1991, 1305.
541 A. Marinetti, P. Le Floch, and F. Mathey, *Organometallics*, 1991, **10**, 1190.
542 R. de Vaumas, A. Marinetti, and F. Mathey, *J. Organomet. Chem.*, 1991, **413**, 411.
543 P. Le Floch, D. Carmichael, and F. Mathey, *Organometallics*, 1991, **10**, 2432.
544 K.M. Doxsee, E.M. Hanawalt, G.S. Shen, T.J.R. Weakley, H. Hope, and C.B. Knobler, *Inorg. Chem.*, 1991, **30**, 3381.
545 A. Marinetti, S. Bauer, L. Ricard, and F. Mathey, *J. Chem. Soc., Dalton Trans.*, 1991, 597.
546 N. Hoa, T. Huy, L. Ricard, and F. Mathey, *Organometallics*, 1991, **10**, 3958.
547 M.L. Sierra, N. Maigrot, C. Charrier, L. Ricard, and F. Mathey, *Organometallics*, 1991, **10**, 2835.
548 M.E. Barr, S.K. Smith, B. Spencer, and L.F. Dahl, *Organometallics*, 1991, **10**, 3983.
549 D.J. Darensbourg, H.P. Wiegreffe, and J.H. Reibenspies, *Organometallics*, 1991, **10**, 6.
550 M. Baudler and T. Etzbach, *Angew. Chem., Int. Ed. Engl.*, 1991, **30**, 580.
551 S.G. Feng, L. Luan, P. White, M.S. Brookhart, J.L. Templeton, and C.G. Young, *Inorg. Chem.*, 1991, **30**, 2582.
552 J.E. Combariza, M. Barfield, and J.H. Enemark, *J. Phys. Chem.*, 1991, **95**, 5463.

INADEQUATE for monosubstituted α_2-$[P_2W_{17}VO_{62}]^{7-}$.[553] N.m.r. data have also been reported for $[\{1,2\text{-}(2\text{-}SC_6H_4S)_2ethane\}Mo(NHNR_2)(NO)]$, ($^{13}C$, ^{14}N, ^{95}Mo),[554] $[[\{1,2\text{-}(2\text{-}SC_6H_4S)_2ethane\}$-$Mo(NO)]_2O]$, ($^{13}C$, ^{14}N),[555] $[WOF_4(H_2NNHR)]$, (^{13}C),[556] $[MN(NPh_2)_3]$, (M = Mo, W; ^{13}C),[557] $[W(NBu^t)_2(OCPh_3)_2]$, (^{13}C),[558] $[Li(THF)_4][W\{N(C_6H_3Pr^i{}_2\text{-}2,6)\}_3Cl]$, ($^{13}C$),[559] $[(Me_3SiO)_2$-$W{=}NCMe_3]$, (^{13}C),[560] [M(5-Bu^t-2-E-pyrimidine)$_4$], (M = Mo, W; E = O, S, Se; ^{13}C),[561] $[Mo_3Pt_2S_4Cl_4(PEt_3)_6]$, ($^{195}Pt$),[562] $[W(S)_2(PMe_3)_4]$, (^{13}C),[563] $[W(O)Cl_2(PMePh_2)(dppe)]$, ($^{13}C$),[564] $[W(Te)_2(PMe_3)_4]$, (^{13}C, ^{125}Te),[565] molybdate and tungstate complexes of perseitol, galactitol, and D-mannitol, (^{13}C),[566] molybdate complexes of alditols and aldoses, (^{95}Mo),[567] $[\{(CySi)_6O_{12}\}_2Mo_2]$, ($^{13}C$, ^{29}Si),[568] $[N_2H_5]_2[Mo_2O_4(C_2O_4)_2(OH_2)_2]$, ($^{13}C$, ^{15}N, ^{17}O, ^{95}Mo),[569] $[W(O)_2Cl_2(MeOCH_2$-$CH_2OMe)]$, (^{13}C),[570] $[PO_4\{M(O)(O_2)_2\}_4]^{3-}$, (M = Mo, W; ^{183}W),[571] $[SiW_{11}O_{40}(SiR_3)_2]^{4-}$, ($^{29}Si$, ^{183}W),[572] α-R-$[SiW_9O_{34}]^{10-}$, (^{29}Si, ^{183}W),[573] $[WM^1{}_3(OH_2)_2(M^2W_9O_{34})_2]^{12-}$, ($M^1$, M^2 = Co, Zn; ^{183}W),[574] α-$[H_2As_2W_{12}O_{48}]^{12-}$, ($^{183}W$),[575] $[Bu^n{}_3NH]_3[PW_{12}O_{40}(DMSO)_3(OH_2)_4]$, ($^{13}C$),[576] α_2-$[P_2W_{17}O_{61}M(OH_2)]^{n-10}$, ($M^{n+}$ = Mn^{3+}, Fe^{3+}, Co^{2+}, Ni^{2+}, Cu^{2+}; ^{183}W),[577] $[(glutathionate)CrO_3]^-$, (^{13}C, ^{17}O),[578] $[MoO(S_2)(S_2CNR_2)_2]$, (^{13}C),[579] $[MoO\{S_2P(OR)_2\}_2(bipy)]$, ($^{13}C$),[580] $[Mo_3S_7(1,2\text{-}O_2C_6H_4)_3]^{2-}$, ($^{13}C$),[581] $[MoOS_3(CuNCS)_n]^{2-}$, (^{13}C, ^{95}Mo),[582] $[W(1,2\text{-}S_2C_6H_4)_2(1\text{-}MeS\text{-}2\text{-}$

553 M. Kozik, R. Acerete, C.F. Hammer, and L.C.W. Baker, *Inorg. Chem.*, 1991, **30**, 4429.
554 D. Sellmann, W. Kern, G. Pöhlmann, F. Knoch, and M. Moll, *Inorg. Chim. Acta*, 1991, **185**, 155.
555 D. Sellmann, B. Seubert, F. Knoch, and M. Moll, *Z. Anorg. Allg. Chem.*, 1991, **600**, 95.
556 S.G. Sakharov, S.A. Zarelus, Yu.V. Kokunov, and Yu.A. Buslaev, *Koord. Khim.*, 1991, **17**, 330 (*Chem. Abstr.*, 1991, **115**, 63 344).
557 Z. Gebeyehu, F. Weller, B. Neumüller, and K. Dehnicke, *Z. Anorg. Allg. Chem.*, 1991, **593**, 99.
558 D.C. Bradley, A.J. Howes, M.B. Hursthouse, and J.D. Runnacles, *Polyhedron*, 1991, **10**, 477.
559 Y.-W. Chao, P.M. Rodgers, D.E. Wigley, S.J. Alexander, and A.L. Rheingold, *J. Am. Chem. Soc.*, 1991, **113**, 6326.
560 D.F. Eppley, P.T. Wolczanski, and G.D. Van Duyne, *Angew. Chem., Int. Ed. Engl.*, 1991, **30**, 584.
561 C.J. Donahue, V.A. Martin, B.A. Schoenfelner, and E.C. Kosinki, *Inorg. Chem.*, 1991, **30**, 1588.
562 T. Saito, T. Tsuboi, Y. Kajitani, T. Yamagata, and H. Imoto, *Inorg. Chem.*, 1991, **30**, 3575.
563 D. Rabinovich and G. Parkin, *J. Am. Chem. Soc.*, 1991, **113**, 5904.
564 S.L. Brock and J.M. Mayer, *Inorg. Chem.*, 1991, **30**, 2138.
565 D. Rabinovich and G. Parkin, *J. Am. Chem. Soc.*, 1991, **113**, 9421.
566 S. Chapelle and J.F. Verchere, *Carbohydr. Res.*, 1991, **221**, 279 (*Chem. Abstr.*, 1991, **115**, 50 105).
567 M. Matulova and V. Bilik, *Chem. Pap.*, 1990, **44**, 703 (*Chem. Abstr.*, 1991, **114**, 102 615).
568 T.A. Budzichowski, S.T. Chacon, M.H. Chisholm, F.J. Feher, and W. Streib, *J. Am. Chem. Soc.*, 1991, **113**, 689.
569 A.N. Startsev, S.A. Shkuropat, O.V. Klimov, M.A. Fedotov, P.E. Kolosov, V.K. Fedorov, S.P. Degtyarev, and D.I. Kochubei, *Koord. Khim.*, 1991, **17**, 229 (*Chem. Abstr.*, 1991, **115**, 20 869).
570 K. Dreisch, C. Andersson, and C. Stålhandske, *Polyhedron*, 1991, **10**, 2417.
571 C. Aubry, G. Chottard, N. Platzer, J.M. Brégeault, R. Thouvenot, F. Chauveau, C. Huet, and H. Ledon, *Inorg. Chem.*, 1991, **30**, 4409.
572 P. Judeinstein, C. Deprun, and L. Nadjo, *J. Chem. Soc., Dalton Trans.*, 1991, 1991.
573 N. Ammari, G. Hervé, and R. Thouvenot, *New J. Chem.*, 1991, **15**, 607.
574 C.M. Tourne, G.F. Tourne, and F. Zonnevijlle, *J. Chem. Soc., Dalton Trans.*, 1991, 143.
575 R. Contant and R. Thouvenot, *Can. J. Chem.*, 1991, **69**, 1488.
576 Y. Gu and J. Ding, *Gaodeng Xuexiao Huaxue Xuebao*, 1990, **11**, 115 (*Chem. Abstr.*, 1991, **114**, 54 744).
577 D.K. Lyon, W.K. Miller, T. Novet, P.J. Domaille, E. Evitt, D.C. Johnson, and R.G. Finke, *J. Am. Chem. Soc.*, 1991, **113**, 7209.
578 S.L. Brauer and K.E. Wetterhahn, *J. Am. Chem. Soc.*, 1991, **113**, 3001.
579 X.F. Yan and C.H. Young, *Aust. J. Chem.*, 1991, **44**, 361.
580 R. Ratnani, G. Srivastava, and R.C. Mehrotra, *Transition Met. Chem. (London)*, 1991, **16**, 204.
581 H. Zimmermann, K. Hegetschweiter, T. Keller, V. Gramlich, H.W. Schmalle, W. Petter, and W. Schneider, *Inorg. Chem.*, 1991, **30**, 4336.
582 A. Beheshti and C.D. Garner, *J. Sci., Islamic Repub. Iran*, 1990, **1**, 270 (*Chem. Abstr.*, 1991, **115**, 196 712).

$SC_6H_4)]^-$, (^{13}C),[583] and [Cl_3W(μ-tetrahydrothiophene)$_2${μ-S(CH_2)$_4$Se}$WCl_3]^-$, (^{77}Se).[584]

Complexes of Mn, Tc, and Re.—On the basis of Fenske-Hall m.o. calculations performed on Mn(I) complexes, the ^{55}Mn chemical shift is dominated by σ_p. The deviation from spherical symmetry of the electron distribution about Mn correlates with the experimental δ(^{55}Mn).[585] The original analysis of ^1H T_1 minima data for [$ReH_5(H_2)${P(C_6H_4Me-4)$_3$}$_2$] gives a H—H bond distance which disagrees with that determined by neutron diffraction. The inclusion of the metal-hydride dipole-dipole contribution produced a better fit of data.[586] ^1H T_1 minima data for [ReH_6{PPh($CH_2CH_2PPh_2)_2$}] gives a H-H distance of 1.17 Å. ^{13}C and ^{31}P n.m.r. spectra were also reported.[587] The variable temperature ^1H T_1 data for [ReH_5{P($OCH_2)_3$CEt}$_{3-n}$($PPh_3)_n$], n = 0 or 1, are consistent with a classical hydride structure. ^{13}C n.m.r. spectra were also reported.[588] The temperature dependence of the ^1H T_1 of the hydride in *mer,trans*-[ReH(CO)$_3$($PPh_3)_2$] has been analysed in order to separate the Re-H and H-H dipole-dipole contributions.[589] N.m.r. data have also been reported for [Mn_2(μ-H)(μ-CyNCHNCy)(CO)$_6$(dppm)], (^{13}C),[590] [Mn_2(CO)$_6$(μ-H){μ-$ClCH_2$SC(S)PCy_3}], (^{13}C),[591] [Tc(η^2-*N,S*-thiourea)H($PMe_3)_4]^+$, (^{13}C),[592] [Re{η^5-2,4-$Me_2C_5H_5$}-$H_2(PPh_3)_2$], (^{13}C),[593] [$H_6Re_4(CO)_{12}]^{2-}$, (^{13}C),[594] [(η^5-$C_5H_5)_2$ReH], (^{13}C),[595] [Re(O)H(η^2-PhC≡CPh)$_2$], (^{13}C),[596] [Re_2(μ-η^1,η^2-C≡CPh)(μ-H)(CO)$_7$(NCMe)], (^{13}C),[597] *cis*-[Mn(CH_2OMe)-(CO)$_4$(PPh_3)], (^{13}C),[598] [(OC)$_5$Mn($CH_2)_3$Mn(CO)$_5$], (^{13}C),[599] [RMn(CO)$_5$], (^{13}C),[600] [(OC)$_5$Mn-CH_2CH=CHCO_2Ph], (^{13}C),[601] [(OC)$_5$MnCH(OSiR^1_3)Me], (^{13}C),[602] *cis, trans*-[Mn(CO)$_2$(CN-2,6-$Me_2C_6H_3$)(COMe)], (^{13}C),[603] [(OC)$_4$Mn{O=C(OEt)C($SnMe_3$)=CCO_2Et}], (^{13}C),[604] [(η^5-C_5H_5)-(OC)$_2$MnPt(μ-C=CHPh)(dppe)], (^{13}C),[605] [(OC)$_4$Fe{μ-CRO}M(CO)$_4$], (M = Mn, Re; ^{13}C),[606] and

583 D. Sellmann, W. Kern, and M. Moll, *J. Chem. Soc., Dalton Trans.*, 1991, 1773.
584 P.M. Boorman, X. Gao, J.F. Fait, and M. Parvez, *Inorg. Chem.*, 1991, **30**, 3886.
585 T.W. Claydon, jun., and B.E. Burston, *New J. Chem.*, 1991, **15**, 713.
586 X.-L. Luo, J.A.K. Howard, and R.H. Crabtree, *Magn. Reson. Chem.*, 1991, **29**, S89.
587 X.-L. Luo and R.H. Crabtree, *J. Chem. Soc., Dalton Trans.*, 1991, 587.
588 X.-L. Luo, D. Michos, and R.H. Crabtree, *Inorg. Chem.*, 1991, **30**, 4286.
589 X.-L. Luo, H. Liu, and R.H. Crabtree, *Inorg. Chem.*, 1991, **30**, 4740.
590 F.J.G. Alonso, M.G. Sanz, and V. Riera, *J. Organomet. Chem.*, 1991, **421**, C12.
591 B. Alvarez, S. García-Granda, Y. Jeannin, D. Miguel, J.A. Miguel, and V. Riera, *Organometallics*, 1991, **10**, 3005.
592 P.L. Watson, J.A. Albanese, J.C. Calabrese, D.W. Ovenall, and R.G. Smith, *Inorg. Chem.*, 1991, **30**, 4638.
593 T.E. Waldman, A.L. Rheingold, and R.D. Ernst, *J. Organomet. Chem.*, 1991, **401**, 331.
594 S.W. Lee and M.G. Richmond, *Inorg. Chem.*, 1991, **30**, 2237.
595 C. Apostolidis, B. Kanellakopulus, R. Maier, J. Rebizant, and M.L. Ziegler, *J. Organomet. Chem.*, 1991, **409**, 243.
596 R.R. Conry and J.M. Mayer, *Organometallics*, 1991, **10**, 3160.
597 S. Top, M. Gunn, G. Jaouen, J. Vaissermann, J.-C. Daran, and J.R. Thornback, *J. Organomet. Chem.*, 1991, **414**, C22.
598 D.H. Gibson, K. Owens, S.K. Mandal, W.E. Sattich, and J.O. Franco, *Organometallics*, 1991, **10**, 1203.
599 E. Lindner and M. Pabel, *J. Organomet. Chem.*, 1991, **414**, C19.
600 E. Lindner, C. Haase, and H.A. Mayer, *Chem. Ber.*, 1991, **124**, 1985.
601 A.P. Masters, J.F. Richardson, and T.S. Sorensen, *Can. J. Chem.*, 1990, **68**, 221.
602 B.T. Gregg, P.K. Hanna, E.J. Crawford, and A.R. Cutler, *J. Am. Chem. Soc.*, 1991, **113**, 384.
603 F.J.G. Alonso, A. Llamazares, V. Riera, M. Vivanco, M.R. Diaz, and S.G. Granda, *J. Chem. Soc., Chem. Commun.*, 1991, 1058.
604 H. Kandler, H.W. Bosch, V. Shklover, and H. Berke, *J. Organomet. Chem.*, 1991, **409**, 233.
605 A.B. Antonova, S.V. Kovalenko, A.A. Johansson, E.D. Korniyets, I.A. Sukhina, A.G. Ginzburg, and P.V. Petrovskii, *Inorg. Chim. Acta*, 1991, **182**, 49.
606 D. Xu, H.D. Kaesz, and S.I. Khan, *Inorg. Chem.*, 1991, **30**, 1341.

$[(\eta^5\text{-}C_5H_4Me)Mn(=CR^1NR^2{}_2)(CO)_2]$, ($^{13}C$).[607]

Axial and equatorial isomers of $[M^1M^2(CO)_9PR_3]$ have been unambiguously identified by ^{31}P n.m.r. spectroscopy.[608] The ^{29}Si-n.m.r. signal of $[(OC)_4Mn(SiMe_2)_2OMe]$ is at remarkably high frequency, δ 115.4. The ^{13}C n.m.r. spectrum was also reported.[609] The ^{13}C n.m.r. spectrum of $[FeMn(\eta^5\text{-}C_5H_5)(CO)_7]$ is a singlet at 200 K.[610] The ^{13}C n.m.r. spectrum of $[Re_6{}^{13}C(^{13}CO)_{19}]^{2-}$ shows complete CO scrambling, even at -85°C.[611] N.m.r. data have also been reported for $[(\eta^5\text{-}C_5H_5)Re(CO)_2(\eta^2\text{-}trans\text{-}MeCH{=}CHMe)]$, ($^{13}C$),[612] $[(\eta^5\text{-}C_5Me_5)Re(\eta^2\text{-benzothiophene})(CO)_2]$, ($^{13}C$),[613] $[(\eta^5\text{-}C_5H_5)Re(NO)(PPh_3)(ClC_6H_5)]^+$, ($^{13}C$),[614] $[Mn(\eta^3\text{-}NR_2CHCHCHEt)(CO)_3]$, ($^{13}C$),[615] $[Re(NAr)(C_3R_3)\{OC(CF_3)_2Me\}_2]$, ($^{13}C$),[616] $[MeRe(O)_2(9,10\text{-}O_2C_{14}H_8)(NC_5H_5)]$, ($^{13}C$),[617] $[MeRe(OCMe_2CMe_2O)(O)_2]$, ($^{13}C$),[618] $[Me_4(O)_2Re_2(\mu\text{-}O)_2]$, ($^{13}C$),[619] $[RhReMe(CO)_4(dppm)_2]^+$, (^{13}C),[620] $[(\eta^5\text{-}C_5H_5)Re(CH_2CH_2CH_2CH_2)(CO)_2]$, ($^{13}C$),[621] $[Re_2(\mu\text{-}O)(\mu\text{-}S)O_2(CH_2Bu^t)_4]$, ($^{13}C$),[622] *mer,trans*-$[Re(CO)_3(PPh_3)_2(CH_2NHCH_2Ph)]$, ($^{13}C$),[623] $[(\eta^5\text{-}C_5H_5)_2ReR]$, ($^{13}C$),[624] $[RReO_3]$, (^{13}C, ^{17}O),[625] $[(OC)_3Fe(\mu\text{-}\eta^4\text{:}\eta^1\text{-}C_6H_7)Re(CO)_5]$, ($^{13}C$),[626] $[(\eta^5\text{-}C_5H_5)Re\{C{=}CH(CH_2)_nCH_2\}(NO)(PPh_3)]$, ($^{13}C$),[627] $[(\eta^5\text{-}C_5Me_5)Re(NO)(PPh_3)C(=O)CH{=}C(OH)M(CO)_4]$, (M = Mn, Re; ^{13}C),[628] $[1,4\text{-}\{(OC)_5Re\}_2C_6H_4]$, ($^{13}C$),[629] $[(\eta^5\text{-}C_5H_4Me)Re(NO)(PPh_3)(C{\equiv}CH)]$, ($^{13}C$),[630] $[(OC)_4(Ph_3P)Re(\mu\text{-}\eta^1\text{:}\eta^1\text{:}\eta^2\text{-}C{\equiv}C)Pt(CO)(PPh_3)Re(CO)_4]$, ($^{13}C$, ^{195}Pt),[631] $[Re(=CHBu^t)(NC_6H_3Pr^i{}_2\text{-}2,6)(O\text{-}2,6\text{-}C_6H_3Cl_2)_2]$, ($^{13}C$),[632] *cis*-$[(\eta^5\text{-}C_5Me_5)(OC)Cl_2Re{=}C(OH)Et]$, ($^{13}C$),[633] $[(\eta^5\text{-}C_6H_6PEt_3)Mn(CO)_3]^+$, ($^{13}C$),[634] $[(\eta^5\text{-}C_6H_6)_2Mn_2(CO)_6]^{2-}$, ($^{13}C$),[635] $[(\eta^5\text{-}$

607 U. Kirchgässner, H. Piana, and U. Schubert, *J. Am. Chem. Soc.*, 1991, **113**, 2228.
608 W.L. Ingham, A.E. Leins, and N.J. Coville, *S. Afr. J. Chem.*, 1991, **44**, 6 (*Chem. Abstr.*, 1991, **115**, 104 669).
609 T. Takeuchi, H. Tobita, and H. Ogino, *Organometallics*, 1991, **10**, 835.
610 A.J.M. Caffyn, M.J. Mays, and P.R. Raithby, *J. Chem. Soc., Dalton Trans.*, 1991, 2349.
611 G. Hsu, S.R. Wilson, and J.R. Shapley, *Inorg. Chem.*, 1991, **30**, 3881.
612 C.P. Casey and C.S. Yi, *Organometallics*, 1991, **10**, 33.
613 M.-G. Choi, M.J. Robertson, and R.J. Angelici, *J. Am. Chem. Soc.*, 1991, **113**, 4005.
614 J.J. Kowalczyk, A.M. Arif, and J.A. Gladysz, *Organometallics*, 1991, **10**, 1079.
615 N.N. Zuniga Villarreal, M.A. Paz-Sandoval, P. Joseph-Nathan, and R.O. Esquivel, *Organometallics*, 1991, **10**, 2616.
616 I.A. Weinstock, R.R. Schrock, and W.M. Davis, *J. Am. Chem. Soc.*, 1991, **113**, 135.
617 J. Takacs, M.R. Cook, P. Kiprof, J.G. Kuchler, and W.A. Herrmann, *Organometallics*, 1991, **10**, 316.
618 W.A. Herrmann, P. Watzlowik, and P. Kiprof, *Chem. Ber.*, 1991, **124**, 1101.
619 W.A. Herrmann, C.C. Romao, P. Kiprof, J. Behm, M.R. Cook, and M. Taillefer, *J. Organomet. Chem.*, 1991, **413**, 11.
620 D.M. Antonelli and M. Cowie, *Organometallics*, 1991, **10**, 2550.
621 E. Lindner and W. Wassing, *Organometallics*, 1991, **10**, 1640.
622 S. Cai, D.M. Hoffman, and D.A. Wierda, *Inorg. Chem.*, 1991, **30**, 827.
623 D.H. Gibson and K. Owens, *Organometallics*, 1991, **10**, 1216.
624 D.M. Heinekey and G.L. Gould, *Organometallics*, 1991, **10**, 2977.
625 W.A. Herrmann, C.C. Romao, R.W. Fischer, P. Kiprof, and C. de M. de Bellefon, *Angew. Chem., Int. Ed. Engl.*, 1991, **30**, 185.
626 B. Niemer, J. Breimair, B. Wagner, K. Polborn, and W. Beck, *Chem. Ber.*, 1991, **124**, 2227.
627 J.J. Kowalczyk, A.M. Arif, and J.A. Gladysz, *Chem. Ber.*, 1991, **124**, 729.
628 J.M. O'Connor, R. Uhrhammer, A.L. Rheingold, and D.M. Roddick, *J. Am. Chem. Soc.*, 1991, **113**, 4530.
629 S.F. Mapolie and J.R. Moss, *Polyhedron*, 1991, **10**, 717.
630 J.A. Ramsden, F. Agbossou, D.R. Senn, and J.A. Gladysz, *J. Chem. Soc., Chem. Commun.*, 1991, 1360.
631 T. Weidmann, V. Weinrich, B. Wagner, C. Robl, and W. Beck, *Chem. Ber.*, 1991, **124**, 1363.
632 M.H. Schofield, R.R. Schrock, and L.Y. Park, *Organometallics*, 1991, **10**, 1844.
633 C.P. Casey, H. Sakaba, and T.L. Underiner, *J. Am. Chem. Soc.*, 1991, **113**, 6673.
634 D.A. Brown, W.K. Glass, and K.M. Kreddan, *J. Organomet. Chem.*, 1991, **413**, 233.
635 R.L. Thompson, S.J. Geib, and N.J. Cooper, *J. Am. Chem. Soc.*, 1991, **113**, 8961.

$C_6Me_5CH_2)Mn(CO)_3]$, (^{13}C),[636] [(η^5-H-naphthalene)$Mn(CO)_3$], (^{13}C),[637] [(η^5-C_5H_5)$Mn(CO)_2$-{$PMe_2(CH_2)_nPMe_2$}], (^{13}C),[638] (36), (^{13}C),[639] [(η^5-C_5H_4Me)$Mn(CO)_2${$\overline{P(BNBu^t{}_2)_2P}$}], ($^{11}B$, ^{13}C),[640] [(η^5-$C_5H_4CH_2CHMeNHMe$)$Mn(CO)_2$], (^{13}C),[641] [$(OC)_3Mn$(η^5-$C_5H_4C{\equiv}C$-η^5-C_5H_4)-$Fe(CO)_2CH_2Ph$], (^{13}C),[642] [(η^5-C_5Me_5)$Tc(CO)_3$], (^{99}Tc)[643] [(η^5-C_5Me_5)$_2Tc_2(CO)_3$], (^{13}C),[644] [(η^5-C_5H_5)$Tc(CO)_3$], (^{13}C),[645] [(η^5-C_5H_5)$Re(NO)(PPh_3)(IR)]^+$, (^{13}C),[646] [(η^5-C_5H_5)Re(NO)-(NH_2Ph)(PPh_3)]$^+$, (^{13}C),[647] [(η^5-C_5H_5)Re(NO)(PPh_3)(NH_nMe_{3-n})]$^+$, (^{13}C, ^{15}N),[648] [(η^5-C_5H_5)-Re(NO)(PPh_3)(OCR^1R^2CN)], (^{13}C),[649] [(η^5-C_5H_5)Re(NO)(PPh_3)($ICH_2CH_2Bu^t$)]$^+$, (^{13}C),[650] [(η^5-C_5H_5)Re(NO)(PPh_3){η^2-O=C(CH_2X)$_2$}]$^+$, (^{13}C),[651] [(η^5-C_5H_4Me)ReO_3], (^{13}C),[652] [(η^5-C_5Me_5)-$Re(CO)_2(PPh_3)$], (^{13}C),[653] [(η^5-C_5Me_5)$_2Re_2(CO)_4$], (^{13}C),[654] [(η^5-C_5Me_5)$(OC)_2Re$(μ_2-η^4-thiophene)$Fe(CO)_3$], (^{13}C),[655] (37), (^{13}C, ^{77}Se),[656] [(η^5-C_5Me_5)Re(NO)(PPh_3)X], (^{13}C),[657] [(η^6-C_6Me_6)$Mn(CO)_2(S_2CH)$], (^{13}C),[658] [$C(NMe_2)_3$][$Mn(CO)_5$], (^{13}C),[659] [Mn_3(μ-CO)$_2$(CO)$_{10}$]$^{3-}$, (^{13}C),[660] [$(OC)_3Mn(CNBu^t)(S_2CPCy_3)]^+$, ($^{13}C$),[661] *fac*-[$(OC)_3M$(dppe)(OR)], ($^{13}C$),[662] [$Mn_2$($\mu$-$S_2CPR_3$)($\mu$-Br)$(CO)_5(PEt_3)$], ($^{13}C$),[663] [$Mn_2$($\mu$-$S_2CPR_3$)$(CO)_6$], ($^{13}C$),[664] [Tc($CNCMe_2$-$CO_2Me$)$_6$]$^+$, ($^{99}Tc$),[665] [$Re(CO)_3(S_2CPCy_3)Br$], ($^{13}C$),[666] [$RhRe(CO)_3$($\mu$-$CO_3$)(dppm)$_2$], ($^{13}C$),[667] [(bipy)$Re(CO)_2${$P(OEt)_3$}$_2$]$^+$, ($^{13}C$),[668] and [$Re_7C(CO)_{21}Ir$($\eta^4$-1,5-$C_8H_{12}$)(CO)]$^{2-}$, ($^{13}C$).[669]

636 D.M. LaBrush, D.P. Eyman, N.C. Baenziger, and L.M. Mallis, *Organometallics*, 1991, **10**, 1026.
637 R.L. Thompson, S. Lee, A.L. Rheingold, and N.J. Cooper, *Organometallics*, 1991, **10**, 1657.
638 A.A. Sorensen and G.K. Yank, *J. Am. Chem. Soc.*, 1991, **113**, 7061.
639 A.J. Arduengo, tert., M. Lattman, H.V.R. Dias, J.C. Calabrese, and M. Kline, *J. Am. Chem. Soc.*, 1991, **113**, 1799.
640 G. Linti and H. Nöth, *Z. Anorg. Allg. Chem.*, 1991, **593**, 124.
641 T.-F. Wang, T.-Y. Lee, Y.-S. Wen, and L.-K. Liu, *J. Organomet. Chem.*, 1991, **403**, 353.
642 C. Lo Sterzo and G. Bocelli, *J. Chem. Soc., Dalton Trans.*, 1991, 1881.
643 W.A. Herrmann, R. Alberto, J.C. Bryan, and A.P. Sattelberger, *Chem. Ber.*, 1991, **124**, 1107.
644 K. Raptis, B. Kanellakopulos, B. Nuber, and M.L. Ziegler, *J. Organomet. Chem.*, 1991, **405**, 323.
645 K. Raptis, E. Dornberger, B. Kanellakopulos, B. Nuber, and M.L. Ziegler, *J. Organomet. Chem.*, 1991, **408**, 61 (*Chem. Abstr.*, 1991, **115**, 71 822).
646 A. Igau and J.A. Gladysz, *Polyhedron*, 1991, **10**, 1903.
647 M.A. Dewey, A.M. Arif, and J.A. Gladysz, *J. Chem. Soc., Chem. Commun.*, 1991, 712.
648 M.A. Dewy, D.A. Knight, D.P. Klein, A.M. Arif, and J.A. Gladysz, *Inorg. Chem.*, 1991, **30**, 4995.
649 D.M. Dalton, C.M. Garner, J.M. Fernández and J.A. Gladysz, *J. Org. Chem.*, 1991, **56**, 6823.
650 A. Igau and J.A. Gladysz, *Organometallics*, 1991, **10**, 2327.
651 D.P. Klein, D.M. Dalton, N.Q. Méndez, A.M. Arif, and J.A. Gladysz, *J. Organomet. Chem.*, 1991, **412**, C7.
652 W.A. Herrmann, M. Taillefor, C. de M. de Bellefon, and J. Behm, *Inorg. Chem.*, 1991, **30**, 3247.
653 A.H. Klahn, C. Leiva, K. Mossert, and X. Zhang, *Polyhedron*, 1991, **10**, 1873.
654 C.P. Casey, H. Sakaba, P.N. Hazin, and D.P. Powell, *J. Am. Chem. Soc.*, 1991, **113**, 8165.
655 M.-G. Choi and R.J. Angelici, *Organometallics*, 1991, **10**, 2436.
656 M.-G. Choi and R.J. Angelici, *J. Am. Chem. Soc.*, 1991, **113**, 5651.
657 D.L. Lichtenberger, A. Rai-Chaudhuri, M.J. Seidel, J.A. Gladysz, S.K. Agbossou, A. Igau, and C.H. Winter, *Organometallics*, 1991, **10**, 1355.
658 S.J. Schauer, D.P. Eyman, R.J. Bernhardt, M.A. Wolff, and L.M. Mallis, *Inorg. Chem.*, 1991, **30**, 570.
659 W. Petz and F. Weller, *Z. Naturforsch., B*, 1991, **46**, 297 (*Chem. Abstr.*, 1991, **115**, 20 942).
660 W. Schatz, H.-P. Neumann, B. Nuber, B. Kanellapulos, and M.L. Ziegler, *Chem. Ber.*, 1991, **124**, 453.
661 D. Miguel, V. Riera, and J.A. Miguel, *J. Organomet. Chem.*, 1991, **412**, 127.
662 S.K. Mandal, D.M. Ho, and M. Orchin, *Inorg. Chem.*, 1991, **30**, 2244.
663 D. Miguel, J.A. Pérez-Martínez, V. Riera, and S. García-Granda, *J. Organomet. Chem.*, 1991, **420**, C12.
664 D. Miguel, V. Riera, J.A. Miguel, M. Gómez, and X. Soláns, *Organometallics*, 1991, **10**, 1683.
665 J.F. Kronauge, A. Davison, A.M. Roseberry, C.E. Costello, S. Maleknia, and A.G. Jones, *Inorg. Chem.*, 1991, **30**, 4265.
666 B. Alvarez, D. Miguel, V. Riera, J.A. Miguel, and S. García-Granda, *Organometallics*, 1991, **10**, 384.
667 D.M. Antonelli and M. Cowie, *Organometallics*, 1991, **10**, 2173.
668 C. Pac, S. Kaseda, K. Ishii, and S. Yanagida, *J. Chem. Soc., Chem. Commun.*, 1991, 787.
669 L. Ma, L.P. Szajek, and J.R. Shapley, *Organometallics*, 1991, **10**, 1662.

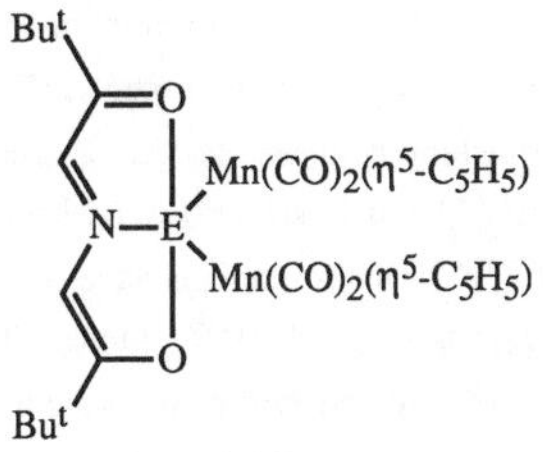

(36)

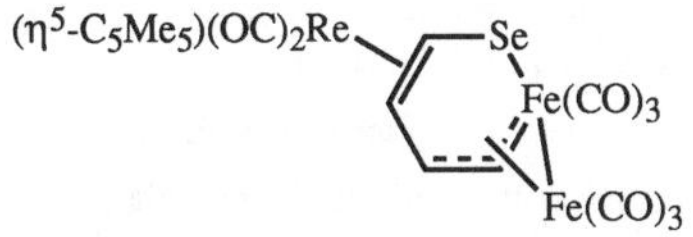

(37)

The sensitivity of a 250 MHz n.m.r. spectrometer for ^{99}Tc has been determined using $[TcO_4]^-$. Over a period of 12 h, a 10^{-7} M concentration can be detected.[670] N.m.r. data have also been reported for $[Mn_7(trien)_2(dien)_2O_4(OAc)_8]^{4+}$, ($^{13}C$),[671] $[Tc(C_8H_5N_2N{=}NH)_3]^+$, ($^{99}Tc$),[672] (38), ($^{13}C$),[673] [TcO-(Schiff base ligand)], (^{13}C),[674] $[HB(pz)_3TcO_3]$, (^{99}Tc),[675] *mer*-$[Tc(NS)Cl_2(PMe_2Ph)_3]$, (^{13}C, ^{99}Tc),[676] $[ORe\{SCH_2CH_2NCH_2CH_2NCMe{=}CHC(O)Me\}]$, ($^{13}C$),[677] *trans*-$[(O)_2Re(3\text{-}ClC_5H_4N)_2(NC_5H_5)_2]^+$, ($^{13}C$),[678] $[Bu^t_2Si(OReO_3)_2]$, and $Bu^t_2SiO_2\{TeCl_2(\mu\text{-}Cl)_2TeCl_2\}$, ($^{13}C$, ^{29}Si, ^{125}Te).[679]

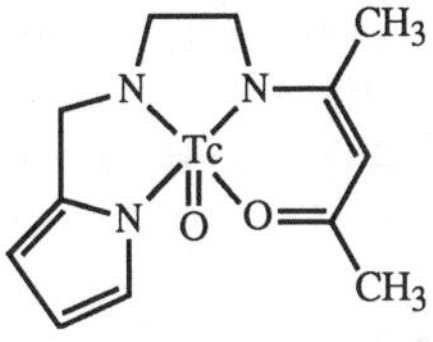

(38)

Complexes of Fe, Ru, and Os.—The structure of $[RuH_4(PPh_3)_3]$ has been studied by n.m.r. spectroscopy. The results indicate rapid oscillations of the hydrogen ligands.[680] The non-classical structure of *cis*-$[P(CH_2CH_2PPh_2)_3RuH(\eta^2\text{-}H_2)]^+$ has been established using variable temperature ^1H and ^{31}P T_1 measurements and J(HD) values.[681] The ^1H n.m.r. spectra of $[(\eta^5\text{-}C_5Me_5)RuH_3(PCy_3)]$ and $[(\{(\eta^5\text{-}C_5Me_5)Ru(PCy_3)\}H(\mu\text{-}H)_2)Cu(\mu\text{-}Cl)]_2$ are AB_2 with temperature dependent coupling constants. ^{13}C n.m.r. spectra were also reported.[682] The hydride ^1H n.m.r. spectrum of

670 M. Findeisen, B. Lorenz, and M. Wahren, *Isotopenpraxis*, 1990, **26**, 520 (*Chem. Abstr.*, 1991, **114**, 54 531).
671 R. Bhula and D.C. Weatherburn, *Angew. Chem., Int. Ed. Engl.*, 1991, **30**, 688.
672 T. Nicholson, A. Mahmood, G. Morgan, A.G. Jones, and A. Davison, *Inorg. Chim. Acta*, 1991, **179**, 53.
673 G.F. Morgan, M. Deblaton, W. Hussein, J.R. Thornback, G. Evrard, F. Durant, J. Stach, U. Abram, and S. Abram, *Inorg. Chim. Acta*, 1991, **190**, 257.
674 S. Liu, S.J. Rettig, and C. Orvig, *Inorg. Chem.*, 1991, **30**, 4915.
675 J.A. Thomas and A. Davidson, *Inorg. Chim. Acta*, 1991, **190**, 231.
676 W. Hiller, R. Hübener, B. Lorenz, L. Kaden, M. Findeisen, J. Stach, and U. Abram, *Inorg. Chim. Acta*, 1991, **181**, 161.
677 U. Abram, S. Abram, J. Stach, R. Wollert, G.F. Morgan, J.R. Thornback, and M. Deblaton, *Z. Anorg. Allg. Chem.*, 1991, **600**, 15.
678 M.S. Ram and J.T. Hupp, *Inorg. Chem.*, 1991, **30**, 130.
679 H.W. Roesky, A. Mazzah, D. Hesse, and M. Noltemeyer, *Chem. Ber.*, 1991, **124**, 519.
680 G.D. Gusev, V.I. Bakhmutov, A.B. Vymenits, and M.E. Vol'pin, *Metalloorg. Khim.*, 1990, **3**, 1432 (*Chem. Abstr.*, 1991, **114**, 113 737); G.D. Gusev, A.B. Vymenits, and V.I. Bakhmutov, *Inorg. Chim. Acta*, 1991, **179**, 195.
681 C. Bianchini, P.J. Perez, M. Peruzzini, F. Zanobini, and A. Vacca, *Inorg. Chem.*, 1991, **30**, 279.
682 T. Arliguie, B. Chaudret, F.A. Jalon, A. Otero, J.A. Lopez, and F.J. Lahoz, *Organometallics*, 1991, **10**, 1888.

$[RuH(H_2)X(PCy_3)_2]$ consists of a broad signal at δ -16.3 with a T_1 minimum of 30 ms.[683] $^1J(^2H\text{-}^1H)$ for *trans*-$[M(H)(\eta^2\text{-}HD)L_2]^+$, M = Fe, Ru, Os, decreases in the order Ru>Fe>Os. ^1H T_1 values were also determined.[684] The T_1 criterion for distinguishing between classical and non-classical hydrides has been assessed using $[MX(\eta^2\text{-}H_2)(dcpe)]^+$,[685] and $[OsH_4\{P(C_6H_4Me\text{-}4)_3\}_3]$.[686] For $[Os(NH_3)_4(H_2)]^{2+}$, $J(^2H\text{-}^1H)$ depends on the solvent.[687] ^1H T_1 measurements have been used to show that the H—H distance in *trans*$[Os(\eta^2\text{-}H_2)H(depe)_2]^+$ is long.[688] ^1H T_1 measurements of hydrides in $[HOs_3(CO)_{10}(O_2CH)]$ and related compounds are dominated by dipole-dipole interaction with other ^1H nuclei in the compound.[689] N.m.r. data have also been reported for $[FeH(dmpe)_2Me]$, (^{13}C),[690] $[FeH(O_2CR)(CO)(PEt_3)_2]$, ($^{13}C$),[691] $[(\eta^5\text{-}C_5H_5)FeH(CO)(MMe_3)_2]$, (M = Si, Sn; ^{13}C),[692] $[HFe_3(CO)_{11}]^-$, (^{13}C),[693] $[HFe_3(CO)_9(C{\equiv}CSiMe_3)]$, ($^{13}C$),[694] $[\{HFe_4(CO)_{12}C\}BXY]$, (^{13}C),[695] $[RuH(\eta^5\text{-}C_8H_{11})_2]^+$, ($^{13}C$),[696] $[RuH(Me_2PCH_2CH_2PMeCH_2CH_2PMeCH_2CH_2PMeC\text{-}H_2)]$, ($^{13}C$),[697] *cis*-$[RuH(OAr)(PMe_3)_4]$, ($^{13}C$),[698] $[(\eta^5\text{-}C_5Me_5)_2Ru_2(\mu\text{-}H)_2(\mu\text{-}O_2CCF_3)_2]$, ($^{13}C$),[699] $[\{\eta^4\text{-}2,5\text{-}Ph_2\text{-}3,4\text{-}(4\text{-}FC_6H_4)_2\text{-cyclopentanone}\}Ru_2H(\mu\text{-}H)(CO)_4]$, ($^{13}C$),[700] $[(dmpm)_2\text{-}Ru_2H_2]$, ($^{13}C$),[701] $[(\eta^5\text{-}C_5Me_5)_2Ru_2(\mu\text{-}D)_2(\mu\text{-}CO)]$, ($^{13}C$),[702] $[H_3Ru_3(COMe)(CO)_7(PPh_2)_3CH]$, ($^{13}C$),[703] $[Ru_3(\mu\text{-}H)(CO)_9(\mu\text{-}MeOCH)(PPh_3)]$, ($^{13}C$),[704] $[Ru_3H_2(CO)_9(\mu_3\text{-}COMe)\{Rh(CO)_2\text{-}(PPh_3)\}]$, ($^{13}C$),[705] $[Ag(PPh_3)_4][\{Ru_3H_2(CO)_9(\mu_3\text{-}COMe)\}_2Ag]$, ($^{13}C$),[706] $[Ru_3Rh_2H_2(CO)_{12}\text{-}(PPh_3)_2]$, ($^{13}C$),[707] $[(\eta^5\text{-}C_5Me_5)Ru(\mu_2\text{-}H)(\mu_2\text{-}SPr^i)\{\eta^2\text{-}\mu_2\text{-}Me_3SiC{\equiv}CC({=}CHSiMe_3)C{\equiv}CSiMe_3\}\text{-}Ru(\eta^5\text{-}C_5Me_5)]$, ($^{13}C$),[708] (39), ($^{13}C$),[709] $[(\mu_2\text{-}H)Ru_3(CO)_9(\mu_3\text{-}1\text{-}4\text{-}\eta^4\text{-dihydrothiophene})]$, ($^{13}C$),[710] $[(\mu\text{-}H)(\mu_3\text{-}\eta^3\text{-}R^1CCR^2CR^3)Ru_3(CO)_9]$, ($^{13}C$),[711] $[Ru_3(\mu\text{-}H)(\mu_3\text{-}2\text{-}NH\text{-}6\text{-}Me\text{-}C_5H_3N)(\mu\text{-}$

683 B. Chaudret, G. Chung, O. Eisenstein, S.A. Jackson, F.J. Lahoz, and J.A. Lopez, *J. Am. Chem. Soc.*, 1991, **113**, 2314.
684 M.T. Bautista, E.P. Cappellani, S.D. Drouin, and R.H. Morris, *J. Am. Chem. Soc.*, 1991, **113**, 4876.
685 A. Mezzetti, A. Del Zotto, P. Rigo, and E. Farnetti, *J. Chem. Soc., Dalton Trans.*, 1991, 1525.
686 P.J. Desrosiers, L. Cai, Z. Lin, R. Richards, and J. Halpern, *J. Am. Chem. Soc.*, 1991, **113**, 4173.
687 Z.W. Li and H. Taube, *J. Am. Chem. Soc.*, 1991, **113**, 8946.
688 K.A. Earl, G. Jia, P.A> Maltby, and R.H. Morris, *J. Am. Chem. Soc.*, 1991, **113**, 3027.
689 S. Aime, M. Cisero, R. Gobetto, D. Osella, and A.J. Arce, *Inorg. Chem.*, 1991, **30**, 1614.
690 L.D. Field, A.V. George, and B.A. Messerle, *J. Chem. Soc., Chem. Commun.*, 1991, 1339.
691 M. Jänicke, H.-U. Hund, and H. Berke, *Chem. Ber.*, 1991, **124**, 719.
692 M. Akita, T. Oku, M. Tanaka, and Y. Moro-oka, *Organometallics*, 1991, **10**, 3080.
693 R.B. King, G.S. Chorghade, N.K. Bhattacharyya, E.M. Holt, and G.J. Long, *J. Organomet. Chem.*, 1991, **411**, 419.
694 J.J. Schneider, M. Nolte, and C. Krüger, *J. Organomet. Chem.*, 1991, **403**, C4.
695 X. Meng, N.P. Rath, T.P. Fehlner, and A.L. Rheingold, *Organometallics*, 1991, **10**, 1986.
696 F. Bouachir, B. Chaudret, F. Dahan, F. Agbossou, and I. Tkatchenko, *Organometallics*, 1991, **10**, 455.
697 L. Dahlenburg, S. Kerstan, and D. Werner, *J. Organomet. Chem.*, 1991, **411**, 457.
698 K. Ohshiro and A. Yamamoto, *Organometallics*, 1991, **10**, 404.
699 H. Suzuki, T. Kakigano, M. Igarashi, M. Tanaka, and Y. Moro-oka, *J. Chem. Soc., Chem. Commun.*, 1991, 283.
700 N. Menashe and Y. Shvo, *Organometallics*, 1991, **10**, 3885.
701 J.F. Hartwig, R.A. Anderson, and R.G. Bergman, *Organometallics*, 1991, **10**, 1710.
702 B.-S. Kang, U. Koelle, and U. Thewalt, *Organometallics*, 1991, **10**, 2569.
703 M.R. Churchill, C.H. Lake, W.G. Feighery, and J.B. Keister, *Organometallics*, 1991, **10**, 2384.
704 J. Evans, P.M. Stroud, and M. Webster, *Acta Crystallogr., Sect. C: Cryst. Struct. Commun.*, 1990, **C46**, 2334 (*Chem. Abstr.*, 1991, **114**, 92 241).
705 J. Evans, P.M. Stroud, and M. Webster, *J. Chem. Soc., Dalton Trans.*, 1991, 2027.
706 J. Evans and P.M. Stroud, *J. Chem. Soc., Dalton Trans.*, 1991, 1351.
707 J. Evans, P.M. Stroud, and M. Webster, *J. Chem. Soc., Dalton Trans.*, 1991, 1017.
708 H. Matsuzaka, Y. Mizobe, M. Nishio, and M. Hidai, *J. Chem. Soc., Chem. Commun.*, 1991, 1011.
709 T. Jenke, H. Stoeckli-Evans, U. Bodensieck, and G. Süss-Fink, *J. Organomet. Chem.*, 1991, **401**, 347.
710 M.-G. Choi, L.M. Daniels, and R.J. Angelici, *Inorg. Chem.*, 1991, **30**, 3647.
711 E. Rosenberg, D.M. Skinner, S. Aime, R. Gobetto, L. Milone, and D. Osella, *Gazz. Chim. Ital.*, 1991, **121**, 313

$PPh_2)_2(CO)_6]$, (^{13}C),[712] $[H_2Ru_6(CO)_{16}(C_6H_4O)]$, ($^{13}C$),[713] $[HRu_4(CO)_{12}B(H)C(Ph)CMeH]$, ($^{11}B$),[714] $[IrOs(H)_2(CO)_3\{\mu_2\text{-}\eta^3\text{-}2\text{-}C_6H_4(Ph)PCH_2PPh_2\}(dppm)]$, ($^{13}C$),[715] $[(\mu\text{-}H)Os_3(\mu\text{-}OCNHCH_2CO_2Et)(NH_2CH_2CO_2Et)(CO)_9]$, ($^{13}C$),[716] and $[Os_9H(CO)_{24}]^-$, (^{13}C).[717]

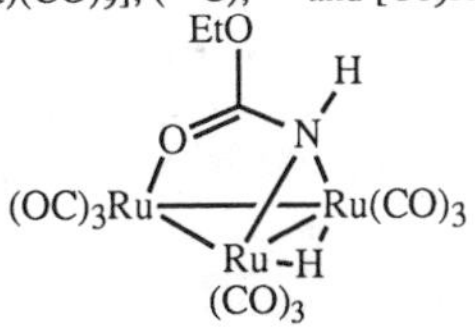

(39)

The ^{29}Si chemical shifts of $[(\eta^5\text{-}C_5H_5)Fe(CO)_2Si_6Me_{11}]$ have been determined. $J(^{29}Si\text{-}^{29}Si)$ was also determined using ^{29}Si INADEQUATE.[718] The ^{119}Sn n.m.r. spectrum of $[(\eta^5\text{-}C_5H_5)(OC)_2Fe\text{-}SnClSnClFe(CO)_2(\eta^5\text{-}C_5H_5)]$ shows $^1J(^{119}Sn\text{-}^{117}Sn)$ of 2981 Hz. The ^{13}C n.m.r. spectrum was also recorded.[719] The ^{13}C n.m.r. spectrum of $[\{(2,4,6\text{-}Me_3C_6H_2)_4porphyrin\}Ru{=}CH{=}CH{=}Ru\text{-}\{(2,4,6\text{-}Me_3C_6H_2)_4porphyrin\}]$ is consistent with carbenoid bonding.[720] The $[AB]_2$ ^{31}P n.m.r. spectrum of $[Ru_2(dmpm)_2(CO)_4\{\mu\text{-}\eta^1\text{:}\eta^1\text{-}C_2(H)_2(CO_2Me)_2\}]$ has been analysed to give $^2J(^{31}P\text{-}^{31}P)$ of 300 Hz. The ^{13}C n.m.r. spectrum was also recorded.[721] N.m.r. data have also been reported for $[FeMe(PMe_3)_3(CO)_2]^+$, (^{13}C),[722] $[(\eta^5\text{-}C_5H_5)Fe(CO)_2CHDCHDCHPhOMe]$, ($^{13}C$),[723] $[(\eta^5\text{-}C_5H_5)M(CO)_2C_6H_{13}]$, (M = Fe, Ru; ^{13}C),[724] $[(\eta^5\text{-}C_5H_5)(OC)_2Fe(CH_2)_nRu(CO)_2(\eta^5\text{-}C_5H_5)]$, ($^{13}C$),[725] $[(\eta^5\text{-}C_5H_5)_2(OC)_2Fe_2(\mu\text{-}CH_2PEt_2)_2]$, ($^{13}C$),[726] $[(1,3\text{–}5\text{–}\eta^4\text{-}C_7H_{10})Fe(CO)_3]$, ($^{13}C$),[727] (40), ($^{13}C$),[728] $[(\eta^5\text{-}C_5H_5)Fe(CO)_2R]$, ($^{13}C$),[729] $[(\eta^5\text{-}C_5H_5)Fe(CO)_2CH(OSiHPh_2)R]$, ($^{13}C$),[730] $[(\eta^5\text{-}C_5H_5)(OC)_2FeCFCHMeCHMe]$, ($^{13}C$),[731] $[(\eta^5\text{-}C_5Me_5)Fe(CO)_2CH{=}CPhC{\equiv}CPh]$, ($^{13}C$),[732]

(*Chem. Abstr.*, 1991, **115**, 208 206).
712 P.L. Andreu, J.A. Cabeza, and V. Riera, *Inorg. Chim. Acta*, 1991, **186**, 225.
713 S. Bhaduri, K. Sharma, H. Khwaja, and P.G. Jones, *J. Organomet. Chem.*, 1991, **412**, 169.
714 S.M. Draper, C.E. Housecroft, A.K. Keep, D.M. Matthews, B.S. Haggerty, and A.L. Rheingold, *J. Organomet. Chem.*, 1991, **410**, C44.
715 R.W. Hilts, R.A. Franchuk, and M. Cowie, *Organometallics*, 1991, **10**, 1297.
716 V.A. Maksakov, V.A. Ershova, A.V. Virovets, Yu.L. Slovokhotov, N.V. Podberezskaya, and Yu.T. Struchkov, *Metalloorg. Khim.*, 1991, **4**, 305 (*Chem. Abstr.*, 1991, **115**, 92 537).
717 A.J. Amoroso, B.F.G. Johnson, J. Lewis, P.R. Raithby, and W.T. Wong, *J. Chem. Soc., Chem. Commun.*, 1991, 814.
718 E. Hengge, M. Eibl, and F. Schrank, *Spectrochim. Acta, Part A*, 1991, **47**, 721 (*Chem. Abstr.*, 1991, **115**, 148 737).
719 Z. Duan, D. Lei, and M.J. Hampden-Smith, *Polyhedron*, 1991, **10**, 2105.
720 N. Rajapakse, B.R. James, and D. Dolphin, *Can. J. Chem.*, 1990, **68**, 2274.
721 K.A. Johnson and W.L. Gladfelter, *Organometallics*, 1991, **10**, 376.
722 J.R. Sowa, jun., V. Zanotti, G. Facchin, and R.J. Angelici, *J. Am. Chem. Soc.*, 1991, **113**, 9185.
723 M. Brookhart and Y. Liu, *J. Am. Chem. Soc.*, 1991, **113**, 939.
724 A. Emeran, M.A. Gafoor, J.K.I. Goslett, Y.H. Liao, L. Pimble, and J.R. Moss, *J. Organomet. Chem.*, 1991, **405**, 237.
725 S.J. Archer, K.P. Finch, H.B. Friedrich, J.R. Moss, and A.M. Crouch, *Inorg. Chim. Acta*, 1991, **182**, 145.
726 D.M. Hester and G.K. Yang, *Organometallics*, 1991, **10**, 369.
727 P. Eilbracht and I. Winkels, *Chem. Ber.*, 1991, **124**, 191.
728 Z. Goldschmidt, E. Genizi, H.E. Gottlieb, H. Berke, H.W. Bosch, and J. Takats, *J. Organomet. Chem.*, 1991, **420**, 419.
729 H.-j. Li and M.M. Turnbull, *J. Organomet. Chem.*, 1991, **419**, 245; B. Giese and G. Thoma, *Helv. Chim. Acta*, 1991, **74**, 1143.
730 M. Akita, O. Mitani, M. Sayama, and Y. Moro-oka, *Organometallics*, 1991, **10**, 1394.
731 T. Omrčen, N.J. Conti, and W.M. Jones, *Organometallics*, 1991, **10**, 913.
732 M. Akita, M. Terada, and Y. Moro-oka, *Organometallics*, 1991, **10**, 2961.

[(η^5-C_5H_5)$(OC)_2$FeC(CH_2OCH_2Ph)=C=CH_2], (^{13}C),[733] [(η^5-C_5Me_5)Fe$(CO)_2${CH=P[Fe$(CO)_4$]-CH($SiMe_3$)$_2$}], (^{13}C),[734] (41), (^{13}C),[735] [$Fe_2(CO)_5$(*P*-dppm){C_2Et_2(CO)C_2Et_2}], (^{13}C),[736] [(η^5-C_5H_5)$(OC)_2$Fe(4-C_5F_4N)], (^{13}C),[737] *trans*-[Fe(C≡CPh)$_2$(dmpe)$_2$], (^{13}C),[738] [(η^5-C_5H_5)Fe$(CO)_2$-(C≡C$SiMe_3$)], (^{13}C),[739] [(η^5-C_5Me_5)Fe$(CO)_2$C≡CFe(η^5-C_5Me_5)$(CO)_2$], (^{13}C),[740] [(η^5-C_5H_5)Fe-(CO)(COMe)(η^2-FC≡CH)], (^{13}C),[741] *R,S,R*-[(η^5-C_5H_5)Fe(CO)(PPh_3)$COCH_2CH$(OH)$C_5H_9O_2$], (^{13}C),[742] [(η^5-C_5H_5)Fe(CO)(PPh_3)C(O)CH_2CH_2C≡CCH_2CH_3], (^{13}C),[743] [{(η^5-C_5H_5)(OC)(R_3P)-MC(O)CH_2CH_2}$_2CH_2$], (M = Fe, Ru; ^{13}C),[744] [{(η^5-C_5H_5)(OC)(R_3P)FeC(O)CH_2}$_2CH_2$], (^{13}C),[745] [(η^5-C_5H_5)$(OC)_2$FeC(O)C(OMe)CH_2-1,2-C_6H_4], (^{13}C),[746] [(η^5-C_5H_5)$(OC)_2$FeC(O)-3-C_5H_4N], (^{13}C),[747] [$(OC)_6Fe_2$(μ-RC=O)$_2$], (^{13}C),[748] [(η^5-C_5H_5)Fe(CO)(PPh_3)(CO_2H)], (^{13}C),[749] [(η^5-C_5H_5)$(OC)_2$Fe=CCMe=CPhCH_2]$^+$, (^{13}C),[750] [Fe(=C=CHCO_2Me)$(CO)_2${P(OMe)$_3$}$_2$],

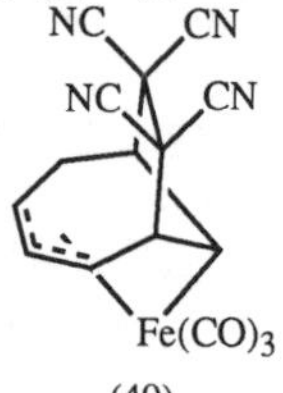

(40)

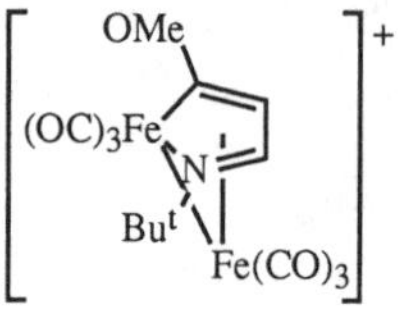

(41)

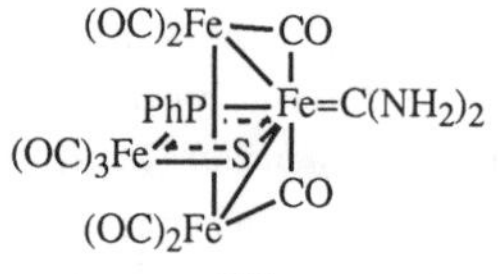

(42)

(^{13}C),[751] [$(OC)_3$Fe=CPhNPhC(CO_2Et)=CMe_2], (^{13}C),[752] (42), (^{13}C),[753] [(η^5-C_5Me_5)Fe(CO)-($PMePh_2$)(CHOMe)]$^+$, (^{13}C),[754] [(η^5-C_5H_5)$_2Fe_2(CO)_2$(μ-CO){μ-C(CN)(OR)}], (^{13}C),[755] [(μ-R^1C≡C)(μ-R^2S)$Fe_2(CO)_6$], (^{13}C),[756] [(η^5-C_5H_5)Fe$(CO)_2SiMe_2GeMe_3$], (^{13}C, ^{29}Si),[757] [(η^5-

733 M.E. Raseta, R.K. Mishra, S.A. Cawood, M.E. Welker, and A.L. Rheingold, *Organometallics*, 1991, **10**, 2936.
734 L. Weber, U. Nolte, P. Jutzi, H.G. Stammler, and B. Neumann, *Chem. Ber.*, 1991, **124**, 989.
735 C.A. Mirkin, K.-L. Lu, T.E. Snead, B.A. Young, G.L. Geoffroy, A.L. Rheingold, and B.S. Haggerty, *J. Am. Chem. Soc.*, 1991, **113**, 3800.
736 R. Giordano, E. Sappa, D. Cauzzi, G. Predieri, A. Tiripicchio, and M.T. Camellini, *J. Organomet. Chem.*, 1991, **412**, C14.
737 R. Chukwu, A.D. Hunter, and B.D. Santarsiero, *Organometallics*, 1991, **10**, 2141.
738 L.D. Field, A.V. George, E.Y. Malouf, I.H.M. Slip, and T.W. Hambley, *Organometallics*, 1991, **10**, 3842.
739 M.P. Gamasa, J. Gimeno, E. Lastra, M. Lanfranchi, and A. Tiripicchio, *J. Organomet. Chem.*, 1991, **405**, 333.
740 M. Akita, M. Terada, S. Oyama, S. Sugimoto, and Y. Moro-oka, *Organometallics*, 1991, **10**, 1561.
741 W. Keim and W. Fischer, *J. Organomet. Chem.*, 1991, **417**, C20.
742 G.J. Bodwell, S.G. Davies, and A.A. Mortlock, *Tetrahedron*, 1991, **47**, 10077.
743 G.J. Bodwell and S.G. Davies, *Tetrahedron Asymmetry*, 1991, **2**, 1075.
744 K.P. Finch, M.A. Gafoor, S.F. Mapolie, and J.R. Moss, *Polyhedron*, 1991, **10**, 963.
745 G.J. Bodwell, S.G. Davies, and S.C. Preston, *J. Organomet. Chem.*, 1991, **402**, C56.
746 S. Tivakornpannarol and W.M. Jones, *Organometallics*, 1991, **10**, 1827.
747 S.G. Davies, A.J. Edwards, R.T. Skerlj, K.H. Sutton, and M. Whittaker, *J. Chem. Soc., Perkin Trans. 1*, 1991, 1027.
748 G. Sundararajan, *Organometallics*, 1991, **10**, 1377.
749 D.H. Gibson, T.-S. Ong, and M. Ye, *Organometallics*, 1991, **10**, 1811.
750 D. Bauer, P. Härter, and E. Herdtweck, *J. Chem. Soc., Chem. Commun.*, 1991, 829.
751 U. Grössmann, H.-U. Hund, H.W. Bosch, H. Schmalle, and H. Berke, *J. Organomet. Chem.*, 1991, **408**, 203.
752 R. Aumann, B. Trentmann, M. Dartmann, and B. Krebs, *Chem. Ber.*, 1991, **124**, 1795.
753 B. Eber, G. Huttner, D. Günauer, W. Imhof, and L. Zsolnai, *J. Organomet. Chem.*, 1991, **414**, 361.
754 M.-J. Tudoret, V. Guerchais, and C. Lapinte, *J. Organomet. Chem.*, 1991, **414**, 373.
755 L. Busetto, M.C. Cassani, V. Zanotti, V.G. Albano, and D. Braga, *J. Organomet. Chem.*, 1991, **415**, 395.
756 D. Seyferth, C.M. Archer, and J.C. Dewan, *Organometallics*, 1991, **10**, 3759.
757 K.H. Pannell and S. Sharma, *Organometallics*, 1991, **10**, 1655.

$C_5H_5)Fe(CO)_2SiMe_2SiMe_2Fe(\eta^5\text{-}C_5H_5)(CO)_2]$, ($^{13}C$, ^{29}Si),[758] $[(Bu^tS)_2Si{=}Fe(CO)_4]$, (^{29}Si),[759] $[Fe_2(\mu\text{-}SiCl_2)_2(CO)_6(PBu^n{}_3)_2]$, ($^{13}C$, ^{29}Si),[760] $[(\eta^5\text{-}C_5H_5)_2Fe(CO)_2\{\mu\text{-}SiBu^t(N_2C_3H_3Me)\}(\mu\text{-}CO)]$, ($^{13}C$, ^{29}Si),[761] (43), (^{13}C, ^{29}Si),[762] $[InCl\{Fe(\eta^5\text{-}C_5H_5)(CO)_2\}_2]$, ($^{13}C$),[763] $[BiCl_2\{Fe(\eta^5\text{-}C_5H_5)(CO)_2\}_2]^-$, ($^{13}C$),[764] $[(Me_3P)_4RuMe\{OC(CMe_2)H\}]$, ($^{13}C$),[765] $[(\eta^5\text{-}C_5H_5)Ru(dppm)S(O)_2\text{-}CH_2X]$, ($^{13}C$),[766] (44), (R = 2,6-$Me_2C_6H_3$; ^{13}C),[767] $[(\eta^5\text{-}C_5Me_5)Ru(CH_2Cl)(NO)Cl]$, ($^{13}C$),[768] $[Ru(CO)_2(CH{=}CHR)Ph(PMe_2Ph)_2]$, ($^{13}C$),[769] $[\{Ru(CMe{=}CHCH{=}NPr^i)(CO)_2(\mu\text{-}X)\}_2]$, ($^{13}C$),[770]

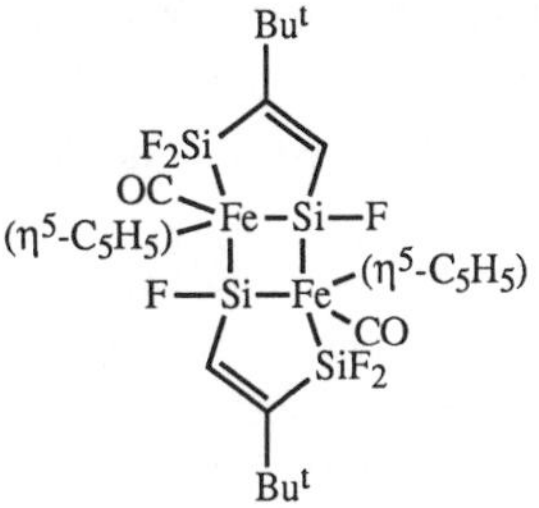

(43)

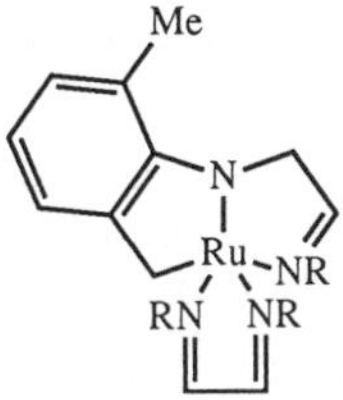

(44)

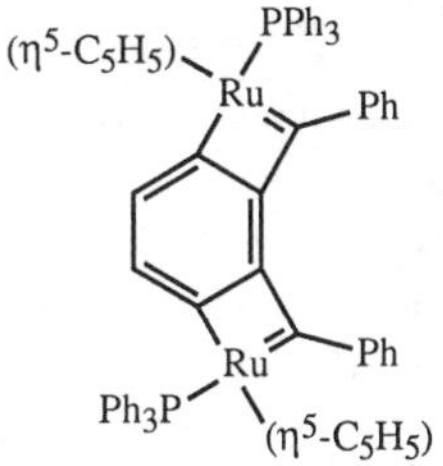

(45)

$[\{HB(pz)_3\}Ru(CO)(PPh_3)C(C{\equiv}CPh){=}CHPh]$, ($^{13}C$),[771] $[\{RuC(CO_2R){=}C(CO_2R)C(CO_2R){=}C\text{-}(CO_2R)\}_2]$, ($^{13}C$),[772] $[Ru\{C_6H_4NHC(O)O\}(PMe_3)_4]$, ($^{13}C$),[773] $[(\eta^6\text{-}C_6Me_6)Ru(C_6H_4CH{=}NPh)\text{-}(PMe_3)]$, ($^{13}C$),[774] (45), ($^{13}C$),[775] $[Ru(C{\equiv}CR)(CO)(PPh_3)_2(NC_5H_5)]^+$, ($^{13}C$),[776] *trans*-$[Ru(C{\equiv}CPh)_2(PMe_3)_4]$, ($^{13}C$),[777] $[(\eta^6\text{-}C_6H_6)Ru(C{\equiv}CPh)Cl(PMe_3)]$, ($^{13}C$),[778] $[(\eta^5\text{-}C_5H_5)Ru(CO)_2\text{-}C{\equiv}CRu(CO)_2(\eta^5\text{-}C_5H_5)]$, ($^{13}C$),[779] $[(\eta^5\text{-}C_5H_5)Ru(C{\equiv}CSnMe_3)(PMe_3)_2]$, ($^{13}C$),[780] $[(\eta^5\text{-}C_5H_5)\text{-}Ru\{C(O)CPh{=}CPhCPh{=}CHPh\}\{P(OMe)_3\}]$, ($^{13}C$),[781] $[(dppe)Ru(CO_2Me)_2(CO)_2]$, ($^{13}C$),[782] $[(\eta^5\text{-}$

758 K.H. Pannell and H. Sharma, *Organometallics*, 1991, **10**, 954.
759 C. Leis, C. Zybill, J. Lachmann, and G. Müller, *Polyhedron*, 1991, **10**, 1163.
760 U. Schubert, M. Knorr, and C. Straßer, *J. Organomet. Chem.*, 1991, **411**, 75.
761 Y. Kawano, H. Tobita, and H. Ogino, *Angew. Chem., Int. Ed. Engl.*, 1991, **30**, 842.
762 K.M. Horng, S.L. Wang, and C.S. Liu, *Organometallics*, 1991, **10**, 631.
763 L.M. Clarkson, N.C. Norman, and L.J. Farrugia, *Organometallics*, 1991, **10**, 1286.
764 W. Clegg, N.A. Compton, R.J. Errington, G.A. Fisher, D.C.R. Hockless, N.C. Norman, and A.G. Orpen, *Polyhedron*, 1991, **10**, 123.
765 J.F. Hartwig, R.G. Bergman, and R.A. Andersen, *Organometallics*, 1991, **10**, 3326.
766 W.A. Schenk and P. Urban, *J. Organomet. Chem.*, 1991, **411**, C27.
767 V. Rosenberger, G. Fendesak, and H. tom Dieck, *J. Organomet. Chem.*, 1991, **411**, 445.
768 J.L. Hubbard, A. Morneau, R.M. Burns, and O.W. Nadeau, *J. Am. Chem. Soc.*, 1991, **113**, 9180.
769 B. Chamberlain and R.J. Mawby, *J. Chem. Soc., Dalton Trans.*, 1991, 2067.
770 W.P. Mul, C.J. Elsevier, M. Van Leijen, and J. Spaans, *Organometallics*, 1991, **10**, 251.
771 N.W. Alcock, A.F. Hill, and R.P. Melling, *Organometallics*, 1991, **10**, 3898.
772 E. Lindner and H. Kühbauch, *J. Organomet. Chem.*, 1991, **403**, C9.
773 J.F. Hartwig, R.G. Bergman, and R.A. Andersen, *J. Am. Chem. Soc.*, 1991, **113**, 6499.
774 G.C. Martin, J.M. Boncella, and E.J. Wucherer, *Organometallics*, 1991, **10**, 2804.
775 M.I. Bruce, G.A. Koutsantonis, M.J. Liddell, and E.R.T. Tiekink, *J. Organomet. Chem.*, 1991, **420**, 253.
776 A.M. Echavarren, J. López, A. Santos, A. Romero, J.A. Hermoso, and A. Vegas, *Organometallics*, 1991, **10**, 2371.
777 S.J. Davies, B.F.G. Johnson, J. Lewis, and P.R. Raithby, *J. Organomet. Chem.*, 1991, **414**, C51.
778 H. Le Bozec, K. Ouzzine, and P.H. Dixneuf, *Organometallics*, 1991, **10**, 2768.
779 G.A. Koutsantonis and J.P. Selegue, *J. Am. Chem. Soc.*, 1991, **113**, 2316.
780 F.R. Lemke, D.J. Szalda, and R.M. Bullock, *J. Am. Chem. Soc.*, 1991, **113**, 8466.
781 M. Crocker, B.J. Dunne, M. Green, and A.G. Orpen, *J. Chem. Soc., Dalton Trans.*, 1991, 1589.
782 J.D. Gargulak, M.D. Noirot, and W.L. Gladfelter, *J. Am. Chem. Soc.*, 1991, **113**, 1054.

$C_5H_5)Ru(=C=CHC_6H_9)(PMe_3)_2]^+$, ($^{13}C$),[783] $[ClRu(=C=C=CR_2)(dppm)_2]^+$, ($^{13}C$),[784] $[\{N(CH_2CH_2PPh_2)_3\}Ru(=C=C=CPh_2)Cl]^+$, ($^{13}C$),[785] $[(\eta^5\text{-}C_5H_5)(Me_3P)_2Ru=C=C(SMe_2)(SMe)]^+$, ($^{13}C$),[786] $[(\eta^6\text{-}C_6Me_6)Ru\{=C(OMe)CH=CMe_2\}Cl(CNR)]$, ($^{13}C$),[787] $[Ru_2(CO)_6\{\mu\text{-}\eta^1\text{:}\eta^1\text{-}PhC=C(PR^1{}_2R^2)CH_2\}(\mu\text{-}PPh_2)]$, ($^{13}C$),[788] $[RuCl_2(HgPh)(NO)\{P(CH_2CH_2CN)_3\}]$, ($^{13}C$),[789] $[Et_4OsO]$, (^{13}C),[790] $[(\eta^6\text{-}1,3,5\text{-}Me_3C_6H_3)OsPr^n(\eta^2\text{-}CH_2=CHMe)Cl]$, ($^{13}C$),[791] $[(\eta^5\text{-}C_5H_5)Os(CH_2SiMe_3)_2(\equiv NBF_3)]$, ($^{13}C$),[792] $[(Me_3SiCH_2)_2Os(O)_2(NC_5H_5)_2]$, ($^{13}C$, ^{17}O),[793] $[(\eta^6\text{-}C_6H_6)Os(CH=CHPh)(PPr^i{}_3)Cl]$, ($^{13}C$),[794] $[Os(C_6F_5)_2(O)_2(NC_5H_5)_2]$, ($^{13}C$, ^{17}O),[795] $[(\eta^6\text{-}1,3,5\text{-}Me_3C_6H_3)Os(=C=CHMe)(PPr^i{}_3)Cl]^+$, ($^{13}C$),[796] $[(\eta^6\text{-}C_6H_6)Os(=C=CHBu^n)(PR_3)I]^+$, ($^{13}C$),[797] $[(\eta^6\text{-}C_6H_6)\overline{Os(=CH_2CH_2CH_2O})(PR_3)I]^+$, ($^{13}C$),[798] and [(tetraphenylporphyrin)Os(=SiR_2)(THF)], (^{29}Si).[799]

The ^1H and ^{13}C n.m.r. spectra of $[Fe(\eta^2\text{-}HCH_2C\equiv CCH_3)(dmpe)_2]^+$ have provided evidence for agostic interaction between the iron and the methyl groups.[800] N.m.r. data have also been reported for $[Fe(\eta^2\text{-}H_2B=CHBRNMe_2)(CO)_4]$, ($^{11}B$, ^{13}C),[801] (46), (^{13}C),[802] $[Fe_2(CO)_6(\mu\text{-}CH=CH_2)(\mu\text{-}CHCCH_2)]$, ($^{13}C$),[803] $[(\eta^5\text{-}C_5H_5)_2Fe_2(CO)_3\{\mu\text{-}C(CN)CNW(CO)_5\}]$, ($^{13}C$),[804] $[Fe_2(CO)_7(Me_3SiC\equiv CNEt_2)]$, ($^{13}C$),[805] $[M_3(CO)_9(\mu_3\text{-}\eta^2\text{-}C\equiv CBu^t)\{\mu_2\text{-}P(C\equiv CBu^t)_2\}]$, (M = Fe, Ru, Os; ^{13}C),[806] $[Fe_2Ir(\mu_3\text{-}C\equiv CPh)(CO)_8(PPh_3)]$, ($^{13}C$),[807] $[Fe_3(CO)_9(\mu_3\text{-}\eta^2\text{-}MeO_2CC\equiv CCO_2Me)]$, ($^{13}C$),[808] $[R^1PFe_3(CO)_9(R^2C\equiv CNEt_2)]$, ($^{13}C$),[809] $[Fe_3(CO)_9(CMe)(Ph_2PC\equiv CR)]^-$, ($^{13}C$),[810] $[(\mu_3\text{-}\eta^2\text{-}C\equiv CBu^t)(CO)_9M^1{}_3(\mu_3\text{-}Hg)M^2(CO)_2(\eta^5\text{-}C_5H_5)]$, ($M^1$ = Ru, Os; M^2 = Fe, Ru; ^{13}C),[811]

783 J.P. Selegue, B.A. Young, and S.L. Logan, *Organometallics*, 1991, **10**, 1972.
784 N. Pirio, D. Touchard, L. Toupet, and P.H. Dixneuf, *J. Chem. Soc., Chem. Commun.*, 1991, 980.
785 A. Wolinska, D. Touchard, P.H. Dixneuf, and A. Romero, *J. Organomet. Chem.*, 1991, **420**, 217.
786 D.C. Miller and R.J. Angelici, *Organometallics*, 1991, **10**, 89.
787 R. Dussel, D. Pilette, P.H. Dixneuf, and W.P. Fehlhammer, *Organometallics*, 1991, **10**, 3287.
788 S.M. Breckenridge, N.J. Taylor, and A.J. Carty, *Organometallics*, 1991, **10**, 837.
789 L. Ballester-Reventos, A. Gutierrez-Alonso, and M.F. Perpiñan-Vielba, *Polyhedron*, 1991, **10**, 1013.
790 K. Rypdal, W.A. Herrmann, S.J. Eder, R.W. Albach, P. Watzlowik, H. Bock, and B. Solouki, *Organometallics*, 1991, **10**, 1331.
791 S. Stahl and H. Werner, *J. Am. Chem. Soc.*, 1991, **113**, 2944.
792 R.W. Marshman, J.M. Shusta, S.R. Wilson, and P.A. Shapley, *Organometallics*, 1991, **10**, 1671.
793 W.A. Herrmann, S.J. Eder, and P. Kiprof, *J. Organomet. Chem.*, 1991, **413**, 27.
794 H. Werner, R. Weinand, W. Knaup, K. Peters, and H.G. Von Schnering, *Organometallics*, 1991, **10**, 3967.
795 W.A. Herrmann, S.J. Eder, and P. Kiprof, *J. Organomet. Chem.*, 1991, **412**, 407.
796 H. Werner, S. Stahl, and W. Kohlmann, *J. Organomet. Chem.*, 1991, **409**, 285.
797 W. Knaup and H. Werner, *J. Organomet. Chem.*, 1991, **411**, 471.
798 H. Werner, W. Knaup, and M. Schulz, *Chem. Ber.*, 1991, **124**, 1121.
799 L.K. Woo, D.A. Smith, and V.G. Young, jun., *Organometallics*, 1991, **10**, 3977.
800 A. Hills, D.L. Hughes, M. Jimenez-Tenorio, G.J. Leigh, C.A. McGeary, A.T. Rowley, M. Bravo, C.E. McKenna, and M.-C. McKenna, *J. Chem. Soc., Chem. Commun.*, 1991, 522.
801 G. Schmid, F. Alraun, and R. Boese, *Chem. Ber.*, 1991, **124**, 2255.
802 N. Sakai, K. Mashima, H. Takaya, R. Yamaguchi, and S. Kozima, *J. Organomet. Chem.*, 1991, **419**, 181.
803 R. Yáñez, J. Ros, and R. Mathieu, *J. Organomet. Chem.*, 1991, **414**, 209.
804 V.G. Albano, S. Bordoni, D. Braga, L. Busetto, A. Palazzi, and V. Zanotti, *Angew. Chem., Int. Ed. Engl.*, 1991, **30**, 847.
805 V. Crocq, J.-C. Daran, Y. Jeanin, B. Eber, and G. Huttner, *Organometallics*, 1991, **10**, 448.
806 B.J. Bobbie, N.J. Taylor, and A.J. Carty, *J. Chem. Soc., Chem. Commun.*, 1991, 1511.
807 M.I. Bruce, G.A. Koutsantonis, and E.R.T. Tiekink, *J. Organomet. Chem.*, 1991, **407**, 391.
808 D. Lentz and M. Reuter, *Chem. Ber.*, 1991, **124**, 773.
809 B. Eber, G. Huttner, Chr. Emmerich, J.C. Daran, B. Heim, and Y. Jeannin, *J. Organomet. Chem.*, 1991, **419**, 43.
810 E. Louattani, J. Suades, and R. Mathieu, *J. Organomet. Chem.*, 1991, **421**, 335.
811 E. Rosenberg, K.I. Hardcastle, M.W. Day, R. Gobetto, S. Hajela, and R. Muftikian, *Organometallics*, 1991, **10**, 203.

$[Au_2Fe_2M(\mu_4\text{-}C{\equiv}CPh)(CO)_7(PPh_3)_3]$, (M = Rh, Ir; ^{13}C),[812] $[Co_3Fe(\eta^5\text{-}C_5H_5)(\mu_4\text{-}\eta^2\text{-}C{=}CHCF_3)(\eta^6\text{-}C_7H_8)(\mu\text{-}CO)_2(CO)_4]$, ($^{13}C$),[813] $[Ru(OH_2)_4\{(\eta^2\text{-}CH_2{=}CHCH_2)_2O\}]^{2+}$, ($^{13}C$),[814] $[(\eta^6\text{-}C_6H_6)Ru(2\text{-}C_6H_4CHSCH_2\text{-}2\text{-}C_6H_4CH_2SCH_2)]^+$, ($^{13}C$),[815] $[(\eta^3{:}\eta^2{:}\eta^3\text{-}C_{12}H_{18})RuCl(OH_2)]^+$, ($^{13}C$),[816] $[(\eta^2\text{-benzyne})Ru(PMe_3)_4]$, ($^{13}C$),[817] $[Ru_2(dmpm)_2(CO)_2(MeO_2CC{\equiv}CCO_2Me)_2]$, ($^{13}C$),[818] $[Ru_3(CO)_8(\mu\text{-}PPh_2)_2(\mu_3\text{-}C_6H_4CO)]$, ($^{13}C$),[819] $[Ru_3(\mu_3\text{-}\eta^2\text{-}CH_3C{\equiv}CCH_3)(CO)_{10}]$, ($^{13}C$),[820] $[Ru_3(CO)_6(Pr^iN{=}CHCH{=}CMe)_2]$, ($^{13}C$),[821] $[Ru_5(\mu_5\text{-}C{\equiv}CPh)(\mu\text{-}PPh_2)(\mu_4\text{-}PPh)(CO)_{13}]$, ($^{13}C$),[822] $[(\eta^2\text{-}C_6H_{10})(\eta^6\text{-}C_6H_3Me_3\text{-}1,3,5)Os(CO)]$, ($^{13}C$),[823] $[\{\eta^6\text{-}1\text{-}MeO\text{-}4\text{-}(MeN{=}CMe)C_6H_4\}\text{-}Os(NH_3)_5]^{2+}$, ($^{13}C$),[824] $[(\eta^2\text{-}2,5\text{-}Me_2\text{-pyrrole})Os(NH_3)_5]^{2+}$, ($^{13}C$),[825] and $[RhOs(CO)_3(\mu\text{-}F_3CC{\equiv}CCF_3)(dppm)_2]^+$, ($^{13}C$),[826] $[(\eta^3\text{-}C_3H_5)Fe(CO)_3P\{{=}C(SiMe_3)_2\}_2]$, ($^{13}C$),[827] (47), ($^{13}C$),[828] $[\{\eta^3\text{-}CH_2C(OSiMe_3)CR^1R^2\}Fe(CO)_2(NO)]$, ($^{13}C$),[829] $[(\eta^3\text{-}C_5H_7)Fe(depe)_2]^+$, ($^{13}C$),[830] (48), ($^{13}C$),[831] $[(\mu_2\text{-}\eta^3{:}\eta^3\text{-}C_4H_5Bu^t)Fe_2(CO)_5(\mu\text{-}PPh_2)]$, ($^{13}C$),[832] $[Fe_2(CO)(\mu\text{-}CO)\{\mu\text{-}\eta^3\text{-}C(O)C_2Me[C(O)R]\}(\eta^5\text{-}C_5H_5)_2]$, ($^{13}C$),[833] $[Ru(\eta^3\text{-}RC_3CHR)\{PhP(OEt)_2\}_4]^+$, ($^{13}C$),[834] *syn, mer*-$[RuCl(\eta^3\text{-}PhC_3CHPh)(Cy_2PCH_2CH_2CH_2)_2PPh]$, ($^{13}C$),[835] $[(\eta^3\text{-}Me_3SiCH_2CHCHCSiMe_3)(CO)_2\text{-}(PPh_3)_2]$, ($^{13}C$),[836] (49), ($^{13}C$),[837] $[\{(\eta^3{:}\eta^3\text{-}CH_2CMeCHCH_2CH_2CHCMeCH_2)RuCl_2\}(dppe)]$,

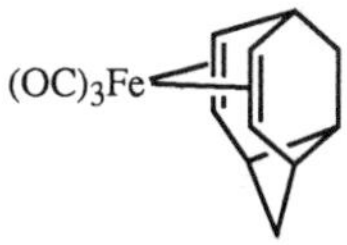

(46)

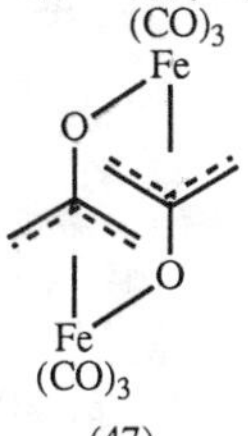

(47)

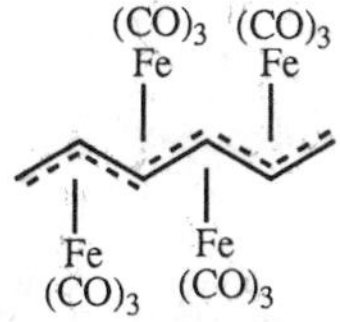

(48)

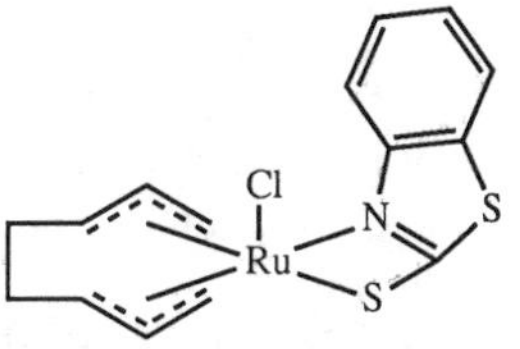

(49)

812 M.I. Bruce, G.A. Koutsantonis, and E.R.T. Tiekink, *J. Organomet. Chem.*, 1991, **408**, 77.
813 R. Rumin, F. Robin, F.Y. Pétillon, K.W. Muir, and I. Stevenson, *Organometallics*, 1991, **10**, 2274.
814 D.V. McGrath, R.H. Grubbs, and J.W. Ziller, *J. Am. Chem. Soc.*, 1991, **113**, 3611.
815 L.R. Hanton and T. Kemmitt, *J. Chem. Soc., Chem. Commun.*, 1991, 700.
816 J.W. Steed and D.A. Tocher, *J. Chem. Soc., Chem. Commun.*, 1991, 1609.
817 J.F. Hartwig, R.G. Bergman, and R.A. Andersen, *J. Am. Chem. Soc.*, 1991, **113**, 3404.
818 K.A. Johnson and W.L. Gladfelter, *J. Am. Chem. Soc.*, 1991, **113**, 5097.
819 J.P.H. Charmant, H.A.A. Dickson, N.J. Grist, J.B. Keister, S.A.R. Knox, D.A.V. Morton, A.G. Orpen, and J.M. Viñas, *J. Chem. Soc., Chem. Commun.*, 1991, 1393.
820 S. Rivomanana, G. Lavigne, N. Lugan, and J.-J. Bonnet, *Organometallics*, 1991, **10**, 2285.
821 W.P. Mul, C.J. Elsevier, W.J.J. Smeets, and A.L. Spek, *Inorg. Chem.*, 1991, **30**, 4152.
822 C.J. Adams, M.I. Bruce, B.W. Skelton, and A.H. White, *J. Organomet. Chem.*, 1991, **420**, 87.
823 H. Werner, S. Stahl, and M. Schulz, *Chem. Ber.*, 1991, **124**, 707.
824 M.E. Kopach, J. Gonzalez, and W.D. Harman, *J. Am. Chem. Soc.*, 1991, **113**, 8972.
825 W.H. Myers, M. Sabat, and W.D. Harman, *J. Am. Chem. Soc.*, 1991, **113**, 6682.
826 R.W. Hilts, R.A. Franchuk, and M. Cowie, *Organometallics*, 1991, **10**, 304.
827 H.J. Metternich and E. Niecke, *Angew. Chem., Int. Ed. Engl.*, 1991, **30**, 312.
828 M. Frey and T.A. Jenny, *J. Organomet. Chem.*, 1991, **421**, 257.
829 K. Itoh, S. Nakanishi, and Y. Otsuji, *Bull. Chem. Soc. Jpn.*, 1991, **64**, 2965.
830 J.R. Bleeke and R.J. Wittenbrink, *J. Organomet. Chem.*, 1991, **405**, 121.
831 M. Iyoda, Y. Kuwatani, M. Oda, K. Tatsumi, and A. Nakamura, *Angew. Chem., Int. Ed. Engl.*, 1991, **30**, 1670.
832 S.M. Breckenridge, S.A. MacLaughlin, N.J. Taylor, and A.J. Carty, *J. Chem. Soc., Chem. Commun.*, 1991, 1718.
833 A. Wong, R.V. Pawlick, C.G. Thomas, D.R. Leon, and L.-K. Liu, *Organometallics*, 1991, **10**, 530.
834 G. Albertin, P. Amendola, S. Antoniutti, S. Ianelli, G. Pelizzi, and E. Bordignon, *Organometallics*, 1991, **10**, 2876.
835 G. Jia and D.W. Meek, *Organometallics*, 1991, **10**, 1444.
836 Y. Wakatsuki, H. Yamazaki, Y. Maruyama, and I. Shimizu, *J. Chem. Soc., Chem. Commun.*, 1991, 261.
837 J.G. Toerien and P.H. van Rooyen, *J. Chem. Soc., Dalton Trans.*, 1991, 1563.

(^{13}C),[838] [$Ru_2(CO)_6(R^1CH_2CCHNR^2{}_2)$], (^{13}C),[839] [$Ru_4(CO)_{10}\{CH_2C{=}CHCH{=}NBu^t\}_2$], (^{13}C),[840] [($\eta^4$-diene)$Fe(CO)_3$], (^{13}C),[841] [($\eta^4$-PhCH=CHCR=CCO)$Fe(CO)_3$], (^{13}C),[842] [{$\eta^4$-PhC=CMeCMe=C($CO_2$Me)$CMe_2$NPh}$Fe(CO)_3$], (^{13}C),[843] [($\eta^4$-1-$R^1$-4-$R^2$-$C_6H_6$)$Fe(CO)_3$], (^{13}C),[844] [($\eta^4$-$O_2$SCH=CHCH=CHCH=CH)$Fe(CO)_3$], (^{13}C),[845] (50), (^{13}C),[846] (51), (^{13}C),[847] (52),

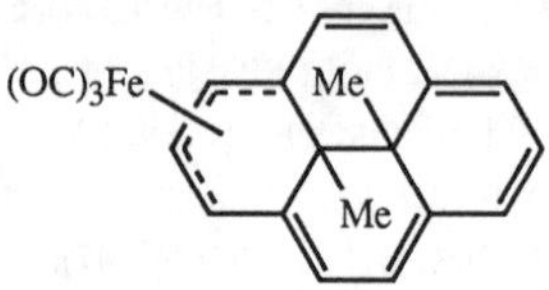

(50)

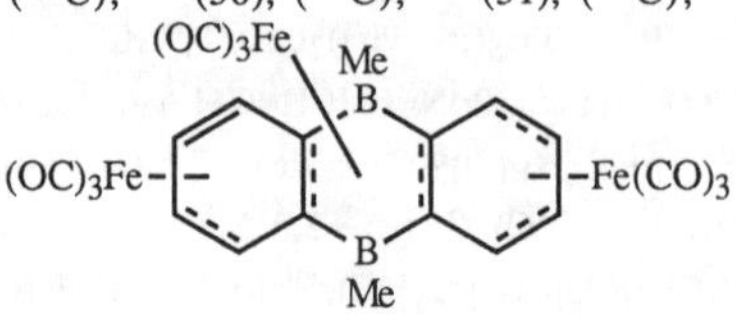

(51)

(^{13}C),[848] [(η^4-PhCH=CHCH=NCHBu^{t}Ph)$Fe(CO)_3$], (^{13}C),[849] [{η^4-Bu^{t}S(O)CH=CHCMe=O}-$Fe(CO)_3$], (^{13}C),[850] [{η^4-$R^1R^2C{=}CR^3C(NEt_2){=}S$}$Fe(CO)_3$], (^{13}C),[851] (53), (^{11}B, ^{13}C),[852] [(η^4-C_4H_6)$Fe(CO)_3$], (^{13}C),[853] [{η^4-(2,4,6-$Me_3C_6H_2$)$_2SiC_3H_4Bu^t$}$Fe(CO)_3$], (^{13}C, ^{29}Si),[854]

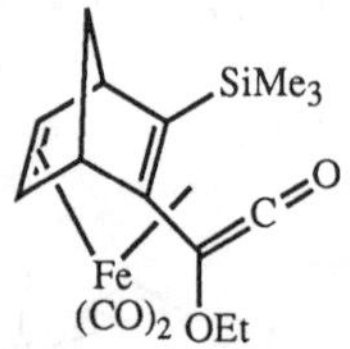

(52)

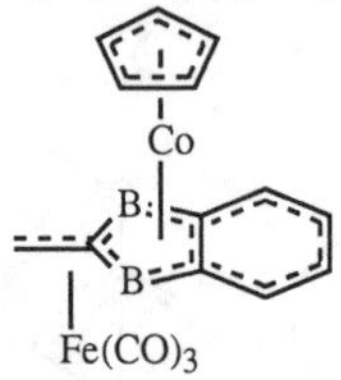

(53)

[(η^6-C_6H_5CN)Ru(η^4-1,5-C_8H_{12})Ru], (^{13}C),[855] [(η^6-1,3,5-$Me_3C_6H_3$)Ru($C_{18}H_{15}$)]$^+$, (^{13}C),[856] [(η^4-2,4-hexadiene)Ru(acac)$_2$], (^{13}C),[857] [(η^5-C_5H_5)(η^4-cyclopentanone)Ru(NCMe)]$^+$, (^{13}C),[858]

838 J.G. Toerien and P.H. van Rooyen, *J. Chem. Soc., Dalton Trans.*, 1991, 2693.
839 W.P. Mul, C.J. Elsevier, L.H. Polm, K. Vrieze, M.C. Zoutberg, D. Heijdenrijk, and C.H. Stam, *Organometallics*, 1991, **10**, 2247.
840 W.P. Mul, C.J. Elsevier, M. van Leijen, K. Vrieze, and A.L. Spek, *Organometallics*, 1991, **10**, 533.
841 W.A. Donaldson, C. Tao, D.W. Bennett, and D.S. Grubisha, *J. Org. Chem.*, 1991, **56**, 4563; R.E. Lehmann and J.K. Kochi, *Organometallics*, 1991, **10**, 190.
842 N.W. Alcock, C.J. Richards, and S.E. Thomas, *Organometallics*, 1991, **10**, 231.
843 R. Aumann, B. Trentmann, C. Krüger, and F. Lutz, *Chem. Ber.*, 1991, **124**, 2595.
844 M.-C.P. Yeh, M.-L. Sun, and S.-K. Lin, *Tetrahedron Lett.*, 1991, **32**, 113.
845 K. Nishino, M. Takagi, T. Kawata, I. Murata, J. Inanaga, and K. Nakasuji, *J. Am. Chem. Soc.*, 1991, **113**, 5059.
846 R.H. Mitchell and P. Zhou, *Angew. Chem., Int. Ed. Engl.*, 1991, **30**, 1013.
847 H. Schulz, H. Pritzkow, and W. Siebert, *Chem. Ber.*, 1991, **124**, 2203.
848 J. Park, S. Kang, D. Whang, and K. Kim, *Organometallics*, 1991, **10**, 3413.
849 K.G. Morris and S.E. Thomas, *J. Chem. Soc., Perkin Trans. 1*, 1991, 97.
850 A. Ibbotson, A.M.Z. Slawin, S.E. Thomas, G.J. Tustin, and D.J. Williams, *J. Chem. Soc., Chem. Commun.*, 1991, 1534.
851 H. Alper and D.A. Brandes, *Organometallics*, 1991, **10**, 2457.
852 Z. Nagy-Magos, A. Feßenbecker, H. Pritzkow, W. Siebert, I. Hyla-Kryspin, and R. Gleiter, *Chem. Ber.*, 1991, **124**, 2677.
853 L. Girard, J.H. MacNeil, A. Mansour, A.C. Chiverton, J.A. Page, S. Fortier, and M.C. Baird, *Organometallics*, 1991, **10**, 3114.
854 W. Ando, T. Yamamoto, H. Saso, and Y. Kabe, *J. Am. Chem. Soc.*, 1991, **113**, 2791.
855 M.A. Bennett, H. Neumann, M. Thomas, X.-q. Wang, P. Pertici, P. Salvadori, and G. Vitulli, *Organometallics*, 1991, **10**, 3237.
856 M.A. Bennett, S. Pelling, G.B. Robertson, and W.A. Wickramasinghe, *Organometallics*, 1991, **10**, 2166.
857 R.D. Ernst, E. Melendez, L. Stahl, and M.L. Ziegler, *Organometallics*, 1991, **10**, 3635.
858 K. Kirchner and H. Taube, *J. Am. Chem. Soc.*, 1991, **113**, 7039.

[{η^4-CH(CO_2Me)=C(CO_2Me)C(CO_2Me)=CH(CO_2Me)}(η^5-C_5H_5)RuI], (^{13}C),[859] [(η^4-MeC=CH-CH=NPr^i)$_2$Ru(CO)$_2$], (^{13}C),[860] and [(η^4-C_4H_6)Ru(η^6-C_6H_6)], (^{13}C).[861]

^{13}C T_1 measurements have been used to study the motion of ferrocene complexes included in cyclodextrins.[862] ^{1}H and ^{13}C n.m.r. spectroscopy has been used to study the structural conformations of ruthenocene and ferrocene containing cryptands.[863] A series of derivatives of organotin ferrocene has been studied by ^{119}Sn n.m.r. spectroscopy and some factors influencing the ^{119}Sn chemical shift were discussed.[864] ^{1}H NOESY has been used to investigate the incorporation of [(η^5-C_5H_5)Fe{η^5-$C_5H_4CH_2NMe_2(CH_2)_n$CONHC$_{10}H_6SO_3$}] into α=cyclodextrin.[865] ^{1}H and ^{13}C n.m.r. spectroscopy has been used to study the configuration of 1,1'-(1,4,10-trioxa-7,13-diazacyclopentadecane-7,13-diyldicarbonyl)ferrocene when complexed with Be^{2+}, Mg^{2+}, Ca^{2+}, Sr^{2+}, and Ba^{2+}.[866] The solution conformation of [(η^5-$C_5H_4PPh_2$)Fe(η^5-C_5H_3-1-PPh_2-2-CHMeX)] has been determined by a combination of two dimensional n.m.r. methods based on ^{1}H, ^{13}C, and ^{31}P observation.[867] The bonding in iron(II) cyclopentadienyl and arene sandwich complexes has been studied. Correlations have been found between ^{13}C and ^{57}Fe n.m.r. and ^{57}Fe Mössbauer data.[868] A correlation between δ(^{13}CO) and the pK_a of the pyridine ligand in [(η^5-C_5H_5)Fe(CO)$_2$(NC$_5H_4$R)]$^+$ has been found.[869] ^{1}H n.m.r. spectroscopic studies of [(η^5-C_5H_5)Ru^{II}(η^5-$C_5H_4C_5H_4$-η^5)Ru^{IV}(η^5-C_5H_5)X]$^+$ suggest that there is rapid electron exchange between the ruthenium atoms.[870] N.m.r. data have also been reported for [(η^5-C_6H_6X)(η^5-C_5H_5)Fe], (^{13}C),[871] [(η^5-C_5H_5)Fe(η^5-C_5H_4R)], (^{13}C),[872] [(η^5-C_5H_5)Fe(η^5-C_5H_4-4-$C_3H_2N_2$)Pd(η^3-1-MeC_3H_4)]$_2$, (^{11}B, ^{13}C),[873] [(η^5-C_5H_5)Fe(η^5-C_5H_4C{NN(C_6H_4-2CO$_2$H)}$_2$H)], (^{13}C),[874] [(η^5-C_5H_5)Fe(η^5-C_5H_4CH=NOH)], (^{13}C),[875] [(η^5-C_5H_5)Fe(η^5-C_5H_4CMe=NNHCOR)], (^{13}C),[876] *trans*-[{(η^5-C_5H_5)Fe(η^5-C_5H_4CHMePPh$_2$)}$_2$-MCl_2], (M = Pd, Pt; ^{13}C),[877] [(η^5-C_5H_5)Fe(η^5-C_5H_4COCH=CHAr)], (^{13}C),[878] [(η^5-C_5H_5)Fe(η^5-C_5H_4COGeR$_3$)], (^{13}C),[879] [(η^5-C_5H_5)Fe{η^5-C_5H_4C(O)NHCH$_2$CH$_2$S(CH$_2$SCH$_2$)$_n$CH$_2$NHC(O)-C_5H_4-η^5}Fe(η^5-C_5H_5)], (^{13}C),[880] [(η^5-C_5H_5)Fe(η^5-C_5H_4SiPh$_3$)], (^{13}C),[881] [(η^5-C_5H_4S-2-*exo*-

859 M.I. Bruce, G.A. Koutsantonis, E.R.T. Tiekink, and B.K. Nicholson, *J. Organomet. Chem.*, 1991, **420**, 271.
860 W.P. Mul, C.J. Elsevier, and J. Spaans, *J. Organomet. Chem.*, 1991, **402**, 125.
861 G.E. Herberich and T.P. Spaniol, *J. Chem. Soc., Chem. Commun.*, 1991, 1457.
862 R.M.G. Roberts and J.F. Warmsley, *J. Organomet. Chem.*, 1991, **405**, 357.
863 C.D. Hall and N.W. Sharpe, *J. Organomet. Chem.*, 1991, **405**, 365.
864 H. Jiang, B. Yan, S. Shi, and G. Xu, *Fenxi Huaxue*, 1991, **19**, 185 (*Chem. Abstr.*, 1991, **114**, 229 102).
865 R. Ishnin and A.E. Kaifer, *J. Am. Chem. Soc.*, 1991, **113**, 8188.
866 C.D. Hall, J.H.R. Tucker, and N.W. Sharpe, *Organometallics*, 1991, **10**, 1727.
867 A. Togni, R.E. Blumer, and P.S. Pregosin, *Helv. Chim. Acta*, 1991, **74**, 1533.
868 A. Houlton, J.R. Miller, R.M.G. Roberts, and J. Silver, *J. Chem. Soc., Dalton Trans.*, 1991, 467.
869 H. Schumann, M. Speiss, W.P. Bosman, J.M.M. Smits, and P.T. Beurskens, *J. Organomet. Chem.*, 1991, **403**, 165.
870 M. Watanabe, H. Sano, and I. Motoyama, *Chem. Lett.*, 1990, 1667 (*Chem. Abstr.*, 1991, **114**, 6785).
871 R.G. Sutherland, C. Zhang, and A. Piórko, *J. Organomet. Chem.*, 1991, **419**, 357.
872 P.D. Beer, E.L. Tite, and A. Ibbotson, *J. Chem. Soc., Dalton Trans.*, 1991, 1691.
873 K. Niedenzu, J. Serwatowski, and S. Trofimenko, *Inorg. Chem.*, 1991, **30**, 524.
874 Y. Gök and H.B. Senturk, *Inorg. Chim. Acta*, 1991, **180**, 53.
875 G. Ferguson, W. Bell, and C. Glidewell, *J. Organomet. Chem.*, 1991, **405**, 229.
876 Z. Tan and X. Wu, *Wuhan Daxue Xuebao Ziran Kexueban*, 1990, 77 (*Chem. Abstr.*, 1991, **114**, 102 302).
877 M. Sawamura, H. Hamashima, and Y. Ito, *Tetrahedron Asymmetry*, 1991, **2**, 593.
878 Á.G. Nagy, P. Sohár, and J. Márton, *J. Organomet. Chem.*, 1991, **410**, 357.
879 H.K. Sharma, F. Cervantes-Lee, and K.H. Pannell, *J. Organomet. Chem.*, 1991, **409**, 321.
880 R.A. Holwerda, T.W. Robison, R.A. Bartsch, and B.P. Czech, *Organometallics*, 1991, **10**, 2652.
881 G.A. Olah, T. Bach, and G.K.S. Prakash, *New J. Chem.*, 1991, **15**, 571.

norbornyl)$_2$Fe], (^{13}C),[882] $[\{(\eta^5\text{-}C_5H_5)Fe(\eta^5\text{-}C_5H_4SC_5H_4\text{-}\eta^5)\}_2Fe]$, (E = S, Se, Te; ^{13}C),[883] $[\{(\eta^5\text{-}C_5H_5)Fe(\eta^5$-4,8-ethano-2,4,6,8-tetrahydro-*s*-indacene-2,6-diyl-$\eta^5)\}_2Fe]$, (^{13}C),[884] $[(\eta^5\text{-}C_5H_5)Fe(\eta^5\text{-}C_5H_3\text{-}2\text{-}PPh_2\text{-}C_5H_3\text{-}2\text{-}PPh_2\text{-}\eta^5)Fe(\eta^5\text{-}C_5H_5)]$, (^{13}C),[885] $[(\eta^5\text{-}C_5H_5)Fe\{\eta^5\text{-}C_5H_3\text{-}1\text{-}PdXL\text{-}2\text{-}C(S)NMe_2\}]$, (^{13}C),[886] $[(\eta^5\text{-}C_5H_4COMe)(\eta^5\text{-}C_5H_4SiMe_3)Fe]$, (^{13}C),[887] $[Fe\{\eta^5\text{-}C_5H_4C(O)O(CH_2)_nOC_6H_4\text{-}4\text{-}C_6H_4\text{-}4\text{-}CN\}_2]$, (^{13}C),[888] ferrocene containing polyphosphate esters, (^{13}C),[889] $[\{\eta^5\text{-}C_5H_4CH_2S(CH_2CH_2O)_5CH_2CH_2SCH_2C_5H_4\text{-}\eta^5\}Fe]$, (^{13}C),[890] $[\{\eta^5\text{-}C_5H_4CH_2\text{-}S(CH_2)_mS(CH_2)_nS(CH_2)_mSCH_2C_5H_4\text{-}\eta^5\}Fe]$, and its Pt(II) complexes, (^{13}C, ^{195}Pt),[891] $[\{\eta^5\text{-}C_5H_3(SiMe_2SiMe_2)_2C_5H_3\text{-}\eta^5\}M]$, (M = Fe, Co$^+$, Sn, Pb; ^{13}C, ^{29}Si, ^{59}Co, ^{119}Sn, ^{207}Pb),[892] $[\{\eta^5\text{-}1\text{-}Bu^t\text{-}Bu^tC(O)C_5H_3\}_2Fe]$, (^{13}C),[893] $[(\eta^5\text{-}C_5Me_4CH{=}CH_2)_2Fe]$, (^{13}C),[894] $[(\eta^5\text{-}C_4Me_4NH)(\eta^5\text{-}C_4Me_4NMe)Fe]^+$, (^{11}B, ^{13}C),[895] $[(\eta^5\text{-}2,5\text{-}Bu^t_2C_5H_3)_2Fe]$, (^{13}C),[896] $[(\eta^5\text{-}C_5H_5)\{\eta^5\text{-}(Me_3SiO)_3\text{-}C_3P_2\}Fe]$, (^{13}C),[897] $[(\eta^5\text{-}2,5\text{-}Me_2C_4H_2Sb)_2Fe]$, (^{13}C),[898] $[(\eta^5\text{-}C_5H_5)Fe\{\eta^6\text{-}C_6(CH_2CH_2Ph)_6\}]^+$,

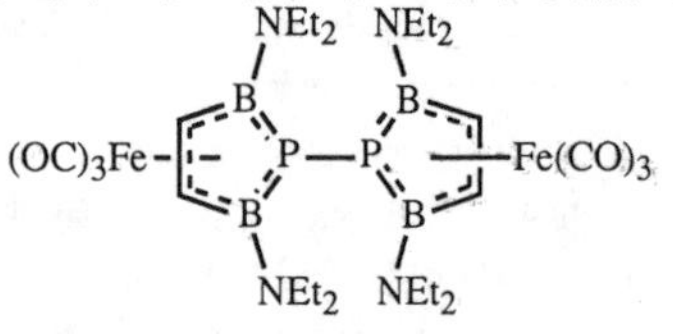

(54)

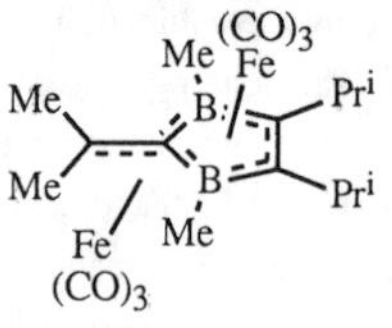

(55)

(^{13}C),[899] $[(\eta^5\text{-}C_5H_5)(\eta^6\text{-}C_6H_5X)Fe]^+$, (^{13}C),[900] $[(\eta^5\text{-}C_5H_5)(\eta^6\text{-arene})Fe]^+$, (^{13}C),[901] $[(\eta^5\text{-}C_5H_5)(\eta^6\text{-arene})M]^+$, (M = Fe, Ru; ^{13}C),[902] $[(\eta^5\text{-}C_5Ph_5)(\eta^6\text{-}C_6Me_{6-n}H_n)Fe]^+$, (^{13}C),[903] (54), (^{11}B, ^{13}C),[904] (55), (^{11}B, ^{13}C),[905] $[(\eta^5\text{-}C_5Me_5)Rh\{\eta^5\text{-}C_4Me_4Fe(CO)_3\}]$, (^{13}C),[906] $[(\eta^5\text{-}C_5Me_5)Fe(CO)_2\text{-}$

882 M. Herberhold, O. Nuyken, and T. Pöhlmann, *J. Organomet. Chem.*, 1991, **405**, 217.
883 M. Herberhold, H.-D. Brendel, O. Nuyken, and T. Pöhlmann, *J. Organomet. Chem.*, 1991, **413**, 65.
884 H. Atzkern, B. Huber, F.H. Köhler, G. Müller, and R. Müller, *Organometallics*, 1991, **10**, 238.
885 M. Sawamura, A. Yamauchi, T. Takegawa, and Y. Ito, *J. Chem. Soc., Chem. Commun.*, 1991, 874.
886 M. Nonoyama and K. Hamamura, *J. Organomet. Chem.*, 1991, **407**, 271.
887 A.F. Cunningham, jun., *J. Am. Chem. Soc.*, 1991, **113**, 4864.
888 J. Bhatt, B.M. Fung, and K.M. Nicholas, *J. Organomet. Chem.*, 1991, **413**, 263.
889 K. Kishore, P. Kannan, and K. Iyanar, *J. Polym. Sci., Part A: Polym. Chem.*, 1991, **29**, 1039 (*Chem. Abstr.*, 1991, **115**, 30 041).
890 S.J. Rao, C.I. Milberg, and R.C. Petter, *Tetrahedron Lett.*, 1991, **32**, 3775.
891 M. Sato, H. Asano, and S. Akabori, *J. Organomet. Chem.*, 1991, **401**, 363.
892 P. Jutzi, R. Krallmann, G. Wolf, B. Neumann, and H.-G. Stammler, *Chem. Ber.*, 1991, **124**, 2391.
893 W. Bell and C. Glidewell, *J. Organomet. Chem.*, 1991, **411**, 251.
894 M. Ogasa, D.T. Mallin, D.W. Macomber, M.D. Rausch, R.D. Rogers, and A.N. Rollins, *J. Organomet. Chem.*, 1991, **405**, 41.
895 N. Kuhn, A. Kuhn, and E.-M. Lampe, *Chem. Ber.*, 1991, **124**, 997.
896 N. Kuhn, K. Jendral, R. Boese, and D. Bläser, *Chem. Ber.*, 1991, **124**, 89.
897 E. Niecke and D. Schmidt, *J. Chem. Soc., Chem. Commun.*, 1991, 1659.
898 A.J. Ashe, tert., T.R. Diephouse, J.W. Kampf, and S.M. Al-Tawell, *Organometallics*, 1991, **10**, 2068.
899 M.J. Zaworotko, K.C. Sturge, L. Nunez, and R.D. Rogers, *Organometallics*, 1991, **10**, 1806.
900 K. Bambridge and R.M.G. Roberts, *J. Organomet. Chem.*, 1991, **401**, 125.
901 R.C. Cambie, S.J. Janssen, P.S. Rutledge, and P.D. Woodgate, *J. Organomet. Chem.*, 1991, **420**, 387; J.-R. Hamon, P. Hamon, S. Sinbandhit, P. Guenot, and D. Astruc, *J. Organomet. Chem.*, 1991, **413**, 243.
902 R.C. Cambie, S.A. Coulson, L.G. Mackay, S.J. Janssen, P.S. Rutledge, and P.D. Woodgate, *J. Organomet. Chem.*, 1991, **409**, 385.
903 L.D. Field, T.W. Hambley, P.A. Lay, C.M. Lindall, and A.F. Masters, *J. Chem. Soc., Dalton Trans.*, 1991, 1499.
904 M. Drieß, P. Frankhauser, H. Pritzkow, and W. Siebert, *Chem. Ber.*, 1991, **124**, 1497.
905 A. Feßenbecker, M. Enders, H. Pritzkow, W. Siebert, I. Hyla-Kryspin, and R. Gleiter, *Chem. Ber.*, 1991, **124**, 1505.
906 S. Luo, A.E. Ogilvy, T.B. Rauchfuss, A.L. Rheingold, and S.R. Wilson, *Organometallics*, 1991, **10**, 1002.

{$PP(C_6H_2Bu^t{}_3$-2,4,6)CHC(O)NRC(O)CH}], (^{13}C),[907] [$(\eta^5$-$C_5Me_5)Fe(CO)_2$-$P(=PC_6H_2Me_3$-2,4,6)OC(OR)=$NNCO_2R$], (^{13}C),[908] (56), (^{13}C),[909] [$(\eta^5$-$C_5Me_5)Fe(CO)_2P$-$(N=CPh_2)_2$], (^{13}C),[910] (57), (^{13}C),[911] [$(\eta^5$-$C_5Me_5)Fe(CO)_2P(SiMe_3)P=C(SiMe_3)_2$], (^{13}C),[912] [$(\eta^5$-$C_5Me_5)M(CO)_2PCl_2$], (M = Fe, Ru; ^{13}C),[913] (58), (^{13}C),[914] [$(\eta^5$-$C_5H_5)Fe(dppe)(NCR)]^+$,

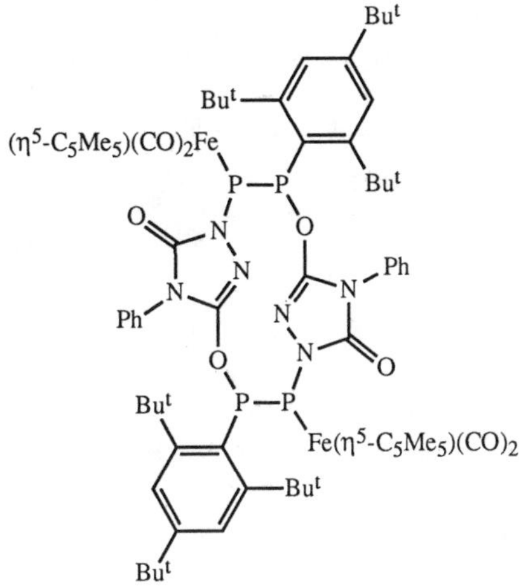

(56)

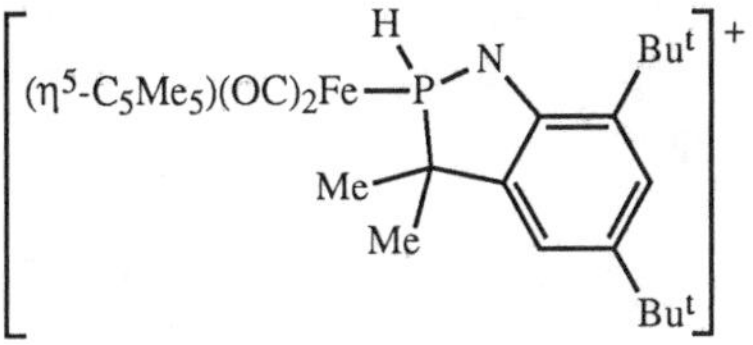

(57)

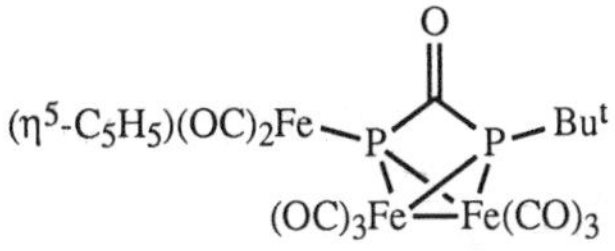

(58)

(^{13}C),[915] [$(\eta^5$-$C_5H_5)Fe(CO)_2(ER_3)]^+$, (E = N, P, As, Sb, Bi; ^{13}C),[916] [$(\eta^5$-$C_5H_5)Fe(CO)_2$-$(EMe_2)]^+$, (M = S, Se, Te; ^{13}C),[917] [$(\eta^5$-$C_5H_5)Fe(CO)_2SC(O)Me$], (^{13}C),[918] [$(\eta^5$-$C_5H_5)Fe(CO)$(4-Ph_2P-dibenzothiophene)I], (^{13}C),[919] [$(\eta^5$-$CH_2CMeCHCMeO)_2Ru$], (^{13}C),[920] [$(\eta^5$-$C_5H_5)Ru(\eta^5$-

907 L. Weber, M. Frebel, A. Müller, and H. Bögge, *Organometallics*, 1991, **10**, 1130.
908 L. Weber, H. Bastian, A. Müller, and H. Bögge, *Organometallics*, 1991, **10**, 2.
909 L. Weber, H. Bastian, R. Boese, and H.-G. Stammler, *J. Chem. Soc., Chem. Commun.*, 1991, 1778.
910 L. Weber, U. Sonnenberg, H.G. Stammler, and B. Neumann, *Z. Naturforsch.*, B, 1991, **46**, 714 (*Chem. Abstr.*, 1991, **115**, 232 459).
911 J. Hein and E. Niecke, *J. Chem. Soc., Chem. Commun.*, 1991, 48.
912 L. Weber, R. Kirchhoff, R. Boese, and H.-G. Stammler, *J. Chem. Soc., Chem. Commun.*, 1991, 1293.
913 L. Weber and U. Sonnenberg, *Chem. Ber.*, 1991, **124**, 725.
914 L. Weber and H. Schumann, *Chem. Ber.*, 1991, **124**, 265.
915 H. Schumann, L. Eguren, and J.W. Ziller, *J. Organomet. Chem.*, 1991, **408**, 361.
916 H. Schumann and L. Eguren, *J. Organomet. Chem.*, 1991, **403**, 183.
917 H. Schumann and J.W. Ziller, *J. Organomet. Chem.*, 1991, **410**, 365.
918 K.R. Powell, W.J. Elias, and M.E. Welker, *J. Organomet. Chem.*, 1991, **407**, 81.
919 L.V. Dunkerton, C.C. Hinckley, J. Tyrrell, and P.D. Robinson, *Report*, 1989, DOE/PC/88861-T18; Order No. DE90010591, 31 pp. Avail. NTIS. From *Energy Res. Abstr.*, 1990, **15**, Abstr. No. 29968 (*Chem. Abstr.*, 1991, **115**, 12 181).
920 T. Schmidt and R. Goddard, *J. Chem. Soc., Chem. Commun.*, 1991, 1427.

indenyl-η^6)Ru(η^5-C_5H_5)]$^+$, (^{13}C),[921] [(η^5-C_5Me_5)$_2$$Ru_2$($\eta^6$;$\eta^6$-PhC≡CPh)]$^{2+}$, (^{13}C),[922] [($\eta^5$-$C_5Me_5$)Ru($\eta^6$-$C_8H_{10}$)]$^+$, (^{13}C),[923] [($\eta^5$-$C_5Me_5$)Ru($\eta^6$-quinoline)]$^+$, (^{13}C),[924] [($\eta^5$-$C_5H_5$)Ru($\eta^6$-arene)]$^+$, (^{13}C),[925] [($\eta^5$-$C_5H_5$)Ru($NH_2$)($Cy_2PCH_2CH_2PCy_2$)], (^{13}C),[926] [($\eta^5$-$C_5H_5$)Ru($NH_2Bu^t$)-($PPh_3$){P(OMe)$_3$}]$^+$, (^{13}C),[927] [($\eta^5$-$C_5H_5$)RuCl($PPh_3$){P(OMe)$_3$}], (^{13}C),[928] [($\eta^5$-$C_5H_5$)Ru-($PMe_3$)$_2${S(Me)C≡CMe}]$^+$, (^{13}C),[929] [($\eta^5$-$C_5H_5$)RuCl($Cy_2PCH_2CH_2PCy_2$)], (^{13}C),[930] [($\eta^5$-$C_5Me_4CH_2Cl$)$RuCl_3$($SMe_2$)], (^{13}C),[931] [$RuRh_3$($\eta^5$-$C_5Me_5$)($\eta^5$-$C_5H_5$)$_3$($\mu_3$-CO)$_3$]$^+$, (^{13}C),[932] [$CoRu_5$($\mu_4$-PPh)($\mu_4$-C≡CPh)($\mu$-$PPh_2$)(CO)$_{12}$($\eta^5$-$C_5H_5$)], (^{13}C),[933] [($\eta^5$-$C_5Me_5$)Os{$\eta^7$-$C_5Me_3$-($CH_2$)$_2$}], (^{13}C),[934] [($\eta^6$-1,3,5-$Me_3C_6H_3$)$_2$Fe]$^{2+}$, (^{13}C),[935] [($\eta^6$-1-Me-4-$Pr^iC_6H_4$)Ru($\mu$-pz)$_2$($\mu$-CO)Rh(dppb)]$^+$, (^{13}C),[936] [($\eta^6$-$C_6H_6$)Ru($NC_6H_3Pr^i_2$-2,6)]$_2$, (^{13}C),[937] [($\eta^6$-$C_6H_6$)Ru{($C_6F_5$)$_2$P-$CH_2CH_2P$($C_6F_5$)$_2$}], (^{13}C),[938] [($\eta^6$-1-Me-4-$Pr^iC_6H_4$)Ru{O=P(OEt)$_2$-$\mu$}$_3$Co($\eta^5$-$C_5H_5$)]$^+$, (^{13}C),[939] [($\eta^6$-arene)$RuCl_2$($SR_2$)], (^{13}C),[940] [($\eta^6$-$C_6H_6$)Os(=N=$CPh_2$)($PPr^i_3$)]$^+$, (^{13}C),[941] and [($\eta^6$-1-Me-4-$Pr^iC_6H_4$)Os(=$NC_6H_3Me_2$-2,6)], (^{15}N).[942]

The bonding in *trans*-[Fe(CO)$_3$(PR_3)$_2$] has been investigated by i.r., ^{31}P n.m.r., and ^{57}Fe Mössbauer spectroscopy.[943] Unexpected ^{13}C n.m.r. shifts are observed for one bis-*ansa* iron(II) CO porphyrin system and are considered to be inconsistent with a linear Fe—CO unit perpendicular to the porphyrin plane.[944] *Ab initio* calculations have been performed that demonstrate an essentially linear correlation between ν(CO) and δ(^{13}C) or δ(^{17}O) of metal carbonyls. The results were applied to a wide range of carbonmonoxyhaem proteins.[945] A series of mixed metal carbonyl clusters containing

921 A.R. Kudinov, P.V. Petrovskii, Yu.T. Struchkov, A.I. Yanovskii, and M.I. Rybinskaya, *J. Organomet. Chem.*, 1991, **421**, 91.
922 X.D. He, B. Chaudret, F. Dahan, and Y.-S. Huang, *Organometallics*, 1991, **10**, 970.
923 D. Rondon, B. Chaudret, X.-D. He, and D. Labroue, *J. Am. Chem. Soc.*, 1991, **113**, 5671.
924 R.H. Fish, R.H. Fong, A. Tran, and E. Baralt, *Organometallics*, 1991, **10**, 1209.
925 R.C. Cambie, L.G. Mackay, P.S. Rutledge, M. Tercel, and P.D. Woodgate, *J. Organomet. Chem.*, 1991, **409**, 263.
926 F.L. Joslin, M.P. Johnson, J.T. Mague, and D.M. Roundhill, *Organometallics*, 1991, **10**, 41.
927 F.L. Joslin, M.P. Johnson, J.T. Mague, and D.M. Roundhill, *Organometallics*, 1991, **10**, 2781.
928 F.L. Joslin, J.T. Mague, and D.M. Roundhill, *Organometallics*, 1991, **10**, 521; H. Lehmkuhl, R. Schwickardi, G. Mehler, C. Krüger, and R. Goddard, *Z. Anorg. Allg. Chem.*, 1991, **606**, 141.
929 D.C. Miller and R.J. Angelici, *Organometallics*, 1991, **10**, 79.
930 F.L. Joslin, J.T. Mague, and D.M. Roundhill, *Polyhedron*, 1991, **10**, 1713.
931 C. Wei, F. Aigbirhio, H. Adams, N.A. Bailey, P.D. Hempstead, and P.M. Maitlis, *J. Chem. Soc., Chem. Commun.*, 1991, 883.
932 A.R. Kudinov, D.V. Muratov, M.I. Rybinskaya, P.V. Petrovskii, A.V. Mironov, T.V. Timofeeva, Yu.L. Slovokhotov, and Yu.T. Struchkov, *J. Organomet. Chem.*, 1991, **414**, 97.
933 C.J. Adams, M.I. Bruce, B.W. Skelton, and A.H. White, *J. Organomet. Chem.*, 1991, **420**, 95.
934 A.Z. Kreindlin, E.I. Fedin, P.V. Petrovskii, M.I. Rybinskaya, R.M. Minyaev, and R. Hoffmann, *Organometallics*, 1991, **10**, 1206.
935 R.E. Lehmann and J.K. Kochi, *J. Am. Chem. Soc.*, 1991, **113**, 501.
936 D. Carmona, J. Ferrer, F.J. Lahoz, L.A. Oro, J. Reyes, and M. Esteban, *J. Chem. Soc., Dalton Trans.*, 1991, 2811.
937 T.P. Kee, L.Y. Park, J. Robbins, and R.R. Schrock, *J. Chem. Soc., Chem. Commun.*, 1991, 121.
938 J.D. Koola and D.M. Roddick, *J. Am. Chem. Soc.*, 1991, **113**, 1450.
939 R.S. Tanke, E.M. Holt, and R.H. Crabtree, *Inorg. Chem.*, 1991, **30**, 1714.
940 M. Gaye, B. Demerseman, and P.H. Dixneuf, *J. Organomet. Chem.*, 1991, **411**, 263.
941 T. Daniel, M. Müller, and H. Werner, *Inorg. Chem.*, 1991, **30**, 3118.
942 R.I. Michelman, R.A. Andersen, and R.G. Bergman, *J. Am. Chem. Soc.*, 1991, **113**, 5100.
943 H. Inoue, T. Takei, G. Heckmann, and E. Fluck, *Z. Naturforsch., B*, 1991, **46**, 682 (*Chem. Abstr.*, 1991, **115**, 104 644).
944 B. Boitrel, A. Lecas-Nawrocka, and E. Rose, *Tetrahedron Lett.*, 1991, **32**, 2129.
945 J.D. Augspurger, C.E. Dykstra, and E. Oldfield, *J. Am. Chem. Soc.*, 1991, **113**, 2447.

cobalt and ruthenium atoms has been investigated by ^{59}Co and ^{99}Ru n.m.r. spectroscopy.[946] The ^{13}C n.m.r. spectrum of $[Ru_3(CO)_{10}(\mu\text{-}CO)(\mu\text{-}CNCF_3)]$ shows only one signal for ^{13}CO.[947] N.m.r. data have also been reported for $[(OC)_4FePH\{N(SiMe_3)_2\}NHSiMe_3$, (^{13}C),[948] $[Fe(RPPRCH{=}PR)(CO)_3]$, (R = 2,4,6-$Bu^t{}_3C_6H_2$; ^{13}C),[949] $[\{(Pr^i{}_2NP)_2OCR^1R^2\}Fe_2(CO)_6]$, (^{13}C),[950] $[\{(Pr^i{}_2NP)_2HCOCHPMe_3\}Fe_2(CO)_6]$, (^{13}C),[951] $[Fe_3(CO)_8(CNPh)(\mu_3\text{-}S)_2]$, (^{13}C),[952] $[(\mu_3\text{-}CO)_2\{(\eta^5\text{-}C_5Me_5)Ni\}_2(OC)_3Fe]$, (^{13}C),[953] $[Fe_4(CO)_9(PR)E]$, (E = PR, Se, Te; ^{13}C, ^{125}Te),[954] $[Fe_3CuI(CO)_9(CCO)]^{2-}$, (^{13}C),[955] $[\{Fe_2(CO)_8\}\{\mu_4\text{-}\eta^2\text{-}Cu_2(Cy_2PCH_2CH_2PCy_2)\}]^{2-}$, (^{13}C),[956] $[Fe_4(CO)_{12}BHAu_2(AsPh_3)_2]$, (^{11}B),[957] $[(Bu^t{}_4\text{-phthalocyanine})Ru(CNBu^t)_2]$, (^{13}C),[958] $[RuI(CO)_2\{PhP(CH_2CH_2PPh_2)_2\}]$, (^{13}C),[959] $[(Ph_3P)(OC)_2Ru(SEt)_3Na(THF)]_2$, (^{13}C),[960] $[(OC)_4(Ph_3P)_2Ru_2(\mu\text{-}SEt)_4(\mu_3\text{-}SEt)_2Na_2(THF)_2]$, (^{13}C),[961] $[Ru_6(\mu_4\text{-}Hg)(\mu_3\text{-}2\text{-}NH_2C_5H_4N)_2\text{-}(CO)_{18}]$, (^{13}C),[962] $[Ru_{10}N(CO)_{24}]$, (^{14}N),[963] $[\{Ru_6C(CO)_{16}\}_2Hg]^{2-}$, (^{13}C),[964] $[Os_3(CO)_{11}$(1-Ph-3,4-Me_2-phosphole)], (^{13}C),[965] $[Os_{18}Hg_3C_2(CO)_{42}]^{2-}$, (^{13}C),[966] and $[Os_{20}(CO)_{40}]^{2-}$, (^{13}C).[967]

Complete ^{1}H and ^{13}C n.m.r. assignments of Ru(II) complexes of two new 2-pyridyl quinoline ligands have been made.[968] The field and temperature dependence of the ^{31}P T_1 in $[Ru_{55}(PBu^t{}_3)_{12}Cl_{20}]$ follows the power law $T_1^{-1} \propto T^nB^{-m}$.[969] The statistical incorporation of $^{13}C_2$ into C_{60} has been investigated by ^{13}C n.m.r. spectroscopy after conversion to $[C_{60}(OsO_4)\text{-}(NC_5H_4Bu^t\text{-}4)]$.[970] N.m.r. data have also been reported for substituted (1,2-naphthalocyaninato)iron compounds, (^{13}C),[971] (59), (^{13}C, ^{119}Sn),[972] (tetraethylphthalocyaninato)iron(II) compounds,

946 P. Braunstein, J. Rosé, P. Granger, and T. Richert, *Magn. Reson. Chem.*, 1991, **29**, S31.
947 D. Lentz, R. Marschall, and E. Hahn, *Chem. Ber.*, 1991, **124**, 777.
948 N. Dufour, J.-P. Majoral, A.-M. Caminade, R. Choukroun, and Y. Dromzée, *Organometallics*, 1991, **10**, 45.
949 J. Hein, M. Nieger, and E. Niecke, *Organometallics*, 1991, **10**, 13.
950 R.B. King, N.K. Bhattacharyya, and E.M. Holt, *J. Organomet. Chem.*, 1991, **421**, 247.
951 R.B. King, N.K. Bhattacharyya, and E.M. Holt, *Inorg. Chem.*, 1991, **30**, 1174.
952 H.G. Raubenheimer, L. Linford, G. Kruger, and A. Lombard, *J. Chem. Soc., Dalton Trans.*, 1991, 2795.
953 J.J. Schneider, R. Goddard, and C. Krüger, *Organometallics*, 1991, **10**, 665.
954 B. Eber, D. Buchholz, G. Huttner, Th. Fässler, W. Imhof, M. Fritz, J.C. Jochims, J.C. Daran, and Y. Jeanin, *J. Organomet. Chem.*, 1991, **401**, 49.
955 A.S. Gunale, M.P. Jensen, C.L. Stern, and D.F. Shriver, *J. Am. Chem. Soc.*, 1991, **113**, 1458.
956 H. Deng and S.G. Shore, *Organometallics*, 1991, **10**, 3486.
957 C.E. Housecroft, M.S. Shongwe, A.L. Rheingold, and P. Zanello, *J. Organomet. Chem.*, 1991, **408**, 7.
958 M. Hanack and P. Vermehren, *Chem. Ber.*, 1991, **124**, 1733.
959 G. Jia and D.W. Meek, *Inorg. Chim. Acta*, 1991, **178**, 195.
960 P.G. Jessop, S.J. Rettig, C.-L. Lee, and B.R. James, *Inorg. Chem.*, 1991, **30**, 4617.
961 P.G. Jessop, S.J. Rettig, and B.R. James, *J. Chem. Soc., Chem. Commun.*, 1991, 773.
962 P.L. Andreu, J.A. Cabeza, A. Llamazares, V. Riera, C. Bois, and Y. Jeannin, *J. Organomet. Chem.*, 1991, **420**, 431.
963 P.J. Bailey, G.C. Conole, B.F.G. Johnson, J. Lewis, M. McPartlin, A. Moule, and D.A. Wilkinson, *Angew. Chem., Int. Ed. Engl.*, 1991, **30**, 1706.
964 B.F.G. Johnson, W.-L. Kwik, J. Lewis, P.R. Raithby, and V.P. Saharan, *J. Chem. Soc., Dalton Trans.*, 1991, 1037.
965 A.J. Deeming, N.I. Powell, A.J. Arce, Y. De Sanctis, and J. Manzur, *J. Chem. Soc., Dalton Trans.*, 1991, 3381.
966 L.H. Gade, B.F.G. Johnson, J. Lewis, M. McPartlin, T. Kotch, and A.J. Lees, *J. Am. Chem. Soc.*, 1991, **113**, 8698.
967 A.J. Amoroso, L.H. Gade, B.F.G. Johnson, J. Lewis, P.R. Raithby, and W.-T. Wong, *Angew. Chem., Int. Ed. Engl.*, 1991, **30**, 107.
968 P. Finacciaro, A. Mamo, and C. Tringali, *Magn. Reson. Chem.*, 1991, **29**, 1165.
969 M.P.J. Van Staveren, H.B. Brom, L.J. De Jongh, and G. Schmid, *Z. Phys. D: At., Mol. Clusters*, 1991, **20**, 333 (*Chem. Abstr.*, 1991, **115**, 63 165).
970 J.M. Hawkins, S. Loren, A. Meyer, and R. Nunlist, *J. Am. Chem. Soc.*, 1991, **113**, 7770; J.M. Hawkins, A. Meyer, S. Loren, and R. Nunlist, *J. Am. Chem. Soc.*, 1991, **113**, 9394.
971 M. Hanack, G. Renz, J. Strähle, and S. Schmid, *J. Org. Chem.*, 1991, **56**, 3501.
972 Y.Z. Voloshin, N.A. Kostromina, A.Y. Nazarenko, and E.V. Polshin, *Inorg. Chim. Acta*, 1991, **185**, 83.

(^{13}C),[973] [{HO(3-pyridyl)$_3$}$_2$Fe]$^{2+}$, (^{13}C),[974] macrobicyclic boron containing iron(II) cyclooctanedione dioximates, (^{11}B, ^{13}C),[975] (60), (^{13}C),[976] [(bipy)$_2$Ru{bipy-5,5'-(4-C$_5$H$_5$NMe)$_2$}]$^{4+}$, (^{13}C),[977] [Ru(O)$_2$(bipy)(SCH$_2$CHRCO$_2$)], (^{13}C),[978] Ru(II) complexes of

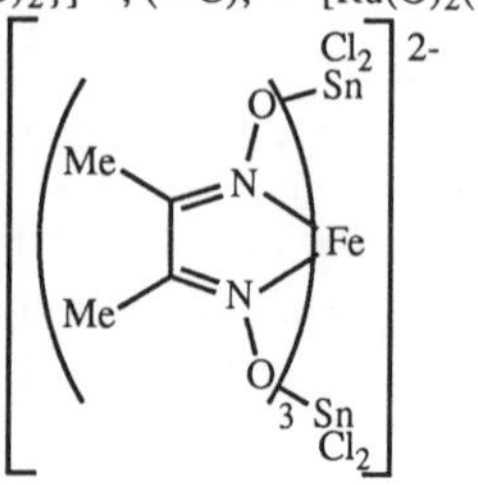

(59)

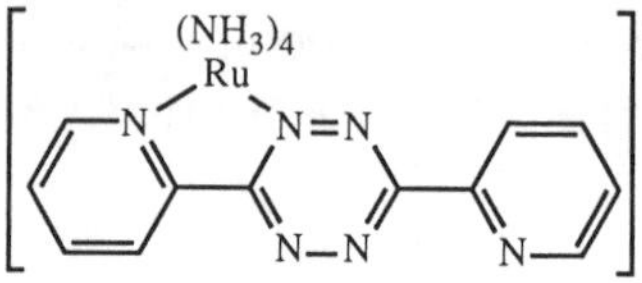

(60)

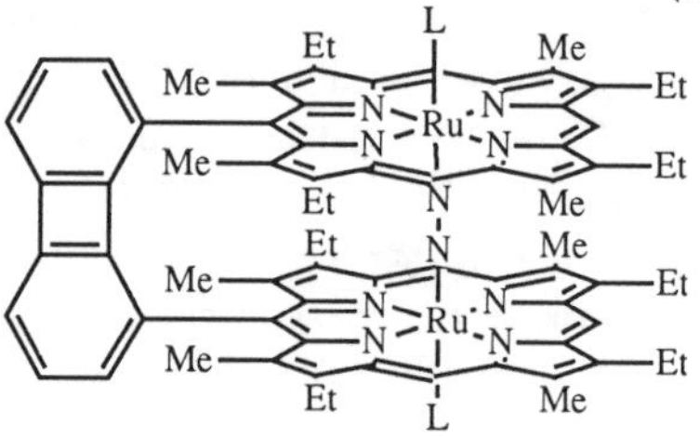

(61)

6-(*N*-pyrazolyl)-2,2-bipyridyls, (^{13}C),[979] (61), (L = 3-Bu^t-4-Ph-imidazole, ^{15}N),[980] [pzB(pz)$_3$Ru-(NCPh)$_2$Cl], (^{13}C),[981] [Ru(edta)(adenosine)]$^-$, (^{13}C),[982] [Ru(2-O$_2$CC$_5$H$_4$N)$_3$], (^{13}C),[983] [RuCl$_2$(4-MeC$_6$H$_4$CN)$_4$], (^{13}C),[984] [RuCl$_2${HB(3,5-Me$_2$C$_3$HN$_2$)$_3$}NO], (^{11}B, ^{13}C),[985] [RuX$_3$(8-hydroxyquinolinate)(NO]$^-$, (^{13}C),[986] [M(N)(O$_2$CCH$_2$CH$_2$S)$_2$]$^-$, (^{13}C),[987] [Os(NAr)$_2$(O$_2$CMe)$_2$-(PMe$_2$Ph)], (^{13}C),[988] [Ru{P(CH$_2$CH$_2$PPh$_2$)$_3$}(O$_2$CMe)]$^+$, (^{13}C),[989] [Ru$_2$(μ-O)(μ-O$_2$CMe)$_2$(O$_2$C-Me)$_2$(PPh$_3$)$_2$], (^{13}C),[990] [Ru(S$_2$C$_6$H$_4$)$_2$(PMe$_3$)$_2$], (^{13}C),[991] H$_n$Ru$_4$O$_6$–(OH$_2$)$_{12}$]$^{(4+n)+}$, (^{17}O),[992]

973 A. Beck, K.M. Mangold, and M. Hanack, *Chem. Ber.*, 1991, **124**, 2315.
974 R.K. Boggess, A.H. Lamson, and S. York, *Polyhedron*, 1991, **10**, 2791.
975 Yu.Z. Voloshin, N.A. Kostromina, A.Yu. Nazarenko, V.N. Shuman, and E.V. Pol'shin, *Ukr. Khim. Zh. (Russ. Ed.)*, 1990, **56**, 1238 (*Chem. Abstr.*, 1991, **115**, 104 659).
976 J.E.B. Johnson, C. de Groff, and R.R. Ruminski, *Inorg. Chim. Acta*, 1991, **187**, 73.
977 K. Bierig, R.J. Morgan, S. Tysoe, H.D. Gafney, T.C. Strekas, and A.D. Baker, *Inorg. Chem.*, 1991, **30**, 4898.
978 W.S. Bigham and P.A. Shapley, *Inorg. Chem.*, 1991, **30**, 4093.
979 A.J. Downard, G.E. Honey, and P.J. Steel, *Inorg. Chem.*, 1991, **30**, 3733.
980 J.P. Collman, J.E. Hutchison, M.A. Lopez, R. Guilard, and R.A. Reed, *J. Am. Chem. Soc.*, 1991, **113**, 2794.
981 M. Onishi, K. Ikemoto, and K. Hiraki, *Inorg. Chim. Acta*, 1991, **190**, 157.
982 B.T. Khan and K. Annapoorna, *Polyhedron*, 1991, **10**, 2465.
983 M.C. Barral, R. Jiménez-Aparicio, E.C. Royer, M.J. Saucedo, F.A. Urbanos, E. Gutiérrez-Puebla, and C. Ruíz-Valero, *J. Chem. Soc., Dalton Trans.*, 1991, 1609.
984 C.M. Duff and G.A. Heath, *J. Chem. Soc., Dalton Trans.*, 1991, 2401.
985 M. Onishi, *Bull. Chem. Soc. Jpn.*, 1991, **64**, 3039.
986 E. Miki, K. Harada, Y. Kamata, M. Umehara, K. Mizumachi, T. Ishimori, M. Nakahara, M. Tanaka, and T. Nagai, *Polyhedron*, 1991, **10**, 583.
987 J.J. Schwab, E.C. Wilkinson, S.R. Wilson, and P.A. Shapley, *J. Am. Chem. Soc.*, 1991, **113**, 6124.
988 M.H. Schofield, T.P. Kee, J.T. Anhaus, R.R. Schrock, K.H. Johnson, and W.M. Davis, *Inorg. Chem.*, 1991, **30**, 3595.
989 C. Bianchini, E. Farnetti, P. Frediani, M. Graziani, M. Peruzzini, and A. Polo, *J. Chem. Soc., Chem. Commun.*, 1991, 1336.
990 M.C. Barral, R. Jiménez-Aparicio, E.C. Royer, F.A. Urbanos, A. Monge, and C. Ruiz-Valero, *Polyhedron*, 1991,

$[RuCl_2(1,5\text{-dithiacyclooctane 1-oxide})_2]$, ($^{13}C$),[993] $[(OsO_2)_2(\text{tartrate})_3]^{4-}$, ($^{13}C$),[994] and $[Fe(S_2C_6H_4)_2]^{2-}$, (^{13}C).[995]

Complexes of Co, Rh, and Ir.—Unusually short minimum T_1 values, 40 to 43 ms at 200 MHz, have been observed for $[CoH(dppe)_2]$, $[CoH_2(dppe)_2]^+$, and $[CoH(CO)(PPh_3)_3]$. The relaxation data were interpreted taking into account the proton-metal dipole-dipole interactions.[996] (^1H, ^{103}Rh) and (^1H, ^{103}Rh){^2H} inverse correlation has been applied to $[RhH_2(H_2)\{(N_2C_3HMe_2\text{-}3,5)_3BH\}]$. It was shown that the large isotope shift can be used to separate the one dimensional ^1H and $^1H\{^2H\}$ n.m.r. spectra.[997] INEPT has been used to determine ^{15}N and ^{103}Rh n.m.r. data on $[RhH_2(PPh_3)_2(bipy)]^+$, and the stereochemistry was determined.[998] N.m.r. data have also been reported for $[(\eta^5\text{-}C_5H_4SiEt_3)RhH_2(SiEt_3)_2]$, ($^{13}C$, ^{29}Si, ^{103}Rh),[999] $[(C_6H_4O_2B)RhHCl(PPr^i_3)_2]$, ($^{11}B$),[1000] $\{RhH(PBu^n_3)_3C{\equiv}CC_6H_4\text{-}4\text{-}C{\equiv}C\}_n$, ($^{13}C$),[1001] $[(\eta^5\text{-}C_5H_5)RhH\{C_6H_3(CF_3)_2\text{-}2,5\}(PMe_3)]$, ($^{13}C$),[1002] $[RhH(SnCl_3)_5]^{3-}$, (^{119}Sn),[1003] $[C(PPh_2{=}O)_3IrH_2(SiMe_2Ph)_2]$, ($^{13}C$),[1004] $[IrH(CH{=}CHCH{=}CHO)(PEt_3)_3]$, ($^{13}C$),[1005] $[(\eta^5\text{-}C_5Me_5)IrH(\eta^4\text{-}1,5\text{-}C_8H_{12})]^+$, ($^{13}C$),[1006] $[(\eta^5\text{-}C_5Me_5)IrH(OEt)(PPh_3)]$, ($^{13}C$),[1007] and $[HIr(\eta^2\text{-}C_8H_{14})(NH_3)_4]^+$, ($^{13}C$).[1008]

The Co—N(1,5,6-Me_3-benzimidazole) bond lengths of $[RCo(dmgH)_2(1,5,6\text{-}Me_3\text{-benzimidazole})]$ show a fairly linear relation with the electronic parameter of the axial R group, derived from the ^{13}C n.m.r. spectra of their pyridine analogues. Steric effects were also discussed.[1009] The ^{19}F (CF_3) chemical shift of $[(CF_3)_2CFCo(dmgH)_2L]$ are largely independent of L, but the ^{19}F (CF) chemical shifts correlate linearly with the ligand basicity.[1010] $[MeRh(CO)_2I_3]^-$ has been observed as an intermediate in the reaction of $[Rh(CO)_2I_2]^-$ with MeI.[1011] Extensive use of COSY has been made to establish the assignment of the ^1H connectivity in (62).[1012] N.m.r. data have also been reported for HCF_2-cobinamides, (^{13}C),[1013] $[(\eta^2\text{-}C{\equiv}CPh)Co_2(CO)_6MeSi(OR)_2]$, ($^{13}C$),[1014] $[CH_2{=}CHCH_2CH_2\text{-}$

10, 113.
991 D. Sellmann, M. Geck, F. Knoch, and M. Moll, *Inorg. Chim. Acta*, 1991, **186**, 187.
992 A. Patel and D.T. Richens, *Inorg. Chem.*, 1991, **30**, 3789.
993 B.W. Arbuckle, P.K. Bharadwaj, and W.K. Musker, *Inorg. Chem.*, 1991, **30**, 440.
994 C.F. Edwards and W.P. William, *Polyhedron*, 1991, **10**, 61.
995 D. Sellmann, M. Geck, and M. Moll, *J. Am. Chem. Soc.*, 1991, **113**, 5259.
996 D.G. Gusev, A.B. Vymenits, and V.I. Bakhmutov, *Inorg. Chem.*, 1991, **30**, 3116.
997 D. Nanz, W. von Philipsborn, U.E. Bucher, and L.M. Venanzi, *Magn. Reson. Chem.*, 1991, **29**, S38.
998 B.T. Heaton, C. Jacob, W. Heggie, P.R. Page, and I. Villax, *Magn. Reson. Chem.*, 1991, **29**, S21.
999 S.B. Duckett and R.N. Perutz, *J. Chem. Soc., Chem. Commun.*, 1991, 28.
1000 S.A. Westcott, N.J. Taylor, T.B. Marder, R.T. Baker, N.J. Jones, and J.C. Calabrese, *J. Chem. Soc., Chem. Commun.*, 1991, 304.
1001 H.B. Fyfe, M. Mlekuz, D. Zargarian, N.J. Taylor, and T.B. Marder, *J. Chem. Soc., Chem. Commun.*, 1991, 188.
1002 S.T. Belt, L. Dong, S.B. Duckett, W.D. Jones, M.G. Partridge, and R.N. Perutz, *J. Chem. Soc., Chem. Commun.*, 1991, 266.
1003 I. Hall and K.R. Koch, *Polyhedron*, 1991, **10**, 1721.
1004 R.S. Tanke and R.H. Crabtree, *Organometallics*, 1991, **10**, 415.
1005 J.R. Bleeke, T. Haile, and M.Y. Chiang, *Organometallics*, 1991, **10**, 19.
1006 J.R. Sowa, jun., and R.J. Angelici, *J. Am. Chem. Soc.*, 1991, **113**, 2537.
1007 D.S. Glueck, L.J.N. Winslow, and R.G. Bergman, *Organometallics*, 1991, **10**, 1462.
1008 R. Koelliker and D. Milstein, *Angew. Chem., Int. Ed. Engl.*, 1991, **30**, 706.
1009 X. Solans, M. Gomez, and C. López, *Transition Met. Chem. (London)*, 1991, **16**, 176.
1010 P.J. Toscano, H. Brand, S. Geremia, L. Randaccio, and E. Zangrando, *Organometallics*, 1991, **10**, 713.
1011 A. Haynes, B.E. Mann, D.J. Gulliver, G.E. Morris, and P.M. Maitlis, *J. Am. Chem. Soc.*, 1991, **113**, 8567.
1012 E.C. Constable, R.P.G. Henney, and D.A. Tocher, *J. Chem. Soc., Dalton Trans.*, 1991, 2335.
1013 K.L. Brown, X. Zou, and L. Salmon, *Inorg. Chem.*, 1991, **30**, 1949.
1014 U. Lay and H. Lang, *J. Organomet. Chem.*, 1991, **418**, 79.

$CH_2Co(dmgH)_2(NC_5H_4Bu^t\text{-}4)]$, ($^{13}C$),[1015] $[RCo(dmgH)_2(NC_5H_5)]$, (^{13}C),[1016] [RCo(diphenylglyoximato)$_2$(NC$_5$H$_5$)], (^{13}C),[1017] (adeninylpropyl)cobalamin, (^{13}C),[1018] alkylcobalamin, (^{13}C),[1019] $[Co_2(CO)_6(\mu\text{-}CO)\{\mu\text{-}C(CH_2OSiMe_3)(OSiPh_3)\}]$, ($^{13}C$),[1020] (63), ($^{13}C$, ^{59}Co),[1021] $[(\eta^5\text{-}C_5H_5)\text{-}Co\{PhC{=}CPhAlH(NEt_3)CPh{=}CPh\}(PPh_3)]$, ($^{13}C$, ^{27}Al),[1022] $[(\eta^5\text{-}C_5H_5)Co(CCl{=}CClCCl{=}CCl)\text{-}(PPh_3)]$, ($^{13}C$),[1023] $[EtCH(OH)CH_2COCo(CO)_4]$, (^{13}C),[1024] [MeC(O)Co(1,4,8,11-tetra-azatetradecane)(OH$_2$)]$^{2+}$, (^{13}C),[1025] $[(\eta^5\text{-}C_5H_5)Co\{{=}C(NHPr)NMeC{=}S\}(PMe_3)]^+$, ($^{13}C$),[1026] (64), (R = C≡CR; ^{13}C),[1027] $[Me_2Si(SiMe_2SiMe_2)_2SiMeCo(CO)_3(PPh_3)]$, ($^{13}C$, ^{29}Si),[1028] $[Me\{(\eta^2\text{-}C{\equiv}CPh)Co_2(CO)_6\}SiClCo(CO)_4]$, ($^{13}C$),[1029] (65), ($^{13}C$),[1030] $[Co_2(CO)_6(\mu\text{-}CO)\{\mu\text{-}Me_2Ge(CH_2)_n\text{-}GeMe_2\}]$, ($^{13}C$),[1031] $[Co(SnMe_3)(CO)_3L]$, (^{13}C, ^{119}Sn),[1032] $[Co(SnCl_mBr_{3\text{-}m})(SnCl_nBr_{3\text{-}n})(CO)_3]$,

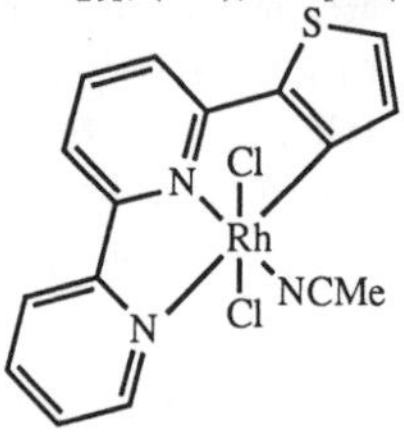

(62)

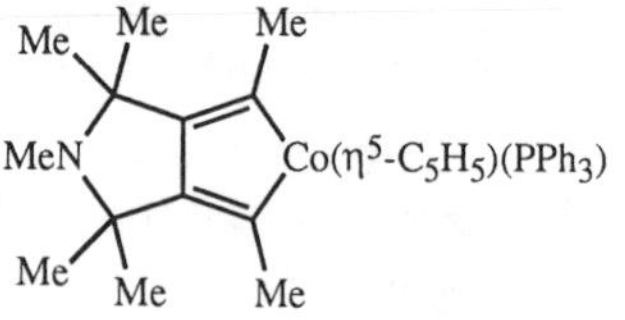

(63)

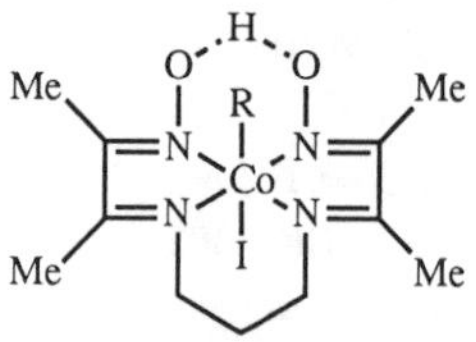

(64)

(^{119}Sn),[1033] $[(\eta^5\text{-}C_5Me_5)_2Rh_2(\mu\text{-}CH_2)_2MeL]^+$, ($^{13}C$),[1034] $[(\eta^5\text{-}C_5Me_5)Rh(CO)Br(CH_3)]$, ($^{13}C$),[1035] $[Rh_2(\mu\text{-}dppm)_2(\mu\text{-}CO)(CH_3)_2]$, ($^{13}C$),[1036] $[(\eta^5\text{-}C_5H_5)Rh(CH_2PPr^i_3)(\eta^2\text{-}C_2H_4)]$, ($^{13}C$),[1037] $[Rh(CH_2PPh_2{=}NC_6H_4Me\text{-}4)(\eta^4\text{-}C_8H_{12})]$, ($^{13}C$, ^{103}Rh),[1038] [EtO$_2$CCHIRh{(4-

1015 J. Hartung and B. Giese, *Chem. Ber.*, 1991, **124**, 387.
1016 B.P. Branchaud and G.-X. Yu, *Organometallics*, 1991, **10**, 3795.
1017 C. López, S. Alvarez, M. Font-Bardia, and X. Solans, *J. Organomet. Chem.*, 1991, **414**, 245.
1018 T.G. Pagano, L.G. Marzilli, M.M. Flocco, C. Tsai, H.L. Carrell, and J.P. Glusker, *J. Am. Chem. Soc.*, 1991, **113**, 531.
1019 Y.W. Alelyunas, P.E. Fleming, R.G. Finke, T.G. Pagano, and L.G. Marzilli, *J. Am. Chem. Soc.*, 1991, **113**, 3781.
1020 A. Sisak, A. Sironi, M. Moret, C. Zucchi, F. Ghelfi, and G. Pályi, *J. Chem. Soc., Chem. Commun.*, 1991, 176.
1021 Z. Zhou, L.P. Battaglia, G.P. Chiusoli, M. Costa, M. Nardelli, C. Pelizzi, and G. Predieri, *J. Organomet. Chem.*, 1991, **417**, 51.
1022 A.A. Aradi, F.E. Hong, and T.P. Fehlner, *Organometallics*, 1991, **10**, 2726.
1023 K. Sünkel, *Chem. Ber.*, 1991, **124**, 2449.
1024 J. Kreisz, F. Ungváry, A. Sisak, and L. Markó, *J. Organomet. Chem.*, 1991, **417**, 89.
1025 A. Bakac and J.H. Espenson, *J. Chem. Soc., Chem. Commun.*, 1991, 1497.
1026 H. Werner and B. Strecker, *J. Organomet. Chem.*, 1991, **413**, 379.
1027 B. Giese, M. Zehnder, M. Neuburger, and F. Trach, *J. Organomet. Chem.*, 1991, **412**, 415.
1028 E. Hengge and M. Eibl, *Organometallics*, 1991, **10**, 3185.
1029 H. Lang, U. Lay, and L. Zsolnai, *J. Organomet. Chem.*, 1991, **417**, 377.
1030 P. Dufour, M.-J. Menu, M. Dartiguenave, Y. Dartiguenave, and J. Dubac, *Organometallics*, 1991, **10**, 1645.
1031 J. Barrau and N.B. Hamida, *Inorg. Chim. Acta*, 1991, **178**, 141.
1032 C. Loubser, J.L.M. Dillen, and S. Lotz, *Polyhedron*, 1991, **10**, 2535.
1033 K.M. Mackay, B.K. Nicholson, M. Service, and R.A. Thomson, *Main Group Met. Chem.*, 1990, **13**, 197 (*Chem. Abstr.*, 1991, **114**, 258 420).
1034 J. Martinez, H. Adams, N.A. Bailey, and P.M. Maitlis, *J. Organomet. Chem.*, 1991, **405**, 393.
1035 D. Monti, M. Bassetti, G.J. Sunley, P.D. Ellis, and P.M. Maitlis, *Organometallics*, 1991, **10**, 4015.
1036 K.W. Kramarz, T.C. Eisenschmid, D.A. Deutsch, and R. Eisenberg, *J. Am. Chem. Soc.*, 1991, **113**, 5090.
1037 H. Werner, O. Schippel, J. Wolf, and M. Schulz, *J. Organomet. Chem.*, 1991, **417**, 149.
1038 P. Imhoff, S.C.A. Nefkens, C.J. Elsevier, K. Goubitz, and C.H. Stam, *Organometallics*, 1991, **10**, 1421.

$MeC_6H_4)_4$-porphyrin}], (^{13}C),[1039] [(η^5-$C_5Me_5)_2Rh_2(CO)_2(\mu$-$CMe_2)$], (^{13}C),[1040] [RhBr(CCl_3)Cl-(CO)($PMe_3)_2$], (^{13}C),[1041] [(PhHC=CH)Rh(dmgH)$_2$(PPh_3)], (^{13}C),[1042] [MeC($CH_2PPh_3)_3$-Rh(CH=CHCH=CH)Cl], (^{13}C),[1043] [(η^5-C_5Me_5)Rh(CH=CHCH=CHS)(PMe_3)], (^{13}C),[1044] [(Et_3)(Bu^{t}NC)$_2$Rh{C=C($SiMe_3$)N=NNBu^t}], (^{13}C),[1045] [(η^5-C_5Me_5)RhPh(PMe_2Ph)X], (^{13}C),[1046] [Rh(C_6H_4-2-CH_2-2-C_5H_4N)($PEt_3)_2Cl_2$], (^{13}C),[1047] [Rh(C_6H_4-2-$C_3H_3N_2)_2Cl]_2$, (^{13}C),[1048] [{EtC(O)}RhCl_2($PPh_3)_2$], (^{13}C),[1049] [(RC≡C)Rh(=C=CHR)(PPr$^i_3)_2$], (^{13}C),[1050] [Rh{=C=C-($SiMe_3)_2$}(PPr$^i_3)_2$Cl], (^{13}C, ^{29}Si),[1051] [(Pr$^i_3$P)$_2$ClRh=C=CHC_6H_4CH=C=RhCl(PPr$^i_3)_2$], (^{13}C),[1052] [(η^5-C_5H_5)Rh{=C(CHR)N(tosyl)}(PPr$^i_3$)]$^+$, (^{13}C),[1053] [(η^5-$C_5H_4CH_2$-2-$C_4H_7O)_2Rh_2$-(CO)$_2$], (^{13}C),[1054] [Rh(HgPh)(O_2CCF_3)(quinoline)(CO){P(C_6H_4OMe-4)$_3$}], (^{13}C),[1055]

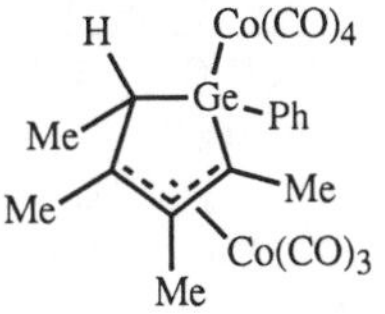

(65)

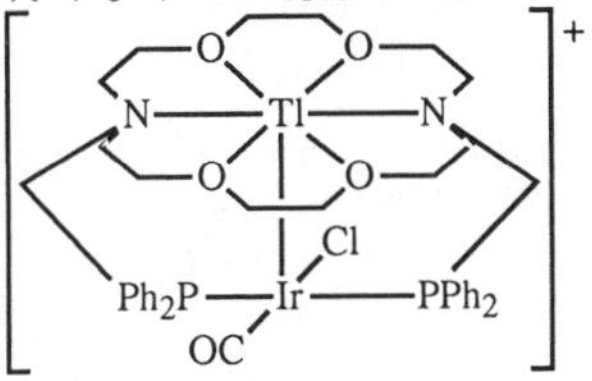

(66)

[Rh($SnCl_3$)(η^2-$C_2H_4)_2(PR_3)_2$], (^{119}Sn),[1056] [{η^2-MeC(O)CH=CH_2}$_3$Rh$SnCl_3$], (^{13}C, ^{119}Sn),[1057] [Rh(CO)X($PEt_3)_2(TeF_3)]^+$, (^{13}C),[1058] [(η^5-C_5Me_5)Ir(CH_3)(CH_2Ph)(PPh_3)], (^{13}C),[1059] [{ClPt-($PEt_3)_2CH_2C_5H_4$-η^5}Ir(CH_3)(Ph)(CO)], (^{13}C),[1060] [Ir(CH_3)I(PH_2Ph){N($SiMe_2CH_2PPh_2)_2$}], (^{13}C),[1061] [(η^5-C_5Me_5)Ir($CH_2CH_2CH_2$)(η^2-C_2H_4)], (^{13}C),[1062] [Ir(CH_2CH=CHCH=CH_2)($PEt_3)_3$],

1039 J. Maxwell and T. Kodadek, *Organometallics*, 1991, **10**, 4.
1040 R.C. Stevens, J.S. Ricci, T.F. Koetzle, and W.A. Herrmann, *J. Organomet. Chem.*, 1991, **412**, 425.
1041 C.J. Cable, H. Adams, N.A. Bailey, J. Crosby, and C. White, *J. Chem. Soc., Chem. Commun.*, 1991, 165.
1042 D. Steinborn, U. Sedlak, and M. Dargatz, *J. Organomet. Chem.*, 1991, **415**, 407.
1043 C. Bianchini, A. Meli, M. Peruzzini, A. Vacca, and F. Vizza, *Organometallics*, 1991, **10**, 645.
1044 W.D. Jones and L. Dong, *J. Am. Chem. Soc.*, 1991, **113**, 559.
1045 E. Deydier, M.-J. Menu, M. Dartiguenave, and Y. Dartiguenave, *J. Chem. Soc., Chem. Commun.*, 1991, 809.
1046 W.D. Jones and V.L. Kuykendall, *Inorg. Chem.*, 1991, **30**, 2615.
1047 K. Hiraki, M. Onishi, and H. Kishino, *Bull. Chem. Soc. Jpn.*, 1991, **64**, 1695.
1048 P.J. Steel, *J. Organomet. Chem.*, 1991, **408**, 395.
1049 J.A. Miller and J.A. Nelson, *Organometallics*, 1991, **10**, 2958.
1050 M. Schäfer, J. Wolf, and H. Werner, *J. Chem. Soc., Chem. Commun.*, 1991, 1341.
1051 D. Schneider and H. Werner, *Angew. Chem., Int. Ed. Engl.*, 1991, **30**, 700.
1052 H. Werner, T. Rappert, and J. Wolf, *Isr. J. Chem.*, 1990, **30**, 377.
1053 H. Werner, U. Brekau, and M. Dziallas, *J. Organomet. Chem.*, 1991, **406**, 237.
1054 T.E. Bitterwolf, K.A. Lott, and A.J. Rest, *J. Organomet. Chem.*, 1991, **408**, 137.
1055 M. Cano, J.V. Heras, M.A. Lobo, M. Martinez, E. Pinilla, E. Gutierrez, and M.A. Monge, *Polyhedron*, 1991, **10**, 187.
1056 V. García and M.A. Garralda, *Inorg. Chim. Acta*, 1991, **180**, 177.
1057 I.P. Kovalev, Yu.N. Kolmogorov, Yu.A. Strelenko, A.V. Ignatenko, M.G. Vinogradov, and G.I. Nikishin, *J. Organomet. Chem.*, 1991, **420**, 125.
1058 E.A.V. Ebsworth, J.H. Holloway, and P.G. Watson, *J. Chem. Soc., Chem. Commun.*, 1991, 1443.
1059 D.S. Glueck and R.G. Bergman, *Organometallics*, 1991, **10**, 1479.
1060 J.A. Miguel-Garcia, H. Adams, N.A. Bailey, and P.M. Maitlis, *J. Organomet. Chem.*, 1991, **413**, 427.
1061 M.D. Fryzuk, K. Joshi, R.K. Chadha, and S.J. Rettig, *J. Am. Chem. Soc.*, 1991, **113**, 8724.
1062 J.B. Wakefield and J.M. Stryker, *J. Am. Chem. Soc.*, 1991, **113**, 7057.

(^{13}C),[1063] $[(\eta^5\text{-}C_5H_5)_2Ir_2(\mu\text{-}\eta^2\text{-}CH{=}CHBu^t)_2]$, (^{13}C),[1064] $[Ir\{C(=CHBu^t)CH{=}CHBu^t\}\text{-}Ph(PMe_3)_3]$, (^{13}C),[1065] $[Ir(C_6H_4Me\text{-}2)(HgC_6H_4Me\text{-}2)(PPh_3)Cl]$, (^{13}C),[1066] and (66), (^{13}C).[1067]

Coherence pathways and "inverse" spectroscopy of I_nS_m spin systems have been applied to $(^{31}P,^{130}Rh)\text{-}\{^1H\}$ in $[(\eta^4\text{-norbornadiene})_2Rh_2(\mu\text{-}4,5\text{-}Me_2\text{-phosphabenzene-2-2'-pyridyl})]^{2+}$.[1068] The n.m.r. spectra of $[Rh\{R_2P(CH_2)_nPR_2\}_2]^+$ and $[Rh(\eta^4\text{-diene})\{R_2P(CH_2)_nPR_2\}]^+$ have been measured by inverse $(^{31}P,^{103}Rh)\text{-}\{^1H\}$ observation. The ^{103}Rh chemical shift range is -1350 to 200.[1069] ^1H COSY n.m.r. spectroscopy has been used to assign the ^1H n.m.r. spectrum of $[(\eta^4\text{-}1,5\text{-}C_8H_{12})Rh(\mu\text{-8-aza-9-methyladenine-H})]_2$.[1070] N.m.r. data have also been reported for $[Co(CO)_{4-n}(\eta^2\text{-alkene})_n]^-$, (^{13}C),[1071] $[\{(\eta^3\text{-}C_3H_5)Co(CO)_2\}_2(dppe)]$, (^{13}C),[1072] $[(\eta^5\text{-}C_5H_5)Co(\eta^2\text{-}C,C\text{-}R^1N{=}C{=}CR^2R^3)(PMe_3)]$, (^{13}C),[1073] (67), (^{13}C),[1074] (68), (^{13}C),[1075] $[(\mu\text{-}PhC_2Ph)Co_2(CO)_5(P_2Ph_4)]$, (^{13}C),[1076] $[(\mu\text{-}PhC_2R)Co_2(CO)_6]$, (^{13}C),[1077] $[\{Co_2(CO)_6(\mu\text{-}PhC_2)\}_2]$, (^{13}C),[1078] $[\{Co_2(CO)_6(\mu\text{-}PhC_2)\}_2PPh]$, (^{13}C),[1079] $[\{\mu\text{-}S(CH_2CH_2S)_2C_2CH_2\}Co_2(CO)_6]$, (^{13}C),[1080] $[Ph_2Si\{OCH_2C_2[Co_2(CO)_6]CH_2O\}_2SiPh_2]$, (^{13}C),[1081] $[Co_2Rh_2(CO)_7(\mu\text{-}CO)(\mu_4\text{-}\eta^2\text{-}HC{\equiv}CBu^n)(PPh_3)]$, (^{13}C),[1082] $[Co_4(CO)_8L_2(RC{\equiv}CR)]$, (^{13}C),[1083] $[(\eta^2\text{-}C_2H_4)_2Rh(1,4,7\text{-trithia-cyclononane})]^+$, (^{13}C),[1084] $[(\eta^5\text{-}C_5H_5)_2Rh_2(\mu\text{-}CO)\{\mu\text{-}\eta^2{:}\eta^2\text{-}CH_2{=}CMeC(CF_3){=}CHCF_3\}]$, (^{13}C),[1085] $[Ir_2(CO)_2(\mu\text{-}I)(\mu\text{-}MeO_2CC{\equiv}CCO_2Me)(dppm)_2]^+$, (^{13}C),[1086] $[(\eta^3\text{-}C_3H_5)Co(PPr^i{}_2CH_2\text{-}CH_2PPr^i{}_2)]$, (^{13}C),[1087] $[(\eta^5\text{-}C_5Me_5)Rh(\eta^3\text{-}MeO_2CCHCHCHCH_2CO_2Me)_2]$, (^{13}C),[1088] $[(\eta^5\text{-}C_5H_5)Co(\eta^4\text{-diene})]$, (^{13}C),[1089] $[(\eta^5\text{-}C_5Me_5)Co(\eta^4\text{-}C_7H_7C_7H_7\text{-}\eta^4)Co(\eta^5\text{-}C_5Me_5)]$, (^{13}C),[1090] $[(\eta^5\text{-}$

1063 J.R. Bleeke, D. Boorsma, M.Y. Chiang, T.W. Clayton, jun., T. Haile, A.M. Beatty, and Y.-F. Xie, *Organometallics*, 1991, **10**, 2391.
1064 A. Nessel, O. Nürnberg, J. Wolf, and H. Werner, *Angew. Chem., Int. Ed. Engl.*, 1991, **30**, 1006.
1065 H.E. Selnau and J.S. Merola, *J. Am. Chem. Soc.*, 1991, **113**, 4008.
1066 W.R. Roper and G.C. Saunders, *J. Organomet. Chem.*, 1991, **409**, C19.
1067 A.L. Balch, F. Neve, and M.M. Olmstead, *J. Am. Chem. Soc.*, 1991, **113**, 2995.
1068 D. Nanz and W. von Philipsborn, *J. Magn. Reson.*, 1991, **92**, 560.
1069 J.M. Ernsting, C.J. Elsevier, W.G.J. de Lange, and K. Timmer, *Magn. Reson. Chem.*, 1991, **29**, S118.
1070 W.S. Sheldrick and B. Günther, *J. Organomet. Chem.*, 1991, **402**, 265.
1071 F. Ungvary, *Organometallics*, 1991, **10**, 3053.
1072 C. Loubser, H.M. Roos, and S. Lotz, *J. Organomet. Chem.*, 1991, **402**, 393.
1073 B. Strecker, G. Hörlin, M. Schulz, and H. Werner, *Chem. Ber.*, 1991, **124**, 285.
1074 P. Magnus and T. Pitterna, *J. Chem. Soc., Chem. Commun.*, 1991, 541.
1075 P. Magnus and M. Davies, *J. Chem. Soc., Chem. Commun.*, 1991, 1522.
1076 A.J.M. Caffyn, M.J. Mays, G.A. Solan, D. Braga, P. Sabatino, G. Conole, M. McPartlin, and H.R. Powell, *J. Chem. Soc., Dalton Trans.*, 1991, 3103.
1077 H. Lang and L. Zsolnai, *Z. Naturforsch., B*, 1990, **45**, 1529 (*Chem. Abstr.*, 1991, **114**, 102 350).
1078 B.F.G. Johnson, J. Lewis, P.R. Raithby, and D.A. Wilkinson, *J. Organomet. Chem.*, 1991, **408**, C9.
1079 H. Lang and L. Zsolnai, *Chem. Ber.*, 1991, **124**, 259.
1080 A. Gelling, J.C. Jeffery, D.C. Povey, and M.J. Went, *J. Chem. Soc., Chem. Commun.*, 1991, 349.
1081 R.H. Cragg, J.C. Jeffery, and M.J. Went, *J. Chem. Soc., Dalton Trans.*, 1991, 137.
1082 I. Ojima, N. Clos, R.J. Donovan, and P. Ingallina, *Organometallics*, 1991, **10**, 3211.
1083 D. Osella, M. Ravera, C. Nervi, C.E. Housecroft, P.R. Raithby, P. Zanello, and F. Laschi, *Organometallics*, 1991, **10**, 3253.
1084 A.J. Blake, M.A. Halcrow, and M. Schröder, *J. Chem. Soc., Chem. Commun.*, 1991, 253.
1085 R.S. Dickson and B.C. Greaves, *J. Chem. Soc., Chem. Commun.*, 1991, 1300.
1086 B.A. Vaartstra, J. Xiao, J.A. Jenkins, R. Verhagen, and M. Cowie, *Organometallics*, 1991, **10**, 2708.
1087 M.D. Fryzuk, J.B. Ng, S.J. Rettig, J.C. Huffman, and K. Jonas, *Inorg. Chem.*, 1991, **30**, 2437.
1088 M. Brookhart and S. Sabo-Etienne, *J. Am. Chem. Soc.*, 1991, **113**, 2777.
1089 R. Boese, J. Rodriguez, and K.P.C. Vollhardt, *Angew. Chem., Int. Ed. Engl.*, 1991, **30**, 993.
1090 R.P. Aggarwal, N.G. Connelly, B.J. Dunne, M. Gilbert, and A.G. Orpen, *J. Chem. Soc., Dalton Trans.*, 1991, 1.

$C_5H_5)Co(\eta^4-R_4C_4H_2)]$, ($^{13}C$),[1091] (69), (R = CO_2Me; ^{13}C),[1092] $[(\eta^5-C_5Me_5)Co(\eta^4-C_5Me_5H)]$, ($^{13}C$),[1093] $[(\eta^5-C_5H_5)Co(\eta^4-C_4H_4SO_2)]$, ($^{13}C$),[1094] $[(\eta^4$-norbornadiene)Rh{$Ph_2P(CH_2)_nPMe_3$}-Cl]$^+$, (^{13}C),[1095] $[(\eta^4$-norbornadiene)Rh(bipy)]$^+$, (^{13}C),[1096] $[(\eta^4-1,5-C_8H_{12})Rh(4,4'-Pr^i_2-2,2'$-bioxazoline)]$^+$, ($^{13}C$),[1097] $[(\eta^4-1,5-C_8H_{12})Rh\{S(-2-C_5H_4N)_2\}]^+$, ($^{13}C$),[1098] $[(\eta^4-1,5-C_8H_{12})_2Rh_2\{\mu-S(CH_2)_3NMe_2CH_2Ph\}_2]^{2+}$, ($^{13}C$),[1099] $[(\eta^5-C_5H_5)Rh(\eta^4-Ph_2Bu^tC_3P)]$, ($^{13}C$),[1100] $[(\eta^5-C_5H_5)Ir(\eta^4-1,5-C_8H_{12})]$, ($^{13}C$),[1101] $[Ir(\eta^4-1,5-C_8H_{12})$(R-benzimidazole)Cl], ($^{13}C$),[1102] and *fac*-$[Ir\{\eta^4-C(CH_2)_3\}\{N(SiMe_2CH_2PPh_2)_2\}]$, ($^{13}C$).[1103]

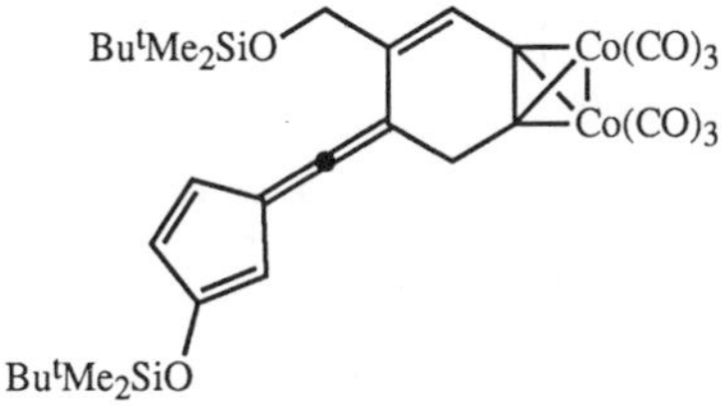

(67)

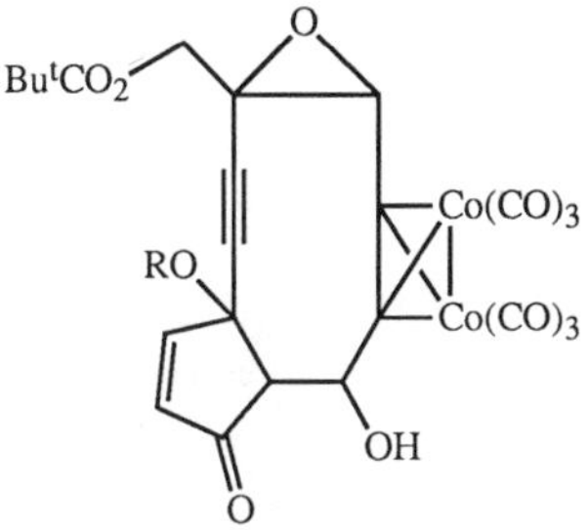

(68)

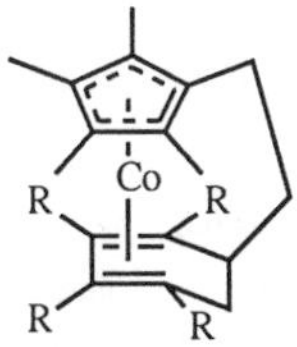

(69)

The effects of substituents on ^{103}Rh chemical shifts for a series of $[(\eta^5-C_5H_4R)Rh(CO)_2]$ have been investigated. The effect of electron-donating and -withdrawing substituents on the ^{103}Rh n.m.r. shift was qualitatively consistent with the traditional model of d_π electron back-donation.[1104] N.m.r. data have also been reported for $[1-\{(\eta^5-2,3,4,5-Ph_4C_4B)(\eta^5-C_5H_5)Co\}_2O]$, ($^{11}B$),[1105] $[(\eta^5-$

1091 V.L. Shirokii, V.A. Knizhnikov, A.V. Mosin, M.G. Novikova, N.A. Maier, and Yu.A. Ol'dekop, *Metalloorg. Khim.*, 1991, **4**, 1059 (*Chem. Abstr.*, 1991, **115**, 289 672).
1092 J. Okuda, K.H. Zimmermann, and E. Herdtweck, *Angew. Chem., Int. Ed. Engl.*, 1991, **30**, 430.
1093 J.J. Schneider, R. Goddard, S. Werner, and C. Krüger, *Angew. Chem., Int. Ed. Engl.*, 1991, **30**, 1124.
1094 R. Albrecht and E. Weiss, *J. Organomet. Chem.*, 1991, **413**, 355.
1095 E. Renaud, R.B. Russell, S. Fortier, S.J. Brown, and M.C. Baird, *J. Organomet. Chem.*, 1991, **419**, 403.
1096 P. Čudić, B. Klaić, Z. Raza, D. Šepac, and V. Sunjic, *Tetrahedron*, 1991, **47**, 5295.
1097 M. Onishi and K. Isagawa, *Inorg. Chim. Acta*, 1991, **179**, 155.
1098 G. Tresoldi, P. Piraino, E. Rotondo, and F. Faraone, *J. Chem. Soc., Dalton Trans.*, 1991, 425.
1099 J.C. Bayón, P. Esteban, J. Real, C. Claver, A. Polo, A. Ruiz, and S. Castillón, *J. Organomet. Chem.*, 1991, **403**, 393.
1100 P. Binger, J. Haas, A.T. Herrmann, F. Langhauser, and C. Krüger, *Angew. Chem., Int. Ed. Engl.*, 1991, **30**, 310.
1101 L.P. Szajek and J.R. Shapley, *Organometallics*, 1991, **10**, 2512.
1102 Y.S. Ramaswamy, R. Halesha, N.M.N. Gowda, and G.K.N. Reddy, *Indian J. Chem., Sect. A*, 1991, **30**, 393 (*Chem. Abstr.*, 1991, **115**, 92 539).
1103 M.D. Fryzuk, K. Joshi, and S.J. Rettig, *Organometallics*, 1991, **10**, 1642.
1104 P.B. Graham, M.D. Rausch, K. Taeschler, and W. Von Philipsborn, *Organometallics*, 1991, **10**, 3049.
1105 F.-E. Hong, T.P. Fehlner, and A.L. Rheingold, *Organometallics*, 1991, **10**, 1213.

$C_5H_4COMe)Co(\eta^2$-tcne)($CNMeCH_2CH_2NMe$)], (^{13}C),[1106] (70), (^{13}C),[1107] [(η^5-C_5H_5)RhCl(μ-Ph_2Ppy)Pt(CO)Cl], (^{13}C),[1108] [(η^5-C_5H_4-neomenthyl)$RhCl_2]_2$, (^{13}C),[1109] [(η^5-$C_5H_4PPh_2$)M(CO)$]_2^{2+}$, (M = Rh, Ir; ^{13}C),[1110] [(η^5-C_5Me_5)Rh($C_4H_3N_2C_4H_3N_2$)], (^{13}C),[1111] [(η^5-C_5Me_5)Rh($NC_5H_5)_3]^{2+}$, (^{13}C),[1112] [(η^5-C_5Me_5)RhCl(en)$]^+$, (^{13}C),[1113] [(η^5-C_5Me_5)M(dibenzothiophene)Cl_2], (M = Rh, Ir; ^{13}C),[1114] [(η^5-C_5Me_5)Ir(NBu^t)], (^{13}C, ^{15}N),[1115] [(η^5-C_5Me_5)Ir(4,4'-R_2-bipy)Cl$]^+$, (^{13}C),[1116] and [(η^5-C_5Me_5)Ir(PMe_3)Se_4], (^{13}C).[1117]

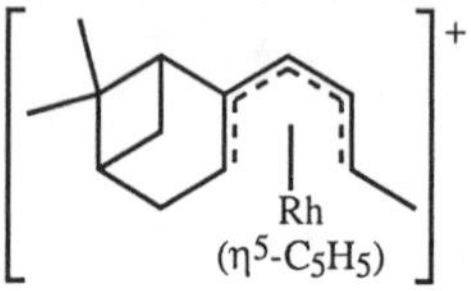

(70)

The ^{59}Co n.m.r. signal of [$Co_2(CO)_8$] in supercritical CO_2 is about 6 times sharper than in C_6D_6. The ^{13}C n.m.r. signal was also recorded.[1118] The ^{31}P n.m.r. spectrum of [$M_3(\mu$-$PPh_2)_3(CO)_nL_2$], M = Rh, Ir, has been analysed.[1119] N.m.r. data have also been reported for [Co_2{μ-P(C≡CPh)($OC_6H_2Bu^t_3$-2,4,6)}$_2(CO)_6$], (^{13}C),[1120] [H_2B(3,5-$Me_2C_3HN_2)_2$Rh(CN-neopentyl)$_2$], (^{13}C),[1121] [$M_2(CO)_4${μ-3,5-$(O_2C)_2C_3HN_2$}], (M = Rh, Ir; ^{13}C),[1122] *trans* -[Rh(CO)(OCRNMe_2)-($PPh_3)_2$], (^{13}C),[1123] [Rh(CO)(acac)(1,3,2-diazaphosphorinane)], (^{13}C),[1124] [{Bu^t_3P(OC)Rh}$_2$(μ-Cl)(μ-SR)], (^{13}C),[1125] *cis*-[{(Bu^t_3As)(CO)Rh}$_2$(μ-Cl)(μ-SR)], (^{13}C),[1126] [($Ph_3P)_2$Pt{CH[C(O)-Ph]$_2$S}Rh(CO)Cl], (^{13}C),[1127] and [Ir(CO)(2-PhC_5H_4N)($PPh_3)_2]^+$, (^{13}C).[1128]

^{59}Co n.m.r. spectroscopy has been used to study the interactions of [$Co(CN)_6]^{3-}$ and [$Co(NH_3)_6]^{3+}$ with nematic liquid crystals.[1129] The linkage isomers of [$Co(NH_3)_5$(succinamide)] have been

1106 S. Delgado, C. Moreno, and M.J. Macazaga, *Polyhedron*, 1991, **10**, 725.
1107 A. Salzer, H. Schmalle, R. Stauber, and S. Streiff, *J. Organomet. Chem.*, 1991, **408**, 403.
1108 S. Lo Schiavo, E. Rotondo, G. Bruno, and F. Faraone, *Organometallics*, 1991, **10**, 1613,
1109 P.A. Schofield, H. Adams, N.A. Bailey, E. Cesarotti, and C. White, *J. Organomet. Chem.*, 1991, **412**, 273.
1110 X. He, A. Maisonnat, F. Dahan, and R. Poilblanc, *Organometallics*, 1991, **10**, 2443.
1111 M. Ladwig and W. Kaim, *J. Organomet. Chem.*, 1991, **419**, 233.
1112 R.H. Fish, E. Baralt, and H.-S. Kim, *Organometallics*, 1991, **10**, 1965.
1113 G. Carcia, G. Sánchez, I. Romero, I. Solano, M.D. Santana, and G. López, *J. Organomet. Chem.*, 1991, **408**, 241.
1114 K.M. Rao, C.L. Day, R.A. Jacobsen, and R.J. Angelici, *Inorg. Chem.*, 1991, **30**, 5046.
1115 D.S. Glueck, J. Wu, F.J. Hollander, and R.G. Bergman, *J. Am. Chem. Soc.*, 1991, **113**, 2041.
1116 R. Ziessel, *Angew. Chem., Int. Ed. Engl.*, 1991, **30**, 844.
1117 M. Herberhold, G.-X. Jin, and A.L. Rheingold, *Chem. Ber.*, 1991, **124**, 2245.
1118 J.W. Rathke, R.J. Klingler, and T.R. Krause, *Organometallics*, 1991, **10**, 1350.
1119 D.E. Berry, J. Browning, K. Dehghan, K.R. Dixon, N.J. Meanwell, and A.J. Phillips, *Inorg. Chem.*, 1991, **30**, 396.
1120 H. Lang, M. Leise, and L. Zsolnai, *J. Organomet. Chem.*, 1991, **410**, 379.
1121 W.D. Jones and E.T. Hessell, *Inorg. Chem.*, 1991, **30**, 778.
1122 J.C. Bayón, G. Net, P. Esteban, P.G. Rasmussen, and D.F. Bergstrom, *Inorg. Chem.*, 1991, **30**, 4771.
1123 L. Song and P.J. Stang, *Inorg. Chim. Acta*, 1991, **188**, 107.
1124 A.L. Zavalishina, S.S. Dorogotovtsev, A.T. Teleshev, E.I. Orzhekovskaya, A.T. Bekker, and E.E. Nifant'ev, *Zh. Obshch. Khim.*, 1990, **60**, 2634 (*Chem. Abstr.*, 1991, **115**, 63 347).
1125 M. Eisen, P. Weitz, S. Shtelzer, J. Blum, H. Schumann, B. Gorella, and F.H. Görlitz, *Inorg. Chim. Acta*, 1991, **188**, 167.
1126 H. Schumann, B. Gorella, M. Eisen, and J. Blum, *J. Organomet. Chem.*, 1991, **412**, 251.
1127 J. Fawcett, W. Henderson, R.D.W. Kemmitt, and D.R. Russell, *J. Chem. Soc., Dalton Trans.*, 1991, 2595.
1128 F. Neve, M. Ghedini, G. De Munno, and A. Crispini, *Organometallics*, 1991, **10**, 1143.
1129 M. Iida and A.S. Tracey, *J. Phys. Chem.*, 1991, **95**, 7891.

characterised by ^1H and ^{13}C n.m.r. spectroscopy.[1130] In (71), L = $[SO_3]^{2-}$, a weak agostic interaction has been observed between a C—H and the cobalt, reducing $^1J(^{13}C—^1H)$ from 129 to 122 Hz.[1131] The solvent dependence of the ^{13}C and ^{59}Co n.m.r. shifts for *trans*- and *cis*-cobalt polyamine-*N*-polycarboxylates has been explained in terms of the angular overlap model.[1132] ^1H and ^{13}C n.m.r. spectroscopy has been used to assign the exchangeable acetate group in vitamin B_{12}.[1133] The ^1H, ^{13}C, and ^{15}N n.m.r. signals from dicyanocabalamin have been fully assigned.[1134] ^{13}C n.m.r. spectra has been used to study the biosynthesis of vitamin B_{12}.[1135] ^{15}N n.m.r. spectroscopy has been used to investigate linkage isomerism in $[(H_3N)_5Rh(glycinato)]^{2+}$.[1136] N.m.r. data have also been reported for $[(H_3N)_5Co(imidazole)]^{3+}$, ($^{13}C$),[1137] $[(H_3N)_5Co\{NH_2CH_2CH_2NH_2CH_2$-$(C_5H_4N)H\}]^{5+}$, ($^{13}C$),[1138] $[(H_3N)_5Co\{N(CHO)_2\}]^{2+}$, ($^{13}C$),[1139] $[(H_3N)_4CoNC_5H_2$-2-C(O)O-4-CO_2Et-6-OH], (^{13}C),[1140] $[(H_3N)_3Co$(L-malate)], (^{13}C, ^{59}Co),[1141] *mer*-$[Co(NH_2CH_2CH_2N{=}CH$-$CH_2NH_2)(en)Cl]^{2+}$, (^{13}C),[1142] [Co(NO)(Schiff base)], (^{15}N, ^{59}Co),[1143] $[(en)_2Co\{O_2CC{=}NCH_2CH(OH)CH_2\}]^{2+}$, ($^{13}C$),[1144] $[(en)_2Co(NMeCOCO_2)]^+$, (^{13}C),[1145] $[(acac)_2$-$Co(ClCH_2CH_2NHCH_2CH_2NHCH_2CH_2Cl)]^+$, ($^{13}C$),[1146] Co(III) complexes of *N*-β-alanyl-(*S*)-aspartate, (^{13}C),[1147] $[\{N(CH_2CH_2CH_2NH_2)_3\}Co(\mu\text{-}O)_2SO_2]^+$, ($^{13}C$),[1148] $[\{1,4\text{-}(H_2NCH_2CH_2$-$CH_2)_2$-1.4.7-diazacyclononane$\}CoCl]^{2+}$, ($^{13}C$),[1149] *trans*-$[Co(NO_2)_2$(1,4,7,10-tetra-azacycloheptadecane)$]^+$, (^{13}C),[1150] *trans*-$[Co(aziridine)_4(NO_2)_2]^+$, ($^{13}C$),[1151] $[\{1\text{-}[(O_2C)_2C(NH_2)]$-1,4,7-triazacyclononane$\}Co]^+$, ($^{13}C$),[1152] *cis*-$[Co(acac)_2(NC_5H_5)_2]^+$, ($^{13}C$),[1153] $[(-)_{589}$-*anti*(*N*)-Δ-*cis*(*N*),*cis*(*O*)-Λ-*cis*(*N*),*cis*(*O*)-di-μ-(OH)-(*S*-arginine)$_4Co_2]^{4+}$, (^{13}C),[1154] *cis*-$[Co(histamine)_2$-$L^1L^2]^{3+}$, (^{13}C),[1155] *cis*-β-[Co{*N*-(2-pyridylmethyl)-2-(2-aminoethylthioacetamide)(amino acidate)$]^+$,

[1130] P.M. Angus and W.G. Jackson, *Inorg. Chem.*, 1991, **30**, 4806.
[1131] W.E. Broderick, K. Kanamori, R.D. Willett, and J.I. Legg, *Inorg. Chem.*, 1991, **30**, 3875.
[1132] S. Kaizaki, N. Koine, and N. Sakagami, *Bull. Chem. Soc. Jpn.*, 1991, **64**, 2058.
[1133] A.I. Scott, N.J. Stolowich, B.P. Atshaves, P. Karuso, M.J. Warren, H.J. Williams, M. Kajiwara, K. Kurumaya, and T. Okazaki, *J. Am. Chem. Soc.*, 1991, **113**, 9891.
[1134] K.L. Brown, H.B. Brooks, B.D. Gupta, M. Victor, H.M. Marques, D.C. Scooby, W.J. Goux, and R. Timkovich, *Inorg. Chem.*, 1991, **30**, 3430.
[1135] D. Thibaut, F. Blanche, L. Debussche, F.J. Leeper, and A.R. Battersby, *Proc. Natl. Acad. Sci. U.S.A.*, 1990, **87**, 8800 (*Chem. Abstr.*, 1991, **114**, 96 892).
[1136] T.G. Appleton and M.R. Cox, *Magn. Reson. Chem.*, 1991, **29**, S80.
[1137] A.G. Blackman, D.A. Buckingham, C.R. Clark, and J. Simpson, *J. Chem. Soc., Dalton Trans.*, 1991, 3031.
[1138] M. Gajhede, A. Hammershøi, and L.K. Skov, *Acta Chem. Scand.*, 1991, **45**, 474.
[1139] A.G. Blackman, D.A. Buckingham, C.R. Clark, and J. Simpson, *Inorg. Chem.*, 1991, **30**, 1635.
[1140] J. MacB. Harrowfield, A.M. Sargeson, and P.O. Whimp, *Inorg. Chem.*, 1991, **30**, 1732.
[1141] S. Bunel, C. Ibarra, V. Calvo, A. Blaskó, C.A. Bunton, and N.L. Keder, *Polyhedron*, 1991, **10**, 2495.
[1142] W.G. Jackson and R.J. Walsh, *Inorg. Chem.*, 1991, **30**, 4813.
[1143] L.F. Larkworthy and S.K. Sengupta, *Inorg. Chim. Acta*, 1991, **179**, 157.
[1144] A. Hammershøi, R.M. Hartshorn, and A.M.M. Sargeson, *J. Chem. Soc., Dalton Trans.*, 1991, 621.
[1145] L. Grøndahl, A. Hammershøi, R.M. Hartshorn, and A.M. Sargeson, *Inorg. Chem.*, 1991, **30**, 1800.
[1146] D.C. Ware, W.R. Wilson, W.A. Denny, and C.E.F. Rickard, *J. Chem. Soc., Chem. Commun.*, 1991, 1171.
[1147] T. Ama, R. Maki, H. Kawaguchi, and T. Yasui, *Bull. Chem. Soc. Jpn.*, 1991, **64**, 459.
[1148] M. Banaszczyk, J.-J. Lee, and F.M. Menger, *Inorg. Chem.*, 1991, **30**, 1972.
[1149] D.G. Fortier and A. McAuley, *J. Chem. Soc., Dalton Trans.*, 1991, 101.
[1150] M. Tsuchimoto, M. Kita, and J. Fujita, *Bull. Chem. Soc. Jpn*, 1991, **64**, 2554.
[1151] D.C. Ware, B.G. Sim, K.G. Robinson, W.A. Denny, P.J. Brothers, and G.R. Clark, *Inorg. Chem.*, 1991, **30**, 3750.
[1152] T. Kojima, R. Kuroda, S. Yano, and M. Hidai, *Inorg. Chem.*, 1991, **30**, 3580.
[1153] J.G. Wardeska, *Synth. React. Inorg. Metal-Org. Chem.*, 1991, **21**, 669.
[1154] P.N. Radivojša, N. Juranić, M.B. Ćelap, K. Toriumi, and K. Saito, *Polyhedron*, 1991, **10**, 2717.
[1155] E. Danilczuk, *Pol. J. Chem.*, 1990, **64**, 453 (*Chem. Abstr.*, 1991, **115**, 269 090).

(^{13}C),[1156] [Co_3(aminoethanolate)$_6$], (^{59}Co),[1157] [Co(Schiff base)Br_2], (^{13}C),[1158] [Co($PhCH_2$-CH_2NCMeCXCMeO)$_3$], (^{13}C),[1159] *trans*-(*O*)-[Co{*N*,*N*-ethylene-bis(D-penicillamine)}], (^{13}C),[1160] (72), (^{13}C),[1161] [Ar_4porphyrinRhCl], (^{13}C),[1162] [{M($NH_2CH_2CH_2S$)$_3$}$_4Zn_4O$]$^{6+}$, (M = Rh, Ir; ^{13}C),[1163] [M{Zn(S_2CNEt_2)$_3$}$_2$], (M = Co^{III}, Pd^{II}, Ag^{I}, Cd^{II}, Tl^{I}; ^{13}C),[1164] and [Co(Se_2CNR_2)$_3$], (^{59}Co, ^{77}Se).[1165]

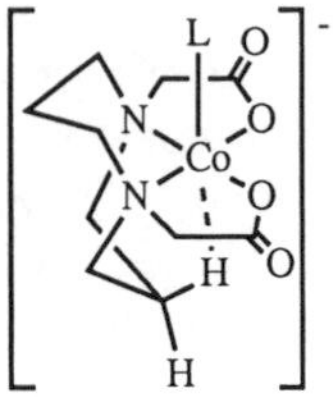

(71)

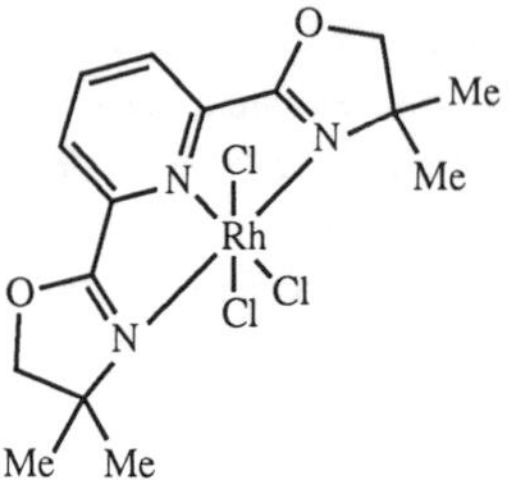

(72)

Complexes of Ni, Pd, and Pt.—Irradiation of the hydride in [(Bu^t_2PH)PdH(μ-PBu^t_2)$_2$Pd-($PHBu^t_2$)] produces a n.O.e. in a ^{31}P n.m.r. signal.[1166] N.m.r. data have also been reported for [($Pr^i_2PCH_2CH_2CH_2PPr^i_2$)$_2Pd_2$($\mu$-H)$_2$($\mu$-CO)], ($^{13}C$),[1167] [($Pr^i_2PCH_2CH_2CH_2PPr^i_2$)$_2Pd_2$($\mu$-H)$_2$].$LiBEt_4$, ($^{11}B$),[1168] *cis*-PtH($SiMePh_2$)($PPh_3$)$_2$], ($^{29}Si$, ^{195}Pt),[1169] [PtH($PPh_2CH_2CH_2$-$SiMe_2Cl$)$_2$Cl], (^{195}Pt),[1170] [Pt_3(μ_3-H){μ_3-Au(PPh_3)}(μ-dppm)]$^{2+}$, (^{195}Pt),[1171] and [PtH{Au(PPh_3)}$_8$]$^+$, (^{195}Pt).[1172]

Agostic interactions in [Ni(CH_2CH_2-H)($Bu^t_2PCH_2CH_2PBu^t_2$)] have been studied by ^{13}C n.m.r. spectroscopy.[1173] Phase sensitive ^1H NOESY has been used on [Pd(C_6H_4-2-CH_2NMe_2)-(quinoline)NO_3] and related palladium and platinum compounds. M-H-C interactions were

1156 P.J. Toscano, K.A. Belsky, T.C. Hsieh, T. Nicholson, and J. Zubieta, *Polyhedron*, 1991, **10**, 977.
1157 T.Sh. Kapanadze, A.P. Gulya, V.M. Novotortsev, O.G. Ellert, V.M. Shcherbakov, Yu.V. Kokunov, and Yu.A. Bushaev, *Koord. Khim.*, 1991, **17**, 934 (*Chem. Abstr.*, 1991, **115**, 196 791); T.Sh. Kapanadze, Yu.E. Gorbunova, Yu.V. Kokunov, and Yu.A. Buslaev, *Mendeleev Commun.*, 1991, 55 (*Chem. Abstr.*, 1991, **115**, 293 504).
1158 M. Mohapatra, V. Chakravortty, and K.C. Dash, *Synth. React. Inorg. Metal-Org. Chem.*, 1991, **21**, 275.
1159 S. Abdul Samath, N. Raman, K. Jeyasubramanian, and S.K. Ramalingam, *Polyhedron*, 1991, **10**, 1687.
1160 K.-i. Okamoto, N. Fushimi, T. Konno, and J. Hidaka, *Bull. Chem. Soc. Jpn*, 1991, **64**, 2635.
1161 H. Nishiyama, M. Horihata, T. Hirai, S. Wakamatsu, and K. Itoh, *Organometallics*, 1991, **10**, 2706.
1162 S. Licoccia, M. Paci, P. Tagliatesta, R. Paolesse, S. Antonaroli, and T. Boschi, *Magn. Reson. Chem.*, 1991, **29**, 1084.
1163 T. Konno, K.-i. Okamoto, and J. Hidaka, *Inorg. Chem.*, 1991, **30**, 2253.
1164 N.K. Singh, C. Kaw, and N. Singh, *Bull. Chem. Soc. Jpn.*, 1991, **64**, 1440.
1165 W. Dietzsch, N.V. Duffy, G.A. Katsoulos, and B. Olk, *Inorg. Chim. Acta*, 1991, **184**, 89.
1166 A. Albinati, F. Lianza, M. Pasquali, M. Sommovigo, P. Leoni, P.S. Pregosin, and H. Rüegger, *Inorg. Chem.*, 1991, **30**, 4690.
1167 M. Portnoy, F. Frolow, and D. Milstein, *Organometallics*, 1991, **10**, 3960.
1168 M.D. Fryzuk, B.R. Lloyd, G.K.B. Clentsmith, and S.J. Rettig, *J. Am. Chem. Soc.*, 1991, **113**, 4332.
1169 C. Müller and U. Schubert, *J. Organomet. Chem.*, 1991, **405**, C1.
1170 S.L. Grundy, R.D. Holmes-Smith, S.R. Stobart, and M.A. Williams, *Inorg. Chem.*, 1991, **30**, 3333.
1171 N.C. Payne, R. Ramachandran, G. Schoettel, J.J. Vittal, and R.J. Puddephatt, *Inorg. Chem.*, 1991, **30**, 4048.
1172 J.J. Bour, P.P.J. Schlebos, R.P.F. Kanters, M.F.J. Schoondergang, H. Addens, A. Overweg, and J.J. Steggerda, *Inorg. Chim. Acta*, 1991, **181**, 195.
1173 F.M. Conroy-Lewis, L. Mole, A.D. Redhouse, S.A. Litster, and J.L. Spencer, *J. Chem. Soc., Chem. Commun.*, 1991, 1601.

discussed.[1174] The ^1H, ^{13}C, ^{31}P, and ^{195}Pt n.m.r. spectra of compounds such as (73) have been reported. $^3J(^{195}Pt\text{-}^{195}Pt)$ was shown to be between 800 and 1100 Hz.[1175] The order parameter has been determined for $\{Pt(PBu^n_3)_2C{\equiv}CC{\equiv}CPt(PBu^n_3)_2C{\equiv}CC_6H_4C{\equiv}C\}$ using ^2H n.m.r. spectroscopy.[1176] ^1H n.m.r. n.O.e. measurements have been used to determine the conformation of $[Pt(\eta^3\text{-}CH_2CMeCH_2)(\eta^2\text{-}CH_2{=}CHPh)(C_6F_5)]$.[1177] Two-dimensional $^{13}C\text{-}^1H$, $^{31}P\text{-}^1H$, and $^{195}Pt\text{-}^1H$ shift correlations have been used to determine the relative signs of coupling constants for *trans*-$[(Bu^n_3P)_2Pt(C{\equiv}CH)_2]$ and *cis*-$[(Et_2PCH_2CH_2PEt_2)Pt(C{\equiv}CH)_2]$.[1178] N.m.r. data have also been reported for $[NiMe(O_2CCHPh_2)(PMe_3)_2]$, ($^{13}C$),[1179] $[(2,6\text{-}Pr^i_2C_6H_3)N{=}CHCH{=}NC_6H_3Pr^i_2\text{-}2,6\}NiCH_2CH_2CF_2CF_2]$, ($^{13}C$),[1180] $[Ni\{C_8H_{14}C(O)O\}(Cy_2PCH_2CH_2\text{-}2\text{-}C_5H_4N)]$, ($^{13}C$),[1181] $[Ni\{C_8H_{12}C(O)O\}(Cy_2PCH_2CH_2\text{-}2\text{-}C_5H_4N)]$, ($^{13}C$),[1182] $[Ni\{CHCH[C(O)NPh]CH_2CH_2O\}\text{-}(dppe)]$, ($^{13}C$),[1183] (74), ($^{13}C$),[1184] (75), ($^{13}C$),[1185] $[9\text{-anthracyl-}10\text{-}Ni(PPh_3)_2Cl]$, ($^{13}C$),[1186]

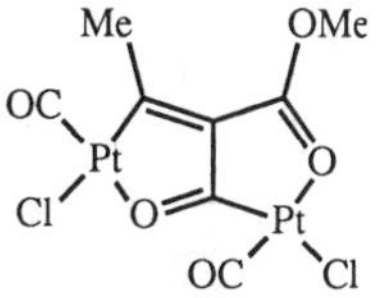

(73)

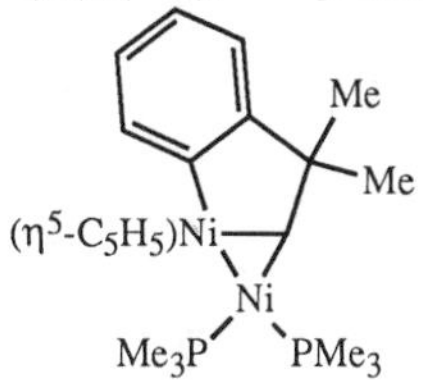

(74)

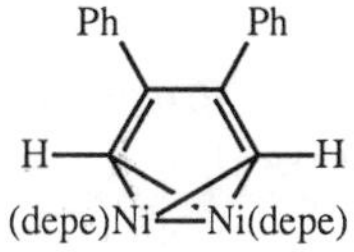

(75)

$[NiX\{2,6\text{-}(Et_2NCH_2)_2C_6H_3\}]$, ($^{13}C$),[1187] $[EtPd(PMe_3)_2(SPh)]$, (^{13}C),[1188] $[Pd(CHMeCHMeN\text{-}Me_2)(en)]^+$, ($^{13}C$),[1189] (76), ($^{13}C$),[1190] $[(dppe)Pd\{C_5H_{10}C({=}O)Me\}]^+$, ($^{13}C$),[1191] $[M\{C(CO_2Me)_2\text{-}CH_2\text{-}2\text{-}(8\text{-hydroxyquinoline})ML]$, (M = Pd, Pt; ^{13}C),[1192] $[Pd\{C_4H_2OC(NMe_2){=}Se\}X]_2$, ($^{13}C$),[1193] $[Pd(C_6H_4\text{-}2\text{-}CH{=}NPr^i)(PMe_3)_3]^+$, ($^{13}C$),[1194] $[(Bu^i_2S)(2,6\text{-}Me_2C_6H_3)Pd(\mu\text{-}MeCO_2)_2Pd(\mu\text{-}$

1174 P.S. Pregosin and F. Wombacher, *Magn. Reson. Chem.*, 1991, **29**, S106.
1175 C. Allevi, M. Bonfa, L. Garlaschelli, and M.C. Malatesta, *Gazz. Chim. Ital.*, 1990, **120**, 711 (*Chem. Abstr.*, 1991, **114**, 122 672).
1176 A. Abe, N. Kimura, and S. Tabata, *Mocromolecules*, 1991, **24**, 6238 (*Chem. Abstr.*, 1991, **115**, 233 305).
1177 H. Kurosawa, K. Miki, N. Kasai, and I. Ikeda, *Organometallics*, 1991, **10**, 1607.
1178 B. Wrackmeyer, *Z. Naturforsch., B*, 1991, **46**, 35 (*Chem. Abstr.*, 1991, **114**, 207 456).
1179 H.-F. Klein, T. Wiemer, M. Dartiguenave, and Y. Dartiguenave, *Inorg. Chim. Acta*, 1991, **189**, 35.
1180 W. Schröder, W. Bonrath, and K.R. Pörschke, *J. Organomet. Chem.*, 1991, **408**, C25.
1181 H. Hoberg, A. Ballesteros, and A. Sigan, *J. Organomet. Chem.*, 1991, **403**, C19.
1182 H. Hoberg and A. Ballesteros, *J. Organomet. Chem.*, 1991, **411**, C11.
1183 H. Hoberg and M. Nohlen, *J. Organomet. Chem.*, 1991, **412**, 225.
1184 E. Carmona, E. Gutiérrez-Puebla, A. Monge, M. Paneque, and M.L. Poveda, *J. Chem. Soc., Chem. Commun.*, 1991, 148.
1185 C.J. Lawrie, H.E. Dankosh, and B.K. Carpenter, *J. Organomet. Chem.*, 1991, **411**, C7.
1186 Z.-h. Zhou and T. Yamamoto, *J. Organomet. Chem.*, 1991, **414**, 119.
1187 J.A.M. van Beck, G. van Koten, M.J. Ramp, N.C. Coenjaarts, D.M. Grove, K. Goubitz, M.C. Zoutberg, C.H. Stam, W.J.J. Smeets, and A.L. Spek, *Inorg. Chem.*, 1991, **30**, 3059.
1188 K. Osakada, Y. Ozawa, and A. Yamamoto, *J. Chem. Soc., Dalton Trans.*, 1991, 759.
1189 L. Zhang and K. Zetterberg, *Organometallics*, 1991, **10**, 3806.
1190 C.-S. Li, C.-H. Cheng, F.-L. Liao, and S.-L. Wang, *J. Chem. Soc., Chem. Commun.*, 1991, 710.
1191 F. Ozawa, T. Hayashi, H. Koide, and A. Yamamoto, *J. Chem. Soc., Chem. Commun.*, 1991, 1469.
1192 A. Yoneda, G.R. Newkome, and K.J. Theriot, *J. Organomet. Chem.*, 1991, **401**, 217.
1193 M. Nonoyama and K. Nonoyama, *Polyhedron*, 1991, **10**, 2265.
1194 L.A. Villanueva, K. Abboud, and J.M. Boncella, *Organometallics*, 1991, **10**, 2969.

$MeCO_2)_2Pd(2,6\text{-}Me_2C_6H_3)(SBu^i{}_2)]$, ($^{13}C$),[1195] (77), ($^{13}C$),[1196] $[M\{C_6H_4C(NHR^1)=NR^2\}Cl]_2$, (M = Pd, Pt; ^{13}C),[1197] $[X(R^1{}_3P)_2PdC{\equiv}CC(=NR^2)C(=NR^2)Pd(PR^1{}_3)_2X]$, ($^{13}C$),[1198] $[(OC)Pd\{Au\text{-}(PPh_3)\}_8]^{2+}$, ($^{13}C$),[1199] $[PtMe_3\{S_2P(OR)_2\}]_2$, (^{13}C),[1200] $[Me_2Pt(C_6H_4\text{-}CH=NR)Cl(SMe_2)]$, ($^{13}C$, ^{195}Pt),[1201] $[MePt(CH_2CH_2CH_2CH_2)I(dppm)]$, ($^{13}C$, ^{195}Pt),[1202] $[MePt(Ph_2PCH_2CH_2)_3P]$, (^{195}Pt),[1203] $[(acac)PtMe(PPh_3)]$, (^{13}C),[1204] (78), (^{13}C),[1205] $[(R^1{}_3P)_2Pr(CH_2)_2CHR^2]$, ($^{13}C$),[1206]

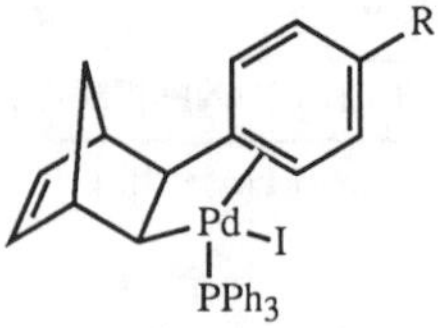
(76)

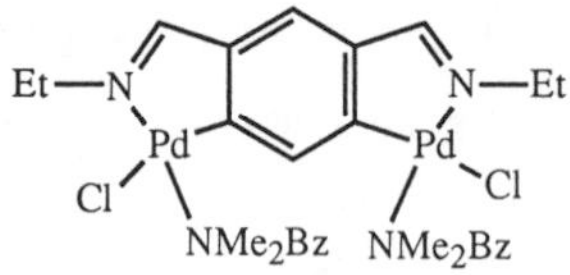
(77)

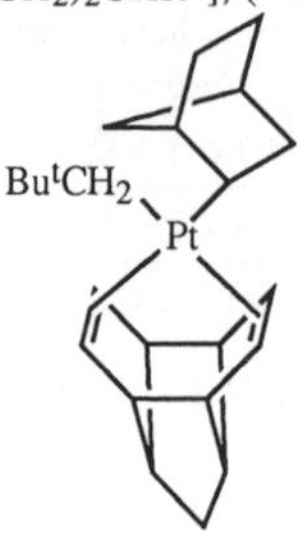
(78)

$[(C_5H_5N)_2Pt\{(CH_2)_2CHR\}Cl_2]$, ($^{13}C$),[1207] $[Pt(CH_2CH_2CPh=O)(CO)Cl]$, ($^{13}C$),[1208] $[(2,6\text{-}Me_2C_6H_3N=CHCH=NC_6H_3Me_2\text{-}2,6)Pt\{CH_2C(=CMe_2)C=CH_2CMe_2OO\}]$, ($^{13}C$),[1209] $[(Cy_2PCH_2\text{-}CH_2PCy_2)Pt(CH_2SiMe_3)\{In(CH_2SiMe_3)_2\}]$, ($^{13}C$),[1210] $[(\eta^3\text{-}C_3H_5)PtCH_2C_6H_4P(C_6H_4OMe\text{-}2)_2]$, ($^{13}C$),[1211] $[Pt\{\sigma,\sigma\text{-}C_8H_{12}(PPh_3)\}(dppe)]^{2+}$, ($^{13}C$),[1212] $[Pt(PEt_3)_2\{\mu\text{-}\eta^1{:}\eta^3\text{-}C(Ph)(H)C(Ph)C=C\text{-}(Ph)\}Pt(PEt_3)_2(C{\equiv}CPh)]^+$, ($^{13}C$),[1213] (79), ($^{13}C$),[1214] $[Pt(SnPh_2CH_2CH_2PPh_2)Ph(PPh_3)]$, ($^{119}Sn$, ^{195}Pt), [1215] *trans*-$[PhPt(NHCO_2H)(PCy_3)_2]$, (^{13}C, ^{195}Pt),[1216] (80), (^{13}C),[1217] (81), (^{13}C),[1218]

1195 Y. Fuchita, M. Kawakami, and K. Shimoke, *Polyhedron*, 1991, **10**, 2037.
1196 S. Chakladar, P. Paul, K. Venkatsubramanian, and K. Nag, *J. Chem. Soc., Dalton Trans.*, 1991, 2669.
1197 J. Barker, N.D. Cameron, M. Kilner, M.M. Mahmoud, and S.C. Wallwork, *J. Chem. Soc., Dalton Trans.*, 1991, 3435.
1198 K. Onitsuka, H. Ogawa, T. Joh, S. Takahashi, Y. Yamamoto, and H. Yamazaki, *J. Chem. Soc., Dalton Trans.*, 1991, 1531.
1199 L.N. Ito, A.M.P. Felicissimo, and L.H. Pignolet, *Inorg. Chem.*, 1991, **30**, 988.
1200 R. Visalakshi and V.K. Jain, *Transition Met. Chem. (Weinhein, Ger.)*, 1990, **15**, 278.
1201 C.M. Anderson, M. Crespo, M.C. Jennings, A.J. Lough, G. Ferguson, and R.J. Puddephatt, *Organometallics*, 1991, **10**, 2672.
1202 M. Rashidi and R.J. Puddephatt, *J. Organomet. Chem.*, 1991, **412**, 445.
1203 M. Peter, M. Probst, P. Peringer, and E.P. Müller, *J. Organomet. Chem.*, 1991, **410**, C29.
1204 H. Jin and K.J. Cavell, *J. Organomet. Chem.*, 1991, **419**, 259.
1205 T.R. Lee, D.A. Wierda, and G.M. Whitesides, *J. Am. Chem. Soc.*, 1991, **113**, 8745.
1206 C. Cartagna, R. Galarini, A. Musco, and R. Santi, *Organometallics*, 1991, **10**, 3956.
1207 J.O. Hoberg and P.W. Jennings, *Organometallics*, 1991, **10**, 8.
1208 K. Ikura, I. Ryu, A. Ogawa, N. Sonoda, S. Harada, and N. Kasai, *Organometallics*, 1991, **10**, 528.
1209 H. Tom Dieck, G. Fendesak, and C. Mung, *Polyhedron*, 1991, **10**, 255.
1210 R.A. Fischer and J. Behm, *J. Organomet. Chem.*, 1991, **413**, C10.
1211 M. Sano and Y. Nakamura, *J. Chem. Soc., Dalton Trans.*, 1991, 417.
1212 S. Fallis, G.K. Anderson, and N.P. Rath, *Organometallics*, 1991, **10**, 3180.
1213 E. Baralt, C.M. Lukehart, A.T. McPhail, and D.R. McPhail, *Organometallics*, 1991, **10**, 516.
1214 M.C. Aversa, P. Bonaccorsi, M. Cusumano, P. Giannetto, and D. Minniti, *J. Chem. Soc., Dalton Trans.*, 1991, 3431.
1215 C. Müller and U. Schubert, *Chem. Ber.*, 1991, **124**, 2181.
1216 S. Park, A.L. Rheingold, and D.M. Roundhill, *Organometallics*, 1991, **10**, 615.
1217 G.S. Hanan, J.E. Kickham, and S.J. Loeb, *J. Chem. Soc., Chem. Commun.*, 1991, 893.
1218 P.S. Pregosin, F. Wombacher, A. Albinati, and F. Lianzi, *J. Organomet. Chem.*, 1991, **418**, 249.

$[(Cy_2PCH_2CH_2PCy_2)\overline{PtC_6H_4\text{-}2\text{-}Si}(SiMe_3)_3]$, ($^{29}Si$),[1219] *o*-platinated complexes of 4,6-Ph_2-5-PhO-1,2-dihydropyrimidine-2-one, (^{195}Pt),[1220] (82), (^{13}C),[1221] $[Pt\{Au(PPh_3)\}\{P(CH_2CH_2PPh_2)_3\}]^+$, ($^{195}Pt$),[1222] $[(OC)(Ph_3P)Pt\{Au(PPh_3)\}_5]^+$, ($^{13}C$),[1223] $[Pt(CuCl)\{Au(PPh_3)\}_8]^{2+}$, ($^{195}Pt$),[1224] $[Pt(PPh_3)\{Au(PPh_3)\}_6(AuX)_3]$, ($^{13}C$, ^{195}Pt),[1225] $[(dppe)\overline{PtSiMe_2CH_2CH_2SiMe_2}]$, ($^{29}Si$, ^{195}Pt),[1226] and $[Pt_3(\mu_3\text{-}SnF_3)_2(\mu\text{-dppm})_3]$, ($^{195}Pt$).[1227]

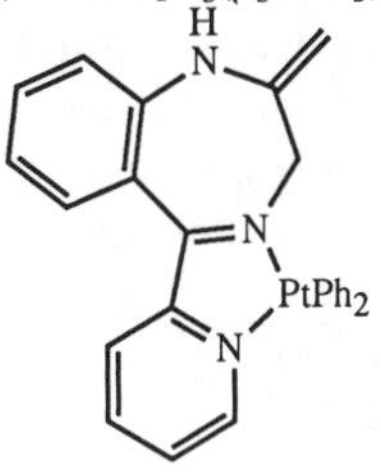

(79)

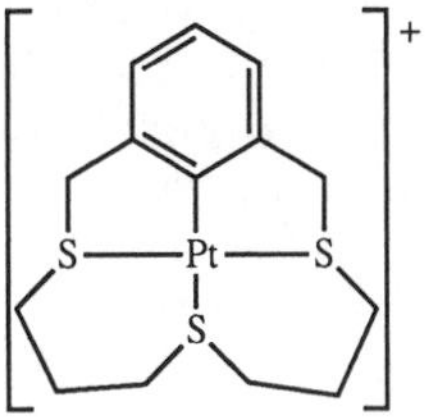

(80)

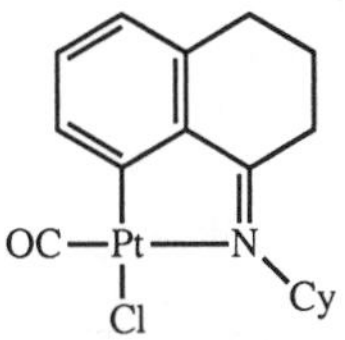

(81)

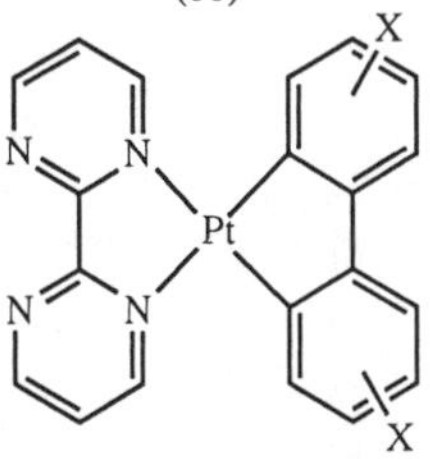

(82)

Two-dimensional NOESY studies have been performed on $[Pd(\eta^3\text{-}C_{10}H_{15})\{(R)(+)\text{-BINAP}\}]^+$ to study the relative spatial orientation of the phenyl and allyl moieties.[1228] The ^{195}Pt-^{31}P-$\{^1H\}$ heteronuclear multiple quantum correlation spectrum of $[Pt_3(CO)_3(PPh_2Pr^i)_3]$ has been measured and used to carry out spectra editing of the different isotopomers with respect to ^{195}Pt.[1229] N.m.r. data have also been reported for $[M\{\eta^2\text{-}\overline{C(CO_2Me)=C(CO_2Me)CH_2C}HCH=CH\}(PPh_3)_2]$, (M = Ni, Pt; ^{13}C),[1230] $[(\mu\text{-}\eta^2{:}\eta^2\text{-}C_7H_{12})\{Ni(\eta^2{:}\eta^2\text{-}C_7H_{12})\}_2]$, ($^{13}C$),[1231] [Ni{$\eta^3$,$\eta^2$,$\eta^2$-dideca-2(*E*),6(*E*),10(*Z*)-trien-1-yl}], (^{13}C),[1232] $[PtCl(en)(\eta^2\text{-}C_2H_4)][PtCl_3(\eta^2\text{-}C_2H_4)]$, ($^{195}Pt$),[1233] $[(\eta^2\text{-}C_2H_4)PtCl_2L]$, {L

1219 L.S. Chang, M.P. Johnson, and M.J. Fink, *Organometallics*, 1991, **10**, 1219.
1220 L.F. Krylova and L.D. Dikanskaya, *Metalloorg. Khim.*, 1991, **4**, 752 (*Chem. Abstr.*, 1991, **115**, 232 484).
1221 H.-A. Brune, R. Hohenadel, G. Schmidtberg, and U. Ziegler, *J. Organomet. Chem.*, 1991, **402**, 179.
1222 M. Peter, P. Peringer, and E.P. Müller, *J. Chem. Soc., Dalton Trans.*, 1991, 2459.
1223 L.N. Ito, A.M.P. Felcissimo, and L.H. Pignolet, *Inorg. Chem.*, 1991, **30**, 387.
1224 M.F.J. Schoondergang, J.J. Bour, P.P.J. Schlebos, A.W.P. Vermeer, W.P. Bosman, J.M.M. Smits, P.T. Beurskens, and J.J. Steggerda, *Inorg. Chem.*, 1991, **30**, 4704.
1225 M.F.J. Schoondergang, J.J. Bour, G.P.F. Van Strijdonck, P.P.J. Schlebos, W.P. Bosman, J.M.M. Smits, P.T. Beurskens, and J.J. Steggerda, *Inorg. Chem.*, 1991, **30**, 2048.
1226 U. Schubert and C. Müller, *J. Organomet. Chem.*, 1991, **418**, C6.
1227 M.C. Jennings, G. Schoettel, S. Roy, R.J. Puddephatt, G. Douglas, L. Manojlović-Muir, and K.W. Muir, *Organometallics*, 1991, **10**, 580.
1228 H. Rüegger, R.W. Kunz, C.J. Ammann, and P.S. Pregosin, *Magn. Reson. Chem.*, 1991, **29**, 197.
1229 H. Rüegger and D. Moskau, *Magn. Reson. Chem.*, 1991, **29**, S11.
1230 H. Choi, J.W. Hershberger, A.R. Pinhas, and D.M. Ho, *Organometallics*, 1991, **10**, 2930.
1231 B. Proft, K.-R. Pörschke, F. Lutz, and C. Krüger, *Chem. Ber.*, 1991, **124**, 2667.
1232 R. Taube, J.-P. Gehrke, P. Boehme, and K. Scherzer, *J. Organomet. Chem.*, 1991, **410**, 403.
1233 N. Farrell, S.G. de Almeida, and Y. Qu, *Inorg. Chim. Acta*, 1991, **178**, 209.

= (83); E = O, NH; ^{13}C},[1234] *cis*-[$Pt(\eta^2-C_2H_4)(DMSO)Cl_2$], ($^{13}C$),[1235] [$Pt\{\eta^2-CH_2=CHCH_2CH(CO_2Me)NH_2\}Cl(PBu^n_3)]^+$, ($^{13}C$),[1236] [$Pt(\eta^2$-alkene)$(PPh_3)_2$], ($^{13}C$, ^{195}Pt),[1237] [$\{O(SiMe_2CH=CH_2-\eta^2)_2Pt(\eta^2-CH_2=CHSiMe_2)\}_2O$], ($^{13}C$, ^{29}Si, ^{195}Pt),[1238] [$Pt(\eta^2$-alkene)(amino acid)Cl], (^{13}C, ^{195}Pt),[1239] [$(\eta^2-C_{10}H_{14})Pt(PPh_3)_2$], ($^{13}C$),[1240] [$(\eta^2-MeO_2CCH=CH-CO_2Me)Pt(PCy_3)_2$], ($^{13}C$),[1241] [$Pt_3(\mu_3-\eta^2-HC\equiv CH)(CO)(\mu$-dppm$)_3]^{2+}$, ($^{13}C$, ^{195}Pt),[1242] [$\{(Et_3P)_2Pt\}_6C_{60}$], (^{13}C),[1243] [$(\eta^3-C_3H_5)Ni(\eta^5-C_5Me_5)$], ($^{13}C$),[1244] [$(\eta^3-Bu^tMe_2SiOCHCHCH_2)-NiCl]_2$, ($^{13}C$),[1245] [$(\eta^3-C_3H_5)Pd(O_2CH)(PMePh_2)$], ($^{13}C$),[1246] [$Pd(\eta^3$-allyl){(*S*)-(*N*-$PPh_2$)(2-$Ph_2POCH_2$)pyrrolidine}]$^+$, ($^{13}C$),[1247] [$(\eta^3-C_6H_8O_2CCH_3)Pd$(phen)]$^+$, ($^{13}C$),[1248] [$Pt(\eta^3-CH_2CR-CH_2)(PR_3)_2]^+$, ($^{13}C$),[1249] [$(\eta^5-C_5H_4Me)Ni(PPh_3)X$], ($^{13}C$),[1250] (84), (R = Bu^t; ^{13}C),[1251]

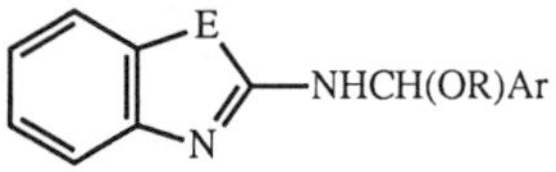

(83)

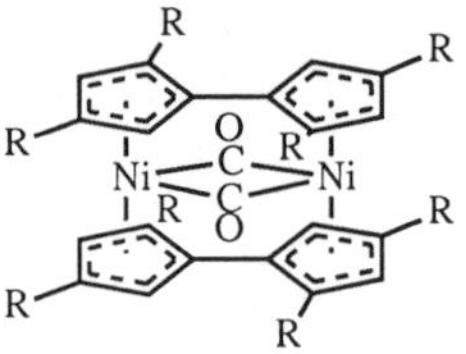

(84)

[$Ni(CO)(PCl=NC_6H_2Bu^t_3$-2,4,6)$(PPh_3)_2$], (^{13}C),[1252] *syn*-[$Pd_2Cl_2(\mu-CO)\{\mu-(Ph_2P)_2CHMe\}_2$], ($^{13}C$),[1253] [{Pd(dppm)$(O_2CCF_3)\}_2(\mu$-CO)], ($^{13}C$),[1254] [$Ni(CNSiMe_3)_4$], ($^{13}C$),[1255] *cis*-[$PtX_2(CO)(\eta^2-C_2H_4)$], ($^{13}C$),[1256] *trans*-[$\{(Et_3P)_2PtPh\}_2(\mu-CO)_2$], ($^{13}C$),[1257] and [$Pt_3(CO)(CNMe)(\mu$-dppm$)_3]^{2+}$, ($^{13}C$, ^{195}Pt).[1258]

^{13}C T_1 measurements have been used to study motion of

1234 A. Albinati, C. Arz, H. Berger, and P.S. Pregosin, *Inorg. Chim. Acta*, 1991, **190**, 119.
1235 V.Yu. Kukushkin, V.K. Belsky, V.E. Konovalov, G.A. Kirakosyan, L.V. Konovalov, A.I. Moiseev, and V.M. Tkachuk, *Inorg. Chim. Acta*, 1991, **185**, 143.
1236 I. Zahn, K. Polborn, B. Wagner, and W. Beck, *Chem. Ber.*, 1991, **124**, 1065.
1237 F. Caruso, M. Camalli, G. Pellizer, F. Asaro, and M. Lenarda, *Inorg. Chim. Acta*, 1991, **181**, 167.
1238 P.B. Hitchcock, M.F. Lappert, and N.J.W. Warhurst, *Angew. Chem., Int. Ed. Engl.*, 1991, **30**, 438.
1239 L.E. Erickson, G.S. Jones, J.L. Blanchard, and K.J. Ahmed, *Inorg. Chem.*, 1991, **30**, 3147.
1240 K. Komatsu, H. Kamo, R. Tsuji, H. Masuda, and K. Takeuchi, *J. Chem. Soc., Chem. Commun.*, 1991, 71.
1241 H. Rüegger, *Magn. Reson. Chem.*, 1991, **29**, 113.
1242 L. Manojlovic-Muir, K.W. Muir, M. Rashidi, G. Schoettel, and R.J. Puddephatt, *Organometallics*, 1991, **10**, 1719.
1243 P.J. Fagan, J.C. Calabrese, and B. Malone, *J. Am. Chem. Soc.*, 1991, **113**, 9409.
1244 H. Lehmkuhl, J. Näser, G. Mehler, T. Keil, F. Danowski, R. Benn, R. Mynott, G. Schroth, B. Gabor, C. Krüger, and P. Betz, *Chem. Ber.*, 1991, **124**, 441.
1245 J.R. Johnson, P.S. Tulley, P.B. Mackenzie, and M. Sabat, *J. Am. Chem. Soc.*, 1991, **113**, 6172.
1246 M. Oshima, I. Shimizu, A. Yamamoto, and F. Ozawa, *Organometallics*, 1991, **10**, 1221.
1247 E. Cesarotti, M. Grassi, L. Prati, and F. Demartin, *J. Chem. Soc., Dalton Trans.*, 1991, 2073.
1248 A. Vitagliano, B. Åkermark, and S. Hansson, *Organometallics*, 1991, **10**, 2592.
1249 S.A. Benyunes, L. Brandt, M. Green, and A.W. Parkins, *Organometallics*, 1991, **10**, 57.
1250 L. Ballester, S. Perez, A. Gutierrez, M.F. Perpiñan, E. Gutierrez-Puebla, A. Monge, and C. Ruiz, *J. Organomet. Chem.*, 1991, **414**, 411.
1251 P. Jutzi, J. Schnittger, W. Wieland, B. Neumann, and H.-G. Stammler, *J. Organomet. Chem.*, 1991, **415**, 425.
1252 J. Hein, E. Niecke, M.F. Meidine, B.F.T. Passos, and J.F. Nixon, *J. Chem. Soc., Chem. Commun.*, 1991, 41.
1253 G. Bensenyei, C.L. Lee, Y. Xie, and B.R. James, *Inorg. Chem.*, 1991, **30**, 2446.
1254 D.J. Wink, B.T. Creagan, and S. Lee, *Inorg. Chim. Acta*, 1991, **180**, 183.
1255 E. Bessler, B.M. Barbosa, W. Hiller, and J. Weidelin, *Z. Naturforsch., B*, 1991, **46**, 490.
1256 H. Alper, Y. Huang, D.B. Dell'Amico, F. Calderazzo, N. Pasqualetti, and C.A. Verachini, *Organometallics*, 1991, **10**, 1665.
1257 I. Torresan, R.A. Michelin, A. Marsella, A. Zanardo, F. Pinna, and G. Strukul, *Organometallics*, 1991, **10**, 623.
1258 R.J. Puddephatt, M. Rashidi, and J.J. Vittal, *J. Chem. Soc., Dalton Trans.*, 1991, 2835.

[Ni{OCHRCH[C(O)C_6H_4X]CMe=NCH_2CH_2N=CMeCH[C(O)C_6H_4X]CHRO}].[1259] Samples of urine from patients treated with carboplatin have been analysed by ^{1}H n.m.r. spectroscopy.[1260] ^{13}C-{^{1}H} n.m.r. spectroscopy has been used to investigate the effect on an oligodeoxyribonucleotide of intrastrand cross-linking by a platinum anticancer drug.[1261] ^{1}H and ^{13}C n.m.r. spectroscopy has been used to study the binding of Ni^{2+} and Zn^{2+} to some heparin monosaccharides.[1262] N.m.r. data have also been reported for [Ni(acac-H)($NHEt_2$)], (^{13}C),[1263] $[NiL]^{2+}$, {L = (85); ^{13}C},[1264] [Ni(2,6-$Pr^i_2C_6H_3$N=CHCH=N$C_6H_3Pr^i_2$-2,6)$_2$], (^{13}C),[1265] [Ni{HC(CPhNC_6H_4NCPh)$_2$CH}], (^{13}C),[1266] [1,3-C_6H_4{CH_2NMeCMe=C(CH=N$CH_2CMe_2CH_2$M=CH)$_2$C=CMeNMeCH_2}Ni]$^{2+}$, (^{13}C),[1267] $[NiL]^{2+}$, (L = pyridine-2-azo-*p*-phenyltetramethylguanidine, ^{13}C),[1268] [M(salicylaldehyde thiobenzoylhydrazone)(NC_5H_5)], (M = Ni, Zn; ^{13}C),[1269] (86), (^{13}C),[1270] [Ni{$SCH_2CH(NH_2)$-CO_2Et}$_2$], (^{13}C),[1271] [Ni{$SCMe_2CH(NH_2)CO_2$}$_2$]$^{2-}$, (^{13}C),[1272] [{HN($CH_2CH_2NHCH_2CH_2NH$-CH_2CH_2)$_2$NH}Pd_2Br_2]$^{2+}$, (^{13}C),[1273] (87), (^{13}C),[1274] *cis*-[M{4-$R^1C_6H_4$C(O)NC(S)N(CH_2R^2)$_2$}$_2$],

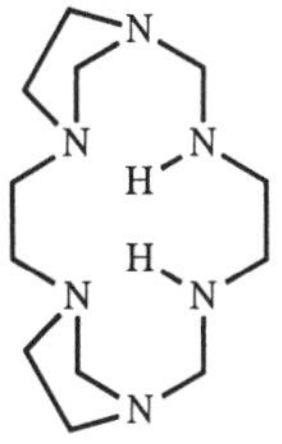

(85)

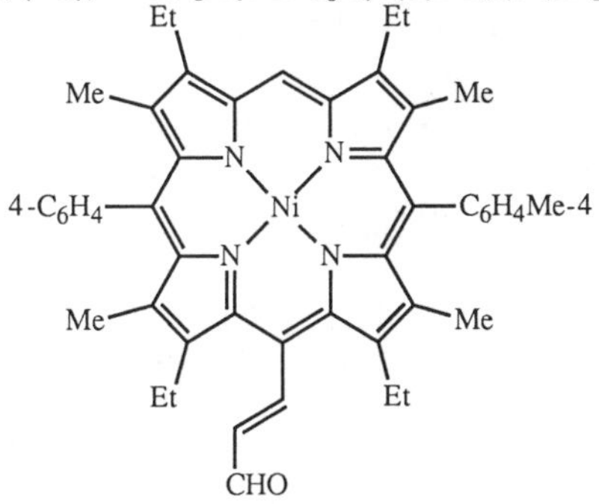

(86)

(M = Pd, Pt; ^{13}C, ^{195}Pt),[1275] Pd, Pt, Cd, and Hg complexes of isoorotic and 2-thioisoorotic acids, (^{13}C),[1276] Pd(II) and Pt(II) complexes of substituted pyrimidines, (^{13}C),[1277] [$MX_2(PMe_2py)_2$], (M =

1259 M. Fujiwara, Y. Ichiki, S. Kinoshita, and H. Wakita, *Polyhedron*, 1991, **10**, 2833.
1260 J.D. Ranford, P.J. Sadler, K. Balmanno, and D.R. Newell, *Magn. Reson. Chem.*, 1991, **29**, S125.
1261 S. Mukundan, jun., Y. Xu, G. Zon, and L.G. Marzilli, *J. Am. Chem. Soc.*, 1991, **113**, 3021.
1262 D.M. Whitfield and Bibudhendra Sarkar, *J. Inorg. Biochem.*, 1991, **41**, 157.
1263 M.T. Toshev, V.G. Yusupov, K.N. Zelenin, L.A. Khorseeva, V.V. Alekseev, Kh.B. Dustov, G.G. Aleksandrov, Z.R. Ashurov, and N.A. Parpiev, *Koord. Khim.*, 1991, **17**, 61 (*Chem. Abstr.*, 1991, **115**, 20 838).
1264 M.P. Suh, S.-G. Kang, V.L. Goedken, and S.-H. Park, *Inorg. Chem.*, 1991, **30**, 365.
1265 W. Bonrath, K.R. Pörschke, R. Mynott, and C. Krüger, *Z. Naturforsch., B*, 1990, **45**, 1647 (*Chem. Abstr.*, 1991, **114**, 102 378).
1266 R.P. Hotz, S.T. Purrington, P.J. Hochgesang, P. Singh, and R.D. Bereman, *Synth. React. Inorg. Metal-Org. Chem.*, 1991, **21**, 253.
1267 P.A. Padolik, A.J. Jircitano, N.W. Alcock, and D.H. Busch, *Inorg. Chem.*, 1991, **30**, 2713.
1268 K.T. Leffek and A. Jarczewski, *Canad. J. Chem.*, 1991, **69**, 1238.
1269 K.N. Zelenin, L.A. Khorseeva, M.T. Toshev, V.V. Alekseev, and Kh.B. Dustov, *Zh. Obshch. Khim.*, 1990, **60**, 2549 (*Chem. Abstr.*, 1991, **115**, 40 743).
1270 M.J. Gunter, B.C. Robinson, J.M. Gulbis, and E.R.T. Tiekink, *Tetrahedron*, 1991, **47**, 7853.
1271 N. Baidya, D. Ndreu, M.M. Olmstead, and P.K. Mascharak, *Inorg. Chem.*, 1991, **30**, 2448.
1272 N. Baidya, M.M. Olmstead, and P.K. Mascharak, *Inorg. Chem.*, 1991, **30**, 3967.
1273 A. McAuley, T.W. Whitcombe, and M.J. Zaworotko, *Inorg. Chem.*, 1991, **30**, 3513.
1274 R.J.P. Corriu, G. Bolin, J.J.E. Moreau, and C. Vernhet, *J. Chem. Soc., Chem. Commun.*, 1991, 211.
1275 K.R. Koch and M.C. Matoetoe, *Magn. Reson. Chem.*, 1991, **29**, 1158.
1276 F. Hueso-Ureña, M.N. Moreno-Carretero, J.M. Salas-Peregrín, and G. Alvarez de Cienfuegos-López, *J. Inorg.*

Pd, Pt; ^{13}C),[1278] [$PdCl_2$(D,L-selenamethionine)], (^{13}C),[1279] (88), (^{13}C),[1280]

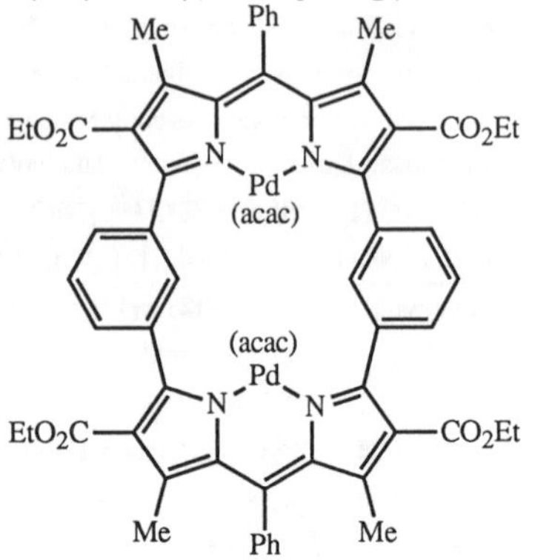

(87)

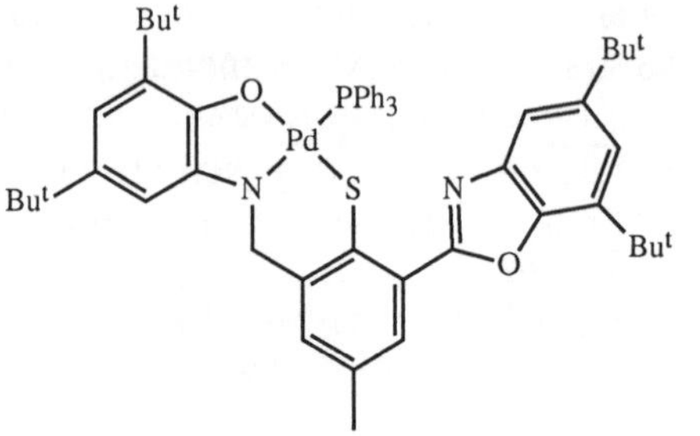

(88)

[Pd{$MeSCH_2CH_2CH(CO_2H)NH_2$}($N_2C_4H_3SH$)Cl]$^+$, (^{13}C),[1281] *cis*-[Pt$(NH_3)_2$(cytosine)Cl]$^+$, (^{13}C),[1282] [Pt$(NH_3)_2$(6-aminouracil)$_2$]$^{2+}$, (^{13}C),[1283] [$Pt_2(NH_3)_4(C_4H_6NO)_2(NO_2)(NO_3)$]$^{2+}$, ($^{13}C$, ^{195}Pt),[1284] *cis*-[PtCl$(NH_3)_2${3-(2-aminoethoxy)estrone}]$^+$, (^{195}Pt),[1285] [$PtCl_2(NH_2)(NO_2$-imidazole)], (^{195}Pt),[1286] [{$PtCl_2(NH_3)$}$_2$(5'-guanosine monophosphate)], (^{195}Pt),[1287] *trans*-[Pt_2(μ-$NH_2)_2(Ph_2PO)_2(PMePh_2)_2$], ($^{15}N$),[1288] [Pt$(NH_2C_5H_9)_2(O_2CR)_2$], ($^{13}C$, ^{195}Pt),[1289] *cis*-[$PtCl_2(PBu^n_3)$(amine)], (^{15}N),[1290] [Pt(9,10-Me_2-phen)(en)]$^{2+}$-nucleotide complexes, (^{13}C),[1291] [Pt($HOCH_2CH_2NHCH_2CH_2NHCH_2CH_2OH$)(oxalate)], ($^{13}C$, ^{195}Pt),[1292] [Pt$(OH)_2$(tmeda)], (^{13}C, ^{195}Pt),[1293] *trans*(Cl,NO_2)-[Pt(bipy)($ONNCH_2CH_2NH_2$)Cl(NO_2)]$^+$, (^{13}C),[1294] [Cl_2Pt{$NH_2CHCH(NH_2)CH_2SCHC_4H_8CO_2H$}], ($^{13}C$),[1295] *cis*-[$PtCl_2${anthraquinonyl-NH$(CH_2)_n$-

Biochem., 1991, **43**, 17.
1277 B.T. Khan and S.M. Zakeeruddin, *Transition Met. Chem. (London)*, 1991, **16**, 119.
1278 T. Suzuki, M. Kita, K. Kashiwabara, and J. Fujita, *Bull. Chem. Soc. Jpn.*, 1990, **63**, 3434.
1279 A.A. Isab and A.R.A. Al-Arfaj, *Transition Met. Chem. (London)*, 1991, **16**, 304.
1280 B.F. Hoskins, R. Robson, G.A. Williams, and J.C. Wilson, *Inorg. Chem.*, 1991, **30**, 4160.
1281 B.T. Khan, J. Bhatt, K. Najmuddin, S. Shamsuddin, and K. Annapoorna, *J. Inorg. Biochem.*, 1991, **44**, 55.
1282 K.I. Yakovlev, A.I. Stetsenko, G.M. Alekseeva, A.F. Imsyrova, A.L. Konovalova, N.S. Kamaletdinov, and T.Yu. Glazkova, *Khim.-Farm. Zh.*, 1991, **25**, 48 (*Chem. Abstr.*, 1991, **115**, 104 663).
1283 S.A. D'yachenko, E.A. Semenova, N.A. Smorygo, D.N. Bochkov, and B.A. Ivin, *Koord. Khim.*, 1991, **17**, 510 (*Chem. Abstr.*, 1991, **115**, 104 635).
1284 T. Abe, H. Moriyama, and K. Matsumoto, *Inorg. Chem.*, 1991, **30**, 4198.
1285 J. Altman, T. Castrillo, W. Beck, G. Bernhardt, and H. Schönenberger, *Inorg. Chem.*, 1991, **30**, 4085.
1286 F.D. Rochon, P.C. Kong, R. Melanson, K.A. Skov, and N. Farrell, *Inorg. Chem.*, 1991, **30**, 4531.
1287 Y. Qu and N. Farrell, *J. Am. Chem. Soc.*, 1991, **113**, 4851.
1288 N.W. Alcock, P. Bergamini, T.J. Kemp, P.G. Pringle, S. Sostero, and O. Traverso, *Inorg. Chem.*, 1991, **30**, 1594.
1289 A.R. Khokhar and Q. Xu, *J. Coord. Chem.*, 1990, **22**, 53.
1290 H.A. Al-Lohedan, I.M. Al-Najjar, and Z.A. Issa, *Orient. J. Chem.*, 1989, **5**, 129 (*Chem. Abstr.*, 1991, **114**, 177 278).
1291 A. Odani, H. Masuda, O. Yamauchi, and S.-i. Ishiguro, *Inorg. Chem.*, 1991, **30**, 4484.
1292 Q. Xu, A.R. Khokhar, and J.L. Bear, *Inorg. Chim. Acta*, 1991, **178**, 107.
1293 F.D. Rochon and A. Morneau, *Magn. Reson. Chem.*, 1991, **29**, 120.
1294 O.N. Ardianova, M.I. Gel'fman, E.V. Kovaleva, O.N. Evstafeva, I.F. Golovaneva, and N.V. Asadulina, *Zh. Neorg. Khim.*, 1990, **35**, 3090 (*Chem. Abstr.*, 1991, **115**, 148 945).
1295 A.F. Noels, N. Nihant, and A.J. Hubert, *Bull. Soc. Chim. Belg.*, 1991, **100**, 497.

$NH(CH_2)_2NH_2\}]$, (^{195}Pt),[1296] [Pt(aspartato)(1,2-cyclohexanediamine)], (^{13}C, ^{195}Pt),[1297] $[Pt(O_2CR)_2$(1,2-cyclohexanediamine)], (^{13}C, ^{195}Pt),[1298] $[Pt\{(O_2C)_2CHR\}$(1,2-cyclohexanediamine)], (^{13}C, ^{195}Pt),[1299] $[Pt(O_2C$-neodecyl$)_2$(1,2-cyclohexanediamine)], (^{13}C, ^{195}Pt),[1300] *trans*-$[PtCl_2\{E$-N(=PPh_3)C(Ph)=$CHCO_2R\}$(NCPh)], (^{13}C),[1301] [Pt{RHNC(S)NN=CHR}Cl], (^{13}C, ^{195}Pt),[1302] [PtCl{2-(2-pyridyl)-4,5-dimethylphosphinine-HOMe}$(PMe_3)]^+$, (^{13}C, ^{195}Pt),[1303] $[Pt(PPh_2$-2-py$)_3]$, (^{195}Pt),[1304] *cis*-$[PtCl_2(N=CPhOCH_2)_2]$, (^{13}C),[1305] aminoalkanol Pt(II) complexes, (^{195}Pt),[1306] anticancer platinum pyrimidine greens, (^{13}C, ^{195}Pt),[1307] $[Pt\{O_2CC(=CHPh)N$-C(O)Me$\}(PMePh_2)_2]$, (^{13}C),[1308] $[Pt(NO_2)_4]^{2-}$, (^{15}N, ^{195}Pt),[1309] $[(NO_2)Pt(\mu\text{-}O)_3Pt_3(NO_2)_6]^{5-}$, ($^{15}N$, ^{195}Pt),[1310] Pt(II)-penicillin complexes, (^{13}C),[1311] $[Pt(S_2{}^{15}N_2H)Cl(PMe_2Ph)]$, ($^{15}N$),[1312] $[Pt\{PSi(C_6H_2Me_3\text{-}2,4,6)_2Si(C_6H_2Me_3\text{-}2,4,6)_2P\}(PPh_3)_2]$, ($^{29}Si$),[1313] Pt(II) complexes of dithiolate ligands, (^{195}Pt),[1314] [Pt(2,4-pentanedione-3-selenato)$(PPh_3)_2]$, (^{13}C),[1315] $[Pt\{PhP(CH_2CH_2\text{-}PPh_2)_2\}\{Ph_2PCH_2P(E)Ph_2\}]^{2+}$, (E = S, Se; ^{77}Se, ^{195}Pt),[1316] $[PtL_2]$, {L = (89); ^{195}Pt},[1317] $[Pt\{P[=C(SiMe_3)_2]_2\}Cl(PEt_3)_2]$, ($^{13}C$),[1318] [{indolyl$(CH_2)_2$NHC(O)$(CH_2)_2$C(O)NH$(CH_2)_2$-$C_6H_3O_2\}Pt(PPh_3)_2]$, ($^{13}C$, ^{195}Pt),[1319] $[Pt\{Se_2C(CMeO)_2\}(PMe_2Ph)_2]$, ($^{13}C$),[1320] $[Pd_2Cl_2(\mu\text{-}Cl)(\mu\text{-}SCH_2CH_2CH_2Cl)(PMe_3)_2]$, ($^{13}C$),[1321] *trans*-$[(NC)_2Pd(\mu\text{-dppm})_2AgX]$, ($^{13}C$),[1322] $[PtCl_5(PEt_3)]^-$,

1296 D. Gibson, K.F. Gean, J. Katzhendle, R. Beb-Shoshan, A. Ramu, and I. Ringel, *J. Med. Chem.*, 1991, **34**, 414 (*Chem. Abstr.*, 1991, **114**, 54 735).
1297 A.H. Talebian, D. Bensely, A. Ghiorghis, C.F. Hammer, P.S. Schein, and D. Green, *Inorg. Chim. Acta*, 1991, **179**, 281.
1298 A.R. Khokhar, Q. Xu, and S. Al-Baker, *J. Coord. Chem.*, 1991, **24**, 77.
1299 A. Talebian, D. Bensely, D. Green, and P.S. Schein, *J. Coord. Chem.*, 1990, **22**, 165.
1300 A.R. Khokhar, S. Al-Baker, T. Brown, and R. Perez-Soler, *J. Med. Chem.*, 1991, **34**, 325 (*Chem. Abstr.*, 1991, **114**, 35 414).
1301 J. Vicente, M.-T. Chicote, J. Fernández-Baeza, F.J. Lahoz, and J.A. López, *Inorg. Chem.*, 1991, **30**, 3617.
1302 K.R. Koch, *J. Coord. Chem.*, 1991, **22**, 289.
1303 B. Schmid, L.M. Venanzi, A. Albinati, and F. Mathey, *Inorg. Chem.*, 1991, **30**, 4693.
1304 Y. Xie and B.R. James, *J. Organomet. Chem.*, 1991, **417**, 277.
1305 R.A. Michelin, R. Bertani, M. Mozzon, G. Bombieri, F. Benetollo, and R.J. Angelici, *Organometallics*, 1991, **10**, 1751.
1306 A.R. Khokhar, Q. Xu, R.A. Newman, and Z.H. Siddik, *J. Inorg. Biochem.*, 191, **43**, 57.
1307 T. Uemura, T. Shimura, H. Nakanishi, T. Tomohiro, Y. Nagawa, and H. Okuno, *Inorg. Chim. Acta*, 1991, **181**, 11.
1308 R.D.W. Kemmitt, S. Mason, and D.R. Russell, *J. Organomet. Chem.*, 1991, **415**, C9.
1309 T.G. Appleton, K.J. Barnham, J.R. Hall, and M.T. Mathieson, *Inorg. Chem.*, 1991, **30**, 2751.
1310 V.I. Privalov, V.V. Lapkin, V.P. Tarasov, and Yu.A. Buslaev, *Mendeleev Commun.*, 1991, 59 (*Chem. Abstr.*, 1991, **115**, 21 500); V.I. Privalov, V.V. Lapkin, V.P. Tarasov, and Yu.A. Buslaev, *Appl. Magn. Reson.*, 1991, **1**, 445 (*Chem. Abstr.*, 1991, **115**, 221 502).
1311 T. Grochowski and K. Samochocka, *Polyhedron*, 1991, **10**, 1473.
1312 P.F. Kelly, I.P. Parkin, R.N. Sheppard, and J.D. Woollins, *Heteroat. Chem.*, 1991, **2**, 301 (*Chem. Abstr.*, 1991, **115**, 196 528).
1313 A.D. Fanta, M. Driess, D.R. Powell, and R. West, *J. Am. Chem. Soc.*, 1991, **113**, 7806.
1314 R. Colton and V. Tedesco, *Inorg. Chem.*, 1991, **30**, 2451.
1315 S. Yamazaki, *Chem. Lett.*, 1991, 1299 (*Chem. Abstr.*, 1991, **115**, 221 823).
1316 A. Handler, P. Peringer, and E.P. Müller, *Polyhedron*, 1991, **10**, 2471.
1317 M.J. Baker, K.N. Harrison, A.G. Orpen, P.G. Pringle, and G. Shaw, *J. Chem. Soc., Chem. Commun.*, 1991, 803.
1318 H.-J. Metternich, E. Niecke, J.F. Nixon, R. Bartsch, P.B. Hitchcock, and M.F. Meidine, *Chem. Ber.*, 1991, **124**, 1973.
1319 H.C. Apfelbaum, J. Blum, and F. Mandelbaum-Shavit, *Inorg. Chim. Acta*, 1991, **186**, 243.
1320 S. Yamazaki, T. Ueno, and M. Hojo, *Bull. Chem. Soc. Jpn.*, 1991, **64**, 1404.
1321 J.H. Yamamoto, G.P.A. Yap, and C.M. Jensen, *J. Am. Chem. Soc.*, 1991, **113**, 5060.
1322 S. Lu, *Jiegou Huaxue*, 1991, **10**, 70 (*Chem. Abstr.*, 1991, **115**, 293 492).

(^{195}Pt),[1323] [$PtCl_2(PEt_3)$(PPhCPh=CPh)], (^{13}C),[1324] [Pt_2Cl_2(μ-X)(μ-TeAr)$(PR_3)_2$], (^{13}C, ^{125}Te, ^{195}Pt),[1325] [$Pt_2X^1{}_2$(μ-X^2)(μ-SePh)$(PR_3)_2$], (^{195}Pt),[1326] [Ni{Sb($OCMe_2CMe_2O$)$_2$}$_2$], (^{13}C),[1327] [Ni_3{$SCH_2CH(CH_3)S$}$_4$], (^{13}C),[1328] [Ni(*rac*-$MeSCH_2CH_2PPhCH_2CH_2PPhCH_2CH_2SMe$)]$^{2+}$, (^{13}C),[1329] [$M(Se_2C_2S_2CS)_2$]$^{2-}$, (M = Ni, Pd, Pt, Zn, Cd, Hg; ^{77}Se),[1330] [$Ni_4Se_4(Se_3)_5(Se_4)$]$^{4-}$, (^{77}Se),[1331] [Pd{RCMeC(CO_2Et)=NOH}$_2$], (^{13}C),[1332] $M^1{}_6[M^2(OH_2)(HIO_6)_2].nH_2O$, ($M^1$ = Na, K; M^2 = Pd, Pt; ^{127}I, ^{195}Pt),[1333] [M(2,6-Me_2-4*H*-pyran-4-thion)$_2X_2$], (M = Pd, Pt; ^{13}C),[1334] [$Pd_2Te_4O_{24}H_2$]$^{10-}$, [$Pt(OH)_2(HTeO_6)_2$]$^{8-}$, (^{125}Te, ^{195}Pt),[1335] [Pt{$(O_2C)_2CHMe$}$_2$]$^{2-}$, (^{13}C, ^{195}Pt),[1336] *cis*-[$PtCl_2${MeS(O)R}$_2$], (^{13}C, ^{195}Pt),[1337] and [Pt{S($CH_2CH_2SCH_2CH_2NHCO$)$_2$-CMe_2}(OH_2)]$^{2+}$, (^{13}C).[1338]

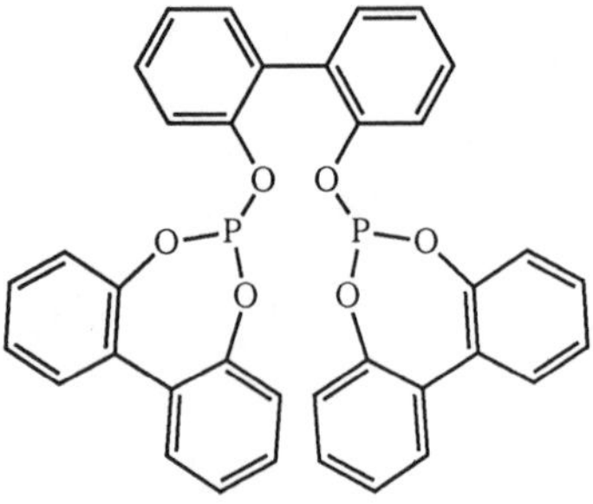

(89)

Complexes of Cu, Ag, and Au.—The six-coordinate central carbon atom in [{(4-$Me_2NC_6H_4$)PPh_2Au}$_6{}^{13}C$]$^{2+}$ has a ^{13}C chemical shift of 137.27.[1339] $^2J(^{13}C,^{13}C)$ is *ca.* 20 Hz for [RCu(^{13}CN)]$^-$ and this has been taken as confirmation of a linear ion.[1340] N.m.r. data have also been reported for [{$(Ph_2PCH_2)_2NCH_2CH_2N(CH_2PPh_2)_2$}$Cu_2(BH_4)_4$], (^{11}B, ^{13}C),[1341] [$Bu^n{}_3SnCu(CN)$-

1323 A. Albinati, W. Kaufmann, and L.M. Venanzi, *Inorg. Chim. Acta*, 1991, **188**, 145.
1324 S.S. Al Juaid, D. Carmichael, P.B. Hitchcock, A. Marinetti, F. Mathey, and J.F. Nixon, *J. Chem. Soc., Dalton Trans.*, 1991, 905.
1325 V.K. Jain and S. Kannan, *J. Organomet. Chem.*, 1991, **418**, 349.
1326 V.K. Jain and S. Kannan, *J. Organomet. Chem.*, 1991, **405**, 265.
1327 A.K.S. Gupta, R. Bohra, and R.C. Mehrotra, *Synth. React. Inorg. Metal-Org. Chem.*, 1991, **21**, 445.
1328 H. Liu, D. Daxu, X. Chen, and B. Kang, *Bopuxue Zazhi*, 1990, **7**, 231 (*Chem. Abstr.*, 1991, **114**, 16 291).
1329 T. Kitagawa, M. Kita, K. Kashiwabara, and J. Fujita, *Bull. Chem. Soc. Jpn.*, 1991, **64**, 2942.
1330 B. Olk and R.-M. Olk, *Z. Anorg. Allg. Chem.*, 1991, **600**, 89.
1331 J.M. McConnachie, M.A. Ansari, and J.A. Ibers, *J. Am. Chem. Soc.*, 1991, **113**, 7078.
1332 Y.S. Jun, M.H. Lee, I.W. Kim, and S.H. Kim, *J. Korean Chem. Soc.*, 1991, **35**, 368 (*Chem. Abstr.*, 1991, **115**, 221 807).
1333 W. Levason, M.D. Spicer, and M. Webster, *J. Coord. Chem.*, 1991, **23**, 67.
1334 G. Faraglia, F. Barbaro, and S. Sitran, *Transition Met. Chem. (Weinhein, Ger.)*, 1990, **15**, 242.
1335 W. Levason, M.D. Spicer, and M. Webster, *Inorg. Chem.*, 1991, **30**, 967.
1336 S.O. Dunham, R.D. Larsen, and E.H. Abbott, *Inorg. Chem.*, 1991, **30**, 4328.
1337 L. Antolini, U. Folli, D. Iarossi, L. Schenetti, and F. Taddei, *J. Chem. Soc., Perkin Trans. 2*, 1991, 955.
1338 E. Kimura, Y. Kurogi, T. Tojo, M. Shionoya, and M. Shiro, *J. Am. Chem. Soc.*, 1991, **113**, 4857.
1339 H. Schmidbaur, B. Brachthäuser, and O. Steigelmann, *Angew. Chem., Int. Ed. Engl.*, 1991, **30**, 1488.
1340 S.H. Bertz, *J. Am. Chem. Soc.*, 1991, **113**, 5470.
1341 M.M.T. Khan, P. Paul, and K. Venkatasubramanian, *Polyhedron*, 1991, **10**, 1827.

Li], (^{13}C),[1342] [Cu{3,5-$Me_2C_3HN_2C(S)S$}(CNR)(PPh_3)], (^{13}C),[1343] (90), (^{13}C),[1344] [(PhC≡C)-Au(PPh_3)], (^{13}C),[1345] [Au{C(NC_5H_{10})$_2$}Cl], (^{13}C),[1346] [dppe($AuSiMe_3$)$_2$], (^{29}Si),[1347] [TlAu-($C_5H_4PPh_2$)$_2$]$_2$, (^{13}C),[1348] and [AuBr(CO)], (^{13}C).[1349]

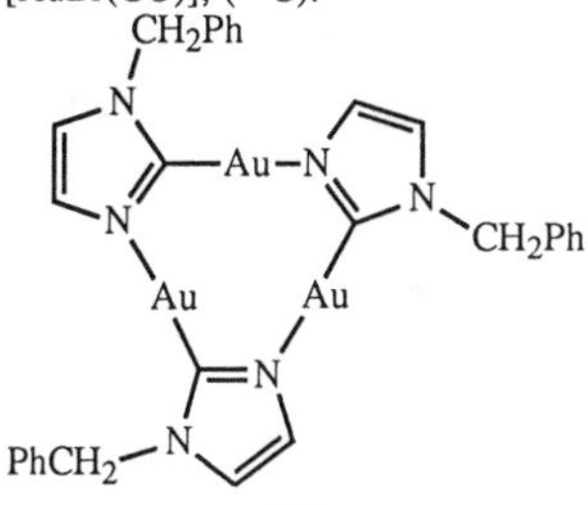

(90)

$AgNO_3$-NCMe complexes have been studied in thermotropic liquid crystals by n.m.r. spectroscopy.[1350] ^1H-{^{109}Ag} heteronuclear multiple quantum coherence transfer experiments have been used to determine metal to cysteine connectivities in silver substituted yeast metallothionein.[1351] The ^{31}P and ^{63}Cu n.m.r. spectra of [Cu(dmpe)$_2$]$_2$$^{2+}$ shows $^1J(^{63}Cu-^{31}P)$.[1352] N.m.r. data have also been reported for [Cu(PPh_3)(heterocyclic thione)Cl], (^{13}C),[1353] [Cu_2{(Ph_2PCH_2)$_2NCH_2CH_2$-N(CH_2PPh_2)$_2$}Cl_2], (^{13}C),[1354] [Cu_2{(Ph_2PCH_2)$_2NCH_2CH_2N$(CH_2PPh_2)$_2$}X_2], (^{11}B, ^{13}C),[1355] [M_n{N(CH_2CH_2N=CH-6-C_5H_4N-2-2'-C_5H_4N-2-CH=NCH_2CH_2)$_3$N}]$^{n+}$, (M = Cu, n = 2; M = Ag, n = 3; ^{13}C),[1356] [Cu_2(2-Me_3SiN-6-MeC_5H_3N)$_2$], (^{13}C, ^{29}Si),[1357] [Ag{MeN($CH_2CH_2SCH_2$-$CH_2SCH_2CH_2$)$_2$NMe}]$^+$, (^{13}C),[1358] [Ag{μ-P(OCBu^t=CH)$_2$N}$_4$]$^+$, (^{13}C),[1359] [M(thiophene-2-carbaldehyde)$_2$]$^+$, (M = Cu, Ag; ^{109}Ag),[1360] [Au{H_2C($CH_2NHCH_2CH_2NHCH_2$)$_2CH_2$}]$^{3+}$,

1342 R.D. Singer, M.W. Hutzinger, and A.C. Oehlsclager, *J. Org. Chem.* 1991, **56**, 4933.
1343 G.A. Ardizzoia, M. Angaroni, G. LaMonica, M. Moret, and N. Masciocchi, *Inorg. Chim. Acta*, 1991, **185**, 63.
1344 F. Bonati, A. Burini, B.R. Pietroni, and B. Bovio, *J. Organomet. Chem.*, 1991, **408**, 271.
1345 M.L. Ganadu, L. Naldini, G. Crisponi, and V. Nurchi, *Spectrochim. Acta, Part A*, 1991, **47**, 615 (*Chem. Abstr.*, 1991, **115**, 63 131).
1346 W.P. Fehlhammer and W. Finck, *J. Organomet. Chem.*, 1991, **414**, 261.
1347 H. Piana, H. Wagner, and U. Schubert, *Chem. Ber.*, 1991, **124**, 63.
1348 G.K. Anderson and N.P. Rath, *J. Organomet. Chem.*, 1991, **414**, 129.
1349 D. Belli Dell'Amico, F. Calderazzo, P. Robino, and A. Segre, *Gazz. Chim. Ital.*, 1991, **121**, 51 (*Chem. Abstr.*, 1991, **115**, 40 758); D.B. Dell'Amico, F. Calderazzo, P. Robino, and A. Segre, *J. Chem. Soc., Dalton Trans.*, 1991, 3017.
1350 G.A.N. Gowda, R.G. Weiss, and C.L. Khetrapal, *Liq. Cryst.*, 1991, **10**, 659 (*Chem. Abstr.*, 1991, **115**, 246 391).
1351 S.S. Narula, R.K. Mehra, D.R. Winge, and I.M. Armitage, *J. Am. Chem. Soc.*, 1991, **113**, 9354.
1352 B. Mohr, E.E. Brooks, N. Rath, and E. Deutsch, *Inorg. Chem.*, 1991, **30**, 4541.
1353 M.S. Hussain, A.R. Al-Arfaj, and M.L. Hossain, *Transition Met. Chem. (Weinberg, Ger.)*, 1990, **15**, 264.
1354 M.M.T. Khan, P. Paul, K. Venkatasubramanian, and S. Purohit, *Inorg. Chim. Acta*, 1991, **183**, 229.
1355 M.M.T. Khan, P. Paul, K. Venkatasubramanian, and S. Purohit, *J. Chem. Soc., Dalton Trans.*, 1991, 3405.
1356 J. de Mendoza, E. Mesa, J.-C. Rodriguez-Ubis, P. Vásquez, F. Vögtle, P.-M. Windscheif, K. Rissanen, J.-M. Lehn, D. Lilienbaum, and R. Ziessel, *Angew. Chem., Int. Ed. Engl.*, 1991, **30**, 1331.
1357 L.M. Engelhardt, G.E. Jacobsen, W.C. Patalinghug, B.W. Skelton, C.L. Raston, and A.H. White, *J. Chem. Soc., Dalton Trans.*, 1991, 2859.
1358 A.J. Blake, G. Reid, and M. Schröder, *J. Chem. Soc., Dalton Trans.*, 1991, 615.
1359 A.J. Arduengo, tert., H.V.R. Dias, and J.C. Calabrese, *J. Am. Chem. Soc.*, 1991, **113**, 7071.
1360 J.F. Modder, J.M. Ernsting, K. Vrieze, M. De Wit, C.H. Stam, and G. Van Koten, *Inorg. Chem.*, 1991, **30**, 1208.

(^{13}C),[1361] [(Me_3P)Au(3'-N_3-3'-desoxythymidine)], (^{13}C),[1362] [{Au(PPh_3)}$_3$NR]$^+$, (^{13}C),[1363] [M(PR_3)(2,5,8-trithia[9]-*o*-benzophane)]$^+$, (M = Cu, Ag; ^{13}C),[1364] [$Cu_2(OC_4Cl_4OH)_2(dppm)_2$], (^{13}C),[1365] [($Bu^n_3P$)Ag($SPPh_2$)$_3$C], (^{13}C),[1366] [(8-mercaptotheophyllinato-*S*)Au(PPh_3)], (^{13}C),[1367] [Au(2-thiouracil)(PEt_3)], (^{13}C),[1368] [Ag(S_2CAr)(PPh_3)$_2$], (^{13}C),[1369] [(Ph_3P)Au($SePPh_3$)]$^+$, (^{77}Se),[1370] [(Ph_3P)Au{SeC(NH_2)$_2$}]$^+$, (^{77}Se),[1371] [Cu($OCEt_3$)], (^{13}C),[1372] [Cu{MeCH=CHC(O)-NHC(S)NHR}$_2$Cl], (^{13}C),[1373] and [Ag(Se_4)]$^+$, (^{77}Se),[1374]

Complexes of Zn, Cd, and Hg.—A review entitled 'Mercury(II)-thiolate chemistry and the mechanism of the heavy metal biosensor MerR' contains ^{199}Hg n.m.r. spectroscopy.[1375]

Low temperature ^{113}Cd n.m.r. spectra of [Cd(BH_4)$_2$] show $^1J(^{113}Cd\text{-}^1H)$ = 273 Hz.[1376] ^{1}H, ^{13}C, ^{29}Si n.m.r. data for [{(Me_3Si)$_n$$CH_{3-n}$}$_2$Zn] are strongly dependent on *n* and indicate a flattening of the α-carbon atom with substitution.[1377] A symposium report on the n.m.r. spectra of oriented organomercury compounds has appeared.[1378] The measurement of ^{199}Hg n.m.r. spectra of methylmercury species by the heteronuclear multiple quantum coherence indirect proton detection method has been applied to the [Hg(CH_3)]$^+$-thiol system.[1379] ^{19}F chemical shifts have been determined for [R_nMOC_6H_3-2,6-X_2-4-F], R_nM = PhHg, Ph_3Sn, Et_3Sn, Ph_3Pb. The introduction of *o*-halo atoms can increase the electron donating ability of the R_nMO groups due to intramolecular coordination and polarization of the metal—O bond.[1380] N.m.r. data have also been reported for [Me_2Zn(Bu^tN=CHCH=NBu^t)], (^{13}C),[1381] [EtZn{OC(OEt)=CHNEt$_2$}], (^{13}C),[1382] [Et_2Zn(OR)]$^-$,

1361 E. Kimura, Y. Kurogi, and T. Takahashi, *Inorg. Chem.*, 1991, **30**, 4117.
1362 T. Pill, K. Polborn. A. Kleinschmidt, V. Erfle, W. Breu, H. Wagner, and W. Beck, *Chem. Ber.*, 1991, **124**, 1541.
1363 A. Grohmann, J. Riede, and H. Schmidbaur, *J. Chem. Soc., Dalton Trans.*, 1991, 783.
1364 B. De Groot, G.R. Giesbrecht, S.J. Loeb, and G.K.H. Shimizu, *Inorg. Chem.*, 1991, **30**, 177.
1365 T.A. Annan, J.E. Kickham, and D.G. Tuck, *Can. J. Chem.*, 1991, **69**, 251.
1366 S.O. Grim, S.A. Sangokoya, A.L. Rheingold, W. McFarlane, I.J. Colquhoun, and R.D. Gilardi, *Inorg. Chem.*, 1991, **30**, 2519.
1367 E. Colacio, A. Romerosa, J. Ruiz, P. Roman, J.M. Gutierrez-Zomilla, A. Vegas, and M. Martinez-Ripoll, *Inorg. Chem.*, 1991, **30**, 3743.
1368 C.S.W. Harker, E.R.T. Tiekink, and M.W. Whitehouse, *Inorg. Chim. Acta*, 1991, **181**, 23.
1369 A.M.M. Lanfredi, F. Ugozzoli, F. Asaro, G. Pellizer, N. Marsich, and A. Camus, *Inorg. Chim. Acta*, 1991, **190**, 71.
1370 P.G. Jones and C. Thöne, *Inorg. Chim. Acta*, 1991, **181**, 291.
1371 P.G. Jones and C. Thöne, *Chem. Ber.*, 1991, **124**, 2725.
1372 A.P. Purdy, C.F. George, and J.H. Callahan, *Inorg. Chem.*, 1991, **30**, 2812.
1373 J. Cernak, J. Chomic, P. Kutschy, D. Surcinova, and M. Dzurilla, *Inorg. Chim. Acta*, 1991, **181**, 85.
1374 S.P. Huang and M.G. Kanatzidis, *Inorg. Chem.*, 1991, **30**, 1455.
1375 J.G. Wright, M.J. Natan, F.M. MacDonnell, D.M. Ralston, and T.V. O'Halloran, *Prog. Inorg. Chem.*, 1990, **38**, 323 (*Chem. Abstr.*, 1991, **114**, 198 319).
1376 H. Nöth and M. Thomann, *Z. Naturforsch., B*, 1990, **45**, 1482 (*Chem. Abstr.*, 1991, **114**, 73 954).
1377 M. Westerhausen, B. Rademacher, and W. Poll, *J. Organomet. Chem.*, 1991, **421**, 175.
1378 C.L. Khetrapal, *Adv. Organomet., Proc. Indo-Sov. Symp. "Organomet. Chem.", 1st. 1988*, (Pub. 1989), 199. Ed. D.V.S. Jain (*Chem. Abstr.*, 1991, **114**, 207 401).
1379 J.M. Robert and D.L. Rabenstein, *Anal. Chem.*, 1991, **63**, 2674.
1380 V.M. Pachevskaya, L.S. Golovchenko, B.A. Kvasov, E.I. Fedin, and D.N. Kravtsov, *Metalloorg. Khim.*, 1991, **4**, 889 (*Chem. Abstr.*, 1991, **115**, 183 481).
1381 M. Kaupp, H. Stoll, H. Preuss, W. Kaim, T. Stahl, G. van Koten, E. Wissing, W.J.J. Smeets, and A.L. Spek, *J. Am. Chem. Soc.*, 1991, **113**, 5606.
1382 F.H. Van der Steen, J. Boersma, A.L. Spek, and G. Van Koten, *Organometallics*, 1991, **10**, 2467.

(^{13}C),[1383] (91), (^{13}C),[1384] [BrZnCH$_2$R], (^{13}C),[1385] [$R_2M_4(O_2CNEt_2)_6$], (M = Zn, Cd; ^{13}C),[1386] [$R^1Zn(S_2CNR^2{}_2)$], (^{13}C),[1387] [$(Ph_3Sn)_2Zn(PPh_3)_2$], (^{119}Sn),[1388] [Me_2Cd(bipy)], (^{113}Cd),[1389] [Me_2Cd(dioxan)], (^{13}C, ^{113}Cd),[1390] [$Cd(CF_3)\{CF_2OC(O)CF_3\}$], ($^{113}Cd$),[1391] [$Cd(n\text{-}C_6F_{13})_2$], ($^{13}C$),[1392] (92), ($R_nM$ = MeHg, PhHg, Me_2Tl, Ph_2Tl; ^{13}C),[1393] [$(\eta\text{-}C_5H_5)Cd\{N(SiMe_3)_2\}]_2$, ($^{13}C$),[1394] [RHg(rhodaninato)], [R_2Tl(rhodaninato)], (^{13}C, ^{199}Hg, ^{205}Tl),[1395] [$RHgS_2PPh_2$], (^{13}C, ^{199}Hg),[1396] [$BrHg\{CH_2C(CH_3)OCHROOC(CH_3)_2\}$], ($^{13}C$),[1397] [RHgBr], ($^{13}C$),[1398] [$Bu^tHgP\{\mu\text{-}SiBu^t(C_6H_2Me_3\text{-}2,4,6)\}_2PHgBu^t$], ($^{29}Si$),[1399] [$(4\text{-}RC_6H_4)HgL$], ($^{13}C$),[1400] [$Hg(CF_3)_2$], [$M(CF_3)_3$], M = Tl, Bi, and [$Ga(CF_3)_4]^-$, ($^{13}C$, ^{71}Ga).[1401]

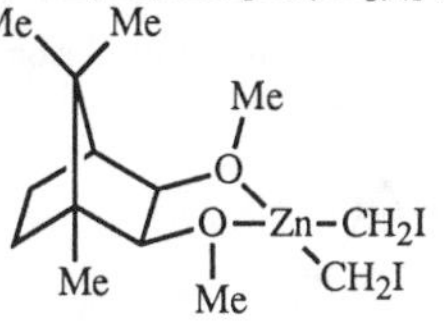

(91)

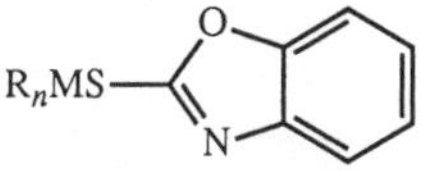

(92)

The binding of [$Hg(^{13}CN)]^+$ to protein thiol groups has been investigated by ^{13}C n.m.r. spectroscopy.[1402] ^1H n.m.r. spectroscopy has been used to show that (5-pyridyl-10,15,20-Ph_3-porphyrin)Zn aggregates with pyridine to Zn coordination.[1403] ^{13}C n.m.r. spectroscopy has been used to study the coordination of several 4-acyl-5-pyrazolone derivatives to Zn^{2+}, Cd^{2+}, Sn^{2+}, and Pb^{2+}.[1404] Two-dimensional ^1H-^{13}C correlation spectroscopy has been used to assign the signals of the bleomycin-Zn^{2+} complex.[1405] Two-dimensional ^{113}Cd-^1H HOESY of some cadmium complexes has been reported.[1406] ^{113}Cd n.m.r. spectroscopy has been used to investigate the interaction of

[1383] R.M. Fabicon, M. Parvez, and H.G. Richey, jun., *J. Am. Chem. Soc.*, 1991, **113**, 1412.
[1384] S.E. Denmark, J.P. Edwards, and S.R. Wilson, *J. Am. Chem. Soc.*, 1991, **113**, 723.
[1385] F. Lambert, B. Kirschleger, and J. Villiéras, *J. Organomet. Chem.*, 1991, **405**, 273.
[1386] M.B. Hursthouse, M.A. Malik, M. Motevalli, and P. O'Brien, *J. Chem. Soc., Chem. Commun.*, 1991, 1690.
[1387] M.B. Hursthouse, M.A. Malik, M. Motevalli, and P. O'Brien, *Organometallics*, 1991, **10**, 730.
[1388] H.F. Klein, J. Montag, and U. Zucha, *Inorg. Chim. Acta*, 1990, **177**, 43.
[1389] M.J. Almond, M.P. Beer, M.G.B. Drew, and D.A. Rice, *Organometallics*, 1991, **10**, 2072.
[1390] M.J. Almond, M.P. Beer, M.G.B. Drew, and D.A. Rice, *J. Organomet. Chem.*, 1991, **421**, 129.
[1391] W. Tyrra and D. Naumann, *Can. J. Chem.*, 1991, **69**, 327.
[1392] D. Naumann, K. Glinka, and W. Tyrra, *Z. Anorg. Allg. Chem.*, 1991, **594**, 95.
[1393] M.V. Castaño, H. Calvo, A. Sánchez, J.S. Casas, J. Sordo, Y.P. Mascarenhas, and C. de O.P. Santos, *J. Organomet. Chem.*, 1991, **417**, 327.
[1394] C.C. Cummins, R.R. Schrock, and W.M. Davis, *Organometallics*, 1991, **10**, 3781.
[1395] N. Playá, A. Macías, J.M. Varela, A. Sánchez, J.S. Casas, and J. Sordo, *Polyhedron*, 1991, **10**, 1465.
[1396] J. Zukerman-Schpector, E.M. Vazquez-Lopez, A. Sanchez, J.S. Casas, and J. Sordo, *J. Organomet. Chem.*, 1991, **405**, 67.
[1397] A.J. Bloodworth and A. Shah, *J. Chem. Soc., Chem. Commun.*, 1991, 947.
[1398] A.J. Bloodworth, K.H. Chan, C.J. Cooksey, and N. Hargreaves, *J. Chem. Soc., Perkin Trans. 1*, 1991, 1923.
[1399] M. Drieß, H. Pritzkow, and M. Reisgys, *Chem. Ber.*, 1991, **124**, 1923.
[1400] V.K. Ahluwalia, J. Kaur, B.S. Ahuja, and G.S. Sodhi, *J. Inorg. Biochem.*, 1991, **42**, 147 (*Chem. Abstr.*, 1991, **115**, 29 505).
[1401] D. Naumann, W. Strauss, and W. Tyrra, *J. Organomet. Chem.*, 1991, **407**, 1.
[1402] S. Bagger, K. Breddam, and B.R. Byberg, *J. Inorg. Biochem.*, 1991, **42**, 97 (*Chem. Abstr.*, 1991, **115**, 45 069).
[1403] E.B. Fleischer and A.M. Shachter, *Inorg. Chem.*, 1991, **30**, 3763.
[1404] T.I. Takedo, T. Nakamura, S. Umetani, and M. Matsui, *Bunseki Kagaku*, 1990, **39**, 559 (*Chem. Abstr.*, 1991, **114**, 73 826).
[1405] D. Williamson, I.J. McLennan, A. Bax, M.P. Gamesik, and J.D. Glickson, *J. Biomol. Struct. Dyn.*, 1990, **8**, 375 (*Chem. Abstr.*, 1991, **114**, 24 603).
[1406] X. Yan, J. Fan, X. Xu, D. Wang, G. Nie, and B. Qian, *Magn. Reson. Chem.*, 1991, **29**, 1092.

$CdCl_2$ with cyanocobalamin.[1407] N.m.r. data have also been reported for $[\{HB(C_3H_3N_2)_3\}_2Zn]$, $[\{HB(C_3H_3N_2)_3\}PbCl]$, ($^{13}C$),[1408] $[\{HB(3\text{-}Bu^t\text{-}5\text{-}MeC_3HN_2)_3\}Zn(OH)]$, ($^{13}C$),[1409] [Zn(1,6,11-triazacyclopentadecane)Cl]$^+$, (^{13}C),[1410] $[M(MeNHCH_2CH_2NHCH_2CH_2NHCH_2CH_2NHCH_2CH_2\text{-}NHMe)]^{2+}$, (M = Zn, Cd; ^{13}C),[1411] $[Zn\{H_2NC(Me)(CH_2NHCH_2CH_2NHCH_2)_2CMeNH_2\}]^{2+}$, ($^{13}C$),[1412] $[EtOC(O)CH(NEt_2)CH(NHBu^t)CH{=}NBu^tZnCl_2]$, ($^{13}C$),[1413] $[\{(Me_3Si)_2N\}\text{-}Zn(O{=}CMeCH_2CMe_2O)]_2$, ($^{13}C$),[1414] $[M\{2\text{-furyl-}C(O)NHN{=}CMeC_6H_4\text{-}2\text{-}NH_2\}_2Cl_2]$, (M = Zn, Cd, Hg; ^{13}C),[1415] Zn^{2+}-thymulin complex, (^{13}C),[1416] Zn^{2+}-dodecamer oligodeoxyribonucleotide, (^{13}C),[1417] (93), (^{13}C),[1418] $[Zn(2\text{-}SC_5H_4N)_2(phen)]$, ($^{13}C$),[1419] $[Cd\{O{=}CMeCH{=}CMeNH(CH_2)_n\text{-}NH_2\}_2]^{2+}$, ($^{13}C$),[1420] Cd^{2+} complexed by borate esters of 2-amino-2-deoxy-D-gluconate, (^{11}B, ^{13}C,

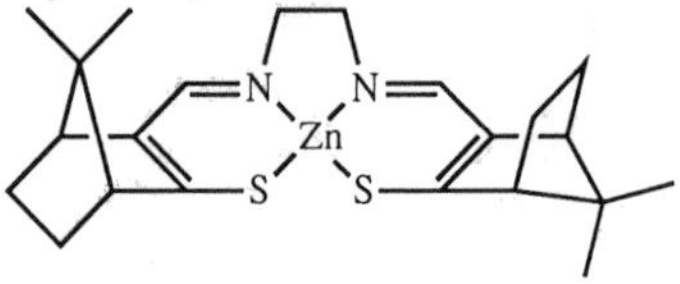

(93)

^{113}Cd),[1421] [(pyropheophorbidato-a methyl ester)Cd], (^{13}C),[1422] Cd substituted superoxide dismutase, (^{113}Cd),[1423] Cd^{2+} bound to ovotransferrin, (^{13}C, ^{113}Cd),[1424] a polypeptide-Cd^{2+} cluster, (^{113}Cd),[1425] $[M(\text{L-ascorbate})_2]$, (M = Zn, Cd; ^{13}C),[1426] $[Zn(SeC_6H_2Bu^t_3\text{-}2,4,6)_2(O{=}CHC_4H_4OMe\text{-}4)]_2$, ($^{13}C$),[1427] Cd^{2+}-$MeOCH{=}CH_2$-alkaline phosphatase, (^{113}Cd),[1428] $[Cd(xanthosinato)_2(OH_2)_6]$,

1407 M.D. Couce, J.M. Varela, A. Sánchez, J.S. Casas, J. Sordo, and M. López-Rivadulla, *J. Inorg. Biochem.*, 1991, **41**, 1.
1408 G.G. Lobbia, F. Bonati, and P. Cecchi, *Synth. React. Inorg. Metal-Org. Chem., 1991, 21*, 1141.
1409 R. Alsfasser, S. Trofimenko, A. Looney, G. Parkin, and H. Vahrenkamp, *Inorg. Chem.*, 1991, **30**, 4098.
1410 K. Ida, H. Okawa, H. Sakiyama, M. Kodera, S. Kida, and I. Murase, *Inorg. Chim. Acta*, 1991, **184**, 197.
1411 J. Aragó, A. Bencini, A. Bianchi, E. Garcia-España, M. Micheloni, P. Paoletti, J.A. Ramirez, and A. Rodriguez, *J. Chem. Soc., Dalton Trans.*, 1991, 3077.
1412 P.V. Bernhardt, G.A. Lawrance, M. Maeder, M. Rossignoli, and T.W. Hambley, *J. Chem. Soc., Dalton Trans.*, 1991, 1167.
1413 F.H. van der Steen, H. Kleijn, A.L. Spek, and G. van Koten, *J. Org. Chem.*, 1991, **56**, 5868.
1414 S.C. Goel, M.Y. Chiang, and W.E. Buhro, *J. Am. Chem. Soc.*, 1991, **113**, 7069.
1415 B. Singh and A.K. Srivastav, *Synth. React. Inorg. Metal-Org. Chem.*, 1991, **21**, 457.
1416 Manh Thong Cung, M. Marraud, P. Lefrancier, M. Dardenne, J.F. Bach, and J.P. Laussac, *J. Biol. Chem.*, 1988, **263**, 5574 (*Chem. Abstr.*, 1991, **114**, 36 345).
1417 X. Jia, G. Zon, and L.G. Marzilli, *Inorg. Chem.*, 1991, **30**, 228.
1418 L. Casella, M. Gullotti, R. Pagliarin, and M. Sisti, *J. Chem. Soc., Dalton Trans.*, 1991, 2527.
1419 M.L. Duran, J. Romero, J.A. Garcia-Vazquez, R. Castro, A. Castineiras, and A. Sousa, *Polyhedron*, 1991, **10**, 197.
1420 M. Kwiatkowski, E. Kwiatkowski, A. Olechnowicz, and B. Kościuszko, *Polyhedron*, 1991, **10**, 945.
1421 J. van Haveren, J.A. Peters, J.G. Batelaan, A.P.G. Kieboom, and H. van Bekkum, *J. Chem. Soc., Dalton Trans.*, 1991, 2649.
1422 J. Iturraspe and A. Gossauer, *Helv. Chim. Acta*, 1991, **74**, 1713.
1423 P. Kofod, R. Bauer, E. Danielsen, E. Larsen, and M.J. Bjerrum, *Eur. J. Biochem.*, 1991, **198**, 607 (*Chem. Abstr.*, 1991, **115**, 88 235).
1424 M. Sola, *Eur. J. Biochem*, 1990, **194**, 349 (*Chem. Abstr.*, 1991, **114**, 57 767).
1425 P.L. Gadhavi, A.L. Davis, J.F. Povey, J. Keeler, and E.D. Laue, *FEBS Lett.*, 1991, **281**, 223 (*Chem. Abstr.*, 1991, **114**, 243 071).
1426 H.A. Tajmir-Riahi, *J. Inorg. Biochem.*, 1991, **42**, 47.
1427 M. Bochmann, K.J. Webb, M.B. Hursthouse, and M. Mazid, *J. Chem. Soc., Chem. Commun.*, 1991, 1735.
1428 J. Afflitto, K.A. Smith, M. Patel, A. Esposito, E. Jensen, and A. Gaffar, *Pharm. Res.*, 1991, **8**, 1384 (*Chem. Abstr.*, 1991, **115**, 263 050).

(^{13}C),[1429] [Hg(thiomalate)$_2$]$^{4-}$, [Pb(thiomalate)]$^-$, (^{13}C),[1430] [M(EC$_6$H$_2$Bu$^t_3$-2,4,6)$_2$], (M = Zn, Cd; E = S, Se; ^{13}C, ^{77}Se),[1431] [M{(2-benzothiazolyl)$_2$CH$_2$}$_2$], (M = Zn, Ni, Hg; ^{13}C, ^{15}N),[1432] [M(4-acetylpyridinthiosemicarbazone)$_2$X$_2$], (M = Zn, Cd, Hg; ^{13}C),[1433] [M(Et$_2$NCS$_2$)X], (M = Zn, Cd, Hg; ^{14}N, ^{113}Cd, ^{199}Hg),[1434] [M(S$_2$CNEt$_2$)$_2$(phen)], (M = Zn, Cd, Hg; ^{14}N, ^{113}Cd),[1435] [Zn(seleno-1,3-dithiole-2-selenolate)$_2$]$^{2-}$, (^{13}C, ^{77}Se),[1436] Cd^{2+}-metallothionein, (^{113}Cd),[1437] [Cd(TeC$_6$H$_2$R$_3$-2,4,6)$_2$], (^{13}C, ^{125}Te),[1438] [Hg(S_m)(S_n)]$^{2-}$, (^{199}Hg),[1439] [Hg(EC$_6$H$_2$R$_3$-2,4,6)$_2$], (E = S, Se; ^{77}Se, ^{199}Hg),[1440] [Hg([15]aneS$_5$)]$^{2+}$, (^{13}C),[1441] [Hg(2,6-Me$_2$4*H*-pyran-4-thione)$_2$X$_2$], (^{13}C),[1442] and [Hg{SeCF(CF$_3$)$_2$}$_2$], Me$_3$SiHg{SeCF(CF$_3$)$_2$}$_2$, (^{13}C, ^{29}Si, ^{77}Se).[1443]

3 Dynamic Systems

This section is in three main parts: (i) 'Fluxional Molecules', dealing with rate processes involving no molecular change, (ii) 'Equilibria', dealing with the use of n.m.r. spectroscopy to measure the position of equilibria and ligand-exchange reactions, including solvation, and (iii) 'Course of Reactions', dealing with the use of n.m.r. spectroscopy to monitor the course of reactions. Each section is ordered by the Periodic Table.

Four relevant reviews have appeared: 'Quantitative investigations of molecular stereodynamics by one- and two-dimensional methods',[1444] 'Some perspectives of 1,1'-bis(diphenylphosphino)ferrocene as a ligand complex', which contains n.m.r. fluxionality,[1445] 'Equilibria studies in solution', which reviews the advantages and disadvantages of the n.m.r. technique for the determination of equilibrium constants in solution,[1446] and 'Nuclear magnetic relaxation in polyelectrolyte solution'.[1447]

Two papers have appeared which are broadly based. The permutational character of the *dsd*

1429 M. Quirós, J.M. Salas, M.P. Sánchez, J.R. Alabart, and R. Faure, *Inorg. Chem.*, 1991, **30**, 2916.
1430 M. Maeda, K. Okada, K. Wakabayashi, K. Honda, K. Ito, and Y. Kinjo, *J. Inorg. Biochem.*, 1991, **42**, 37.
1431 M. Bochmann, K.J. Webb, M.B. Hursthouse, and M. Mazid, *J. Chem. Soc., Dalton Trans.*, 1991, 2317.
1432 A. Abbotto, V. Alanzo, S. Bradamante, G.A. Pagani, C. Rizzoli, and G. Calestani, *Gazz. Chim. Ital.*, 1991, **121**, 365 (*Chem. Abstr.*, 1991, **115**, 269 225).
1433 P. Souza, L. Sanz, V. Fernandez, A. Arquero, E. Gutierrez, and A. Monge, *Z. Naturforsch., B*, 1991, **46**, 767 (*Chem. Abstr.*, 1991, **115**, 148 995).
1434 S.M. Zemskova, I.M. Oglezneva, M.A. Fedotov, S.A. Gromilov, and S.V. Larionov, *Sib. Khim. Zh.*, 1991, 115 (*Chem. Abstr.*, 1991, **115**, 221 737).
1435 S.M. Zemskova, I.M. Oglezneva, M.A. Fedotov, L.K. Glukhikh, and S.V. Larionov, *Izv. Sib. Otd. Akad. Nauk SSSR, Ser. Khim. Nauk*, 1990, 89 (*Chem. Abstr.*, 1991, **114**, 258 472)
1436 R.M. Olk, B. Olk, J. Rohloff, and E. Hoyer, *Z. Chem.*, 1990, **30**, 445 (*Chem. Abstr.*, 1991, **114**, 220 125).
1437 M.J. Cismowski, S.S. Narula, I.M. Armitage, M.L. Chernaik, and P.C. Huang, *J. Biol. Chem.*, 1991, **266**, 24 390 (*Chem. Abstr.*, 1991, **115**, 250 602).
1438 M. Bochmann, A.P. Coleman, K.J. Webb, M.B. Hursthouse, and M. Mazid, *Angew. Chem., Int. Ed. Engl.*, 1991, **30**, 973.
1439 T.D. Bailey, R.M.H. Banda, D.C. Craig, I.G. Dance, I.N.L. Ma, and M.L. Scudder, *Inorg. Chem.*, 1991, **30**, 187.
1440 M. Bochmann and K.J. Webb, *J. Chem. Soc., Dalton Trans.*, 1991, 2325.
1441 A.J. Blake, E.C. Pasteur, G. Reid, and M. Schröder, *Polyhedron*, 1991, **10**, 1545.
1442 G. Faraglia, Z. Guo, and S. Sitran, *Polyhedron*, 1991, **10**, 351.
1443 A. Haas, C. Limberg, and M. Spehr, *Chem. Ber.*, 1991, **124**, 423.
1444 K.G. Orrell, V. Šik, and D. Stephenson, *Prog. Nucl. Magn. Reson. Spectrosc.*, 1990, **22**, 141.
1445 T.S.A. Hor and L.T. Phang, *Bull. Singapore Natl. Inst. Chem.*, 1990, **18**, 29 (*Chem. Abstr.*, 1991, **115**, 173 243).
1446 A.I. Popov, *Pract. Spectrosc.*, 1991, **11**, 485 (*Chem. Abstr.*, 1991, **114**, 130 232).
1447 J.C. Leyte, *Makromol. Chem., Macromol. Symp.*, 1990, **34**, 81 (*Chem. Abstr.*, 1991, **114**, 63 082).

mechanisms in heptacoordinate chemistry has been examined and applied to n.m.r. line-shape analysis.[1448] The estimation of the error of the determination of the equilibrium constant by n.m.r. spectroscopy has been discussed.[1449]

Fluxional Molecules.—*Lithium.* ^{6}Li n.m.r. spectroscopy has been used to demonstrate the exchange of three ^{6}Li signals from [(4-$R^1C_6H_4CHCHR^2CHR^2CHC_6H_4R^1$-4)$Li_2$] and $\Delta G^\ddagger$ determined. The ^{13}C n.m.r. spectrum was also reported.[1450]

Sodium. $\Delta G^\ddagger$ for amide rotation in 1-acetyl-1-aza-4,7,10,13-tetraoxacyclopentadecane is 73 ± 1 kJ mol^{-1} but decreases to 54 ± 1 kJ mol^{-1} in the Na^+ complex.[1451]

Magnesium, Calcium, and Barium. ^{2}H n.m.r. spectroscopy has been used to study the dynamic processes occurring in the Mg^{2+} and Ca^{2+} complexes with 2-propylenediaminetetraacetic acid.[1452] The fluxionality of [Ca(Bu^t-calix[8]arene-H_2)$(DMF)_4$] has been investigated by ^{1}H n.m.r. spectroscopy.[1453] ^{1}H and ^{13}C n.m.r. spectroscopy has been used to demonstrate restricted rotation in [(η^5-$C_5Pr^i_4H$)$_2$M], M = Ca, Ba, with $\Delta G^\ddagger \leq 11.1$ kcal mol^{-1}, M = Ca, and 11.1 ± 0.5 kcal mol^{-1}, M = Ba.[1454]

Lutetium. ^{1}H and ^{13}C n.m.r. spectroscopy has been used to demonstrate role exchange between the η^3- and η^5-dienyl forms in [Lu(η^3-2,4-$Me_2C_5H_5$)(η^5-2,4-$Me_2C_5H_5$)$_2$].[1455]

Titanium. The ^{1}H and ^{13}C n.m.r. spectra of [(η^5-C_5Me_4H)$_2$Ti(C_6H_4Me-4)$_2$] show restricted rotation of the Ti—C_6H_4Me bond with $\Delta G^\ddagger$ ~60.5 kJ mol^{-1}.[1456] In the ^{13}C n.m.r. spectrum of [(η^5-2,4-$Bu^t_2C_5H_5$)$_2$Ti], there is restricted rotation of the staggered dienyl ligands with $\Delta G^\ddagger = 15.5$ kcal mol^{-1}. The ^{13}C n.m.r. spectrum of [(η^1-2,4-$Bu^t_2C_5H_5$)$_2$Zn] was also recorded.[1457] The dynamic ^{1}H n.m.r. spectrum of [(η^5-C_5H_5)$_2$Ti(μ-$AsMe_2AsMe_2$)$_2$Ti(η^5-C_5H_5)$_2$] shows a temperature dependent ring inversion with $\Delta G^\ddagger$ ~13 kcal mol^{-1}.[1458] The fluxionality of [(η^5-2-R^1-4-$R^2C_5H_2CH_2CH_2C_5H_2$-2-R^1-4-R^4-η^5)$TiCl_2$] has been studied by ^{1}H and ^{13}C n.m.r. spectroscopy.[1459] According to variable temperature ^{1}H and ^{13}C n.m.r. spectroscopy, [PhTi(OCH_2CH_2)$_3$N] is a dimer undergoing a fluxional gearing motion at room temperature, but becoming monomeric on warming.[1460] Variable temperature ^{1}H, ^{13}C, and ^{31}P n.m.r. spectroscopy has been used to study Ti—N bond rotation in

1448 J. Brocas and M. Bauwin, *J. Math. Chem.*, 1991, **6**, 281 (*Chem. Abstr.*, 1991, **115**, 190 786).
1449 V.T. Panyushkin, A.B. Vashchuk, and S.N. Bolotin, *Koord. Khim.*, 1991, **17**, 1042 (*Chem. Abstr.*, 1991, **115**, 190 921).
1450 P. Schade, T. Schäfer, K. Müllen, D. Bender, K. Knoll, and K. Bronstert, *Chem. Ber.*, 1991, **124**, 2833.
1451 G.C. Saunders, *Polyhedron*, 1991, **10**, 1877.
1452 R. Song, Y. Ba, G. Zhang, Y. Li, and Z. Qui, *Wuli Huaxue Xuebao*, 1991, **7**, 102 (*Chem. Abstr.*, 1991, **114**, 151 408).
1453 J.M. Harrowfield, M.I. Ogden, W.R. Richmond, and A.H. White, *J. Chem. Soc., Dalton Trans.*, 1991, 2153.
1454 R.A. Williams, K.F. Tesh, and T.P. Hanusa, *J. Am. Chem. Soc.*, 1991, **113**, 4843.
1455 H. Schumann and A. Dietrich, *J. Organomet. Chem.*, 1991, **401**, C33.
1456 P. Courtot, R. Pichon, J.Y. Salaun, and L. Toupet, *Can. J. Chem.*, 1991, **69**, 661.
1457 R.D. Ernst, J.W. Freeman, P.N. Swepston, and D.R. Wilson, *J. Organomet. Chem.*, 1991, **402**, 17.
1458 P. Gowik and Th. Klapötke, *J. Organomet. Chem.*, 1991, **404**, 349.
1459 S. Collins, Y. Hong, R. Ramachandran, and N.J. Taylor, *Organometallics*, 1991, **10**, 2349.
1460 W.M.P.B. Menge and J.G. Verkade, *Inorg. Chem.*, 1991, **30**, 4628.

$[Ti(NR_2)_3Cl]$.[1461] The ^1H n.m.r. spectrum of $[Ti(R,R\text{-}Et_2tartrate)(OBu^t)_2]_2$ shows exchange of the inequivalent CH protons and $\Delta G^\ddagger$, $\Delta H^\ddagger$, and $\Delta S^\ddagger$ were determined. The ^{13}C and ^{17}O n.m.r. spectra were also reported.[1462]

Zirconium, and Hafnium. Variable temperature ^1H, ^{11}B, and ^{31}P n.m.r. spectroscopy has shown that $[M_2H_3(BH_4)_5(PMe_3)_2]$, M = Zr, Hf, is fluxional.[1463] The ^1H n.m.r. spectrum of $[(CH_2\text{-}1\text{-}tetrahydroindenyl\text{-}\eta^5)_2Zr(CH_3)(\eta^2\text{-}C_6H_5)BPh_3]$ is dynamic due to motion on the coordinated phenyl ring with $\Delta G^\ddagger$ ~12.1 kcal mol^{-1}. The ^{13}C n.m.r. spectrum was also reported.[1464] The variable temperature ^1H n.m.r. spectrum of $[(CH_2\text{-}1\text{-}indenyl\text{-}\eta^5)_2ZrC(SiMe_3{=}CMe_2)]^+$ shows restricted rotation due to metal interaction with $\Delta G^\ddagger = 10.7 \pm 0.2$ kcal mol^{-1}. ^{13}C and ^{29}Si n.m.r. spectra were also reported.[1465] $[(\eta^5\text{-}C_5H_5)_2\overline{Zr(CH_2CH{=}CHN}Ph)]$ is fluxional due to ring inversion with $\Delta G^\ddagger$ = 46.8 ± 0.2 kJ mol^{-1}. ^{13}C n.m.r. spectra were also recorded.[1466] ^1H n.m.r. spectroscopy has been used to measure $\Delta G^\ddagger$ = 7 kcal mol^{-1} for the exchange

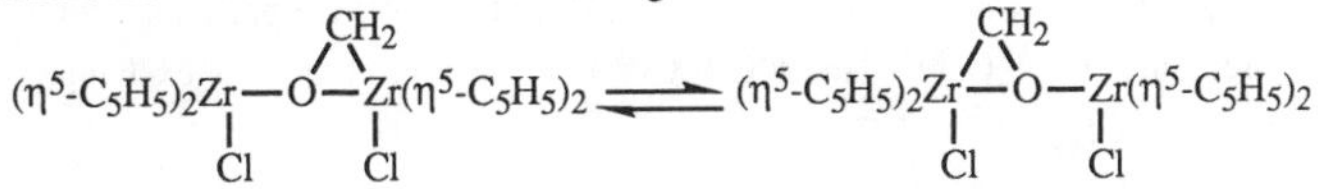

The ^{13}C n.m.r. spectrum was reported.[1467] A similar dynamic process was observed for $[(\eta^5\text{-}C_5Me_5)ClZr(\mu\text{-}\eta^5\text{-}C_5H_4C_5H_4\text{-}\eta^5)(\mu\text{-}OCH_2)ZrCl(\eta^5\text{-}C_5Me_5)]$.[1468] Variable temperature ^1H, ^{13}C, and ^{31}P n.m.r. spectroscopy has been used to investigate the fluxionality of (94), involving exchange of PPh_2 coordination to the CMe group.[1469] $[(\eta^5\text{-}C_5H_5)Hf(\eta^3\text{-}1,1,2\text{-}Me_3C_3H_2)(\eta^4\text{-}2,3\text{-}Me_2C_4H_4)]$ shows three separate dynamic processes in the ^1H n.m.r. spectrum.[1470] Variable temperature ^1H n.m.r. spectra of $[(\eta^5\text{-}C_5Me_5)(\eta^3\text{-}1,2\text{-}Me_2C_3H_3)ZrX_2]$ and $[(\eta^5\text{-}C_5Me_5)(\eta^2\text{-}CH_2PPh_2)(\eta^4\text{-}2,3\text{-}Me_2C_4H_4)Zr]$ show dynamic processes.[1471] Allyl exchange occurs in $[(\eta^5\text{-}C_5H_5)Hf(\eta^3\text{-}1,1,2\text{-}Me_3C_3H_2)Br_2]$ with $\Delta G^\ddagger{}_{220}$ = 40.9 ± 1 kJ mol^{-1}.[1472] ^1H n.m.r. spectroscopy has been used to determine $\Delta G^\ddagger$ =16.5 ± 0.3 kcal mol^{-1} for ring inversion in (95). The ^{13}C n.m.r. spectrum was also

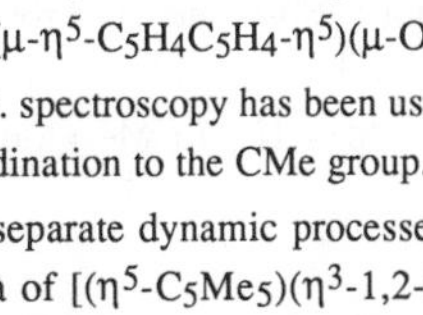

(94)

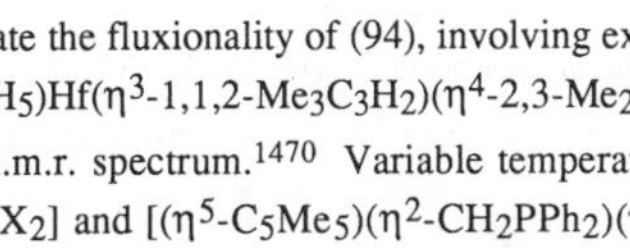

(95)

[1461] D.G. Dick, R. Rousseau, and D.W. Stephan, *Can. J. Chem.*, 1991, **69**, 357.
[1462] M.G. Finn and K.B. Sharpless, *J. Am. Chem. Soc.*, 1991, **113**, 113.
[1463] J.E. Gozum and G.S. Girolami, *J. Am. Chem. Soc.*, 1991, **113**, 3829.
[1464] A.D. Horton and J.H.G. Frijns, *Angew. Chem., Int. Ed. Engl.*, 1991, **30**, 1152.
[1465] A.D. Horton and A.G. Orpen, *Organometallics*, 1991, **10**, 3910.
[1466] J.M. Davis, R.J. Whitby, and A. Jaxa-Chamiec, *J. Chem. Soc., Chem. Commun.*, 1991, 1743.
[1467] G. Erker, M. Mena, and M. Bendix, *J. Organomet. Chem.*, 1991, **410**, C5.
[1468] C.J. Curtis and R.C. Haltiwanger, *Organometallics*, 1991, **10**, 3220.
[1469] W. Tikkanen and J.W. Ziller, *Organometallics*, 1991, **10**, 2266.
[1470] T.J. Prins, B.E. Hauger, P.J. Vance, M.E. Wemple, D.A. Kort, J.P. O'Brien, M.E. Silver, and J.C. Huffman, *Organometallics*, 1991, **10**, 979.
[1471] P.J. Vance, T.J. Prins, B.E. Bauger, M.E. Wemple, L.M. Pederson, D.A. Kort, M.R. Kannisto, S.J. Geerligs, R.S. Kelly, J.J. McCandless, J.C. Huffman, and D.G. Peters, *Organometallics*, 1991, **10**, 917.
[1472] B.E. Hauger, P.J. Vance, T.J. Prins, M.E. Wemple, D.A. Kort, M.E. Silver, and J.C. Huffman, *Inorg. Chim. Acta*, 1991, **187**, 91.

recorded.[1473] Ring inversion has also been studied in (96) by ^1H and ^{13}C n.m.r. spectroscopy.[1474] Restricted rotation has been demonstrated by ^1H and ^{13}C n.m.r. spectroscopy in $[\{\eta^5\text{-}1,2,4\text{-}(Me_3Si)_3C_5H_2\}_2MCl_2]$, M = Zr, Hf, with $\Delta G^\ddagger$ = 11.0 and 11.3 kcal mol^{-1} respectively.[1475] Dynamic processes in $[(\eta^5\text{-}C_5H_5)_2Zr(\text{catecholate})]$ have been studied by ^1H and ^{13}C n.m.r. spectroscopy and $\Delta G^\ddagger$ determined.[1476]

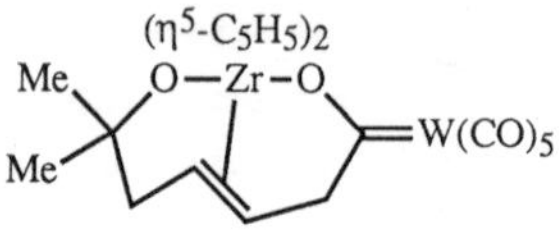

(96)

Vanadium. Restricted rotation of the dienyl in $[(\eta^5\text{-}C_5H_7)V(CO)_2(PMe_3)_2]$ has been investigated by ^1H and ^{13}C n.m.r. spectroscopy and $\Delta G^\ddagger$ = 45.2 kJ mol^{-1}.[1477] Activation parameters of the exchange between two types of glycinate groups in $[VO_2\{N(CH_2CO_2)_3\}]^{2-}$ have been determined from ^{13}C n.m.r. measurements.[1478] ^{13}C, ^{51}V, and ^{13}C EXSY have been used to show exchange of the CH_2 groups in $[VO_2\{O_2CCH_2NHC(CH_2OH)_3\}]$.[1479]

Niobium and Tantalum. ^1H and ^{13}C n.m.r. measurements for cyclo-octatetraene role exchange in $[(\eta^5\text{-}C_5H_4Me)Nb(\eta^3\text{-}C_8H_8)(\eta^4\text{-}C_8H_8)]$ have given $\Delta G^\ddagger$ = 76 ± 2 kJ mol^{-1}.[1480] $\Delta G^\ddagger$ for rotation about the M—M bond in $[(\eta^5\text{-}C_5H_5)_2M^1M^2(\eta^5\text{-}C_5H_5)(CO)_3]$, M^1 = Nb, Ta, M^2 = Mo, W, has been determined and the ^{13}C n.m.r. spectrum recorded.[1481]

Chromium, Molybdenum and Tungsten. The variable temperature ^1H n.m.r. spectrum of $[W(PMe_3)_3H_5Li]_4$ shows one hydride signal at 323 K and three at low temperature and $\Delta G^\ddagger$ was determined. ^{13}C, ^{119}Sn, and ^{183}W n.m.r. spectra were recorded for $[W(PMe_3)_3H_5(SnBu^n{}_3)]$.[1482] The ^1H (hydride) signal of $[WH(CO)_3\{PhP(CH_2CH_2PPh_2)_2\}]^+$ is a double doublet at low temperature and a triplet at high.[1483] A similar behaviour has been observed for $[WH(OTf)(CO)_3(PCy_3)_2]$.[1484]

In (97), the two methyl groups exchange with $\Delta G^\ddagger$ = 80.2 ± 5 kJ mol^{-1}. The ^{13}C and ^{29}Si n.m.r. spectra were also recorded.[1485] Variable temperature ^1H and ^{13}C n.m.r. spectra of $[(\eta^5\text{-}C_5H_5)M(NO)(\eta^1\text{-}CH_2Ph)(\eta^3\text{-}CH_2Ph)]$, M = Mo, W, show role exchange between the two benzyl

1473 G. Erker, R. Pfaff, C. Krüger, and S. Werner, *Organometallics*, 1991, **10**, 3559.
1474 G. Erker, F. Sosna, P. Betz, S. Werner, and C. Krüger, *J. Am. Chem. Soc.*, 1991, **113**, 564.
1475 C.H. Winter, D.A. Dobbs, and X.-X. Zhou, *J. Organomet. Chem.*, 1991, **403**, 145.
1476 G. Erker and R. Noe, *J. Chem. Soc., Dalton Trans.*, 1991, 685.
1477 W.-J. Lin, G.-H. Lee, S.-M. Peng, and R.-S. Liu, *Organometallics*, 1991, **10**, 2519.
1478 M.H. Lee and K. Schaumburg, *Bull. Korean Chem. Soc.*, 1990, **11**, 399 (*Chem. Abstr.*, 1991, **114**, 34 653).
1479 D.C. Crans, P.M. Ehde, P.K. Shin, and L. Petersson, *J. Am. Chem. Soc.*, 1991, **113**, 3728.
1480 G.E. Herberich, U. Englert, and P. Roos, *Chem. Ber.*, 1991, **124**, 2663.
1481 D. Perrey, J.C. Leblanc, C. Moïse, and J. Martin-Gil, *J. Organomet. Chem.*, 1991, **412**, 363.
1482 A. Berry, M.L.H. Green, J.A. Bandy, and K. Prout, *J. Chem. Soc., Dalton Trans.*, 1991, 2185.
1483 V. Zanotti, V. Rutar, and R.J. Angelici, *J. Organomet. Chem.*, 1991, **414**, 177.
1484 L.S. Van der Sluys, K.A. Kubat-Martin, G.J. Kubas, and K.G. Caulton, *Inorg. Chem.*, 1991, **30**, 306.
1485 R. Probst, C. Leis, S. Gamper, E. Herdtweck, C. Zybill, and N. Auner, *Angew. Chem., Int. Ed. Engl.*, 1991, **30**, 1132.

groups.[1486] $[Mo\{C(O)CH_3\}(S_2CH)(CO)(PMe_3)_2]$ shows agostic interaction between the Mo and acetyl methyl. ^1H, ^{13}C, and ^{31}P n.m.r. spectra show a fast equilibrium between two degenerate ground state structures.[1487]

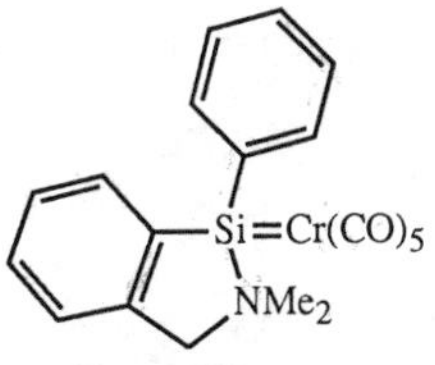

(97)

$[\{HB(C_3H_3N_2)_3\}Mo(CO)_2(\eta^2\text{-}CPh{=}CHCH{=}NMeBu^t)]^+$ is fluxional due to rapid inversion at the vinyl β-carbon. The ^{13}C n.m.r. spectrum was recorded.[1488] The fluxional behaviour of $[(\eta^5\text{-}C_5H_5)Mo\{{=}C(CF_3)C(CF_3)PR_3\}(\eta^2\text{-}CF_3C{\equiv}CCF_3)]$, $[(\eta^5\text{-}C_5H_5)Mo(SC_6F_5)(PMe_2Ph)\{\eta^4\text{-}C_4(CF_3)_4\}]$, and $[(\eta^5\text{-}C_5H_5)MoF\{\eta^3\text{-}C(CF_2)C(CF_3){=}C(CF_3)C(CF_3)SC_6F_5\}L]$ has been studied by ^{19}F n.m.r. spectroscopy.[1489] There is a dynamic process which equilibrates the two sides of the fulvene signals of (98) with $\Delta G^\ddagger = 15.5 \pm 0.12$ kcal mol^{-1} and a second $\Delta G^\ddagger = 17.6 \pm 1.0$ kcal mol^{-1} which coalesce the two phenyl groups. The ^{13}C n.m.r. spectrum was also recorded.[1490] The molecular mobility of $Co_2(CO)_6$-alkyne-estradiol and $[(\eta^5\text{-}C_5H_5)_2Mo(CO)_4(\text{alkyne-estradiol})]$ complexes has been studied by ^{13}C line width analysis.[1491] The ^1H and ^{13}C n.m.r. spectra of $[(\eta^5\text{-}C_5H_5)_2Mo_2(CO)_4(\mu\text{-}C{\equiv}CPh)]^-$ show two cyclopentadienyl signals at low temperature and one at high and $\Delta G^\ddagger = 51 \pm 2$ kJ mol^{-1}.[1492] ^1H and ^{13}C n.m.r. spectroscopy have provided evidence for the fluxional behaviour of (99)].[1493] The ^1H n.m.r. spectrum of the ethyl group of $[(\eta^5\text{-}C_5Me_5)Mo(\eta^2\text{-}EtN{=}CMe)(CO)_2]^+$ is ABX_3 at low temperature and A_2X_3 at high temperature, and $\Delta G^\ddagger = 17.2$ kcal

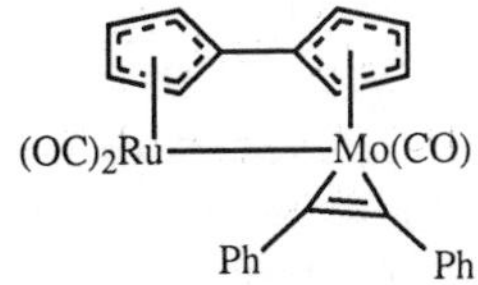

(98)

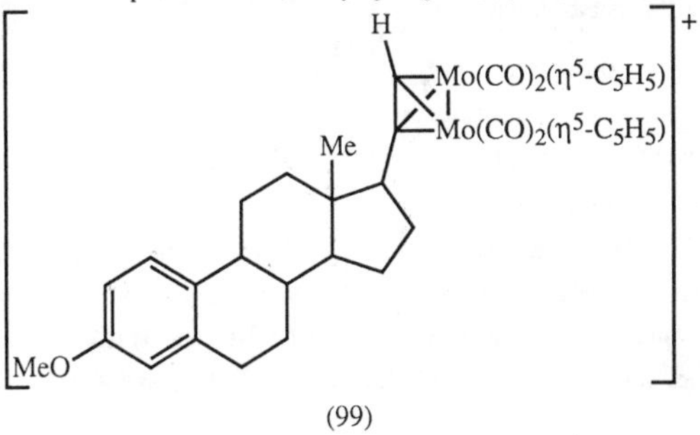

(99)

1486 P. Legzdins, E.C. Philips, J. Trotter, V.C. Yee, F.W.B. Einstein, and R.H. Jones, *Organometallics*, 1991, **10**, 986.
1487 E. Carmona, L. Contreras, M.L. Poveda, and L.J. Sanchez, *J. Am. Chem. Soc.*, 1991, **113**, 4322.
1488 A.S. Gamble, P.S. White, and J.L. Templeton, *Organometallics*, 1991, **10**, 693.
1489 N.M. Agh-Atabay, L.J. Canoira, L. Carlton, and J.L. Davidson, *J. Chem. Soc., Dalton Trans.*, 1991, 1175.
1490 R. Boese, M.A. Huffman, and K.P.C. Vollhardt, *Angew. Chem., Int. Ed. Engl.*, 1991, **30**, 1463.
1491 D. Osella, E. Stein, G. Jaouen, and P. Zanello, *J. Organomet. Chem.*, 1991, **401**, 37.
1492 S.F.T. Froom, M. Green, R.J. Mercer, K.R. Nagle, A.G. Orpen, and R.A. Rodrigues, *J. Chem. Soc., Dalton Trans.*, 1991, 3171.
1493 C. Cordier, M. Gruselle, G. Jaouen, V.I. Bakhmutov, M.V. Galakhov, L.L. Troitskaya, and V.I. Sokolov, *Organometallics*, 1991, **10**, 2303.

mol^{-1}. The ^{13}C n.m.r. spectrum was also recorded.[1494] The ^{1}H n.m.r. spectrum of (100) shows dynamic behaviour. The ^{13}C, ^{31}P, and ^{95}Mo n.m.r. spectra were also recorded.[1495] The fluxionality of $[(\eta^5\text{-}C_5H_5)M\{\eta^3\text{-}C(CF_3)C(CF_3)CR^1{=}CR^2(SR)\}(\eta^2\text{-}CF_3C{\equiv}CCF_3)]$, M = Mo, W, has been studied by ^{19}F n.m.r. spectroscopy. The lowest energy process was ascribed to inversion of the

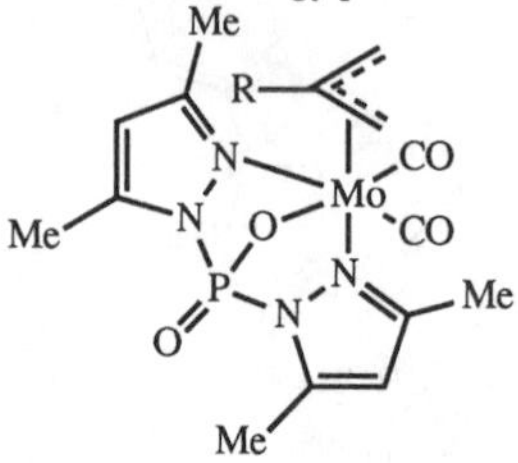

(100)

sulphur on the β-carbon atom and the second due to rotation of the CF_3 attached to the β-carbon.[1496] Temperature dependent ^{1}H and ^{13}C n.m.r. spectra have demonstrated the fluxionality in Scheme 1.[1497] The barrier to alkyne rotation in $[W(\eta^2\text{-}PhC{\equiv}CPh)_3(SnR_3)]^-$ has been determined and the ^{13}C and ^{119}Sn n.m.r. spectra recorded.[1498] The barrier for 2-butyne rotation in $[W(CO)(PMe_3)_2(S_2C\text{-}NC_4H_8)(\eta^2\text{-}MeC{\equiv}CMe)]^+$ is 66.9 ± 1 kJ mol^{-1}. The ^{13}C n.m.r. spectrum was recorded.[1499] Similarly in $[WI(CO)(NCCH_2\text{-}3\text{-}C_4H_3S)(dppm)(\eta^2\text{-}MeC{\equiv}CMe)]^+$, the barrier for 2-butyne rotation is 51.3 kJ mol^{-1}, and the ^{13}C n.m.r. spectrum was recorded.[1500] ^{1}H and ^{13}C n.m.r. spectroscopy has shown that $[(\eta^5\text{-}C_5Me_5)(OC)_2W\{\eta^2\text{-}C(NEt)Me\}]$ is fluxional.[1501] ^{1}H n.m.r. spectroscopy has been used to determine the rate of rotation of the nitrile in $[(bipy)W(\eta^2\text{-}MeC{\equiv}N)(PMe_3)_2Cl]$. The ^{13}C n.m.r. spectrum was also determined.[1502]

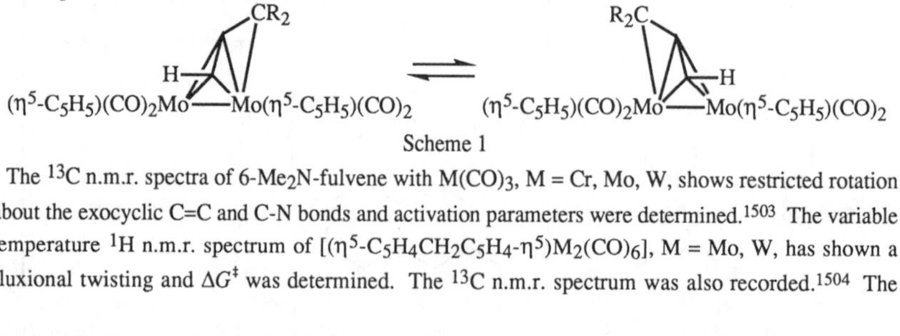

Scheme 1

The ^{13}C n.m.r. spectra of 6-Me_2N-fulvene with $M(CO)_3$, M = Cr, Mo, W, shows restricted rotation about the exocyclic C=C and C-N bonds and activation parameters were determined.[1503] The variable temperature ^{1}H n.m.r. spectrum of $[(\eta^5\text{-}C_5H_4CH_2C_5H_4\text{-}\eta^5)M_2(CO)_6]$, M = Mo, W, has shown a fluxional twisting and $\Delta G^{\ddagger}$ was determined. The ^{13}C n.m.r. spectrum was also recorded.[1504] The

1494 A.C. Filippou, W. Grünleitner, C. Völkl, and P. Kiprof, *J. Organomet. Chem.*, 1991, **413**, 181.
1495 V.S. Joshi, V.K. Kale, K.M. Sathe, A. Sarkar, S.S. Tavale, and C.G. Suresh, *Organometallics*, 1991, **10**, 2898.
1496 L. Carlton, N.M. Agh-Atabay, and J.L. Davidson, *J. Organomet. Chem.*, 1991, **413**, 205.
1497 M.V. Galakhov, V.I. Bakhmutov, I.V. Barinov, and O.A. Reutov, *J. Organomet. Chem.*, 1991, **421**, 65; I.V. Barinov, O.A. Reutov, A.V. Polyakov, A.I. Yanovsky, Yu.T. Struchkov, and V.I. Sokolov, *J. Organomet. Chem.*, 1991, **418**, C24.
1498 D.J. Wink and N.J. Cooper, *Organometallics*, 1991, **10**, 494.
1499 P.K. Baker and K.R. Flower, *J. Organomet. Chem.*, 1991, **405**, 299; P.K. Baker and K.R. Flower, *Polyhedron*, 1991, **10**, 711.
1500 P.K. Baker, M.G.B. Drew, S. Edge, and S.D. Ridyard, *J. Organomet. Chem.*, 1991, **409**, 207.
1501 A.C. Filippou, W. Grünleitner, and P. Kiprof, *J. Organomet. Chem.*, 1991, **410**, 175.
1502 J. Barrera, M. Sabat, and W.D. Harman, *J. Am. Chem. Soc.*, 1991, **113**, 8178.
1503 V.I. Bakhmutov, M.V. Galakhov, B.N. Strunin, and V.A. Nikanorov, *Metalloorg. Khim.*, 1990, **3**, 1329 (*Chem. Abstr.*, 1991, **114**, 122 612).
1504 T.E. Bitterwolf and A.L. Rheingold, *Organometallics*, 1991, **10**, 3856.

variable temperature ^{13}C n.m.r. spectra of [(η^5-$C_5H_nMe_{5-n}$)$MCo_2(CO)_8CCO_2Pr^i$], M = Mo, W, have revealed the existence of two interconverting rotamers.[1505] ^1H and ^{31}P variable temperature n.m.r. spectra have shown fluxionality for [(η^5-C_5H_7)$W(CO)_2$(dmpe)]$^+$.[1506] The variable temperature ^1H and ^{13}C n.m.r. spectra of [(η^5-C_5Me_5)Mo(NO)(CH_2Ph)Cl] have shown exchange of the *o*-hydrogens.[1507] The fluxionality of the ruthenium carbonyls in [(η^5-C_5H_5)$WRu_3(CO)_9$(μ_3-COMe)(C=CH-Ph)] has been studied by variable temperature ^{13}C n.m.r. spectroscopy.[1508] Restricted rotation in some complexes of the type [(η^6-arene)$M(CO)_2L$], M = Cr, Mo, W, has been studied by ^{13}C n.m.r. spectroscopy.[1509]

The ^{13}C n.m.r. spectra of [($Bu^t_2PNSNPBu^tPh$)$W(CO)_4$] shows restricted rotation about the P—C bond. $^1J(^{31}P$-$^{15}N)$ was determined.[1510] In [$Cr(CO)_5${P(C_6H_4Me-2)$_3$}] and [$Fe(CO)_4${P(C_6H_4Me-2)$_3$}], the barrier to intramolecular CO exchange has been determined by ^{13}C n.m.r. spectroscopy.[1511] The barrier to B—N rotation in [$Cr(CO)_4L$], [(η^4-1,5-C_8H_{12})NiL], and [NiL_2], L = $C_4H_4BNPr^i_2$, has been determined and the ^{11}B and ^{13}C n.m.r. spectra recorded.[1512] The n.m.r. spectra of [$Cr(CO)_3$(1,4,7-trithiacyclononane)] indicates fluxionality.[1513] The fluxionality of [$MI_2(CO)_3L_2$], M = Mo, W, has been studied by ^{13}C and ^{31}P n.m.r. spectroscopy.[1514] The ^{13}C n.m.r. spectrum of [$WI(CO)_3(PPh_3)$(acac)] shows one CO signal at room temperature and two at -70°C.[1515] The ^{13}C n.m.r. spectrum of [{HB(3,5-$Me_2C_3HN_2$)$_3$}$W(CO)_2I$] shows the trigonal twist with $\Delta G^\ddagger$ = 13.9 kcal mol^{-1}.[1516]

^1H n.m.r. spectroscopy shows NH/NH_2 exchange in [$M_2(OBu^t)_4(HNPh)_2(H_2NPh)_2$], M = Mo, W.[1517] The ^1H n.m.r. spectrum of [$M_2(OCPh_3)_2(NMe_2)_4$], M = Mo, W, is temperature dependent showing a mixture of interconverting *anti* and *gauche* rotamers, and $\Delta G^\ddagger$ was determined for M—N rotation barriers.[1518] The ^1H n.m.r. spectrum of *anti*-[$W_2(GePh_3)_2(NMe_2)_4$] shows methyl exchange with $\Delta G^\ddagger$ = 74 kJ mol^{-1}.[1519] The barrier to rotation about the $MeOC_6H_4$—CHO bond in [{HC(py)$_3$}$W(NO)_2$(η^1-4-anisaldehyde)] has been determined from ^1H and ^{13}C n.m.r. spectra.[1520]

Manganese and Rhenium. ^1H and ^{31}P n.m.r. spectroscopy has been used to examine exchange in

1505 K.A. Sutin, L. Li, C.S. Frampton, B.G. Sayer, and M.J. McGlinchey, *Organometallics*, 1991, **10**, 2362.
1506 S.L. Wu, C.Y. Cheng, S.L. Wang, and R.S. Liu, *Inorg. Chem.*, 1991, **30**, 311.
1507 N.H. Dryden, P. Legzdins, J. Trotter, and V.C. Yee, *Organometallics*, 1991, **10**, 2857.
1508 Y. Chi, S.H. Chuang, L.K. Liu, and Y.S. Wen, *Organometallics*, 1991, **10**, 2485.
1509 G. Hunter, R.L. MacKay, P. Kremminger, and W. Weissensteiner, *J. Chem. Soc., Dalton Trans.*, 1991, 3349; J.A. Chudek, G. Hunter, R.L. MacKay, P. Kremminger, and W. Weissensteiner, *J. Chem. Soc., Dalton Trans.*, 1991, 3337; K.V. Kilway and J.S. Siegel, *J. Am. Chem. Soc.*, 1991, **113**, 2332; M. Mailvaganam, C.S. Frampton, S. Top, B.G. Sayer, and M.J. McGlinchey, *J. Am. Chem. Soc.*, 1991, **113**, 1177.
1510 M. Herberhold, S.M. Frank, and B. Wrackmeyer, *J. Organomet. Chem.*, 1991, **410**, 159.
1511 J.A.S. Howell, M.G. Palin, P. McArdle, D. Cunningham, Z. Goldschmidt, H.E. Gottlieb, and D. Hezroni-Langerman, *Inorg. Chem.*, 1991, **30**, 4683.
1512 G.E. Herberich, M. Negele, and H. Ohst, *Chem. Ber.*, 1991, **124**, 25.
1513 H.J. Kim, Y. Do, H.W. Lee, J.H. Jeong, and Y.S. Sohn, *Bull. Korean Chem. Soc.*, 1991, **12**, 257 (*Chem. Abstr.*, 1991, **115**, 125 617).
1514 P.K. Baker, M.A. Beckett, and L.M. Severs, *J. Organomet. Chem.*, 1991, **409**, 213.
1515 P.K. Baker and D.ap. Kendrick, *Polyhedron*, 1991, **10**, 433.
1516 M.H. Chisholm, I.P. Parkin, W.E. Streib, and K.S. Folting, *Polyhedron*, 1991, **10**, 2309.
1517 S.G. Feng, C.C. Philipp, A.S. Gamble, P.S. White, and J.L. Templeton, *Organometallics*, 1991, **10**, 3504.
1518 M.H. Chisholm, I.P. Parkin, J.C. Huffman, E.M. Lobkovsky, and K. Folting, *Polyhedron*, 1991, **10**, 2839.
1519 M.H. Chisholm, I.P. Parkin, and J.C. Huffman, *Polyhedron*, 1991, **10**, 1215.
1520 J.W. Faller and Y. Ma, *J. Am. Chem. Soc.*, 1991, **113**, 1579.

$[(PCy_2H)ReH_5(\mu\text{-}PCy_2)RhH(PCy_2H)]$ and related compounds and $\Delta G^{\ddagger}$ was determined.[1521] Restricted rotation about the Re—P bonds in $[ReCl(H_2)(PMe_2Ph_2)_4]$ has been examined. A T_1 minimum of 0.092 s at 400 MHz has been determined for the H_2 group.[1522]

Variable temperature ^1H n.m.r. spectra of $[ReH(\eta^2\text{-}C_2H_4)_2(PMe_2Ph)_3]$ show olefin rotation with $\Delta G^{\ddagger}$ = 12.2 kcal mol^{-1}. The ^{13}C n.m.r. spectrum was reported.[1523] Variable temperature ^1H n.m.r. spectra of $[Re_3(\mu\text{-}H)_3(\mu\text{-}\eta^2\text{-}CHNR)(CO)_9]^-$ have shown that the triple bridging formimidoyl group undergoes two types of dynamic processes causing partial or total equalisation of the hydrides and consisting, presumably, of rotations of this ligand on the Re_3 face.[1524] The dynamic processes operative in the three isomers of $[Re_2Pt(\mu\text{-}H)_2(CO)_8(PPh_3)_2]$ have been studied by ^1H and ^{31}P n.m.r. spectroscopy including EXSY. ^{13}C n.m.r. spectra were also recorded.[1525]

^1H n.m.r. spectroscopy shows that $[(2,4,6\text{-}Bu^t_3C_6H_2O)(\eta^1\text{-}C_5Me_5)P{=}Mn(CO)_4]$ is fluxional due to C_5Me_5 sigmatropic shifts with $\Delta G^{\ddagger} = 60.5 \pm 0.8$ kJ mol^{-1}. The ^{13}C and ^{55}Mn n.m.r. spectra were also recorded.[1526] The ^1H n.m.r. spectrum of $[(OC)_5ReCH_2CH_2Mn(CO)_5]$ shows CH_2 exchange, while the spectrum of *cis*-$[\{(OC)_5ReCH_2CH_2\}_2Os(CO)_4]$ is $[AB]_2$.[1527] The ^{13}C n.m.r. spectrum of $[(OC)_4ReCPhCPhCPh]$ shows exchange *via* $[(OC)_4Re(\eta^3\text{-}C_3Ph_3)]$, making the phenyls equivalent.[1528]

^1H and ^{13}C n.m.r. spectroscopy has been used to show that $[(EtC{\equiv}CEt)(O)Re(\mu\text{-}O)(\mu\text{-}EtC{\equiv}CEt)Re(EtC{\equiv}CEt)_2]$ is fluxional.[1529] The barrier to allene rotation in $[(\eta^5\text{-}C_5Me_5)Re(CO)_2(\eta^2\text{-}allene)]$ has been determined.[1530]

The fluxional behaviour of *cis*-$[(OC)_4Os\{\mu\text{-}\eta^1\text{:}\eta^5\text{-}C_6H_6Mn(CO)_3\}_2]$ has been studied and $\Delta G^{\ddagger}$ = 49 kJ mol^{-1} for 1,2-shifts.[1531] Exchange of the diastereotopic germanium phenyl groups in $[(\eta^5\text{-}C_5H_5)Re(GePh_2OTf)(NO)(PPh_3)]$ has been investigated by ^{13}C n.m.r. spectroscopy and $\Delta G^{\ddagger}_{268}$ = 12.6 kcal mol^{-1}.[1532] The ^1H and ^{13}C n.m.r. spectra of $[(\eta^5\text{-}C_5H_5)Re(CO)(PPh_3)(O{=}CEt_2)]$ show that the rhenium is moving between the lone pairs of the ketone with $\Delta G^{\ddagger} = 7.0 \pm 0.1$ kcal mol^{-1}.[1533] Methyl exchange in $[(\eta^5\text{-}C_5H_5)Re(NO)(PPh_3)(SMe_2)]^+$ occurs with $\Delta G^{\ddagger}$ = 9.6 kcal mol^{-1}.[1534] ^1H n.m.r. spectroscopy has shown inversion of sulphur in $[ReX(CO)_3(1\text{-}MeS\text{-}2\text{-}Ph_2PC_6H_4)]$ and $[ReX(CO)_3\{PhP(C_6H_4\text{-}2\text{-}SMe)_2\}]$.[1535]

Iron, Ruthenium, and Osmium. The skeletal rearrangement of $[H_2Fe\{P(CH_2CH_2CH_2PMe_2)_3\}]$ has been investigated by magnetisation transfer.[1536] The intramolecular rearrangement of

1521 R.T. Baker, T.E. Glassman, D.W. Ovenall, and J.C. Calabrese, *Isr. J. Chem.*, 1991, **31**, 33.
1522 F.A. Cotton and R.L. Luck, *Inorg. Chem.*, 1991, **30**, 767.
1523 S. Komiya and A. Baba, *Organometallics*, 1991, **10**, 3105.
1524 T. Beringhelli, G. D'Alfonso, A. Minoja, G. Ciani, M. Moret, and A. Sironi, *Organometallics*, 1991, **10**, 3131.
1525 T. Beringhelli, G. D'Alfonso, and A.P. Minoja, *Organometallics*, 1991, **10**, 394.
1526 H. Lang, M. Leise, and C. Emmerich, *J. Organomet. Chem.*, 1991, **418**, C9.
1527 J. Breimair, B. Niemer, K. Raab, and W. Beck, *Chem. Ber.*, 1991, **124**, 1059.
1528 C. Löwe, V. Shklover, and H. Berke, *Organometallics*, 1991, **10**, 3396.
1529 E. Spaltenstein and J.M. Mayer, *J. Am. Chem. Soc.*, 1991, **113**, 7744.
1530 J.M. Zhuang and D. Sutton, *Organometallics*, 1991, **10**, 1516.
1531 B. Niemer, J. Breimair, T. Voelkel, B. Wagner, K. Polborn, and W. Beck, *Chem. Ber.*, 1991, **124**, 2237.
1532 K.E. Lee, A.M. Arif, and J.A. Gladysz, *Organometallics*, 1991, **10**, 751.
1533 D.M. Dalton and J.A. Gladysz, *J. Chem. Soc., Dalton Trans.*, 1991, 2741.
1534 N. Quiros Mendez, A.M. Arif, and J.A. Gladysz, *Organometallics*, 1991, **10**, 2199.
1535 E.W. Abel, D. Ellis, K.G. Orrell, and V. Šik, *Polyhedron*, 1991, **10**, 1603.
1536 L.D. Field, N. Bampos, and B.A. Messerle, *Magn. Reson. Chem.*, 1991, **29**, 36; M.V. Baker, L.D. Field, and

$[FeH(SiR_3)(CO)_4]$ has been studied by variable temperature ^{13}C n.m.r. spectroscopy.[1537] In $[FeH(Et_2PCH_2CH_2PEt_2)_2(BH_4)]$ and $[FeH(Pr_2PCH_2CH_2PPr_2)_2(BH_4)]$ there is rapid BH_4 rotation.[1538] $[\{P(CH_2CH_2PPh_2)_3\}RuH(\eta^2\text{-}H_2)]^+$ is dynamic with a 2:1 hydride signal at -85 °C and one signal at 60 °C. The non-classical structure was established by ^1H and ^{31}P n.m.r. data, T_1 measurements, and $J(^2H\text{-}^1H)$ values.[1539] $\Delta G^\ddagger$ for cyclopentadienyl rotation in $[\{\eta^5\text{-}C_5H_4C(C_6H_4Me\text{-}4)_3\}RuH_2(PPh_3)_2]^+$ has been determined as 13.3 kcal mol^{-1}.[1540] The ^1H n.m.r. spectrum of $[(Ph_3P)_2(H)Ru(\mu\text{-}H)_3Ru(PPh_3)_3]$ shows four hydride signals at -90 °C and three exchange at room temperature.[1541] Variable temperature ^1H, ^{13}C, and ^{31}P n.m.r. studies indicate that ancillary ligands affect the barriers to aryl ring rotation in $[(Me_3P)_4RuH(NHAr)]$ and $[(Me_3P)_4RuH(OAr)]$.[1542] The fluxionality of (101) has been studied. The ^1H n.m.r. spectrum shows two hydride signals at low temperature which exchange with $\Delta G^\ddagger = 39.2$ kJ mol^{-1}. The ^{13}C n.m.r. spectrum shows five ^{13}CO signals at 175 K and one signal at 219 K.[1543] The volume of activation for hydride fluxionality on transition-metal clusters has been determined by analysis of ^1H n.m.r. line shapes at pressures up to 200 MPa. The exchange of the two hydrides of $[(\mu\text{-}H)_2Ru_3(\mu_3\text{-}CHCO_2Me)(CO)_9]$ occurs with $\Delta V^\ddagger = +4.1 \pm 0.3$ cm^3 mol^{-1}, while for $[H(\mu\text{-}H)Os_3(CO)_{10}(PPh_3)]$, $\Delta V^\ddagger = -0.8 \pm 0.4$ cm^3 mol^{-1}.[1544] Variable temperature ^{13}C n.m.r. spectroscopy has been used to study carbonyl exchange in $[HRu_3(CO)_{11}]^-$.[1545] The $^1H\{^{31}P\}$ spectra of $[(\eta^6\text{-}C_6H_6)Os(PR_3)H_3]^+$ show fluxionality, and at low temperature large hydride-hydride coupling constants are observed.[1546] ^{13}C EXSY has been used to study carbonyl exchange in $[Os_3Pt(\mu\text{-}H)_2(CO)_9(PR_3)_2]$ and $\Delta G^\ddagger$ determined.[1547] Variable temperature ^1H and ^{13}C n.m.r. spectroscopy has been used to study intramolecular rotation of the Os_3 triangle in $[PtOs_5(CO)_{15}(\mu_3\text{-}S)(\mu\text{-}H)_6]$ and $\Delta G^\ddagger$ determined as 12.3 kcal mol^{-1}.[1548] ^{13}C EXSY has been used to study carbonyl exchange in $[HOs_6(CO)_{17}]^-$.[1549]

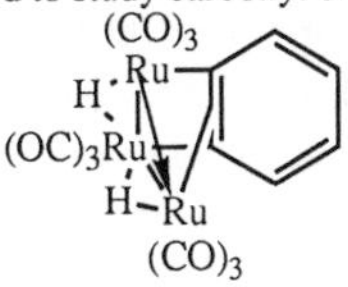

(101)

Variable temperature ^{13}C n.m.r. spectroscopy has been used to determine $\Delta G^\ddagger$ for rotation about the P—Fe bond in $[(\eta^5\text{-}C_5H_5)Fe(CO)(PPh_3)(\{C(O)Me\}]$ as 10.3 ± 0.1 kcal mol^{-1}.[1550] Restricted rotation of the C—NH_2 bond in (102) has been studied, and the ^{13}C n.m.r. spectrum recorded.[1551]

D.J. Young, *Appl. Organomet. Chem.*, 1990, **4**, 551 (*Chem. Abstr.*, 1991, **114**, 94 080).
1537 J.W. Connolly, M.J. Hatlee, A.H. Cowley, and P.R. Sharp, *Polyhedron*, 1991, **10**, 841.
1538 M.V. Baker and L.D. Field, *Appl. Organomet. Chem.*, 1990, **4**, 543 (*Chem. Abstr.*, 1991, **114**, 94 079).
1539 C. Bianchini, P.J. Perez, M. Peruzzini, F. Zanobini, and A. Vacca, *Inorg. Chem.*, 1991, **30**, 279.
1540 O.B. Ryan, K.-T. Smith, and M. Tilset, *J. Organomet. Chem.*, 1991, **421**, 315.
1541 L.S. Van Den Sluys, G.J. Kubas, and K.G. Caulton, *Organometallics*, 1991, **10**, 1033.
1542 J.F. Hartwig, R.A. Andersen, and R.G. Bergman, *Organometallics*, 1991, **10**, 1875.
1543 E. Bonfantini and P. Vogel, *J. Chem. Soc., Chem. Commun.*, 1991, 1334.
1544 J.B. Keister, U. Frey, D. Zbinden, and A.E. Merbach, *Organometallics*, 1991, **10**, 1497.
1545 G. Süss-Fink, M. Langenbahn, and F. Neumann, *Organometallics*, 1991, **10**, 815.
1546 D.M. Heinekey and T.G.P. Harper, *Organometallics*, 1991, **10**, 2891.
1547 L.J. Farrugia and S.E. Rae, *Organometallics*, 1991, **10**, 3919.
1548 R.D. Adams, M.P. Pompeo, and W. Wu, *Inorg. Chem.*, 1991, **30**, 2899.
1549 B.F. Johnson, J. Lewis, M.A. Pearsall, and L.G. Scott, *J. Organomet. Chem.*, 1991, **402**, C27.
1550 S.G. Davies, A.E. Derome, and J.P. McNally, *J. Am. Chem. Soc.*, 1991, **113**, 2854.
1551 B. Eber, G. Huttner, L. Zsolnai, and W. Imhof, *J. Organomet. Chem.*, 1991, **402**, 221.

Windshield wiper fluxionality of the (μ-CH=CH$_2$) ligand has been observed for [Fe$_2$(CO)$_4$(μ-CH=CH$_2$)(μ-PPh$_2$)(μ-dppm)] by ^{31}P n.m.r. spectroscopy. The ^{13}C n.m.r. spectrum was also recorded.[1552] The variable temperature ^{13}C n.m.r. spectrum of [Fe$_2$(CO)$_6$(μ-σ,π-MeC=CH)(μ-EtS)] shows one CO signal at room temperature and two at -60 °C.[1553] The dynamic behaviour of [Fe(CO)$_3${μ-Si(OMe)$_3$}(μ-dppm)Rh(CO)] has been studied.[1554] The variable temperature ^{13}C, ^{19}F, and ^{31}P n.m.r. spectra of [Fe(SiF$_2$CH=CBu^{t}SiF$_2$)(CO)$_3$(PR$_3$)] have been analysed and interpreted as a trigonal twist.[1555] Exchange of the OMe groups in (103) occurs with $\Delta G^{\ddagger}$ = 65 ± 2 kJ mol^{-1}. The ^{29}Si and ^{119}Sn n.m.r. spectra were recorded.[1556] ^{1}H and ^{13}C n.m.r. studies of [Os$_2$(μ-CH=CHPh)-(μ-Br)(CO)$_5$(PPh$_3$)] indicate that the ethenyl group undergoes a rapid $\pi\rightarrow\sigma,\sigma\rightarrow\pi$ rearrangement.[1557] The variable temperature ^{1}H and ^{13}C n.m.r. spectra of [Os$_3$(μ-H)(μ-Pr^{n}N=CEt)(CO)$_8$(PPh$_3$)] show exchange of the hydrides between isomers.[1558]

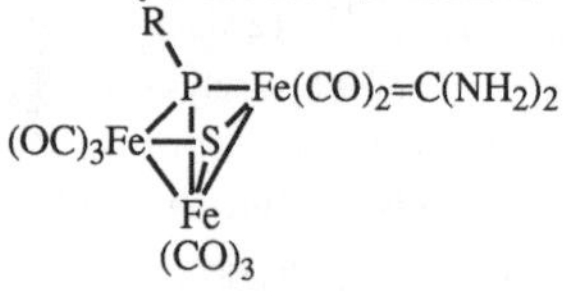

(102)

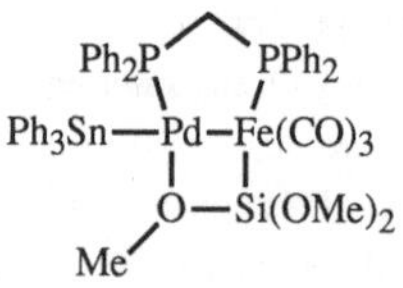

(103)

Carbonyl scrambling in [Ru$_3$(CO)$_9$(μ-CO)(C$_2$H$_2$)] has been studied by ^{13}C n.m.r. spectroscopy.[1559] Variable temperature ^{13}C and ^{31}P n.m.r. spectroscopy has been used to establish solution structures and conformer populations for a variety of cyclic and acyclic [(η^4-diene)Fe(CO)$_2$L] complexes.[1560] ^{13}C n.m.r. spectroscopy has been used to study P—C rotation in [(η^4-C$_6$H$_8$)Fe(CO)$_2${P(C$_6$H$_4$Me-2)$_3$}] and $\Delta G^{\ddagger}$ determined.[1561]

^{13}C T_1 values have been reported for a series of ferrocene derivatives and related sandwich complexes, and used to determine the relative importance of overall molecular reorientation and internal rotation.[1562] Variable temperature ^{1}H and ^{13}C n.m.r. studies on [(η^5-1,3-R$_2$C$_5$H$_3$)$_2$M], M = Fe, Ru, have yielded $\Delta G^{\ddagger}$ values.[1563] Temperature dependent ^{1}H and ^{13}C n.m.r. spectra have shown that *syn*-[(η^5-C$_5$H$_5$)Fe(η^5-tetrahydro-4,4,8,8-Me$_4$-4,8-disila-*s*-indacenyl)] is non-rigid and undergoes a 1,2-silatropic shift with E_a = 49.2 ± 2.9 kJ mol^{-1}.[1564] Activation barriers have been determined for

1552 G. Hogarth, *J. Organomet. Chem.*, 1991, **407**, 91.
1553 D. Seyferth, C.M. Archer, D.P. Ruschke, M. Cowie, and R.W. Hilts, *Organometallics*, 1991, **10**, 3363.
1554 P. Braunstein, M. Knorr, E. Villarroya, A. DeCian, and J. Fischer, *Organometallics*, 1991, **10**, 3714.
1555 C.Y. Lee, Y. Wang, and C.S. Liu, *Inorg. Chem.*, 1991, **30**, 3893.
1556 P. Braunstein, M. Knorr, H. Piana, and U. Schubert, *Organometallics*, 1991, **10**, 828.
1557 S.H. Chuang, Y. Chi, F.L. Liao, S.L. Wang, S.M. Peng, G.H. Lee, J.C. Wu, and K.M. Horng, *J. Organomet. Chem.*, 1991, **410**, 85.
1558 M. Day, D. Espitia, K.I. Hardcastle, S.E. Kabir, E. Rosenberg, R. Gobetto, L. Milone, and D. Osella, *Organometallics*, 1991, **10**, 3550.
1559 S. Aime, R. Gobetto, L. Milone, D. Osella, L. Violano, A.J. Arce, and Y. De Sanctis, *Organometallics*, 1991, **10**, 2854.
1560 J.A.S. Howell, G. Walton, M.-C. Tirvengadum, A.D. Squibb, M.G. Palin, P. McArdle, D. Cunningham, Z. Goldschmidt, H. Gottlieb, and G. Strul, *J. Organomet. Chem.*, 1991, **401**, 91.
1561 J.A.S. Howell, M.G. Palin, M.-C. Tirvengadum, D. Cunningham, P. McArdle, Z. Goldschmidt, and H.E. Gottlieb, *J. Organomet. Chem.*, 1991, **413**, 269.
1562 R.M.G. Roberts and J.F. Warmsley, *J. Organomet. Chem.*, 1991, **405**, 347.
1563 E.W. Abel, N.J. Long, K.G. Orrell, A.G. Osborne, and V. Šik, *J. Organomet. Chem.*, 1991, **403**, 195.
1564 M. Fritz, J. Hiermeier, N. Hertkorn, F.H. Köhler, G. Müller, G. Reber, and O. Steigelmann, *Chem. Ber.*, 1991, **124**, 1531.

C—N bond rotation in ferrocenecarboxamides.[1565] Dynamic n.m.r. studies have shown that $\Delta G^{\ddagger}$ of the bridge reversal fluxion in $[Fe(\eta^5\text{-}C_5H_4Te)_2S]$, E = S, Se, Te, are 56.3, 55.4, and 51.8 kJ mol^{-1} respectively.[1566] Bridge reversal in $[M(\eta^5\text{-}C_5H_4E^1)_2E^2]$, E^1, E^2 = S, Se, M = Ru, Os, has been studied using EXSY, and $\Delta G^{\ddagger}$ determined.[1567] Bridge reversal in $[Fe(\eta^5\text{-}C_5H_4E)_2CH_2]$ has been investigated by ^{13}C n.m.r. spectroscopy and $\Delta G^{\ddagger}$ determined.[1568] The dynamics of $[(\eta^5\text{-}C_5H_5)_2Fe_2(CN)(CO)_3]^-$ have been investigated by ^1H n.m.r. spectroscopy.[1569] The interconversion of *cis*-$[(\eta^5\text{-}C_5H_5)_2Fe_2(CO)_2(\mu\text{-}CO)(\mu\text{-}CNPh)]$ and *trans*-$[(\eta^5\text{-}C_5H_5)_2Fe_2(CO)_2(\mu\text{-}CO)(\mu\text{-}CNPh)]$ has $\Delta G^{\ddagger}$ of 14.8 kcal mol^{-1}.[1570] Restricted rotation of the phenyl groups in $[(\eta^5\text{-}C_5Ph_4H)_2Ru]$ has been studied by ^1H n.m.r. spectroscopy.[1571] The variable temperature ^1H n.m.r. spectrum of $[\{\eta^5\text{-}C_5Ph_4C_6H_3(OMe)_2\text{-}2,5\}Ru(CO)_2Br]$ shows restricted rotation.[1572] The ^1H dynamic n.m.r. spectrum of $[(\eta^5\text{-}C_5H_4Me)_4Ru_4S_4]^{2+}$ shows exchange of cyclopentadienyl groups with $\Delta G^{\ddagger}$ = 52 kJ mol^{-1}.[1573]

Dynamic ^{13}C n.m.r. spectra have indicated restricted rotation of the Bu^t group and hindered CO scrambling for (104) and (105).[1574] The fluxionality of $[Fe_3(CO)_{10}L(CNCF_3)]$ has been studied by ^{13}C n.m.r. spectroscopy.[1575] The ^{13}C T_1 values and n.O.e.s have been measured for the phenyl groups of $[FeCo_2(CO)_9(\mu\text{-}PPh)]$ and the rate of phenyl rotation determined.[1576] Carbonyl scrambling in $[FeCo_2(CO)_9(\mu\text{-}PPhH)]^-$ has been investigated by ^{13}C n.m.r. spectroscopy.[1577] The ^{13}C n.m.r. spectrum of $[Ru_2(CO)_6(\mu\text{-}SePh)_2]$ shows one CO signal at room temperature and three at -40 °C.[1578] At -48 °C, the ^{13}C n.m.r. spectrum of $[(\eta^5\text{-}C_5H_5)(OC)Ir\{Os(CO)_4\}_2]$ exhibits five carbonyl signals in the ratio 2:2:1:2:2 which exchange at higher temperatures with $\Delta G^{\ddagger}$ = 14.2 kcal mol^{-1}. In contrast, $\Delta G^{\ddagger}$ = 7.4 kcal mol^{-1} for $[(\eta^5\text{-}C_5Me_5)(OC)Ir\{Os(CO)_4\}_2]$.[1579] ^{13}C NOESY has been used to study exchange in $[Os_6(CO)_{18\text{-}n}(PPh_3)_n]$.[1580]

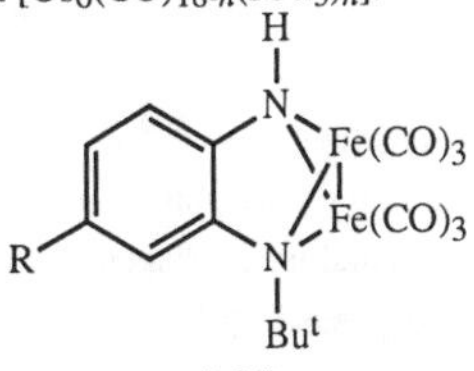

(104)

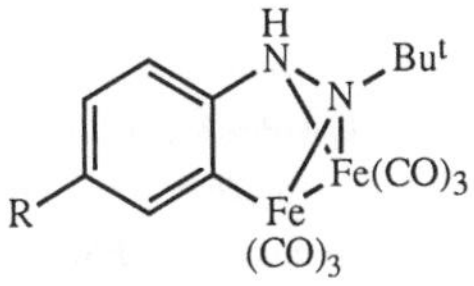

(105)

1565 R.C. Petter and S.J. Rao, *J. Org. Chem.*, 1991, **56**, 2932.
1566 E.W. Abel, K.G. Orrell, A.G. Osborne, V. Šik, and W. Guoxiong, *J. Organomet. Chem.*, 1991, **411**, 239.
1567 E.W. Abel, N.J. Long, K.G. Orrell, A.G. Osborne, and V. Šik, *J. Organomet. Chem.*, 1991, **419**, 375.
1568 M. Herberhold and P. Leitner, *J. Organomet. Chem.*, 1991, **411**, 233.
1569 W.P. Fehlhammer, A. Schroeder, F. Schoder, J. Fuchs, A. Vökl, B. Boyadjiev, and S. Schrölkamp, *J. Organomet. Chem.*, 1991, **411**, 405.
1570 A.R. Manning, G. McNally, and P. Soye, *Inorg. Chim. Acta*, 1991, **180**, 103.
1571 R.J. Hoobler, J.V. Adams, M.A. Hutton, T.W. Francisco, B.S. Haggerty, A.L. Rheingold, and M.P. Castellani, *J. Organomet. Chem.*, 1991, **412**, 157.
1572 S.B. Colbran, D.C. Craig, W.M. Harrison, and A.E. Grimley, *J. Organomet. Chem.*, 1991, **408**, C33.
1573 E.J. Houser, J. Amarasekera, T.B. Rauchfuss, and S.R. Wilson, *J. Am. Chem. Soc.*, 1991, **113**, 7440.
1574 H. Kisch, P. Reißer, and F. Knoch, *Chem. Ber.*, 1991, **124**, 1143.
1575 D. Lentz and R. Marschall, *Organometallics*, 1991, **10**, 1487.
1576 P. Yuan, M.G. Richmond, and M. Schwartz, *Inorg. Chem.*, 1991, **30**, 588.
1577 M.J. Don and M.G. Richmond, *Inorg. Chem.*, 1991, **30**, 1703.
1578 P.L. Andreu, J.A. Cabeza, D. Miguel, V. Riera, M.A. Villa, and S. Garcia-Granda, *J. Chem. Soc., Dalton Trans.*, 1991, 533.
1579 A. Riesen, F.W.B. Einstein, A.K. Ma, R.K. Pomeroy, and J.A. Shipley, *Organometallics*, 1991, **10**, 3629.
1580 B.F.G. Johnson, J. Lewis, M.-A. Pearsall, and L.G. Scott, *J. Organomet. Chem.*, 1991, **413**, 337.

The rates of rotation of 2-Me-imidazole when coordinated to a Fe(III) porphyrin have been studied by n.m.r. spectroscopy.[1581] ^1H n.m.r. spectroscopy has been used to study the rotational mobility in myoglobin.[1582] ^{31}P n.m.r. spectroscopy has been used to study axial-equatorial exchange of the PCy_2 groups in [{$PhP(CH_2CH_2PCy_2)_2$}$RuCl_2$]. The ^{13}C n.m.r. spectrum was also reported.[1583] ^{31}P n.m.r. spectroscopy has been used to show that *fac*-[$Ru(O_2CMe)_2(PPh_3)_3$] is fluxional, being ABC at 203 K and a singlet at 308 K, with $\Delta G^{\ddagger} = 48.7$ kJ mol^{-1}.[1584]

Cobalt. ^1H n.m.r. spectroscopy has been used to examine the exchange in Scheme 2. The ^{13}C n.m.r. spectrum was also recorded.[1585] Phenyl group rotation in [$(\eta^5\text{-}C_5H_5)_2Co_3(\mu\text{-CO})(CO)_3(\mu\text{-CPh})$] has been studied using ^{13}C T_1 measurements.[1586] The reorientational dynamics of [$Co_2Ni(\eta^5\text{-}C_5H_5)(CO)_6(\mu_3\text{-CPh})$] have also been studied by ^{13}C T_1 measurements.[1587] The ^1H n.m.r. spectrum of [$(\eta^5\text{-}C_5H_5)_3Co_3(\eta^6\text{-}C_6H_5CH{=}CHMe)$] shows one cyclopentadienyl signal at room temperature and three at 22 K. The ^{13}C n.m.r. spectrum was also recorded.[1588] Carbonyl exchange in [$(4\text{-}MeOC_6H_4)_2CHCCo_3(CO)_9$] has been studied by ^{13}C n.m.r. spectroscopy.[1589]

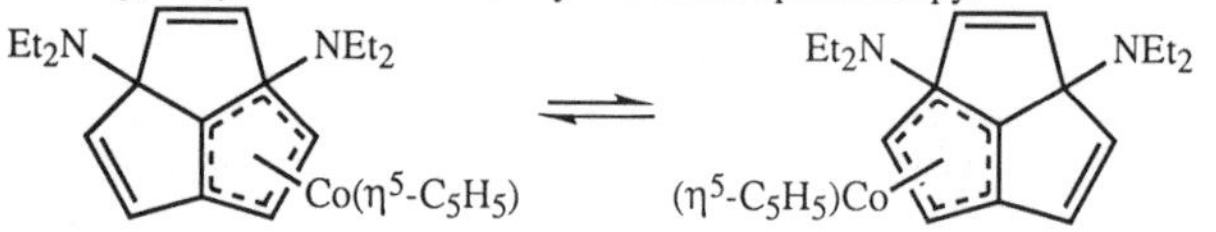

Scheme 2

Rotation of the *N*-Me-imidazole coordinated to Co(III) porphyrins has been investigated using ^{13}C T_1 measurements. ^{59}Co relaxation was also examined.[1590] ^1H two-dimensional EXSY has been applied to Co(II) ion transfer through [12]crown-4 rings.[1591]

Rhodium and Iridium. $^1H\{^{31}P\}$ and $^1H\{^{103}Rh\}$ n.m.r. spectroscopy has been used to determine the structure of [$MeC(CH_2PPhMe)_3RhH_2\{Au(PPh_3)\}_3$]$^+$. Variable temperature n.m.r. studies have revealed three dynamic processes, rapid rotation of $\{Au(PPh_3)\}_3$, multiple exchange of diastereotopic hydrides, and slow rotation of the *RRS/SSR*-triphos unit in one diastereomer.[1592] The reversible arm-off dissociation of the tripodal ligand $MeC(CH_2PPh_2)_3$ in [$HRh(CO)\{MeC(CH_2PPh_2)_3\}$] under H_2/CO has been studied by high pressure i.r. and n.m.r. spectroscopy.[1593] Variable temperature ^1H n.m.r. spectroscopy of [$\{MeC(CH_2PPh_2)_3\}IrH_2(\eta^2\text{-}C_2H_4)$] shows exchange between the hydride and olefin.[1594] ^1H, ^2H, and ^{15}N n.m.r. spectroscopy has been used to study exchange between the

1581 M. Nakamura and N. Nakamura, *Chem. Lett.*, 1991, 627 (*Chem. Abstr.*, 1991, **115**, 148 902).
1582 G.N. LaMar, J.B. Hauksson, L.B. Dugad, P.A. Liddell, N. Venkataramana, and K.M. Smith, *J. Am. Chem. Soc.*, 1991, **113**, 1544.
1583 G. Jia, I.-m. Lee, D.W. Meek, and J.C. Gallucci, *Inorg. Chim. Acta*, 1991, **177**, 81.
1584 E. Lindner, R. Fawzi, W. Hiller, A. Carvill, and M. McCann, *Chem. Ber.*, 1991, **124**, 2691.
1585 G. Schröder, H. Butenschön, R. Boese, T. Lendvai, and A. de Meijere, *Chem. Ber.*, 1991, **124**, 2423.
1586 P. Yuan, M.G. Richmond, and M. Schwartz, *Inorg. Chem.*, 1991, **30**, 679.
1587 P. Yuan, M.J. Don, M.G. Richmond, and M. Schwartz, *Inorg. Chem.*, 1991, **30**, 3704.
1588 H. Wadepohl, K. Büchner, M. Herrmann, and H. Pritzkow, *Organometallics*, 1991, **10**, 861.
1589 M.F. D'Agostino, C.S. Frampton, and M.J. McGlinchey, *Organometallics*, 1991, **10**, 1383.
1590 L. Cassidei, H. Bang, J.O. Edwards, and R.G. Lawler, *J. Phys. Chem.*, 1991, **95**, 7186.
1591 F.L. Dickert, M. Feigl, W. Gumbrecht, H.U. Meissner, and P.P. Otto, *Z. Phys. Chem. (Munich)*, 1991, **170**, 185 (*Chem. Abstr.*, 1991, **115**, 190 825).
1592 D. Imhof, H. Rüegger, L.M. Venanzi, and T.R. Ward, *Magn. Reson. Chem.*, 1991, **29**, S73.
1593 G. Kiss and I.T. Horvath, *Organometallics*, 1991, **10**, 3798.
1594 P. Barbaro, C. Bianchini, A. Meli, M. Peruzzini, A. Vacca, and F. Vizza, *Organometallics*, 1991, **10**, 2227.

hydride and NH_3 protons in $[(\eta^2\text{-}C_8H_{14})IrH(NH_3)_3L]^+$.[1595]

Analysis of the variable temperature ^1H, ^{13}C, and ^{31}P n.m.r. spectra of compounds such as $[\{MeC(CH_2PPh_2)_3\}Rh(\eta^2\text{-}C_2H_4)(NCMe)]^+$ suggests that these cations are fluxional with preliminary arm-off dissociation of one of the three phosphorus groups.[1596] The effect of high pressure on ethylene rotation in $[(\eta^5\text{-}C_5H_5)Rh(\eta^2\text{-}C_2H_4)(\eta^2\text{-}C_2F_4)]$ has been studied.[1597] Alkene exchange kinetics of $[(\eta^2\text{-alkene})_2Rh(acac)]$ have been studied by ^1H, ^{13}C, and ^{103}Rh n.m.r. spectroscopy.[1598] ^{13}C and ^{31}P n.m.r. spectroscopy has been used to study the fluxionality of $[\{MeC(CH_2PPh_2)_3\}Rh(\mu\text{-}Cl)(\mu\text{-}\eta^2,\eta^2\text{-}C_2H_2)Rh\{(Ph_2PCH_2)_3CMe\}]^+$.[1599] Dynamic processes in $[(\eta^5\text{-}C_5Me_5)_2M_2(CO)_2\{\mu\text{-}CR(OMe)\}]$, M = Co, Rh, have been studied by ^{13}C n.m.r. spectroscopy.[1600] The indenyl effect on ethylene rotation in $[(\eta^5\text{-}C_5H_5)Ir(\eta^2\text{-}C_2H_4)]$ and $[(\eta^5\text{-indenyl})Ir(\eta^2\text{-}C_2H_4)]$ has been studied. The free energies of activation are 5 to 6 kcal mol^{-1} less for the indenyl complexes than for the corresponding cyclopentadienyl complexes.[1601]

Exchange in Rh(I)-diene-bisphosphine-Cl complexes has been studied by ^1H, ^{13}C, and ^{31}P n.m.r. spectroscopy.[1602] Variable temperature ^{31}P n.m.r. spectra of $[(\eta^4\text{-}1,5\text{-}C_8H_{12})Rh\{C[P(S)Ph_2]_3\}]$ give $\Delta G^{\ddagger} = 46 \pm 2$ kJ mol^{-1} for exchange of coordinated and non-coordinated sulphur. The ^{13}C n.m.r. spectrum was also recorded.[1603] A similar study has been performed on $[(\eta^4\text{-}1,5\text{-}C_8H_{12})\{C[P(S)Ph_2][P(O)Ph_2]_2\}]$.[1604] Exchange of olefin in $[(\eta^4\text{-}1,5\text{-}C_8H_{12})Rh\{Ph_2P(2\text{-}C_5H_4N)\}Cl]$ has been examined and attributed to the nitrogen of the pyridine coordinating, followed by a Berry pseudorotation. The ^{13}C n.m.r. spectrum was also recorded.[1605] The fluxionality of (106) has been investigated by ^{13}C n.m.r. spectroscopy.[1606] The ^{13}C and ^{31}P n.m.r. spectra of

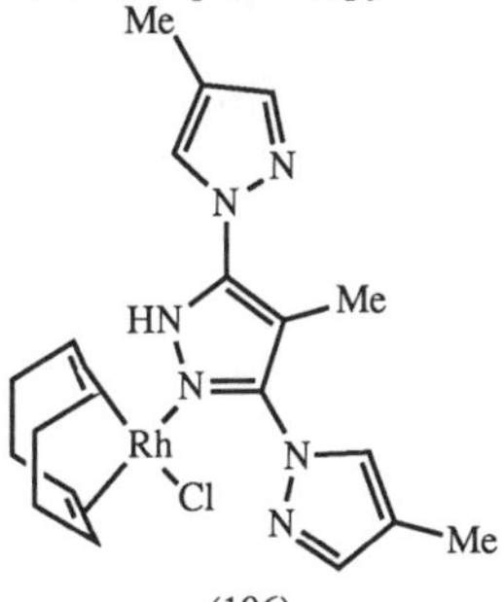

(106)

1595 R. Koelliker and D. Milstein, *J. Am. Chem. Soc.*, 1991, **113**, 8524.
1596 D.J. Rauscher, E.G. Thaler, J.C. Huffman, and K.G. Caulton, *Organometallics*, 1991, **10**, 2209.
1597 J. Jonas and X. Peng, *Ber. Bunsenges. Phys. Chem.*, 1991, **95**, 243.
1598 B. Åkermark, J. Glaser, L. Öhrström, and K. Zetterberg, *Organometallics*, 1991, **10**, 733.
1599 C. Bianchini, D. Masi, A. Meli, M. Peruzzini, A. Vacca, F. Laschi, and P. Zanello, *Organometallics*, 1991, **10**, 636.
1600 A.G. Avent, S.A. Benyunes, P.A. Chaloner, N.G. Getts, and P.B. Hitchcock, *J. Chem. Soc., Dalton Trans.*, 1991, 1417.
1601 L.P. Szajek, R.J. Lawson, and J.R. Shapley, *Organometallics*, 1991, **10**, 357.
1602 G. Szalontai, P. Sándor, and J. Bakos, *Magn. Reson. Chem.*, 1991, **29**, 449.
1603 J. Browning, K.R. Dixon, R.W. Hilts, N.J. Meanwell, and F. Wang, *J. Organomet. Chem.*, 1991, **410**, 389.
1604 S.O. Grim, P.B. Kettler, and J.B. Thoden, *Organometallics*, 1991, **10**, 2399.
1605 E. Rotondo, G. Battaglia, C.G. Arena, and F. Faraone, *J. Organomet. Chem.*, 1991, **419**, 399.
1606 R.M. Claramunt, C. Lopez, D. Sanz, J. Elguero, D. Carmona, M. Esteban, L.A. Oro, and M. Begtrup, *J. Organomet. Chem.*, 1991, **412**, 259.

$[\{Ph_2P(S)C(PPh_2S)_2\}Ir(\eta^4\text{-}1,5\text{-}C_8H_{12})]$ show P=S exchange.[1607] ^{31}P n.m.r. spectroscopy has been used to show phosphorus exchange in $[\{MeC(CH_2PPh_2)_3\}Ir(\eta^4\text{-}C_6H_6)]^+$ and $\Delta H^\ddagger$ and $\Delta S^\ddagger$ were determined.[1608] The fluxionality of ^{13}CO enriched $[CH_2\{(\eta^5\text{-}C_5H_4)Ir(CO)\}_2(\mu\text{-}CO)]$ has been studied by ^{13}C n.m.r. spectroscopy.[1609] The ^{13}C n.m.r. spectrum of $[Ir_6(CO)_{15}\{Au(PPh_3)\}]^{2-}$ consists of one signal at 50 °C and four signals at -90 °C.[1610]

The energy of breaking the Rh—Rh bond in $[(\text{tetramesitylporphyrin})Rh]_2$ has been determined.[1611] The ^1H and ^{31}P n.m.r. spectra of *trans-mer*-$[IrCl_2(PMe_3)_3(PMe_2Ph)]^+$ show restricted rotation about the Ir-PMe_2Ph bond.[1612]

Nickel, Palladium and Platinum. $\Delta H^\ddagger$, $\Delta S^\ddagger$, and $\Delta G^\ddagger$ have been determined for the rearrangement of $[(Ph_3P)_3MH]^+$. The suggested mechanism is *via* a tetrahedral intermediate/transition state, even though $\Delta S^\ddagger$ = -30 e.u.[1613] Chemical exchange in $[H_3Pt_2(dppe)]^+$ has been studied using T_1, T_2, and HD isotopic perturbation.[1614] ^1H, ^{13}C, and ^{31}P n.m.r. spectroscopy has been used to show that $[PtEt\{1,2\text{-}(Bu^t_2PCH_2)_2C_6H_4\}]^+$, $[PtEt(Bu^t_2PCH_2CH_2CH_2PBu^t_2)]^+$, and $[PtH(\eta^2\text{-}C_2H_4)(Bu^t_2P\text{-}CH_2CH_2PBu^t_2)]^+$ undergo two fluxional processes; agostic methyl group rotation and exchange between the hydride-ethene and ethyl forms of the compounds.[1615] The non-rigidity of *cis*-$[PtH(SR)(Ph_2PCH_2)_3CMe]$ has been studied.[1616] $[Au(PPh_3)\{Au(PPh_3)\}_7]^{2+}$, $[Pt(H)(PPh_3)\text{-}\{Au(PPh_3)\}_7]^{2+}$, and $[Pt(Ag)\{Au(PPh_3)\}_8]^{3+}$ have been shown to be fluxional.[1617] Variable temperature ^1H, ^{13}C, ^{31}P, and ^{195}Pt n.m.r. spectra have shown that $[Pt_4(\mu\text{-}H)(\mu\text{-}CO)(\mu\text{-}dppm)_3(dppm\text{-}P)]^+$ is fluxional and an analysis of the coupling constants has yielded mechanistic information.[1618]

$P(C_6H_4F\text{-}4)_3$ exchange in $[(2\text{-}MeC_6H_4)Ni\{P(C_6H_4F\text{-}4)_3\}_2(O_2CC_6H_4F\text{-}4)]$ has been studied by ^{19}F and ^{31}P n.m.r. spectroscopy.[1619] Hindered rotation of the ethene in $[(\eta^2\text{-}C_2H_4)_2NiSnL\text{-}\{CH(SiMe_3)_2\}_2]$ has been investigated by ^1H and ^{13}C n.m.r. spectroscopy.[1620] Variable temperature ^1H n.m.r. spectroscopy has been used to study the non-rigidity of $[PtXMe_3\{(2\text{-}MeSC_6H_4)PPh_2\}]$.[1621] ^1H, ^{125}Te, and ^{195}Pt n.m.r. spectroscopy, including EXSY, has been used

1607 S.O. Grim, P.B. Kettler, and J.S. Merola, *Inorg. Chim. Acta*, 1991, **185**, 57.
1608 C. Bianchini, K.G. Caulton, C. Chardon, O. Eisenstein, K. Folting, T.J. Johnson, A. Meli, M. Peruzzini, D.J. Rauscher, W.E. Streib, and F. Vizza, *J. Am. Chem. Soc.*, 1991, **113**, 5127.
1609 T.E. Bitterwolf, A. Gambaro, F. Gottardi, and G. Valle, *Organometallics*, 1991, **10**, 1416.
1610 R. Della Pergola, F. Demartin, L. Garlaschelli, M. Manassero, S. Martinengo, N. Masciocchi, and M. Sansoni, *Organometallics*, 1991, **10**, 2239.
1611 B.B. Wayland, S. Ba, and A.E. Sherry, *J. Am. Chem. Soc.*, 1991, **113**, 5305.
1612 A.J. Deeming, S. Doherty, and J.E. Marshall, *Polyhedron*, 1991, **10**, 1857.
1613 A.R. Siedle, R.A. Newmark, and W.B. Gleason, *Inorg. Chem.*, 1991, **30**, 2005.
1614 S. Aime, R. Gobetto, A.L. Bandini, G. Banditelli, and G. Minghetti, *Inorg. Chem.*, 1991, **30**, 316.
1615 L. Mole, J.L. Spencer, N. Carr, and A.G. Orpen, *Organometallics*, 1991, **10**, 49.
1616 F. Cecconi, P. Innocenti, S. Midollini, S. Moneti, A. Vacca, and J.A. Ramirez, *J. Chem. Soc., Dalton Trans.*, 1991, 1129.
1617 R.P.F. Kanters, P.P.J. Schlebos, J. Bour, J.J. Steggerda, W.E.J.R. Maas, and R. Janssen, *Inorg. Chem.*, 1991, **30**, 1709 (*Chem. Abstr.*, 1991, **114**, 198 150).
1618 G. Douglas, L. Manojlovic-Muir, K.W. Muir, M.C. Jennings, B.R. Lloyd, M. Rashidi, G. Schoettel, and R.J. Puddephatt, *Organometallics*, 1991, **10**, 3927.
1619 G.I. Drogunova, L.S. Isaeva, A.S. Peregudov, and D.N. Kravtsov, *Metalloorg. Khim.*, 1991, **4**, 659 (*Chem. Abstr.*, 1991, **115**, 92 584).
1620 C. Pluta, K.R. Pörschke, R. Mynott, P. Betz, and C. Krüger, *Chem. Ber.*, 1991, **124**, 1321.
1621 E.W. Abel, D. Ellis, K.G. Orrell, and V. Šik, *J. Chem. Res., Synop.*, 1991, 222.

to study tellurium inversion in [$PtIMe_3$\{MeTe(CH_2)$_3$TeMe\}].[1622] The dynamics of [(η^2-C_2H_4)PtMeCl(η^4-3,7-dimethylenebicyclo[3.3.1]nonane)] have been analysed using ^{1}H, ^{13}C, and ^{195}Pt n.m.r. spectroscopy.[1623] Restricted rotation about the Pt—C(Ar) bond in [Pt(η^1-$CH_2CH{=}CH_2$)($C_6H_3Cl_2$-2,5)(dppe)] has been investigated by ^{1}H n.m.r. spectroscopy.[1624] The temperature dependent ^{1}H n.m.r. spectrum of [Fe(η^5-$C_5H_4PPh_2$)$_2$Pt(CH_2SiMe_2Ph)$_2$] shows exchange with $\Delta G^‡$ = ~54 kJ mol^{-1}.[1625]

$\Delta G^‡$ has been determined for hindered C—N rotation in [(η^2-C_2H_4)$PtCl_2$(guanine)].[1626] ^{1}H n.m.r. spectroscopy has been used to study propene rotation in [(η^2-$CH_2{=}CHMe$)PtX_2(2,7-Me_2-phen)].[1627] The fluxionality of [(bicyclo[2.2.1]hept-2-yl)Pt\{$R_2P(CH_2)_nPR_2$\}]$^+$, which contains an agostic bond, has been investigated.[1628]

The ^{1}H n.m.r. spectrum of [(η^3-1,3-$Bu^t_2C_5H_5$)(η^5-1,3-$Bu^t_2C_5H_3$)Ni] shows restricted rotation. The ^{13}C n.m.r. spectrum was also recorded.[1629] ^{1}H two dimensional NOESY has been used to determine the structure of [Pd(η^3-2-MeC_3H_4)(bipy)]$^+$, and exchange was examined using EXSY.[1630] Magnetisation transfer has been used to demonstrate the migration of the nickel from one end of the anthracene ligand to the other without exchange with free anthracene in [(η^4-$C_{14}H_{10}$)Ni(PBu^n_3)$_2$].[1631] $\Delta G^‡$ has been determined for ring rotation in [Ni(η^5-C_4H_4BR)$_2$] using ^{1}H, ^{11}B, ^{13}C n.m.r. and spectroscopy.[1632]

Restricted rotation in [Pt(en)(DMSO)(5'-GMP-*N*7)]$^{2+}$ has been studied.[1633] Variable temperature ^{13}C n.m.r. spectra of $HOCH_2CH_2N$\{$CH_2CH_2N(CH_2CH_2OH)CH_2CH_2$\}$NCH_2CH_2OH$ complexes with Pd^{II}, Cd^{II}, and Hg^{II} show exchange between two square antiprisms.[1634] Rotation about the Pt-Pt vector in [$Pt^{III}_2Cl_6$\{HN=C(OH)Bu^t\}$_4$] has $\Delta G^‡$ =15.7 ± 0.5 kcal mol^{-1}.[1635] The fluxionality of [Pd_2(μ-*N*-*S*-η^2-2-SC_5H_4N)$_2Cl_2$(PM_3)$_2$] has been investigated by ^{1}H, ^{13}C and ^{31}P n.m.r. spectroscopy and $\Delta G^‡$, $\Delta H^‡$, and $\Delta S^‡$ determined.[1636] Exchange between coordinated and uncoordinated ether in [Ni(*O,P*-$R_2PCH_2C_6H_9O$)(*P*-$R_2PCH_2C_6H_9O$)Cl]$^+$ has been investigated.[1637] The ^{1}H n.m.r. spectrum of [$PdCl_2$\{($PhCH_2S$)$_2CH_2$\}$_2$] is temperature dependent due to 1,3-shifts of the sulphur ligand. The ^{13}C n.m.r. spectrum was also recorded.[1638] Variable temperature ^{195}Pt n.m.r. spectroscopy has been used to study the stereodynamics of complex thioethers like methionine

1622 E. Abel, K.G. Orrell, S.P. Scanlan, D. Stephenson, T. Kemmitt, and W. Levason, *J. Chem. Soc., Dalton Trans.*, 1991, 591.
1623 L. Mink, M.F. Rettig, and R.M. Wing, *J. Am. Chem. Soc.*, 1991, **113**, 2065.
1624 H. Kurosawa, K. Shiba, K. Ohkita, and I. Ikeda, *Organometallics*, 1991, **10**, 3941.
1625 B.C. Ankianiec and G.B. Young, *Polyhedron*, 1991, **10**, 1411.
1626 L. Cavallo, R. Cini, J. Kobe, L.G. Marzilli, and G. Natile, *J. Chem. Soc., Dalton Trans.*, 1991, 1867.
1627 F.P. Fanizzi, F.P. Intini, L. Maresca, G. Natile, M. Lanfranchi, and A. Tiripicchio, *J. Chem. Soc., Dalton Trans.*, 1991, 1007.
1628 N. Carr, B.J. Dunne, L. Mole, A.G. Orpen, and J.L. Spencer, *J. Chem. Soc., Dalton Trans.*, 1991, 863.
1629 J.J. Schneider, R. Goddard, C. Krüger, S. Werner, and B. Metz, *Chem. Ber.*, 1991, **124**, 301.
1630 A. Albinati, R.W. Kunz, C.J. Ammann, and P.S. Pregosin, *Organometallics*, 1991, **10**, 1800.
1631 A. Stranger, *Organometallics*, 1991, **10**, 2979.
1632 G.E. Herberich, U. Englert, M. Hostalek, and R. Laven, *Chem. Ber.*, 1991, **124**, 17.
1633 E.L.M. Lempers, M.J. Bloemink, and J. Reedijk, *Inorg. Chem.*, 1991, **30**, 201.
1634 P.-A. Pittet, G.S. Laurence, S.F. Lincoln, M.L. Turonek, and K.P. Wainwright, *J. Chem. Soc., Chem. Commun.*, 1991, 1205.
1635 R. Cini, F.P. Fanizzi, F.P. Intini, and G. Natile, *J. Am. Chem. Soc.*, 1991, **113**, 7805.
1636 J.H. Yamamoto, W. Yoshida, and C.M. Jensen, *Inorg. Chem.*, 1991, **30**, 1353.
1637 E. Lindner and J. Dettinger, *Z. Naturforsch., B*, 1991, **46**, 432 (*Chem. Abstr.*, 1991, **115**, 104 648).
1638 Y. Fuchita, H. Maruyama, M. Kawatani, and K. Hiraki, *Polyhedron*, 1991, **10**, 561.

complexed to platinum^{II}.[1639]

Copper. Ligand dynamics have been investigated for Cu^{I} and Zn^{II} complexes of 2,6-{4-(5-methylimidazoyl)CH_2CH_2N=CMe}$_2$pyridine.[1640] The ^{31}P n.m.r. spectrum of [(μ-dppf){Cu(dppf)}$_2$]$^{2+}$ is broad at room temperature, but is described as AB_2 at -25 °C.[1641]

Silver. Exchange between bonded and non-bonded sulphur atoms in [Ag(1,3,6,9,11,14-hexathiacyclohexadecane)]$^+$ has been studied.[1642]

Zinc and Cadmium. Dynamic ^1H n.m.r. spectroscopy has been used to determine $\Delta G^{\ddagger}$ for two mechanisms of zinc intramolecular inversion of tetrahedral coordination.[1643] The fluxionality of [(bipy)Zn(η^1-SCPh=CPhSMe)(η^2-SCPh=CPhSMe)] has been studied by ^1H n.m.r. spectroscopy.[1644] The dynamic enantiomerization of [M{RN=CHCPr^i=C(SH)Ph}$_2$] has been investigated by ^1H and ^{13}C n.m.r. spectroscopy, and $\Delta G^{\ddagger}$ determined.[1645]

Mercury. The variable temperature ^{13}C n.m.r. spectra of [Hg{1,4,8,11-(2-$HOCH_2CH_2$)$_4$-1,4,8,11-tetraazacyclotetradecane}]$^{2+}$ has been used to derive the thermodynamic parameters for pairwise exchange of the hydroxyethyl arms.[1646]

Boron. Restricted P—C bond rotation has been observed in $HC(PPh_2BH_3)_3$ by ^1H, ^{11}B, and ^{13}C n.m.r. spectroscopy.[1647] $\Delta G^{\ddagger}$ for $(PhMe_2P)_2Pt$ rotation on [1,1-$(PhMe_2P)_2$-2-Me-1,2,3-$PtC_2B_8H_9$] is 63 kJ mol^{-1}.[1648] The fluxionality of *closo*-[1,1,1,1-$(THF)_4$-1,2,3-$SmC_2B_9H_{11}$] has been studied by ^{11}B n.m.r. spectroscopy.[1649] Dynamic molecular processes in [$ArBMe_3$]Li(THF) have been investigated by ^1H, ^7Li, ^{11}B, and ^{13}C n.m.r. spectroscopy.[1650] Dynamic ^1H and ^{13}C n.m.r. spectroscopy has been used to determine the rate of sulphur inversion in (107).[1651] Rotational barriers around B—P bonds in compounds such as (2,4,6-$Me_3C_6H_2$)$_3$BPPhLi(OEt_2) have been investigated by ^1H, ^{11}B, and ^{31}P n.m.r. spectroscopy.[1652] The B—N rotational barrier in $Me_2NBPhOCH_2CF_3$ has been determined by ^1H, ^{11}B, and ^{13}C n.m.r. spectroscopy as 16.5 kcal mol^{-1}.[1653]

1639 E. Ratilla, *Report*, 1990, **IS-T-1453**; Order No. DE91000747, 219 pp. Avail. NTIS. From *Energy Res. Abstr.*, 1990, **15**, Abstr. No. 53375 (*Chem. Abstr.*, 1991, **115**, 88 590).
1640 D.K. Coggin, J.A. Gonzalez, A.M. Kook, D.M. Stanbury, and L.J. Wilson, *Inorg. Chem.*, 1991, **30**, 1115.
1641 G. Pilloni, R. Graziani, B. Longato, and B. Corain, *Inorg. Chim. Acta*, 1991, **190**, 165.
1642 B. De Groot and S.J. Loeb, *Inorg. Chem.*, 1991, **30**, 3103.
1643 M.S. Korobov, G.S. Borodkin, L.E. Nivorozhkin, and V.I. Minkin, *Zh. Obshch. Khim.*, 1991, **61**, 929 (*Chem. Abstr.*, 1991, **115**, 288 230).
1644 C. Zhang, R. Chadha, H.K. Reddy, and G.N. Schrauzer, *Inorg. Chem.*, 1991, **30**, 3865.
1645 L.E. Konstantinovskii, R. Ya. Olekhnovich, M.S. Korobov, L.E. Nivorozhkin, and V.I. Minkin, *Zh. Obshch. Khim.*, 1990, **60**, 1930 (*Chem. Abstr.*, 1991, **114**, 114 054); L.E. Konstantinovskii, R.Ya. Olekhnovich, M.S. Korobov, L.E. Nivorozhkin, and V.I. Minkin, *Polyhedron*, 1991, **10**, 771.
1646 P. Clarke, S.F. Lincoln, and K.P. Wainwright, *Inorg. Chem.*, 1991, **30**, 134; P.A. Pittet, G.S. Laurence, S.F. Lincoln, M.L. Turonek, and K.P. Wainwright, *J. Chem. Soc., Chem. Commun.*, 1991, 1205.
1647 H. Schmidbaur, A. Stützer, and E. Herdtweck, *Chem. Ber.*, 1991, **124**, 1095.
1648 J.D. Kennedy, M. Thornton-Pett, B. Ŝtíbr, and T. Jelínek, *Inorg. Chem.*, 1991, **30**, 4484.
1649 M.J. Manning, C.B. Knobler, R. Khattar, and F.M. Hawthorne, *Inorg. Chem.*, 1991, **30**, 2009.
1650 E. Kalbarczyk-Bidelska and S. Pasynkiewicz, *J. Organomet. Chem.*, 1991, **407**, 143.
1651 S. Toyota and M. Oki, *Bull. Chem. Soc. Jpn.*, 1991, **64**, 1563.
1652 D.C. Pestana and P.P. Power, *J. Am. Chem. Soc.*, 1991, **113**, 8426.
1653 S.Y. Shaw and R.H. Neilson, *Inorg. Chem.*, 1991, **30**, 148.

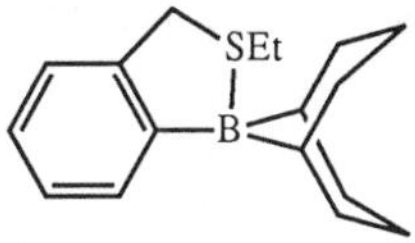

(107)

Aluminium and Gallium. Variable temperature ^1H n.m.r. studies have established that the Al_2S_2 and Ga_2S_2 ring systems in $[R_2MEC_6F_5]_2$, M = Al, Ga; E = O, S, undergo rapid inversion/exchange, which leads to equivalence of the alkyl groups attached to the metal atoms. $^6J(^{19}F-^1H)$ is observed.[1654]

Silicon, Germanium, Tin, and Lead. Long range coupling constants have been used to determine the barriers to rotation about C_{sp^2}—C_{sp^3} bonds for $PhCH_2SiH_3$ and $PhCH_2SiCl_3$ as 7.4 ± 1.6 kJ mol^{-1} and 8.1 kJ mol^{-1}.[1655] Variable temperature ^1H n.m.r. spectroscopy has been used to show axial-equatorial hydrogen exchange in $[H_2Si(OPr^i)_3]^-$. The ^{29}Si n.m.r. spectrum was also recorded.[1656] The ^{13}C n.m.r. spectrum of $[(Me_3Si)_2C{=}CCH(SiMe_3)_2]^+$ shows three methyl signals at -130 °C and two at -100 °C.[1657] Internal rotation in $\{(2,4,6\text{-}Me_3C_6H_2)_2\}C{=}C(OH)SiMe_3$ has $\Delta G^\ddagger = 11.1$ kcal mol^{-1}.[1658] Temperature dependent ^1H and ^{13}C n.m.r. spectroscopy has shown that two enantiomers of (108) undergo a degenerate and a non-degenerate silatropic rearrangement with a barrier of E_a = 48.5 kJ mol^{-1}.[1659] Prototropic rearrangements have been observed for (109), M = Si, using two dimensional ACCORDION experiments. The ^{13}C n.m.r. spectra of (109), M = Si, Ge, Sn, were also recorded.[1660] Inversion of the central ring of (110), R = Me, occurs with $\Delta G^\ddagger = 16$ kcal mol^{-1}. The ^{13}C and ^{29}Si n.m.r. spectra were also recorded.[1661] The molecular dynamics of $(SiMe_2Ar\text{-}SiMe_2OSiMeR)_n$ have been studied using ^{29}Si T_1 measurements.[1662] 1,3-shifts of the organotin group in (109), M = Sn, have been studied.[1663]

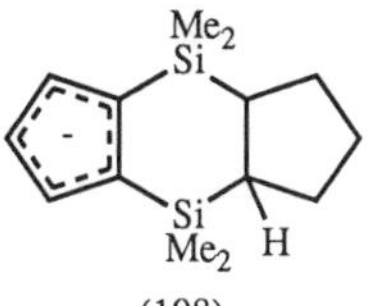

(108)

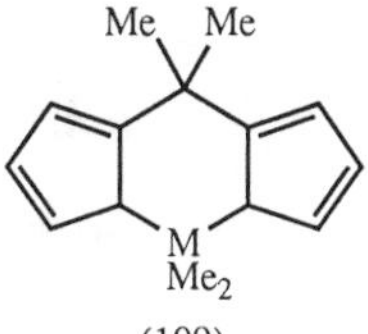

(109)

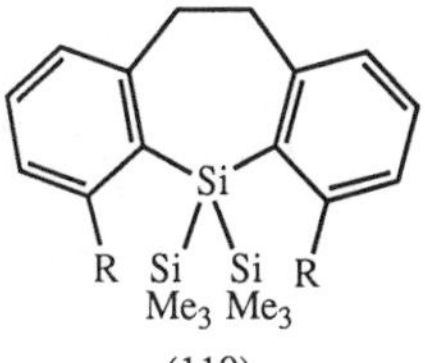

(110)

1654 D.G. Hendershot, R. Kumar, M. Barber, and J.P. Oliver, *Organometallics*, 1991, **10**, 1917.
1655 T. Schaefer, R. Sebastian, and G.H. Penner, *Can. J. Chem.*, 1991, **69**, 496.
1656 R.J.P. Corriu, C. Guérin, B.J.L. Henner, and Q. Wang, *Organometallics*, 1991, **10**, 3574.
1657 H.-U. Siehl, F.-P. Kaufmann, Y. Apeloig, V. Braude, D. Danovich, A. Berndt, and K. Stamatis, *Angew. Chem., Int. Ed. Engl.*, 1991, **30**, 1479.
1658 E.B. Nadler and Z. Rappoport, *Tetrahedron Lett.*, 1991, **32**, 1233.
1659 J. Hiermeier, F.H. Köhler, and G. Müller, *Organometallics*, 1991, **10**, 1787.
1660 I.E. Nifant'ev, V.L. Yarnykh, M.V. Borzov, B.A. Mazurchik, V.I. Mstislavskii, V.A. Roznyatovskii, and Yu.A. Ustynyuk, *Organometallics*, 1991, **10**, 3739.
1661 L.D. Lange, J.Y. Corey, and N.P. Rath, *Organometallics*, 1991, **10**, 3189.
1662 C.W. Chu, L.C. Dickinson, K. Fugishiro, M. Itoh, R.W. Lenz, and J.C.W. Chien, *Macromolecules*, 1991, **24**, 4615 (*Chem. Abstr.*, 1991, **115**, 72 569).
1663 I.E. Nifant'ev, A.K. Shestakova, D.A. Lemenovskii, Yu.L. Slovokhotov, and Yu.T. Struchkov, *Metalloorg. Khim.*, 1991, **4**, 292 (*Chem. Abstr.*, 1991, **115**, 49 839).

The dynamic behaviour of 1,4-$C_6H_4\{OSiMe_2(\eta^1-C_5H_5)\}_2$ has been investigated by variable temperature ^1H n.m.r. spectroscopy.[1664] The temperature dependent ^1H n.m.r. spectra of $R^1R^2Si(\mu-PH)_2SiR^1R^2$ exhibit two dynamic processes, hindered rotation of the aryl groups at silicon and inversion at phosphorus. The ^{29}Si n.m.r. spectrum was also recorded.[1665] The fluxionality of some five coordinate compounds such as $[\overline{CH_2(CH_2)_nSi}(O_2C_6H_4)F]^-$ has been investigated by ^{19}F n.m.r. spectroscopy. The ^{13}C and ^{29}Si n.m.r. spectra were also recorded.[1666] Pseudo-rotation in (111) has been investigated by ^1H n.m.r. spectroscopy.[1667]

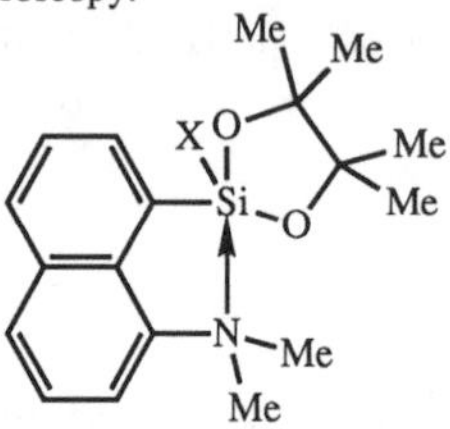

(111)

$[\{H_2B(C_3H_3N_2)_2\}_2Sn]$ shows two dynamic processes, axial-equatorial exchange, and boat-boat flip. The ^{11}B and ^{119}Sn n.m.r. spectra were also recorded.[1668] Intramolecular proton transfer has been investigated for $[Sn(OPr^i)_4(HOPr^i)]$. ^{13}C and ^{119}Sn n.m.r. spectra were also recorded.[1669]

Phosphorus. Axial-equatorial ^{19}F exchange in $[RPF_2(tmeda)]^+$ has been studied by ^{13}C, ^{19}F, and ^{31}P n.m.r. spectroscopy.[1670] Pseudorotation on $(PhO)\overline{P(OCH_2CMe_2CH_2O})$(9,10-$O_2$-phenanthrene) has been investigated.[1671] ^1H and ^{13}C n.m.r. spectroscopy has been used to study Berry pseudorotation in compounds such as $(2,6-Me_2C_6H_3O)\overline{P(OCH_2CMe_2CH_2O})(1-O-2-NHC_6H_4)$,[1672] and $(2,6-Me_2C_6H_3S)\overline{P(OCH_2CMe_2CH_2O})(OC_6F_5)_2$.[1673]

Arsenic, Antimony, and Bismuth. The ^1H n.m.r. spectrum of $Cl\overline{SbCMe{=}CMeCH_2S}$ shows exchange of the CH_2 protons. The ^{13}C n.m.r. spectrum was also recorded.[1674] The fluxionality of [2.2.2]paracyclophane coordinated to AsX_3, SbX_3,and BiX_3 has been studied.[1675] $\Delta G^{\ddagger}$ for Bi—N rotation in $Bi(NMe_2)_3$ is 40.6 kJ mol^{-1}. The ^{13}C n.m.r. spectrum was also recorded.[1676]

1664 H. Plenio, *Chem. Ber.*, 1991, **124**, 2185.
1665 M. Drieß, H. Pritzkow, and M. Reisgys, *Chem. Ber.*, 1991, **124**, 1931.
1666 R.O. Day, C. Streelatha, J.A. Deiters, S.E. Johnson, J.M. Holmes, L. Howe, and R.R. Holmes, *Organometallics*, 1991, **10**, 1758.
1667 F.H. Carre, R.J.P. Corriu, G.F. Lanneau, and Z. Yu, *Organometallics*, 1991, **10**, 1236.
1668 D.L. Reger, S.J. Knox, M.F. Huff, A.L. Rheingold, and B.S. Haggerty, *Inorg. Chem.*, 1991, **30**, 1754.
1669 M.J. Hampden-Smith, T.A. Wark, A. Rheingold, and J.C. Huffman, *Can. J. Chem.*, 1991, **69**, 121.
1670 T. Kaukorat, P.G. Jones, and R. Schmutzler, *Chem. Ber.*, 1991, **124**, 1335.
1671 R.R. Holmes, K.C.K. Swamy, J.M. Holmes, and R.O. Day, *Inorg. Chem.*, 1991, **30**, 1052.
1672 R.O. Day, K.C.K. Swamy, L. Fairchild, J.M. Holmes, and R.R. Holmes, *J. Am. Chem. Soc.*, 1991, **113**, 1627.
1673 J. Hans, R.O. Day, L. Howe, and R.R. Holmes, *Inorg. Chem.*, 1991, **30**, 3132.
1674 R.A. Fisher, R.B. Nielsen, W.M. Davis, and S.L. Buchwald, *J. Am. Chem. Soc.*, 1991, **113**, 165.
1675 T. Probst, O. Steigelmann, J. Riede, and H. Schmidbaur, *Chem. Ber.*, 1991, **124**, 1089.
1676 W. Clegg, N.A. Compton, R.J. Errington, G.A. Fisher, M.E. Green, D.C.R. Hockless, and N.C. Norman, *Inorg. Chem.*, 1991, **30**, 4680.

Selenium and Tellurium. Molecular reorientation of SeF_6 and TeF_6 has been studied using ^{19}F T_1 measurements.[1677]

Equilibria.—*Solvation Studies of Ions.* The ^1H line width of $H_2{}^{17}O$ in strong electrolytes has been measured and exchange rates determined.[1678] ^1H n.m.r. spectroscopy has been used to study ionic solvation in water-DMSO and water-DMF mixtures.[1679] Water-oil microemulsions have been studied as model systems for the study of water confined in microenvironments.[1680] The solvation of alkali metal fluorides, nitrates, perchlorates, $[ScF_6]^{3-}$, $[GaF_6]^{3-}$, and $[InF_6]^{3-}$ has been studied by ^{19}F, ^{45}Sc, ^{71}Ga, ^{115}In, and ^{133}Cs n.m.r. spectroscopy.[1681]

Group Ia. Organo-lithium ion pairing in THF/HMPA has been investigated by ^6Li, ^7Li, ^{13}C, and ^{31}P n.m.r. spectroscopy.[1682] Isotope effects on the reorientational motion of hydrated water molecules in alkali metal bromide dilute aqueous solutions have been studied by ^1H and ^{17}O n.m.r. spectroscopy.[1683] The dynamic properties of water molecules coordinated to Li^+ and F^- in undercooled LiCl and KF solutions have been investigated by n.m.r. spectroscopy.[1684] Alkali metal-water interactions have been studied in acetonitrile.[1685] Aqueous solutions containing NaCl and KCl at various concentrations have been studied by ^{17}O n.m.r. spectroscopy.[1686] ^2H and ^{17}O T_1 measurements on D_2O in dilute solutions of KF, KCl, KBr, and KI have been made and reorientation motion of hydrated water molecules determined.[1687] The linewidths and chemical shifts of ^{14}N have been measured for NaSCN and NaN_3 in a variety of solvents. The linewidth dependence on the solvent was attributed to ion-solvent interaction.[1688] ^{23}Na T_1 and T_2 measurements have been made on sodium dodecyl sulphate in some aqueous and non-aqueous systems.[1689] The standard chemical potential for Li^+ in 2-methyltetrahydrofuran has been determined from ^7Li T_1 measurements.[1690]

Magnesium and Calcium. ^{25}Mg n.m.r. spectra of $Mg(ClO_4)_2$ solution have been measured in water-acetone and water-acetonitrile.[1691] Cation exchange in the system calcium(II) or magnesium(II), complexing agent, zeolite-NaA has been investigated and cation-anion coordination in mixed solvents

1677 C.J. Jameson, A.K. Jameson, and R.J. Terry, *J. Phys. Chem.*, 1991, **95**, 2982 (*Chem. Abstr.*, 1991, **114**, 171 640).
1678 J. Diratsaoglu, S. Hauber, H.G. Hertz, and K.J. Müller, *Z. Phys. Chem. (Munich)*, 1990, **168**, 13 (*Chem. Abstr.*, 1991, **114**, 193 419).
1679 V.A. Abakshin, O.V. Eliseeva, and G.A. Krestov, *Zh. Neorg. Khim.*, 1990, **35**, 2324 (*Chem. Abstr.*, 1991, **114**, 50 632).
1680 D. Senatra, L. Lendinara, and M.G. Giri, *Prog. Colloid Polym. Sci.*, 1991, **84**, 122 (*Chem. Abstr.*, 1991, **115**, 287 749).
1681 G.A. Kirakosyan, S.V. Loginov, T.V. Galuzina, V.P. Tarasov, and Yu.A. Buslaev, *Zh. Neorg. Khim.*, 1990, **35**, 2306 (*Chem. Abstr.*, 1991, **114**, 50 631).
1682 H.J. Reich and J.P. Borst, *J. Am. Chem. Soc.*, 1991, **113**, 1835.
1683 A. Shimizu and Y. Taniguchi, *Bull. Chem. Soc. Jpn.*, 1991, **64**, 221.
1684 E.W. Lang, W. Fink, and H. Radkowitsch, *NATO ASI Ser., Ser. C*, 1991, **329**, 393 (*Chem. Abstr.*, 1991, **115**, 143 932).
1685 J.F. O'Brien, D.W. Johnson, K. Kuehner, T.R. Brewer, and R.G. Keil, *Inorg. Chim. Acta*, 1991, **183**, 113.
1686 A. Higashiyama, Y. Yamamoto, and R. Chujo, *Bull. Chem. Soc. Jpn.*, 1991, **54**, 285.
1687 A. Shimizu and Y. Taniguchi, *Bull. Chem. Soc. Jpn.*, 1991, **64**, 1613.
1688 S.C.F. Au-Yeung, *J. Magn. Reson.*, 1991, **92**, 10.
1689 A. Ceglie, G. Colafemmina, M. Della Monica, L. Burlamacchi, and M. Monduzzi, *J. Colloid Interface Sci.*, 1991, **146**, 363 (*Chem. Abstr.*, 1991, **115**, 190 421).
1690 A.I. Mishustin, N.F. Tovmash, and A.V. Plakhotnik, *Zh. Fiz. Khim.*, 1991, **65**, 1411 (*Chem. Abstr.*, 1991, **115**, 100 579).
1691 K. Miura, H. Matsuda, S. Kikuchi, and H. Fukui, *J. Chem. Soc., Faraday Trans.*, 1991, **87**, 837.

studied by using ^{35}Cl n.m.r. spectroscopy.[1692]

Scandium. The extraction of scandium by di-2-ethylhexyl phosphoric acid has been studied by ^{31}P and ^{45}Sc n.m.r. spectroscopy.[1693]

Lanthanides. Variable pressure ^{17}O n.m.r. kinetics have been applied to lanthanide solvates.[1694] ^{1}H, ^{2}H, and ^{13}C relaxation measurements have been applied to the solvation of Gd(III) chelates.[1695]

Thorium and Uranium. The kinetics of water exchange of $UO_2(NO_3)_2(OH_2)_2$ in acetone-water have been studied by ^{1}H n.m.r. spectroscopy.[1696] The solvate complexes formed during the extraction of $UO_2(NO_3)_2$ and $Th(NO_3)_4$ by tributylphosphate have been studied by ^{1}H, ^{13}C, ^{15}N, and ^{31}P n.m.r. spectroscopy.[1697] The solvation of $UO_2(O_3SC_6H_5)_2$ in water-tributyl phosphate has been studied by dynamic n.m.r. spectroscopy.[1698]

Chromium. The solvation of $[Cr(SCN)_6]^{3-}$ has been studied using ^{1}H n.m.r. line broadening.[1699]

Manganese. The relative amounts of free and complexed manganese ions have been determined using ^{1}H T_1 and T_2 measurements.[1700] A ^{1}H T_1 and T_2 study has been used to investigate the complexation of Mn^{II} with carboxy containing ligands. ^{13}C n.m.r. spectra were also recorded.[1701] Water coordination to manganese(II) tetrakis(4-sulphophenyl)porphine has been investigated.[1702]

Iron. Water exchange on $[Fe(edta)(OH_2)]^-$,[1703] and the pressure dependence of DMF exchange on $[Fe\{1,2\text{-}[(O_2CCH_2)_2N]_2C_6H_4\}]^-$ have been determined using ^{17}O n.m.r. spectroscopy.[1704]

Ruthenium. Acetonitrile exchange on $[(\eta^5\text{-}C_5H_5)Ru(NCMe)_3]^+$ and $[(\eta^6\text{-}C_6H_6)Ru(NCMe)_3]^{2+}$ has been studied by ^{1}H n.m.r. spectroscopy as a function of pressure.[1705]

Cobalt. ^{13}C and ^{59}Co n.m.r. spectroscopy has been used to study the solvation of compounds such as $[Co(edta)]^-$.[1706] The kinetics of water exchange in Co^{2+} or Cu^{2+} ethylenediaminedisuccinic acid-water have been studied by ^{1}H n.m.r. spectroscopy.[1707]

Rhodium. A variable pressure ^{17}O n.m.r. study of water exchange on $[Rh(OH_2)_6]^{3+}$ has been

1692 A.R. Ebaid, A.K. Barakat, N.A. Khalifa, A.T. Kandil, L.M. Maesen, and H. Van Bekkum, *Recl. Trav. Chim. Pays-Bas*, 1991, **110**, 374 (*Chem. Abstr.*, 1991, **115**, 190 607).

1693 V.Yu. Korovin, S.B. Randarevich, S.V. Bodaratskii, V.V. Trachevskii, and I.V. Pastukhova, *Koord. Khim.*, 1991, **17**, 561 (*Chem. Abstr.*, 1991, **114**, 254 814).

1694 L. Helm and A.E. Merbach, *Eur. J. Solid State Inorg. Chem.*, 1991, **28**, 245 (*Chem. Abstr.*, 1991, **114**, 235 894).

1695 S. Aime, M. Botta, and G. Ermondi, *J. Magn. Reson.*, 1991, **92**, 572.

1696 M.E.D.G. Azenha, H.D. Burrows, S.J. Formosinho, V.M.S. Gil, and M. da G. Miguel, *J. Coord. Chem.*, 1991, **24**, 15.

1697 A.M. Rozen, A.S. Nikiforov, N.A. Kartasheva, N.V. Neumoev, V.S. Markov, V.E. Yushmanov, and P.Yu. Shkarin, *Dokl. Akad. Nauk SSSR*, 1990, **315**, 1140 (*Chem. Abstr.*, 1991, **115**, 143 860).

1698 E.O. Ashevskaya, L.L. Shcherbakova, and V.A. Shcherbakov, *J. Radioanal. Nucl. Chem.*, 1990, **143**, 307 (*Chem. Abstr.*, 1991, **114**, 130 305); V.A. Shcherbakov, L.L. Shcherbakova, and E.O. Ashevskaya, *Dokl. Akad. Nauk SSSR*, 1991, **316**, 943 (*Chem. Abstr.*, 1991, **115**, 122 054).

1699 A. Wakisaka, H. Sakuragi, and K. Tokumaru, *J. Chem. Soc., Faraday Trans.*, 1991, **87**, 2167.

1700 K.E. Kellar and N. Foster, *Anal. Chem.*, 1991, **63**, 2919.

1701 V.G. Shtyrlin, S.A. Bikchantaeva, V.V. Trachevskii, and A.V. Zakharov, *Koord. Khim.*, 1991, **17**, 676 (*Chem. Abstr.*, 1991, **115**, 36 443).

1702 G. Hernandez and R.G. Bryant, *Bioconjugate Chem.*, 1991, **2**, 394 (*Chem. Abstr.*, 1991, **115**, 227 238).

1703 M. Mizuno, S. Funahashi, N. Nakasuka, and M. Tanaka, *Inorg. Chem.*, 1991, **30**, 1550.

1704 M. Mizuno, S. Funahashi, N. Nakasuka, and M. Tanaka, *Bull. Chem. Soc. Jpn.*, 1991, **64**, 1988.

1705 W. Luginbuehl, P. Zbinden, P.A. Pittet, T. Armbruster, H.B. Buergi, A.E. Merbach, and A. Ludi, *Inorg. Chem.*, 1991, **30**, 2350.

1706 T. Taura, *Bull. Chem. Soc. Jpn.*, 1991, **64**, 2362.

1707 N.N. Tananaeva, V.V. Strashko, and R.V. Tikhonova, *Ukr. Khim. Zh. (Russ. Ed.)*, 1990, **56**, 1139 (*Chem. Abstr.*, 1991, **114**, 235 819).

reported.[1708]

Nickel. The solvation of [Ni(1,4,8,11-Me_4-1,4,8,11-tetraazacyclotetradecane)]$^{2+}$ by H_2O and DMSO has been studied by ^1H n.m.r. spectroscopy.[1709]

Palladium and Platinum. The aquation of [$Pd(O_2CCH_3)_2$], [$Pd(OH)_4$]$^{2-}$, and [$Pd(O_4C_2)_2$]$^{2-}$ has been studied by ^{17}O n.m.r. spectroscopy.[1710] Solvent exchange on [PdL_4]$^{2+}$ has been studied by high pressure ^1H n.m.r. spectroscopy.[1711] ^{13}C and ^{195}Pt n.m.r. spectroscopy has been used to investigate the solvolysis of [$Pt_2(NH_3)_4$(3,3-Me_2-glutarimidate)$_2$]$^{2+}$ in D_2O and DMSO.[1712]

Copper. The dynamics of water molecules in Cu^{2+} aqueous solution have been investigated by ^{17}O n.m.r. spectroscopy. The influence of the Jahn-Teller distortion on scalar relaxation has been examined.[1713] The solvation of Cu^{2+} in the presence of an amine oxidase has been studied by ^1H T_1 measurements.[1714]

Silver. The solvent self diffusion coefficients and ^{109}Ag chemical shifts in $AgNO_3$-CH_3CN-H_2O mixtures have been studied.[1715]

Zinc. ^{67}Zn n.m.r. spectroscopy has been used to study the structure of $ZnCl_2$ in water, and species such as [$ZnCl_n(OH_2)_m$]$^{2-n}$ identified.[1716]

Cadmium. The dissociation of $CdCl_2$ into [$Cd(DMF)_6$]$^{2+}$, [$CdCl(DMF)_5$]$^+$, and [$CdCl_3$]$^-$ has been investigated by ^{113}Cd n.m.r. spectroscopy. When [BH_4]$^-$ is added, it coordinates, and the complexation was also studied by ^{11}B and ^{113}Cd n.m.r. spectroscopy.[1717]

Gallium. The extraction of Ga^{III} from aqueous HCl by a $(BuO)_3PO$ solid extractant has been studied by ^{31}P and ^{71}Ga n.m.r. spectroscopy.[1718]

Silicon. ^2H and ^{14}N n.m.r. spectroscopy has been used to study the interaction of CD_3CN with octadecylsilica surfaces.[1719]

Chlorine. A new method of division has been used for the viscosity B-coefficient and has been extended to partitioning of the n.m.r. B'-coefficient both in aqueous and non-aqueous systems. The method was applied to ions and the hydration number of Cl^- determined.[1720]

Ionic Equilibria. Group Ia. ^6Li, ^7Li, and ^{13}C T_1 and $T_{1\rho}$ measurements have been made to study ion pair dynamics and the structure of lithium fluorenide.[1721] Exchange between Ph_nMLi_m, M = I, Te,

1708 G. Laurenczy, I. Rappaport, D. Zbinden, and A.E. Merbach, *Magn. Reson. Chem.*, 1991, **29**, S45.
1709 E. Iwamoto, J. Nishimoto, T. Yokoyama, K. Yamamoto, and T. Kumamaru, *J. Chem. Soc., Faraday Trans.*, 1991, **87**, 1537.
1710 M.A. Fedotov, S.Yu. Troitskii, and V.A. Likholobov, *Koord. Khim.*, 1990, **16**, 1675 (*Chem. Abstr.*, 1991, **114**, 134 960).
1711 N. Hallinan, V. Besancon, M. Forster, G. Elbaze, Y. Ducommun, and A.E. Merbach, *Inorg. Chem.*, 1991, **30**, 1112.
1712 H. Urata, H. Moriyama, and K. Matsumoto, *Inorg. Chem.*, 1991, **30**, 3914.
1713 D.H. Powell, L. Helm, and A.E. Merbach, *J. Chem. Phys.*, 1991, **95**, 9258.
1714 D.M. Dooley, M.A. McGuirl, C.E. Cote, P.F. Knowles, I. Singh, M. Spiller, R.D. Brown, tert., and S.H. Koenig, *J. Am. Chem. Soc.*, 1991, **113**, 754.
1715 B. Wittekopf, N. Weiden, and K.G. Weil, *Ber. Bunsenges. Phys. Chem.*, 1991, **95**, 520.
1716 J.A. Tossell, *J. Phys. Chem.*, 1991, **95**, 366.
1717 G. Linti, H. Nöth, and M. Thomann, *Z. Naturforsch., B*, 1990, **45**, 1463 (*Chem. Abstr.*, 1991, **114**, 74 205).
1718 V. Yu. Korovin, S.B. Randarevich, Yu.N. Pogorelov, and A.V. Zhuravleva, *Zh. Neorg. Khim.*, 1991, **36**, 205 (*Chem. Abstr.*, 1991, **115**, 167 682).
1719 E.H. Ellison and D.B. Marshall, *J. Phys. Chem.*, 1991, **95**, 808.
1720 M.M. Bhattacharyya, *Bull. Chem. Soc. Jpn.*, 1991, **64**, 644.
1721 I. Sethson, B. Eliasson, and U. Edlund, *Magn. Reson. Chem.*, 1991, **29**, 1012.

and Hg, has been studied by ^{7}Li and ^{13}C n.m.r. spectroscopy.[1722] Complexation of Li^{+}, Na^{+}, K^{+}, Cs^{+}, Ag^{+}, and Tl^{+} with 4,7,13,16-tetraoxa-1,10-diazabicyclo[8.8.2]eicosane has been studied by ^{23}Na n.m.r. spectroscopy.[1723] Complexation of Li^{+} and Ag^{+} by 4,7,13,16-tetraoxa-1,10-diazabicyclo[8.8.2]eicosane has been studied in seven solvents by ^{7}Li n.m.r. spectroscopy.[1724] ^{6}Li, ^{15}N, and ^{31}P n.m.r. spectroscopy has been used to investigate the structure of $\overline{LiNCMe_2CH_2CH_2CH_2C}Me_2$ and $Li(NPr^{i}_2)$ in the presence of $(Me_2N)_3PO$.[1725] ^{7}Li T_1 measurements and ^{13}C and ^{31}P chemical shifts have been used to study the binding of Li^{+} and Mg^{2+} to ATP and ADP.[1726] The complexation of Li^{+}, Na^{+}, K^{+}, Cs^{+}, Cd^{2+}, and Pb^{2+} by azamacrocyclic ether-bislactones and amine-bislactam has been investigated by ^{1}H, ^{7}Li, ^{23}Na, and ^{133}Cs n.m.r. spectroscopy.[1727] ^{7}Li n.m.r. spectroscopy has been used to study the interaction of Li^{+} with (112).[1728] ^{7}Li n.m.r. spectroscopy has been used to study the interaction of Li^{+} with

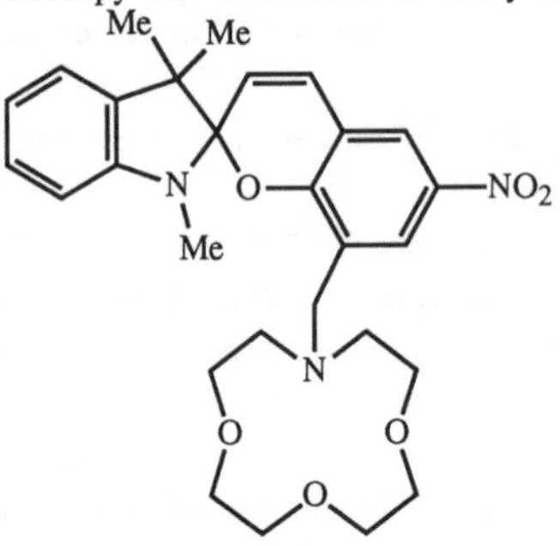

(112)

phospholipids.[1729] The coordination of Li^{+} to monobenzene-15-crown-5 has been studied using two-dimensional ^{7}Li exchange spectroscopy.[1730] ω-Lithium poly(styrene)sulphonates have been studied using ^{1}H and ^{13}C n.m.r. spectroscopy.[1731] Tissue lithium concentration has been measured by ^{7}Li n.m.r. spectroscopy in patients with bipolar disorder.[1732] The Li^{+}-crown ether stability constant has been determined by ^{7}Li n.m.r. spectroscopy in a molten salt at room temperature.[1733] Li^{+} and Ca^{2+} binding to amino acids has been investigated using ^{1}H, ^{13}C, and ^{17}O n.m.r. spectroscopy.[1734]

1722 H.J. Reich, D.P. Green, and N.H. Phillips, *J. Am. Chem. Soc.*, 1991, **113**, 1414.
1723 S.F. Lincoln and A.K.W. Stephens, *Inorg. Chem.*, 1991, **30**, 3529; S.F. Lincoln and T. Rodopoulos, *Inorg. Chim. Acta*, 1991, **190**, 223.
1724 A. Abou-Hamdan and S.F. Lincoln, *Inorg. Chem.*, 1991, **30**, 462.
1725 F.E. Romesberg, J.H. Gilchrist, A.T. Harrison, D.J. Fuller, and D.B. Collum, *J. Am. Chem. Soc.*, 1991, **113**, 5751.
1726 A. Abraha, D.E. Mota de Freitas, M. Margarida, C.A. Castro, and C.F.G.C. Geraldes, *J. Inorg. Biochem.*, 1991, **42**, 191.
1727 A. Zhang, S. Wang, and S. Qin, *Huaxue Xuebao*, 1991, **49**, 164 (*Chem. Abstr.*, 1991, **115**, 16 400).
1728 K. Kimura, T. Yamashita, and M. Yokoyama, *J. Chem. Soc., Chem. Commun.*, 1991, 147.
1729 J.F.M. Post and D.A. Wilkinson, *J. Colloid Interface Sci.*, 1991, **143**, 174 (*Chem. Abstr.*, 1991, **114**, 171 965).
1730 K.M. Briere, H.D. Dettman, and C. Detellier, *J. Magn. Reson.*, 1991, **94**, 600.
1731 R.P. Quirk and J. Kim, *Macromolecules*, 1991, **24**, 4515 (*Chem. Abstr.*, 1991, **115**, 72 400).
1732 L. Gyulai, S.W. Wicklund, R. Greenstein, M.S. Bauer, P. Ciccione, P.C. Whybrow, J. Zimmerman, G. Kovachich, and W. Alves, *Biol. Psychiatry*, 1991, **29**, 1161 (*Chem. Abstr.*, 1991, **115**, 126 343).
1733 E.M. Eyring, *Report*, 1990, **Order No. AD-A218819**, 6 pp. Avail. NTIS. From *Gov. Rep. Announce. Index (U.S.)*, 1990, **90**, Abstr. No. 033 224 (*Chem. Abstr.*, 1991, **114**, 151 382).
1734 H. Schmidbaur, P. Kiprof, O. Kumberger, and J. Riede, *Chem. Ber.*, 1991, **124**, 1083.

Chelation of 2-substituted 1-lithioxides has been studied by ^{1}H, ^{6}Li, and two-dimensional ^{6}Li-^{1}H HOESY n.m.r. spectroscopy.[1735] Lithium exchange in $[Li_4(OC_6H_3Me_2\text{-}3,5)_2][ClO_4]_2$ has been studied by ^{6}Li, ^{7}Li, ^{13}C, and ^{35}Cl n.m.r. spectroscopy, including ^{6}Li EXSY.[1736] The structure and chemical equilibria of lithium β-diketonates have been studied by ^{13}C n.m.r. spectroscopy.[1737] Ionic conductivity in CF_3SO_3Li-polyethylene oxide has been investigated using ^{7}Li and ^{19}F n.m.r. spectroscopy.[1738]

^{2}H and ^{23}Na n.m.r. relaxation in a hexagonal lyotropic liquid crystal has been investigated.[1739] ^{23}Na n.m.r. spectroscopy has been used to study the complexation of Na^+ by (113).[1740] The stability of 18-crown-6 ether with Na^+ and K^+ in CD_3OD and D_2O has been estimated from ^{13}C T_1 measurements.[1741] A ^{23}Na n.m.r. study of the complexation of $NaBPh_4$ by [5,5]-dibenzo-30-crown-10 in nitromethane, acetonitrile, acetone, and pyridine has been reported.[1742] ^{13}C n.m.r. titration experiments have been used to investigate the interaction between S(4-benzo-15-crown-5)$_2$ and metals such as Na^+, K^+, Cu^+, and Ag^+.[1743] Complexation of $NaBPh_4$ by poly(propylene oxide) has been investigated by ^{23}Na n.m.r. spectroscopy.[1744] The dissociation kinetics of calixarene ester-Na^+ complexes have been determined using ^{23}Na n.m.r. spectroscopy.[1745] ^{13}C, ^{17}O, and ^{23}Na n.m.r. spectroscopy has been used to study the influences of water, sucrose, and sodium chloride on starch gelatinization.[1746] The interaction of Na^+ and K^+ with κ-carrageenan has been studied by n.m.r. spectroscopy.[1747] The competitive displacement of Mg^{2+} by Na^+ and K^+ from ribosomal RNA has been followed by ^{25}Mg n.m.r. spectroscopy.[1748] ^{23}Na n.m.r. studies of the interaction of alkali metal ions with the calcium binding sites of calcium ATPase[1749] and with concentrated DNA solutions at low supporting electrolyte concentration[1750] have been reported. The intracellular Na^+ concentration in dog kidney cortical tubules has been monitored by ^{23}Na n.m.r. spectroscopy using dysprosium tripolyphosphate as a shift reagent.[1751] A correction to an earlier report of a theoretical n.m.r. lineshape study of a two-site exchange model of Na^+ ions in an intracellular environment has been published.[1752] A method has been developed for the absolute quantification of intracellular Na^+

1735 M.A. Nichols, A.T. McPhail, and E.M. Arnett, *J. Am. Chem. Soc.*, 1991, **113**, 6222.
1736 L.M. Jackman, E.F. Rakiewicz, and A.J. Benesi, *J. Am. Chem. Soc.*, 1991, **113**, 4101.
1737 Y. Araki, A. Iwase, S. Kudo, T.M. Suzuki, and T. Yokoyama, *Bull. Chem. Soc. Jpn.*, 1991, **64**, 2931.
1738 N. Boden, S.A. Leng, and I.M. Ward, *Solid State Ionics*, 1991, **45**, 261 (*Chem. Abstr.*, 1991, **115**, 19 408).
1739 P.O. Quist, B. Halle, and I. Furo, *J. Chem. Phys.*, 1991, **95**, 6945.
1740 P. Clarke, S.F. Lincoln, and E.R.T. Tiekink, *Inorg. Chem.*, 1991, **30**, 2747.
1741 C. Erk, *Thermochim. Acta*, 1991, **180**, 317 (*Chem. Abstr.*, 1991, **115**, 36 543).
1742 K.M. Briere, H.P. Graves, T.S. Rana, L.J. Maurice, and C. Detellier, *J. Phys. Org. Chem.*, 1990, **3**, 435 (*Chem. Abstr.*, 1991, **115**, 7972).
1743 P.D. Beer, C.G. Crane, and M.G.B. Drew, *J. Chem. Soc., Dalton Trans.*, 1991, 3235.
1744 Y.S. Pak, K.J. Adamic, S.G. Greenbaum, M.C. Wintersgill, J.J. Fontanella, and C.S. Coughlin, *Solid State Ionics*, 1991, **45**, 277 (*Chem. Abstr.*, 1991, **115**, 20 728).
1745 T. Jin and K. Ichikawa, *J. Phys. Chem.*, 1991, **95**, 2601.
1746 P. Chinachoti, V.A. White, L. Lo, and T.R. Stengle, *Cereal Chem.*, 1991, **68**, 238 (*Chem. Abstr.*, 1991, **115**, 134 387).
1747 V.P. Yur'ev, A.L. Blumenfel'd, E.E. Braudo, and V.B. Tolstoguzov, *Colloid Polym. Sci.*, 1991, **269**, 850 (*Chem. Abstr.*, 1991, **115**, 210 571).
1748 S.S. Reid and J.A. Cowan, *J. Am. Chem. Soc.*, 1991, **113**, 673.
1749 I.M. Timonin, S.N. Dvoryantsev, V.V. Petrov, E. Ruuge, and D.O. Levitskii, *Biochim. Biophys. Acta*, 1991, **1066**, 43 (*Chem. Abstr.*, 1991, **115**, 67 408).
1750 T.E. Strzelecka and R.L. Rill, *Biopolymers*, 1990, **30**, 803 (*Chem. Abstr.*, 1991, **114**, 20 451).
1751 H. Ammann, Y. Boulanger, and P. Vinay, *Magn. Reson. Med.*, 1990, **16**, 368 (*Chem. Abstr.*, 1991, **115**, 131 051).
1752 E. Berggren and P.O. Westlund, *Biophys. J.*, 1990, **58**, 813 (*Chem. Abstr.*, 1991, **114**, 38 630).

using triple quantum filtered ^{23}Na n.m.r. spectroscopy.[1753] ^{23}Na n.m.r. imaging of foods has been reported.[1754] The dynamics of the phosphate component in water solutions of (1-x)$NaPO_3$-xNaF glasses have been studied using ^{19}F and ^{31}P n.m.r. spectroscopy.[1755]

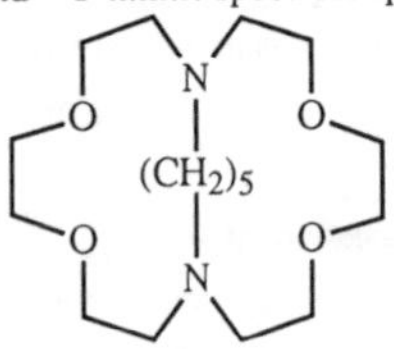

(113)

A potassium naphthalide solution has been studied by ^{39}K n.m.r. spectroscopy as a function of temperature and concentration and complexation was examined.[1756] A pulsed field gradient spin-echo n.m.r. study of self-diffusion in isotropic micellar solutions containing KO_2CR and Me_4Si has been published.[1757] ^{13}C n.m.r. spectroscopy has been used to investigate the interaction of (114) with KX, where the K^+ goes into the polyether, and the X^- coordinates to boron.[1758] The complexation of N,N'-dicarboxymethyl macrocyclic ether-bislactones and amine-bislactam with K^+, Cd^{2+}, and Pb^{2+} has been studied by ^1H n.m.r. spectroscopy.[1759] ^{23}Na, ^{39}K, ^{87}Rb, and ^{133}Cs n.m.r. spectroscopy has been used to study the interaction of group Ia cations with furcellaran gels.[1760]

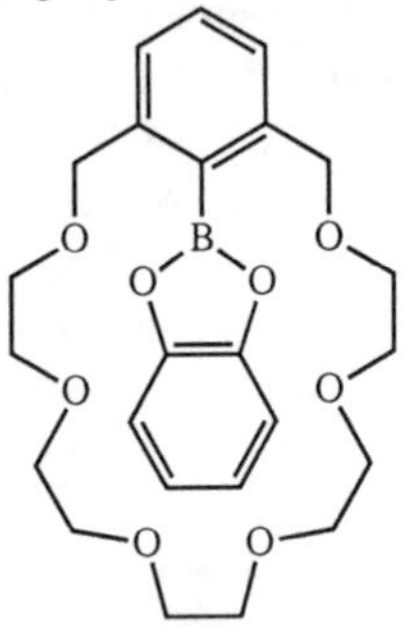

(114)

^{87}Rb T_1 and T_2 measurements have been used to study the interaction of Rb^+ with ι-carrageenan.[1761] ^{87}Rb n.m.r. spectroscopy has been used to measure K^+ fluxes across membranes in intact tissues.[1762] A ^{133}Cs competitive study of complex formation between Rb^+ and Cs^+ with crown

1753 J.L. Allis, A.M.L. Seymour, and G.K. Radda, *J. Magn. Reson.*, 1991, **93**, 71.
1754 N. Ishida, T. Kobayashi, H. Kano, S. Nagai, and H. Ogawa, *Agric. Biol. Chem.*, 1991, **55**, 2195 (*Chem. Abstr.*, 1991, **115**, 230 684).
1755 N.I. Yumashev, A.A. Pronkin, A.A. Al'in, and L.V. Yumasheva, *Fiz. Khim. Stekla*, 1991, **17**, 197 (*Chem. Abstr.*, 1991, **115**, 15 987).
1756 M. Sokól, J. Grobelny, and M. Kowalczuk, *Magn. Reson. Chem.*, 1991, **29**, 1204.
1757 M.A. Desando, G. Lahajnar, I. Zupancic, and L.W. Reeves, *J. Mol. Liq.*, 1990, **47**, 171 (*Chem. Abstr.*, 1991, **114**, 129 820).
1758 M.T. Reetz, C.M. Niemeyer, and K. Harms, *Angew. Chem., Int. Ed. Engl.*, 1991, **30**, 1472.
1759 A. Zhang, *Huaxi Yike Daxue Xuebao*, 1991, **22**, 216 (*Chem. Abstr.*, 1991, **115**, 241 136).
1760 S.F. Tanner, V.J. Morris, and P.S. Belton, *Int. J. Biol. Macromol.*, 1990, **12**, 302 (*Chem. Abstr.*, 1991, **114**, 143 862).
1761 L. Piculell and C. Rochas, *Carbohydr. Res.*, 1990, **208**, 127.
1762 M.C. Steward, Y. Seo, M. Murakami, and H. Watari, *Proc. R. Soc. London, Ser. B*, 1991, **243**, 115 (*Chem. Abstr.*, 1991, **115**, 178 511).

ethers in acetonitrile has been reported.[1763]

Group IIa. ^1H n.m.r. spectroscopy has been used to study alkaline earth complexes with 18-crown-6.[1764] ^{13}C n.m.r. measurements of the complexation to cryptand C222 have been used to determine alkaline earth metal mixtures.[1765] ^1H pulsed gradient n.m.r. self-diffusion studies of the barium ethyl(octyl)phosphate-water system have been reported.[1766] K^+ and Ca^{2+} binding to calbindin D_{9k} has been studied by ^{39}K and ^{43}Ca n.m.r. spectroscopy.[1767] The interaction of Mg^{2+}, Al^{3+}, and F^- with G protein α subunits has been investigated by ^{19}F and ^{31}P n.m.r. spectroscopy.[1768] ^{31}P n.m.r. spectroscopy has been used to study the interaction of Mg^{2+} with inositol 1,4,5-bisphosphate.[1769] A ^{31}P n.m.r. study of the mechanism of, and salt effect on, magnesium ATP exchange has been reported.[1770] Magnesium binding to ATP and related compounds has been determined by total line shape analysis of ^{25}Mg n.m.r. spectra.[1771] ^{25}Mg n.m.r. spectroscopy has been used to investigate the coordination of Mg^{2+} to 5S rRNA.[1772]

Scandium. Scandium extraction by (2-Et-hexyl)$_2$phosphate has been studied by ^{31}P n.m.r. spectroscopy[1773] and by ^{31}P and ^{45}Sc n.m.r. spectroscopy.[1774]

Yttrium, Lanthanum, and the Lanthanides. ^1H and ^{15}N n.m.r. spectroscopy has been used to study the extraction mechanism of Ce^{IV} by a primary amine.[1775] The complexation of Dy^{3+} with α-alanine has been studied by ^1H n.m.r. spectroscopy.[1776] EDTA denticity for La^{3+} and $[VO_2]^+$ in the presence of Gd^{3+} has been studied by ^{13}C n.m.r. line broadening.[1777] ^1H and ^{13}C n.m.r. shift and relaxation measurements have shown that the lanthanide complexes of $\{Pr^nNHC(O)CH_2N(CH_2COOH)CH_2CH_2\}_2NCH_2COOH$ occur in isomeric forms.[1778] ^1H and ^{13}C T_1 measurements have been used to investigate complexes of $HO_2CCH_2N\{CH_2CH_2N(CH_2CO_2H)_2\}_2$ with La^{3+}, Pr^{3+}, and Eu^{3+}.[1779]

Exchange in $[Y(OSiPh_3)_3(OPBu^n{}_3)]$ and related complexes has been studied by ^{31}P n.m.r. spectroscopy.[1780] The complexation of Er^{3+} by $[NO_3]^-$ in aqueous solution has been studied by

1763 R.T. Streeper and S. Khazaeli, *Polyhedron*, 1991, **10**, 221.
1764 M.K. Amini and M. Shamsipur, *J. Phys. Chem.*, 1991, **95**, 9601.
1765 L.A. Fedorov, H. Yuan, M. Zuo, and L. Shen, *Bopuxue Zazhi*, 1991, **8**, 73 (*Chem. Abstr.*, 1991, **115**, 105 005).
1766 T. Yoshida, K. Miyagai, S. Aoki, K. Taga, and H. Okabayashi, *Colloid Polym. Sci.*, 1991, **269**, 713 (*Chem. Abstr.*, 1991, **115**, 143 810).
1767 S. Linse, C. Johansson, P. Brodin, T. Grundstroem, T. Drakenberg, and S. Forsén, *Biochemistry*, 1991, **30**, 154 (*Chem. Abstr.*, 1991, **114**, 19 644).
1768 T. Higashijima, M.P. Graziano, H. Suga, M. Kainosho, and A.G. Gilman, *J. Biol. Chem.*, 1991, **266**, 3396 (*Chem. Abstr.*, 1991, **114**, 159 376).
1769 A.M. White, M.A. Varney, S.P. Watson, S. Rigby, C. Liu, J.G. Ward, C.B. Reese, H.C. Graham, and R.J.P. Williams, *Biochem. J.*, 1991, **278**, 759 (*Chem. Abstr.*, 1991, **115**, 180 364).
1770 O.E. Bishop, R. Miles, and B.E. Smith, *Biochim. Biophys. Acta*, 1990, **1019**, 276 (*Chem. Abstr.*, 1991, **114**, 2892).
1771 J.A. Cowan, *Inorg. Chem.*, 1991, **30**, 2740.
1772 J.A. Cowan, *J. Am. Chem. Soc.*, 1991, **113**, 675.
1773 V.V. Shatalov, V.A. Lagutenkov, N.M. Meshcheryakov, and O.V. Chesalina, *Zh. Neorg. Khim.*, 1991, **36**, 199 (*Chem. Abstr.*, 1991, **115**, 264 463).
1774 V.Yu. Korovin, S.B. Randarevich, S.V. Bodaratskii, and V.V. Trachevskii, *Zh. Neorg. Khim.*, 1990, **35**, 2404 (*Chem. Abstr.*, 1991, **114**, 50 549).
1775 X. Zhang and D. Li, *Fenxi Huaxue*, 1991, **19**, 353 (*Chem. Abstr.*, 1991, **115**, 121 258).
1776 A.R. Mustafina, F.V. Devyatov, S.G. Abdullina, S.G. Vul'fson, and Yu.I. Sal'nikov, *Izv. Vyssh. Uchebn. Zaved., Khim. Khim. Tekhnol.*, 1990, **33**, 3 (*Chem. Abstr.*, 1991, **114**, 172 521).
1777 K. Popov, A. Vendilo, and N. Dyatlova, *Magn. Reson. Chem.*, 1991, **29**, 301.
1778 C.F.G.C. Geraldes, A.M. Urbano, M.C. Alpoim, M.A. Hoefnagel, and J.A. Peters, *J. Chem. Soc., Chem. Commun.*, 1991, 656.
1779 S. Aime and M. Botta, *Inorg. Chim. Acta*, 1991, **177**, 101.
1780 M.J. McGeary, P.S. Coan, K. Folting, W.E. Streib, and K.G. Caulton, *Inorg. Chem.*, 1991, **30**, 1723.

n.m.r. spectroscopy.[1781] ^{13}C n.m.r. spectroscopy has been used to investigate the complexation of Gd^{3+} by citrate.[1782] ^{1}H n.m.r. spectroscopy has been used to study the dynamic structure of lanthanide nitate complexes with 18-crown-6.[1783] ^{1}H n.m.r. titrations have been used to investigate complexation between β-cyclodextrin and β-benzyloxy-α-propionic acid derivatives of dota and dpta complexes of Gd^{3+}.[1784] Inner and outer sphere complexes of Y^{3+} with halide ions have been studied by ^{89}Y n.m.r. spectroscopy.[1785]

Uranium. The dynamics in the $[UO_2(CO_3)_3]^{4-}$-$[CO_3]^{2-}$ system have been studied using ^{13}C n.m.r. line-broadening.[1786] The complexes formed during U^{VI} extraction by (2-Et-hexyl)$_2$phosphate have been studied by ^{31}P n.m.r. spectroscopy.[1787] The UO_2F_2-H_2O-Me_2CO system has been studied by ^{1}H n.m.r. spectroscopy.[1788]

Titanium. ^{1}H, ^{13}C, and ^{27}Al n.m.r. spectroscopy has been used to study the interaction of $[(\eta^5-C_5H_5)_2Ti(CH_2SiMe_3)Cl]$ with $AlCl_3$ and provided evidence for the formation of $[(\eta^5-C_5H_5)_2Ti(CH_2SiMe_3)][AlCl_4]$.[1789] Nuclear magnetic relaxation has been used to study equilibria between Ti^{III}/Ti^{IV} or V^{III}/V^{IV} and tartrate.[1790]

Zirconium and Hafnium. The mechanism of extraction and separation of Zr^{IV} and Hf^{IV} by petroleum sulphoxide from hydrochloric acid has been studied by n.m.r. spectroscopy.[1791]

Vanadium. Complex formation between vanadate and amino acids and simple peptides has been studied by ^{51}V n.m.r. spectroscopy.[1792] ^{51}V n.m.r. spectroscopy has been used to study the formation and decomposition of peroxyvanadate complexes.[1793] The V^{V}-oxalate system has been investigated using ^{13}C, ^{17}O, and ^{51}V n.m.r. spectroscopy.[1794] Coordination site exchange in $[VO_2(C_2O_4)_2]^{3-}$ has been studied using ^{13}C n.m.r. spectroscopy.[1795] ^{17}O and ^{51}V n.m.r. spectroscopy has been used to investigate aqueous molybdovanadates.[1796] The effects of vanadate and vanadate complexes on the rates of exchange of phosphoryl groups catalysed by phosphoglucomutase have been determined by ^{31}P n.m.r. spectroscopy.[1797] Complex formation between vanadate and nucleosides has been studied by ^{1}H and ^{51}V n.m.r. spectroscopy.[1798] Vanadate interactions with bovine Cu,Zn-superoxide dismutase have been studied by ^{51}V n.m.r.

1781 N.V. Utyaganov, S.G. Vul'fson, F.V. Devyatov, and Yu.I. Sal'nikov, *Zh. Neorg. Khim.*, 1990, **35**, 2877 (*Chem. Abstr.*, 1991, **114**, 151 518).
1782 G.E. Jackson and J. du Toit, *J. Chem. Soc., Dalton Trans.*, 1991, 1463.
1783 S.P. Babailov, Yu.G. Kriger, and S.P. Gabuda, *Izv. Akad. Nauk SSSR, Ser. Khim.*, 1990, 2661 (*Chem. Abstr.*, 1991, **114**, 134 676).
1784 S. Aime, M. Botta, M. Panero, M. Grandi, and F. Uggeri, *Magn. Reson. Chem.*, 1991, **29**, 923.
1785 R. Takahashi and S. Ishiguro, *J. Chem. Soc., Faraday Trans.*, 1991, **87**, 3379.
1786 E. Brucher, J. Glaser, and I. Toth, *Inorg. Chem.*, 1991, **30**, 2239.
1787 V.A. Lagutenkov, M.V. Mikeev, and N.M. Meshcheryakov, *Koord. Khim.*, 1991, **17**, 1436 (*Chem. Abstr.*, 1991, **115**, 264 472).
1788 A.A. Lychev, V.A. Mikhalev, and D.N. Suglobov, *Radiokhimiya*, 1990, **32**, 25 (*Chem. Abstr.*, 1991, **114**, 89 301).
1789 J.J. Eisch, K.R. Caldwell, S. Werner, and C. Krüger, *Organometallics*, 1991, **10**, 3417.
1790 A.N. Glebov, S.M. Shavaleeva, O.Yu. Tarasov, and Yu. I. Sal'nikov, *Zh. Neorg. Khim.*, 1991, **36**, 1760 (*Chem. Abstr.*, 1991, **115**, 167 710).
1791 Q. Wang and Z. Zhang, *Huagong Yejin*, 1990, **11**, 317 (*Chem. Abstr.*, 1991, **114**, 172 488).
1792 J.S. Jaswal and A.S. Tracey, *Can. J. Chem.*, 1991, **69**, 1600.
1793 J.S. Jaswal and A.S. Tracey, *Inorg. Chem.*, 1991, **30**, 3718.
1794 P.M. Ehde, L. Pettersson, and J. Glaser, *Acta Chem. Scand.*, 1991, **45**, 998.
1795 M.H. Lee and K. Schaumburg, *Magn. Reson. Chem.*, 1991, **29**, 865.
1796 O.W. Howarth, L. Pettersson, and I. Andersson, *J. Chem. Soc., Dalton Trans.*, 1991, 1799.
1797 G.L. Mendz, *Arch. Biochem. Biophys.*, 1991, **291**, 201 (*Chem. Abstr.*, 1991, **115**, 274 377).
1798 A.S. Tracey and C.H. Leon-Lai, *Inorg. Chem.*, 1991, **30**, 3200.

spectroscopy.[1799]

Niobium. The formation of Nb^V alkoxides has been studied by n.m.r. spectroscopy.[1800]

Chromium. ^{13}C n.m.r. spectroscopy has been used to examine the complexation of $[(OC)_3Cr\{\eta^6\text{-}C_6H_5SiMe(OCH_2CH_2)_4O\}]$ to Li^+, Na^+, and K^+.[1801] ^1H and ^{13}C n.m.r. spectroscopy has been used to study outer sphere coordination of $[Cr(tropolonate)_3]$.[1802] The binding of Cr^{III} complexes to phosphate groups of ATP has been studied by ^{31}P n.m.r. spectroscopy.[1803]

Molybdenum and Tungsten. ^{11}B n.m.r. spectroscopy has been used to investigate the complexation of BEt_3 to $[(\eta^5\text{-}C_5H_5)WMe(CHO)(CO)_2]^-$. The ^{13}C n.m.r. spectrum was also given.[1804] Self exchange reactions of $[(\eta^5\text{-}C_5H_5)M(CO)_3X]/[(\eta^5\text{-}C_5H_5)M(CO)_3]^-$, M = Mo, W, couples have been measured by ^1H n.m.r. spectroscopy.[1805] The extraction of tungsten by long-chain aliphatic amines from fluoride solutions has been studied by ^{19}F n.m.r. spectroscopy.[1806] The complexation of D-glucaric acid with $[MoO_4]^{2-}$ and $[WO_4]^{2-}$ has been investigated using ^1H, ^{13}C, and ^{95}Mo n.m.r. spectroscopy.[1807] ^1H n.m.r. spectroscopy has been used to study the complexation of molybdate to salicylhydroxamic acid.[1808] ^{31}P n.m.r. spectroscopy has been used to investigate the effect of acidification on $[H_3PMo_{11}O_{39}]^{7-}$.[1809] Using ^{17}O, ^{31}P, and ^{183}W n.m.r. spectroscopy, the effect of pH on isopoly- and heteropolytungstates has been studied.[1810] A ^{17}O n.m.r. study of the composition of mixed molybdotungstate complexes in weakly acidic aqueous solutions has been studied.[1811]

Manganese. The effect of Mn^{2+} on ^{35}Cl linewidths of Cl^- binding to photosystem II has been investigated.[1812]

Technetium. Complexation in the $Sn^{II}\text{-}HO_2CCH_2N\{CH_2CH_2N(CH_2CO_2H)_2\}_2\text{-}[TcO_4]^-$ system has been studied by ^1H n.m.r. spectroscopy.[1813]

Rhenium. Variable temperature ^1H n.m.r. spectroscopy has provided information on the equilibrium between $[Re_3(\mu\text{-}H)_4(CO)_9\{C(CH_3)O\}]^{2-}$ and $[Re_3(\mu\text{-}H)_3(CO)_9\{CH(CH_3)O\}]^{2-}$.[1814] ^1H and ^{13}C n.m.r. spectra have provided evidence for an equilibrium between $[(\eta^5\text{-}$

1799 L. Wittenkeller, A. Abraha, R. Ramasamy, D.M. de Freitas, L.A. Theisen, and D.C. Crans, *J. Am. Chem. Soc.*, 1991, **113**, 7872.

1800 D.J. Eichorst, D.A. Payne, and A.N.A. Wragg, *Ceram. Trans.*, 1990, **11**, 375 (*Chem. Abstr.*, 1991, **114**, 10 960).

1801 M. Moran, C.M. Casado, I. Cuadrado, and J. Losada, *Inorg. Chim. Acta*, 1991, **185**, 33.

1802 A.N. Kitaigorodskii and U. Edlund, *Acta Chem. Scand.*, 1991, **45**, 534.

1803 A. Kortenkamp and P. O'Brien, *Carcinogenesis (London)*, 1991, **12**, 921 (*Chem. Abstr.*, 1991, **115**, 24 148).

1804 J.T. Gauntlett, B.E. Mann, M.J. Winter, and S. Woodward, *J. Chem. Soc., Dalton Trans.*, 1991, 1427.

1805 C.L. Schwarz, R.M. Bullock, and C. Creutz, *J. Am. Chem. Soc.*, 1991, **113**, 1225.

1806 M.A. Medkov, A.A. Smol'kov, B.N. Chernyshov, V.V. Kon'shin, and N.M. Laptash, *Zh. Neorg. Khim.*, 1990, **35**, 2421 (*Chem. Abstr.*, 1991, **114**, 50 550).

1807 M.L. Ramos, M.M. Caldeira, and V.M.S. Gil, *Inorg. Chim. Acta*, 1991, **180**, 219.

1808 Kh.T. Sharipov, G.A. Alimova, and N.N. Tananaeva, *Koord. Khim.*, 1991, **17**, 467 (*Chem. Abstr.*, 1991, **114**, 254 837).

1809 L.A. Combs-Walker and C.L. Hill, *Inorg. Chem.*, 1991, **30**, 4016.

1810 O.M. Kulikova, R.I. Maksimovskaya, S.M. Kulikov, and I.V. Kozhevnikov, *Izv. Akad. Nauk SSSR, Ser. Khim.*, 1991, 1726 (*Chem. Abstr.*, 1991, **115**, 246 499).

1811 R.I. Maksimovskaya, S.M. Maksimov, A.A. Blokhin, and V.P. Taushkanov, *Zh. Neorg. Khim.*, 1991, **36**, 1011 (*Chem. Abstr.*, 1991, **115**, 125 518).

1812 T. Wydrzynski, F. Baumgart, F. MacMillan, G. Renger, and T. Vaenngard, *Curr. Res. Photosynth., Proc. Int. Conf. Photosynth., 8th*, 1990, **1**, 749 (*Chem. Abstr.*, 1991, **115**, 68 650).

1813 V.V. Sergeev, T.A. Babushkina, A.V. Nikitina, R.A. Stukan, and G.E. Kodina, *Koord. Khim.*, 1991, **17**, 1215 (*Chem. Abstr.*, 1991, **115**, 269 277).

1814 T. Beringhelli, G. D'Alfonso, M. Freni, G. Ciani, M. Moret, and A. Sironi, *J. Organomet. Chem.*, 1991, **412**, C4.

C_5H_5)Re(NO)(PPh_3)($SiMe_2Cl$)] and $AlCl_3$ to give [(η^5-C_5H_5)Re(NO)(PPh_3)($SiMe_2$)][$AlCl_4$]. Other complexes such as [(η^5-C_5H_5)Re($NOBCl_3$)(PPh_3)($SiMe_2Cl$)] were formed and ^{11}B, ^{13}C, and ^{27}Al n.m.r. spectra were recorded.[1815]

Iron. Exchange between isomers of [(η^5-C_5H_5)(OC)(R_3P)Fe=CHMe]$^+$ has been investigated by ^1H and ^{13}C n.m.r. spectroscopy.[1816] ^1H, ^{13}C, and ^{29}Si has been used to study electron exchange between the ferrocene and ferrocinium forms of *syn*-[(η^5-C_5H_5)Fe(η^5-4,4,8,8-Me_4tetrahydro-4,8-disila-*s*-indacenediyl-η^5)Fe(η^5-C_5H_5)].[1817] The complexation of alkyldimethyl(ferrocenylmethyl)-ammonium salts by cyclodextrin has been studied by ^1H n.m.r. spectroscopy.[1818]

NOESY has been used to study (tetraphenylporphyrin)Fe^{III} complexes as a function of time and exchange of imidazole ligands.[1819] The nature of the aqueous solution axial ligands of [Fe{(1-Me-pyridinium-4-yl)$_4$porphyrin}]$^{5+}$ has been examined as a function of pH.[1820] ^1H two dimensional n.m.r. spectroscopy has been used to study chemical exchange in cytochrome c.[1821] The formation of Fe^{III}-edta complexes at high pH and ionic strength has been studied using nuclear magnetic relaxation dispersion measurements.[1822] The complexation of Fe^{II} to keratin hydrolysate has been studied by ^{13}C n.m.r. spectroscopy.[1823]

The site selection interaction of Fe^{III} and Cu^{II} with saccharides has been investigated by ^{13}C n.m.r. spectroscopy.[1824] The kinetics of electron exchange between [Fe(1,4,7-trithiacyclononane)$_2$]$^{2+}$ and [Fe(1,4,7-trithiacyclononane)$_2$]$^{3+}$ have been studied using ^1H and ^{13}C n.m.r. spectroscopy.[1825] The temperature and concentration dependence of the ^{27}Al n.m.r. spectra of $FeCl_3$-$AlCl_3$ have been reported.[1826]

Ruthenium. The pK_a of [(η^5-C_5H_5)RuH_2(dppe)]$^+$ has been determined by ^1H and ^{31}P n.m.r. spectroscopy.[1827] ^1H n.m.r. spectroscopy has been used to investigate the *cis-trans* equilibrium in [(η^3;η^3-$CH_2CMeCHCH_2CH_2CHCMeCH_2$)Ru($NCMe)_2$Cl]$^+$. The ^{13}C n.m.r. spectrum was also reported.[1828] ^1H n.m.r. spectroscopy has been used to investigate electron exchange between [(η^5-C_5H_5)$_2$Ru] and [(η^5-C_5H_5)$_2$Ru]$^+$.[1829] Facile Cl^-/[OH]$^-$ in [(η^6-C_6H_6)Ru(μ-Cl)(μ-$C_3H_3N_2$)$_2$Ru(η^6-C_6H_6)]$^+$ has been studied by ^1H n.m.r. spectroscopy.[1830] High pressure ^{13}C and ^{17}O n.m.r. spectroscopy has been used to show that [Ru(OH_2)$_6$]$^{2+}$ reacts quantitatively with CO (50 bar) in H_2O

1815 K.E. Lee, A.M. Arif, and J.A. Gladysz, *Chem. Ber.*, 1991, **124**, 309.
1816 M. Brookhart, Y. Liu, E.W. Goldman, D.A. Timmers, and G.D. Williams, *J. Am. Chem. Soc.*, 1991, **113**, 927.
1817 H. Atzkern, J. Hiermeier, F.H. Köhler, and A. Steck, *J. Organomet. Chem.*, 1991, **408**, 281.
1818 R. Isnin, C. Salam, and A.E. Kaifer, *J. Org. Chem.*, 1991, **56**, 35.
1819 F.A. Walker and U. Simonis, *J. Am. Chem. Soc.*, 1991, **113**, 8652.
1820 M.A. Ivanca, A.G. Lappin, and W.R. Scheidt, *Inorg. Chem.*, 1991, **30**, 711.
1821 Y. Feng, A.J. Wand, H. Roder, and S.W. Englander, *Biophys. J.*, 1991, **59**, 323 (*Chem. Abstr.*, 1991, **114**, 139 232).
1822 G. Hernandez, J. Rogalskyj, and R.G. Bryant, *J. Coord. Chem.*, 1991, **24**, 9 (*Chem. Abstr.*, 1991, **115**, 196 543).
1823 M. Hoshino, J.C. Song, T. Yoshida, and T. Uryu, *Kobunshi Ronunshu*, 1991, **48**, 341 (*Chem. Abstr.*, 1991, **115**, 50 769).
1824 K. Araki, *Kenkyu Hokoku - Asahi Garasu Zaidan*, 1990, **57**, 173 (*Chem. Abstr.*, 1991, **115**, 136 507).
1825 H. Doine and T.W. Swaddle, *Can. J. Chem.*, 1990, **68**, 2228.
1826 A.P. Shut'ko, L.I. Butchenko, Z.Z. Rozhkova, and V.N. Kvasko, *Zh. Fiz. Khim.*, 1990, **64**, 2372 (*Chem. Abstr.*, 1991, **114**, 50 579).
1827 G. Jia and R.H. Morris, *J. Am. Chem. Soc.*, 1991, **113**, 875.
1828 D.N. Cox, R.W.H. Small, and R. Roulet, *J. Chem. Soc., Dalton Trans.*, 1991, 2013.
1829 M. Watanabe and H. Sano, *Chem. Lett.*, 1991, 555 (*Chem. Abstr.*, 1991, **115**, 49 924).
1830 W.S. Sheldrick and H.S. Hagen-Eckhard, *J. Organomet. Chem.*, 1991, **410**, 73.

to form $[Ru(CO)(OH_2)_5]^{2+}$.[1831] Concentrated solutions of $[Ru(NO)(NO_2)_4(OH)]^{2-}$ and $[Ru(NO)(NO_2)_2(OH_2)(OH)]$ have been studied by ^{14}N, ^{15}N, ^{17}O, and ^{99}Ru.[1832]

Osmium. The kinetics of electron self-exchange of $[Os(CN)_6]^{4-}/[Os(CN)_6]^{3-}$ have been investigated by ^{13}C n.m.r. spectroscopy.[1833]

Cobalt. The coordination of amino acids to some water soluble Co^{III} porphyrins has been studied by ^1H n.m.r. spectroscopy.[1834] The association constants for ion pairs between cobalt sepulchrate and various counter anions have been studied by ^{59}Co n.m.r. spectroscopy.[1835] Co^{II}, Ni^{II}, Cu^{II}, Zn^{II}, Cd^{II}, Hg^{II}, and Pb^{II} complexation by 15 and 18 membered diaza crown ethers has been investigated by ^1H and ^{13}C n.m.r. spectroscopy.[1836] ^1H and ^{13}C n.m.r. spectroscopy has been used to investigate imidazole binding to Co^{II} substituted human carbonic anhydrase.[1837] The electron self-exchange rate constant for $[Co\{MeC(CH_2SCH_2CH_2SCH_2)_3CMe\}]^{2+}/[Co\{MeC(CH_2SCH_2CH_2SCH_2)_3CMe\}]^{3+}$ has been studied by ^1H and ^{13}C n.m.r. spectroscopy.[1838]

Iridium. The equilibrium between $[IrCl_6]^{3-}$ and $[NO_2]^-$ has been studied using ^{15}N n.m.r. spectroscopy.[1839]

Nickel. Ni(II) carbonic anhydrase and its adducts with $[NO_3]^-$, $[MeCO_2]^-$, $[NCO]^-$, and $[N_3]^-$ have been investigated by ^1H n.m.r. spectroscopy and binding constants determined. T_1 was determined and the assignments made with the aid of n.O.e. measurements.[1840] ^1H n.m.r. spectroscopy has been used to study biphasic extraction of Ni^{2+}.[1841] Enhanced Ni^{II} chelation by gem-dimethyl substituted macrocyclis thioethers has been studied using ^1H n.m.r. spectroscopy.[1842]

Palladium and Platinum. The binding constants for amines to $[PdMe(dmpe)L]^+$ have been determined by ^{31}P n.m.r. spectroscopy.[1843] Complexation between $[PdCl_4]^{2-}$ and $[SnCl_3]^-$ has been studied by ^{119}Sn n.m.r. spectroscopy.[1844] ^1H and ^2H n.m.r. spectroscopy has been used to study the Pd^{II} catalysed equilibrium[1845]

$$Me_2C{=}CHC(CD_3)(CH_3)OH + MeOH \rightleftharpoons MeO(H_3C)_2CCH{=}C(CD_3)(CH_3) + H_2O$$

The reaction of *cis*-$[Pt(NH_3)_2(OH_2)_2]^{2+}$ with creatinine has been studied by ^{13}C and ^{195}Pt n.m.r. spectroscopy.[1846] ^1H, ^{15}N, and ^{195}Pt n.m.r. spectroscopy has been used to study the reaction of *cis*-$[Pt\{1\text{-}(2''\text{-}HOCH_2CH_2)\text{-}2\text{-}Me\text{-}5\text{-}nitroimidazole\}_2Cl_2]$ with guanosine monophosphate and some

1831 G. Laurenczy, L. Helm, A. Ludi, and A.E. Merbach, *Helv. Chim. Acta*, 1991, **74**, 1236.
1832 M.A. Fedotov and A.V. Belyaev, *Koord. Khim.*, 1991, **17**, 103 (*Chem. Abstr.*, 1991, **114**, 177 237).
1833 D.H. Macartney, *Inorg. Chem.*, 1991, **30**, 3337.
1834 E. Mikros, F. Gaudemer, and A. Gaudemer, *Inorg. Chem.*, 1991, **30**, 1806.
1835 J. Sotomayor, H. Santos, and F. Pina, *Can. J. Chem.*, 1991, **69**, 567.
1836 J. Kim, C.J. Yoon, H.B. Park, and S.J. Kim, *J. Korean Chem. Soc.*, 1991, **35**, 119 (*Chem. Abstr.*, 1991, **115**, 100 431).
1837 C. Luchinat, R. Monnanni, and M. Sola, *Inorg. Chim. Acta*, 1991, **177**, 133.
1838 P. Osvath, A.M. Sargeson, B.W. Skelton, and A.H. White, *J. Chem. Soc., Chem. Commun.*, 1991, 1036.
1839 A.B. Venediktov, S.V. Korenev, M.A. Fedotov, and A.V. Belysaev, *Koord. Khim.*, 1990, **16**, 1400 (*Chem. Abstr.*, 1991, **114**, 74 109).
1840 J.M. Moratal, M.-J. Martinez-Ferrer, A. Donaire, J. Castells, J. Salgado, and H.R. Jiménez, *J. Chem. Soc., Dalton Trans.*, 1991, 3393.
1841 C. Tondre and D. Canet, *J. Phys. Chem.*, 1991, **95**, 4810.
1842 J.M. Desper, S.H. Gellman, R.E. Wolf, jun., and S.R. Cooper, *J. Am. Chem. Soc.*, 1991, **113**, 8663.
1843 A.L. Seligson and W.C. Trogler, *J. Am. Chem. Soc.*, 1991, **113**, 2520.
1844 R.Kh. Karymova, L. Ya. Al't, I.A. Agapov, and V.K. Duplyakin, *Zh. Neorg. Khim.*, 1991, **36**, 469.
1845 C.M. Dumiao, J.W. Francis, and P.M. Henry, *Organometallics*, 1991, **10**, 1400.
1846 C.F.G.C. Geraldes, M. Aragon-Salgado, and J. Martin-Gil, *Polyhedron*, 1991, **10**, 799.

oligonucleotides.[1847] The formation of $[PtCl_n(NO_2)_{4-n}]^{2-}$ from $[PtCl_4]^{2-}$ and $[NO_2]^-$ has been studied by ^{15}N and ^{195}Pt n.m.r. spectroscopy.[1848] The interaction between Pd^{II} and L-citrulline has been investigated by ^1H and ^{13}C n.m.r. spectroscopy.[1849] A variable pressure ^1H n.m.r. study of ligand exchange on $[M(S_2C_4H_4)_2]^{2+}$, M = Pd, Pt, and $[Pt(SMe_2)_4]^{2+}$ has been investigated and $\Delta H^\ddagger$, $\Delta S^\ddagger$, and $\Delta V^\ddagger$ determined.[1850]

Copper. Low temperature ^1H, ^{13}C, ^{29}Si, and ^{119}Sn n.m.r. spectroscopy has been used to study the species present when CuCN, MeLi, and R_3MLi, M = Si, Sn, are mixed.[1851] A similar ^1H, ^{13}C, and ^{29}Si n.m.r. study has been made of the CuCN, MeLi, and $PhMe_2SiLi$ system.[1852] ^1H nuclear magnetic relaxation has been used to study Cu^{II} binding to NH_3, $H_2NCH_2CH_2NH_2$, and triethylene tetraammine.[1853] The electron self-exchange rate of $[Cu\{(5\text{-MeimidH})_2DAP\}]^+/[Cu\{5\text{-}(MeimidH)_2DAP\}]^{2+}$ has been determined as a function of temperature by dynamic n.m.r. spectroscopy.[1854] Equilibrium constants for the binding of pyridine and similar ligands to $[Cu(O_2CCH_3)_2]$ have been determined by ^1H n.m.r. spectroscopy.[1855] The rate constants of ligand exchange reactions of the complexes of Cu^{II} with amino acids have been determined by ^1H n.m.r. line-broadening studies.[1856] The binding of Cu^{2+} to heparin has been studied by ^1H and ^{13}C n.m.r. spectroscopy.[1857] The value of the self-exchange electron transfer rate constant for $[Cu([15]aneS_5]^+/[Cu([15]aneS_5]^{2+}$ has been determined using ^1H line broadening. ^{13}C n.m.r. data were also reported.[1858]

Silver. The binding of Ag^+, Hg^{2+}, and Pb^{2+} to several thiacrown ethers has been studied using ^{13}C T_1 measurements.[1859]

Zinc, Cadmium, and Mercury. ^{11}B and ^{113}Cd n.m.r. spectroscopy has been used to study the CdI_2-MBH_4 system.[1860] ^1H and ^{13}C n.m.r. spectroscopy has been used to investigate the formation of $[RZnL][AlR_4]$ from R_2Zn, R_3Al, and 1,4,8,11-Me_4-1,4,8,11-diazacyclotetradecane or 211 cryptand.[1861] The formation constants for the complexation of $[MeHg]^+$ with captopril have been determined by ^1H and ^{13}C n.m.r. spectroscopy.[1862] ^1H and ^{13}C n.m.r. spectroscopy has been used to study complex formation between (115) and $ZnCl_2$.[1863] ^1H n.m.r. spectroscopy has been used to investigate the equilibrium between $Zn^{2+}$with alaninehydroxamic acid, leucine, sarcosine, and

1847 F.M. Macdonald and P.J. Sadler, *Magn. Reson. Chem.*, 1991, **29**, S52.
1848 M.A. Fedotov, S.V. Korenev, and A.V. Belyaev, *Koord. Khim.*, 1990, **16**, 1272 (*Chem. Abstr.*, 1991, **114**, 74 190).
1849 M.L. Ganadu, V. Leoni, G. Crisponi, and V. Nurchi, *Polyhedron*, 1991, **10**, 333.
1850 U. Frey, S. Elmroth, B. Moullet, L.I. Elding, and A.E. Merbach, *Inorg. Chem.*, 1991, **30**, 5033.
1851 S. Sharma and A.C. Oehlschlager, *J. Org. Chem.*, 1991, **56**, 770.
1852 R.D. Singer and A.C. Oehlschlager, *J. Org. Chem.*, 1991, **56**, 3510.
1853 L.S. Szczepaniak and R.G. Bryant, *Inorg. Chim. Acta*, 1991, **184**, 7.
1854 D.K. Coggin, J.A. Gonzalez, A.M. Kook, C. Bergman, T.D. Brennan, W.R. Scheidt, D.M. Stanbury, and L.J. Wilson, *Inorg. Chem.*, 1991, **30**, 1125.
1855 A.L. Abuhijleh and I.Y. Ahmed, *Polyhedron*, 1991, **10**, 793.
1856 Y. Li, Q. Zhang, and R. Chen, *Wuli Huaxue Xuebao*, 1991, **7**, 82 (*Chem. Abstr.*, 1991, **115**, 240 952).
1857 R.N. Rej, K.R. Holme, and A.S. Perlin, *Carbohydr. Res.*, 1990, **207**, 143 (*Chem. Abstr.*, 1991, **114**, 102 636).
1858 A.M.Q. Vande Linde, K.L. Juntunen, O. Mols, M.B. Ksebati, L.A. Ochrymowycz, and D.B. Rorabacher, *Inorg. Chem.*, 1991, **30**, 5037.
1859 G. Wu, W. Jiang, J.D. Lamb, J.S. Bradshaw, and R.M. Izatt, *J. Am. Chem. Soc.*, 1991, **113**, 6538.
1860 H. Nöth and M. Thomann, *Z. Naturforsch., B*, 1990, **45**, 1472 (*Chem. Abstr.*, 1991, **114**, 74 134).
1861 R.M. Fabicon, A.D. Pajerski, and H.G. Richey, jun., *J. Am. Chem. Soc.*, 1991, **113**, 6680.
1862 A.A. Isab, *J. Chem. Soc., Dalton Trans.*, 1991, 449.
1863 C. Bolm, K. Weickhardt, M. Zehnder, and T. Ranff, *Chem. Ber.*, 1991, **124**, 1173.

histidine.[1864] The interaction of *N*-(phenylsulphonyl)glycine and *N*-(tolylsulphonyl)glycine with Zn^{2+} and Cd^{2+} has been studied by ^1H n.m.r. spectroscopy.[1865] A ^1H n.m.r. titration of (116) with Zn^{2+}, Cd^{2+}, and Hg^{2+} gave evidence for 1:1 and 1:2 species.[1866] The ^1H n.m.r. temperature dependence of Zn^{II} and Cd^{II} complexes of 1,2-propanediaminetetraacetic acid has been studied.[1867] The dissociation of $[Zn\{S_2P(OR)_2\}_2]$ has been studied by ^{31}P n.m.r. spectroscopy.[1868]

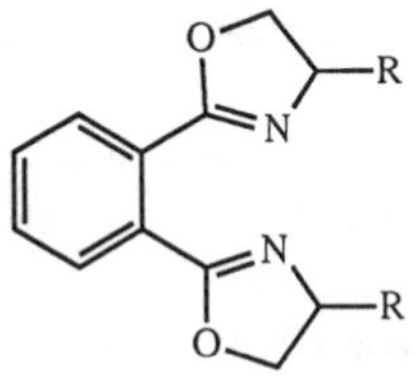

(115)

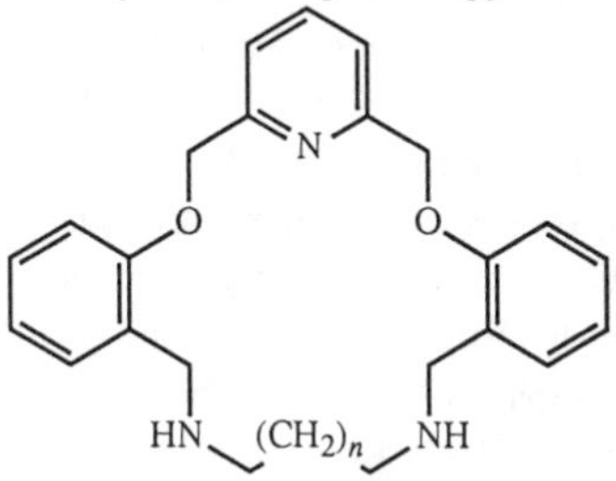

(116)

Boron. ^1H ROESY has been used to investigate ion pairing between (117) and $[BH_4]^-$.[1869] The binding constants between *closo*-$[3,3,3\text{-}(OC)_3\text{-}3,1,2\text{-}MC_2B_9H_{11}]^-$, M = Mn, Re, with α- and β-cyclodextrin have been determined by ^1H n.m.r. spectroscopy.[1870] The formation constants of peroxyborates have been determined by ^{11}B n.m.r. spectroscopy.[1871] ^{11}B n.m.r. spectroscopy has been used to investigate equilibria between borate and three phenolic carboxylic acids to give borate esters.[1872] ^{11}B and ^{31}P n.m.r. spectroscopy has been used to study borate complexing with hydroxyethylidenediphosphonic acid.[1873] Aluminosilicate and borosilicate species have been identified in solution by ^{11}B, ^{27}Al, and ^{29}Si n.m.r. spectroscopy.[1874]

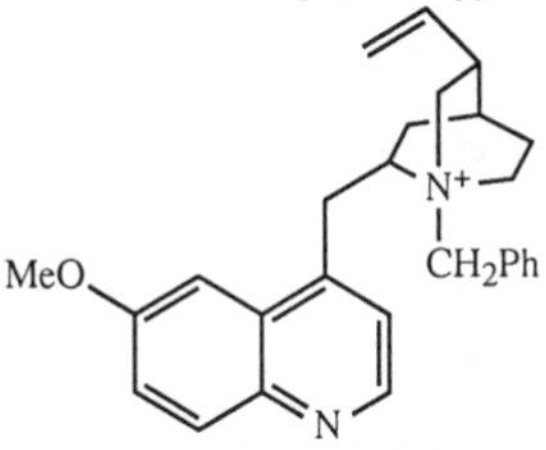

(117)

Aluminium. The composition of solutions prepared from $LiAlH_4$ and $AlCl_3$ has been studied using

1864 B. Kurzak, H. Koslowski, and P. Decock, *J. Inorg. Biochem.*, 1991, **41**, 71.
1865 G.G. Battistuzzi, M. Borsari, L. Menabue, M. Saladini, and M. Sola, *Inorg. Chem.*, 1991, **30**, 498.
1866 H. Adams, N.A. Bailey, D.E. Fenton, I.G. Ford, S.J. Kitchen, M.G. Williams, P.A. Tasker, A.J. Leong, and L.F. Lindoy, *J. Chem. Soc., Dalton Trans.*, 1991, 1665.
1867 F. Li and R. Song, *Magn. Reson. Chem.*, 1991, **29**, 735.
1868 Q. Lu, X. Yang, and S. Zhu, *Yingyong Huaxue*, 1990, **7**, 54 (*Chem. Abstr.*, 1991, **114**, 65 426).
1869 T.C. Pochapsky, P.M. Stone, and S.S. Pochapsky, *J. Am. Chem. Soc.*, 1991, **113**, 1460.
1870 P.A. Chetcuti, P. Moser, and G. Rihs, *Organometallics*, 1991, **10**, 2895.
1871 B.N. Chernyshov, *Zh. Neorg. Khim.*, 1990, **35**, 2341 (*Chem. Abstr.*, 1991, **114**, 74 056); B.N. Chernyshov, A.Yu. Vaseva, and O.V. Brovkina, *Zh. Neorg. Khim.*, 1991, **36**, 460 (*Chem. Abstr.*, 1991, **114**, 220 165).
1872 C.F. Bell, B.C. Gallagher, K.A.K. Lott, E.L. Short, and L. Walton, *Polyhedron*, 1991, **10**, 613.
1873 N.A. Kostromina, I.N. Tret'yakova, L.B. Novikova, and V.V. Trachevskii, *Ukr. Khim. Zh. (Russ. Ed.)*, 1991, **57**, 227 (*Chem. Abstr.*, 1991, **115**, 121 284); N.A. Kostromina, I.N. Tret'yakova, L.B. Novikova, and V.V. Trachevskii, *Ukr. Khim. Zh. (Russ. Ed.)*, 1990, **56**, 899 (*Chem. Abstr.*, 1991, **114**, 130 381).
1874 R.F. Mortlock, A.T. Bell, and C.J. Radke, *J. Phys. Chem.*, 1991, **95**, 372.

^{27}Al n.m.r. spectroscopy.[1875] The complexation of Al^{III} by $HOCH_2CH_2OH$ and $H_2NCH_2CH_2NH_2$ in the presence of F^- has been investigated using ^{19}F n.m.r. spectroscopy.[1876] ^{13}C n.m.r. spectroscopy has been used to study the complexing of Al^{III} by diethylenetriaminopentaacetate.[1877] ^{27}Al n.m.r. spectroscopy has been used to investigate the binding of $[Al(OH)_4]^-$ and $[Al(OH_2)_6]^{3+}$ in mixed amphiphilic liquid crystalline bilayer systems.[1878] The hydrolysis of aluminates in aqueous peroxide solutions has been studied by ^{19}F and ^{27}Al n.m.r. spectroscopy.[1879] Some new aluminium polyoxycations have been characterised by ^{27}Al n.m.r. spectroscopy.[1880] The composition of alumina sol-gels has been studied using ^{27}Al n.m.r. spectroscopy.[1881] ^{27}Al n.m.r. spectroscopy has been used to carry out speciation on Al^{3+} with acetate and oxalate.[1882]

The binding isomerism of Al^{3+}-phosphinate, -phosphite, and -phosphate complexes has been investigated by ^{31}P n.m.r. spectroscopy.[1883] The interaction of Al^{3+} and phosphate[1884] and cyclopolyphosphates[1885] has been studied by ^{27}Al and ^{31}P n.m.r. spectroscopy. The binding of $[AlF_4]^-$ to nucleoside diphosphates has been investigated using ^1H and ^{19}F n.m.r. spectroscopy.[1886]

Equilibria in sodium aluminosilicate solution have been investigated using ^{27}Al n.m.r. spectroscopy.[1887] The effect of silicate ratio on the distribution of silicate and aluminosilicate anions has been investigated by ^{27}Al and ^{29}Si n.m.r. spectroscopy.[1888] The incorporation of aluminium into silicate anions has been investigated by ^{29}Si n.m.r. spectroscopy.[1889] Retarding and accelerating effects of Al^{3+} on the growth of polysilicic acid particles have been studied by ^{27}Al n.m.r. spectroscopy.[1890] Fluorine-, silicon-, and phosphorus-containing complexes in wet-process phosphoric acid have been investigated using ^{19}F and ^{31}P n.m.r. spectroscopy.[1891]

N.m.r. spectroscopy has been used to study the effect of dilution with water on the speciation of polymeric aluminium chloride.[1892] The removal of protons from ambient-temperature chloroaluminate

1875 V.A. Mazin, V.A. Kazakov, and V.N. Titova, *Elektrokhimiya*, 1991, **27**, 417 (*Chem. Abstr.*, 1991, **114**, 255 779).
1876 S.P. Petrosyants and E.R. Buslaeva, *Koord. Khim.*, 1991, **17**, 161 (*Chem. Abstr.*, 1991, **115**, 20 864).
1877 K.I. Popov, N.E. Kovaleva, A.G. Vendilo, and N.M. Dyatlova, *Zh. Neorg. Khim.*, 1991, **36**, 1021 (*Chem. Abstr.*, 1991, **115**, 58 289).
1878 M. Iida and A.S. Tracey, *Langmuir*, 1991, **7**, 202.
1879 V.V. Kon'shin and B.N. Chernyshov, *Koord. Khim.*, 1990, **16**, 1314 (*Chem. Abstr.*, 1991, **114**, 172 500).
1880 G. Fu, L.F. Nazar, and A.D. Bain, *Chem. Mater.*, 1991, **3**, 602 (*Chem. Abstr.*, 1991, **115**, 55 393).
1881 L.F. Nazar, D.G. Napier, D. Lapham, and E. Epperson, *Mater. Res. Soc. Symp. Proc.*, 1990, **180**, 117 (*Chem. Abstr.*, 1991, **115**, 285 483).
1882 F. Thomas, A. Masion, J.Y. Bottero, J. Rouiller, F. Genevrier, and D. Boudot, *Environ. Sci. Technol.*, 1991, **25**, 1553 (*Chem. Abstr.*, 1991, **115**, 98 828).
1883 Q. Feng and H. Waki, *Polyhedron*, 1991, **10**, 659.
1884 M.A. Fedotov, V.P. Shmachkova, and N.S. Kotsarenko, *Izv. Sib. Otd. Akad. Nauk SSSR, Ser. Khim. Nauk*, 1990, 80 (*Chem. Abstr.*, 1991, **114**, 215 680).
1885 Q. Feng, H. Waki, and G. Kura, *Polyhedron*, 1991, **10**, 1527.
1886 D.J. Nelson and R.B. Martin, *J. Inorg. Biochem.*, 1991, **43**, 37.
1887 B. Fahlke, W. Wieker, D. Mueller, K. Kintscher, W. Roscher, P. Knop, G. Nemitz, W. Hoese, and H. Fuertig, **Ger. (East) DD 279 661** (Cl. C01B33/28), 13 Jun 1990, Appl. 297 086, 05 Dec 1986 (*Chem. Abstr.*, 1991, **114**, 12 807).
1888 R.F. Mortlock, A.T. Bell, A.K. Chakraborty, and C.J. Radke, *J. Phys. Chem.*, 1991, **95**, 4501.
1889 R.F. Mortlock, A.T. Bell, and C.J. Radke, *J. Phys. Chem.*, 1991, **95**, 7847.
1890 T. Yokoyama, Y. Takahashi, and T. Tarutani, *J. Colloid Interface Sci.*, 1991, **141**, 559 (*Chem. Abstr.*, 1991, **114**, 172 135).
1891 V.M. Norwood, tert. and J.J. Kohler, *Fert. Res.*, 1991, **28**, 221 (*Chem. Abstr.*, 1991, **115**, 135 031).
1892 C. Yao, Y. Sun, P. Zhao, and L. Fei, *Huanjing Huaxue*, 1991, **10**, 1 (*Chem. Abstr.*, 1991, **115**, 141 904); C. Yao, Y. Sun, P. Zhao, and L. Fei, *Water Treat.*, 1991, **6**, 163 (*Chem. Abstr.*, 1991, **115**, 239 195).

ionic liquids has been investigated by ^{1}H and ^{13}C n.m.r. spectroscopy.[1893] ^{17}O n.m.r. spectroscopy has been used to assess the behaviour of oxide in the molten salts composed of $AlCl_3$ and 1-Et-3-Me imidazolium chloride.[1894]

Gallium and Indium. Gallium, indium, and yttrium complexes of 1,4,7-triazacyclononane-1,4,7-triacetic acid have been studied by ^{13}C and ^{71}Ga n.m.r. spectroscopy.[1895] The ^{71}Ga n.m.r. spectrum of hydrolysed solutions of Ga^{3+} shows a signal at δ 173.2 relative to $[Ga(OH_2)_6]^{3+}$ which was ascribed to a Ga_{13} species analogous to Al_{13}.[1896]

The extraction of indium with 1,10-(1-Ph-3-Me-5-HO-4-pyrazolyl)-1,10-decanedione and $(C_8H_{17})_3P$ has been investigated using ^{31}P n.m.r. spectroscopy.[1897]

Thallium. The ^{1}H n.m.r. spectrum of $[(tetraphenylporphyrin)Tl(O_2CCH_3)]$ shows acetate exchange of 32.65 s^{-1} at -57 °C. The ^{13}C n.m.r. spectrum was also reported.[1898]

Silicon. Bond ordering in silicate glasses has been investigated by n.m.r. spectroscopy.[1899] The effects of $[R_4N]^+$ on the equilibrium distribution of silicate oligomers in aqueous alkaline silicate solutions have been investigated by ^{29}Si n.m.r. spectroscopy.[1900]

Germanium. The complexation of Ge^{4+} by edta has been studied by ^{13}C n.m.r. spectroscopy in the presence of Gd^{3+}.[1901]

Tin. ^{1}H and ^{119}Sn n.m.r. spectroscopy has been used to investigate base hydrolysis of $MeSnCl_3$ and $BuSnCl_3$.[1902] Equilibria of some Sn^{II} complexes of sulphate, chloride, perchlorate, and fluoride have been studied by ^{119}Sn n.m.r. spectroscopy.[1903]

Nitrogen. ^{1}H n.m.r. saturation transfer measurements have been applied to $[^{15}NH_4]^+$ in H_2O-D_2O and the secondary isotope effect for proton dissociation determined.[1904] Cation-cation association of $[Me_4N]^+$ has been investigated using ^{1}H and ^{2}H T_1 and self-diffusion measurements.[1905]

Phosphorus. The equilibrium

$$HF + HPO_3H_2 \rightleftharpoons H_2O + FHPO_2H$$

has been investigated by n.m.r. spectroscopy.[1906]

Bismuth. ^{1}H and ^{13}C n.m.r. spectroscopy has been used to investigate the formation of Bi^{III}-citrate complexes.[1907]

1893 M.A.M. Noel, P.C. Trulove, and R.A. Osteryoung, *Anal. Chem.*, 1991, **63**, 2892; S.G. Park, P. Trulove, R.T. Carlin, and R.A. Osteryoung, *Proc. - Electrochem. Soc.*, 1990, **90**, 290.
1894 R.A. Osteryoung, *Report*, 1990, **AFOSR-TR-90-0084**; Order No. AD-A217742, 39 pp. Avail. NTIS. From *Gov. Rep. Announce. Index (U.S.)*, 1990, **90**, Abstr. No. 029,938 (*Chem. Abstr.*, 1991, **114**, 236 485).
1895 C. Broan, J.P. Cox, A.S. Craig, R. Kataky, D. Parker, A. Harrison, A.M. Randall, and G. Ferguson, *J. Chem. Soc., Perkin Trans. 2*, 1991, 87.
1896 L.F. Nazar, S.W. Liblong, and X.T. Yin, *J. Am. Chem. Soc.*, 1991, **113**, 5889.
1897 A. Tayeb, G.J. Goetz-Grandmont, and J.P. Brunette, *Monatsh. Chem.*, 1991, **122**, 453.
1898 J.-C. Chen, H.-S. Jang, and J.-H. Chen, *Polyhedron*, 1991, **10**, 2069.
1899 S.J. Gurman, *J. Non-Cryst. Solids*, 1990, **125**, 151 (*Chem. Abstr.*, 1991, **114**, 12 610).
1900 W.M. Hendricks, A.T. Bell, and C.J. Radke, *J. Phys. Chem.*, 1991, **95**, 9513; W.M. Hendricks, A.T. Bell, and C.J. Radke, *J. Phys. Chem.*, 1991, **95**, 9519; E. Blen, E. Lippmaa, M. Magi, G.I. Agafonov, and V.I. Korneev, *Zh. Prikl. Khim. (Leningrad)*, 1990, **63**, 1636 (*Chem. Abstr.*, 1991, **114**, 16 767).
1901 K.I. Popov, I.I. Seifullina, and T.P. Batalova, *Koord. Khim.*, 1991, **17**, 452 (*Chem. Abstr.*, 1991, **114**, 254 836).
1902 S.J. Blunden and R. Hill, *Inorg. Chim. Acta*, 1991, **177**, 219.
1903 Yu.V. Kokunov and I.E. Rakov, *Koord. Khim.*, 1990, **16**, 1616 (*Chem. Abstr.*, 1991, **114**, 130 383).
1904 C.L. Perrin and R.E. Engler, *J. Phys. Chem.*, 1991, **95**, 8431.
1905 M. Holz and K.J. Patil, *Ber. Bunsenges. Phys. Chem.*, 1991, **95**, 107.
1906 J.W. Larson, *Polyhedron*, 1991, **10**, 1695.
1907 E. Asato, W.L. Driessen, R.A.G. de Graaff, F.B. Hulsbergen, and J. Reedijk, *Inorg. Chem.*, 1991, **30**, 4210.

Tellurium. ^{125}Te n.m.r. spectroscopy has been used to study exchange and redox reactions of organotellurium(IV) dithiolate and organotellurium(II) complexes. ^{13}C n.m.r. spectra were also recorded.[1908] Hydrolysis equilibria of $[TeCl_6]^{2-}$ in HCl have been measured by ^{125}Te n.m.r. spectroscopy.[1909]
Halides. The anion binding properties of poly(vinylpyrrolidone) have been investigated by ^{35}Cl, ^{81}Br, and ^{127}I n.m.r. spectroscopy.[1910] Chloride binding to sarcoplasmic reticulum membranes has been studied by ^{35}Cl n.m.r. spectroscopy.[1911] A ^{35}Cl and ^{37}Cl n.m.r. study of chloride binding to the etythrocyte anion transport protein has been published.[1912]

Equilibria among Uncharged Species. Lithium. ^{7}Li, ^{11}B and ^{13}C n.m.r. spectroscopy has been used to demonstrate an equilibrium between 1-Me_2B-2-$Me_2NCH_2C_6H_4$ and $[LiMe]_4$ to give species such as (118).[1913]

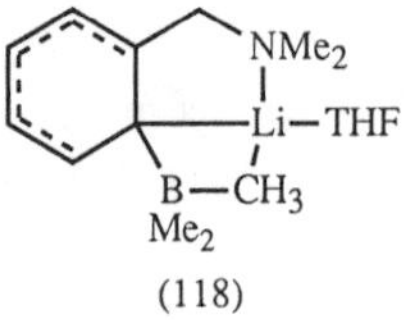

(118)

Magnesium. The Schlenk equilibrium has been investigated by ^{1}H and ^{13}C n.m.r. spectroscopy for some aryl magnesium compounds.[1914] ^{1}H and ^{13}C n.m.r. spectroscopy has been used to determine the rate of exchange of free and coordinated ether on PhMgBr:CuI reagent.[1915]
Lanthanides. Two-dimensional ^{13}C n.m.r. spectroscopy has been used to demonstrate $\eta^3 \leftrightarrow \eta^4$ equilibria in compounds formed from isoprene and the $[(CF_3CO_2)LnCl]$-EtOH-Bu^i_2AlH catalyst.[1916]
Uranium. The complexation of (119) with neutral molecules has been assessed by ^{1}H n.m.r. titration.[1917]
Titanium and Zirconium. The mechanism of interaction of alkylzirconocene dichlorides with methylalumoxane in the presence of ethene has been studied by ^{1}H and ^{13}C n.m.r. spectroscopy.[1918] Variable temperature ^{1}H and ^{31}P n.m.r. spectroscopy has been used to demonstrate a monomer$\leftrightarrow$dimer equilibrium for $[(\eta^5\text{-}C_5H_5)_2Ti(PEt_2)]$. The ^{13}C n.m.r. spectrum of $[(\eta^5\text{-}C_5H_5)_2Zr(PEt_2)]_2$ was also reported.[1919] Ligand exchange on adducts of $[TiCl_3(OR)]$ with ketones

1908 A.M. Bond, D. Dakternieks, R. Di Goacomo, and A.F. Hollenkamp, *Organometallics*, 1991, **10**, 3310.
1909 J.B. Milne, *Can. J. Chem.*, 1991, **69**, 987.
1910 J.D. Song, R. Ryoo, and M.S. Jhon, *Macromolecules*, 1991, **24**, 1727 (*Chem. Abstr.*, 1991, **114**, 165 420).
1911 I.R. Vetter, H. Hanssum, and H.G. Baeumert, *Biochem. Biophys. Acta*, 1991, **1067**, 9 (*Chem. Abstr.*, 1991, **115**, 226 537).
1912 W.S. Price, P.W. Kuchel, and B.A. Cornell, *Biophys. Chem.*, 1991, **40**, 329 (*Chem. Abstr.*, 1991, **115**, 153 345).
1913 E. Kalbarczyk-Bidelska and S. Pasynkiewicz, *J. Organomet. Chem.*, 1991, **417**, 1.
1914 P.R. Markies, R.M. Altink, A. Villena, O.S. Akkerman, F. Bickelhaupt, W.J.J. Smeets, and A.L. Spek, *J. Organomet. Chem.*, 1991, **402**, 289.
1915 S.K. Nahar, H.P. Nur, M.T. Rahman, M. Nilsson, and T. Olsson, *J. Organomet. Chem.*, 1991, **408**, 261.
1916 X. Zhang, F. Pei, X. Li, Y. Jin, J. Ding, and S. Zhang, *Gaofenzi Xuebao*, 1990, 391 (*Chem. Abstr.*, 1991, **114**, 247 830).
1917 A.R. Van Doorn, M. Bos, S. Harkema, J. Van Eerden, W. Verboom, and D.N. Reinboudt, *J. Org. Chem.*, 1991, **56**, 2371.
1918 L.A. Nekhaeva, B.A. Krentsel, V.P. Mar'in, I.M. Khrapova, V.L. Khodzhaeva, A.I. Mikaya, and S.I. Ganicheva, *Neftekhimiya*, 1991, **31**, 209 (*Chem. Abstr.*, 1991, **115**, 115 144).
1919 D.G. Dick and D.W. Stephan, *Can. J. Chem.*, 1991, **69**, 1146.

has been studied by ^{1}H and ^{13}C n.m.r. spectroscopy.[1920]

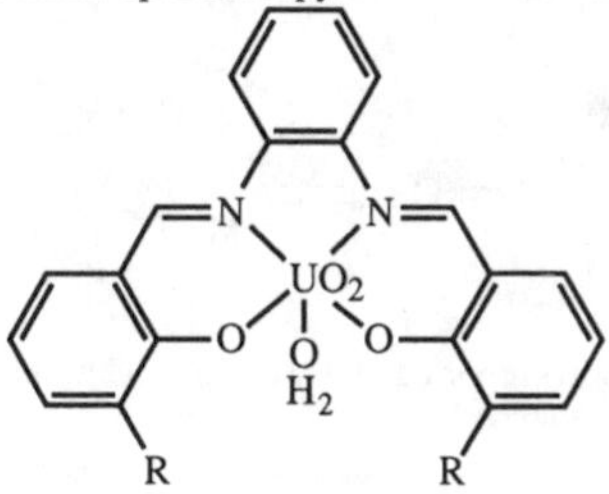

(119)

Vanadium. ^{1}H, ^{13}C and ^{51}V n.m.r. spectroscopy has been used to study the interaction of [$VOCl_2(OR)$] with ketones.[1921]

Niobium. ^{1}H n.m.r. spectroscopy has been used to demonstrate the equilibrium

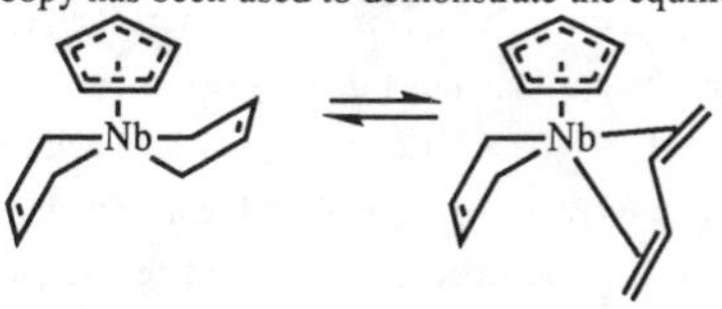

and $\Delta G^{\ddagger}$ was determined as 80 ± 2 kJ mol^{-1}. The ^{13}C n.m.r. spectrum was also reported.[1922]

Chromium. Metal-metal bond energies in [$(\eta^5\text{-}C_5H_5)Cr(CO)_2L]_2$, L = CO, $P(OMe)_3$, have been determined by n.m.r. spectroscopy.[1923]

Molybdenum. The exchange reactions of *trans*-[$Mo(CO_2)_2(PMe_3)_4$] with chelating phosphines and isocyanides have been studied by ^{1}H, ^{13}C, and ^{31}P n.m.r. spectroscopy.[1924] ^{1}H n.m.r. spectroscopy has indicated exchange between *anti*- and *gauche*-isomers of [$Mo_2X_2(PMe_2Ph)_3$].[1925] Isomerism in the [$Mo_2(\mu\text{-}O_2CCF_3)_4$]-bipy system has been investigated by ^{1}H and ^{19}F n.m.r. spectroscopy.[1926]

Tungsten. Variable temperature ^{1}H and ^{31}P n.m.r. spectroscopy has been used to demonstrate the equilibrium

$$[(RO)_3W(\mu\text{-}H)(\mu\text{-}OR)_2W(OR)_2L] \rightleftharpoons [W_2(\mu\text{-}H)(OR)_7] + L$$

^{13}C n.m.r. spectra were also reported.[1927] The aggregation of [$(\eta^5\text{-}C_5Me_5)Me_3W{=}NNLi_2$] has been studied by ^{1}H, ^{7}Li, ^{13}C, and ^{15}N n.m.r. spectroscopy.[1928]

Manganese. The formation of Mn^{III}, Co^{II}, or Fe^{III} porphyrin molecular complexes with π-donors such as anthracene has been studied by ^{1}H n.m.r. spectroscopy.[1929]

Iron and Ruthenium. Hydrogen binding in complexes such as [$MH(\eta^2\text{-}H_2)(R_2PCH_2CH_2PR_2)$], M =

[1920] B. Bachand and J.D. Wuest, *Organometallics*, 1991, **10**, 2015.
[1921] Phan Viet Minh Tan, V. Sharma, and J.D. Wuest, *Inorg. Chem.*, 1991, **30**, 3026.
[1922] G.E. Herberich, U. Englert, K. Linn, P. Roos, and J. Runsink, *Chem. Ber.*, 1991, **124**, 975.
[1923] L.Y. Goh and Y.Y. Lim, *J. Organomet. Chem.*, 1991, **402**, 209.
[1924] E. Carmona, A.K. Hughes, M.A. Munoz, D.M. O'Hare, P.J. Perez, and M.L. Poveda, *J. Am. Chem. Soc.*, 1991, **113**, 9210.
[1925] J.C. Gordon, H.D. Mui, R. Poli, and K.J. Ahmed, *Polyhedron*, 1991, **10**, 1667.
[1926] J.H. Matonic, S.J. Chen, S.P. Perlepes, K.R. Dunbar, and G. Christou, *J. Am. Chem. Soc.*, 1991, **113**, 8169.
[1927] S.T. Chacon, M.H. Chisholm, K. Folting, M.J. Hampden-Smith, and J.C. Huffman, *Inorg. Chem.*, 1991, **30**, 3122.
[1928] T.E. Glassman, A.H. Liu, and R.R. Schrock, *Inorg. Chem.*, 1991, **30**, 4723.
[1929] O.A. Chamaeva and A.N. Kitaigorodskii, *Izv. Akad. Nauk SSSR, Ser. Khim.*, 1990, 1755 (*Chem. Abstr.*, 1991, **114**, 89 398).

Fe, Ru, has been investigated by ^1H n.m.r. spectroscopy.[1930] ^{31}P n.m.r. spectroscopy has been used to investigate the equilibrium

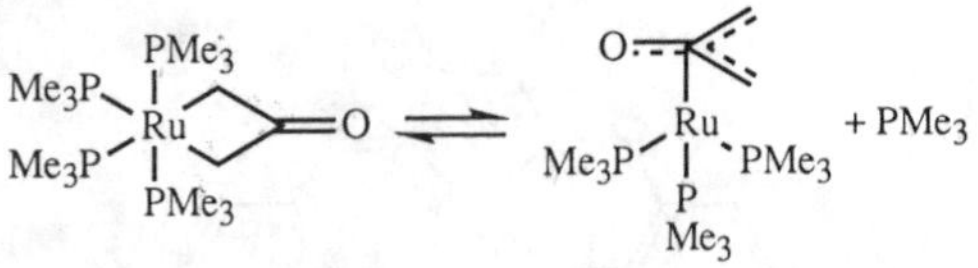

The ^{13}C n.m.r. spectrum was also reported.[1931] ^1H and ^{13}C n.m.r. spectroscopy has been used to investigate the equilibrium[1932]

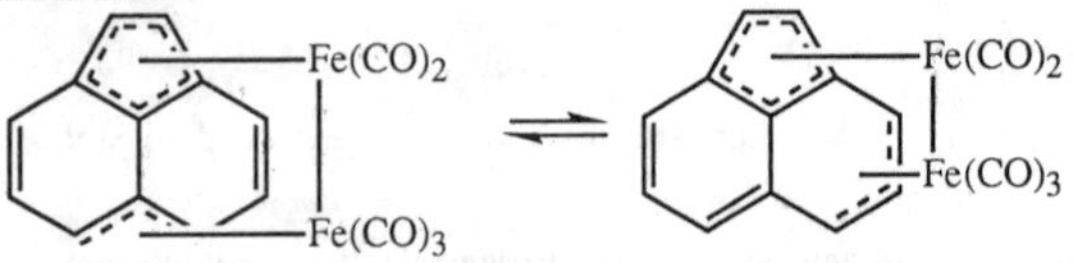

The ^1H n.m.r. spectrum of $[(\eta^5\text{-}C_5Me_5)RuCl_2]_2$ is very temperature dependent, moving from δ 1.6 at 109 K to δ 8 at 350 K. This was attributed to a diamagnetic↔paramagnetic equilibrium. The ^{13}C n.m.r. spectrum was reported.[1933] The ^1H and ^{13}C n.m.r. spectra of $[(\eta^6\text{-}p\text{-cymene})RuCl(\mu\text{-}pz)_2Ir(CO)_2]$ show interconversion of two isomers.[1934] The association energy of $[Fe\{N(SiMe_3)_2\}_2]$ has been investigated by ^1H n.m.r. spectroscopy.[1935] Pyridine coordination to [Fe(BAE)] has been investigated by ^1H n.m.r. spectroscopy.[1936]

Cobalt. The *mer*↔*fac* isomerization of $[Co(H_2NCH_2CH_2O)_3]$ has been studied by ^{59}Co n.m.r. spectroscopy.[1937]

Rhodium. The equilibrium

$$3[(\eta^5\text{-}C_5Me_5)Rh(SR_f)_2] \rightleftharpoons [(\eta^5\text{-}C_5Me_5)_2Rh_2(\mu\text{-}SR_f)_3][(\eta^5\text{-}C_5Me_5)Rh(SR_f)_3]$$

has been studied using ^{19}F n.m.r. spectroscopy.[1938] The fragmentation and recombination reactions of $[Rh_4(CO)_{12\text{-}n}L_n]$ under a pressure of CO or CO-H_2 have been studied by ^{13}C and ^{31}P n.m.r. spectroscopy.[1939] ^{31}P n.m.r. spectroscopy has been used to study the equilibrium between $[Rh_2(4\text{-}MeC_6H_4NCHNC_6H_4Me\text{-}4)_2(O_2CCF_3)_2(OH_2)_2]$ and $2\text{-}Ph_2PC_5H_4N$.[1940]

Iridium. The coordination of H_2,[1941] HCl,[1942] and H_2O[1943] to $[IrHCl_2(PR_3)_2]$ has been studied by

1930 M.T. Bautista, E.P. Cappellani, S.D. Drouin, R.H. Morris, C.T. Schweitzer, A. Sella, and J. Zubkowski, *J. Am. Chem. Soc.*, 1991, **113**, 4876.
1931 J.F. Hartwig, R.G. Bergman, and R.A. Andersen, *Organometallics*, 1991, **10**, 3344.
1932 H. Nagashima, T. Fukahori, and K. Itoh, *J. Chem. Soc., Chem. Commun.*, 1991, 786.
1933 U. Kölle, J. Kossakowski, N. Klaff, L. Wesemann, U. Englert, and G.E. Herberich, *Angew. Chem., Int. Ed. Engl.*, 1991, **30**, 690.
1934 D. Carmona, J. Ferrer, A. Mendoza, F.J. Lahoz, J. Reyes, and L.A. Oro, *Angew. Chem., Int. Ed. Engl.*, 1991, **30**, 1171.
1935 M.M. Olmstead, P.P. Power, and S.C. Shoner, *Inorg. Chem.*, 1991, **30**, 2547.
1936 H.Y. Liu, B. Scharbert, and R.H. Holm, *J. Am. Chem. Soc.*, 1991, **113**, 9529.
1937 T.Sh. Kapanadze, M.A. Elerdashvili, A.P. Gulya, and Yu.A. Buslaev, *Zh. Neorg. Khim.*, 1991, **36**, 1212 (*Chem. Abstr.*, 1991, **115**, 143 658).
1938 J.J. Garcia, H. Torrens, H. Adams, N.A. Bailey, and P.M. Maitlis, *J. Chem. Soc., Chem. Commun.*, 1991, 74.
1939 D.T. Brown, T. Eguchi, B.T. Heatón, J.A. Iggo, and R. Whyman, *J. Chem. Soc., Dalton Trans.*, 1991, 677.
1940 E. Rotondo, G. Bruno, F. Nicolò, S. Lo Schiavo, and P. Piraino, *Inorg. Chem.*, 1991, **30**, 1195.
1941 D.G. Gusev, V.I. Bakhmutov, V.V. Grushin, and M.E. Vol'pin, *Inorg. Chim. Acta*, 1991, **177**, 115.
1942 D.G. Gusev, A.B. Vymenits, and V.I. Bakhmutov, *Mendeleev Commun.*, 1991, 24 (*Chem. Abstr.*, 1991, **115**, 221 904).
1943 V.I. Bakhmutov, D.G. Gusev, A.B. Vymenits, V.V. Grushin, and M.E. Vol'pin, *Metalloorg. Khim.*, 1991, **4**, 164 (*Chem. Abstr.*, 1991, **114**, 172 385).

^1H and ^2H n.m.r. spectroscopy. Rapid intermolecular ligand exchange for *trans*-[Ir(CO)X(PR_3)$_2$] has been studied by ^{31}P n.m.r. spectroscopy.[1944]

Nickel. Self-aggregation of Ni^{II} and Zn^{II} 2-vinylphylloerythrins has been studied by ^1H n.m.r. spectroscopy.[1945] Ligand exchange in [Ni(dppe)X_2] has been investigated using ^{31}P n.m.r. spectroscopy. Large shielding anisotropy of the ^{31}P shielding tensor has been found.[1946]

Palladium. ^{31}P n.m.r. spectroscopy has been used to investigate PPh_3 complexation to [Pd_2(dba)$_3$].[1947]

Platinum. The reversible coordination of alkenes to *trans*-[PtH(SnX_3)(PR_3)$_2$] has been investigated by ^1H and ^{31}P n.m.r. spectroscopy.[1948] ^{31}P n.m.r. spectroscopy has been used to demonstrate the equilibrium[1949]

$$cis\text{-[Pt(COCOPh)(PPh}_3\text{)(dppe)]Cl} \rightleftharpoons cis\text{-[Pt(COCOPh)Cl(dppe)]} + \text{PPh}_3$$

Cis-trans isomerism of some square-planar Pt^{II} nitroimidazole complexes has been studied using ^1H, ^{13}C, and ^{195}Pt n.m.r. spectroscopy.[1950] ^{31}P, ^{77}Se, and ^{195}Pt n.m.r. spectroscopy has been used to investigate the interaction of [Pt(S_2COPr^n)$_2$] with a variety of bidentate ligands.[1951]

Copper. The interaction of $PhCH_2CH_2CH_2Ph$ with macromolecule-CuCl-$AlCl_3$ complexes has been studied using ^{13}C n.m.r. spectroscopy.[1952] Fluxionality and interconversion of species of the type [Cu_4(C_6H_4Me-2)$_4$(SMe_2)$_2$] have been investigated by ^1H and ^{13}C n.m.r. spectroscopy.[1953] ^1H and ^{13}C n.m.r. spectroscopy has been used to determine $\Delta G^{\ddagger} = 14.1 \pm 0.5$ kcal mol^{-1} for exchange of hfacH with [Cu(η^4-1,5-C_8H_{12})(hfac)].[1954] ^1H n.m.r. line broadening has been used to investigate the interaction of [Cu(salicylideneglycinate)(OH_2)] with cytidine and cytosine.[1955] Exchange between conformers of [Cu($SC_6H_3CHR^1NMe_2$-2-R^2-3)]$_3$ has been investigated using ^1H and ^{13}C n.m.r. spectroscopy.[1956]

Zinc. Dimerization equilibria of [Zn{$S_2P(OR)_2$}$_2$] in various solvents have been studied by ^{31}P n.m.r. spectroscopy.[1957]

Mercury. ^1H and ^{13}C n.m.r. spectroscopy has been used to study the interaction of $HgCl_2$ and Hg(O_2CCH_3)$_2$ with purine and pyrimidine nucleosides.[1958] Exchange equilibria between R^1SH and R^2HgSR^3, $R^2_3SnSR^3$, and $R^2_3PbSR^3$ have been investigated using ^{19}F n.m.r. spectroscopy.[1959]

Boron and Aluminium. Ligand exchange in adducts of ethylaluminium hydrides has been investigated

1944 J.S. Thompson and J.D. Atwood, *J. Am. Chem. Soc.*, 1991, **113**, 7429.
1945 R.J. Abraham, A.E. Rowan, K.E. Mansfield, and K.M. Smith, *J. Chem. Soc., Perkin Trans. 2*, 1991, 515.
1946 P.S. Jarrett and P.J. Sadler, *Inorg. Chem.*, 1991, **30**, 2098.
1947 V. Farina and B. Krishnan, *J. Am. Chem. Soc.*, 1991, **113**, 9585.
1948 A.B. Permin and V.S. Petrosyan, *Appl. Organomet. Chem.*, 1990, **4**, 111 (*Chem. Abstr.*, 1991, **114**, 6780).
1949 Y.-J. You, J.-T. Chen, M.-C. Cheng, and Y. Wang, *Inorg. Chem.*, 1991, **30**, 3621.
1950 F.M. Macdonald and P.J. Sadler, *Polyhedron*, 1991, **10**, 1443.
1951 R. Colton and V. Tedesco, *Inorg. Chim. Acta*, 1991, **183**, 161.
1952 H. Hirai, *J. Macromol. Sci., Chem.*, 1990, **A27**, 1293 (*Chem. Abstr.*, 1991, **114**, 45 448).
1953 B. Lenders, D.M. Grove, G. Van Koten, W.J.J. Smeets, P. Van der Sluis, and A.L. Spek, *Organometallics*, 1991, **10**, 786.
1954 K.M. Chi, H.-K. Shin, M.J. Hampden-Smith, E.N. Duesler, and T.T. Kodas, *Polyhedron*, 1991, **10**, 2293.
1955 I. Samasundaram, M.K. Kommiya, and M. Palaniandavar, *J. Chem. Soc., Dalton Trans.*, 1991, 2083.
1956 D.M. Knotter, H.L. Van Maanen, D.M. Grove, A.L. Spek, and G. Van Koten, *Inorg. Chem.*, 1991, **30**, 3309.
1957 P.G. Harrison, P. Brown, M.J. Hynes, J.M. Kiely, and J. McManus, *J. Chem. Res. (S)*, 1991, 174.
1958 B.T. Khan, *Proc. Indo-Sov. Symp. "Organomet. Chem.", 1st. 1988*, (Pub. 1989), 316. Ed. D.V.S. Jain (*Chem. Abstr.*, 1991, **114**, 164 697).
1959 A.S. Peregudov, E.M. Rokhlina, E.I. Fedin, and D.N. Kravtsov, *Metalloorg. Khim.*, 1990, **3**, 1351 (*Chem. Abstr.*, 1991, **115**, 29 502).

by ^1H n.m.r. spectroscopy.[1960] Complexes of secondary and tertiary acetates with BCl_3, BBr_3, and $EtAlCl_2$ have been studied using ^1H n.m.r. spectroscopy.[1961] ^{13}C, ^{17}O, and ^{27}Al n.m.r. spectroscopy has been used to investigate complex formation between $AlCl_3$ and RCOCl.[1962] An equilibrium between monomer and trimer of [Al(acac)Cl_2] has been investigated using ^{13}C and ^{27}Al n.m.r. spectroscopy.[1963] A simple linear fitting procedure for the determination of first-order rate constants from two-dimensional exchange spectra in rapidly relaxing systems has been presented and applied to halogen exchange between BCl_3 and BBr_3.[1964]

Tin. The complexation of DMSO to 2,2'-(3-Me_2ClSn-naphthyl)$_2$ has been investigated by ^{13}C and ^{119}Sn n.m.r. spectroscopy.[1965] The coordination of dialkyl phosphonates with tin and organotin chlorides has been studied by ^1H, ^{31}P, and ^{119}Sn n.m.r. spectroscopy.[1966] ^{119}Sn n.m.r. spectroscopy has been used to examine *cis-trans* isomerism of $[Sn(\mu\text{-}Cl)\{\overline{NCMe_2(CH_2)_3CMe_2}\}]_2$ with $\Delta G^{\ddagger}{}_{192}$ = 32.7 kJ mol^{-1}.[1967] A ^{13}C, ^{119}Sn, and ^{121}Sb n.m.r. study of Lewis acid-base interaction of anhydrous $SnCl_4$ and $SbCl_5$ with some common solvents has been reported.[1968]

Phosphorus. Cis-trans isomerism of $R^1R^2C(CH_2Se)_2CHP(O)(PMe)_2$ has been investigated by ^1H and ^{13}C n.m.r. spectroscopy.[1969] The interconversion of isomers of P_3Se_4X has been investigated using ^{31}P and ^{77}Se n.m.r. spectroscopy.[1970]

Selenium. Equilibria in S_2Cl_2-Se_2Cl_2 and $Se_{8-n}S_n$ systems have been studied using ^{77}Se n.m.r. spectroscopy.[1971]

Tellurium. ^{125}Te n.m.r. spectroscopy has been used to study ligand exchange reactions of organotellurides.[1972]

Course of Reactions.—*Lithium.* A ^{13}C n.m.r. method has been developed for the quantitative determination of radical anions in relatively high concentration and applied to the reaction of PhLi with CO.[1973]

Uranium. The hydrolysis of uranyl salts has been investigated by ^{17}O n.m.r. spectroscopy,[1974] which has also been used to monitor the oxidation of $[NO_2]^-$ by $[U(DMSO)_8]^{4+}$.[1975] ^1H CIDNP has

1960 D.M. Frigo, P.J. Reuvers, D.C. Bradley, H. Chudzynska, H.A. Meinema, J.G. Kraaijkamp, and K. Timmer, *Chem. Mater.*, 1991, **3**, 1097 (*Chem. Abstr.*, 1991, **115**, 256 261).
1961 K. Matyjaszewski and C.H. Lin, *J. Polym. Sci., Part A: Polym. Chem.*, 1991, **29**, 1439 (*Chem. Abstr.*, 1991, **115**, 184 013).
1962 F. Bigi, G. Casnati, G. Sartori, and G. Predieri, *J. Chem. Soc., Perkin Trans. 2*, 1991, 1319.
1963 J. Lewiński, S. Pasynkiewicz, and J. Lipkowski, *Inorg. Chim. Acta*, 1991, **178**, 113.
1964 E.F. Derose, J. Castillo, D. Saulys, and J. Morrison, *J. Magn. Reson.*, 1991, **93**, 347.
1965 R. Krishnamurti, H.G. Kuivila, N.S. Shaik, and J. Zubieta, *Organometallics*, 1991, **10**, 423.
1966 P.N. Nagar, *J. Indian Chem. Soc.*, 1990, **67**, 703 (*Chem. Abstr.*, 1991, **115**, 8947).
1967 R.W. Chorley, P.B. Hitchcock, B.S. Jolly, M.F. Lappert, and G.A. Lawless, *J. Chem. Soc., Chem. Commun.*, 1991, 1302.
1968 X. Mao and G. Xu, *Wuji Huaxue Xuebao*, 1990, **6**, 95 (*Chem. Abstr.*, 1991, **115**, 100 512).
1969 M. Mikolajczyk, M. Mikina, P. Graczyk, M.W. Wieczorek, and G. Bujacz, *Tetrahedron Lett.*, 1991, **32**, 4189.
1970 H.-P. Baldus, R. Blachnik, P. Loennecke, and B.W. Tattershall, *J. Chem. Soc., Dalton Trans.*, 1991, 2643.
1971 R. Steudel, B. Plinke, D. Jensen, and F. Baumgart, *Polyhedron*, 1991, **10**, 1037.
1972 R.U. Kirss and D.W. Brown, *Organometallics*, 1991, **10**, 3597.
1973 N.S. Nudelman, F. Doctorovich, and G. Amorin, *Tetrahedron Letts.*, 1990, **31**, 2533.
1974 R.B. King, C.M. King, and R. Garber, *Mater. Res. Soc. Symp. Proc.*, 1990, **180**, 1083 (*Chem. Abstr.*, 1991, **115**, 265 098).
1975 Y.Y. Park, Y. Ikeda, M. Harada, and H. Tomiyasu, *Chem. Lett.*, 1991, 1329 (*Chem. Abstr.*, 1991, **115**, 143 714).

been observed during the photodecomposition of uranyl carboxylates.[1976]

Zirconium. The interaction of $[(\eta^5\text{-}C_5H_5)_2ZrCl_2]$ with methylalumoxane has been studied by ^1H, ^{13}C, ^{27}Al, and ^{29}Si n.m.r. spectroscopy.[1977] The thermal rearrangement of $[(\eta^5\text{-}C_5H_5)_2ZrNBu^tC(CH_2SiMe_2CH_2)\text{=}CNBu^t]$ has been followed by ^1H n.m.r. spectroscopy and the ^{13}C n.m.r. spectrum recorded.[1978]

Vanadium. The reaction of $[P_2W_{17}O_{61}]^{10-}$ or $[P_2W_{18}O_{62}]^{6-}$ with $[VO_3]^-$ has been followed by n.m.r. spectroscopy.[1979]

Molybdenum and Tungsten. ^1H n.m.r. spectroscopy has been used to monitor the reaction of $[(\eta^5\text{-}C_5Me_5)Mo(CO)_3H]$ with DMSO,[1980] and the interconversion of $[(Bu^tCH_2)_3W{\equiv}CSiMe_3]$ and $[(Bu^tCH_2)_2(Me_3SiCH_2)W{\equiv}CBu^t]$. The ^{13}C n.m.r. spectra were also recorded.[1981] The reaction in the $[MoOCl_5]^-\text{-}SnCl_2$ system in DMF has been studied by ^{119}Sn n.m.r. spectroscopy.[1982] ^1H n.m.r. spectroscopy has been used to study the rate of decomposition of *trans*-$[Mo\{CHBu^tCH(C_6F_5)O\}(NAr)(OBu^t)_2]$. The ^{13}C n.m.r. spectrum was also recorded.[1983] ^1H n.m.r. spectroscopy has been used to monitor the deuteriation of benzothiophene, catalysed by $[\{(\eta^5\text{-}C_5H_4Me)Mo\}_2(S_2CH_2)(\mu\text{-}S)(\mu\text{-}SH)]^+$.[1984] The mechanism of ^{13}CO incorporation into *fac*-$[\{1,5\text{-}(Ph_2PCH_2)_2\text{-}3\text{-oxabicyclo}[3.3.0]\text{octane}\}Mo(CO)_3]$ has been investigated by ^{13}C n.m.r. spectroscopy.[1985] The reaction of $[W(CO)_5(OC_6H_3Ph_2\text{-}2,6)]^-$ with CO_2 to give $[W(CO)_5(O_2COC_6H_3Ph_2\text{-}2,6)]^-$ has been monitored by ^{13}C n.m.r. spectroscopy in a sapphire n.m.r. tube.[1986] ^{31}P n.m.r. spectroscopy has been used to monitor the reaction of $[Pt(\eta^2\text{-}C_2H_4)(PPh_3)_2]$ with $[W(CO)_5\{PPhC(OEt)\text{=}CPhCHPh\}]$.[1987] The reaction of 2-naphthylmethylmercaptan with $[W(CO)_6]$ has been monitored by ^1H n.m.r. spectroscopy.[1988] Terminal ligand substitution reactions of $[Mo_3(\mu_3\text{-}CMe)(\mu_3\text{-}O)(\mu\text{-}O_2CMe)_6L_3]^+$ and $[Mo_3(\mu_3\text{-}O)(\mu\text{-}O_2CMe)_6(OH_2)_3]^{2+}$ have been studied using ^1H, ^{13}C, ^{95}Mo, and ^{183}W n.m.r. spectroscopy.[1989] ^{19}F n.m.r. spectroscopy has been used to investigate ligand exchange reactions of $[(Mo_6Cl_8)F_6]^{2-}$ and $[(Mo_6Cl_8)X_6]^{2-}$ in MeCN. The ^{95}Mo chemical shifts depend linearly on the electronegativity of X.[1990]

Manganese. The kinetics of the reaction of $[HMn(CO)_5]$ with $[(OC)_5MnCH_2CH_2CH\text{=}CMePh]$ have

1976 S.V. Rykov, I.V. Khudyakov, E.D. Skakovskii, H.D. Burrows, S.J. Formosinho, and M. da G.M. Miguel, *J. Chem. Soc., Perkin Trans. 2*, 1991, 835.
1977 L.A. Nekhaeva, G.N. Bondarenko, S.V. Rykov, A.I. Nekhaev, B.A. Krentsel, V.P. Mar'in, L.I. Vyshinskaya, I.M. Khrapova, A.V. Polonskii, and N.N. Korneev, *J. Organomet. Chem.*, 1991, **406**, 139.
1978 F.J. Berg and J.L. Petersen, *Organometallics*, 1991, **10**, 1599.
1979 K.Zh. Serikpaeva and A.K. Il'yasova, *Izv. Akad. Nauk Kaz. SSR, Ser. Khim.*, 1990, 3 (*Chem. Abstr.*, 1991, **114**, 74 043).
1980 G.J. Kubas, G. Kiss, and C.D. Hoff, *Organometallics*, 1991, **10**, 2870.
1981 K.G. Caulton, M.H. Chisholm, W.E. Streib, and Z. Xue, *J. Am. Chem. Soc.*, 1991, **113**, 6082.
1982 N.D. Chichirova, V.P. Tikhonova, I.Yu. Sal'nikov, V.V. Trachevskii, and A.S. Khramov, *Zh. Neorg. Khim.*, 1991, **36**, 980 (*Chem. Abstr.*, 1991, **115**, 63 389).
1983 G.C. Bazan, R.R. Schrock, and M.B. O'Regan, *Organometallics*, 1991, **10**, 1062.
1984 L. Lopez, G. Godziela, and M.R. DuBois, *Organometallics*, 1991, **10**, 2660.
1985 M.R. Mason and J.G. Verkade, *J. Am. Chem. Soc.*, 1991, **113**, 6309.
1986 D.J. Darensbourg, B.L. Mueller, C.J. Bischoff, S.S. Chojnacki, and J.H. Reibenspies, *Inorg. Chem.*, 1991, **30**, 2418.
1987 F.A. Ajulu, S.S. Al-Juaid, D. Carmichael, P.B. Hitchcock, M.F. Meidine, J.F. Nixon, F. Mathey, and N.H.T. Huy, *J. Organomet. Chem.*, 1991, **406**, C20.
1988 C.T. Ng and T.Y. Luh, *J. Organomet. Chem.*, 1991, **412**, 121.
1989 K. Nakata, A. Nagasawa, N. Soyama, Y. Sasaki, and T. Ito, *Inorg. Chem.*, 1991, **30**, 1575.
1990 K. Harder, G. Peters, and W. Preetz, *Z. Anorg. Allg. Chem.*, 1991, **598**, 139.

been studied by ^{1}H n.m.r. spectroscopy. The ^{13}C n.m.r. spectrum was also reported.[1991] ^{1}H n.m.r. spectroscopy has been used to monitor the reaction[1992]

[(octaethylporphyrin)Mn≡N] + [(tetraphenylporphyrin)MnCl] $\rightleftharpoons$ [(octaethylporphyrin)MnCl] + [(tetraphenylporphyrin)Mn≡N]

Rhenium. ^{1}H n.m.r. spectroscopy has been used to monitor the thermolysis of $[(\eta^5\text{-}C_5H_5)Re(=C_5H_8)(NO)(PPh_3)]^+$ to give $[(\eta^5\text{-}C_5H_5)Re(\eta^2\text{-}C_5H_8)(NO)(PPh_3)]^+$.[1993]

Iron. The reaction of $[(Ph_2EtP)_3FeH_3Cu(PEtPh_2)]$ with $^{13}CO_2$ has been monitored by ^{13}C n.m.r. spectroscopy.[1994] ^{59}Co n.m.r. spectroscopy has been used to study site selectivity of the reactions between $[HMCo_3(CO)_{12}]$, M = Fe, Ru, and various two-electron donor ligands.[1995] The reaction of $[(\eta^5\text{-}C_5H_5)Fe(CO)_2C(O)Me]$ with R_3SiH to give $[(\eta^5\text{-}C_5H_5)Fe(CO)_2CHMeOSiR_3]$ has been monitored by ^{1}H and ^{13}C n.m.r. spectroscopy.[1996] ^{1}H n.m.r. spectroscopy has been used to follow the photolysis of $[(\eta^5\text{-}C_5H_5)Fe(CO)_2SiMe_2SiMe_2Fe(\eta^5\text{-}C_5H_5)(CO)_2]$ to give $[\{(\eta^5\text{-}C_5H_5)Fe(CO)\}_2(\mu\text{-}CO)\{\mu\text{-}SiMe(SiMe_3)\}]$. The ^{13}C and ^{29}Si n.m.r. spectra were also measured.[1997] ^{31}P n.m.r. spectroscopy has been used to monitor the reaction of (120) with $Li[BEt_3H]$ and the ^{13}C n.m.r. spectrum was recorded.[1998] The pyrolysis of $[Fe_3(CO)_{10}(L)(\mu_2\text{-}CNCF_3)]$ has been investigated by ^{1}H, ^{13}C, and ^{19}F n.m.r. spectroscopy.[1999] Horse radish peroxidase catalyzed oxidation of $[SCN]^-$ by H_2O_2 has been studied by ^{15}N n.m.r. spectroscopy.[2000] ^{1}H n.m.r. spectroscopy has been used to monitor the reaction of $[Fe_4S_4LCl]^{2-}$, LH_3 = (121), with $[4\text{-}MeC_6H_4S]^-$, 1,4,7-triazanonane, or Bu^tCN.[2001]

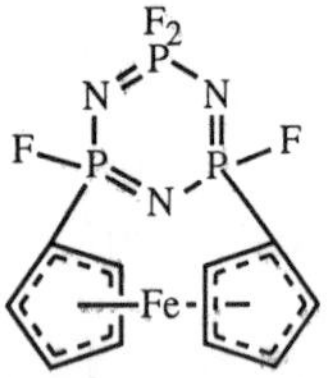

(120)

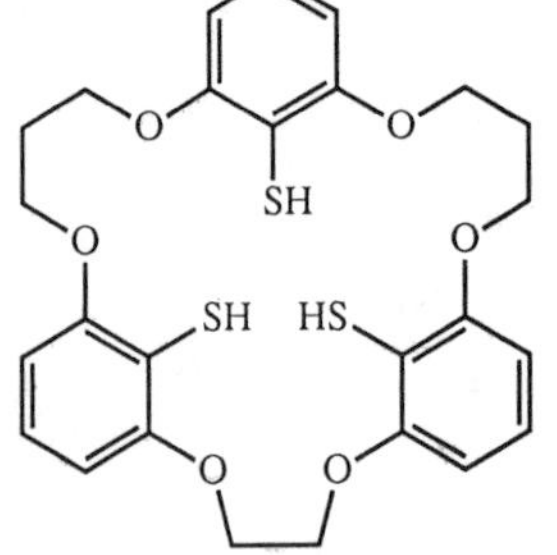

(121)

Ruthenium. ^{1}H n.m.r. spectroscopy has been used to monitor the thermolysis of *cis*-$[(Me_3P)_4RuH(CH_2Ph)]$ in the presence of PMe_3. The ^{13}C n.m.r. spectrum was also reported.[2002] ^{2}H n.m.r. spectroscopy has been used to study the reaction of $[(\eta^5\text{-}C_5Me_5)RuBr_2(\eta^3\text{-}CH_2CRCH_2)]$

1991 R.M. Bullock and B.J. Rappoli, *J. Am. Chem. Soc.*, 1991, **113**, 1659.
1992 L.K. Woo, J.G. Goll, D.J. Czapla, and J.A. Hays, *J. Am. Chem. Soc.*, 1991, **113**, 8478.
1993 C. Roger, G.S. Bodner, W.G. Hatton, and J.A. Gladysz, *Organometallics*, 1991, **10**, 3266.
1994 L.S. Van Der Sluys, M.M. Miller, G.J. Kubas, and K.G. Caulton, *J. Am. Chem. Soc.*, 1991, **113**, 2513.
1995 P. Braunstein, J. Rose, P. Granger, J. Raya, S.E. Bouaoud, and D. Grandjean, *Organometallics*, 1991, **10**, 3686.
1996 P.K. Hanna, B.T. Gregg, and A.R. Cutler, *Organometallics*, 1991, **10**, 31.
1997 K. Ueno, N. Hamashima, M. Shimoi, and H. Ogino, *Organometallics*, 1991, **10**, 959.
1998 H.R. Allcock, W.D. Coggio, I. Manners, and M. Parvez, *Organometallics*, 1991, **10**, 3090.
1999 D. Lentz and R. Marschall, *Chem. Ber.*, 1991, **124**, 497.
2000 S. Modi, D.V. Behere, and S. Mitra, *Biochim. Biophys. Acta*, 1991, **1080**, 45 (*Chem. Abstr.*, 1991, **115**, 250 908).
2001 M.A. Whitener, G. Peng, and R.H. Holm, *Inorg. Chem.*, 1991, **30**, 2411.
2002 J.F. Hartwig, R.A. Anderson, and R.G. Bergman, *J. Am. Chem. Soc.*, 1991, **113**, 6492.

with $BrMg(CH_2)_4MgBr$ to give $[(\eta^5\text{-}C_5Me_5)Ru(\eta^4\text{-}C_4H_6)(\eta^1\text{-}CH_2CRCH_2)]$. The ^{13}C n.m.r. spectrum was also recorded.[2003] ^1H and ^{13}C n.m.r. spectroscopy has been used to follow ligand replacement on $[(\eta^5\text{-}C_5H_5)Ru(CNMe)_3]^+$.[2004] The *cis*-$[Ru(bipy)_2(CO)X]^+$ catalysed homogeneous reduction of CO_2 has been monitored by ^{13}C n.m.r. spectroscopy.[2005] High-pressure ^{13}C and ^{17}O n.m.r. spectra have shown that $[Ru(OH_2)_6]^{2+}$ reacts quantitatively with CO in H_2O to form $[Ru(CO)(OH_2)_5]^{2+}$.[2006]

Osmium. Thermal reactions of $[H(\mu\text{-}H)Os_3(CO)_{10}(CNR)]$ have been investigated by ^1H n.m.r. spectroscopy.[2007] ^1H n.m.r. spectroscopy has been used to follow the reaction of $[Os_3(CO)_8(\mu\text{-}OMe)(\mu_3\text{-}\eta^2\text{-}MeCHCNMe_2)(\mu\text{-}H)H]$ with CO.[2008]

Cobalt. ^{31}P n.m.r. spectroscopy has been used to monitor the rate of conversion of $[\{P(CH_2CH_2PPh_2)_3\}CoH(C{\equiv}CPh)]^+$ to $[\{P(CH_2CH_2PPh_2)_3\}Co(C{=}CHPh)]^+$. The ^{13}C n.m.r. spectrum was also reported.[2009] ^1H and ^{13}C n.m.r. spectroscopy has been used to monitor the decomposition of $[Co(CH_2Ph)\{H(ON{=}CHCH{=}NCH_2)_2CH_2\}]$.[2010] ^1H n.m.r. spectroscopy has been used to monitor the reaction of $GeMeH_3$[2011] and $GeMeH_2GeH_3$[2012] with $[Co_2(CO)_8]$. ^1H n.m.r. spectroscopy has been used to monitor CO loss from $[(\overline{CH_2CMe{=}CMeCH_2Ge})\{Co(CO)_4\}_2]$. The ^{13}C n.m.r. spectrum was also reported.[2013] ^{13}C chemical shifts are sensitive to a secondary $^1H/^2H$ isotope shift and have been used to monitor proton exchange at amines coordinated to Co^{III}.[2014] $^1H/^2H$ exchange at $[Co(pyridoxylideneglycinate)_2]^-$ has been monitored by n.m.r. spectroscopy.[2015] The reactions of 2-hydroxyiminopropionates with Co^{2+} or Ni^{2+} have been followed by ^{13}C n.m.r. spectroscopy.[2016]

Rhodium. The reaction of $[(\eta^5\text{-}C_5Me_5)RhH_2(PR_3)]$ with D_2 has been monitored by ^1H and ^2H n.m.r. spectroscopy.[2017]

Iridium. ^2H n.m.r. spectroscopy has been used to monitor the reaction of (122) with D_2O.[2018] ^1H n.m.r. spectroscopy has been used to monitor the oxidative addition of EtI to *trans*-$[Ir(CO)CH_3\{P(C_6H_4Me\text{-}4)_3\}_2]$.[2019] ^1H n.m.r. spectroscopy has been used to follow the photolysis of $[(\eta^5\text{-}C_5Me_5)IrCl_2(PCy_3)]$, and the products investigated by ^{13}C n.m.r. spectroscopy.[2020] ^{31}P

2003 H. Nagashima, Y. Michino, K.-i. Ara, and T. Fukahori, *J. Organomet. Chem.*, 1991, **406**, 189.
2004 R.H. Fish, H.S. Kim, and R.H. Fong, *Organometallics*, 1991, **10**, 770.
2005 R. Ziessel, *NATO ASI Ser., Ser. C*, 1990, **314**, 79 (*Chem. Abstr.*, 1991, **114**, 100 839).
2006 G. Laurenczy, L. Helm, A. Ludi, and A.E. Merbach, *Helv. Chim. Acta*, 1991, **74**, 1236.
2007 E.V. Anslyn, M. Green, G. Nicola, and E. Rosenberg, *Organometallics*, 1991, **10**, 2600.
2008 R.D. Adams, M.P. Pompeo, and J.T. Tanner, *Organometallics*, 1991, **10**, 1068.
2009 C. Bianchini, M. Peruzzini, A. Vacca, and F. Zanobini, *Organometallics*, 1991, **10**, 3697.
2010 B.E. Daikh and R.G. Finke, *J. Am. Chem. Soc.*, 1991, **113**, 4160.
2011 S.G. Anema, S.K. Lee, K.M. Mackay, B.K. Nicholson, and M. Service, *J. Chem. Soc., Dalton Trans.*, 1991, 1201.
2012 S.G. Anema, S.K. Lee, K.M. Mackay, L.C. McLeod, B.K. Nicholson, and M. Service, *J. Chem. Soc., Dalton Trans.*, 1991, 1209.
2013 D. Lei, M.J. Hampden-Smith, J.W. Garvey, and J.C. Huffman, *J. Chem. Soc., Dalton Trans.*, 1991, 2449.
2014 W.G. Jackson, *Inorg. Chem.*, 1991, **30**, 1570.
2015 A.G. Sykes, R.D. Larsen, J.R. Fischer, and E.H. Abbot, *Inorg. Chem.*, 1991, **30**, 2911.
2016 V.V. Skopenko, R.D. Lampeka, T.Yu. Sliva, and D.I. Stakhov, *Ukr. Khim. Zh. (Russ. Ed.)*, 1990, **56**, 675 (*Chem. Abstr.*, 1991, **114**, 54 773).
2017 W.D. Jones, V.L. Kuykendall, and A.D. Selmeczy, *Organometallics*, 1991, **10**, 1577.
2018 A.C. Albeniz, D.M. Heinekey, and R.H. Crabtree, *Inorg. Chem.*, 1991, **30**, 3632.
2019 J.S. Thompson and J.D. Atwood, *Organometallics*, 1991, **10**, 3525.
2020 D.A. Freedman and K.R. Mann, *Inorg. Chem.*, 1991, **30**, 836.

n.m.r. spectroscopy has been used to monitor the oxidative addition of CH_3COCl to *trans*-$[Ir(CO)(OH)\{P(C_6H_4Me\text{-}4)_3\}_2]$.[2021]

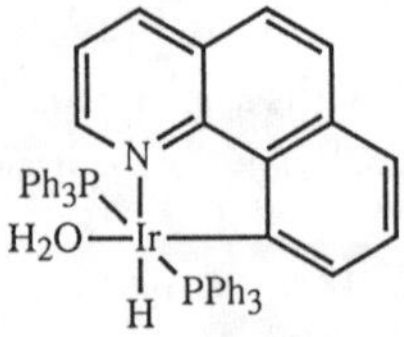

(122)

Nickel. ^{15}N n.m.r. spectroscopy has been used to investigate the protonation of $[Ni(S_2{}^{15}N_2H)_2]$.[2022]
Palladium. The isomerization of $F_3CC(CD_3)(OH)CH{=}C(CH_3)CF_3$ catalysed by $[PdCl_4]^{2-}$ has been monitored by ^{13}C n.m.r. spectroscopy.[2023] ^1H n.m.r. spectroscopy has been used to monitor the decomposition of $[PdMe_2X(CH_2Ph)(bipy)]$ to give, selectively, ethane.[2024] The processes occurring after dissolution of $[Me_3N]_3[Pd(SnCl_3)_5]$ in $MeNO_2$ have been studied by ^{119}Sn n.m.r. spectroscopy.[2025] The mechanism of catalysis of the Arbuzov reaction of non-activated aryl halides by $[Pd(PPh_3)_3]$ has been investigated by n.m.r. spectroscopy.[2026]
Platinum. ^1H n.m.r. spectroscopy has been used to monitor the isomerization of $[Me_2(HO)Pt\{N(CH_2CO_2)_2\}]^-$. The ^{13}C n.m.r. spectrum was also recorded.[2027] ^1H n.m.r. spectroscopy has been used to monitor the hydrolysis of *cis*-$[PtMe_2(DMSO)_2]$.[2028] *Cis-trans* isomerism of *cis*-$[Pt(CH_2Bu^t)Cl(PEt_3)_2]$ has been monitored by ^1H and ^{31}P n.m.r. spectroscopy.[2029] ^1H n.m.r. spectroscopy has been used to monitor the reaction between $Me_2C{=}CMeOSO_2CF_3$ and $[(\eta^2\text{-}C_2H_4)Pt(PPh_3)_2]$ to give $[Pt(CMe{=}CMe_2)(OTf)(PPh_3)_2]$ and $[(\eta^3\text{-}1,1,3\text{-}Me_3C_3H_2)Pt(PPh_3)_2]^+$. The ^{13}C n.m.r. spectrum was also reported.[2030]

The rate of displacement of guanosine by cyanide from *cis*-$[Pt(NH_3)_2(guanosine)_2]^{2+}$ has been determined using ^{13}C n.m.r. spectroscopy.[2031] ^{195}Pt n.m.r. spectroscopy has been used to study human metabolism of *cis*-$[PtCl_2(Pr^iNH_2)_2]$.[2032] ^{195}Pt n.m.r. spectroscopy has been used to investigate the species present during the hydrolysis of the cys-gly peptide bond in the presence of $[PtCl_4]^{2-}$.[2033] ^{31}P n.m.r. studies of the reaction of *cis*-$[PtCl_2(PMe_2Ph)_2]$ with $S_4N_4/S_4{}^{15}N_4$ show complete ^{15}N scrambling.[2034]
Copper. The reaction of $[(PhMe_2Si)_3CuLi_2]$, $[PhMe_2SiMeCuCNLi_2]$, $[(Me_3Sn)_3CuLi_2]$ and

2021 J.S. Thompson, S.L. Randall, and J.D. Atwood, *Organometallics*, 1991, **10**, 3906.
2022 M.L. Calatayud, J.A. Ramirez, and J. Faus, *J. Chem. Soc., Dalton Trans.*, 1991, 2995.
2023 J.W. Francis and P.M. Henry, *Organometallics*, 1991, **10**, 3498.
2024 A.J. Canty, P.R. Traill, B.W. Skelton, and A.H. White, *J. Organomet. Chem.*, 1991, **402**, C33.
2025 R.Kh. Karymova, L.Ya. Al't, I.A. Agapov, and V.K. Duplyakin, *Zh. Neorg. Khim.*, 1991, **36**, 464 (*Chem. Abstr.*, 1991, **115**, 20 888).
2026 V.V. Sentemov, E.A. Krasil'nikova, and I.V. Berdnik, *Zh. Obshch. Khim.*, 1991, **61**, 374 (*Chem. Abstr.*, 1991, **115**, 92 415).
2027 T.G. Appleton, R.D. Berry, J.R. Hall, and J.A. Sinkinson, *Inorg. Chem.*, 1991, **30**, 3860.
2028 D. Minniti and M.F. Parisi, *Inorg. Chim. Acta*, 1991, **188**, 127.
2029 G. Alibrandi, L.M. Scolaro, and R. Romeo, *Inorg. Chem.*, 1991, **30**, 4007.
2030 Z. Zhong, R.J. Hinkle, A.M. Arif, and P.J. Stang, *J. Am. Chem. Soc.*, 1991, **113**, 6196.
2031 M.M. Jones and J.A. Beaty, *Inorg. Chem.*, 1991, **30**, 1584.
2032 L. Pendyala, B.S. Krishnan, J.R. Walsh, A.V. Arakali, J.W. Cowens, and P.J. Creaven, *Cancer Chemother. Pharmacol.*, 1989, **25**, 10 (*Chem. Abstr.*, 1991, **114**, 156 524).
2033 I.E. Burgeson and N.M. Kostić, *Inorg. Chem.*, 1991, **30**, 4299.
2034 C.W. Allen, P.F. Kelly, and J.D. Woollins, *J. Chem. Soc., Dalton Trans.*, 1991, 1343.

$[Me_3SnMeCu(CN)Li_2]$ with cyclohex-2-en-1-one has been investigated using ^{13}C n.m.r. spectroscopy.[2035]

Boron. ^{11}B n.m.r. spectroscopy has been used to monitor the reaction of $[H_2BCONHEt]_2$ with NH_2Me. The ^{13}C n.m.r. spectrum was also recorded.[2036] ^{11}B n.m.r. spectroscopy has been used to investigate F^- as a cage opening agent of *closo*-carboranes such as 1,6-$C_2B_4H_6$, 2,4-$C_2B_5H_7$, 1,2-$C_2B_{10}H_{12}$, and 1,7-$C_2B_{10}H_{12}$.[2037] The formation of $RB(CMe{=}CPr^i)_2BPr^i$ has been monitored by ^1H, ^{11}B, ^{13}C, and ^{119}Sn n.m.r. spectroscopy.[2038] ^{11}B n.m.r. spectroscopy has been used to monitor the photolysis of $[Ph_3BC_6H_4Me\text{-}4]^-$,[2039] and the irradiation of $[Ph_3BC{\equiv}CPh]^-$.[2040]

^{11}B n.m.r. spectroscopy has been used to investigate the reaction of $Me_2NCH_2PhBCl_3$ with KF to yield $Me_2NCH_2PhBCl_{3-n}F_n$ and $[BF_4]^-$.[2041] The solvolysis of $[(phen)BF_2]^+$ has been monitored by ^{11}B and ^{19}F n.m.r. spectroscopy.[2042] The decomposition of inorganic boron peroxide compounds in aqueous solution has been studied using ^{11}B n.m.r. spectroscopy.[2043] The polymerization of isobutylene, initiated by 1,3,5-$(MeOCMe_2)_3C_6H_3/BCl_3$ has been investigated using ^{11}B n.m.r. spectroscopy.[2044]

Aluminium. ^{27}Al n.m.r. spectroscopy has been used to follow the hydrolysis of $[Al(OPr^i)_3]$.[2045] The formation of alumina-silica gels has been investigated using n.m.r. spectroscopy.[2046] ^1H and ^{13}C n.m.r. spectroscopy has been used to investigate the reaction of Me_2NPh with $AlCl_3$ melts.[2047] ^{27}Al n.m.r. spectroscopy has been used to study the thermal decomposition of basic aluminium chlorides.[2048]

Silicon. The reaction between Me_6Si_2 and quinones has been studied by CIDNP.[2049] The kinetics of photooxidation of Bu^tMe_2SiOR have been monitored by ^1H and ^{13}C n.m.r. spectroscopy.[2050] The reaction of $MeCO_2H$ in $(MeCO)_2O$ to give $RSiCl_2(O_2CMe)$ has been investigated by ^1H n.m.r. spectroscopy. The ^{29}Si n.m.r. spectrum was also recorded.[2051] The kinetics of polymerization of $PhSi(OMe)_3$ have been investigated by ^{29}Si n.m.r. spectroscopy.[2052] The polymerization of $Si(OR)_4$ has been studied using ^{29}Si n.m.r. spectroscopy.[2053] A similar study has been performed using ^1H

2035 S. Sharma and A.C. Oehlschlager, *Tetrahedron*, 1991, **47**, 1177.
2036 M.R.M.D. Charandabi, D.A. Feakes, M.L. Ettel, and K.W. Morse, *Inorg. Chem.*, 1991, **30**, 2433.
2037 H. Tomita, H. Luu, and T. Onak, *Inorg. Chem.*, 1991, **30**, 812.
2038 B. Wrackmeyer and G. Kehr, *Polyhedron*, 1991, **10**, 1497.
2039 J.D. Wilkey and G.B. Schuster, *J. Am. Chem. Soc.*, 1991, **113**, 2149.
2040 M.A. Kropp, M. Baillargeon, K.M. Park, K. Bhamidapaty, and G.B. Schuster, *J. Am. Chem. Soc.*, 1991, **113**, 2155.
2041 J. Atchekzaî, A. Ouassas, C. R'kha, B. Bonnetot, H. Mongeot, and B. Frange, *Synth. React. Inorg. Metal-Org. Chem.*, 1991, **21**, 1133.
2042 K.R. Koch and S. Madelung, *Polyhedron*, 1991, **10**, 2221.
2043 B.N. Chernyshov, A.Yu. Vaseva, O.V. Brovkina, and E.V. Pashnina, *Zh. Prikl. Khim. (Leningrad)*, 1991, **64**, 22 (*Chem. Abstr.*, 1991, **115**, 167 530).
2044 Z. Zsuga, T. Kelen, and J. Borbely, *Polym. Bull. (Berlin)*, 1991, **26**, 417 (*Chem. Abstr.*, 1991, **115**, 280 641).
2045 P. Monsef-Mirzai, P.M. Watts, W.R. McWhinnie, and H.W. Gibbs, *Inorg. Chim. Acta*, 1991, **188**, 205.
2046 W.G. Fahrenholtz, S.L. Hietala, D.M. Smith, A.J. Hurd, C.J. Brinker, and W.L. Earl, *Mater. Res. Soc. Symp. Proc.*, 1990, **180**, 229 (*Chem. Abstr.*, 1991, **115**, 259 243).
2047 S.-G. Park, P.C. Trulove, R.T. Carlin, and R.A. Osteryoung, *J. Am. Chem. Soc.*, 1991, **113**, 3334.
2048 P. Brand, D. Mueller, and W. Gessner, *Cryst. Res. Technol.*, 1990, **25**, 951 (*Chem. Abstr.*, 1991, **114**, 54 818).
2049 M. Igarashi, T. Ueda, M. Wakasa, and Y. Sakaguchi, *J. Organomet. Chem.*, 1991, **421**, 9.
2050 W. Adam and X. Wang, *J. Org. Chem.*, 1991, **56**, 4737.
2051 K. Käppler, A. Porzel, U. Scheim, and K. Rühlmann, *J. Organomet. Chem.*, 1991, **402**, 155.
2052 L.W. Kelts and T.E. Long, *Polym. Prepr. (Am. Chem. Soc., Div. Polym. Chem.)*, 1990, **31**, 701 (*Chem. Abstr.*, 1991, **114**, 144 072).
2053 F. Brunet, B. Cabane, M. Dubois, and B. Perly, *J. Phys. Chem.*, 1991, **95**, 945; F. Devreux, J.P. Boilot, F.

n.m.r. spectroscopy.[2054] ^{29}Si n.m.r. spectroscopy has been used to study the hydrolysis of $Mg(OMe)_2$-$Si(OMe)_4$.[2055] ^{29}Si n.m.r. spectroscopy has been used to study the role of pressure on the hydrolysis and condensation kinetics of $Si(OMe)_4$ under neutral conditions.[2056] The formation of lithium silicate gels from the hydrolysis of $Si(OEt)_4$ in the presence of $LiNO_3$ has been followed by ^{29}Si n.m.r. spectroscopy.[2057] The gelation process of organically modified silicates has been investigated by ^{29}Si n.m.r. spectroscopy.[2058] Incipient polymerization of silica in acid-catalyzed $Si(OMe)_4$ systems has been studied by ^{29}Si n.m.r. spectroscopy.[2059]

Tin. The hydrolysis of $[SnMe_3(OH_2)_2]^+$ has been investigated by ^{119}Sn n.m.r. spectroscopy.[2060]

Lead. ^{207}Pb n.m.r. spectroscopy has been used to study the reactivity of $Pb_6O_4(OEt)_4\{Nb(OEt)_5\}_4$.[2061]

Phosphorus. The hydrolysis of poly(dichlorophosphazene) has been monitored by ^{31}P n.m.r. spectroscopy.[2062]

Arsenic. The reaction between Me_2AsNMe_2 and secondary amines has been monitored by ^1H and ^{13}C n.m.r. spectroscopy.[2063]

Selenium. The kinetics of the reaction of the thiolate form of D-penicillamine with bis(D-penicillamine)selenide have been studied by ^1H n.m.r. spectroscopy.[2064] ^{77}Se n.m.r. spectroscopy has been used to monitor the decomposition of 1,2,3,4,5-Se_5S_2 to 1,2,3,4,5,6-Se_6S_2 and 1,2,3,4-Se_4S_2.[2065]

Tellurium. ^1H n.m.r. spectroscopy has been used to follow the loss of H_2O_2 from (123), E = O, S, Se, Te.[2066]

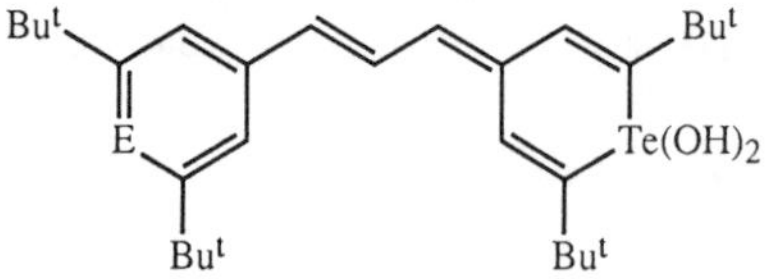

(123)

Xenon. ^1H and ^{19}F CIDNP is observed in the reaction of XeF_2 with dinitromethyl compounds.[2067]

Chaput, and A. Lecomte, *Mater. Res. Soc. Symp. Proc.*, 1990, **180**, 211 (*Chem. Abstr.*, 1991, **115**, 285 665); T.N.M. Bernards, M.J. Van Bommel, and A.H. Boonstra, *J. Non-Cryst. Solids*, 1991, **134**, 1 (*Chem. Abstr.*, 1991, **115**, 240 969).

2054 G.H. Bogush, C.J. Brinker, P.D. Majors, and D.M. Smith, *Mater. Res. Soc. Symp. Proc.*, 1990, **180**, 491 (*Chem. Abstr.*, 1991, **115**, 261 521).

2055 K.E. Yeager and J.M. Burlitch, *Chem. Mater.*, 1991, **3**, 387.

2056 D.W. Hua, Y. Masuda, and J. Jonas, *J. Mol. Liq.*, 1991, **48**, 233.

2057 M. Smaihi, D. Petit, J.P. Boilot, P. Bergez, and A. Lecomte, *Adv. Ceram.*, 1990, **27**, 23 (*Chem. Abstr.*, 1991, **114**, 10 837).

2058 Y.J. Chung, S.J. Ting, and J.D. Mackenzie, *Mater. Res. Soc. Symp. Proc.*, 1990, **180**, 981 (*Chem. Abstr.*, 1991, **115**, 285 680).

2059 J.J. Van Beek, D. Seykens, J.B.H. Jansen, and R.D. Schuiling, *J. Non-Cryst. Solids*, 1991, **134**, 14 (*Chem. Abstr.*, 1991, **115**, 184 076).

2060 M.J. Hynes, J.M. Keely, and J. McManus, *J. Chem. Soc., Dalton Trans.*, 1991, 3427.

2061 L.G. Hubert-Pfalzgraf, R. Papiernik, M.C. Massiani, and B. Septe, *Mater. Res. Soc. Symp. Proc.*, 1990, **180**, 393 (*Chem. Abstr.*, 1991, **115**, 285 490).

2062 D.G. Gabler and J.F. Haw, *Macromolecules*, 1991, **24**, 4218 (*Chem. Abstr.*, 1991, **115**, 30 065).

2063 C.J. Thomas, L.K. Krannich, and C.L. Watkins, *Synth. React. Inorg. Metal-Org. Chem.*, 1991, **21**, 427.

2064 D.L. Rabenstein, T.M. Scott, and W. Guo, *J. Org. Chem.*, 1991, **56**, 4176.

2065 P. Pekonen, Y. Hiltunen, R.S. Laitinen, and T.A. Pakkanen, *Inorg. Chem.*, 1991, **30**, 3679.

2066 M.R. Detty, *Organometallics*, 1991, **10**, 702.

2067 A.E. Trubitsyn, I.V. Tselinskii, A.A. Mel'nikov, M.B. Shcherbinin, M.B. Taraban, V.I. Mar'yasova, and T.V.

4 Paramagnetic Complexes

In this section, compounds of *d*-block transition elements will be considered first and then those of the lanthanide and actinide elements. Papers concerning the use of paramagnetic complexes as 'shift reagents' are usually omitted.

Five reviews have appeared: 'Elucidation of the composition and structure of paramagnetic coordination compounds in solution from nuclear-spin-relaxation data',[2068] 'Fundamental concepts of n.m.r. in paramagnetic systems. Part I. The isotropic shift',[2069] 'Fundamental concepts of n.m.r. in paramagnetic systems. Part II. Relaxation Effects',[2070] 'The electronic structure of paramagnetic polynuclear metal clusters in proteins studied through proton n.m.r. spectroscopy',[2071] and 'Graph method for determination of stability constants of lanthanide shift reagent complexes from n.m.r. data'.[2072]

Hydroxy-aryl metal chelates have been examined for diagnostic n.m.r. imaging.[2073]

The Transition Metals.—The magnetic properties of transition metal complexes in solution have been studied using ^{1}H n.m.r. spectroscopy.[2074] N.m.r. studies of the intermediate states of SR calcium ATPase using Co^{III}, Cr^{III}, and Rh^{III} nucleotides have been reported.[2075]

Titanium. The ^{13}C n.m.r. spectrum of $[(\eta^5\text{-}C_5Me_5)_2TiMe]$ has been reported.[2076]

Vanadium. The ^{1}H n.m.r. spectra of $[(\eta^5\text{-}C_5H_5)VX_3]$[2077] and $[V(NCH_2CH_2N{=}CHC_6H_4\text{-}2\text{-}O)_3]$[2078] have been reported.

Chromium. Ion pairing between Cr^{III} complexes and sulphonate anions has been studied using ^{1}H broadening.[2079] The ^{1}H and ^{13}C n.m.r. spectrum of (124) has been reported.[2080]

Molybdenum and Tungsten. ^{1}H n.m.r. spectra have been reported for *cis-mer*-$[MoOCl_2(PMe_2Ph)_3]$,[2081] *mer*-$[MoCl_3(PMe_3)_3]$,[2082] *mer-trans*-$MoCl_3(THF)_2(PR_3)]$,[2083] $[Mo_2X_6$-

2068 K.I. Zamaraev and V.M. Nekipelov, Zh. Strukt. Khim., 1990, 31, 111 (*Chem. Abstr.*, 1991, **114**, 130 237).
2069 J.D. Satterlee, *Concepts Magn. Reson.*, 1990, **2**, 69 (*Chem. Abstr.*, 1991, **115**, 148 733).
2070 J.D. Satterlee, *Concepts Magn. Reson.*, 1990, **2**, 119 (*Chem. Abstr.*, 1991, **115**, 125 213).
2071 L. Banci, I. Bertini, F. Briganti, and C. Luchinat, *New J. Chem.*, 1991, **25**, 467.
2072 V.D. Buikliskii, S.L. Bel'skaya, and V.T. Panyushkin, *Koord. Khim.*, 1991, **17**, 135 (*Chem. Abstr.*, 1991, **114**, 130 241).
2073 R.B. Lauffer and S.K. Larsen, ***PCT Int. Appl. WO 91 03,200***, Cl. GO1N1/00, 21 Mar 1991, US Appl. 399,737, 28 Aug 1989; 45 pp. (*Chem. Abstr.*, 1991, **115**, 274 765).
2074 S.G. Vul'fson, A.N. Glebov, O.Yu. Tarasov, and Yu.I. Sal'nikov, *Dokl. Akad. Nauk SSSR*, 1990, **314**, 386 (*Chem. Abstr.*, 1991, **114**, 54 521).
2075 T.A. Kuntzweiler and C.M. Grisham, *Soc. Gen. Physiol. Ser.*, 1991, **46**, 231 (*Chem. Abstr.*, 1991, **115**, 274 509).
2076 G.A. Luinstra, L.C. Ten Cate, H.J. Heeres, J.W. Pattiasina, A. Meetsma, and J.H. Teuben, *Organometallics*, 1991, **10**, 3227.
2077 D.B. Morse, T.B. Rauchfuss, and S.R. Wilson, *Inorg. Chem.*, 1991, **30**, 775.
2078 K. Ramesh and R. Mukherjee, *J. Chem. Soc., Dalton Trans.*, 1991, 3259.
2079 M. Iida and H. Yokoyama, *Bull. Chem. Soc. Jpn.*, 1991, **64**, 128.
2080 H. Atzkern, J. Hiermeier, B. Kanellakopulos, F.H. Köhler, G. Müller, and O. Steigelmann, *J. Chem. Soc., Chem. Commun.*, 1991, 997.
2081 P.J. Desrochers, K.W. Nebesny, M.J. LaBarre, S.E. Lincoln, T.M. Loehr, and J.H. Enemark, *J. Am. Chem.*

$(PR_3)_4]$,[2084] and $[M_2Fe_6Se_8(SEt)_9]^{3-}$, M = Mo, W.[2085] ^{183}W n.m.r. spectra have been reported for α-$[Co^{II}W_{12}O_{40}]^{8-}$.[2086]

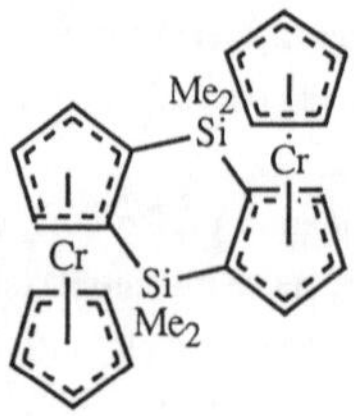

(124)

Manganese. [{(4-sulphatophenyl)$_4$porphyrin}Mn] has been evaluated for magnetic resonance imaging of tumours.[2087] The effect of Gd complexes with 1,2-cyclohexanediaminetetraacetic acid, iminodiacetic acid, and diethylenetriaminepentaacetic acid and $[Mn(edta)(OH_2)]^{2-}$ on the ^{13}C line widths of $[VO_2(edta)]^{3-}$ has been studied.[2088] ^1H n.m.r. spectra have been reported for $[Mn\{N(CH_2N{=}CHC_6H_3Me\text{-}2\text{-}O)_3\}]$,[2089] [Mn(hfac)$_2$(2-Pr^i-4,4,5,5-Me$_4$-4,4-dihydro-1*H*-imidazoline-1-oxyl-3-oxide)], (including ^{19}F),[2090] $[Mn_4O_2(O_2CR)_7(pic)_2]^-$,[2091] $[M(OC_6H_2Bu^t_3\text{-}2,4,6)_2]_2$, M = Mn, Fe,[2092] and $[Mn^{III}(salicylhydroxamate)Cl_{0.5}(DMF)]_4$.[2093] Mn^{2+} induced relaxation has been used to investigate its absorption into roots.[2094] Analysis of low-resolution, low-field n.m.r. relaxation data with the Pade-Laplace method has been applied to $MnCl_2$ doped water samples.[2095]

Rhenium. ^1H n.m.r. data have been reported for *mer*-$[ReCl_3(dppm)\{Ph_2PCH_2P(O)Ph_2\}]$,[2096] and $[ReFe_3S_4(SEt)_4(dmpe)]^-$.[2097]

Iron. The magnetic moment of [Fe(6,13-Me$_2$-1,4,8,11-tetraazacyclotetradecane-6,13-diamine)]$^{3+}$ has been determined using the Evans' method.[2098] Anti-ferromagnetic coupling in [Fe$_2${*N,N,N',N'*-(2-benzimidazolyl)$_4$-2-HO-1,3-diaminopropane}(μ-OH)(NO$_3$)$_2$]$^{2+}$ has been studied using ^1H n.m.r.

Soc., 1991, **113**, 9193.
2082 K. Yoon, G. Parkin, and A.L. Rheingold, *J. Am. Chem. Soc.*, 1991, **113**, 1437.
2083 R. Poli and J.C. Gordon, *Inorg. Chem.*, 1991, **30**, 4550.
2084 R. Poli and H.D. Mui, *Inorg. Chem.*, 1991, **30**, 65.
2085 M.A. Greaney, C.L. Coyle, R.S. Pilato, and E.I. Stiefel, *Inorg. Chim. Acta*, 1991, **189**, 81.
2086 N. Casan-Pastor, P. Gomez-Romero, G.B. Jameson, and L.C.W. Baker, *J. Am. Chem. Soc.*, 1991, **113**, 5658.
2087 D.A. Place, P.J. Faustino, P.C.M. Van Zijl, A. Chesnick, and J.S. Cohen, *Invest. Radiol.*, 1990, **25**, S69 (*Chem. Abstr.*, 1991, **114**, 139 054).
2088 K.I. Popov, A.G. Vendilo, and N.M. Dyatlova, *Dokl. Akad. Nauk SSSR*, 1990, **314**, 660 (*Chem. Abstr.*, 1991, **114**, 73 840).
2089 K. Ramesh, D. Bhuniya, and R. Mukherjee, *J. Chem. Soc., Dalton Trans.*, 1991, 2917.
2090 F. Ferraro, D. Gatteschi, R. Sessoli, and M. Corti, *J. Am. Chem. Soc.*, 1991, **113**, 8410.
2091 E. Libby, J.K. McCusker, E.A. Schmitt, K. Folting, D.N. Hendrickson, and G. Christou, *Inorg. Chem.*, 1991, **30**, 3486.
2092 R.A. Bartlett, J.J. Ellison, P.P. Power, and S.C. Shoner, *Inorg. Chem.*, 1991, **30**, 2888.
2093 M.S. Lah and V.L. Pecoraro, *Inorg. Chem.*, 1991, **30**, 878.
2094 S. Ratkovic and Z. Vucinic, *Plant Physiol. Biochem. (Paris)*, 1990, **28**, 617 (*Chem. Abstr.*, 1991, **115**, 88 583).
2095 C. Tellier, M. Guillou-Charpin, D. Le Botlan, and F. Pelissolo, *Magn. Reson. Chem.*, 1991, **29**, 164.
2096 X.L.R. Fontaine, E.H. Fowles, T.P. Layzell, B.L. Shaw, and M. Thornton-Pett, *J. Chem. Soc., Dalton Trans.*, 1991, 1519.
2097 S. Ciurli and R.H. Holm, *Inorg. Chem.*, 1991, **30**, 743.
2098 P.V. Bernhardt, P. Comba, T.W. Hambley, and G.A. Lawrance, *Inorg. Chem.*, 1991, **30**, 942.

spectroscopy.[2099] ^{1}H n.m.r. spectroscopy has been used to study the spin-crossover transition in $[Fe(4,4'\text{-bi-}1,2,4\text{-triazole})_2(SCN)_2]$.[2100] ^{1}H n.m.r. data have also been reported for $[CH_2(CH_2C_5H_4\text{-}\eta^5)_2Fe]$,[2101] $[(\eta^5\text{-}C_5Me_5)Fe(dppe)X]^+$, (^{13}C),[2102] $[Fe\{2,6\text{-}(3\text{-}R^1\text{-}5\text{-}R^2C_3HN_2\text{-}CH_2)_2C_5H_3N\}_2]^{2+}$,[2103] $[Fe\{N(CH_2\text{-}2\text{-}C_5H_4N)_3\}(1,2\text{-}O_2C_6H_2Bu^t_2\text{-}3,5)]^+$,[2104] $[Fe_2\{1,2\text{-}[2\text{-}OC_6H_4C(O)NH]_2C_6H_4\}_2(N\text{-Me-imidazolyl})_2]$,[2105] [Fe{pentane-2,4-dione(*S*-methylisothiosemicarbazone)$_2$}I],[2106] $[Fe(O\text{-}2\text{-}C_6H_4CH{=}NC_6H_4\text{-}2\text{-}S)_2]^+$,[2107] (125), (^{13}C),[2108] $[Fe_2O\{(NC_5H_4\text{-}2\text{-}CH_2)_2NCH_2CO_2\}_2(OCH_2Ph)]^+$,[2109] and $[Fe_4O_2(O_2CR)_7(bipy)_2]^+$.[2110]

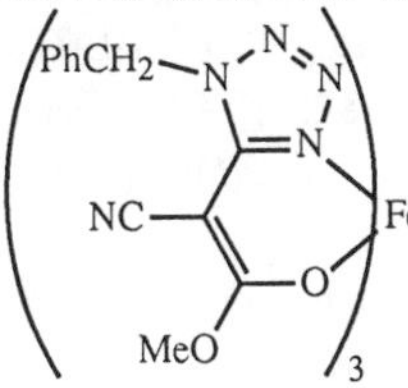

(125)

^{1}H n.m.r. spectroscopy has been used to study some haem peptides encapsulated in aqueous detergent molecules.[2111] ^{1}H and ^{13}C T_1 measurements on Fe(III) tetraphenylporphyrin-imidazole complexes have been used to study the relationship between paramagnetic relaxation rate and electron-nucleus distance.[2112] Two-dimensional COSY experiments have been carried out on a variety of paramagnetic iron complexes of porphyrins and chlorins.[2113] ^{1}H n.m.r. shifts and ^{13}C T_1 measurements have been reported for various π-acceptors in solutions containing paramagnetic Co^{II} and Fe^{III} porphyrins.[2114] ^{1}H n.m.r. hyperfine shift pattern has been suggested as a probe for ligation state in high-spin ferric haemoproteins.[2115] The optimum conditions for applying ^{1}H COSY and TOCSY to paramagnetic ferric protohaemin complexes have been determined.[2116] Very long-range isotope shifts have been observed in the 500 MHz ^{1}H n.m.r. spectra of several deuteriated

2099 B.A. Brennan, Q. Chen, C. Juarez-Garcia, A.E. True, C.J. O'Connor, and L. Que, jun., *Inorg. Chem.*, 1991, **30**, 1937.
2100 A. Ozarowski, Y. Shunzhong, B.R. McGarvey, A. Mislankar, and J.E. Drake, *Inorg. Chem.*, 1991, **30**, 3167.
2101 T.-Y. Dong, H.-M. Lin, M.-Y. Hwang, T.-Y. Lee, L.-H. Tseng, S.-M. Peng, and G.-H. Lee, *J. Organomet. Chem.*, 1991, **414**, 227.
2102 C. Roger, P. Hamon, L. Toupet, H. Rabaa, J.Y. Saillard, J.R. Hamon, and C. Lapinte, *Organometallics*, 1991, **10**, 1045.
2103 S. Mahapatra, N. Gupta, and R. Mukherjee, *J. Chem. Soc., Dalton Trans.*, 1991, 2911.
2104 H.G. Jang, D.D. Cox, and L. Que, *J. Am. Chem. Soc.*, 1991, **113**, 9200.
2105 A. Stassinopoulos, G. Schulte, G.C. Papaefthymiou, and J.P. Caradonna, *J. Am. Chem. Soc.*, 1991, **113**, 8686.
2106 V.M. Leovac, R. Herak, B. Prelesnik, and S.R. Niketic, *J. Chem. Soc., Dalton Trans.*, 1991, 2295.
2107 J.W. Pyrz, X. Pan, D. Britton, and L. Que, jun., *Inorg. Chem.*, 1991, **30**, 3461.
2108 R.W. Saalfrank, C.-J. Lurz, K. Schobert, O. Struck, E. Bill, and A.X. Trautwein, *Angew. Chem., Int. Ed. Engl.*, 1991, **30**, 1494.
2109 S. Ménage and L. Que, jun., *New J. Chem.*, 1991, **15**, 431.
2110 J.K. McCusker, J.B. Vincent, E.A. Schmitt, M.L. Mino, K. Shin, D.K. Coggin, P.M. Hagen, J.C. Huffman, G. Christou, and D.N. Hendrickson, *J. Am. Chem. Soc.*, 1991, **113**, 3012.
2111 S. Mazumdar, O.K. Medhi, and S. Mitra, *Inorg. Chem.*, 1991, **30**, 700; S. Mazumdar, *J. Chem. Soc., Dalton Trans.*, 1991, 2091.
2112 Y. Yamamoto, N. Nanai, R. Chûjô, and Y. Inoue, *Bull. Chem. Soc. Jpn.*, 1991, **64**, 3199.
2113 K.A. Keating, J.S. De Ropp, G.N. La Mar, A.L. Balch, F.Y. Shiau, and K.M. Smith, *Inorg. Chem.*, 1991, **30**, 3258.
2114 A.N. Kitaigorodskii and U. Edlund, *Magn. Reson. Chem.*, 1991, **29**, 693.
2115 K. Rajarathnam, G.N. La Mar, M.L. Chiu, S.G. Sligar, J.P. Singh, and K.M. Smith, *J. Am. Chem. Soc.*, 1991, **113**, 7886.
2116 I. Bertini, F. Capozzi, C. Luchinat, and P. Turano, *J. Magn. Reson.*, 1991, **95**, 244.

haemins.[2117] A systematic approach towards the complete assignment of ^{13}C resonances for horse ferrocytochrome has been described.[2118] N.m.r. data have also been reported for some porphyrins,[2119] (^{19}F),[2120] (^{13}C),[2121] ferricytochrome,[2122] and (2,3,7,8,12,13,17,18-Me_8-corrolato)Fe(III) with axial ligands.[2123]

Iron oxide particles have been examined as contrast enhancing agents in magnetic resonance imaging.[2124] ^1H chemical shifts and relaxation times have been determined for aqueous solutions of $[Fe(OH_2)_6]^{3+}$.[2125] N.m.r. data have also been reported for $[Fe_2O(OAc)_2\{P(methylimidazol-yl)_3\}]$,[2126] $[Fe_6E_6(SEt_3)_4Cl_2]$, E = S, Se,[2127] $[Fe_4S_4L]^{2-}$, L = (126),[2128] esters of amino acids bound to $[Fe_4S_4]^{2+}$ clusters,[2129] $[Fe_4S_6(SEt)_4]^{4-}$,[2130] a complex formed from $[Fe_4S_4(SPh)_4]^{2-}$ with ATP, (^{31}P),[2131] iron-sulphur proteins,[2132] and $[M(SePh)_4]^{2-}$, M = Fe, Co, Ni.[2133]

Ruthenium. The ^1H n.m.r. isotropic shift data of $[Ru(tetraphenylporphyrin)Br_2]$ are consistent with π-donation from X.[2134]

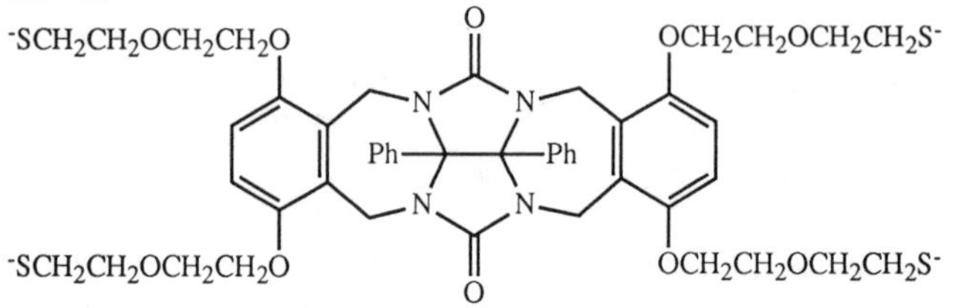

(126)

2117 C.J. Medforth, F.Y. Shiau, G.N. La Mar, and K.M. Smith, *J. Chem. Soc., Chem. Commun.*, 1991, 590.
2118 Y. Gao, J. Boyd, and R.J.P. Williams, *Eur. J. Biochem.*, 1990, **194**, 355 (*Chem. Abstr.*, 1991, **114**, 77 299).
2119 E.T. Kintner and J.H. Dawson, *Inorg. Chem.*, 1991, **30**, 4892; S. Mosseri, J.C. Mialocq, B. Perly, and P. Hambright, *J. Phys. Chem.*, 1991, **95**, 4659; E.P. Sullivan, jun., J.D. Grantham, C.S. Thomas, and S.H. Strauss, *J. Am. Chem. Soc.*, 1991, **113**, 5264; P.W. Crawford and M.D. Ryan, *Inorg. Chim. Acta*, 1991, **179**, 25; A. Malek, L. Latos-Grazynski, T.J. Bartczak, and A. Zadlo, *Inorg. Chem.*, 1991, **30**, 3222; R. Guilard, I. Perrot, A. Tabard, P. Richard, C. Lecomte, Y.H. Liu, and K.M. Kadish, *Inorg. Chem.*, 1991, **30**, 27.
2120 A. Nanthakumar and H.M. Goff, *Inorg. Chem.*, 1991, **30**, 4460; K.M. Kadish, A. Tabard, W. Lee, Y.H. Liu, C. Ratti, and R. Guilard, *Inorg. Chem.*, 1991, **30**, 1542.
2121 B. Garcia, C.H. Lee, A. Blasko, and T.C. Bruice, *J. Am. Chem. Soc.*, 1991, **113**, 8118.
2122 S.J. Moench, T.M. Shi, and J.D. Satterlee, *Eur. J. Biochem.*, 1991, **197**, 631 (*Chem. Abstr.*, 1991, **115**, 3513); R. Timkovich, *Inorg. Chem.*, 1991, **30**, 37; L. Banci, I. Bertini, P. Turano, J.C. Ferrer, and A.G. Mauk, *Inorg. Chem.*, 1991, **30**, 4510.
2123 S. Licoccia, M. Paci, R. Paolesse, and T. Boschi, *J. Chem. Soc., Dalton Trans.*, 1991, 461.
2124 S. Frank, I. Stuermer, and P.C. Lauterbur, *Nuklearmedizin, Suppl. (Stuttgart)*, 1990, **26**, 48 (*Chem. Abstr.*, 1991, **115**, 274 722); H.G. Gundersen, T. Bach-Gansmo, E. Holtz, A.K. Fahlvik, and A. Berg, *Invest. Radiol.*, 1990, **25**, S67 (*Chem. Abstr.*, 1991, **114**, 139 053).
2125 Q. Lin, H. Yuan, and W. Liu, *Chin. Sci. Bull.*, 1990, **35**, 1754 (*Chem. Abstr.*, 1991, **114**, 155 650).
2126 R.H. Beer, W.B. Tolman, S.G. Bott, and S.J. Lippard, *Inorg. Chem.*, 1991, **30**, 2082.
2127 B.S. Snyder, M.S. Reynolds, R.H. Holm, G.C. Papaefthymiou, and R.B. Frankel, *Polyhedron*, 1991, **10**, 203.
2128 C.F. Martens, H.L. Blonk, T. Bongers, J.G.M. van der Linden, G. Beurskens, P.T. Beursgens, J.M.M. Smits, and R.J.M. Nolte, *J. Chem. Soc., Chem. Commun.*, 1991, 1623.
2129 D.J. Evans and G.J. Leigh, *J. Inorg. Biochem.*, 1991, **42**, 25.
2130 S.A. Al-Ahmad, J.W. Kampf, R.W. Dunham, and D. Coucouvanis, *Inorg. Chem.*, 1991, **30**, 1163.
2131 Y. Wu, D. Zeng, C. Li, L. Hong, G. Ling, and Q. Cai, *Wuli Huaxue Xuebao*, 1991, **7**, 400 (*Chem. Abstr.*, 1991, **115**, 153 871).
2132 I. Bertini, F. Briganti, C. Luchinat, A. Scozzafava, and M. Sola, *J. Am. Chem. Soc.*, 1991, **113**, 1237; 7084.
2133 J.M. McConnachie and J.A. Ibers, *Inorg. Chem.*, 1991, **30**, 1770.
2134 M. Ke, C. Sishta, B.R. James, D. Dolphin, J.W. Sparapany, and J.A. Ibers, *Inorg. Chem.*, 1991, **30**, 4766.

Cobalt. The temperature dependent ^1H shifts of $[Co_3(\eta^5\text{-}C_5H_4Me)_3(\mu_3\text{-}S)_2]$ have been examined and the shifts analysed.[2135] N.m.r. data have also been reported for $[(\eta^5\text{-}C_5H_5)Co\{P(O)(OEt)_2\}_3UCl_2\text{-}(\eta^5\text{-}C_5H_5)]$,[2136] $[Co^{II}Cl(N\text{-ethyletiopyrinato})]$,[2137] $[\{HB(3\text{-}Pr^i\text{-}4\text{-}Br)_3\}_2Co]$,[2138] Co^{II} and Ni^{II} complexes with the enaminoketone of the imidazolidine nitroxyl radical,[2139] Co^{II} salen complexes,[2140] Co^{II} complexes with α-amino acids,[2141] $[Co_3(S_2C_6H_4)_3(PBu^n{}_3)_3]$,[2142] and $[Me_3NCH_2CONH_2]_2\text{-}[Co(SPh)_4]$.[2143]

Nickel. A methylnickel(II) derivative of coenzyme F430 has been detected by ^2H n.m.r. spectroscopy.[2144] N.m.r. data have also been reported for $[Ni(en)_2]^{2+}$ complex of $(N{=}CMeCMe{=}N)_n$,[2145] $[Ni(terpy)Cl_2]$,[2146] $[Ni(H_2NCH_2CH_2NCMe{=}CHCMe{=}O)(imidazole)]$,[2147] $[Ni(8\text{-hydroxyquinoline})_2]$,[2148] and (127), ($^{13}C$).[2149]

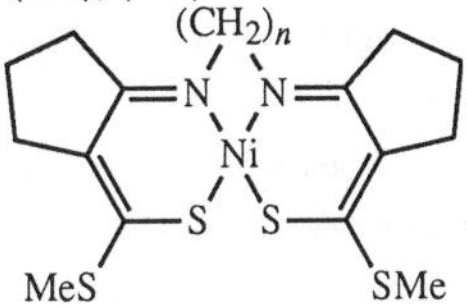

(127)

Copper. N.m.r. data have been reported for $[\{HB(3,5\text{-}Me_2C_3HN_2)_3\}Cu(PPh_3)]$.[2150]

Complexes of the Lanthanides and Actinides.—*Lanthanides.* Europium (*S*,*S*)-ethylenediamine-*N*,*N*'-disuccinate has been proposed as a chiral shift reagent for use in aqueous solution.[2151] A Gd^{3+} complex of 2,3-dihydroxy-*p*-xylylene-diaminotetraacetic acid has been used as a water soluble relaxation reagent.[2152] N.m.r. data have also been reported for $[(\eta^5\text{-}C_5Me_5)_2CeCH_2\text{-}Ph]$, ($^{13}C$),[2153] $[(\eta^5\text{-}C_5H_5)_3Nd(NC_5H_4Me\text{-}4)]$,[2154] $[(\eta^5\text{-}C_5Me_5)_2Sm(Bu^tN{=}CHCH{=}NBu^t)]$,

2135 C.R. Pulliam, J.B. Thoden, A.M. Stacy, B. Spencer, M.H. Englert, and L.F. Dahl, *J. Am. Chem. Soc.*, 1991, **113**, 7398.
2136 D. Baudry, M. Ephritikhine, W. Kläui, M. Lance, M. Nierlich, and J. Vigner, *Inorg. Chem.*, 1991, **30**, 2333.
2137 K. Konishi, T. Sugino, T. Aida, and S. Inoue, *J. Am. Chem. Soc.*, 1991, **113**, 6487.
2138 J.C. Calabrese, P.J. Domaille, S. Trofimenko, and G.J. Long, *Inorg. Chem.*, 1991, **30**, 2795.
2139 A.M. Atskanov, A.V. Podopelov, V.I. Ovcharenko, K.E. Vostrikova, L.B. Volodarskii, A.K. Mikitaev, and R.Z. Sagdeev, *Sib. Khim. Zh.*, 1991, 57 (*Chem. Abstr.*, 1991, **115**, 83 945).
2140 C.P. Cheng, Y.S. Hwang, and R.S. Hwang, *J. Chin. Chem. Soc. (Taipei)*, 1990, **37**, 457 (*Chem. Abstr.*, 1991, **114**, 16 298).
2141 V.P. Tikhonov, *Koord. Khim.*, 1991, **17**, 1094 (*Chem. Abstr.*, 1991, **115**, 269 164).
2142 B. Kang, J. Peng, M. Hong, D. Wu, X. Chen, L. Weng, X. Lei, and H. Liu, *J. Chem. Soc., Dalton Trans.*, 1991, 2897.
2143 M.A. Walters, J.C. Dewan, C. Min, and S. Pinto, *Inorg. Chem.*, 1991, **30**, 2656.
2144 S.-K. Lin and B. Jaun, *Helv. Chim. Acta*, 1991, **74**, 1725.
2145 W.B. Euler, *Polyhedron*, 1991, **10**, 859.
2146 N. Baidya, M. Olmstead, and P.K. Mascharak, *Inorg. Chem.*, 1991, **30**, 929.
2147 J.-P. Costes and J.-P. Laurent, *Inorg. Chim. Acta*, 1990, **177**, 277.
2148 G.E. Jackson and L.G. Scott, *S. Afr. J. Chem.*, 1990, **43**, 75 (*Chem. Abstr.*, 1991, **114**, 239 166).
2149 E.M. Martin, R.D. Bereman, and P. Singh, *Inorg. Chem.*, 1991, **30**, 957.
2150 N. Kitajima, T. Koda, S. Hashimoto, T. Kitagawa, and Y. Moro-oka, *J. Am. Chem. Soc.*, 1991, **113**, 5664.
2151 J. Kido, Y. Okamoto, and H.G. Brittain, *J. Org. Chem.*, 1991, **56**, 1412.
2152 F. Shi, Y. Xiao, G. Tang, and J. Ni, *Bopuxue Zazhi*, 1990, **7**, 433 (*Chem. Abstr.*, 1991, **115**, 125 362).
2153 M. Booij, A. Meetsma, and J.H. Teuben, *Organometallics*, 1991, **10**, 3246.
2154 H. Reddmann, H. Schultze, H.D. Amberger, G.V. Shalimoff, and N.M. Edelstein, *J. Organomet. Chem.*, 1991,

(^{13}C),[2155] [2,4,6-$(F_3C)_3C\{N(SiMe_3)\}_2Nd(\mu\text{-}Cl)_2Li(THF)_2$],[2156] Eu^{3+} complex of bis(homooxa)-*p*-Bu^t-calix[4]arene,[2157] lanthanide chelates as magnetic resonance imaging contrast agents,[2158] lanthanide shift reagents for anionic cyanine dyes,[2159] Eu^{3+} and Y^{3+} complexes of a series of *N,N'*-bis(carboxymethyl) macrocyclic ether bis(lactones), (^{13}C),[2160] lanthanide alaninate complexes, (^{13}C),[2161] gly-L-leu complexes with Ho^{3+} and Yb^{3+}, (^{13}C),[2162] AMP, ADP, and ATP complexes of Pr^{3+} and Nd^{3+}, (^{31}P),[2163] and salicylaldehyde (*N*-benzoyl)glycyl hydrazone complexes of Y^{3+}, Pr^{3+}, Nd^{3+}, Sm^{3+}, Eu^{3+}, Gd^{3+}, and Dy^{3+}, (^{13}C).[2164]

^1H T_1 has been measured for aqueous solutions of MCl_3, M = Pr, Nd, Tb, Dy.[2165] A novel method for water signal suppression in ^1H n.m.r. spectra has been proposed and used for Dy^{III} complexes which cause a marked decrease of water T_2.[2166] The spin matrix density method has been used to describe the -105 °C ^1H n.m.r. spectrum of acetone solutions of $Yb(O_2CCH_3)_3$.[2167] The solvate states of Tb^{III}, Ho^{III}, Er^{III}, Tm^{III}, and Yb^{III} have been studied in aqueous DMSO by ^1H magnetic relaxation.[2168] The extent of inner-shell ion-pair formation of Er^{3+} with nitrate ion in aqueous mixtures has been studied by ^{15}N n.m.r. spectroscopy.[2169] The interaction of Gd^{3+} with dihexyl-*N,N*-diethylcarbamoylmethylene phosphonate has been studied.[2170] The hyperfine interaction constants of lanthanide-phosphorus have been determined for $[LnCl_2(HMPA)_4]^+$ by ^1H and ^{31}P n.m.r. spectroscopy.[2171] The effects of diamagnetic salts on lanthanide induced paramagnetic shifts of the n.m.r. signals of organic substrates have been examined.[2172] Lanthanide induced ^{17}O n.m.r. shifts of diastereotopic sulphonyl oxygen atoms have provided a basis for determining equilibrium constants.[2173] A series of lanthanide acetate and haloacetate complexes have been studied by ^1H,

411, 331.
2155 A. Recknagel, M. Noltemeyer, and F.T. Edelmann, *J. Organomet. Chem.*, 1991, **410**, 53.
2156 A. Recknagel, F. Knösel, H. Gornitzka, M. Noltemeyer, F.T. Edelmann, and U. Behrens, *J. Organomet. Chem.*, 1991, **417**, 363.
2157 Z. Asfari, J.M. Harrowfield, M.I. Ogden, J. Vicens, and A.H. White, *Angew. Chem., Int. Ed. Engl.*, 1991, **30**, 854.
2158 C.A. Chang, *Eur. J. Solid State Inorg. Chem.*, 1991, **28**, 237 (*Chem. Abstr.*, 1991, **114**, 253 919).
2159 I.V. Komarov, A.V. Turov, S.V. Popov, Yu.L. Slominskii, and M.Yu. Kornilov, *Teor. Eksp. Khim.*, 1991, **27**, 197 (*Chem. Abstr.*, 1991, **115**, 210 193).
2160 R.C. Holz, C.A. Chang, and W. DeW. Horrocks, jun., *Inorg. Chem.*, 1991, **30**, 3270.
2161 Y. Xu, K. Yao, J. Chen, and L. Shen, *Huaxue Xuebao*, 1990, **48**, 884 (*Chem. Abstr.*, 1991, **114**, 16 331).
2162 J. Ren, C. Niu, F. Pei, W. Wang, and J. Ni, *Bopuxue Zazhi*, 1991, **8**, 89 (*Chem. Abstr.*, 1991, **115**, 92 940).
2163 S.N. Misra, *Adv. Organomet. Proc. Indo-Sov. Symp. "Organomet. Chem.", 1st. 1988*, (Pub. 1989), 228 (*Chem. Abstr.*, 1991, **115**, 178 496).
2164 G. Singh and T.R. Rao, *Spectrochim. Acta, Part A*, 1991, **47**, 727 (*Chem. Abstr.*, 1991, **115**, 173 365).
2165 A.S. Rao, R. Srinivasa, and V. Arulmozhi, *Indian J. Pure Appl. Phys.*, 1991, **29**, 586 (*Chem. Abstr.*, 1991, **115**, 148 746).
2166 S. Aime, M. Botta, L. Barbero, F. Uggeri, and F. Fedeli, *Magn. Reson. Chem.*, 1991, **29**, S85.
2167 V.D. Buikliskii, V.Ya. Kavun, and B.N. Chernyshov, *Koord. Khim.*, 1991, **17**, 424 (*Chem. Abstr.*, 1991, **115**, 83 940).
2168 F.V. Devyatov, V.F. Safina, and Yu.I. Sal'nikov, *Izv. Vyssh. Uchebn. Zaved., Khim. Khim. Tekhnol.*, 1990, **33**, 59 (*Chem. Abstr.*, 1991, **114**, 50 653).
2169 A. Fratiello, V. Kubo-Anderson, S. Azimi, E. Marinez, D. Matejka, R. Perrigan, and B. Yao, *J. Solution Chem.*, 1991, **20**, 893.
2170 N.V. Jarvis, *Solvent Extr. Ion Exch.*, 1991, **9**, 697 (*Chem. Abstr.*, 1991, **115**, 241 131).
2171 V.V. Skopenko, V.M. Amirkhanov, A.V. Turov, and V.V. Trachevskii, *Koord. Khim.*, 1991, **17**, 124 (*Chem. Abstr.*, 1991, **114**, 176 906).
2172 V.D. Buikliskii, V.T. Banyushkin, and I.I. Soldatenko, *Zh. Obshch. Khim.*, 1990, **60**, 2413 (*Chem. Abstr.*, 1991, **114**, 246 719).
2173 T.A. Powers and S.A. Evans, jun., *Tetrahedron Lett.*, 1990, **31**, 5835; T.A. Powers, S.A. Evans, jun., K. Pandiarajan, and J.C.N. Benny, *J. Org. Chem.*, 1991, **56**, 5589.

^{13}C, and ^{17}O n.m.r. spectroscopy.[2174] The paramagnetic ion-induced relaxation rate enhancements of ^{6}Li in adducts of Li^{+} and $[Ln(triphosphate)_2]^{7-}$ in aqueous solution have shown that up to seven monovalent counterions can coordinate in the second coordination sphere. The pseudocontact shift of ^{7}Li n.m.r. shift data suggests that in the second coordination sphere some preference of the counterions for the axial region opposite the water ligand may exist.[2175] N.m.r. data have also been reported for $[Ln(PW_{11}O_{39})_2]^{11-}$, ($^{17}O$, ^{31}P, ^{183}W),[2176] $[(Bu^t_3CO)_3Nd(THF)]$,[2177] lanthanide complexes of 5-sulphosalicylic acid, (^{13}C),[2178] and complexes of vanadyl diethylenetriaminepentaacetate with Gd cyclohexanediaminetetraacetate, (^{13}C).[2179]

Actinides. The n.m.r. chemical shift for a $5f^1$ system in an intermediate crystal field environment of octahedral symmetry has been calculated.[2180] N.m.r. data have also been reported for $[(\eta^5\text{-}C_5H_4SiMe_3)_3UH]$,[2181] $[(\eta^5\text{-}C_5H_5)_3UR_2]^-$,[2182] $[(\eta^3\text{-}C_3H_5)_2U(OCBu^t_3)_2]$,[2183] $[(\eta^5\text{-}C_5H_5)Co\{P(O)(OEt)_2\}_3U(\eta^5\text{-}C_5H_5)Cl_2]$,[2184] $[NC_5H_3C(2,6\text{-}CH_2C_5H_4\text{-}\eta^5)_2UCl_2]$,[2185] $[\{(\eta^5\text{-}C_5H_4SiMe_3)_3U\}_2(\mu\text{-}O)]$, ($^{13}C$),[2186] $[(\eta^5\text{-}C_5H_4SiMe_3)_3U(N_3)U(\eta^5\text{-}C_5H_4SiMe_3)_3]^+$,[2187] $[(\eta^5\text{-indenyl})_3U(THF)]$,[2188] and $[L_2U(OCPH_2CPh_2O)_2]$.[2189]

5 Solid-state N.M.R. Spectroscopy

This section consists of three main parts: 'Motion in Solids', 'Structure of Solids', and 'Molecules Sorbed Onto Solids'.

A number of reviews have appeared, entitled 'Solid-state nuclear magnetic resonance',[2190] 'Solid-state n.m.r.',[2191] 'Solid state n.m.r. of spin $\frac{1}{2}$ nuclei',[2192] 'High resolution ^{19}F n.m.r. of solids',[2193]

2174 D.F. Mullica and G.A. Wilson, *Inorg. Chim. Acta*, 1991, **177**, 209.
2175 R. Ramasamy, D. Mota de Freitas, C.F.G.C. Geraldes, and J.A. Peters, *Inorg. Chem.*, 1991, **30**, 3188.
2176 M.A. Fedotov, B.Z. Pertsikov, and D.K. Danovich, *Koord. Khim.*, 1991, **17**, 234 (*Chem. Abstr.*, 1991, **114**, 258 309).
2177 M. Wedler, J.W. Gilje, U. Pieper, D. Stalke, M. Noltemeyer, and F.T. Edelmann, *Chem. Ber.*, 1991, **124**, 1163.
2178 J. Ren, F, Pei, and W. Wang, *Chin. J. Chem.*, 1990, 512 (*Chem. Abstr.*, 1991, **115**, 173 158).
2179 K.I. Popov, E.O. Tolkacheva, and N.M. Dyatlova, *Zh. Neorg. Khim.*, 1991, **36**, 1017 (*Chem. Abstr.*, 1991, **115**, 40 519).
2180 K.H. Lee, *J. Magn. Reson.*, 1991, **92**, 247.
2181 J.-C. Berthet, J.-F. Le Maréchal, and M. Ephritikhine, *J. Chem. Soc., Chem. Commun.*, 1991, 360.
2182 C. Villiers and M. Ephritikhine, *New J. Chem.*, 1991, **15**, 559.
2183 C. Baudin, D. Baudry, M. Ephritikhine, M. Lance, A. Navaza, M. Nierlich, and J. Vigner, *J. Organomet. Chem.*, 1991, **415**, 59.
2184 M. Wedler, J.W. Gilje, M. Noltemeyer, and F.T. Edelmann, *J. Organomet. Chem.*, 1991, **411**, 271.
2185 G. Paolucci, R.D. Fischer, F. Benetollo, R. Seraglia, and G. Bombieri, *J. Organomet. Chem.*, 1991, **412**, 327.
2186 J.-C. Berthet, J.-F. Le Maréchal, M. Nierlich, M. Lance, J. Vigner, and M. Ephritikhine, *J. Organomet. Chem.*, 1991, **408**, 335.
2187 J.C. Berthet, M. Lance, M. Nierlich, J. Vigner, and M. Ephritikhine, *J. Organomet. Chem.*, 1991, **420**, C9.
2188 X. Jemine, J. Goffart, S. Bettonville, and J. Fuger, *J. Organomet. Chem.*, 1991, **415**, 363.
2189 C. Villiers, R. Adams, M. Lance, M. Nierlich, J. Vigner, and M. Ephritikhine, *J. Chem. Soc., Chem. Commun.*, 1991, 1144.
2190 P.M. Henrichs and J.M. Hewitt, *Phys. Methods Chem. (2nd Ed.)*, 1990, **5**, 345 (*Chem. Abstr.*, 1991, **114**, 176 648).
2191 F. Horii, *Kobunshi*, 1991, **39**, 888 (*Chem. Abstr.*, 1991, **114**, 198 123).
2192 R.A. Wind, *Pract. Spectrosc.*, 1991, **11**, 125 (*Chem. Abstr.*, 1991, **114**, 176 643).
2193 R.K. Harris and P. Jackson, *Chem. Rev.*, 1991, **91**, 1427.

'N.m.r. of quadrupolar nuclei in the solid state',[2194] 'Solid state n.m.r. of quadrupolar nuclei',[2195] 'Studies of heterogeneous catalysis using high-resolution solid state n.m.r.'[2196] 'Analysis of solid inorganic phosphates',[2197] 'Solid-state nuclear magnetic resonance spectroscopy of metal coordination complexes and organometallics',[2198] 'Solid state ^{13}C n.m.r. of metal carbonyls',[2199] 'High resolution ^{31}P solid state m.a.s. n.m.r. in phosphorus transition metal compounds',[2200] 'N.m.r. spectroscopy of solid electrolytes',[2201] 'High frequency and solid state n.m.r. techniques for the study of ionically conductive glasses',[2202] 'Relationship between nuclear spin relaxation and diffusion in glasses',[2203] 'Nuclear spectroscopic investigations of high temperature superconductor ceramics',[2204] 'Nuclear magnetic resonance for oxide superconductors',[2205] 'Nuclear resonance of high-temperature superconductors',[2206] 'Present status of the theoretical study of high T_c oxides',[2207] 'A μSR and n.m.r. study of the magnetic phases of bismuth based high T_c superconductors',[2208] N.m.r. studies of the electronic and magnetic properties of high T_c materials',[2209] 'Electronic structure of the high T_c superconductors',[2210] '^{205}Tl n.m.r. high T_c superconductors: hyperfine interactions of localised and delocalised electrons',[2211] '^{63}Cu n.q.r. and n.m.r. studies in yttrium barium copper oxide superconductors',[2212] 'Silver sodalites: novel optically responsive nanocomposites',[2213] 'The Al_2SiO_5 polymorphs. Aluminium-silicon disorder in sillimanite',[2214] 'Chemical structural analysis of zeolites by solid state n.m.r.',[2215] 'Zeolite crystallography. Structure determination in the absence of conventional single-crystal data' which contains m.a.s. n.m.r. spectroscopy,[2216] 'One- and two-dimensional high resolution solid state n.m.r. studies of zeolite lattice structures',[2217] 'Basic principles

2194 F. Taulelle, *NATO ASI Ser., Ser. C*, 1990, **322**, 393 (*Chem. Abstr.*, 1991, **114**, 113 529).
2195 J.L. Dye, A.S. Ellaboudy, and J. Kim, *Pract. Spectrosc.*, 1991, **11**, 217 (*Chem. Abstr.*, 1991, **114**, 176 644).
2196 V.M. Mastikhin and K.I. Zamaraev, *Appl. Magn. Reson.*, 1990, **1**, 295 (*Chem. Abstr.*, 1991, **115**, 215 972).
2197 T. Umegaki, *Zairyo Gijutsu*, 1991, **9**, 135 (*Chem. Abstr.*, 1991, **115**, 221 954).
2198 A. Yamasaki, *Coord. Chem. Rev.*, 1991, **109**, 107.
2199 B.E. Hanson, *Met.-Met. Bonds Clusters Chem. Catal., [Proc. Ind.-Univ. Coop. Chem. Program], 7th 1989*, (Pub. 1990), 231. Ed. J.P. Fackler, Plenum, N.Y. (*Chem. Abstr.*, 1991, **115**, 148 495).
2200 R. Gobetto, *Mater. Chem. Phys.*, 1991, **29**, 221 (*Chem. Abstr.*, 1991, **115**, 293 163).
2201 V.M. Buznik, *Ionnye Rasplavy Tverd. Elektrolity*, 1990, **5**, 58 (*Chem. Abstr.*, 1991, **115**, 40 215).
2202 J.H. Kennedy, Z. Zhang, and H. Eckert, *Mater. Res. Soc. Symp. Proc.*, 1991, **210**, 611 (*Chem. Abstr.*, 1991, **115**, 221 379).
2203 H. Jain, O. Kanert, and K.L. Ngai, *Diffus. Defect Data, Pt. A*, 1991, **75**, 163 (*Chem. Abstr.*, 1991, **115**, 104 287).
2204 A.K. Zhetbaev, *Izv. Akad. Nauk SSR, Ser. Fiz.*, 1990, **54**, 1739 (*Chem. Abstr.*, 1991, **114**, 16 115).
2205 M. Takigawa, *Kotai Butsuri*, 1990, **25**, 723 (*Chem. Abstr.*, 1991, **114**, 113 505).
2206 A.A. Gippiks, E.A. Kravchenko, and V.V. Moshchalkov, *Zh. Neorg. Khim.*, 1991, **36**, 568 (*Chem. Abstr.*, 1991, **114**, 219 781).
2207 H. Ebisawa, *Kagaku Kogyo*, 1990, **41**, 205 (*Chem. Abstr.*, 1991, **114**, 33 638).
2208 R. De Renzi, G. Guildi, P. Carretta, G. Calestani, and S.F.J. Cox, *IOP Short Meet. Ser.*, 1989, **22**, 11 (*Chem. Abstr.*, 1991, **115**, 195 421).
2209 H. Alloul, *Physica B (Amsterdam)*, 1991, **169**, 51 (*Chem. Abstr.*, 1991, **115**, 147 251).
2210 T.M. Rice, F. Mila, and F.C. Zhang, *Philos. Trans. R. Soc. London, Ser. A*, 1991, **334**, 459 (*Chem. Abstr.*, 1991, **115**, 142 476).
2211 M. Mehring, F. Hentsch, and N. Winzek, *Springer Ser. Solid-State Sci.*, 1990, **99**, 184 (*Chem. Abstr.*, 1991, **115**, 104 299).
2212 D. Brinkmann, *Springer Ser. Solid-State Sci.*, 1990, **99**, 195 (*Chem. Abstr.*, 1991, **115**, 104 300).
2213 G.A. Ozin, A. Stein, G.D. Stucky, and J.P. Godber, *Inclusion Phenom. Mol. Recognit., [Proc. Int. Symp.], 5th 1988*, (Pub. 1990), 379 (*Chem. Abstr.*, 1991, **115**, 211 228).
2214 D.M. Kerrick, *Rev. Mineral.*, 1990, **22**, 187 (*Chem. Abstr.*, 1991, **114**, 167 933).
2215 S. Nakata, *Kagaku Kogyo*, 1990, **41**, 354 (*Chem. Abstr.*, 1991, **114**, 198 127).
2216 L.B. McCusker, *Acta Crystallogr., Sect. A: Found. Crystallogr.*, 1991, **47**, 297 (*Chem. Abstr.*, 1991, **115**, 82 451).
2217 C.A. Fyfe, Y. Feng, H. Grondey, G.T. Kokotailo, and H. Gies, *Chem. Rev.*, 1991, **91**, 1525.

and recent results of ^{1}H magic angle spinning and pulsed field gradient nuclear magnetic resonance studies on zeolites',[2218] 'Solid state n.m.r. spectroscopy applied to zeolites',[2219] 'Recent advances in techniques for characterising zeolite structures',[2220] 'Solid-state n.m.r. studies of molecular sieve catalysts',[2221] 'The application of n.m.r. to silicates and aluminosilicates',[2222] 'Crystal chemistry of clay minerals' contains ^{29}Si m.a.s. n.m.r. spectroscopy,[2223] 'Solid state nuclear magnetic resonance - new ways of looking at age-old materials' which is about clays,[2224] 'N.m.r. of adsorbed species and surfaces',[2225] 'Solid state n.m.r. techniques for the study of surface phenomena',[2226] 'N.m.r. of adsorbed molecules used as probes for surface investigation',[2227] 'N.m.r. spectroscopy of dynamics at catalytic surfaces',[2228] 'Motion of organic species occluded or sorbed within zeolites',[2229] 'High resolution solid state n.m.r. studies of the effect of temperature and adsorbed organic molecules on zeolite lattice structures',[2230] 'N.m.r. studies on molecular motion of hydrocarbons in different faujasites',[2231] '^{129}Xe n.m.r. of adsorbed xenon: a method for study of zeolites and metal modified zeolites',[2232] and 'N.m.r. spectroscopy of xenon in confined spaces: chlathrates, intercalates, and zeolites'.[2233]

Motion in Solids.—Nuclear spin relaxation and atomic motion in inorganic glasses have been investigated using ^{2}H n.m.r. spectroscopy.[2234] N.m.r. spectroscopy has been used to study the microporosity of hydrophilic materials.[2235] The movement of alkali ions in inorganic glasses has been examined, paying particular attention to lithium in lithium triborate glasses.[2236] ^{2}H n.m.r. studies of solid and liquid hydrates of LiOH, NaOH, KOH, and CsOH indicate that the $^2H^+$ is mobile.[2237] Lithium mobility in Li_4GeO_4, Li_2CrO_4, and Li_2WO_4 has been studied by ^{7}Li n.m.r. spectroscopy.[2238]

2218 H. Pfeifer, D. Freude, and J. Kaerger, *Stud. Surf. Sci. Catal.*, 1991, **65**, 89 (*Chem. Abstr.*, 1991, **115**, 196 366).
2219 G. Engelhardt, *Stud. Surf. Sci. Catal.*, 1991, **58**, 285 (*Chem. Abstr.*, 1991, **115**, 104 298).
2220 D.E.W. Vaughan, M.M.J. Treacy, and J.M. Newsam, *NATO ASI Ser., Ser. B*, 1990, **221**, 99 (*Chem. Abstr.*, 1991, **114**, 219 969).
2221 J. Klinowski, *Chem. Rev.*, 1991, **91**, 1459.
2222 T. Maekawa and H. Maekawa, *Kobutsugaku Zasshi*, 1990, **19**, 339 (*Chem. Abstr.*, 1991, **115**, 53 491).
2223 M. Sato, *Seramikkusu*, 1991, **26**, 288 (*Chem. Abstr.*, 1991, **115**, 194 369).
2224 K.J.D. MacKenzie, *Chem. N.Z.*, 1991, **55**, 30 (*Chem. Abstr.*, 1991, **115**, 261 435).
2225 J.B. Nagy, *NATO ASI Ser., Ser. C*, 1990, **322**, 371 (*Chem. Abstr.*, 1991, **115**, 99 854).
2226 J.P. Ansermet, C.P. Slichter, and J.H. Sinfelt, *Prog. Nucl. Magn. Reson. Spectrosc.*, 1990, **22**, 401 (*Chem. Abstr.*, 1991, **115**, 240 328).
2227 J. Fraissard, *Stud. Surf. Sci. Catal.*, 1990, **57**, B201 (*Chem. Abstr.*, 1991, **114**, 69 816).
2228 T.M. Duncan, *Shokubai*, 1990, **32**, 510 (*Chem. Abstr.*, 1991, **114**, 172 233).
2229 J.M. Newsam, B.G. Silbernagel, M.T. Melchior, T.O. Brun, and F. Trouw, *Inclusion Phenom. Mol. Recognit., [Proc. Int. Symp.], 5th 1988*, (Pub. 1990), 325. Ed. J.L. Atwood (*Chem. Abstr.*, 1991, **115**, 162 277).
2230 C.A. Fyfe, G.T. Kokotailo, H. Gies, and H. Strobl, *Inclusion Phenom. Mol. Recognit., [Proc. Int. Symp.], 5th 1988*, (Pub. 1990), 299. Ed. J.L. Atwood (*Chem. Abstr.*, 1991, **115**, 162 276).
2231 H.T. Lechert, W.D. Basler, and M. Jia, *NATO ASI Ser., Ser. B*, 1990, **221**, 183 (*Chem. Abstr.*, 1991, **115**, 79 045).
2232 M.A. Springuel-Huet and J. Fraissard, *Pet. Tech.*, 1991, **361**, 90 (*Chem. Abstr.*, 1991, **115**, 259 220).
2233 C. Dybowski, N. Bansal, and T.M. Duncan, *Annu. Rev. Phys. Chem.*, 1991, **42**, 433 (*Chem. Abstr.*, 1991, **115**, 293 162).
2234 O. Kanert, J. Steinert, H. Jain, and K.L. Ngai, *J. Non-Cryst. Solids*, 1991, **131**, 1001 (*Chem. Abstr.*, 1991, **115**, 97 581); G. Williams, *J. Non-Cryst. Solids*, 1991, **131**, 1 (*Chem. Abstr.*, 1991, **115**, 99 821).
2235 V.V. Zhizhenkov and E.A. Egorov, *Granitsa Razdela, Prochn. i Razrushenie Kompozition. Mater*, 1989, 137. From *Ref. Zh., Khim*, 1991, Abstr. No 2M61 (*Chem. Abstr.*, 1991, **115**, 264 135).
2236 H. Jain, *J. Non-Cryst. Solids*, 1991, **131**, 961 (*Chem. Abstr.*, 1991, **115**, 147 465).
2237 Z. Borkowska, U. Frese, E. Kriegsmann, U. Stimming, M. Gavish, K.J. Adamic, and S.G. Greenbaum, *Ber. Bunsenges. Phys. Chem.*, 1991, **95**, 1033.
2238 E.I. Burmakin and A.A. Lakhtin, *Izv. Akad. Nauk SSSR, Neorg. Mater.*, 1991, **27**, 837 (*Chem. Abstr.*, 1991, **115**, 40 521).

^{7}Li T_1 and $T_{1\rho}$ values have been measured in $(1-x)GeO_2.{}_xLi_2O$ glasses and Li^+ jumps observed.[2239] β-radiation detected n.m.r. has been used to observe ^{8}Li and ^{12}B dynamics in $Li_2O.3B_2O_3$ glasses.[2240] Temperature dependent ^{7}Li n.m.r. spectroscopy has been used to study lithium motion in Li_2CO_3 based lead borate glasses. The ^{11}B n.m.r. spectra were also examined.[2241] The magnitude of F^- conduction in xLiF.(95-x)ZrF_4.5LaF_3 glasses has been examined using ^{19}F n.m.r. spectroscopy.[2242] Ionic transport in LiBr mixed with alumina has been investigated using ^{7}Li T_1, $T_{1\rho}$, and T_2 measurements.[2243] ^{19}F n.m.r. spectroscopy has been used to study ionic mobility in $Na_{1-x}Y_xF_{1+2x}$.[2244] ^{23}Na n.m.r. spectra of NaCl single crystal have been used to investigate the deformation induced atomic diffusion.[2245] Order-disorder in $RbCaF_3$ has been investigated using ^{87}Rb n.m.r. spectroscopy.[2246] The effects of tunnelling of hydrogen on the ^{1}H n.m.r. line shape in $Rb_{1-x}(NH_4)_xH_2PO_4$ have been studied.[2247] The T_1 of ^{87}Rb in $Rb_{0.56}(ND_4)_{0.44}$ is due to the thermally activated jumping of D in O—D····O bonds. This results in a large isotope effect.[2248] ^{1}H n.m.r. spectroscopy has been used to investigate proton diffusion in CsH_3O_2.[2249] Tunnelling and T_1 of hydrogen dissolved in scandium metal has been investigated.[2250] The phase change in $(Pr^iOH)_2Ce_2(OPr^i)_8$ has been investigated by ^{2}H n.m.r. spectroscopy.[2251]

^{1}H n.m.r. spectroscopy has been used to study hydrogen diffusion in $ZrV_2H_x(D)_x$ and $HfV_2H_x(D)_x$.[2252] N.m.r. spectroscopy has been used to study hydrogen mobility in $ZrCr_2H_x$.[2253] ^{13}C n.m.r. spectroscopy has been used to study stereochemical non-rigidity of $[(\eta^5-C_5H_5)_2Ti(\eta^1-C_5H_5)_2]$. $E_a = 33.2\pm1.0$ kJ mol^{-1} and A = 2.9 x 10^{10} s^{-1}.[2254] Dynamic processes such as diene flip and ring reorientation have been investigated in compounds such as $[(\eta^5-C_5H_4Bu^t)_2ZrC_4H_6]$ using

2239 O. Kanert, M. Kloke, R. Küchler, S. Ruckstein, and H. Jain, *Ber. Bunsenges. Phys. Chem.*, 1991, **95**, 1061.
2240 A. Schirmer, P. Heitjans, B. Bader, P. Freilaender, H.J. Stoeckmann, and H. Ackermann, *J. Phys.: Condens. Matter*, 1991, **3**, 4323 (*Chem. Abstr.*, 1991, **115**, 55 149).
2241 B. Wang, S.P. Szu, and M. Greenblatt, *J. Non-Cryst. Solids*, 1991, **134**, 249 (*Chem. Abstr.*, 1991, **115**, 237 814).
2242 Y. Kawamoto, R. Kanno, and J. Fujiwara, *J. Mater. Sci. Lett.*, 1991, **10**, 804 (*Chem. Abstr.*, 1991, **115**, 140 851).
2243 J.H. Strange, S.M. Rageb, and R.C.T. Slade, *Philos. Mag. A*, 1991, **64**, 1159 (*Chem. Abstr.*, 1991, **115**, 292 158).
2244 A.D. Toshmatov, F.L. Aukhadeev, D.N. Terpilovskii, V.A. Dudkin, S.L. Korableva, and L.D. Livanova, *Fiz. Tverd. Tela (Leningrad)*, 1990, **32**, 2563 (*Chem. Abstr.*, 1991, **115**, 19 525); A.D. Toshmatov, F.L. Aukhadeev, D.N. Terpilovskii, V.A. Dudkin, R.Sh. Zhdanov, S.L. Korableva, Sh.I. Yagudin, and L.D. Livanova, *Paramagnit. Rezonans*, 1990, 107. From *Ref. Zh., Fiz. (A-Zh.)*, 1991, Abstr. No. 5N703 (*Chem. Abstr.*, 1991, **115**, 292 098).
2245 K. Detemple, O. Kanert, K.L. Murty, and J.T.M. De Hosson, *Phys. Rev. B; Condens. Matter*, 1991, **44**, 1988 (*Chem. Abstr.*, 1991, **115**, 146 962).
2246 D. Ledue and J. Teillet, *Phys. Status Solidi A*, 1991, **126**, 335 (*Chem. Abstr.*, 1991, **115**, 245 582).
2247 S. Dattagupta, B. Tadic, R. Pirc, and R. Blinc, *Phys. Rec. B: Condens. Matter*, 1991, **44**, 4387 (*Chem. Abstr.*, 1991, **115**, 173 182).
2248 J. Dolinsek, *Wiss. Beitr. -Martin-Luther-Univ. Halle-Wittenberg*, 1990, 87 (*Chem. Abstr.*, 1991, **115**, 148 717).
2249 R.E. Lechner, H.J. Bleif, H. Dachs, R. Marx, M. Stahn, and I. Anderson, *Solid State Ionics*, 1991, **46**, 25 (*Chem. Abstr.*, 1991, **115**, 142 892).
2250 I. Svare, D.R. Torgeson, and F. Borsa, *Phys. Rev. B: Condens. Matter*, 1991, **43**, 7448 (*Chem. Abstr.*, 1991, **114**, 239 154).
2251 B.A. Huggins, P.D. Ellis, P.S. Gradeff, B.A. Vaartstra, K. Yunlu, J.C. Huffman, and K.G. Caulton, *Inorg. Chem.*, 1991, **30**, 1720.
2252 A.V. Skripov, M.Yu. Belyaev, S.V. Rychkova, and A.P. Stepanov, *J. Phys.: Condens. Matter*, 1991, **3**, 6277 (*Chem. Abstr.*, 1991, **115**, 190 283).
2253 A.V. Skripov, M.Yu. Belyaev, and A.P. Stepanov, *Solid State Commun.*, 1991, **78**, 909 (*Chem. Abstr.*, 1991, **115**, 173 165).
2254 S.J. Heyes and C.M. Dobson, *J. Am. Chem. Soc.*, 1991, **113**, 463.

^{13}C n.m.r. spectroscopy.[2255] The fluxionality of $[(\eta^8\text{-}C_8H_8)(\eta^4\text{-}C_8H_8)Zr]$ has been investigated using solid state ^{13}C n.m.r. spectroscopy.[2256] Proton dynamics in $TiS_2\text{-}NH_3$ intercalates have been investigated.[2257] Lithium motion in pure and Ta doped $LiZr_2(PO_4)_3$ has been investigated using ^7Li and ^{31}P n.m.r. spectroscopy.[2258] N.m.r. spectroscopy has been used to study ionic mobility in xerogel polyvanadates.[2259] ^1H and ^{19}F n.m.r. spectra have been used to study the dynamics of the ionic groups in $(NH_4)_2ZrF_6$ and $(NH_4)_2HfF_6$.[2260] Na^+ motion and electron hopping in β-$Na_{0.33}V_2O_5$ have been investigated.[2261]

^{13}C and ^{31}P c.p. m.a.s. n.m.r. studies of the polytopal ligand rearrangement processes in $[WH_6(PMe_3)_3]$ have been reported.[2262] Internal rotation about the $C^2\text{-}C^3$ bond has been observed in the ^{13}C c.p. m.a.s. n.m.r. spectrum of $[M_2(\eta^3\text{-}HC^1C^2C^3Me_2)(CO)_4(\eta^5\text{-}C_5H_5)_2]^+$, M = Mo, W.[2263] The dynamic behaviour in the solid state of $[(\eta^6\text{-}C_6Me_6)Cr(CO)_3]$, $[(\eta^6\text{-}1,2,3\text{-}C_6Me_3H_3)Cr(CO)_3]$, and $[(\eta^6\text{-}1,2,4,6\text{-}C_6Me_4H_2)Cr(CO)_3]$ has been examined using variable temperature ^1H T_1 measurements. The lowest energy process is methyl rotation and C_6Me_6 reorientation next.[2264] Variable temperature ^{13}C c.p. m.a.s. n.m.r. spectroscopy has been used to deduce $\Delta G^\ddagger$ = *ca* 60 kJ mol^{-1} for $Cr(CO)_3$ rotation in (128).[2265]

Transport properties in the $Li_xMo_2O_4$,[2266] $Li_xW_3O_9F$,[2267] and LiW_2O_9F systems[2268] have been

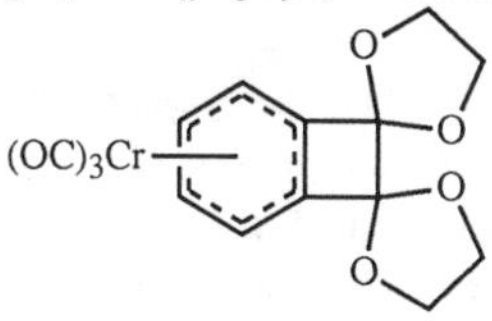

(128)

investigated using ^7Li n.m.r. spectroscopy.

Ferrocene enchlathrated in deoxycholic acid[2269] and thiourea[2270] has been investigated using ^{13}C and ^2H n.m.r. spectroscopy. A kinetic analysis of the ^{13}C c.p. m.a.s. n.m.r. spectrum of

2255 D. Braga, F. Grepioni, and E. Parisini, *Organometallics*, 1991, **10**, 3735.
2256 D.M. Rogers, S.R. Wilson, and G.S. Girolami, *Organometallics*, 1991, **10**, 2419.
2257 P.F. McMillan, V. Cajipe, P. Molinie, M.F. Quinton, V. Gourlaouen, and P. Colombet, *Chem. Mater.*, 1991, **3**, 796 (*Chem. Abstr.*, 1991, **115**, 166 961).
2258 D. Peit, F. Chaput, and J.P. Boilot, *Mater. Sci. Monogr.*, 1991, **66C**, 2275 (*Chem. Abstr.*, 1991, **115**, 238 022).
2259 A.V. Dmitriev, T.A. Denisova, S.G. Arsenov, V.L. Volkov, and R.N. Pletnev, *Zh. Neorg. Khim.*, 1991, **36**, 1786 (*Chem. Abstr.*, 1991, **115**, 219 819).
2260 V.Ya. Kavun, V.I. Sergienko, B.N. Chernyshov, B.V. Bukvetskii, N.A. Dedenko, N.G. Bakeeva, and L.N. Ignat'eva, *Zh. Neorg. Khim.*, 1991, **36**, 1004 (*Chem. Abstr.*, 1991, **114**, 257 636).
2261 J.C. Badot, N. Baffier, and A. Fourrier-Lamer, *J. Non-Cryst. Solids*, 1991, **131**, 1227 (*Chem. Abstr.*, 1991, **115**, 124 823).
2262 S.J. Heyes, M.L.H. Green, and C.M. Dobson, *Inorg. Chem.*, 1991, **30**, 1930.
2263 M.V. Galakhov, V.I. Bakhmutov, and V.I. Barinov, *Magn. Reson. Chem.*, 1991, **29**, 506.
2264 S. Aime, D. Braga, R. Gobetto, F. Grepioni, and A. Orlandi, *Inorg. Chem.*, 1991, **30**, 951.
2265 H.G. Wey, P. Betz, I. Topalović, and H. Butenschön, *J. Organomet. Chem.*, 1991, **411**, 369.
2266 S. Colson, J.M. Tarascon, S. Szu, and L.C. Klein, *Mater. Res. Soc. Symp. Proc.*, 1991, **210**, 405 (*Chem. Abstr.*, 1991, **115**, 219 812).
2267 M. Menetrier, S.H. Chang, K.S. Suh, J. Senegas, J.P. Chaminade, and C. Delmas, *Mater. Res. Soc. Symp. Proc.*, 1991, **210**, 345 (*Chem. Abstr.*, 1991, **115**, 215 372).
2268 C. Delmas, S.H. Chang, M. Menetrier, K.S. Suh, J. Senegas, and J.P. Chaminade, *Solid State Ionics*, 1990, **40-41(Pt. 2)**, 560 (*Chem. Abstr.*, 1991, **114**, 54 657).
2269 F. Imashiro, N. Kitazaki, D. Kuwahara, T. Nakai, and T. Terao, *J. Chem. Soc., Chem. Commun.*, 1991, 85.
2270 S.J. Heyes, N.J. Clayden, and C.M. Dobson, *J. Phys. Chem.*, 1991, **95**, 1547.

$[Os_3(CO)_8(\eta^2-C_2H_4)(\mu_3:\eta^2:\eta^2:\eta^2-C_6H_6)]$ has indicated the presence of four intramolecular processes.[2271] Dynamic processes in $[MRu_3(CO)_9(C_2Bu^t)(PPh_3)]$, M = Cu, Ag, have been investigated by ^{31}P n.m.r. spectroscopy.[2272] The fluxionality of $RuCl_2$ adsorbed on a polysiloxane supported ether phosphine has been investigated by ^{31}P n.m.r. spectroscopy.[2273] Hydrogen tunnelling in PdH_x has been studied.[2274] Variable temperature ^{127}I m.a.s. n.m.r. spectroscopy has been used to study Ag^+ diffusion in β-AgI.[2275]

The reorientation of $[(D_3C)_3NH][BPh_4]$ has been investigated using ^2H n.m.r. spectroscopy.[2276] Anion diffusion in the poly(*p*-phenylene)-tetrafluoroborate system has been studied using ^1H and ^{19}F n.m.r. spectroscopy.[2277] Solids containing $[PCl_4]^+$, $[PBr_4]^+$, and $[BX_4]^-$ have been investigated by ^{11}B and ^{31}P n.m.r. spectroscopy.[2278] Cationic self-diffusion in ionic plastic phases of $TlNO_2$ and $TlNO_3$ in TlNCS has been studied using ^{203}Tl and ^{205}Tl n.m.r. spectroscopy.[2279] ^1H n.m.r. spectroscopy has been used to investigate the dynamics of the cation in Me_4NTlCl_4 and Et_4NTlCl_4.[2280]

The solid state dynamics of C_{60} and C_{70} have been studied using ^{13}C n.m.r. spectroscopy.[2281] ^{133}Cs n.m.r. spectroscopy has been used to study cation mobility in zeolite A.[2282] N.m.r. diffusion experiments have been used to study anisotropic diffusion in ZSM-5 type zeolites.[2283]

Motion in dimethyl siloxane networks has been investigated by ^2H n.m.r. spectroscopy.[2284] ^1H n.m.r. spectroscopy has been used to study the dynamics of hydrogen-containing groups in the structure of hydrosodalite.[2285] Ionic diffusion in alkali silicate glass has been investigated.[2286] Slow atomic motions close to the glass transition in $K_2Si_4O_9$ have been studied using two-dimensional ^{29}Si n.m.r. spectroscopy.[2287] Molecular diffusion and n.m.r. relaxation of water in unsaturated porous silica glass have been investigated.[2288] The structure and dynamics of chemically modified silica gels

2271 S.J. Heyes, C.M. Dobson, M.A. Gallop, B.F.G. Johnson, and J. Lewis, *Inorg. Chem.*, 1991, **30**, 3850.
2272 S.S.D. Brown, I.D. Salter, D.J. Smith, N.J. Clayden, and C.M. Dobson, *J. Organomet. Chem.*, 1991, **408**, 439.
2273 E. Lindner, A. Bader, and H.A. Mayer, *Z. Anorg. Allg. Chem.*, 1991, **598**, 235.
2274 R.L. Armstrong, *Bull. Magn. Reson.*, 1990, **12**, 116 (*Chem. Abstr.*, 1991, **115**, 221 494).
2275 G.W. Wagner, *Solid State Ionics*, 1991, **47**, 143 (*Chem. Abstr.*, 1991, **115**, 246 285).
2276 M.L.H. Gruwel, *J. Chem. Soc., Faraday Trans.*, 1991, **87**, 1715.
2277 V.G. Shteinberg, B.A. Shumm, A.F. Zueva, I.N. Erofeev, and O.N. Efimov, *Synth. Met.*, 1991, **41**, 321 (*Chem. Abstr.*, 1991, **115**, 61 980).
2278 K.B. Dillon, R.K. Harris, P.N. Gates, A.S. Muir, and A. Root, *Spectrochim. Acta, Part A*, 1991. **47**, 831 (*Chem. Abstr.*, 1991, **115**, 173 174).
2279 Y. Furukawa, H. Nagase, R. Ikeda, and D. Nakamura, *Bull. Chem. Soc. Jpn.*, 1991, **64**, 3105.
2280 M. Lenck, S.Q. Dou, and A. Weiss, *Z. Naturforsch., A: Phys. Sci.*, 1991, **46**, 777 (*Chem. Abstr.*, 1991, **115**, 219 539).
2281 R. Tycko, G. Dabbagh, R.C. Haddon, D.C. Douglass, A.M. Mujsce, M.L. Kaplan, and A.R. Kortan, *Mater. Res. Soc. Symp. Proc.*, 1991, **206**, 703 (*Chem. Abstr.*, 1991, **115**, 263 834); R.D. Johnson, C.S. Yannoni, J. Salem, G. Meijer, and D.S. Bethune, *Mater. Res. Soc. Symp. Proc.*, 1991, **206**, 715 (*Chem. Abstr.*, 1991, **115**, 263 836).
2282 M.K. Ahn and L.E. Iton, *J. Phys. Chem.*, 1991, **95**, 4496.
2283 J. Kärger, *J. Phys. Chem.*, 1991, **95**, 5558 (*Chem. Abstr.*, 1991, **115**, 36 051).
2284 C.D. Poon, E.T. Samulski, and A.I. Nakatani, *Makromol. Chem., Macromol. Symp.*, 1990, **40**, 109 (*Chem. Abstr.*, 1991, **115**, 9922).
2285 V.A. Detinich, *Kristallografiya*, 1991, **36**, 662 (*Chem. Abstr.*, 1991, **115**, 82 577).
2286 G.N. Greaves, S.J. Gurman, C.R.A. Catlow, A.V. Chadwick, and S. Houde-Walter, *Report*, 1990, **DL/SCI/P-705E**; Order No. PB91-122184, 16 pp. Avail. NTIS. From *Gov. Rep. Announce Index (U.S.)*, 1991, **91**, Abstr. No. 111 411 (*Chem. Abstr.*, 1991, **115**, 261 514).
2287 I. Farnan and J.F. Stebbins, *J. Non-Cryst. Solids*, 1990, **124**, 207 (*Chem. Abstr.*, 1991, **114**, 30 522).
2288 F. D'Orazio, S. Bhattacharja, W.P. Halperin, K. Eguchi, and T. Mizusaki, *Phys. Rev. B: Condens. Matter*, 1990, **42**, 9810 (*Chem. Abstr.*, 1991, **114**, 54 530).

have been investigated using ^{13}C and ^{29}Si c.p. m.a.s. n.m.r. spectroscopy.[2289] Molecular motions in Me_3MCl, M = Sn, Pb, have been investigated by measuring the second ^{1}H moment as a function of temperature.[2290] Fluoride ion diffusion and proton motion in MSn_2F_5, M = $[NH_4]^+$, Rb^+, Cs^+,[2291] and $N_2H_5Sn_3F_7$,[2292] have been investigated using ^{1}H and ^{19}F T_1, $T_{1\rho}$, and T_2 measurements. Anionic transport in PbF_2 has been investigated using ^{19}F n.m.r. spectroscopy.[2293]

Proton tunnelling in ammonium compounds has been studied.[2294] ^{1}H and ^{19}F second moment and T_1 measurements on $(NH_4)_3AlF_6$ have confirmed that electrical conductivity is attributable to self diffusion of $[NH_4]^+$.[2295] Motion in $(ND_4)_2SnCl_6$ below 60 K has been studied using ^{2}H T_1 measurements.[2296] Molecular motions and phase transitions in $MeNH_3PbX_3$ have been investigated by ^{1}H T_1 measurements.[2297] ^{1}H n.m.r. studies of molecular motions of $[Bu^tNH_3]^+$ in $[Bu^tNH_3]_2[TeX_6]$ have been reported.[2298] ^{1}H n.m.r. spectroscopy has been used to study cationic motion in solid $[Bu^tNH_3]Cl$ and $[Bu^tNH_3]Br$.[2299] The molecular dynamics of $(N_2H_5)_2SO_4$ have been investigated by ^{1}H T_1 and second moment measurements.[2300] The dynamics of $[Me_4N]^+$ in sodalite have been investigated by ^{2}H and ^{13}C n.m.r. spectroscopy.[2301] A novel ionic plastic phase of $[Me_4N][NCS]$ has been investigated using ^{1}H T_1 and $T_{1\rho}$ measurements.[2302] The temperature dependence of ^{1}H T_1 of $[Me_4N]_2[PbCl_6]$ shows two minima at 190 and 115 K.[2303]

The activation energy for ion reorientation in Na_2DPO_3 has been determined.[2304] Methyl group rotation in Me_4SbI has been investigated.[2305] Two-dimensional solid state ^{31}P n.m.r. spectroscopy has been used to study diluent motion in a phosphate containing glass.[2306]

Structure of Solids.——The basic equations for the solid state n.m.r. line shapes of quadrupole nuclei have been presented and used in microcomputer programs for the simulation of spectra with or

2289 K. Albert and E. Bayer, *J. Chromatogr.*, 1991, **544**, 345 (*Chem. Abstr.*, 1991, **115**, 20 742).
2290 D. Zhang, S.Q. Dou, and A. Weiss, *Z. Naturforsch., A*, 1991, **46**, 337 (*Chem. Abstr.*, 1991, **115**, 20 726).
2291 K. Hirokawa, H. Kitahara, Y. Furukawa, and D. Nakamura, *Ber. Bunsenges. Phys. Chem.*, 1991, **95**, 651.
2292 W. Granier, T. Ala, S. Vilminot, J.P. Battut, J. Dupuis, and S. Soudani, *Solid State Ionics*, 1991, **44**, 159 (*Chem. Abstr.*, 1991, **114**, 176 951).
2293 S.G. Bakhvalov, V.M. Buznik, V.A. Vopilov, I.S. Kernasyuk, and A.N. Matsulev, *Zh. Neorg. Khim.*, 1990, **35**, 3013 (*Chem. Abstr.*, 1991, **115**, 39 382).
2294 M. Punkkinen, A.H. Vuorimaki, and E.E. Ylinen, *Physica B (Amsterdam)*, 1990, **160**, 338 (*Chem. Abstr.*, 1991, **114**, 34 626).
2295 Y. Furukawa, A. Sasaki, and D. Nakamura, *Solid State Ionics*, 1990, **42**, 223 (*Chem. Abstr.*, 1991, **114**, 15 554).
2296 L.P. Ingman, E. Koivula, M. Punkkinen, E.E. Ylinen, and Z.T. Lalowicz, *Physica B (Amsterdam)*, 1990, **162**, 281 (*Chem. Abstr.*, 1991, **114**, 34 637).
2297 Q. Xu, T. Eguchi, H. Nakayama, N. Nakamura, and M. Kishita, *Z. Naturforsch., A*, 1991, **46**, 240 (*Chem. Abstr.*, 1991, **114**, 237 988).
2298 H. Ishida, S. Inada, N. Hayama, D. Nakamura, and R. Ikeda, *Ber. Bunsenges. Phys. Chem.*, 1991, **95**, 866.
2299 H. Ishida, S. Inada, N. Hayama, D. Nakamura, and R. Ikeda, *Z. Naturforsch., A*, 1991, **46**, 265 (*Chem. Abstr.*, 1991, **114**, 258 310).
2300 K. Ganesan, R. Damle, and J. Ramakrishna, *Spectrochim. Acta, Part A*, 1990, **46**, 1549 (*Chem. Abstr.*, 1991, **114**, 16 332).
2301 C.J.J. Den Ouden, K.P. Datema, F. Visser, M. Mackay, and M.F.M. Post, *Zeolites*, 1991, **11**, 418 (*Chem. Abstr.*, 1991, **115**, 79 682).
2302 T. Tanabe, D. Nakamura, and R. Ikeda, *J. Chem. Soc., Faraday Trans.*, 1991, **87**, 987.
2303 Y. Furukawa, Y. Baba, S. Gima, M. Kaga, T. Asaji, R. Ikeda, and D. Nakamura, *Z. Naturforsch., A*, 1991, **46**, 809 (*Chem. Abstr.*, 1991, **115**, 221 526).
2304 J.D. Trudeau, J.L. Schwartz, and T.C. Farrar, *J. Phys. Chem.*, 1991, **95**, 9788.
2305 G. Burbach and A. Weiss, *Z. Naturforsch., A*, 1991, **46**, 759 (*Chem. Abstr.*, 1991, **115**, 221 525).
2306 C. Zhang, P. Wang, A.A. Jones, P.T. Inglefield, and R.T. Kambour, *Macromolecules*, 1991, **24**, 338 (*Chem. Abstr.*, 1991, **114**, 24 929).

without magic-angle spinning.[2307] Novel measurements of nuclear spin cross-relaxation in metal hydrides have been reported.[2308] N.m.r. spectroscopy has been used to determine spatially resolved pore size information.[2309] Considerations and examples for a uniform behaviour of spin-lattice relaxation in glasses have been discussed.[2310]

^{6}Li m.a.s. n.m.r. spectroscopy has been proposed as a powerful probe of lithium containing materials. It was compared with ^{7}Li n.m.r. spectroscopy.[2311] The ^{14}N nuclear magnetic shielding tensor has been determined in single crystals of Li_3N and KNO_3.[2312] High resolution solid state ^{7}Li n.m.r. spectra of $[Li(12\text{-crown-}4)_2][Li(\{N(SiMe_3)_2\}_2SPh)_2]$ have been reported.[2313] ^{6}Li, ^{7}Li and ^{133}Cs n.m.r. spectroscopy has been used to study macroscopically oriented hydrated Li- and Cs-DNA fibres.[2314] Polymer electrolytes based on composites of poly{bis(methoxyethoxy)phosphazene} and poly(ethylene oxide) have been studied using ^{7}Li n.m.r. spectroscopy.[2315] ^{13}C and ^{6}Li solid state n.m.r. spectroscopy has been used to characterise the morphology and dynamics of $LiClO_4$/poly(ethylene oxide) electrolytes.[2316] ^{19}F and ^{31}P n.m.r. spectroscopy has been used to study the structural state of fluorine in $(1\text{-}x)LiPO_{3}.{}_{x}LiF$.[2317] The ^{7}Li n.m.r. linewidths and T_1 for poly(propylene glycol) complexed with LiO_3SCF_3 have been reported.[2318] Relaxation processes in ionic conductive chalcogenide glasses have been studied by n.m.r. spectroscopy.[2319] Polarization transport of the β-active ^{8}Li nuclei in the spatially disordered ^{6}Li system of LiF crystals has been carried out using the β n.m.r. technique.[2320] Ion-ion interactions of Li^+ in $LiAlCl_4$ and $LiGaCl_4$ in SO_2 have been investigated using ^{7}Li n.m.r. spectroscopy.[2321]

A simple procedure for the determination of the quadrupole interaction parameters and isotropic chemical shifts from m.a.s. n.m.r. spectra of half-integer spin nuclei in solids has been applied to ^{23}Na and ^{27}Al.[2322] Satellite transitions in m.a.s. n.m.r. spectra of quadrupole nuclei have been applied to ^{17}O, ^{23}Na, and ^{27}Al m.a.s. n.m.r. spectroscopy.[2323] Quadrupolar perturbed ^{23}Na and ^{35}Cl n.m.r. spectroscopy has been shown to be a powerful method to investigate the Na and Cl sites in

2307 J.P. Amoureux, C. Fernandez, and P. Granger, *NATO ASI Ser., Ser. C*, 1990, **322**, 409 (*Chem. Abstr.*, 1991, **114**, 176 897).
2308 D.B. Baker, M.S. Conradi, R.E. Norberg, D.R. Torgeson, and R.G. Barnes, *J. Less-Common Met.*, 1991, **172**, 379 (*Chem. Abstr.*, 1991, **115**, 293 311).
2309 B. Ewing, P.J. Davis, P.D. Majors, G.P. Drobny, D.M. Smith, and W.L. Earl, *Stud. Surf. Sci. Catal.*, 1991, **62**, 709 (*Chem. Abstr.*, 1991, **115**, 240 736).
2310 M. Gruene and W. Mueller-Warmuth, *Ber. Bunsenges. Phys. Chem.*, 1991, **95**, 1068.
2311 S.P. Bond, A. Gelder, J. Homer, W.R. McWhinnie, and M.C. Perry, *J. Mater. Chem.*, 1991, **1**, 327 (*Chem. Abstr.*, 1991, **115**, 221 495).
2312 T.J. Bastow and S.N. Stuart, *Chem. Phys. Lett.*, 1991, **180**, 305 (*Chem. Abstr.*, 1991, **115**, 40 550).
2313 F. Pauer, J. Rocha, and D. Stalke, *J. Chem. Soc., Chem. Commun.*, 1991, 1477.
2314 L. Einarsson, L. Nordenskioeld, A. Rupprecht, I. Furo, and T.C. Wong, *J. Magn. Reson.*, 1991, **93**, 34.
2315 K.J. Adamic, S.G. Greenbaum, K.M. Abraham, M. Alamgir, M.C. Wintersgill, and J.J. Fontanella, *Chem. Mater.*, 1991, **3**, 534 (*Chem. Abstr.*, 1991, **114**, 229 974).
2316 J.F. O'Gara, G. Nazri, and D.M. MacArthur, *Solid State Ionics*, 1991, **47**, 87 (*Chem. Abstr.*, 1991, **115**, 216 268).
2317 N.I. Yumashev, A.A. Pronkin, A.A. Il'in, and L.V. Yumasheva, *Fiz. Khim. Stekla*, 1991, **17**, 210 (*Chem. Abstr.*, 1991, **115**, 15 988).
2318 S.H. Chung, K.R. Jeffrey, and J.R. Stevens, *J. Chem. Phys.*, 1991, **94**, 1803.
2319 A. Pradel and M. Ribes, *J. Non-Cryst. Solids*, 1991, **131**, 1063 (*Chem. Abstr.*, 1991, **115**, 103 587).
2320 Yu.G. Abov, M.I. Bulgakov, S.P. Borovlev, A.D. Gul'ko, V.M. Garochkin, F.S. Dzheparov, S.V. Stepanov, S.S. Trostin, and V.E. Shestopal, *Zh. Eksp. Teor. Fiz.*, 1991, **99**, 962 (*Chem. Abstr.*, 1991, **114**, 258 334).
2321 T.J. Lee, P.C. Yao, and G.T.K. Fey, *J. Power Sources*, 1991, **45**, 143 (*Chem. Abstr.*, 1991, **115**, 95 866).
2322 G. Engelhardt and H. Koller, *Magn. Reson. Chem.*, 1991, **29**, 941.
2323 J. Skibsted, N.C. Nielsen, H. Bildsoe, and H.J. Jakobsen, *J. Magn. Reson.*, 1991, **95**, 88.

$Na(CN)_xCl_{1-x}$ and $Na_xK_{1-x}CN$.[2324] The ^{13}C chemical shifts of [Na(15-crown-5)(SCN)] have been analysed in terms of torsional angle influences.[2325] A high resolution n.m.r. spectroscopic study of the central transition of $I = \frac{3}{2}$ quadrupolar nuclei has been applied to ^{23}Na in $NaNO_2$ and Na_2SO_4 powders.[2326] Quadrupole broadening of the n.m.r. satellites of ^{23}Na has been applied to single crystals of $NaNO_2$.[2327] Local structure of polyethylene glycol-NaI solid polyelectrolyte has been studied using ^{13}C and ^{23}Na n.m.r. spectroscopy.[2328] The techniques of n.m.r. imaging have been developed for simple inorganic solids and applied to mechanically induced defects in NaCl.[2329] Acoustic n.m.r. measurements have been reported for ^{23}Na in NaCl crystals irradiated with 20 MHz longitudinal acoustic waves.[2330]

The ^{39}K nuclear quadrupole coupling parameters have been measured in 16 common inorganic potassium salts.[2331] A ^{39}K n.m.r. study of the antiferroelectric phase transition in KSCN has been reported.[2332] The temperature variation of the ^2H and ^{39}K nuclear quadrupole coupling constants has been measured for KOH and KOD.[2333] The phase transitions in NH_4, Cs, K, and Rb perbromates have been studied using n.m.r. spectroscopy.[2334] The addition of a small amount of KBr to modify the susceptibility of materials has been used to assess bulk susceptibility using ^{13}C c.p. m.a.s. n.m.r. spectroscopy. This was demonstrated with the addition of $MnSO_4.H_2O$ and vermiculite.[2335] The effect of laser irradiation on the T_1 in KCl doped with Cs has been studied.[2336]

A two-dimensional ^{87}Rb n.m.r. separation of inhomogeneous from homogeneous lineshapes in a proton glass DRADP has been reported.[2337] A previous report of the n.m.r. determination of the Edwards-Anderson order parameter in $Rb_{1-x}(ND_4)_xD_2PO_4$ has been corrected.[2338] The temperature dependence of T_1 in CsSCN has been examined.[2339] The crystalline electride formed from caesium metal and 15-crown-5 contains sandwiched Cs^+ trapped electrons according to ^{133}Cs n.m.r. measurements.[2340] Local electric field gradients in $CsClO_4$ have been studied using ^{35}Cl and ^{133}Cs n.m.r. spectroscopy.[2341]

2324 W. Wiotte, J. Petersson, R. Blinc, and S. Elschner, *Phys. Rev. B: Condens. Matter*, 1991, **43**, 12751 (*Chem. Abstr.*, 1991, **115**, 148 721).
2325 G.W. Buchanan, S. Mathias, Y. Lear, and C. Bensimon, *Can. J. Chem.*, 1991, **69**, 404.
2326 G. Li and X. Wu, *Wuli Xuebao*, 1990, **39**, 1848 (*Chem. Abstr.*, 1991, **114**, 176 903).
2327 D.G. Hughes and L. Pandey, *J. Magn. Reson.*, 1991, **93**, 361.
2328 Y. Ueda, H. Ohki, N. Nakamura, and H. Chihara, *Bull. Chem. Soc. Jpn.*, 1991, **64**, 2416.
2329 B.H. Suits, *Report*, 1989, **ARO-A23808.3-MS-F**; Order No. AD-A217073, 10 pp. Avail. NTIS. From *Gov. Rep. Announce. Index (U.S.)*, 1990, **90**, Abstr. No. 023,328 (*Chem. Abstr.*, 1991, **114**, 175 216).
2330 G. Floridi, R. Lamanna, and P. Diodati, *Acoust. Lett.*, 1991, **15**, 71 (*Chem. Abstr.*, 1991, **115**, 293 325).
2331 T.J. Bastow, *J. Chem. Soc., Faraday Trans.*, 1991, **87**, 2453 (*Chem. Abstr.*, 1991, **115**, 125 391).
2332 R. Blinc, J. Seliger, T. Apih, J. Dolinsek, I. Zupancic, O. Plyushch, A. Fuith, W. Schranz, H. Warhanek et al., *Phys. Rev. B: Condens. Matter*, 1991, **43**, 569 (*Chem. Abstr.*, 1991, **114**, 93 097).
2333 T.J. Bastow, S.L. Segel, and K.R. Jeffrey, *Solid State Commun.*, 1991, **78**, 565 (*Chem. Abstr.*, 1991, **115**, 62 415).
2334 V.P. Tarasov, V.I. Privalov, K.S. Gavrichev, V.E. Gorbunov, Yu.K. Gusev, and Yu.A. Buslaev, *Koord. Khim.*, 1990, **16**, 1603 (*Chem. Abstr.*, 1991, **114**, 112 219).
2335 T.K. Pratum, *J. Magn. Reson.*, 1991, **91**, 581.
2336 L.L. Buishvili and I.I. Topchyan, *Fiz. Tverd. Tela (Leningrad)*, 1990, **32**, 2638 (*Chem. Abstr.*, 1991, **114**, 239 142).
2337 J. Dolinsek, *J. Magn. Reson.*, 1991, **92**, 312.
2338 S. Chen and D.C. Ailion, *Phys. Rev. B.: Condens. Matter*, 1991, **43**, 8672 (*Chem. Abstr.*, 1991, **114**, 239 158).
2339 Y. Furukawa, R. Ikeda, and D. Nakamura, *J. Phys. Soc. Jpn.*, 1991, **60**, 2100 (*Chem. Abstr.*, 1991, **115**, 125 372).
2340 S.B. Dawes, J.L. Eglin, K.J. Moeggenborg, J. Kim, and J.L. Dye, *J. Am. Chem. Soc.*, 1991, **113**, 1605.
2341 V.P. Tarasov, M.A. Meladze, G.A. Kirakosyan, A.E. Shvelashvili, and Yu.A. Buslaev, *Phys. Status Solidi B*,

Beryllophosphate and zincophosphate zeolite X type molecular sieves have been investigated using ^{31}P n.m.r. spectroscopy.[2342] The species formed when MgH_2 is mixed with $[(\eta^5-C_5H_5)_2ZrMe_2]$ on a carrier to form a Ziegler catalyst have been investigated.[2343] $[M(\eta^5-C_5H_4Bu^t)_2]$, M = Mg, Ca, Sr, Ba, has been characterised using ^{13}C n.m.r. spectroscopy.[2344] ^{13}C c.p. m.a.s. n.m.r. spectroscopy has been used to show that compounds such as $[Mg(anthracene)(THF)_2]$ have Mg-C9,10 interactions.[2345] The effects of the additions of CaO, SrO, BaO, or Li_2O to a synthetic albite feldspar glass have been studied using ^{23}Na, ^{27}Al, and ^{29}Si m.a.s. n.m.r. spectroscopy.[2346] Ferromagnetism of the ^{19}F nuclear spins in a $KMgF_3$ single crystal has been observed by n.m.r. spectroscopy in the microkelvin range.[2347] A theoretical explanation of the spin-lattice coupling coefficients for Ni^{2+} ions in $KMgF_3$ and $KZnF_3$ has been published.[2348] The Ca, Sr, and Ba salts of $[anthracene]^{2-}$ have been studied by ^{13}C c.p. m.a.s. n.m.r. spectroscopy.[2349] The ^{35}Cl and ^{81}Br electric field gradient tensors of tris(sarcosine) calcium chloride and bromide have been determined.[2350] ^{27}Al n.m.r. spectroscopy has been used to study $Ca_8(Al_{12}O_{24})(WO_4)_2$ and $Sr_8(Al_{12}O_{24})(CrO_4)_2$.[2351] A new technique for solid n.m.r. imaging has been applied to ^{31}P imaging in solid bone.[2352] The conversion of calcium phosphate bioceramics into living bone has been studied by ^1H and ^{31}P n.m.r. spectroscopy.[2353] Ionic disorder in fluorite crystals has been investigated by n.m.r. spectroscopy.[2354] ^{19}F and ^{31}P n.m.r. spectroscopy has been used to investigate glasses from $Ba(PO_3)_2$-CaF_2-AlF_3.[2355] The n.m.r. spectrum of $SrVO_3$ has been reported.[2356] Disordered $Ca_{1-x}Th_xF_{2+2x}$ has been studied using ^{19}F n.m.r. spectroscopy.[2357]

Superconducting films in the Bi-Sr-Ca-Cu-O systems have been investigated by ^1H n.m.r. spectroscopy.[2358] A study of the ^{17}O n.m.r. spectra of the normal phase of $(Bi,Pb)_2Sr_2Ca_2Cu_3O_y$ has permitted the signals due to CuO_2 having pyramidal and square planar coordination to be distinguished.[2359] Oxygen sites in $Bi_2Sr_2Ca_{n-1}Cu_nO_{4+2n}$ have been characterised by ^{17}O n.m.r.

1991, **167**, 271 (*Chem. Abstr.*, 1991, **115**, 221 515).
2342 W.T.A. Harrison, T.E. Gier, K.L. Moran, J.M. Nicol, H. Eckert, and G.D. Stucky, *Chem. Mater.*, 1991, **3**, 27 (*Chem. Abstr.*, 1991, **114**, 73 947).
2343 R. Benn and W. Herrmann, *Angew. Chem., Int. Ed. Engl.*, 1991, **30**, 426.
2344 M.G. Gardiner, C.L. Raston, and C.H.L. Kennard, *Organometallics*, 1991, **10**, 3680.
2345 W.M. Brooks, C.L. Raston, R.E. Sue, F.J. Lincoln, and J.J. McGinnity, *Organometallics*, 1991, **10**, 2098.
2346 A.P.S. Mandair and W.R. McWhinnie, *Polyhedron*, 1991, **10**, 55.
2347 P. Bonamour, V. Bouffard, C. Fermon, M. Goldman, J.F. Jacquinot, and G. Saux, *Phys. Rev. Lett.*, 1991, **66**, 2810 (*Chem. Abstr.*, 1991, **115**, 63 123).
2348 W. Zheng, *Philos. Mag. Lett.*, 1990, **62**, 115 (*Chem. Abstr.*, 1991, **114**, 34 490).
2349 H. Bönnemann, B. Bogdanovic, R. Brinkman, N. Egeler, R. Benn, I. Topalovic, and K. Seevogel, *Main Group Met. Chem.*, 1990, **13**, 341 (*Chem. Abstr.*, 1991, **115**, 159 219).
2350 T. Erge, D. Michel, J. Petersson, and F. Engelke, *Phys. Status Solidi A*, 1991, **123**, 325 (*Chem. Abstr.*, 1991, **114**, 176 956).
2351 J.J. Van der Klink, W.S. Veeman, and H. Schmid, *J. Phys. Chem.*, 1991, **95**, 1508.
2352 L. Li, *Phys. Med. Biol.*, 1991, **36**, 199 (*Chem. Abstr.*, 1991, **114**, 202 591).
2353 J.L. Miquel, L. Facchini, A.P. Legrand, X. Marchandise, P. Lecouffe, M. Chanavaz, M. Donazzan, C. Rey, and J. Lemaitre, *Clin. Mater.*, 1990, **5**, 115 (*Chem. Abstr.*, 1991, **114**, 192 516).
2354 M.T. Hutchings and W. Hayes, *Report*, 1989, **AERE-R-13544**; Order No. N90-20862, 116 pp. Avail. NTIS. From *Gov. Rep. Announce. Index (U.S.)*, 1990, **90**, Abstr. No. 049,807 (*Chem. Abstr.*, 1991, **114**, 154 277).
2355 U. Baerenwald, M. Dubiel, W. Matz, and D. Ehrt, *J. Non-Cryst. Solids*, 1991, **130**, 171 (*Chem. Abstr.*, 1991, **115**, 97 587).
2356 M. Onoda, H. Ohta, and H. Nagasawa, *Solid State Commun.*, 1991, **79**, 281 (*Chem. Abstr.*, 1991, **115**, 221 427).
2357 K.S. Suh, J. Senegas, J.M. Reau, and P. Hagenmuller, *Phys. Status Solidi B*, 1991, **164**, 375 (*Chem. Abstr.*, 1991, **115**, 103 150).
2358 S. Katayama and M. Sekine, *J. Mater. Res.*, 1991, **6**, 36 (*Chem. Abstr.*, 1991, **114**, 219 154).
2359 A. Trokiner, L. Le Noc, J. Schneck, A.M. Pougnet, R. Mellet, J. Primot, H. Savary, Y.M. Gao, and S. Aubry,

spectroscopy.[2360] Nuclear relaxation of $Bi_2Sr_2CaCu_2O_8$ has been investigated.[2361] The diamagnetization of the high T_c oxide superconductor Pb-doped Bi-Sr-Ca-Cu-O powder in the vortex state has been studied by measuring the n.m.r. spectrum of the surface material coated on the sample.[2362] ^{43}Ca and ^{205}Tl n.m.r. spectroscopy has been used to study structural disorder in $Bi_2Sr_2CaCu_2O_8$ and $Tl_2Ba_2CaCu_2O_8$.[2363] The ^{19}F n.m.r. spectra of $Bi_2Sr_2CaCu_2O_{8-y}F_y$ and $Bi_2Sr_2Ca_2Cu_3O_{10-y}F_y$ have been reported.[2364] The ^{17}O T_1 and Knight shift in $Tl_2Ba_2CaCu_2O_{6+x}$, $La_{1.85}Sr_{0.15}CuO_{4-x}$, and $Ba_{0.6}K_{0.4}BiO_3$ have been measured.[2365] ^{205}Tl n.m.r. spectra have been measured for the high T_c superconductor thallium barium calcium copper oxide.[2366] N.m.r. data have also been reported for the La-Sr-Ba-Cu-O system,[2367] (^{17}O, ^{63}Cu, ^{205}Tl),[2368] (^{17}O, ^{63}Cu),[2369] (^{63}Cu),[2370] (^{139}La),[2371] the La-Sr-Cu-O-F system, (^{19}F),[2372] the Y-Ba-Cu-O system,[2373] (^{17}O,

Phys. Rec. B: Condens. Matter, 1991, **44**, 2426 (*Chem. Abstr.*, 1991, **115**, 147 816).

2360 R. Dupree, Z.P. Han, A.P. Howes, D.M. Paul, M.E. Smith, and S. Male, *Physica C (Amsterdam)*, 1991, **175**, 269 (*Chem. Abstr.*, 1991, **115**, 39 672); A. Trokiner, L. Le Noc, R. Mellet, A.M. Pougnet, D. Morin, Y.M. Gao, J. Primot, and J. Schneck, *Phase Transitions*, 1991, **30**, 147 (*Chem. Abstr.*, 1991, **115**, 245 292).

2361 R.E. Walstedt, R.F. Bell, and D.B. Mitzi, *Phys. Rev. B: Condens. Matter*, 1991, **44**, 7760 (*Chem. Abstr.*, 1991, **115**, 246 381).

2362 Y. Maniwa, H. Sato, T. Mituhashi, K. Mizoguchi, I. Shiozaki, and K. Kume, *Jpn. J. Appl. Phys., Part 2*, 1991, **30**, L572 (*Chem. Abstr.*, 1991, **114**, 219 916).

2363 V.P. Tarasov, V.I. Privalov, V.I. Ozhogin, A.Yu. Yakubovskii, L.D. Shustov, S.V. Verkhovskii, Yu.I. Zhdanov, K.N. Mikhalev, B.A. Aleksashin, et al., *Sverkhprovodimost: Fiz., Khim., Tekh.*, 1991, **4**, 133 (*Chem. Abstr.*, 1991, **115**, 39 667).

2364 V.M. Buznik, V.E. Volkov, A.I. Livshits, and N.P. Fokina, *Zh. Neorg. Khim.*, 1990, **35**, 2299 (*Chem. Abstr.*, 1991, **114**, 54 522).

2365 L. Reven, J. Shore, S. Yang, T. Duncan, D. Schwartz, J. Chung, and E. Oldfield, *Phys. Rev. B: Condens. Matter*, 1991, **43**, 10466 (*Chem. Abstr.*, 1991, **114**, 258 333).

2366 J.C. Jol, D. Reefman, H.B. Brom, T. Zetterer, D. Hahn, H.H. Otto, and K.F. Renk, *Physica C (Amsterdam)*, 1991, **175**, 12 (*Chem. Abstr.*, 1991, **114**, 257 834); Y.Q. Song, M. Lee, W.P. Halperin, L.M. Tonge, and T.J. Marks, *Phys. Rev. B: Condens. Matter*, 1991, **44**, 914 (*Chem. Abstr.*, 1991, **115**, 63 162); L. Mihaly, K. Tompa, I. Bakonyi, P. Banki, E. Zsoldos, S. Pekker, G. Oszlanyi, and G. Hutiray, *Prog. High Temp. Supercond.*, 1988, **14**, 1227 (*Chem. Abstr.*, 1991, **114**, 16 281); M. Mehring, F. Hentsch, and N. Winzek, *NATO ASI Ser., Ser. B*, 1991, **246**, 87 (*Chem. Abstr.*, 1991, **115**, 196 531); Yu.I. Zhdanov, K.N. Mikhalev, B.A. Aleksashin, S.V. Verkhovskii, K.A. Okulova, V.I. Voronin, L.D. Shustov, A.Yu. Yakubovskii, and A.I. Akimov, *Sverkhprovodimost: Fiz., Khim., Tekh.*, 1990, **3**, 194 (*Chem. Abstr.*, 1991, **114**, 34 629).

2367 H. Monien, P. Monthoux, and D. Pines, *Phys. Rev. B: Condens. Matter*, 1991, **43**, 275 (*Chem. Abstr.*, 1991, **114**, 113 745).

2368 Y. Kitaoka, K. Fujiwara, K. Ishida, T. Kondo, and K. Asayama, *J. Magn. Magn. Mater.*, 1990, **90**, 619 (*Chem. Abstr.*, 1991, **114**, 175 588).

2369 K. Asayama, G.Q. Zheng, Y. Kitaoka, K. Ishida, and K. Fujiwara, *Physica C (Amsterdam)*, 1991, **178**, 281 (*Chem. Abstr.*, 1991, **115**, 173 172); K. Ishida, Y. Kitaoka, G.Q. Zheng, and K. Asayama, *J. Phys. Soc. Jpn.*, 1991, **60**, 3516 (*Chem. Abstr.*, 1991, **115**, 268 078).

2370 M.A. Kennard, Y. Song, K.R. Poeppelmeier, and W.P. Halperin, *Chem. Mater.*, 1991, **3**, 672 (*Chem. Abstr.*, 1991, **115**, 63 118); T. Imai, K. Yoshimura, T. Uemura, H. Yasuoka, and K. Kosuge, *J. Phys. Soc. Jpn.*, 1990, **59**, 3846 (*Chem. Abstr.*, 1991, **114**, 73 845); S. Ohsugi, Y. Kitaoka, K. Ishida, and K. Asayama, *J. Phys. Soc. Jpn.*, 1991, **60**, 2351 (*Chem. Abstr.*, 1991, **115**, 125 385); Y.Q. Song, M.A. Kennard, M. Lee, K.R. Poeppelmeier, and W.P. Halperin, *Phys. Rev. B: Condens. Matter*, 1991, **44**, 7159 (*Chem. Abstr.*, 1991, **115**, 221 524).

2371 A.V. Zalesskii, V.G. Krivenko, A.P. Levanyuk, I.E. Lipinski, Yu.G. Metlin, and T.A. Sivokon, *Sverkhprovodimost: Fiz., Khim., Tekh.*, 1990, **3**, 202 (*Chem. Abstr.*, 1991, **114**, 54 514).

2372 E.F. Kukovitskii, R.G. Mustafin, and G.B. Teitel'baum, *Physica B (Amsterdam)*, 1991, **169**, 643 (*Chem. Abstr.*, 1991, **115**, 221 493).

2373 M.C. Nuss, P.M. Mankiewich, M.L. O'Malley, E.H. Westerwick, and P.B. Littlewood, *Phys. Rev. Lett.*, 1991, **66**, 3305 (*Chem. Abstr.*, 1991, **115**, 62 278); Y. Maniwa, T. Mituhashi, K. Mizoguchi, and K. Kume, *Physica C (Amsterdam)*, 1991, **175**, 401 (*Chem. Abstr.*, 1991, **115**, 39 673); F. Marsiglio, *Phys. Rev. B: Condens. Matter*, 1991, **44**, 5373 (*Chem. Abstr.*, 1991, **115**, 195 690); M. Solanki-Moser, D.P. Tunstall, and W.J. Webster, *Supercond. Sci. Technol.*, 1990, **3**, 464 (*Chem. Abstr.*, 1991, **114**, 93 778); H. Luetgemeier, *J. Magn. Magn. Mater.*, 1990, **90**, 633 (*Chem. Abstr.*, 1991, **114**, 176 795); K.N. Shrivastava, *Phys. Rep.*, 1991, **200**, 51 (*Chem. Abstr.*, 1991, **114**, 239 051); L.D. Rotter, Z. Schlesinger, R.T. Collins, F. Holtzberg, C. Field, U.W. Welp, G.W. Crabtree, J.Z. Liu, Y. Fang, et al., *Phys. Rev. Lett.*, 1991, **67**, 2741 (*Chem. Abstr.*, 1991, **115**, 290 006); H. Niki, T. Higa, S. Tomiyoshi, M. Omori, T. Kajitani, T. Sato, T. Shinohara, T. Suzuki, K.

^{63}Cu, ^{89}Y),[2374] (^{17}O, ^{63}Cu),[2375] (^{17}O, ^{89}Y),[2376] (^{17}O),[2377] (^{63}Cu),[2378] (^{89}Y),[2379] (^{135}Ba),[2380] $YBa_2Cu_3O_{7-x}F_x$, (^{19}F),[2381] $Tl_2Ba_2CuO_{6+y}$, (^{63}Cu, ^{205}Tl),[2382] and $BaCo_xTi_xFe_{12-2x}O_{19}$, ($^{57}Fe$).[2383]

Yagasaki, and R. Igei, *J. Magn. Magn. Mater.*, 1990, **90**, 672 (*Chem. Abstr.*, 1991, **114**, 219 111); E.A. Eremina, A.B. Yaroslavtsev, N.N. Oleinikov, A.V. Strelkov, and Yu.D. Tret'yakov, *Sverkhprovodimost: Fiz., Khim., Tekh.*, 1989, **2**, 13 (*Chem. Abstr.*, 1991, **114**, 134 987)

2374 S.E. Barrett, J.A. Martindale, D.J. Durand, C.H. Pennington, C.P. Slichter, T.A. Friedmann, J.P. Rice, and D.M. Ginsberg, *Phys. Rev. Lett.*, 1991, **66**, 108 (*Chem. Abstr.*, 1991, **114**, 54 550); H. Monien, D. Pines, and M. Takigawa, *Phys. Rev. B: Condens Matter*, 1991, **43**, 258 (*Chem. Abstr.*, 1991, **114**, 113 744).

2375 E. Lippmaa, E. Joon, I. Heinmaa, A. Miller, V. Miidel, R. Stern, and S. Vija, *Prog. High Temp. Supercond.*, 1990, **21**, 242 (*Chem. Abstr.*, 1991, **114**, 16 286); M. Takigawa, A.P. Reyes, P.C. Hammel, J.D. Thompson, R.H. Heffner, Z. Fisk, and K.C. Ott, *Phys. Rev. B: Condens. Matter*, 1991, **43**, 247 (*Chem. Abstr.*, 1991, **114**, 113 633); A.Yu. Zavidonov and M.V. Eremin, *Sverkhprovodimost: Fiz., Khim., Tekh.*, 1990, **3**, 2700 (*Chem. Abstr.*, 1991, **115**, 83 950); C. Berthier, Y. Berthier, P. Butaud, M. Horvatic, Y. Kitaoka, and P. Segransan, *NATO ASI Ser., Ser. B*, 1991, **246**, 73 (*Chem. Abstr.*, 1991, **115**, 196 530); H. Monien, *NATO ASI Ser., Ser. B*, 1991, **246**, 111 (*Chem. Abstr.*, 1991, **115**, 196 532); M. Takigawa, *NATO ASI Ser., Ser. B*, 1991, **246**, 61 (*Chem. Abstr.*, 1991, **115**, 196 529).

2376 T. Ohno, T. Kanashiro, and K. Mizuno, *J. Phys. Soc. Jpn.*, 1991, **60**, 2040 (*Chem. Abstr.*, 1991, **115**, 63 146); C. Berthier, Y. Berthier, P. Butaud, M. Horvatic, and P. Segransan, *Springer Ser. Solid-State Sci.*, 1990, **99**, 209 (*Chem. Abstr.*, 1991, **115**, 104 465); H. Alloul, *Springer Ser. Solid-State Sci.*, 1990, **99**, 201 (*Chem. Abstr.*, 1991, **115**, 104 464).

2377 F. Barriquand, P. Odier, and D. Jerome, *Physica C (Amsterdam)*, 1991, **177**, 230 (*Chem. Abstr.*, 1991, **115**, 83 964); I.A. Goidenko and M.V. Eremin, *Sverkhprovodimost: Fiz., Khim., Tekh.*, 1991, **4**, 451 (*Chem. Abstr.*, 1991, **115**, 103 886); Y. Kohori, T. Sugata, T. Kohara, and Y. Oda, *J. Magn. Magn. Mater.*, 1990, **90**, 667 (*Chem. Abstr.*, 1991, **114**, 176 922); J.H. Brewer, R.F. Kiefl, J.F. Carolan, P. Dosanjh, W.N. Hardy, S.R. Kreitzman, Q. Li, T.M. Riseman, P. Schleger, et al., *Hyperfine Interact.*, 1991, **63**, 177 (*Chem. Abstr.*, 1991, **114**, 176 010).

2378 T. Machi, I. Tomeno, T. Miyatake, N. Koshizuka, S. Tanaka, T. Imai, and H. Yasuoka, *Physica C (Amsterdam)*, 1991, **173**, 32 (*Chem. Abstr.*, 1991, **114**, 155 551); O.M. Vyaselev, P.A. Komonovich, and I.F. Shchegolev, *Sverkhprovodimost: Fiz., Khim., Kekh.*, 1991, **4**, 55 (*Chem. Abstr.*, 1991, **114**, 258 331); K. Ishida, Y. Kitaoka, T. Yoshitomi, N. Ogata, T. Kamino, and K. Asayama, *Physica C (Amsterdam)*, 1991, **179**, 29 (*Chem. Abstr.*, 1991, **115**, 172 112); K. Szunyogh, G.H. Schadler, P. Weinberger, and R. Monnier, *Physica B (Amsterdam)*, 1990, **163**, 259 (*Chem. Abstr.*, 1991, **114**, 34 630); D.E. MacLaughlin, A.P. Reyes, M. Takigawa, P.C. Hammel, R.H. Heffner, J.D. Thompson, and J.E. Crow, *Physica B (Amsterdam)*, 1991, **171**, 245 (*Chem. Abstr.*, 1991, **115**, 148 576); D. Brinkmann, *Prog. High Temp. Supercond.*, 1990, **21**, 230 (*Chem. Abstr.*, 1991, **114**, 16 285); R.A. Brand, C. Sauer, H. Luetgemeier, P.M. Meuffels, and W. Zinn, *Hyperfine Interact.*, 1990, **55**, 1229 (*Chem. Abstr.*, 1991, **114**, 15 239); P. Mendels, H. Alloul, J.F. Marucco, J. Arabski, and G. Collin, *Physica C (Amsterdam)*, 1990, **171**, 429 (*Chem. Abstr.*, 1991, **114**, 93 777); C.H. Pennington and C.P. Slichter, *Phys. Rev. Lett.*, 1991, **66**, 381 (*Chem. Abstr.*, 1991, **114**, 73 852); M. Matsumura, H. Yamagata, Y. Yamada, K. Ishida, Y. Kitaoka, K. Asayama, H. Takagi, H. Iwabuchi, and S. Uchida, *J. Magn. Magn. Mater.*, 1990, **90**, 661 (*Chem. Abstr.*, 1991, **114**, 198 175); H. Luetgemeier, *Springer Ser. Solid-State Sci.*, 1990, **99**, 214 (*Chem. Abstr.*, 1991, **115**, 104 466); A.P. Reyes, D.E. MacLaughlin, M. Takigawa, P.C. Hammel, R.H. Heffner, J.D. Thompson, and J.E. Crow, *Phys. Rev. B: Condens. Matter*, 1991, **43**, 2989 (*Chem. Abstr.*, 1991, **114**, 176 901); M. Takigawa, J.L. Smith, and W.L. Hults, *Phys. Rev. B: Condens. Matter*, 1991, **44**, 7764 (*Chem. Abstr.*, 1991, **115**, 246 382); M. Mali, I. Mangelschots, H. Zimmermann, and D. Brinkmann, *Physica C (Amsterdam)*, 1991, **175**, 581 (*Chem. Abstr.*, 1991, **115**, 40 531); O.A. Anikeenok, M.V. Eremin, R.Sh. Zhdanov, V.V. Naletov, M.P. Rodionova, and M.A. Teplov, *Pis'ma Zh. Eksp. Teor. Fiz.*, 1991, **54**, 154 (*Chem. Abstr.*, 1991, **115**, 196 562).

2379 L.L. Buishvili and G.I. Mamniashvili, *Sverkhprovodimost: Fiz., Khim., Tekh.*, 1989, **2**, 75 (*Chem. Abstr.*, 1991, **114**, 113 731); H. Alloul, T. Ohno, H. Casalta, J.F. Marucco, P. Mendels, J. Arabski, G. Collin, and M. Mehbod, *Physica C (Amsterdam)*, 1990, **171**, 419 (*Chem. Abstr.*, 1991, **114**, 113 586); H.B. Brom and H. Alloul, *Physica C (Amsterdam)*, 1991, **177**, 297 (*Chem. Abstr.*, 1991, **115**, 83 878); T. Ohno, H. Alloul, P. Mendels, G. Collin, and J.F. Marucco, *J. Magn. Magn. Mater.*, 1990, **90**, 657 (*Chem. Abstr.*, 1991, **114**, 219 803); R. Dupree, Z.P. Han, D.M. Paul, T.G.N. Babu, and C. Greaves, *Physica C (Amsterdam)*, 1991, **179**, 311 (*Chem. Abstr.*, 1991, **115**, 196 561); D.P. Tunstall and W.J. Webster, *Supercond. Sci. Technol.*, 1991, **4**, S406 (*Chem. Abstr.*, 1991, **115**, 19 807).

2380 H. Luetgemeier, V. Florent'ev, and A. Yakubovski, *Springer Ser. Solid-State Sci.*, 1990, **99**, 222 (*Chem. Abstr.*, 1991, **115**, 83 938).

2381 H. Pan, B.C. Gerstein, H.R. Loeliger, and T.A. Vanderah, *Appl. Magn. Reson.*, 1990, **1**, 101 (*Chem. Abstr.*, 1991, **115**, 148 714).

2382 Y. Kitaoka, K. Fujiwara, K. Ishida, K. Asayama, Y. Shimakawa, T. Manako, and Y. Kubo, *Physica C (Amsterdam)*, 1991, **179**, 107 (*Chem. Abstr.*, 1991, **115**, 173 180).

2383 V.S. Abramov, V.B. Tyutyunnik, V.P. Pashchemo, A.G. Titenko, and V.I. Arkharov, *Dokl. Akad. Nauk SSSR*, 1990, **312**, 638 (*Chem. Abstr.*, 1991, **114**, 16 294).

The local magnetic field at Y in $Y_6Fe_{23}H_x$ has been studied using ^{89}Y n.m.r. spectroscopy.[2384] A transverse field μSR study of $LaH_{2.75}$ has been reported.[2385] ^1H T_1 measurements have been used to study LaH_x.[2386] ^1H T_1 and ^{139}La Knight shift measurements have been used to study the metal-nonmetal transition in the LaH_2-LaH_3 system.[2387] ^1H n.m.r. spectroscopy has been used to study crystal field levels in PrH_y,[2388] and SmH_y.[2389] A ^{13}C c.p. m.a.s. n.m.r. spectrum of $[Lu\{CH(SiMe_3)_2\}_3(\mu\text{-}Cl)]K$ shows asymmetry due to potassium coordination.[2390] N.m.r. data have also been reported for $(Nd_{1-x}Y_x)_2Co_{14}B$, (^{59}Co),[2391] $SmRh_3B_2$, (^{11}B),[2392] $Y_2Co_{17}C_x$, (^{89}Y),[2393] CeM_2Si_2, (M = Ru, Cu; ^{29}Si, ^{63}Cu),[2394] $Sc_xTi_{1-x}N_x$,[2395] Y-Si-Al-O-N glass, (^{15}N),[2396] La-Si-Al-O-N glass, (^{15}N),[2397] $[Y(O_2CCH_3)_3.4H_2O]$, (^{13}C),[2398] $[Tb(O_3SOEt)_3.9H_2O]$,[2399] La_2CuO_4,[2400] (^{63}Cu),[2401] (^{139}La),[2402] $Nd_{2-x}Ce_xCuO_2$,[2403] (^{63}Cu, ^{65}Cu),[2404] $Ln_{2-x}Ce_xCuO_2$, Ln = Nd, Sm, Pr, (^{63}Cu),[2405] $Pr_{2-x}Ce_xCuO_2$, (^{65}Cu, ^{141}Pr),[2406] $La_2Cu_{0.97}Zr_{0.03}O_4$, ($^{139}La$),[2407] $Sm_3Fe_5O_{12}$,[2408]

2384 V.A. Vasil'kovskii, M.I. Bartashevich, A.A. Gorlenko, and N.M. Kovtun, *Fiz. Tverd. Tela (Leningrad)*, 1990, **32**, 3089 (*Chem. Abstr.*, 1991, **115**, 62 878).

2385 M.R. Chowdhury, G.A. Styles, E.F.W. Seymour, S.F.J. Cox, C.A. Scott, and G.H. Eaton, *IOP Short Meet. Ser.*, 1989, **22**, 77 (*Chem. Abstr.*, 1991, **115**, 148 694).

2386 C.T. Chang, *Report*, 1989, **IS-T-1378**; Order No. DE89007571, 172 pp. Avail. NTIS. From *Energy Res. Abstr.*, 1989, **14**, Abstr. No. 17810 (*Chem. Abstr.*, 1991, **114**, 218 878).

2387 R.G. Barnes, C.T. Chang, M. Belhoul, D.R. Torgeson, R.J. Schoenberger, B.J. Beaudry, and E.F.W. Seymour, *J. Less-Common Met.*, 1991, **172**, 411 (*Chem. Abstr.*, 1991, **115**, 140 002).

2388 M. Belhoul, R.J. Schoenberger, D.R. Torgeson, and R.G. Barnes, *J. Less-Common Met.*, 1991, **172**, 366 (*Chem. Abstr.*, 1991, **115**, 246 373).

2389 O.J. Zogal, S. Idziak, M. Drulis, and K. Niedzwiedz, *Phys. Status Solidi B*, 1991, **167**, K55 (*Chem. Abstr.*, 1991, **115**, 196 566).

2390 C.J. Schaverien and J.B. van Mechelen, *Organometallics*, 1991, **10**, 1704.

2391 E. Jedryka, M. Wojcik, P. Panissod, A.T. Pedziwiatr, and M. Slępowronski, *J. Appl. Phys.*, 1991, **69**, 6043 (*Chem. Abstr.*, 1991, **115**, 246 368).

2392 T. Ohno, Y. Kishimoto, T. Yamada, Y. Michihiro, and T. Kanashiro, *J. Magn. Magn. Mater.*, 1990, **90**, 547 (*Chem. Abstr.*, 1991, **114**, 198 256).

2393 C. Kapusta, M. Rosenberg, K.V. Rao, Z. Han, T.H. Jacobs, and K.H.J. Buschow, *J. Less-Common Met.*, 1991, **169**, L5 (*Chem. Abstr.*, 1991, **115**, 40 264).

2394 H. Nakamura, T. Iwai, Y. Kitaoka, and K. Asayama, *J. Magn. Magn. Mater.*, 1990, **90**, 490 (*Chem. Abstr.*, 1991, **114**, 176 920).

2395 B.N. Dudkin and E.V. Vanchikova, *Zh. Obshch. Khim.*, 1991, **61**, 521 (*Chem. Abstr.*, 1991, **115**, 148 590).

2396 D. Kruppa, R. Dupree, and M.H. Lewis, *Mater. Lett.*, 1991, **11**, 195 (*Chem. Abstr.*, 1991, **115**, 76 919).

2397 D.P. Thompson, M.J. Leach, and R.K. Harris, *C-MRS Int. Symp. Proc. 1990*, (Pub. 1991), **2**, 435 (*Chem. Abstr.*, 1991, **115**, 213 332).

2398 F. Ribot, P. Toledano, and C. Sanchez, *Inorg. Chim. Acta*, 1991, **185**, 239.

2399 L.K. Aminov, A.G. Volodin, A.V. Egorov, V.V. Naletov, M.S. Tagirov, M.A. Teplov, and G. Feller, *Appl. Magn. Reson.*, 1990, **1**, 113 (*Chem. Abstr.*, 1991, **115**, 125 361).

2400 S. Chakravarty, M.P. Gelfand, P. Kopietz, R. Orbach, and M. Wollensak, *Phys. Rev. B: Condens. Matter*, 1991, **43**, 2796 (*Chem. Abstr.*, 1991, **114**, 155 665).

2401 T. Kohara, K. Ueda, Y. Kohori, and Y. Oda, *J. Magn. Magn. Mater.*, 1990, **90**, 669 (*Chem. Abstr.*, 1991, **114**, 176 923).

2402 V.A. Borodin, V.D. Doroshev, Yu.M. Ivanchenko, M.M. Savosta, and A.E. Filippov, *Pis'ma Zh. Eksp. Teor. Fiz.*, 1990, **52**, 1073 (*Chem. Abstr.*, 1991, **114**, 54 544).

2403 K. Kobayashi, Y. Goto, S. Matsushima, and G. Okada, *Jpn. J. Appl. Phys., Part 2*, 1991, **30**, L1106 (*Chem. Abstr.*, 1991, **115**, 62 300).

2404 S. Kambe, H. Yasuoka, H. Takagi, S. Uchida, and Y. Tokura, *J. Phys. Soc. Jpn.*, 1991, **60**, 400 (*Chem. Abstr.*, 1991, **114**, 176 952); S.M. Doyle, R. Dupree, A.P. Howes, D.M. Paul, and M.E. Smith, *Bull. Mater. Sci.*, 1991, **14**, 619 (*Chem. Abstr.*, 1991, **115**, 220 268).

2405 K. Kumagai, M. Abe, S. Tanaka, Y. Maeno, and T. Fujita, *J. Magn. Magn. Mater.*, 1990, **90**, 675 (*Chem. Abstr.*, 1991, **114**, 176 924).

2406 O.N. Bakharev, A.G. Volodin, A.V. Egorov, M.S. Tagirov, and M.A. Teplov, *Pis'ma Zh. Eksp. Teor. Fiz.*, 1990, **52**, 1012 (*Chem. Abstr.*, 1991, **114**, 16 318).

2407 V.L. Matukhin, V.N. Anashkin, A.I. Pogorel'tsev, E.F. Kukovitskii, and I.A. Safin, *Sverkhprovodimost: Fiz., Khim., Tekh.*, 1991, **4**, 511 (*Chem. Abstr.*, 1991, **115**, 148 735).

2408 M.I. Kurkin and V.V. Serikov, *Fiz. Met. Metalloved.*, 1991, 103 (*Chem. Abstr.*, 1991, **115**, 40 512).

and $Er_xY_{3-x}Fe_5O_{12}$, (^{57}Fe).[2409]

The influence of synthesis parameters on the structural properties of mixed oxides derived from $Si(OEt)_4$ and [$Ti(OPr^i)_4$] precursors has been studied by ^{29}Si n.m.r. spectroscopy.[2410] The strong metal support interaction state of ruthenium/titania has been investigated by ^{1}H n.m.r. spectroscopy.[2411] The temperature and frequency dependence of ^{7}Li, ^{19}F, and ^{23}Na T_1 values have been studied in fluorozirconate glasses.[2412] N.m.r. data have also been reported for $[(\eta^5\text{-}C_5H_5)(\eta^5\text{-}C_5Me_5)ZrH_2]_2$,[2413] ZrH_x with Mn, Cr, Fe impurities,[2414] $ZrNiH_x$,[2415] Zr_2PdH_x,[2416] polyzirconocarbosilane, (^{13}C, ^{29}Si),[2417] titanite, (^{29}Si),[2418] $BaZrO_3$, $SrZrO_3$, (^{91}Zr),[2419] $M^1M^2O_3$, (M^1 = Ca, Sr, Ba; M^2 = Ti, Zr; ^{47}Ti, ^{49}Ti, ^{91}Zr, ^{137}Ba),[2420] vanadium-titanium oxide catalysts, (^{51}V),[2421] zirconium aluminate fibres, (^{27}Al),[2422] $M_xGa_8Ga_{8+x}Ti_{16-x}O_{56}$, (^{71}Ga, ^{87}Rb, ^{133}Cs),[2423] [$Ti(OPr^i)_4$]-[$Pb(O_2CCH_3)_2$] sol-gel, (^{1}H, ^{13}C),[2424] [$\{(OC)_3Mn(\eta^5\text{-}C_5H_3MeCH_2)\}_2NH$] intercalated into α-$Zr(HPO_4)_2.H_2O$, (^{1}H, ^{13}C),[2425] $K_2TiSnO_2(PO_4)_2$, (^{119}Sn),[2426] [$Zr(O_3PCH_2CH_2COOH)_2$], (^{13}C),[2427] [$Zr(HPO_4)_2.H_2O$], (^{31}P),[2428] $MTi_2(PO_4)_3$, (M = Li, Na, K; ^{31}P),[2429] titanium-oxo-acetate polymers, (^{13}C),[2430] [$Zr(O_3PCH_2CH_2COCl)_2$], (^{13}C, ^{31}P),[2431] composites of α-

2409 V.A. Borodin, V.D. Doroshev, T.N. Tarasenko, M.M. Savosta, and P. Novak, *J. Phys.: Condens. Matter*, 1991, **3**, 5881 (*Chem. Abstr.*, 1991, **115**, 221 507).
2410 K.L. Walther, A. Wokaun, B.E. Handy, and A. Baiker, *J. Non-Cryst. Solids*, 1991, **134**, 47 (*Chem. Abstr.*, 1991, **115**, 216 016).
2411 C.C.A. Riley, P. Jonsen, P. Meehan, J.C. Frost, K.J. Packer, and J.P.S. Badyal, *Catal. Today*, 1991, **9**, 121 (*Chem. Abstr.*, 1991, **114**, 235 775).
2412 S. Estalji, O. Kanert, J. Steinert, H. Jain, and K.L. Ngai, *Phys. Rev. B: Condens Matter*, 1991, **43**, 7481 (*Chem. Abstr.*, 1991, **114**, 239 155).
2413 H.G. Woo, J.F. Walzer, and T.D. Tilley, *Macromolecules*, 1991, **24**, 6863 (*Chem. Abstr.*, 1991, **115**, 256 799).
2414 J.W. Han, D.R. Torgeson, and R.G. Barnes, *Phys. Rev. B: Condens. Matter*, 1990, **42**, 7710 (*Chem. Abstr.*, 1991, **114**, 34 649).
2415 P.V. Afanas'ev and V.V. Lunin, *Dokl. Akad. Nauk SSSR*, 1990, **315**, 892 (*Chem. Abstr.*, 1991, **114**, 193 469).
2416 D.B. Baker, E.K. Jeong, M.S. Conradi, R.E. Norberg, and R.C. Bowman, jun., *J. Less-Common Met.*, 1991, **172**, 373 (*Chem. Abstr.*, 1991, **115**, 293 310).
2417 F. Babonneau and G.D. Soraru, *J. Eur. Ceram. Soc.*, 1991, **8**, 29 (*Chem. Abstr.*, 1991, **115**, 213 223); F. Babonneau, P. Barre, J. Livage, and M. Verdaguer, *Mater. Res. Soc. Symp. Proc.*, 1990, **180**, 1035 (*Chem. Abstr.*, 1991, **115**, 285 508).
2418 F.C. Hawthorne, L.A. Groat, M. Raudsepp, N.A. Ball, M. Kimata, F.D. Spike, R. Gaba, N.M. Halden, G. Lumpkin, *et al.*, *Am. Mineral.*, 1991, **76**, 370 (*Chem. Abstr.*, 1991, **115**, 53 554).
2419 J.S. Hartman, F.P. Koffyberg, and J.A. Ripmeester, *J. Magn. Reson.*, 1991, **91**, 400.
2420 J.J. Fitzgerald, S.S. Han, S.F. Dec, M.F. Davis, C.E. Bronnimann, and G.E. Maciel, *NIST Spec. Publ.*, 1991, **804**, 173 (*Chem. Abstr.*, 1991, **115**, 148 719).
2421 G. Centi, D. Pinelli, F. Trifiro, D. Ghoussoub, M. Guelton, and L. Gengembre, *J. Catal.*, 1991, **130**, 238 (*Chem. Abstr.*, 1991, **115**, 36 384).
2422 P.S. Kalinin, D.B. Dorzhiev, V.N. Zueva, and V.E. Khazanov, *Neorg. Mater.*, 1991, **27**, 1550 (*Chem. Abstr.*, 1991, **115**, 268 924).
2423 Y. Onoda, M. Watanabe, Y. Fujiki, S. Yoshikado, T. Ohachi, and I. Taniguchi, *Solid State Ionics*, 1990, **40**, 147 (*Chem. Abstr.*, 1991, **114**, 16 301).
2424 S.D. Ramamurthi and D.A. Payne, *Mater. Res. Soc. Symp. Proc.*, 1990, **180**, 79 (*Chem. Abstr.*, 1991, **115**, 285 481).
2425 C.F. Lee and M.E. Thompson, *Inorg. Chem.*, 1991, **30**, 4.
2426 S.J. Crennell, J.J. Owen, A.K. Cheetham, J.A. Kaduk, and R.H. Jarman, *Eur. J. Solid State Inorg. Chem.*, 1991, **28**, 397 (*Chem. Abstr.*, 1991, **114**, 258 405).
2427 D.A. Burwell and M.E. Thompson, *Chem. Mater.*, 1991, **3**, 730 (*Chem. Abstr.*, 1991, **115**, 63 274).
2428 H. Benhamza, P. Barboux, A. Bouhaouss, F.A. Josien, and J. Livage, *J. Mater. Chem.*, 1991, **1**, 681 (*Chem. Abstr.*, 1991, **115**, 246 530).
2429 A.M. Korduban, I.P. Napibin, E.A. Pashchenko, A.V. Podenezhko, N.S. Slobodyanik, N.V. Stus, N.V. Stus, V.V. Trachevskii, and A.P. Shpak, *Dokl. Akad. Nauk Ukr. SSR, Ser. B: Geol., Khim. Biol. Nauki*, 1990, 49 (*Chem. Abstr.*, 1991, **115**, 120 334).
2430 S. Doeuff, M. Henry, and C. Sanchez, *Mater. Res. Bull.*, 1990, **25**, 1519 (*Chem. Abstr.*, 1991, **114**, 134 819).
2431 D.A. Burwell and M.E. Thompson, *Chem. Mater.*, 1991, **3**, 14 (*Chem. Abstr.*, 1991, **114**, 73 956).

$Zr(HPO_4)_2.H_2O$ and Al_2O_3, (^{31}P),[2432] $Li_2Zr_6Cl_{15}Mn$, and $LiZr_6Cl_{15}Fe$, (^7Li).[2433]

Enhanced ^1H T_1 values have been measured over a range of frequencies and temperatures in the Nb-H, Nb-V-H, and Ta-H metal-hydrogen systems.[2434] The synthesis of silicon-tantalum-carbide ceramics from tantalum alkoxide modified polycarbosilane has been studied by ^{13}C and ^{29}Si n.m.r. spectroscopy.[2435] The effect of the anisotropic magnetic shift on the polycrystalline first order quadrupole lineshapes for non coincident principal axes has been examined and applied to ^{51}V in V_2O_5.[2436] $(VO)_2P_2O_7$ has been proposed as an internal standard in high temperature ^{31}P n.m.r. spectroscopy.[2437] The sol-gel route to Nb_2O_5 has been studied using ^{13}C n.m.r. spectroscopy.[2438] ^{51}V n.m.r. spectroscopy has been used to study the mechanisms of the structural distortions in solids, using MVF_6 as an example.[2439] N.m.r. data have also been reported for vanadium-oxygen compounds, (^{51}V),[2440] V_2O_5, (^{51}V),[2441] $H_2V_{12}O_{31-y}$, (^{51}V),[2442] $LiVO_2$, (^{51}V),[2443] $Na_xV_2O_5$, (^{23}Na, ^{51}V),[2444] $NbVO_5$, (^{51}V, ^{93}Nb),[2445] $x(Co_3O_4).(1-x)(V_2O_5)$,[2446] Na_2O-B_2O_3-V_2O_5 glasses, (^{51}V),[2447] PbO-V_2O_5-B_2O_3 glasses,[2448] $H_2V_{12}O_{31-\delta}.nH_2O$, ($^{51}V$),[2449] $[NH_4]_3[VO_2(C_2O_4)_2]$.-$3H_2O$, (^{13}C),[2450] vanadium containing silicalite, (^{29}Si, ^{51}V),[2451] β-$VOPO_4$, (^{31}P),[2452] Bi_2O_3-V_2O_5

2432 R.C.T. Slade and J.A. Knowles, *Solid State Ionics*, 1991, **46**, 45 (*Chem. Abstr.*, 1991, **115**, 82 899).
2433 J. Zhang and J.D. Corbett, *Inorg. Chem.*, 1991, **30**, 431.
2434 L.R. Lichty, J.W. Han, D.R. Torgeson, R.G. Barnes, and E.F.W. Seymour, *Phys. Rev. B: Condens. Matter*, 1990, **42**, 7734 (*Chem. Abstr.*, 1991, **114**, 54 524).
2435 K. Thorne, E. Liimatta, and J.D. Mackenzie, *J. Mater. Res.*, 1991, **6**, 2199 (*Chem. Abstr.*, 1991, **115**, 238 070).
2436 P.W. France, *J. Magn. Reson.*, 1991, **92**, 30.
2437 H. Pan and B.C. Gerstein, *J. Magn. Reson.*, 1991, **92**, 618.
2438 P. Griesmar, G. Papin, C. Sanchez, and J. Livage, *Chem. Mater.*, 1991, **3**, 335 (*Chem. Abstr.*, 1991, **114**, 148 655).
2439 S.G. Kozlova, N.K. Moroz, and S.P. Gabuda, *Zh. Strukt. Khim.*, 1990, **31**, 143 (*Chem. Abstr.*, 1991, **114**, 92 273).
2440 M. Nabavi, F. Taulelle, C. Sanchez, and M. Verdaguer, *J. Phys. Chem. Solids*, 1990, **51**, 1375 (*Chem. Abstr.*, 1991, **114**, 111 204).
2441 M. Nabavi, C. Sanchez, and J. Livage, *Philos. Mag. B*, 1991, **63**, 941 (*Chem. Abstr.*, 1991, **115**, 36 098).
2442 A.V. Dmitriev, S.G. Arsenov, and V.L. Volkov, *J. Inclusion Phenom. Mol. Recognit. Chem.*, 1991, **11**, 89 (*Chem. Abstr.*, 1991, **115**, 268 933).
2443 M. Onoda, T. Naka, and H. Nagasawa, *J. Phys. Soc. Jpn.*, 1991, **60**, 2550 (*Chem. Abstr.*, 1991, **115**, 173 084).
2444 N.A. Zhuravlev, A.V. Dmitriev, A.A. Lakhtin, A.G. Maksimov, V.L. Volkov, and R.N. Pletnev, *Zh. Strukt. Khim.*, 1990, **31**, 59 (*Chem. Abstr.*, 1991, **114**, 258 314); N.A. Zhuravlev, A.V. Dmitriev, and P. Ya Novak, *Fiz. Tverd. Tela (Leningrad)*, 1990, **32**, 2899 (*Chem. Abstr.*, 1991, **114**, 258 312).
2445 J. Davis, D. Tinet, J.J. Fripiat, J.M. Amarilla, B. Casal, and E. Ruiz-Hitzky, *J. Mater. Res.*, 1991, **6**, 393 (*Chem. Abstr.*, 1991, **114**, 155 672).
2446 L. Stanescu, E. Burzo, T. Farcas, E. Lazar, and A. Mateiu, *Rev. Roum. Phys.*, 1990, **35**, 395 (*Chem. Abstr.*, 1991, **115**, 58 203).
2447 S. Xu and X. Wu, *Bopuxue Zazhi*, 1990, **7**, 345 (*Chem. Abstr.*, 1991, **114**, 190 796).
2448 H. Doweidar, I.A. Gohar, A.A. Megahed, and G. El-Damrawi, *Solid State Ionics*, 1991, **46**, 275 (*Chem. Abstr.*, 1991, **115**, 124 508).
2449 A.V. Dmitriev, S.G. Arsenov, T.A. Denisov, and V.L. Volkov, *Zh. Neorg. Khim.*, 1991, **36**, 2158 (*Chem. Abstr.*, 1991, **115**, 293 309).
2450 M.H. Lee, *Bull. Korean Chem. Soc.*, 1991, **12**, 243 (*Chem. Abstr.*, 1991, **115**, 83 958).
2451 M.S. Rigutto and H. Van Bekkum, *Appl. Catal.*, 1991, **68**, L1 (*Chem. Abstr.*, 1991, **114**, 151 252); J. Kornatowski, M. Sychev, V. Goncharuk, and W.H. Baur, *Stud. Surf. Sci. Catal.*, 1991, **65**, 581 (*Chem. Abstr.*, 1991, **115**, 293 434).
2452 J. Li, M.E. Lashier, G.L. Schrader, and B.C. Gerstein, *Appl. Catal.*, 1991, **73**, 83 (*Chem. Abstr.*, 1991, **115**, 100 207); N.H. Batis, H. Batis, A. Ghorbel, J.C. Vedrine, and J.C. Volta, *J. Catal.*, 1991, **128**, 248 (*Chem. Abstr.*, 1991, **114**, 172 258).

system, (^{51}V),[2453] (Li,H)NbO_3, (^{7}Li),[2454] MgO doped $LiNbO_3$,[2455] (Pb,Mg)Nb_2O_6, (^{93}Nb),[2456] $[Al_{1.67}(OH)_4(OH_2)_{1.67}][Ca_2Nb_3O_{10}]$, ($^{27}Al$, ^{93}Nb),[2457] $\dot{A}lNbO_4$, (^{27}Al),[2458] $Na_4Nb(PO_4)_3$, (^{31}P),[2459] and $LiTaO_3$, (^{7}Li).[2460]

Deuterium quadrupole coupling constants and asymmetry parameters in a bridging hydride have been determined for $[Et_4N][^2HM_2(CO)_{10}]$, M = Cr, W.[2461] The ^{31}P chemical shift anisotropy, scalar coupling constants to ^{55}Mn, ^{95}Mo, ^{97}Mo and ^{183}W, and the ^{55}Mn quadrupole coupling constants have been determined for compounds such as $[(\eta^5\text{-}C_5H_5)M(CO)_2\{(CH_2)_nPPh_2\}]$, M = Cr, Mo, W, and $[(OC)_4Mn\{(CH_2)_nPPh_2\}]$.[2462] The hydration properties of alkali metal ions incorporated in cubic ammonium molybdate have been studied by ^1H n.m.r. spectroscopy.[2463] N.m.r. data have also been reported for hydrated sodium molybdenum bronzes,[2464] $[\{H_2C(3,5\text{-}Me_2C_3HN_2)_2\}Mo(CO)_4]$, ($^{13}C$),[2465] $[NPr^n_4][Cr(C^{15}N)(CO)_5]$, ($^{13}C$, ^{15}N),[2466] $MoO_3\text{-}Al_2O_3$ catalysts, (^{95}Mo),[2467] $Li(Na)_xMo_2O_4$, (^{7}Li, ^{23}Na),[2468] $MoOPO_4$, (^{31}P),[2469] and $Cd_{1-x}In_xCr_2Se_4$.[2470]

The mechanism of Li^+ insertion into $\lambda\text{-}MnO_2$ has been studied by n.m.r. spectroscopy.[2471] Double quantum n.m.r. of oriented nuclei has been observed for $^{54}Mn\text{-}Mn(O_2CCH_3)_2.4H_2O$ and $^{54}Mn\text{-}MnCl_2.4H_2O$.[2472] The ^{31}P n.m.r. spectra for undoped and Mn- and Ni-doped hopeite crystals, $Zn_{2.4}Mn_{0.4}Ni_{0.2}(PO_4)_2.4H_2O$ have been measured.[2473] N.m.r. data have also been reported for $Zn_{0.97}Mn_{0.03}S$,[2474] (Mn,Zn)F_2, (^{19}F),[2475] (Co,Mn)$Cl_2.2H_2O$,[2476] $[Me_4N][MnCl_3]$,[2477]

2453 F.D. Hardcastle, I.E. Wachs, H. Eckert, and D.A. Jefferson, *J. Solid State Chem.*, 1991, **90**, 194 (*Chem. Abstr.*, 1991, **114**, 215 600).
2454 Y. Kanazaki, R. Chitrakar, T. Ohsaka, M. Abe, H. Uyama, and O. Matsumoto, *Yoyuen oyobi Koon Kagaku*, 1991, **34**, 45 (*Chem. Abstr.*, 1991, **115**, 267 247).
2455 L.J. Hu, Y.H. Chang, C.S. Chang, S.J. Yang, M.L. Hu, and W.S. Tse, *Mod. Phys. Lett. B*, 1991, **5**, 789 (*Chem. Abstr.*, 1991, **115**, 169 336).
2456 V.V. Laguta, M.D. Glinchuk, I.P. Bykov, A.N. Titov, and E.M. Andreev, *Fiz. Tverd. Tela (Leningrad)*, 1990, **32**, 3132 (*Chem. Abstr.*, 1991, **114**, 237 992).
2457 S. Hardin, D. Hay, M. Millikan, J.V. Sanders, and T.W. Turney, *Chem. Mater.*, 1991, **3**, 977 (*Chem. Abstr.*, 1991, **115**, 164 695).
2458 P.G. Pries de Oliveira, F. Lefebvre, M. Primet, J.G. Eon, and J.C. Volta, *J. Catal.*, 1991, **130**, 293 (*Chem. Abstr.*, 1991, **115**, 36 385).
2459 S. Prabakar and K.J. Rao, *J. Solid State Chem.*, 1991, **91**, 186 (*Chem. Abstr.*, 1991, **114**, 198 267).
2460 P.P. Man, *Mol. Phys.*, 1990, **69**, 337.
2461 A.J. Kim, F.R. Fronczek, L.G. Butler, S. Chen, and E.A. Keiter, *J. Am. Chem. Soc.*, 1991, **113**, 9090.
2462 E. Lindner, R. Fawzi, H.A. Mayer, K. Eichele, and K. Pohmer, *Inorg. Chem.*, 1991, **30**, 1102.
2463 A.J. Khan, Y. Kanzaki, and M. Abe, *J. Chem. Soc., Faraday Trans.*, 1991, **87**, 2669.
2464 N. Sotani, K. Eda, and M. Kunitomo, *J. Solid State Chem.*, 1990, **89**, 123 (*Chem. Abstr.*, 1991, **114**, 54 651).
2465 V.S. Joshi, A. Sarkar, and P.R. Rajamohanan, *J. Organomet. Chem.*, 1991, **409**, 341.
2466 E. Bär, J. Fuchs, D. Rieger, F. Aguilar-Parrilla, H.-H. Limbach, and W.P. Fehlhammer, *Angew. Chem., Int. Ed. Engl.*, 1991, **30**, 88.
2467 J.C. Edwards and P.D. Ellis, *Langmuir*, 1991, **7**, 2117 (*Chem. Abstr.*, 1991, **115**, 190 686).
2468 S. Colson, S.P. Szu, L.C. Klein, and J.M. Tarascon, *Solid State Ionics*, 1991, **46**, 283 (*Chem. Abstr.*, 1991, **115**, 147 480).
2469 L. Lezama, K.S. Suh, G. Villeneuve, and T. Rojo, *Solid State Commun.*, 1990, **76**, 449 (*Chem. Abstr.*, 1991, **114**, 16 346).
2470 N.M. Kovtun, V.Ya. Mitrofanov, V.K. Prokopenko, A.Ya. Fishman, and A.A. Shemyakov, *Fiz. Nizk. Temp. (Kiev)*, 1991, **17**, 110 (*Chem. Abstr.*, 1991, **115**, 83 941).
2471 Y. Kanzaki, A. Taniguchi, and M. Abe, *J. Electrochem. Soc.*, 1991, **138**, 333 (*Chem. Abstr.*, 1991, **114**, 73 939).
2472 M. Le Gros, A. Kotlicki, and B.G. Turrell, *Phys. Lett. A*, 1991, **154**, 75 (*Chem. Abstr.*, 1991, **114**, 176 960).
2473 N. Sato, *J. Mater. Sci. Lett.*, 1991, **10**, 115 (*Chem. Abstr.*, 1991, **114**, 176 899).
2474 V. Bindilatti, T.Q. Vu, and Y. Shapira, *Solid State Commun.*, 1991, **77**, 423 (*Chem. Abstr.*, 1991, **114**, 198 209).
2475 M. Itoh and H. Yasuoka, *J. Magn. Magn. Mater.*, 1990, **90**, 359 (*Chem. Abstr.*, 1991, **114**, 176 917).
2476 H. Kubo, K. Zenmyo, T. Hamasaki, H. Deguchi, and K. Takeda, *Fukuoka Kogyo Daigaku Erekutoronikusu Kenkyusho Shoho*, 1990, **7**, 5 (*Chem. Abstr.*, 1991, **115**, 20 713); H. Kubo, K. Zenmyo, T. Hamasaki, H.

$[Me_4N][Re_3Cl_{10}(OH_2)_2].2H_2O$, ($^{13}C$),[2478] and $CsMnI_3$, (^{134}Cs).[2479]

The ^{13}C n.m.r. chemical shielding tensor of the bridging methylene unit in *cis*-[(η^5-$C_5H_5)_2Fe_2(CO)_2(\mu$-$CO)(\mu$-$CH_2)$] has been determined by ^{13}C c.p. m.a.s. n.m.r. spectroscopy.[2480] Solid state ^{13}C n.m.r. spectroscopy has been used to quantify the degree of asymmetry of bonding for the semi-bridging carbonyl groups in iron carbonyl complexes.[2481] N.m.r. spectroscopy has been used to study the ferrite formation in the $Ca(OH)_2$-Fe_2O_3 and $CaCO_3$-Fe_2O_3 systems.[2482] N.m.r. imaging has been used to study domain structure in $FeBO_3$.[2483] N.m.r. data have also been reported for some ferrocene amides, (^{13}C),[2484] [$Fe(\eta^5$-$C_5H_4PPh_2)_2Au(Ph_2PC_5H_4$-$\eta^5)Fe(\eta^5$-$C_5H_5)$], ($^{13}C$),[2485] $^{17}O_2$ containing picket fence porphyrin, (^{17}O),[2486] [Fe(bistriazolyl)$(SeCN)_2$], (^1H),[2487] martensite,[2488] Fe_2O_3,[2489] pentasil-type iron silicates, (^{29}Si),[2490] GeFeAPO-5 molecular sieve, (^{27}Al, ^{31}P),[2491] and $FeNi_2S_4$.[2492]

Solid state ^2H n.m.r. spectra of single crystals of 2*H*-SnS_2 intercalated with [$Co(\eta^5$-$C_5D_5)_2$] indicate that the guest molecules are ordered within the host layers with their principal axes parallel to the host layer.[2493] N.m.r. data have also been reported for [$Rh(C_6H_2Me_3$-2,4,6$)_3$], (^{13}C),[2494] [(η^3-$C_3H_5)_2RhCl]_2$, (^{13}C),[2495] cobalt dispersed on various catalyst supports,[2496] CoAPO-5 molecular sieve, (^{31}P),[2497] and Co_3S_4, (^{59}Co).[2498]

^{113}Cd m.a.s. n.m.r. spectroscopy has been used to investigate CN group disorder in

Deguchi, and K. Takeda, *J. Magn. Magn. Mater.*, 1990, **90**, 355 (*Chem. Abstr.*, 1991, **114**, 176 916).
2477 H. Benner, J.A. Holyst, and J. Loew, *Europhys. Lett.*, 1991, **14**, 383 (*Chem. Abstr.*, 1991, **114**, 176 950).
2478 B. Jung, G. Meyer, and E. Herdtweck, *Z. Anorg. Allg. Chem.*, 1991, **604**, 27.
2479 S. Maegawa, N. Fujiwara, T. Kohmoto, and T. Goto, *J. Magn. Magn. Mater.*, 1990, **90**, 271 (*Chem. Abstr.*, 1991, **114**, 176 915).
2480 A.J. Kim, M.I. Altbach, and L.G. Butler, *J. Am. Chem. Soc.*, 1991, **113**, 4831.
2481 G.E. Hawkes, K.D. Sales, S. Aime, R. Gobetto, and L.Y. Lian, *Inorg. Chem.*, 1991, **30**, 1489.
2482 Yu.S. Yusfin, N.F. Pashkov, G.V. Shcheblykin, and A.N. Kapranov, *Izv. Vyssh. Uchebn. Zaved., Chern. Metall.*, 1991, 4 (*Chem. Abstr.*, 1991, **114**, 251 085).
2483 Kh.G. Bogdanova, V.A. Golenishchev-Kutuzov, L.I. Medvedev, and M.M. Shakirzyanov, *Fiz. Tverd. Tela (Leningrad)*, 1991, **33**, 379 (*Chem. Abstr.*, 1991, **115**, 83 965).
2484 M.C. Grossel, M.R. Goldspink, J.A. Hriljac, and S.C. Weston, *Organometallics*, 1991, **10**, 851,
2485 A. Houlton, R.M.G. Roberts, J. Silver, and R.V. Parish, *J. Organomet. Chem.*, 1991, **418**, 269.
2486 E. Oldfield, H.C. Lee, C. Coretsopoulos, F. Adebodun, K.D. Park, S. Yang, J. Chung, and B. Phillips, *J. Am. Chem. Soc.*, 1991, **113**, 8680.
2487 A. Ozarowski, Y. Shunzhong, B.R. McGarvey, A. Mislankar, and J.E. Drake, *Inorg. Chem.*, 1991, **30**, 3167.
2488 O.G. Bakharev, V.G. Gavrilyuk, and V.M. Nadutov, *Fiz. Met. Metalloved.*, 1990, 196 (*Chem. Abstr.*, 1991, **114**, 47 207).
2489 T. Bouhacina, G. Ablart, J. Pescia, and Y. Servant, *Solid State Commun.*, 1991, **78**, 573 (*Chem. Abstr.*, 1991, **115**, 83 922).
2490 Y. Ohno, Y. Mitsuhashi, Y. Mizutani, and Y. Fukui, *Sekiyu Gakkaishi*, 1991, **34**, 366 (*Chem. Abstr.*, 1991, **115**, 104 590).
2491 S. Han, L. Wang, and G. Yu, *Gaodeng Xuexiao Huaxue Xuebao*, 1990, **11**, 1171 (*Chem. Abstr.*, 1991, **114**, 177 060).
2492 A.Ya. Kesler, S.G. Smirnov, K.V. Pokholok, and B.N. Viting, *Izv. Akad. Nauk SSSR, Neorg. Mater.*, 1991, **27**, 1162 (*Chem. Abstr.*, 1991, **115**, 215 296).
2493 C. Grey, J.S.O. Evans, D. O'Hare, and S.J. Heyes, *J. Chem. Soc., Chem. Commun.*, 1991, 1380.
2494 R.S. Hay-Motherwell, S.U. Koschmieder, G. Wilkinson, B. Hussain-Bates, and M.B. Hursthouse, *J. Chem. Soc., Dalton Trans.*, 1991, 2821.
2495 S.A. King, D. Van Engen, H.E. Fischer, and J. Schwartz, *Organometallics*, 1991, **10**, 1195.
2496 A.N. Murty, *Report*, 1990, **DOE/PC/79917-T6**; Order No. DE91007443, 82 pp. Avail. NTIS. From *Energy Res. Abstr.*, 1991, **16**, Abstr. No. 12304 (*Chem. Abstr.*, 1991, **115**, 235 992).
2497 S. Han and J. Zhou, *Shiyou Huagong*, 1990, **19**, 673 (*Chem. Abstr.*, 1991, **114**, 231 438).
2498 H. Nishihara, T. Kanomata, T. Kaneko, and H. Yasuoka, *J. Appl. Phys.*, 1991, **69**, 4618 (*Chem. Abstr.*, 1991, **115**, 40 524).

$Cd(NH_3)_2Ni(CN)_4.2C_6H_6$ and related compounds.[2499] ^{13}C, ^{31}P, and ^{195}Pt c.p. m.a.s. n.m.r. spectra have been measured on $[PtX_2(PR_3)_2]$ and the PR_3 groups are found to be inequivalent.[2500] N.m.r. data have also been reported for aged Pd^3H_x, (^3He),[2501] bis{(cyclooctane-1,5-diyl)bis(pyrazol-1-yl)borato}$_2$Ni, (^1H, ^{13}C),[2502] {PtCl(PBu$^t_3$)(PBu$^t_2$CMe$_2$CH$_2$)}, (^{31}P),[2503] [Ni(η^2-O=CC=C=O)(PPh$_3$)$_2$], (^{13}C, ^{31}P),[2504] [Ni(en)$_2$(NO$_2$)(ClO$_4$)],[2505] [Ni(cyclohexanediamine)$_2$X]X, (^{13}C),[2506] $BaNi_2(PO_4)_2$, (^{31}P),[2507] and (perylene)$_2$[M(maleonitriledithiolate)$_2$], (M = Pt, Au, ^1H).[2508]

The quadrupole moment of ^{186}Au has been determined.[2509] The ^{31}P c.p. m.a.s. n.m.r. spectrum of [Ag(1-Ph-3,4-Me$_2$phosphole)$_3$X] shows $^1J(^{109}Ag-^{31}P)$.[2510] N.m.r. data have also been reported for (methylenedithiotetrathiafulvalene)$_2$Au(CN)$_2$,[2511] [Cu(bipy)$_2$][PF$_6$], (^{15}N),[2512] [(Ph$_3$P)(4-MeC$_5$H$_4$N)CuCl], (^{31}P),[2513] phenylcyanamidocopper(I) complexes, (^{31}P),[2514] dicyanoquinonediimine complexes of Cu,[2515] κ-(BEDT-TTF)$_2$Cu(NCS)$_2$, (^1H)[2516] [(Ph$_3$P)$_3$CuF].4PPh$_3$.4MeOH, (^{31}P),[2517] [Cu(1-Ph-3,4-Me$_2$phosphole)$_3$X], (^{31}P),[2518] [Au(1-Ph-dibenzophosphole)$_3$ClBr$_2$],

2499 S. Nishikiori, C.I. Ratcliffe, and J.A. Ripmeester, *Canad. J. Chem.*, 1990, **68**, 2270.
2500 R.K. Harris, I.J. McNaught, P. Reams, and K.J. Packer, *Magn. Reson. Chem.*, 1991, **29**, S60.
2501 G.C. Abell and D.F. Cowgill, *Phys. Rev. B: Condens. Matter*, 1991, **44**, 4178 (*Chem. Abstr.*, 1991, **115**, 221 511).
2502 S. Trofimenko, F.B. Hulsbergen, and J. Reedijk, *Inorg. Chim. Acta*, 1991, **183**, 203.
2503 X. Han, H. Ruegger, and J. Sonderegger, *Chin. Sci. Bull.*, 1991, **36**, 382 (*Chem. Abstr.*, 1991, **115**, 173 161).
2504 A.K. List, M.R. Smith, tert., and G.L. Hillhouse, *Organometallics*, 1991, **10**, 361.
2505 M. Chiba, Y. Ajiro, H. Kikuchi, T. Kubo, and T. Morimoto, *Phys. Rev. B: Condens. Matter*, 1991, **44**, 2838 (*Chem. Abstr.*, 1991, **115**, 148 743); N. Fujiwara, T. Goto, T. Kohmoto, and S. Maegawa, *J. Magn. Magn. Mater.*, 1990, **90**, 229 (*Chem. Abstr.*, 1991, **114**, 176 913).
2506 R. Ikeda, T. Tamura, and M. Yamashita, *Chem. Phys. Lett.*, 1990, **173**, 466 (*Chem. Abstr.*, 1991, **114**, 73 838).
2507 P. Gaveau, J.P. Boucher, L.P. Regnault, and Y. Henry, *J. Appl. Phys.*, 1991, **69**, 6228 (*Chem. Abstr.*, 1991, **115**, 40 527).
2508 R.T. Henriques, C. Bourbonnais, P. Wzietek, D. Kongeter, J. Vioron, and D. Jerome, *Synth. Met.*, 1991, **42**, 2339 (*Chem. Abstr.*, 1991, **115**, 104 497).
2509 B. Hinfurtner, E. Hagn, E. Zech, and R. Eder, *Phys. Rev. Lett.*, 1991, **67**, 812 (*Chem. Abstr.*, 1991, **115**, 144 146).
2510 S. Attar, N.W. Alcock, G.A. Bowmaker, J.S. Frye, W.H. Bearden, and J.H. Nelson, *Inorg. Chem.*, 1991, **30**, 4166.
2511 K. Kanoda, Y. Kobayashi, T. Takahashi, T. Inukai, and G. Saito, *Phys. Rev. B: Condens. Matter*, 1990, **42**, 8678 (*Chem. Abstr.*, 1991, **114**, 93 772).
2512 S. Kitagawa, M. Munakata, D. Deguchi, and T. Fujito, *Magn. Reson. Chem.*, 1991, **29**, 566.
2513 P.C. Healy, A.K. Whittaker, J.D. Kildea, B.W. Skelton, and A.H. White, *Aust. J. Chem.*, 1991, **44**, 729.
2514 E.W. Ainscough, E.N. Baker, M.L. Brader, A.M. Brodie, S.L. Ingham, J.M. Waters, J.V. Hanna, and P.C. Healy, *J. Chem. Soc., Dalton Trans.*, 1991, 1243.
2515 U. Langohr, J.U. Von Schuetz, H.C. Wolf, H. Meixner, and S. Huenig, *Synth. Met.*, 1991, **42**, 1855 (*Chem. Abstr.*, 1991, **115**, 104 490).
2516 T. Takahashi, K. Kanoda, K. Akiba, K. Sakao, M. Watabe, K. Suzuki, and G. Saito, *Synth. Met.*, 1991, **42**, 2005 (*Chem. Abstr.*, 1991, **115**, 147 752); T. Klutz, U. Haeberlen, D. Schweitzer, and H.J. Keller, *Synth. Met.*, 1991, **42**, 2115 (*Chem. Abstr.*, 1991, **115**, 83 969); T. Klutz, U. Haeberlen, and D. Schweitzer, *J. Phys.: Condens. Matter*, 1990, **2**, 10417 (*Chem. Abstr.*, 1991, **114**, 113 755).
2517 P.C. Healy, J.V. Hanna, J.D. Kildea, B.W. Skelton, and A.H. White, *Aust. J. Chem.*, 1991, **44**, 427.
2518 S. Attar, G.A. Bowmaker, N.W. Alcock, J.S. Frye, W.H. Bearden, and J.H. Nelson, *Inorg. Chem.*, 1991, **30**, 4743.

(^{31}P),[2519] CuO, (^{63}Cu),[2520] (^{17}O, ^{63}Cu),[2521] tobermorite from CuO and silicic acid, (^{29}Si),[2522] $[Cu(O_2CR)_2]_2$, (^{13}C),[2523] adducts of 1,5,9,13-tetraselenacyclohexadecane with $Cu(O_3SCF_3)$ and $Hg(CN)_2$, (^{77}Se),[2524] $AgZnPO_4$, (^{31}P),[2525] K_2CuF_4, (^{63}Cu, ^{65}Cu),[2526] and $K_2CuCl_4.2H_2O$, (^{1}H).[2527]

Framework ordering in solid $Cd(CN)_2$ has been studied using ^{113}Cd n.m.r. spectroscopy.[2528] Anisotropies in the ^{31}P chemical shift tensor and the ^{199}Hg-^{31}P spin-spin coupling tensor for a series of mercury phosphines[2529] and phosphonates[2530] have been determined. $Zn\{S_2P(OEt)_2\}_2$ has been proposed as a compound to use to set the magic angle for ^{31}P m.a.s. n.m.r. spectroscopy.[2531] N.m.r. data have also been reported for a L-carnosine Zn^{II} complex, (^{13}C, ^{15}N),[2532] (BEDT-TTF)$_4Hg_{2.78}Cl_8$,[2533] $CdGeP_2$, (^{31}P, ^{113}Cd),[2534] $[Cd(PR_3)_2X_2]$, (^{31}P, ^{113}Cd),[2535] $CdGeAs_2$, (^{113}Cd),[2536] Zn salt of poly(methacrylic acid), (^{13}C),[2537] CdO-B_2O_3-SiO_2, (^{11}B),[2538] Cd_2SnO_4, (^{111}Cd),[2539] $[Zn\{S_2P(OR)_2\}_2]$, (^{31}P),[2540] $[Cd(S_2CNEt_2)_2]$, (^{113}Cd),[2541] $[Cd\{S_2C{=}C(CN)_2\}_2]^{2-}$, ($^{113}Cd$),[2542] $Hg_{1-x}Cd_xTe$, (^{125}Te),[2543] $[M(SR)_n]^{2-n}$, (M = Cd, Hg; ^{113}Cd, ^{199}Hg),[2544]

2519 S. Attar, J.H. Nelson, W.H. Bearden, N.W. Alcock, L. Solujić, and E.B. Milosavljević, *Polyhedron*, 1991, **10**, 1939.
2520 R.G. Graham, D. Fowler, J.S. Lord, P.C. Riedi, and B.M. Wanklyn, *Phys. Rev. B: Condens. Matter*, 1991, **44**, 7091 (*Chem. Abstr*., 1991, **115**, 246 377); L. Cristofolini, G. Amoretti, C. Bucci, P. Carretta, R. De Renzi, G. Guidi, F. Licci, and G. Calestani, *Physica C (Amsterdam)*, 1991, **181**, 121 (*Chem. Abstr*., 1991, **115**, 245 583).
2521 J.P. Lu, Q. Si, J.H. Kim, and K. Levin, *Physica C (Amsterdam)*, 1991, **179**, 191 (*Chem. Abstr*., 1991, **115**, 173 181).
2522 Y. Okada, M. Shimoda, T. Mitsuda, and H. Toraya, *Onoda Kenkyu Hokoku*, 1990, **42**, 199 (*Chem. Abstr*., 1991, **115**, 98 039).
2523 G.C. Campbell, J.H. Reibenspies, and J.F. Haw, *Inorg. Chem*., 1991, **30**, 171.
2524 R.J. Batchelor, F.W.B. Einstein, I.D. Gay, J.H. Gu, and B.M. Pinto, *J. Organomet. Chem*., 1991, **411**, 147.
2525 M. Andratschke, *Z. Anorg. Allg. Chem*., 1991, **601**, 103.
2526 M. Fujii, *J. Phys. Soc. Jpn*., 1990, **59**, 4449 (*Chem. Abstr*., 1991, **114**, 54 463).
2527 P.J. Kang, J.O. Kwag, and I. Yu, *Sae Mulli*, 1990, **30**, 225 (*Chem. Abstr*., 1991, **114**, 155 647).
2528 S. Nishikiori, C.I. Ratcliffe, and J.A. Ripmeester, *J. Chem. Soc., Chem. Commun*., 1991, 735.
2529 W.P. Power, M.D. Lumsden, and R.E. Wasylishen, *J. Am. Chem. Soc*., 1991, **113**, 8257.
2530 W.P. Power, M.D. Lumsden, and R.E. Wasylishen, *Inorg. Chem*., 1991, **30**, 2997.
2531 A. Kubo and C.A. McDowell, *J. Magn. Reson*., 1991, **92**, 409.
2532 T. Matsukura, T. Takahashi, Y. Nishimura, T. Ohtani, M. Sawada, and K. Shibata, *Chem. Pharm. Bull*., 1990, **38**, 3140.
2533 A.V. Skiprov, A.P. Stepanov, I. Heinmaa, A. Vainrub, R.N. Lyubovskaya, and M.Z. Aldoshina, *Physica C (Amsterdam)*, 1990, **172**, 340 (*Chem. Abstr*., 1991, **114**, 93 785).
2534 D. Franke, R. Maxwell, D. Lathrop, and H. Eckert, *J. Am. Chem. Soc*., 1991, **113**, 4822.
2535 J.M. Kessler, J.H. Reeder, R. Vac, C. Yeung, J.H. Nelson, J.S. Frye, and N.W. Alcock, *Magn. Reson. Chem*., 1991, **29**, S94.
2536 D.R. Franke and H. Eckert, *J. Phys. Chem*., 1991, **95**, 331.
2537 M. Asada, N. Asada, A. Toyoda, I. Ando, and H. Kurosu, *J. Mol. Struct*., 1991, **244**, 237.
2538 D.J. Cha and S.J. Chung, *Sae Mulli*, 1990, **30**, 544 (*Chem. Abstr*., 1991, **114**, 148 593).
2539 K.J.D. MacKenzie, C.M. Cardile, and R.H. Meinhold, *J. Phys. Chem. Solids*, 1991, **52**, 969 (*Chem. Abstr*., 1991, **115**, 195 222).
2540 P.G. Harrison and P. Brown, *Wear*, 1991, **148**, 123 (*Chem. Abstr*., 1991, **115**, 211 528).
2541 O.F.Z. Kahn and P. O'Brien, *Polyhedron*, 1991, **10**, 325.
2542 H.Y. Li and E.L. Amma, *Inorg. Chim. Acta*, 1991, **177**, 5.
2543 H.M. Vieth, S. Vega, N. Yellin, and D. Zamir, *J. Phys. Chem*., 1991, **95**, 1420.
2544 R.A. Santos, E.S. Gruff, S.A. Koch, and G.S. Harbison, *J. Am. Chem. Soc*., 1991, **113**, 469.

$[Hg(S_2COR)_2]_n$, (^{13}C),[2545] K_2ZnCl_4, (^{39}K),[2546] Rb_2ZnCl_4, (^{35}Cl),[2547] $(NH_4)_2ZnCl_4$, (^{14}N),[2548] and $Cd_{1-x}Bi_xF_{2x}$, (^{19}F).[2549]

^2H nuclear spin dipolar dephasing in quadrupole echo-experiments has been applied to $(D_3C)_3NHBPh_4$.[2550] N.m.r. data have also been reported for $NaBH_4$, $Na_2B_{12}H_{12}$, (^{11}B),[2551] $R_4NB_3H_8$, (^{11}B),[2552] preceramic organoboron-silicon polymers,[2553] boron carbide solid solution, (^{13}C),[2554] and boron in a silicon matrix, (^{11}B).[2555]

The state of boron nitrides in a matrix of dispersed silica matrixes has been studied using ^{11}B n.m.r. spectroscopy.[2556] A method has been presented which retrieves structural information from m.a.s. n.m.r. spectra in glasses broadened by second order quadrupole effects and was applied to ^{11}B m.a.s. n.m.r. spectra of a borate glass.[2557] The acidic sites in aluminium modified porous glasses have been characterised by ^{27}Al n.m.r. spectroscopy.[2558] ^{11}B n.m.r. spectroscopy has been used to study the removal of boron from boron containing pentasil type zeolites.[2559] The thermal stability of boron containing ZSM-5 molecular sieve has been investigated using ^{11}B n.m.r. spectroscopy.[2560] Some silicon ceramics which contain some boron have been investigated by ^{11}B n.m.r. spectroscopy.[2561] N.m.r. data have also been reported for boron nitrides, (^{11}B),[2562] Li_2O-Na_2O-B_2O_3-SiO_2 glasses, (^6Li, ^7Li, ^{29}Si),[2563] $Li_2B_4O_7$, (^7Li, ^{11}B),[2564] $28Li_2O$-$16Li_2Cl_2$-$56B_2O_3$, (^7Li),[2565] $xNa_2O.yB_2O_3$,

2545 A.M. Hounslow and E.R.T. Tiekink, *J. Crystallogr. Spectrosc. Res.*, 1991, **21**, 133 (*Chem. Abstr.*, 1991, **114**, 257 316).
2546 B. Topic, U. Haeberlen, R. Blinc, and S. Zumer, *Phys. Rev. B: Condens. Matter*, 1991, **43**, 91 (*Chem. Abstr.*, 1991, **114**, 73 356).
2547 G. Papavassiliou, F. Milia, R. Blinc, and S. Zumer, *Solid State Commun.*, 1991, **77**, 891 (*Chem. Abstr.*, 1991, **115**, 20 717).
2548 D. Michel, B. Mueller, J. Petersson, A. Trampert, and R. Walisch, *Phys. Rev. B: Condens. Matter*, 1991, **43**, 7507 (*Chem. Abstr.*, 1991, **114**, 239 156).
2549 K.S. Suh, J. Senegas, J.M. Reau, and P. Hagenmuller, *J. Solid State Chem.*, 1991, **93**, 469 (*Chem. Abstr.*, 1991, **115**, 219 918).
2550 M.L.H. Gruwel, *J. Phys. Chem.*, 1991, **95**, 10 109.
2551 B. Gruner, V. Prochazka, J. Subrt, S. Hermanek, and R. Pospech, *Eur. J. Solid State Inorg. Chem.*, 1991, **28**, 597 (*Chem. Abstr.*, 1991, **115**, 173 423).
2552 B. Bonnetot, H. Mongeot, B. Frange, and A. Ouassas, *Eur. J. Solid State Inorg. Chem.*, 1991, **28**, 547 (*Chem. Abstr.*, 1991, **115**, 124 120).
2553 M.T.S. Hsu, T.S. Chen, and S.R. Riccitiello, *J. Appl. Polym. Sci.*, 1991, **42**, 851 (*Chem. Abstr.*, 1991, **114**, 127 591).
2554 T.L. Aselage and D. Emin, *AIP Conf. Proc.*, 1991, **231**, 177 (*Chem. Abstr.*, 1991, **115**, 171 087).
2555 D.Y. Han and H. Kessemeier, *Phys. Rev. Lett.*, 1991, **67**, 346 (*Chem. Abstr.*, 1991, **115**, 83 967).
2556 I.P. Beletskii, L.A. Bondar, V.V. Brei, V.E. Klimenko, and A.A. Chuiko, *Teor. Eksp. Khim.*, 1991, **27**, 90 (*Chem. Abstr.*, 1991, **115**, 79 557).
2557 P. Mustarelli, R. Riccardi, S. Scotti, and M. Villa, *Phys. Chem. Glasses*, 1991, **32**, 129 (*Chem. Abstr.*, 1991, **115**, 83 944).
2558 I. Tschistowskaja and F. Janowski, *Chem. Tech. (Leipzig)*, 1991, **43**, 252 (*Chem. Abstr.*, 1991, **115**, 143 525).
2559 B. Unger, K.-P. Wendlandt, H. Toufar, W. Schwieger, K.H. Bergk, E. Brunner, and W. Reschetilowski, *J. Chem. Soc., Faraday Trans.*, 1991, **87**, 3099.
2560 S.B. Hong, Y.S. Uh, S.I. Woo, and J.K. Lee, *J. Chem. Eng.*, 1991, **8**, 1 (*Chem. Abstr.*, 1991, **115**, 74 544).
2561 D. Seyferth, H. Plenio, W.S. Rees, jun., and K. Buchner, *Front. Organosilicon Chem., [Proc. Int. Symp. Organosilicon Chem.], 9th*, 1990 (Pub. 1991), 15 (*Chem. Abstr.*, 1991, **115**, 285 398).
2562 P.S. Marchetti, D. Kwon, W.R. Schmidt, L.V. Interrante, and G.E. Maciel, *Chem. Mater.*, 1991, **3**, 482 (*Chem. Abstr.*, 1991, **114**, 233 570).
2563 P.J. Bray, J.F. Emerson, D. Lee, S.A. Feller, D.L. Bain, and D.A. Feil, *J. Non-Cryst. Solids*, 1991, **129**, 240 (*Chem. Abstr.*, 1991, **114**, 239 172).
2564 Yu.N. Ivanov, Ya.V. Burak, and K.S. Aleksandrov, *Fiz. Tverd. Tela (Leningrad)*, 1990, **32**, 3379 (*Chem. Abstr.*, 1991, **115**, 40 536).
2565 M. Grüne and W. Müller-Warmuth, *Ber. Bunsenges. Phys. Chem.*, 1991, **95**, 1068.

(^{11}B),[2566] $K_{4.5}B_{4.5}(O_2)_7(OH)_4$, ($^1H$),[2567] alkaline earth boroaluminate crystals and glasses,[2568] (^{11}B, ^{27}Al),[2569] K_2O-B_2O_3-Al_2O_3 glasses, (^{11}B),[2570] gels from $Si(OEt)_4$ and H_3BO_3, (^{11}B, ^{29}Si),[2571] aluminium borates, (^{11}B, ^{27}Al),[2572] sodium borosilicate glasses with a low silica content,[2573] boron-zeolites and borosilicates, (^1H, ^{11}B, ^{23}Na, ^{29}Si),[2574] boron containing ZSM-12, (^{11}B),[2575] Li_2Cl_2-B_2O_3-SiO_2 xerogels, (^7Li, ^{11}B, ^{29}Si),[2576] Li_2O-B_2O_3-Al_2O_3-SiO_2, (^{11}B),[2577] SrO-B_2O_3-Al_2O_3-TiO_2, (^{11}B, ^{27}Al),[2578] Li_2O-B_2O_3-SiO_2,[2579] B_2O_3-SiO_2-PbO, Al_2O_3-SiO_2-PbO, (^{11}B, ^{27}Al, ^{29}Si),[2580] SiO_2 supported BPO_4, (^{31}P),[2581] SAPO-5, BAPO-5, LiAPO-5, (^{11}B, ^{29}Si),[2582] $Li_{6+x}(B_{10}S_{18})S_x$, ($^7Li$),[2583] cyclic boron-sulphur and boron-selenium compounds, (^{10}B, ^{11}B),[2584] B_2S_3-Li_2S-LiI glasses, (^{11}B),[2585] and $[Bu_4N][BF_4]$, (^1H, ^{19}F).[2586]

The formation of high strength alumina fibres has been followed by ^{13}C n.m.r. spectroscopy.[2587] The incorporation of aluminium and gallium into a zeolite has been monitored by ^{27}Al, ^{29}Si, and ^{71}Ga n.m.r. spectroscopy.[2588] N.m.r. data have also been reported for AlN from precursors, (^{13}C, ^{27}Al),[2589] Al_2O_3 fibres, (^{27}Al),[2590] aluminium gels, (^{27}Al),[2591] the product from $Al(OPr^i)_3$

2566 S. Xu and X. Wu, *Wuli Xuebao*, 1990, **39**, 714 (*Chem. Abstr.*, 1991, **114**, 87 027).
2567 V.M. Kalenik, S.A. Zakharova, Yu.L. Martynyuk, and B.N. Chernyshov, *Izv. Akad. Nauk SSSR, Neorg. Mater.*, 1991, **27**, 1097 (*Chem. Abstr.*, 1991, **115**, 147 035).
2568 B.C. Bunker, R.J. Kirkpatrick, and R.K. Brow, *J. Am. Ceram. Soc.*, 1991, **74**, 1425 (*Chem. Abstr.*, 1991, **115**, 55 136).
2569 B.C. Bunker, R.J. Kirkpatrick, R.K. Brow, G.L. Turner, and C. Nelson, *J. Am. Ceram. Soc.*, 1991, **74**, 1430 (*Chem. Abstr.*, 1991, **115**, 55 137).
2570 M.S. Shim, H.K. Yang, M.J. Kang, M.S. Kim, H.L. Song, S.J. Chung, H.T. Kim, and D.J. Cha, *Sae Mulli*, 1990, **30**, 734 (*Chem. Abstr.*, 1991, **114**, 252 291).
2571 H. Asaoka, *J. Mol. Catal.*, 1991, **68**, 301.
2572 N.A. Ovramenko, O.F. Zakharchenko, A.S. Litovchenko, V.V. Trachevskii, V.I. Shutova, and F.D. Ovcharenko, *Dokl. Akad. Nauk SSSR*, 1990, **312**, 1145 (*Chem. Abstr.*, 1991, **114**, 93 907).
2573 H. Doweidar and M.S. Meikhail, *Phys. Chem. Glasses*, 1990, **31**, 239 (*Chem. Abstr.*, 1991, **114**, 67 555).
2574 J.C. Edwards, C.L. O'Young, and P.J. Giammatteo, *Prepr. - Am. Chem. Soc., Div. Pet. Chem.*, 1991, **36**, 285 (*Chem. Abstr.*, 1991, **115**, 235 664).
2575 Y. Zhao, S. Xiang, and H. Li, *Cuihua Xuebao*, 1990, **11**, 323 (*Chem. Abstr.*, 1991, **114**, 54 664).
2576 B. Wang, S. Szu, M. Tsai, M. Greenblatt, and L.C. Klein, *Solid State Ionics*, 1991, **47**, 297 (*Chem. Abstr.*, 1991, **115**, 244 901).
2577 M.S. Shim, S.J. Moon, M.J. Gang, D.S. Byun, M.S. Kim, S.J. Chung, H.T. Kim, and D.J. Cha, *Sae Mulli*, 1990, **30**, 557 (*Chem. Abstr.*, 1991, **114**, 148 594).
2578 T. Huebert, U. Banach, K. Witke, and P. Reich, *Phys. Chem. Glasses*, 1991, **32**, 58 (*Chem. Abstr.*, 1991, **114**, 212 481).
2579 Y. Tang and Z. Jiang, *Guangxue Xuebao*, 1990, **10**, 1107 (*Chem. Abstr.*, 1991, **115**, 140 836).
2580 S. Prabakar and K.J. Rao, *Philos. Mag. B*, 1991, **64**, 401 (*Chem. Abstr.*, 1991, **115**, 246 386).
2581 S. Sato, M. Hasegawa, T. Sodesawa, and F. Nozaki, *Bull. Chem. Soc. Jpn.*, 1991, **64**, 516.
2582 S. Qiu, W. Tian, W. Pang, T. Sun, and D. Jiang, *Zeolites*, 1991, **11**, 371 (*Chem. Abstr.*, 1991, **115**, 63 264).
2583 P. Zum Hebel, B. Krebs, M. Gruene, and W. Müller-Warmuth, *Solid State Ionics*, 1990, **43**, 133 (*Chem. Abstr.*, 1991, **114**, 113 890).
2584 R. Conrady-Pigorsch, W. Müller-Warmuth, G. Schetlik, M. Wienkenhöver, and B. Krebs, *Ber. Bunsenges. Phys. Chem.*, 1991, **95**, 453.
2585 K.S. Suh, A. Hojjaji, G. Villeneuve, M. Menetrier, and A. Levasseur, *J. Non-Cryst. Solids*, 1991, **128**, 13 (*Chem. Abstr.*, 1991, **114**, 254 337).
2586 B. Szafranska, Z. Pajak, and A. Kozak, *Z. Naturforsch., A*, 1991, **46**, 545 (*Chem. Abstr.*, 1991, **115**, 83 960).
2587 J.C.W. Chien, Y. Yang, J.R. Martinez, S. Dong, V. Telluri, and K. Jakus, *J. Polym. Sci., Part A: Polym. Chem.*, 1991, **29**, 495 (*Chem. Abstr.*, 1991, **114**, 127 610); T. Yogo and H. Iwahara, *J. Mater. Sci.*, 1991, **26**, 5292 (*Chem. Abstr.*, 1991, **115**, 213 370).
2588 M.A. Camblor, J.A. Martens, P.J. Grobet, and P.A. Jacobs, *Stud. Surf. Sci. Catal.*, 1991, **65**, 613 (*Chem. Abstr.*, 1991, **115**, 293 435).
2589 N. Hashimoto, H. Yoden, K. Nomura, and S. Deki, *Nippon Seramikkusu Kyokai Gakujutsu Ronbunshi*, 1991, **99**, 751 (*Chem. Abstr.*, 1991, **115**, 213 302).
2590 T. Nishio and T. Fujiki, *Nippon Kagaku Kaishi*, 1991, 1346 (*Chem. Abstr.*, 1991, **115**, 261 707); R.J. Lussier, *J. Catal.*, 1991, **129**, 225 (*Chem. Abstr.*, 1991, **114**, 215 438); H.W. Lee, G.D. Kim, H.J. Jung, and C.E. Kim, *Yoop Hakhoechi*, 1991, **28**, 60 (*Chem. Abstr.*, 1991, **115**, 164 750); W.J. DeSisto, Y.Y. Qian, C.

hydrolysis, (^{13}C, ^{27}Al),[2592] Al_2O_3 films, (^{27}Al),[2593] γ-$Al(OH)_3$, (^{27}Al),[2594] AlO(OH), (^{27}Al),[2595] aluminium hydroxide chlorides, (^{27}Al),[2596] $3CaO.Al_2O_3$, (^{27}Al),[2597] $CaO.10Al_2O_3$, (^{27}Al),[2598] high-Al_2O_3 cement, (^{27}Al),[2599] vanadium contaminated aluminas and aluminosilicate gels, (^{51}V),[2600] ruby, (^{27}Al),[2601] Pb^{2+} in β-alumina, (^{207}Pb),[2602] products from the reaction of γ-Al_2O_3 with SO_2, (^{27}Al),[2603] and halogen promoted γ-alumina, (^{27}Al).[2604]

Silicon-aluminium-carbon-oxygen ceramic precursors have been characterised by ^{13}C, ^{27}Al, and ^{29}Si n.m.r. spectroscopy.[2605] The state of OH groups before their destruction in kaolinite has been studied by ^1H and ^{27}Al n.m.r. spectroscopy.[2606] The ^{29}Si m.a.s. n.m.r. spectrum of Al_2SiO_5 consists of two signals.[2607] The localization of hydrogen atoms in $Na_8Al_6Si_6O_{24}(OH)_2.2H_2O$ has been achieved using ^1H n.m.r. spectroscopy.[2608] N.m.r. data have also been reported for aluminosilicate gels, (^{27}Al, ^{29}Si),[2609] laponite gels, (^7Li, ^{23}Na),[2610] Al_2SiO_5, (^{27}Al),[2611]

Hannigan, J.O. Edwards, and R. Kershaw, *Report*, 1990, **Tr-3**; Order No. AD-A217570, 9 pp. Avail. NTIS. From *Gov. Rep. Announce. Index (U.S.)*, 1990, **90**, Abstr. No. 024,613 (*Chem. Abstr.*, 1991, **114**, 48 119).

2591 T.E. Wood, A.R. Siedle, J.R. Hill, R.P. Skarjune, and C.J. Goodbrake, *Mater. Res. Soc. Symp. Proc.*, 1990, **180**, 97 (*Chem. Abstr.*, 1991, **115**, 285 482).

2592 M. Inoue, H. Kominami, and T. Inui, *J. Chem. Soc., Dalton Trans.*, 1991, 3331.

2593 Y. Kobayashi, S. Nakata, and Y. Kurokawa, *Sekiyu Gakkaishi*, 1991, **34**, 197 (*Chem. Abstr.*, 1991, **114**, 148 728).

2594 R.C.T. Slade, J.C. Southern, and I.M. Thompson, *J. Mater. Chem.*, 1991, **1**, 563 (*Chem. Abstr.*, 1991, **115**, 293 307).

2595 B.S. Hemingway, R.A. Robie, and J.A. Apps, *Am. Mineral.*, 1991, **76**, 445 (*Chem. Abstr.*, 1991, **115**, 53 558).

2596 R. Bertram, D. Mueller, W. Gessner, H. Goerz, and S. Schoenherr, *Z. Chem.*, 1990, **30**, 416 (*Chem. Abstr.*, 1991, **114**, 257 277).

2597 J. Skibsted, H. Bildsoe, and H.J. Jakobsen, *J. Magn. Reson.*, 1991, **92**, 669.

2598 J.A.M. Van Hoek, F.J.J. Van Loo, R. Metselaar, J.W. De Haan, and A.J. Van den Berg, *Solid State Ionics*, 1991, **45**, 93 (*Chem. Abstr.*, 1991, **114**, 235 700).

2599 W. Gessner, S. Moehmel, J. Kieser, and M. Haewecker, *Calcium Aluminate Cem., Proc. Int. Symp.*, 1990, 52. Ed. R.J. Mangabhai (*Chem. Abstr.*, 1991, **114**, 252 466); D.J. Greenslade and D.J. Williamson, *Calcium Aluminate Cem., Proc. Int. Symp.*, 1990, 81. Ed R.J. Mangabhai (*Chem. Abstr.*, 1991, **115**, 14 158).

2600 P.S. Iyer, H. Eckert, M.L. Occelli, and J.M. Stencel, *ACS Symp. Ser.*, 1991, **452**, 242 (*Chem. Abstr.*, 1991, **115**, 12 063).

2601 A. Szabo, T. Muramoto, and R. Kaarli, *Phys. Rev. B: Condens. Matter*, 1990, **42**, 7769 (*Chem. Abstr.*, 1991, **114**, 16 317).

2602 B.M. Sass, B.H. Suits, and D. White, *Mater. Res. Soc. Symp. Proc.*, 1991, **210**, 553 (*Chem. Abstr.*, 1991, **115**, 221 509).

2603 V.P. Shmachkova, N.S. Kotsarenko, E.M. Moroz, and V.M. Mastikhin, *Kinet. Katal.*, 1991, **32**, 916 (*Chem. Abstr.*, 1991, **115**, 240 928).

2604 J. Thomson, G. Webb, and J.M. Winfield, *J. Mol. Catal.*, 1991, **67**, 117.

2605 F. Babonneau, G.D. Soraru, K.J. Thorne, and J.D. Mackenzie, *J. Am. Ceram. Soc.*, 1991, **74**, 1725 (*Chem. Abstr.*, 1991, **115**, 140 915).

2606 A.S. Litovchenko, O.D. Ishutina, and A.M. Kalinichenko, *Phys. Status Solidi A*, 1991, **123**, K57 (*Chem. Abstr.*, 1991, **114**, 193 426).

2607 J.S. Hartman and B.L. Sherriff, *J. Phys. Chem.*, 1991, **95**, 7575.

2608 V.A. Detinich, V.Yu. Galitskii, and B.N. Grechushnikov, *Dokl. Akad. Nauk SSSR*, 1990, **314**, 625 (*Chem. Abstr.*, 1991, **114**, 33 506).

2609 W.G. Fahrenholtz, S.L. Hietala, P. Newcomer, N.R. Dando, D.M. Smith, and C.J. Brinker, *J. Am. Ceram. Soc.*, 1991, **74**, 2393 (*Chem. Abstr.*, 1991, **115**, 285 541).

2610 J. Grandjean and P. Laszlo, *J. Magn. Reson.*, 1991, **92**, 404.

2611 L.B. Alemany, D. Massiot, B.L. Sherriff, M.E. Smith, and F. Taulelle, *Chem. Phys. Lett.*, 1991, **177**, 301 (*Chem. Abstr.*, 1991, **114**, 198 268); S.F. Dec, J.J. Fitzgerald, J.S. Frye, M.P. Shatlock, and G.E. Maciel, *J. Magn. Reson.*, 1991, **93**, 403.

$Na_8(AlSiO_4)_6(NO_2)_2$, (^{23}Na, ^{29}Si),[2612] (^{29}Si),[2613] $Na_8(AlSiO_4)_6(MO_4)_2$, (M = Mn, Cl; ^{29}Si),[2614] $Na_8(AlSiO_4)_6(NO_3)_2$, (^{29}Si),[2615] $KMg_3(Si_3AlO_{10})F_2$, (^{29}Si),[2616] Na_2O-Al_2O_3-SiO_2 glasses, (^{29}Si),[2617] and Al_2O_3-SiO_2 glasses, (^{27}Al, ^{29}Si).[2618]

The structure and thermal transformations of allophanes have been studied by ^{27}Al and ^{29}Si n.m.r. spectroscopy.[2619] ^{27}Al and ^{29}Si m.a.s. n.m.r. spectroscopy has been used to study the aluminium-silicon distribution in synthetic $K_xMg_2Al_{4+x}Si_{5-x}O_{18}$ corderites.[2620] ^{27}Al and ^{29}Si n.m.r. spectroscopy has been used to examine the rehydration of metakaolinite to kaolinite.[2621] The effects of a $LiNO_3.3H_2O$ mineralizer on the thermal stability and phase transformations of kaolinite have been studied by ^{7}Li, ^{27}Al, and ^{29}Si n.m.r. spectroscopy.[2622] The evolution of structural changes during flash calcination of kaolinite has been studied using ^{27}Al and ^{29}Si n.m.r. spectroscopy.[2623] Structural changes induced on mullite precursors by thermal treatment have been investigated by ^{27}Al m.a.s. n.m.r. spectroscopy.[2624] ^{29}Si n.m.r. spectroscopy has been used to investigate the condensation in mullite powders formed from $Al(OR)_3$ and $Si(OR)_4$.[2625] A two-dimensional *J*-scaled ^{29}Si COSY spectrum of highly siliceous mordenite has been reported.[2626] The short-range-order of the Si,Al cation distribution over the tetrahedral sheets of phyllosilicates has been studied from a quantitative analysis of ^{29}Si n.m.r. spectra.[2627] Dealumination and aluminium intercalation of vermiculite has been investigated by ^{27}Al and ^{29}Si n.m.r. spectroscopy.[2628] N.m.r. data have also been reported for amorphous silica-aluminas, (^{29}Si),[2629] silica-alumina matrices, (^{27}Al),[2630] Silica Springs allophane, (^{27}Al, ^{29}Si),[2631] amblygonite-montebrasite series, (^{31}P),[2632] sodium beidellite, (^{29}Si),[2633] North Sea

[2612] P.B. Kempa, G. Engelhardt, J.C. Buhl, J. Felsche, G. Harvey, and C. Baerlocher, *Zeolites*, 1991, **11**, 558 (*Chem. Abstr.*, 1991, **115**, 171 364).
[2613] M.T. Weller, S.M. Dodd, and M.R.M. Jiang, *J. Mater. Chem.*, 1991, **1**, 11 (*Chem. Abstr.*, 1991, **114**, 239 225).
[2614] M.T. Weller and K.E. Haworth, *J. Chem. Soc., Chem. Commun.*, 1991, 734.
[2615] J.C. Buhl, *Thermochim. Acta*, 1991, **189**, 75 (*Chem. Abstr.*, 1991, **115**, 269 031).
[2616] F.D. Duldulao and J.M. Burlitch, *Chem. Mater.*, 1991, **3**, 772 (*Chem. Abstr.*, 1991, **115**, 164 680).
[2617] H. Maekawa, T. Maekawa, K. Kawamura, and T. Yokokawa, *J. Phys. Chem.*, 1991, **95**, 6822.
[2618] R.K. Sato, P.F. McMillan, P. Dennison, and R. Dupree, *J. Phys. Chem.*, 1991, **95**, 4483.
[2619] K.J.D. MacKenzie, M.E. Bowden, and R.H. Meinhold, *Clays Clay Miner.*, 1991, **39**, 337 (*Chem. Abstr.*, 1991, **115**, 236 413).
[2620] J. Senegas, A.R. Grimmer, D. Muller, P.T.D. Mercurio, and B. Frit, *J. Mater. Sci.*, 1991, **26**, 5053 (*Chem. Abstr.*, 1991, **115**, 267 250).
[2621] J. Rocha and J. Klinowski, *J. Chem. Soc., Chem. Commun.*, 1991, 582; J. Rocha, J.M. Adams, and J. Klinowski, *J. Solid State Chem.*, 1990, **89**, 260 (*Chem. Abstr.*, 1991, **114**, 66 047).
[2622] J. Rocha, J. Klinowski, and J.M. Adams, *J. Mater. Sci.*, 1991, **26**, 3009 (*Chem. Abstr.*, 1991, **115**, 34 144).
[2623] R.C.T. Slade and T.W. Davies, *J. Mater. Chem.*, 1991, **1**, 361 (*Chem. Abstr.*, 1991, **115**, 141 011).
[2624] J. Sanz, I. Sobrados, A.L. Cavalieri, P. Pena, S. De Aza, and J.S. Moya, *J. Am. Ceram. Soc.*, 1991, **74**, 2398 (*Chem. Abstr.*, 1991, **115**, 285 542).
[2625] T. Heinrich, F. Raether, O. Spormann, and J. Fricke, *J. Appl. Crystallogr.*, 1991, **24**, 788 (*Chem. Abstr.*, 1991, **115**, 188 278).
[2626] W. Kolodsiejski, P.J. Barrie, H. He, and J. Klinowski, *J. Chem. Soc., Chem. Commun.*, 1991, 961.
[2627] C.P. Herrero and J. Sanz, *J. Phys. Chem. Solids*, 1991, **52**, 1129 (*Chem. Abstr.*, 1991, **115**, 219 409).
[2628] J.B. D'Espinose de la Caillerie and J.J. Fripiat, *Clays Clay Miner.*, 1991, **39**, 270 (*Chem. Abstr.*, 1991, **115**, 95 996).
[2629] C. Doremieux-Morin, C. Martin, J.M. Bregeault, and J. Fraissard, *Appl. Catal.*, 1991, **77**, 149 (*Chem. Abstr.*, 1991, **115**, 288 173).
[2630] W.C. Cheng and K. Rajagopalan, *ACS Symp. Ser.*, 1991, **452**, 198 (*Chem. Abstr.*, 1991, **114**, 210 168).
[2631] C.W. Childs, R.L. Parfitt, and R.H. Newman, *Clay Miner.*, 1990, **25**, 329 (*Chem. Abstr.*, 1991, **114**, 9737).
[2632] L.A. Groat, M. Raudsepp, F.C. Hawthorne, T.S. Ercit, B.L. Sherriff, and J.S. Hartman, *Am. Mineral.*, 1990, **75**, 992 (*Chem. Abstr.*, 1991, **114**, 27 322).
[2633] J.T. Kloprogge, A.M.J. Van der Eerden, J.B.H. Jansen, and J.W. Geus, *Geol. Mijnbouw*, 1990, **69**, 351 (*Chem. Abstr.*, 1991, **115**, 162 310).

Jurassic illite/smectite, (^{27}Al, ^{29}Si),[2634] cordierite, (^{27}Al, ^{29}Si),[2635] mordenite, (^{29}Si),[2636] mullite, (^{27}Al, ^{29}Si),[2637] phlogopite, (^{27}Al, ^{29}Si),[2638] silicalite,[2639] aluminosilicate sodalites, (^{7}Li, ^{23}Na, ^{27}Al),[2640] and tobermorites, (^{27}Al, ^{29}Si).[2641]

^{1}H n.m.r. spectroscopy has been used to investigate the acidity of zeolites.[2642] Using silicon-aluminium distribution in natural zeolites from ^{27}Al and ^{29}Si n.m.r. spectra, it has been shown that Loewenstein's rule is invariably adhered to.[2643] The temperature dependent T_1 values of ^{17}O, ^{23}Na, ^{27}Al and ^{71}Ga have been determined in a series of hydrated zeolites.[2644] ^{27}Al n.m.r. spectroscopy has been used to determine the ratio of framework to non-framework aluminium in zeolites.[2645] High resolution solid-state n.m.r. investigations of the lattice structures of zeolite catalysts and sorbents have been reported.[2646] ^{27}Al and ^{29}Si n.m.r. spectroscopy has been used to study the transformation of high silica gmelinite into analcime.[2647]

Two commercial hydrocracking catalysts have been examined by ^{27}Al and ^{29}Si n.m.r. spectroscopy.[2648] ^{27}Al-{^{1}H} cross polarization and ultra high speed ^{27}Al m.a.s. n.m.r. spectroscopy have been used to characterize some ultra-stabilized zeolite Y.[2649] The effect of non-framework aluminium ions on the hydrostability of ultra-stabilized zeolite Y has been studied by ^{27}Al m.a.s. n.m.r. spectroscopy.[2650] The interaction of zeolite alumina with matrix silica in ultra-stabilized zeolites Y has been investigated using ^{27}Al n.m.r. spectroscopy.[2651] ^{29}Si n.m.r. spectroscopy has been used to study the hydrothermal transformation of natural clinoptilolite into zeolites Y and P_1.[2652] The effect of Fe^{3+} on T_1 and T_2 of ^{29}Si has been investigated for zeolite Y.[2653] Solid state n.m.r. measurements

2634 H. Lindgreen, H. Jacobsen, and H.J. Jakobsen, *Clays Clay Miner.*, 1991, **39**, 54 (*Chem. Abstr.*, 1991, **114**, 210 916).
2635 U. Selvaraj, S. Komarneni, and R. Roy, *J. Am. Ceram. Soc.*, 1990, **73**, 3663 (*Chem. Abstr.*, 1991, **114**, 67 647).
2636 A. Xie, C. You, C. Bi, Y. Hu, and Y. Qiu, *Cuihua Xuebao*, 1991, **12**, 353 (*Chem. Abstr.*, 1991, **115**, 264 101).
2637 L.H. Merwin, A. Sebald, H. Rager, and H. Schneider, *Phys. Chem. Miner.*, 1991, **18**, 47 (*Chem. Abstr.*, 1991, **115**, 75 460).
2638 S. Circone, A. Navrotsky, R.J. Kirkpatrick, and C.M. Graham, *Am. Mineral.*, 1991, **76**, 1485 (*Chem. Abstr.*, 1991, **115**, 236 389).
2639 G. Bellussi, A. Carati, M.G. Clerici, and A. Esposito, *Stud. Surf. Sci. Catal.*, 1991, **63**, 421 (*Chem. Abstr.*, 1991, **115**, 190 716).
2640 N.C. Nielsen, H. Bildsoe, H.J. Jakobsen, and P. Norby, *Zeolites*, 1991, **11**, 622 (*Chem. Abstr.*, 1991, **115**, 196 541).
2641 M. Tsuji, S. Komarneni, and P. Malla, *J. Am. Ceram. Soc.*, 1991, **74**, 274 (*Chem. Abstr.*, 1991, **115**, 95 361).
2642 H.G. Karge, *Stud. Surf. Sci. Catal.*, 1991, **65**, 133 (*Chem. Abstr.*, 1991, **115**, 287 686).
2643 A. Alberti, *Stud. Surf. Sci. Catal.*, 1991, **60**, 107 (*Chem. Abstr.*, 1991, **115**, 186 855).
2644 J. Haase, K.D. Park, K. Guo, H.K.C. Timken, and E. Oldfield, *J. Phys. Chem.*, 1991, **95**, 6996.
2645 D. Zhou, J. Wang, Z. Guo, L. Ma, and D. Ding, *Huaxue Wuli Xuebao*, 1991, **4**, 152 (*Chem. Abstr.*, 1991, **115**, 288 046).
2646 C.A. Fyfe, G.T. Kokotailo, H. Gies, and H. Strobl, *NATO ASI Ser., Ser. C*, 1990, **322**, 425 (*Chem. Abstr.*, 1991, **114**, 151 199).
2647 P.N. Joshi, A. Thangaraj, and V.P. Shiralkar, *Zeolites*, 1991, **11**, 164 (*Chem. Abstr.*, 1991, **114**, 196 709).
2648 L. Flanagan, *Report*, 1990, **IS-T-1518**; Order No. DE91000666, 67 pp. Avail. NTIS. From *Energy Res. Abstr.*, 1990, **15**, Abstr. No. 53304 (*Chem. Abstr.*, 1991, **115**, 12 069).
2649 L. Kellberg, M. Linsten, and H.J. Jakobsen, *Chem. Phys. Lett.*, 1991, **182**, 120 (*Chem. Abstr.*, 1991, **115**, 125 390).
2650 A. Yoshida, Y. Adachi, and K. Inoue, *Zeolites*, 1991, **11**, 549 (*Chem. Abstr.*, 1991, **115**, 173 268); A. Yoshida, K. Inoue, and Y. Adachi, *Zeolites*, 1991, **11**, 223 (*Chem. Abstr.*, 1991, **114**, 193 308).
2651 A. Corma, M. Grande, V. Fornes, S. Cartlidge, and M.P. Shatlock, *Appl. Catal.*, 1990, **66**, 45 (*Chem. Abstr.*, 1991, **114**, 50 269).
2652 C. De las Pozas-Del Rio, E. Reguera-Ruiz, C. Diaz-Aguila, and R. Roque-Malherbe, *J. Solid State Chem.*, 1991, **94**, 215 (*Chem. Abstr.*, 1991, **115**, 246 528).
2653 A. Thangaraj and S. Ganapathy, *Indian J. Chem., Sect A*, 1990, **29**, 1080 (*Chem. Abstr.*, 1991, **114** 155 658).

support the claim aluminium is reinserted into the framework of steamed samples of zeolite Y upon treatment with highly alkaline solutions.[2654] Short-range order of the silicon/aluminium distribution on the faujasite framework has been investigated using ^{29}Si n.m.r. data.[2655] Siliceous faujasitic zeolites produced by direct synthesis or by secondary synthesis have been compared using ^{27}Al and ^{29}Si n.m.r. spectroscopy.[2656] The role of extra framework aluminium in the reactions of 2-methylpentane in ultra-stabilized zeolite Y has been studied by n.m.r. spectroscopy.[2657] ^{29}Si n.m.r. spectroscopy has been used to study long versus short range silicon/aluminium ordering in zeolites X and Y.[2658] Fast m.a.s. n.m.r. spectroscopy with ^{1}H-^{27}Al cross polarization has been used to quantitatively monitor several Al species in dealuminated zeolite Y.[2659] The synthesis of high-silica zeolite Y has been monitored by ^{29}Si n.m.r. spectroscopy.[2660] The realumination of dealuminated zeolite Y has been studied using ^{29}Si n.m.r. spectroscopy.[2661] The dealumination of zeolite Y has been studied using ^{29}Si n.m.r. spectroscopy.[2662] ^{23}Na m.a.s. n.m.r. spectroscopy has been used to study the acid-base precursor chemistry in zeolite Y.[2663] Cation location in Na,La zeolite Y has been investigated using ^{23}Na nutation n.m.r. spectroscopy.[2664] Evidence for slow tumbling of Na^+ in hydrated NaY zeolite has been obtained from ^{23}Na n.m.r. spectroscopy.[2665] ^{29}Si n.m.r. spectroscopy has been used to study alkali exchanged NaY zeolites.[2666] GaAlY zeolite has been studied using ^{27}Al, ^{29}Si, and ^{71}Ga n.m.r. spectroscopy.[2667] Zeolite catalysts have been studied using ^{27}Al and ^{29}Si n.m.r. spectroscopy.[2668] The role of the amorphous matrix in the hydrothermal ageing of fluid cracking catalysts has been investigated using ^{29}Si n.m.r. spectroscopy.[2669]

^{27}Al and ^{29}Si n.m.r. spectroscopy has been used to examine the synthesis of zeolite NaA from metakaolinite.[2670] The role of an aluminium-tertiary alkanolamine chelate in the synthesis of large crystal zeolite NaA has been studied using ^{27}Al n.m.r. spectroscopy.[2671] An unusual high sodium natural chabazite and a variety of aluminium rich chabazite like species have been characterized by ^{27}Al and ^{29}Si m.a.s. n.m.r. spectroscopy.[2672] ^{27}Al n.m.r. spectroscopy has been used to investigate the

2654 P.J. Barrie, L.F. Gladden, and J. Klinowski, *J. Chem. Soc., Chem. Commun.*, 1991, 592 (*Chem. Abstr.*, 1991, **115**, 16 081).
2655 C.P. Herrero, *J. Phys. Chem.*, 1991, **95**, 3282.
2656 J. Dwyer, K. Karim, W.J. Smith, N.E. Thompson, R.K. Harris, and D.C. Apperley, *J. Phys. Chem.*, 1991, **95**, 8826.
2657 G.R. Bamwenda, W.A. Groten, and B.W. Wojciechowski, *Stud. Surf. Sci. Catal.*, 1991, **68**, 753 (*Chem. Abstr.*, 1991, **115**, 258 652).
2658 C.P. Herrero, L. Utrera, and R. Ramirez, *Chem. Phys. Lett.*, 1991, **183**, 199 (*Chem. Abstr.*, 1991, **115**, 190 512); C.P. Herrero, *J. Chem. Soc., Faraday Trans.*, 1991, **87**, 2837.
2659 J. Rocha and J. Klinowski, *J. Chem. Soc., Chem. Commun.*, 1991, 1121.
2660 A. Yoshida, *Nippon Kogaku Kaishi*, 1991, 110 (*Chem. Abstr.*, 1991, **114**, 105 132).
2661 Z. Zhang, X. Liu, Y. Xu, and R. Xu, *Zeolites*, 1991, **11**, 232 (*Chem. Abstr.*, 1991, **114**, 193 309).
2662 G.M. Telbiz, A.I. Prilipko, and I.V. Mishin, *Stud. Surf. Sci. Catal.*, 1991, **65**, 563 (*Chem. Abstr.*, 1991, **115**, 216 010).
2663 L. McMurray, A.J. Holmes, A. Kuperman, G.A. Ozin, and S. Ozkar, *J. Phys. Chem.*, 1991, **95**, 9448 (*Chem. Abstr.*, 1991, **115**, 240 497).
2664 C.F. Lin and K.J. Chao, *J. Phys. Chem.*, 1991, **95**, 9411 (*Chem. Abstr.*, 1991, **115**, 240 485).
2665 R. Challoner and R.K. Harris, *Zeolites*, 1991, **11**, 265 (*Chem. Abstr.*, 1991, **114**, 193 361).
2666 J.Y. Chern and K.J. Chao, *J. Chin. Chem. Soc. (Taipei)*, 1991, **38**, 123 (*Chem. Abstr.*, 1991, **114**, 193 284).
2667 J. Dwyer and K. Karim, *J. Chem. Soc., Chem. Commun.*, 1991, 905.
2668 S. Yue and M. Gu, *Shiyou Huagong*, 1991, **20**, 21 (*Chem. Abstr.*, 1991, **115**, 32 017).
2669 P. Gelin and T. Des Couriers, *Appl. Catal.*, 1991, **72**, 179 (*Chem. Abstr.*, 1991, **115**, 36 365).
2670 J. Rocha, J. Klinowski, and J.M. Adams, *J. Chem. Soc., Faraday Trans.*, 1991, **87**, 309.
2671 M. Morris, A. Sacco, jun., A.G. Dixon, and R.W. Thompson, *Zeolites*, 1991, **11**, 178 (*Chem. Abstr.*, 1991, **114**, 196 560).
2672 K.A. Thrush and S.M. Kuznicki, *J. Chem. Soc., Faraday Trans.*, 1991, **87**, 1031.

stability of the tetrahedral aluminium sites in zeolite β.[2673] Structural disorder in zeolite L has been investigated by ^{29}Si m.a.s. n.m.r. spectroscopy.[2674] The synthesis and properties of zeolite Linde Q have been studied using ^{27}Al and ^{29}Si n.m.r. spectroscopy.[2675] The structure and location of the organic base in zeolite EU-1 have been determined using ^{13}C c.p. m.a.s. n.m.r. spectroscopy.[2676] The synthesis of iron-silicate analogues of mordenite has been followed by m.a.s. n.m.r. spectroscopy.[2677]

ZSM-5 zeolites synthesized without a template have been characterized using ^{23}Na, ^{27}Al, and ^{29}Si n.m.r. spectroscopy.[2678] Secondary synthesis on modified pentasil zeolites has been studied using ^{27}Al n.m.r. spectroscopy.[2679] ^{13}C m.a.s. n.m.r. spectroscopy has been used to study the role of pyrrolidine in various high silica zeolites.[2680] The mechanism of ZSM-5 formation has been studied using ^{29}Si n.m.r. spectroscopy.[2681] The influence of synthesis variables on the formation of ZSM-5 has been investigated using ^{27}Al n.m.r. spectroscopy.[2682] The Bu_4N template effect during the formation of ZSM-5 has been studied using ^{13}C n.m.r. spectroscopy.[2683] The hydroxonium ion HZSM-5 catalysts have been studied using both broad line and high resolution ^{1}H, ^{13}C, ^{27}Al, and ^{29}Si n.m.r. spectroscopy.[2684] ^{13}C, ^{27}Al, and ^{29}Si n.m.r. spectroscopy has been used to study coked zeolites, HZSM-5.[2685] Three-dimensional Si-O-Si connectivities in the monoclinic form of ZSM-5 have been investigated using two-dimensional ^{29}Si m.a.s. n.m.r. INADEQUATE experiments.[2686] The synthesis of ZSM-48 in the presence of 1,8-diaminooctane and hexamethonium bromide has been analysed by ^{13}C n.m.r. spectroscopy.[2687] Organic bases have been located and Brønsted acid sites identified in zeolites using ^{14}N n.m.r. spectroscopy.[2688] Structural defects in highly siliceous MFI

2673 E. Bourgeat-Lami, P. Massiani, F. Di Renzo, F. Fajila, and T. Des Courieres, *Catal. Lett.*, 1990, **5**, 265 (*Chem. Abstr.*, 1991, **114**, 155 797); E. Bourgeat-Lami, P. Massiani, F. Di Renzo, P. Espiau, F. Fajula, and T. Des Courieres, *Appl. Catal.*, 1991, **72**, 139 (*Chem. Abstr.*, 1991, **115**, 36 363).

2674 K. Tsutsumi, A. Shiraishi, K. Nishimiya, M. Kato, and T. Takaishi, *Stud. Surf. Sci. Catal.*, 1991, **60**, 141 (*Chem. Abstr.*, 1991, **115**, 287 774).

2675 K.J. Andries, B. De Wit, P.J. Grobet, and H.J. Bosmans, *Zeolites*, 1991, **11**, 116 (*Chem. Abstr.*, 1991, **114**, 177 061); K.J. Andries, H.J. Bosmans, and P.J. Grobet, *Zeolites*, 1991, **11**, 124 (*Chem. Abstr.*, 1991, **114**, 196 776).

2676 A. Thangaraj, P.R. Rajamohan, S. Ganapathy, and P. Ratnasamy, *Zeolites*, 1991, **11**, 69 (*Chem. Abstr.*, 1991, **114**, 133 285).

2677 A.J. Chandwadkar, R.N. Bhat, and P. Ratnasamy, *Zeolites*, 1991, **11**, 42 (*Chem. Abstr.*, 1991, **114**, 113 870).

2678 Z. Guo, J. Wang, K. Wang, D. Zhou, S. Xiang, and H. Li, *Gaodeng Xuexiao Huaxue Xuebao*, 1990, **11**, 800 (*Chem. Abstr.*, 1991, **114**, 89 083).

2679 W. Reschetilowski, W.D. Einicke, B. Meier, E. Brunner, and H. Ernst, *Stud. Surf. Sci. Catal.*, 1991, **65**, 529 (*Chem. Abstr.*, 1991, **115**, 293 432).

2680 R. Kumar, A. Thangaraj, and P. Rajmohanan, *Indian J. Chem., Sect. A*, 1990, **29**, 1083 (*Chem. Abstr.*, 1991, **114**, 151 002); A. Thangaraj, R. Kumar, and P. Ratnasamy, *Zeolites*, 1991, **11**, 573 (*Chem. Abstr.*, 1991, **115**, 143 269).

2681 C.D. Chang and A.T. Bell, *Catal. Lett.*, 1991, **8**, 305 (*Chem. Abstr.*, 1991, **115**, 18 850).

2682 G.N. Rao, P.N. Joshi, V.P. Shiralkar, B.S. Rao, and A.N. Kotasthane, *Indian J. Technol.*, 1990, **28**, 697 (*Chem. Abstr.*, 1991, **115**, 143 564); R. Challoner and R.K. Harris, *J. Magn. Reson.*, 1991, **94**, 288.

2683 T. Song, L. Li, and R. Xu, *Bopuxue Zazhi*, 1990, **7**, 321 (*Chem. Abstr.*, 1991, **114**, 134 799).

2684 P. Batamack, C. Dorémieux-Morin, J. Fraissard, and D. Freude, *J. Phys. Chem.*, 1991, **95**, 3790; E. Brunner, H. Ernst, D. Freude, T. Froehlich, M. Hunger, and H. Pfeifer, *J. Catal.*, 1991, **127**, 34 (*Chem. Abstr.*, 1991, **114**, 50 278).

2685 H. Ernst, D. Freude, M. Hungar, and H. Pfeifer, *Stud. Surf. Sci. Catal.*, 1991, **65**, 397 (*Chem. Abstr.*, 1991, **115**, 216 008).

2686 C.A. Fyfe, H. Grondey, Y. Feng, and G.T. Kokotailo, *Chem. Phys. Lett.*, 1990, **173**, 211 (*Chem. Abstr.*, 1991, **114**, 53 177).

2687 G. Giordano, N. Dewale, Z. Gabelica, J.B. Nagy, and E.G. Derouane, *Appl. Catal.*, 1991, **71**, 79 (*Chem. Abstr.*, 1991, **115**, 20 806).

2688 A. Thangaraj, P.R. Rajamohanan, P.M. Suryavanshi, and S. Ganapathy, *J. Chem. Soc., Chem. Commun.*, 1991,

zeolites have been studied using ^{29}Si n.m.r. spectroscopy.[2689] N.m.r. data have also been reported for AlNa-ZSM-5 (^{1}H, ^{27}Al),[2690] Fe-ZSM-5, (^{27}Al, ^{29}Si),[2691] Ni-ZSM-5,[2692] Al-ZSM-5, (^{27}Al),[2693] (^{27}Al, ^{29}Si),[2694] Ga-ZSM-5, (^{27}Al, ^{29}Si, ^{71}Ga),[2695] (^{1}H, ^{13}C),[2696] HP-ZSM-5, (^{27}Al, ^{31}P),[2697] P-ZSM-5, (^{1}H, ^{27}Al),[2698] ZSM-11, (^{29}Si),[2699] Fe-ZSM-12, (^{29}Si),[2700] H-ZSM-20, (^{27}Al),[2701] MFI type zeolites containing fluoride, (^{19}F),[2702] and Ti-rich MFI zeolites, (^{29}Si).[2703]

The structure of synthetic fraiponite has been studied using ^{29}Si n.m.r. spectroscopy.[2704] Variations in the interlayer cation sites of clay minerals have been studied by ^{133}Cs n.m.r. spectroscopy.[2705] ^{113}Cd n.m.r. spectra of Cd^{2+} exchanged clays have been studied and the chemical shift tensor determined.[2706] The formation of β'-Sialon from montmorillonite-polyacrylonitrile composite by carbothermal reduction has been studied by ^{27}Al and ^{29}Si n.m.r. spectroscopy.[2707]

High alumina glasses in the $CaO\text{-}Al_2O_3\text{-}SiO_2$ system have been investigated by ^{27}Al n.m.r. spectroscopy.[2708] ^{29}Si m.a.s. n.m.r. spectroscopy has been used to study the structural changes of the silicate ions in tobermorite during carbonation of autoclaved lightweight aerated concrete.[2709] Leaching effects on silicate polymerization in lead and zinc doped Portland cement have been studied using ^{29}Si n.m.r. spectroscopy.[2710] Cement hydration has been studied using ^{1}H n.m.r. spectroscopy.[2711] Alkaline earth aluminosilicate glasses have been studied using ^{27}Al and ^{29}Si n.m.r.

493.

2689 J.M. Chezeau, L. Delmotte, J.L. Guth, and Z. Gabelica, *Zeolites*, 1991, **11**, 598 (*Chem. Abstr.*, 1991, **115**, 171 154).

2690 T. Romotowski, J. Komorek, V.M. Mastikhin, and A.V. Nosov, *Zeolites*, 1991, **11**, 491 (*Chem. Abstr.*, 1991, **115**, 79 685).

2691 W. Shen, Y. Yu, Y. Yen, Y. Ding, and C. Hu, *Chin. Chem. Lett.*, 1991, **2**, 249 (*Chem. Abstr.*, 1991, **115**, 125 350); G. Vorbeck, M. Richter, R. Fricke, B. Parlitz, E. Schreier, K. Szulzewsky, and B. Zibrowius, *Stud. Surf. Sci. Catal.*, 1991, **65**, 631 (*Chem. Abstr.*, 1991, **115**, 216 011).

2692 S. Yao, S. Qiu, W. Pang, T. Sun, and D. Jiang, *Chin. J. Chem.*, 1991, **9**, 109 (*Chem. Abstr.*, 1991, **115**, 246 511).

2693 B. Staudte, M. Hunger, and M. Nimz, *Zeolites*, 1991, **11**, 837 (*Chem. Abstr.*, 1991, **115**, 264 125).

2694 S. Zheng, G. Xie, B. Zhang, F. Shi, and K. Zheng, *Fudan Xuebao, Ziran Kexueban*, 1990, **29**, 273 (*Chem. Abstr.*, 1991, **115**, 161 555).

2695 J. Luo, C. Tan, and Q. Dong, *Cuihua Xuebao*, 1990, **11**, 462 (*Chem. Abstr.*, 1991, **114**, 220 003).

2696 R. Challoner, R.K. Harris, S.A.I. Barri, and M.J. Taylor, *Zeolites*, 1991, **11**, 827 (*Chem. Abstr.*, 1991, **115**, 264 123).

2697 G. Lischke, R. Eckelt, H.G. Jerschkewitz, B. Parlitz, E. Schreier, W. Storek, B. Zibrowius, and G. Oehlmann, *J. Catal.*, 1991, **132**, 229 (*Chem. Abstr.*, 1991, **115**, 190 744).

2698 W. Reschetilowski, B. Meier, M. Hunger, B. Unger, and K.P. Wendlant, *Angew. Chem., Int. Ed. Engl.*, 1991, **30**, 686.

2699 C.A. Fyfe, Y. Feng, H. Grondey, G.T. Kokotailo, and A. Mar, *J. Phys. Chem.*, 1991, **95**, 3747.

2700 W. Wang and W. Pang, *Cuihua Xuebao*, 1991, **12**, 156 (*Chem. Abstr.*, 1991, **115**, 16 235).

2701 Y. Sun, P.J. Chu, and J.H. Lunsford, *Langmuir*, 1991, **7**, 3027 (*Chem. Abstr.*, 1991, **115**, 288 147).

2702 L. Delmotte, M. Soulard, F. Guth, A. Seive, A. Lopez, and J.L. Guth, *Zeolites*, 1990, **10**, 778 (*Chem. Abstr.*, 1991, **114**, 34 650).

2703 A. Thangaraj, R. Kumar, S.P. Mirajkar, and P. Ratnasamy, *J. Catal.*, 1991, **130**, 1 (*Chem. Abstr.*, 1991, **115**, 36 375).

2704 N. Takahashi, M. Tanaka, and T. Satoh, *Nippon Kagaku Kaishi*, 1991, 962 (*Chem. Abstr.*, 1991, **115**, 77 112).

2705 C.A. Weiss, jun., R.J. Kirkpatrick, and S.P. Altaner, *Am. Mineral.*, 1990, **75**, 970 (*Chem. Abstr.*, 1991, **114**, 65 994).

2706 D. Tinet, A.M. Faugere, and R. Prost, *J. Phys. Chem.*, 1991, **95**, 8804.

2707 T. Bastow, S.G. Hardin, and D.W. Turney, *J. Mater. Sci.*, 1991, **26**, 1443 (*Chem. Abstr.*, 1991, **114**, 169 718).

2708 R.K. Sato, P.F. McMillan, P. Dennison, and R. Dupree, *Phys. Chem. Glasses*, 1991, **32**, 149 (*Chem. Abstr.*, 1991, **115**, 140 858).

2709 Y. Ikeda, Y. Yasuike, and Y. Takashima, *Nippon Seramikkusu Kyokai Gakujutsu Ronbunshi*, 1991, **99**, 423 (*Chem. Abstr.*, 1991, **115**, 14 214).

2710 J.D. Ortego, Y. Barroeta, F.K. Cartledge, and H. Akhter, *Environ. Sci. Technol.*, 1991, **25**, 1171 (*Chem. Abstr.*, 1991, **114**, 213 689).

2711 R. Rumm, H. Haranczyk, H. Peemoeller, and M.M. Pintar, *Cem. Concr. Res.*, 1991, **21**, 391 (*Chem. Abstr.*,

spectroscopy.[2712] The hydrothermal reaction of a rhyolitic composition glass has been studied.[2713] The behaviour of the silicate anion in cement and cement hydrate has been studied by ^{29}Si m.a.s. n.m.r. spectroscopy.[2714] ^{29}Si n.m.r. spectroscopy has been used to compare the hydration processes of tricalcium silicate and β-dicalcium silicate.[2715] ^{19}F and ^{27}Al n.m.r. spectroscopy has provided evidence for five- and six-coordinate aluminium fluoride complexes in fluorine bearing aluminosilicate glasses.[2716] Monodispersed mesoporous catalyst matrices for fluid catalytic cracking have been characterized by ^{27}Al n.m.r. spectroscopy.[2717]

^{27}Al, ^{29}Si, and ^{31}P m.a.s. n.m.r. spectroscopy has been used to investigate aluminosilicate, silicophosphate, and aluminosilicophosphate gels and the evolution of crystalline structures on heating the gels.[2718] ^{27}Al and ^{31}P n.m.r. spectroscopy has been used to investigate $AlPO_4$-8 and its transformation from VPI-5 and its interaction with water.[2719] The spin-echo double resonance technique has been applied to ^{27}Al and ^{31}P in $AlPO_4$-5. It is possible to detect that ^{31}P is coupled to ^{27}Al.[2720] Solid state n.m.r. spectroscopy has been used to study the effects of ion sorption in $AlPO_4$-5.[2721] ^{27}Al and ^{31}P n.m.r. spectroscopy has been used to study the transformation of VPI-5/MCM-9 into $AlPO_4$-8/SAPO-8.[2722] The incorporation of silicon into the framework of SAPO-5,[2723] and SAPO-37,[2724] has been studied by ^{1}H, ^{27}Al, ^{29}Si, and ^{31}P n.m.r. spectroscopy. N.m.r. data have also been reported for aluminium metaphosphate, (^{31}P),[2725] aluminium orthophosphate, (^{1}H, ^{27}Al, ^{31}P),[2726] amorphous aluminium phosphates, (^{27}Al, ^{29}Si),[2727] VPI-5, (^{27}Al, ^{29}Si),[2728] (^{31}P),[2729]

1991, **114**, 170 055).

2712 C.I. Merzbacher, B.L. Sherriff, J.S. Hartman, and W.B. White, *J. Non-Cryst. Solids*, 1990, **124**, 194 (*Chem. Abstr.*, 1991, **114**, 30 521).

2713 W.H.A. Yang and R.J. Kirkpatrick, *Am. Mineral.*, 1990, **75**, 1009 (*Chem. Abstr.*, 1991, **114**, 46 796).

2714 Y. Okada, H. Aibasaki, and K. Hitotsuya, *Semento, Konkurito Ronunshu*, 1989, 14 (*Chem. Abstr.*, 1991, **114**, 29 065).

2715 Y. Tong, H. Du, and L. Fei, *Cem. Concr. Res.*, 1991, **21**, 509 (*Chem. Abstr.*, 1991, **115**, 55 602); Y. Tong, H. Du, and L. Fei, *Cem. Concr. Res.*, 1990, **20**, 986 (*Chem. Abstr.*, 1991, **114**, 11 224).

2716 S.C. Kohn, R. Dupree, M.G. Mortuza, and C.M.B. Henderson, *Am. Mineral.*, 1991, **76**, 306 (*Chem. Abstr.*, 1991, **114**, 147 362).

2717 W.A. Wachter, **U.S. US 5,051,385**, Cl. 502-64; B01J29/06, 24 Sep 1991, US Appl. 215,163, 05 Jul 1988; 30 pp. (*Chem. Abstr.*, 1991, **115**, 259 767).

2718 S. Prabakar, K.J. Rao, and C.N.R. Rao, *J. Mater. Res.*, 1991, **6**, 592.

2719 M. Stoecker, D. Akporiaye, and K.P. Lillerud, *Appl. Catal.*, 1991, **69**, L7 (*Chem. Abstr.*, 1991, **114**, 151 406).

2720 E.R.H. Van Eck, R. Janssen, W.E.J.R. Maas, and W.S. Veeman, *Chem. Phys. Lett.*, 1990, **174**, 428 (*Chem. Abstr.*, 1991, **114**, 54 525).

2721 R.H. Meinhold and N.J. Tapp, *Zeolites*, 1991, **11**, 401 (*Chem. Abstr.*, 1991, **115**, 16 089).

2722 L. Maistriau, Z. Gabelica, E.G. Derouane, E.T.C. Vogt, and J. Van Oene, *Zeolites*, 1991, **11**, 583 (*Chem. Abstr.*, 1991, **115**, 173 269).

2723 B. Zibrowius, E. Loeffler, G. Finger, E. Sonntag, M. Hunger, and J. Kornatowski, *Stud. Surf. Sci. Catal.*, 1991, **65**, 537 (*Chem. Abstr.*, 1991, **115**, 293 433).

2724 P.P. Man, M. Briend, M.J. Peltre, A. Lamy, P. Beaunier, and D. Barthomeuf, *Zeolites*, 1991, **11**, 563 (*Chem. Abstr.*, 1991, **115**, 143 268).

2725 G. Palavit, P. Vast, and R. Chi, *Calorim. Anal. Therm.*, 1988, **19**, P14.1 (*Chem. Abstr.*, 1991, **114**, 198 550).

2726 J. Sanz, J.M. Campelo, and J.M. Marinas, *J. Catal.*, 1991, **130**, 642 (*Chem. Abstr.*, 1991, **115**, 58 066).

2727 I.L. Mudrakovskii, V.P. Shmachkova, N.S. Kotsarenko, and V.M. Mastikhin, *Kinet. Katal.*, 1991, **32**, 1207 (*Chem. Abstr.*, 1991, **115**, 293 322); S. Prabakar, K.J. Rao, and C.N.R. Rao, *Mater. Res. Bull.*, 1991, **26**, 805 (*Chem. Abstr.*, 1991, **115**, 118 987).

2728 L.B. McCusker, C. Baerlocher, E. Jahn, and M. Buelow, *Zeolites*, 1991, **11**, 308 (*Chem. Abstr.*, 1991, **115**, 19 083); S. Yao, S. Qiu, W. Pang, L. Wu, L. Li, and Q. Xin, *Chem. Res. Chin. Univ.*, 1991, **7**, 11 (*Chem. Abstr.*, 1991, **115**, 143 132).

2729 J.P. van Braam Houckgeest, B. Kraushaar-Czarnetzki, R.J. Dogterom, and A. de Groot, *J. Chem. Soc., Chem. Commun.*, 1991, 666; M.E. Davis, B.D. Murray, and M. Narayana, *ACS Symp. Ser.*, 1990, **437**, 48 (*Chem. Abstr.*, 1991, **114**, 112 042).

(^{27}Al),[2730] SAPO-5, (^{29}Si),[2731] $AlPO_4$-8, (^{27}Al),[2732] (^{29}Si),[2733] FeMnAPO-5, (^{27}Al, ^{29}Si),[2734] $AlPO_4$-11, (^{27}Al),[2735] $AlPO_4$-14, (^{27}Al, ^{31}P),[2736] $AlPO_4$-14A,[2737] $AlPO_4$-21, $AlPO_4$-25, (^{27}Al),[2738] and SAPO-37, (^{27}Al, ^{29}Si, ^{31}P).[2739]

^{31}P n.m.r. spectroscopy has been used to study the P_2S_5-Al_2S_3-Li_2S system.[2740] ^{27}Al n.m.r. spectroscopy has been used to study polymeric aluminium chloride.[2741] ^{27}Al n.m.r. spectroscopy has been used to study premelting phenomena in crystals of $NaAlCl_4$.[2742]

^{69}Ga and ^{71}Ga n.m.r. spectroscopy has been used to study the gallium analogue of ZSM-5.[2743] Tetrahedral site ordering in synthetic gallium albite has been studied by ^{29}Si m.a.s. n.m.r. spectroscopy.[2744] GeAPO-5 molecular sieve has been characterised by ^{27}Al and ^{31}P n.m.r. spectroscopy.[2745]

Solid-state ^{13}C n.m.r. spectroscopy has been used to determine the bond lengths in C_{60}.[2746] The local coordination of carbon atoms in amorphous carbon has been studied by ^{1}H and ^{13}C n.m.r. spectroscopy.[2747] ^{13}C n.m.r. spectroscopy has been used to characterise diamond thin films grown in a low pressure hot-filament reactor.[2748] The two-dimensional metallic hydrogen lattice in K-H-graphite systems has been characterised by ^{1}H n.m.r. spectroscopy.[2749] ^{11}B, ^{23}Na, ^{27}Al, ^{29}Si, and ^{31}P n.m.r. spectroscopy has been used to characterise the inorganic components of coals and combustion residues.[2750] ^{13}C n.m.r. spectroscopy has been used to study the carbonaceous residues from meteorites.[2751] The attempted preparation of carbon nitride from carbon-nitrogen pyrolyzates

2730 P.J. Grobet, A. Samoson, H. Geerts, J.A. Martens, and P.A. Jacobs, *J. Phys. Chem.*, 1991, **95**, 9620; J.O. Perez, P.J. Chu, and A. Clearfield, *J. Phys. Chem.*, 1991, **95**, 9994.

2731 R. Wang, C.F. Lin, Y.S. Ho, L.J. Leu, and K.J. Chao, *Appl. Catal.*, 1991, **72**, 39.

2732 J. Rocha, X. Liu, and J. Klinowski, *Chem. Phys. Lett.*, 1991, **182**, 531 (*Chem. Abstr.*, 1991, **115**, 196 546).

2733 D. Zhao, W. Pang, X. Yao, J. Qiu, and Z. Ye, *Gaodeng Xuexiao Huaxue Xuebao*, 1991, **12**, 304 (*Chem. Abstr.*, 1991, **115**, 269 033).

2734 S. Han, G. Yu, and M. Zhou, *Hilin Daxue Ziran Kexue Xuebao*, 1990, 110 (*Chem. Abstr.*, 1991, **114**, 209 903).

2735 P.J. Barrie, M.E. Smith, and J. Klinowski, *Chem. Phys. Lett.*, 1991, **180**, 6 (*Chem. Abstr.*, 1991, **115**, 63 126).

2736 B. Zibrowius, U. Lohse, and J. Richter-Mendau, *J. Chem. Soc., Faraday Trans.*, 1991, **87**, 1433.

2737 M. Goepper and J.L. Guth, *Zeolites*, 1991, **11**, 477 (*Chem. Abstr.*, 1991, **115**, 125 486).

2738 R. Jelinek, B.F. Chmelka, Y. Wu, P.J. Grandinetti, A. Pines, P.J. Barrie, and J. Klinowski, *J. Am. Chem. Soc.*, 1991, **113**, 4097.

2739 A.F. Ojo, J. Dwyer, J. Dewing, and K. Karim, *J. Chem. Soc., Faraday Trans.*, 1991, **87**, 2679.

2740 J.H. Kennedy, C. Schaupp, H. Eckert, and M. Ribes, *Solid State Ionics*, 1991, **45**, 21 (*Chem. Abstr.*, 1991, **114**, 197 165).

2741 C. Yao, X. Yan, Y. Sun, P. Zhao, and L. Fei, *Huanjing Kexue Xuebao*, 1989, **9**, 488 (*Chem. Abstr.*, 1991, **114**, 253 747).

2742 B. Krebs, H. Greiwing, C. Brendel, and F. Taulelle, *Inorg. Chem.*, 1991, **30**, 981.

2743 A.P.M. Kentgens, C.R. Bayense, J.H.C. Van Hooff, J.W. De Haan, and L.J.M. Van de Ven, *Chem. Phys. Lett.*, 1991, **176**, 399 (*Chem. Abstr.*, 1991, **114**, 155 667).

2744 B.L. Sherriff, M.E. Fleet, and P.C. Burns, *J. Solid State Chem.*, 1991, **94**, 52 (*Chem. Abstr.*, 1991, **115**, 219 520).

2745 S. Han, L. Wang, and G. Yu, *Jilin Daxue Ziran Kexue Xuebao*, 1990, 85 (*Chem. Abstr.*, 1991, **115**, 246 510).

2746 C.S. Yannoni, P.B. Bernier, D.S. Bethune, G. Meijer, and J.R. Salem, *J. Am. Chem. Soc.*, 1991, **113**, 3190.

2747 H. Pan, M. Pruski, B.C. Gerstein, F. Li, and J.S. Lannin, *Phys. Rev. B: Condens. Matter*, 1991, **44**, 6741 (*Chem. Abstr.*, 1991, **115**, 240 147).

2748 K.M. McNamara, K.K. Gleason, and M.W. Geis, *Mater. Res. Soc. Symp. Proc.*, 1990, **162**, 207 (*Chem. Abstr.*, 1991, **115**, 146 890).

2749 T. Enoki, K. Nakazawa, K. Suzuki, S. Miyajima, T. Chiba, Y. Iye, H. Yamamoto, and H. Inokuchi, *J. Less-Common Met.*, 1991, **172**, 20 (*Chem. Abstr.*, 1991, **115**, 186 259).

2750 P. Burchill, O.W. Howarth, and B.J. Sword, *Fuel*, 1991, **70**, 361 (*Chem. Abstr.*, 1991, **114**, 231 823).

2751 T. Murae, A. Masuda, and T. Takahashi, *Proc. NIPR Symp. Antarct. Meteorites*, 1990, **3**, 211 (*Chem. Abstr.*, 1991, **114**, 250 989).

has been followed by ^{13}C n.m.r. spectroscopy.[2752] The reaction of C_nXF_m with SO_2 has been investigated using ^{19}F n.m.r. spectroscopy.[2753] N.m.r. data have also been reported for lithium potassium graphite oxides,[2754] graphite oxide, (^{13}C),[2755] and graphite fluorides, (^{19}F).[2756]

Hydrogen bonding in hydrogenated amorphous silicon has been investigated by ^{1}H[2757] and ^{2}H[2758] n.m.r. spectroscopy. A new way to graft SiH_4 to silica gel using $[Ru_3(CO)_{12}]$ as a catalyst has been studied using ^{29}Si n.m.r. spectroscopy.[2759] ^{29}Si, ^{119}Sn, and ^{207}Pb n.m.r. spectra have been used to study $[(\eta^5\text{-}C_5Me_5)_2M]$, M = Si, Sn, Pb.[2760] The formation of SiC from polymethylsilane,[2761] siloxanes,[2762] and a vinylic polysilane[2763] has been studied using ^{13}C and ^{29}Si n.m.r. spectroscopy. ^{13}C and ^{29}Si n.m.r. spectroscopy has been used to characterize sites in silicon carbide polytypes.[2764] ^{13}C n.m.r. spectroscopy has been used to study structural defects in amorphous and microcrystalline SiC.[2765] The ^{29}Si chemical shift tensor of the silicon atom in fluorosilatrane has been determined.[2766] The doublet structure observed in the ^{2}H n.m.r. spectra of deformed dimethyl silicone rubber networks has been interpreted in terms of strain-dependent crosslink fluctuations.[2767] The crystalline conformation of $(Me_2SiO)_n$ has been investigated using ^{13}C and ^{29}Si n.m.r. spectroscopy.[2768] The phase behaviour of $(R_2SiO)_n$ has been investigated. Diffusive rotation of the chain segments around the molecule along the axis has been indicated by ^{29}Si n.m.r. spectroscopy.[2769] The structure of two different hybrid siloxane-oxide systems has been studied by ^{29}Si n.m.r. spectroscopy.[2770] N.m.r. data have also been reported for polymethylsilsesquioxanes, (^{29}Si),[2771] $(ViHSiNH)_x$, (^{13}C,

2752 L. Maya, D.R. Cole, and E.W. Hagaman, *J. Am. Ceram. Soc.*, 1991, **74**, 1686 (*Chem. Abstr.*, 1991, **115**, 140 914).
2753 V.M. Paasonen, A.S. Nazarov, V.M. Grankin, and I.I. Yakovlev, *Zh. Neorg. Khim.*, 1990, **35**, 2205 (*Chem. Abstr.*, 1991, **114**, 54 656).
2754 A.F. Manukhin, A.K. Tsvetnikov, L.A. Matveenko, V.M. Buznik, and A.N. Matsulaev, *Zh. Neorg. Khim.*, 1991, **36**, 1956 (*Chem. Abstr.*, 1991, **115**, 221 635).
2755 M. Mermoux, Y. Chabre, and A. Rousseau, *Carbon*, 1991, **29**, 469 (*Chem. Abstr.*, 1991, **114**, 148 796).
2756 A. Hannwi, M. Daoud, D. Djurado, J.C. Cousseins, Z. Fawal, and J. Dupuis, *Synth. Met.*, 1991, **44**, 75 (*Chem. Abstr.*, 1991, **115**, 148 747).
2757 S.E. Ready, J.B. Boyce, N.M. Johnson, J. Walker, and K.S. Stevens, *Mater. Res. Soc. Symp. Proc.*, 1990, **192**, 127 (*Chem. Abstr.*, 1991, **114**, 93 795).
2758 J. Bodart, P. Santos-Filho, and R.E. Norberg, *J. Non-Cryst. Solids*, 1989, **114**, 825 (*Chem. Abstr.*, 1991, **114**, 150 814).
2759 G. Felix, F. Meiouet, A.P. Legrand, H. Hommel, and H. Taibi, *J. Chromatogr. Sci.*, 1991, **29**, 457 (*Chem. Abstr.*, 1991, **115**, 240 601).
2760 B. Wrackmeyer, A. Sebald, and L.H. Merwin, *Magn. Reson. Chem.*, 1991, **29**, 260.
2761 Z.F. Zhang, F. Babonneau, R.M. Laine, Y. Mu, J.F. Harrod, and J.A. Rahn, *J. Am. Ceram. Soc.*, 1991, **74**, 670 (*Chem. Abstr.*, 1991, **114**, 169 698).
2762 R.B. Taylor and G.A. Zank, *Polym. Prepr. (Am. Chem. Soc., Div. Polym. Chem.)*, 1991, **32**, 586 (*Chem. Abstr.*, 1991, **115**, 213 205).
2763 W.R. Schmidt, L.V. Interrante, R.H. Doremus, T.K. Trout, P.S. Marchetti, and G.E. Maciel, *Chem. Mater.*, 1991, **3**, 257 (*Chem. Abstr.*, 1991, **114**, 148 651).
2764 M. O'Keeffe, *Chem. Mater.*, 1991, **3**, 332 (*Chem. Abstr.*, 1991, **114**, 154 264); D.C. Apperley, R.K. Harris, G.L. Marshall, and D.P. Thompson, *J. Am. Ceram. Soc.*, 1991, **74**, 777 (*Chem. Abstr.*, 1991, **114**, 233 585).
2765 S.J. Ting, C.J. Chu, and J.D. Mackenzie, *Mater. Res. Soc. Symp. Proc.*, 1990, **192**, 529 (*Chem. Abstr.*, 1991, **115**, 20 714).
2766 J.H. Iwamiya and G.E. Maciel, *J. Magn. Reson.*, 1991, **92**, 590.
2767 M.G. Brereton, *Macromolecules*, 1991, **24**, 6160 (*Chem. Abstr.*, 1991, **115**, 234 383).
2768 F.C. Schilling, M.A. Gomez, and A.E. Tonelli, *Macromolecules*, 1991, **24**, 6552 (*Chem. Abstr.*, 1991, **115**, 257 064).
2769 M. Moller, S. Siffrin, G. Koegler, and D. Oelfin, *Makromol. Chem., Macromol. Symp.*, 1990, **34**, 171 (*Chem. Abstr.*, 1991, **114**, 103 146).
2770 S. Dire, L. Bois, F. Babonneau, and G. Carturan, *Polym. Prepr. (Am. Chem. Soc., Div. Polym. Chem.)*, 1991, **32**, 501 (*Chem. Abstr.*, 1991, **115**, 188 172).
2771 R.M. Laine, J.A. Rahn, K.A. Youngdahl, and J.F. Harrod, *Report*, 1990, **TR-21**; Order No. AD-A222643, 13 pp. Avail. NTIS. From *Gov. Rep. Announce. Index (U.S.)*, 1990, **90**, Abstr. No. 050,814 (*Chem. Abstr.*, 1991,

^{29}Si),[2772] $(Me_2HSi)_8Si_8O_{20}$, (^{29}Si),[2773] boron doped β-SiC, (^{11}B),[2774] molded SiC products, (^{13}C),[2775] preceramic precursors and pyrolyzed products,[2776] di-*n*-hexylsilylene/di-*n*-pentylsilylene copolymers, (^{29}Si),[2777] poly(di-*n*-hexyl)silanes, (^{29}Si),[2778] copoly(Me_2Si-Ph_2Si), (^{29}Si),[2779] silicon containing polyacetylenes, (^{13}C),[2780] Si_3N_4/SiC ceramics, (^{13}C, ^{29}Si),[2781] Si_3N_4, (^{14}N, ^{29}Si),[2782] β'-Sialon ceramics, (^{29}Si),[2783] silicone rubber, (^{13}C, ^{29}Si),[2784] dimethylsilicones reinforced with *in situ* precipitated silica, (^{29}Si),[2785] a phosphine functionalized polysiloxane, (^{29}Si, ^{31}P),[2786] and siloxane-thioether copolymers, (^{13}C).[2787]

The impact of acidic/hydrothermal treatment on pore structural and chromatographic properties of porous silicas has been investigated with ^{29}Si n.m.r. spectroscopy.[2788] The *in situ* pore structure analysis during ageing and drying of silica gels has been performed using n.m.r. spectroscopy.[2789] The influence of preparation parameters on pore structure of silica gels prepared from $Si(OEt)_4$ has been studied using ^{29}Si n.m.r. spectroscopy.[2790] ^{29}Si n.m.r. spectroscopy has been used to study cation binding to sol-gel silicates.[2791] N.m.r. imaging has been used to characterise the pore structure of silica gel prepared from $Si(OEt)_4$.[2792] N.m.r. data have also been reported for Si_2O_3, (^{29}Si),[2793]

114, 233 518).

2772 N.S.C.K. Yive, R. Corriu, C. Leclercq, P.H. Mutin, and A. Vioux, *New J. Chem.*, 1991, **15**, 85.

2773 D. Hoebbel, I. Pitsch, and D. Heidemann, *Z. Anorg. Allg. Chem.*, 1991, **592**, 207.

2774 D.J. O'Donnell, T.T.P. Cheung, G.F. Schuette, and M. Sardashti, *J. Am. Ceram. Soc.*, 1991, **74**, 2025 (*Chem. Abstr.*, 1991, **115**, 140 958).

2775 K. Shimada, T. Sawaki, S. Watanabe, A. Nakaishi, and M. Ogasawara, *Jpn. Kokai Tokkyo Koho JP 02,293,379 [90,293,379]* (Cl. C04B35/56), 04 Dec 1990, Appl. 89/109,177, 01 May 1989; 10 pp. (*Chem. Abstr.*, 1991, **115**, 77 216).

2776 F. Babonneau, J. Livage, and R.M. Laine, *Polym. Prepr. (Am. Chem. Soc., Div. Polym. Chem.)*, 1991, **32**, 579 (*Chem. Abstr.*, 1991, **115**, 213 203).

2777 H. Frey, K. Matyjaszewski, M. Moeller, and D. Oelfin, *Colloid Polym. Sci.*, 1991, **269**, 442 (*Chem. Abstr.*, 1991, **115**, 72 623).

2778 H. Frey, M. Moeller, and K. Matyjaszewski, *Synth. Met.*, 1991, **42**, 1571 (*Chem. Abstr.*, 1991, **115**, 115 204).

2779 T. Takayama and I. Ando, *J. Mol. Struct.*, 1991, **243**, 101.

2780 G. Costa, A. Grosso, M.C. Sacchi, P.C. Stein, and L. Zetta, *Macromolecules*, 1991, **24**, 2858 (*Chem. Abstr.*, 1991, **114**, 208 255).

2781 D.M. Narsavage, L.V. Interrante, P.S. Marchetti, and G.E. Maciel, *Chem. Mater.*, 1991, **3**, 721 (*Chem. Abstr.*, 1991, **115**, 76 994); R.M. Stewart, N.R. Dando, D. Seyferth, and A.J. Perrotta, *Polym. Prepr. (Am. Chem. Soc., Div. Polym. Chem.)*, 1991, **32**, 569 (*Chem. Abstr.*, 1991, **115**, 213 199).

2782 A.C. Olivieri and G.R. Hatfield, *J. Magn. Reson.*, 1991, **94**, 535.

2783 G.D. Soraru, A. Ravagni, R. Campostrini, and F. Babonneau, *J. Am. Ceram. Soc.*, 1991, **74**, 2220 (*Chem. Abstr.*, 1991, **115**, 213 275).

2784 S. Hayashi and K. Hayamizu, *Bull. Chem. Soc. Jpn.*, 1991, **64**, 1386.

2785 L. Garrido, J.E. Mark, C.C. Sun, J.L. Ackerman, and C. Chang, *Macromolecules*, 1991, **24**, 4067 (*Chem. Abstr.*, 1991, **115**, 30 906).

2786 E. Lindner, A. Bader, and H.A. Mayer, *Inorg. Chem.*, 1991, **30**, 3783.

2787 E. Klemm, U. Beil, and R. Maertin, *Acta Polym.*, 1991, **42**, 20 (*Chem. Abstr.*, 1991, **114**, 144 124).

2788 K.K. Unger, K.D. Lork, B. Pfleiderer, K. Albert, and E. Bayer, *J. Chromatogr.*, 1991, **556**, 395 (*Chem. Abstr.*, 1991, **115**, 215 627).

2789 D.M. Smith, P.J. Davis, and C.J. Brinker, *Mater. Res. Soc. Symp. Proc.*, 1990, **180**, 235 (*Chem. Abstr.*, 1991, **115**, 285 668).

2790 B. Handy, K.L. Walther, A. Wokaun, and A. Baiker, *Stud. Surf. Sci. Catal.*, 1991, **63**, 239 (*Chem. Abstr.*, 1991, **115**, 190 707).

2791 J. Sanchez and A. McCormick, *Chem. Mater.*, 1991, **3**, 320 (*Chem. Abstr.*, 1991, **114**, 148 803).

2792 D.M. Smith, R. Deshpande, C.J. Brinker, W.L. Earl, B. Ewing, and P.J. Davis, *Prepr. - Am. Chem. Soc., Div. Pet. Chem.*, 1991, **36**, 489 (*Chem. Abstr.*, 1991, **115**, 285 344).

2793 V. Belot, R.J.P. Corriu, D. Leclercq, P. Lefevre, P.H. Mutin, A. Vioux, and A.M. Flank, *J. Non-Cryst. Solids*, 1991, **127**, 207 (*Chem. Abstr.*, 1991, **114**, 198 341).

opal and other hydrous silicas, (^{29}Si),[2794] and base-catalyzed silica gels.[2795]

^{29}Si chemical shifts in a wide variety of silicates have been related to a parameter, *P*, which takes into account the electronegativity, the structural description of the silicate units, and the cation.[2796] The room temperature setting process in compacts of various silicate and non-silicate particles bonded with Na silicate has been studied by ^{29}Si n.m.r. spectroscopy.[2797] Bond ordering in silicate glasses has been examined using n.m.r. data.[2798] ^{29}Si n.m.r. studies of $K_2Si_4O_9$ have been used to show five-coordination at silicon.[2799] The effect of a magnetic field on the structure formation of alkali metal silicate glasses has been studied using ^{29}Si n.m.r. spectroscopy.[2800] Pressure-induced silicon coordination in alkali oxide-silica melts has been studied by ^{23}Na and ^{29}Si n.m.r. spectroscopy.[2801] The bonding states of Na^+ in tetrasilicic sodium fluor mica have been investigated by ^{23}Na n.m.r. spectroscopy.[2802] ^{29}Si m.a.s. n.m.r. spectroscopy has been used to monitor the double chain/triple chain intergrowths in hydrous silicates.[2803] ^{29}Si has been used to study six-coordinated silicon at high pressure in silicate minerals.[2804] The formation of silicate anions in 80 year old concrete has been studied by ^{29}Si n.m.r. spectroscopy.[2805] The hydration process of β-dicalcium silicate has been followed by ^{29}Si n.m.r. spectroscopy.[2806] The ferroelastic-t-incommensurate transition in Sr_2SiO_4 has been investigated by ^{29}Si n.m.r. spectroscopy.[2807] N.m.r. data have also been reported for octadecasil, (^{13}C, ^{19}F, ^{29}Si),[2808] alkali silicate glasses, (^{29}Si),[2809] magadiite, (^{29}Si),[2810] $K_2MgSi_5O_{12}$, (^{29}Si),[2811] $MgSiO_3$, (^{29}Si),[2812] $Mg_3Si_4O_{10}(OH)_2$, (^{29}Si),[2813] some silicates, (^{17}O,

2794 S.J. Adams, G.E. Hawkes, and E.H. Curzon, *Am. Mineral.*, 1991, **76**, 1863 (*Chem. Abstr.*, 1991, **115**, 283 701).
2795 G.J. Garvey and B.E. Smith, *Mater. Res. Soc. Symp. Proc.*, 1990, **180**, 223 (*Chem. Abstr.*, 1991, **115**, 285 667).
2796 S. Prabakar, K.J. Rao, and C.N.R. Rao, *Chem. Phys. Lett.*, 1991, **183**, 176 (*Chem. Abstr.*, 1991, **115**, 196 551); B.L. Sherriff, H.D. Grundy, and J.S. Hartman, *Eur. J. Mineral.*, 1991, **3**, 751 (*Chem. Abstr.*, 1991, **115**, 283 683).
2797 K.J.D. Mackenzie, I.W.M. Brown, P. Ranchod, and R.H. Meinhold, *J. Mater. Sci.*, 1991, **26**, 763 (*Chem. Abstr.*, 1991, **114**, 107 651).
2798 S.J. Gurman, *Report*, 1989, **DK/SCI/P-657T**; Order No. PB90-209073, 16 pp. Avail. NTIS. From *Gov. Rep. Announce. Index (U.S.)*, 1990, **90**, Abstr. No. 038,904 (*Chem. Abstr.*, 1991, **114**, 215 168).
2799 J.F. Stebbins, *Nature (London)*, 1991, **351**, 638.
2800 G.V. Zhmykhov, V.G. Pitsyuga, L.G. Pryadko, and V.I. Borul'ko, *Fiz. Khim. Stekla*, 1991, **17**, 345 (*Chem. Abstr.*, 1991, **115**, 261 506).
2801 X. Xue, J.F. Stebbins, M. Kanzaki, P.F. McMillan, and B. Poe, *Am. Mineral.*, 1991, **76**, 8 (*Chem. Abstr.*, 1991, **114**, 168 135).
2802 M. Soma, A. Tanaka, H. Seyama, S. Hayashi, and K. Hayamizu, *Clay Sci.*, 1990, **8**, 1 (*Chem. Abstr.*, 1991, **115**, 196 668).
2803 J. Rocha, M.D. Welch, and J. Klinowski, *J. Am. Chem. Soc.*, 1991, **113**, 7100.
2804 J.F. Stebbins and M. Kanzaki, *Science (Washington, D.C., 1883-)*, 1991, **251**, 294 (*Chem. Abstr.*, 1991, **114**, 106 239).
2805 Y. Okada, M. Manada, and S. Fujii, *Semento, Konkurito Ronbunshu*, 1990, 58 (*Chem. Abstr.*, 1991, **115**, 77 367).
2806 Y. Tong, H. Du, and L. Fei, *Cem. Concr. Res.*, 1991, **21**, 355 (*Chem. Abstr.*, 1991, **114**, 170 052).
2807 B.L. Phillips, R.J. Kirkpatrick, and J.G. Thompson, *Phys. Rev. B: Condens. Matter*, 1991, **43**, 13280 (*Chem. Abstr.*, 1991, **115**, 38 986).
2808 P. Caullet, J.L. Guth, J. Hazm, J.M. Lamblin, and H. Gies, *Eur. J. Solid State Inorg. Chem.*, 1991, **28**, 345 (*Chem. Abstr.*, 1991, **115**, 40 631).
2809 H. Maekawa, T. Maekawa, K. Kawamura, and T. Yokokawa, *J. Non-Cryst. Solids*, 1991, **127**, 53 (*Chem. Abstr.*, 1991, **114**, 193 023).
2810 G. Scholzen, K. Beneke, and G. Lagaly, *Z. Anorg. Allg. Chem.*, 1991, **597**, 183.
2811 S.C. Kohn, R. Dupree, M.G. Mortuza, and C.M.B. Henderson, *Phys. Chem. Miner.*, 1991, **18**, 144 (*Chem. Abstr.*, 1991, **115**, 171 392).
2812 R.J. Kirkpatrick, D. Howell, B.L. Phillips, X.D. Cong, E. Ito, and A. Navrotsky, *Am. Mineral.*, 1991, **76**, 672 (*Chem. Abstr.*, 1991, **114**, 232 133).
2813 K. Urabe, I. Kenmoku, K. Kawabe, and Y. Izumi, *J. Chem. Soc., Chem. Commun.*, 1991, 867.

(^{29}Si),[2814] CaO-SiO_2, (^{29}Si),[2815] lead phosphosilicate glasses, (^{29}Si, ^{31}P),[2816] and Li_2S-SiS_2.[2817]

The ^{119}Sn n.m.r. spectrum of solid R_3SnF shows that the tin is five-coordinate with the ^{119}Sn equally coupled to ^{19}F nuclei.[2818] N.m.r. data have also been reported for R_2SnCl(oxinate), (^{1}H, ^{13}C, ^{15}N, ^{119}Sn),[2819] substituted 2,2-di-*n*-butyl-4-oxobenzo-1,3,2-dioxastannins, (^{13}C, ^{119}Sn),[2820] $(Me_2SnE)_3$, (^{119}Sn, ^{125}Te),[2821] lithium germanate glasses, (^{7}Li),[2822] metal hydrogen phosphates, (M = Ti, Zr, Sn; ^{31}P, ^{119}Sn),[2823] $PbHPO_4$, (^{31}P),[2824] $Sn(OSO_2F)_2$, (^{119}Sn),[2825] $[M^1\{M^2(OBu^t)_3\}_2]$, (M^1 = Ce, Sn, Ob; M^2 = Mg, Ca, Sr, Ba, Cd, Eu),[2826] $CsSnCl_3$, $CsPbCl_3$, (^{1}H, ^{119}Sn, ^{133}Cs),[2827] and $MSnF_4$, (M = Ba, Sr, Pb; ^{19}F).[2828]

^{15}N chemical shift standards for solid state have been discussed and the chemical shifts of some $[NH_4]^+$ salts reported.[2829] The ^{31}P chemical shift anisotropy of $P(CN)_3$ is -96 ± 5 p.p.m.[2830] Anisotropy of J (^{31}P-^{31}P) in homonuclear spin-pair systems in rotating solids has been examined for 1,2-(2,4,6-$Bu^t{}_3C_6H_2)_2$diphosphene.[2831] Forbidden n.m.r. transitions between singlet and triplet states in $I = \frac{1}{2}$ pair systems in the rotating solids have been observed in the ^{31}P n.m.r. spectrum for 1,2-(2,4,6-$Bu^t{}_3C_6H_2)_2$diphosphene. $^{1}J(^{31}P\text{-}^{31}P)$ = 577 ± 15 Hz.[2832] The ^{31}P n.m.r. spectrum of $P(CH_2Cl)Cl_4$ shows that it is $[P(CH_2Cl)Cl_3][P(CH_2Cl)Cl_5]$.[2833] N.m.r. data have also been reported for Ph_3PE, (E = lone pair, O, S, Se; ^{13}C, ^{31}P),[2834] and Ph_nPCl_{5-n}, (^{31}P).[2835]

^{15}N and ^{31}P c.p. m.a.s. n.m.r. spectra of $[2,4,6\text{-}Bu^t{}_3C_6H_2N{\equiv}P][AlCl_4]$ have been used to determine the shift tensors.[2836] The ^{31}P chemical shift tensors of three polycrystalline and one

2814 K.T. Mueller, Y. Wu, B.F. Chmelka, J. Stebbins, and A. Pines, *J. Am. Chem. Soc.*, 1991, **113**, 32.

2815 Y. Okada, T. Ogai, and T. Mitsuda, *Semento, Konkurito Ronbunshu*, 1990, 42 (*Chem. Abstr.*, 1991, **115**, 77 441); G.M.M. Bell, J. Bensted, F.P. Glasser, E.E. Lachowski, D.R. Roberts, and M.J. Taylor, *Adv. Cem. Res.*, 1990, **3**, 23 (*Chem. Abstr.*, 1991, **114**, 170 042); M. Kansaki, J.F. Stebbins, and X. Xue, *Geophys. Res. Lett.*, 1991, **18**, 463 (*Chem. Abstr.*, 1991, **114**, 250 783).

2816 S. Prabakar, K.J. Rao, and C.N.R. Rao, *Mater. Res. Bull.*, 1991, **26**, 285 (*Chem. Abstr.*, 1991, **114**, 235 471).

2817 S.W. Martin, H.K. Patel, F. Borsa, and D. Torgeson, *J. Non-Cryst. Solids*, 1991, **131**, 1041 (*Chem. Abstr.*, 1991, **115**, 83 018).

2818 H. Bai, R.K. Harris, and H. Reuter, *J. Organomet. Chem.*, 1991, **408**, 167.

2819 A. Lyčka, J. Holeček, A. Sebald, and I. Tkáč, *J. Organomet. Chem.*, 1991, **409**, 331.

2820 M. Boualam, R. Willem, J. Gelan, A. Sebald, P. Lelieveld, D. De Vos, and M. Gielen, *Appl. Organomet. Chem.*, 1990, **4**, 335 (*Chem. Abstr.*, 1991, **115**, 8940).

2821 I.D. Gay, C.H.W. Jones, and R.D. Sharma, *J. Magn. Reson.*, 1991, **91**, 186.

2822 O. Kanert, M. Kloke, R. Kuechler, S. Rueckstein, and H. Jain, *Ber. Bunsenges. Phys. Chem.*, 1991, **95**, 1061; R. Kuechler, O. Kanert, S. Rueckstein, and H. Jain, *J. Non-Cryst. Solids*, 1991, **128**, 328 (*Chem. Abstr.*, 1991, **115**, 39 398).

2823 M.J. Hudson and A.D. Workman, *J. Mater. Chem.*, 1991, **1**, 375 (*Chem. Abstr.*, 1991, **115**, 221 496); M.J. Hudson, A.D. Workman, and R.J.W. Adams, *Solid State Ionics*, 1991, **46**, 159 (*Chem. Abstr.*, 1991, **115**, 104 486).

2824 F. Ermark and U. Haeberlen, *J. Phys.: Condens. Matter*, 1991, **3**, 1909 (*Chem. Abstr.*, 1991, **114**, 218 672).

2825 D.C. Adams, T. Birchall, R. Faggiani, R.J. Gillespie, and J.E. Vekris, *Canad. J. Chem.*, 1991, **69**, 2122.

2826 M. Veith, J. Hans, L. Stahl, P. May, V. Huch, and A. Sebald, *Z. Naturforsch., B*, 1991, **46**, 403 (*Chem. Abstr.*, 1991, **115**, 104 605).

2827 S. Sharma, N. Weiden, and A. Weiss, *Z. Naturforsch., A*, 1991, **46**, 329 (*Chem. Abstr.*, 1991, **114**, 257 268).

2828 J.E. Callanan and R.D. Weir, *J. Chem. Thermodyn.*, 1991, **23**, 411 (*Chem. Abstr.*, 1991, **115**, 58 387).

2829 S. Hayashi and K. Hayamizu, *Bull. Chem. Soc. Jpn.*, 1991, **64**, 688.

2830 M.A.H.A. Al-juboori, A. Finch, P.J. Gardner, and C.J. Groombridge, *Polyhedron*, 1991, **10**, 1831.

2831 R. Challoner, T. Nakai, and C.A. McDowell, *J. Magn. Reson.*, 1991, **94**, 433; R. Challoner, T. Nakai, and C.A. McDowell, *J. Chem. Phys.*, 1991, **94**, 7038.

2832 T. Nakai, R. Challoner, and C.A. McDowell, *Chem. Phys. Lett.*, 1991, **180**, 13 (*Chem. Abstr.*, 1991, **115**, 63 127).

2833 K.B. Dillon and T.A. Straw, *J. Chem. Soc., Chem. Commun.*, 1991, 234.

2834 J.A. Davies, S. Dutremez, and A.A. Pinkerton, *Inorg. Chem.*, 1991, **30**, 2380.

2835 M.A.H.A. Al-Juboori, P.N. Gates, and A.S. Muir, *J. Chem. Soc., Chem. Commun.*, 1991, 1270.

2836 R.D. Curtis, M.J. Schriver, and R.E. Wasylishen, *J. Am. Chem. Soc.*, 1991, **113**, 1493.

vitreous polymorphs of P_2O_5 have been determined.[2837] The ^{31}P chemical shift tensor in crystalline phosphoric acid at 123K is axially symmetric with $\Delta\sigma = +60$ p.p.m.[2838] Sonically induced narrowing of solid state n.m.r. spectra has been developed and applied to the ^{31}P n.m.r. spectrum of sodium phosphate and the ^{27}Al n.m.r. spectrum of aluminium sulphate.[2839] The spinning rate dependence of the ^{31}P m.a.s. n.m.r. spectra in condensed phosphates has been measured.[2840] Cation effects have been observed on the ^{31}P m.a.s. n.m.r. chemical shifts of metaphosphate glasses.[2841] A ^{133}Cs n.m.r. study of the ferroelectric and antiferroelectric transitions in CsH_2PO_4 has been reported.[2842] The lineshape of the quadrupole perturbed ^2H n.m.r. spectrum of $Rb_{1-x}(ND_4)_xD_2PO_4$ has been derived using the pseudospin Glauber model.[2843] ^{19}F and ^{31}P T_1 values of Na_2PFO_3 have been determined and chemical shift anisotropies, coupling constants and correlation times determined.[2844] N.m.r. data have also been reported for sulphonated (aryloxy)- and (arylamino)phosphazenes, (^{13}C, ^{31}P),[2845] P-O-N-H system, (^{31}P),[2846] tripolyphosphate cements, (^{31}P),[2847] KH_2AsO_4, (^{75}As),[2848] P_4S_n, (^{31}P),[2849] $P_xAs_yS_z$, (^{31}P),[2850] P_2Se_5, (^{31}P, ^{77}Se),[2851] P_4Se_4, (^{31}P),[2852] P-Se glasses, (^{31}P),[2853] $(TMTSF)_2PF_6$,[2854] and $CsPCl_6$, (^{31}P).[2855]

^1H n.m.r. imaging has been applied to salt-water ice mixtures.[2856] Laser production of a large-spin polarization in frozen ^{129}Xe has been reported.[2857] N.m.r. data have also been reported for 1,3-$[(E_2N_2N)C_6H_4(CN)_2E_2]$, (E = S, Se; ^{77}Se),[2858] bis(tetramethyltetraselenafulvolenium)per-

2837 A.R. Grimmer and G.U. Wolf, *J. Solid State Inorg. Chem.*, 1991, **28**, 221.

2838 A.J. Vila, C.M. Lagier, G. Wagner, and A.C. Olivieri, *J. Chem. Soc., Chem. Commun.*, 1991, 683.

2839 J. Homer, P. McKeown, W.R. McWhinnie, S.U. Patel, and G.J. Tilstone, *J. Chem. Soc., Faraday Trans.*, 1991, **87**, 2253.

2840 S. Hayashi and K. Hayamizu, *Chem. Phys.*, 1991, **157**, 381.

2841 R.K. Brow, C.C. Phifer, G.L. Turner, and R.J. Kirkpatrick, *J. Am. Ceram. Soc.*, 1991, **74**, 1287 (*Chem. Abstr.*, 1991, **115**, 55 131).

2842 P.J. Schuele and V.H. Schmidt, *Ferroelectrics*, 1991, **117**, 35 (*Chem. Abstr.*, 1991, **115**, 39 835).

2843 R. Pirc, B. Tadic, R. Blinc, and R. Kind, *J. Non-Cryst. Solids*, 1991, **131**, 92 (*Chem. Abstr.*, 1991, **115**, 120 264); R. Pirc, B. Tadic, R. Blinc, and R. Kind, *Phys. Rev. B: Condens. Matter*, 1991, **43**, 2501 (*Chem. Abstr.*, 1991, **114**, 155 664).

2844 T.C. Farrar and M.J. Jablonsky, *J. Phys. Chem.*, 1991, **95**, 9159.

2845 H.R. Allcock, R.J. Fitzpatrick, and L. Salvati, *Chem. Mater.*, 1991, **3**, 1120 (*Chem. Abstr.*, 1991, **115**, 233 056).

2846 V. Avotins, A. Vitola, T.N. Miller, E.A. Pashchenko, V.V. Trachevskii, and V.B. Chernogorenko, *Latv. PSR Zinat. Akad. Vestis, Kim. Ser.*, 1990, 703 (*Chem. Abstr.*, 1991, **115**, 125 383).

2847 M.C. Connaway-Wagner, W.G. Klemperer, and J.F. Young, *Chem. Mater.*, 1991, **3**, 5 (*Chem. Abstr.*, 1991, **114**, 67 969).

2848 J.L. Bjorkstam, *Ferroelectrics*, 1991, **117**, 215 (*Chem. Abstr.*, 1991, **115**, 39 843).

2849 M.C. Demarcq, *Ind. Eng. Chem. Res.*, 1991, **30**, 1906 (*Chem. Abstr.*, 1991, **115**, 74 572); T. Bjorholm and H.J. Jakobsen, *J. Am. Chem. Soc.*, 1991, **113**, 27.

2850 L. Koudelka, M. Pisarcik, L.N. Blinov, and M.S. Gutenev, *J. Non-Cryst. Solids*, 1991, **134**, 86 (*Chem. Abstr.*, 1991, **115**, 215 429)

2851 R. Blachnik, H.P. Baldus, P. Lönnecke, and B.W. Tattershall, *Angew. Chem., Int. Ed. Engl.*, 1991, **30**, 605.

2852 G.R. Burns, J.R. Rollo, J.D. Sarfati, and K.R. Morgan, *Spectrochim. Acta, Part A*, 1991, **47**, 811 (*Chem. Abstr.*, 1991, **115**, 104 501).

2853 D. Lathrop and H. Eckert, *Phys. Rev. B: Condens. Matter*, 1991, **43**, 7279 (*Chem. Abstr.*, 1991, **114**, 171 865).

2854 T. Takahashi, T. Harada, Y. Kobayashi, K. Kanoda, K. Suzuki, K. Murata, and G. Saito, *Synth. Met.*, 1991, **43**, 3985 (*Chem. Abstr.*, 1991, **115**, 148 651).

2855 A.S. Muir, *Polyhedron*, 1991, **10**, 2217.

2856 W.A. Edelstein and E.M. Schulson, *J. Glaciol.*, 1991, **37**, 177 (*Chem. Abstr.*, 1991, **115**, 284 025).

2857 G.D. Cates, D.R. Benton, M. Gatzke, W. Happer, K.C. Hasson, and N.R. Newbury, *Phys. Rev. Lett.*, 1990, **65**, 2591 (*Chem. Abstr.*, 1991, **114**, 16 319).

2858 M.P. Andrews, A.W. Cordes, D.C. Douglass, R.M. Fleming, S.H. Glarum, R.C. Haddon, P. Marsh, R.T. Oakley, T.T.M. Palstra, L.F. Schneemeyer, G.W. Trucks, R. Tycko, J.V. Waszczak, K,M. Young, and N.M. Zimmerman, *J. Am. Chem. Soc.*, 1991, **113**, 3559.

chlorate,[2859] fluorosulphate doped polyacetylene,[2860] SF_6 chlathrate hydrate, (^{19}F),[2861] $NH_4IO_3.2HIO_3$, and $KIO_3.2HIO_3$.[2862]

Molecules Sorbed Onto Solids.—This section is divided into two subsections: 'Water sorbed Onto Solids' and 'Atoms and Other Molecules Sorbed Onto Solids'.

Water Sorbed Onto Solids.—The transverse relaxation times of water adsorbed on highly dispersed pyrogenic oxides of silicon, titanium and titanium-silicon have been reported.[2863] The correlation times of water, acetone, dioxane, toluene, DMF, and THF adsorbed on silica gel, alumina or charcoal have been determined.[2864] Water adsorption on thermally expanded graphite has been investigated.[2865] Water in albite glass has been investigated by ^1H n.m.r. spectroscopy.[2866] Water mobility in vermiculites with different exchangeable cations has been investigated by n.m.r. spectroscopy.[2867] ^1H n.m.r. spectroscopy has been used to study water adsorbed on synthetic chrysotile asbestos.[2868] Water molecules adsorbed on Brønsted and Lewis acid sites in zeolites and amorphous silica-aluminas have been studied.[2869] Water adsorption on silica gel has been studied.[2870] Surface induced spin-lattice relaxation of water in tricalcium silicate gels has been investigated.[2871] The order acquisition by clay platelets in a magnetic field has been studied by using the residual quadrupolar splittings of the ^2H and ^{17}O signals of the D_2O solvent.[2872] The effect of paramagnetic relaxation on these systems has also been examined.[2873] The sorption of water, methanol, and ammonia on $AlPO_4$-5 has been studied by ^2H, ^{27}Al, and ^{31}P n.m.r. spectroscopy.[2874] Two-dimensional spin echo and three-dimensional chemical shift imaging techniques have been used

[2859] J.M. Delrieu and N. Kinoshita, *J. Magn. Magn. Mater.*, 1990, **90**, 763 (*Chem. Abstr.*, 1991, **114**, 176 926); J.M. Delrieu and N. Kinoshita, *Synth. Met.*, 1991, **43**, 3947 (*Chem. Abstr.*, 1991, **115**, 124 538).

[2860] K. Mizoguchi, F. Shimizu, K. Kume, and S. Masubuchi, *Synth. Met.*, 1991, **41**, 185 (*Chem. Abstr.*, 1991, **115**, 82 976).

[2861] A.W.K. Khanzada, *Phys. Chem. (Peshawar, Pak.)*, 1989, **8**, 14 (*Chem. Abstr.*, 1991, **115**, 148 715).

[2862] D.F. Baisa, E.D. Chesnokov, and A.I. Shanchuk, *Fiz. Tverd. Tela (Leningrad)*, 1990, **32**, 3295 (*Chem. Abstr.*, 1991, **115**, 40 535).

[2863] V.V. Turov, V.I. Zarko, and A.A. Chuiko, *Ukr. Khim. Zh. (Russ. Ed.)*, 1990, **56**, 1262 (*Chem. Abstr.*, 1991, **114**, 254 468).

[2864] Y. Xue, Y.R. Du, C.H. Ye, and Y.H. Kong, *Appl. Magn. Reson.*, 1991, **1**, 413 (*Chem. Abstr.*, 1991, **115**, 196 537).

[2865] V.V. Turov, K.V. Pogorelyi, L.A. Mironova, and A.A. Chuiko, *Zh. Fiz. Khim.*, 1991, **65**, 170 (*Chem. Abstr.*, 1991, **114**, 172 118).

[2866] V.O. Zavel'skii and N.I. Bezmen, *Geokhimiya*, 1990, 1120 (*Chem. Abstr.*, 1991, **114**, 66 004).

[2867] A.I. Sosikov, N.K. Red'kina, P.P. Kushch, E.V. Summanen, E.A. Dzhavadyan, L.N. Erofeev, and B.A. Rozenberg, *Izv. Akad. Nauk SSSR, Neorg. Mater.*, 1990, **26**, 2148 (*Chem. Abstr.*, 1991, **114**, 69 652).

[2868] S. Ozeki, Y. Masuda, H. Sano, H. Seki, and K. Ooi, *J. Phys. Chem.*, 1991, **95**, 6309.

[2869] M. Hunger, D. Freude, and H. Pfeifer, *J. Chem. Soc., Faraday Trans.*, 1991, **87**, 657; P. Batamack, C. Doremieux-Morin, R. Vincent, and J. Fraissard, *Chem. Phys. Lett.*, 1991, **180**, 545 (*Chem. Abstr.*, 1991, **115**, 100 001).

[2870] U. Zscherpel, E. Brunner, and B. Staudte, *Z. Phys. Chem. (Leipzig)*, 1990, **271**, 931 (*Chem. Abstr.*, 1991, **114**, 109 378).

[2871] F. Milia, Y. Bakopoulos, and L. Miljkovic, *Z. Naturforsch., A*, 1991, **46**, 697 (*Chem. Abstr.*, 1991, **115**, 125 393).

[2872] A. Delville, J. Grandjean, and P. Laszlo, *J. Phys. Chem.*, 1991, **95**, 1383; V.D. Shantarin, V.F. Blashchanitsa, V.S. Voitenko, L.S. Podenko, and M.A. Sigorskii, *Str-vo Glubok. Tazved. Skvazhin v Zap. Sibiri, Tyumen*, 1989, 123. From *Ref. Zh., Geol.*, 1991, Abstr. No. 3D289 (*Chem. Abstr.*, 1991, **115**, 259 539).

[2873] L.S. Podenko and S.I. Dmitriev, *Zh. Prikl. Spektrosk.*, 1991, **55**, 146 (*Chem. Abstr.*, 1991, **115**, 293 316).

[2874] I. Kustanovich and D. Goldfarb, *J. Phys. Chem.*, 1991, **95**, 8818.

to analyse oil-water replacement in limestone.[2875] N.m.r. imaging of water in porous materials has been reported.[2876] A comment on a previous paper[2877] entitled 'Using ^{129}Xe n.m.r. spectroscopy to measure diffusivity of water in microporous zeolite Y' has appeared.[2878]

Atoms and Other Molecules Sorbed Onto Solids.—The ^{1}H n.m.r. spectra of molecules adsorbed on a graphite surface have been simulated mathematically.[2879] Anomalous diffusion in adsorbent-adsorbate systems has been studied by pulsed field gradient n.m.r. spectroscopy.[2880]

Hydrogen chemisorption on Ru/TiO_2 has been investigated by ^{1}H n.m.r. spectroscopy.[2881] The ^{1}H and ^{13}C nuclear spin dynamics in adsorbed hydrogen and ethylene on Ru/SiO_2 and M,Ru/SiO_2, M = Cu, Ag, Au, have been used to probe the surface concentrations of reactive intermediates.[2882] ^{1}H n.m.r. spectroscopy has been used to estimate the extent of metal surface covered by TiO_2 overlayers in H_2 reduced Rh/TiO_2 catalysts.[2883] The chemisorption of deuterium on Rh/SiO_2 has been investigated by ^{2}H n.m.r. spectroscopy.[2884] ^{1}H n.m.r. spectroscopy has been used to study hydrogen chemisorption on Pt/SiO_2.[2885] The supercooling of molecular hydrogen in zeolites has been investigated using ^{1}H T_2 measurements.[2886] Hydrogen bonding and microvoids in hydrogenated amorphous silicon have been studied by ^{1}H n.m.r. spectroscopy.[2887] The ^{2}H and ^{3}H nuclear relaxation of $^{2}H_2$-$^{3}H_2$ adsorbed onto silica aerogel has been reported.[2888]

Nuclear spin relaxation of ^{6}Li and ^{7}Li adsorbed on metal surfaces has been studied.[2889] A similar study has been performed for ^{7}Li and ^{23}Na on tungsten surfaces.[2890] ^{6}Li n.m.r. spectroscopy has been used to investigate Li^{+} adsorbed on a silicon(111) surface.[2891] Zeigler-Natta catalysts supported on $MgCl_2$ have been studied by ^{13}C n.m.r. spectroscopy.[2892] Surface species supported on vanadia-alumina catalysts have been investigated by ^{51}V n.m.r. spectroscopy.[2893] M.a.s. ^{13}C n.m.r.

2875 J.M. Dereppe, C. Moreaux, and K. Schenker, *J. Magn. Reson.*, 1991, **91**, 596; J.J. Dechter, R.A. Komoroski, and S. Ramaprasad, *J. Magn. Reson.*, 1991, **93**, 142.
2876 S.N. Sarkar, E.W. Wooten, and R.A. Komoroski, *Appl. Spectrosc.*, 1991, **45**, 619 (*Chem. Abstr.*, 1991, **115**, 60 702).
2877 N. Bansal and C. Dybowski, *J. Magn. Reson.*, 1990, **89**, 21.
2878 J. Kaerger, *J. Magn. Reson.*, 1991, **93**, 184.
2879 K.V. Pogorelyi, V.V. Turov, L.A. Moronova, and A.A. Chuiko, *Dokl. Akad. Nauk Ukr. SSR*, 1991, 128 (*Chem. Abstr.*, 1991, **115**, 196 542).
2880 J. Kärger and H. Spindler, *J. Am. Chem. Soc.*, 1991, **113**, 7571.
2881 P. Jonsen and C.C.A. Christopher, *J. Chem. Soc., Faraday Trans.*, 1991, **87**, 1917.
2882 M. Pruski, D.K. Sanders, X. Wu, S.J. Hwang, T.S. King, and B.C. Gerstein, *Prepr. Pap. - Am. Chem. Soc., Div. Fuel Chem.*, 1991, **36**, 1839 (*Chem. Abstr.*, 1991, **115**, 182 413).
2883 J.P. Belzunegui, J.M. Rojo, and J. Sanz, *J. Phys. Chem.*, 1991, **95**, 3463.
2884 T.H. Chang, C.P. Cheng, and C.T. Yeh, *J. Phys. Chem.*, 1991, **95**, 5239.
2885 D. Rouabah, R. Benslama, and J. Fraissard, *Chem. Phys. Lett.*, 1991, **179**, 218 (*Chem. Abstr.*, 1991, **114**, 235 790).
2886 M. Rall, J.P. Brison, and N.S. Sullivan, *Phys. Rev. B: Condens. Matter*, 1991, **44**, 9639 (*Chem. Abstr.*, 1991, **115**, 287 608).
2887 M. Zheng, E.J. Vanderheiden, P.C. Taylor, R. Shinar, S. Mitra, and J. Shinar, *Mater. Res. Soc. Symp. Proc.*, 1990, **192**, 657 (*Chem. Abstr.*, 1991, **115**, 20 715).
2888 P.C. Souers, E.M. Fearon, J.D. Sater, E.R. Mapoles, J.R. Gaines, and P.A. Fedders, *J. Vac. Sci. Technol., A*, 1991, **9**, 232 (*Chem. Abstr.*, 1991, **114**, 194 338).
2889 M. Riehl-Chudoba and D. Fick, *Surf. Sci.*, 1991, **251**, 97 (*Chem. Abstr.*, 1991, **115**, 120 826); M. Riehl-Chudoba, U. Memmert, and D. Fick, *Surf. Sci.*, 1991, **245**, 180 (*Chem. Abstr.*, 1991, **114**, 235 644).
2890 M. Bickert, W. Widdra, and D. Fick, *Surf. Sci.*, 1991, **251**, 931 (*Chem. Abstr.*, 1991, **115**, 100 073).
2891 J. Chrost and D. Fick, *Surf. Sci.*, 1991, **251**, 78 (*Chem. Abstr.*, 1991, **115**, 100 047).
2892 P. Sormunen, T. Hjertberg, and E. Iiskola, *Makromol. Chem.*, 1990, **191**, 2663 (*Chem. Abstr.*, 1991, **114**, 7308).
2893 O.B. Lapina, V.M. Mastikhin, L.G. Simonova, and Yu.O. Bulgakova, *J. Mol. Catal.*, 1991, **69**, 61.

spectroscopy has been used to study [$Mo(CO)_6$] and [$W(CO)_6$] adsorbed on zeolites X and Y.[2894] Quantitative ^{95}Mo n.m.r. spectroscopy has been used to investigate the adsorption of oxomolybdate on a γ-Al_2O_3 support.[2895] The interaction of Cs^+ with SiO_2 supported 12-molybdophosphoric acid has been investigated using ^{31}P n.m.r. spectroscopy.[2896] The ^{31}P T_1 of SiO_2 supported 12-molybdophosphoric acid depends on the polyanion:SiO_2 ratio.[2897] ^{29}Si n.m.r. spectroscopy has provided evidence for 12-molybdosilicate in SiO_2 supported catalysts.[2898] ^{1}H n.m.r. spectroscopy has been used to investigate SiO_2 supported group VII monometallic and group VIII-group IV bimetallic catalysts and Al_2O_3 and SiO_2 supported Pt-Re bimetallic catalysts.[2899] ^{13}C n.m.r. spectroscopy has been used to study the complexation of noble metals by a complexing sorbent with 1,3(5) dimethylpyrazole functional groups.[2900] The intercalation of 2-aminoethylferrocene into the layered host lattices MoO_3, 2H-TaS_2, and α-$Zr(HPO_4)_2.H_2O$ has been investigated using ^{13}C n.m.r. spectroscopy.[2901] Supported ruthenium catalysts have been investigated by n.m.r. spectroscopy.[2902] The reversible loading of [(η^6-2,3,4,5-Me_4-thiophene)$Ru(OTf)_2$]$_x$ onto γ-Al_2O_3 has been studied using ^{13}C n.m.r. spectroscopy.[2903] The ^{13}C m.a.s. n.m.r. spectrum of [$Os_3(^{13}CO)_{12}$] supported on SiO_2 has been measured to give both the isotropic chemical shift and the principal components of the chemical shift tensor.[2904] Rhodium olefin complexes, chemisorbed onto γ-Al_2O_3, have been characterized by ^{13}C n.m.r. spectroscopy.[2905] The intercalation of $Rh(PPh)_3$ into clay has been studied by ^{27}Al, ^{29}Si, and ^{31}P n.m.r. spectroscopy.[2906] The adsorption of Cd^{2+} from seawater by humic acids has been studied by ^{13}C n.m.r. spectroscopy.[2907]

Confined geometry effects on the reorientational dynamics of molecular liquids in porous silica glasses have been studied using ^{2}H n.m.r. spectroscopy.[2908] The self-diffusion of hydrocarbons in montmorillonite has been characterised using pulse n.m.r. spectroscopy.[2909] A pulsed field gradient n.m.r. study of diffusion anisotropy in oriented ZSM-5[2910] and NaX[2911] zeolites has been reported.

2894 W.M. Shirley, C.A. Powers, and J.S. Frye, *Inorg. Chem.*, 1991, **30**, 4182.
2895 P. Sarrazin, B. Mouchel, and S. Kasztelan, *J. Phys. Chem.*, 1991, **95**, 7405.
2896 S. Kasztelan, E. Payen, and J.B. Moffat, *J. Catal.*, 1991, **128**, 479 (*Chem. Abstr.*, 1991, **114**, 151 277).
2897 R. Thouvenot, C. Rocchiccioli-Deltcheff, and M. Fournier, *J. Chem. Soc., Chem. Commun.*, 1991, 1352.
2898 R. Thouvenot, M. Fournier, and C. Rocchiccioli-Deltcheff, *J. Chem. Soc., Faraday Trans.*, 1991, **87**, 2829.
2899 X. Wu, *Report*, 1990, **IS-T-1441**; Order No. DE 91000610, 410 pp. Avail. NTIS. From *Energy Res. Abstr.*, 1990, **15**, Abstr. No. 53214 (*Chem. Abstr.*, 1991, **114**, 172 274).
2900 G.V. Myasoedova, I.I. Antokol'skaya, I.L. Krylova, G.R. Ishmiyarova, S.B. Savvin, A.V. Rebrov, L.A. Fedorov, V.K. Belyaeva, I.N. Marov, and V.I. Seraya, *Zh. Anal. Khim.*, 1991, **46**, 1077 (*Chem. Abstr.*, 1991, **115**, 222 092).
2901 K. Chatakondu, C. Formstone, M.L.H. Green, D. O'Hare, J.M. Twyman, and P.J. Wiseman, *J. Mater. Chem.*, 1991, **1**, 205 (*Chem. Abstr.*, 1991, **115**, 20 802).
2902 J.C. Keizenberg, *Report*, 1990, **IS-T-1404**; Order No. DE90011812, 119 pp. Avail. NTIS. From *Energy Res. Abstr.*, 1990, **15**, Abstr. No. 36699 (*Chem. Abstr.*, 1991, **115**, 57 831).
2903 E.A. Ganja, T.B. Rauchfuss, and C.L. Stern, *Organometallics*, 1991, **10**, 270.
2904 T.H. Walter, G.R. Frauenhoff, J.R. Shapley, and E. Oldfield, *Inorg. Chem.*, 1991, **30**, 4732.
2905 S.A. Vierkötter, C.E. Barnes, T.L. Hatmaker, J.E. Penner-Hahn, C.M. Stinson, B.A. Huggins, A. Benesi, and P.D. Ellis, *Organometallics*, 1991, **10**, 3803.
2906 M.M. Taqui Khan, S.A. Samad, M.R.H. Siddiqui, H.C. Bajaj, and G. Ramachandraiah, *Polyhedron*, 1991, **10**, 2729.
2907 M. Sohn and S. Rajski, *Org. Geochem.*, 1990, **15**, 439 (*Chem. Abstr.*, 1991, **114**, 191 928).
2908 G. Liu, Y. Li, and J. Jonas, *J. Chem. Phys.*, 1991, **95**, 6892.
2909 N.K. Dvoyashkin and A.I. Maklakov, *Kolloidn. Zh.*, 1991, **53**, 631 (*Chem. Abstr.*, 1991, **115**, 167 368).
2910 U. Hong, J. Kärger, R. Kramer, H. Pfeifer, G. Seiffert, U. Müller, K.K. Unger, H.B. Lueck, and T. Ito, *Zeolites*, 1991, **11**, 816 (*Chem. Abstr.*, 1991, **115**, 264 122); J. Kärger and H. Pfeifer, *J. Chem. Soc., Faraday Trans.*, 1991, **87**, 1989; K.P. Datema, C.J.J. Den Ouden, W.D. Ylstra, H.P.C.E. Kuipers, M.F.M. Post, and J. Kärger, *J. Chem. Soc., Faraday Trans.*, 1991, **87**, 1935.
2911 U. Hong, J. Kärger, and H. Pfeifer, *J. Am. Chem. Soc.*, 1991, **113**, 4812.

^{1}H, ^{13}C, ^{27}Al, and ^{29}Si n.m.r. spectroscopy has been used to study the coking of weakly hydrothermally dealuminated HZSM-5 zeolites during the cracking of n-hexane.[2912] ^{1}H and ^{2}H n.m.r. spectroscopy has been used to study the adsorption state of hydrocarbons on four typical charcoals.[2913]

^{13}C n.m.r. spectroscopy has been used to study benzenoid aromatics adsorbed on Al_2O_3 and SiO_2. Techniques such as SEFT, J-RESOLVED, DEPT, and COSY were used.[2914] ^{2}H n.m.r. spectroscopy has been used to investigate C_6D_6 adsorbed on boehmite glass.[2915] ^{129}Xe n.m.r. spectroscopy has been used to study the distribution of aromatic molecules in NaY zeolite.[2916] A polemic on an earlier paper[2917] entitled 'The diffusion of benzene in high silica zeolite ZSM-5 studied by PFGNMR and QUENS' has appeared.[2918] The dynamics of FC_6H_5 adsorbed on graphite and BN have been studied using ^{2}H n.m.r. spectroscopy.[2919]

Adsorbed CO and C_2H_4 on colloidal metals have been investigated by ^{13}C n.m.r. spectroscopy.[2920] Caesium induced structural changes of adsorbed ethene on caesium promoted silver catalyst have been studied using ^{13}C n.m.r. spectroscopy.[2921] ^{13}C n.m.r. spectroscopy has been used to investigate the conversion of ethene to coke on supported platinum catalysts.[2922] The interaction of ethene molecules with hydroxyl groups in NaHA and NaH ZSM-5 zeolites has been studied by ^{29}Si n.m.r. spectroscopy.[2923] Hydrolytically stable bonded chromatographic phases, prepared through hydrosilylation of olefins on a hydride-modified silica intermediate, have been investigated by n.m.r. spectroscopy.[2924] Variable temperature m.a.s. ^{13}C n.m.r. spectroscopy has been used to study the reactions of isobutylene in zeolites HY and HZSM-5.[2925] ^{13}C, ^{29}Si, and ^{129}Xe n.m.r. spectroscopy has been used to study coke formation and the deactivation of H-ZSM-5.[2926] ^{2}H n.m.r. spectroscopy has been used to study the formation of ethylidyne from acetylene and hydrogen coadsorbed on platinum.[2927] Acetylene polymerization in a HZSM-5 zeolite has been investigated using ^{13}C n.m.r. spectroscopy.[2928]

Methylamine syntheses from methanol and ammonia on various zeolite catalysts have been studied

2912 H. Ernst and H. Pfeifer, *Z. Phys. Chem. (Leipzig)*, 1990, **271**, 1145 (*Chem. Abstr.*, 1991, **114**, 246 666).
2913 B. Hu, N. Yang, J. Hu, W. Liu, Y. Kong, and Y. Du, *Bopuxue Zazhi*, 1990, **7**, 405 (*Chem. Abstr.*, 1991, **115**, 79 526).
2914 M. Ebener, G. Von Fircks, and H. Günther, *Helv. Chim. Acta*, 1991, **74**, 1296.
2915 J. Fukasawa, C.D. Poon, and E.T. Samulski, *Langmuir*, 1991, **7**, 1727 (*Chem. Abstr.*, 1991, **115**, 99 970).
2916 B.F. Chmelka, J.G. Pearson, S.B. Liu, R. Ryoo, L.C. De Menorval, and A. Pines, *J. Phys. Chem.*, 1991, **95**, 303.
2917 J. Kärger and H. Jobic, *Colloids Surf.*, 1991, **58**, 203 (*Chem. Abstr.*, 1991, **115**, 240 613).
2918 M. Morgan *et al.*, *Colloids Surf.*, 1989, **36**, 209.
2919 B. Boddenberg and V. Grundke, *Z. Naturforsch., A*, 1991, **46**, 211 (*Chem. Abstr.*, 1991, **114**, 193 247).
2920 J.S. Bradley, J. Millar, E.W. Hill, and M. Melchior, *ACS Symp. Ser.*, 1990, **437**, 160 (*Chem. Abstr.*, 1991, **115**, 100 075).
2921 J. Wang and P.D. Ellis, *J. Am. Chem. Soc.*, 1991, **113**, 9675.
2922 M. Pruski, X. Wu, M.W. Smale, B.C. Gerstein, and T.S. King, *Stud. Surf. Sci. Catal.*, 1991, **68**, 699 (*Chem. Abstr.*, 1991, **115**, 186 443).
2923 J. Datka, *Zeolites*, 1991, **11**, 739 (*Chem. Abstr.*, 1991, **115**, 240 611).
2924 J.E. Sandoval and J.J. Pesek, *Anal. Chem.*, 1991, **63**, 2634.
2925 N.D. Lazo, B.R. Richardson, P.D. Schettler, J.L. White, E.J. Munson, and J.F. Haw, *J. Phys. Chem.*, 1991, **95**, 9420.
2926 M.C. Barrage, F. Bauer, H. Ernst, J. Fraissard, D. Freude, and H. Pfeifer, *Catal. Lett.*, 1990, **6**, 201 (*Chem. Abstr.*, 1991, **114**, 254 666).
2927 C.A. Klug, C.P. Slichter, and J.H. Sinfelt, *J. Phys. Chem.*, 1991, **95**, 2119; C.A. Klug, C.P. Slichter, and J.H. Sinfelt, *J. Phys. Chem.*, 1991, **95**, 7033.
2928 C. Pereira, G.T. Kokotailo, and R.J. Gorte, *J. Phys. Chem.*, 1991, **95**, 705.

by ^{27}Al and ^{29}Si n.m.r. spectroscopy.[2929] A spin-lock pulse sequence has been used in solid state ^{1}H n.m.r. spectroscopy to suppress the dominant water signal and to observe very narrow signals of the Pr_2NH template in the hydrated molecular sieve VPI-5.[2930] The intercalation of α-$[M(O_3POH)_2.H_2O]$, M = Zr, Sn, with $HC\equiv CCH_2NH_2$ and $N\equiv CCH_2NH_2$ has been investigated by ^{13}C n.m.r. spectroscopy.[2931] The structure and dynamics of amino functional silanes adsorbed on silica surfaces have been studied using ^{2}H and ^{29}Si n.m.r. spectroscopy.[2932] The intercalation and pillaring of α-$[Zr(O_3POH)_2.H_2O]$ with $H_2N(CH_2)_3Si(OEt)_3$ have been investigated by ^{13}C, ^{29}Si, and ^{31}P n.m.r. spectroscopy.[2933] ^{13}C n.m.r. spectroscopy has been used to study the character of coke deposited on HZSM-5 and HY zeolites in acetone conversion.[2934] The decomposition of THF on MoS_2 and Li_xMoS_2 has been studied by ^{13}C n.m.r. spectroscopy.[2935] The interaction of methylsalicylate with smectic clays has been investigated by ^{1}H[2936] and ^{13}C[2937] n.m.r. spectroscopy. ^{13}C n.m.r. spectroscopy has been used to study kaolinite-acrylic acid intercalation.[2938] ^{1}H n.m.r. spectroscopy has been used to study the chemistry of methanol adsorbed on ZSM-5 and several Si,Al- and Si,Al,P-based molecular sieves,[2939] and H_2S and methanol adsorbed on charcoal.[2940] The results of ^{1}H and ^{13}C n.m.r. studies of the reaction between H_2S and methanol over solid catalysts have been presented.[2941] The behaviour of $CuZr_2(PO_4)_3$ in the decomposition of Pr^iOH has been studied using ^{1}H and ^{31}P n.m.r. spectroscopy.[2942] The chemisorption of CO onto a rhodium electrode has been investigated by ^{13}C n.m.r. spectroscopy.[2943] High resolution ^{13}C n.m.r. spectroscopy has been used to observe the adsorption-desorption process for CO on 70 Å microcrystalline palladium[2944] and platinum[2945] in colloidal suspension. The roles of CO as either an intermediate or a catalyst in the conversion of MeOH to gasoline on HZSM-5 have been investigated using ^{13}C n.m.r. spectroscopy.[2946]

Chemically modified glass surfaces have been characterised by ^{13}C and ^{29}Si n.m.r. spectroscopy.[2947] ^{13}C and ^{29}Si n.m.r. spectroscopy has been used to study alkyldihydrosilanes

2929 K. Segawa and H. Tachibana, *J. Catal.*, 1991, **131**, 482 (*Chem. Abstr.*, 1991, **115**, 182 563).

2930 W. Kolodziejski, J. Rocha, H. He, and J. Klinowski, *Appl. Catal.*, 1991, **77**, L1 (*Chem. Abstr.*, 1991, **115**, 287 886).

2931 J.E. Pillion and M.E. Thompson, *Chem. Mater.*, 1991, **3**, 777 (*Chem. Abstr.*, 1991, **115**, 173 260).

2932 H.J. Kang and F.D. Blum, *J. Phys. Chem.*, 1991, **95**, 9391.

2933 L. Li, X. Liu, Y. Ge, L. Li, and J. Klinowski, *J. Phys. Chem.*, 1991, **95**, 5910.

2934 J. Novakova, L. Kubelkova, V. Bosacek, and K. Mach, *Zeolites*, 1991, **11**, 135 (*Chem. Abstr.*, 1991, **114**, 151 262); V. Bosacek, L. Kubelkova, and J. Novakova, *Stud. Surf. Sci. Catal.*, 1991, **65**, 337 (*Chem. Abstr.*, 1991, **115**, 231 602).

2935 M.W. Juzkow and I.D. Gay, *J. Phys. Chem.*, 1991, **95**, 9911.

2936 P. O'Brien and C.J. Williamson, *Chem. Ind. (London)*, 1990, 752 (*Chem. Abstr.*, 1991, **114**, 30 720).

2937 P. O'Brien and C.J. Williamson, *Polyhedron*, 1991, **10**, 567.

2938 Y. Sugahara, T. Nagayama, K. Kuroda, A. Doi, and C. Kato, *Clay Sci.*, 1991, **8**, 69 (*Chem. Abstr.*, 1991, **115**, 259 248).

2939 M.W. Anderson, J. Klinowski, and P.J. Barrie, *J. Phys. Chem.*, 1991, **95**, 235.

2940 F. Deng, L. Li, H. Sun, and Y. Du, *Wuli Huaxue Xuebao*, 1990, **6**, 648 (*Chem. Abstr.*, 1991, **114**, 109 372).

2941 A.V. Nosov, V.M. Mastikhin, and A.V. Mashkina, *J. Mol. Catal.*, 1991, **66**, 73.

2942 A. Serghini, R. Brochu, M. Ziyad, M. Loukah, and J.C. Védrine, *J. Chem. Soc., Faraday Trans.*, 1991, **87**, 2487.

2943 J.A. Caram and C. Gutierrez, *J. Electroanal. Chem. Interfacial Electrochem.*, 1991, **307**, 99 (*Chem. Abstr.*, 1991, **115**, 80 843).

2944 J.S. Bradley, J.M. Millar, and E.W. Hill, *J. Am. Chem. Soc.*, 1991, **113**, 4016.

2945 J.S. Bradley, J.M. Millar, E.W. Hill, and S. Behal, *J. Catal.*, 1991, **129**, 530.

2946 E.J. Munson, N.D. Lazo, M.E. Moellenhoff, and J.F. Haw, *J. Am. Chem. Soc.*, 1991, **113**, 2783.

2947 K. Albert, B. Pfleiderer, E. Bayer, and R. Schnabel, *J. Colloid Interface Sci.*, 1991, **142**, 35 (*Chem. Abstr.*, 1991, **114**, 193 043).

attached to silica.[2948] ^{13}C n.m.r. spectroscopy has been used to study the adsorption of silyl ethers on $MgCl_2$ supported Ziegler-Natta catalysts.[2949] T_1 measurements have been used to investigate the changes in Brownian dynamics caused by dimethyl siloxane adsorption on spin-labelled silica gel within a boundary region 5 to 15 Å from the adsorbate surface.[2950] Chain relaxation of adsorbed dimethylsiloxane on a γ-Al_2O_3 substrate has been studied by measuring ^{13}C T_1 values.[2951] The structural characteristics of a series of oligomeric bonded phases have been investigated by ^{13}C and ^{29}Si n.m.r. spectroscopy.[2952] Silanized and polymer-coated octadecyl reversed phases have been characterized by ^{29}Si n.m.r. spectroscopy.[2953] Dimethyloctadecylsilyl-modified silica has been studied using ^{13}C n.m.r. lineshapes.[2954] The hydrolysis of (γ-aminopropyl)triethoxysilane silylated imogolite and the formation of a silylated tubular silicate-layered silicate nanocomposite have been investigated by ^{29}Si n.m.r. spectroscopy.[2955] The structure and dynamics of dimethyloctadecylsilyl-modified silica have been investigated using wide line ^{2}H n.m.r. spectroscopy.[2956] Chemically bonded phases for reversed-phase high-performance liquid chromatography have been studied using ^{13}C and ^{29}Si n.m.r. spectroscopy.[2957] The conformations of α-ω diols and *n*-alcohols grafted on silica have been studied by ^{13}C n.m.r. spectroscopy.[2958] Thermally stable phenylpolysiloxane fused silica capillary columns have been evaluated with the aid of ^{13}C n.m.r. spectroscopy.[2959] ^{29}Si cross-polarization dynamics at the poly(vinyl-alcohol)-SiO_2 sol-gel interface have been investigated.[2960] Chemically bonded liquid crystals for high-performance liquid chromatography on silica have been characterised by ^{13}C n.m.r. spectroscopy.[2961]

The reaction of NH_3 with surface species formed from acetone on a HZSM-5 zeolite has been investigated by ^{13}C n.m.r. spectroscopy.[2962] Temperature-programmed desorption of NH_3 from HZSM-5 zeolites has been investigated by ^{1}H n.m.r. spectroscopy.[2963] The role of iron impurities in the formation of electron-accepting sites in HZSM-5 zeolites has been investigated by studying the ^{15}N

2948 F. Meiouet, G. Felix, H. Taibi, H. Hommel, and A.P. Legrand, *Chromatographia*, 1991, **31**, 335 (*Chem. Abstr.*, 1991, **115**, 71 693).

2949 P. Sormunen, T.T. Pakkanen, E. Vahasarja, T.A. Pakkanen, and E. Iiskola, *Stud. Surf. Sci. Catal.*, 1990, **56**, 139 (*Chem. Abstr.*, 1991, **114**, 229 442).

2950 J. Van Alsten, *Macromolecules*, 1991, **24**, 5320 (*Chem. Abstr.*, 1991, **115**, 137 169).

2951 D.W. Ovenall and J. Van Alsten, *Polym. Prepr. (Am. Chem. Soc., Div. Polym. Chem.)*, 1990, **31**, 543 (*Chem. Abstr.*, 1991, **114**, 165 328).

2952 S.O. Akapo and C.F. Simpson, *J. Chromatogr.*, 1991, **557**, 515 (*Chem. Abstr.*, 1991, **115**, 190 630).

2953 M.J.J. Hetem, J.W. De Haan, H.A. Claessens, C.A. Cramers, A. Deege, and G. Schomburg, *J. Chromatogr.*, 1991, **540**, 53 (*Chem. Abstr.*, 1991, **114**, 172 174).

2954 R.C. Zeigler and G.E. Maciel, *J. Phys. Chem.*, 1991, **95**, 7345.

2955 L.M. Johnson and T.J. Pinnavaia, *Langmuir*, 1991, **7**, 2636 (*Chem. Abstr.*, 1991, **115**, 246 504).

2956 R.C. Zeigler and G.E. Maciel, *J. Am. Chem. Soc.*, 1991, **113**, 6349.

2957 B. Buszewski, J. Schmid, K. Albert, and E. Bayer, *J. Chromatogr.*, 1991, **552**, 415 (*Chem. Abstr.*, 1991, **115**, 167 339).

2958 A. Tuel, H. Hommel, A.P. Legrand, H. Balard, M. Sidqi, and E. Papirer, *Colloids Surf.*, 1991, **58**, 17 (*Chem. Abstr.*, 1991, **115**, 215 768).

2959 Q. Wu, M. Hetem, C.A. Cramers, and J.A. Rijks, *J. High Resolut. Chromatogr.*, 1990, **13**, 811 (*Chem. Abstr.*, 1991, **114**, 94 379).

2960 N. Zumbulyadis and J.M. O'Reilly, *Macromolecules*, 1991, **24**, 5294 (*Chem. Abstr.*, 1991, **115**, 137 365).

2961 J.J. Pesek, M.A. Vidensek, and M. Miller, *J. Chromatogr.*, 1991, **556**, 373 (*Chem. Abstr.*, 1991, **115**, 240 554).

2962 Z. Dolejsek, J. Novakova, V. Bosacek, and L. Kubelkova, *Zeolites*, 1991, **11**, 244 (*Chem. Abstr.*, 1991, **114**, 228 285).

2963 B. Hunger, J. Hoffmann, O. Heitzsch, and M. Hunger, *J. Therm. Anal.*, 1990, **36**, 1379 (*Chem. Abstr.*, 1991, **114**, 215 333).

n.m.r. spectra of adsorbed N_2 and N_2O.[2964] By using ^{31}P n.m.r. spinning side band analysis, the PMe_3 complex of La exchanged H-Y zeolites has been analysed.[2965] The adsorption of PMe_3 on a mixture of HY zeolite and $AlCl_3$ results in $AlCl_3$-PMe_3 coordination complexes which have been characterised by ^{27}Al and ^{31}P n.m.r. spectroscopy.[2966] The adsorption of PMe_3 on silica/alumina has been studied using ^{31}P n.m.r. spectroscopy.[2967] SiO_2/Al_2O_3 prepared by chemical vapour deposition has been characterised by ^{29}Si n.m.r. spectroscopy and by the ^{31}P n.m.r. spectrum of chemisorbed PMe_3.[2968] An n.m.r. study of $P(OMe)_3$ adsorbed on silica has been reported.[2969] The adsorption of phosphate ester on $BaTiO_3$ has been investigated by ^{31}P n.m.r. spectroscopy.[2970] The results from a ^{31}P n.m.r. study of phosphate adsorption at a boehmite/aqueous solution interface have been reported.[2971] The binding of $(EtO)_3PO$ to montmorillonite has been investigated by ^{13}C and ^{31}P n.m.r. spectroscopy.[2972]

^{13}C n.m.r. spectroscopy has been used to observe $[Me_3O]^+$ formation from Me_2O on zeolite HZSM-5.[2973] The microdynamic motion of DMSO adsorbed in zeolite NaX has been studied by ^1H n.m.r. spectroscopy.[2974] A two-dimensional *J*-resolved ^{13}C n.m.r. study of DMSO intercalated in kaolinite has been measured at natural abundance.[2975] Poly(*p*-phenylene selenide) doped with SO_3 has been investigated by ^1H n.m.r. spectroscopy.[2976]

Optical pumping has been used to enhance the pulsed n.m.r. signal of ^{129}Xe, allowing the detection of low-pressure xenon gas and xenon adsorbed on powdered solids.[2977] The titration by hydrogen and oxygen of bare or covered platinum surfaces has been followed by ^{129}Xe n.m.r. spectroscopy.[2978] ^{129}Xe n.m.r. spectroscopy has been used to study xenon gas adsorbed on amorphous carbon.[2979] Porous carbon has been analysed with ^{129}Xe n.m.r. spectroscopy.[2980] ^{129}Xe n.m.r. spectroscopy has been used to investigate coal micropores.[2981] Line-broadening effects

2964 V.M. Mastikhin, S.V. Filimonova, I.L. Mudrakovskii, and V.N. Romannikov, *J. Chem. Soc., Faraday Trans.*, 1991, **87**, 2247.
2965 P.J. Chu, R.R. Carvajal, and J.H. Lunsford, *Chem. Phys. Lett.*, 1990, **175**, 407 (*Chem. Abstr.*, 1991, **114**, 113 760).
2966 P.J. Chu, A. De Mallmann, and J.H. Lunsford, *J. Phys. Chem.*, 1991, **95**, 7362.
2967 S. Sato, T. Sodesawa, and F. Nozaki, *Shokubai*, 1990, **32**, 342 (*Chem. Abstr.*, 1991, **114**, 69 842).
2968 S. Sato, T. Sodesawa, F. Nozaki, and H. Shojii, *J. Mol. Catal.*, 1991, **66**, 343.
2969 I.D. Gay, A.J. McFarlan, and B.A. Morrow, *J. Phys. Chem.*, 1991, **95**, 1360.
2970 N. Le Bars, D. Tinet, A.M. Faugere, H. Van Damme, and P. Levitz, *J. Phys. III*, 1991, **1**, 707 (*Chem. Abstr.*, 1991, **115**, 190 561).
2971 W.F. Bleam, P.E. Pfeffer, S. Goldberg, R.W. Taylor, and R. Dudley, *Langmuir*, 1991, **7**, 1702 (*Chem. Abstr.*, 1991, **115**, 100 114).
2972 P. O'Brien, C.J. Williamson, and C.J. Groombridge, *Chem. Mater.*, 1991, **3**, 276 (*Chem. Abstr.*, 1991, **114**, 151 107).
2973 E.J. Munson and J.F. Haw, *J. Am. Chem. Soc.*, 1991, **113**, 6303.
2974 M. Hunger, A.L. Diaz Perez and A. Martinez-D'Alessandro, *Z. Phys. Chem. (Leipzig)*, 1990, **271**, 1009 (*Chem. Abstr.*, 1991, **114**, 69 719).
2975 J. Rocha, W. Kolodziejski, and J. Klinowski, *Chem. Phys. Lett.*, 1991, **176**, 395 (*Chem. Abstr.*, 1991, **114**, 155 666).
2976 W. Czerwinski, L. Kreja, P. Hruszka, J. Jurga, and B. Brycki, *J. Mater. Sci.*, 1991, **26**, 5921 (*Chem. Abstr.*, 1991, **115**, 280 719).
2977 D. Raftery, H. Long, T. Meersmann, P.J. Grandinetti, L. Reven, and A. Pines, *Phys. Rev. Lett.*, 1991, **66**, 584 (*Chem. Abstr.*, 1991, **114**, 155 661).
2978 G.P. Valenca and M. Boudart, *J. Catal.*, 1991, **128**, 447 (*Chem. Abstr.*, 1991, **114**, 151 276).
2979 D.J. Suh, T.J. Park, S.K. Ihm, and R. Ryoo, *J. Phys. Chem.*, 1991, **95**, 3767.
2980 H.C. Foley, N. Bansal, D.S. Lafyatis, and C. Dybowski, *Prepr. - Am. Chem. Soc., Div. Pet. Chem.*, 1991, **36**, 502 (*Chem. Abstr.*, 1991, **115**, 287 814).
2981 C. Tsiao and R.E. Botto, *Energy Fuels*, 1991, **5**, 87 (*Chem. Abstr.*, 1991, **114**, 46 232).

for ^{129}Xe adsorbed in the amorphous state of solid polymers have been investigated.[2982]

^{129}Xe n.m.r. spectroscopy of adsorbed xenon in zeolites has been used to determine void spaces.[2983] ^{129}Xe n.m.r. spectroscopy has been used to study xenon adsorption in NaA zeolite cavities.[2984] ^{129}Xe n.m.r. spectroscopy has been compared with ^{1}H pulsed field gradient n.m.r. measurements for the location of the deposit of carbonaceous compounds on zeolite A.[2985] The thermal stability of zeolite beta has been studied by ^{129}Xe n.m.r. spectroscopy.[2986] The interaction of Xe and zeolite ρ has been studied by ^{129}Xe n.m.r. spectroscopy.[2987]

A palladium cluster supported on X and Y zeolites has been investigated by ^{129}Xe n.m.r. spectroscopy. Xenon is more strongly adsorbed onto the palladium cluster.[2988] ^{129}Xe n.m.r. spectroscopy has been used to study silver exchanged X and Y zeolites.[2989] The origin of the ^{129}Xe n.m.r. line splitting observed in the study of Y zeolite has been investigated by comparing the n.m.r. spectra obtained before and after mixing various Y zeolites.[2990] The relaxation times of xenon adsorbed in Na-Y zeolites have been measured as a function of xenon uptake.[2991] ^{129}Xe n.m.r. spectroscopy has been used to study cation location in Y zeolite.[2992] The distribution of cations in lanthanum exchanged NaY zeolites has been investigated using ^{129}Xe n.m.r. spectroscopy.[2993] The distribution of Pt and adsorbed guest species in zeolites has been studied using ^{129}Xe n.m.r. spectroscopy.[2994] The ^{129}Xe n.m.r. spectrum of NaY-Pt/NaY and NaY-Pt/Al_2O_3 indicates that the n.m.r. peak pattern depends on the mixing method.[2995] ^{129}Xe n.m.r. spectroscopy has been used to study competitive Ar adsorption on coked HY zeolite,[2996] and dealuminated zeolites.[2997] ^{129}Xe and ^{27}Al n.m.r. spectroscopy has been used to study high-silica zeolites synthesized in a non alkaline fluoride medium.[2998] The variation of ^{129}Xe n.m.r. chemical shifts with aluminium content in ZSM-5

2982 A.P.M. Kentgens, H.A. Van Boxtel, R.J. Verweel, and W.S. Veeman, *Macromolecules*, 1991, **24**, 3712 (*Chem. Abstr.*, 1991, **114**, 248 126).

2983 Q. Chen, M.A. Springuel-Huet, and J. Fraissard, *Stud. Surf. Sci. Catal.*, 1991, **65**, 219 (*Chem. Abstr.*, 1991, **115**, 196 558).

2984 A.V. McCormick and B.F. Chmelka, *Mol. Phys.*, 1991, **73**, 603; P.R. Van Tassel, H.T. Davis, and A.V. McCormick, *Mol. Phys.*, 1991, **73**, 1107; B.F. Chmelka, D. Raftery, A.V. McCormick, L.C. De Menorval, R.D. Levine, and A. Pines, *Phys. Rev. Lett.*, 1991, **66**, 580 (*Chem. Abstr.*, 1991, **114**, 129 956).

2985 T. Ito, J. Fraissard, J. Kärger, and H. Pfeifer, *Zeolites*, 1991, **11**, 103 (*Chem. Abstr.*, 1991, **114**, 151 261).

2986 S.B. Liu, J.F. Wu, L.J. Ma, T.C. Tsai, and I. Wang, *J. Catal.*, 1991, **132**, 432 (*Chem. Abstr.*, 1991, **115**, 288 158).

2987 C.J. Tsiao, J.S. Kauffman, D.R. Corbin, L. Abrams, E.E. Carroll, jun., and C. Dybowski, *J. Phys. Chem.*, 1991, **95**, 5586.

2988 J.K. Kim, S.K. Ihm, J.Y. Lee, and R. Ryoo, *J. Phys. Chem.*, 1991, **95**, 8546.

2989 R. Grosse, R. Burmeister, B. Boddenberg, A. Gedeon, and J. Fraissard, *J. Phys. Chem.*, 1991, **95**, 2443; A. Gedeon, R. Burmeister, R. Grosse, B. Boddenberg, and J. Fraissard, *Chem. Phys. Lett.*, 1991, **179**, 191 (*Chem. Abstr.*, 1991, **114**, 215 325).

2990 R. Ryoo, C. Pak, and B.F. Chmelka, *Zeolites*, 1990, **10**, 790 (*Chem. Abstr.*, 1991, **11**, 30 599).

2991 M.L. Smith and C. Dybowski, *J. Phys. Chem.*, 1991, **95**, 4942.

2992 K.J. Chao, S.H. Chen, and S.B. Liu, *Stud. Surf. Sci. Catal.*, 1991, **60**, 123 (*Chem. Abstr.*, 1991, **115**, 269 036).

2993 D.S. Shy, S.H. Chen, J. Lievens, S.B. Liu, and K.J. Chao, *J. Chem. Soc., Faraday Trans.*, 1991, **87**, 2855; Q.J. Chen, T. Ito, and J. Fraissard, *Zeolites*, 1991, **11**, 239 (*Chem. Abstr.*, 1991, **114**, 193 144).

2994 B.F. Chmelka, *Report*, 1989, **LBL-28545**; Order No. DE90013871, 306 pp. Avail. NTIS. From *Energy Res. Abstr.*, 1990, **15**, Abstr. No. 39327 (*Chem. Abstr.*, 1991, **114**, 150 969).

2995 R. Ryoo, C. Pak, D.H. Ahn, L.C. De Menorval, and F. Figueras, *Catal. Lett.*, 1990, **7**, 417 (*Chem. Abstr.*, 1991, **114**, 151 306).

2996 J.T. Miller, B.L. Meyers, and G.J. Ray, *J. Catal.*, 1991, **128**, 436 (*Chem. Abstr.*, 1991, **114**, 151 275).

2997 R.L. Cotterman, D.A. Hickson, S. Cartlidge, C. Dybowski, C. Tsiao, and A.F. Venero, *Zeolites*, 1991, **11**, 27 (*Chem. Abstr.*, 1991, **114**, 129 874).

2998 Q.J. Chen, J.L. Guth, A. Seive, P. Caullet, and J. Fraissard, *Zeolites*, 1991, **11**, 798 (*Chem. Abstr.*, 1991, **115**, 269 039).

and ZSM-11 zeolites has been reported.[2999] Coke formation on ZSM-5 zeolites has been investigated using ^{129}Xe n.m.r. spectroscopy.[3000] ^{13}C and ^{129}Xe n.m.r. spectroscopy has been used to study the role of $[R_4N]^+$ in the synthesis of high-silica zeolites.[3001] Variable temperature ^{129}Xe n.m.r. studies of a pillared montmorillonite have been reported.[3002] ^{129}Xe n.m.r. spectroscopy has been used to characterise $AlPO_4$-8 and related molecular sieves.[3003]

6 Group IIIB Compounds

Boron Hydrides and Carboranes.—^{11}B chemical shifts have been calculated for H_3BNH_3 as a function of B—N distance.[3004] The reactions of μ-$Me_2NB_2H_5$ with Me_2AsNMe_2, $MeAs(NMe_2)_2$, and $As(NMe_2)_3$ have been studied using ^1H, ^{11}B, and ^{13}C n.m.r. spectroscopy and species such as $Me_2NHBH_2NMe_2BH_3$ identified.[3005] N.m.r. data have also been reported for $M[BH_4]$, $M[B_3H_8]$, (M = K, Rb, Cs; ^{11}B),[3006] $MeC(CH_2PMe_2BH_3)_3$, (^{11}B, ^{13}C),[3007] $Me_2NCH_2BH_2NMe_2CH_2BH$-CO_2H, (^{13}C),[3008] $(Pr^iO)_2P(O)CH_2NHBH_2CN$, ($^{11}B$, ^{13}C),[3009] $Me_3BH_2CONR^1R^2$, (^{11}B),[3010] $MeC(CHPPh_2)_2BH_2$, (^{11}B, ^{13}C),[3011] $(EtO)_2(HO)PBH_2X$, (^{11}B, ^{13}C),[3012] $[HBEt_3]^-$, (^{11}B, ^{13}C),[3013] diisopinocamphenylborane, (^{11}B, ^{13}C),[3014] $(C_8H_{14}BH)_2$, (^{11}B, ^{13}C),[3015] $pyPr^iBH$-$C(O)NHEt$, (^{11}B, ^{13}C),[3016] $[Sn\{(pz)_3BH\}_2]$, (^{11}B, ^{13}C, ^{119}Sn),[3017] $[\{\eta^3\text{-}HB(3\text{-}Bu^tpzH)_3\}Cl]^+$, ($^{13}C$),[3018] $R_{3-n}SnX_n(3,5\text{-}Me_2C_3HN_2)_3BH$, ($^{13}C$, ^{119}Sn),[3019] and $HB(1,2\text{-}O_2\text{-}3\text{-}Me\text{-}6\text{-}Bu^tC_6H_2)$, ($^{13}C$).[3020]

^{11}B n.m.r. spectroscopy has been used to investigate the thermolysis of B_6H_{12}.[3021] ^{11}B and ^{13}C T_1 values have been determined for $[B_{12}H_{12}]^{2-}$, $[B_{10}H_{10}]^{2-}$, and $C_2B_{10}H_{12}$, and the quadrupole coupling

2999 S.M. Alexander, J.M. Coddington, and R.F. Howe, *Zeolites*, 1991, **11**, 368 (*Chem. Abstr.*, 1991, **115**, 16 086).
3000 C. Tsiao, C. Dybowski, A.M. Gaffney, and J.A. Sofranko, *J. Catal.*, 1991, **128**, 520 (*Chem. Abstr.*, 1991, **114**, 151 279).
3001 Q. Chen, J.B. Nagy, J. Fraissard, J. El Hage-Al Asswad, Z. Gabelica, E.G. Derouane, R. Aiello, F. Crea, G. Giordano, and A. Nastro, *NATO ASI Ser., Ser. B*, 1990, **221**, 87 (*Chem. Abstr.*, 1991, **114**, 220 006).
3002 P.J. Barrie, G.F. McCann, I. Gameson, T. Rayment, and J. Klinowski, *J. Phys. Chem.*, 1991, **95**, 9416; G. Fetter, D. Tichit, L.C. De Menorval, and F. Figueras, *Appl. Catal.*, 1990, **65**, L1 (*Chem. Abstr.*, 1991, **114**, 12 847).
3003 Q.J. Chen, J. Fraissard, H. Cauffriez, and J.L. Guth, *Zeolites*, 1991, **11**, 534 (*Chem. Abstr.*, 1991, **115**, 143 267); J.A. Martens, E. Feijen, J.L. Lievens, P.J. Grobet, and P.A. Jacobs, *J. Phys. Chem.*, 1991, **95**, 10025; M.J. Annen, M.E. Davis, and B.E. Hanson, *Catal. Lett.*, 1990, **6**, 331 (*Chem. Abstr.*, 1991, **115**, 99 939).
3004 M. Bühl, T. Steinke, P. von R. Schleyer, and R. Boese, *Angew. Chem., Int. Ed. Engl.*, 1991, **30**, 1160.
3005 D.K. Srivastava, L.K. Krannich, and C.L. Watkins, *Inorg. Chem.*, 1991, **30**, 2441.
3006 T.G. Hill, R.A. Godfroid, J.P. White, tert., and S.G. Shore, *Inorg. Chem.*, 1991, **30**, 2952.
3007 H. Schmidbaur, T. Wimmer, J. Lachmann, and G. Müller, *Chem. Ber.*, 1991, **124**, 275.
3008 N.E. Miller, *Inorg. Chem.*, 1991, **30**, 2228.
3009 M. Mittakanti and K.W. Morse, *Inorg. Chem.*, 1991, **30**, 2434.
3010 M.K. Das, P. Mukherjee, and S. Roy, *Bull. Chem. Soc. Jpn.*, 1990, **63**, 3658.
3011 H. Schmidbaur, S. Gamper, C. Paschalidis, O. Steigelmann, and G. Müller, *Chem. Ber.*, 1991, **124**, 1525.
3012 A. Sood, C.K. Sood, I.H. Hall, and B.F. Spielvogel, *Tetrahedron*, 1991, **47**, 6915.
3013 R. Köster, W. Schüßler, R. Boese, and D. Bläser, *Chem. Ber.*, 1991, **124**, 2259.
3014 H.C. Brown, K.S. Bhat, and P.K. Jadhav, *J. Chem. Soc., Perkin Trans. 1*, 1991, 2633.
3015 R. Köster and M. Yalpani, *Pure Appl. Chem.*, 1991, **63**, 387.
3016 W.J. Mills, C.H. Sutton, M.W. Baize, and L.J. Todd, *Inorg. Chem.*, 1991, **30**, 1046.
3017 M.N. Hansen, K. Niedenzu, J. Serwatowska, J. Serwatowski, and K.R. Woodrum, *Inorg. Chem.*, 1991, **30**, 866.
3018 A. Looney, G. Parkin, and A.L. Rheingold, *Inorg. Chem.*, 1991, **30**, 3099.
3019 G.G. Lobbia, F. Bonati, P. Cecchi, A. Lorenzotti, and C. Pettinari, *J. Organomet. Chem.*, 1991, **403**, 317.
3020 K. Burgess, W.A. van der Donk, and M.J. Ohlmeyer, *Tetrahedron Asymmetry*, 1991, **2**, 613.
3021 R. Greatrex, N.N. Greenwood, and S.D. Waterworth, *J. Chem. Soc., Dalton Trans.*, 1991, 643.

constants estimated.[3022] The high frequency ^{11}B n.m.r. shifts for B(10) in the *closo*-B_9H_9X series, the so-called antipodal effect, have been well reproduced by IGLO calculations.[3023] The nuclear quadrupole coupling constants of ^{10}B and ^{11}B in $B_{10}C_2H_{12}$ have been determined.[3024] The structures and molecular dynamics of *ortho*- and *meta*-carboranes have been studied by ^1H and ^{11}B n.m.r. spectroscopy.[3025] ^{11}B, ^{13}C, and ^{199}Hg n.m.r. spectroscopy has been used to show that 1,2-$[HgC_2B_{10}H_{12}]_4$ forms an adduct with LiCl.[3026] N.m.r. data have also been reported for $[H_3B\text{-}\mu_2\text{-}S(B_2H_5)]^-$, ($^{11}B$),[3027] $[\{Ph_2P(CH_2)_nPPh_2\}PdB_3H_7]$, ($^{11}B$),[3028] $[(dppe)PtB_3H_7]$, (^{11}B),[3029] $[Me_4N][B_3H_8]$, (^{11}B),[3030] $B_4H_8(\overline{SCH_2CH_2S})$, ($^{11}B$, ^{13}C),[3031] *conjuncto*-$[B_4H_8Fe_4(CO)_{12}]$, (^{11}B),[3032] $B_5H_8P(SiMe_3)\{CH(OSiMe_3)Bu^t\}$, ($^{11}B$, ^{13}C),[3033] $[\{Bu^t(Me_3SiO)HCPH\}B_5H_8]$, ($^{11}B$, ^{13}C),[3034] $[(\eta^5\text{-indenyl})Fe(Et_2C_2B_4H_4)Ni(Et_2MeC_3B_2Et_2)Co(\eta^5\text{-}C_5H_5)]$, ($^{11}B$, ^{13}C),[3035] $[(\eta^5\text{-}C_5H_5)(OC)_2Fe(B_5H_7PPh_2)]$, ($^{11}B$, ^{13}C),[3036] $[(OC)(Ph_3P)_2HOs(PMe_2Ph)ClPtB_5H_7]$, ($^{11}B$),[3037] $[(\eta^5\text{-}C_5Me_5)Co\{Et_2C_2B_3H_4C(OAc)=CH_2\}]$, ($^{11}B$, ^{13}C),[3038] $[(\eta^5\text{-}C_5Me_5)Co(R_2C_2B_3H_5)Co(R_2C_2B_3H_5)Co(\eta^5\text{-}C_5Me_5)]$, ($^{11}B$, ^{13}C),[3039] $[2,4\text{-}(Me_3Si)_2C_2B_4H_4]^{2-}$, ($^{11}B$, ^{13}C),[3040] $[YCl(THF)\{\eta^5\text{-}(Me_3Si)_2C_2B_4H_4\}_2]^{2-}$, ($^{11}B$, ^{13}C),[3041] [1,1'-*commo*-Cr{2,3-$(Me_3Si)_2$-2,3-$C_2B_4H_4\}_2$], (^{11}B),[3042] $[(\eta^5\text{-}C_4Me_4S)Fe(Et_2C_2B_4H_4)]$, ($^{11}B$),[3043] $[(B_6H_{10})Fe(CO)_4]$, (^{11}B),[3044] [CH_2\{*closo*-1-(η^5-C_5H_4)-Co(2,3-$Et_2C_2B_4H_4$)\}$_2$], (^{11}B),[3045] $[(\eta^5\text{-}C_5Me_5)Co\{(Me_3Si)C_2B_4H_5\}]$, ($^{11}B$),[3046] $[(Et_2C_2B_4H_4)Co(Me_4C_5CH_2Ph)]$, ($^{11}B$),[3047] *closo*-1-$Bu^t$-1-Ga-2,3-$(Me_3Si)_2$-2,3-$C_2B_4H_4$, ($^{11}B$, ^{13}C),[3048] 1-Pr^i-1-In-2,3-$(Me_3Si)_2$-2,3-$C_2B_4H_4$, (^{11}B, ^{13}C),[3049] 1-(2,4,6-$Bu^t_3C_6H_2$)-1-P-2,3-$(Me_3Si)_2$-2,3-

3022 Y.L. Pascal and O. Convert, *Magn. Reson. Chem.*, 1991, **29**, 308.
3023 M. Bühl, P. von R. Schleyer, Z. Havias, D. Hnyk, and S. Heřmánek, *Inorg. Chem.*, 1991, **30**, 3107.
3024 A. Loetz and J. Voitlaender, *J. Chem. Phys.*, 1991, **95**, 3208.
3025 E.C. Reynhardt and S. Froneman, *Mol. Phys.*, 1991, **74**, 61.
3026 X. Yang, C.B. Knobler, and M.F. Hawthorne, *Angew. Chem., Int. Ed. Engl.*, 1991, **30**, 1507.
3027 H. Binder, H. Loos, K. Dermentzis, H. Borrmann, and A. Simon, *Chem. Ber.*, 1991, **124**, 427.
3028 C.E. Housecroft, B.A.M. Shaykh, A.L. Rheingold, and B.S. Haggerty, *Inorg. Chem.*, 1991, **30**, 125.
3029 B.S. Haggerty, C.E. Housecroft, A.L. Rheingold, and B.A.M. Shaykh, *J. Chem. Soc., Dalton Trans.*, 1991, 2175.
3030 D. Yu, G. Yuan, and G. Zhang, *Fenxi Ceshi Tongbao*, 1990, **9**, 5 (*Chem. Abstr.*, 1991, **114**, 143 497).
3031 H. Binder, P. Melidis, S. Söylemez, and G. Heckmann, *J. Chem. Soc., Chem. Commun.*, 1991, 1091.
3032 C.S. Jun, X. Meng, K.J. Haller, and T.P. Fehlner, *J. Am. Chem. Soc.*, 1991, **113**, 3603.
3033 R.W. Miller, K.J. Donaghy, and J.T. Spencer, *Organometallics*, 1991, **10**, 1161.
3034 R.W. Miller, K.J. Donaghy, and J.T. Spencer, *Phosphorus, Sulphur*, 1991, **57**, 287.
3035 A. Fessenbecker, M. Stephan, R.N. Grimes, H. Pritzkow, U. Zenneck, and W. Siebert, *J. Am. Chem. Soc.*, 1991, **113**, 3061.
3036 B.H. Goodreau, R.L. Ostrander, and J.T. Spencer, *Inorg. Chem.*, 1991, **30**, 2066.
3037 J. Bould, J.E. Crook, N.N. Greenwood, and J.D. Kennedy, *J. Chem. Soc., Dalton Trans.*, 1991, 185.
3038 K.W. Piepgrass, J.H. Davis, jun., M. Sabat, and R.N. Grimes, *J. Am. Chem. Soc.*, 1991, **113**, 680.
3039 K.W. Piepgrass, J.H. Davis, jun., M. Sabat, and R.N. Grimes, *J. Am. Chem. Soc.*, 1991, **113**, 681.
3040 N.S. Hosmane, L. Jia, H. Zhang, J.W. Bausch, G.K.S. Prakash, R.E. Williams, and T.P. Onak, *Inorg. Chem.*, 1991, **30**, 3793.
3041 A.R. Oki, H. Zhang, and N.S. Hosmane, *Organometallics*, 1991, **10**, 3964.
3042 A.R. Oki, H. Zhang, J.A. Maguire, N.S. Hosmane, H. Ro, and W.E. Hatfield, *Organometallics*, 1991, **10**, 2996.
3043 K.J. Chase and R.N. Grimes, *Inorg. Chem.*, 1991, **30**, 3957.
3044 L. Barton and D.K. Srivastava, *Organometallics*, 1991, **10**, 2982.
3045 C.A. Plumb, P.J. Carroll, and L.G. Sneddon, *Inorg. Chem.*, 1991, **30**, 4678.
3046 M.A. Benvenuto and R.N. Grimes, *Inorg. Chem.*, 1991, **30**, 2836.
3047 J.H. Davis, jun., M.A. Benvenuto, and R.N. Grimes, *Inorg. Chem.*, 1991, **30**, 1765.
3048 N.S. Hosmane, K.J. Lu, H. Zhang, L. Jia, A.H. Cowley, and M.A. Mardones, *Organometallics*, 1991, **10**, 963.
3049 N.S. Hosmane, K.J. Lu, H. Zhang, A.H. Cowley, and M.A. Mardones, *Organometallics*, 1991, **10**, 392.

$C_2B_4H_4$, (^{11}B, ^{13}C),[3050] *hypho*-$C_3B_4H_{12}$, (^{11}B),[3051] [2-(η^6-C_6Me_6)-*closo*-2,1-$RuSB_8H_8$], (^{11}B),[3052] 6,9-CNB_8H_{13}, (^{11}B),[3053] [5-Cl-6-(η^6-C_6Me_6)-*nido*-6-RuB_9H_{12}], (^{11}B),[3054] *nido*-Et_{10}-2,4,6-$C_4B_{10}H_4$, (^{11}B),[3055] [6,6,6,6-$(OC)_3(Ph_3P)$-6-WB_9H_{13}], (^{11}B),[3056] *closo*-[{$(Ph_3P)_2RhH$}N-$B_{10}H_{11}$], (^{11}B),[3057] [RhCl(7,8-μ-SCH_2CH_2S-$C_2B_9H_{10}$)(7,8-μ-SCH_2CH_2S-$C_2B_9H_9$)], (^{11}B),[3058] [9-Cl-8-(η^5-C_5Me_5)-*nido*-8,7-$IrNB_9H_{11}$], (^{11}B),[3059] [9-Cl-8-(η^5-C_5Me_5)-*nido*-8,7-$IrSB_9H_{10}$], (^{11}B),[3060] [Pd(PPh_3)Cl(7-MeS-8-Me-11-Ph_2P-7,8-$C_2B_9H_{10}$)], (^{11}B),[3061] [10-*endo*-{$(Ph_3P)Au$}-7,8-*nido*-$C_2B_9H_9Me_2$]$^-$, (^{11}B),[3062] [*closo*-3,1,2-$TlC_2B_9H_{11}$]$^-$, (^{11}B),[3063] [$Tl_2B_9H_9C_2Me_2$], (^{11}B),[3064] [$(XC_6H_4)_nB_{12}H_{12-n}$]$^{2-}$, ($^{11}B$, ^{13}C),[3065] [$(X_2C_6H_3)_nB_{12}H_{12-n}$]$^{2-}$, ($^{11}B$, ^{13}C),[3066] [$(Cl_3C_6H_2)_nB_{12}H_{12-n}$]$^{2-}$, ($^{11}B$, ^{13}C),[3067] $NB_{11}H_{12}$, (^{11}B, ^{14}N),[3068] [$B_{11}H_{11}E$]$^-$, (E = As, Sb, Bi; ^{11}B),[3069] [*nido*-7,9-$C_2B_{10}H_{13}$]$^-$, (^{11}B),[3070] $C_6H_9C_2B_{10}H_{11}$, (^{13}C),[3071] 3-allyl-*o*-carborane, (^{11}B, ^{13}C),[3072] $B_{10}H_{10}C_2C_4H_6O$, (^{11}B, ^{13}C),[3073] 9-FSO_3-*o*- and *m*-carboranes, (^{11}B),[3074] 3-H_2N-7,8-dicarbaundecaborate, (^{11}B),[3075] [WCo(μ-CMe)$(CO)_3$(η^4-C_4Me_4)(η^5-$C_2B_9H_9Me_2$)], (^{11}B, ^{13}C),[3076] [(η^5-C_5H_5)MoW(μ-CC_6H_4Me-4)$(CO)_3${η^5-$C_2B_9H_8$($CH_2C_6H_4Me$-4)}Me_2], (^{11}B, ^{13}C),[3077] [W_2-Cu(μ-CC≡CBu^t)$_2$$(CO)_4$($\eta^5$-$C_2B_9H_9Me_2$)$_2$]$^-$, ($^{11}B$, ^{13}C),[3078] [MoW(μ-C_4Me_4){σ,η^5-CH(C_6H_4Me-

3050 N.S. Hosmane, K.J. Lu, A.H. Cowley, and M.A. Mardones, *Inorg. Chem.*, 1991, **30**, 1325.
3051 R. Greatrex, N.N. Greenwood, and M. Kirk, *J. Chem. Soc., Chem. Commun.*, 1991, 1510.
3052 M. Bown, X.L.R. Fontaine, N.N. Greenwood, and J.D. Kennedy, *Z. Anorg. Allg. Chem.*, 1991, **602**, 17.
3053 J. Holub, T. Jelínek, J. Plešek, B. Štíbr, S. Heřmánek, and J.D. Kennedy, *J. Chem. Soc., Chem. Commun.*, 1991, 1389.
3054 M. Bown, X.L.R. Fontaine, N.N. Greenwood, and J.D. Kennedy, *Z. Anorg. Allg. Chem.*, 1991, **598**, 45.
3055 R. Köster, G. Seidel, B. Wrackmeyer, D. Bläser, R. Boese, M. Bühl, and P. von R. Schleyer, *Chem. Ber.*, 1991, **124**, 2715.
3056 P.K. Baker, M.A. Beckett, and L.M. Severs, *Polyhedron*, 1991, **10**, 1663.
3057 H.-P. Hansen, J. Müller, U. Englert, and P. Paetzold, *Angew. Chem., Int. Ed. Engl.*, 1991, **30**, 1377.
3058 F. Teixidor, A. Romerosa, C. Viñas, J. Rius, C. Miravittles, and J. Casabó, *J. Chem. Soc., Chem. Commun.*, 1991, 192.
3059 K. Nestor, X.L.R. Fontaine, J.D. Kennedy, B. B. Štíbr, K. Baše, and M. Thornton-Pett, *Collect. Czech. Chem. Commun.*, 1991, **56**, 1607.
3060 K. Nestor, X.L.R. Fontaine, N.N. Greenwood, J.D. Kennedy, and M. Thornton-Pett, *J. Chem. Soc., Dalton Trans.*, 1991, 2657.
3061 F. Teixidor, J. Casabó, A.M. Romerosa, C. Viñas, J. Rius, and C. Miravitlles, *J. Am. Chem. Soc.*, 1991, **113**, 9895.
3062 J.A.K. Howard, J.C. Jeffery, P.A. Jelliss, T. Sommerfeld, and F.G.A. Stone, *J. Chem. Soc., Chem. Commun.*, 1991, 1664.
3063 M.J. Manning, C.B. Knobler, M.F. Hawthorne, and Y. Do, *Inorg. Chem.*, 1991, **30**, 3589.
3064 P. Jutzi, D. Wegener, and M.B. Hursthouse, *Chem. Ber.*, 1991, **124**, 295.
3065 W. Preetz and R. von Bismarck, *J. Organomet. Chem.*, 1991, **411**, 25.
3066 R. von Bismarck and W. Preetz, *J. Organomet. Chem.*, 1991, **418**, 147.
3067 R. von Bismarck and W. Preetz, *J. Organomet. Chem.*, 1991, **418**, 157.
3068 J. Müller, J. Runsink, and P. Paetzold, *Angew. Chem., Int. Ed. Engl.*, 1991, **30**, 175.
3069 A. Ouassas, C. R'Kha, H. Mongeot, and B. Frange, *Inorg. Chim. Acta*, 1991, **180**, 257.
3070 J. Plesek, B. Stíbr, X.L.R. Fontaine, J.D. Kennedy, S. Heřmánek, and T. Jelínek, *Collect. Czech. Chem. Commun.*, 1991, **56**, 1618.
3071 Q. Huang, H.L. Gingrich, and M. Jones, jun., *Inorg. Chem.*, 1991, **30**, 3254.
3072 J. Li, C.F. Logan, and M. Jones, jun., 1991, **30**, 4866.
3073 T. Ghosh, H.L. Gingrich, C.K. Kam, E.C. Mobraaten, and M. Jones, jun., *J. Am. Chem. Soc.*, 1991, **113**, 1313.
3074 V.N. Lebedev, M.V. Troitskaya, and L.I. Zakharkin, *Metalloorg. Khim.*, 1991, **4**, 956 (*Chem. Abstr.*, 1991, **115**, 208 053).
3075 L.I. Zakharkin, V.A. Ol'shevskaya, D.D. Sulaimankulova, and V.A. Antonivich, *Izv. Akad. Nauk SSSR, Ser. Khim.*, 1991, 1145 (*Chem. Abstr.*, 1991, **115**, 159 220).
3076 N. Carr, J.R. Fernandez, and F.G.A. Stone, *Organometallics*, 1991, **10**, 2718.
3077 S.A. Brew, M.D. Mortimer, and F.G.A. Stone, *J. Chem. Soc., Dalton Trans.*, 1991, 811.
3078 J.-L. Cabioch, S.J. Dossett, S. Hart, M.U. Pilotti, and F.G.A. Stone, *J. Chem. Soc., Dalton Trans.*, 1991, 519.

4)$C_2B_9H_8Me_2$}(η^7-C_7H_7)], (^{11}B, ^{13}C),[3079] [*closo*-3,3,3-$(OC)_3$-3-Ph_3Sn-3,1,2-$MC_2B_9H_{11}$]$^-$, (M = Cr, Mo, W; ^{11}B, ^{119}Sn),[3080] [WFe_2{μ_3-η^3,η^5-CH=C=CBu^{t}C(O)$C_2B_9H_8Me_2$}$(CO)_8$], (^{11}B, ^{13}C),[3081] [ReRh{μ-σ,η^5-$C_2B_9H_7(CH_2C_6H_4Me$-4)}Me_2(η^5-C_5H_4Me)], (^{11}B, ^{13}C),[3082] [*commo*-3,3'-Fe(8-Et_3N-3,1,2-$FeC_2B_9H_{10})_2$], (^{11}B),[3083] [*closo*-3,3-$(OC)_2$-3,1,2-$FeC_2B_9H_{11}$]$^{2-}$, (^{11}B),[3084] [*closo*-3-OC-3-L^1-3-L^2-3,1,2-$FeC_2B_9H_{11}$], {L^1,L^2 = CO, PPh_3, CH_3CN, $P(OMe)_3$; ^{11}B},[3085] [*closo*-3,3-$(Ph_3P)_2$-3-H-3-Cl-3,1,2-$RuC_2B_9H_{11}$], (^{11}B),[3086] [Rh(CO)(PPh_3)(η^5-$C_2B_9H_9Me_2$)]$^-$, (^{11}B, ^{13}C),[3087] [3-(η^5-C_5Me_5)-*closo*-3,1,2-$RhAs_2B_9H_9$], (^{11}B),[3088] [Pd{7,8-μ-[S(CH_2CH_2)S]-$C_2B_9H_{10}$}$_2$], (^{11}B),[3089] [3,3-$(R_3P)_2$-*closo*-3,1,2-$PtAs_2B_9H_9$], (^{11}B, ^{195}Pt),[3090] [Pt(Bu$^t_2$PCH_2CH_2-PBu$^t_2$)(*closo*-$CB_{11}H_{12}$)][$CB_{11}H_{12}$], (^{11}B, ^{13}C),[3091] [9,9-$(Ph_3P)_2$-*arachno*-9,5,6-$PtC_2B_7H_{11}$], (^{11}B),[3092] [Cu(1-$B_{10}H_9N_2)_2$]$^-$, (^{11}B),[3093] [3-Ph_3P-4-Me_2S-3,1,2-*closo*-$CuC_2B_9H_{10}$], (^{11}B),[3094] [$B_{12}H_{12-x}(HgO_2CCF_3)_x$]$^{2-}$, (^{11}B),[3095] [ClE($Me_2C_2B_9H_9$)$AlCl_3$], and [Sn($Me_2C_2B_9H_9$).THF], (E = As, P; ^{11}B, ^{13}C, ^{27}Al, ^{119}Sn).[3096]

Other Compounds of Boron.—The very high frequency shifts observed for ^{11}B resonances of compounds with a boron atom in an interstitial hole have been calculated by the Fenske-Hall methodology.[3097] (129) has $\delta(^{11}B)$ at 135.4, and the ^{13}C n.m.r. spectrum was also recorded.[3098] N.m.r. data have also been reported for (2,3,5,6-$Me_4C_6H)_2$BB=C{B(C_6HMe-2,3,5,6)}$_2$C$(SiMe_3)_2$, (^{11}B, ^{13}C),[3099] (130), (^{11}B, ^{13}C),[3100] $B_6(NEt_2)_6$, (^{11}B, ^{13}C),[3101] and $B_6X_nY_{6-n}$, (X ≠ Y = Cl, Br, I; ^{11}B).[3102]

The ^{11}B and ^{13}C chemical shifts of $B(CH_2CH=CH_2)_3$, $Me_2BC_3H_5$, and $F_2BC_3H_5$ have been compared with those calculated theoretically.[3103] N.m.r. data have also been reported for

3079 S.J. Dossett, I.J. Hart, M.U. Pilotti, and F.G.A. Stone, *J. Chem. Soc., Dalton Trans.*, 1991, 511.
3080 J. Kim, Y. Do, Y.S. Sohn, C.B. Knobler, and M.F. Hawthorne, *J. Organomet. Chem.*, 1991, **418**, C1.
3081 N. Carr, S.J. Dossett, and F.G.A. Stone, *J. Organomet. Chem.*, 1991, **413**, 223.
3082 M.U. Pilotti, F.G.A. Stone, and I. Topaloglu, *J. Chem. Soc., Dalton Trans.*, 1991, 1621.
3083 H.C. Kang, S.S. Lee, C.B. Knobler, and M.F. Hawthorne, *Inorg. Chem.*, 1991, **30**, 2024.
3084 S.S. Lee, C.B. Knobler, and M.F. Hawthorne, *Organometallics*, 1991, **10**, 1054.
3085 S. Lee, C.B. Knobler, and M.F. Hawthorne, *Organometallics*, 1991, **10**, 670.
3086 I.T. Chizhevskii, I.A. Lobanova, V.I. Bregadze, P.V. Petrovskii, A.V. Polyakov, A.I. Yanovskii, and Yu.T. Struchkov, *Metalloorg. Khim.*, 1991, **4**, 957 (*Chem. Abstr.*, 1991, **115**, 208 054).
3087 M.U. Pilotti, I. Topaloglu, and F.G.A. Stone, *J. Chem. Soc., Dalton Trans.*, 1991, 1355.
3088 X.L.R. Fontaine, J.D. Kennedy, M. McGrath, and T.R. Spalding, *Magn. Reson. Chem.*, 1991, **29**, 711.
3089 F. Teixidor, J. Casabó, C. Viñas, E. Sanchez, L. Escriche, and R. Kivekäs, *Inorg. Chem.*, 1991, **30**, 3053.
3090 M. McGrath, T.R. Spalding, X.L.R. Fontaine, J.D. Kennedy, and M. Thornton-Pett, *J. Chem. Soc., Dalton Trans.*, 1991, 3223.
3091 G.S. Mhinzi, S.A. Litster, A.D. Redhouse, and J.L. Spencer, *J. Chem. Soc., Dalton Trans.*, 1991, 2769.
3092 B. Štíbr, K. Baše, T. Jelínek, X.L.R. Fontaine, J.D. Kennedy, and M. Thornton-Pett, *Collect. Czech. Chem. Commun.*, 1991, **56**, 646.
3093 L.L. Ng, B.K. Ng, K. Shelly, C.B. Knobler, and M.F. Hawthorne, *Inorg. Chem.*, 1991, **30**, 4278.
3094 E.J.M. Hamilton and A.J. Welch, *Polyhedron*, 1991, **10**, 471.
3095 A.B. Yakushev, I.B. Sivaev, K.A. Solntsev, and N.T. Kuznetsov, *Koord. Khim.*, 1990, **16**, 867 (*Chem. Abstr.*, 1991, **114**, 54 714).
3096 P. Jutzi, D. Wegener, and M. Hursthouse, *J. Organomet. Chem.*, 1991, **418**, 277.
3097 R. Khattar, T.P. Fehlner, and P.T. Czech, *New J. Chem.*, 1991, **15**, 705.
3098 T. Mennekes, P. Paetzold, and R. Boese, *Angew. Chem., Int. Ed. Engl.*, 1991, **30**, 173.
3099 A. Höfner, B. Ziegler, R. Hunold, P. Willershausen, W. Massa, and A. Berndt, *Angew. Chem., Int. Ed. Engl.*, 1991, **30**, 594.
3100 D. Broom, U. Seebold, M. Noltemeyer, and A. Meller, *Chem. Ber.*, 1991, **124**, 2645.
3101 M. Baudler, K. Rockstein, and W. Oehlert, *Chem. Ber.*, 1991, **124**, 1149.
3102 W. Preetz and M. Stallbaum, *Z. Naturforsch., B*, 1990, **45**, 1113 (*Chem. Abstr.*, 1991, **114**, 74 171).
3103 M. Bühl, P. von R. Schleyer, M.A. Ibrahim, and T. Clark, *J. Am. Chem. Soc.*, 1991, **113**, 2466.

$Bu^tMe_2SiOCH(Pr^i)CHMeCH{=}CH(BEt_2)OC(O)NPr^i_2$, ($^{13}C$),[3104] (131), ($^{11}B$, ^{13}C, ^{119}Sn),[3105]

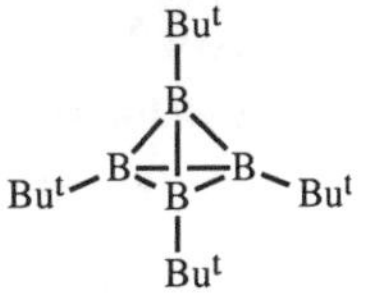
(129)

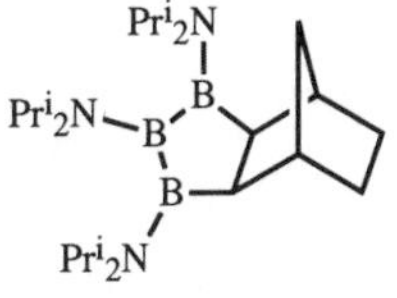
(130)

(132), (^{13}C),[3106] (133), (^{11}B, ^{13}C),[3107] 9-{2-(R_2NCH_2)C_6H_4}-9-borabicyclo[3.3.1]nonane,

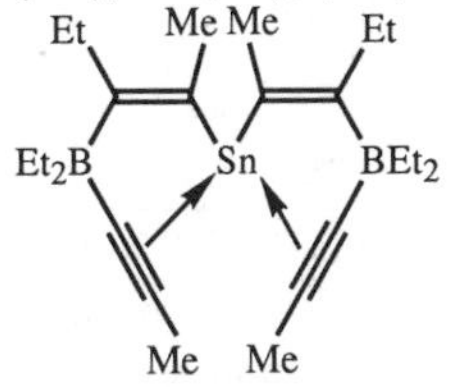
(131)

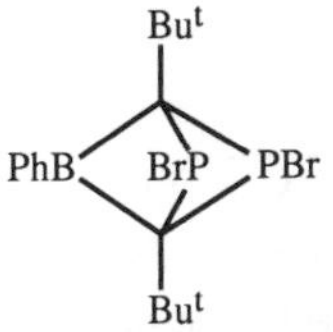
(132)

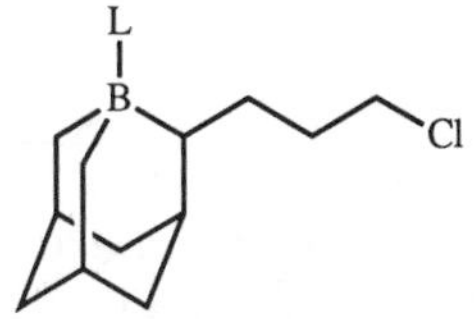
(133)

(^{13}C),[3108] $Me_2SnCMe{=}CEtBEt_2N{=}S{=}NSiMe_3$, ($^{11}B$, ^{13}C, ^{15}N, ^{29}Si, ^{119}Sn),[3109] $Et_2BCEt{=}CR^1$-$SnMe_2OR^2$, (^{11}B, ^{13}C, ^{29}Si, ^{119}Sn),[3110] $Et_2B(\mu\text{-}3,5\text{-}Pr^i_2C_3HN_2)_2BEt_2$, ($^{11}B$, ^{13}C),[3111] $R_2NB\{CH_2C(=CH_2)CH_2\}_2BNR_2$, ($^{11}B$),[3112] (134), ($^{11}B$, ^{13}C, ^{15}N, ^{29}Si),[3113] $R_2B(\mu\text{-pz})_2MEt_2$, $R_2B(\mu\text{-pz})_2SnMe_2Cl$, $R_2Si(\mu\text{-pz})_2$, (M = Al, Ga; ^{11}B, ^{13}C, ^{27}Al, ^{29}Si, ^{119}Sn),[3114] (135),

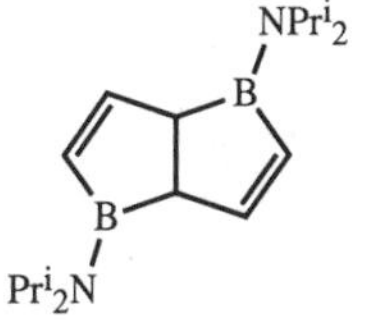
(134)

(135)

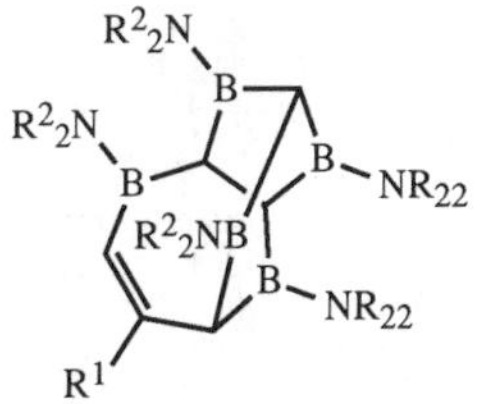
(136)

3104 S. Birkinshaw and P. Kocieński, *Tetrahedron Lett.*, 1991, **32**, 6961.
3105 B. Wrackmeyer, G. Kehr, and R. Boese, *Angew. Chem., Int. Ed. Engl.*, 1991, **30**, 1370.
3106 P. Binger, T. Wettling, R. Schneider, F. Zurmühlen, U. Bergsträsser, J. Hoffmann, G. Maas, and M. Regitz, *Angew. Chem., Int. Ed. Engl.*, 1991, **30**, 206.
3107 Yu.N. Bubnov, T.V. Potapova, and M.E. Gursky, *J. Organomet. Chem.*, 1991, **412**, 311.
3108 S. Toyota and M. Oki, *Bull. Chem. Soc. Jpn.*, 1991, **64**, 1554.
3109 B. Wrackmeyer, S.M. Frank, M. Herberhold, H. Borrmann, and A. Simon, *Chem. Ber.*, 1991, **124**, 691.
3110 B. Wrackmeyer, K. Wagner, A. Sebald, L.H. Merwin, and R. Boese, *Magn. Reson. Chem.*, 1991, **29**, S3.
3111 M. Yalpani, R. Köster, and R. Boese, *Chem. Ber.*, 1991, **124**, 1699.
3112 G.E. Herberich, U. Eigendorf, and C. Ganter, *J. Organomet. Chem.*, 1991, **402**, C17.
3113 A. Meller, C. Böker, U. Seebold, D. Bromm, W. Maringgele, A. Heine, R. Herbst-Irmer, E. Pohl, D. Stalke, M. Noltemeyer, and G.M. Sheldrick, *Chem. Ber.*, 1991, **124**, 1907.
3114 C.H. Dungan, W. Maringgele, A. Meller, K. Niedenzu, H. Nöth, J. Serwatowska, and J. Serwatowski, *Inorg. Chem.*, 1991, **30**, 4799.

(^{13}C),[3115] (136), (^{11}B, ^{13}C),[3116] $(Me_3Sn)_2CR^1CR^2{}_2BR^3NR^4{}_2$, ($^{11}B$, ^{13}C, ^{29}Si, ^{119}Sn),[3117] $[SnN\{B(C_6H_2Me_3\text{-}2,4,6)_2\}]_4$, ($^{11}B$),[3118] $(F_3C)_2BNMe_2C(O)NBu^t$, (^{11}B, ^{13}C),[3119] (137), (^{11}B, ^{13}C, ^{17}O, ^{29}Si),[3120] (138), (^{11}B, ^{13}C, ^{119}Sn),[3121] $(2,4,6\text{-}Me_3C_6H_2)_2B(1\text{-adamantyl})PPPh_2$, ($^{11}B$),[3122] and $EtBCEt_2CR^1(SnMe_3)SiMeR^2O$, ($^{11}B$, ^{13}C, ^{17}O, ^{29}Si, and ^{119}Sn).[3123]

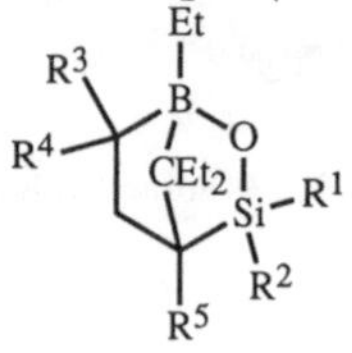

(137)

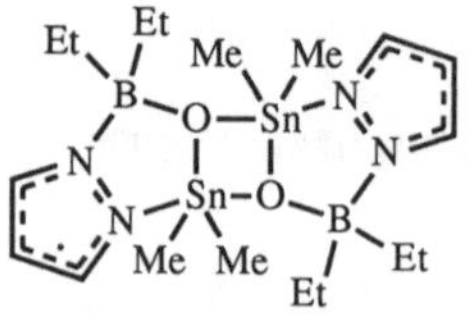

(138)

The conformation of $R^1BOCR^2R^3CH_2CMeR^4NH$[3124] and $Pr^iCH_2BOCHR^1CHMeCHR^2NH$[3125] has been investigated using ^1H and ^{13}C n.m.r. spectroscopy. ^1H n.m.r. spectroscopy has been used to show that the half chair of (139) is kept even at 125 °C, where it forms a nematic phase.[3126] The binding of aryl boronic acids to chymotrypsin and subtilisin has been studied by ^{11}B n.m.r. spectroscopy.[3127] A tetrahedral boronic acid-β-lactamase complex has been observed using ^{11}B n.m.r. spectroscopy.[3128] N.m.r. data have also been reported for $R^1BNR^2SNR^3$ (^{11}B, ^{13}C, ^{15}N, ^{29}Si),[3129] (140), (^{11}B, ^{13}C, ^{14}N),[3130] $(PrBNBu^t)_2$, (^{11}B),[3131] $MeBNRCH{=}CHNR$, (^{11}B, ^{13}C),[3132] α-boranyldiazomethane, (^{11}B),[3133] $Et_3B_3N_3H_2(SiMe_3)$, (^{11}B, ^{13}C),[3134] $BrEt_2B_3N_3Me_3$,

3115 U. Brand, S. Hünig, K. Peters, and H.G. Von Schnering, *Chem. Ber.*, 1991, **124**, 1187.
3116 W. Maringgele, U. Seebold, A. Heine, D. Stalke, M. Noltemeyer, G.M. Sheldrick, and A. Meller, *Organometallics*, 1991, **10**, 2097.
3117 B. Wrackmeyer and K. Wagner, *Chem. Ber.*, 1991, **124**, 503.
3118 H. Chen, R.A. Bartlett, H.V.R. Dias, M.M. Olmstead, and P.P. Power, *Inorg. Chem.*, 1991, **30**, 3390.
3119 A. Ansorge, D.J. Brauer, H. Bürger, F. Dörrenbach, T. Hagen, G. Pawelke, and W. Weuter, *J. Organomet. Chem.*, 1991, **407**, 283.
3120 R. Köster, G. Seidel, and G. Müller, *Chem. Ber.*, 1991, **124**, 1017.
3121 K. Niedenzu, H. Nöth, J. Serwatowska, and J. Serwatowski, *Inorg. Chem.*, 1991, **30**, 3249.
3122 D.C. Pestana and P.P. Power, *Inorg. Chem.*, 1991, **30**, 528.
3123 R. Köster, G. Seidel, and B. Wrackmeyer, *Chem. Ber.*, 1991, **124**, 1003.
3124 A.R. Kalyuskii, V.V. Kuznetsov, Yu.E. Brusilovskii, M.G. Glukhova, A.I. Gren, and V.Ya. Gorbatyuk, *Zh. Org. Khim.*, 1990, **26**, 2498 (*Chem. Abstr.*, 1991, **115**, 159 223).
3125 A.R. Kalyuskii, V.V. Kuznetsov, O.S. Timofeev, and A.I. Gren, *Zh. Obshch. Khim.*, 1990, **60**, 2093 (*Chem. Abstr.*, 1991, **114**, 42 852).
3126 H. Matsubara, T. Tanaka, Y. Takai, M. Sawada, K. Seto, H. Imazaki, and S. Takahashi, *Bull. Chem. Soc. Jpn.*, 1991, **64**, 2103.
3127 J.E. Baldwin, T.D.W. Claridge, A.E. Derome, C.J. Schofield, and B.D. Bradley, *Bioorg. Med. Chem. Lett.*, 1991, **1**, 9 (*Chem. Abstr.*, 1991, **115**, 44 939); S. Zhong, F. Jordan, C. Kettner, and L. Polgar, *J. Am. Chem. Soc.*, 1991, **113**, 9429.
3128 J.E. Baldwin, T.D.W. Claridge, A.E. Derome, B.D. Smith, M. Twyman, and S.G. Waley, *J. Chem. Soc., Chem. Commun.*, 1991, 573.
3129 C.D. Habben, A. Heine, G.M. Sheldrick, D. Stalke, M. Bühl, and P. von R. Schleyer, *Chem. Ber.*, 1991, **124**, 47.
3130 Yu.N. Bubnov, M.E. Gursky, and D.G. Pershin, *J. Organomet. Chem.*, 1991, **412**, 1.
3131 P. Paetzold, *Pure Appl. Chem.*, 1991, **63**, 345.
3132 G. Schmid, J. Lehr, M. Polk, and R. Boese, *Angew. Chem., Int. Ed. Engl.*, 1991, **30**, 1015.
3133 M.-P. Arthur, A. Baceiredo, and G. Bertrand, *J. Am. Chem. Soc.*, 1991, **113**, 5856.
3134 J. Bai, K. Niedenzu, J. Serwatowska, and J. Serwatowski, *Inorg. Chem.*, 1991, **30**, 4631.

(^{11}B, ^{13}C),[3135] $Pr^i{}_4B_4N_4(CH_2Ph)_4$, (^{11}B, ^{13}C),[3136] $B_n(NMe_2)_nR_2$, (^{11}B, ^{13}C),[3137] $Bu^tB{=}NBu^tP$-R, (^{11}B, ^{13}C),[3138] $R^1BOCH_2CHR^2CH_2NR^3$, (^{13}C),[3139] Me_2NBPh_2OBPhO, (^{11}B),[3140] (141), (^{11}B, ^{13}C),[3141] $XBR^1NR^2BR^3X$, (^{11}B, ^{13}C),[3142] $R^1B(QR^2)_2$, (^{11}B),[3143] (142), (^{13}C),[3144] $ICH_2BOCMe_2CMe_2O$, (^{11}B),[3145] $CH_2{=}CHCHXBOCMe_2CMe_2O$, ($^{13}C$),[3146] BuCH=CHB(1,2-$O_2C_6H_4$), ($^{11}B$, ^{13}C),[3147] $PhB(OSiBu^t{}_2O)BPh$, (^{11}B, ^{13}C, ^{29}Si),[3148] 2,5-$\{CH_3C(O)NH\}_2C_6H_3BOCMe_2CMe_2O$, ($^{11}B$),[3149] $RSB(O)_2BPh$, (^{11}B),[3150] $BSi_2O_3Ph_5$, (^{11}B, ^{13}C),[3151] SeSeCEt=CEtB(OR), (^{11}B, ^{13}C, ^{77}Se),[3152] and (143), (^{11}B, ^{13}C).[3153]

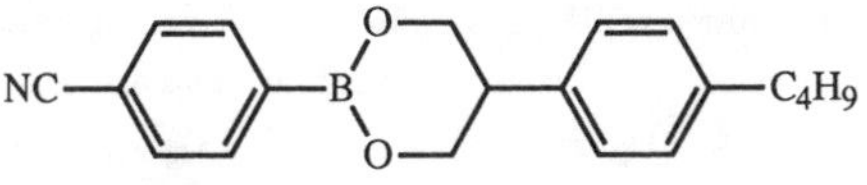

(139)

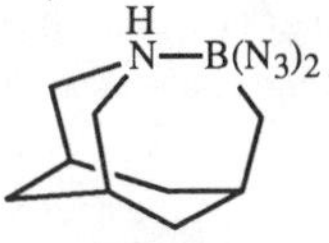

(140)

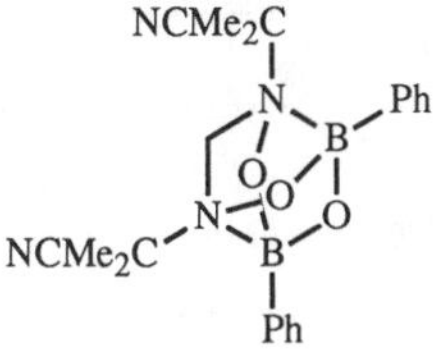

(141)

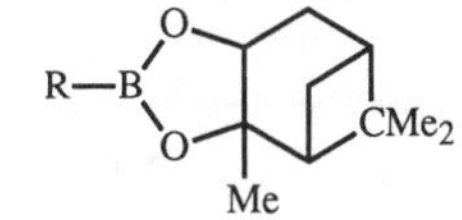

(142)

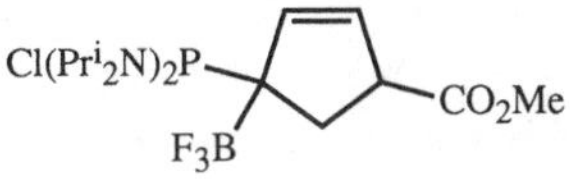

(143)

Boric acid esters in wine have been determined by ^{11}B n.m.r. spectroscopy.[3154] N.m.r. data have also been reported for $\{(Me_3Si)_2NBHNH_2\}_2NH$, ($^{11}B$),[3155] $MeN\{B[N(SiMe_3)_2]N{=}\}S$, ($^{11}B$, ^{13}C),[3156] $(Pr^i{}_2N)_2BN{=}C{=}NSiPr^i{}_3$, ($^{11}B$, ^{13}C, ^{29}Si),[3157] $MeP(O)\{OB(NHBu^t)(NC_5Me_4H_6)\}_2$,

3135 J. Bai and K. Niedenzu, *Inorg. Chem.*, 1991, **30**, 2955.
3136 B. Thiele, P. Schreyer, U. Englert, P. Paetzold, R. Boese, and B. Wrackmeyer, *Chem. Ber.*, 1991, **124**, 2209.
3137 H. Nöth and M. Wagner, *Chem. Ber.*, 1991, **124**, 1963.
3138 R. Streubel, E. Niecke, and P. Paetzold, *Chem. Ber.*, 1991, **124**, 765.
3139 A.R. Kalyuskii, V.V. Kuznetsov, Yu.E. Shapiro, S.A. Bochkor, and A.I. Gren, *Khim. Geterotsikl. Soedin*, 1990, 1424 (*Chem. Abstr.*, 1991, **114**, 122 473).
3140 W. Kliegel, U. Riebe, S.J. Rettig, and J. Trotter, *Can. J. Chem.*, 1991, **69**, 1222.
3141 W. Kliegel, G. Lubkowitz, S.J. Rettig, and J. Trotter, *Can. J. Chem.*, 1991, **69**, 1217.
3142 P. Paetzold, B. Redenz-Stormanns, and R. Boese, *Chem. Ber.*, 1991, **124**, 2435.
3143 M.V. Rangaishenvi, B. Singaram, and H.C. Brown, *J. Org. Chem.*, 1991, **56**, 3286.
3144 X. Wang, *J. Chem. Soc., Chem. Commun.*, 1991, 1515.
3145 A. Whiting, *Tetrahedron Lett.*, 1991, **32**, 1503.
3146 R.W. Hoffmann and J.J. Wolff, *Chem. Ber.*, 1991, **124**, 563.
3147 T.E. Cole, R. Quintanilla, and S. Rodewald, *Organometallics*, 1991, **10**, 3777.
3148 A. Mazzah, A. Haoudi-Mazzah, M. Noltemeyer, and H.W. Roesky, *Z. Anorg. Allg. Chem.*, 1991, **604**, 93.
3149 S.X. Cai and J.F.W. Keana, *Bioconjugate Chem.*, 1991, **2**, 317 (*Chem. Abstr.*, 1991, **115**, 183 402).
3150 V. Chaturvedi and J.P. Tandon, *Main Group Met. Chem.*, 1990, **13**, 259 (*Chem. Abstr.*, 1991, **114**, 228 995).
3151 D.A. Foucher, A.J. Lough, and I. Manners, *J. Organomet. Chem.*, 1991, **414**, C1.
3152 C.D. Habben and S. Pusch, *Z. Anorg. Allg. Chem.*, 1991, **593**, 217.
3153 J.-M. Sotiropoulos, A. Baceiredo, K.H. von Locquenghien, F. Dahan, and G. Bertrand, *Angew. Chem., Int. Ed. Engl.*, 1991, **30**, 1154.
3154 O. Lutz, E. Humpfer, and M. Spraul, *Naturwissenschaften*, 1991, **78**, 67 (*Chem. Abstr.*, 1991, **114**, 162 347).
3155 K.J.L. Paciorek, S.R. Masuda, L.A. Hoferkamp, J.H. Nakahara, and R.H. Kratzer, *Inorg. Chem.*, 1991, **30**, 577.
3156 C.D. Habben, *Z. Anorg. Allg. Chem.*, 1991, **606**, 229.
3157 M.-P. Arthur, H.P. Goodwin, A. Baceiredo, K.B. Dillon, and G. Bertrand, *Organometallics*, 1991, **10**, 3205.

(^{11}B, ^{13}C),[3158] $(OH)_2BNH_2CH_2CH_2C(O)O$, (^{11}B),[3159] (1,2-$C_6H_4O_2$)B(OPr^i)N(=CHPh)N-(=CHPh)B(OPr^i)(1,2-$O_2C_6H_4$), (^{11}B),[3160] $Me_2N(BFBF_2)_2NMe_2$, (^{11}B),[3161] 2-Pr^iNH-1,3-Pr^i_2-4,5-(Pr^iNH)$_2$-1,3,7-diazaborolidine, (^{11}B, ^{13}C),[3162] $Cl_2M(NMeCMe)_2CH$, (M = B, Al; ^{11}B, ^{13}C),[3163] (R^1O)H(O)POBOCH$_2$CHR^{2}CHR^{3}O, (^{11}B, ^{13}C),[3164] borate esters of carbohydrate oximes, (^{11}B, ^{13}C),[3165] $OCH_2CH_2CH_2OBSP(S)(OEt)_2$, ($^{11}B$, ^{13}C),[3166] BF_3 complexes with dimethylglyoxime and similar ligands, (^{11}B),[3167] and [Pr^i_2N][BF_4], (^{11}B, ^{13}C).[3168]

Complexes of Other Group IIIB Elements.—A characteristic U-shaped relationship between the ^{27}Al n.m.r. chemical shift and n has been found for $[Et_{3-n}AlX_n]_2$, together with a correlation between the available ^{27}Al nuclear quadrupole coupling constants and δ for four coordinate aluminium atoms.[3169] N.m.r. data have also been reported for [$H_2Al\{MeN(CH_2CH_2NMe_2)_2\}$][$AlH_4$], ($^{13}C$, ^{27}Al),[3170] $[H_3Al(NMe_2CH_2Ph)]_2$, (^{13}C),[3171] [$(H_3Ga)_2$(tmeda)], (^{13}C),[3172] [$H_5Ga_3\{(NMe\text{-}CH_2)_2\}_2$], ($^{13}C$),[3173] $[Me_2AlC(SiMe_3)_2PMe_2]_2$, ($^{13}C$, ^{27}Al),[3174] [$Me_2AlC(SiMe_3)_2PMe_3LiCl$], ($^7Li$, ^{13}C, ^{27}Al),[3175] [$C_5H_{10}MCH_2CH_2CH_2NMe_2$], (M = Al, Ga; ^{13}C),[3176] (144), (^{13}C),[3177] $[Al_{12}Bu^i_{12}]^{2-}$, (^{13}C),[3178] (145), (X = O, S; ^{13}C),[3179] [$\{(Me_3Si)_2CH(Cl)Al\}_2CH_2$], ($^{13}C$),[3180] [$Al(\eta^5\text{-}C_5Me_5)]_4$, ($^{27}Al$),[3181] [$(Bu^tCH_2)_2MCH_2PPh_2$], (M = In, Ga; ^{13}C),[3182] (146), (M = Ga, In;

3158 H. Nöth and J. Schübel, *Chem. Ber.*, 1991, **124**, 1687.
3159 M.K. Das, N.N. Bandyopadhyay, and S. Roy, *Synth. React. Inorg. Metal-Org. Chem.*, 1991, **21**, 931.
3160 V. Chaturvedi, L. Chaturvedi, and J.P. Tandon, *Main Group Met. Chem.*, 1990, **13**, 253 (*Chem. Abstr.*, 1991, **114**, 228 994).
3161 W. Haubold, U. Kraatz, and W. Einholz, *Z. Anorg. Allg. Chem.*, 1991, **592**, 35.
3162 W. Maringgele, D. Stalke, A. Heine, and A. Meller, *Eur. J. Solid State Inorg. Chem.*, 1991, **28**, 867 (*Chem. Abstr.*, 1991, **115**, 246 729).
3163 N. Kuhn, A. Kuhn, J. Lewandowski, and M. Speis, *Chem. Ber.*, 1991, **124**, 2197.
3164 O.P. Singh, R.K. Mehrotra, and G. Srivastava, *Synth. React. Inorg. Metal-Org. Chem.*, 1991, **21**, 717.
3165 J. Van Haveren, M.H.B. Van den Burg, J.A. Peters, J.G. Batelaan, A.P.G. Kieboom, and H. Van Bekkum, *J. Chem. Soc., Perkin Trans., 2*, 1991, 321.
3166 O.P. Singh, R.K. Mehrotra, and G. Srivastava, *Phosphorus, Sulphur*, 1991, **60**, 147.
3167 K. Dey, A. Gangopadhyay, and A.K. Biswas, *Proc. Natl. Acad. Sci., India, Sect. A*, 1989, **59**, 5 (*Chem. Abstr.*, 1991, **114**, 207 312).
3168 B. Neumüller and F. Gahlmann, *J. Organomet. Chem.*, 1991, **414**, 271.
3169 Z. Cerny, J. Machacek, J. Fusek, S. Hermanek, O. Kriz, and B. Casensky, *J. Organomet. Chem.*, 1991, **402**, 139.
3170 J.L. Atwood, K.D. Robinson, C. Jones, and C.L. Raston, *J. Chem. Soc., Chem. Commun.*, 1991, 1697.
3171 J.L. Atwood, F.R. Bennett, F.M. Elms, C. Jones, C.L. Raston, and K.D. Robinson, *J. Am. Chem. Soc.*, 1991, **113**, 8183.
3172 J.L. Atwood, S.G. Bott, F.M. Elms, C. Jones, and C.L. Raston, *Inorg. Chem.*, 1991, **30**, 3792.
3173 J.L. Atwood, S.G. Bott, C. Jones, and C.L. Raston, *Inorg. Chem.*, 1991, **30**, 4868.
3174 H.H. Karsch, K. Zellner, J. Lachmann, and G. Müller, *J. Organomet. Chem.*, 1991, **409**, 109.
3175 H.H. Karsch, K. Zellner, and G. Müller, *J. Chem. Soc., Chem. Commun.*, 1991, 466.
3176 H. Schumann, U. Hartmann, W. Wassermann, O. Just, A. Dietrich, L. Pohl, M. Hostalek, and M. Lokai, *Chem. Ber.*, 1991, **124**, 1113.
3177 W. Uhl, M. Layh, and W. Massa, *Chem. Ber.*, 1991, **124**, 1511.
3178 W. Hiller, K.-W. Klinkhammer, W. Uhl, and J. Wagner, *Angew. Chem., Int. Ed. Engl.*, 1991, **30**, 179.
3179 U.M. Dzhemilev, A.G. Ibragimov, A.P. Zolotarev, R.R. Muslukhov, and G.A. Tolstikov, *Izv. Akad. Nauk SSSR, Ser. Khim.*, 1990, 2831 (*Chem. Abstr.*, 1991, **114**, 164 318).
3180 W. Uhl and M. Layh, *J. Organomet. Chem.*, 1991, **415**, 181.
3181 C. Dohmeier, C. Robl, M. Tacke, and H. Schnöckel, *Angew. Chem., Int. Ed. Engl.*, 1991, **30**, 564.
3182 O.T. Beachley, jun., M.A. Banks, M.R. Churchill, W.G. Feighery, and J.C. Fettinger, *Organometallics*, 1991, **10**, 3036.

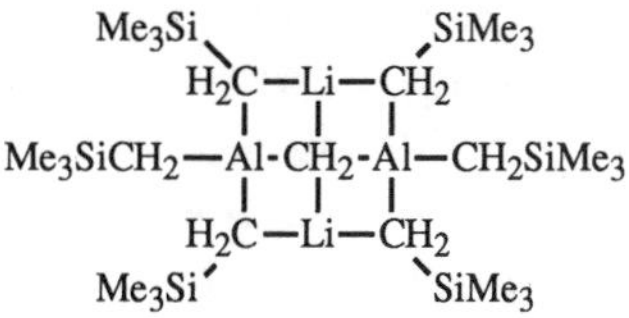

(144)

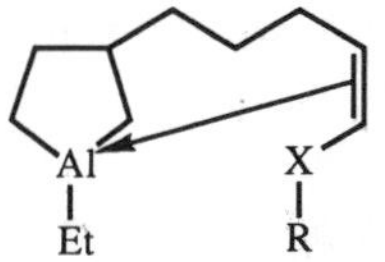

(145)

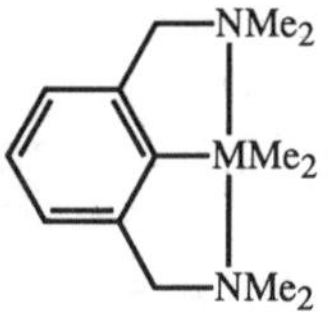

(146)

^{13}C),[3183] (147), (^{13}C, ^{31}P),[3184] (148), (^{29}Si),[3185] [$(PhCH_2)_nInCl_{3-n}$], (^{13}C),[3186] [$(\eta^5$-2,4-$Bu^t_2C_5H_2)_2Tl_2$], (^{13}C),[3187] [$MeAlNR]_3$, (149), (^{13}C, ^{27}Al),[3188] [$R^1_2M(NR^2)_2XR^1$], (M = Al, Ga,

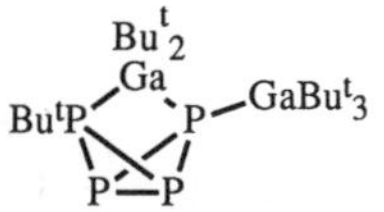

(147)

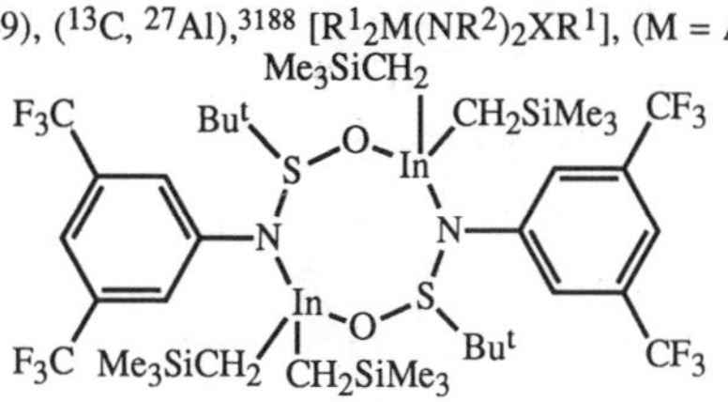

(148)

In, Tl; ^{13}C),[3189] [$Me_2Al(OC_6H_2$-2,6-Bu^t_2-4-Me)(NH_3)], (^{13}C, ^{14}N),[3190] [Me_2-AlO(MeO)C=C(H)N(Bu^t)Me], (^{13}C),[3191] [$R_2Al(OC_6H_2Bu^t_2$-2,6-Me-4)L], (^{13}C),[3192] [Me(Cl)-$AlPMe_2$=C($SiMe_3$)$SiMe_2PMe_2C(SiMe_3)_2$], (^{13}C, ^{27}Al),[3193] [$Me_4Al_2O]_3$, (^{27}Al),[3194] [R_nAl-$(OAr)_{3-n}]_m$, (^{27}Al),[3195] [$R_2M(\mu$-$OC_6H_4EMe)]_2$, (M = Al, Ga; ^{13}C, ^{27}Al),[3196] [$Me_2Al(OSiMe_3)]_2$, (^{13}C, ^{29}Si),[3197] [RAl{HN(CH_2)$_2$NHAlR_2}$_2$], (^{13}C, ^{27}Al),[3198] [$Me_2MNC_4H_4$], (M = Ga, In; ^{13}C),[3199] [$(Me_3SiCH_2)_2GaAs(SiMe_3)_2Ga(CH_2SiMe_3)_2Cl$], (^{13}C),[3200] [$Bu^t_2GaPH_2]_3$, (^{13}C),[3201]

3183 H. Schumann, U. Hartmann, and W. Wassermann, *Chem. Ber.*, 1991, **124**, 1567.
3184 M.P. Power and A.R. Barron, *Angew. Chem., Int. Ed. Engl.*, 1991, **30**, 1153.
3185 U. Dembowski, M. Noltemeyer, J.W. Gilje, and H.W. Roesky, *Chem. Ber.*, 1991, **124**, 1917.
3186 B. Neumüller, *Z. Anorg. Allg. Chem.*, 1991, **592**, 42.
3187 P. Jutzi, J. Schnittger, and M.B. Hursthouse, *Chem. Ber.*, 1991, **124**, 1693.
3188 K.M. Waggoner and P.P. Power, *J. Am. Chem. Soc.*, 1991, **113**, 3385.
3189 D. Kottmair-Maieron, R. Lechler, and J. Weidlein, *Z. Anorg. Allg. Chem.*, 1991, **593**, 111.
3190 M.D. Healy, J.T. Leman, and A.R. Barron, *J. Am. Chem. Soc.*, 1991, **113**, 2776.
3191 F.H. van der Steen, G.P.M. van Mier, A.L. Spek, J. Kroon, and G. van Koten, *J. Am. Chem. Soc.*, 1991, **113**, 5742.
3192 M.D. Healy, J.W. Ziller, and A.R. Barron, *Organometallics*, 1991, **10**, 597.
3193 H.H. Karsch, K. Zellner, and G. Müller, *Organometallics*, 1991, **10**, 2884.
3194 N.N. Korneev, V.S. Kolesov, N.I. Ivanova, A.V. Polonskii, I.M. Khrapova, A.V. Kisin, and N.N. Govorov, *Metalloorg. Khim.*, 1991, **4**, 595 (*Chem. Abstr.*, 1991, **115**, 92 346).
3195 R. Benn, E. Janssen, H. Lehmkuhl, A. Rufińska, K. Angermund, P. Betz, R. Goddard, and C. Krüger, *J. Organomet. Chem.*, 1991, **411**, 37.
3196 D.G. Hendershot, M. Barber, R. Kumar, and J.P. Oliver, *Organometallics*, 1991, **10**, 3302.
3197 R. Mulhaupt, J. Calabrese, and S.D. Ittel, *Organometallics*, 1991, **10**, 3403.
3198 Z. Jiang, L.V. Interrante, D. Kwon, F.S. Tham, and R. Kullnig, *Inorg. Chem.*, 1991, **30**, 995; Z. Jiang, L.V. Interrante, D. Kwon, F.S. Tham, and R. Kullnig, *Report*, 1990, **TR-12**; Order No. AD-A225758, 38 pp. Avail. NTIS. From *Gov. Rep. Announce. Index (U.S.)*, 1991, **91**, Abstr. No. 115,457 (*Chem. Abstr.*, 1991, **115**, 280 096).
3199 K. Locke, J. Weidlein, F. Scholz, N. Bouanah, N. Brianese, P. Zanella, and Y. Gao, *J. Organomet. Chem.*, 1991, **420**, 1.
3200 R.L. Wells, J.W. Pasterczyk, A.T. McPhail, J.D. Johansen, and A. Alvanipour, *J. Organomet. Chem.*, 1991, **407**, 17.
3201 A.H. Cowley, P.R. Harris, R.A. Jones, and C.M. Nunn, *Organometallics*, 1991, **10**, 652.

$[Bu^t_2Ga(\mu\text{-}SH)]_2$, ($^{13}C$),[3202] $[\{Cl(GaPh_2)_2As(SiMe_3)CH_2\}_2CH_2]$, ($^{13}C$),[3203] $[(Me_3SiCH_2)_2InCl]$, (^{13}C),[3204] $[(Me_3SiCH_2)_2InOMe]$, (^{13}C),[3205] $[(Me_3SiCH_2)_2InN(SO_2Me)_2]$, ($^{13}C$),[3206] $[(Me_3Si\text{-}CH_2)(Bu^tCH_2)InCl]$, ($^{13}C$),[3207] $[(2,4,6\text{-}Me_3C_6H_2)_2InBr]$, ($^{13}C$),[3208] $[Me_2Tl$(2-thioorotato)], (^{13}C),[3209] $[(Me_3SiCH_2)_2TlN(SO_2Me)_2]$, ($^{13}C$),[3210] (150), ($^{13}C$),[3211] $[Bu^tGa(PHC_6H_2Bu^t_3\text{-}2,4,6)_2]$, ($^{13}C$),[3212] (151), ($^{13}C$),[3213] $[Bu^tGa(\mu_3\text{-}S)]_4$, ($^{13}C$),[3214] $[(2,4,6\text{-}Ph_3C_6H_2)GaPCy]_3$, ($^{13}C$),[3215] and $[X_2InCH_2X]$, (^{13}C).[3216]

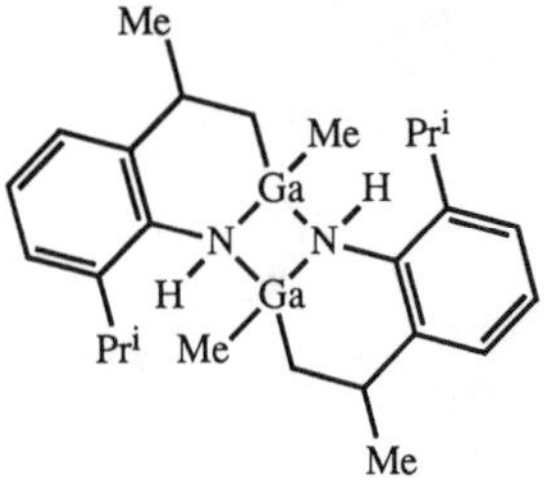

(149)

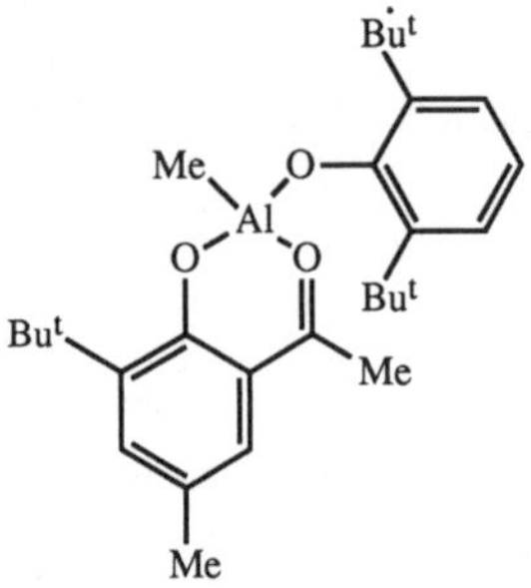

(150)

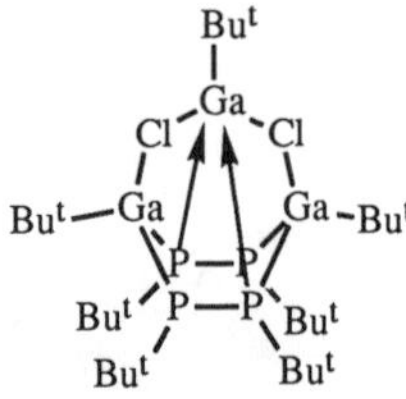

(151)

The solution conformation of the Al^{3+} and Ga^{3+} complexes of ferrichrome and deferriferrichrome has been determined using ^{1}H n.O.e. measurements.[3217] $NH_4Al(SO_4)_2$ has been proposed as a water T_2 reduction agent for use with the C.P.M.G. pulse sequence for water suppression.[3218] The structural state of Al^{3+} in Al_2O_3 containing liquid phases has been characterised using ^{27}Al n.m.r. spectroscopy.[3219] ^{27}Al and ^{29}Si n.m.r. spectroscopy has been used to investigate oligiomers from the $(Pr^iO)_3Al\text{-}Al(NO_3)_3\text{-}Si(OEt)_4\text{-}H_2O$ system.[3220] $[Tl(SCN)]^{2+}$ has been characterised by ^{13}C, ^{14}N, ^{15}N, and ^{205}Tl n.m.r. spectroscopy. $^{2}J(^{205}Tl\text{-}^{13}C)$ is 733 Hz and $^{3}J(^{205}Tl\text{-}^{15}N)$ is 143 Hz, showing

3202 M.B. Power and A.R. Barron, *J. Chem. Soc., Chem. Commun.*, 1991, 1315.
3203 W.K.Holley, J.W. Pasterczyk, C.G. Pitt, and R.L. Wells, *Heteroat. Chem.*, 1990, **1**, 475.
3204 R.L. Wells, L.J. Jones, A.T. McPhail, and A. Alvanipour, *Organometallics*, 1991, **10**, 2345.
3205 T. Douglas and K.H. Theopold, *Inorg. Chem.*, 1991, **30**, 594.
3206 A. Blaschette, A. Michalides, and P.G. Jones, *J. Organomet. Chem.*, 1991, **411**, 57.
3207 O.T. Beachley, jun., J.D. Maloney, M.R. Churchill, and C.H. Lake, *Organometallics*, 1991, **10**, 3568.
3208 J.T. Leman, J.W. Ziller, and A.R. Barron, *Organometallics*, 1991, **10**, 1766.
3209 M.S. Garcia-Tasende, B.E. Rivero, A. Castiñeiras, A. Sanchez, J.S. Casas, J. Sordo, W. Hiller, and J. Strähle, *Inorg. Chim. Acta*, 1991, **181**, 43.
3210 A. Blaschette, P.G. Jones, A. Michalides, and M. Näveke, *J. Organomet. Chem.*, 1991, **415**, 25.
3211 M.B. Power, S.G. Bott, E.J. Bishop, K.D. Tierce, J.L. Atwood, and A.R. Barron, *J. Chem. Soc., Dalton Trans.*, 1991, 241.
3212 D.A. Atwood, A.H. Cowley, R.A. Jones, and M.A. Mardones, *J. Am. Chem. Soc.*, 1991, **113**, 7050.
3213 A.H. Cowley, R.A. Jones, M.A. Mardones, J.L. Atwood, and S.G. Bott, *Angew. Chem., Int. Ed. Engl.*, 1991, **30**, 1141.
3214 A.H. Cowlet, R.A. Jones, P.R. Harris, D.A. Atwood, L. Contreras, and C.J. Burek, *Angew. Chem., Int. Ed. Engl.*, 1991, **30**, 1143.
3215 H. Hope, D.C. Pestana, and P.P. Power, *Angew. Chem., Int. Ed. Engl.*, 1991, **30**, 691.
3216 T.A. Annan, D.G. Tuck, M.A. Khan, and C. Peppe, *Organometallics*, 1991, **10**, 2159.
3217 K.L. Constantine, A. De Marco, M. Madrid, C.L. Brooks, tert., and M. Llinas, *Biopolymers*, 1990, **30**, 239 (*Chem. Abstr.*, 1991, **114**, 2383).
3218 G. Huang and X. Li, *Fenxi Ceshi Tongbao*, 1990, **9**, 66 (*Chem. Abstr.*, 1991, **114**, 176 895).
3219 D. Massiot, F. Taulelle, and J.P. Coutures, *Colloq. Phys.*, 1990, C5-425/C5-431 (*Chem. Abstr.*, 1991, **114**, 28 667).
3220 T. Nishio and Y. Fujiki, *Nippon Seramikkusu Kyokai Gakujutsu Ronbunshi*, 1991, **99**, 654 (*Chem. Abstr.*, 1991, **115**, 164 742).

that sulphur is coordinated to thallium.[3221] N.m.r. data have also been reported for $[PCl(NPr^i_2)_2NPhAlCl_3]$, ($^{13}C$, ^{27}Al),[3222] aluminium porphyrins, (^{27}Al),[3223] $[\{Al(salen)\}_2(\mu\text{-}O)]$, ($^{27}Al$),[3224] *S*-methylisopropylidenehydrazinecarbodithioate, (^{13}C, ^{27}Al),[3225] $[Me_2Si(NMe_2)_2GaCl_3]$, (^{13}C, ^{29}Si),[3226] $[C_6H_{10}(CH_2NHCH_2CEt_2S)_2GaCl]$, ($^{13}C$),[3227] [Al(1-R-3-O-2-Me-4-pyridinone)$_3$], (^{27}Al),[3228] $[SO_4Al_{12}(OH)_{24}(OH_2)_{12}]^{10+}$, ($^{27}Al$),[3229] $[Me_3SnOAl(OPr^i)_2]$, (^{27}Al, ^{119}Sn),[3230] $[Al(OPr^i)_{3-n}(OSiMe_3)_n]$, ($^{27}Al$),[3231] $[Al(OSiMe_3)_4]^-$, (^{13}C, ^{27}Al),[3232] $[M^1(OPCl_2NPCl_2O)_3]$, $[Me_3M^2N\{P(O)Cl_2\}_2]$, (M^1 = Al, Ga, In; M^2 = Si, Sn; ^{13}C, ^{27}Al, ^{29}Si, ^{119}Sn),[3233] [(*trans*-1-$Me_3SiOC_6H_{10}$-2-O)$AlCl_2$], (^{13}C, ^{27}Al),[3234] $[AlBr_3(BuOH)_n]$, (^{13}C, ^{27}Al),[3235] [Tl(1,4,7-trithiacyclononane)]$^+$, (^{13}C),[3236] $[Bu^t_2P{=}PHSiMe_3][AlCl_4]$, ($^{13}C$, ^{27}Al, ^{29}Si),[3237] and TlF, (^{205}Tl).[3238]

7 Group IVB Elements

Two reviews have appeared:- '^{29}Si n.m.r. spectroscopy of trimethylsilyl tags',[3239] and 'Possibilities for studying polar petroleum fractions by means of ^{29}Si n.m.r. spectroscopy'.[3240]

Methods of suppression of acoustic ringing have been examined and applied to base-line roll effects in ^{73}Ge n.m.r. spectroscopy.[3241] The application of Hahn spin-echo extended X-^1H shift correlated (X = ^{13}C, ^{29}Si, ^{119}Sn, ^{207}Pb) two-dimensional n.m.r. spectroscopy has been described. This allows the complete suppression of the central line leaving solely the ^{15}N satellites and leading both to accurate measurement of long-range ^{15}N-^1H coupling constants and relative signs of $J(^{15}N\text{-}^1H)$ and $J(^{15}N\text{-}X)$, and $^{15}N/^{14}N$ isotope effects on ^{119}Sn and ^{207}Pb chemicals shifts.[3242]

3221 J. Blixt, R.K. Dubey, and J. Glaser, *Inorg. Chem.*, 1991, **30**, 2824.
3222 N. Burford, R.E.v.H. Spence, and J.F. Richardson, *J. Chem. Soc., Dalton Trans.*, 1991, 1615.
3223 J.M. DeSimone, M. Stangle, J.S. Riffle, and J.E. McGrath, *Makromol. Chem., Macromol. Symp.*, 1991, **42**, 373 (*Chem. Abstr.*, 1991, **115**, 208 655).
3224 P.L. Gurian, L.K. Cheatham, J.W. Ziller, and A.R. Barron, *J. Chem. Soc., Dalton Trans.*, 1991, 1449.
3225 A.R. Barron and G. Davies, *Heteroat. Chem.*, 1990, **1**, 291 (*Chem. Abstr.*, 1991, **114**, 24 003).
3226 W.R. Nutt, J.S. Blanton, A.M. Boccanfuso, L.A. Silks, tert., A.R. Garber, and J.D. Odom, *Inorg. Chem.*, 1991, **30**, 4136.
3227 L.C. Francesconi, B.L. Liu, J.J. Billings, P.J. Carroll, G. Graczyk, and H.F. Kung, *J. Chem. Soc., Chem. Commun.*, 1991, 94.
3228 L. Simpson, S.J. Rettig, J. Trotter, and C. Orvig, *Can. J. Chem.*, 1991, **69**, 893.
3229 W. He, S. Lin, and Z. Li, *Shuichuli Jishu*, 1989, **15**, 228 (*Chem. Abstr.*, 1991, **114**, 213 589).
3230 P.N. Kapoor, A.K. Bhagi, R.N. Kapoor, and H.K. Sharma, *J. Organomet. Chem.*, 1991, **420**, 321.
3231 K. Folting, W.E. Streib, K.G. Caulton, O. Pomelet, and L.G. Hubert-Pfalzgraf, *Polyhedron*, 1991, **10**, 1639.
3232 M.H. Chisholm, J.C. Huffman, and J.L. Wesemann, *Polyhedron*, 1991, **10**, 1367.
3233 A. Mazzah, H.-J. Gosink, J. Liebermann, and H.W. Roesky, *Chem. Ber.*, 1991, **124**, 753.
3234 V. Sharma, M. Simard, and J.D. Wuest, *Inorg. Chem.*, 1991, **30**, 579; F. Belanger-Gariepy, K. Hoogsteen, V. Sharma, and J.D. Wuest, *Inorg. Chem.*, 1991, **30**, 4140.
3235 M.V. Dzhulai, Yu.S. Bogachev, R.R. Shifrina, V.G. Khutsishvili, N.N. Shapet'ko, and E.N. Gur'yanova, *Zh. Obshch. Khim.*, 1990, **60**, 2322 (*Chem. Abstr.*, 1991, **115**, 40 748).
3236 A.J. Blake, J.A. Greig, and M. Schröder, *J. Chem. Soc., Dalton Trans.*, 1991, 529.
3237 H. Grützmacher and H. Pritzkow, *Angew. Chem., Int. Ed. Engl.*, 1991, **30**, 709.
3238 D. Cho, K. Sangster, and E.A. Hinds, *Phys. Rev. A*, 1991, **44**, 2783 (*Chem. Abstr.*, 1991, **115**, 215 059).
3239 J. Scraml, *Prog. Nucl. Magn. Reson. Spectrosc.*, 1990, **22**, 289 (*Chem. Abstr.*, 1991, **114**, 100 674).
3240 V. Chvalovsky, J. Schraml, and V. Blechta, *God. Sofii. Univ. 'Kliment Okhridski', Khim Fak.*, 1985, (Pub. 1990), **79**, 197 (*Chem. Abstr.*, 1991, **114**, 9196).
3241 A.L. Wilkins, R.A. Thomson, and K.M. Mackay, *Main Group Met. Chem.*, 1990, **13**, 219 (*Chem. Abstr.*, 1991, **115**, 104 460).
3242 E. Kupče and B. Wrackmeyer, *Magn. Reson. Chem.*, 1991, **29**, 351.

Ab initio Hartree-Foch calculations have been used to estimate the ^{13}C shielding tensor for C_{60}.[3243] Theoretical calculations on the STO-3G basis have been used to interpret the ^{13}C n.m.r. spectrum of C_{70}.[3244] ^{13}C n.m.r. data have also been reported for some fullerenes.[3245]

For compounds such as $H_2Si(Cl)SH$, the ^1H and ^{29}Si chemical shifts and coupling constants have been compared with related compounds.[3246] N.m.r. data have also been reported for $(Me_3SiCH_2)_nSiH_{4-n}$, (^{13}C, ^{29}Si),[3247] $(Me_3Si)_3SiSiH_3$, (^{13}C, ^{29}Si),[3248] $(H_2SiCH_2CH_2)_n$, (^{13}C, ^{29}Si),[3249] $EtSiH_2GeH_3$, (^{29}Si),[3250] (152), (^{13}C, ^{29}Si),[3251] $R(PhC{\equiv}C)SiH_2$, (^{13}C, ^{29}Si),[3252] $(PhSiH_2)_3CMe$, (^{13}C, ^{29}Si),[3253] $(CHF_2)SiH_3$, (^{13}C, ^{29}Si),[3254] (153), (^{13}C, ^{29}Si),[3255] $PhSiH_2(SiHPh)_nSiH_2Ph$, (^{29}Si),[3256] $H_nSi_3Ph_{8-n}$, (^{29}Si),[3257] 3-$MeOCH_2$-4-Me_2SiH-pyridine, (^{29}Si),[3258]

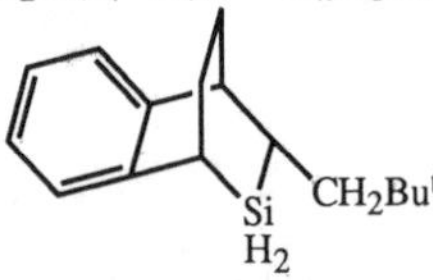

(152)

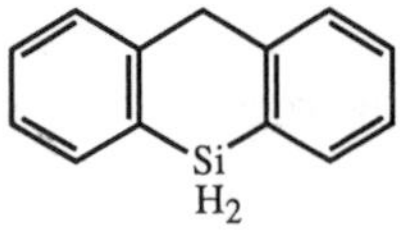

(153)

$Me_2SiHCH_2CHMeCO_2Me$, (^{13}C),[3259] $HMe_2SiSnPh_2Me$, (^{29}Si, ^{119}Sn),[3260] $H(PhMeSi)_nH$, (^{13}C, ^{29}Si),[3261] $Me(PhC{\equiv}C)Si(H)NPr^i{}_2$, ($^{13}C$, ^{29}Si),[3262] $CH_2{=}CHCH_2Si(OSiMe_2H)_3$, ($^{13}C$),[3263] (154), ($^{13}C$),[3264] $HSi(1\text{-}C_6H_4\text{-}2\text{-}SCH_2)_3\text{-}1,3,5\text{-}C_6H_3$, ($^{13}C$),[3265] $[HSi(OR)_4]^-$, (^{29}Si),[3266] (155),

3243 P.W. Fowler, P. Lazzeretti, M. Malagoli, and R. Zanasi, *J. Phys. Chem.*, 1991, **95**, 6404.
3244 J. Baker, P.W. Fowler, P. Lazzeretti, M. Malagoli, and R. Zanasi, *Chem. Phys. Lett.*, 1991, **184**, 182 (*Chem. Abstr.*, 1991, **115**, 215 344).
3245 F. Diederich, R. Ettl, Y. Rubin, R.L. Whetten, R. Beck, M. Alvarez, S. Anz, D. Sensharma, F. Wudli, et al., *Science (Washington, D.C., 1883-)*, 1991, **252**, 548 (*Chem. Abstr.*, 1991, **115**, 63 229); R.L. Whetten, M.M. Alvarez, S.J. Anz, K.E. Schriver, R.D. Beck, F.N. Diederich, Y. Rubin, R. Ettl, C.S. Foote, et al., *Mater. Res. Soc. Symp. Proc.*, 1991, **206**, 639 (*Chem. Abstr.*, 1991, **115**, 246 485); G. Zhennan, Q. Jiuxin, Z. Xihuang, W. Yongqing, Z. Xing, F. Sunqi, and G. Zizhao, *J. Phys. Chem.*, 1991, **95**, 9615; P.-M. Allemand, A. Koch, F. Wudl, Y. Rubin, F. Diederich, M.M. Alvarez, S.J. Anz, and R.L. Whetten, *J. Am. Chem. Soc.*, 1991, **113**, 1050; R.D. Johnson, G. Meijer, J.R. Salem, and D.S. Bethune, *J. Am. Chem. Soc.*, 1991, **113**, 3619; R. Ettl, I. Chao, F. Diederich, and R.L. Whetten, *Nature (London)*, 1991, **353**, 149.
3246 H.G. Horn, B. Toepfer, and M. Hemeke, *Chem.-Ztg.*, 1991, **115**, 15 (*Chem. Abstr.*, 1991, **114**, 258 497).
3247 C.K. Whitmarsh and L.V. Interrante, *J. Organomet. Chem.*, 1991, **418**, 69.
3248 Y. Derouiche and P.D. Lickiss, *J. Organomet. Chem.*, 1991, **407**, 41.
3249 B. Boury, R.J.P. Corriu, D. Leclercq, P.H. Mutin, J.-M. Planeix, and A. Vioux, *Organometallics*, 1991, **10**, 1457.
3250 T. Lobreyer, J. Oeler, and W. Sundermeyer, *Chem. Ber.*, 1991, **124**, 1405.
3251 N. Auner, C. Seidenschwarz, E. Herdtweck, and N. Sewald, *Angew. Chem., Int. Ed. Engl.*, 1991, **30**, 444.
3252 H. Lang and U. Lay, *Z. Anorg. Allg. Chem.*, 1991, **596**, 7.
3253 H. Schmidbaur, J. Zech, D.W.H. Rankin, and H.E. Robertson, *Chem. Ber.*, 1991, **124**, 1953.
3254 H. Bürger, R. Eujen, and P. Moritz, *J. Organomet. Chem.*, 1991, **401**, 149.
3255 Y. van den Winkel, B.L.M. van Baar, F. Bickelhaupt, W. Kulik, C. Sierakowski, and G. Maier, *Chem. Ber.*, 1991, **124**, 185.
3256 J.P. Banovetz, K.M. Stein, and R.M. Waymouth, *Organometallics*, 1991, **10**, 3430.
3257 K. Hassler and U. Katzenbeisser, *J. Organomet. Chem.*, 1991, **421**, 151.
3258 M. Mazhar, S. Ali, and I.A. Zia, *J. Chem. Soc. Pak.*, 1990, **12**, 216 (*Chem. Abstr.*, 1991, **114**, 584).
3259 R. Skoda-Földes, L. Kollár, and B. Heil, *J. Organomet. Chem.*, 1991, **408**, 297.
3260 W. Uhlig, *J. Organomet. Chem.*, 1991, **421**, 189.
3261 J.Y. Corey, X.-H. Zhu, T.C. Bedard, and L.D. Lange, *Organometallics*, 1991, **10**, 924.
3262 H. Lang and U. Lay, *Z. Anorg. Allg. Chem.*, 1991, **596**, 17.
3263 L.J. Mathias and T.W. Carothers, *J. Am. Chem. Soc.*, 1991, **113**, 4043.
3264 M. Ishikawa, Y. Yozuriha, T. Horio, and A. Kunai, *J. Organomet. Chem.*, 1991, **402**, C20.
3265 R.P. L'Esperance, A.P. West, jun., D. Van Engen, and R.A. Pascal, jun., *J. Am. Chem. Soc.*, 1991, **113**, 2672.
3266 R.J.P. Corriu, C. Guérin, B. Henner, and Q. Wang, *Organometallics*, 1991, **10**, 2297.

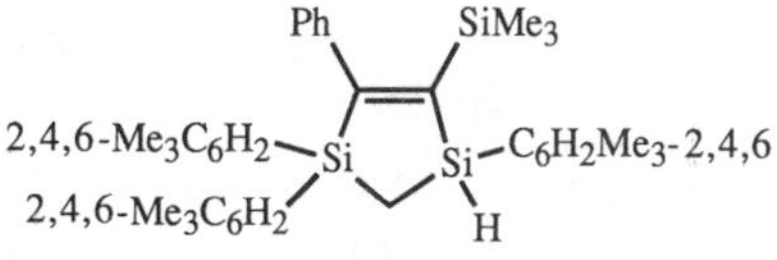

(154)

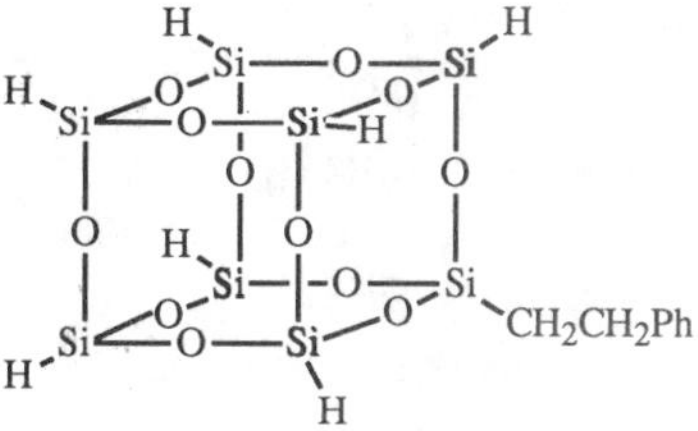

(155)

(^{13}C),[3267] $(Me_3Si)_2Si(H)SSiMe_3$, (^{13}C, ^{29}Si),[3268] $H_nMe_{2-n}\dot{G}e(CH_2)_3\dot{C}H_2$, ($^{13}C$, ^{73}Ge),[3269] and $HMe_2Sn(CH_2)_nSnMe_2H$, (^{13}C, ^{119}Sn).[3270]

The resonance frequencies of ^1H, ^{13}C, and ^{29}Si nuclei in pure Me_4Si have been measured accurately by spinning the liquid sample at the magic angle and the shifts are free from the effects of the bulk magnetic susceptibility and the solvent.[3271] The ^1H, ^{13}C, and ^{29}Si n.m.r. spectra of Me_4Si dissolved in two nematic liquid crystals have been measured.[3272]

The ^1H and ^{13}C n.m.r. spectra of $Me_5C_5GeC(SiMe_3)_3$ show a singlet for the C_5Me_5 group indicating rapid fluxionality.[3273] The signs of coupling constants have been determined for $(Me_3M^1)_3C(M^2Me_3)$, M = Si, Sn, Pb, using ^{13}C two dimensional n.m.r. spectroscopy.[3274] The effects of substituent on ^{13}C, ^{29}Si, and ^{119}Sn chemical shifts in $Me_3SnCR^1{=}C{=}CR^2R^3$, R^1, R^2, R^3 = H, MMe_3, SEt, M = Si, Ge, Sn, have been investigated and J values are additive.[3275] The stereochemical dependence of $^3J(^{119}Sn\text{-}^{15}N)$ has been examined for $Me_3Sn(Me_3M)C{=}C(NEt_3)$-$CH_2OMe$, M = Si, Ge, Sn. The ^1H, ^{13}C, ^{15}N, ^{29}Si, and ^{119}Sn n.m.r. spectra were recorded.[3276] The signs of coupling constants have been determined for $Me_3PbC{\equiv}CR$, R = Me, Me_3M, $C{\equiv}CPbMe_3$, M = Si, Sn, Pb, using ^{13}C two dimensional n.m.r. spectroscopy.[3277] N.m.r. data have also been reported for $[Me_3Si]^+$, (^{13}C, ^{29}Si),[3278] (156), (^{13}C, ^{29}Si),[3279] $P_3N_3(CH_2SiMe_3)_6$, (^{13}C),[3280] $Me_2\dot{S}iCH_2C(CH_2SiMe_3){=}C(CH_2SiMe_3)CH_2\dot{C}Me_2$, ($^{13}C$, ^{29}Si),[3281] $Sn\{CH(SiMe_3)C_6H_4\text{-}2\text{-}NMe_2\}_2$, ($^{119}Sn$),[3282] $(Me_3Si)_2CHSiMe_2CH{=}CH_2$, ($^{13}C$),[3283] $\{(Me_3Si)_2CH\}_2Ge{=}C_{13}H_8$,

3267 G. Calzaferri, D. Herren, and R. Imhof, *Helv. Chim. Acta*, 1991, **74**, 1278.
3268 M. Ballestri, C. Chatgilialoglu, and G. Seconi, *J. Organomet. Chem.*, 1991, **408**, C1.
3269 Y. Takeuchi, K. Tanaka, and T. Harazono, *Bull. Chem. Soc. Jpn.*, 1991, **64**, 91.
3270 T.N. Mitchell and B.S. Bronk, *Organometallics*, 1991, **10**, 936.
3271 K. Hayamizu, *Bull. Chem. Soc. Jpn.*, 1991, **64**, 685.
3272 Y. Hiltunen and J. Jokissaari, *Chem. Phys. Lett.*, 1990, **175**, 585 (*Chem. Abstr.*, 1991, **114**, 113 762).
3273 P. Jutzi, A. Becker, C. Leue, H.G. Stammler, B. Neumann, M.B. Hursthouse, and A. Karaulov, *Organometallics*, 1991, **10**, 3838.
3274 B. Wrackmeyer and H. Zhou, *Spectrochim. Acta, Part A*, 1991, **47**, 849 (*Chem. Abstr.*, 1991, **115**, 208 115).
3275 E. Liepiņš, I. Birğele, E. Lukevics, E.T. Bogoradovskii, and V.S. Zavgorodnii, *J. Organomet. Chem.*, 1991, **402**, 43.
3276 E. Kupče, B. Wrackmeyer, and E. Lukevics, *Magn. Reson. Chem.*, 1991, **29**, 444.
3277 B. Wrackmeyer and K. Horchler von Locquenghien, *Main Group Met. Chem.*, 1990, **13**, 387 (*Chem. Abstr.*, 1991, **115**, 232 385).
3278 J.B. Lambert, L. Kania, W. Schilf, and J.A. McConnell, *Organometallics*, 1991, **10**, 2578.
3279 A.R. Basindale and M. Borbaruah, *J. Chem. Soc., Chem. Commun.*, 1991, 1499.
3280 H.R. Allcock, W.D. Coggio, M. Parvez, and M.L. Turner, *Organometallics*, 1991, **10**, 677.
3281 N. Wiberg, S. Wagner, and G. Fischer, *Chem. Ber.*, 1991, **124**, 1981.
3282 J.T.B.H. Jastrzebski, D.M. Grove, J. Boersma, G. van Koten, and J.-M. Ernsting, *Magn. Reson. Chem.*, 1991, **29**, S25.
3283 D. Seyferth and H. Lang, *Organometallics*, 1991, **10**, 551.

(^{13}C),[3284] (157), (^{13}C),[3285] (158), (^{13}C),[3286] $Me_2SiCHMeCH{=}CHCHMeC(SiMe_3)_2$, $(^{13}C, {}^{29}Si)$,[3287] (159), (^{13}C),[3288] $(Me_3Si)_2CN{=}NN(SiBu^t{}_2Me)SnMe_2$, $(^{13}C, {}^{29}Si, {}^{119}Sn)$,[3289] $(Me_3Si)_3CGeCH(SiMe_3)_2$, $(^{13}C, {}^{29}Si)$,[3290] $SnC(SiMe_3)_2CH_2CH_2C(SiMe_3)_2$, $(^{13}C, {}^{29}Si, {}^{119}Sn)$,[3291] $(Me_3Si)_3CSi(\mu\text{-}E)_2(\mu\text{-}E_2)C(SiMe_3)_3$, (E = S, Se; ^{13}C, ^{29}Si),[3292] $(Me_3Si)_3CTeN{=}C{=}N\text{-}TeC(SiMe_3)_3$, $(^{13}C, {}^{125}Te)$,[3293] $\{(Me_3Si)_3CTe\}_2E$, (E = S, Se; ^{125}Te),[3294] $MeC(O)CH{=}CMeO\text{-}SiMe_2OCBu^t{=}CHSiMe_3$, $(^{13}C, {}^{29}Si)$,[3295] $(Me_3Si)MeSiCPh{=}CSiMe_3$, (^{13}C),[3296] $Me_3SiCH{=}CH\text{-}SiMePhCH_2CMe{=}CMe_2$, (^{13}C),[3297] $(Me_3Si)(Me_3Sn)C{=}CCH_2CH_2CH_2CHPr^n$, (^{13}C),[3298] $Me_3Si\text{-}C({=}CH_2)CH(SnMe_3)(SnBu^n{}_3)$, $(^{29}Si, {}^{119}Sn)$,[3299] $(Me_2N)_3P{=}C(SiMe_3)P(S){=}NSiMe_3$, (^{13}C),[3300] Me_3Si-quinoline, $(^{13}C, {}^{14}N, {}^{15}N, {}^{29}Si)$,[3301] 1-$Me_3Si$-2-$Pr^iCH_2SiMe_2$-3-$MeC_6H_3$, (^{13}C),[3302] 2,3-$(Me_3Si)_2C_6H_4OH$, (^{29}Si),[3303] 1-$CH_2{=}CMeCHMeCH_2GeMe_2$-2-$Me_3SiC_6H_4$, (^{13}C),[3304] 1,2-$(Me_3Si)_2$-2-hexene, (^{29}Si),[3305] and Me_3SiR, (^{13}C).[3306]

3284 M. Lazraq, C. Couret, J. Escudie, and J. Satge, *Polyhedron*, 1991, **10**, 1153.
3285 J. Okuda, E. Herdtweck, and E.M. Zeller, *Chem. Ber.*, 1991, **124**, 1575.
3286 J. Ohshita, H. Ohsaki, M. Ishikawa, A. Tachibana, Y. Kurosaki, T. Yamabe, and A. Minato, *Organometallics*, 1991, **10**, 880.
3287 N. Wiberg, G. Fischer, and S. Wagner, *Chem. Ber.*, 1991, **124**, 769.
3288 E. Niecke, R. Streubel, and M. Nieger, *Angew. Chem., Int. Ed. Engl.*, 1991, **30**, 90.
3289 N. Wiberg and S.-K. Vasisht, *Angew. Chem., Int. Ed. Engl.*, 1991, **30**, 93.
3290 P. Jutzi, A. Becker, H.G. Stammler, and B. Neumann, *Organometallics*, 1991, **10**, 1647.
3291 M. Kira, R. Yauchibara, R. Hirano, C. Kabuto, and H. Sakurai, *J. Am. Chem. Soc.*, 1991, **113**, 7785.
3292 H. Yoshida, Y. Kabe, and W. Ando, *Organometallics*, 1991, **10**, 27.
3293 W. Fimml and F. Sladky, *Chem. Ber.*, 1991, **124**, 1131.
3294 C. Köllemann and F. Stadky, *Organometallics*, 1991, **10**, 2101.
3295 G. Maas, M. Alt, K. Schneider, and A. Fronda, *Chem. Ber.*, 1991, **124**, 1295.
3296 J. Ohshita and M. Ishikawa, *J. Organomet. Chem.*, 1991, **407**, 157.
3297 M. Ishikawa, Y. Nishimura, and H. Sakamoto, *Organometallics*, 1991, **10**, 2701.
3298 S.A. Rao and P. Knochel, *J. Am. Chem. Soc.*, 1991, **113**, 5735.
3299 T.N. Mitchell and U. Schneider, *J. Organomet. Chem.*, 1991, **407**, 319.
3300 U. Krüger, H. Pritzkow, and H. Grützmacher, *Chem. Ber.*, 1991, **124**, 329.
3301 E. Lukevics, E. Liepins, I. Segal, and M. Fleisher, *J. Organomet. Chem.*, 1991, **406**, 283.
3302 M. Ishikawa and H. Sakamoto, *J. Organomet. Chem.*, 1991, **414**, 1.
3303 H. Jancke, R. Wolff, and M. Lauterbach, *Z. Chem.*, 1990, **30**, 376 (*Chem. Abstr.*, 1991, **115**, 8867).
3304 K.L. Bobbitt, V.M. Maloney, and P.P. Gaspar, *Organometallics*, 1991, **10**, 2772.
3305 T. Suzuki, P.Y. Lo, and G.A. Lawless, *J. Chem. Soc., Dalton Trans.*, 1991, 439.
3306 Y. Ito, M. Suginome, T. Matsuura, and M. Murakami, *J. Am. Chem. Soc.*, 1991, **113**, 8899; A.K. Saxena, C.S. Bisaria, S.K. Saini, and L.M. Pande, *Synth. React. Inorg. Metal-Org. Chem.*, 1991, **21**, 401; H. Monti, P. Piras, M. Afshari, and R. Faure, *J. Mol. Struct.*, 1991, **243**, 31; Q. Xie, S. Hu, Z. Shan, F. Zhang, C. Lin, and C. Huang, *Huaxue Xuebao*, 1991, **49**, 81 (*Chem. Abstr.*, 1991, **114**, 247 360); R.L. Wells, A.P. Purdy, and C.G. Pitt, *Phosphorus Sulphur*, 1991, **57**, 1; S. Okamoto, T. Yoshino, and F. Sato, *Tetrahedron: Asymmetry*, 1991, **2**, 35; S. Kobayashi, J. Kadokawa, and H. Uyama, *Macromolecules*, 1991, **24**, 4475 (*Chem. Abstr.*, 1991, **115**, 50 371); R.S. Archibald, D.P. Chinnery, A.D. Fanta, and R. West, *Organometallics*, 1991, **10**, 3769; W.E. Billups, G.-A. Lee, B.E. Arney, jun., and K.H. Whitmire, *J. Am. Chem. Soc.*, 1991, **113**, 7980; Y.T. Jeon, C.-P. Lee, and P.S. Mariano, *J. Am. Chem. Soc.*, 1991, **113**, 8847; N. Tokitoh, T. Matsumoto, H. Suzuki, and R. Okazaki, *Tetrahedron Lett.*, 1991, **32**, 2049; K.-T. Kang, S.S. Kim, and J.C. Lee, *Tetrahedron Lett.*, 1991, **32**, 4341; N. Choi, Y. Kabe, and W. Ando, *Tetrahedron. Lett.*, 1991, **32**, 4573; S. Ebeling, D. Matthies, and D. McCarthy, *Phosphorus, Sulphur*, 1991, **60**, 265; N. Lage, S. Masson, and A. Thuiller, *J. Chem. Soc., Perkin Trans.1*, 1991, 3389; H. Nakahira, I. Ryu, M. Ikebe, N. Kambe, and N. Sonoda, *Angew. Chem., Int. Ed. Engl.*, 1991, **30**, 177; H.-U. Reißig and C. Hippeli, *Chem. Ber.*, 1991, **124**, 115; K. Kadei and F. Vögtle, *Chem. Ber.*, 1991, **124**, 909; M. Conrads and J. Mattay, *Chem. Ber.*, 1991, **124**, 1425; B.M. Trost, T.A. Grese, and D.M.T. Chan, *J. Am. Chem. Soc.*, 1991, **113**, 7350; B.M. Trost and T.A. Grese, *J. Am. Chem. Soc.*, 1991, **113**, 7363; M.C. Pirrung, N. Krishnamurthy, D.S. Nunn, and A.T. McPhail, *J. Am. Chem. Soc.*, 1991, **113**, 4910; K. Burgess, W.A. van der Donk, M.B. Jarstfer, and M.H. Ohlmeyer, *J. Am. Chem. Soc.*, 1991, **113**, 6139; F. Babudri, V. Fiandanese, G. Marchese, and F. Naso, *J. Chem. Soc., Chem. Commun.*, 1991, 237; A.K. Jhingan and T. Meehan, *Tetrahedron*, 1991, **47**, 1621; B. Mergardt, K. Weber, G. Adwidjaja, and E. Schaumann, *Angew. Chem., Int. Ed. Engl.*, 1991, **30**, 1687; P. Jankowski, S. Marczak, M. Masnyk, and J. Wicha, *J. Organomet. Chem.*, 1991, **403**, 49; D. Mesnard and L. Miginiac, *J. Organomet.*

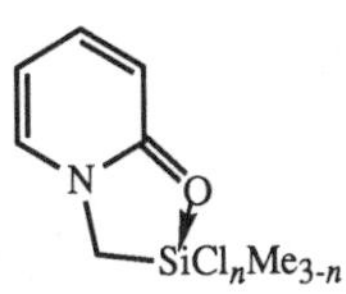

(156)

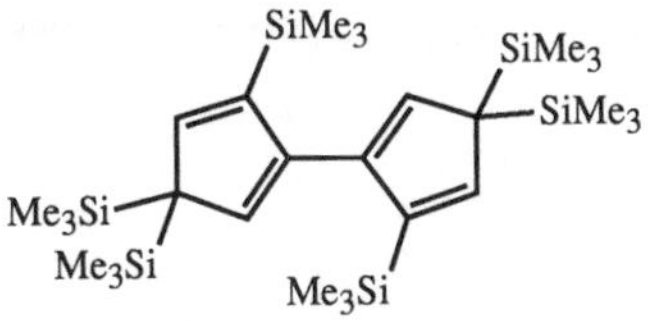

(157)

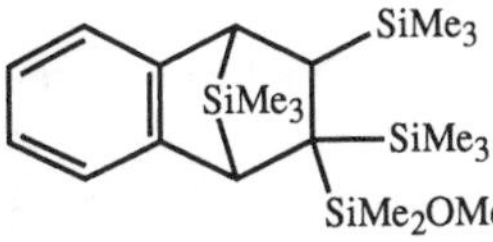

(158)

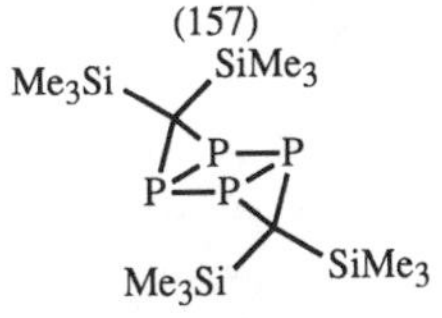

(159)

The ^{13}C n.m.r. spectra of α,ω-diphenylpermethylated oligosilanes indicate the highest electron density on the *ipso*-carbon atoms.[3307] N.m.r. data have also been reported for $\{C_6H_8(CH_2)_2SiMe_2\}_8Si_8O_{20}$, ($^{13}C$, ^{29}Si),[3308] copolymers of 1,3-Br_2-2,2-dihexyltetramethyltrisilane, (^{29}Si),[3309] copolymer of 1,1-$Me_2$1-silacyclopent-3-ene and 1,1-$Ph_2$1-silacyclopent-3-ene, (^{13}C, ^{29}Si),[3310] poly(benzyldimethylvinylsilane), (^{13}C),[3311] poly{carbo(dimethyl)silanes}, (^{13}C, ^{29}Si),[3312] (160), (^{13}C),[3313] copolymers of $(4\text{-ClCOC}_6H_4)_2SiMe_2$ and $(3\text{-ClCOC}_6H_4)_2SiMe_2$ with 2,5-$(4\text{-}H_2NC_6H_4)_2$-3,4-Ph_2-thiophene and $(4\text{-}H_2NC_6H_4)_2O$, (^{13}C),[3314] (161), (^{13}C),[3315] $PhSiMe_2CMeHI$, (^{13}C),[3316] $PhMe_2SiC(SiMe_2)_3CSiMe_2Ph$, ($^{13}C$, ^{29}Si),[3317] $(2,4,6\text{-}Me_3C_6H_2)Me_2SiR$, ($^{13}C$),[3318] Me_2SiPhR, (^{13}C),[3319] Me_2SiBu^tR, (^{13}C),[3320] $Me_2\overline{SiCH=CHCH=CHCH=CH}$, ($^{13}C$),[3321] (162),

Chem., 1991, **403**, 299; E. Block, M. Gernon, H. Kang, G. Ofori-Akai, and J. Zubieta, *Inorg. Chem.*, 1991, **30**, 1736; N. Lage, S. Masson, and A. Thuiller, *J. Chem. Soc., Perkin Trans. 1*, 1991, 2269; P. Zanirato, *J. Chem. Soc., Perkin Trans. 1*, 1991, 2789; M. Frasch, W. Sundermeyer, and J. Waldi, *Chem. Ber.*, 1991, **124**, 1805; R. Askani and J. Hoffmann, *Chem. Ber.*, 1991, **124**, 2307; B. Mauzé and L. Miginiac, *J. Organomet. Chem.*, 1991, **411**, 69; A. Fürstner, G. Kollegger, and H. Weidmann, *J. Organomet. Chem.*, 1991, **414**, 295; M. Grignon-Dubois, M. Fialeix, and C. Biran, *Can. J. Chem.*, 1991, **69**, 2014.

3307 K.E. Ruehl and K. Matyjaszewski, *J. Organomet. Chem.*, 1991, **410**, 1.

3308 I. Pitsch, D. Hoebbel, H. Jancke, and W. Hiller, *Z. Anorg. Allg. Chem.*, 1991, **596**, 63.

3309 R. West, R. Menescal, T. Asuke, and C.H. Yuan, *Polym. Prepr. (Am. Chem. Soc. Div. Polym. Chem.)*, 1990, **31**, 248 (*Chem. Abstr.*, 1991, **114**, 186 458).

3310 Q. Zhou, G. Manuel, and W.P. Weber, *Report*, 1990, **Order No. AD-A221686**, 5 pp. Avail. NTIS. From *Gov. Rep. Announce. Index (U.S.)*, 1990, **90**, Abstr. No. 045,720 (*Chem. Abstr.*, 1991, **114**, 165 479).

3311 J. Oku, T. Hasegawa, T. Kawakita, Y. Kondo, and M. Takaki, *Macromolecules*, 1991, **24**, 1253 (*Chem. Abstr.*, 1991, **114**, 123 164).

3312 K.B. Wagener and D.W. Smith, jun., *Macromolecules*, 1991, **14**, 6073 (*Chem. Abstr.*, 1991, **115**, 233 020).

3313 T.C. Bedard, J.Y. Corey, L.D. Lange, and N.P. Rath, *J. Organomet. Chem.*, 1991, **401**, 261.

3314 T.S. Jahnke, D.J. Walker, and S.S. Mohite, *Polym. Prepr. (Am. Chem. Soc., Div. Polym. Chem.)*, 1991, **32**, 310 (*Chem. Abstr.*, 1991, **115**, 280 914).

3315 H. Sakamoto and M. Ishikawa, *J. Organomet. Chem.*, 1991, **418**, 305.

3316 H. Hengelsberg, R. Tacke, K. Fritsche, C. Syldatk, and F. Wagner, *J. Organomet. Chem.*, 1991, **415**, 39.

3317 W. Ando, H. Yoshida, K. Kurishima, and M. Sugiyama, *J. Am. Chem. Soc.*, 1991, **113**, 7790.

3318 H. Müller, U. Weinzierl, and W. Seidel, *Z. Anorg. Allg. Chem.*, 1991, **603**, 15.

3319 M.A. Sparks and J.S. Panek, *J. Org. Chem.*, 1991, **56**, 3431; R. Liao, Q. Xie, and Y. Wu, *Gaodeng Xuexiao Huaxue Xuebao*, 1990, **11**, 1072 (*Chem. Abstr.*, 1991, **115**, 49 790); I. Fleming, Y. Landais, and P.R. Raithby, *J. Chem. Soc., Perkin Trans. 1*, 1991, 715.

3320 S.G. Davies and M.R. Shipton, *J. Chem. Soc., Perkin Trans. 1*, 1991, 501; B.A. Keay and J.-L.J. Bontront, *Can. J. Chem.*, 1991, **69**, 1326; S.R. Angle and D.O. Arnaìz, *Tetrahedron Lett.*, 1991, **32**, 2327.

3321 Y. Nakadaira, R. Sato, and H. Sakurai, *Organometallics*, 1991, **10**, 435.

(^{13}C,^{29}Si),[3322] $Me_2Si\{ArC(Me)=NN=C(NH_2)O\}_2$, ($^{13}C$),[3323] $R^1R^2MeSiSiMeR^3R^4$, (^{13}C),[3324]

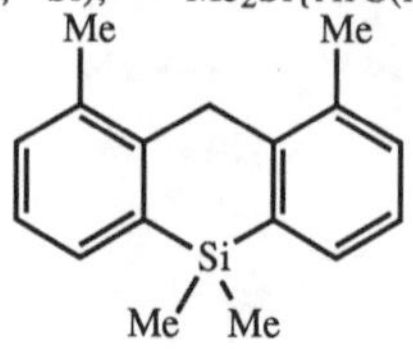

(160)

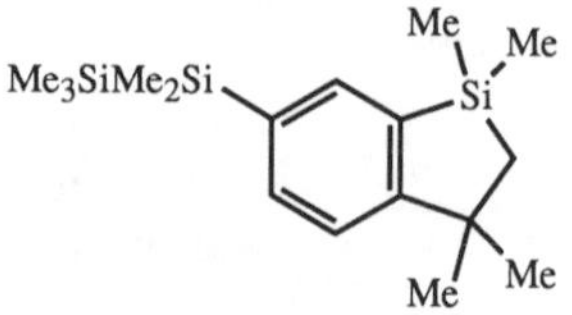

(161)

(162)

$MePh_2SiCH=CH_2$, (^{13}C),[3325] difluorocyclopropanated polymer of 1-Me-1-Ph-1-sila-*cis*-pent-3-ene, (^{13}C, ^{29}Si),[3326] and poly(1-Me-1-Ph-1-silapentane), (^{13}C, ^{29}Si).[3327]

^{14}N n.m.r. spectroscopy has been used to distinguish between nitrile imides, *e.g.*, $Pr^i_3SiC\equiv N$-$NSiPr^i_3$ and isomeric diazo compounds, *e.g.*, $Pr^i_3SiCHN_2$.[3328] Benzoyl derivatives of SeR, TeR, and I show higher frequency ^{17}O signals than those of the lighter elements on the same group of the Periodic Table whereas $ArCOGeR_3$ is at as high a frequency as $ArCOSiR_3$. This behaviour was discussed in terms of π-bond order and electronic excitation energy.[3329] N.m.r. data have also been reported for (163), (^{13}C, ^{29}Si),[3330] $Ph_3SiCH=CHPh$, (^{13}C),[3331] $(2,4,6\text{-}Me_3C_6H_2)N=C$-$(SiBu^t_2)_2C=N(C_6H_2Me_3\text{-}2,4,6)$, ($^{13}C$, ^{29}Si),[3332] Et_3SiR, (^{13}C),[3333] Pr^i_3SiR, (^{13}C),[3334] $Ph_3MCH_2CH=CH_2$, (M = Si, Ge, Sn; ^{13}C),[3335] (164), (^{13}C),[3336] (165), {R = (166); ^{13}C},[3337]

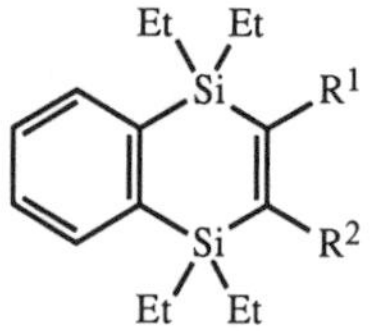

(163)

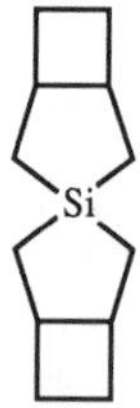

(164)

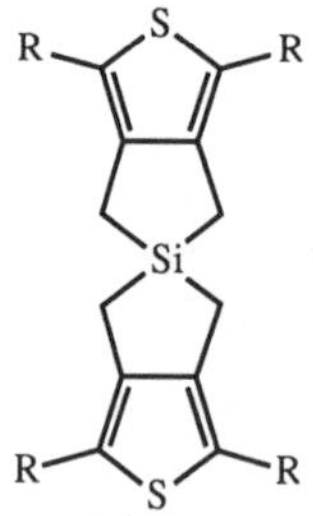

(165)

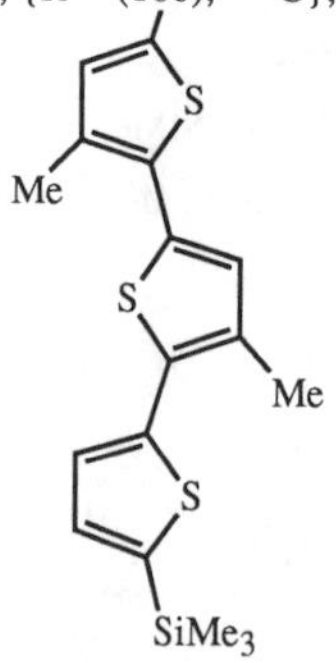

(166)

3322 A. Sekiguchi, I. Maruki, K. Ebata, C. Kabuto, and H. Sakurai, *J. Chem. Soc., Chem. Commun.*, 1991, 341.
3323 D. Singh and R.V. Singh, *Main Group Met. Chem.*, 1990, **13**, 309 (*Chem. Abstr.*, 1991, **114**, 207 338).
3324 A. Kunai, T. Kawakami, E. Toyoda, and M. Ishikawa, *Organometallics*, 1991, **10**, 893.
3325 B. Marciniec and C. Pietraszuk, *J. Organomet. Chem.*, 1991, **412**, 301.
3326 H.S.J. Lee and W.P. Weber, *Polym. Prepr. (Am. Chem. Soc., Div. Polym. Chem.)*, 1990, **31**, 424 (*Chem. Abstr.*, 1991, **114**, 165 028).
3327 X. Liao and W.P. Weber, *Polym. Bull. (Berlin)*, 1991, **25**, 621 (*Chem. Abstr.*, 1991, **115**, 50 457).
3328 K.H. von Locquenghien, R. Réau, and G. Bertrand, *J. Chem. Soc., Chem. Commun.*, 1991, 1192.
3329 H. Dahn and P. Péchy, *J. Chem. Soc., Perkin Trans. 2*, 1991, 1721.
3330 M. Ishikawa, H. Sakamoto, and T. Tabuchi, *Organometallics*, 1991, **10**, 3173.
3331 L.N. Lewis, G. Karen, G.L. Bryant, jun., and P.E. Donahue, *Organometallics*, 1991, **10**, 3750.
3332 M. Weidenbruch, B. Brand-Roth, S. Pohl, and W. Saak, *Polyhedron*, 1991, **10**, 1147.
3333 I. Ojima, P. Ingallina, R.J. Donovan, and N. Clos, *Organometallics*, 1991, **10**, 38.
3334 Y. Rubin, C.B. Knobler, and F. Diederich, *Angew. Chem., Int. Ed. Engl.*, 1991, **30**, 698.
3335 G. Hagen and H. Mayr, *J. Am. Chem. Soc.*, 1991, **113**, 4954.
3336 K. Nakanishi, K. Mizuno, and Y. Otsuji, *J. Chem. Soc., Chem. Commun.*, 1991, 91.
3337 J.M. Tour, R. Wu, and J.S. Schumm, *J. Am. Chem. Soc.*, 1991, **113**, 7064.

$Ph_2Si(CH_2CH{=}CHCH_2)_2SiPh_2$, ($^{13}C$, ^{29}Si),[3338] poly(1,1-Ph_2-1-sila-cyclopent-3-ene), (^{13}C, ^{29}Si),[3339] poly(1-Ph-1-sila-*cis*-pent-3-ene), (^{13}C, ^{29}Si),[3340] $Bu^t{}_2SiCH_2CH_2$, (^{13}C, ^{29}Si),[3341] $Ph_2SiCHMeCHCH{=}CHMe$, (^{13}C, ^{29}Si),[3342] (2,4,6-$Me_3C_6H_2$)$_2$SiCHMeCEt, (^{13}C, ^{29}Si),[3343] (adamantyl)$_2$SiCHMeCR, (^{13}C, ^{29}Si),[3344] (167), (^{13}C),[3345] (168), (^{13}C),[3346] Ph_2Bu^tSiR, (^{13}C),[3347] (2,4,6-$Me_3C_6H_2$)$_2$SiCR^{1}R^{2}OSi($C_6H_2Me_3$-2,4,6)$_2$, (^{29}Si),[3348] Ph_3SiR, (^{13}C),[3349] (2,4,6-$Me_3C_6H_2$)$_2$SiGe($C_6H_2Me_3$-2,4,6)$_2$Ge($C_6H_2Me_3$-2,4,6)$_2$, (^{13}C, ^{29}Si),[3350] $(Me_3Si)_3SiC(O)$-NMe_2, (^{13}C, ^{29}Si),[3351] $[(Me_3Si)_3SiGeCl_2]_2$, (^{13}C, ^{29}Si),[3352] $(Me_3Si)_2SiCMePhCMePhSi(SiMe_3)_2$, ($^{13}C$, ^{29}Si),[3353] (169), (^{13}C),[3354] $(Me_3SiSiMe_2)_2SiMePh$, (^{13}C),[3355] $(Me_3Si)_2SiCH_2CH{=}CHOC$-(adamantyl)($OSiMe_3$), ($^{13}C$, ^{29}Si),[3356] $(Me_3Si)_2SiCH_2CH{=}\dot{C}(OR)C$(adamantyl)($OSiMe_3$), ($^{13}C$, ^{29}Si),[3357] $(Me_3Si)_3SiR$, (^{13}C),[3358] $Me_3SiOSiMe(SiMe_3)CH(SiMe_3)$adamantyl, ($^{13}C$, ^{29}Si),[3359] $OCH_2CH_2CH_2CH{=}CHSiMePhSiMe_3$, ($^{13}C$),[3360] (1-naphthyl)$_2SiMeSiMe_3$, ($^{13}C$),[3361] $Si_2Cl_nMe_{6-n}$, (^{29}Si),[3362] $ClCH_2SiMe_2SiMe_2SiMe_2CH_2Cl$, ($^{13}C$, ^{29}Si),[3363] $Me_2Si(SiMe_2SiMe_2)_2SiMeSiMe_3$, ($^{29}Si$),[3364] $Me(SiMe_2)_nMe$, (^{13}C, ^{29}Si),[3365] $Me_2SiSiMe_2SiMe_2CH_2CPh{=}CPhCH_2$, ($^{13}C$, ^{29}Si),[3366] 4-$FC_6H_4SiMe_2SiMe_2$-4-$C_6H_4CH{=}C(CN)_2$, (^{13}C, ^{29}Si),[3367] 4-$Me_2NC_6H_4SiMe_2SiMe_2$-4-C_6H_4-

3338 J.T. Anhaus, W. Clegg, S.P. Collingwood, and V.C. Gibson, *J. Chem. Soc., Chem. Commun.*, 1991, 1720.
3339 D.A. Stonich and W.P. Weber, *Polym. Bull. (Berlin)*, 1991, **25**, 629 (*Chem. Abstr.*, 1991, **115**, 30 013).
3340 X. Liao, Y.H. Ko, G. Manuel, and W.P. Weber, *Polym. Bull. (Berlin)*, 1991, **25**, 63 (*Chem. Abstr.*, 1991, **114**, 229 542).
3341 P. Boudjouk, E. Black, and R. Kumarathasan, *Organometallics*, 1991, **10**, 2095.
3342 S. Zhang and R.T. Conlin, *J. Am. Chem. Soc.*, 1991, **113**, 4272.
3343 S. Zhang, P.E. Wagenseller, and R.T. Conlin, *J. Am. Chem. Soc.*, 1991, **113**, 4278.
3344 D.H. Pae, M. Xiao, M.Y. Chiang, and P.P. Gaspar, *J. Am. Chem. Soc.*, 1991, **113**, 1281.
3345 H. Brunner and F. Prester, *J. Organomet. Chem.*, 1991, **411**, C1.
3346 I.N. Jung, B.R. Yoo, M.E. Lee, and P.R. Jones, *Organometallics*, 1991, **10**, 2529.
3347 Y. Rubin, S.S. Lin, C.B. Knobler, J. Anthony, A.M. Boldi, and F. Diederich, *J. Am. Chem. Soc.*, 1991, **113**, 6943.
3348 A.D. Fanta, D.J. De Young, J. Belzner, and R. West, *Organometallics*, 1991, **10**, 3466.
3349 M. Taniguchi, K. Oshima, and K. Utimoto, *Tetrahedron Lett.*, 1991, **32**, 2783.
3350 K.M.Baines and J.A. Cooke, *Organometallics*, 1991, **10**, 3419.
3351 S.S. Al-Juaid, Y. Derouiche, P.B. Hitchcock, P.D. Lickiss, and A.G. Brook, *J. Organomet. Chem.*, 1991, **403**, 293.
3352 S.P. Mallela and R.A. Geanangel, *Inorg. Chem.*, 1991, **30**, 1480.
3353 J. Ohshita, Y. Masaoka, and M. Ishikawa, *Organometallics*, 1991, **10**, 3775.
3354 J. Ohshita, H. Ohsaki, M. Ishikawa, A. Tachibana, Y. Kurosaki, T. Yamabe, T. Tsukihara, K. Takahashi, and Y. Kiso, *Organometallics*, 1991, **10**, 2685.
3355 A. Kunai, T. Kawakami, E. Toyoda, and M. Ishikawa, *Organometallics*, 1991, **10**, 2001.
3356 A.G. Brook, S.S. Hu, W.J. Chatterton, and A.J. Lough, *Organometallics*, 1991, **10**, 2752.
3357 A.G. Brook, S.S. Hu, A.K. Saxena, and A.J. Lough, *Organometallics*, 1991, **10**, 2758.
3358 P. Arya, M. Lesage, and D.D.M. Wayner, *Tetrahedron Lett.*, 1991, **32**, 2853.
3359 A.G. Brook, P. Chiu, J. McClenaghnan, and A.J. Lough, *Organometallics*, 1991, **10**, 3292.
3360 K. Takaki, H. Sakamoto, Y. Nishimura, Y. Sugihara, and M. Ishikawa, *Organometallics*, 1991, **10**, 888.
3361 J. Ohshita, H. Ohsaki, and M. Ishikawa, *Organometallics*, 1991, **10**, 2695.
3362 R. Lehnert, M. Hoeppner, and H. Kelling, *Z. Anorg. Allg. Chem.*, 1990, **591**, 209.
3363 T. Kobayashi and K.H. Pannell, *Organometallics*, 1991, **10**, 1960.
3364 E. Hengge and P.K. Jenkner, *Z. Anorg. Allg. Chem.*, 1991, **606**, 97.
3365 J. Maxka, L.M. Huang, and R. West, *Organometallics*, 1991, **10**, 656.
3366 T. Shimizu, K. Shimizu, and W. Ando, *J. Am. Chem. Soc.*, 1991, **113**, 354.
3367 G. Mignani, A. Krämer, G. Puccetti, I. Ledoux, J. Zyss, and G. Soula, *Organometallics*, 1991, **10**, 3656.

$CH{=}C(CN)_2$, (^{13}C, ^{29}Si),[3368] polymers with SiSiGe sequences, (^{29}Si),[3369] $(PhMeSi)_6$, (^{13}C, ^{29}Si),[3370] $MeFSi(SiMePhSiMePh)_2SiFMe$, ($^{29}Si$),[3371] and $[(4\text{-}MeC_6H_4)_2Si]_n$, ($n$ = 4, 5; ^{13}C, ^{29}Si).[3372]

(167)

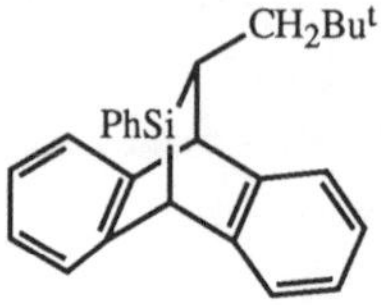
(168)

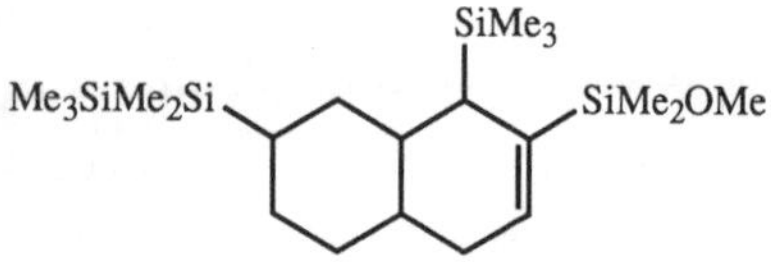
(169)

^1H, ^{13}C, and ^{73}Ge n.m.r. spectroscopy has been used to study $Me_{4-n}Ge(C{\equiv}CH)_n$. The origin of the chemical shifts could not simply be attributed to the electron densities as estimated by MNDO calculation.[3373] Two dimensional ^1H-^{13}C correlation has been used for some organotin derivatives to determine the relative signs of $J(^{119}Sn\text{-}^{13}C)$ and $J(^{119}Sn\text{-}^1H)$.[3374] ^1H, ^{13}C, ^{31}P, and ^{119}Sn n.m.r. data have been presented for vinyl phosphines in which the vinyl group bears Me_3Sn residues. The magnitudes of $^3J(^{31}PC{=}C^{119}Sn)$, $^4J(^{31}PC{=}CSn^{13}C)$ and $^5J(^{31}PC{=}CSnCH)$ are determined by through space coupling when P and C are *cis*.[3375] ^{119}Sn n.m.r. spectroscopy has been used to determine the *cis-trans* ratio of $Bu^n{}_3SnCH_2CH{=}CHOMe$.[3376] ^{13}C T_1 measurements have been used to study motion of $(n\text{-}C_8H_{17})_4Sn$ as a function of temperature, pressure and frequency.[3377] For Ph_4Sn, Ph_3SnR and Ph_2SnR_2, the linear relationship

$$\left|{}^1J(^{119}Sn\text{-}^{13}C)\right| = (15.56 \pm 0.84)\theta - (1160 \pm 101)$$

has been obtained, where θ is the C-Sn-C angle.[3378] N.m.r. data have also been reported for $Me_2M(C_5HMe_4)_2$, (M = Ge, Sn; ^{13}C),[3379] (170), (^{13}C),[3380] $(2,4,6\text{-}Me_3C_6H_2)_nGeF_{4-n}$, ($^{13}C$),[3381] organotin complexes of Schiff bases, (^{13}C, ^{119}Sn),[3382] 10-X-9-Me_3Sn-triptycene, (^{13}C, ^{119}Sn),[3383] $Me_3SnC(O)CH{=}CH_2$, (^{13}C, ^{119}Sn),[3384] $Me_3SnCH{=}CMePPh_2$, (^{119}Sn),[3385] $Me_3SnCH{=}CCH_2$,

3368 G. Mignani, M. Barzoukas, J. Zyss, G. Soula, F. Balegroune, D. Grandjean, and D. Josse, *Organometallics*, 1991, **10**, 3660.
3369 H. Isaka, M. Fujiki, M. Fujino, and N. Matsumoto, *Macromolecules*, 1991, **24**, 2647 (*Chem. Abstr.*, 1991, **114**, 186 233).
3370 J. Maxka, F.K. Mitter, D.R. Powell, and R. West, *Organometallics*, 1991, **10**, 660.
3371 W. Uhlig and A. Tzschach, *Z. Chem.*, 1990, **30**, 254 (*Chem. Abstr.*, 1991, **114**, 6594).
3372 T. Nakano, Y. Ochi-Ai, and Y. Nagai, *Chem. Express*, 1990, **5**, 857 (*Chem. Abstr.*, 1991, **114**, 122 482).
3373 E. Liepiņš, M.V. Petrova, E.T. Bogoradovskii, and V.S. Zavgorodnii, *J. Organomet. Chem.*, 1991, **410**, 287.
3374 A. Lyčka, J. Jirman, and J. Holeček, *Magn. Reson. Chem.*, 1991, **29**, 1212.
3375 T.N. Mitchell, K. Heesche-Wagner, and H.J. Belt, *Magn. Reson. Chem.*, 1991, **29**, 78.
3376 K. Koeber, J. Gore, and J.-M. Vatele, *Tetrahedron Lett.*, 1991, **32**, 1187.
3377 D.G. Gillies, S.J. Matthews, and L.H. Sutcliffe, *Magn. Reson. Chem.*, 1991, **29**, 1221.
3378 J. Holeček, K. Handlir, M. Nadvornik, and A. Lyčka, *Z. Chem.*, 1990, **30**, 265 (*Chem. Abstr.*, 1991, **114**, 24 058).
3379 H. Schumann, L. Esser, J. Loebel, A. Dietrich, D. Van der Helm, and X. Ji, *Organometallics*, 1991, **10**, 2585.
3380 J.-D. Andriamizaka, C. Couret, J. Escudié, A. Laporterie, G. Manuel, and M. Regitz, *J. Organomet. Chem.*, 1991, **419**, 57.
3381 P. Rivière, M. Rivière-Baudet, A. Castel, D. Desor, and C. Abdennadher, *Phosphorus Sulphur*, 1991, **61**, 189.
3382 P. Dixit, J.P. Tandon, R.B. Goyal, and J.P. Agnihotri, *Main Group Met. Chem.*, 1990, **13**, 407 (*Chem. Abstr.*, 1991, **115**, 159 292).
3383 W. Adcock and V.S. Iyer, *Magn. Reson. Chem.*, 1991, **29**, 381.
3384 A. Ricci, A. Degl'Innocenti, A. Capperucci, G. Reginato, and A. Mordini, *Tetrahedron Lett.*, 1991, **32**, 1899.
3385 T.N. Mitchell and K. Heesche, *J. Organomet. Chem.*, 1991, **409**,

(^{119}Sn),[3386] $MeCH{=}C(SnMe_3)CO_2Me$, (^{13}C, ^{119}Sn),[3387] (171), (R = $SnMe_3$, $SbMe_4$; ^{13}C, ^{29}Si, ^{119}Sn),[3388] $[Ph_4SnMe]^-$, (^{13}C, ^{119}Sn),[3389] $Bu^n{}_3SnR$, (^{13}C),[3390] $Bu^n{}_3SnC(OSiMe_2R^1){=}CR^2R^3$, ($^{119}Sn$),[3391] $Bu^n{}_3SnCH_2CHR^1CHR^2CH_2SiMe_3$, ($^{13}C$, ^{119}Sn),[3392] $Bu^n{}_3SnCH{=}CHCH_2N(SiMe_3)_2$, ($^{13}C$, ^{119}Sn),[3393] (172), (^{13}C),[3394] (173), (^{13}C),[3395] $(2,4,6\text{-}Pr^i{}_3C_6H_2)SnBu^t{}_2SnBu^t{}_2(C_6H_2Pr^i{}_3\text{-}2,4,6)$, ($^{13}C$),[3396] $[Bu^t{}_2Sn\{CH(PPh_3)\}_2SnBu^t{}_2][BF_4]_2$, ($^{11}B$, ^{13}C, ^{119}Sn),[3397] (8-NMe_2-naphthyl)$SnPhR^1R^2$, (^{13}C, ^{119}Sn),[3398] Ph_3SnR, (^{13}C, ^{119}Sn),[3399] (174), (R = 2,6-$Et_2C_6H_3$; ^{13}C, ^{119}Sn),[3400] $[Sn(C_6H_3Et_2\text{-}2,6)]_{10}$, ($^{119}Sn$),[3401] $\{2,4,6\text{-}(F_3C)_3C_6H_2\}_2Sn$, ($^{13}C$, ^{119}Sn),[3402] (3,5-$Bu^t{}_2C_6H_3CMe_2CH_2)_6Pb_2$, ($^{13}C$),[3403] $\{2,4,6\text{-}(F_3C)_3C_6H_2\}_2Pb$ (^{13}C, ^{207}Pb),[3404] and (4-$MeC_6H_4)_6M_2$, (M = Sn, Pb; ^{13}C, ^{119}Sn, ^{207}Pb).[3405]

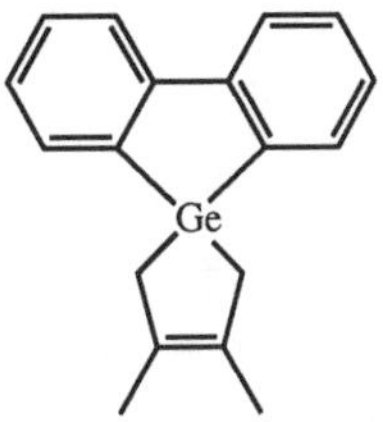
(170)

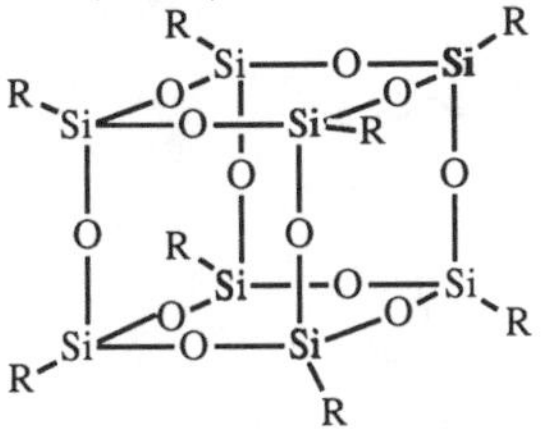

(171)

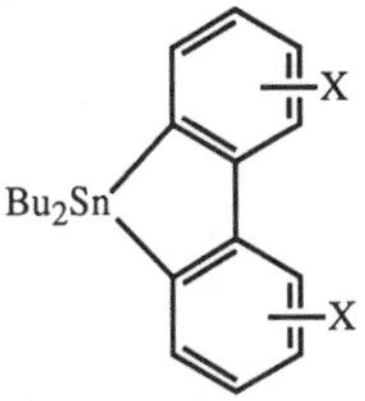

(172)

$Se\{N(SiMe_3)_2\}_2$, (^{77}Se),[3406] $Bu^tSiMe_2NLiNLiSiMe_2Bu^t$, ($^{13}C$, ^{29}Si),[3407] $PhMe_2SiNMe_2$,

3386 T.N. Mitchell and U. Schneider, *J. Organomet. Chem.*, 1991, **405**, 195.
3387 J.C. Cochran, K.M. Terrence, and H.K. Phillips, *Organometallics*, 1991, **10**, 2411.
3388 F.J. Feher and K.J. Weller, *Inorg. Chem.*, 1991, **30**, 880.
3389 A.J. Ashe, tert., L.L. Lohr, and S.M. Al-Taweel, *Organometallics*, 1991, **10**, 2424.
3390 M.A. Tius, J. Gomez-Galeno, X.-q. Gu, and J.H. Zaidi, *J. Am. Chem. Soc.*, 1991, **113**, 5775; J.R. McCarthy, D.P. Matthews, D.M. Stemerick, E.W. Huber, P. Bey, B.J. Lippert, R.D. Snyder, and P.S. Sunkara, *J. Am. Chem. Soc.*, 1991, **113**, 7439; Y. Guindon, J.-F. Lavallée, M. Llinas-Brunet, G. Horner, and J. Rancourt, *J. Am. Chem. Soc.*, 1991, **113**, 9701; J.K. Stille, H.Su, D.H. Hill, P. Schneider, M. Tanaka, D.L. Morrison, and L.S. Hegedus, *Organometallics*, 1991, **10**, 1993; M. Yoshitake, M. Yamamoto, S. Kohmoto, and K. Yamada, *J. Chem. Soc., Perkin Trans. 1*, 1991, 2157; M. Yoshitake, M. Yamamoto, S. Kohmoto, and K. Yamada, *J. Chem. Soc., Perkin Trans. 1*, 1991, 2161.
3391 J.-B. Verlhac, M. Pereyre, and H. Shin, *Organometallics*, 1991, **10**, 3007.
3392 Y. Tsuji and Y. Obora, *J. Am. Chem. Soc.*, 1991, **113**, 9368.
3393 R.J.P. Corriu, G. Bolin, and J.J.E. Moreau, *Tetrahedron Lett.*, 1991, **32**, 4121.
3394 H.-A. Brune, R. Hohenadel, G. Schmidtberg, and U. Ziegler, *J. Organomet. Chem.*, 1991, **402**, 171.
3395 M. Weidenbruch, K. Schäfers, J. Schlaefke, K. Peters, and H.G. von Schnering, *J. Organomet. Chem.*, 1991, **415**, 343.
3396 M. Weidenbruch, J. Schlaefke, K. Peters, and H.G.von Schnering, *J. Organomet. Chem.*, 1991, **414**, 319.
3397 H. Grützmacher and H. Pritzkow, *Organometallics*, 1991, **10**, 938.
3398 J.T.B.H. Jastrzebski, J. Boersma, P.M. Esch, and G. Van Koten, *Organometallics*, 1991, **10**, 930.
3399 S.M.S.V. Doidge-Harrison, I.W. Nowell, P.J. Cox, R.A. Howie, O.J. Taylor, and J.L. Wardell, *J. Organomet. Chem.*, 1991, **401**, 273; K.G. Penman, W. Kitching, and G. Tagliavini, *Organometallics*, 1991, **10**, 1320.
3400 L.R. Sita and I. Kinoshita, *J. Am. Chem. Soc.*, 1991, **113**, 5070.
3401 L.R. Sita and I. Kinoshita, *J. Am. Chem. Soc.*, 1991, **113**, 1856.
3402 H. Gruetzmacher, H. Pritzkow, and F.T. Edelmann, *Organometallics*, 1991, **10**, 23.
3403 R. Okazaki, K. Shibata, and N. Tokitoh, *Tetrahedron Lett.*, 1991, **32**, 6601.
3404 S. Brooker, J.-K. Buijink, and F.T. Edelmann, *Organometallics*, 1991, **10**, 25.
3405 C. Schneider and M. Dräger, *J. Organomet. Chem.*, 1991, **415**, 349.
3406 A. Haas, J. Kasprowski, K. Angermund, P. Beyz, C. Krüger, Y.-H. Tsay, and S. Werner, *Chem. Ber.*, 1991, **124**, 1895.
3407 C. Drost, U. Klingebiel, and M. Noltemeyer, *J. Organomet. Chem.*, 1991, **414**, 307.

(^{29}Si),[3408] Me_3SiCl(benzothiazoline), (^{13}C),[3409] (176), (^{13}C, ^{29}Si),[3410] $Bu^t_2\overline{SiNRCH=CHNRSi}$-$Bu^t_2$, ($^{13}C$),[3411] $(Bu^t_2SiFNH)_2SiBu^t_2$, (^{13}C, ^{29}Si),[3412] (2,4,6-$Me_3C_6H_2)_2\overline{SiNPhNPhSi}(C_6H_2Me_3$-2,4,6$)_2$, ($^{13}C$, ^{29}Si),[3413] (2,4,6-$Me_3C_6H_2)_3GeNHR$, (^{13}C),[3414] (2,4,6-$Me_3C_6H_2)_2Ge$(fluorenyl)

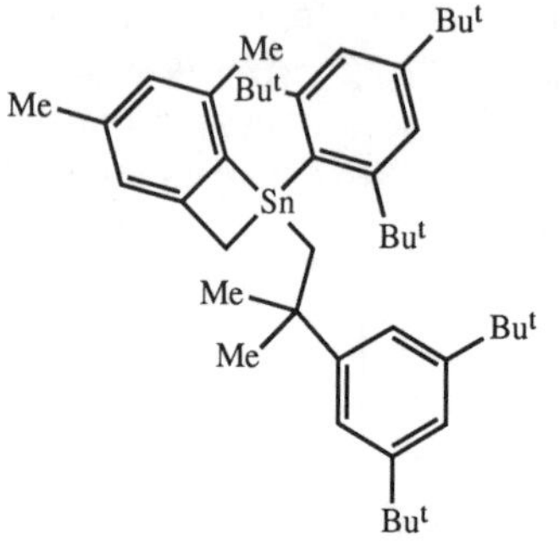

(173)

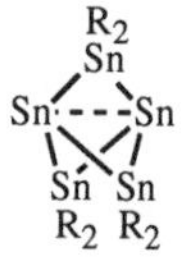

(174)

$N=CPh_2$, (^{13}C),[3415] $Me_3SnN=C=NC(O)CH_2OR$, (^{13}C),[3416] $MeC(O)N(SnMe_3)(OSiMe_3)$, (^{13}C, ^{15}N, ^{29}Si),[3417] $Me_2\overline{Sn(C_6H_4\text{-}2\text{-}CHMeN}Me_2)Br$, ($^{119}Sn$),[3418] Ph_3SnCl complexes with some Schiff bases, (^{13}C),[3419] $(Me_2Bu^tSi)_4P_2$, (^{13}C),[3420] (2,4,6-$Pr^i_3C_6H_2)_2Si=PSiMe_2Bu^t$, ($^{29}Si$),[3421] and $Bu^t_2FSiPHSiBu^t_3$, (^{13}C, ^{29}Si).[3422]

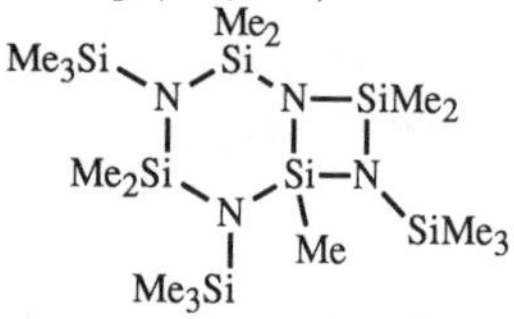

(175)

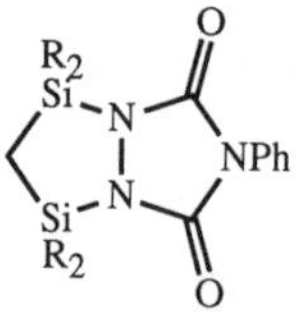

(176)

A selective variant of the standard INEPT experiment has been applied to the assignment of ^{29}Si n.m.r. lines in trimethylsilylated compounds.[3423] The protonated XH heteroatom in oil fractions has

3408 R.J.P. Corriu, D. Leclercq, P.H. Mutin, J.M. Planeix, and A. Vioux, *J. Organomet. Chem.*, 1991, **406**, C1.
3409 K. Singh and J.P. Tandon, *Indian J. Chem., Sect. A*, 1991, **30**, 283 (*Chem. Abstr.*, 1991, **114**, 247 357).
3410 W. Ando, M. Kako, and T. Akasaka, *J. Am. Chem. Soc.*, 1991, **113**, 6286.
3411 M. Weidenbruch, A. Lesch, and K. Peters, *J. Organomet. Chem.*, 1991, **407**, 31.
3412 T. Kottke, U. Klingebiel, M. Noltemeyer, U. Pieper, S. Walter, and D. Stalke, *Chem. Ber.*, 1991, **124**, 1941.
3413 A. Sakakibara, Y. Kabe, T. Shimizu, and W. Ando, *J. Chem. Soc., Chem. Commun.*, 1991, 43.
3414 M. Rivière-Baudet, P. Rivière, A. Castel, A. Morère, and C. Abdennhader, *J. Organomet. Chem.*, 1991, **409**, 131.
3415 M. Lazraq, J. Escudié, C. Couret, J. Satgé, and M. Soufiaoul, *Organometallics*, 1991, **10**, 1140.
3416 L. Jäger, *Z. Anorg. Allg. Chem.*, 1991, **600**, 181.
3417 J. Schraml, H.-M. Boldhaus, F. Erdt, E.W. Krahé, and C. Bliefert, *J. Organomet. Chem.*, 1991, **406**, 299.
3418 J.T.B.H. Jastrzebski, J. Boersma, and G. van Koten, *J. Organomet. Chem.*, 1991, **413**, 43.
3419 A. Kumari, R.V. Singh, and J.P. Tandon, *Indian J. Chem., Sect. A*, 1991, **30**, 468 (*Chem. Abstr.*, 1991, **115**, 71 769).
3420 H. Westermann and M. Nieger, *Inorg. Chim. Acta*, 1991, **177**, 11.
3421 M. Drieß, *Angew. Chem., Int. Ed. Engl.*, 1991, **30**, 1022.
3422 N. Wiberg and H. Schuster, *Chem. Ber.*, 1991, **124**, 93.
3423 V. Blechta and J. Schraml, *Collect. Czech. Chem. Commun.*, 1991, **56**, 258.

been analysed by ^{29}Si n.m.r. spectroscopy after silylation.[3424] The reactivity of $[Me_3SiO]^-$ and $\{(CH_2{=}CH)Me_2Si\}_2O$ has been correlated with ^{17}O n.m.r. chemical shifts.[3425] N.m.r. data have also been reported for $Me_3SiOSiMe_2R$, (^{13}C),[3426] $HCF_2CF_2CH_2OCH_2CH_2CH_2SiMe(OSiMe_3)_2$, ($^{13}C$),[3427] $Me_2(Me_3SiO)SiNH_2$, (^{29}Si),[3428] $Si_8O_{20}(SiMe_3)_8$, (^{29}Si),[3429] Me_3SiOCN_3PhP, (^{13}C),[3430] $(RO)_2P(OSiMe_3)$, (^{13}C),[3431] $\{(4\text{-}MeC_6H_4)_2PO\}_nSiMe_{4-n}$, ($^{13}C$),[3432] Me_3SiOR, (^{13}C),[3433] $Me_2SiR(OSO_2CF_3)$, {R = $C{\equiv}CPh$, $GePh_3$, $SnPh_3$, PPh_2, $N(SiMe_3)_2$; ^{13}C, ^{29}Si},[3434] $Me_2Si(C{\equiv}CPh)(OSO_2CF_3)$, ($^{13}C$),[3435] $Me_2RSiOPr^i$, (^{13}C),[3436] $Me_2(4\text{-}RC_6H_4)SiO_3H$, ($^{13}C$, ^{29}Si),[3437] 1,4-$(HOSiMe_2)_2C_6H_4$, (^{13}C, ^{29}Si),[3438] $\{(C_6H_{13})Me_2Si\}_2O$, (^{29}Si),[3439] $(Pr^iO)Me$-$SiCH_2CH_2CH_2CHCHBu^nOH$, (^{13}C),[3440] 3,5-$(NO_2)_2C_6H_3CO_2CH_2SiMePhBu^t$, ($^{13}C$),[3441] {SiMe-$PhSiMe(OSO_2CF_3)\}_n$, ($^{13}C$, ^{29}Si),[3442] Bu^tMe_2SiOR, (^{13}C),[3443] Bu^tPh_2SiOR, (^{13}C, ^{29}Si),[3444]

3424 C.G. Béguin, C. Bennouna, H. Bitar, A. Dahbi, and L. Léna, *Magn. Reson. Chem.*, 1991, **29**, 576.

3425 L.G. Klapshina, Yu.A. Kurskii, E. Liepiņš, L.M. Terman, and S.V. Shaulova, *Metalloorg. Khim.*, 1991, **4**, 34 (*Chem. Abstr.*, 1991, **114**, 207 340).

3426 S.A. Swint and M.A. Buese, *J. Organomet. Chem.*, 1991, **402**, 145.

3427 G. Sonnek, C. Rabe, G. Schmaucks, R. Kaden, and I. Lehms, *J. Organomet. Chem.*, 1991, **405**, 179.

3428 E. Popowski, I. Hillert, H. Kelling, and H. Jancke, *Z. Anorg. Allg. Chem.*, 1991, **601**, 133.

3429 I. Hasegawa, *Polyhedron*, 1991, **10**, 1097.

3430 Y.K. Rodi, L. Lopez, C. Malavaud, M.T. Boisdon, and J. Barrans, *J. Chem. Soc., Chem. Commun.*, 1991, 23.

3431 A.P. Avdeenko, A.A. Tolmachev, V.V. Pirozhenko, and E.I. Gol'farb, *Zh. Obsch. Khim.*, 1990, **60**, 2272 (*Chem. Abstr.*, 1991, **115**, 29 475).

3432 K.M. Cooke, T.P. Kee, A.L. Langton, and M. Thornton-Pett, *J. Organomet. Chem.*, 1991, **419**, 171.

3433 W. Adam, L. Hadjiarapoglou, V. Jäger, J. Klicić, B. Seidel, and X. Wang, *Chem. Ber.*, 1991, **124**, 2361; R.W. Hoffmann and M. Julius, *Ann.*, 1991, 811; J.A. Faunce, B.A. Grisso, and P.B. Mackenzie, *J. Am. Chem. Soc.*, 1991, **113**, 3418; R.G. Salomon, N.D. Sachinvala, S. Roy, B. Basu, S.R. Raychaudhuri, D.B. Miller, and R.B. Sharma, *J. Am. Chem. Soc.*, 1991, **113**, 3085; A.G. Myers, P.M. Harrington, and E.Y. Kuo, *J. Am. Chem. Soc.*, 1991, **113**, 694; N. Chatani, Y. Kajikawa, H. Nishimura, and S. Murai, *Organometallics*, 1991, **10**, 21; C.M. Hettrick and W.J. Scott, *J. Am. Chem. Soc.*, 1991, **113**, 4903; C.J. Hawker, R. Lee, and J.M.J. Fréchet, *J. Am. Chem. Soc.*, 1991, **113**, 4583; H. Xia, Y. Zou, and R. Pan, *Gaodeng Xuexiao Huaxue Xuebao*, 1991, **12**, 699 (*Chem. Abstr.*, 1991, **115**, 280 631); R. Tripathy, P.J. Carroll, and E.R. Thornton, *J. Am. Chem. Soc.*, 1991, **113**, 7630; D.A. Evans, R.P. Polniaszek, K.M. DeVries, D.E. Guinn, and D.J. Mathre, *J. Am. Chem. Soc.*, 1991, **113**, 7613; R.W. Hoffmann and M. Bewersdorf, *Chem. Ber.*, 1991, **124**, 1259; R.W. Hoffmann, M. Bewersdorf, M. Krüger, W. Mikolaiski, and R. Stürmer, *Chem. Ber.*, 1991, **124**, 1243; A.K. Beck and D. Seebach, *Chem. Ber.*, 1991, **124**, 2897; H. Poleschner and M. Heydenreich, *Magn. Reson. Chem.*, 1991, **29**, 1231; B.H. Kim and J.Y. Lee, *Tetrahedron Asymmetry*, 1991, **2**, 1359; K.I. Sutowardoyo and D. Sinou, *Tetrahedron Asymmetry*, 1991, **2**, 437; H.M.L. Davies, T.J. Clark, and G.F. Kimmer, *J. Org. Chem.*, 1991, **56**, 6440.

3434 W. Uhlig, *J. Organomet. Chem.*, 1991, **409**, 377.

3435 W. Uhlig, *Z. Anorg. Allg. Chem.*, 1991, **603**, 109.

3436 D. Enders and S. Nakai, *Chem. Ber.*, 1991, **124**, 219.

3437 B. Plesnicar, J. Cerkovnik, J. Koller, and F. Kovac, *J. Am. Chem. Soc.*, 1991, **113**, 4946.

3438 G.N. Babu and R.A. Newmark, *Macromolecules*, 1991, **24**, 4503.

3439 T. Suzuki and I. Mita, *J. Organomet. Chem.*, 1991, **414**, 311.

3440 K. Matsumoto, Y. Takeyama, K. Oshima, and K. Utimoto, *Tetrahedron Lett.*, 1991, **32**, 4545.

3441 R. Tacke, S. Brakmann, F. Wuttke, J. Fooladi, C. Syldatk, and D. Schomburg, *J. Organomet. Chem.*, 1991, **403**, 29.

3442 W. Uhlig, *J. Organomet. Chem.*, 1991, **402**, C45.

3443 P.K. Somers, T.J. Wandless, and S.L. Schriber, *J. Am. Chem. Soc.*, 1991, **113**, 8045; W. Adam and X. Wang, *J. Org. Chem.*, 1991, **56**, 7244; M. Lautens and P. Chiu, *Tetrahedron Lett.*, 1991, **32**, 4827; K.C. Nicolaou, C.A. Veale, C.-K. Hwang, J. Hutchinson, C.V.C. Prasad, and W.W. Ogilvie, *Angew. Chem., Int. Ed. Engl.*, 1991, **30**, 299; N. Nakanishi, J. Tsujikawa, N. Hara, T. Tanimoio, and K. Koizumi, *Yakugaku Zazzhi*, 1990, **110**, 477 (*Chem. Abstr.*, 1991, **114**, 6992); T. Matsumoto, M. Katsuki, H. Jona, and K. Suzuki, *J. Am. Chem. Soc.*, 1991, **113**, 6982; C.M. Hudson, M.R. Marzabadi, K.D. Moeller, and D.G. New, *J. Am. Chem. Soc.*, 1991, **113**, 7372; S.E. Denmark and B.R. Henke, *J. Am. Chem. Soc.*, 1991, **113**, 2177; J. Mulzer, H.M. Kirstein, J. Buschmann, C. Lehmann, and P. Luger, *J. Am. Chem. Soc.*, 1991, **113**, 910; D. Seebach, T. Maetzke, W. Petter, B. Klötzer, and D.A. Plattner, *J. Am. Chem. Soc.*, 1991, **113**, 1781; M.J. Pregel and E. Buncel, *Can. J. Chem.*, 1991, **69**, 130; M.W. Anderson, B. Hildebrandt, G. Dahmann, and R.W. Hoffmann, *Chem. Ber.*, 1991, **124**, 2127.

3444 C.M. Dreef-Tromp, E.M.A. van Dam, H. van den Elst, J.E. van den Boogaart, G.A. van der Marel, and J.H. van

Et_3SiOR, (^{13}C),[3445] Pr^i_3SiOR, (^{13}C),[3446] (177), (^{13}C, ^{29}Si),[3447] $R^1{}_nSi(OR^2)_{4-n}$, (^{13}C),[3448] (4-FC_6H_4)(C_6H_{11})$Si(OH)CH_2CH_2CH_2NR_2$, (^{13}C, ^{29}Si),[3449] $R^1R^2R^3MO_2CR^4$, (M = Si, Ge, Sn;

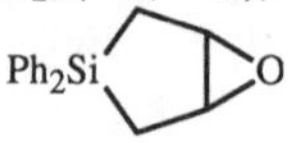

(177)

^{13}C, ^{17}O, ^{29}Si, ^{119}Sn),[3450] $Bu^tSi(OCH_2CH_2CH_2)_3$-2,4,6-C_6H_3, (^{13}C),[3451] $(Me_3CCH_2)Ph_2SiOR$, (^{13}C),[3452] (2,4,6-$Me_3C_6H_2)_2Si(OEt)CHPh_2$, ($^{13}C$, ^{29}Si),[3453] (2,4,6-$Me_3C_6H_2)_2\overline{SiESi}(C_6H_2Me_3$-2,4,6$)_2$, (E = O, S, Se, Te; ^{29}Si, ^{77}Se, ^{125}Te),[3454] and $Ph_2RSiOMe$, (^{13}C).[3455]

^1H and ^{13}C n.m.r. spectroscopy has been used to characterise the products from the reaction of $(Bu^n{}_3Sn)_2O$ with $ArCO_2H$.[3456] N.m.r. data have also been reported for (2,4,6-$Me_3C_6H_2)_2Ge$-(fulvenyl)$OCMe{=}CH_2$, (^{13}C),[3457] $R^1{}_nSn(O_2CCH_2S_2CNR^2{}_2)_{4-n}$, ($^{13}C$),[3458] $R_2\overline{Sn(CH_2)_nSO}_3$, ($^{13}C$, ^{119}Sn),[3459] (178), (^{13}C),[3460] (179), (^{119}Sn),[3461] poly($Bu^n{}_3Sn$-methacrylate), (^{13}C),[3462] $Bu^n{}_3Sn$

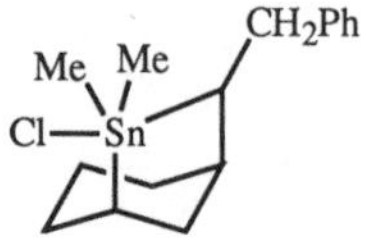

(178)

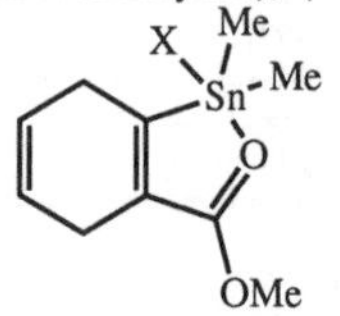

(179)

containing antifouling polymers, (^{13}C),[3463] tin in reaction materials where $Bu^n{}_3SnF$ has been used as an aldol catalyst, (^{119}Sn),[3464] $Bu^n{}_3SnOAc$, (^{13}C, ^{119}Sn),[3465] $Bu^n{}_3SnO_2CR$, (^{13}C, ^{119}Sn),[3466]

Boom, *Recl. Trav. Chim. Pays-Bas*, 1991, **110**, 378; I.W.J. Still and W. Daoquan, *Phosphorus Sulphur*, 1991, **62**, 83.

3445 D.L.J. Clive, C. Zhang, K.S.K. Murthy, W.D. Hayward, and S. Daigneault, *J. Org. Chem.*, 1991, **56**, 6447; L.D. Quin, J.C. Kisalus, J.J. Skolimowski, and N.S. Rao, *Phosphorus Sulphur*, 1990, **54**, 1.

3446 R. Munschauer and G. Maas, *Angew. Chem., Int. Ed. Engl.*, 1991, **30**, 306.

3447 Y.T. Park, G. Manuel, R. Bau, D. Zhao, and W.P. Weber, *Organometallics*, 1991, **10**, 1586.

3448 M. Harkonen, J.V. Seppala, and T. Vaananen, *Makromol. Chem.*, 1991, **192**, 721 (*Chem. Abstr.*, 1991, **114**, 207 910).

3449 R. Tacke, K. Mahner, C. Strohmann, B. Forth, E. Mutschler, T. Friebe, and G. Lambrecht, *J. Organomet. Chem.*, 1991, **417**, 339.

3450 I. Zicmane, E. Liepiņš, L.M. Ignatovich, and E. Lukevics, *J. Organomet. Chem.*, 1991, **417**, 355.

3451 R. Damrauer, J.A. Hankin, and R.C. Haltiwanger, *Organometallics*, 1991, **10**, 3962.

3452 M.H. Hopkins, L.E. Overman, and G.M. Rishton, *J. Am. Chem. Soc.*, 1991, **113**, 5354.

3453 K. Kabeta, D.R. Powell, J. Hanson, and R. West, *Organometallics*, 1991, **10**, 827.

3454 R.P.-K. Tan, G.R. Gillette, D.R. Powell, and R. West, *Organometallics*, 1991, **10**, 546.

3455 B.R. Yoo, M.E. Lee, and I.N. Jung, *J. Organomet. Chem.*, 1991, **410**, 33.

3456 M.G. Muralidhara and V. Chandrasekhar, *Indian J. Chem., Sect. A*, 1991, **30**, 487 (*Chem. Abstr.*, 1991, **115**, 92 453).

3457 M. Lazraq, C. Couret, J. Escudie, J. Satgé, and M. Dräger, *Organometallics*, 1991, **10**, 1771.

3458 S.W. Ng and V.G.K. Das, *J. Organomet. Chem.*, 1991, **409**, 143.

3459 A. Bernstein and H. Weichmann, *Z. Anorg. Allg. Chem.*, 1991, **603**, 41.

3460 M. Ochiai, S. Iwaki, Y. Nagao, S. Kitagawa, and M. Munakata, *Tetrahedron Lett.*, 1991, **32**, 4945.

3461 U. Kolb, M. Dräger, and B. Jousseaume, *Organometallics*, 1991, **10**, 2737.

3462 D. Xu, Y. Wang, Z. Wu, and Z. Han, *Gongneng Gaofenzi Xuebao*, 1990, **3**, 48 (*Chem. Abstr.*, 1991, **115**, 30 196); Z. Han, S. Li, Z. Zhu, and Q. Zhu, *Gaofenzi Xuebao*, 1989, 600 (*Chem. Abstr.*, 1991, **114**, 164 933).

3463 R.R. Joshi, J.R. Dharia, and S.K. Gupta, *Inorg. Met.-Containing Polym. Mater., [Proc. Am. Chem. Soc. Int. Symp.] 1989*, (Pub. 1990), 381 (*Chem. Abstr.*, 1991, **115**, 262 972).

3464 S. Kobayashi, H. Uchiro, Y. Fujishita, I. Shiina, and T. Mukaiyama, *J. Am. Chem. Soc.*, 1991, **113**, 4247.

3465 H.-V. Pham, B. Rusterholz, E. Pretsch, and W. Simon, *J. Organomet. Chem.*, 1991, **403**, 311.

3466 Q. Xie, S. Li, S. Zhang, D. Zhang, Z. Zhang, and J. Hu, *Huaxue Xuebao*, 1991, **49**, 723 (*Chem. Abstr.*, 1991, **115**, 208 118).

$Bu^n{}_3SnO_3SAr$, (^{119}Sn),[3467] $[Bu^n{}_3SnO_2CC_6H_4SO_3]^-$, ($^{13}C$),[3468] $(Bu^n{}_3Sn)_2O$ derivatives of phenolic compounds, (^{119}Sn),[3469] $ClBu_2Sn(CH_2)_3O(CH_2CH_2O)_2Me$, ($^{13}C$, ^{119}Sn),[3470] $R^1{}_3SnOP(O)(H)(OR^2)$, (^{13}C, ^{119}Sn),[3471] $N(CH_2CHRO)_3Ge(OR^2)$, $R^1{}_3SnOR^2$, (^{13}C),[3472] and $(C_6H_{11})_3SnO_2CCH_2OR$, ($^{13}C$).[3473]

The thiocarbonyl ^{13}C chemical shifts of $Pr^iCS_2MMe_3$, M = C, Si, Ge, Sn, Pb, correlate linearly with the Allred-Rochow electronegativity of the metals and their π-π^* transition energies except for the silyl esters.[3474] The temperature dependence of the ^{119}Sn chemical shifts of $Bu^n{}_3SnSR$ has been examined. The ^{13}C chemical shifts were also recorded.[3475] Mercaptans in petroleum fractions have been characterized by ^{119}Sn n.m.r. spectroscopy after reaction with R_3Sn reagents.[3476] N.m.r. data have also been reported for Bu^tMe_2SiSR, (^{13}C),[3477] $Ph_nGe(S_2COR)_{4-n}$, (^{13}C),[3478] and (2,4,6-$Pr^i{}_3C_6H_2)_2SnTeSn(C_6H_2Pr^i{}_3$-2,4,6$)_2$, ($^{13}C$, ^{119}Sn, ^{125}Te).[3479]

^{29}Si n.m.r. spectroscopy has been used to infer the coordination of silicon in (180). The ^{13}C n.m.r. spectra were also recorded.[3480] N.m.r. data have also been reported for 4-$PhC_6H_4SiPr^i{}_2Cl$, (^{13}C),[3481] (fluorenyl)$_3$GeF, (^{13}C),[3482] Me_3SnCl, (^{119}Sn),[3483] R_3SnCl, (^{13}C),[3484] $[Me_3SnCl_2]^-$, (^{13}C, ^{119}Sn),[3485] 2-{2,4-$(NO_2)_2C_6H_3N$=N}$C_6H_4SnMe_2Cl$, (^{13}C, ^{119}Sn),[3486] $RCH(OH)CH_2C(=CH_2)SnMe_2Cl$, ($^{13}C$, ^{119}Sn),[3487] $\{(Me_3Si)_2CH\}_2SnXCR_2H$, ($^{13}C$, ^{119}Sn),[3488] and $RSnMe_2I$, (^{13}C).[3489]

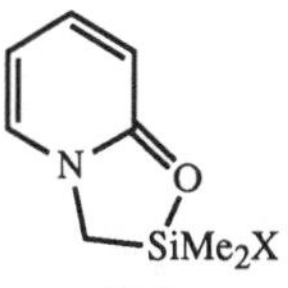

(180)

3467 F. Thunecke and R. Borsdorf, *J. Prakt. Chem.*, 1991, **333**, 489.
3468 S.W. Ng, V.G.K. Das, and E.R.T. Tiekink, *J. Organomet. Chem.*, 1991, **411**, 121.
3469 E. Kolehmainen, J. Paasivirta, R. Kauppinen, T. Otollinen, S. Kasa, and R. Herzschuh, *Int. J. Environ. Anal. Chem.*, 1991, **43**, 19 (*Chem. Abstr.*, 1991, **114**, 214 013).
3470 F. Ferkous, D. Messadi, B. De Jeso, M. Degueil-Castaing, and B. Maillard, *J. Organomet. Chem.*, 1991, **420**, 315.
3471 E. Nietzschmann and K. Kellner, *Phosphorus Sulphur*, 1991, **55**, 73.
3472 M. Nasim, L.I. Livantsova, G.S. Zaitseva, and J. Lorberth, *J. Organomet. Chem.*, 1991, **403**, 85.
3473 Q. Xie and J. Zheng, *Youji Huaxue*, 1991, **11**, 82.
3474 S. Kato, A. Hori, M. Mizuta, T. Kalada, K. Fujieda, and Y. Ikebe, *J. Organomet. Chem.*, 1991, **420**, 13.
3475 F. Thunecke, D. Schulze, and R. Borsdorf, *Z. Chem.*, 1990, **30**, 444 (*Chem. Abstr.*, 1991, **114**, 185 659).
3476 E. Rafii, M. Nagassoum, R. Faure, R. Foon, L. Lena, and J. Metzger, *Fuel*, 1991, **70**, 132 (*Chem. Abstr.*, 1991, **114**, 65 411).
3477 G.A. Kraus and B. Andersh, *Tetrahedron Lett.*, 1991, **32**, 2189.
3478 J.E. Drake, A.G. Mislankar, and M.L.Y. Wong, *Inorg. Chem.*, 1991, **30**, 2174.
3479 A. Schäfer, M. Weidenbruch, W. Saak, S. Pohl, and H. Marsmann, *Angew. Chem., Int. Ed. Engl.*, 1991, **30**, 834.
3480 A.R. Bassindale and M. Borbaruah, *J. Chem. Soc., Chem. Commun.*, 1991, 1501.
3481 J. Anthony and F. Diederich, *Tetrahedron Lett.*, 1991, **32**, 3787.
3482 G. Anselme, J. Escudié, C. Couret, and J. Satgé, *J. Organomet. Chem.*, 1991, **403**, 93.
3483 W. Linert, A. Sotriffer, and V. Gutmann, *J. Coord. Chem.*, 1990, **22**, 21.
3484 T.N. Mitchell, K. Kwetkat, and B. Godry, *Organometallics*, 1991, **10**, 1633.
3485 S.E. Johnson, K. Polborn, and H. Nöth, *Inorg. Chem.*, 1991, **30**, 1410.
3486 W.P. Neumann and C. Wicenec, *J. Organomet. Chem.*, 1991, **420**, 171.
3487 T.N. Mitchell, U. Schneider, and K. Heesche-Wagner, *J. Organomet. Chem.*, 1991, **411**, 107.
3488 G. Anselme, C. Couret, J. Escudié, S. Richelme, and J. Satgé, *J. Organomet. Chem.*, 1991, **418**, 321.
3489 A.J. Lucke and D.J. Young, *Tetrahedron Lett.*, 1991, **32**, 807.

The temperature and pressure dependence of self diffusion of $(Me_2SiO)_4$ and $(Me_2SiNH)_3$ has been examined.[3490] Chain motion at the $(Me_2SiO)_n$-filler interface has been studied by ^1H and ^2H n.m.r. spectroscopy.[3491] The orientation ability of thermotropic liquid crystalline comb-like siloxane polymers has been studied by ^1H n.m.r. spectroscopy.[3492] ^{13}C n.m.r. spectroscopy has been used to investigate the local dynamics in non-oriented mesomorphic polysiloxanes.[3493] Ph_2SiCl_2 has been used as a dimerization agent to convert enantiomeric alcohols into a mixture of meso and DL diastereomeric silyl acetals which were distinguished by ^{13}C n.m.r. spectroscopy.[3494] ^2H n.m.r. spectroscopy has been used to study C_6D_6 in a liquid crystalline polysiloxane.[3495] N.m.r. data have also been reported for $PhN(SiMe_2OSiMe_2)_2O$, (^{29}Si),[3496] (181), (^{29}Si),[3497] (182), (^{13}C),[3498] $Me_2SiClN{=}PR^1R^2R^3$, (^{13}C, ^{29}Si, ^{31}P),[3499] $O(R_2SiOSiR_2)_2Se$, (^{13}C, ^{29}Si, ^{77}Se),[3500] $(MeO_2CHMeCH_2SiO)_n$, (^{13}C),[3501] polysiloxane starburst polymers, (^{13}C, ^{29}Si),[3502] polysiloxanes, (^{29}Si),[3503] 5,6-benzo-1,3,2-dioxa(dithia)silepins, (^{13}C),[3504] silarylene-siloxane copolymers, (^{13}C, ^{29}Si),[3505] fluorinated polysiloxanes, (^{13}C, ^{29}Si),[3506] (183), (^{13}C, ^{29}Si),[3507] $MeSi(OMe)_2(CH_2$-$NC_4H_8)$, (^{13}C, ^{29}Si),[3508] $NC(CH_2)_2SiMe(OCH_2CH_2)_2NMe$, ($^{13}C$),[3509] polyorganosiloxane alkali metal alcoholates, (^{29}Si),[3510] $(Bu^t_2SiFNBu^t)_2S_2$, (^{13}C, ^{29}Si),[3511] (184), (^{13}C, ^{29}Si),[3512]

3490 A. Greiner-Schmid, M. Has, and H.D. Luedemann, Z. *Naturforsch., A*, 1990, **45**, 1281 (*Chem. Abstr.*, 1991, **114**, 192 980).
3491 V.M. Litvinov and H.W. Spiess, *Makromol. Chem., Macromol. Symp.*, 1991, **44**, 33 (*Chem. Abstr.*, 1991, **115**, 137 433).
3492 H. Roth and B. Kruecke, *Mol. Cryst. Liq. Cryst.*, 1990, **193**, 223 (*Chem. Abstr.*, 1991, **114**, 92 293); H. Roth and B. Kruecke, *Acta Polym.*, 1991, **42**, 140 (*Chem. Abstr.*, 1991, **115**, 72 582).
3493 H. Ouladi, F. Laupretre, P. Sergot, L. Monnerie, M. Mauzac, and H. Richard, *Macromolecules*, 1991, **24**, 1800 (*Chem. Abstr.*, 1991, **114**, 108 173).
3494 X. Wang, *Tetrahedron Lett.*, 1991, **32**, 3651.
3495 M. Schulz, A. Van der Est, E. Roessler, G. Kossmehl, and H.M. Vieth, *Macromolecules*, 1991, **24**, 5040 (*Chem. Abstr.*, 1991, **115**, 93 391).
3496 Z. Lasocki and M. Witekowa, *J. Organomet. Chem.*, 1991, **408**, 27.
3497 A.B. Zachernyuk, V.B. Isaev, B.D. Lavrukhin, and A.A. Zhdanov, *Zh. Obshch. Khim.*, 1990, **60**, 2107 (*Chem. Abstr.*, 1991, **114**, 81 934).
3498 D. Singh, R.B. Goyal, and R.V. Singh, *Appl. Organomet. Chem.*, 1991, **5**, 45 (*Chem. Abstr.*, 1991, **114**, 185 610).
3499 W. Wolfsberger, *Chem.-Ztg.*, 1991, **115**, 89 (*Chem. Abstr.*, 1991, **115**, 71 697).
3500 P. Boudjouk, S.R. Bahr, and D.P. Thompson, *Organometallics*, 1991, **10**, 778.
3501 M. Hong, N. Qiu, and J. Hu, *Xiamen Daxue Xuebao, Ziran Kexueban*, 1990, **29**, 300 (*Chem. Abstr.*, 1991, **115**, 257 298).
3502 A. Morikawa, M. Kakimoto, and Y. Imai, *Macromolecules*, 1991, **24**, 3469 (*Chem. Abstr.*, 1991, **114**, 247 907).
3503 M.J. Van Bommel, T.N.M. Bernards, and A.H. Boonstra, *J. Non-Cryst. Solids*, 1991, **128**, 231 (*Chem. Abstr.*, 1991, **115**, 55 142); J.M. Bellama, S.R. Meyer, and R.E. Pellenbarg, *Appl. Organomet. Chem.*, 1991, **5**, 107 (*Chem. Abstr.*, 1991, **115**, 21 184).
3504 I.Kh. Shakirov, L.V. Egorova, V.V. Klochkov, L.K. Yuldasheva, P.P. Chernov, A.R. Safiullin, N.D. Ibragimova, and R.P. Arshinova, *Zh. Obshch. Khim.*, 1991, **61**, 147 (*Chem. Abstr.*, 1991, **115**, 92 359).
3505 R.A. Newmark and G.N. Babu, *Macromolecules*, 1991, **24**, 4510 (*Chem. Abstr.*, 1991, **115**, 72 635).
3506 B. Boutevin and B. Youssef, *Macromolecules*, 1991, **24**, 629 (*Chem. Abstr.*, 1991, **114**, 63 995).
3507 R.J.P. Corriu, G.F. Lanneau, and V.D. Mehta, *J. Organomet. Chem.*, 1991, **419**, 9.
3508 R. Tacke, J. Sperlich, C. Strohmann, and G. Mattern, *Chem. Ber.*, 1991, **124**, 1491.
3509 V.V. Belyaeva, E.I. Brodskaya, G.A. Kuznetsova, V.P. Baryshok, T.V. Kashik, and M.G. Voronkov, *Metalloorg. Khim.*, 1991, **4**, 808 (*Chem. Abstr.*, 1991, **115**, 208 073).
3510 O.V. Kononov, S.Ya. Lazarev, V.D. Lobkov, and L.V. Osetrova, *Vysokomol. Soedin., Ser. B*, 1991, **33**, 330 (*Chem. Abstr.*, 1991, **115**, 208 677).
3511 U. Klingebiel, F. Pauer, G.M. Sheldrick, and D. Stalke, *Chem. Ber.*, 1991, **124**, 2651.
3512 R. Corriu, G. Lanneau, and C. Priou, *Angew. Chem., Int. Ed. Engl.*, 1991, **30**, 1130.

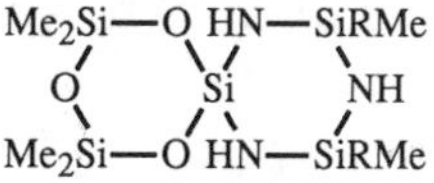
(181)

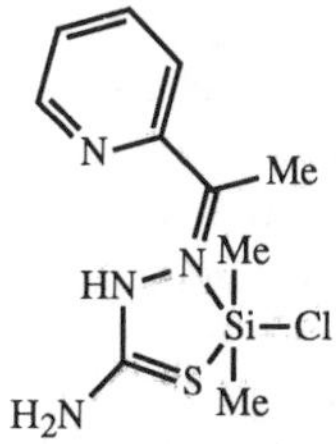
(182)

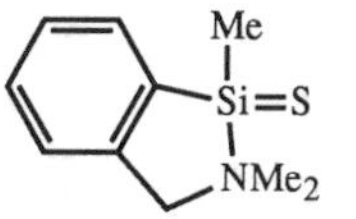
(183)

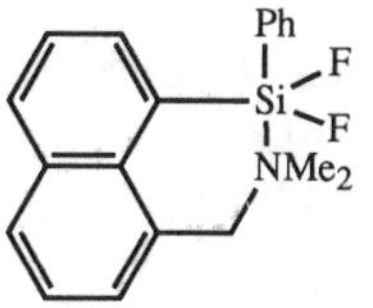
(184)

$Bu^t_2Si(PHPh)_2$, (^{29}Si),[3513] $P(SiBu^t_2F)_2Li(THF)_2$, (^{13}C, ^{29}Si),[3514] Ph_2SiL_2, (L_2 = *N,S*-chelate; ^{13}C),[3515] $Cl_2SiCR{=}CRCHCH_2Bu^t$, (^{13}C, ^{29}Si),[3516] (2,4,6-$Bu^t_3C_6H_2$)P=$SiBu^t$($C_6H_2Bu^t_3$-2,4,6), (^{13}C, ^{29}Si),[3517] and $(Bu^t_2SiOH)_2$, (^{29}Si).[3518]

^1H n.O.e. measurements have been used to show that R_2Sn(8-quinolinolate) and R_2Sn(8-quinolinethiolate) are distorted *cis*-octahedral. The ^{13}C T_1 values were also reported.[3519] N.m.r. data have also been reported for [Et_2Ge(3-N-2-MeO_2C-thiophene)]$_2$, (^{13}C),[3520] $CH_2CMe{=}CMeCH_2Ge$-$(NBu^t)_2SiMe_2$, (^{13}C),[3521] (fulvenyl)$_2Ge(OMe)_2$, (^{13}C),[3522] (2,4,6-$Me_3C_6H_2)_2GeOCHPhP$-($C_6H_2Pr^i_3$-2,4,6), (^{13}C),[3523] $R^1R^2GeSe_2C_6H_3R^3$, (^{77}Se),[3524] $Me_2SnCl_2(OCH_2C_3H_3N_2)$, ($^{119}Sn$),[3525] $Me_2SnCl_2(HOC_{10}H_6CH{=}NC_6H_4OMe)_2$, ($^{13}C$),[3526] (185), ($^{13}C$, ^{119}Sn),[3527] Me_2Sn-$(O_2CCBr_3)_2$, (^{13}C, ^{119}Sn),[3528] $Et_2Sn(O_2CC_4H_3S)_2$, (^{13}C, ^{119}Sn),[3529] $[ClMe_2SnS_2C_2S_2CO]^-$, (^{13}C, ^{119}Sn),[3530] $Bu^n_2Sn\{OP(O)(H)OEt\}_2$, ($^{13}C$, ^{119}Sn),[3531] $R^1_2Sn(O_2CR^2)_2$, (^{119}Sn),[3532] [{Bu^n_2Sn-(OC_6H_4OMe-2)}$_2$O]$_2$, (^{13}C, ^{119}Sn),[3533] $Bu^n_2Sn(O_2CR)_2$, (^{13}C, ^{17}O, ^{119}Sn),[3534] ($H_2NC_6H_4$-$CO_2)_2SnBu^n_2$, (^{13}C, ^{17}O, ^{119}Sn),[3535] [$Bu_2Sn(O_2CC_6H_3XY)_2$], (^{13}C, ^{119}Sn),[3536] (MeO_2CCH_2-

3513 W. Uhlig, *Z. Anorg. Allg. Chem.*, 1991, **601**, 125.
3514 U. Klingebiel, M. Meyer, U. Pieper, and D. Stalke, *J. Organomet. Chem.*, 1991, **408**, 19.
3515 D. Singh and R.V. Singh, *Phosphorus Sulphur*, 1991, **61**, 57.
3516 N. Auner, C. Seidenschwarz, and E. Herdtweck, *Angew. Chem., Int. Ed. Engl.*, 1991, **30**, 1151.
3517 Y. Van den Winkel, H.M.M. Bastiaans, and F. Bickelhaupt, *J. Organomet. Chem.*, 1991, **405**, 183.
3518 R. West and E.K. Pham, *J. Organomet. Chem.*, 1991, **403**, 43.
3519 I. Tkáč, J. Holeček, and A. Lyčka, *J. Organomet. Chem.*, 1991, **418**, 311.
3520 M. Rivière-Baudet and A. Morère, *Phosphorus Sulphur*, 1991, **62**, 211.
3521 G. Billeb, K. Bootz, W.P. Neumann, and G. Steinhoff, *J. Organomet. Chem.*, 1991, **406**, 303.
3522 G. Anselme, C. Couret, J. Escudié, and J. Satgé, *Synth. React. Inorg. Metal-Org. Chem.*, 1991, **21**, 229.
3523 H. Ranaivonjatovo, J. Escudié, C. Couret, and J. Satgé, *J. Organomet. Chem.*, 1991, **415**, 327.
3524 P. Tavares, P. Meunier, B. Gautheron, G. Dousse, and H. Lavayssiere, *Phosphorus Sulphur*, 1991, **55**, 249.
3525 B. Salgado, E. Freijanes, A.S. Gonzáles, J.S. Casas, J. Sordo, U. Casellato, and R. Graziani, *Inorg. Chim. Acta*, 1991, **185**, 137.
3526 S.-G. Teoh, S.-B. Teo, G.-Y. Yeap, and J.-P. Declercq, *Polyhedron*, 1991, **10**, 2683.
3527 A. Jain, S. Saxena, and A.K. Rai, *Indian J. Chem., Sect. A*, 1991, **30**, 881 (*Chem. Abstr.*, 1991, **115**, 232 392).
3528 R. Parkash, M. Meena, and S. Singh, *Bull. Chem. Soc. Jpn.*, 1991, **64**, 1443.
3529 C. Vatsa, V.K. Jain, T. Kesavadas, and E.R.T. Tiekink, *J. Organomet. Chem.*, 1991, **410**, 135.
3530 S.M.S.V. Doidge-Harrison, R.A. Howie, J.T.S. Irvine, G. Spencer, and J.L. Wardell, *J. Organomet. Chem.*, 1991, **414**, C5.
3531 K. Kellner, L. Rodewald, and K. Schenzel, *Phosphorus Sulphur*, 1991, **55**, 65.
3532 S.J. Blunden, R. Hill, and S.E. Sutton, *Appl. Organomet. Chem.*, 1991, **5**, 159 (*Chem. Abstr.*, 1991, **115**, 114 664).
3533 C. Vatsa, V.K. Jain, T.K. Das, and E.R.T. Tiekink, *J. Organomet. Chem.*, 1991, **418**, 329.
3534 J. Holeček, A. Lyčka, M. Nádvorník, and K. Handlír, *Collect. Czech. Chem. Commun.*, 1991, **56**, 1908.
3535 S.P. Narula, S.K. Bharadwaj, H.K. Sharma, Y. Sharda, and G. Mairesse, *J. Organomet. Chem.*, 1991, **415**, 203.
3536 A. Meriem, R. Willem, M. Biesemans, B. Mahieu, D. de Vos, P. Lelieveld, and M. Gielen, *Appl. Organomet. Chem.*, 1991, **5**, 195 (*Chem. Abstr.*, 1991, **115**, 136 249).

$CH_2)_2Sn(S_2CNMe_2)Cl$, (^{13}C, ^{119}Sn),[3537] $[MeN(CH_2CH_2CH_2)_2SnS]_n$, ($^{119}Sn$),[3538] $Sn\{CH(SiMe_3)C_6H_4NMe_2\text{-}2\}_2$, ($^{13}C$),[3539] $\{2\text{-}R^1NHC_6H_4CO_2SnR^2{}_2\}_2O$, ($^{13}C$, ^{119}Sn),[3540] R_2Sn(pyridine-2-carboxylate)$_2$, (^{13}C, ^{119}Sn),[3541] $[\{R_2Sn(O_2CBu^t)\}_2O]_2$, ($^{13}C$, ^{119}Sn),[3542] $[Bu_2Sn(O_2CC_6H_4S\text{-}2\text{-})_2]$, ($^{13}C$, ^{119}Sn),[3543] (8-Me_2N-1-naphthyl)$_2SnX_2$, (^{13}C, ^{119}Sn),[3544] $[2,4,6\text{-}\{(Me_3Si)_2CH\}_3C_6H_2]_2SnSe_4$, ($^{13}C$, ^{77}Se),[3545] and $(2,4,6\text{-}Pr^i{}_3C_6H_2)_2Sn(\mu\text{-}Se)_2Sn(C_6H_2Pr^i{}_3\text{-}2,4,6)_2$, ($^{13}C$, ^{77}Se, ^{119}Sn).[3546]

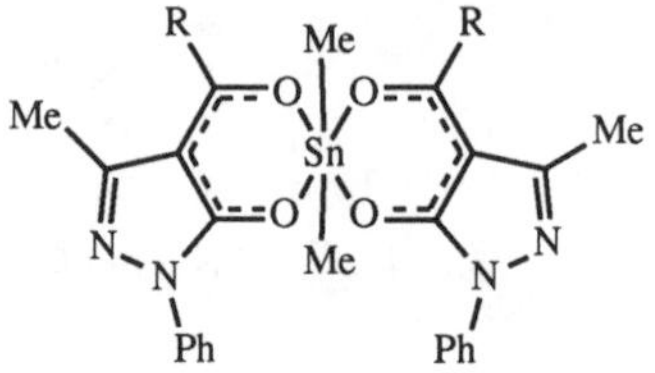

(185)

^{13}C and ^{29}Si n.m.r. spectroscopy has been used to study the polycondensation of heteropolysiloxanes.[3547] N.m.r. data have also been reported for methoxysilyl terminated butadienes, (^{13}C, ^{29}Si),[3548] a highly branched polycarbosilane, (^{13}C, ^{29}Si),[3549] $C_4H_8NHCH_2CH_2Si(O_2C_{10}H_6)_2$, ($^{13}C$, ^{29}Si),[3550] (171), ($R = C_6H_{13}$; ^{13}C),[3551] $(EtO)_3SiCHMeCH_2Si(OEt)_3$, ($^{13}C$),[3552] $(Pr^iSi)_4(PH)_6$, (^{29}Si),[3553] $(C_6H_{11})_7Si_7O_{12}E$, (E = P, As, Sb; ^{13}C, ^{29}Si),[3554] $N(CH_2CHRO)_3Si\overline{CHCH_2O}$, ($^{13}C$),[3555] $[(C_{10}H_6O_2)_2SiPh]^-$, (^{13}C, ^{29}Si),[3556] oligophenylethoxysilanes, (^{29}Si),[3557] $Cl_3GeCH(C_6H_4R)CH_2CO_2H$, ($^{13}C$),[3558] *trans*-$Bu^t{}_2$-(1-Ph-3-Me-4-PhCO-pyrazolon-5-ato)$_2Sn$, (^{13}C,

3537 O.-S. Jung, J.H. Jeong, and Y.S. Sohn, *Organometallics*, 1991, **10**, 2217.
3538 B.M. Schmidt, M. Draeger, and K. Jurkschat, *J. Organomet. Chem.*, 1991, **410**, 43.
3539 J.T.B.H. Jastrzebski, P. Van der Schaaf, J. Boersma, and G. Van Koten, *New J. Chem.*, 1991, **15**, 301.
3540 A. Meriem, M. Biesemans, R. Willem, B. Mahieu, D. De Vos, P. Lelieveld, and M. Geilen, *Bull. Soc. Chim. Belg.*, 1991, **100**, 367.
3541 C. Vatsa, V.K. Jain, and T. Kesavadas, *Indian J. Chem., Sect. A*, 1991, **30**, 451 (*Chem. Abstr.*, 1991, **115**, 49 843).
3542 C. Vatsa, V.K. Jain, T. Kesavadas, and E.R.T. Tiekink, *J. Organomet. Chem.*, 1991, **408**, 157.
3543 M. Boualam, R. Willem, M. Biesemans, B. Mahieu, and M. Gielen, *Heteroat. Chem.*, 1991, **2**, 447 (*Chem. Abstr.*, 1991, **115**, 256 305).
3544 J.T.B.H. Jastrzebski, P.A. Van der Schaaf, J. Boersma, G. Van Koten, M. De Wit, Y. Wang, D. Heijdenrijk, and C.H. Stam, *J. Organomet. Chem.*, 1991, **407**, 301.
3545 N. Tokitoh, Y. Matsuhashi, and R. Okazaki, *Tetrahedron Lett.*, 1991, **32**, 6151.
3546 A. Schäfer, M. Weidenbruch, W. Saak, S. Pohl, and H. Marsmann, *Angew. Chem., Int. Ed. Engl.*, 1991, **30**, 963.
3547 M.P. Besland, C. Guizard, N. Hovnanian, A. Larbot, L. Cot, J. Sanz, I. Sobrados, and M. Gregorkiewitz, *J. Am. Chem. Soc.*, 1991, **113**, 1982.
3548 W.E. Lindsell, K. Radha, and I. Soutar, *Polym. Int.*, 1991, **25**, 1 (*Chem. Abstr.*, 1991, **115**, 30 887).
3549 C.K. Whitmarsh and L.V. Interrante, *Organometallics*, 1991, **10**, 1336.
3550 C. Strohmann, R. Tacke, G. Mattern, and W.F. Kuhs, *J. Organomet. Chem.*, 1991, **403**, 63.
3551 D. Herren, H. Bürgy, and G. Calzaferri, *Helv. Chim. Acta*, 1991, **74**, 24.
3552 B. Marciniec, H. Maciejewski, and J. Mirecki, *J. Organomet. Chem.*, 1991, **418**, 61.
3553 M. Baudler, W. Oehlert, and K.F. Tebbe, *Z. Anorg. Allg. Chem.*, 1991, **598**, 9.
3554 F.J. Feher and T.A. Budzichowski, *Organometallics*, 1991, **10**, 812.
3555 M. Nasim, L.I. Livantsova, D.P. Krut'ko, G.S. Zaitseva, J. Lorberth, and M. Otto, *J. Organomet. Chem.*, 1991, **402**, 313.
3556 K.C.K. Swamy, C. Sreelatha, R.O. Day, J. Holmes, and R.R. Holmes, *Inorg. Chem.*, 1991, **30**, 3126.
3557 V.V. Severnyi, M.A. Karyugin, A.V. Kisin, and I.B. Sokol'skaya, *Zavod. Lab.*, 1990, **56**, 41 (*Chem. Abstr.*, 1991, **114**, 103 164).
3558 L. Sun, Y. Wu, and M. Bai, *Youji Huaxue*, 1991, **11**, 303 (*Chem. Abstr.*, 1991, **115**, 136 246).

^{119}Sn),[3559] $RSi(OCH_2CH_2)_3N$, (^{13}C, ^{29}Si),[3560] $[BuSn(O)(O_2CR)]_6$, (^{13}C, ^{119}Sn),[3561] [BuSn(S)-{S=C(Ph)CHCPh=O}]$_2$, (^{13}C, ^{119}Sn),[3562] $ROOCCH_2CH_2Sn(S_2CNMe_2)(E^1CH_2CH_2)_2E^2$, ($E^1$, E^2 = O, S, NMe; ^{119}Sn),[3563] $(C_7H_{13})_2Si_7O_{12}SnMe$, ($^{13}C$, ^{29}Si),[3564] and (186), (^{13}C, ^{119}Sn).[3565]

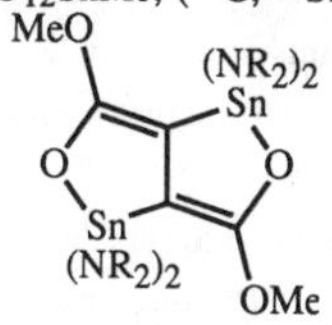

(186)

The ^{29}Si chemical shift of $N_3Si(NHCH_2CH_2)_3N$ is anomolously low relative to the Cl-Si analogue. The ^{13}C n.m.r. data were also reported.[3566] The effect of deprotonation on the ^{29}Si n.m.r. shielding for the series $Si(OH)_4$ to $[SiO_4]^{4-}$ has been calculated.[3567] ^{207}Pb n.m.r. spectroscopy has been used to monitor the condensation of lead alkoxides.[3568] ^1H and ^{13}C n.m.r. spectroscopy has been used to study solid and liquid crystalline Pb^{II} soaps.[3569] The ^{19}F T_1 of SiF_4 in O_2 has been measured as a function of density, temperature and magnetic field.[3570] N.m.r. data have also been reported for Sn^{IV} protoporphyrin IX complex of equine myoglobin, (^{119}Sn),[3571] $(acac)_2Si(NCS)_2$, (^{13}C, ^{29}Si),[3572] $[Si(maltolate)_3]^+$, (^{29}Si),[3573] $(MeO)_2Si\{OCO(CF_2)_4H\}_2$, ($^{13}C$),[3574] $Sn\{(OPPh_2)_2N\}_2X_2$, (^{13}C, ^{119}Sn),[3575] $Sn(S_2COR)_2ClX$, (^{13}C, ^{119}Sn),[3576] and $[Pb(EPh)_3]^-$, (^{207}Pb).[3577]

8 Group VB Elements

A review has appeared:- 'Organometallic chemistry of diphosphazene ligands', which contains n.m.r. spectroscopy.[3578]

T_1 has been measured as a function of temperature for ^{14}N in N_2 gas in mixtures with Ar, Kr, Xe,

3559 C. Pettinari, G. Rafaiani, G.G. Lobbia, A. Lorenzotti, F. Bonati, and B. Bovio, *J. Organomet. Chem.*, 1991, **405**, 75.
3560 R.J. Garant, L.M. Daniels, S.K. Das, M.N. Janakiraman, R.A. Jacobson, and J.G. Verkade, *J. Am. Chem. Soc.*, 1991, **113**, 5728.
3561 V.B. Mokal, V.K. Jain, and E.R.T. Tiekink, *J. Organomet. Chem.*, 1991, **407**, 173.
3562 N. Seth, V.D. Gupta, G. Linti, and H. Nöth, *Chem. Ber.*, 1991, **124**, 83.
3563 O.S. Jung, J.H. Jeong, and Y.S. Sohn, *Organometallics*, 1991, **10**, 761.
3564 F.J. Feher, T.A. Budzichowski, R.L. Blanski, K.J. Weller, and J.W. Ziller, *Organometallics*, 1991, **10**, 2526.
3565 P.B. Hitchcock, H.A. Jasim, M.F. Lappert, W.P. Leung, A.K. Rai, and R.E. Taylor, *Polyhedron*, 1991, **10**, 1203.
3566 J. Woning and J.G. Verkade, *Organometallics*, 1991, **10**, 2259.
3567 J.A. Tossell, *Phys. Chem. Miner.*, 1991, **17**, 654 (*Chem. Abstr.*, 1991, **114**, 189 259).
3568 R. Papiernik, L.G. Hubert-Pfalzgraf, and M.C. Massiani, *Polyhedron*, 1991, **10**, 1657.
3569 G. Feio, H.D. Burrows, C.F.G.C. Geraldes, and T.J.T. Pinheiro, *Liq. Cryst.*, 1991, **9**, 417 (*Chem. Abstr.*, 1991, **114**, 198 275).
3570 C.J. Jameson, A.K. Jameson, and J.K. Hwang, *J. Chem. Phys.*, 1991, **94**, 172 (*Chem. Abstr.*, 1991, **114**, 113 757).
3571 R.S. Deeb and D.H. Peyton, *J. Biol. Chem.*, 1991, **266**, 3728 (*Chem. Abstr.*, 1991, **114**, 180 663).
3572 S.P. Narula, R. Shankar, B. Kaur, and S. Soni, *Polyhedron*, 1991, **10**, 2463.
3573 D.F. Evans and C.Y. Wong, *Polyhedron*, 1991, **10**, 1131.
3574 N. Yoshino, S.-i. Tominaga, and H. Hirai, *Bull. Chem. Soc. Jpn.*, 1991, **64**, 2735.
3575 R.O. Day, R.R. Holmes, A. Schmidpeter, K. Stoll, and L. Howe, *Chem. Ber.*, 1991, **124**, 2443.
3576 S. Sharma, R. Bohra, and R.C. Mahrotra, *Synth. React. Inorg. Metal-Org. Chem.*, 1991, **21**, 741.
3577 S.H. Blanchard, P.A.W. Dean, V. Manivannan, R.S. Srivastava, and J.J. Vittal, *J. Fluorine Chem.*, 1991, **51**, 93 (*Chem. Abstr.*, 1991, **114**, 229 061).
3578 S.S. Krishnamurthy, *Proc.-Indian Acad. Sci., Chem. Sci.*, 1990, **102**, 283 (*Chem. Abstr.*, 1991, **114**, 6548).

CO, CO_2, HCl, CH_4, CF_4, and SF_6 mixtures. The relaxation is dominated by the quadrupolar mechanism.[3579] The ^{15}N shielding for NH_3 has been calculated.[3580] The temperature dependence of the ^{31}P shielding in PH_3 has been remeasured and calculated.[3581] The ^{31}P nucleus of $[HPO_3]^-$ has been found to relax by chemical shift anisotropy, dipole-dipole, and spin-rotation interactions.[3582] N.m.r. data have also been reported for $[H_2N{=}S{=}NH_2]^{2+}$, (^{14}N),[3583] $Bu^tCH_2AsH_2$, and $Me_2GaAs(CH_2Bu^t)_2$, (^{13}C).[3584]

A simple, rapid, and non-destructive method for the detection and determination of phosphines and phosphine oxides in solution by ^1H and ^{31}P n.m.r. spectroscopy has been proposed.[3585] The ^{31}P n.m.r. spectrum of $[P_3CH_2]^-$ has been analysed as $A[BX]_2$.[3586] N.m.r. data have also been reported for $R^1{}_2R^2P{=}CCl_2$, (^{13}C),[3587] $\overline{CH_2(CH_2NMe)_2P}Pr^i(OLi)$, ($^6Li$, ^{13}C),[3588] $[(Pr^i{}_2N)_2P(S)\text{-}\overline{C{=}NN\{P(NPr^i{}_2)_2Me\}CH(OEt)C}H_2]^+$, ($^{13}C$),[3589] $[(Me_2N)_3PC{\equiv}P]^+$, (^{13}C),[3590] (2,4,6-$Bu^t{}_3C_6H_2$)-$\overline{PC(=CCl_2)P}(C_6H_2Bu^t{}_3\text{-}2,4,6)$, ($^{13}C$),[3591] poly(chlorocarbophosphazene), (^{13}C),[3592] $P_6(C_5Me_5)_4$, (^{13}C),[3593] $P_{12}Pr_4$, (^{13}C),[3594] $C_4Me_4AsAsC_4Me_4$, (^{13}C),[3595] $HOCH_2CH_2CH_2AsO_3H_2$, (^{13}C),[3596] (187), (M = As, Sb, Bi; ^{13}C),[3597] $Ph_3Sb(OCH_2CH_2)_2NH$, (^{13}C),[3598] $Ph_2Sb(OCMeCH{=}CMeNR)$,

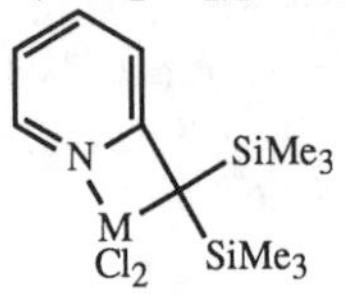

(187)

(^{13}C),[3599] Ph_2SbSCN, (^{14}N),[3600] and Ph_2BiCF_3, (^{13}C).[3601]

3579 C.J. Jameson, A.K. Jameson, and M.A. ter Horst, *J. Chem. Phys.*, 1991, **95**, 5799.
3580 C.J. Jameson, A.C. de Dios, and A.K. Jameson, *J. Chem. Phys.*, 1991, **95**, 1069.
3581 C.J. Jameson, A.C. de Dios, and A.K. Jameson, *J. Chem. Phys.*, 1991, **95**, 9042.
3582 C.L. Tsai, W.S. Price, Yu.C. Chang, B.C. Perng, and L.P. Hwang, *J. Phys. Chem.*, 1991, **95**, 7546.
3583 A. Haas and T. Mischo, *Z. Anorg. Allg. Chem.*, 1991, **606**, 191.
3584 J.C. Pazik, C. George, and A. Berry, *Inorg. Chim. Acta*, 1991, **187**, 207.
3585 T. Kupka, A. Wawer, J.O. Dziegielewski, and P.S. Zacharias, *Fresenius. J. Anal. Chem.*, 1991, **339**, 253 (*Chem. Abstr.*, 1991, **114**, 177 676).
3586 M. Baudler and J. Hahn, *Z. Naturforsch., B*, 1990, **45**, 1139 (*Chem. Abstr.*, 1991, **114**, 6637).
3587 E.A. Romanenko, A.P. Marchenko, G.N. Koidan, and A.M. Pinchuk, *Zh. Obshch. Khim.*, 1990, **60**, 2229 (*Chem. Abstr.*, 1991, **115**, 8932).
3588 S.E. Denmark, P.C. Miller, and S.R. Wilson, *J. Am. Chem. Soc.*, 1991, **113**, 1468.
3589 M. Granier, A. Baceiredo, G. Bertrand, V. Huch, and M. Veith, *Inorg. Chem.*, 1991, **30**, 1161.
3590 U. Fleischer, H. Grützmacher, and U. Krüger, *J. Chem. Soc., Chem. Commun.*, 1991, 302.
3591 M. Yoshifuji, K. Toyota, H. Yoshimura, K. Hirotsu, and A. Okamoto, *J. Chem. Soc., Chem. Commun.*, 1991, 124.
3592 H.R. Allcock, S.M. Coley, I. Manners, O. Nuyken, and G. Renner, *Macromolecules*, 1991, **24**, 2024 (*Chem. Abstr.*, 1991, **114**, 164 996).
3593 P. Jutzi, R. Kroos, A. Mueller, H. Bogge, and M. Penk, *Chem. Ber.*, 1991, **124**, 75.
3594 M. Baudler, H. Jachow, and K.F. Tebbe, *Z. Anorg. Allg. Chem.*, 1991, **593**, 9.
3595 S.C. Sendlinger, B.S. Haggerty, A.L. Rheingold, and K.H. Theopold, *Chem. Ber.*, 1991, **124**, 2453.
3596 G.M. Tsivgoulis, D.N. Sotiropoulos, and P.V. Ioannou, *Phosphorus Sulphur*, 1991, **57**, 189.
3597 C. Jones, L.M. Engelhardt, P.C. Junk, D.S. Hutchings, W.C. Patalinghug, C.L. Raston, and A.H. White, *J. Chem. Soc., Chem. Commun.*, 1991, 1560.
3598 D. Kraft and M. Wieber, *Z. Anorg. Allg. Chem.*, 1991, **605**, 137.
3599 R. Gupta, Y.P. Singh, and A.K. Rai, *Indian J. Chem., Sect. A*, 1991, **30**, 541 (*Chem. Abstr.*, 1991, **115**, 114 662).
3600 G.E. Forster, I.G. Southerington, M.J. Begley, and D.B. Sowerby, *J. Chem. Soc., Chem. Commun.*, 1991, 54.
3601 S. Pasenock, D. Naumann, and W. Tyrra, *J. Organomet. Chem.*, 1991, **417**, C47.

The valence tautomerism of $[P_7]^{3-}$ has been studied by ^{31}P COSY measurements.[3602] N.m.r. data have also been reported for $(Pr^i{}_2N)_2P(N)_2PCl(NPr^i{}_2)$, ($^{13}C$),[3603] $N_3P_3(NMe_2)_3(NCS)_3$, (^{13}C),[3604] $N_3P_3(OPh)_5NH(CH_2)_3SiMe_2H$, ($^{13}C$),[3605] $N_3P_3(OMe)_{6-n}(OC_6H_4Me\text{-}4)_n$, ($^{13}C$),[3606] $N_3P_3Cl_4\{(OCH_2CH_2)_2O\}$, ($^{13}C$),[3607] $(PhO)_2P\{N(SO_3)P(OPh)_2\}_2NSO_3$, ($^{13}C$),[3608] $N_3P_3Cl_4\{(OCH_2)_2C(CO_2Et)_2\}$, ($^{13}C$),[3609] $(N_4P_4Cl_7)_2$(spermidine), (^{13}C),[3610] $N_4P_4Cl_6(OCH_2)_2CMe_2$, (^{13}C),[3611] $N_4P_4Cl_6(OCH_2)_2C(CO_2Et)_2$, ($^{13}C$),[3612] $\{NPCl_2N{=}P\{N{=}P(OPh)_3\}_2N{=}PCl_2\}_n$, ($^{13}C$),[3613] $\{S(O)Cl{=}NP(OPh)_2{=}NP(OPh)_2{=}N\}_n$, ($^{13}C$),[3614] and $Sb(OR)_5NH_3$, (^{13}C).[3615]

^{15}N n.m.r. spectroscopy has been used to determine the ^{18}O enrichment of $[^{15}NO_3]^-$ using an isotope shift of 0.056 p.p.m. from $[^{15}N(^{16}O)_3]^-$ to $[^{15}N(^{16}O)_2(^{18}O)]^-$.[3616] A semiempirical formula for the calculation of the ^{31}P chemical shift has been derived from bond polarization theory. The influence of structure and coordination on the ^{31}P chemical shift of phosphates was examined.[3617] N.m.r. data have also been reported for $[SNS]^+$, (^{14}N),[3618] $Sb(OR)_3$, (^{13}C),[3619] $(Pr^iO)Sb(OCH_2CH_2O)$, (^{13}C),[3620] $(Pr^iO)Sb(OCH_2CH_2)_2E$, (E = O, S; ^{13}C),[3621] $Bi(OCH_2CH_2OMe)_3$, (^{13}C),[3622] $[R^1CSNSCR^2]^+$, (^{13}C, ^{14}N),[3623] $[C_7H_{10}S_3N_3]^+$, (^{13}C, ^{14}N),[3624] and $[SbCl_6]^-$, (^{121}Sb).[3625]

9 Compounds of Groups VI, Neon and Xenon

Water quality has been assessed from ^{17}O line width measurements.[3626] ^{77}Se n.m.r.

3602 M. Baudler and J. Hahn, *Z. Naturforsch., B*, 1990, **45**, 1279 (*Chem. Abstr.*, 1991, **114**, 34 727).
3603 H. Rolland, E. Ocando-Mavarez, P. Potin, J.P. Majoral, and G. Bertrand, *Inorg. Chem.*, 1991, **30**, 4095.
3604 H.R. Allcock, J.S. Rutt, and M. Parvez, *Inorg. Chem.*, 1991, **30**, 1776.
3605 H.R. Allcock, C.J. Nelson, and W.D. Coggio, *Organometallics*, 1991, **10**, 3819.
3606 S. Karthikeyan and S.S. Krishnamurthy, *J. Chem. Soc., Dalton Trans.*, 1991, 299.
3607 R.A. Shaw and S. Ture, *Phosphorus Sulphur*, 1991, **57**, 103; A.H. Alkubaisi, S.R. Contractor, H.G. Parkes, L.S. Shaw, and R.A. Shaw, *Phosphorus Sulphur*, 1991, **56**, 143.
3608 E. Montoneri, G. Ricca, M. Gleria, and M.C. Gallazzi, *Inorg. Chem.*, 1991, **30**, 150.
3609 A.H. Alkubaisi and R.A. Shaw, *Phosphorus Sulphur*, 1991, **55**, 49.
3610 A. Kilic, Z. Kilic, and R.A. Shaw, *Phosphorus Sulphur*, 1991, **57**, 111.
3611 H.A. Al-Madfa and R.A. Shaw, *Phosphorus Sulphur*, 1991, **55**, 59.
3612 A. Kilic and R.A. Shaw, *Phosphorus Sulphur*, 1991, **57**, 95.
3613 D.C. Ngo, J.S. Rutt, and H.R. Allcock, *J. Am. Chem. Soc.*, 1991, **113**, 5075.
3614 M. Liang and I. Manners, *J. Am. Chem. Soc.*, 1991, **113**, 4044.
3615 T. Athar, R. Bohra, and R.C. Mehrotra, *J. Indian Chem. Soc.*, 1990, **67**, 535 (*Chem. Abstr.*, 1991, **114**, 113 980).
3616 D.W. Johnson and D.W. Margerum, *Inorg. Chem.*, 1991, **30**, 4845.
3617 U. Sternberg, F. Pietrowski, and W. Priess, *Z. Phys. Chem. (Munich)*, 1990, **168**, 115 (*Chem. Abstr.*, 1991, **114**, 155 673).
3618 M.V.F. Brooks, T.S. Cameron, F. Grein, S. Parsons, J. Passmore, and M.J. Schriver, *J. Chem. Soc., Chem. Commun.*, 1991, 1079.
3619 K.V. Kobakhidze, L.V. Shmelev, E.E. Grinberg, A.V. Kessinikh, and A.A. Efremov, *Vysokochist. Veshchestva*, 1990, 142 (*Chem. Abstr.*, 1991, **114**, 16 316).
3620 A.K. Sen Gupta, R. Bohra, R.C. Mehrotra, and K. Das, *Main Group Met. Chem.*, 1990, **13**, 321 (*Chem. Abstr.*, 1991, **115**, 246 562).
3621 A.K. Sen Gupta, R. Bohra, and R.C. Mehrotra, *Indian J. Chem., Sect. A*, 1991, **30**, 588 (*Chem. Abstr.*, 1991, **115**, 125 624).
3622 M.-C. Massiani, R. Papiernik, L.G. Hubert-Pfalzgraf, and J.-C. Daran, *Polyhedron*, 1991, **10**, 437.
3623 S. Parsons, J. Passmore, M.J. Schriver, and X. Sun, *Inorg. Chem.*, 1991, **30**, 3342.
3624 A. Apblett, T. Chivers, A.W. Cordes, and R. Vollmerhaus, *Inorg. Chem.*, 1991, **30**, 1392.
3625 G.R. Willey and M. Ravindran, *Inorg. Chim. Acta*, 1991, **183**, 167; G.R. Willey, H. Collins, and M.G.B. Drew, *J. Chem. Soc., Dalton Trans.*, 1991, 961.
3626 Y. Fukazawa, *Kenkyu Hokoku - Kanagawa-ken Kogyo Shikensho*, 1990, 99 (*Chem. Abstr.*, 1991, **115**, 98 724).

characterisation of ^{77}Se labelled ovine erythrocyte glutathione peroxidase[3627] and bovine haemoglobin[3628] has been described. ^{77}Se n.m.r. spectroscopy has been used to detect chirality using chiral oxazolidine-2-selones as reagents.[3629] The ^{77}Se n.m.r. spectrum of (188) shows two ^{77}Se n.m.r. signals due to diastereomers.[3630] N.m.r. data have also been reported for $[HSe]^-$, $[Se_n]^{2-}$, (^{77}Se),[3631] $[HE]^-$, $[E_n]^{2-}$, (^{77}Se, ^{123}Te, ^{125}Te),[3632] $(Ph_3C)_2S_5$, (^{13}C),[3633] $[Me_2SCF_3]^+$, (^{13}C),[3634] MeSeR, (^{13}C),[3635] MeSeC(S)R, (^{13}C),[3636] $MeSeC_6H_3$-2-CO_2H-6-NO_2, (^{13}C, ^{77}Se),[3637] $Se(CH_2C{\equiv}CH)_2$, (^{13}C),[3638] (189), (^{13}C),[3639] organic selenide and diselenide compounds, (^{13}C),[3640]

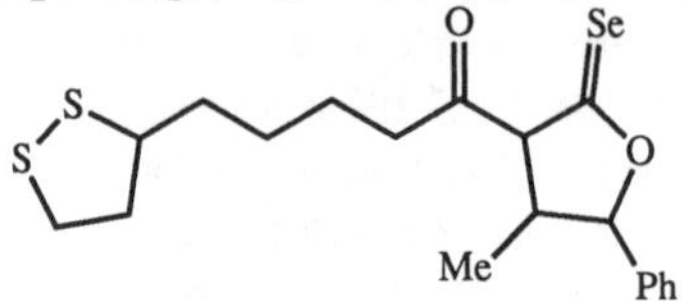

(188)

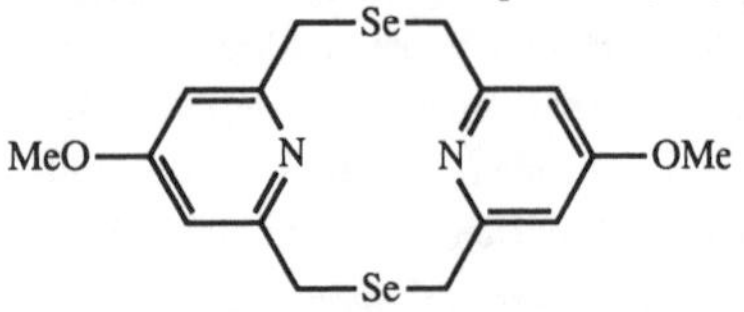

(189)

$RE(CH_2)_nCO_2Me$, (E = Se, Te; ^{13}C),[3641] RSePh, (^{13}C),[3642] $RSeC_6H_4CF_3$-4, (^{13}C),[3643] $R_FC(Se)_2CF_R$, (^{13}C, ^{77}Se),[3644] $Te(CF_2)_2Se$, (^{13}C, ^{77}Se, ^{125}Te),[3645] PhSeCH=CHCH=CHOMe, (^{77}Se),[3646] (190), (^{13}C, ^{77}Se),[3647] $PhE(C_6H_{11})$, (E = Se, Te; ^{13}C, ^{77}Se, ^{125}Te),[3648] (191), (^{13}C, ^{77}Se),[3649] PhSeR, (^{13}C),[3650] *5H,7H*-dibenzo[*b,g*][1,5]diselenocin, (^{13}C, ^{77}Se),[3651] (192), (^{13}C, ^{15}N),[3652] R_4Te, (^{13}C, ^{125}Te),[3653] $MeTeCH_2Ph$, (^{125}Te),[3654] 4-$ROC_6H_4TeCH_2CH_2OH$, (^{13}C),[3655]

3627 P. Gettins and B.C. Crews, *J. Biol. Chem.*, 1991, **266**, 4804 (*Chem. Abstr.*, 1991, **114**, 202 418).
3628 P. Gettins and S.A. Wardlaw, *J.Biol. Chem.*, 1991, **266**, 3422 (*Chem. Abstr.*, 1991, **114**, 202 862).
3629 L.A. Silks, tert., J. Peng, J.D. Odom, and R.B. Dunlap, *J. Org. Chem.*, 1991, **56**, 6733.
3630 L.A. Silks, tert., J. Peng, J.D. Odom, and R.B. Dunlap, *J. Chem. Soc., Perkin Trans. 1*, 1991, 2491.
3631 J. Cusick and I. Dance, *Polyhedron*, 1991, **10**, 2629.
3632 M. Bjoergvinsson and G.J. Schrobilgen, *Inorg. Chem.*, 1991, **30**, 2540.
3633 R. Steudel, S. Förster, and J. Albertsen, *Chem. Ber.*, 1991, **124**, 2357.
3634 R. Minkwitz and V. Gerhard, *Z. Anorg. Allg. Chem.*, 1991, **591**, 143.
3635 T. Kataoka, M. Yoshimatsu, H. Shimizu, and M. Hori, *Tetrahedron Lett.*, 1991, **32**, 105.
3636 M. Khalid, J.-L. Ripoll, and Y. Vallée, *J. Chem. Soc., Chem. Commun.*, 1991, 964.
3637 C. Lambert and L. Christiaens, *Tetrahedron*, 1991, **47**, 9053.
3638 R. Gleiter, S. Rittinger, and H. Langer, *Chem. Ber.*, 1991, **124**, 357.
3639 J. Breitenbach, M. Nieger, and F. Vögtle, *Chem. Ber.*, 1991, **124**, 2583.
3640 K. Tokami, A. Mizuno, and K. Takagi, *Kenkyu Kiyo - Wakayama Kogyo Koto Senmon Gakko*, 1989, **24**, 52 (*Chem. Abstr.*, 1991, **114**, 5691).
3641 L.K. Jie, S.F. Marcel, Y.K. Cheung, S.H. Chau, and B.F.Y. Yan, *J. Chem. Soc., Perkin Trans. 2*, 1991, 501.
3642 M. Tingoli, M. Tiecco, D. Chianelli, R. Balducci, and A. Temperini, *J. Org. Chem.*, 1991, **56**, 6809; M. Tiecco, L. Testaferri, M. Tingoli, D. Chianelli, and D. Bartoli, *J. Org. Chem.*, 1991, **56**, 4529.
3643 H. Kuniyasu, A. Ogawa, S.-I. Miyazaki, I. Ryu, N. Kambe, and N. Sonoda, *J. Am. Chem. Soc.*, 1991, **113**, 9796.
3644 R. Boese, A. Haas, and M. Spehr, *Chem. Ber.*, 1991, **124**, 51.
3645 R. Boese, A. Haas, and C. Limberg, *J. Chem. Soc., Chem. Commun.*, 1991, 1378.
3646 M.D. Fryzuk, G.S. Bates, and C. Stone, *J. Org. Chem.*, 1991, **56**, 7201.
3647 S. Ogawa, S. Sato, T. Erata, and N. Furukawa, *Tetrahedron Lett.*, 1991, **32**, 3179.
3648 H. Duddeck, P. Wagner, and A. Biallass, *Magn. Reson. Chem.*, 1991, **29**, 248.
3649 H. Fujihara, H. Mima, T. Erata, and N. Furukawa, *J. Chem. Soc., Chem. Commun.*, 1991, 98.
3650 H. Ali and J.E. van Lier, *J. Chem. Soc., Perkin Trans. 1*, 1991, 269.
3651 H. Fujihara, Y. Ueno, J.J. Chiu, and N. Furukawa, *Chem. Lett.*, 1991, 1649 (*Chem. Abstr.*, 1991, **115**, 289 566).
3652 A.L. Nivorozhkin, L.E. Nivorozhkin, L.E. Konstantinovskii, and V.I. Minkin, *Mendeleev Commun.*, 1991, 78 (*Chem. Abstr.*, 1991, **115**, 207 826).
3653 R.W. Gedridge, jun., K.T. Higa, and R.A. Nissan, *Organometallics*, 1991, **10**, 286.
3654 R.U. Kiras, D.W. Brown, K.T. Higa, and R.W. Gedridge, jun., *Organometallics*, 1991, **10**, 3589.
3655 A.K. Singh and S. Thomas, *Polyhedron*, 1991, **10**, 2065.

α-tellurocarbonyl compounds, (^{125}Te),[3656] (193), (^{13}C, ^{125}Te),[3657] RClC=CHTeC_6H_4OMe-4, (^{13}C),[3658] (194), (^{13}C),[3659] PhTeC(CH_2OH)=CHTePh, (^{13}C),[3660] and (2-HO-5-MeC_6H_3)$_2$Te, (^{13}C).[3661]

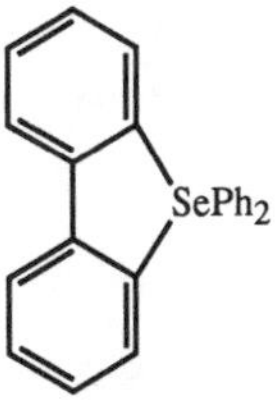

(190)

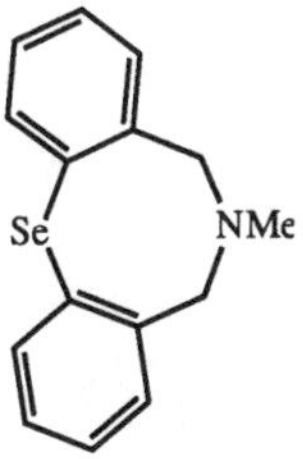

(191)

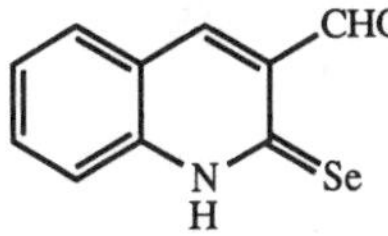

(192)

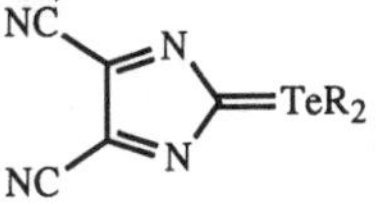

(193)

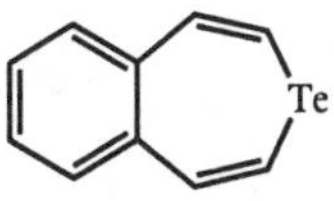

(194)

^{1}J(^{77}Se-^{77}Se) has been measured for R^{1}SeSeR^2. A correlation of ^{1}J(^{77}Se-^{77}Se) with the average ^{77}Se shielding has been found for perhalomethyl diselenides and organyl diselenides. ^{1}J(^{77}Se-^{13}C) was also reported.[3662] N.m.r. data have also been reported for (195), (^{13}C, ^{77}Se),[3663] (196), (^{77}Se),[3664] P($O_2C_6Cl_4$)(SePh), (^{13}C),[3665] PhTeNPr$^i_2$, (^{13}C, ^{125}Te),[3666] (197), (^{15}N, ^{125}Te),[3667] (2,4,6-Bu$^t_3$$C_6H_2$)SeSe($C_6H_2Bu^t_3$-2,4,6), (^{13}C, ^{77}Se),[3668] (198), (^{13}C),[3669] (199), (^{13}C, ^{77}Se),[3670] Me_3TeS_2COR, (^{13}C),[3671] R$^1_2$Te{S_2P(OR^2)$_2$}$_2$, (^{13}C, ^{125}Te),[3672] (200), (^{13}C, ^{125}Te),[3673] R$^1_3$TeS_2P(OR^2)$_2$, (^{13}C),[3674] Ph_3TeSR, (^{13}C, ^{125}Te),[3675] Ph_3Te(S_2CPR_2), (^{13}C),[3676] $F_5SCF_2C(CF_3)_3$, (^{13}C),[3677] ($BrCH_2CH_2$)$_2$SeBr_2, (^{13}C),[3678] $CH_3C(O)CH_2Te(Cl)_2C_6H_4$OMe-4,

3656 L.A. Silks, tert., J.D. Odom, and R.B. Dunlap, *Synth. Commun.*, 1991, **21**, 1105 (*Chem. Abstr.*, 1991, **115**, 182 222).
3657 G. Tappeiner, B. Bildstein, and F. Sladky, *Chem. Ber.*, 1991, **124**, 699.
3658 J.V. Comasseto, H.A. Stefani, A. Chieffi, and J. Zukerman-Schpector, *Organometallics*, 1991, **10**, 845.
3659 H. Sashida, H. Kurahashi, and T. Tsuchiya, *J. Chem. Soc., Chem. Commun.*, 1991, 802.
3660 A. Ogawa, K. Yokoyama, H. Yokoyama, R. Obayashi, N. Kambe, and N. Sonoda, *J. Chem. Soc., Chem. Commun.*, 1991, 1748.
3661 A.K. Singh, S. Thomas, and B.L. Khandelwal, *Polyhedron*, 1991, **10**, 2693.
3662 W. Gombler, *Magn. Reson. Chem.*, 1991, **29**, 73.
3663 H. Fujihara, H. Mima, M. Ikemori, and N. Furukawa, *J. Am. Chem. Soc.*, 1991, **113**, 6337.
3664 B.E. Maryanoff and M.C. Rebarchak, *J. Org. Chem.*, 1991, **56**, 5203.
3665 T.A. Annan, Z. Tian, and D.G. Tuck, *J. Chem. Soc., Dalton Trans.*, 1991, 19.
3666 T. Murai, K. Nonomura, K. Kimura, and S. Kato, *Organometallics*, 1991, **10**, 1095.
3667 V.I. Minkin, I.D. Sadekov, A.A. Maksimenko, O.E. Kompan, and Yu.T. Struchkov, *J. Organomet. Chem.*, 1991, **402**, 331.
3668 W.-W. du Mont, S. Kubiniok, L. Lange, S. Pohl, W. Saak, and I. Wagner, *Chem. Ber.*, 1991, **124**, 1315.
3669 H. Fujihara, Y. Ueno, J.-J. Chiu, and N. Furukawa, *J. Chem. Soc., Chem. Commun.*, 1991, 1052.
3670 H. Fujihara, M. Yabe, J.-J. Chiu, and N. Furukawa, *Tetrahedron Lett.*, 1991, **32**, 4345.
3671 M. Wieber and S. Rohse, *Phosphorus Sulphur*, 1991, **55**, 79.
3672 T.N. Srivastava, J.D. Singh, and S.K. Srivastava, *Phosphorus Sulphur*, 1991, **55**, 117.
3673 H. Fujihara, T. Ninoi, R. Akaishi, T. Erata, and N. Furukawa, *Tetrahedron Lett.*, 1991, **32**, 4537.
3674 M. Wieber and S. Rohse, *Phosphorus Sulphur*, 1991, **55**, 91.
3675 M. Wieber and S. Rohse, *Z. Anorg. Allg. Chem.*, 1991, **592**, 202.
3676 M. Wieber and S. Rohse, *Phosphorus Sulphur*, 1991, **55**, 85.
3677 H.-S. Huang, H. Roesky, and R.J. Lagow, *Inorg. Chem.*, 1991, **30**, 789.
3678 S. Akabori, Y. Takanohashi, S. Aoki, and S. Sato, *J. Chem. Soc., Perkin Trans. 1*, 1991, 3121.

(^{13}C),[3679] $(C_6H_{10}Cl)Te(Cl)_2C_6H_4OMe$-4, ($^{13}C$),[3680] (201), ($^{13}C$),[3681] $O(SiMe_2CH_2)_2TeI_2$, (^{13}C),[3682] and $Ph_2TeBr(S_2CNR_2)$, (^{13}C, ^{125}Te).[3683]

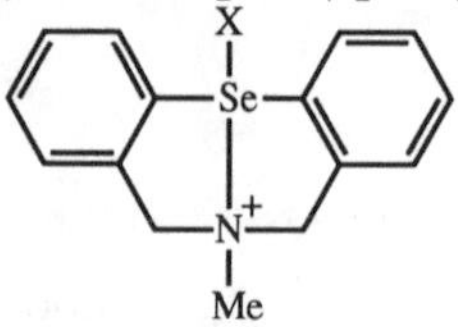

(195)

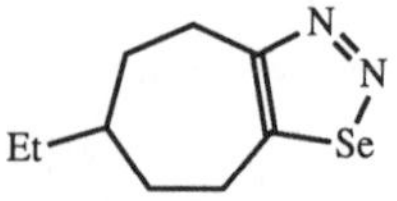

(196)

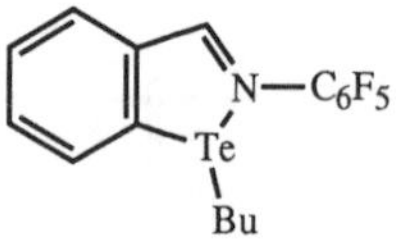

(197)

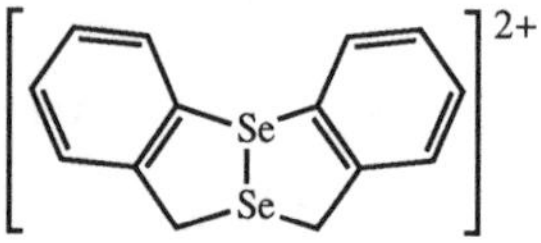

(198)

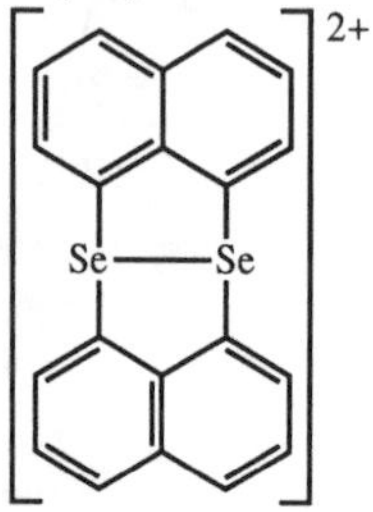

(199)

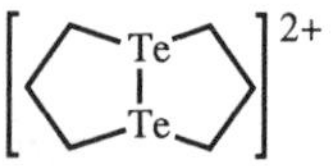

(200)

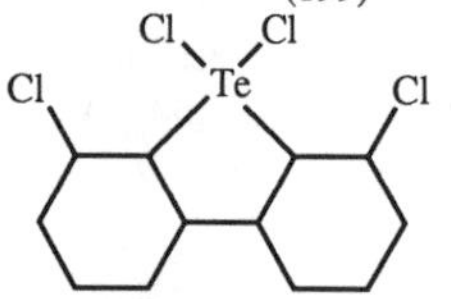

(201)

The ^{77}Se and ^{31}P T_1 values have been reported for $Bu^t{}_3PSe$.[3684] Electronic and steric effects in Ar_3PSe compounds have been discussed with the assistance of $^1J(^{77}Se-^{31}P)$.[3685] The reaction of 13 aliphatic ketones with $TeCl_4$ has been studied using ^1H, ^{13}C, and ^{125}Te n.m.r. spectroscopy.[3686] The reaction of $[TeF_5]^-$ with diols has been studied using ^{19}F and ^{125}Te n.m.r. spectroscopy.[3687] N.m.r. data have also been reported for tri and tetra tertiary phosphine selenides, (^{13}C, ^{77}Se),[3688] (202), (^{13}C, ^{77}Se),[3689] $\{(Pr^i{}_2N)_2P\}_2E$, (E = S, Se; ^{13}C, ^{77}Se),[3690] $Bu^tSSeOSBu^t$, (^{13}C, ^{77}Se),[3691]

3679 H.A. Stefani, A. Chieffi, and J.V. Comasseto, *Organometallics*, 1991, **10**, 1178.
3680 M.A. Malik, M.E.S. Ali, F.J. Berry, J. Kaur, M. Rowshani, and B.C. Smith, *Inorg. Chim. Acta*, 1991, **180**, 251.
3681 M. Albeck and T. Tamary, *J. Organomet. Chem.*, 1991, **420**, 35.
3682 A.Z. Al-Rubaie, S. Uemura, and H. Masuda, *J. Organomet. Chem.*, 1991, **410**, 309.
3683 J.H.E. Bailey, J.E. Drake, and M.L.Y. Wong, *Can. J. Chem.*, 1991, **69**, 1948.
3684 G.H. Penner, *Can. J. Chem.*, 1991, **69**, 1054.
3685 J. Malito and E.C. Alyea, *Phosphorus Sulphur*, 1990, **54**, 95.
3686 D.H. O'Brien, K.J. Irgolic, and C.K. Huang, *Heteroat. Chem.*, 1990, **1**, 215 (*Chem. Abstr.*, 1991, **114**, 5777).
3687 Yu.V. Kokunov, V.M. Afanas'ev, M.P. Gustyakova, and Yu.A. Buslaev, *Koord. Khim.*, 1991, **17**, 652 (*Chem. Abstr.*, 1991, **115**, 125 536).
3688 R. Colton and T. Whyte, *Aust. J. Chem.*, 1991, **44**, 525.
3689 F. Krech, K. Issleib, A. Zschunke, C. Mügge, and S. Skvorcov, *Z. Anorg. Allg. Chem.*, 1991, **594**, 66.
3690 H. Westermann, M. Nieger, and E. Niecke, *Chem. Ber.*, 1991, **124**, 13.
3691 J.L. Kice, D.M. Wilson, and J.M. Espinola, *J. Org. Chem.*, 1991, **56**, 3520.

$SeSX_2$, (^{77}Se),[3692] $[TeSe_3]^{2-}$, (^{77}Se, ^{125}Te),[3693] $[TeF_5]^-$, (^{125}Te),[3694] and $[TeF_7]^-$, (^{125}Te).[3695]

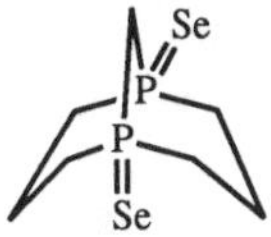

(202)

The ^{21}Ne n.m.r. spectrum of ^{21}Ne enriched gas dissolved in liquid crystalline ZLI1167 has been used to investigate the temperature dependence of the ^{21}Ne quadrupole coupling and electric field gradient.[3696] ^{21}Ne n.m.r. shieldings in solvents have been measured and correlated with ^{129}Xe chemical shifts.[3697] The quadrupole couplings have been reported for ^{83}Kr and ^{131}Xe in some mixed liquid crystals based on ZLI1167 and EBBA.[3698] The n.m.r. signal of low pressure ^{129}Xe has been greatly enhanced with linearly polarized laser radiation.[3699] T_1 of ^{131}Xe in ^{87}Rb atoms has been measured.[3700] ^{129}Xe gas to solution chemical shifts show a linear relation to ^{13}C chemical shifts of methane in the same solvents but are 27.1 times more sensitive.[3701] The anisotropy of the ^{129}Xe shielding tensor for xenon gas dissolved in the nematic and smectic phases has been studied.[3702] The n.m.r. relaxation of ^{131}Xe in quadrupole solvents has been studied.[3703] The ^{19}F and ^{129}Xe n.m.r. spectra of planar $[XeF_5]^-$ have been reported.[3704]

10 Appendix

This appendix contains a list of papers in which the use of nuclei other than ^1H, ^{19}F, and ^{31}P, has been described. The nuclei are ordered by increasing atomic number and mass.

^2H 20, 21, 88, 164, 281, 286, 684, 687, 997, 1176, 1452, 1539, 1595, 1687, 1695, 1719, 1739, 1845, 1905, 1941-1943, 2003, 2014, 2015, 2017, 2018, 2144, 2234, 2237, 2251, 2271, 2276, 2284, 2296, 2301, 2333, 2461, 2493, 2550, 2758, 2767, 2843, 2872, 2874, 2884, 2888, 2908, 2913, 2915, 2919, 2927, 2932, 2956, 3491, and 3495.

^3H 2888.

3692 J. Milne, *J. Chem. Soc., Chem. Commun.*, 1991, 1048.
3693 M. Bjorgvinsson, J.F. Sawyer, and G.J. Schrobilgen, *Inorg. Chem.*, 1991, **30**, 4238.
3694 Yu.V. Kokunov, Yu.E. Gorbunova, V.M. Afamas'ev, V.N. Petrov, R.L. Davidovich, and Yu.A. Buslav, *J. Fluorine Chem.*, 1990, **50**, 285 (*Chem. Abstr.*, 1991, **114**, 114 041).
3695 K.O. Christe, J.C.P. Sanders, G.J. Schrobilgen, and W.W. Wilson, *J. Chem. Soc., Chem. Commun.*, 1991, 837; A.-R. Mahjoub and K. Seppelt, *J. Chem. Soc., Chem. Commun.*, 1991, 840.
3696 P. Ingman, J. Jokisaari, O. Pulkkinen, P. Diehl, and O. Muenster, *Chem. Phys. Lett.*, 1991, **182**, 253 (*Chem. Abstr.*, 1991, **115**, 148 745).
3697 P. Diehl, O. Muenster, and J. Jokisaari, *Chem. Phys. Lett.*, 1991, **178**, 147 (*Chem. Abstr.*, 1991, **114**, 198 278).
3698 P. Ingman, J. Jokisaari, and P. Diehl, *J. Magn. Reson.*, 1991, **92**, 163.
3699 X. Zeng, C. Wu, M. Zhao, S. Li, L. Li, X. Zhang, Z. Liu, and W. Liu, *Chem. Phys. Lett.*, 1991, **182**, 538 (*Chem. Abstr.*, 1991, **115**, 196 547); X. Zeng, C. Wu, M. Zhao, S. Li, L. Li, X. Zhang, Z. Liu, and W. Liu, *Yuanzi Yu Fenzi Wuli Xuebao*, 1990, **7**, 1636 (*Chem. Abstr.*, 1991, **115**, 40 511).
3700 Z. Liu, X. Sun, X. Zeng, and Q. He, *Chin. Phys. Lett.*, 1990, **7**, 388 (*Chem. Abstr.*, 1991, **114**, 176 893).
3701 P. Diehl, R. Ugolini, N. Suryaprakash, and J. Jokisaari, *Magn. Reson. Chem.*, 1991, **29**, 1163.
3702 O. Muenster, J. Jokisaari, and P. Diehl, *Mol. Cryst. Liq. Cryst.*, 1991, **206**, 179 (*Chem. Abstr.*, 1991, **115**, 196 555).
3703 A. Dejaegere, M. Luhmer, M.L. Stien, and J. Reisse, *J. Magn. Reson.*, 1991, **91**, 362.
3704 K.O. Christe, E.C. Curtis, D.A. Dixon, H.P. Mercier, J.C.P. Sanders, and G.J. Schrobilgen, *J. Am. Chem. Soc.*, 1991, **113**, 3351 (*Chem. Abstr.*, 1991, **114**, 198 360).

^{3}He 21 and 2501.

^{6}Li 36, 39, 40, 62, 63, 64, 68, 94, 389, 1450, 1682, 1721, 1725, 1735, 1736, 2175, 2311, 2314, 2316, 2320, 2563, 2889, 2891, and 3588.

^{7}Li 41, 42, 48, 57, 70, 71, 73, 74-76, 78, 79, 91, 93, 94, 389, 1650, 1682, 1690, 1721, 1722, 1724, 1726-1730, 1732, 1733, 1736, 1738, 1913, 1928, 2175, 2238, 2239, 2241, 2243, 2258, 2266-2268, 2311, 2313-2315, 2318, 2321, 2412, 2433, 2454, 2460, 2468, 2563-2565, 2576, 2583, 2610, 2622, 2640, 2822, 2889, 2890, and 3175.

^{8}Li 2240 and 2320.

^{9}Be 117.

^{10}B 2584 and 3024.

^{11}B 20, 24, 73, 131, 166, 169, 170, 174, 184, 334, 339, 466, 640, 714, 801, 852, 873, 895, 904, 905, 957, 975, 985, 1000, 1105, 1168, 1341, 1355, 1421, 1463, 1512, 1632, 1647, 1649, 1650, 1652, 1653, 1668, 1717, 1804, 1815, 1860, 1871-1874, 1913, 2036-2044, 2241, 2278, 2392, 2538, 2551, 2552, 2555-2557, 2559-2562, 2564, 2566, 2570-2572, 2574-2578, 2580, 2582, 2584, 2585, 2780, 2774, 3004-3007, 3009-3017, 3021-3070, 3072-3103, 3105, 3107, 3109-3114, 3116-3123, 3127-3138, 3140-3143, 3145, 3147-3168, and 3397.

^{12}B 2240.

^{13}C 12, 18, 20, 22, 23-25, 28, 29-32, 35, 37-41, 43, 45-57, 60, 68-70, 72, 73, 75-78, 82, 91, 92, 94, 95, 96, 98, 100-116, 119-122, 124-126, 129, 130, 132-144, 146, 148-163, 165, 167, 168, 170-210, 212, 213, 216-245, 248-259, 261-267, 270-285, 287-294, 296-299, 305, 306, 309-311, 318, 319, 325, 327, 329, 332-334, 337-342, 345-362, 364-412, 415-479, 481-510, 513-551, 554-561, 563-566, 568-570, 576, 578-583, 587, 588, 590-607, 609-642, 644-664, 666-669, 671, 673, 674, 676-679, 682, 690-713, 715-717, 719-758, 760-863, 866-869, 871-941, 944, 945, 946-956, 958-962, 964-968, 970-979, 981-991, 993-995, 999, 1001, 1002, 1004-1009, 1013-1032, 1034-1055, 1057-1067, 1071-1103, 1106-1118, 1120-1128, 1130, 1132-1135, 1137-1142, 1144-1156, 1158-1164, 1167, 1173, 1175, 1178-1202, 1204-1214, 1216-1218, 1221, 1223, 1225, 1230-1232, 1234-1259, 1261-1284, 1289, 1291-1295, 1297-1303, 1305, 1307, 1308, 1311, 1315, 1318-1322, 1324, 1325, 1327-1329, 1332, 1334, 1336-1346, 1348, 1349, 1353-1359, 1361-1369, 1372, 1373, 1377, 1381-1387, 1390, 1392-1398, 1400-1402, 1404, 1405, 1408-1422, 1424, 1426, 1427, 1429-1433, 1436, 1438, 1441-1443, 1450, 1454-1457, 1459-1462, 1464-1469, 1473-1482, 1485-1488, 1490-1495, 1497-1505, 1507-1512, 1514-1516, 1520-1523, 1525, 1526, 1528, 1529, 1532, 1533, 1537, 1542, 1543, 1545, 1547-1553, 1555, 1557-1564, 1568, 1574-1580, 1583, 1585-1590, 1596, 1598-1600, 1602-1607, 1609, 1610, 1615, 1618, 1620, 1623, 1629, 1632, 1634, 1636, 1638, 1645-1647, 1650, 1651, 1653, 1657, 1659, 1661, 1666, 1669, 1670, 1672-1674, 1676, 1682, 1695, 1697, 1701, 1706, 1712, 1721, 1722, 1726, 1731, 1734, 1736, 1737, 1741, 1743, 1746, 1758, 1765, 1777-1779, 1782, 1786, 1789, 1794, 1795, 1801, 1802, 1804, 1807, 1815-1817, 1823-1825, 1828, 1831, 1833, 1836-1838, 1846, 1849, 1851, 1852, 1857-1859, 1861-1863, 1877, 1893, 1895, 1898, 1901, 1907, 1908, 1913-1916, 1918-1922, 1924, 1927, 1928, 1931-1934, 1939, 1950, 1952-1954, 1956, 1958, 1962, 1963, 1965, 1968, 1969, 1973, 1977, 1978, 1981,

2491, 2558, 2569, 2572, 2578, 2580, 2588-2599, 2601, 2603-2606, 2609, 2611, 2618-2624, 2627, 2628, 2630, 2631, 2634, 2635, 2637, 2638, 2640, 2641, 2643-2651, 2656, 2659, 2667, 2668, 2670-2673, 2675, 2678, 2679, 2682, 2684, 2685, 2690, 2691, 2693-2695, 2697, 2698, 2701, 2707, 2708, 2712, 2716-2720, 2722-2724, 2726-2730, 2732, 2734-2736, 2738, 2739, 2741, 2742, 2745, 2750, 2839, 2874, 2906, 2912, 2929, 2966, 2998, 3096, 3114, 3169, 3170, 3174, 3175, 3181, 3188, 3193-3196, 3198, 3219, 3220, 3222-3225, 3228-3235, and 3237.

^{29}Si 18, 20, 41, 46, 58, 72, 75, 78, 108, 109, 123, 129, 149, 154, 167, 191, 207, 231, 247, 273, 309, 322, 335, 366, 568, 572, 573, 609, 679, 718, 757-762, 799, 854, 892, 999, 1028, 1051, 1169, 1219, 1226, 1238, 1313, 1347, 1357, 1377, 1399, 1443, 1465, 1485, 1556, 1656, 1661, 1662, 1665, 1666, 1817, 1851, 1852, 1874, 1888, 1889, 1900, 1977, 1997, 2051-2053, 2055-2059, 2223, 2287, 2289, 2346, 2394, 2410, 2417, 2418, 2435, 2451, 2490, 2522, 2563, 2571, 2574, 2576, 2580, 2582, 2588, 2605, 2607, 2609, 2612-2623, 2625-2629, 2631, 2633-2638, 2641, 2643, 2647, 2648, 2652, 2653, 2655, 2656, 2658, 2660-2662, 2666-2670, 2672, 2674, 2675, 2678, 2681, 2684-2686, 2689, 2691, 2694, 2695, 2699, 2700, 2703, 2704, 2707, 2709, 2710, 2712, 2714, 2715, 2718, 2723, 2724, 2727, 2728, 2731, 2733, 2734, 2739, 2744, 2750, 2759-2766, 2768-2773, 2777-2779, 2781-2786, 2788, 2790, 2791, 2793, 2794, 2796, 2797, 2799-2801, 2803-2816, 2898, 2906, 2912, 2923, 2926, 2929, 2932, 2933, 2947, 2948, 2952, 2953, 2957, 2960, 2968, 3109, 3110, 3113, 3114, 3117, 3120, 3123, 3129, 3148, 3157, 3185, 3197, 3220, 3226, 3233, 3237, 3239, 3240, 3242, 3246-3258, 3260-3262, 3266, 3268, 3271, 3272, 3275, 3276, 3278, 3279, 3281, 3287, 3289-3292, 3295, 3299, 3301, 3303, 3305, 3308-3310, 3312, 3317, 3322, 3326, 3327, 3330, 3332, 3338-3344, 3348, 3350-3353, 3356, 3357, 3359, 3362-3372, 3388, 3407, 3408, 3410, 3412, 3413, 3417, 3421-3424, 3428, 3429, 3434, 3437-3439, 3442, 3444, 3447, 3449, 3450, 3453, 3454, 3480, 3496, 3497, 3499, 3500, 3502, 3503, 3505-3508, 3510-3514, 3516-3518, 3547-3550, 3553, 3554, 3556, 3557, 3560, 3564, 3566, 3567, 3572, and 3573.

^{35}Cl 1692, 1736, 1812, 1910-1912, 2324, 2341, 2350, and 2547.

^{37}Cl 1912.

^{39}K 59, 89, 90, 1756, 1760, 1767, 2331-2333, and 2546.

^{43}Ca 1767 and 2363.

^{45}Sc 1681, 1693, and 1774.

^{47}Ti 246 and 2420.

^{49}Ti 246 and 2420.

^{51}V 294, 295, 300-308, 312, 314-324, 326, 1479, 1792-1794, 1796, 1798, 1799, 1921, 2421, 2436, 2439-2445, 2447, 2449, 2451, 2453, 2600, and 2893.

^{55}Mn 585, 1526, and 2462.

^{57}Fe 868, 2383, and 2409.

^{59}Co 11, 892, 946, 1021, 1118, 1129, 1132, 1141, 1143, 1157, 1165, 1590, 1706, 1835, 1937, 1995, 2391, and 2498.

^{63}Cu 1352, 2212, 2368-2370, 2374, 2375, 2378, 2382, 2394, 2401, 2404, 2405, 2520, 2521, and 2526.

^{65}Cu 2404, 2406, and 2526.

^{67}Zn 1716.

^{69}Ga 2743.

^{71}Ga 1401, 1681, 1718, 1895, 1896, 2423, 2588, 2644, 2667, 2695, and 2743.

^{73}Ge 3241, 3269, and 3373.

^{75}As 2848.

^{77}Se 214, 215, 520, 584, 656, 1165, 1316, 1330, 1331, 1370, 1371, 1374, 1431, 1436, 1440, 1443, 1951, 1970, 1971, 2065, 2524, 2851, 2858, 3152, 3406, 3454, 3500, 3524, 3545, 3546, 3627-3632, 3637, 3644-3649, 3651, 3662-3664, 3668, 3670, 3684, 3685, and 3688-3693.

^{81}Br 1910 and 2350.

^{83}Kr 3698.

^{87}Rb 1760-1762, 2246, 2248, 2337, and 2423.

^{89}Y 129, 148, 1785, 2374, 2376, 2379, 2384, and 2393.

^{91}Zr 2419 and 2420.

^{93}Nb 293, 313, 328, 2445, 2456, and 2457.

^{95}Mo 477, 520, 552, 554, 567, 569, 582, 1495, 1807, 1989, 1990, 2462, 2467, and 2895.

^{97}Mo 2462.

^{99}Tc 643, 665, 670, 672, 675, and 676.

^{99}Ru 946 and 1832.

^{103}Rh 997, 998, 999, 1038, 1069, 1104, 1592, and 1598.

^{109}Ag 510, 1351, 1360, 1715, and 2510.

^{111}Cd 2539.

^{113}Cd 1376, 1389-1391, 1406, 1407, 1421, 1423-1425, 1428, 1434, 1435, 1437, 1717, 1860, 2499, 2528, 2534-2537, 2539, 2541, 2542, 2544, and 2706.

^{115}In 1681.

^{117}Sn 20, 42, 281, and 719.

^{119}Sn 20, 42, 108, 137, 183, 281, 332, 363, 719, 864, 892, 972, 1003, 1032, 1033, 1056, 1057, 1215, 1388, 1482, 1498, 1556, 1668, 1669, 1844, 1851, 1902, 1903, 1965-1968, 1982, 2025, 2038, 2060, 2426, 2760, 2818-2821, 2823, 2825, 2827, 3017, 3019, 3080, 3096, 3105, 3109, 3110, 3114, 3117, 3121, 3123, 3230, 3233, 3242, 3260, 3270, 3275, 3276, 3282, 3289, 3291, 3299, 3374-3376, 3378, 3382-3389, 33913393, 3397-3402, 3405, 3418, 3450, 3459, 3461, 3464-3467, 3469-3471, 3475, 3476, 3479, 3483, 3485-3488, 3525, 3527-3538, 3540-3544, 3546,3559, 3561-3563, 3565, 3571, 3575, and 3576.

^{121}Sb 1968 and 3625.

^{123}Te 3632.

^{125}Te 565, 679, 954, 1325, 1335, 1438, 1622, 1908, 1909, 1972, 2543, 2821, 3293, 3294, 3454, 3479, 36323645, 3648, 3653, 3654, 3656, 3657, 3666, 3667, 3672, 3673, 3675, 3683, 3686, 3687, and 3693-3695.

^{127}I 1333, 1910, and 2275.

^{129}Xe 2232, 2233, 2857, 2878, 2916, 2926, 2977-3003, 3697, 3698, 3699, 3701, 3702, and 3704.

^{131}Xe 3698, 3700, and 3703.

^{133}Cs 20, 61, 67, 75, 97, 1681, 1727, 1760, 1763, 2282, 2314, 2340, 2341, 2423, 2705, 2827, and 2842.

^{134}Cs 2479.

^{137}Ba 2420.

^{139}La 2371, 2387, 2402, and 2407.

^{141}Pr 2406.

^{171}Yb 137, 145, and 147.

^{183}W 322, 324, 330, 331, 343, 520, 553, 571-575, 577, 1482, 1810, 1989, 2086, 2176, and 2462.

^{195}Pt 480, 562, 631, 891, 1169-1172, 1175, 1175, 1178, 1201-1203, 1215, 1216, 1220, 1222, 1224-1227, 1229, 1233, 1237-1239, 1242, 1258, 1275, 1284-1287, 1289, 1292, 1293, 1296-1300, 1302-1304, 1306, 1307, 1309, 1310, 1314, 1316, 1317, 1319, 1323, 1325, 1326, 1333, 1335-1337, 1618, 1622, 1623, 1639, 1712, 1846-1848, 1950, 1951, 2032, 2033, 2500, and 3090.

^{186}Au 2509.

^{199}Hg 1375, 1379, 1395, 1396, 1434, 1439, 1440, 2529, 2530, 2544, and 3026.

^{203}Tl 2279.

^{205}Tl 1395, 2211, 2279, 2363, 2366, 2368, 2382, 3221, and 3238.

^{207}Pb 892, 2061, 2602, 2760, 3242, 3404, 3405, 3568, and 3577.

2
Nuclear Quadrupole Resonance Spectroscopy

BY K. B. DILLON

1 Introduction

This chapter reports on the pure nuclear quadrupole resonance (n.q.r.) spectra of quadrupolar ($I > \frac{1}{2}$) nuclei in inorganic or organometallic solids. There have been some reports of n.q.r. imaging this year, and this technique may become of increasing importance. The literature has again been dominated by n.q.r. studies on high-temperature superconductors, particularly of copper nuclei. N.q.r. and n.m.r. studies of phase transitions in disordered and ordered crystals have been reviewed[1], as has n.q.r. in intermetallic compounds.[2] Investigations by n.q.r. of intramolecular complexes formed by halogen-containing organic compounds of Si, Ge and Sn have been surveyed,[3] and pressure studies of molecular dynamics and phase transitions, mainly involving ^{35}Cl n.q.r., have been reviewed.[4] Recent ^{63}Cu n.q.r. and n.m.r. data for the high-T_c superconductors $La_{2-x}Sr_xCuO_4$, $YBa_2Cu_3O_7$ and $Tl_2Ba_2CuO_{6+y}$ have been reviewed,[5] as have n.q.r. and n.m.r. studies of high-T_c superconductors in the period 1987-1989.[6] Four more limited reviews, mainly concerned with results obtained in the authors' laboratories, have appeared, on ^{139}La and ^{63}Cu n.q.r. spin-lattice relaxation data for $La_{2-x}Sr_xCuO_4$ ($0 \leq x \leq 0.3$),[7,8] copper n.q.r. and n.m.r. of Y-Ba-Cu-O superconductors,[9] and ^{63}Cu n.q.r. (together with ^{17}O and ^{205}Tl n.m.r.) of high-T_c superconductors.[10]

The normal format is adopted in the more detailed sections, i.e. results for main group elements (Groups II, III, V and VII), followed by those for transition metals and lanthanides.

2 Main Group Elements

2.1 Group II (Barium-135) - N.q.r. (and n.m.r.) of ^{135}Ba nuclei in $YBa_2Cu_3O_y$ has been observed at 4.2K from a sample isotopically enriched with ^{135}Ba.[11] The pure n.q.r. line occurred at 27.0(1) MHz. The results confirmed the electric field gradient (e.f.g.) calculations by Ambrosch-Draxl et al. Some ^{63}Cu data were also obtained.

2.2 Group III (Boron-10 and -11, and Aluminium-27) - ^{11}B (and ^{10}B in some cases) n.q.r. measurements have been reported for crystalline B_2O_3, vitreous B_2O_3, which showed two distinct boron sites, polycrystalline lithium diborate ($Li_2O.2B_2O_3$) and H_3BO_3.[12] Where the ^{10}B frequencies were measurable, values of the asymmetry parameter η were calculated. Boron n.m.r. and n.q.r. studies of glass structures have been described, including results for normal and ^{10}B-enriched vitreous B_2O_3, $Li_2O.B_2O_3$ and $2Na_2O.B_2O_3$ (the last two compounds unenriched).[13] Two boron sites were found in $Li_2O.B_2O_3$, probably corresponding to two types of chain unit. An as yet unexplained response was seen in the 700 kHz region. The n.q.r. of $2Na_2O.B_2O_3$ indicated the presence of pyroborate and orthoborate units, in a ratio of about 3:1.

The n.q.r. of ^{10}B and ^{11}B nuclei in p-$B_{10}C_2H_{12}$ has been measured at 77K by a double resonance technique, enabling e^2Qq/h and η to be evaluated.[14] Ab initio calculations of the e.f.g. tensors at the boron nuclei in o-, m- and p-$B_{10}C_2H_{12}$ were performed, and the results agreed well with experimental data. No reorientational jumps of the carborane molecules at 77K were apparent on the n.q.r. timescale. The technique of double resonance with spin mixing by continuous coupling (DRCC) has been used for observation of ^{10}B and ^{11}B n.q.r. from $BaO.B_2O_3.5H_2O$ and $BaO.B_2O_3.4H_2O$.[15] The boron quadrupole coupling data were discussed in terms of the crystal structure.

All the ^{27}Al n.q.r. frequencies (both transitions) have been measured for three different forms of crystalline Al_2SiO_5, i.e. andalusite (gem quality sample), sillimanite (both at 77K), and kyanite (at room temperature), enabling e^2Qq/h and η to be calculated.[16] The results were compared with those from n.m.r. studies of single crystals. In general the parameters derived from the n.q.r. data were more accurate.

2.3 Group V (Nitrogen-14, Arsenic-75, Antimony-121 and -123, and Bismuth-209) - The temperature (T)-dependence of ^{14}N signals for $K_4M(CN)_8.2H_2O$ (M = Mo or W) has been studied from 77 to 293K.[17] Twelve lines were found, six for each transition, in agreement with the crystal structure. Two of the six nitrogen sites showed a differing T-dependence from the others. The lack of matching between lines from similar sites in these isostructural compounds was attributed to variation in the cyanide charge distribution produced by differences in orbital overlap, as a result of relativistic effects present in the tungsten atom. The construction of a sensitive Fourier-transform spectrometer based on a d.c. SQUID (superconducting quantum interference device) amplifier has been described, suitable for obtaining low frequency (below 200 kHz) n.m.r. and n.q.r.[18,19] It was used to obtain ^{14}N n.q.r. signals from NH_4ClO_4, among other applications.

^{14}N n.q.r. (the ν^- line) has been obtained from a thin layer sample of $NaNO_2$ at room temperature by using a meanderline, or zig-zag, surface coil.[20] The results were compared with those obtained using a solenoid of comparable dimensions. The sensitivity of n.q.r. detection using surface coils was shown to be greatly enhanced by using fast-pulsing techniques. The advantages and limitations of the method were discussed. The induction and echo signals have been seen for the first time in the effective field of a multipulse n.q.r. train, from a single crystal sample of $NaNO_2$ at 77K.[21] It was demonstrated experimentally that these effects could be utilised for n.q.r. tomography. The Stark effect on the ν^- line of the ^{14}N n.q.r. from ferroelectric $NaNO_2$ has been investigated at 77K in both powder and single crystal samples, the latter with either a single domain or a multidomain structure.[22] The line shifts observed in the single domain crystals were closely related to the change in polarisation at the ^{14}N site, caused by the applied electric field. This could be explained qualitatively in terms of the point charge model, on the assumption that $NaNO_2$ was an ideal ionic crystal. No shifts were seen in the resonances from the other samples, but the lines were broadened. The results for the multidomain and powder samples could be explained quite well by using the first moment and second moment, respectively, of the n.q.r.

line in a single domain crystal under an external electric field. Attempts have been made to observe ^{14}N n.q.r. by double resonance techniques from large protein complexes, such as carbon monoxy haemoglobin.[23] Success had not been achieved as yet, but some of the problems were discussed, and the apparatus was being modified and improved in an effort to overcome these difficulties.

An n.q.r. imaging procedure has been described, analogous to n.m.r. rotating frame zeugmatography.[24] No magnetic field gradients were used, but the necessary r.f. gradients were produced by surface coils, enabling objects larger than the coil diameter to be investigated. The technique was illustrated by using cylindrical As_2O_3 samples, partially with a sandwich structure, to which different T gradients had been applied. It was expected to be particularly suitable for detecting gradients of physical parameters that influence the resonance position, such as stress or pressure and temperature, and for obtaining the spatial distribution of the chemical composition. This approach has been extended to the recording of two-dimensional n.q.r. images, and again illustrated using As_2O_3 powder, which allowed the spatial distribution of the As nuclei to be depicted.[25] The advantages and disadvantages of the method were discussed. ^{75}As n.q.r., including the resonance frequency at 4.2K and the T-dependence of T_1^{-1} between 2 and 200K, has been measured for Yb_4As_3.[26] The relaxation rate followed the law T_1T = constant in a narrow T range between 10 and 20K, and became almost independent of T below 4K, suggesting that the ground state properties of Yb_4As_3 were not so simple as to be described by the Fermi liquid picture. The theoretical basis for the use of pulsed methods in very low T n.q.r. (or n.m.r.) spin-echo experiments has been discussed, including the necessity for minimising sample heating, and a method was established to give good thermal anchoring of the sample.[27] This involved using pulses much shorter than the standard $\pi/2$—π pulses. The expression for the echo amplitude was confirmed by experiments on a powder sample of high-purity arsenic. Preliminary T_1 measurements suggested that the Korringa relationship holds down to 150mK in arsenic.

The compounds $LH[Sb_2F_7]$ (L = 1,2,4-triazole) and $L'H[SbF_4]$ (L' = 3-, 4- or 5-amino-1,2,4-triazole) have been prepared[28] from SbF_3 and acidic solutions of L or L'. They were characterised by $^{121,123}Sb$ n.q.r. at 77K, and i.r. spectroscopy. Two distinct Sb sites were found in $LH[Sb_2F_7]$. SbF_5 reacted with H_2SO_4 and H_2SeO_4 to give $SbF(OH)(HSO_4)$ and $SbF(SeO_4).H_2O$ respectively.[29] The products were characterised by $^{121,123}Sb$ n.q.r. at 77K, among other techniques. A theoretical treatment has been developed for the T-dependence of the ^{123}Sb n.q.r. parameters in $CsSbClF_3$, where Bayer theory could not explain the T-dependence of the e.f.g. asymmetry.[30] Good agreement with the experimental results was obtained.

The T-dependence from 77-300K of the ^{121}Sb and ^{35}Cl n.q.r. parameters has been determined for $[PhNH_3]_nSbCl_{3+n}$ (n = 1,2 or 3).[31] A good correlation was found between the ^{121}Sb quadrupolar coupling constant and the charge on the anion. The T-dependence of the ^{35}Cl n.q.r. frequencies for the compounds with n = 2 or 3 showed the characteristic pattern of 3-centre, 4-electron bonds. From the n.q.r. results, the anion for n = 3 was expressed as $[SbCl_5]^{2-}\ Cl^-$, rather than $[SbCl_6]^{3-}$. The T-dependence of ^{121}Sb and ^{81}Br n.q.r. for 4,4'-bipyridinium pentabromoantimonate(III) from 77 to 300K has been measured.[32] No signals could be detected from the analogous 2,2'-bipyridinium compound, although the crystal structures of both compounds were determined. The Sb n.q.r. parameters showed only a weak T-dependence.

The ^{81}Br spectrum consisted of five lines over a wide frequency range, from 34 to 113 MHz at 77K. The Br resonances were assigned from the crystal structure. Bonding in the anion was discussed, together with possible models to explain the observed T-dependence of the bromine frequencies. The use of multifrequency pulsed n.q.r. for detecting second-order spin-lattice coupling mechanisms, and for determining transition probabilities, has been discussed theoretically, and illustrated experimentally by measurements on a highly pure sample of Sb.[33] The quadrupolar mechanism was shown to contribute 5-10% to the total relaxation process at 77K. Above the Debye temperature the importance of quadrupolar coupling increased dramatically, and the results suggested the probable emergence of a two-phonon relaxation mechanism.

In a preliminary report, the ^{209}Bi n.q.r. spectrum has been recorded for the antiferromagnetic region of $Bi_2Sr_2YCu_2O_{8+\delta}$.[34] A very different e.f.g. distribution was found from that in $Bi_2Sr_2CaCu_2O_8$, which is at the superconducting end of the phase diagram. The results were consistent with the idea that charge transfer takes place in the Bi layers.

2.4 Group VII (Chlorine-35 and -37, Bromine -79 and -81, and Iodine -127) - The ^{35}Cl n.q.r. frequencies at 293K have been measured for a single crystal of $CsHg_2Cl_5$; the X-ray structure was also determined.[35] Five resonances were observed between 11 and 23 MHz. The two lowest frequency lines were assigned to Hg-Cl-Hg bridging bonds. A good correlation was found between bond length and frequency for the three higher frequency resonances and for those of molecular $HgCl_2$. The electronic structures and dynamics of some dichloromethylene-phosphoranes $X(YZ)P{=}CCl_2$ and -phosphines $XP{=}CCl_2$ have been probed by ^{35}Cl n.q.r., including relaxation time measurements.[36] Literature values for ^{13}C and ^{31}P chemical shifts were also used in the discussion. Analysis of the data led to a unified approach to the electronic structures of the compounds as simple quasi-alternant systems. Features in the structural dynamics were caused by differences in the P-C multiple bond structure, and by steric interactions between substituents on the P and C atoms. The ^{35}Cl n.q.r. frequencies at 77K have been measured for a series of 3-substituted 1-dichlorophosphoryl adamantanes.[37] The results were correlated with the inductive constants of the substituents on adamantane. The mechanism of increased conductivity of the adamantane nucleus was examined within the framework of a quasi-π-electron approach.

The ^{35}Cl frequencies for some n-alkoxy- and n-alkyl-chlorosilanes have been correlated with the Taft induction constants.[38] In $Me(CH_2)_nOSiCl_3$ and $Me(CH_2)_nSiCl_3$ the ^{35}Cl n.q.r. frequency was found to oscillate between even and odd numbers of CH_2 groups. The X-ray structures of NMe_4TlCl_4 and NEt_4TlCl_4 have been determined at room temperature.[39] The compounds underwent phase transitions at 239 and 222K, respectively. No ^{35}Cl signals were observed for the high temperature phase in either case, attributed to the anion dynamics. The T-dependence of the ^{35}Cl n.q.r. was studied in the low temperature phase from 77K to T_c, the NMe_4 compound giving two lines and the NEt_4 compound having a four line spectrum. Possible structures in this phase were discussed. The dynamics of the cation were investigated by ^1H n.m.r. ^{35}Cl n.q.r. spin-lattice relaxation times (T_{1Q}) have been measured as a function of T from 77 to 320K for $[NMe_4]_2[MCl_6]$ (M = Pb, Sn or Te).[40] Some ^1H n.m.r. relaxation data (T_{1H}) were also obtained for the Pb compound, and were already available for the others. In the room temperature phases, T_{1Q} was modulated strongly by the protonic motions of the cation, suggesting that these took place between disordered orientations. The T-dependence for the Pb complex, where T_{1Q} stemmed from cationic motion responsible for a deep T_{1H} minimum, differed from that for the Sn and Te compounds, where T_{1Q} was determined by cationic motion giving rise to shallow T_{1H} minima. The results suggested that the Me groups in the respective complexes take up different orientations in the crystals.

^{35}Cl n.q.r. frequencies have been measured at ca. 77, 210 and 290K for M_2PdCl_6 (M = K, Rb, Cs or NH_4), and in the range 150-350K (the observation limits) for the clathrate $[NMe_4]_2[PdCl_6].xCl_2$ ($x \leq 1$).[41] The crystal structure of the clathrate was also determined. A similar complex $[NMe_4]_2[SnCl_6].Cl_2$ was prepared, and shown to be isomorphous with the palladium compound. The ^{35}Cl n.q.r. of the Pd clathrate was as expected for the compound without Cl_2 inclusion, indicating that the interaction was weak. Signals for the Cl_2 molecule could not be detected, possibly because of motion. The X-ray data demonstrated that the result of inclusion was expansion of the unit cell. T-dependence studies on ^{35}Cl n.q.r. from $(NH_4)_2ZnCl_4$, together with T_1 measurements for ^{35}Cl and ^{1}H nuclei, have shown the existence of two polymorphic modifications of the compound, with different phase transition sequences between 266 and 406K.[42] Two new low temperature phase transitions, at 55 and 80K, were also observed. The experimental data were discussed in terms of the structures of the various phases. ^{35}Cl frequencies and T_1 values have been measured for a single crystal of Rb_2ZnCl_4 at various temperatures close to the paraelectric-incommensurate transition temperature (29.40°C).[43] The results could be explained by the theory of spin-lattice relaxation via large scale fluctuations of the modulation wave in incommensurate systems. At lower temperatures the conventional theory of relaxation via phason and amplitudon fluctuations applied. T-dependence measurements between 288 and 318K of the ^{35}Cl frequency in mixed $[Rb_{1-x}(NH_4)_x]_2ZnCl_4$ crystals, with a controlled amount of impurity, have shown the existence of a temperature region where incommensurate inhomogeneous broadening of the n.q.r. spectra was averaged out.[44] As a result, the characteristic line splitting started at a temperature below T_c, an effect enhanced by the presence of impurities. The experimental evidence suggested that impurities might prevent the onset of incommensurate long-range order in incommensurate insulators.

Two ^{35}Cl n.q.r. signals have been recorded at both 77 and 182K for $SnCl_4.(Et_2O)_2$.[45] Separate η values were also determined at 182K by Fourier analysis of the spectrum of slow beats of the spin echo envelope in a weak magnetic field, but only an average value was obtained at 77K. The crystal structure of the compound confirmed the trans arrangement of the ligands. Intramolecular H-Cl contacts were deduced to cause significant deviations from axial symmetry for the electron distribution at the ^{35}Cl nuclei. Two ^{35}Cl resonances have been observed for 2-MeO-5-$MeC_6H_3SnCl_3$, the higher frequency line being more intense.[46] This line was attributed to two Cl atoms of the $SnCl_3$ group, symmetrically placed relative to the plane of the aromatic ring, and the other resonance to the third Cl. The results were in agreement with the crystal structure, which was determined. ^{35}Cl n.q.r. data (including some from the literature) for 5-coordinate (tbp) and 6-coordinate (ψ-octahedral) complexes of $SnCl_4$ have been analysed.[47] Some new measurements were reported, including η values obtained from Fourier analysis of the spectrum of the spin echo pulse envelope in a weak magnetic field.

Measurements of η enabled axial ($\eta \sim 0$) and equatorial (η 0.05-0.16) chlorines to be differentiated in octahedral complexes, while cis and trans isomers could also be distinguished. The main differences between ψ–octahedral and tbp complexes were due to the relatively higher population of the 3p σ-orbitals of the equatorial Sn-Cl bonds in octahedral species, indicating a greater degree of charge transfer than for the corresponding bonds in tbp systems.

Complex formation between $SnCl_4$ and the aryl dichlorophosphates 2-$XC_6H_4OPOCl_2$ (X = H, F, Cl or Br) has been investigated by means of ^{35}Cl n.q.r. at 77K.[48] When X=H, a cis-octahedral 1:2 complex was formed, regardless of the reactant ratio. The other ligands formed 1:1 tbp complexes from equimolar quantities of reactants, and octahedral 1:2 complexes from 1:2 reactant ratios. Complexation caused an increase in frequency for Cl bound to P, and a decrease in frequency for Cl bound to Sn, related to changes in electron density in the bonds. The ^{35}Cl spectra at 77K of RR'CO-$SnCl_4$ (1:1 or 2:1) systems have shown the formation of tbp or octahedral complexes, depending on the ratios of the components and the nature of the ketone.[49] Mixtures of spatial isomers, or mixtures of complexes with different coordination numbers for tin, were often observed, and changes in the spectra with time sometimes indicated a transition to another spatial configuration. Features of the electron distribution in the tbp complexes were discussed. Calculations on Ph_2PbCl_2, which has a polymeric chain structure with all chlorines bridging, have suggested that the ^{35}Cl n.q.r. frequency should be around 5 to 6 MHz.[50]

The pressure-(p) dependence up to 5 kbar and T-dependence from 77-300K have been determined for ^{35}Cl signals from $NaClO_3$ and $Ba(ClO_3)_2.H_2O$.[51] The torsional frequency of the ClO_3^- ion and its variation with p and T were evaluated from the n.q.r. frequency, by using the harmonic approximation. The p effect on internal motions was smaller in $Ba(ClO_3)_2.H_2O$ than in $NaClO_3$. At 77K the p coefficient of frequency was very small for $NaClO_3$, but larger for the Ba compound, in keeping with the observation that the torsional frequency in $NaClO_3$ was unaffected by p at this temperature, whereas it increased at the rate of 12 cm^{-1} $kbar^{-1}$ in $Ba(ClO_3)_2.H_2O$. Recent n.q.r. measurements at 293 ± 2K, made with a pulsed spectrometer for $KClO_3$-$NaClO_3$ mixtures in varying mol %, have shown two signals for the individual components.[52] The results did not agree with earlier CW measurements. Line broadening was still observed relative to the spectra of the pure compounds, and possible causes were discussed. A modification has been devised for a simple n.q.r. spectrometer to improve the stability and sensitivity, involving the insertion of a phase shifter with operational amplifiers between the oscilloscope and audio-oscillator.[53] This enabled the ^{35}Cl signal from $NaClO_3$ to be observed on the oscilloscope screen at 295K. A precision marginal oscillator n.q.r. spectrometer has been described that operates in the range 14-35 MHz.[54] Results were presented for $NaClO_3$ and $KClO_3$ at room temperature, and for $KClO_3$ at various temperatures from 77 to 393K. The quadrupolar coupling constants for both chlorine isotopes were obtained with high precision.

An n.q.r. imaging experiment has been conducted on the ^{35}Cl n.q.r. from a powder sample of $NaClO_3$ at room temperature, by using various perturbing Zeeman fields.[55] The results were described, and the advantages and limitations of the method were discussed. The technique was expected to become an important tool for obtaining the spatial distributions of n.q.r.-sensitive quantities, such as T and p in disordered solids. A new method of performing Zeeman-perturbed n.q.r. experiments, called Zeeman-incremental spin-echo envelope modulation, or ZISEEM, has been described, and its use illustrated via ^{35}Cl n.q.r. of both single crystal and polycrystalline samples of $NaClO_3$ at 77K.[56] For single crystal samples, the method could be used to eliminate problems associated with T_2 delay. For polycrystalline samples, however, the decay of the Zeeman oscillations was determined mainly by powder-averaging effects rather than T_2 processes, giving a marginal gain only over previous methods. The advantages and disadvantages of the technique were discussed, including application to multiple-quantum experiments. The n.q.r. frequency of the ^{35}Cl nucleus in a single $NaClO_3$ crystal has been measured as a function of isothermal uniaxial stress along the (100), (110) and (111) crystal directions near room temperature, in the p range 0-4 MPa.[57] A linear and directional behaviour was observed, with negative slopes corresponding to the (100) and (110) directions, but positive for the (111) direction. A model was developed to allow the linear terms of the static and dynamic contributions to the variation in e.f.g. at the nucleus to be obtained. These were deduced to be of the same order of magnitude. A theoretical treatment has been given for broadband excitation in pure n.q.r. of I = 3/2 nuclei.[58] Experiments were carried out on powder samples of $HgCl_2$, $NaClO_3$ and C_6Cl_5OH, to verify broadband excitation of n.q.r. composite pulses.

^{35}Cl frequencies at 77K have been measured for ReX_2Cl_{12} (X = Se or Te), Re_2XCl_{12} (X = S or Se), and for the γ and β forms of $ReCl_4$.[59] The ^{79}Br n.q.r. spectrum of $ReBr_3$ was also recorded. The ReX_2Cl_{12} complexes were found to have the structure $ReCl_6(XCl_3)_2$, while Re_2SCl_{12} could be represented as $Re_2Cl_9(SCl_3)$. The Se analogue of the latter compound had a more complex structure, with two crystallographically non-equivalent $SeCl_3$ groups. The results for $ReBr_3$ indicated that it was not isostructural with $ReCl_3$, which is a trinuclear cluster Re_3Cl_9. Incommensurate (I) systems have been studied by n.q.r. and n.m.r. techniques, with emphasis on T_1 measurements.[60] The results included n.q.r. data for ^{35}Cl nuclei in Rb_2ZnCl_4, and ^{79}Br nuclei in β-$ThBr_4$. The phason and amplitudon contributions to T_1^{-1} were determined separately in each case. The phason gaps were evaluated, and the results compared with those from other techniques. The T-dependence of the lineshape for $ThBr_4$ was established over the entire I phase, enabling the order parameter to be determined. Molecular motions and the known phase transitions in $MeNH_3PbX_3$ (X = Cl, Br or I) have been investigated via the T-dependence of the halogen n.q.r. frequencies from 77 to 300K, and the ^1H spin-lattice relaxation times.[61] In the highest temperature tetragonal and cubic phases, isotropic reorientation of the cations took place, with $E_a \sim 11$ kJ mol^{-1}. T_1 measurements in the lowest temperature (orthorhombic) phases indicated that the cations underwent correlated C_3 reorientations in the chloride and iodide, whereas correlated and uncorrelated reorientations were excited in the bromide. The activation energy was derived in each case. T_1 in the low temperature region of the orthorhombic phases was shown to be governed by rotational tunnelling of the cations. Spin-lattice relaxation for I = 3/2 nuclei has been considered theoretically, together with n.q.r. detection methods, and instrumental improvements were suggested as a result.[62] The T-dependence of the resonance frequency and T_1 was examined in a series of model compounds, and a predictive model for the T-dependence was developed. Relaxation time measurements were used to analyse molecular motions in a series of main group halides.

The crystal structure of $HgBr_2$ has been refined, and a correlation observed between the ^{81}Br n.q.r. frequency and the Hg-Br bond lengths.[63] A similar relationship between frequency and bond length was found for some bromomercurates, showing that n.q.r. provides an effective method for determining the structural pattern in such compounds. ^{79}Br and ^{81}Br frequencies at 77K have been measured for various germanes containing the group $Br_3GeCEC(O)X$, where E = C or N.[64] From the results the Ge atom in $Br_3GeCH_2\overline{NCO(CH_2)_nCH_2}$ (n = 2 or 3) was deduced to be pentacoordinate, as a result of intramolecular interaction with the carbonyl oxygen. A possible similar although weaker interaction was detected in $Br_3GeCH_2CH(Me)COOMe$, but not in the other compounds studied. The alternative suggestion of dimerisation appeared less likely. The ^{79}Br n.q.r. spectrum at various temperatures from 4.2 to 293K has been recorded for the orthorhombic phase of $(NbBr_5)_2$, for which Nb data had been previously reported.[65] It consisted of two singlet lines near 110 MHz and two doublet lines near 60 MHz. This pattern closely resembled the ^{35}Cl spectrum of the high temperature modification of $NbCl_5$, suggesting that the latter compound might also be orthorhombic. Structural possibilities for the $(NbBr_5)_2$ dimer were discussed.

The known phase transitions in $CsPbBr_3$ have been corroborated by T-dependence studies of the ^{81}Br n.q.r. from 77 to 420K.[66] Above 406K (T_c) the compound has a cubic Perovskite-type structure, giving rise to a single resonance. At T_c this split into two components, one of double intensity, for the tetragonal phase II. The phase transition to the low temperature orthorhombic phase, expected at 316K, could not be detected by n.q.r., although one line did show a change of slope. Two resonances in a 2:1 intensity ratio were found from this phase also. The results were compared with those for other compounds of the type ABX_3. The T-dependence of the ^{79}Br and ^{81}Br resonances from enH_2CdBr_4, $enH_2(CdBr_3)_2.H_2O$ and $(enH_2)_2CdBr_6$, where en = $H_2NCH_2CH_2NH_2$, has been measured from 77 to 388K.[67] The crystal structures of the last two compounds were also determined. The complex enH_2CdBr_4 underwent a phase transition at ca. 340K (increasing T), and 320K (decreasing T). A three line spectrum was observed for ^{81}Br below this transition, and a single resonance from the higher T phase. No phase transitions were apparent for $enH_2(CdBr_3)_2.H_2O$, and the spectrum showed three ^{81}Br(or ^{79}Br) resonances at all temperatures. The structure was characterised by chains of $[Cd_4Br_{12}]^{4-}$ units, built up by edge-connected octahedra $CdBr_6$, with the corners of two neighbouring octahedra forming part of a bridging tetrahedron $CdBr_4$, alternately cis and trans on the chain. The compound $(enH_2)_2CdBr_6$ was formulated as $[(enH_2)^{2+}]_2\ [CdBr_4]^{2-}$, $2Br^-$ from the crystal structure. It gave a four-line ^{81}Br spectrum, three between 40 and 70 MHz and the fourth as low as 10 MHz, attributed to H-bonded Br^-. Some resonances had anomalous T-coefficients in the last two compounds, and all could be assigned from the crystal structures.

The T-dependence of the ^{79}Br or ^{127}I n.q.r. (as appropriate) has been determined in the range 77-420K for some ortho-substituted anilinium bromides and iodides $2\text{-}RC_6H_4NH_3^+\ X^-$ (X = Br or I; R = Cl, CN, Et, NH_2 and $NH_3^+\ X^-$).[68] Phase transitions were detected in $2\text{-}EtC_6H_4NH_3^+\ Br^-.\frac{1}{2}H_2O$ (T_c 164K), $2\text{-}EtC_6H_4NH_3^+I^-$ (T_c 214K), $[1,2\text{-}C_6H_4(NH_3)_2]^{2+}\ [Br]_2^-$

and its deutero-analogue with ND_3 groups (T_c 209K), and $[1,2\text{-}C_6H_4(NH_3)_2]^{2+}$ $[I]_2^-$ (T_c 173K). The results were discussed, and the data compared with those reported previously for para-substituted anilinium halides. ^{127}I n.q.r. frequencies (both transitions) have been measured at 77K for $(Me_nC_5H_{5-n})Fe(CO)_2$ I for n =1-5, enabling e^2Qq/h and η to be evaluated.[69] Data were also given for n = 2, 3 or 5 at 296K. The quadrupolar coupling constants decreased slightly with n, as expected. The terminal iodine atom was relatively weakly affected by a change in the number of Me groups, however, unlike the Mn or Re derivatives, where the central atom quadrupolar coupling constants were very sensitive to changes in the Cp ring. The T-dependence of the ^{127}I n.q.r. parameters for Cs_2ZnI_4 from 400-77K has revealed phase transitions at 120, 108 and 96K.[70] Other physical techniques were also used, and the presence of an incommensurate phase between 120 and 108K was deduced. Anomalies became apparent above 120K, provisionally interpreted as due to the formation of dynamic clusters of phase transition precursors. The structure of the compound in the various phases and the nature of the phase transitions were discussed. The p-dependence of the ^{127}I n.q.r. parameters and the phase transition temperature T_c (214K at 1 bar) has been determined for $NH_4IO_3.2HIO_3$ up to 4.5 kbar.[71] The phase transition was deduced to be mixed (order-disorder/displacement), and due to the antipolar ordering of the protons of bifurcated H-bonds, leading to doubling of the unit cell. The observed decrease of T_c with rise in pressure was explained on the basis of the Ising model, by the change of structure of the H-bonds under pressure. Other results for Group VII elements are described in the sub-sections on Group V[31,32] and Copper-63 and -65.[72]

3 Transition Metals and Lanthanides

3.1 Vanadium-51 - The pure n.q.r. of ^{51}V in KVO_3, $NaVO_3$ and NH_4VO_3 has been recorded at both room temperature and 77K, enabling e^2Qq/h and η to be evaluated.[73] These values were considerably more accurate than those obtained previously by n.m.r. Multiquantum transitions were observed in some instances for large η values. A sensitive n.q.r. spectrometer was constructed for this work, which uses 35 g samples.

3.2 Manganese-55 - The ^{55}Mn frequencies at 77K have been measured for a series of complexes $Mn(CO)_3R$, where R = $C_5H_2Me_3$, C_5HMe_4, C_5Me_5 or C_5Cl_5.[74] The results were compared with previous data for R = C_5H_4Me. The ^{35}Cl n.q.r. frequency was also found for the chloro-derivative, together with room temperature data for ^{55}Mn in this compound. The changes in the quadrupolar coupling constant were deduced to be primarily controlled by the resonance characteristics of the substituents on the Cp ring.

3.3 Copper-63 and -65 - The T-dependence of the ^{63}Cu (and ^{65}Cu) n.q.r. frequencies between 77 and 300K has been determined for seven copper(I) compounds $[CuL_2]X$, where L is a hindered pyridine (2,6-Me_2py or 2,4,6-Me_3py) and X is the counter-ion.[75] The crystal structures of five of the complexes were also obtained. All of the complexes had essentially linear N-Cu-N fragments, but they could be separated into two distinct types, one with a dihedral angle of 51°-61° between the two pyridine rings, and the other with a dihedral angle near 0°, giving a planar structure. A good correlation was observed between the Cu-N bond length and the ^{63}Cu n.q.r. frequency. The Cu-N bond was significantly longer in the planar complexes, and the resonance frequency lower. The T-dependence of the ^{63}Cu (and ^{65}Cu) n.q.r. from 77-300K for the copper (I) polyhalide anionic complexes $[CuBr_2]^-$, $[CuBr(Cl)]^-$, $[Cu_2Br_4]^{2-}$, $[Cu_2Br_5]^{3-}$ and $[Cu_2I_4]^{2-}$ showed no evidence of phase transitions.[72] All contained one type of Cu atom only, apart from one of the phases of $[PPh_4]_2$ $[Cu_2I_4]$, where the two Cu atoms were inequivalent. Bromine frequencies at 77 and 298K were also recorded for $[Cu_2Br_4]^{2-}$ and $[Cu_2Br_5]^{3-}$. The Zeeman ^{63}Cu n.q.r. spectrum of a single crystal of $[NPr_4]_2$ $[Cu_2I_4]$ gave η as 0.262 (1), with the z axis of the e.f.g. perpendicular to the plane of the anion and the x axis along the terminal Cu-I bond. The results were discussed in terms of the electronic structures of the complexes. The e.f.g. at the ^{63}Cu nucleus in Cu_2O and CuO has been evaluated as a function of p (up to 12 kbar) by means of the n.q.r. frequency.[76] The p-coefficients were given for Cu_2O at 77 and 295K, and for CuO at 295K. The results indicated that Cu_2O was not a purely ionic crystal, as often assumed in the analysis of the e.f.g. in high-T_c superconductors, and that covalent bonding was important in CuO. It was suggested that meaningful p-dependence measurements below 10 kbar should prove possible for high-T_c superconductors.

Several reports have again appeared on copper n.q.r. studies of high-T_c superconductors; they are grouped, as far as possible, by research teams. ^{63}Cu n.q.r. has been observed from the heavy-fermion system $CeCu_2Si_2$.[77,78] In superconducting specimens ($T_c \simeq$ 0.6K), a phase transition was detected at ca. 0.9K. Zero- and longitudinal-field muon spin relaxation measurements suggested a static magnetic ordering below 0.8K, probably in an incommensurate spin-density-wave state, rather than a spin-glass state, from the n.q.r. data. The exotic transition observed in n.q.r. was expected to be closely associated with this magnetic ordering. Results from a non-superconducting sample were consistent with extensive disorder in the Cu site occupation, such that the spin-density wave might not exist. The superconducting transition temperature T_c could be enhanced by the presence of a spin-density wave, so the superconductivity would be lost if this was depressed. ^{63}Cu n.q.r. at room temperature has been recorded for $YBa_2Cu_3O_{7-\delta}$.[79] The normal two resonances were observed, and their intensities supported the assignment of the lower frequency line to Cu(I) and the higher frequency line to Cu(II) sites.

Copper n.q.r. and n.m.r. measurements at 4.2 and 1.2K have been reported for the compounds MBa_2Cu_3Oy, where M = Y, or a rare earth metal heavier than Pr (which do not disturb the superconductivity), and for M = Pr, which destroys the superconductivity even at full oxygen loading.[80,81] Magnetic ordering was found only at the Cu(2) sites in each instance, either at low oxygen concentration or for M = Pr. In the Y and Pr compounds, the e.f.g. tensors at the Cu(1) sites with two, three or four oxygen neighbours were in agreement, indicating the same valence state for the Y and Pr in these compounds. Various physical techniques, including the ^{63}Cu n.q.r. of the Cu(1) site at 300, 100 and 4.2K, have been used to study a magnetic transition which doubles the magnetic unit cell along the c-axis in antiferromagnetic $Y_{1-u}Gd_uBa_2(Cu_{1-x}Fe_x)_3O_6$ (x = 0 or 0.01).[82] The hyperfine field distribution at Fe substituted for Cu(1) sites was attributed to inhomogeneities in Fe-Fe coordination, i.e. to small Fe clusters. From analysis of n.m.r. measurements on ^{63}Cu nuclei at 505K in the paramagnetic phase of $YBa_2Cu_3O_6$, the n.q.r. frequency of the plane Cu(2) sites has been derived.[83] ^{63}Cu n.q.r. data have been obtained for powder samples of $YBa_2Cu_3O_{7-\delta}$, previously oriented by mechanically vibrating the powder in a d.c. magnetic field of 8T at room temperature.[84] The T-dependence of the frequency showed a non-monotonic behaviour near T_c, possibly indicative of a ferroelectric or antiferroelectric transition, as also suggested by ultrasonic measurements. In the superconducting state the lineshape of the n.q.r. signals depended on the magnetic field applied during the cooling procedure, or after zero-field cooling. The results yielded information about the trapped flux.

Copper n.q.r. and n.m.r. data have been reported for $La_{2-x}Sr_xCuO_4$ ($0.10 \leq x \leq 0.20$).[85] For each composition, two distinct Cu sites were detected, the relative occupation of which correlated well with the Sr doping. From T-dependence n.m.r. studies, both sites were deduced to have the same electronic spin susceptibility and hyperfine field, and to occur in the superconducting phase with the same T_c. They had distinct orbital frequency shifts and electric field gradients, however. The results were unexpected in view of the crystallographic structure. The T-dependence of the n.q.r. (and n.m.r.) of Cu ions in $La_{1.84}Sr_{0.16}CuO_4$ has been found between 300 and 5K.[86] Analysis of the spectra showed two electronically distinct Cu sites, with uniaxial symmetry of the e.f.g. tensors, and the principal axis along the crystal c-axis. This result constrained possible structure models. N.m.r. and n.q.r. studies of ^{63}Cu T_1 values in $YBa_2Cu_3O_7$ below T_c have suggested that the orbital pairing in the superconducting state is d-like, and that at low temperatures and strong fields the fluxoid cores may dominate the relaxation mechanism.[87]

The T-dependence of T_1^{-1} for ^{63}Cu nuclei in $YBa_2Cu_4O_8$ has been determined over a wide range up to 680K.[88] Some Knight shift measurements were also reported. In the normal state, T_1^{-1} for the planar Cu(2) sites increased rapidly with T from T_c up to ca. 250K. Above this temperature the T-dependence was approximately linear. The behaviour of T_1^{-1} for the chain Cu(1) sites appeared quite similar to that of the chain sites in $YBa_2Cu_3O_7$, suggesting that superlinear T-dependence was a characteristic of the one-dimensional chain structure. The T-dependence of T_1^{-1} has similarly been measured from 4.2 to 340K for ^{63}Cu nuclei in $Y_{0.925}Ca_{0.075}Ba_2Cu_4O_{8+\delta}$ (T_c 90K); results were compared with those for $YBa_2Cu_4O_8$.[89] The increase in T_c upon substitution of Ca^{2+} for Y^{3+} could be explained by an increase in hole concentration at the Cu-O_2 plane, and by depression of the antiferromagnetic spin correlations. Copper n.q.r. (and n.m.r.) has been observed at 1.3K from $Pb_2Sr_2Y_{1-x}Ca_xCu_3O_{8+\delta}$, with $x = 0$ or 0.5.[90] For the non-superconducting sample ($x = 0$), n.q.r. was detected from the Cu(1) sites, and antiferromagnetic nuclear resonance from the Cu(2) sites. Two sets of n.q.r. signals were observed for the superconducting sample ($x = 0.5$), and assigned to Cu(1) (higher frequency) and Cu(2) (lower frequency) sites, indicating the disappearance of antiferromagnetic ordering. The T-dependence of T_1^{-1} for the n.q.r. signals from the superconducting sample was also obtained, and found to be similar to those seen in other high-T_c superconducting oxides.

The effect of quenching on various physical parameters, including Cu n.q.r. at 1.4 K and T_2 at 1.4 and 4.2K, has been investigated for $Bi_2Sr_2CaCu_2O_x$.[91] Results were given for samples quenched from 873K (T_c ~90K) and 573K (T_c ~70K). The resonances were very broad, but, contrary to expectation, the quenching effect on T_2 was not large. The n.q.r. frequencies of the Cu(1) and Cu(2) sites in $RBa_2Cu_3O_6$ (antiferromagnetic) and $RBa_2Cu_3O_7$ (superconducting), where R is a rare earth, have been correlated with the ionic radii of R.[92] A linear relationship was found except for the Cu(2) signals when R = Pr or Nd, where hybridisation between the 4f and CuO_2 states became important. The Cu n.q.r. of the Cu(1) site, and the antiferromagnetic n.m.r. of the Cu(2) site, have been investigated as a function of excess oxygen, y, in $Pb_2Sr_2(Y_{1-x}Ca_x)Cu_3O_{8+y}$.[93] From the results the electronic configuration of Cu(1) was deduced to change from $3d^{10}$ to $3d^9$ with oxygen-doping, producing extreme line broadening. Various physical methods, including Cu n.q.r., have been used to study anomalies due to the low-T structural transition around x = 0.12 in $La_{2-x}Ba_xCuO_4$.[94] The n.q.r. spectra were extensively broadened for $x < 0.14$, showing that the local charged states of the CuO_2 plane were largely distributed in the low T phase. T_2^{-1} was also suppressed for $0.11 \leq x \leq 0.14$. N.q.r. and n.m.r. measurements at 1.4 K have been made for Cu nuclei in $La_2CuO_{4+\delta}$ and $(La_{1-x}Sr_x)_2CuO_{4+\delta}$, including samples treated in a high pressure of oxygen gas.[95] The results indicated that oxygen-loaded $La_2CuO_{4+\delta}$ was composed of a non-superconducting antiferromagnetic phase poor in oxygen, and a superconducting oxygen-rich phase, where the electronic state around Cu was very similar to that in the Sr-doped samples. The effects of oxygen-loading on the Sr-doped sample were not distinct up to the partial pressure of 400 bar used in this work. ^{63}Cu n.q.r., including the T-dependence of T_1^{-1}, has been recorded for $Bi_2Sr_2CuO_6$.[96] At 1.3K the spectrum contained a broad line around 25-35 MHz; the broadening was attributed to atomic modulation. T_1^{-1} was proportional to T below ca. 10K, and saturated above 15K. No anomaly associated with superconductivity appeared, indicating the presence of a large fraction of the normal phase in this system. The Cu n.q.r. spectrum of superconducting $La_2CuO_{4+\delta}$ at 1.3K showed signals around 33.1 and 36.0 MHz.[97] The signal intensity of oxygen-loaded samples increased with the degree of oxygen loading. T_1 measurements for the ^{63}Cu nuclei indicated that the paramagnetic phase was in a superconducting state. Some La n.m.r. measurements were also described. The n.q.r. for the Cu(1) and Cu(2) sites in $YBa_2(Cu_{1-x}Fe_x)_3O_{7-y}$ has been studied as a function of x.[98] Signals could be detected up to x = 0.06 at 100K. The results indicated that Fe substituted predominantly at Cu(1) sites. The x dependence of the integrated intensities suggested that none of the Cu sites located in the four unit cells around the Fe impurity contributed to the observations, i.e. they were wiped out.

To obtain information on an anomalous peak at 35K, seen in the T-dependence plot of T_2^{-1} for ^{63}Cu nuclei from air-annealed $YBa_2Cu_3O_{7-\delta}$, the carrier concentration was changed by annealing in oxygen, and by substituting La for Ba.[99]. On annealing with O_2, the 35K peak was suppressed, and another small peak at 50K appeared. When 5% substitution of La for Ba was carried out and the sample annealed in O_2, which could reduce the carrier concentration, the spectrum was similar to that of the original air-annealed sample. The anomalous peak was deduced to be related to the carrier concentration, but not to the dynamics of oxygen vacancies in the Cu(1) chain. The T-dependence of the ^{63}Cu n.q.r. frequency and T_1 for $LuInCu_4$ has been determined, and compared with previous results for $YbInCu_4$.[100] For the Lu compound, T_1 followed the Korringa law over the whole T range studied, indicating relaxation through the conduction electrons. A different T-dependence plot was observed for the Yb compound near the transition temperature T_v (ca. 40K), with a large contribution to the relaxation deduced to arise from the 4f electrons. There was no transition in $LuInCu_4$ at this temperature. The T-dependence of T_1^{-1} for Cu nuclei in $YbAgCu_4$ and the isostructural $LuAgCu_4$ has been investigated.[101] For $YbAgCu_4$ the relaxation rate decreased markedly with decreasing T, consistent with delocalisation of the 4f electrons. The 4f spin relaxation rate deduced from T_1 showed the typical T-dependence of a dense Kondo system, i.e. it had a minimum around 110 ± 20K, considered to be a characteristic temperature of the system. The p-dependence of the ^{63}Cu n.q.r. frequency and T_c has been found for $La_{1.85}Sr_{0.15}CuO_4$ and $YBa_2Cu_4O_8$ under high pressure up to 1.3 GPa.[102,103] The frequency decreased with increasing p in the La compound, while that for the CuO_2 plane sites increased with p in $YBa_2Cu_4O_8$. Charge transfer was concluded to take place under pressure, and a model proposed to account for the results. The ratio of the hole density of Cu, n(Cu), to the hole number, n(o), in each in-plane oxygen, was considered to be of prime importance in governing T_c, with the enhancement of T_c under pressure due to the change of this value to the most favourable one, ~1.0.

The superconducting properties of Zn-doped $YBa_2Cu_3O_7$ have been investigated by ^{63}Cu n.q.r. and n.m.r., including T_1 measurements.[104] The T_1T constant law held at low temperature, far below T_c of 79K and 68K for Zn contents of 0.01 and 0.02 respectively, providing clear evidence for gapless superconductivity with a finite density of states at the Fermi level. The rapid increase of T_1^{-1} just below T_c was almost unaffected by doping. The properties of $Tl_2Ba_2CuO_{6+y}$ have been similarly studied, including ^{63}Cu n.q.r. for T_c = 72, 40 and 0K.[105] The value of T_1 (and the Knight shift) below T_c was almost the same as in $YBa_2Cu_3O_7$, although the behaviour above T_c was strongly modified, in the sense that the q-dependence of the antiferromagnetic spin fluctuation became less distinct on doping. The results could be well interpreted by a "gapless d-wave" model, with a finite density of states at the Fermi surface. Extensive T_1 measurements as a function of T for ^{63}Cu nuclei in $La_{2-x}Sr_xCuO_4$ ($0.075 \le x \le 0.20$) have been reported.[106,107,108] Below T_c, a rapid decrease of T_1^{-1} became apparent, independent of the Sr content. For x = 0.20, T_1^{-1} was not saturated, but still T-dependent even in the low temperature region. Hence saturation did not arise from trivial impurities, but rather from an intrinsic enhancement of T_1^{-1}, caused by the localised nature of the d spins.[108] The relaxation behaviour in both the superconducting and normal states was discussed. Copper n.q.r. and n.m.r. techniques have been used to establish the magnetic and superconducting phase diagram of $Ce_{1-x}Th_xCu_{2.2}Si_2$.[109] Static magnetic ordering, with T_N = 2.4K and 4.5K for x = 0.08 and 0.12 respectively, was shown by broadening of the n.q.r. (and n.m.r.) spectra, and a specific heat anomaly. For lightly-doped compounds with $x \le 0.064$ these effects were not evident, but there was a dramatic reduction in signal intensity below a temperature ranging from 0.9 to 1.3K, depending on the Th content. The magnetic transition in undoped and lightly Th-doped $CeCu_2Si_2$ samples differed from the static magnetic ordering in heavily-doped compounds.

Crystal field gradient calculations have been used to identify all the ^{63}Cu n.q.r. lines from $YBa_2Cu_3O_{7-\delta}$ with different δ.[110] The Cu electrons were found to be localised, but the oxygen π-orbital holes were widely distributed. T_1 and T_2 values were measured for $\delta = 0.05$ and 0.18, and a band structure model was proposed to explain the results. The spin-spin relaxation of Cu nuclei in $YBa_2Cu_3O_{6.95}$ and $YbBa_2Cu_3O_{6.9}$ has been studied at the n.q.r. frequencies in weak magnetic fields at 4.2K.[111] The structural features seen in the relaxation of $^{63}Cu(2)$ nuclei for both compounds were explained as resulting from a spin-spin interaction of the resonating Cu(2) nuclei with $Cu^{2+}(2)$ centres localised nearby. It was suggested that the copper n.q.r. is observed in Zhang-Rice singlet states. ^{63}Cu (and ^{65}Cu) n.q.r. has been detected for the phases $Bi_{2.17}Sr_{1.86}CuO_z$ and $Bi_{4.53}Sr_{7.58}(Cu_{0.88}Al_{0.12})_5O_7$.[112] The n.q.r. spectra were discussed with relation to the Cu lattice sites. Differences in spin-lattice relaxation and the T-dependence of the spin-echo effect were found between these phases. ^{63}Cu n.q.r. at 4.2K has been detected, using the method of spin-echo amplitude measurement in a weak magnetic field, from $YBa_2Cu_3O_{6.1}$ samples treated in I_2 vapour, and from $YBa_2Cu_3O_y$ (y = 6.9, 6.5 or 6.1).[113] Two resonances were seen from the iodinated samples, at 31.4 and 23.9 MHz, and the assignments to Cu(2) and Cu(1) sites respectively were confirmed. The results supported the conclusion that iodination did not give rise to the orthorhombic-2 phase of $YBa_2Cu_3O_{6.5}$. ^{63}Cu n.q.r. has been recorded from $YBa_2Cu_3O_{7-\delta}$ ($0.05 \le \delta \le 0.50$) by registering the spin-echo signal.[114] The T-dependence of T_1 and T_2 was also found. The evolution of the spectra, which varied with δ, was discussed in connection with the manifestation of ordering in the neighbourhood of the Cu(1) atoms. N.q.r. studies of the ^{63}Cu nuclei, and Mössbauer spectra of ^{119}Sn nuclei, have been reported for $YBa_2Cu_{3-x}Sn_xO_{7-\delta}$.[115] The Cu resonance lines were broadened by the presence of tin, and an additional low intensity line was found in the Cu(1) region for x = 0.06, possibly indicating short-range order in the arrangement of the Cu and Sn atoms. The introduction of the impurity caused no significant effect on T_1. Cu n.q.r. spectra as a function of x have been recorded for the normal state of $YBa_2(Cu_{1-x}Zn_x)_3O_{7-\delta}$.[116] The intensities of the signals for both the Cu(1) and Cu(2) sites decreased with increasing x. Analysis of the data indicated that the Zn atoms occupy Cu(2) sites, and that their influence extended to the neighbouring four Cu atoms in the same plane, and to the Cu(1) site just below the Zn atom. The effect of Zn doping on the superconductivity was compared with that of Ni doping, and the mechanism for the disappearance of superconductivity caused by Zn substitution was discussed. Other results for ^{63}Cu have been described in the sub-section on Group II.[11]

3.4 Zirconium-91 - The ^{91}Zr n.q.r. signal for the $\frac{3}{2} \leftrightarrow \frac{5}{2}$ transition in $ZrSiO_4$ has been located at room temperature (6.142 MHz, linewidth ca. 70 kHz).[117] T_1 was also measured as 0.57 ± 0.07s. The experiments required 6,000 or more transients, and a delay time between pulse pairs of at least 1s, to get a good signal-to-noise ratio.

3.5 Lanthanum-139 - The nuclear spin relaxation rate for ^{139}La nuclei ($\frac{5}{2} \leftrightarrow \frac{7}{2}$ transition) has been measured as a function of T in $La_{2-x}Sr_xCuO_4$ for x = 0.12.[118] The results provided evidence for a structural phase transition at ca. 7K, below T_c. This conclusion was supported by ultrasonic experiments. The nuclear relaxation of ^{139}La nuclei from 1.7 to 77K has been recorded for $La_2Cu_{0.97}Zn_{0.03}O_4$.[119] The data indicated that zinc doping largely conserved the anomalous T-dependence of T_1^{-1} and T_2^{-1} seen for the undoped material. The primary effect of Zn doping was a noticeable reduction in relaxation rates between 5 and 15K. A method has been devised for separating the magnetic and quadrupolar contributions to the spin-lattice relaxation rate in n.q.r. relaxation, without completely solving the master equations.[120] It was illustrated experimentally for ^{139}La relaxation ($\frac{5}{2} \leftrightarrow \frac{7}{2}$ transition) in $La_{2-x}Sr_xCuO_4$. The T-dependence of the n.q.r. frequency and linewidth from 77-325K for ^{139}La nuclei ($\frac{3}{2} \leftrightarrow \frac{5}{2}$ transition) in a nearly stoichiometric single crystal of $La_2CuO_{4+\delta}$ showed no anomaly at 150K, suggesting no order-disorder phase transition at this temperature.[121] The results did not agree with a perturbed angular correlation study of ^{111}In-doped $La_{1-x}Sr_xCuO_{4+\delta}$, where a structural phase transition in the intrinsic material was suggested. Substantial structural changes in the metallic phase of $La_2CuO_{4+\delta}$ have been detected below 220K by ^{139}La n.q.r. and n.m.r. spectroscopy.[122] Results were reported for crystals with T_c 38 and 28K. Upon cooling the T_c 38K sample from 220 to 100K, the n.q.r. line shifted down in frequency by over 1 MHz, broadened from ca. 30 kHz to over 1 MHz, and developed a double peak structure. A similar spectrum was obtained from the T_c 28K sample at 75K, except that the double peak structure was absent. The results emphasised the loss of positional order and non-trivial structural change on cooling. The double peak showed the presence of two distinct La sites in the metallic phase at low temperature.

3.6 Praseodymium-141 - Two-pulse and three-pulse stimulated n.q.r. echoes have been measured from Pr^{3+} in $YAlO_3$, to determine the dephasing times.[123] The results were compared with the theory of Hu and Hartmann. The apparent discrepancy between the two-pulse and three-pulse fits was attributed to fluctuations caused by different host nuclei.

3.7 Rhenium-185 and -187 - The rhenium n.q.r. spectra for $(Me_nC_5H_{5-n})Re(CO)_3$ (n = 1-5) have been recorded at 77K.[124] Results were also given for the pentamethyl Cp derivative at 273K. The quadrupolar coupling constant increased monotonically as Me groups were added to the Cp ring, whereas the η value changed irregularly, with a maximum for the penta-methyl compound. The crystal structure of the latter was also determined. Bonding in the compounds was discussed in the light of the results.

References

1. F. Borsa and A. Rigamonti, *Top. Curr. Phys.*, 1991, **45** (Struct. Phase Transitions II), 83.
2. V.L. Matukhin and I.A. Safin, *Radiospektroskopiya Kondensir. Sred*, M. 1990, 127; from *Ref. Zh., Fiz.* (A-Zh) 1991, Abstr. No. 3S292; *Chem. Abstr.*, 1991, **115**, 125377.
3. V.P. Feshin and G.A. Polygalova, *J. Organomet. Chem.*, 1991, **409**, 1.
4. M. Maćkowiak, *Bull. Magn. Reson.*, 1991, **12**, 141.
5. K. Asayama, G.-Q. Zheng, Y. Kitaoka, K. Ishida and K. Fujiwara, *Physica C*, 1991, **178**, 281.
6. A.A. Gippius, E.A. Kravchenko and V.V. Moshchalkov, *Russ. J. Inorg. Chem.*, 1991, **36**, 318.
7. F. Borsa and A. Rigamonti, *NASA Conf. Publ.*1991, **3100** (AMSAHTS 90: Adv. Mater. Sci. Appl. High Temp. Supercond.), 399.
8. A. Rigamonti, *Springer Ser. Solid-State Sci.*, 1990, **99** (Electron. Prop. High-T_c Supercond. Relat. Compd.), 218.
9. D. Brinkmann, *Springer Ser. Solid-State Sci.*, 1990, **99** (Electron. Prop. High-T_c Supercond. Relat. Compd.), 195.
10. Y. Kitaoka, K. Fujiwara, K. Ishida, T. Kondo and K. Asayama, *J. Magn. Magn. Mater.*, 1990, **90-91**, 619.
11. H. Lütgemeier, V. Florentiev and A. Yakubovski, *Springer Ser. Solid-State Sci.*, 1990, **99** (Electron. Prop. High-T_c Supercond. Relat. Compd.), 222.

12. P.J. Bray, S. Gravina and D. Lee, *AIP Conf. Proc.*, 1991, **231** (Boron-Rich Solids), 271.
13. P.J. Bray, J.F. Emerson, D. Lee, S.A. Feller, D.L. Bain and D.A. Feil, *J. Non-Cryst. Solids*, 1991, **129**, 240.
14. A. Lötz and J. Voitländer, *J. Chem. Phys.*, 1991, **95**, 3208.
15. G. Molchanov, V. Anferov, J. Svirksts and H. Gode, *Latv. PSR Zinat. Akad. Vestis, Kim. Ser.*, 1990, (2), 192.
16. D. Lee and P.J. Bray, *J. Magn. Reson.*, 1991, **94**, 51.
17. J. Murgich, I. Bonalde, A. Díaz, and J.A. Abanero, *J. Magn. Reson.*, 1991, **93**, 47.
18. N.-Q. Fan, *Diss. Abstr. Int. B*, 1991, **52**, 2115.
19. N.-Q. Fan and J. Clarke, *Rev. Sci. Instrum.*, 1991, **62**, 1453.
20. M.L. Buess, A.N. Garroway and J.B. Miller, *J. Magn. Reson.*, 1991, **92**, 348.
21. V.L. Ermakov, R. Kh. Kurbanov, D. Ya. Osokin and V.A. Shagalov, *JETP Lett.*, 1991, **54**, 466.
22. K.T. Han and S.H. Choh, *J. Korean Phys. Soc.*, 1991, **24**, 159.
23. H.S. Panth, L. Rowan and E. Bergmann, *Speculations Sci. Technol.*, 1990, **13**, 228.
24. E. Rommel, D. Pusiol, P. Nickel and R. Kimmich, *Meas. Sci. Technol.*, 1991, **2**, 866.
25. P. Nickel, E. Rommel, R. Kimmich and D. Pusiol, *Chem. Phys. Lett.*, 1991, **183**, 183.
26. H. Nakamura, S. Hirooka, Y. Kitaoka, K. Asayama, O. Nakamura, T. Suzuki and T. Kasuya, *Physica B*, 1991, **171**, 242.
27. I.P. Goudemond, J.M. Keartland and M.J.R. Hoch, *J. Low-Temp. Phys.*, 1991, **82**, 369.
28. R.L. Davidovich, L.A. Zemnukhova, G.A. Fedorishcheva, T.G. Ermakova and V.A. Lopyrev, *Koord. Khim.*, 1990, **16**, 1319; *Chem. Abstr.*, 1991, **114**, 74105.
29. R.L. Davidovich, L.A. Zemnukhova, G.A. Fedorishcheva, A.A. Udovenko and M.F. Eiberman, *Koord. Khim.*, 1990, **16**, 926; *Chem. Abstr.*, 1990, **113**, 203816.
30. V.N. Rykovanov, R. Sh. Lotfullin, L.A. Zemnukhova, A.A. Boguslavskii and R.L. Davidovich, *Phys. Status Solidi B*, 1991, **165**, K13.
31. T. Okuda, Y. Hashimoto, H. Terao, K. Yamada and S. Ichiba, *J. Mol. Struct.*, 1991, **245**, 103.
32. H. Ishihara, S.-Q. Dou and A. Weiss, *Ber. Bunsenges. Phys. Chem.*, 1991, **95**, 659.
33. J.M. Keartland, G.C.K. Folscher and M.J.R. Hoch, *Phys. Rev. B*, 1991, **43**, 8362.
34. R. De Renzi, G. Guidi, C. Bucci and R. Tadeschi, *Hyperfine Interact.*, 1990, **63**, 295.
35. V.I. Pakhomov, A.V. Goryunov, I.N. Ivanova-Korfini, A.A. Boguslavskii and R. Sh. Lotfullin, *Russ. J. Inorg. Chem.*, 1991, **36**, 800.
36. E.A. Romanenko, A.P. Marchenko, G.N. Koidan and A.M. Pinchuk, *J. Gen. Chem. USSR*, 1990, **60**, 1992.
37. E.A. Romanenko, E.É. Lavrova, R.I. Yurchenko and O.M. Voitsekhovskaya, *Theoret. Exptl. Chem.*, 1991, **27**, 116.
38. A.F. Babkin, I.P. Biryukov and Yu. I. Khudobin, *Russ. J. Phys. Chem.*, 1990, **64**, 1010.
39. M. Lenck, S.-Q. Dou and A. Weiss, *Z. Naturforsch.*, 1991, **46a**, 777.
40. Y. Furukawa, Y. Baba, S.-e. Gima, M. Kaga, T. Asaji, R. Ikeda and D. Nakamura, *Z. Naturforsch.*, 1991, **46a**, 809.
41. P. Storck and A. Weiss, *Z. Naturforsch.*, 1991, **46b**, 1214.
42. A.M. Kotelevets, S.V. Pogrebnyak and E.D. Chesnokov, *Bull. Acad. Sci. USSR, Phys. series*, 1990, **54**(6), 174.

43. G. Papavassiliou, F. Milia, R. Blinc and S. Žumer, *Solid State Commun.*, 1991, **77**, 891.
44. F. Milia, G. Papavassiliou and A. Anagnostopoulos, *Phys. Rev. B*, 1991, **43**, 4464.
45. A.V. Yatsenko, L.A. Aslanov, M. Yu. Burtsev and E.A. Kravchenko, *Russ. J. Inorg. Chem.*, 1991, **36**, 1147.
46. S.N. Gurkova, A.I. Gusev, N.V. Alekseev, V.P. Feshin, I.M. Lazarev, G.V. Dolgushin and M.G. Voronkov, *Organomet. Chem. in the USSR*, 1988, **1**, 685.
47. É.A. Kravchenko, V.G. Morgunov, M. Yu. Burtsev and Yu. A. Buslaev, *J. Gen. Chem. USSR*, 1990, **60**, 1737.
48. V.P. Feshin, G.V. Dolgushin, I.M. Lazarev and M.G. Voronkov, *Koord. Khim. (Engl. transl.)*, 1990, **16**, 555.
49. V.P. Feshin, G.V. Dolgushin, I.M. Lazarev and M.G. Voronkov, *Koord. Khim. (Engl. transl.)*, 1990, **16**, 656.
50. E.M. Berksoy and M.A. Whitehead, *J. Organomet. Chem.*, 1991, **410**, 293.
51. T.V. Krishnamoorthy, P.K. Babu and J. Ramakrishna, *J. Mol. Struct.*, 1991, **249**, 377.
52. J. Haleš and J. Kasprzak, *Acta Phys. Slovaca*, 1991, **41**, 65.
53. H. Nabeshima, *Phys. Rep. Kumamoto Univ.*, 1990, **8**, 67.
54. F. Dehui, L. Dehai, Z. Xiaoli and Z. Bide, *Chinese J. Magn. Reson.*, 1991, **8**, 123.
55. S. Matsui, K. Kose and T. Inouye, *J. Magn. Reson.*, 1990, **88**, 186.
56. G.J. Bowden, J.M. Cadogan, J. Khachan and R.P. Starrett, *J. Magn. Reson.*, 1991, **93**, 603.
57. R.C. Zamar and A.H. Brunetti, *J. Phys.: Condens. Matter*, 1991, **3**, 2401.
58. A. Ramamoorthy and P.T. Narasimhan, *Mol. Phys.*, 1991, **73**, 207.
59. E.V. Bryukhova, V.L. Kolesnichenko, S.I. Kuznetsov and N.I. Timoshchenko, *Bull. Acad. Sci. USSR, Div. Chem. Sci.*, 1991, **40**, 175.
60. S. Chen, *Diss. Abstr. Int. B*, 1991, **51**, 2432.
61. Q. Xu, T. Eguchi, H. Nakayama, N. Nakamura and M. Kishita, *Z. Naturforsch.*, 1991, **46a**, 240.
62. A.A. Koukoulos, *Diss. Abstr. Int . B*, 1991, **51**, 4363.
63. V.I. Pakhomov, A.V. Goryunov, I.N. Ivanova-Korfini, A.A. Boguslavskii and R. Sh. Lotfullin, *Russ. J. Inorg. Chem.*, 1990, **35**, 1407.
64. V.P. Feshin, P.A. Nikitin, G.A. Polygalova, T.K. Gar, O.A. Dombrova and N.A. Viktorov, *Organomet. Chem. in the USSR*, 1991, **4**, 258.
65. H. Sekiya, N. Okubo and Y. Abe, *Phys. Lett. A*, 1991, **152**, 495.
66. S. Sharma, N. Weiden and A. Weiss, *Z. Naturforsch.*, 1991, **46a**, 329.
67. V.G. Krishnan, S.-q. Dou, H. Paulus and A. Weiss, *Ber. Bunsenges. Phys. Chem.*, 1991, **95**, 1256.
68. J. Hartmann and A. Weiss, *Z. Naturforsch.*, 1991, **46a**, 367.
69. G.K. Semin, E.V. Bryukhova, S.I. Kuznetsov, I.R. Lyatifov and Sh. N. Abdulova, *Organomet. Chem. in the USSR*, 1991, **4**, 346.
70. I.P. Aleksandrova, S.V. Primak, E.V. Shemetov and A.I. Kruglik, *Sov. Phys. Solid State*, 1991, **33**, 758.
71. D.F. Baisa, A.I. Barabash, A.I. Shanchuk and E.A. Shadchin, *Sov. Phys. Crystallogr.*, 1991, **36**, 685.
72. S. Ramaprabhu and E.A.C. Lucken, *J. Chem. Soc. Dalton Trans.*, 1991, 2615.
73. D. Mao, P.J. Bray and G.L. Petersen, *J. Amer. Chem. Soc.*, 1991, **113**, 6812.
74. G.K. Semin, E.V. Bryukhova, S.I. Kuznetsov, I.R. Lyatifov and G.M. Dzhafarov, *Organomet. Chem. in the USSR*, 1988, **1**, 507.

75. A. Habiyakare, E.A.C. Lucken and G. Bernardinelli, *J. Chem. Soc. Dalton Trans.*, 1991, 2269.
76. R.G. Graham, P.C. Riedi and B.M. Wanklyn, *J. Phys. : Condens. Matter*, 1991, **3**, 135.
77. C. Tien, *Phys. Rev. B*, 1991, **43**, 83.
78. C. Tien, *Chinese J. Phys. (Taipei)*, 1990, **28**, 527.
79. L.-A. Zheng, J.-S. Ma, Y.-J. Gu, X.-W. Wu, Y.-Q. Yu and S.-P. Xue, *Chinese Sci. Bull.*, 1990, **35**, 1878.
80. H. Lütgemeier, *Springer Ser. Solid-State Sci.*, 1990, **99** (Electron. Prop. High-T_c Supercond. Relat. Compd.), 214.
81. H. Lütgemeier, *J. Magn. Magn. Mater.*, 1990, **90-91**, 633.
82. L. Bottyán, H. Lütgemeier, J. Dengler, S. Pekkev, A. Rockenbauer, A. Jánossy and D.L. Nagy, *Springer Ser. Solid-State Sci.*, 1990, **99** (Electron. Prop. High-T_c Supercond. Relat. Compd.), 230.
83. M. Mali, I. Mangelschots, H. Zimmermann and D. Brinkmann, *Physica C*, 1991, **175**, 581.
84. H. Riesemeier, H. Kamphausen, E.-W. Scheidt, G. Stadermann, K. Lüders and V. Müller, *Springer Ser. Solid-State Sci.*, 1990, **99** (Electron. Prop. High-T_c Supercond. Relat. Compd.), 225.
85. M.A. Kennard, Y. Song, K.R. Poeppelmeier and W.P. Halperin, *Chem. Mater.*, 1991, **3**, 672.
86. Y.-Q. Song, M.A. Kennard, M. Lee, K.R. Poeppelmeier and W.P. Halperin, *Phys. Rev. B*, 1991, **44**, 7159.
87. J.A. Martindale, S.E. Barrett, C.A. Klug, K.E. O'Hara, S.M. DeSoto, C.P. Slichter, T.A. Friedmann and D.M. Ginsberg, *Physica C*, 1991, **185-9**, 93.
88. T. Machi, I. Tomeno, T. Miyatake, N. Koshizuka and S. Tanaka, *Physica C*, 1991, **173**, 32.
89. T. Machi, I. Tomeno, T. Miyatake, K. Tai, N. Koshizuka, S. Tanaka and H. Yasuoka, *Physica C*, 1991, **185-9**, 1147.
90. M. Yoshikawa, K. Yoshimura, T. Imai, Y. Ueda, H. Yasuoka and K. Kosuge, *J. Phys. Soc. Jpn.*, 1991, **60**, 37.
91. T. Ishida, K. Koga, S. Nakamura, Y. Iye, K. Kanoda, S. Okui, T. Takahashi, T. Oashi and K. Kumagai, *Physica C*, 1991, **176**, 24.
92. T. Takatsuka, K.-i. Kumagai, H. Nakajima and A. Yamanaka, *Physica C*, 1991, **185-9**, 1071.
93. O. Okada, T. Oashi, K.-i. Kumagai, T. Noje, Y. Koike and Y. Saito, *Physica C*, 1991, **185-9**, 1075.
94. K.-i. Kumagai, H. Matoba, N. Wada, M. Okaji and K. Nera, *J. Phys. Soc. Jpn.*, 1991, **60**, 1448.
95. T. Kohara, K. Ueda, Y. Kohori and Y. Oda, *J. Magn. Magn. Mater.*, 1990, **90-91**, 669.
96. Y. Kohori, K. Ueda and T. Kohara, *Physica C*, 1991, **185-9**, 1187.
97. T. Kohara, K. Ueda, Y. Kohori and Y. Oda, *Physica C*, 1991, **185-9**, 1189.
98. N. Matsumura, H. Yamagata and Y. Oda, *Physica C*, 1991, **185-9**, 1135.
99. M. Tei, H. Matsuzawa, H. Sakamoto, K. Mizoguchi and K. Kume, *Physica C*, 1991, **185-9**, 1227.
100. K. Nakajima, H. Nakamura, Y. Kitaoka, K. Asayama, K. Yoshimura and T. Nitta, *J. Magn. Magn. Mater.*, 1990, **90-91**, 581.
101. H. Nakamura, K. Nakajima, Y. Kitaoka, K. Asayama, K. Yoshimura and T. Nitta, *Physica B*, 1991, **171**, 238.
102. G.-Q. Zheng, E. Yanase, K. Ishida, Y. Kitaoka, K. Asayama, Y. Kodama, R. Tanaka, S. Nakamichi and S. Endo, *Solid State Commun.*, 1991, **79**, 51.
103. G.-Q. Zheng, E. Yanase, K. Ishida, Y. Kitaoka, K. Asayama, Y. Kodama and S. Endo, *Physica C*, 1991, **185-9**, 767.

104. K. Ishida, Y. Kitaoka, T. Yoshitomi, N. Ogata, T. Kamino and K. Asayama, *Physica C*, 1991, **179**, 29.

105. K. Fujiwara, Y. Kitaoka, K. Ishida, K. Asayama, Y. Shimakawa, T. Manako and Y. Kubo, *Physica C*, 1991, **184**, 207.

106. Y. Kitaoka, K. Ishida, S. Ohsugi, K. Fujiwara and K. Asayama, *Physica C*, 1991, **185-9**, 98.

107. S. Ohsugi, Y. Kitaoka, K. Ishida and K. Asayama, *J. Phys. Soc. Jpn.*, 1991, **60**, 2351.

108. S. Ohsugi, Y. Kitaoka, K. Ishida and K. Asayama, *Physica C*, 1991, **185-9**, 1099.

109. Y. Kitaoka, H. Nakamura, T. Iwai, K. Asayama, O. Ahlheim, C. Geibel, C. Schank and F. Steglich, *J. Phys. Soc. Jpn.*, 1991, **60**, 2122.

110. A. Yu. Zavidonov, M.V. Eremin, O.N. Bakharev, A.V. Egorov, V.V. Naletov, M.S. Tagirov and M.A. Teplov, *Sverkhprovodimost : Fiz., Khim., Tekh.*, 1990, **3**, 1597; *Chem. Abstr.*, 1991, **114**, 113754.

111. O.A. Anikeenok, M.V. Eremin, R. Sh. Zhdanov, V.V. Naletov, M.P. Rodionova and M.A. Teplov, *JETP Lett.*, 1991, **54**, 149.

112. A.V. Zalesskii, I.E. Lipinski, E.M. Smirnovskaya, T.A. Khimich and A.A. Bush, *Sverkhprovodimost : Fiz., Khim., Tekh.*, 1990, **3**, 2719; *Chem. Abstr.*, 1991, **115**, 83951.

113. V.L. Matukhin, I.A. Safin, V.N. Anashkin, O.V. Zharikov and Yu.A. Ossipyan, *Solid State Commun.*, 1991, **79**, 1063.

114. A.M. Bogdanovich, Yu. I. Zhdanov, K.N. Mikhalev, B.A. Aleksashin, S.V. Verkhovskiy and V.V. Serikov, *Phys. Met. Metall.*, 1990, **70**, 94.

115. A.V. Strekalovskaya, Yu. I. Zhdanov, Ye. A. Shabunin, K.N. Mikhalev, N.P. Filippova, A.M. Sorkin, S.V. Verkhovskiy, V.A. Tsurin, V.L. Kozhevnikov, A.M. Bogdanovich and V.V. Serikov, *Phys. Met. Metall.*, 1990, **70**, 208.

116. H. Yamagata, K. Inada and M. Matsumura, *Physica C*, 1991, **185-9**, 1101.

117. B.B. Carlisle, R.J.C. Brown and T.J. Bastow, *J. Phys. : Condens. Matter*, 1991, **3**, 3675.

118. T. Goto, T. Nomoto, T. Hanaguri, T. Shinohara, T. Sato and T. Fukase, *J. Phys. Soc. Jpn.*, 1991, **60**, 3581.

119. V.L. Matukhin, V.N. Anashkin, A.I. Pogorel'tsev, E.F. Kukovitskii and I.A. Safin, *Supercond.*, 1991, **4**, 435.

120. T. Rega, *J. Phys. : Condens. Matter*, 1991, **3**, 1871.

121. D.E. MacLaughlin, P.C. Hammel, J.P. Vithayathil, P.C. Canfield, Z. Fisk, R.H. Heffner, A.P. Reyes, J.D. Thompson and S.-W. Cheong, *Phys. Rev. Lett.*, 1991, **67**, 525.

122. P.C. Hammel, E.T. Ahrens, A.P. Reyes, R.H. Heffner, P.C. Canfield, S.-W. Cheong, Z. Fisk and J.E. Schirber, *Physica C*, 1991, **185-9**, 1095.

123. L.E. Erickson, *Phys. Rev. B*, 1991, **43**, 12723.

124. G.K. Semin, E.V. Bryukhova, S.I. Kuznetsov, L.G. Kuz'mina, T.L. Khotsyanova, I.R. Lyatifov and Sh. N. Abdulova, *Organomet. Chem. in the USSR*, 1991, **4**, 414.

3
Rotational Spectroscopy

BY J. H. CARPENTER

1 Introduction

The number of reported papers involving rotational analysis of high-resolution spectra in all regions of the electromagnetic spectrum has risen significantly this year. The largest fractional increases are in the sections for weakly bound complexes, diatomics and triatomics. Spectroscopic study of the structure and dynamics of weakly bound complexes is giving considerable insight into the initial stages of the reactions between molecules.

Reviews of interest include ones on the study of molecular spectroscopy by *ab initio* methods[1], photochemical processes in weakly bound complexes[2], the structure and spectra of negative ions[3], the spectroscopy of jet-cooled ions and radicals[4], multiphoton excitation and mass-selective ion detection for neutral and ion spectroscopy[5], the structure and spectroscopy of the alkali hydride diatomic molecules and their ions[6], improved fits for the rovibrational constants of electronic states of N_2 and O_2[7], the resonance rotational Raman effect[8], tunable far-infrared spectrometers[9], and interstellar molecules.[10]

2 Weakly Bound Complexes

Unless otherwise stated, the complexes in this section were produced in supersonic jets.

The sub-Doppler ultraviolet (UV) spectrum by mass-selected Resonance-enhanced Multiphoton Ionisation (REMPI) of the complex of benzene with neon could be rotationally resolved.[11] This indicated that the neon lay on the C_6 axis of benzene at 346 pm from the plane. There was a decrease of 4 pm on going from the ground S_0 state to the S_1 excited state. A similar study[12] of benzene.Kr and benzene.Xe gave rotational constants and mean centre of mass distances (R_{cm}) which complemented the work on the Ne and Ar complexes.

Three bands were observed[13] in the laser-induced fluorescence (LIF) in the *B*-*X* transition of NeCN. These indicated a mean centre of mass distance (R_{cm}) of 379(7) pm, and a barrier to internal rotation of 17.2(1.0) cm^{-1}, similar to NeN_2 and ArN_2. The rotationally resolved electronic spectrum[14] of NeOH and NeOD associated with the *A*-*X* transition of OH gave R_{cm} = 373(4) pm for NeOH, a value close to that in ArOH where the rare gas van der Waals (vdW) radius is larger but the interaction is stronger.

The Fourier Transform microwave (FTMW) spectrum[15] of $Ar.CH_3CN$ revealed that it is an asymmetric rotor with nearly free internal rotation of the methyl group. The C_3 axis of CH_3CN is almost perpendicular to the R_{cm} vector, with R_{cm} = 365.05(7) pm. The dipole moment of 3.802(5) D lies along the *b*-axis. The vdW stretching force constant was determined from the centrifugal distortion constants. In the FTMW spectra[16] of Ar with CH_2=CHF, CH_2=CF_2 and

$CHF=CF_2$ the argon sits over the FCCH, FCF and FCCF chains respectively; it thus lies at the periphery of the molecule in contact with the largest number of heavy atoms. From the rotationally resolved electronic spectrum[17] of the complex of argon with *trans*-stilbene, a "substitution" position for the argon was obtained using the complex and the uncomplexed molecule. The argon lies above the plane of one of the rings but displaced by more than 100 pm from its centre, a result not expected from atom-atom pair potential calculations. A rotational contour analysis of the fluorescence excitation spectrum[18] of the complex of argon with *para*-difluorobenzene is consistent with Ar being above the centre of the ring at 350(50) pm in the ground S_0 state but 10(4) pm closer in the S_1 state. The dipole moment of the complex of argon with furan was found from Stark shifts in its FTMW spectrum.[19] At 0.701(6) D it is greater than the 0.685(11) D of furan itself. This contrasts with the Ar.pyrrole complex where the dipole decreases on complexation; the results are explained by the different induced dipoles on the argon. Various Ar_n.AT clusters (where AT = 3-amino-*s*-tetrazine) were observed in the LIF spectra[20] of mixtures of Ar and AT; those for Ar.AT and Ar_2.AT were rotationally resolved and indicated that the argon atoms were on the C_6 axis (above and below in the case of Ar_2.AT) at about 330 pm from the ring. The Ar_3.AT spectra could not be interpreted because of overlap with higher clusters. The complexes of 4-(N,N-dimethylamino)benzonitrile with argon and H_2O (as well as the uncomplexed molecule) were studied using rotational band contour analysis of their LIF spectra.[21] In the argon complex the *c*-type band with Q branch is consistent with the assumption that the argon lies above the ring carbon bonded to the nitrile group. In the H_2O complex the absence of a Q-branch implies a *b*-type band with the H_2O not above the ring as predicted; it probably lies in the plane with the oxygen of the water interacting with the triple bond of CN.

The infrared (IR) tunable-diode laser (TDL) spectrum[22] of $Ar.CO_2$ at 2376 cm^{-1} is a combination band of $\nu_{as}(CO_2)$ with the intermolecular bending vibration. The large positive inertial defect implies a large vibrational amplitude; the average structure in the combination band has R_{cm} = 360 pm and θ(OCAr) = 76.9°, somewhat different from the almost identical values in the ground and $\nu_{as}(CO_2)$ vibrational states. The millimetre-wave (mmw) spectrum[23] of the first excited Π state of Ar.HCN shows it to be highly non-rigid with a large zero-point amplitude in the bending vibration and a contraction of R_{cm} on excitation. The vibrationally averaged angle between R_{cm} and the molecular axis is 63°, with the Ar tilted towards the hydrogen, so the complex changes from approximately linear Ar.HCN in the ground state to T-shaped in the excited state. An extension[24] of previous FTMW work on $\chi_{aa}(^{14}N)$ and $\chi_{aa}(^{83}Kr)$ of Kr.HCN showed both values increasing from J=0 to J=7.

The optothermal IR spectrum[25] of $Ar.NH_3$ revealed a Π-Σ band near $^qR(0,0)$ of ν_2 of NH_3. The high resolution enabled the *l*-type coupling constant q, as well as other rotational constants, to be well determined. Two new bands were seen in the vibration-rotation-tunnelling (VRT) spectra[26] of $Ar.NH_3$ in the far infrared (FIR). They were assigned and analysed to give an effective angular potential for the complex with a global minimum near the T-shaped configuration (angle between R_{cm} and C_3 axis = 99°) with unequal higher minima at 0° and 180°. A mmw study[27] of $Ar.NH_3$ showed similar Σ and Π states in which the inversion tunnelling was virtually unchanged in the Σ state from free NH_3 but nearly quenched in the Π state.

FTMW spectroscopy was used to obtain the first pure rotational spectrum[28] of ArOH and ArOD. In addition to rotational constants, quadrupole coupling constants for ^{2}D and the hyperfine constant for ^{1}H were determined; these gave an average angle between R_{cm} and O-H of 59° for ArOH and 52° for ArOD, both larger than in Ar.HF. Stimulated emission pumping (SEP) of the vdW bending and stretching mode frequencies in the ground electronic state of ArOH showed a Renner-Teller splitting of the rotational levels.[29] Rotational, fine and hyperfine structure were resolved[30] in the LIF of $\tilde{A}^2\Sigma^+$-$\tilde{X}^2\Pi$ of ArOH and ArOD. The variation of R_{cm} in the $\tilde{A}$ and $\tilde{X}$ states indicate that whereas the $\tilde{X}$ state is a true vdW complex, the $\tilde{A}$ state has incipient chemical bonding and is Ar.HO rather than Ar.OH. The submillimetre-wave (submmw) spectrum[31] of Ar.D_2O was explained in terms of nearly free internal rotation. States not yet detected in Ar.H_2O were observed in the FIR spectra[32] of Ar.D_2O and Ar.HOD, while five new VRT states (two Σ and three Π) in H_2O were also studied[33] by FIR and analysed, indicating strong angular-radial coupling. The ν_3 = 0-1 transitions of Ar.H_2O in the NIR region[34] were added to FIR data to give an angular potential energy surface with barriers to internal rotation of 19 cm^{-1} in plane and 33 cm^{-1} out of plane, only slightly higher than in the ground state.

The FTMW rotation-tunnelling spectrum[35] of Ar.SO_2 showed *a*- and *c*-type transitions; in addition to rotational constants the Coriolis coupling between rotation and tunnelling was determined; the results suggested that, of two possible structures, that in which the Ar is above the SO_2 plane with the angle between Ar--S and R_{cm} = 54.4° and that between the *a*-axis and the C_2 axis of SO_2 = 121.0° is the more likely.

In the NIR spectrum[36] of Ne.DF four beautifully simple bands involving ν(DF) in combination with internal rotations and vdW stretching vibrations were seen. An improved potential, with the global minimum of -86 cm^{-1} at linear Ne--D-F, a secondary minimum of -55 cm^{-1} at linear Ne--F-D and a saddle point of -39 cm^{-1} near perpendicular geometry, was determined.As the lowest bound state is about 4 cm^{-1} above the saddle, the rotation of DF is only slightly hindered. The internal rotor dynamics of Ne.DCl and Ar.DCl were investigated using IRTDL spectroscopy.[37] The relative intensities of the fundamental DCl stretch and its combination with the Σ and Π bends were very different; in Ne.DCl the combination is much stronger than the fundamental (implying almost free rotation of the DCl), whereas in Ar.DCl the motion is a restricted libration. Analysis[38] of three low-lying vdW vibrations of Ar.HF gave rotational constants, and vibrational frequencies which were all about 5 cm^{-1} less those found in combination with ν(HF). This contrasts with Ar.HCl where the values were almost identical, and is consistent with the previously noted strengthening of the vdW bond on HF excitation. The FTMW spectra[39] of excited vdW states of Ar.HCl and Ar.DCl gave rotational and quadrupole constants which showed that the hydrogen is predominantly away from the argon atom. Anomalous values of centrifugal distortion constants in the Σ stretch vibration implied strong Coriolis perturbation by the Π bending state. Such coupling was investigated in a FIR study[40] of Ar.HCl; the effective rotational constants in the Σ bending state agree with the theoretical prediction that the structure is Ar--ClH rather than Ar--HCl, but the deperturbed constants do not. The discrepancy is attributed to large-amplitude vibrational averaging. Several rovibrational bands of Ar.DCl were analysed in its IRTDL spectrum.[41] The two-laser optical-optical double resonance (OODR) spectra[42] of Kr.Cl_2 and Xe.Cl_2 showed incomplete rotational resolution but were consistent with T- shaped complexes.

The complexes of Hg with CO_2 and with OCS were shown by FTMW spectroscopy to be T-shaped like the corresponding rare-gas complexes. In Hg.CO_2 the vdW force constants obtained[43] from centifugal distortion constants were considerably larger than in Ar.CO_2, as was the dipole moment of 0.107(3) D. A similar analysis was performed[44] for Hg.OCS, and the *eQq* value for ^{201}Hg was small as expected (and as observed for ^{83}Kr in Kr.OCS).

The FIR spectra[45] of the H_2 and D_2 dimers were seen at 20 K in a static long pathlength cell as lines accompanying the corresponding pure rotational transitions of the monomers. They were analysed using a close-coupled formalism between two virtually free internal rotors. A similar FTIR study[46] of the CO.H_2 and CO.D_2 complexes in the ν(CO) region indicated a small red shift on complexation. A preliminary assignment was performed on the basis of a T-shaped "triatomic" with free internal rotation of the H_2 and the CO motion between semirigid rotor and free internal rotor limits.

A linear B--B--HCl structure was inferred from the FTMW spectra[47] of B_2H_6.HCl. Tunnelling was observed in some isotopomers but not those containing ^{10}B^{11}B or DCl. The two tunnelling states were separately analysed, and it was suggested that the motion is a reorientation of B_2H_6 by 180° in the bridging plane coupled with a gear-like counter-rotation of the HCl by 360°.

The N_2.HCCH complex has a low dipole moment but its FTMW spectrum[48] was measured and the *eQq* values for ^{14}N in several isotopomers could be determined. By correcting these for vibrational averaging an effective *eQq* value for free N_2 of -5.01(13) MHz was estimated. The corresponding FTMW spectra[49] of CO.HCCH could be analysed using a simple linear semirigid rotor model. No OC--HCCD (only OC--DCCH) was seen; similar observations on species containing ^{13}C suggested that there was a zero-point energy difference of 5-15 cm^{-1} between the conformations rather than nearly free rotation of the HCCH. The linear structure of OC.HCCH was also confirmed in the rovibrational spectra[50] at 3300 cm^{-1} and the R_{cm} distances in the ^{12}CO and ^{13}CO complexes showed that the structure is HCCH--CO rather than HCCH--OC. The HCCH.NNO complex was shown to have a planar parallel structure from its IR spectrum[51] obtained with a F-centre laser. R_{cm} = 330.7 pm and the geometry is very similar to that of HCCH.CO_2.

The complex between benzene and $^{15}N_2$ was measured by FTMW spectroscopy.[52] Two internal rotation states imply almost free rotation of the N_2 parallel to the ring in the complex. The R_{cm} value of 349.8 pm is close to that of C_6H_6.$^{14}N_2$ found in the UV.

A reinvestigation of the FTMW spectrum[53] of eight isotopomers of $(SO_2)_2$ disclosed a high-barrier tunnelling motion similar to that in $(H_2O)_2$. The structure has a plane of symmetry with one molecule straddling it and the other in the plane; the angle between the C_2 axes of the SO_2 molecules is 113.5° since they point in opposite directions. The dipole moment was almost exactly along the *a*-axis. The FTMW spectra of complexes of ozone with ethene[54] and ethyne[55] shed some light on the processes leading up to the formation of ozonides. The structures of both complexes are similar, with the terminal oxygen atoms close to the C-C bond and the two molecules roughly parallel. Tunnelling splittings are observed, corresponding to rotation of the organic molecule by 180° around a C_2 axis. *Ab initio* calculations suggest that in both cases the complex is in a shallow minimum separated by the transition state from the ozonide.

The $HCCH.SO_2$ complex was shown[56] by FTMW to have a mirror plane; both molecules straddle the plane with the internuclear axis of HCCH perpendicular to the C_2 axis of SO_2. This C_2 axis is tilted 14° from the perpendicular to R_{cm} with the sulfur closer to the HCCH. The structure is similar to that of $C_2H_4.SO_2$ but differs from the ozone complex.

Several FTMW studies of complexes of SO_2 with nitrogen-containing organic compounds showed that they all have an essentially L-shaped structure with the sulfur closest to the nitrogen. In $Me_3N.SO_2$ the crystal structure was consistent with the gas-phase structure,[57] with the SO_2 plane about 75° from the C_3 axis of the Me_3N, but the gas phase r(NS) of 226(3) pm shortens to 205(1) pm in the crystal. The $Me_2NH.SO_2$ complex was found[58] to contain no symmetry plane. The nitrogen lone pair points toward SO_2, nearly perpendicular to the SO_2 plane with one methyl group lying in the plane that contains the N and S atoms and the C_2 axis of SO_2. The dipole moment and ^{14}N quadrupole coupling constants imply a charge transfer of less than 0.25e from N to S, similar to that in $Me_3N.SO_2$. In the complex between pyridine and SO_2 the plane of symmetry of the molecule is perpendicular to the planes of both monomers; the pyridine ring lies at 15(5)° and the SO_2 plane at 81(5)° to the N--S bond.[59] The relative intensities of a- and c-type transitions in the optothermal IR spectra[60] of $SO_2.HCN$ rule out all structures except that in which the SO_2 plane is perpendicular to the NCH axis and S is closest to N. A similar study[60] of $SO_2.HF$ showed substantial broadening of the lines.

The linear isomer of HCN.HCCH is favoured when neon rather than argon was used as carrier gas. In a FTMW study[61] of five isotopomers the R_{cm} and average bending angles of the monomers were determined; the force constant obtained from the D_J implies a well depth of 660 cm^{-1}. The first IR spectra[62] of linear SCO.HCN were obtained by optothermal spectroscopy using a F-centre laser. Both ν_1 and its hot band with ν_7 were observed; the rotational constants confirm the structure deduced from previous FTMW work. High resolution IR spectra[63] of the NH_3 umbrella motion in $NH_3.HCN$ were obtained by optothermal detection of radiation from a CO_2 laser tuned by microwave sidebands. The blue shift of 91.8 cm^{-1} from ν_2 of NH_3 implies a decrease in vdW bonding energy in the excited state, while a decrease in B implies a concomitant extension of the H- bond.

All three symmetries of tunnelling states were observed in the VRT spectra[64] of $(NH_3)_2$ in the FIR. The spectra were consistent with the potential minimum at a structure in which the N atoms are closest to each other with their lone pairs at 49° and 65° to the N--N vector but pointing in opposite directions. This lack of H-bonding is consistent with NH_3 not being a proton donor in any of its binary complexes.

The molecular beam electric resonance (MBER) spectra[65] of $NH_3.NCCN$ and $(NCCN)_2$ show that both have a heavy atom C_{2v} structure, with the C_3 axis of NH_3 or the internuclear axis of the second NCCN perpendicular to the NCCN. In $NH_3.NCCN$ there is free internal rotation of

the hydrogen atoms which point away from the NCCN. In $(NCCN)_2$ only *a*-type transitions are seen and the polar structure differs from that of the non-polar $(CO_2)_2$ dimer.
A REMPI study[66] of complexes of benzene with one and two H_2O molecules is mainly vibronic, but rotational band contours indicate that in $C_6H_6.H_2O$ the water lies on the C_6 axis with the hydrogens towards the ring and free internal rotation. In $C_6H_6.(H_2O)_2$, one H_2O is in a similar position but the other lies outside the ring at about the same distance from its plane. Rotational constants were obtained for $OC.H_2O$ and $OC.D_2O$ from rotation-tunnelling bands in the submmw region.[67] Tunnelling splitting was seen in the FTMW spectra[68] of $PF_3.H_2O$ and $PF_3.D_2O$ but not PF_3.HDO. Rotational constants were obtained and the dipole moment measured. The complex has C_s symmetry with the H atoms and two F atoms straddling the mirror plane, H_2O over the PF_2 face and OH and PF in pseudo-eclipsed configuration. Two tunnelling states were also observed in the FTMW spectrum[69] of $O_3.H_2O$; only one could be fitted to a semirigid rotor model, implying a low barrier to internal rotation. The complex has C_s symmetry with O_3 straddling the plane but H_2O in the plane with one of its hydrogens close to the terminal O atoms of O_3 which are tilted towards the water. This contrasts with $SO_2.H_2O$ where it is the sulfur which is closest to the water, and the H atoms are at 90° to the plane with definitely no H-bonding. A submmw study[70] of $(H_2O)_2$ extended and improved previous work. The FTMW spectrum[71] of H_2C=CH-C≡CH.HF reveals that HF attaches to the triple bond in a T-shaped structure which is within 9° of planarity, unlike the HCl complex. This difference is explained by changes in steric interactions between the species. (H,F) spin-spin coupling and some previously unobserved *c*-type transitions were observed in the FTMW spectrum[72] of $(CH_2)_2$O.HF. Inversion doubling about the pyramidal oxygen was still not seen but the spin-spin coupling constants imply that O--H-F may be non-linear. The corresponding sulfur complex, $(CH_2)_2$S.HF, was also characterised by FTMW spectroscopy,[73] using a fast mixing nozzle with <10 μs between mixing and spectroscopy. Both *a*- and *b*-type transitions were observed, so the complex has only C_s symmetry, with the HBr H-bonded to a pyramidal sulfur (as in H_2S.HF) and again some evidence for a bent H-bond. In the optothermal spectra[74] of HCN.HX (X = Cl, Br, I) around 3300 cm^{-1} the lines were sharp, showing no vibrational predissociation. Rovibrational constants were found, giving *r*(N-H) = 208.89 (HCl), 208.70 (DCl), 215.36 (HBr) and 230.14 pm (HI). A FTIR study[75] at -54°C and -36°C of a static sample of H^{13}CN.HF enabled an isotopically enriched sample to be used. All four stretches were observed as fundamentals and/or combinations. Lifetimes in the ν_1 and ν_3 states were J dependent. The structure of NNO.HCl from the FTMW spectrum[76] is a planar asymmetric top with the HCl and NNO approximately parallel, Cl nearest to the central N and H nearest to O but, unusually, Cl is nearer to NNO than is H. The ground state rotational constants from a FTMW study[77] of PH_3.HI show it is a symmetric top H_3P--HI with R_{cm} = 438.22 pm and considerable oscillation of the monomers. There is no transfer of the hydrogen from HI to PH_3.
Several studies of clusters of three or more species have been reported. Two-colour pump-probe spectroscopy was used for the $\tilde{B}$-$\tilde{X}$ excitation spectrum[78] of He_2Cl_2. Unlike $HeCl_2$, the rotational structure could not be analysed using a semirigid rotor model; the complex is an extremely floppy liquid-like cluster. Three bands in the UV spectrum[79] of $C_6H_6.Ar_2$ were

rotationally resolved to give rotational constants for ground (S_0) and excited (S_1) states. The complex studied was unambiguously determined to have an Ar atom on each side of the ring in a D_{6h} configuration. The average distance of the Ar atoms from the ring is 357.7(1) pm in S_0, identical to that in C_6H_6.Ar which indicates no interaction between the two Ar atoms which each have little effect on the structure of the ring. Another isomer of $C_6H_6.Ar_2$ was found in a REMPI study;[80] fitting of theoretical structures to observed rotational contours suggested that this has both atoms on the same side of the ring with one above the centre and the other to one side. The pump-probe spectrum[81] of Ar_2Cl_2 showed a distorted tetrahedron with $r(Ar_2\text{-}Cl_2)$ = 390(50) pm and r(Ar-Ar) = 410(60) pm, while that for Ar_3Cl_2 was inconclusive but suggested that both "belt" (linear bent Ar_3 over Cl_2) and "triangular" (equilateral Ar_3 over Cl_2) forms may be present. The red shift of ν(HF) in the IR difference frequency laser (DFL) spectra[82] of Ar_nHF (n=1-4) as well as the changes in structure were investigated. The incremental red shift decreased as n increased; for Ar_3HF and Ar_4HF the vibrational frequencies were halfway between the free HF and HF in Ar matrix values. Two studies[83,84] of the FIR spectra of Ar_2HCl reported VRT behaviour. The change in R_{cm} between the ground state and the Σ bending vibration at 39.5 cm^{-1} showed that the T-shaped structure of the ground state (with Ar--Ar as the horizontal and HCl as the vertical and H closer to Ar_2) changes to one in which Cl is closer to Ar_2 in the excited vibrational state. A second band at 37.2 cm^{-1} gives b-type transitions, implying axis-switching from a to b as the vibrational state changes.

The FTMW spectra[85] of eight isotopomers of $HCN(CO_2)_2$ yielded a structure of C_2 symmetry in which the slipped parallel $(CO_2)_2$ dimer has HCN along its C_2 axis with N nearest to $(CO_2)_2$. The r(C--C) distance of the $(CO_2)_2$ part is 7.7 pm shorter than in the $(CO_2)_2$ dimer itself. A splitting into doublets is attributed to an inversion of the cluster by counter-rotation of the CO_2 units by 140° from their equilibrium positions 20.3° out of the ac-plane. Another FTMW study,[86] of $(H_2O)_2.CO_2$, also showed doublet splitting. The structure is cyclical with a six-membered ring (C=O--H-O--H-O), this being related to the structures of the $(H_2O)_2$ and $CO_2.H_2O$ dimers. The dipole moment and inertial defects imply an essentially planar structure although some of the hydrogen atoms may be out of plane. The rotational coherence spectra[87] of the cluster of 1-naphthol with two H_2O molecules and its deuterated analogue gave rotational constants consistent with a cyclic trimer similar to $(H_2O)_3$. The plane of the benzene ring lies approximately perpendicular to that of the O atoms.

3 Triatomic ions and molecules

In a NIR study[88] undertaken to aid the search for higher overtones of H_3^+ on Jupiter and other extraterrestrial locations, four lines belonging to the $3\nu_2(l=1)$ overtone of H_3^+ at 6880 cm^{-1} were assigned on the basis of the temperature dependence of their intensities and calculations from first principles.

The first complete rotational analysis of the spectrum of a metal dihalide was performed on ν_3 and hot bands in the FTIR emission spectrum[89] of BeF_2. Enough vibrational states were analysed to give an equilibrium structure: r(BeF) = 137.29710(95) pm. The $\tilde{A}^2\Pi$-$\tilde{X}^2\Sigma$ transitions of CaOH and CaOD showed strong Fermi resonances in the LIF spectrum;[90] the $\tilde{A}^2\Pi_{3/2}$ (10^00) of CaOH and the $\tilde{A}^2\Pi_{1/2}$ (10^00) of CaOD were severely perturbed by the (02^00) state. The $\tilde{B}^2\Sigma^+$-$\tilde{X}^2\Sigma^+$ transition of CaOD in the ground vibrational states was analysed[91] to give r_0 values which fit well with CaOH and other X(II)OH species. In the corresponding transition of BaOH the upper level was severely perturbed, but deperturbed rotational constants were obtained.[92] In addition it was shown that the $\tilde{B}$-$\tilde{A}$ relaxation was collisional rather than radiative.

The MW spectra of the unstable species FBS and BrBS showed they were linear. The fluorine species[93] gave r_s(BS) = 160.6(3) pm (same as ClBS) and r_1(BF) = 128.4(3) pm. For the bromine molecule[94] r_s(BS) = 160.8(2) pm, r_s(BBr) = 183.1(2) pm, and the bromine eQq values imply a π-bond character of 25.5% and ionic character of 14.5% in the B-Br bond.

The LIF spectrum[95] of jet-cooled Al_2O at 38250 cm^{-1} shows partially resolved PQR structure, assigned to a $^1\Pi_u$-$^1\Sigma_g^+$ transition of a linear species with $B \sim 0.1087$ cm^{-1} and hence r(AlO) = 164 pm.

The IR DFL spectrum[96] of C_2D at 2800 cm^{-1} showed two bands: the (11^10)-(00^00) vibrational band and a perturbed vibronic transition $^2\Sigma^+$-$^2\Sigma^+$ from $\tilde{X}$(00^01). Avoided crossing were seen in the FIR laser magnetic resonance (LMR) spectrum[97] of HCO, giving a full set of rotational and Zeeman parameters.

A new triplet band system, $\tilde{b}^3\Pi_g$-$\tilde{a}^3\Pi_u$, of C_3 was observed[98] at 6500 cm^{-1} in both emission (FTIR) and absorption (laser). The rovibrational constants gave $r_0(\tilde{b})$ = 128.6 pm, $r_0(\tilde{a})$ = 129.8 pm (compared with $r_0(\tilde{X}^1\Sigma_g^+)$ = 127.8 pm, $r_0(\check{A}^1\Pi_u)$ = 130.5 pm, the latter having the same configuration as $\tilde{a}$). Two mmw transitions[99] identified CCO for the first time in space (having been previously characterised in the laboratory). The IR TDL spectrum[100] of ν_1 of CCN gave B and D_J values for the two spin-orbit components of the $\tilde{X}^2\Pi$ state. A mmw study[101] of NCS, formed in a hollow cathode discharge in CS_2 and N_2, gave greatly improved B and D_J values compared with previous optical work, as well as other parameters. The first high-resolution IR study[102] of ICN, of the ν(CN) stretch at 2180 cm^{-1}, gave improved rotational constants and a band origin 9 cm^{-1} lower than the previously reported value.

Several studies[103-107] of CO_2 in the IR region gave information on intensities and dipole moment matrix elements, pressure shifts and accurate frequencies. Similar studies[108-113] on OCS included an investigation[113] of vibrational energy transfer from excited N_2; under some circumstances a population inversion between (04^00) and (03^10) was observed. The FTIR and mmw spectra[114] of several monoisotopomers of OCSe enabled B for ν_2 and ν_3, as well as B_0, D_0, ΔB, ΔD (and q where applicable) to be determined. Stimulated emission pumping (SEP) to a rotationally resolved state of v' = 22 of the ground electronic state of CS_2 was accomplished by three-level polarisation spectroscopy[115] *via* the intermediate $\tilde{a}^3A_2$(0,10,0) state. Perturbations give parallel and perpendicular bands in which the molecule behaves like an asymmetric top with θ(SCS) = 147.6°. Radiation decay from the same $\tilde{a}^3A_2$(0,10,0) state showed[116] two lifetimes varying with J; this was attributed to mixing of the three spin components by strong rotational coupling.

Analysis of the band contours in the LIF spectra[117] of CFCl gave r_v and θ_v values for the $\tilde{A}$ and $\tilde{X}$(000) states. On excitation the angle increases from 107.6° to 128.1°, while *r*(CF) increases from 130.7 to 132.7 pm and *r*(CCl) decreases from 171.4 to 165.2 pm. The *K*-structure was clearly resolved in the LIF excitation spectrum[118] of jet-cooled CCl_2. The differences in rotational constants (*A*'-*B*') and (*A*"-*B*") were determined. Another study[119] of the same $\tilde{A}^1B_1$-$\tilde{X}^1A_1$ transition produced an excited state structure: r_0(CCl) = 165.2 pm and θ_0(ClCCl) = 131.4°, an increase of 22.1° compared with the $\tilde{X}$ state.

The LIF of SiH_2, formed in a free-jet expansion after photolysis of $C_6H_5SiH_3$, gave some rotational lines which were not analysed.[120] A similar study[121] of SiC_2 reported several new bands as well as a reassignment of the $\tilde{A}^1B_2$-$\tilde{X}^1A_1$ transition. The large variation of A''_{eff} and ΔG_v with vibrational state showed that the $\tilde{X}$ state is very anharmonic (unlike the $\tilde{A}$ state) with pseudorotation of Si around C_2. The first observation[122] of HNSi in the gas phase was of the ν_1 band by FTIR emission; rovibrational constants were determined.

K-subband heads were seen in the LIF spectrum[123] of the $\tilde{A}^1A''$-$\tilde{X}^1A'$ transition of HGeBr. Assuming bondlengths from similar species, the rotational constants imply an increase in the bond angle from 103(3)° in $\tilde{X}$ to 112(2)° in $\tilde{A}$.

Rotational and hyperfine constants of ND_2 were found from its MW spectrum.[124] An anomalously large ^{14}N spin-rotation constant was attributed to interactions of $\tilde{X}^2B_1$ with the low-lying $\tilde{A}^2A_1$ state. The LIF and dispersed emission spectra[125] of NH_2 were successfully explained by a semirigid bender Renner-Teller hamiltonian.

A hollow cathode discharge in a C_2N_2/He mixture produced CNC^+ whose IR TDL spectrum[126] for the *n*v2 (*n*=0-3) transitions gave rovibrational constants. The *l*-type doubling constant implied a bending wavenumber between 400 and 530 cm^{-1}, much higher than in the isoelectronic C_3. The first high-resolution observations of N_3 (from photolysis of HN_3) were reported in the LIF spectra[127] of the $\tilde{A}^2\Sigma_u$-$\tilde{X}^2\Pi_g$ transitions. The linewidths implied a lifetime between 0.37 and 20 ns in the *A* state. A pulsed FTMW study[128] of $^{15}N_2O$ found T_1 and T_2 for the *J* = 1-0 transition, and the pressure broadening caused by a variety of gases. The dipole moment of NO_2 in the $\tilde{A}^2B_2$ and $\tilde{X}^2A_1$ states was found to vary with rovibronic state.[129] The 'dark' $\tilde{C}^2A_2$ state of NO_2 was probed by OODR.[130,131] Assignments gave rovibrational constants and *r*(NO) = 140 pm, θ(ONO) = 102° in this state. A visible excitation study[132] of jet-cooled NO_2 between 16300 and 18502 cm^{-1} gave significant improvements over previous work. The analysis showed chaotic behaviour of the vibronic levels and irregular rotational behaviour dur to rovibronic interactions. The FIR spectrum[133] of pure isotopomers of ONCl enabled improvements to the rotational constants to be made. A fairly complete analysis was possible of the rotational and Zeeman behaviour of NF_2 in a magic doublet rotational Zeeman spectroscopic study.[134]

A number of studies of water and its isotopomers in pure rotational, vibrotational and rovibronic spectra have been reported.[135-144] Several transitions of H_2O^+ have been assigned.[145] An extension to previous IR TDL spectra of HO_2 gave an improved set of rovibrational constants.[146] Line strengths in the $2\nu_1$ band of HO_2 have been measured.[147] The FIR LMR of FOO near 330 GHz showed many level anticrossing features and enabled Zeeman parameters to be determined.[148] Interest in atmospheric ozone has led to a number of FTIR studies.[149-154] Most of the transitions observed in the FTIR spectrum[155] of ν_1 of OClO were doubled by spin-rotation interaction. No Fermi resonance with $2\nu_2$ was seen, but there was a weak local perturbation between the two vibrational states.

A report of the laser excitation spectrum[156] of $NiCl_2$ at 360 nm is mainly vibronic but intensity alternation in one band implied it was due to $Ni^{35}Cl_2$. The $\tilde{A}^1A'$-$\tilde{X}^1A'$ laser excitation spectrum[157] of CuOH and CuOD at 628.5 nm showed some perturbations. From the effective rotational constants a partial substitution structure: r(CuO) = 177.48(32) pm, r(OH) = 103.48(40) pm and θ(CuOH) = 111.0(1.6)°. The results imply that some covalent bonding between Cu^+ and OH^- occurs.

4 Tetra-atomic ions and molecules

In contrast to the ESR spectrum in an argon matrix at 12 K, no evidence for a bent structure was seen in the first high-resolution spectrum reported for C_4. An IR TDL study[158] of $\nu_3(\sigma_u)$ of the jet-cooled species gave ν_3, B'', B' and D'. FTIR spectroscopy was used[159] to monitor C_3O formed by pyrolysis of fumaroyl dichloride. Rovibrational constants were obtained for both states in ν_1 and seven associated hot bands. A static FTMW Zeeman study[160] of the J=1-0 transition of CNCN gave magnetic susceptibility and molecular electric quadrupole moment values. The improved rotational constants obtained from the submmw spectrum[161] of COF_2 were used to re-fit the ν_2 band. A FTIR study[162] of ν_4 of COF_2 showed that ν_4 interacts with $2\nu_3$ which is 4 cm^{-1} below. The rovibrational constants allow accurate predictions for high-resolution atmospheric studies. An equilibrium structure was obtained from rotational constants from a FTIR study[163] of ν_3, $\nu_2+\nu_3$ and ν_1-ν_5 of FCCF together with some theoretical values: r_e(CC) = 118.65 pm, r_e(CF) = 128.32 pm.

The pure inversion and inversion-rotation spectra[164] of $^{15}ND_3$ in the FIR region gave ground state rotational constants. Two studies of the $\tilde{B}^1E''$ state of NH_3 have been reported: in one,[165] a small but significant Jahn-Teller effect was found by IRODR and jet spectroscopy; in the other,[166] zero-energy kinetic energy PES and REMPI were used to characterise the state as well as determining the ionisation energy of NH_3 and ν_2 of NH_3^+. Another study[167] of NH_3^+ formed in a glow discharge gave band origins and rovibrational constants and showed that the ground state is planar $^2A''$ as for CH_3. In comparison with CH_3, however, NH_3^+ is more rigid and closer to a harmonic oscillator. Measurements of the absolute intensities of rovibrational lines in ν_2 and ν_4 of PH_3 gave values for the transition dipole matrix elements.[168] A simple local mode model accounted for the vibrational dependence of the rotational constants and the offdiagonal vibrational operators found in an analysis of the FTIR spectrum[169] of the first overtone stretching region of SbH_3.

The MW spectrum[170] of H_2NO in its $\tilde{X}^2B_1(v=0)$ ground state showed a-type R-branches with hyperfine structure. The molecule is essentially planar C_{2v} with the unpaired electron in a $p\pi$ orbital perpendicular to the plane. An assumed r(NH) of 101(1) pm gave r(NO) = 128.0(4) pm and θ(HNH) = 122.7(2.2)°. The N-O distance decreases in the series HNO < H_2NO < H_2NOH. The ground state rotational constants obtained from the FTIR spectrum[171] of ν_4 of DN_3 agree with the microwave values, and effective upper state rotational constants were found for each K_a value measured. A simultaneous fit was not possible, because of interaction with other vibrational states. The new observation[172] of *b*-type transitions as well as *a*-type in the FTIR of ν_4 of *cis*-HNO_2 enabled the transition dipole moment to be located at 16(1)° to the *a*-axis. In the *trans* isomer only *a*-type transitions were seen, implying that the transition dipole lies within 5° of the *a*-axis. Three previously unobserved bands were seen in the FTIR spectrum[173] of NO_3 between 1900 and 2200 cm^{-1}. The analysis yielded rovibrational constants from which it was inferred that they were $5\nu_4$, $\nu_1+3\nu_4$ and a hot band. The splittings due to A_1/A_2 interaction for K=3 states of NF_3 were observed in a laser IR study[174] of ν_1 at 1032 cm^{-1}. No significant difference in *eQq* was found between ground state and ν_1 values.

Velocity modulation was used in the IR spectroscopy[175] of H_3O^+ in the ν_3 region. Both ν_3 and $\nu_2+\nu_3-\nu_2$ were analysed; some inversion splitting was seen. In a study of H_2O_2, the molecule was pumped by IR to $\nu_3+\nu_5$ or $\nu_2+\nu_5$ and then excited with visible radiation above the dissociation threshold.[176] The low-resolution spectra showed parallel bands of a near prolate symmetric top, but higher resolution revealed clumps of lines. The linewidths indicated a lifetime of less than 35 ps. A complete molecular structure of HSSH was obtained for the first time from the rotational spectra in the mmw and FIR regions.[177] The effects of the torsional vibration were removed to give r(SS) = 205.64(1) pm, r(SH) = 134.21(2) pm, θ(SSH) = 97.88(5)° and a dihedral angle of 90.3(2)°. Because of large changes of direction of inertial axes on isotopic substitution the dihedral angle could be confirmed from relative intensity measurements. The MW spectrum[178] of the corresponding 'chain' isomer FSSF of F_2S_2 yielded r_0, r_z and r_e structures (r_e(SS) = 188.89 pm, r_e(SF) = 162.95 pm, θ(SSF) = 108.264°, dihedral angle = 87.526°) and, in addition, harmonic force constants were obtained from the centrifugal distortion constants. The centrifugally induced pure rotational spectrum of SO_3 was observed in a static FTMW study.[179] From the ΔK=3 transitions could be found both B and C as well as centrifugal distortion constants. These gave r_0(SO) = 142.04 pm and, using calculated α_1 and experimental α_2, α_3 and α_4, an equilibrium r_e(SO) = 141.74 pm.

5 Penta-atomic ions and molecules

The IR TDL spectrum[180] at 2086 cm^{-1} in a mixture of C_2H_2 and He subjected to a hollow cathode discharge was consistent with its being the ν_3 vibration of linear C_5. Rovibrational constants including B = 2548.7 MHz were obtained. A variety of techniques were used to study the vibrational spectrum of CH_4. The first rotationally-resolved magnetic-field-induced vibrational circular dichroism spectrum[181] was of ν_4 of CH_4; the rotational g-value found was of the same sign but 40% higher than the ground state value. Symmetry-selective cooling was inferred from anomalous intensities in the jet-cooled IR TDL spectrum[182] of ν_4 of CH_4, giving direct evidence that nuclear spin state is conserved during the collision process whereas rotational relaxation is fast. The five vibrational states in the ν_1/ν_3 fundamental region were studied with FTIR and Raman spectroscopy.[183] Intracavity laser absorption was used[184] to observe the weak v=5 stretching overtone of CH_4 in the visible region. A supersonic slit jet was used to obtain simplification of the spectrum.

A FTIR study[185] of $2\nu_1$ of SiH_4, when combined with previous results for $n\nu_1$ vibrations, showed a strikingly rapid decrease of B value with n as a result of local mode effects. A similar effect was seen[186] for GeH_4; here the relation between the interaction parameters is very close to the local mode predictions for $3\nu_1$ and approaching it for $2\nu_1$. Monoisotopic $^{116}SnH_4$ was studied by FTIR and stimulated Raman scattering in the ν_1/ν_3 region.[187]

The perturbation of ν_1 and ν_4 in the FTIR spectrum[188] of $SiH_3{}^{79}Br$ meant that several different fits of rovibrational constants were required. From the mmw spectrum[189] of GeH_3F the r_0, r_z and r_e structures were found: r_e(GeF) = 173.0945(40) pm, r_e(GeH) = 151.451(13) pm and θ_e(FGeH) = 106.073(17)°. The hyperfine structure in the rotational spectrum of GeH_3Cl, studied in a static FTMW spectrometer[190], gave quadrupole and spin-rotation parameters for both chlorine isotopes and ^{73}Ge; the eQq for ^{73}Ge was identical to that in GeH_3F. The high-resolution FTIR spectrum[191] of $^{74}GeH_3{}^{79}Br$ between 2000 and 2250 cm^{-1} showed weak Coriolis resonance between ν_1 and ν_4 and their corresponding hot bands. In the overtone region around 4150 cm^{-1} there were strong local perturbations in $^{74}GeH_3{}^{79}Br$ but only a weak Coriolis resonance in the corresponding $^{70}GeH_3I$ spectra.[192] The upper states are close to the local mode limit. By studying $\nu_2+\nu_3$ of CF_3I at low temperature (230 K), it was possible to resolve K structure of the R(J) manifolds in an IR TDL investigation.[193] Rovibrational (including Coriolis) constants were determined.

Both mmw[194] and IR DFL[195] spectra of $HNCCN^+$ were reported. Rovibrational constants were obtained for ν_1 and its hot band from ν_7; a local perturbation was seen around $J' = 25$. The B_0 and D_J values extracted from the mmw data were more precise than those from the IR spectra. The IR TDL spectrum[196] of FCCCN, formed by a dc discharge through perfluorobenzonitrile, gave band origins, B_v and D_J values for several vibrational states. A rotational analysis of ν_3 of F_2CN_2 in a FTIR study[197] gave much more precise rotational constants than previous two-photon studies but the structure obtained was essentially identical.

The FT FIR spectrum[198] of OCCCO yielded band origins and rotational constants for 18 bands with ν_7 = 0-8. A simplified spectrum and some new features were seen in the jet-cooled FTIR spectrum of ν_3 of OCCCO in a system designed for low throughput rather than low temperature so that unstable species could be investigated.[199] Several hot bands of ν_7 were also seen. A comprehensive rovibrational analysis of the FTIR spectrum[200] of OCCCS gave B_e = 1408.9652(56) MHz and ν_7 = 77.63094(20) cm^{-1}. The first experimental determination of the structure of SCCCS was obtained from its FTIR spectrum.[201] The molecule is linear with r(CS) = 155.26(10) pm and r(CC) = 127.67(1.00) pm. Isotopic shift data allowed the assignment of ν_3 to ν_{as}(CC) and ν_4 to ν_{as}(CC).

Two studies of FTIR spectra of DNO_3 were reported. The ν_8 vibration at around 750 cm^{-1} showed no perturbations,[202] and no torsional splittings were seen[203] in the ν_9 out-of-plane vibration at 350 cm^{-1}.

The use of a pyrolysis nozzle attached to the pulsed jet of a FTMW spectrometer enabled the unstable molecule CH_2=PCl to be studied on a 10 μs timescale.[204] The ^{31}P and chlorine spin-rotation hyperfine splittings were observed. Many level anticrossing signals were seen in the laser Stark spectrum[205] of POF_3. These and laser-MWDR gave precise values of the dipole moment as a function of J and K.

Several vibrations of $FClO_3$ were studied by FTIR and mmw spectroscopy.[206] The A_0 constant was obtained and r_0 and r_e structures determined, the latter using assumptions making it not very accurate. The r_0 structure was r_0(ClO) = 140.5 pm, r_0(ClF) = 160.4 pm and θ_0(OClO) = 115.3°, in reasonable agreement with electron diffraction (ED) values.

A Moret-Bailly hamiltonian was used to analyse the $\nu_3(T_2)$ FTIR spectrum[207] of RuO_4. The simple PQR structure was very congested when natural abundance samples were used, so monoisotopic $^{102}RuO_4$ and $^{102}RuO_4$ were used. Coincidences with N_2O and CO_2 lasers were predicted for use with laser isotope separation studies. The first observations of the Q-branch of ν_3 of OsO_4 at sub-Doppler resolution were made by both Lamb-dip and infrared-radiofrequency double resonance.[208] Tetrahedral fine structure was seen and analysed.

6 Molecules containing six or more atoms

The mmw spectrum[209] at 505°C of $NaBH_4$ showed that it is a C_{3v} symmetric top. A tridentate structure (with three hydrogens bridges between Na and B) was found with r(Na--B) = 230.8(6) pm, $r(BH_b)$ = 128(10) pm, $r(BH_t)$ = 124(10) pm and $\theta(H_bBH_t)$ = 111°. A static FTMW study[210] of BH_3NH_3 and BH_3ND_3 showed a splitting in the h_3 spectrum but not the d_3. This was attributed to two different vibrational states as in $(NH_3)_2$. Rotational and hyperfine constants were found; the eQq values were about twice those in the solid state. A similar study[211] of BH_2NH_2 gave an eQq_{cc} value higher than in most other boron compounds because of the p←p (B←X) π-bonding possible here. Some centrifugal distortion constants as well as quadrupole constants were obtained from the FTMW spectrum[212] of BF_2NH_2. The first identification in the gas phase of the classical donor acceptor complex BF_3NH_3 has been reported.[213] The J=1-0 transitions of four isotopomers were seen in a pulsed jet FTMW study. The observed frequencies, quadrupole hyperfine structure and small internal rotation splitting were consistent with the molecule being a C_{3v} symmetric top with r(BN) = 159(3) pm, almost identical to the solid-state value.

The ν_4 and ν_5 vibrations of C_7 were studied by IR TDL in a supersonic beam,[214,215] and simultaneously analysed. For ν_4 and the ground state the negative D_J and anomalous H_J values suggest that C_7 has a large amplitude anharmonic bending motion as in C_3 but not C_5. There was thus some evidence for alternation of rigidity in linear C_{2n+1} clusters, although little direct evidence was seen in ν_5.

This reviewer could not resist the temptation to include C_{60}, even though the FTIR emission spectrum shows little evidence of rotational structure.[216] There is possibly some PQR structure on a band at 570.3 cm^{-1} which has a similar width of 11-13 cm^{-1} at 1065 K as other bands observed.

The first detection of HC_8CN in the laboratory, in a supersonic jet by FTMW,[217] gave transitions for J'' = 7-24; the B_0 obtained agrees with the less precise astronomical value, and D_J could also be determined.

On heating dimethylsilacyclobutane to 1000°C, the unstable molecule $(CH_3)_2Si{=}CH_2$ was formed and its FTMW spectrum studied.[218] The heavy atom effective structure was r_0(Si=C) = 169.2(3) pm, r_0(C-Si) = 186.8(3) pm and θ(C-Si-C) = 111.4(2)°, with no appreciable tilting of the methyl groups. Analysis of the internal rotation gave V_3 = 4.20(7) kJ mol^{-1}.

In the v_t=2-0 torsional spectrum of N_2H_4, studied by FT FIR, a global fitting to rotational and torsional parameters was performed.[219] Large changes were seen in some of the *trans* tunnelling parameters for the v_t = 0, 1, 2 states; there was no evidence for *cis* tunnelling.

The dipole moment and structure of Me_3CCP were obtained from a MW study.[220] Assumptions about the methyl group gave $r_0(C_aP)$ = 154.3(3) pm, $r_0(C_bC_a)$ = 147.8(3) pm, $r_0(C_cC_b)$ = 154.4(3) pm and $\theta_0(C_aC_bC_c)$ = 109.2(1)°, where C_c are the methyl carbons. Two series of vibrational satellites were seen, one being the CCP bend and the other the degenerate methyl torsional mode. The pure rotational spectrum of PF_5, allowed by a centrifugally induced dipole moment, was observed for the first time, in a FTMW spectrometer[221] at -65°C. No splitting due to Berry pseudorotation was observed. As well as centrifugal distortion constants, the rotational constant difference (*A-B*) was obtained; by combining this with the *B* value from high-resolution IR studies a r_0 structure was determined: $r_0(PF_a)$ = 157.46 pm, $r_0(PF_e)$ = 153.43 pm, in agreement with ED and *ab initio* values.

The structures of SF_5CCH and $SF_5CH{=}CH_2$ were found in a combined MW and ED study.[222] In both, $r(SF_e)$ was longer than $r(SF_a)$ with $r(SF_{mean})$ = 157.4 pm in SF_5CCH and 158.1 pm in $SF_5CH{=}CH_2$. This fits in with the trend from SF_5CH_3 of an increase in bondlength with decreasing electronegativity of the attached group. Similar trends were seen in $\theta(F_aSF_e)$ as expected by VSEPD theory and in *r*(SC) because of the hybridisation of the carbon atom. A similar combined MW and ED study[223] of SF_5CN gave a C_{4v} structure with $r(SF_a)$ = 155.8(6) pm, $r(SF_e)$ = 156.6(6) pm, *r*(SC) = 176.5 pm, *r*(CN) = 115.2(5) pm and $\theta(F_aSF_e)$ = 90.1(2)°. Several excited vibrational states of SF_5CN in the MW spectrum were also analysed.

The first rotationally resolved spectra of $Cr(CO)_6$, $Mo(CO)_6$ and $W(CO)_6$ were obtained from jet cooled samples by IR TDL spectroscopy.[224] The ν_2 CO stretching vibrations were studied; these gave simple PQR structure. A FTMW study of *cyclo*-$C_5H_5Co(CO)_2$ showed splittings and additional transitions due to ^{59}Co quadrupole coupling and to hindered rotation. Analysis of the data gave the tenfold barrier to internal rotation as 28(7) cm^{-1}, much lower than for ferrocene in the solid state but similar to V_6 in CF_3NO_2. The rotational constants implied θ(OC-Co-CO) = 98(5)°. A similar study[226] of $Co(CO)_3NO$ gave B_0, D_J and eQq_{cc} for both ^{59}Co and ^{14}N. Data from the FTIR spectrum[227] of $\nu_1+\nu_3$ of UF_6 with a resolution of 0.008 cm^{-1} agrees with and supplements previous work.

Table: Diatomic Molecules and Ions

Mol.	Ref	Technique	State(s)	Information derived
a) 1-electron species				
HD^+	228	RFIRDR	X	hyperfine structure (hfs)
b) 2-electron species				
H_2	229	IRTDL,DFL	X	Rot. in crystalline p-H_2
	230	XUV Stark	Rydberg	Stark map
HD	231	FTFIR	X	Rot. consts
D_2	232	IRDFL	X	Relax. in solid
Li_2	233	NIR	$2^1\Sigma_g^+$	Dunham coeffs, RKR curve
	234	Pol. sp.	$b^3\Pi_u$	spin-rot consts
	235	Fl. exc.	$C^1\Pi_g$-$X^1\Sigma_g^+$	Dunham coeffs, RKR curve
NaLi	236	FTIR + LIF	$X^1\Sigma^+$	Dunham coeffs, D_e
Na_2	237	LIF	$A^1\Sigma_u^+$	Perturbations
	238	UV exc. sp.	$B^1\Pi_u$	Linewidths, p.e. curve
	239	OODR	$2^3\Delta_g$	Dunham coeffs, r_e=353.654 pm
	240	OODR	$4^1\Sigma_g^+$	Dunham coeffs
	241	OODR	$2^1\Pi$	Dunham coeffs, RKR curve
	242	OODR	$4^1\Sigma_g^+$	Mol. consts, perturb.
	243	FTIR em.	(3s,3d)$^1\Delta_g$	Eq. consts, r_e=341.8 pm
NaK	244	OODR	$B^1\Pi$-$X^1\Sigma^+$	Eq. consts, RKR
NaRb	245	Pol. sp.	$B^1\Pi$-$X^1\Sigma^+$	Mol consts, RKR
K_2	246	LIF, FTIR	$X^1\Sigma_g^+$	Dunham coeffs, D_e
c) 3-electron species				
BeH,BeD	247	Abs., em.	3d Rydberg	Mol. consts, r_e=132.5 pm(BeH)
SrD	248	IRTDL	$X^2\Sigma$	Dunham coeffs, mass-indep. params
BaD	248	IRTDL	$X^2\Sigma$	Dunham coeffs, mass-indep. params
d) 4-electron species				
BH,BD	249	FTVIS	$A^1\Pi$-$X^1\Sigma^+$	Dunham coeffs
GaD	250	IRTDL	$X^1\Sigma$	Dunham coeffs, mass-indep. params
InD	250	IRTDL	$X^1\Sigma$	Dunham coeffs, mass-indep. params
TlD	250	IRTDL	$X^1\Sigma$	Dunham coeffs, mass-indep. params
e) 5-electron species				
CH,CD	251	Em. sp.	$B^2\Sigma^-$-$X^2\Pi$	Rot. consts
CH	252	Em. sp.	$A^2\Delta$-$X^2\Pi$,$B^2\Sigma^-$-$X^2\Pi$	Mol. consts
GeH	253	IRTDL	$X^2\Pi_{1/2}$	Mol. consts
	254	IRTDL	$X^2\Pi_{1/2}$	Vib-rot. & spin-rot. consts
	255	LMR	$X^2\Pi$	trans. dipole mom.
SnH	256	LMR,IRTDL	$X^2\Pi$	Mol. + hf consts
SnD	257	IRTDL	$X^2\Pi$	Mol. consts
TiH	258	LIF Stark	$^4\Gamma_{5/2}$-$X^4(\Omega=3/2)$	Assignment, dip. moment
f) 6-electron species				
Al_2	259	LIF	$E^3\Sigma_g$-$X^3\Pi_u$,$F^3\Sigma_g$-$X^3\Pi_u$	Mol. consts, $r_0(X)$=270.7 pm
			$G^3\Pi_g$-$X^3\Pi_u$	$r_e(G)$=278.8 pm
NH	260	REMPI-PES	$g^1\Delta$-$a^1\Delta$	Mol. consts, I.E.
NH,ND	261	REMPI	$d^1\Sigma^+$-$a^1\Delta$	Mol. consts
PH	262	emission	$b^1\Sigma^+$-$X^3\Sigma^-$	Trans. dipole moms

Table: Diatomic Molecules and Ions (continued)

Mol.	Ref	Technique	State(s)	Information derived
BiH	263	IRTDL	$X^1\Sigma$	Dunham coeffs, mass-independ. params
g) 7-electron species				
OH	264	FIR	$X^2\Pi$	Press. broadening
	265	OODR	$A^2\Sigma$-$X^2\Pi$	Lifetimes
	266	REMPI	$D^2\Sigma^-,3^2\Sigma^-$	Rot. consts
	267	REMPI	$D^2\Sigma^-$	Mol. consts
	268		$X^2\Pi,A^2\Sigma^+,B^2\Sigma^+,C^2\Sigma^+$	Mol. consts
SH	269	FTIR em.	$X^2\Pi$	Intensities
GeH	255	LMR	$X^2\Pi$	Dipole moment
TeD	270	LMR	$X^2\Pi$	Mol. consts
HF^+	271	Photofrag. sp.	$A^2\Sigma^+$-$X^2\Pi$	Dunham coeffs, eQq
CrH	272	LMR	$X^6\Sigma^+$	Mol. & hf consts
	273	Faraday-LMR	$X^6\Sigma^+$	Mol. consts
h) 8-electron species				
HF	274	FTIR em.	$X^1\Sigma$	Rot. consts
LiF	274	FTIR em.	$X^1\Sigma$	Dunham coeffs
$Be^{18}O$	275	Emiss. sp	$B^1\Sigma^+$-$X^1\Sigma^+$	Rot. & eq. consts
MgO	276	IRTDL	$X^1\Sigma^+$	Rot. consts
	277	Exc. sp.	$d^3\Delta$-$a^3\Pi$,$d^3\Delta$-$A^1\Pi$	Mol. consts, $r_e(D)$=186.06 pm,$r_e(d)$=187.10 pm
	278	Exc. sp.	$B^1\Sigma$-$a^3\Pi$,$D^1\Delta$-$a^3\Pi$	Mol. consts, $r_e(A)$=186.87 pm,$r_e(a)$=186.36 pm
AlN	279	LIF jet	$C^3\Pi$-$X^3\Pi$	Mol. consts. perturb., predissoc.
SiC	280	LIF jet	$C^3\Pi$-$X^3\Pi$	Mol. consts.
	281	LIF jet	$C^3\Pi$-$X^3\Pi$	Mol. consts., rad. lifetimes
	282	LIF jet	$C^3\Pi$-$X^3\Pi$	Mol. consts., $r_e(C)$=191.9 pm, $r_e(X)$=171.82 pm
N_2^{2+}	283	VUV em.	$D^1\Sigma_u$-$X^1\Sigma_g$	Intensities
	284	Photofrag. sp.	$^3\Pi_g$-$^3\Sigma_u$	Mol. consts., r_0''=109.406 pm, r_0'=120.78 pm
HF	285	FTFIR	$X^1\Sigma$	Rot. consts
	286	IR heterodyne	$X^1\Sigma$	Freqs, mol. consts
HCl	287	REMPI jet	various	Mol. consts
	288	FIR	$X^1\Sigma$	Press. broadening
	289	REMPI	$F^1\Delta,E,V$	Intens., Λ-doublets
HBr	290	FIR	$X^1\Sigma$	Mol. & hf consts
HI	291	NIRTDL	$X^1\Sigma$	Linestrengths
	292	NIRTDL	$X^1\Sigma$	hf consts
ArH^+	293	FTIR em.	$X^1\Sigma$	Intensities, dipole moms
XeH^+,XeD^+	294	submmw	$X^1\Sigma$	hf consts
MnH	295	Fluor. sp.	$A^7\Pi$-$X^7\Sigma^+$	hf consts
MnD	296	IRTDL	$X^7\Sigma^+$	Dunham + hf consts
i) 9-electron species				
NaKr	297	LIF, abs.	$A^2\Pi$-$X^2\Sigma$	r_e, D_e, p.e. curve
CaF	298	LIF	$B'^2\Delta$	Mol. consts
BaF	299	LIF	$E^2\Pi$-$A^2\Pi$	Eff. mol. consts
BaCl	300	LIF	$C^2\Pi$-$A'^2\Delta$	Mol. consts

Table: Diatomic Molecules and Ions (continued)

Mol.	Ref	Technique	State(s)	Information derived
BaI	301	OODR+LIF	$C^2\Pi$-$X^2\Sigma$	Mol. consts
	302	LIF	$C^2\Pi$-$X^2\Sigma$	Mol. consts (high J)
AlS	303	mmw	$X^2\Sigma$	Rot. consts, r_0 = 203.1515 pm
CN	304	MW	$X^2\Sigma$	hf consts, perturb. $A^2\Pi$
	305	FTIR	$X^2\Sigma$	Dunham coeffs
N_2^+	306	IRTDL	$A^2\Pi_u$-$X^2\Sigma_g$	Eq. consts
	307	Laser sp.	$A^2\Pi_u$-$X^2\Sigma_g$	Mol. consts
$^{15}NO^{2+}$	308	Em. photo.	$B^2\Sigma$-$X^2\Sigma$	Mol. consts
j) 10-electron species				
AlBr	309	Em. photo.	$A^1\Pi$-$X^1\Sigma$	Rot.& eq. consts
GaF	310	FTIR em.	$X^1\Sigma$	Dunham coeffs
GaCl	311	Em. photo.	$a^3\Pi$-$X^1\Sigma$	Eq. consts
YF	312	MWODR	$X^1\Sigma$	Rot. consts
	313	LIF+ em. sp.	$C^1\Sigma$-$X^1\Sigma$,$G^1\Pi$-$X^1\Sigma$	Mol. consts
YCl	314	FTNIR em.	$B^1\Pi$-$X^1\Sigma$	Mol. consts, $r_e(B)$ = 247.0 pm, $r_e(X)$ = 238.3 pm
InCl	315	Abs. photo.	$C^1\Pi$-$X^1\Sigma$	Mol. consts
TlI	316		$A0^+$-$X0^+$	hf consts
CO	317	MW + IR	$X^1\Sigma$	Mol. consts
	318	FTFIR	$X^1\Sigma$	Rot. consts for $^{12}C^{18}O$
	319	IR heterodyne	$X^1\Sigma$	Press. broadening
	320	FTIR solar sp.	$X^1\Sigma$	Dunham coeffs
	321	FTIR	$X^1\Sigma$	Intensities
	322	Exc. sp.	$D^1\Delta$-$X^1\Sigma$	Mol. consts, perturb.
	323	Abs. photo.	$b^3\Sigma$-$a^3\Pi$~$a'^3\Sigma$	Mol. consts, perturb.
	324	Optogalv. sp.	$L'^1\Pi$-$B^1\Sigma$	Mol. consts
	325	REMPI jet	np Rydberg	Mol. consts
CS	326	FTIR em.	$d^3\Delta$-$a^3\Pi$	Assignment, Λ-doubling, perturb.
SiO	327	mmw sp.	$X^1\Sigma$	Dunham coeffs
SnS	328	IRTDL	$X^1\Sigma$	Dunham coeffs, mass-independ. params
N_2	329	REMPI	$a''^1\Sigma_g$-$X^1\Sigma_g$	Mol. consts
	330	FTIR em.	$w^1\Delta_u$-$a^1\Pi_g$	Rot. consts, r_e'=126.852 pm
	331	REMPI	$a''^1\Sigma_g$-$X^1\Sigma_g$	Mol. consts
	332	IR solar	$X^1\Sigma_g$	O & S, v=1-0 quad. sp., Dunham coeffs
	333	NIRDFL	Rydberg	Rot consts
NO^+	334	IRTDL	$X^1\Sigma$	Mol. consts
	335	FTIR em.	$b^3\Pi$-$a^3\Sigma$	Mol. consts, perturb.
	336	REMPI+PES	$X^1\Sigma$	Assignments
PO^+	337	MW sp.	$X^1\Sigma$	Rot. constants, r_e=142.49927 pm
ScF	338	LIF Stark	$C^1\Sigma$-$X^1\Sigma$	Dipole mom.
TiO	339	LIF beam	$E^3\Pi$-$X^3\Delta$	Mol. consts
	340	LIF	$C^3\Delta$-$X^3\Delta$,$B^3\Pi$-$X^3\Delta$	Mol. consts
k) 11-electron species				
SiF	341	LIF	$D(2)^2\Pi$	Mol. constants, lifetimes
NO	342	mmw, FIR	$X^2\Pi$	Rot. + spin consts
	343	LMR	$X^2\Pi$	Press. broadening

Table: Diatomic Molecules and Ions (continued)

Mol.	Ref	Technique	State(s)	Information derived
	344	REMPI	$A^2\Sigma$-$X^2\Pi$	Assignments, intensities
	345	FTIR em.	3d Rydberg	Mol. consts
	346	REMPI	*n*f Rydberg	Zeeman params
	347	REMPI	$C^2\Pi$,$X^2\Pi$	Zeeman params
	348	LIF+REMPI	$A^2\Sigma$-$X^2\Pi$	Parity assignment of Λ-doublets in $X^2\Pi_{1/2}$
	349	4-wave mix.	$B^2\Pi$-$A^2\Sigma$-$X^2\Pi$	Frequencies, intensities
NS	350	REMPI	$F^2\Delta$-$X^2\Pi$	Mol. consts
SO^+	351	MW sp.	$X^2\Pi$	Mol. consts
VO	352	FTMW	$X^4\Sigma^-$	Dipole mom.
NbO	352	FTMW	$X^4\Sigma^-$	Dipole mom.
NiH	353	LMR	$^2\Delta_{5/2}$-$^2\Delta_{3/2}$	Frequencies, g_J
	354	LMR	$X^2\Delta$	Rot + hf consts, Zeeman, r_e=146.94493 pm
	355	LIF	various	Assignments, mol. consts
CeF	356	LIF	Ω'=9/2-Ω"=7/2	Mol. & hf consts, perturb.
l) 12-electron species				
$^{16}O^{18}O$	357	MW sp.	$X^3\Sigma_g^-$	Mol. + spin consts
O_2	358	LIF	*B*-*X*	Predissoc.
	359	REMPI	$d^1\Pi_g$-$X^3\Sigma_g^-$	Assignment, perturb.
	360	Exc. sp.	$c^1\Sigma_u^-$,$A^3\Sigma_u^+$,$A'^3\Delta_u$	Lifetimes
SO	361	LMR	$X^3\Sigma^-$	Rot. + spin consts
	362	REMPI	Rydberg	Mol. consts
S_2	363	Abs. photo.	Rydberg	Mol. constants, r_0
Te_2	364	LIF	*B*-*X*	Mol. consts, RKR
CrMo	365	REMPI jet	$A^1\Sigma$-$X^1\Sigma$	Mol. consts, r_e'=182.64 pm, r_e"=181.82 pm
MoO	366	REMPI jet	various	Mol. constants, r_0
AgD	250	IRTDL	$X^1\Sigma$	Dunham coeffs, mass-indep. params
m) 13-electron species				
Cl_2^+	367	Em. photo. jet	$A^2\Pi_u$-$X^2\Pi_g$	Mol. consts
BrO	368	FTIR	$X^2\Pi_{3/2}$	Mol. consts
MnO	369	LIF	$A^6\Sigma^+$-$X^6\Sigma^+$	Rot. + hf params, perturb.
ZnD	248	IRTDL	$X^2\Sigma$	Dunham coeffs, mass-indep. params
CdD	248	IRTDL	$X^2\Sigma$	Dunham coeffs, mass-indep. params
n) 14-electron species				
Cl_2	370	Laser sp.	*B*-*X*	Assignment
	371	OODR	$0_g^-(^3P_1)$,B'	Dunham coeffs
	372	OODR	$2_g(^3P_2)$	Dunham coeffs, RKR
	373	OODR	$A^3\Pi_u$-$X^1\Sigma_g$	Dunham coeffs, RKR, perturb.
Br_2	374	LIF	$B^3\Pi(0_u^+)$	Mol. consts
	375	pol. sp.	$B^3\Pi(0_u^+)$	hf + spin consts
	376	OODR	$0_g^+(^1D)$	Dunham coeffs, RKR
	377	OODR	0_u^+,1_u,$2_u(^3P_2)$	Mol. consts, RKR, perturb.
IF	378	Em. sp. jet	$D'2$-$A'2$,$E0^+$-$X0^+$	Mol. consts
ICl	379	LIF jet Stark	$A^3\Pi$-$X^1\Sigma$	Spatial orientation
	380	LIF+MWODR	$A^3\Pi$-$X^1\Sigma$	Rot. + hf consts
	381	Em. photo.	$G[1(^3P_1)]$-$A^3\Pi$	Mol. consts, $r_e(G)$ = 322.94 pm

Table: Diatomic Molecules and Ions (continued)

Mol.	Ref	Technique	State(s)	Information derived
	382	Em. photo.	$f[0^+(^3P_0)]$-$X^1\Sigma$	Mol. consts
I_2	383	Sat. sp. jet	*B-X*	hf + spin consts
	384	LIF rot. coh. sp.	*B-X*	Rot. constants
	385	Calc.	*B-X*	Frequencies, intensities
	386	Exc. sp. jet	$1_g(^3P_2)$-*A*	Mol. consts
	387	Exc. sp. jet	*D'-A'*	Rot. consts
PtO	388	FTNIR em.	$A\,^3\Pi$-$X^3\Sigma^-$	Mol. consts
o) Species with more than 14 electrons				
$ArKr^+$	389	Em. sp. jet	617-646 nm	Mol. consts, characterisation, Ω-doubling
Ne_2	390	LIF jet	Rydberg	Rot. consts
	391	LIF jet	Rydberg	Rot. energies & consts, perturb.
NiF	392		$^2\Pi_{3/2}$-$^2\Delta_{5/2}$, $^2\Pi_{3/2}$-$^2\Sigma$	Reanalysis
ZnI	393	Em. photo.	*C-X*,*B-X*	Mol. consts
Cu_2	394	LIF	*A*,*B*,*C*,*G*,*J-X*	Mol. consts, $r_e(C)$=226 pm, $r_e(J)$=215 pm
CuAg	395	REMPI jet	*A*,*A'*,*B'-X*	Mol. consts, $r_0(X)$=237.35 pm, r_e
CuAu	396	REMPI jet	various	Mol. consts, $r_0(X)$=233.02 pm, $r_e(a)$=242.8 pm
Ag_2	397	LIF jet	*A-X*	Assignment, $r_0(X)$=253.35 pm, $r_e(A)$=265.5 pm
	398	Abs. photo.	*B-X*	Mol. consts, $r_e(B)$=257.18 pm, $r_e(X)$=246.92 pm
AgAu	399	REMPI jet	*A-X*,*B-X*	Assignments
Au_2	399	REMPI jet	*a-X* + others	Rot. consts, $r_e(B')$=257.0 pm
SmO	400	LIF		Rot. consts, Ω-doubling

References

1. C.W. Bauschlicher Jr and S.R. Langhoff, *Chem. Rev.*, 1991, **91**, 701.
2. M. Takayanagi and I. Hanazaki, *Chem. Rev.*, 1991, **91**, 1193.
3. D.R. Bates, *Adv. Atom. Mol. Opt. Phys.*, 1990, **27**, 1.
4. P.C. Engelking, *Chem. Rev.*, 1991, **91**, 399.
5. U. Boesl, *J. Phys. Chem.*, 1991, **95**, 2949.
6. W.C. Stwalley, W.T. Zemke and S.C. Yang, *J. Phys. Chem. Ref. Data*, 1991, **20**, 153.
7. R.R. Laher and F.R. Gilmore, *J. Phys. Chem. Ref. Data*, 1991, **20**, 685.
8. L.D. Ziegler, Y.C. Chung, P.G. Wang and Y.P. Zhang, *J. Phys. Chem.*, 1991, **94**, 3394.
9. G.A. Blake, K.B. Laughlin, R.C. Cohen, K.L. Busarow, D-H. Gwo, C.A. Schmuttenmaer, D.W. Steyert and R.J. Saykally, *Rev. Sci. Instrum.*, 1991, **62**, 1693, 1701.
10. J. Lequeux and E. Roueff, *Phys. Reports*, 1991, **200**, 241.
11. Th. Weber, E. Riedle, H.J. Neusser and E.W. Schlag, *J. Mol. Struct.*, 1991, **249**, 69.
12. Th. Weber, E. Riedle, H.J. Neusser and E.W. Schlag, *Chem. Phys. Lett.*, 1991, **183**, 77.
13. Y. Lin and M.C. Heaven, *J. Chem. Phys.*, 1991, **94**, 5765.
14. Y. Lin, S.K. Kulkarni and M.C. Heaven, *J. Phys. Chem.*, 1991, **94**, 1720.
15. R.S. Ford, R.D. Suenram, G.T. Fraser, F.J. Lovas and K.R. Leopold, *J. Chem. Phys.*, 1991, **94**, 5306.
16. Z. Kisiel, P.W. Fowler and A.C. Legon, *J. Chem. Phys.*, 1991, **95**, 2283.
17. B.B. Champagne, D.F. Plusquellic, J.F. Pfanstiel, D.W. Pratt, W.M. van Herpen and W.L. Meerts, *Chem. Phys.*, 1991, **156**, 251.
18. M-C. Su, H-K. O and C.S. Parmenter, *Chem. Phys.*, 1991, **156**, 261.
19. J.J. Oh, K.W. Hillig II, R.L. Kuczkowski and R.K. Bohn, *J. Phys. Chem.*, 1991, **94**, 4453.
20. J.C. Alfano, S.J. Martinez III and D.H. Levy, *J. Chem. Phys.*, 1991, **94**, 1673.
21. O. Kajimoto, H. Yokoyama, Y. Ohshima and Y. Endo, *Chem. Phys. Lett.*, 1991, **179**, 455.
22. S.W. Sharpe, D. Reifschneider, C. Wittig and R.A. Beaudet, *J. Chem. Phys.*, 1991, **94**, 233.
23. A.L. Cooksy, S. Drucker, J. Faeder, C.A. Gottlieb and W. Klemperer, *J. Chem. Phys.*, 1991, **95**, 3017.

24. T.C. Germann, T. Emilsson and H.S. Gutowsky, *J. Chem. Phys.*, 1991, **95**, 6302.
25. G.T. Fraser, A.S. Pine and W.A. Kreiner, *J. Chem. Phys.*, 1991, **94**, 7061.
26. C.A. Schmuttenmaer, R.C. Cohen, J.G. Loeser and R.J. Saykally, *J. Chem. Phys.*, 1991, **95**, 9.
27. E. Zwart, H. Linnartz, W.L. Meerts, G.T. Fraser, D.D. Nelson Jr and W. Klemperer, *J. Chem. Phys.*, 1991, **95**, 793.
28. Y. Ohshima, M. Iida and Y. Endo, *J. Chem. Phys.*, 1991, **95**, 7001.
29. M.T. Berry, M.R. Brustein, M.I. Lester, C.Chakravarty and D.C. Clary, *Chem. Phys. Lett.*, 1991, **178**, 301.
30. B-C. Chang, L. Yu, D. Cullin, B. Rehfuss, J. Williamson, T.A. Miller, W.M. Fawzy, X. Zheng, S.Fei and M. Heaven, *J. Chem. Phys.*, 1991, **95**, 7086.
31. E. Zwart and W.L. Meerts, *Chem. Phys.*, 1991, **151**, 407.
32. S. Suzuki, R.E. Bumgarner, P.A. Stockman, P.G. Green and G.A. Blake, *J. Chem. Phys.*, 1991, **94**, 824.
33. R.C. Cohen and R.J. Saykally, *J. Chem. Phys.*, 1991, **95**, 7891.
34. R. Lascola and D.J. Nesbitt, *J. Chem. Phys.*, 1991, **95**, 7917.
35. L.H. Coudert, K. Matsumara and F.J. Lovas, *J. Mol. Spectrosc.*, 1991, **147**, 46.
36. C.M. Lovejoy and D.J. Nesbitt, *J. Chem. Phys.*, 1991, **94**, 209.
37. M.D. Schuder, D.D. Nelson Jr and D.J. Nesbitt, *J. Chem. Phys.*, 1991, **94**, 5796.
38. M.A. Dvorak, S.W. Reeve, W.A. Burns, A. Grushow and K.R. Leopold, *Chem. Phys. Lett.*, 1991, **185**, 399.
39. C. Chuang and H.S. Gutowsky, *J. Chem. Phys.*, 1991, **94**, 86.
40. S.W. Reeve, M.A. Dvorak, D.W. Firth and K.R. Leopold, *Chem. Phys. Lett.*, 1991, **181**, 259.
41. Z. Wang, A. Quiñones, R.R. Lucchese and J.W. Bevan, *J. Chem. Phys.*, 1991, **95**, 3175.
42. C.R. Bieler, K.E. Spence and K.C. Janda, *J. Phys. Chem.*, 1991, **95**, 5058.
43. M. Iida, Y. Ohshima and Y. Endo, *J. Chem. Phys.*, 1991, **95**, 4772.
44. M. Iida, Y. Ohshima and Y. Endo, *J. Chem. Phys.*, 1991, **94**, 6989.
45. A.R.W. McKellar and J. Schaefer, *J. Chem. Phys.*, 1991, **95**, 3081.
46. A.R.W. McKellar, *Chem. Phys. Lett.*, 1991, **186**, 58.
47. C. Chuang, T.D. Klots, R.S. Ruoff, T. Emilsson and H.S. Gutowsky, *J. Chem. Phys.*, 1991, **95**, 1552.
48. A.C. Legon, A.L. Wallwork and P.W. Fowler, *Chem. Phys. Lett.*, 1991, **184**, 175.
49. A.C. Legon, A.L. Wallwork, J.W. Bevan and Z. Wang, *Chem. Phys. Lett.*, 1991, **180**, 57.
50. M.D. Marshall, J. Kim, T.A. Hu, L.H. Sun and J.S. Muenter, *J. Chem. Phys.*, 1991, **94**, 6334.
51. T.A. Hu, L.H. Sun and J.S. Muenter, *J. Chem. Phys.*, 1991, **95**, 1537.
52. Y. Ohshima, H. Kohguchi and Y. Endo, *Chem. Phys. Lett.*, 1991, **184**, 21.
53. A. Taleb-Bendiab, K.W. Hillig II and R.L. Kuczkowski, *J. Chem. Phys.*, 1991, **94**, 6956.
54. C.W. Gillies, J.Z. Gillies, R.D. Suenram, F.J. Lovas, E. Kraka and D. Cremer, *J. Amer. Chem. Soc.*, 1991, **113**, 2412.
55. J.Z. Gillies, C.W. Gillies, F.J. Lovas, K. Matsumura, R.D. Suenram, E. Kraka and D. Cremer, *J. Amer. Chem. Soc.*, 1991, **113**, 6408.
56. A.M. Andrews, K.W. Hillig II, R.L. Kuczkowski, A.C. Legon and N.W. Howard, *J. Chem. Phys.*, 1991, **94**, 6947.
57. J.J. Oh, M.S. LaBarge, J. Matos, J.W. Kampf, K.W. Hillig II and R.L. Kuczkowski, *J. Amer. Chem. Soc.*, 1991, **113**, 4732.
58. J.J. Oh, K.W. Hillig II and R.L. Kuczkowski, *J. Phys. Chem.*, 1991, **95**, 7211.
59. J.J. Oh, K.W. Hillig II and R.L. Kuczkowski, *J. Amer. Chem. Soc.*, 1991, **113**, 7480.
60. D.C. Dayton and R.E. Miller, *J. Phys. Chem.*, 1991, **94**, 6641.
61. A.I. Jaman, T.C. Germann, H.S. Gutowsky, J.D. Augspurger and C.E. Dykstra, *Chem. Phys.*, 1991, **154**, 281.
62. D.C. Dayton, M.D. Marshall and R.E. Miller, *J. Chem. Phys.*, 1991, **95**, 785.
63. G.T. Fraser, A.S. Pine, W.A. Kreiner and R.D. Suenram, *Chem. Phys.*, 1991, **156**, 523.
64. M. Havenith, R.C. Cohen, K.L. Busarow, D-H. Gwo, Y.T. Lee and R.J. Saykally, *J. Chem. Phys.*, 1991, **94**, 4776.
65. I.I. Suni, S. Lee and W. Klemperer, *J. Phys. Chem.*, 1991, **95**, 2859.
66. A.J. Gotch, A.W. Garrett, D.L. Severance and T.S. Zwier, *Chem. Phys. Lett.*, 1991, **178**, 121.
67. R.E. Bumgarner, S. Suzuki, P.A. Stockman, P.C. Green and G.A. Blake, *Chem. Phys. Lett.*, 1991, **176**, 123.
68. M.S. LaBarge, A.M. Andrews, A. Taleb-Bendiab, K.W. Hillig II and R.K. Bohn, *J. Phys. Chem.*, 1991, **95**, 3523.
69. J.Z. Gillies, C.W. Gillies, R.D. Suenram, F.J. Lovas, T. Schmidt and D. Cremer, *J. Mol. Spectrosc.*, 1991, **146**, 493.
70. E. Zwart, J.J. ter Meulen, W.L. Meerts and L.H. Coudert, *J. Mol. Spectrosc.*, 1991, **147**, 27.
71. Z. Kisiel, P.W. Fowler and A.C. Legon, *Chem. Phys. Lett.*, 1991, **176**, 446.
72. A.C. Legon, A.L. Wallwork and D.J. Millen, *Chem. Phys. Lett.*, 1991, **178**, 279.
73. A.C. Legon, A.L. Wallwork and H.E. Warner, *J.C.S. Faraday Trans.*, 1991, **87**, 3327.
74. P.A. Block and R.E. Miller, *J. Mol. Spectrosc.*, 1991, **147**, 359.
75. A. Quiñones, R.S. Ram and J.W. Bevan, *J. Chem. Phys.*, 1991, **95**, 3980.

76. D.J. Pauley, M.A. Roehrig, L. Adamowicz, J.C. Shea, S.T. Haubrich and S.G. Kukolich, *J. Chem. Phys.*, 1991, **94**, 899.
77. N.W. Howard, A.C. Legon and G.J. Luscombe, *J.C.S. Faraday Trans.*, 1991, **87**, 507.
78. W.D. Sands, C.R. Bieler and K.C. Janda, *J. Chem. Phys.*, 1991, **95**, 729.
79. Th. Weber and H.J. Neusser, *J. Chem. Phys.*, 1991, **94**, 7689.
80. M. Schmidt, M. Mons, J. Le Calvé, P. Millié and C. Cossart-Magos, *Chem. Phys. Lett.*, 1991, **183**, 69.
81. C.R. Bieler, D.D. Evard and K.C. Janda, *J. Phys. Chem.*, 1991, **94**, 7452.
82. A. McIlroy, R. Lascola, C.M. Lovejoy and D.J. Nesbitt, *J. Phys. Chem.*, 1991, **95**, 2636.
83. M.J. Elrod, D.W. Steyert and R.J. Saykally, *J. Chem. Phys.*, 1991, **94**, 58.
84. M.J. Elrod, D.W. Steyert and R.J. Saykally, *J. Chem. Phys.*, 1991, **95**, 3182.
85. H.S. Gutowsky, J. Chen, P.J. Hadjuk and R.S. Ruoff, *J. Phys. Chem.*, 1991, **94**, 7774.
86. K.I. Peterson, R.D. Suenram and F.J. Lovas, *J. Chem. Phys.*, 1991, **94**, 106.
87. L.L. Connell, S.M. Ohline, P.W. Joireman, T.C. Corcoran and P.M. Felker, *J. Chem. Phys.*, 1991, **94**, 4668.
88. S.S. Lee, B.F. Ventrudo, D.T. Cassidy, T. Oka, S. Miller and J. Tennyson, *J. Mol. Spectrosc.*, 1991, **145**, 222.
89. C.I. Frum, R. Engleman Jr and P.F. Bernath, *J. Chem. Phys.*, 1991, **95**, 1435.
90. J.A. Coxon, M. Li and P.I. Presunka, *J. Mol. Spectrosc.*, 1991, **150**, 33.
91. R.A. Hailey, C.N. Jarman, W.T.M.L. Fernando and P.F. Bernath, *J. Mol. Spectrosc.*, 1991, **147**, 40.
92. T. Gustavsson, C. Alcaraz, J. Berlande, J. Cuvellier, J-M. Mestdagh, P. Meynardier, P. de Pujo, O. Sublemontier and J-P. Vesticot, *J. Mol. Spectrosc.*, 1991, **145**, 210.
93. T.A. Cooper, S. Firth and H.W. Kroto, *J.C.S. Faraday Trans.*, 1991, **87**, 1499.
94. T.A. Cooper, S. Firth and H.W. Kroto, *J.C.S. Faraday Trans.*, 1991, **87**, 1.
95. M. Cai, C.C. Carter, T.A. Miller and V.E. Bondybey, *J. Chem. Phys.*, 1991, **95**, 73.
96. W-B. Yan, H.E. Warner and T. Amano, *J. Chem. Phys.*, 1991, **94**, 1712.
97. J.M. Brown, H.E. Radford and T.J. Sears, *J. Mol. Spectrosc.*, 1991, **148**, 20.
98. H. Sasada, T. Amano, C. Jarman and P.F. Bernath, *J. Chem. Phys.*, 1991, **94**, 2401.
99. M. Ohishi, H. Suzuki, S-I. Ishikawa, C. Yamada, H. Kanamori, W.M. Irvine, R.D. Brown, P.D. Godfrey and N. Kaifu, *Astrophys. J.*, 1991, **380**, L39.
100. M. Fehér, C. Salud and J.P. Maier, *J. Mol. Spectrosc.*, 1991, **145**, 246; **150**, 280.
101. T. Amano and T. Amano, *J. Chem. Phys.*, 1991, **95**, 2275.
102. C. Degli Esposti and L. Favero, *J. Mol. Spectrosc.*, 1991, **147**, 84.
103. L.P. Giver and C. Chackerian Jr, *J. Mol. Spectrosc.*, 1991, **148**, 80.
104. L. Rosenmann, S. Langlois, C. Delaye and J. Taine, *J. Mol. Spectrosc.*, 1991, **149**, 167.
105. L. Jörissen, W. Hohe and W. Kreiner, *IEEE J. Quant. Elec.*, 1991, **27**, 1090.
106. Q. Kou and G. Guelachvili, *J. Mol. Spectrosc.*, 1991, **148**, 324.
107. A. Groh, D. Goddon, M. Schneider, W. Zimmermann and W. Urban, *J. Mol. Spectrosc.*, 1991, **146**, 161.
108. L.S. Masukidi, J.G. Lahaye and A. Fayt, *J. Mol. Spectrosc.*, 1991, **148**, 281.
109. T.L. Tan and E.C. Looi, *J. Mol. Spectrosc.*, 1991, **148**, 262.
110. A.G. Maki, J.S. Wells and J.B. Burkholder, *J. Mol. Spectrosc.*, 1991, **147**, 173.
111. G. Blanquet, P. Coupe, F. Derie and J. Walrand, *J. Mol. Spectrosc.*, 1991, **147**, 543.
112. M.D. Vanek, J.S. Wells and A.G. Maki, *J. Mol. Spectrosc.*, 1991, **147**, 398.
113. L.B. Favero, M.C. Righetti and P.G. Favero, *Chem. Phys.*, 1991, **155**, 107.
114. H. Bürger, M. Litz, H. Willner, M. LeGuennec, G. Wlodarczak and J. Demaison, *J. Mol. Spectrosc.*, 1991, **146**, 220.
115. H.T. Liou, P. Dan, H. Yang and J-Y. Yuh, *Chem. Phys. Lett.*, 1991, **176**, 109.
116. H.T. Liou, H. Yang, N.C. Wang and R.W. Joy, *Chem. Phys. Lett.*, 1991, **178**, 80.
117. R. Schlachta, G.M. Lask and V.E. Bondybey, *J.C.S. Faraday Trans.*, 1991, **87**, 2407.
118. Q. Lu, Y. Chen, D. Wang, Y. Zhang, S. Yu, C. Chen, M. Koshi, H. Matsui, S. Koda and X. Ma, *Chem. Phys. Lett.*, 1991, **178**, 517.
119. D.J. Clouthier and J. Karolczak, *J. Chem. Phys.*, 1991, **94**, 1.
120. H. Ishikawa and O. Kajimoto, *J. Mol. Spectrosc.*, 1991, **150**, 610.
121. T.J. Butenhoff and E.A. Rohlfing, *J. Chem. Phys.*, 1991, **95**, 1.
122. M. Elhanine, R. Farrenq and G. Guelachvili, *J. Chem. Phys.*, 1991, **94**, 2529.
123. H. Ito, E. Hirota and K. Kuchitsu, *Chem. Phys. Lett.*, 1991, **177**, 235.
124. M. Kanada, S. Yamamoto and S. Saito, *J. Chem. Phys.*, 1991, **94**, 3423.
125. R.N. Dixon, S.J. Irving, J.R. Nightingale and M. Vervloet, *J.C.S. Faraday Trans.*, 1991, **87**, 2121.
126. M. Fehér, C. Salud and J.P. Maier, *J. Chem. Phys.*, 1991, **94**, 5377.
127. T. Haas and K-H. Gericke, *Ber. Bunsenges. Phys. Chem.*, 1991, **95**, 1289.
128. H-W. Nicolaisen and H. Mäder, *Mol. Phys.*, 1991, **73**, 349.
129. S. Heitz, R. Lampka, D. Weidauer and A. Hese, *J. Chem. Phys.*, 1991, **94**, 2532.
130. H. Nagai, K. Aoki, T. Kusumoto, K. Shibuya and K. Obi, *J. Phys. Chem.*, 1991, **95**, 2718.
131. K. Shibuya, T. Kusumoto, H. Nagai and K. Obi, *J. Chem. Phys.*, 1991, **95**, 720.

132. A. Delon, R. Jost and M. Lombardi, *J. Chem. Phys.*, 1991, **95**, 5701.
133. J.R. Durig, T.J. Geyer, Y.H. Kim, V.F. Kalasinsky and J.K. McDonald, *J. Mol. Struct.*, 1991, **244**, 103.
134. U.E. Frank and W. Huttner, *Chem. Phys.*, 1991, **152**, 261.
135. R. Bhattacharjee, J.S. Muenter and M.D. Marshall, *J. Mol. Spectrosc.*, 1991, **145**, 302.
136. T. Amano and F. Scappini, *Chem. Phys. Lett.*, 1991, **182**, 93.
137. S.L. Shostak, W.L. Ebenstein and J.S. Muenter, *J. Chem. Phys.*, 1991, **94**, 5875.
138. J.C. Pearson, T. Anderson, E. Herbst, F.C. DeLucia and P. Helminger, *Astrophys. J.*, 1991, **379**, L41.
139. R.A. Toth, *J. Opt. Soc. Amer. B*, 1991, **8**, 2236.
140. O.N. Ulenikov and A.S. Zhilyakov, *J. Mol. Spectrosc.*, 1991, **146**, 79.
141. C.P. Rinsland, M.A.H. Smith, V. Malathy Devi and D.C. Benner, *J. Mol. Spectrosc.*, 1991, **150**, 173.
142. J-P. Chevillard, J-Y. Mandin, J-M. Flaud and C. Camy-Peyret, *Canad. J. Phys.*, 1991, **69**, 1286.
143. R.D. Gilbert, M.S. Child and J.W.C. Johns, *Mol. Phys.*, 1991, **74**, 473.
144. R.G. Tonkyn, R. Wiedmann, E.R. Grant and M.G. White, *J. Chem. Phys.*, 1991, **95**, 7033.
145. B. Das and J.W. Farley, *J. Chem. Phys.*, 1991, **95**, 8809.
146. D.D. Nelson Jr and M.S. Zahniser, *J. Mol. Spectrosc.*, 1991, **150**, 527.
147. T.J. Johnson, F.G. Wienhold, J.P. Burrows, G.W. Harris and H. Burkhard, *J. Phys. Chem.*, 1991, **95**, 6499.
148. U. Bley, P.B. Davies, M. Grantz, T.J. Sears and F. Temps, *Chem. Phys.*, 1991, **152**, 281.
149. A. Perrin, A-M. Vasserot, J-M. Flaud, C. Camy-Peyret, V.M. Devi, M.A.H. Smith, C.P. Rinsland, A. Barbe, S. Bouazza and J-J. Plateaux, *J. Mol. Spectrosc.*, 1991, **149**, 519.
150. M.A.H. Smith, C.P. Rinsland and V.M. Devi, *J. Mol. Spectrosc.*, 1991, **147**, 142.
151. J.J. Plateaux, S. Bouazza and A. Barbe, *J. Mol. Spectrosc.*, 1991, **146**, 314.
152. M.N. Spencer and C. Chackerian Jr, *J. Mol. Spectrosc.*, 1991, **146**, 135.
153. C.P. Rinsland, M.A.H. Smith, V.M. Devi, A. Perrin, J-M. Flaud and C. Camy-Peyret, *J. Mol. Spectrosc.*, 1991, **149**, 474.
154. A. Barbe, S. Bouazza, J.J. Plateaux, J.M. Flaud and C. Camy-Peyret, *J. Mol. Spectrosc.*, 1991, **150**, 255.
155. J. Ortigoso, R. Escribano, J.B. Burkholder, C.J. Howard and W.J. Lafferty, *J. Mol. Spectrosc.*, 1991, **148**, 346.
156. L.R. Zink, F.J. Grieman, J.M. Brown, T.R. Gilson and I.R. Beattie, *J. Mol. Spectrosc.*, 1991, **146**, 225.
157. C.N. Jarman, W.T.M.L. Fernando and P.F. Bernath, *J. Mol. Spectrosc.*, 1991, **145**, 151.
158. J.R. Heath and R.J. Saykally, *J. Chem. Phys.*, 1991, **94**, 3271.
159. D. McNaughton, D. McGilvery and F. Shanks, *J. Mol. Spectrosc.*, 1991, **149**, 458.
160. A. Klesing, D.H. Sutter and F. Stroh, *J. Mol. Spectrosc.*, 1991, **148**, 149.
161. E.A. Cohen and W. Lewis-Bevan, *J. Mol. Spectrosc.*, 1991, **148**, 378.
162. C. Camy-Peyret, J-M. Flaud, A. Goldman, F.J. Murcray, R.D. Blatherwick, F.S. Bonomo, D.G. Murcray and C.P. Rinsland, *J. Mol. Spectrosc.*, 1991, **149**, 481.
163. H. Burger, W. Schneider, S. Sommer, W. Thiel and H. Willner, *J. Chem. Phys.*, 1991, **95**, 5660.
164. L. Fusina, M. Carlotti, G. DiLonardo, S.N. Murzin and O.N. Stepanov, *J. Mol. Spectrosc.*, 1991, **147**, 71.
165. J.M. Allen, M.N.R. Ashfold, R.J. Stickland and C.M. Western, *Mol. Phys.*, 1991, **74**, 49.
166. W. Habenicht, G. Reiser and K. Muller-Dethlefs, *J. Chem. Phys.*, 1991, **95**, 4809.
167. S.S. Lee and T. Oka, *J. Chem. Phys.*, 1991, **94**, 1698.
168. R.J. Kshirsagar, K. Singh, R. D'Cunha, V.A. Job, D. Papousek, J.F. Ogilvie and L. Fusina, *J. Mol. Spectrosc.*, 1991, **149**, 152.
169. M. Halonen, L. Halonen, H. Burger and P. Moritz, *J. Chem. Phys.*, 1991, **95**, 7099.
170. H. Mikami, S. Saito and S. Yamamoto, *J. Chem. Phys.*, 1991, **94**, 3415.
171. J. Bendtsen and F.M. Nicolaisen, *J. Mol. Spectrosc.*, 1991, **145**, 123.
172. I. Kleiner, J.M. Guilmot, M. Carleer and M. Herman, *J. Mol. Spectrosc.*, 1991, **149**, 341.
173. K. Kawaguchi, T. Ishiwata, I. Tanaka and E. Hirota, *Chem. Phys. Lett.*, 1991, **180**, 436.
174. W. Hohe and W.A. Kreiner, *J. Mol. Spectrosc.*, 1991, **150**, 28.
175. W.C. Ho, C.J. Pursell and T. Oka, *J. Mol. Spectrosc.*, 1991, **149**, 530.
176. X. Luo and T.R. Rizzo, *J. Chem. Phys.*, 1991, **94**, 889.
177. J. Behrend, P. Mittler, G. Winnewisser and K.M.T. Yamada, *J. Mol. Spectrosc.*, 1991, **150**, 99.
178. R.W. Davis and S. Firth, *J. Mol. Spectrosc.*, 1991, **145**, 225.
179. V. Meyer, D.H. Sutter and H. Dreizler, *Z. Naturforsch. A*, 1991, **46**, 710.
180. N. Moazzen-Ahmadi, S.D. Flatt and A.R.W. McKellar, *Chem. Phys. Lett.*, 1991, **186**, 291.
181. B. Wang, R.K. Yoo, P.V. Croatto and T.A. Keiderling, *Chem. Phys. Lett.*, 1991, **180**, 339.
182. M. Hepp, G. Winnewisser and K.M.T. Yamada, *J. Mol. Spectrosc.*, 1991, **146**, 181.
183. J.M. Jouvard, B. Lavorel, J.P. Champion and L.R. Brown, *J. Mol. Spectrosc.*, 1991, **150**, 201.
184. A. Campargne, M. Chevenier and F. Stoeckel, *Chem. Phys. Lett.*, 1991, **183**, 153.
185. Q. Xi, H. Qian, H. Ma and L. Halonen, *Chem. Phys. Lett.*, 1991, **177**, 261.
186. Q. Zhu, H. Qian and B.A. Thrush, *Chem. Phys. Lett.*, 1991, **186**, 436.
187. A. Tabyaoui, B. Lavorel, G. Pierre and H. Burger, *J. Mol. Spectrosc.*, 1991, **148**, 100.
188. F. Lattanzi, C. Di Lauro and H. Burger, *Mol. Phys.*, 1991, **72**, 575.

189. M. Le Guennec, W. Chen, G. Wlodarczak, J. Demaison, R. Eujen and H. Burger, *J. Mol. Spectrosc.*, 1991, **150**, 493.
190. G. Wlodarczak, N. Heineking and H. Dreizler, *J. Mol. Spectrosc.*, 1991, **147**, 252.
191. R. Escribano, G. Graner and H. Burger, *J. Mol. Spectrosc.*, 1991, **146**, 83.
192. H. Burger and G. Graner, *J. Mol. Spectrosc.*, 1991, **149**, 491.
193. A. Baldacci, S. Giorgianni, P. Stoppa, A. DeLorenzi and S. Ghersetti, *J. Mol. Spectrosc.*, 1991, **147**, 208.
194. T. Amano and F. Scappini, *J. Chem. Phys.*, 1991, **95**, 2280.
195. H.E. Warner and T. Amano, *J. Mol. Spectrosc.*, 1991, **145**, 66.
196. M. Niedenhoff, K.M.T. Yamada, G. Winnewisser, K. Tanaka and T. Okabayashi, *J. Mol. Spectrosc.*, 1991, **145**, 290.
197. H. Sieber, H.J. Neusser, F. Stroh and M. Winnewisser, *J. Mol. Spectrosc.*, 1991, **148**, 453.
198. J. Vander Anwera, J.W.C. Johns and O.L. Polyansky, *J. Chem. Phys.*, 1991, **95**, 2299.
199. A.D. Walters, M. Winnewisser, K. Lattner and B.P. Winnewisser, *J. Mol. Spectrosc.*, 1991, **149**, 542.
200. F. Holland and M. Winnewisser, *J. Mol. Spectrosc.*, 1991, **149**, 45.
201. F. Holland and M. Winnewisser, *J. Mol. Spectrosc.*, 1991, **147**, 496.
202. T.L. Tan, E.C. Looi, K.T. Lua, A.G. Maki, J.W.C. Johns and M. Noel, *J. Mol. Spectrosc.*, 1991, **149**, 425.
203. T.L. Tan, E.C. Looi, K.T. Lua, A.G. Maki, J.W.C. Johns and M. Noel, *J. Mol. Spectrosc.*, 1991, **150**, 486.
204. A.C. Legon and D. Stephenson, *J.C.S. Faraday Trans.*, 1991, **87**, 3325.
205. K. Tanaka, K. Someya and T. Tanaka, *Chem. Phys.*, 1991, **152**, 229.
206. K. Burczyk, H. Burger, M. Le Guennec, G. Wlodarczak and J. Demaison, *J. Mol. Spectrosc.*, 1991, **148**, 65.
207. M. Snels, M.P. Sassi and M. Quack, *Mol. Phys.*, 1991, **72**, 145.
208. L. Ricci, F.S. Pavone, M. Prevedelli, L.R. Zink, M. Inguscio, F. Scappini and M.P. Sassi, *J. Chem. Phys.*, 1991, **94**, 2509.
209. Y. Kawashima, C. Yamada and E. Hirota, *J. Chem. Phys.*, 1991, **94**, 7707.
210. K. Vormann and H. Dreizler, *Z. Naturforsch. A*, 1991, **46**, 1060.
211. K. Vormann, H. Dreizler, J. Doose and A. Guarnieri, *Z. Naturforsch. A*, 1991, **46**, 770.
212. K. Vormann and H. Dreizler, *Z. Naturforsch. A*, 1991, **46**, 909.
213. A.C. Legon and H.E. Warner, *J.C.S. Chem. Commun.*, 1991, 1397.
214. J.R. Heath and R.J. Saykally, *J. Chem. Phys.*, 1991, **94**, 1724.
215. J.R. Heath, A. Van Orden, E. Kuo and R.J. Saykally, *Chem. Phys. Lett.*, 1991, **182**, 17.
216. C.I. Frum, R. Engleman Jr, H.G. Hedderich, P.F. Bernath, L.D. Lamb and D.R. Huffman, *Chem. Phys. Lett.*, 1991, **176**, 504.
217. M. Iida, Y. Ohshima and Y. Endo, *Astrophys. J.*, 1991, **371**, L45.
218. H.S. Gutowsky, J. Chen, P.J. Hadjuk, J.D. Keen, C. Chuang and T. Emilsson, *J. Amer. Chem. Soc.*, 1991, **113**, 4747.
219. N. Ohashi and W.B. Olson, *J. Mol. Spectrosc.*, 1991, **145**, 383.
220. A.D. Couch and A.P. Cox, *J.C.S. Faraday Trans.*, 1991, **87**, 9.
221. C. Styger and A. Bauder, *J. Mol. Spectrosc.*, 1991, **148**, 479.
222. P. Zylka, D. Christen, H. Oberhammer, G.L. Gard and R.J. Terjeson, *J. Mol. Struct.*, 1991, **249**, 285.
223. J. Jacobs, G.S. McGrady, H. Willner, D. Christen, H. Oberhammer and P. Zylka, *J. Mol. Struct.*, 1991, **245**, 275.
224. J-R. Burie, P.B. Davies, G.M. Hansford, N.A. Martin, J. Gang and D.K. Russell, *Mol. Phys.*, 1991, **74**, 919.
225. M.A. Roehrig, Q-Q. Chen, S.T. Haubrich and S.G. Kukolich, *Chem. Phys. Lett.*, 1991, **183**, 84.
226. S.G. Kukolich, M.A. Roehrig, S.T. Haubrich and J.A. Shea, *J. Chem. Phys.*, 1991, **94**, 191.
227. R.N. Mulford, *J. Mol. Spectrosc.*, 1991, **147**, 260.
228. A. Carrington, I.R. McNab, C.A. Montgomerie-Leach and R.A. Kennedy, *Mol. Phys.*, 1991, **72**, 735.
229. M-C. Chan, S.S. Lee, M. Okumura and T. Oka, *J. Chem. Phys.*, 1991, **95**, 88.
230. H.H. Fielding and T.P. Softley, *Chem. Phys. Lett.*, 1991, **185**, 199.
231. L. Ulivi, P. DeNatale and M. Inguscio, *Astrophys. J.*, 1991, **378**, L29.
232. M-C. Chan, L-W. Xu, C.M. Gabrys and T. Oka, *J. Chem. Phys.*, 1991, **95**, 9404.
233. C. He, L.P. Gold and R.A. Bernheim, *J. Chem. Phys.*, 1991, **95**, 7947.
234. W-H. Jeng, X. Xie, L.P. Gold and R.A. Bernheim, *J. Chem. Phys.*, 1991, **94**, 928.
235. K. Ishikawa, S. Kubo and H. Kato, *J. Chem. Phys.*, 1991, **95**, 8803.
236. C.E. Fellows, *J. Chem. Phys.*, 1991, **94**, 5855.
237. H. Knockel, T. Johr, H. Richter and E. Tiemann, *Chem. Phys.*, 1991, **152**, 399.
238. H. Richter, H. Knockel and E. Tiemann, *Chem. Phys.*, 1991, **157**, 217.
239. T-J. Whang, A.M. Lyyra, W.C. Stwalley and L. Li, *J. Mol. Spectrosc.*, 1991, **149**, 505.
240. H. Wang, T-J. Whang, A.M. Lyyra, L. Li and W.C. Stwalley, *J. Chem. Phys.*, 1991, **94**, 4756.
241. T-J. Whang, H. Wang, A.M. Lyyra, L. Li and W.C. Stwalley, *J. Mol. Spectrosc.*, 1991, **145**, 112.
242. L. Li, A.M. Lyyra, W.C. Stwalley, M. Li and R.W. Field, *J. Mol. Spectrosc.*, 1991, **147**, 215.
243. R.F. Barrow, C. Amiot, J. Verges, J. d'Incan, C. Effantin and A. Bernard, *Chem. Phys. Lett.*, 1991, **183**, 94.

244. S. Kasahara, M. Baba and H. Katô, *J. Chem. Phys.*, 1991, **94**, 7713.
245. Y-C. Wang, M. Kajitani, S. Kasahara, M. Baba, K. Ishikawa and H. Katô, *J. Chem. Phys.*, 1991, **95**, 6229.
246. C. Amiot, *J. Mol. Spectrosc.*, 1991, **147**, 370.
247. C. Clerbaux and R. Colin, *Mol. Phys.*, 1991, **72**, 471.
248. H. Birk, R-D. Urban, P. Polomsky and H. Jones, *J. Chem. Phys.*, 1991, **94**, 5435.
249. W.T.M.L. Fernando and P.F. Bernath, *J. Mol. Spectrosc.*, 1991, **145**, 392.
250. R-D. Urban, H. Birk, P. Polomsky and H. Jones, *J. Chem. Phys.*, 1991, **94**, 2523.
251. A. Para, *J. Phys. B*, 1991, **24**, 3179.
252. P.F. Bernath, C.R. Brazier, T. Olsen, R. Hailey, W.T.M.L. Fernando, C. Woods and J.L. Hardwick, *J. Mol. Spectrosc.*, 1991, **147**, 16.
253. Y. Akiyama, K. Tanaka and T. Tanaka, *J. Chem. Phys.*, 1991, **94**, 3280.
254. M. Petri, U. Simon, W. Zimmermann, W.Urban, J.P. Towle and J.M. Brown, *Mol. Phys.*, 1991, **72**, 315.
255. S.H. Ashworth and J.M. Brown, *Chem. Phys. Lett.*, 1991, **182**, 73.
256. W. Zimmermann, U. Simon, M. Petri and W. Urban, *Mol. Phys.*, 1991, **74**, 1287.
257. U. Simon, M. Petri, W. Zimmermann, G. Huhn and W. Urban, *Mol. Phys.*, 1991, **73**, 1051.
258. T.C. Steimle, J.E. Shirley, B. Simard, M. Vasseur and P. Hackett, *J. Chem. Phys.*, 1991, **95**, 7179.
259. M.F. Cai, C.C. Carter, T.A. Miller and V.E. Bondybey, *Chem. Phys.*, 1991, **155**, 233.
260. E. de Beer, M. Born, C.A. de Lange and N.P.C. Westwood, *Chem. Phys. Lett.*, 1991, **186**, 40.
261. M.N.R. Ashfold, S.G. Clement, J.D. Howe and C.M. Western, *J.C.S. Faraday Trans.*, 1991, **87**, 2515.
262. G. DiStefano, M. Lenzi, G. Piciacchia and A. Ricci, *Chem. Phys. Lett.*, 1991, **185**, 212.
263. R-D. Urban, P. Polomsky and H. Jones, *Chem. Phys. Lett.*, 1991, **181**, 485.
264. K.V. Chance, D.A. Jennings, K.M. Evenson, M.D. Vanek, I.G. Nolt, J.V. Radostitz and K. Park, *J. Mol. Spectrosc.*, 1991, **146**, 375.
265. J.A. Gray and R.L. Farrow, *J. Chem. Phys.*, 1991, **95**, 7054.
266. E. de Beer, M.P. Koopmans, C.A. de Lange, Y. Wang and W.A. Chupka, *J. Chem. Phys.*, 1991, **94**, 7634.
267. M. Collard, P. Kerwin and A. Hodgson, *Chem. Phys. Lett.*, 1991, **179**, 422.
268. J.A. Coxon, A.D. Sappey and R.A. Copeland, *J. Mol. Spectrosc.*, 1991, **145**, 41.
269. A. Benidar, R. Farrenq, G. Guelachvili and C. Chackerian Jr, *J. Mol. Spectrosc.*, 1991, **147**, 383.
270. J.P. Towle and J.M. Brown, *Mol. Phys.*, 1991, **74**, 465.
271. P.C. Cosby, H. Helm and M. Larzilliere, *J. Chem. Phys.*, 1991, **94**, 92.
272. S.M. Corkery, J.M. Brown, S.P. Beaton and K.M. Evenson, *J. Mol. Spectrosc.*, 1991, **149**, 257.
273. K. Lipus, E. Bachem and W. Urban, *Mol. Phys.*, 1991, **73**, 1041.
274. H.G. Hedderich, C.I. Frum, R. Engleman Jr and P.F. Bernath, *Canad J. Chem.*, 1991, **69**, 1659.
275. A. Antić-Jovanović, D.S. Pešić, V. Bojović and N. Vukelić, *J. Mol. Spectrosc.*, 1991, **145**, 403.
276. S. Civiš, H.G. Hedderich and C.E. Blom, *Chem. Phys. Lett.*, 1991, **176**, 489.
277. B. Bourguignon and J. Rostas, *J. Mol. Spectrosc.*, 1991, **146**, 437.
278. P.C.F. Ip, K.J. Cross, R.W. Field, J. Rostas, B. Bourguignon and J. McCombie, *J. Mol. Spectrosc.*, 1991, **146**, 409.
279. M. Ebben and J.J. ter Meulen, *Chem. Phys. Lett.*, 1991, **177**, 229.
280. M. Ebben, M. Drabbels and J.J. ter Meulen, *Chem. Phys. Lett.*, 1991, **176**, 404.
281. M. Ebben, M. Drabbels and J.J. ter Meulen, *J. Chem. Phys.*, 1991, **95**, 2292.
282. T.J. Butenhoff and E.A. Rohlfing, *J. Chem. Phys.*, 1991, **95**, 3839.
283. D. Cossart, C. Cossart-Magos and F. Launay, *J.C.S. Faraday Trans.*, 1991, **87**, 2525.
284. D.M. Szaflarski, A.S. Mullin, K. Yokoyama, M.N.R. Ashfold and W.C. Lineberger, *J. Phys. Chem.*, 1991, **95**, 2122.
285. H.G. Hedderich, K. Walker and P.F. Bernath, *J. Mol. Spectrosc.*, 1991, **149**, 314.
286. D. Goddon, A. Groh, H.J. Hanses, M. Schneider and W. Urban, *J. Mol. Spectrosc.*, 1991, **147**, 392.
287. D.S. Green, G.A. Bickel and S.C. Wallace, *J. Mol. Spectrosc.*, 1991, **150**, 303, 354, 388.
288. K. Park, K.V. Chance, I.G. Nolt, J.V. Radostitz, M.D. Vanek, D.A. Jennings and K.M. Evenson, *J. Mol. Spectrosc.*, 1991, **147**, 521.
289. Y. Xie, P.T.A. Reilly, S. Chilukuri and R.J. Gordon, *J. Chem. Phys.*, 1991, **95**, 854.
290. G. Di Lonardo, L. Fusina, P. De Natale, M. Inguscio and M. Prevedelli, *J. Mol. Spectrosc.*, 1991, **148**, 86.
291. H. Riris, C.B. Carlisle, D.E. Cooper, L-G. Wang, T.F. Gallagher and R.H. Tipping, *J. Mol. Spectrosc.*, 1991, **146**, 381.
292. F. Matsushima, S. Kakihata and K. Takagi, *J. Chem. Phys.*, 1991, **94**, 2408.
293. P.A. Martin and G. Guelachvili, *Chem. Phys. Lett.*, 1991, **180**, 344.
294. K.A. Peterson, R.H. Petrmichl, R.L. McClain and R.C. Woods, *J. Chem. Phys.*, 1991, **95**, 2352.
295. T.D. Varberg, R.W. Field and A.J. Merer, *J. Chem. Phys.*, 1991, **95**, 1563.
296. R-D. Urban and H. Jones, *Chem. Phys. Lett.*, 1991, **178**, 295.
297. R. Brühl, J. Kapetanakis and D. Zimmermann, *J. Chem. Phys.*, 1991, **94**, 5865.
298. J. d'Incan, C. Effantin, A. Bernard, J. Vergès and R.F. Barrow, *J. Phys. B*, 1991, **24**, L71.

299. C. Effantin, A. Bernard, J. d'Incan, E. Andrianavalona and R.F. Barrow, *J. Mol. Spectrosc.*, 1991, **145**, 456.
300. C. Amiot and J. Vergès, *Chem. Phys. Lett.*, 1991, **185**, 310.
301. C.A. Leach, J.R. Waldeck, C. Noda, J.S. McKillop and R.N. Zare, *J. Mol. Spectrosc.*, 1991, **146**, 465.
302. D. Zhao, P.H. Vaccaro, A.A. Tsekouras, C.A. Leach and R.N. Zare, *J. Mol. Spectrosc.*, 1991, **148**, 226.
303. S. Takano, S. Yamamoto and S. Saito, *J. Chem. Phys.*, 1991, **94**, 3355.
304. H. Ito, K. Kuchitsu, S. Yamamoto and S. Saito, *Chem. Phys. Lett.*, 1991, **186**, 539.
305. S.P. Davis, M.C. Abrams, M.L.P. Rao and J.W. Brault, *J. Opt. Soc. Amer. B*, 1991, **8**, 198.
306. D.T. Cramb, M.C.L. Gerry, F.W. Dalby and I. Ozier, *Chem. Phys. Lett.*, 1991, **178**, 115.
307. B. Lindgren, P. Royen and M. Zackrisson, *J. Mol. Spectrosc.*, 1991, **146**, 343.
308. D. Cossart and C. Cossart-Magos, *J. Mol. Spectrosc.*, 1991, **147**, 471.
309. H. Bredohl, I. Dubois, E. Mahieu and F. Melen, *J. Mol. Spectrosc.*, 1991, **145**, 12.
310. H. Uehara, K. Horiai, H. Nakagawa and H. Suguro, *Chem. Phys. Lett.*, 1991, **178**, 553.
311. E. Mahieu, I. Dubois and H. Bredohl, *J. Mol. Spectrosc.*, 1991, **150**, 477.
312. J.E. Shirley, W.L. Barclay Jr, L.M. Ziurys and T.C. Steimle, *Chem. Phys. Lett.*, 1991, **183**, 363.
313. L.A. Kaledin and E.A. Shenyavskaya, *Mol. Phys.*, 1991, **72**, 1203.
314. J. Xin, G. Edvinsson and L. Klynning, *J. Mol. Spectrosc.*, 1991, **148**, 59.
315. W.E. Jones and T.D. McLean, *J. Mol. Spectrosc.*, 1991, **150**, 195.
316. H. Bovensmann, H. Knöckel and E. Tiemann, *Mol. Phys.*, 1991, **73**, 813.
317. A. Le Floch, *Mol. Phys.*, 1991, **72**, 133.
318. P. De Natale, M. Inguscio, C.R. Orza and L.R. Zink, *Astrophys. J.*, 1991, **370**, L53.
319. A.J. Mannucci, *J. Chem. Phys.*, 1991, **95**, 7795.
320. R. Farrenq, G. Guelachvili, A.J. Sauval, N. Grevesse and C.B. Farmer, *J. Mol. Spectrosc.*, 1991, **149**, 375.
321. D. Bailly, C. Rossetti, F. Thibault and R.Le Doucen, *J. Mol. Spectrosc.*, 1991, **148**, 329.
322. B.A. Garetz, C. Kittrell and A.C. Le Floch, *J. Chem. Phys.*, 1991, **94**, 843.
323. T. Rytel, *J. Mol. Spectrosc.*, 1991, **145**, 420.
324. S. Sekine, S. Iwata and C. Hirose, *Chem. Phys. Lett.*, 1991, **180**, 173.
325. N. Hosoi, T. Ebata and M. Ito, *J. Phys. Chem.*, 1991, **95**, 4182.
326. J-I. Choe, Y-M. Rho, S-M. Lee, A.C. Le Floch and S.G. Kukolich, *J. Mol. Spectrosc.*, 1991, **149**, 185.
327. R. Mollaaghababa, C.A. Gottlieb, J.M. Vrtilek and P. Thaddeus, *Astrophys. J.*, 1991, **368**, L19.
328. H. Birk and H. Jones, *Chem. Phys. Lett.*, 1991, **181**, 245.
329. T.F. Hanisco and A.C. Kummel, *J. Phys. Chem.*, 1991, **95**, 8565.
330. F. Roux and F. Michaud, *J. Mol. Spectrosc.*, 1991, **149**, 441.
331. K.R. Lykke and B.D. Kay, *J. Chem. Phys.*, 1991, **95**, 2252.
332. C.P. Rinsland, R. Zander, A. Goldman, F.J. Murcray, D.G. Murcray, M.R. Gunson and C.B. Farmer, *J. Mol. Spectrosc.*, 1991, **148**, 274.
333. H. Kanamori, S. Takashima and K. Sakurai, *J. Chem. Phys.*, 1991, **95**, 80.
334. W.C. Ho, I. Ozier, D.T. Cramb and M.C.L. Gerry, *J. Mol. Spectrosc.*, 1991, **149**, 559.
335. K.P. Huber and M. Vervloet, *J. Mol. Spectrosc.*, 1991, **146**, 188.
336. M. Takahashi, H. Ozeki and K. Kimura, *Chem. Phys. Lett.*, 1991, **181**, 255.
337. R.H. Petrmichl, K.A. Peterson and R.C. Woods, *J. Chem. Phys.*, 1991, **94**, 3504.
338. B. Simard, M. Vasseur and P.A. Hackett, *Chem. Phys. Lett.*, 1991, **176**, 303.
339. B. Simard and P.A. Hackett, *J. Mol. Spectrosc.*, 1991, **148**, 128.
340. T. Gustavsson, C. Amiot and J. Vergès, *J. Mol. Spectrosc.*, 1991, **145**, 56.
341. M. Ebben, M. Versluis and J.J. ter Meulen, *J. Mol. Spectrosc.*, 1991, **149**, 329.
342. A.H. Saleck, K.M.T. Yamada and G. Winnewisser, *Mol. Phys.*, 1991, **72**, 1135.
343. Z. Liu, Y. Liu, F. Li, J. Li, K. He, B. Gong and Y. Chen, *Chem. Phys. Lett.*, 1991, **183**, 340.
344. Y. Luo, Y.D. Cheng, H. Ågren, R. Maripuu, W. Seibt, L. Ohlund, P. Ejeklint, B. Carman, K.Z. Xing, Y. Achiba and K. Siegbahn, *Chem. Phys.*, 1991, **153**, 473.
345. A. Bernard, C. Effantin, J. d'Incan, C. Amiot and J. Vergès, *Mol. Phys.*, 1991, **73**, 221.
346. S. Guizard, N. Shafizadeh, M. Horani and D. Gauyacq, *J. Chem. Phys.*, 1991, **94**, 7046.
347. S. Guizard, N. Shafizadeh, D. Chapoulard, M. Horani and D. Gauyacq, *Chem. Phys.*, 1991, **156**, 509.
348. F.H. Geuzebroek, M.G. Tenner, A.W. Kleyn, H. Zacharias and S. Stolte, *Chem. Phys. Lett.*, 1991, **187**, 520.
349. K. Tsukiyama, M. Tsukakoshi and T. Kasuya, *J. Chem. Phys.*, 1991, **94**, 883.
350. M. Barnes, J. Baker, J.M. Dyke and R. Richter, *Chem. Phys. Lett.*, 1991, **185**, 433.
351. T. Amano, T. Amano and H.E. Warner, *J. Mol. Spectrosc.*, 1991, **146**, 519.
352. R.D. Suenram, G.T. Fraser, F.J. Lovas and C.W. Gillies, *J. Mol. Spectrosc.*, 1991, **148**, 114.
353. E. Bachem, W. Urban and Th. Nelis, *Mol. Phys.*, 1991, **73**, 1031.
354. T. Nelis, S.P. Beaton, K.M. Evenson and J.M. Brown, *J. Mol. Spectrosc.*, 1991, **148**, 462.
355. S. Adakkai Kadavathu, R. Scullman, R.W. Field, J.A. Gray and M. Li, *J. Mol. Spectrosc.*, 1991, **147**, 448.
356. Y. Azuma, W.J. Childs and K.L. Menningen, *J. Mol. Spectrosc.*, 1991, **145**, 413.
357. M. Mizushima and S. Yamamoto, *J. Mol. Spectrosc.*, 1991, **148**, 447.

358. X. Yang, A.M. Wodtke and L. Hüwel, *J. Chem. Phys.*, 1991, **94**, 2469.
359. A. Sur, R.S. Friedman and P.J. Miller, *J. Chem. Phys.*, 1991, **94**, 1705.
360. J. Wildt, G. Bednarek, E.H. Fink and R.P. Wayne, *Chem. Phys.*, 1991, **156**, 497.
361. A. S-C. Cheung, J.P.L. Li, T.W. Wong, M. Li and H.E. Radford, *J. Mol. Spectrosc.*, 1991, **150**, 285.
362. M. Barnes, J. Baker, J.M. Dyke, M. Feher and A. Morris, *Mol. Phys.*, 1991, **74**, 689.
363. Y.V. Chandri and C.G. Mahajan, *J. Mol. Spectrosc.*, 1991, **145**, 308.
364. O. Babaky and K. Hussein, *Canad. J. Phys.*, 1991, **69**, 57.
365. E.M. Spain, J.M. Behm and M.D. Morse, *Chem. Phys. Lett.*, 1991, **179**, 411.
366. Y.M. Hamrick, S. Taylor and M.D. Morse, *J. Mol. Spectrosc.*, 1991, **146**, 274.
367. J.C. Choi and J.L. Hardwick, *J. Mol. Spectrosc.*, 1991, **145**, 371.
368. J.J. Orlando, J.B. Burkholder, A.M.R.P. Bopegedera and C.J. Howard, *J. Mol. Spectrosc.*, 1991, **145**, 278.
369. A.G. Adam, Y. Azuma, H. Li, A.J. Merer and T. Chandrakumar, *Chem. Phys.*, 1991, **152**, 391.
370. A. del Olmo, C. Domingo, J.M. Orza and D. Bermejo, *J. Mol. Spectrosc.*, 1991, **145**, 323.
371. T. Ishiwata, Y. Kasai and K. Obi, *J. Chem. Phys.*, 1991, **95**, 60.
372. J-H. Si, T. Ishiwata and K. Obi, *J. Mol. Spectrosc.*, 1991, **147**, 334.
373. T. Ishiwata, A. Ishiguro and K. Obi, *J. Mol. Spectrosc.*, 1991, **147**, 300, 321.
374. E. Martinez, F.J. Basterrechea, J. Albaladejo, F. Castaño and P. Puyuelo, *J. Phys. B*, 1991, **24**, 2765.
375. P. Liu, J. Kieckhäfer and E. Tiemann, *J. Mol. Spectrosc.*, 1991, **150**, 521.
376. T. Ishiwata, O. Nakamura and K. Obi, *J. Mol. Spectrosc.*, 1991, **150**, 262.
377. T. Ishiwata, T. Hara, K. Obi and I. Tanaka, *J. Phys. Chem.*, 1991, **95**, 2763.
378. A.R. Hoy, K.J. Jordan and R.H. Lipson, *J. Phys. Chem.*, 1991, **95**, 611.
379. B. Friedrich and D.R. Herschbach, *Nature*, 1991, **353**, 412; *Z. Phys. D*, 1991, **18**, 153.
380. J.R. Johnson, T.J. Slotterback, D.W. Pratt, K.C. Janda and C.M. Western, *J. Phys. Chem.*, 1991, **94**, 5661.
381. J.B. Hudson, L.J. Sauls, P.C. Tellinghuisen and J. Tellinghuisen, *J. Mol. Spectrosc.*, 1991, **148**, 50.
382. M.A. Stepp, M.A. Kremer, P.C. Tellinghuisen and J. Tellinghuisen, *J. Mol. Spectrosc.*, 1991, **146**, 169.
383. A. Levinger and Y. Prior, *J. Chem. Phys.*, 1991, **94**, 1664.
384. D.M. Willberg, J.J. Breen, M. Gutmann and A.H. Zewail, *J. Phys. Chem.*, 1991, **95**, 7136.
385. S. Gerstenkorn, P. Luc and R.J. LeRoy, *Canad. J. Phys.*, 1991, **69**, 1299.
386. X. Zheng, S. Fei, M.C. Heaven and J. Tellinghuisen, *J. Mol. Spectrosc.*, 1991, **149**, 399.
387. J. Tellinghuisen, S. Fei, X. Zheng and M.C. Heaven, *Chem. Phys. Lett.*, 1991, **176**, 373.
388. C.I. Frum, R. Engleman Jr and P.F. Bernath, *J. Mol. Spectrosc.*, 1991, **150**, 566.
389. F. Holland, K.P. Huber, A.R. Hoy and R.H. Lipson, *J. Mol. Spectrosc.*, 1991, **145**, 164.
390. S.B. Kim, D.J. Kane, J.G. Eden and M.L. Ginter, *J. Chem. Phys.*, 1991, **94**, 145.
391. D.J. Kane, S.B. Kim, J.G. Eden and M.L. Ginter, *J. Chem. Phys.*, 1991, **95**, 3877.
392. C. Dufour, P. Carette and B. Pinchemel, *J. Mol. Spectrosc.*, 1991, **148**, 303.
393. K.J. Jordan and R.H. Lipson, *J. Phys. Chem.*, 1991, **95**, 7204.
394. R.H. Page and C.S. Gudeman, *J. Chem. Phys.*, 1991, **94**, 39.
395. G.A. Bishea, N. Marak and M.D. Morse, *J. Chem. Phys.*, 1991, **95**, 5618.
396. G.A. Bishea, J.C. Pinegar and M.D. Morse, *J. Chem. Phys.*, 1991, **95**, 5630.
397. B. Simard, P.A. Hackett, A.M. James and P.R.R. Langridge-Smith, *Chem. Phys. Lett.*, 1991, **186**, 415.
398. D.S. Pesic and B.R. Vujisic, *J. Mol. Spectrosc.*, 1991, **146**, 516.
399. G.A. Bishea and M.D. Morse, *J. Chem. Phys.*, 1991, **95**, 5646.
400. G. Bujin and C. Linton, *J. Mol. Spectrosc.*, 1991, **147**, 120.

4
Characteristic Vibrations of Compounds of Main Group Elements

BY G. DAVIDSON

1. Group I

The hot bands ($2\nu_2$, $l = 0 \leftarrow n_2$), ($2\nu_2$, $l = 0 \leftarrow \nu_2$) and ($\nu_1 + \nu_2 \leftarrow \nu_1$) were observed and assigned for H_3^+.[1] The H_3^+ overtone $2\nu_2$ ($l = 2$) $\leftarrow$ 0 was observed at about 2μm.[2]

Analysis of the high-resolution gas-phase FTIR spectra of LiH and LiD gave the following vibrational values: ^{7}LiH 1405.50936(266) cm^{-1}, ^{6}LiH 1420.11754(270) cm^{-1}, ^{7}LiD 1054.93684(179) cm^{-1} and ^{6}LiD 1074.33199(182) cm^{-1}.[3] Non-empirical M.O. L.C.A.O. calculations for $Li_2(NC)_2$ gave calculated vibrational wavenumbers for the C_{2h} and C_{2v} isomers. Matrix-isolation IR of vapours above LiCN showed that both isomers were present.[4]

MgO-doped crystals of $LiNbO_3$ gave Raman bands at 151 and 115 cm^{-1}, due to Li^+ and Mg^{2+} translations respectively.[5] The high-pressure Raman spectra of LiOH showed that there is a phase transition at approx. 7 kbar.[6] ^{6}Li/^{7}Li substitution was used to identify the predominantly νLi-O modes in Li_2MO_3, where M = Zr or Hf. All were in the range 350 - 400 cm^{-1}.[7] Similar experiments on $LiGaTiO_4$ showed that the LiO_4 modes were mainly in the range 400 - 450 cm^{-1}, although there were contributions from Li-O stretching in features at higher wavenumbers.[8] Lithium isotope shifts in the vibrational spectra of Li_2MO_3, where M = Ti or Sn, are consistent with the presence of lithium atoms with both octahedral and tetrahedral coordination.[9]

Infrared spectra of CO_3^{2-}-bearing aluminosilicate minerals gave bands below 380 cm^{-1} due to vibrations of NaO_n and CaO_n polyhedra associated with dissociated CO_3^{2-}.[10] The complexes [Na-15-crown-5][V(NX)FCl_3(CH_3CN)], where X = Cl, Br or I, contain a bridging unit (1), for which νNa-F is near 230 cm^{-1}, and νNa-Cl near 172 cm^{-1}.[11]

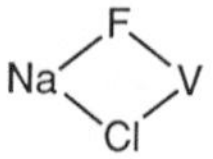

(1)

The Raman spectra of single crystals of K-Zn and K-Ga priderites, i.e. titanates, contain a band at about 78 cm^{-1}, due to motion of the potassium ions.[12] High-temperature IR spectra of CsOH and CsOD vapours include bands due to νCs-O (monomer) near 360 cm^{-1} and νCs-O (dimer) near 230 cm^{-1}.[13]

2 Group II

Localised vibrational modes of $^{9}Be_{In}$ acceptors and $^{28}Si_{In}$ donors are seen at 435 and 359 cm^{-1} respectively in InAs, with corresponding lines at 414 and 316 cm^{-1} in InSb.[14] The linear ions $[X\text{-}Be\text{-}X]^{4-}$, where X = P, As or Sb, have been prepared. Their vibrational assignments are summarised in Table 1.[15]

Table 1 Vibrational assignments for $[X\text{-}Be\text{-}X]^{4-}$ (/cm^{-1})

X =	P	As	Sb
$\nu_3\ (\Pi_u)$	340	310	255
$\nu_1\ (\Sigma_g)$	379	222	-
$\nu_2\ (\Sigma_u)$	860	775	700

The IR and Raman spectra of beryllium iodates and periodates gave some indication of Be-O stretching wavenumbers.[16] *Ab initio* calculations on MX_2, where M = Be, Mg, Ca, Sr or Ba; X = F, Cl, Br or I, gave reasonable agreement with available experimental data on these molecules.[17]

Ab initio calculations gave predicted vibrational wavenumbers for Mg_3 (ν_1 96, ν_2 104 cm^{-1}) and Mg_4 (ν_1 184, ν_2 143, ν_3 167 cm^{-1}).[18] $[Mg(MeCN)_6]_2[Bi_4Cl_{16}]$ gave IR features with contributions from νMg-N at 276 and 250 - 255 cm^{-1}.[19] FTIR spectra of films of chlorophyll *a* and chlorophyll *b* show several bands in the region 200 - 400 cm^{-1} with contributions from νMg-N and νMg-O.[20]

High-resolution IR spectroscopy of MgO showed that the fundamental in the $X^1\Sigma^+$ (ground) state is at 774.73886(21) cm^{-1}, with the first hot-band at 764.43509(30) cm^{-1}. In the $B^1\Sigma^+$ state the band centre is at 814.56832(83) cm^{-1}.[21] $CdMgIn_2O_5$ has IR and Raman spectra for which $^{24}Mg/^{26}Mg$ isotopic substitution show that νMg-O bands are at 375 and 230 cm^{-1}.[22]

$\{M[N(SiMe_3)_2]_2\}_2$ have $\nu_{as}MN_2$ at 410 cm^{-1} (M = Ca) or 378 cm^{-1} (Sr).[23] The IR spectrum of Pb-doped Bi/Sr/Ca/Cu oxide (2223) contains a band due to the Ca-O stretch at 254 cm^{-1} (a_{2u} symmetry).[24] Isotopic ($^{40}Ca/^{44}Ca$) shifts were used to identify Ca-O modes in [bis(glycylglycinato)calcium chloride]$_n$. IR bands were observed at 345, 267 and 242 cm^{-1}, with Raman bands at 305, 158 and 111 cm^{-1}.[25] The IR spectra of CaF_2 in various matrices gave values for ν_3 of the ^{40}Ca and ^{44}Ca isotopomers, e.g. in Ne at 583.2 (^{40}Ca) and 570.9 (^{44}Ca) cm^{-1}. The results were used to estimate the FCaF angle (143° in Ne from the above figures).[26]

Diode laser IR spectra of SrD and BaD gave harmonic wavenumbers for the $^2\Sigma$ (ground) state: 1205.6072(31) cm^{-1} (Sr) and 1169.6790(25) cm^{-1} (Ba).[27] The single-phase, Pb-doped Bi/Sr/Ca/Cu oxide has νSr-O at 254 cm^{-1}.[28]

Raman spectra of single crystallites of $Tl_2Ba_2Ca_{n-1}Cu_nO_{2n+4-\delta}$, where n = 1, 2 or 3, show modes associated with vibrations of bridging O(2) in the BaO planes (near 485 cm^{-1}) and O(3) in TlO planes (near 600 cm^{-1}).[29] Changes in Ba-sensitive modes along the series $Y_{1-x}Eu_xBa_2Cu_3O_7$, where $0 \leqslant x \leqslant 1$, were followed by Raman spectroscopy.[30] A normal coordinate analysis of Ba_3WO_6 suggests that vibrations in the range 71 - 150 cm^{-1} involve the barium octahedra and cubo-octahedra.[31]

3 Group III

3.1 Boron :- Isotopic shifts (H/D), $^{10}B/^{11}B$) were measured for the B-H stretching mode of B implanted in silicon.[32] Characteristic IR bands for BH_4^- and $B_nH_n^{2-}$ were assigned for the new mixed salts $MB_nH_n.MBH_4$, where M = K, Rb or Cs, n = 10 or 12.[33] (2) has νBH at 2400 cm^{-1}, νNH at 3410 cm^{-1}, νC=O at 1570 cm^{-1} and δNH at 1520 cm^{-1}.[34]

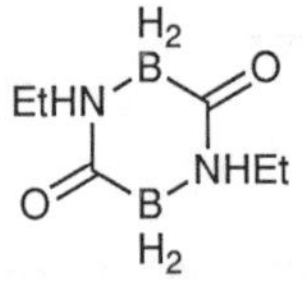

(2)

Vibrational spectra and normal coordinate analyses have been reported for the following: $(^{11}B_n{}^{10}B_{6-n})X_6^{2-}$, where X = Cl or Br;[35] $B_6X_nY_{6-n}^{2-}$, where n = 1 - 5, X ≠ Y = Cl, Br or I;[36] and $(^{11}B_n{}^{10}B_{6-n})I_6^{2-}$, where n = 0 - 6, including pairs of geometric isomers for n = 2, 3 or 4.[37] Analysis of the normal vibrations of $B_{12}H_{12}^{2-}$ showed that the dicarbollide anion acts as a rigid pseudo-atom in skeletal modes of complexes with Sn^{II}.[38]

Hot-filament-activated CVD of boron nitride produces polycrystalline hexagonal boron nitride.[39] Microwave-plasma-enhanced chemical vapour deposition from $NaBH_4$ + NH_3/H_2 produces BN films at 60 torr. These have Raman bands at about 1080 and 1310 cm^{-1}, due to cubic BN.[40]

The reaction products of the high-temperature pyrolysis of B_2H_6 and NH_3 were trapped and identified at low temperatures. All 11 IR-active modes of $H_2B{=}NH_2$ were assigned, e.g. νB=N 1365 (^{10}B) and 1334 (^{11}B) cm^{-1}.[41] Similar experiments with diborane and amines gave data for $H_2B{=}NMe_2$ and for the first time, $H_2B{=}N(Me)H$.[42] Characteristic νBN wavenumbers, and other modes, were identified from the IR spectra of (3), where R = Me or Et, R^1, R^2, R^3 = H, Me or Et.[43]

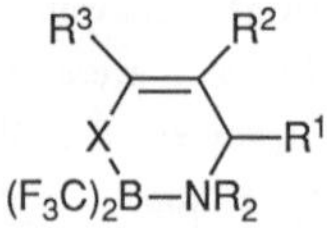

(3)

The Raman spectra of boron arsenide and boron phosphide have been obtained, and assigned with the aid of ^{10}B and ^{11}B substitution.[44]

IR and Raman spectra of $(CH_3)_2B(OCH_3)$ in the solid and gas phases, together with Raman data for the liquid, show that there is only one conformer in all three phases, of

C_s symmetry. A complete assignment was proposed for this conformer.[45] νB-O in $B(OOH)_4^-$ gives a broad IR band at 1070 - 1090 cm^{-1}.[46] A new study of the FTIR spectrum of gaseous $CsBO_2$ gave the assignments summarised in Table 2. For $Cs^{10}BO_2$, ν_1 is at 1995 cm^{-1}, with ν_5 unshifted at 232.8 cm^{-1}.[47]

Table 2 Vibrational assignments for $CsBO_2$ (/cm^{-1})

ν_1	νB=O	1933
ν_2	νB-O	1045
ν_3	δ_{ip}OB=O	632
ν_4	δ_{oop}OB=O	588
ν_5	νCs-O	232.8

Vapour-phase IR and Raman spectra show that the metaboric acid cyclic trimer is in equilibrium with monomeric orthoboric acid and water. At 350°C approximately equal amounts of $(HBO_2)_3$ and H_3BO_3 are present.[48] Modes due to νBO_4^{5-} were identified by boron isotopic shifts in MBO_4, where M = Nb or Ta. These show greater similarity to SiO_4^{4-} than to $B(OH)_4^-$ modes, even though f B-O is less than fSi-O in $M'SiO_4$, where M = Zr or Hf.[49]

Characteristic Raman bands were established for 3 distinct types of borate crystals based solely on metaborate triangular units. These are (i) degenerate intra-annular bonds, D_{3h}; (ii) distorted rings with alternating intra-annular bonds, C_{3h}; and (iii) chains.[50] IR and Raman spectra were used to probe the structural units present in the borate frameworks of alkali-metal-containing borate glasses.[51] Pressure-induced changes in borate glasses were followed by IR spectroscopy.[52]

The IR spectra of alkali borate glasses show that $B_2O_5^{4-}$ and BO_3^{3-} units dominate the structure at high alkali concentrations.[53] IR and Raman spectra of crystalline LiB_3O_5 were assigned using factor group analysis. νB-O correlates with the non-linear optical efficiency of the material.[54] The IR and Raman spectra of β-BaB_2O_4 were assigned in terms of site symmetry.[55] An IR band at 920 cm^{-1} in mixed alkali borosilicate glasses, e.g. $Li_4B_{10}O_{17}$, containing Cr_2O_3 and/or MnO_2, was attributed to a pentaborate unit.[56] IR and Raman spectra were used to follow the devitrification of glassy lithium metaborate.[57]

IR and Raman spectra of glasses in the SrO-B_2O_3-Al_2O_3-TiO_2 system suggested that a wide range of borate species was present.[58] The Raman spectra of glasses in the system xPbO.(100-x)B_2O_3, where $22 \leqslant x \leqslant 85$ show that the boron changes from three- to four coordination with increasing PbO concentration. The BO_4 units are incorporated into boroxol rings as penta- and diborate units.[59] FTIR spectra of V_2O_5/B_2O_3 thin films show that the B_2O_3 component has a boroxol ring structure.[60]

IR spectra were used to identify the structural units in thioborate systems, e.g.

$Na_2S.B_2S_3$, $Na_2S.3B_2S_3$ and $2Na_2S.B_2S_3$.[61]

3.2 Aluminium- $(Cp^*Al)_4$, where $Cp^* = C_5Me_5$, has a Raman band at 377 cm^{-1}, assigned to the breathing mode of the Al_4 tetrahedron..[62] The IR spectra of $(CH_3)_2AlH$, $(CD_3)_2AlH$ and $(CH_3)_2AlD$ in the vapour-phase show that the molecules exist primarily as dimers, with smaller amounts of trimer.[63] There is IR evidence for SiH_3AlH in 1:10 silane : argon matrices containing Al atoms, after photolysis at 400 nm.[64]

Co-deposition of Al and CO_2 in Ar matrices forms $AlCO_2$ molecules. There appear to be two geometrical isomers, the low-temperature form has C_s symmetry, with inequivalent C-O bonds, while the high-temperature form has a ring structure, with symmetrical interaction of Al with the two oxygen atoms.[65] The vapour phase IR spectra of monomeric $AlMe_3$ and $AlMe_3$- d_9 have been reported for the first time, together with data on the gallium analogue. Normal coordinate analyses and *ab initio* calculations were used to confirm the assignments. The data were consistent with free rotation of the methyl groups.[66] (4) has νAl-C at 588 cm^{-1}, $\nu AlBr_t$ at 422 cm^{-1}, while νAl-Br-Al are at 333 and 381 cm^{-1}.[67] IR spectra of mixtures of $Me_3N.AlH_3$ and Me_3Ga show the existence of rapid exchange reactions, leading to the accumulation of methyl groups on the aluminium.[68]

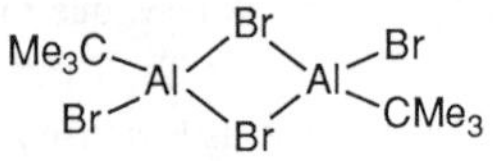

(4)

The IR spectrum of $[\{(Me_3Si)_2N\}_2AlNH_2]_2$ is consistent with the presence of an approximately square four-membered ring. $[\{(Me_3Si)_2N\}_2Al(NH_2)_2]_3Al$, on the other hand, contains a central six-coordinate aluminium, joined to three 4-coordinate aluminium atoms via NH_2 bridges.[69] Novel compounds have been prepared which contain planar trigonal ions, $[ME_3]^{6-}$, where M = Al or Ga, E = Sb; M = In, E = As. Vibrational assignments for these are summarised in Table 3.[70-1]

Table 3 Vibrational assignments for $[ME_3]^{6-}$, where M = Al, Ga, E = Sb; M = In, E = As (/cm^{-1}).

	$AsSb_3^{6-}$	$GaSb_3^{6-}$	$InAs_3^{6-}$
ν_1 (a_1')	132	128	171
ν_2 (a_2")	181	110	99
ν_3 (e')	293/318	192/210	197/215

The IR and Raman spectra of Al-N_2O-Ar and Al-O_2-Ar systems have been recorded. The first shows that mainly AlO, AlO_2 and Al_2O_2 are present, with 3 isomeric forms of the last species.[72] Raman spectra were used to characterise basic aluminium chloride species formed during electrolysis of $AlCl_3$, e.g. $[Al_n(OH)_{2.5n}(H_2O)_m]^{+n/2}$, with aluminium in an octahedral environment.[73] IR and Raman data of aluminosilicate glasses showed that specific structural features can be related to bands in well-defined areas of the IR and/or Raman spectra.[74]

A melt with the composition $NaAl_2OCl_5$ has an IR spectrum which fits with that calculated for a species $[Al_4O_2Cl_{10}]^{2-}$.[75] Characteristic bands of AlO_4, AlO_5 and AlO_6 units were assigned from the IR spectra of MAl_2O_4, $MAl_{12}O_{19}$ and $MMgAl_{10}O_{17}$, where M = Ca, Sr or Ba.[76] The IR and Raman spectra of single crystals of natural garnets containing $Al_2Si_3O_{12}^{6-}$ have been reported. Assignments were proposed for all 17 IR and 25 Raman modes.[77]

Vibrational spectra of $(N_2H_5)_2AlF_5.H_2O$ are consistent with the presence of isolated octahedral anions, $[AlF_5(H_2O)]^{2-}$.[78] The IR spectra of fluoroaluminates formed in the system NaF-AlF_3-H_2O showed the presence of $NaAlF_4.0.83H_2O$, $Na_3Al_3F_{14}.0.5H_2O$ and $Na_3AlF_6.0.167H_2O$.[79]

The Raman spectrum of $EtAlCl_2$ in 1-methyl-3-butylimidazolium chloride solution shows the formation of $EtAlCl_3^-$, $Et_3Al_2Cl_5^-$ and $Et_3Al_3Cl_7^-$.[80] The IR spectra of the co-condensation products of Al, Ga or In with Br_2 in Ar matrices gave evidence for the formation of MBr, MBr_2 (M = Al, Ga or In) and $AlBr_3$. Data for MBr_2 molecules were consistent with C_{2v} symmetry.[81]

3.3 Gallium:- The vibrational spectra of $Ga_2I_4L_2$, where L = THF, py or other donor ligands, all have νGaGa near 140 cm^{-1}, i.e. all have an "ethane-like" structure.[82] Diode laser spectra gave the following harmonic vibrational wavenumbers: GaD 1599.9832(17) cm^{-1}, InD 1475.8597(18) cm^{-1} and TlD 1394.6590(52) cm^{-1}.[83] $EtGaH_2$, as its NMe_3 adduct, has νGaH_2 at 1820 cm^{-1}, δGaH_2 at 752 cm^{-1}. Analogous bands for Et_2GaH are at 1798 cm^{-1} (νGaH) and 820 cm^{-1} (δGaH).[84]

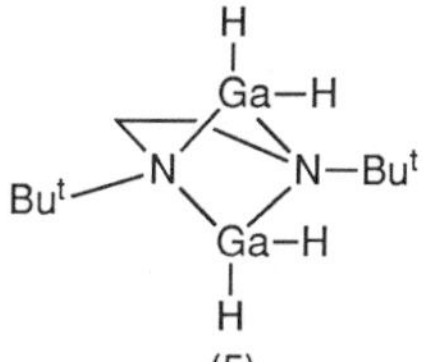

(5)

νGaH bands in (5) are seen at 1870 and 1920 cm^{-1}.[85] The IR spectra of digallane, as vapour, or in Ar or N_2 matrices, have been reported. The following assignments were given, based on Ar matrix data: νGaH_t (b_{2u}) 2015/1996 cm^{-1}, (b_{3u}) 1985/1968 cm^{-1}; νGaH_{br} (b_{3u}) 1283 - 1234 cm^{-1}, (b_{1u}) 1218 - 1195 cm^{-1}; ρGaH_2 (b_{2u}) 773/761 cm^{-1}; δGaH_2 (b_{3u}) 676-666 cm^{-1}; and ρGaH_2 (b_{1u}) 659 - 648 cm^{-1}.[86] Skeletal mode assignments were proposed for the new compounds $(^iC_3H_7)GaX_2$ and $(^iC_3H_7)_2GaX$, where X = Cl, Br or I.[87]

Some vibrational assignments for $Me_3P.GaCl_3$, made with aid of $^{69}Ga/^{71}Ga$ isotopic shifts, are summarised in Table 4.[88] Me_3Ga and AsH_3, in Ar or N_2 matrices, form a 1:1 complex, of C_{3v} symmetry. There was IR evidence for the same species in the gas-phase at 295K.[89]

Table 4 Some vibrational assignments for $Me_3P.GaCl_3$ ($/cm^{-1}$)

a_1	νGaP	371.3	e	νGaCl	377.6
	νGaCl	339.4		$\delta GaCl_3$	142.9
	$\delta GaCl_3$	133.2		ρGaCl	94.0

The Raman spectra of barium gallosilicate glasses at low SiO_2 concentrations are dominated by a Ga-O-Ga stretching mode near 530 cm^{-1}.[90] The IR and Raman spectra of $PbO-Bi_2O_3-Ga_2O_3$ glasses contain bands at 130, 400, 550 and 650 cm^{-1}, which were assigned to νPb-O or νBi-O, νGa-O-Pb or νGa-O-Bi, νGa-O-Ga and $\nu Ga\text{-}O_t$ respectively.[91] Single crystals of $A_x[Ga_8Ga_{8+x}Ti_{6-x}O_{56}]$, where A = K, Rb or Cs, give IR bands near 400 and 475 cm^{-1}, said to be due to Ga(Ti)-O stretching and Ga(Ti)-O-Ga(Ti) bending respectively.[92]

The IR spectra of glasses in the $Na_2S-Ga_2S_3$ system contain characteristic vibrations of tetrahedral GaS_4 units, including sulphur-bridge stretching of corner- and edge-shared tetrahedra.[93] $Ga[Se(2,4,6\text{-}^tBu_3C_6H_2)]_3$ has νGa-Se at 256 cm^{-1}.[94]

IR emission spectra of GaF at 1000°C gave a value for the harmonic stretching wavenumbers of 622.367(11) cm^{-1}.[95] *Ab initio* m.o. calculations on $GaClH_2$, $GaCl_2H$ and $GaCl_3$, together with their dimers, gave a set of predicted wavenumbers. Agreement with experiment was better for Ga_2Cl_6 than for $Ga_2Cl_2H_4$.[96]

3.4 Indium- IR and Raman spectra for (6) gave these assignments: νIn-F 347 cm^{-1} (R = CH_2Ph), 385 cm^{-1} (mesityl); νIn-C 459. 437 cm^{-1} (CH_2Ph), 540 cm^{-1} (mesityl). The cation $[^iPr_2In]^+$ has νIn-C at 519 cm^{-1} (IR only) and 473 cm^{-1} (IR and Raman).[97] IR and Raman assignments for (7), where X = CH_2Ph or Cl, include the following: νIn-Cl ring modes: 225, 206 cm^{-1} (CH_2Ph), 222, 206 cm^{-1} (Cl); νIn-C(Cl) terminal 453 cm^{-1} (X = CH_2Ph; C), 448 cm^{-1} (X = Cl, C) and 297 cm^{-1} (X = Cl, Cl).[98]

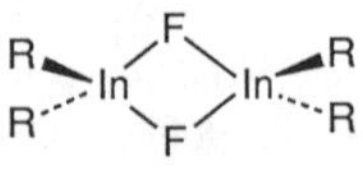

(6)

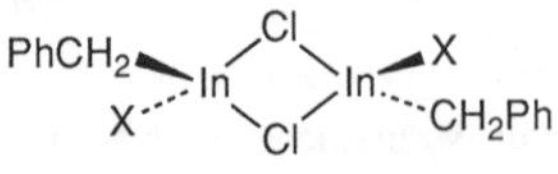

(7)

Raman and IR spectra of $ZnIn_2S_4$, $Zn_2In_2S_5$ and $Zn_3In_2S_6$ show that the indium is four-coordinate in all cases.[99] Raman spectra of concentrated aqueous solutions of $InCl_3$ or $InBr_3$ show that almost linear X-In-X units are present, together with coordinated water molecules.[100]

3.5 Thallium:- The low-wavenumber IR and Raman spectra of TlSPh, TlS-t-C_4H_9 and $TlSC_7H_7$ contain bands due to ν_{as} and ν_s of TlS_3 and TlS_4 polyhedra near 300 and 150 - 170 cm^{-1} respectively.[101]

Ph-C$(NSiMe_3)_2TlCl_2$ and [Ph-C$(NSiMe_3)_2TlCl_3]^-$ both have νTl-Cl near 270 cm^{-1}, i.e. both contain terminal thallium atoms only.[102] νTl-X skeletal mode assignments were given for $TlCl_3$.2L, $TlBrCl_2$.3L, $TlICl_2$(L)(MeCN), $TlBr_3$(L)(MeCN), $TlBr_2$Cl.2L, $TlIBr_2$(L)(MeCN), TlI_3(L)(MeCN) and TlIBrCl.L, where L = 2-ethylimidazole.[103]

4 Group IV

4.1 Carbon- The Raman spectrum of diamond shows linear dependence of the 1332 cm^{-1} band for all-^{12}C diamond with ^{13}C content, falling to 1282 cm^{-1} for the all-^{13}C form.[104] The Raman fundamental mode of diamond can be used as a pressure sensor in a diamond anvil cell at high temperatures.[105]

There is IR evidence for several C_n clusters isolated in argon at 12 K., including linear C_n, where n = 3, 5, 7, 8 and 9.[106] The ν_2 bending vibration of C_3 was found by high-resolution far IR to be at 63.416529(40) cm^{-1}.[107] *Ab initio* m.o. calculations on C_3^+ gave predicted vibrational spectra for several possible geometries.[108]

The matrix IR spectrum of C_3O gave the following fundamental wavenumbers: ν_5 ?120 cm^{-1}, ν_4 580.0 cm^{-1}, ν_3 939.1 cm^{-1}, ν_2 1907.0 cm^{-1} and ν_1 2242.9 cm^{-1}.[109] The FTIR spectrum of transient gaseous C_3O contained ν_1 at 2257.219530(666) cm^{-1}.[110]

C_4 trapped in argon gave an IR band at 1699.8 cm^{-1}, assigned as a combination of νC=C with the C_4 bending mode. This is consistent with a *cis* -bent structure for C_4.[111] Separate results on C_4 confirmed this assignment, with direct observation of the C_4 bending mode at 172.4 cm^{-1}.[112] *Ab initio* results on C_4 suggest that a matrix IR band at 1284 cm^{-1} may belong to a cyclic form of the molecule.[113]

M.o. calculations on carbon clusters C_2 - C_{10} also suggested some assignments for observed matrix IR data.[114-5] Several modes of C_7 were assigned from high-resolution diode-laser spectroscopy.[116-7] The ν_6 (σ_u) stretching fundamental of C_9 was shown to be at 2014.3383(10) cm^{-1}.[118] Theoretical IR spectra were calculated for 22 polyhedral carbon clusters C_n, with n ranging from 20 to 240.[119]

There have, of course, been a large numbers of papers referring to C_{60} and other fullerenes. The most complete vibrational assignment for C_{60} is summarised in Table 5.[120] Other vibrational data, generally agreeing with the above, have also been published, including a number of unassigned bands of C_{70}.[121-8] The IR spectrum of C_{60} at high pressures shows that there is a phase transition at about 14 kbar.[129] The resonance Raman spectrum of C_{60} films show the existence of extra bands. These may be due to solid state effects or possibly the existence of isomers of lower symmetry.[130] SERS for C_{60} on a gold surface gave evidence for lowering of symmetry and perturbation of the electronic structure, compared to the free molecule.[131] Emission FTIR spectra of C_{60} at high temperatures shows that it stable to about 600°C in the absence of oxygen. In the presence of O_2, oxidation starts at about 200°C.[132]

$C_{60}H_{36}$ has IR bands due to νCH at 2925 and 2855 cm^{-1}, with nC=C at 1620 cm^{-1} and other bands at 1459, 1400 and 675 cm^{-1}.[133] The calculated vibrational spectrum of $C_{60}H_{60}$ gave a match between 6 wavenumbers and an unidentified astronomical emission.[134]

Ab initio M.O. calculations on the HCS radical gave calculated wavenumbers for νOH (3104 cm^{-1}), νCS (1165 cm^{-1}) and the bending mode (871 cm^{-1}).[135] The high-resolution FTIR spectrum of CHF_3 shows that ν_1 (νCH) is highly perturbed by a combination band to give three parallel-type bands.[136] ν_5 (δCF_2) for HCF_2Cl is centred at 596.373031(98) cm^{-1}.[137]

Table 5 Vibrational Assignments for C_{60} (Symmetry I_h) (/cm^{-1})

Infrared	Raman	Assignments
-	273	h_g squashing
-	437	h_g
-	496	a_g breathing
527	-	t_{1u}
577	-	t_{1u}
-	710	hg
-	774	h_g
-	1099	?h_g
1183	-	t_{1u}
-	1250	?h_g
-	1428	h_g
1428	-	t_{1u}
-	1470	h_g
-	1575	h_g

Very precise measurements have been reported for $\nu_3+\nu_8$ of H_2CCS(2250.27179(17) cm^{-1}), ν_1 of HDCCS (2283.052116(98) cm^{-1}) and ν_1 of D_2CCS (2237.477635(81) cm^{-1}).[138]

High-resolution FTIR data have been obtained for $^{12}CH_3F$: ν_3 1048.610701(10) cm^{-1}; $2\nu_3$ 2081.380386(26) cm^{-1}; $\nu_3+\nu_6$ 2221.805006(14) cm^{-1}; and ν_6 1182.674392(17) cm^{-1}.[139-40]

A number of fundamentals of CD_2HF were identified from high-resolution IR measurements.[141]

Absolute IR intensities were measured for rovibrational lines belonging to ν_2 and ν_5 of $^{12}CH_3{}^{35}Cl$.[142] Numerous combination bands of $CH_3{}^{35}Cl$ were analysed from the high-resolution FTIR in the range 4250 - 4600 cm^{-1}.[143] ν_3 and ν_6 mode band centres were found with high precision for $CD_3{}^{35}Cl$ and $CD_3{}^{37}Cl$.[144] Similar analyses were performed for several overtone and combination bands of $CH_3{}^{79}Br$; ν_6 of $^{13}CH_3I$;[146] and combination bands of CH_2D_2 in the range 2200 - 2500 cm^{-1}.[147]

The ν_3 band of CNC^+ is centred at 1974.07172(65) cm^{-1}.[148] ν_1 for CCN was found from tunable diode laser IR spectroscopy to be at 1924.0406(27) cm^{-1}.[149] The same technique for protonated cyano-acetylene, HC_3N^+, showed that ν_3 (νCN) was at 2315.1425(4) cm^{-1}.[150] The following fundamentals were observed for $^1H^{13}C^{14}N$-$^1H^{19}F$: ν_1 (νHF) 3716.100(3) cm^{-1}; ν_2 (νCH) 3292.1029(3) cm^{-1}; ν_3 (νCN) 2088.9209(7) cm^{-1}; and ν_4 167.407(5) cm^{-1}.[151] The high-resolution FTIR spectrum of F-C≡C-F has been reported for the first time.[152] The ν_3 band (νCN) of ICN is centred at 2179.18441(6) cm^{-1}.[153] IR data were reported and largely assigned for cyanoformyl fluoride, N≡C-C(=O)F, e.g. νCN 2257 cm^{-1}, νC=O 1857 cm^{-1}. νCF and νCC modes were heavily mixed.[154]

The IR spectrum of PhC≡P shows νC≡P at 1565 cm^{-1}, This corresponds to a C≡P stretching force constant of 812 N m^{-1}, compared to 889 N m^{-1} in MeC≡P, and indicating that there is some conjugation of C≡P with the Ph ring.[155] An *ab initio*calculation of the vibrational wavenumbers of the dimer $(HCP)_2$ has been made.[156]

Revised vibrational constants gave harmonic vibrational wavenumbers of 2169.81 cm^{-1} for CO.[157] There is IR evidence for CO^+, *trans*-OCCO $^+$ and *trans* -OCCO$^-$ in neon matrices.[158] The band centre for ν_4 of $O{=}CF_2$ is at $1243.2660_9 \pm 0.00014$ cm^{-1}.[159]

The IR spectrum of $[C(O)F]_2O_2$ suggests that there is only one conformer in the gas-phase, with ν_sC=O at 1902 cm^{-1}, ν_{as}C=O at 1929 cm^{-1}. This is probably the *syn*, *syn*form (8).[160] ν_6 (νC-O) of peroxyformic acid, HC(=O)O-OH, was shown to have a band centre at 1124.9853 cm^{-1}.[161] The IR and Raman spectra of monothioformic acid, HC(=O)SH, show the presence of two stable planar conformations. The *trans*form is the more stable in the vapour phase.[162]

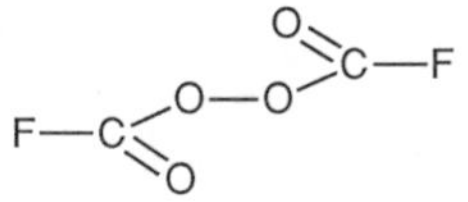

(8)

High-resolution FTIR data for O=C=C=C=S gave the following band centres: ν_1 2257.52915(20) cm^{-1}; ν_2 1961.96697(20) cm^{-1}; ν_3 1279.31516(20) cm^{-1}; and a low-lying CCC bending mode (ν_7) at 77.63094(20) cm^{-1}.[163] The O=C=S molecule has been investigated extensively, both in matrices and the gas-phase.[164-9]

Absolute IR line intensities were measured for the $(31^10) \leftarrow (00^00)$ band of $^{12}C^{16}O_2$, centred at 4416 cm^{-1}.[170] A CO_2 band was seen at 2376.49133(20) cm^{-1} in CO_2.Ar.[171] There is IR evidence for CO_4^- species produced from a Ne/CO_2/O_2 mixture. The data were used for a normal coordinate analysis, and the results were consistent with an $O_2C...O_2^-$ structure of C_s symmetry.[172] Band centres for ν_3 modes for isotopomers of S=C=C=C=S have been measured with high precision.[173]

The molecule FC≡CF has been prepared and fully characterised for the first time. The IR spectrum contained ν_3 (Σ^+_g) at 1349 cm^{-1}. Combination bands produced an estimate of 787 cm^{-1} for ν_2.[174] IR and Raman spectra of solid $CFCl_3$ showed a number of solid-state effects.[175] IR spectra of CF_3X, where X = H, D, Cl or I, yielded detailed assignments, especially for CF_3Cl in liquid argon solution.[176] Some combination bands of CF_3Cl were analysed from high-resolution FTIR data.[177] Raman and IR spectra of solid $CClF_3$ showed no evidence for phase transitions between 20K and the melting point. Internal and external modes were all assigned.[178]

$\nu_2+\nu_3$ for CF_3I is at 1028.2124(10) cm^{-1}.[179] The band centre for ν_1 (ν_sCF_2) of $CF_2{}^{35}Cl_2$ was found at 1101.3769(63) cm^{-1}.[180] The IR and Raman spectra of $CFBr_3$ and $CBrCl_3$ were assigned under C_{3v} symmetry, and normal coordinate analyses were carried out.[181] $(ClCH_2)_2Cl^+$ gave characteristic IR bands at 3070, 3068, 2980, 1233, 1030, 870, 796 and 780 cm^{-1}.[182]

4.2 Silicon- IR and Raman spectra of Ph_3SiSiH_3, $Ph_2HSiSiHPh_2$, PhH_2SiSiH_2Ph and their Si-deuteriated derivatives, all have νSiSi in the range 427 - 523 cm^{-1}.[183] IR (vapour) and Raman (liquid) spectra were reported for Si_2H_5Cl, giving a full assignment for the first time.[184] Detailed vibrational assignments were given for $H_2ISiSiIH_2$, HX_2SiSiX_2H, where X = Br or I, and their deuteriates, as well as for X_3SiSiH_3.[185]

The IR and Raman spectra of $Me_2(MeO)SiH$ in the gas phase show that the main conformer is *cis*.[186] νSiH in binary hydrogenated silicon materials is near 2100 cm^{-1}, consistent with an SiH group attached to relatively large Si monocrystals.[187] The FTIR spectra of H_2O or D_2O on porous silica gave evidence for the formation of SiH/SiD and SiOH/SiOD groups.[188] The FTIR spectra of NH_3 adsorbed on silicon surfaces showed the formation of SiH, $SiNH_2$ and Si_2NH surface species.[189] The IR spectra of SiN_x:H films can be reproduced by the superposition of two components at around 2000 cm^{-1} (νSiH) and 2100 cm^{-1} (νSiN).[190]

The IR and Raman spectra of gaseous and solid $(CH_2{=}CH)SiH_2Cl$ show that only the *gauche* form is present in the solid phase, but *cis* and *gauche* forms are both present in the gas phase.[191] There is IR evidence for Si-Si bond breaking on the adsorption of Si_2H_6 on Si(111) at 120K, forming SiH_3 units. Above 250K, Si-H bond breaking also occurs.[192]

High-resolution FTIR data were reported for $H_3Si^{79}Br$, 2100 and 2330 cm^{-1}, giving ν_1 at 2198.844(66) cm^{-1}, $\nu_2+2\nu_6$ 2190.73(28) cm^{-1}, and $\nu_5+2\nu_6$ 2211.55 cm^{-1}. Combination of these results with earlier data yielded precise values for modes ν_2 to ν_6 also.[193] A high resolution study has been made of the ν_2/ν_4 dyad of $^{29}SiH_4$ and $^{30}SiH_4$ in the range 877 - 954 cm^{-1},[194] and also of the (2000) band of SiH_4, with its band centred at 4309.3485(96) cm^{-1}.[195]

IR spectra have been obtained for β-SiC particles formed by chemical-vapour deposition.[196] The FTIR spectrum of SiC_2 vapour contains ν_1 (a_1) at 1741.3 cm^{-1}, ν_2 (a_1) at 824.3 cm^{-1}, and ν_3 (b_2) at 160.4 cm^{-1} (all figures for $^{28}Si^{12}C_2$).[197]

The IR and Raman spectra of $(CD_3)_3SiC{\equiv}CD$ have been assigned.[198] IR and Raman spectra of solid and gaseous $(CH_2{=}CH)Si(CH_3)H_2$, and the Raman spectrum of the liquid have been reported. *Cis* and *gauche* conformers were found in all phases, with the former as the more stable.[199] Complete vibrational assignments and normal coordinate analyses were made from Raman data for $(CH_3)_{4-n}Si(CD_3)_n$, where n = 0 - 4, $(CH_2D)_{4-n}Si(CH_3)_n$ and $(CHD_2)_{4-n}Si(CD_3)_n$, where n = 0 - 3.[200]

N_2 adsorption on to silane-derived silicon powders leads to conversion of Si-H at the surface to SiN, with a characteristic νSi-N band at about 835 cm^{-1}.[201] The IR spectra of pentacoordinate silicon derivatives show that intramolecular Si←N, Si←O and Si←S bonds produce shifts of up to 100 cm^{-1} in modes adjacent to the silicon coordination sphere.[202] ν_{as}SiNSi bands are in the range 905 - 983 cm-1 in $Me_2[(Me_3Si)_2N]SiH$ and related species.[203] Gas-phase electron diffraction of $Me_3EN{=}C{=}NEMe_3$, where E = Si or Ge, show that both are linear, contradicting earlier vibrational data.[204] IR and Raman spectra of *trans*$SiF_4(NH_3)_2$ could be assigned in terms of local D_{4h} symmetry for SiF_4 and $D_{\infty h}$ for SiN_2.[205] Several IR bands in the range 120 - 500 cm^{-1} for $(^iPrSi)_4(PH)_6$ were assigned to vibrations of the Si_4P_6 skeleton.[206]

H_2O on silicon atoms adsorbed on Ni(001) dissociates to form Si-OH units, with νSi-OH 1015 cm^{-1}, δSi-O-H 800 cm^{-1}. Note that the νSi-OH mode is at higher wavenumbers than on elemental Si.[207] *Ab initio* m.o. calculations gave predicted values for vibrational wavenumbers of H_4SiO_4.[208] A detailed assignment was made from the IR and Raman spectra of $Si(OCH_3)_4$ and $Si(OCF_3)_4$.[209] $(Cl_2SiO)_3$ and $(Cl_2SiO)_4$ gave very similar vibrational spectra. They are also similar to the spectrum of $(PNCl_2)_3$, but very different from that of puckered $(PNCl_2)_4$, confirming the X-ray results which show that the Si_3O_3 and Si_4O_4 rings are both planar.[210]

Silica gel films on single-crystal silicon substrate were characterised by the wavenumbers of Si-O-Si modes.[211] The IR spectra of several pillared clay minerals give characteristic Si-O stretches and bends in the ranges 1000 - 1200 cm^{-1}, 400 - 500 cm^{-1} respectively.[212] Characteristic Si-O stretches and external lattice modes were found for specific feldspar minerals.[213] The IR spectra of various silicate glasses were used to identify the effects of hydration. Changes were seen in both νSi-O and νSi-O-Si regions.[214] A force field has been calculated for zeolites which reproduces accurately the structure and dynamics of silica sodalite.[215] The Raman spectra of the faujasitic family of zeolites show a dependence on the Si/Al ratio.[216]

Raman spectra were used to determine the structures of SiO_4 units in Na_2O - SiO_2 and Li_2O - SiO_2 systems.[217] The Raman spectra of depolymerised peralkaline alumino-silicate melts in M_2O - Al_2O_3 - SiO_2, where M = Li, Na or K, were analysed in terms of SiO_2, SiO_3^{2-} and $Si_2O_5^{2-}$ units.[218] The Raman spectra of $Na_2O.(2\text{-}x)SiO_2.xGeO_2$ glasses, where x is in the range 0 - 2, and $Na_2O.SiO_2.xGeO_2$, where x lies between 0 and 1. The bridge stretching modes involving Si-O-Si, Si-O-Ge and Ge-O-Ge units are coupled to give only one Raman band as the Si/Ge ratio is varied.[219] IR and Raman spectra of glasses $(10\text{-}x)K_2O\text{-}50SiO_2\text{-}(x/3)La_2O_3$, where x = 0, 6 or 7.5, and related systems were assigned in terms of ν_sSiO of SiO_4 tetrahedra with 1, 2 or 4 bridging oxygen atoms.[220] Characteristic SiO_4 stretching and bending modes were found for silicate minerals Mg_2SiO_4, $[(Mg_{0.88}Fe_{0.12})]_2SiO_4$, Fe_2SiO_4 and $CaMgSiO_4$.[221]

Gels of the composition $2MgO.2Al_2O_3.5SiO_2$ can be differentiated in terms of their IR spectra in the νSi-O region.[222] IR spectra of gels in the systems $CaO\text{-}M_2O_3\text{-}SiO_2$, where M = Al or Cr, have been studied. They show the presence of Si-O-M units in the polymers.[223] The IR and Fourier transform Raman spectra of SiO_2/TiO_2 glasses gave evidence for the formation of Si-O-Ti bonds, especially during the latter stages of densification at temperatures in the range 615 - 1000°C.[224]

The FTIR emission spectrum of SiS gave the following harmonic wavenumbers: $^{28}Si^{32}S$ 749.645653(64) cm^{-1}; $^{28}Si^{34}S$ 739.29690(15) cm^{-1}; $^{29}Si^{32}S$ 742.71729(12) cm^{-1}; $^{30}Si^{32}S$ 736.20727(15) cm^{-1}.[225] Another report gave slightly different values from diode laser measurements.[226] Far-IR and Raman spectra of A_2BSiS_4, where A = Cu or Ag; B = Zn, Cd, Hg, Mn, Fe or Co, show a characteristic band of the SiS_4 group at approximately 500 cm^{-1}.[227]

IR and Raman spectra of gaseous and solid cyclopropyltrifluorosilane have been reported and fully assigned.. The SiF_3 torsional mode gave a broad IR band in the gas phase, centred at 46 cm^{-1}.[228] *Ab initio* calculations on the 1:1 adduct of SiF_4 and NH_3 gave calculated wavenumbers in good agreement with experimental values, and confirm the latter.[229] νSi-F in *trans*-SiF_4L_2, where L = pyridine or a substituted pyridine, show that increased basicity of L gave decreased νSi-F, i.e. *cis*-weakening of the Si-F bonds by stronger Si←N.[230]

Si-Cl stretching bands from chloromethylsilanes adsorbed on SiO_2 show the presence of $\text{-Si(Me)}Cl_2$ and =Si(Me)Cl from $MeSiCl_3$. Me_2SiCl_2 gave =Si(Me)Cl species.[231] Matrix IR spectra of species condensed from the systems Si - Cl_2 and Si - Cl_2 - H_2 in equilibrium with solid silicon show the presence of $SiCl_4$, $SiCl_3$, $SiCl_2$ (from Si - Cl_2); HCl, $SiCl_4$, $SiCl_3$, $SiCl_2$, $SiHCl_3$ and SiH_2Cl_2 (Si - Cl_2 - H_2).[232] New vibrational assignments are summarised in Table 6 for SiX_2, where X = Cl or Br.[233]

Table 6 New vibrational assignments for SiX_2 ($/cm^{-1}$)

	ν_1	ν_2	ν_3
X = Cl	513	195	502
Br	404	130	400

4.3 Germanium:- Two reports have been made of diode laser spectroscopy of GeH, in the $X^2\Pi_{1/2}$ state, with ν_o at 1831.8451(12) cm^{-1}.[234-5] IR spectra of atomic hydrogen adsorbed on a Ge(111) surface show that GeH_3, GeH_2 and GeH units are formed below 150 K, but only GeH above 400 K.[236] High-resolution FTIR spectra were recorded for monoisotopic $H_3{}^{74}Ge^{79}Br$ (in regions of ν_1, ν_4[237] and $2\nu_1$ and $\nu_1+\nu_4$[238]). The latter work also included data on $H_3{}^{70}GeI$.

The IR spectrum of oxygen atoms in Ge showed ν_3 of "quasimolecular" Ge_2O.[239] Anodisation of a germanium film using a high-frequency oxygen plasma produced GeO_2 units, with νGeO at 870 cm^{-1}.[240] Structural changes with pressure for GeO_2 glass were followed by Raman spectroscopy at pressures up to 56 GPa.[241] IR spectra of silicate and germanate minerals yielded an empirical formula for predicting νGe-O from νSi-O of the related structure.[242] The pressure and temperature dependences of the Raman bands of Ca_2GeO_4 were measured to 2.7 GPa and 1000 K.[243] Raman bands due to $GeO_4{}^{4-}$ and $Ge_2O_7{}^{6-}$ units were seen for amorphous $Pb_3Ge_3O_{11}$.[244]

The IR spectrum of $Me_2Ge{=}S$ in an argon matrix contained νGe=S at 518 cm^{-1}.[245] Matrix-isolated $S{=}GeCl_2$ gives νGe=S at 580.34 cm^{-1}, and $\nu GeCl_2$ at 404.02 and 439.94 cm^{-1}. Data were also reported involving $^{72}Ge/^{74}Ge$ and $^{35}Cl/^{37}Cl$ isotopic shifts.[246] νGe-S bands were assigned for the phenylgermanium *O*-alkyl dithiocarbamates PhGe-$[S_2CO(^iPr)]_3$, $Ph_2Ge[S_2CO(R)]_2$, where R = ^iPr or ^nBu, and $Ph_3Ge[S_2CO(R)]$, where R = Me, Et, ^iPr or ^nBu.[247] Glasses $Ge_{20}Bi_7Se_{73}$ and $Ge_{20}B_{13}Se_{87}$ both give νGe-Se near 250 cm^{-1}, and νBi-Se/νBi-Ge near 160 cm^{-1}.[248]

$SmGeF_6$ has $GeF_6{}^{2-}$ bands which are very similar to those of $BaGeF_6$, i.e. there is an isolated anion.[249] IR spectra of the Ge-I_2 system contained a band at 230 cm^{-1} for argon-matrix isolated GeI_2, and another at 246 cm^{-1} for GeI.[250]

4.4 Tin and Lead- The IR diode laser spectrum of SnH ($X^2\Pi_{1/2}$) gives ν_0 (for ^{120}SnH) at 1655.49247(30) cm^{-1}.[251] The equivalent wavenumber for SnD is 1191.51494(10) cm^{-1}.[252] The following precise values have been obtained for $^{116}SnH_4$: ν_1 1908.102734(18) cm^{-1}, ν_3 1905.834093(18) cm^{-1}.[253]

The IR and Raman spectra of $MeSn[N(SiMe_3)_2]_3$ contain νSnC at 508 cm^{-1}, νSnN at 390/380/364 cm^{-1}.[254] $R_2Sn(H_2PO_2)$, where R = Me or Ph, show only one IR band due to νSnC_2, i.e. the SnC_2 units are linear.[255] The IR spectrum of $MeO_2CCH_2CH(CO_2Me)$-CH_2SnCl_3 in solution is characteristic of five-coordinate tin.[256] The molecular structure of Me_3SnCl in various donor solvents was studied by far-IR. νSnCl is linearly related to the solvent donor number.[257]

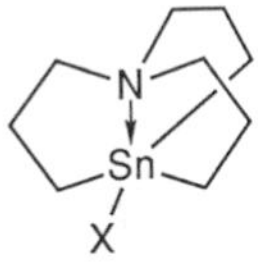

(9)

IR and Raman spectra were assigned for (9), where X = Cl, Br, I or Me. νSnN lay in the range 355 - 384 cm^{-1}, with νSnX 228 cm^{-1} (Cl) or 484 cm^{-1} (Me).[258] $SnCl_4$.L, where L

= 1,4-diaryl-2,3-dimethyl-1,4-diazabutadiene, has IR and Raman spectra consistent with C_{2v} symmetry for the $SnCl_4N_2$ unit.[259]

νSnO bands were assigned for $Sn(O^tBu)_4$ and $Sn(O^tBu\text{-}d_9)_4$. They were at 605, 568 cm^{-1} respectively.[260] νSnOSn bands were seen in the range 620 - 685 cm^{-1} in $\{[R_2Sn(O_2CCH_2SPh)]_2O\}_2$, where R = Me, Et, ^{n}Pr, ^{n}Bu or ^{n}Oct.[261] Table 7 lists the vibrational constants for isotopic forms of SnS obtained by high-resolution IR.[262]

Table 7 Vibrational constants for $^nSn^{32}S$ ($/cm^{-1}$)

n =	124	485.56842(42)
	122	486.38398(42)
	120	487.22679(20)
	119	487.65729(21)
	118	488.09645(28)
	117	488.54083(20)
	116	488.99384(31)

νSnX bands are consistent with *cis* geometry for $SnX_4(detu)_2$, where X = F or Cl, detu = 1,3-diethylthiourea.[263] Raman spectra have been measured for $SnCl_2$ in the temperature range 10 - 290 K. Low temperatures give much better resolution and allowed previously unseen modes to be detected. A band at 18 cm^{-1} is thought to be a lattice mode arising from the vibration of chains of pyramidal $SnCl_3$ groups against each other.[264] Me_3SnCl in donor solvents shows a shift in νSnCl which is linearly related to the solvent donor number.[265] The Raman spectrum of $PCl_4^+SnCl_5^-$ is consistent with D_{3h} symmetry for the anion, despite the C_{2v} symmetry indicated by X-ray diffraction.[266]

νPbO is in the range 413 - 460 cm^{-1} in $M_2[Pb(OH)_6]$, where M = K, Rb or Cs.[267] The IR and Raman spectra of $(C_3H_7NH_3)_2[PbCl_4]$ show that there are phase transitions at about 110 and 170 K.[268] IR and Raman spectra of glasses (65-x)PbO-xPbF_2-PbI_2-35SiO_2, where x = 0, 10, 15, 20 or 25, show νPbI near 140 cm^{-1}, decreasing in intensity with increased PbF_2 concentration.[269]

5. Group V

5.1 Nitrogen :- Stimulated Raman spectroscopy gave accurate spectroscopic constants for N_2 (fundamental and first hot bands): $\nu_{0(0\rightarrow1)}$ 2329.9116(5) cm^{-1}; $\nu_{0(1\rightarrow2)}$ 2301.2517(13) cm^{-1}.[270] N_4^+ is formed by microwave excitation of Ne/N_2 mixtures. In matrices it gives a characteristic IR band at 2237.6 cm^{-1}. Nitrogen shifts confirm that the ion has a linear, centrosymmetric ground state structure.[271]

High-resolution FTIR of difluorodiazirine, (10), shows ν_3 at 805.14 cm^{-1}.[272] IR spectra of 1:1 adducts of N_3F with BF_3, AsF_5 and HF all show an increase in νN(β)N(γ), and a decrease in νN(α)N(β) compared to free F-N(α)N(β)N(γ).[273] A partial analysis of the ν_4 band of DN_3 (950 - 1050 cm^{-1}) shows that strong perturbation by other bands makes a complete analysis impossible.[274] ν_1 (νNH) of $HNCCN^+$ was found to be at 3448.2710(1) cm^{-1}.[275] The FTIR emission spectrum of HNSi shows that ν_1 (νNH) is at 3588.43933(19) cm^{-1}.[276]

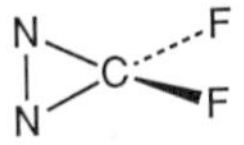

(10)

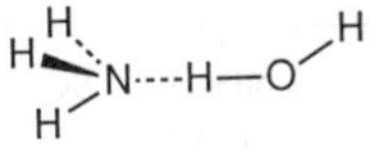

(11)

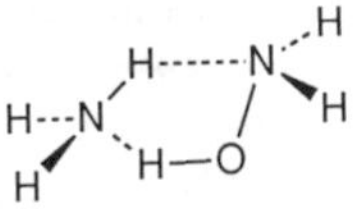

(12)

Diode laser spectra of NH_3^+ show that $\nu_2 \leftarrow 0$ is at 903.3898 cm^{-1}, and $2\nu_2 \leftarrow \nu_2$ at 939.771 cm^{-1}.[277] The IR spectrum of H_2O and NH_3 co-deposited in an Ar or a N_2 matrix is consistent with a linearly hydrogen-bonded 1:1 complex, (11).[278] *Ab initio* m.o. calculations gave a theoretical vibrational spectrum for this adduct.[279] The IR spectrum of the matrix-isolated 1:1 complex of NH_2OH and NH_3 is consistent with structure (12).[280] The nature of hydrogen bonding involving isotopically dilute NH_3D^+ ions in NH_4VO_3 and $(NH_4)_2V_6O_{13}$ has been elucidated using IR data.[281-2]

νNH modes were asigned for both submolecules in the $(NH_3)_2$ dimer.[283] There was IR evidence for an ammonia dimer in the form $H_3N...H\text{-}NH_2$, but with a very weak hydrogen-bonding interaction.[284] Gas-phase vibrational/internal rotational spectra were reported and assigned for the ammoniated ammonium ions $NH_4^+(NH_3)_n$, where n = 1 - 10.[285] FTIR at high resolution shows that the overtone of the torsional mode of N_2H_4 is centred at 670.34902(12) cm^{-1}.[286] A new very detailed normal coordinate analysis has been carried out for trimethylamine, using Raman and IR data from NMe_3, $^{15}NMe_3$, $NMe_2(CD_3)$, $NMe_2(CDH_2)$, $NMe(CD_3)_2$, $N(CD_3)_3$ and $^{15}N(CD_3)_3$.[287]

Precise values have been reported for the vibrational constants of NO^+ (several isotopomers). Evidence was also found for NO^-, $ONNO^+$ and $ONNO^-$ in Ne/NO mixtures subjected to a microwave discharge and then deposited at about 5 K.[288-9] IR and Raman specra were reported and assigned for the new nitryl cations $ON(Cl)F^+$ (νN=O 1695 cm^{-1}, νNF 801 cm^{-1}, νNCl 570 cm^{-1}) and $ON(CF_3)F^+$ (νN=O 1673 cm^{-1}, νNF 843 cm^{-1}).[290] The high-resolution IR spectrum of $^{16}O^{14}N^{35}Cl$ showed that $3\nu_1$ is at 5292.1089(3) cm^{-1}.[291] FTIR spectra of matrix-isolated NOCl show that dimers and trimers are formed at concentrations above 0.1%.[292]

Matrices resulting from microwave discharge and condensation of Ne/N_2O mixtures give IR specra showing the presence of N_2O^+ and NNO_2^-.[293] High-resolution FTIR spectra of the ν_3 and $\nu_1+\nu_3$ modes are centred at 1582.1006(8) cm^{-1} and 2858.7087(8) cm^{-1} respectively.[294] The ν_4 fundamental of HNO_2 is at 790.117045(32) cm^{-1} (*trans*-isomer) or 851.943054(32) cm^{-1} (*cis*-isomer).[295] The ν_2 and ν_3 fundamentals of the NO_3 radical were identified at 762 cm^{-1}, 1492 cm^{-1} respectively.[296] The high-resolution FTIR spectrum of

DNO_3 contains ν_8 at 762.8738 ± 0.0001 cm^{-1}.[297]

The IR spectra of solid phases formed from HNO_3 and H_2O gave evidence for the formation of a wide range of hydrate species.[298] The vibrational spectra of MNO_3, where M = Rb or Cs, show that the NO_3^- modes are very little affected by intramolecular coupling.[299] The photolysis of nitric acid in argon at 12 K gave a number of bands due to peroxynitrous acid, HOONO.[300]

Raman data for solid N_2O_4 have been reported, showing mixtures of N_2O_4 and $NO^+NO_3^-$ on fast deposition, but only molecular N_2O_4 on slower deposition. The presence of two crystalline and one amorphous phases of $NO^+NO_3^-$ was also reported.[301-2]

IR and Raman data were reported for NSCl (gas and solid). The gas phase data agree with earlier results. The solid has νNS about 75 cm^{-1} higher than in the gas, while νSCl disappears. Thus the solid form can be represented as NS^+Cl^-.[303] The photolysis of NCl_3 in argon matrices produces NCl, and, in the presence of NCl_3/Cl_2 aggregates, NCl_2.[304]

5.2 Phosphorus- Detailed assignments were proposed from IR and Raman spectra for P_{11}^{3-} and As_{11}^{3-}, on the basis of D_3 symmetry. Normal coordinate analyses showed very extensive mixing of modes.[305]

The IR spectra of $EtPH_2$ and $EtPD_2$ show the presence of both *trans* and (higher energy) *gauche* conformers.[306] The Raman (liquid and solid) and IR (vapour and solid) spectra of cyclohexylphosphine gave evidence for axial conformers as well as the more stable equatorial form. Detailed assignments were proposed, using PH_2 and PD_2 shifts.[307] FTIR data have been reported for D_{3h} PH_3F_2, giving the assignments shown in Table 8.[308]

Table 8 Vibrational assignments for PH_3F_2 (/cm^{-1})

ν_5	e'	2488.48
ν_3	a_2''	1266.15
ν_8	e''	(1194)*
ν_6	e'	965.45
ν_4	a_2''	755.986
ν_2	a_1'	(597)*
ν_7	e'	(341)*

* estimated from overtones and combinations

The IR and Raman spectra of R_2NPX_2, where R = Me or Et; X = F, Cl or Br, gave detailed vibrational assignments. $\nu_{as}PX_2$ and νCN bands are at lower wavenumbers than in $MePX_2$, Me_3N respectively, because of the decreased strength of the P-X and C-N bonds owing to the increase in P-N bond order from p_π-d_π interactions.[309] Characteristic νP=N and νP-N modes were assigned for $P_3N_3Cl_3(NCN^iPr)_3$, $P_3N_3Cl_4(NCNPh)_2$ and $P_3N_3(NCNPh)_6$.[310] The IR spectra of K^+, Rb^+ and NH_4^+ salts of $[P_2N_3SO_2(NH_2)_4]^-$ show that they are hydrolysed with replacement of three NH_2 groups by OH to form (13).[311]

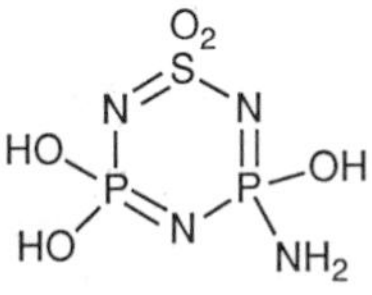

(13)

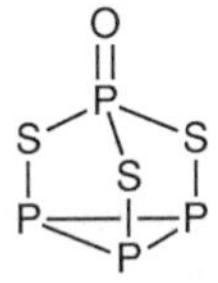

(14)

The Raman spectra of As-P-S glasses, $(As_2S_3)_{1-x}(P_2S_5)_x$, where x = 0 to 0.7, all show a strong band at 418 cm^{-1} associated with the breathing vibration of the $As_2P_2S_4$ rings.[312]

IR data have been used to characterise the products of the following reactions in solid argon at 12 K: P_2/O_2,[313] and P_2/O_3.[314] Several species were identified by IR after the depolymerisation of P_4O_{10} in matrices, e.g. PO_2, P_2O_4, P_2O_5, P_4O_9 and P_4O_8.[315]

IR and Raman spectra of EtP(O)(OEt) showed that there were at least 4 conformers in liquid and solution, but only the the more polar conformers in the solid.[316] νPO wavenumbers of phosphates depend mainly on the P-O distance, but IR and Raman intensities depend on the symmetry of the vibration. A given pattern of bands can often identify a particular structure type.[317]

IR and Raman spectra showed that addition of $MgCaSrBaAl_2F_{14}$ to $Ba(PO_3)_2$ shows that the metaphosphate structure changes to pyro- and orthophosphates.[318] Fast ion-conducting AgI-$AgPO_3$ glasses were characterised by changes in the positions, intensities and band widths of νOPO and νPOP modes, near 1140, 680 cm^{-1} respectively.[319] The following phosphates were studied by IR and Raman spectroscopy: $Na_4P_2O_7.10H_2O$;[320] polycrystalline α-$CaNa_2P_2O_7.4H_2O$, $Ca(NH_4)NaP_2O_7.3H_2O$, $Cd(NH_4)NaP_2O_7.3H_2O$;[321] $Te(OH)_6.2Na_3P_3O_9.6H_2O$;[322] $Ca(NH_4)_3P_3O_{10}.2H_2O$, $Ca_2(NH_4)P_3O_{10}.2.5H_2O$;[323] $K_3H_2P_3O_{10}.H_2O$;[324] $MNa_2P_4O_{12}$, where M = Ca or Sr;[325] and $M_6P_6O_{18}.6H_2O$, where M = Cs or Rb.[326]

The Raman spectrum of S=PCl_2F has been recorded and assigned under C_s symmetry.[327] Matrix-IR spectra of isomers of P_4S_3O gave evidence for isomeric forms with terminal oxygen at the apex, and basal phosphorus atoms, (14). UV photolysis suggested the formation of P_4S_3O (oxo-bridged) and $P_4S_3O_2$.[328] The species formed by an argon discharge on P_4/S_8 included PS, PS_2 and P_2S.[329] Several molecular species were also identified on condensation of P_4S_{10} in argon matrices.[330] Vibrational spectra were reported and assigned for $Na_4P_2S_7$, $Na_2FeP_2S_7$ and $Ag_4P_2S_7$.[331]

The Raman spectrum of P_4Se_4 shows strong bands at 350 and 185 cm^{-1}. The data are consistent with the presence of the selenium analogue of α-P_4S_4, i.e. D_{2h} symmetry. Phase changes in P_4Se_3 were also monitored by Raman spectroscopy.[332]

Raman data were reported for $[PCl_nBr_{4-n}]^+$, where $0 \leqslant n \leqslant 4$. The following characteristic wavenumbers were found: n = 3, 374 cm^{-1}; 2, 327 cm^{-1}; 1, 280 cm^{-1}.[333]

5.3 Arsenic :*Ab initio* m.o. calculations have been made for the IR spectra of As_2O and As_4O.[334] Hydrogen-implanted GaAs has an IR feature at 2029 cm^{-1} due to νAs-H.[335]

Assignments from the matrix IR spectrum of AsH_2F are summarised in Table 9.[336] *Ab initio* SCF-MO calculations for AsH_nF_{3-n}, where n = 0 - 3, and AsH_nF_{5-n}, where n = 0 - 5, gave quite good agreement with experiment for known species (AsH_3, AsF_3 and AsF_5), suggesting that the remaining calculated wavenumbers may be reliable.[337]

Table 9 Vibrational assignments for matrix-isolated AsH_2F (/cm^{-1})

			AsH_2F	AsHDF	AsD_2F
ν_5	(a")	νAsH/D	2117	2113	1526
ν_1	(a')	νAsH/D	2108	1522	1518
ν_2	(a')	δAsH_2	984	-	-
ν_3	(a')	$\delta AsH_2/D_2$	842	-	606
ν_4	(a')	νAsF	649	654	662

νMN modes were assigned for $CH_3SCN.MF_5$, i.e. 270 cm^{-1} (M = As) or 275 cm^{-1} (Sb); and for $CH_3CN.MF_5$, i.e. 270 cm^{-1} (M = As) or 271 cm^{-1} (Sb).[338]

The IR and Raman spectra of $Ca_4As_2O_9$ and $Ca_3(AsO_4)_2$ are almost identical.[339] ν_s and ν_{as} AsOAs modes have been assigned in $Mg_2As_2O_7$ and $Mg(AsO_3)_2$; surprisingly, ns modes are absent from Raman spectra.[340]

There is evidence for ion-ion interaction in acetone or acetonitrile solutions of $LiAsF_6$, from ν_1 of the anion.[341] The Raman spectrum of unstable $AsBr_4^+AsF_6^-$ has been obtained at -196°C, giving the expected characteristic bands. Decomposition was followed by Raman spectroscopy, showing the formation of AsF_3 and Br_2 (*via*$AsBr_3$).[342]

5.4 Antimony and Bismuth- Sb-O doped SiO_2 films in the system $Si(OEt)_4$-$SbCl_3$-O_2 shows νSbO at 960 cm^{-1}.[343] The vibrational spectrum of α-Sb_2O_4 has been assigned in terms of the site symmetry of SbO_6 groups present in the lattice.[344] $MO.Sb_2O_3.nH_2O$, where M = Mg(II), Ca(II) or Pb(II), $n \leqslant 7$, all have νSbO near 580 cm^{-1}.[345] The IR and Raman spectra of crystalline $Sb_4O_5X_2$, where X = Cl or Br, have been reported and assigned, and a normal coordinate analysis performed.[346] νSbO modes have also been assigned for $Ln_3Sb_6O_{12}$, where Ln = Pr, Nd, Sm, Eu, Gd, Tb, Dy, Ho, Er, Tm and Yb.[347] The IR and Raman spectra of $Na(SbF)PO_4$, $Na(SbF)AsO_4$, $NH_4(SbF)PO_4$ and $NH_4(SbF)$-

AsO_4 show the presence of $[(SbF)XO_4]^-$ layers, where X = P or As, formed by sharing 4 corners between XO_4 tetrahedra and $SbFO_4$ pseudo-octahedra.[348]

νSbX modes in SbX_3(15-crown-5), where X = F, Cl, Br or I, are shifted to higher wavenumbers than in SbX_3 for X = F or I, but to lower wavenumbers for X = Cl or Br.[349] The vibrational spectrum of solid $SbCl_3$ shows the existence of an equilibrium between $SbCl_3$ and $Sb^+ + SbCl_6^-$.[350] Assignments have been proposed from the FTIR spectrum of $[Cl_3Sb(\mu\text{-}S)SbCl_3]^{2-}$ in the range 100 - 500 cm^{-1}.[351]

There has been a re-interpretation of the diode laser spectrum of BiH ($^1\Sigma$), giving the vibrational constants 1699.5216(37) cm^{-1} for ^{209}BiH, 1205.4329(22) cm^{-1} for ^{209}BiD.[352]

The fine structure of the emission spectrum of BiO shows ω_e at 688.4 ± 1 cm^{-1} in the $X_2{}^2P_{3/2}$ (ground) state.[353] νBiO modes were assigned in $Bi(ClO_4)_3$ (435 cm^{-1}), $Bi(ClO_4)_4^-$ and $Bi(ClO_4)_5^{2-}$ (420 cm^{-1}).[354] The Raman spectra of bismuth vanadates show that at high Bi : V ratios $BiVO_4$ tetrahedra are present.[355]

Emission spectra of BiF, BiCl, BiBr and BiI have been observed for the $X_2{}^1 \rightarrow X_1{}^{0+}$ transition. The following vibrational constants were obtained from the fine structure: 543.0558(24) cm^{-1} (F), 327.44(15) cm^{-1} (^{35}Cl), 220.38(21) cm^{-1} (^{79}Br) and 168.68(6) cm^{-1} (I).[356]

6. Group VI

6.1 Oxygen :- There is Raman evidence for a new form of O_2, i.e.ζ-O_2, at 20 K, 25 GPa.[357] There have been several reports of high-resolution FTIR studies on various isotopic forms of O_3.[358-62]

The temperature and pressure dependence of the ν_1 band of water vapour were measured to temperatures of 1200 K using inverse Raman spectroscopy.[363] IR data on the νOD band of isotopically dilute HDO in $M(ClO_4)_2$ aqueous solutions, where M = Mg or Ni, gave evidence for HDO molecules interacting with the anion, in the cation coordination sphere, in solvent-shared ion-pairs and in the bulk solvent.[364] ν_2 (δ) of H_2O, HDO and D_2O were measured for liquid water with deuterium contents of up to 99.9%. The peak positions of the three bands vary linearly with molar fractions of the three isotopic species.[365] There have been several contradictory attempts to analyse the Raman spectrum of liquid water in terms of a limited number of distinct components.[366-8] The near-IR spectrum of H_2O in a range of diluting solvents was discussed in terms of the number of hydrogen-bonded OH groups in relation to the proton acceptor nature of the diluent.[369]

ν_2 wavenumbers of 1711 cm^{-1} and 1480/1490 cm^{-1} were found for H_2O decoupled in amorphous D_2O and for HDO decoupled in H_2O/D_2O respectively.[370] The IR spectrum of T_2O ice at 90 K contains these bands: ν_3, 2104 cm^{-1}; ν_1, 1988 cm^{-1}; ν_2, 1023 cm^{-1}.[371]

6.2 Sulphur and Selenium Rings and Chains :- A high-resolution study has been made of the torsional-rotational spectrum of HSSH near 417 cm^{-1}.[372] The IR and Raman spectra of CF_3SSCH_3 and CF_3SSCF_3 are assignable on the assumption that the C-S-S-C dihedral angle is approximately 90°.[373]

Ab initio m.o. calculations gave predicted wavenumbers for the most stable isomers of $H_2S_2O_2$, i.e. HO-S-S-OH, O=S(OH)-SH and $S=S(OH)_2$.[374] The IR and Raman spectra of dithionite (as the NEt_4^+ salt) show that the anion adopts a staggered centrosymmetric conformation in the solid, and in both aqueous and non-aqueous solutions. This behaviour of $S_2O_4^{2-}$ contrasts with the situation where there is a smaller counterion, where the eclipsed conformation is preferred. A normal coordinate analysis shows extensive mixing of some internal coordinates.[375] $SeSCl_2$ has νSCl at 449 cm^{-1}, and νSeS 414 cm^{-1}, consistent with a polymeric, chain formulation.[376]

Matrix-FTIR spectra of S_3 and S_4 in argon show that S_3 has a C_{2v} structure like that of ozone, while S_4 exists in several forms.[377] The new cation $[S(SH)_3]^+$ has been prepared, as $SbCl_6^-$ or AsF_6^- salts. νSH is at 2505 cm^{-1} (νSD 1820 cm^{-1}), and νSS 495 and 477 cm^{-1}. The analogue $[S(SCl)_3]^+$ has mixed νSCl and νSSS bands in the range 543 - 426 cm^{-1}.[378]

Vibrational assignments have been proposed for S_4^{2+}, with data on $^{32}S/^{34}S$ isotopic shifts. Normal coordinate analyses were carried out on several isotopic forms.[379] *Ab initio* calculations were reported for S_4, which gave some agreement with experimental data for some of the possible isomers.[380]

$[PPh_4][SAsS_7]$ has νAsS at 450 cm^{-1} (exocyclic), 344, 324 cm^{-1} (As-cyclic-S), with νSS appearing as weak IR bands near 480 cm^{-1}.[381] Se_nS_{8-n} species have νSeSe near 260 cm^{-1}, νSeS 340 - 380 cm^{-1} and νSS 430 - 470 cm^{-1}.[382]

The IR and Raman spectra of gaseous and solid $CH_3SeSeCF_3$ and $CH_3SeSeCF_3$ have been recorded and assigned, together with the Raman spectra of the liquids. All 3 phases for each contain only one conformer.[383] Several vibrational assignments have been proposed for $Se_3Br_8^{2-}$, $Se_4Br_{14}^{2-}$, $Se_5Br_{12}^{2-}$;[384] and $[Se_3X_{13}]^-$, where X = Cl or Br.[385] The new selenium nitride, (15), gave the following assignments: $\nu_{as}SeN_2$ 828 cm^{-1}; ν_sSeN_2 792 cm^{-1}; ν_sSeN 584 cm^{-1}; $\nu_{as}SeN$ 554, 571 cm^{-1}; $\nu_{as}Se\text{-}Se_2$ 352 cm^{-1} and $\nu_{as}Se\text{-}Se_2$ 325 cm^{-1}.[386] $[K(2,2,2\text{-crypt})]_2Se_7$ has characteristic IR bands of Se_7^{2-}: νSeSe 283, 250, 233 cm^{-1}; deformation 140 cm^{-1}.[387]

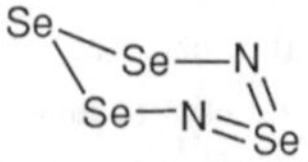

(15)

6.3 Other Sulphur and Selenium Compounds :- The high-pressure Raman spectrum of H_2S reveals a phase transition at about 11 GPa.[388] The vibrational spectra of liquid H_2S and H_2Se have been recorded up to the third overtone region.[389] There is IR evidence for weak hydrogen-bonding between H_2X, where X = S or Se, and HCl, Me_2O and NMe_3 in liquid krypton solutions.[390] Several assignments of IR bands have been made for sulphinic acid, HSO_2H, produced by photolysis of H_2S/SO_2 mixtures in argon matrices at 12 K., e.g. νSH 2591 cm^{-1}, νSO 1209 cm^{-1}.[391]

Partial vibrational assignments have been proposed for CH_3SF_3, CD_3SF_3;[392] CF_3SCD_3,[393] $Me_2SCF_3^+$, $MeS(CF_3)_2^+$;[394] $MeSX_2^+$, Me_2SX^+, where X = Cl or Br;[395] $CH_3S(CN)X^+$ and $(NC)_2SX^+$, where X = F, Cl, Br or I.[396] Fourteen of the fifteen fundamentals have been assigned for CF_3SCN, under C_s symmetry.[397] A full vibrational assignment has been proposed, and a normal coordinate analysis performed for SF_5CN.[398] A vibrational assignment for $S(CN)_3^+$ includes the following: ν_sSC_3 638 cm^{-1}, $\nu_{as}SC_3$ 568 cm^{-1} and ν_sCN 2242 cm^{-1}.[399]

A high-resolution study of the FTIR spectrum of $^{34}S^{16}O_2$ gave a band centre for ν_2 of 513.53866(3) cm^{-1}.[400] A detailed assignment has been proposed for solid SO_2, including isotopic data, and a consideration of both intra- and intermolecular modes.[401] CARS gave evidence for multiphoton excitation in ν_1, but not in ν_2 for SO_2.[402] FTIR data have been given for a number of adducts of SO_2.[403-4] The IR spectrum of $HOSO_2$ in solid argon includes νOH at 3539.8 cm^{-1}, $\nu_{as}SO_2$ 1309.2 cm^{-1}, ν_sSO_2 1097.2 cm^{-1} and νSO 759.3 cm^{-1}.[405]

IR and Raman spectra of SO_2ClF and SO_2BrF have been assigned in terms of C_s symmetry.[406] There is IR evidence for an $H_2O.SO_3$ complex from the matrix-isolation photolysis of SO_2, O_3 and H_2O.[407] The Raman spectrum of SO_3Cl^- is consistent with C_{3v} symmetry.[408] IR and Raman spectra were used to confirm the structure of (16).[409]

The vibrational spectra of crystalline MSO_4, where M = Ba, Sr or Pb, have been re-interpreted in terms of a vibrational unit cell (*Immd*) which is half of the size of the crystallographic unit cell.[410] FTIR was used to determine the isotopic compostion of $S^{16}O_x{}^{18}O_{4-x}{}^{2-}$ formed by the oxidation of FeS_2 under $^{18}O_2$ in $H_2{}^{16}O$ solutions.[411] IR spectra were used to identify structural units in aluminoborosilicate and aluminophosphate glasses containing less than 7% SO_3 (as Na_2SO_4). The SO_4^{2-} ions do not interact with the structural framework.[412]

Matrix-isolation IR studies gave evidence for M_2SeO_4, where M = Na, K, Rb or Cs, of D_{2d} symmetry, M_2SeO_3 (C_s) and $MSeO_2$ (C_{2v}).[413] The IR and Raman spectra of $[Ph_4P][Se_2Cl_6]$ contain the following features: ν_sSeCl_t 333 cm^{-1}, $\nu_sSeCl_{br}Se$ 178 cm^{-1}.[414] $SeBr_2$ gives Raman bands at 291, 265 and 105 cm^{-1}.[415]

6.4 Tellurium :- FT Raman spectra of R_2Te, where R = Me, Et or ^tBu, are all consistent with CTeC skeletons of C_{2v} symmetry.[416] The IR spectra of $Ph_3TeX.SnCl_4$ and $2Ph_3Te$-$X.SnCl_4$, where X = Cl, Br or I, all contain bands typical of Ph_3Te^+.[417]

The IR spectrum of tetragonal TeO_2 is consistent with a rutile-type structure.[418] High-pressure Raman spectra of TeO_2 show that there are phase changes at 1, 4.5, 11 and 22 GPa.[419] IR and Raman spectra of $KIO_3.Te(OH)_6$ show that the TeO_6 unit occupies a site of C_1 symmetry.[420] In $Te(OH)_6.2CO(NH_2)_2$, however, they occupy C_2 sites.[421]

Characteristic cation and anion bands were assigned for $[TeCl_3(15\text{-crown-}5)][TeCl_5]$, and $[TeCl_3(15\text{-crown-}5)]_2[TeCl_6]$.[422] Skeletal mode assignments show that the anion in $PPh_4[TeCl_4(OH)].H_2O$ or $K[TeCl_4(OH)].0.5H_2O$ is monomeric, with νTeCl 287,

255, 239 cm^{-1}. In $K[TeCl_4(OH)]$, however, polymerisation is indicated, with Cl bridging, and $\nu TeCl_t$ at 292, 358 cm^{-1}, $\nu TeCl_{br}$ 158, 187 cm^{-1}. Both types have νTeO ca. 630 cm^{-1}.[423] IR spectra were reported, with some assignments, for $[H_9O_4]_2[TeBr_6]$ and $[H_9O_4][Te_3Br_{13}]$.[424]

7. Group VII

An extremely precise measurement has been made of the vibrational fundamental of HF: 3961.4224986(372) cm^{-1}.[425] *Ab initio* calculations have been carried out to estimate the wavenumbers of intramolecular stretching bands of $(HF)_2$, $(DF)_2$ and (HF)(DF).[426] A number of complexes containing HF have been studied: Ar_nHF, n = 1 - 4;[427] (OCO)HF;[428] B...$(HF)_2$, for a variety of bases (B);[429] UF_6-HF and UF_6-FH.[430] The high-resolution FTIR spectrum of $FClO_3$ contains the following bands: ν_2 716.809727(8) cm^{-1}; ν_3 549.87885(6) cm^{-1}; ν_5 590.314776(20) cm^{-1}; and ν_6 405.60507(4) cm^{-1}.[431]

Resonance Raman effects were found for $^{35}Cl_2$ with 413.1 nm or UV excitation.[432] The Raman spectra of Me_2X^+, where X = Cl, Br or I, include: $\nu_{as}CX$ 647 cm^{-1} (Cl), 560 cm^{-1} (Br) or 526 cm^{-1} (I); ν_sCX 608 cm^{-1} (Cl), 545 cm^{-1} (Br) and 515 cm^{-1} (I).[433]

The concentration dependence of the depolarised Raman scattering of HCl and DCl in liquid CO_2 solutions has been followed.[434] Matrix-FTIR spectra have been obtained for a number of isotopomers of $(HCl)_3$, giving assignments for valence vibrations, in- and out-of-plane librations.[435] DCl in Ne has νDCl at 2091.3717(4) cm^{-1}, and in Ar 2089.4180(2) cm^{-1}.[436] The ν_1 fundamental and the $\nu_1+\nu_2^1$ intermolecular combination band of the van der Waals species Ar-DCl have been probed by diode laser spectroscopy.[437] It was possible to differentiate between WF_6-ClH and WF_6-HCl in terms of their νHCl values: 2862.1 cm^{-1}, 2866.2 cm^{-1} respectively.[438]

The high-resolution FTIR spectrum in the ν_1 region of OClO shows that the band centre is at 945.59226(2) cm^{-1} ($^{35}ClO_2$) or 939.60242(3) cm^{-1} ($^{37}ClO_2$).[439] The high-pressure Raman spectra of $NaClO_3$, $NaBrO_3$ and $KBrO_3$ showed that there are phase changes at 24, 36 and 19 kbar respectively.[440] The IR spectra of $HClO_4/H_2O$ gaseous mixtures show that the $HClO_4$ bands are identical to those of anhydrous perchloric acid. There is, therefore, no formation of $H_2O.HClO_4$ or other associates.[441]

The low-temperature Raman spectrum of $Br_3^+AsF_6^-$ contains the following bands: $\nu_{as}Br_3$ 297 cm^{-1}, ν_sBr_3 293 cm^{-1}, and δBr_3 124 cm^{-1}.[442] The Raman bands seen for the cation in $Br_5^+AsF_6^-$ are consistent with the C_{2h} form (16).[443]

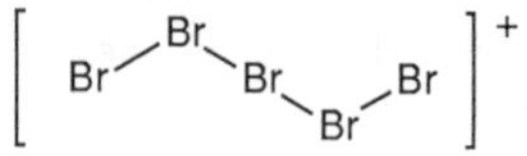

(16)

New, and more precise vibrational data have been reported for BrO: (^{79}BrO) 723.41420 cm^{-1}, (^{81}BrO) 721.92715(7) cm^{-1}.[444] The high-resolution FTIR spectrum of

BrO_3F gave the following band centres: ν_1 876.81436(2) cm^{-1}; ν_2 609.18143(1) cm^{-1}; and ν_3 360.75556(3) cm^{-1}, all for the ^{79}Br isotopomer.[445]

Vibrational assignments for $CH_3ICF_3^+$ include νI-CH_3 496 cm^{-1}; νI-CF_3 264 cm^{-1} and δCIC 119 cm^{-1}.[446] The IR and Raman spectra of $MH_2(IO_3)_3$, where M = K or NH_4, show band splittings due to distortion of the IO_4 tetrahedra.[447] There have been reports of the preparation, for the first time, of IF_6O^-, TeF_6O^{2-} (C_{5v}), TeF_7^- (D_{5h}), IF_8^- and TeF_8^{2-} (D_{4d}). The assignments for IF_6O^- and IF_8^- are summarised in Tables 10 and 11 respectively.[448-9]

Table 10 Vibrational assignments for IF_6O^- (C_{5v}; /cm^{-1})

a_1	ν_1	νI=O	873
	ν_2	νIF_{ax}	649
	ν_3	$\nu_s IF_5$	584
	ν_4	$\delta_{umbrella}$	359
e_1	ν_5	$\nu_{as} IF_5$	585
	ν_6	δ_{wag}I=O	457
	ν_7	$\delta_{wag} IF_{ax}$	405
	ν_8	$\delta_{as} IF_{5i.p}$	260
e_2	ν_9	$\nu_{as} IF_5$	530
	ν_{10}	$\delta_{sciss} IF_{5i.p}$	341

Table 11 Vibrational assignments for IF_8^- (D_{4d}; /cm^{-1})

a_1’	ν_1	$\nu_s IF_{2ax}$	660
	ν_2	$\nu_s IF_5$	590
a_2”	ν_3	$\nu_{as} IF_{2ax}$	595
	ν_4	$\delta_{umbrella} IF_5$	587
e_1’	ν_5	$\nu_{as} IF_5$	550
	ν_6	$\delta_{sciss} IF_{2ax}$	463
	ν_7	$\delta_{as} IF_{5i.p}$	410
e_1”	ν_8	$\delta_{wag} IF_{2ax}$	410
e_2’	ν_9	$\nu_{as} IF_5$	380
	ν_{10}	$\delta_{sciss} IF_5$	314

8. Group VIII

The lowest Σ-bending vibration of Ar-$^{14}NH_3$ was found to be at 26.470633(77) cm^{-1}.[450]

Pentagonal-planar XeF_5^- has been prepared for the first time. The structure was shown by X-ray diffraction and confirmed by IR and Raman spectra. Vibrational assignments, under D_{5h} symmetry, are summarised in Table 12.[451]

Table 12 Vibrational assignments of XeF_5^- (D_{5h}; /cm^{-1})

a_1'	ν_1	ν_s in-plane	502
a_2''	ν_2	δ_s out-of-plane	274
e_1'	ν_3	ν_{as}	400 - 550
	ν_4	δ_{as} in-plane	290
e_2'	ν_5	ν_{as}	423
	ν_6	δ_{as} in-plane	377

REFERENCES

1. M.G.Bawendi, B.D.Rehfuss and T.Oka, *J. Chem. Phys.*, 1990, **93**, 6200.
2. L.W.Xu, C.Gabrys and T.Oka, *J. Chem. Phys.*, 1990, **93**, 6210.
3. A.G.Maki, W.B.Olson and G.Thompson, *J. Mol. Spectrosc.*, 1990, **144**, 257.
4. V.G.Solomonik, *Zh. Fiz. Khim.*, 1990, **64**, 2691.
5. L.J.Hu, Y.H.Chang, C.S.Chang, S.J.Yang, M.L.Hu and W.S.Tse, *Mod. Phys. Lett., B*, 1991, **5**, 789.
6. D.M.Adams and J.Haines, *J. Phys. Chem.*, 1991, **95**, 7064.
7. L.V.Golubeva, N.V.Porotnikov, O.I.Kondratov, and K.I.Petrov, *Russ. J. Inorg. Chem.*, 1990, **35**, 1480.
8. L.V.Golubeva, N.V.Porotnikov, O.I.Kondratov, and K.I.Petrov, *Russ. J. Inorg. Chem.*, 1991, **36**, 102.
9. L.V.Golubeva, N.V.Porotnikov, O.I.Kondratov, and K.I.Petrov, *Russ. J. Inorg. Chem.*, 1990, **35**, 1028.
10. W.R.Taylor, *Eur. J. Mineral.*, 1990, **2**, 547.
11. W.Massa, S.Wocaldo, S.Lutz and K.Dehnicke, *Z. anorg. allg. Chem.*, 1990, **589**, 79.
12. M.Ishii, Y.Fujiki and T.Ohsaka, *Solid State Ionics*, 1990, **40-41**, 150.
13. R.J.M.Konings, A.S.Booij and E.H.P.Cordfunke, *Vib. Spectrosc.*, 1991, **1**, 383.
14. R.Addinall, R.Murray, R.C.Newman, J.Wagner, S.D.Parker, R.L.Williams, R.Droopad, A.G.DeOliveira, I.Ferguson and R.A.Stradling, *Semicond. Sci. Technol.*, 1991, **6**, 147.
15. M.Somer, M.Hartweg, K.Peters, T.Popp and H.G.von Schnering, *Z. anorg. allg. Chem.*, 1991, **595**, 217.
16. M.Maneva, M.Georgiev, N.Lange and H.D.Lutz, *Z. Naturforsch., B*, 1991, **46b**, 795.
17. M.Kaupp, P.v.R.Schleyer, H.Stoll and M.Preuss, *J. Am. Chem. Soc.*, 1991, **113**, 6012.
18. T.J.Lee, A.P.Rendell and P.R.Taylor, *J. Chem. Phys.*, 1990, **93**, 6636.
19. G.R.Willey, H.Collins and M.G.B.Drew, *J. Chem. Soc., Dalton Trans.*, 1991, 961.
20. H.Tajmir-Riahi, G.Wang and R.M.Leblanc, *Photochem. Photobiol.*, 1991, **54**, 265.
21. S.Civiš, H.G.Hedderich and C.E.Blom, *Chem. Phys. Lett.*, 1991, **176**, 489.
22. A.K.Vazhnov and K.I.Petrov, *Russ. J. Inorg. Chem.*, 1990, **35**, 1187.
23. M.Westerhausen, *Inorg. Chem.*, 1991, **30**, 96.
24. L.Zhong, J.Cai, Y.Liu, H.Tang, Y.Sun and J.Du, *Int. J. Millimeter Waves*, 1991, **12**, 239.
25. J.Odo, M.Arima, A.Iwado, N.Motohashi, Y.Tanaka and Y.Saito, *Chem. Pharm. Bull.*, 1991, **39**, 251.
26. I.R.Beattie, P.J.Jones and N.A.Young, *Inorg. Chem.*, 1991, **30**, 2250.
27. H.Birk, R.D.Urban, P.Polomsky and H.Jones, *J. Chem. Phys.*, 1991, **94**, 5435.
28. L.Zhong, J.Cai and H.Tang, *Proc. SPIE - Int. Soc. Opt. Eng.*, 1990, **1514**, 110.
29. K.Matsuishi, Y.Q.Wang, Y.Y.Sun, J.G.Lion, P.H.Hor, M.Gorman and C.W.Chu, *Proc. SPIE - Int. Soc. Opt. Eng.*, 1990, **1336**, 93.
30. I.S.Yang, G.Burns, F.H.Dacol, J.F.Bringley and S.S.Trail, *Physica, C*, 1990, **171**, 31.
31. V.L.Balashov, A.A.Kharlanov, O.I.Kondratov and V.V.Fomichev, *Russ. J. Inorg. Chem.*, 1991, **36**, 254.
32. G.D.Watkins, W.B.Fowler, G.G.Deleo, M.Stavola, D.M.Kozuch, S.J.Pearton and J.Lopata, *Mater. Res. Soc., Symp. Proc.*, 1990, **163**, 367.
33. O.A.Kanaeva, N.T.Kuznetsov and K.A.Solntsev, *Russ. J. Inorg. Chem.*, 1990, **35**, 1421.
34. M.R.M.D.Charandabi, D.A.Feakes, M.L.Ettel and K.W.Morse, *Inorg. Chem.*, 1991, **30**, 2433.
35. J.Thesing, J.Baurmeister, W.Preetz, D.Thery and H.G.von Schnering, *Z. Natuforsch., B*, 1991, **46b**, 800.
36. J.Thesing, M.Stallbaum and W.Preetz, *Z. Naturforsch., B*, 1991, **46b**, 602.

37. J.Thesing, W.Preetz and J.Baurmeister, *Z. Naturforsch., B*, 1991, **46b**, 19.
38. B.A.Kolesov, K.V.Pradenov, S.G.Vasil'eva and O.V.Volkov, *Izv. Sib. Otd. Akad. Nauk SSSR, Ser. Khim. Nauk.*, 1990, 64.
39. R.R.Rye, *J. Vac. Sci. Technol., A*, 1991, **9**, 1099.
40. H.Saitoh and W.A.Yarborough, *Appl. Phys. Lett.*, 1991, **58**, 2228.
41. J.D.Carpenter and B.S.Ault, *J. Phys., Chem.*, 1991, **35**, 3502.
42. J.D.Carpenter and B.S.Ault, *J. Phys., Chem.*, 1991, **35**, 3507.
43. A.Ansorge, D.J.Brauer, H.Bürger, F.Dörrenbach, T.Hagen, G.Pawelke and W.Weuter, *J. Organometal. Chem.*, 1990, **396**, 253.
44. D.R.Taillant, T.L.Aselage and D.Emin, *AIP Conf. Proc.*, 1991, **231**, 301.
45. E.J.Stampf, J.D.Odom, S.V.Saari, Y.H.Kim, M.M.Bergama and J.R.Durig, *J. Mol. Struct.*, 1990, **239**, 113.
46. K.V.Titova and E.I.Kollakova, *Russ.J. Inorg. Chem.*, 1991, **36**, 493.
47. R.J.M.Konings, A.S.Booij and E.H.P.Cordfunke, *J. Mol. Spectrosc.*, 1991, **145**, 451.
48. T.R.Gilson, *J. Chem. Soc., Dalton Trans.*, 1991, 2463.
49. A.M.Heyns, K.J.Range and M.Widenauer, *Spectrochim. Acta, A*, 1990, **46A**, 1621.
50. G.D.Chryssikos, J.A.Kapoutsis, A.P.Patsis and E.I.Kamitsos, *Spectrochim. Acta, A*, 1991, **47A**, 1117.
51. E.I.Kamitsos and G.D.Chryissikos, *J. Mol. Struct.*, 1991, **247**, 1.
52. Z.Zhang and N.Soga, *Phys. Chem. Glasses*, 1991, **32**, 142.
53. C.Liu and C.A.Angell, *J. Chem. Phys.*, 1990, **93**, 7378.
54. Q.Y.Shang, B.S.Hudson and C.Huang, *Spectrochim. Acta, A*, 1991, **47A**, 291.
55. S.Hong, *Proc. SPIE - Int. Soc. Opt. Eng.*, 1991, **1437**, 194.
56. I.A.Gohar, W.Schmitz, P.Paufler and R.M.Elshazly, *Phys. Stat. Solidi, A*, 1990, **122**, 469.
57. G.D.Chryssikos, E.I.Kamitsos, A.P.Patsis, M.S.Bitsis and M.A.Karakassides, *J. Non-Cryst. Solids*, 1990, **126**, 42.
58. T.Hübert, U.Banach, K.Witke and P.Reich., *Phys. Chem. Glasses*, 1991, **32**, 58.
59. B.N.Meera, A.K.Sood, N.Chandrabhas and J.Ramakrishna, *J. Non-Cryst. Solids*, 1990, **126**, 224.
60. G.A.Khan and C.A.Hogarth, *J. Mater. Sci.*, 1991, **26**, 799.
61. S.W.Martin and D.R.Bloyer, *J. Am. Ceram. Soc.*, 1991, **74**, 1003.
62. C.Dohmeier, C.Robl, M.Tacke and H.Schnöckel, *Angew. Chem., Int. Ed. Engl.*, 1991, **30**, 564.
63. A.S.Grady, S.G.Puntambekar and D.K.Russell, *Spectrochim Acta, A*, 1991, **47A**, 47.
64. M.A.Lefcourt and G.A.Ozin, *J. Phys. Chem.*, 1991, **95**, 2623.
65. A.M.Le Quéré, C.Su and L.Manceron, *J. Phys. Chem.*, 1991, **95**, 3031.
66. G.A.Atiyah, A.S.Grady, D.K.Russell and T.A.Claxton, *Spectrochim. Acta, A*, 1991, **47A**, 467.
67. W.Uhl and J.E.O.Schnepf, *Z. anorg. allg. Chem.*, 1991, **595**, 225.
68. A.S.Grady, R.D.Markwell, D.K.Russell and A.C.Jones, *J. Cryst. Growth*, 1990, **106**, 239.
69. K.J.L.Paciorek, J.H.Nakahara, L.A.Hoferkamp, C.George, J.L.Flippen-Anderson, R.Gilardi and W.R.Schmidt, *Chem. Mater.*, 1991, **3**, 82.
70. M.Somer, K.Peters, T.Popp and H.G.von Schnering, *Z. anorg. allg. Chem.*, 1991, **597**, 201.
71. W.Blase, G.Cordier, K.Peters, M.Somer and H.G.von Schnering, *Angew. Chem., Int. Ed. Engl.*, 1991, **30**, 326.
72. G.V.Chertikhin, L.V.Serebrennikov and V.F.Shevel'kov, *Zh. Fiz. Khim.*, 1991, **65**, 1078.
73. L.M.Sharygin, A.V.Korenkova, S.M.Vovk and E.I.Zlokazova, *Russ. J. Inorg. Chem.*, 1991, **36**, 171.
74. W.R.Taylor, *Proc. Ind. Acad. Sci., Earth Planet. Sci.*, 1990, **99**, 99.

75. M.A.Einarsrud, E.Rytter and M.Ystenes, *Vib. Spectrosc.*, 1990, **1**, 61.
76. S.V.Kuznetsov, N.T.Kuznetsov, Yu.I.Krasilov and I.V.Krumins, *Russ. J. Inorg. Chem.*, 1991, **36**, 251.
77. A.M.Hofmeister and A.Chopelas, *Phys. Chem. Miner.*, 1991, **17**, 503.
78. S.Milicev and A.Rahten, *Eur. J. Solid State Inorg. Chem.*, 1991, **28**, 557.
79. A.S.Korubitsyn, V.P.Kondakov, T.A.Permyakova, T.A.Ust'yantseva, I.A.Leont'eva and Z.I.Shishkina, *Izv. Akad. Nauk SSSR, Neorg. Mater.*, 1991, **27**, 108.
80. B.Gilbert, Y.Chauvin and I.Guibard, *Vib. Spectrosc.*, 1991, **1**, 219.
81. E.D.Samsonova, S.B.Osin and V.F.Shevel'kov, *Vestn. Mosk. Univ., Ser. 2: Khim.*, 1991, **32**, 122.
82. J.C.Beamish, A.Boardman and I.J.Worrall, *Polyhedron*, 1991, **10**, 95.
83. R.D.Urban, H.Birk, P.Polomsky and H.Johns, *J. Chem. Phys.*, 1991, **94**, 2523.
84. A.S.Grady, R.D.Markwell and D.K.Russell, *J. Chem. Soc., Chem. Comm.*, 1991, 14.
85. M.J.Henderson, C.H.L.Kennard, C.L.Raston and G.Smith, *J. Chem. Soc., Chem. Comm.*, 1990, 1203.
86. C.R.Pulham, A.J.Downs, M.J.Goode, D.W.H.Rankin and H.E.Robertson, *J. Am. Chem. Soc.*, 1991, **113**, 5149.
87. G.G.Hoffmann and R.Fischer, *Z. anorg. allg. Chem.*, 1990, **590**, 181.
88. D.L.W.Kwoh and R.C.Taylor, *Spectrochim. Acta, A*, 1991, **47A**, 409.
89. E.A.Piocos and B.S.Ault, *J. Phys. Chem.*, 1991, **95**, 6827.
90. C.M.Shaw and J.E.Shelby, *Phys. Chem. Glasses*, 1991, **32**, 48.
91. F.Miyaji and S.Sakka, *J. Non-Cryst. Solids*, 1991, **134**, 77.
92. Y.Fujiki, M.Tatanabe, T.Sasaki and S.Takenouchi, *Nippon Seramikkusu Kyokai Gakujutsu Ronbunshi*, 1990, **98**, 1245 (*Chem. Abs.*, 1991, **114**, 92212).
93. S.Barnier, M.Palazzi, M.Massot and C.Julien, *Solid State Ionics*, 1990, **44**, 81.
94. K.Ruhlandt-Senge and P.P.Power, *Inorg. Chem.*, 1991, **30**, 3683.
95. H.Uehara, K.Horiai, K.Nakagawa and H.Suguro, *Chem. Phys. Lett.*, 1991, **178**, 553.
96. B.J.Buke, T.P.Hamilton and H.F.Schaffer, *Inorg. Chem.*, 1991, **30**, 4275.
87. B.Neumüller and F.Gaulmann, *J. Organometal. Chem.*, 1991, **414**, 271.
98. B.Neumüller, *Z. anorg. allg. Chem.*, 1991, **592**, 42.
99. S.I.Radautsan, N.N.Syrbu, V.E.L'vin and F.G.Donika, *Fiz. Tekh. Poluprovodn., (Leningrad)*, 1990, **24**, 1592 (*Chem. Abs.*, 1991, **114**, 52192).
100. M.A.Marques, M.A.Sousa Oliveira, J.R.Rodrigues, R.M.Cavagnat and J.Devaure, *J. Chem. Soc., Farad. Trans.*, 1990, **86**, 3883.
101. B.Krebs and A.Brömmelhaus, *Z. anorg. allg. Chem.*, 1991, **595**, 167.
102. H.Borgholte, K.Dehnicke, H.Goesmann and D.Fenske, *Z. anorg. allg. Chem.*, 1991, **600**, 7.
103. M.R.Bermejo, M.D.Varela, B.Fernandez, M.I.Fernandez and M.Gayoso, *An. Quim.*, 1990, **86**, 408.
104. C.J.Chu, M.P.D'Evelyn, R.H.Hauge and J.L.Margrave, *J. Mater. Res.*, 1990, **5**, 2405.
105. A.Tardieu, F.Cansell and J.P.Petitet, *J. Appl. Phys.*, 1990, **68**, 3243.
106. J.Szczepanski and M.Vala, *J. Phys. Chem.*, 1991, **95**, 2792.
107. C.A.Schuttenmaer, R.C.Cohen, N.Pugliano, J.R.Heath, A.L.Cooksy, K.L.Busarow and R.J.Saykally, *Science*, 1990, **249**, 897.
108. J.M.L.Martin, J.P.Francois and R.Gijbels, *J. Chem. Phys.*, 1990, **93**, 5037.
109. P.Botschwina and H.P.Reisenauer, *Chem. Phys. Lett.*, 1991, **183**, 217.
110. D.McNaughton, D.McGilvery and F.Shanks, *J. Mol. Spectrosc.*, 1991, **149**, 458.
111. L.N.Shen, P.A.Withey and W.R.M.Graham, *J. Chem. Phys.*, 1991, **94**, 2395.
112. P.A.Withey, L.N.Shen and W.R.M.Graham, *J. Chem. Phys.*, 1991, **95**, 820.
113. J.M.L.Martin, J.P.Francois and R.Gijbels, *J. Chem. Phys.*, 1991, **94**, 3753.

114. J.M.L.Martin, J.P.Francois and R.Gijbels, *J. Comput. Chem.*, 1991, **12**, 52.
115. J.M.L.Martin, J.P.Francois and R.Gijbels, *J. Chem. Phys.*, 1990, **93**, 8850.
116. J.R.Heath, R.A.Sheeks, A.L.Cooksy and R.J.Saykally, *Science*, 1990, **249**, 895.
117. J.R.Heath and R.J.Saykally, *J. Chem. Phys.*, 1991, **94**, 1724.
118. J.R.Heath and R.J.Saykally, *J. Chem. Phys.*, 1990, **93**, 8392.
119. D.Bakowies and W.Thiel, *Chem. Phys.*, 1991, **151**, 309.
120. D.S.Bethune, G.Meijer, W.C.Tang, H.J.Rosen, W.G.Golden, H.Seki, C.A.Brown., *Chem. Phys. Lett.*, 1991, **179**, 181.
121. W.Kretschmer, L.D.Lamb, K.Fostiropoulos and D.R.Huffman, *Nature*, 1990, **347**, 354.
122. H.Ajie, M.M.Alvarez, S.J.Anz, R.D.Beck, F.Diederich, K.Fostiropoulos, D.R.Huffman, W.Krätschmer, Y.Rubin, K.E.Schriver, D.Sensharma and R.L.Whetten, *J. Phys. Chem.*, 1990, **94**, 8630.
123. J.P.Hare, T.J.Dennis, H.W.Kroto, R.Taylor, A.W.Allaf, S.Balm and D.R.M.Walton, *J. Chem. Soc., Chem. Comm.*, 1991, 412.
124. D.S.Bethune, G.Meijer, W.C.Tang and H.J.Rosen, *Chem. Phys. Lett.*, 1990, **174**, 219.
125. C.I.Frum, R.R.Engleman, H.G.Hedderich, P.F.Bernath, L.D.Lamb and D.R.Huffman, *Chem. Phys. Lett.*, 1991, **176**, 504.
126. T.J.Dennis, J.P.Hare, H.W.Kroto, R.Taylor, D.R.M.Walton and P.J.Hendra, *Spectrochim. Acta, A*, 1991, **47A**, 1289.
127. G.B.Adams, J.B.Page, O.F.Sankey, K.Sinha, J.Menendez and D.R.Huffman, *Phys. Rev., B*, 1991, **33**, 4052.
128. D.M.Cox, S.Behal, M.Disko, S.M.Gorun, M.Greaney, C.S.Hsu, E.B.Kollin, J.Millar, J.Robbins, W.Robbins, R.D.Sherwood and P.Tindall, *J. Am. Chem. Soc.*, 1991, **113**, 2940.
129. V.Huang, D.F.R.Gilson and I.S.Butler, *J. Phys. Chem.*, 1991, **95**, 5723.
130. K.Sinha, J.Menendez, G.B.Adams, J.B.Page, O.F.Sankey, L.D.Lamb and D.R.Huffman, *Proc. SPIE - Int. Soc. Opt. Eng.*, 1991, **1437**, 32.
131. R.L.Garrell, T.M.Herne, C.A.Szafranski, F.Diederich, F.Ettl and R.L.Whetten, *J. Am. Chem. Soc.*, 1991, **113**, 6302.
132. A.M.Vassallo, L.S.K.Pang, P.A.Cole-Clarke and M.A.Wilson, *J. Am. Chem. Soc.*, 1991, **113**, 7820.
133. R.E.Haufler, J.Conceicao, L.P.F.Chibante, Y.Chai, N.E.Byrne, S.Flanagan, M.M.Haley, S.C.O'Brien, C.Pan, Z.Xiao, W.E.Billups, M.A.Ciufolini, R.H.Hauge, J.L.Margrave, L.J.Wilson, R.F.Curl and R.E.Smalley, *J. Phys. Chem.*, 1990, **94**, 8634.
134. A.Webster, *Nature,* 1991, **352**, 412.
135. J.Senekowitsch, S.Carter, P.Rosmus and H.J.Werner, *Chem. Phys.*, 1990, **147**, 281.
136. C.J.Pursell and D.P.Weliky, *J. Mol. Spectrosc.*, 1990, **143**, 251.
137. A.Gambi, P.Stoppa, S.Giorganni, A.De Lorenzi, R.Visinoni and S.Ghersetti, *J. Mol. Spectrosc.*, 1991, **145**, 29.
138. C.N.Jarman and H.W.Kroto, *J. Chem. Soc., Farad. Trans.*, 1991, **87**, 1815.
139. D.Papousek, J.F.Ogilvie, S.Civis and M.Winnewisser, *J. Mol. Spectrosc.* 1991, **149**, 109.
140. D.Papousek, R.Tesar, R.Pracna, S,Civiš, M.Winnewisser, S.P.Belov and M.Yu.Tret'yakov, *J. Mol. Spectrosc.*, 1991, **147**, 279.
141. D.Luckhaus and M.Quack, *Chem. Phys. Lett.*, 1991, **180**, 524.
142. F.Cappellani, G.Restelli and G.Tarrago, *J. Mol. Spectrosc.*, 1991, **146**, 326.
143. H.Najib, N.Bensari-Zizi, H.Bürger, G.Guelachvili and C.Alamichel, *Mol. Phys.*, 1990, **70**, 849.
144. J.Dupre-Maquaire, J.Dupre, F.Meyer, C.Meyer and M.Koivusaari, *J. Mol. Spectrosc.*, 1991, **146**, 369.
145. M.Ouahman, H.Najib and N.Bensari-Zizi, *Spectrochim. Acta, A*, 1991, **47A**, 493.
146. S.Alanko, *J. Mol. Spectrosc.*, 1991, **147**, 406.

147. O.N.Ulenikov, A.B.Malikova, G.A.Shevchenko, G.Guelachvili and M.Morillon-Chapey, *J. Mol. Spectrosc.*, 1991, *149*, 160.
148. M.Feher, C.Salud and J.P.Maier, *J. Chem. Phys.*, 1991, **94**, 5377.
149. M.Feher, C.Salud and J.P.Maier, *J. Mol. Spectrosc.*, 1991, **145**, 246.
150. K.Kawaguchi, M.Kajita, K.Tanaka and E.Hirota, *J. Mol. Spectrosc.*, 1990, **144**, 451.
151. A.Quiniones, R.S.Ram and J.W.Bevan, *J. Chem. Phys.*, 1991, **95**, 3980.
152. M.Niedenhoff, K.M.T.Yamada, G.Winnewisser, K.Tanaka and T.Okabayashi, *J. Mol. Spectrosc.*, 1991, **145**, 290.
153. C.Degli Esposti and L.Favero, *J. Mol. Spectrosc.*, 1991, **147**, 84.
154. W.J.Balfour, S.G.Fougere and D.Klapstein, *Spectrochim. Acta, A*, 1991, **47A**, 1127.
155. K.Ohno and H.Matsuura, *J. Mol. Struct.*, 1991, **242**, 303.
156. J.S.Craw and W.B.De Olmeida, *Chem. Phys. Lett.*, 1991, **177**, 517.
157. A.Le Floch, *Mol. Phys.*, 1991, **72**, 133.
158. W.E.Thompson and M.E.Jacox, *J. Chem., Phys.*, 1991, **95**, 735.
159. C.Camy-Peyret, J.M.Flaud, A.Goldman, F.J.Murcray, R.D.Blatherwick, F.S.Bonomo, D.G.Murray and C.P.Rinsland, *J. Mol. Spectrosc.*, 1991, **149**, 481.
160. H.G.Mack, C.O.Della Vedova and H.Oberhammer, *Angew. Chem., Int. Ed. Engl.*, 1991, **30**, 1145.
161. A.Baudet, J.Dommen, H.Hollenstein, D.Luckhaus and M.Quack, *J. Mol. Spectrosc.*, 1990, **143**, 268.
162. C.O.Della Vedova, *J. Raman Spectrosc.*, 1991, **22**, 291.
163. F.Holland and M.Winnewisser, *J. Mol. Spectrosc.*, 1991, **149**, 45.
164. V.I.Lang and J.S.Winn., *J. Chem. Phys.*, 1991, **94**, 5270.
165. L.S.Masukidi, J.G.Lahaye and A.Fayt, *J. Mol. Spectrosc.*, 1991, **148**, 281.
166. A.M.Tolonen, V.M.Horneman and S.Alanko, *J. Mol. Spectrosc.*, 1990, **144**, 18.
167. A.G.Maki, J.S.Wells and D.A.Jennings, *J. Mol. Spectrosc.*, 1990, **144**, 224.
168. T.L.Tan and E.C.Looi, *J. Mol. Spectrosc.*, 1991, **148**, 262.
169. T.L.Tan, E.C.Looi and K.T.Lua, *J. Mol. Spectrosc.*, 1991, **148**, 265.
170. L.P.Giver and C.Chackerian, *J. Mol. Spectrosc.*, 1991, **148**, 80.
171. S.W.Sharpe, D.Reifschneider, C.Wittig and R.A.Beaudet, *J. Chem. Phys.*, 1991., **94**, 233.
172. M.E.Jacox and W.E.Thompson, *J. Phys. Chem.*, 1991, **95**, 2781.
173. F.Holland and M.Winnewisser, *J. Mol. Spectrosc.*, 1991, **147**, 496.
174. H.Bürger and S.Sommer, *J. Chem. Soc., Chem. Comm.*, 1991, 456.
175. L.M.LeBlanc and A.Anderson, *J. Raman Spectrosc.*, 1991, **22**, 355.
176. L.A.Zhigula, T.D.Kolomiitsova and S.M.Melikova, *Mol. Spektrosk.*, 1990, **8**, 198.
177, S.Giorgianni, P.Stoppa, A.Baldacci, A.Gambi and S.Ghersetti, *J. Mol. Spectrosc.*, 1991, **150**, 184.
178. L.M.LeBlanc and A.Anderson, *J. Raman Spectrosc.*, 1990, **21**, 693.
179. A.Baldacci, S.Giorgianni, P.Stoppa, A.De Lorenzi and S.Ghersetti, *J. Mol. Spectrosc.*, 1991, **147**, 208.
180. S.Giorgianni, A.Gambi, A.Baldacci, A.De Lorenzi and S.Ghersetti, *J. Mol. Spectrosc.*, 1990, **144**, 230.
181. S.Durai, V.N.Kumar and S.Mohan, *Orient. J. Chem.*, 1989, **5**, 223,
182, H.Vančik, K.Percač and D.E.Sunko, *J. Am. Chem. Soc.*, 1990, **112**, 7418.
183. K.Hassler and M.Pöschl, *Monatsh.*, 1990, **121**, 365.
184. A.Ben Altabef and R.Escribano, *Spectrochim. Acta, A*, 1991, **47A**, 455.
185. K.Hassler and M.Pöschl, *Spectrochim. Acta, A*, 1991, **47A**, 439.
186. O.A.Rusaeva, M.V.Bovtun, L.V.Khristenko, Yu.A.Pentin and A.V.Yarkov, *Zh. Fiz. Khim.*, 1990, **64**, 2070.
187. S.Furukawa, *J. Phys.: Condens. Matter*, 1990, **2**, 9209.

188, P.Gupta, A.C.Dillon, A.S.Bracker and S.M.George, *Surf. Sci.*, 1991, **245**, 360.
189. A.C.Dillon, P.Gupta, M.B.Robinson, A.S.Bracker and S.M.George, *J. Vac. Sci. Technol., A*, 1991, **9**, 2222.
190. S.Hasegawa, M.Matsuda and Y.Kurata, *Appl. Phys. Lett.*, 1990, **57**, 2211.
191. J.R.Durig, J.F.Sullivan, G.A.Guirgis and M.A.Qtaitat, *J. Phys. Chem.*, 1991, **95**, 1563.
192. K.J.Uram and U.Jansson, *Surf. Sci.*, 1991, **249**, 105.
193. F.Lattanzi, C.Di Lauro and H.Bürger, *Mol. Phys.*, 1991, **72**, 575.
194. H.Prinz, W.A.Kreinber and G.Pierre, *Can. J. Phys.*, 1990, **68**, 551.
195. Q.Zhu, H.Qian, H.Ma and L.Halonen, *Chem. Phys. Lett.*, 1991, **177**, 261.
196. L.Chen, T.Goto and T.Hirai, *J. Mater. Sci.*, 1990, **25**, 4273.
197. J.D.Presilla-Marquez, W.R.M.Graham and R.A.Shepherd, *J. Chem. Phys.*, 1990, **93**, 5424.
198. V.S.Nikitin, M.V.Polyakova, I.I.Baburina, A.V.Belyakov, E.T.Bogoradovskii and V.S.Zavgorodnii, *Spectrochim. Acta, A*, 1990, **46A**, 1669.
199. J.R.Durig, J.F.Sullivan and M.A.Qtaitat, *J. Mol. Struct.*, 1991, **243**, 239.
200. D.Fischer, M.Klostermann and K.L.Öhme, *J. Raman Spectrosc.*, 1991, **22**, 19.
201. B.W.Sheldon and J.S.Haggerty, *J. Am. Ceram. Soc.*, 1991, **74**, 1417.
202. Yu.L.Frolov and M.G.Voronkov, *Metalloorg. Khim.*, 1990, **3**, 1038.
203. E.Popowski, P.Kosse, H.Kelling and H.Jancke, *Z. anorg. allg. Chem.*, 1991, **594**, 179.
204. A.Hammel, H.V.Volden, A.Haaland, J.Weidlein and R.Reischmann, *J. Organometal. Chem.*, 1991,**408**, 35.
205. A.I.Popov, M.D.Val'kovskii, V.F.Sukhoverkhov, N.A.Chumaevskii, A.V.Sakharov, V.O.Gel'mbol'dt and A.A.Ennan, *Russ. J. Inorg. Chem.*, 1991, **36**, 208.
206. M.Baudler, W.Oelhert and K.-F.Tebbe, *Z.anorg. allg. Chem.*, 1991, **598-9**, 9.
207. M.McGonigal and V.M.Bermudez, *Surf. Sci.*, 1991, **241**, 357.
208. W.B.De Almeida and P.J.O'Malley, *J. Mol. Struct.*, 1991, **246**, 179.
209. I.S.Ignat'ev, A.N.Lazarev, T.F.Tenisheva and B.F.Shchegolev, *J. Mol. Struct.*, 1991, **244**, 193.
210. U.Wannagat, G.Bogedain, A.Schervan, H.C.Marsmann, D.J.Brauer, H.Bürger, F.Durrenbach, G.Pawele, C.Krüger and K.H.Claus, *Z. Naturforsch., B*, 1991, **46b**, 931.
211. R.M.Almeida and C.G.Pantano, *Proc. SPIE - Int. Soc. Opt. Eng.*, 1990, **1328**, 329.
212. Y.Hao, Y.Zhang, L.Tao and L.Zheng, *Cuihua Xuebao*, 1990, **11**, 394. (*Chem. Abs.*, 1990, **113**, 236391).
213. T.P.Mernagh, *J. Raman Spectrosc.*, 1991, **22**, 453.
214. R.D.Hosung and R.H.Doremus, *J. Mater. Res.*, 1990, **5**, 2209.
215. J.B.Nicholas, A.J.Hopfinger, F.E.Trouw and L.E.Iton, *J. Am. Chem. Soc.*, 1991, **113**, 4792.
216. P.K.Dutta and J.Twu, *J. Phys. Chem.*, 1991, **95**, 2498.
217. V.N.Bykov, V.N.Anfilogov, I.B.Bobylev and O.A.Berezikova, *Rasplavy*, 1990, 31 (*Chem. Abs.*, 1991, **114**, 48028).
218. B.O.Mysen, *Am. Mineral.*, 1990, **75**, 120.
219. T.Furukawa and W.B.White, *J. Chem. Phys.*, 1991, **95**, 776.
220. A.J.G.Ellison and P.C.Hess, *J. Non-Cryst. Solids*, 1991, **127**, 247.
221. A.Chopelas, *Am. Mineral.*, 1991, **76**, 1101.
222. T.Fukui, C.Sakurai and M.Okuyama, *Nippon Kagaku Kaishi*, 1991, 281 (*Chem. Abs.*, 1991, **114**, 212612).
223. J.Carda, M.I.Burguete, G.Monros, P.Escribano and J.Alarcon, *J. Eur. Ceram. Soc.*, 1990, **6**, 97.
224. C.C.Perry, X.Li and D.N.Waters, *Spectrochim. Acta, A*, 1991, **47A**, 1487.
225. C.I.Frum, R.Englemann and P.F.Bernath, *J. Chem. Phys.*, 1990, **93**, 5457.
226. H.Birk and H.Jones, *Chem. Phys. Lett.*, 1990, **175**, 536.

227. M.Himmrich and H.Haeuseler, *Spectrochim. Acta, A*, 1991, **47A**, 933.
228. T.S.Little, M.Qtaitat, M.Dakkouri and A.Dakkouri, *J. Raman Spectrosc.*, 1990, **21**, 591.
229. J.Hu, L.J.Schaad and B.A.Hess, *J. Am. Chem. Soc.*, 1991, **113**, 1463.
230. V.O.Gel'mbol'dt, *Russ. J. Inorg. Chem.*, 1990, **35**, 1611.
231. C.P.Tripp and M.L.Hair, *Langmuir*, 1991, **7**, 923.
232. J.Leitner, J.Mikulec and C.Černý, *Coll. Czech. Chem. Commun.*, 1990, **55**, 2432.
233. G.Gereshikov, N.Yu.Subbotina and M.Hargittai, *J. Mol. Spectrosc.*, 1990, **143**, 293.
234. Y.Akiyama, K.Tanaka and T.Tanaka, *J. Chem. Phys.*, 1991, **94**, 3280.
235. M.Petri, U.Simon, W.Zimmermann, W.Urban, J.P.Towle and M.Brown, *Mol. Phys.*, 1991, **72**, 315.
236. J.E.Crowell and G.Lu, *J. Electron Spectrosc. Relat. Phenom.*, 1990, **54-5**, 1045.
237. R.Escribano, G.Graner and H.Bürger, *J. Mol. Spectrosc.*, 1991, **146**, 83.
238. H.Bürger and G.Graner, *J. Mol. Spectrosc.*, 1991, **149**, 491.
239. L.I.Khirunenko, V.I.Shakhovtsov, V.K.Shinkarenko and F.M.Vorobkaol, *Fiz. Tekhn. Poluprovodn. (Leningrad)*, 1990, **24**, 1051.
240. C.Liu, F.Xu and L.Liu, *J. Phys.: Condens. Matter*, 1991, **3**, 1293.
241. D.J.Durben and G.H.Wolf, *Phys. Rev., B*, 1991, **43**, 2355.
242. M.Madon and G.D.Price, *J. Geophys. Res.*, 1989, **94**, 15687.
243. P.Gillet, F.Guyot and J.M.Malezieux, *Phys. Earth Planet. Inter.*, 1989, **58**, 141.
244. G.Lan, H.Sun, Z.Yin, J.Wang and H.Wang, *Phys. Stat. Solidi, B*, 1991, **164**, 39.
245. V.N.Khabashesku, S.E.Boganov, P.S.Zuev, O.M.Nefedov, J.Tamas, A.Gomory and I.Beseyei, *J. Organometal. Chem.*, 1991, **402**, 161.
246. R.Kuppe and H.Schnöckel, *Z.anorg. allg. Chem.*, 1991, **592**, 179.
247. J.E.Drake, A.G.Mislankar and M.L.Y.Wong, *Inorg. Chem.*, 1991, **30**, 2174.
248. L.Tichy, H.Ticha, A.Pačesova and J.Petzlt, *J. Non-Cryst. Solids*, 1991, **128**, 191.
249. C.G.Barraclough, R.W.Cockman and T.A.O'Donnell, *Inorg. Chem.*, 1991, **30**, 343.
250. S.A.Zaitsev, S.B.Osin, V.A.Koryazhkin and V.F.Shevel'kov, *Vestn, Mosk. Univ., Ser. 2: Khim.*, 1990, **31**, 128.
251. U.Simon, M.Petri, W.Zimmermann and W.Urban, *Mol. Phys.*, 1990, **71**, 1163.
252. U.Simon, M.Petri, W.Zimmermann, G.Huhn and W.Urban, *Mol. Phys.*, 1991, **73**, 1051.
253. A.Tabyaoui, B.Lavorel, G.Pierre and H.Burger, *J. Mol. Spectrosc.*, 1991, **148**, 100.
254. M.Rannenberg, J.Weidlein and A.Obermeyer, *Z. Naturforsch., B*, 1991, **46b**, 459.
255. A.S.Sall and L.Diop, *Inorg. Chim. Acta*, 1990, **171**, 53.
256. O.S.Jung, J.H.Jeong and Y.S.Sohn, *J.Organometal. Chem.*, 1990, **397**, 17.
257. W.Linert, V.Gutmann and A.Sotriffer, *Vib. Spectrosc.*, 1990, **1**, 199.
258. K.Schenzel, A.Kolbe and P.Reich, *Monatsh.*, 1990, **121**, 615.
259. A.Alvarez-Valdés, M.J.Camazón, M.C.Navarro-Ranninger and T.Torres, *Z. Naturforsch., B*, 1990, **45b**, 1043.
260. M.J.Hampden-Smith, T.A.Wark, A.Rheingold and J.C.Huffman, *Can. J. Chem.*, 1991, **69**, 121.
261. G.K.Sandhu, N.Sharma and E.R.T.Tiekink, *J. Organometal. Chem.*, 1991, **403**, 119.
262. H.Birk and H.Jones, *Chem. Phys. Lett.*, 1991, **181**, 245.
263. D.Tudela, M.A.Khan and J.J Zuckerman, *J. Chem. Soc., Dalton Trans.*, 1991, 999.
264. M.H.Kuok and L.H.Lim, *J. Raman Spectrosc.*, 1990, **21**, 675.
265. W.Linert, A.Sotriffer and V.Gutmann, *J. Coord. Chem.*, 1990, **22**, 21.
266. D.Rohm, C.J.H.Schutte and J.Shamir, *S. Afr. J. Chem.*, 1990, **43**, 110.
267. B.N.Ivanov-Emin, A.M.Il'inets, B.E.Zaitsev, A.V.Kostrikin, F.M.Spiridonov and V.P.Dolganev, *Russ. J. Inorg. Chem.*, 1990, **35**, 1301.
268. Y.Abid, M.Kamoun, A.Daoud and F.Romain, *J. Raman Spectrosc.*, 1990, **21**, 709.
269. S.H.Morgan, D.O.Henderson and R.H.Magruder, *J. Non-Cryst. Solids*, 1991, **128**, 146.

270. A.Tabyaoui, B.Lavorel, G.Millot, R.Saint-Loup, R.Chaux and H.Berger, *J. Raman Spectrosc.*, 1990, **21**, 809.
271. W.E.Thompson and M.E.Jacox, *J. Chem. Phys.*, 1990, **93**, 3856.
272. H.Sieber, H.J.Neusser, F.Stroh and W.Winnewisser, *J. Mol. Spectrosc.*, 1991, **148**, 453.
273. G.Schatte and H.Willner, *Z. Naturforsch., B*, 1991, **46b**, 483.
274. J.Bendtsen and F.M.Nicolaisen, *J. Mol. Spectrosc.*, 1991, **145**, 123.
275. H.E.Warner and T.Amano, *J. Mol. Spectrosc.*, 1991, **145**, 66.
276. M.Elhanine, R.Farrenq and G.Guelachvili, *J. Chem. Phys.*, 1991, **94**, 2529.
277. S.S.Lee and T.Oka, *J. Chem. Phys.*, 1991, **94**, 1698.
278. G.A.Yeo and T.A.Ford, *Spectrochim. Acta, A*, 1991, **47A**, 485.
279. G.A.Yeo and T.A.Ford, *Can. J. Chem.*, 1991, **69**, 632.
280. G.A.Yeo and T.A.Ford, *Spectrochim. Acta, A*, 1991, **47A**, 919.
281. D.De Waal, A.M.Heyns, K.J.Range and C.Eglmeier, *Spectrochim. Acta, A*, 1990, **46A**, 1639.
282. D.De Waal, A.M.Heyns, K.J.Range and C.Eglmeier, *Spectrochim. Acta, A*, 1990, **46A**, 1649.
283. J.P.Perchard, R.B.Bohn and L.Andrews, *J. Phys. Chem.*, 1991, **95**, 2707.
284. A.J.Barnes, *J. Mol. Struct.*, 1990, **237**, 19.
285. J.M.Price, H.W.Crofton and Y.T.Lee, *J. Phys. Chem.*, 1991, **95**, 2182.
286. N.Ohashi and W.B.Olson, *J. Mol. Spectrosc.*, 1991, **145**, 383.
287. L.V.Serbinovskaya, L.V.Kim, Kh.Kh.Muldagaliev and O.V.Agashkin, *Zh. Fiz. Khim.*, 1991, **65**, 1831.
288. W.C.Ho, I.Ozier, D.T.Cramb and M.C.L.Gerry, *J. Mol. Spectrosc.*, 1991, **149**, 559.
289. M.E.Jacox and W.E.Thompson, *J. Chem. Phys.*, 1990, **93**, 7609.
290. R.Minkwitz, D.Bernstein, H.Preut and P.Sartori, *Inorg. Chem.*, 1991, **30**, 2157.
291. C.Alamichel and J.Verges, *Spectrochim. Acta, A*, 1991, **47A**, 915.
292. L.H.Jones and B.I.Swanson, *J. Phys. Chem.*, 1991, **95**, 86.
293. M.E.Jacox, *J. Chem. Phys.*, 1990, **93**, 7622.
294. Y.Hamada, *J. Mol. Struct.*, 1991, **242**, 367.
295. I.Kleiner, J.M.Guilmot, M.Carleer and M.Herman, *J. Mol. Spectrosc.*, 1991, **149**, 341.
296. N.Kawaguchi, T.Ishiwata, I.Tanaka and E.Hirota, *Chem. Phys. Lett.*, 1991, **180**, 436.
297. T.L.Tan, E.C.Looi, K.T.Lua, A.G.Maki, J.W.C.Johns and M.Noel, *J. Mol. Spectrosc.*, 1991, **149**, 425.
298. R.H.Smith, M.T.Leu and L.F.Keyser, *J. Phys. Chem.*, 1991, **95**, 5924.
299. S.F.A.Kettle, E.H.J.Lugwisha and L.J.Norrby, *Can. J. Appl. Spectrosc.*, 1990, **35**, 91.
300. B.M.Cheng, J.W.Lee and Y.P.Lee, *J. Phys. Chem.*, 1991, **95**, 2814.
301. A.Givan and A.Loewenschuss, *Struct. Chem.*, 1990, **1**, 579.
302. A.Givan and A.Loewenschuss, *J. Chem. Phys.*, 1990, **93**, 7592.
303. E.Rost and W.Sawodny, *Spectrochim. Acta, A*, 1990, **46A**, 1793.
304. J.V.Gilbert and L.J.Smith, *J. Phys. Chem.*, 1991, **95**, 7278.
305. H.G.von Schnering, M.Somer, G.Kliche, W.Hönle, T.Meyer, J.Wolf, L.Ohse and P.B.Kempa, *Z. anorg. allg. Chem.*, 1991, **601**, 13.
306. J.R.Durig, C.M.Whang and R.J.Harlan, *Vib. Spectrosc.*, 1990, **1**, 19.
307. T.H.Pai and V.F.Kalasinsky, *J. Raman Spectrosc.*, 1990, **21**, 607.
308. H.Beckers, J.Breidung, H.Bürger, R.Kuna, A.Rahner, W.Schneider and W.Thiel, *J. Chem. Phys.*, 1990, **93**, 4603.
309. Sh.Sh.Nabiev and V.D.Klimov, *Russ. J. Inorg. Chem.*, 1991, **36**, 713.
310. L.Jäger, S.Ahmed and H.Köhler, *Z. anorg. allg. Chem.*, 1990, **591**, 118.
311. I.A.Rozanov, L.Ya.Medvedeva and I.V.Goeva, *Russ. J. Inorg. Chem.*, 1990, **35**, 1416.
312. L.Koudelka, M.Pisarcik, L.N.Blinov and M.S.Gutenev, *J. Non-Cryst. Solids*, 1991, **134**,

86.
313. M.McCluskey and L.Andrews, *J. Phys. Chem.*, 1991, **95**, 2679.
314. M.McCluskey and L.Andrews, *J. Phys. Chem.*, 1991, **95**, 2988.
315. M.McCluskey and L.Andrews, *J. Phys. Chem.*, 1991, **95**, 3545.
316. S.A.Katsyuba, N.I.Monakheva, L.Kh.Ashrafullina, R.R.Shagidullin and S.A.Terent'eva, *Zh. Prikl. Spektrosk.*, 1991, **54**, 912.
317. A.Rulmont, R.Cahay, M.Liegeois-Duyckaerts and P.Tarte, *Eur. J. Solid State Inorg. Chem.*, 1991, **28**, 207.
318. V.D.Khalilev, K.G.Karapetyan, V.L.Bogdanov, E.B.Nasyrova and O.V.Yanush, *Fiz. Khim. Stekla*, 1990, **16**, 529.
319. M.C.R.Shastry and K.J.Rao, *Spectrochim. Acta, A*, 1990, **46A**, 1581.
320. D.Philip, B.L.George and G.Aruldhas, *J. Raman Spectrosc.*, 1990, **21**, 523.
321. I.H.Joe, G.Aruldhas and G.Keresztury, *J. Raman Spectrosc.*, 1991, **22**, 537.
322. D.Philip, G.Aruldhas, X.Mathew, V.U.Nayar, N.Boudjada and B.Lambert-Andron, *J. Raman Spectrosc.*, 1991, **22**, 45.
323. L.S.Ivashkevich, V.A.Lyutsko, T.N.Galkova and G.G.Shvarkova, *Russ. J. Inorg. Chem.*, 1990, **35**, 1025.
324. L.S.Ivashkevich and V.A.Lyutsko, *Russ. J. Inorg. Chem.*, 1990, **35**, 1630.
325. N.Santha and V.U.Nayar, *J. Raman Spoectrosc.*, 1990, **21**, 517.
326. S.Abraham and G.Aruldhas, *J. Raman Spectrosc.*, 1991, **22**, 245.
327. S.Mohan, *Ind. J. Pure Appl. Phys.*, 1990, **28**, 599.
328. Z.Mielke, L.Andrews, K.A.Nguyen and M.S.Gordon, *Inorg. Chem.*, 1990, **29**, 5096.
329. Z.Mielke, G.D.Brabson and L.Andrews, *J. Phys. Chem.*, 1991, **95**, 75.
330. L.Andrews, G.C.Reynolds, Z.Mielke and M.McCluskey, *Inorg. Chem.*, 1990, **29**, 5222.
331. F.Menzel, L.Ohse and W.Brockner, *Heteroatom. Chem.*, 1990, **1**, 357.
332. G.R.Burns, J.R.Rollo, J.D.Sarfrati and K.R.Morgan, *Spectrochim. Acta, A*, 1991, **47A**, 811.
333. K.B.Dillon, R.K.Harris, P.N.Gates, A.S.Muir and A.Root, *Spectrochim. Acta, A*, 1991, **47A**, 831.
334. S.A.Jarrett-Sprague and I.H.Hillier, *Chem. Phys.*, 1990, **148**, 325.
335. H.J.Stein, *Nucl. Instrum. Methods Phys. Res., Sect. B*, 1991, **B59-60**, 1106.
336. L.Andrews and T.C.McInnis, *Inorg. Chem.*, 1991, **30**, 2990.
337. J.Breidung, W.Thiel and A.Komornicki, *Inorg. Chem.*, 1991, **30**, 1067.
338. R.Minkwitz, M.Koch, J.Nowicki and H.Borrmann, *Z. anorg. allg. Chem.*, 1990, **590**, 93.
339. B.K.Kasenov, I.V.Ashlyaeva and A.Z.Beilina, *Dokl. Akad. Nauk SSSR*, 1990, **314**, 452.
340, I.V.Ashlyaeva, B.K.Kasenov and A.Z.Beilina, *Russ. J. Inorg. Chem.*, 1991, **36**, 611.
341. I.S.Perelygin and G.P.Mikhailov, *Zh. Prikl. Spektrosk.*, 1991, **55**, 319.
342. T.Klapotke and J.Passmore, *J. Chem. Soc., Dalton Trans.*, 1990, 3815.
343. N.Vlahovici, C.Pavelescu and I.Kleps, *J. Mater. Sci. Lett.*, 1991, **10**, 920.
344. I.L.Botto, I.B.Schalamuk, S.J.Ametrano and R.E.De Barrio, *An. Asoc. Quim. Argent.*, 1990, **78**, 195.
345. N.M.Palekha, V.P.Karlov, G.N.Butuzov and V.V.Klimov, *Russ. J. Inorg. Chem.*, 1990, **35**, 1095.
346. N.A.Mazhenov, Z.M.Muldakhmetov, B.Z.Nurgaliev and S.Kh.Fazylov, *Izv. Akad. Nauk SSSR, Neorg. Mater.*, 1990, **26**, 1952.
347. I.L.Botto, E.J.Baran, C.Cascales, I.Rasines and R.Saez Puche, *J. Phys. Chem. Solids*, 1991, **52**, 431.
348. K.Holz, F.Obst and R.Mattes, J. *Solid State Chem.*, 1991, **90**, 353.
349. M.Schäfer, J.Pebler, B.Borgsen, F.Weller and K.Dehnicke, *Z. Naturforsch., B*, 1990, **45b**, 1243.

350. H.Friedewold and W.Brockner, *Z. Naturforsch., A*, 1991, **46a**, 595.
351. Z.Lin, D.Wu and L.Zhang, *Jiegou Huaxue*, 1990, **9**, 291 (*Chem. Abs.*, 1991, **114**, 255968).
352. R.D.Urban, P.Polomsky and H.Jones, *Chem. Phys. Lett.*, 1991, **181**, 485.
353. E.H.Fink, K.D.Setzer, D.A.Ramsay and M.Vervloet, *Chem. Phys. Lett.*, 1991, **179**, 103.
354. T.A.Ivanova, V.P.Babaeva and V.Ya.Rosolovskii, *Russ. J. Inorg. Chem.*, 1990, **35**, 1398.
355. F.D.Hardcastle, I.E.Wachs, H.Eckert and D.A.Jefferson, *J. Solid State Chem.*, 1991, **90**, 194.
356. E.H.Fink, K.D.Setzer, D.A.Ramsay and H.Vervloet, *Chem. Phys. Lett.*, 1991, **179**, 95.
357. W.B.Carter, D.Schiferl, M.L.Lowe and D.Gonzales, *J. Phys. Chem.*, 1991, **95**, 2516.
358. A.Perrin, A.M.Vasserot, J.M.Flaud, C.Camy-Peyret, V.M.Devi, M.A.H.Smith, C.P.Rinsland, A.Barbe, S.Bouazza and J.J.Plateaux, *J. Mol. Spectrosc.*, 1991, **149**, 519.
359. V.M.Devi, A.Perrin, J.M.Flaud, C.Camy-Peyret, C.P.Rinsland and M.A.H.Smith, *J. Mol. Spectrosc.*, 1990, **143**, 381.
360. C.P.Rinsland, M.A.H.Smith, V.M.Devi, A.Perrin, J.M.Flaud and C.Camy-Peyret, *J. Mol. Spectrosc.*, 1991, **149**, 474.
361. A.Perrin, A.M.Vasserot, J.M.Flaud, C.Camy-Peyret, C.P.rinsland, M.A.H.Smith and V.M.Devi, *J. Mol. Spectrosc.*, 1990, **143**, 311.
362. A.Barbe, S.Bouazza, J.J.Plateaux, J.M.Flaud and C.Camy-Peyret, *J. Mol. Spectrosc.*, 1991, **150**, 255.
363. D.A.Greenhalgh and L.A.Rahn, *J. Raman Spectrosc.*, 1990, **21**, 847.
364. O.Kristiansson and J.Lindgren, *J. Phys. Chem.*, 1991, **95**, 1488.
365. M.Falk, *J. Raman Spectrosc.*, 1990, **21**, 563.
366. D.V.Luu, L.Cambon and M.Mathlouthi, *J. Mol. Struct.*, 1990, **237**, 411.
367. M.C.Shivaglal and S.Singh, *J. Mol. Liq.*, 1990, **46**, 297.
368. M.Pavlović, G.Baranović and D.Lovreković, *Spectrochim. Acta, A*, 1991, **47A**, 897.
369. A.Burneau, *J. Mol. Liq.*, 1990, **46**, 99.
370. J.P.Devlin, *J. Mol. Struct.*, 1990, **224**, 33.
371. I.Kanesaka, H.Hayashi, M.Kita and K.Kawai, *J. Chem. Phys.*, 1990, **93**, 6113.
322. G.Winnewisser and K.M.T.Yamada, *Vib. Spectrosc.*, 1991, **1**, 263.
373. J.R.Durig and M.M.Berganat, *Struct. Chem.*, 1990, **1**, 561.
374. K.Miaskiewicz and R.Steudel, *J. Chem. Soc., Dalton Trans.*, 1991, 2395.
375. W.C.Hodgeman, J.B.Weinrach and D.W.Bennett, *Inorg. Chem.*, 1991, **30**, 1611.
376. J.Milne, *J. Chem. Soc., Chem. Comm.*, 1991, 1048.
377. G.D.Brabson, Z.Mielke and L.Andrews, *J. Phys. Chem.*, 1991, **95**, 79.
378. R.Minkwitz, R.Krause, H.Härtner and W.Sawodny, *Z. anorg. allg. Chem.*, 1991, **593**, 137.
379. R.Minkwitz, J.Nowicki, W.sawodny and K.Härtner, *Spectrochim. Acta, A*, 1991, **47A**, 151.
380. G.E.Quelch, H.F.Schaefer and C.J.Marsden., *J. Am. Chem. Soc.*, 1990, **112**, 8719.
381. B.Siewert and V.Müller, *Z. anorg. allg. Chem.*, 1991, **595**, 211.
382. R.Steudel, B.Plinke, D.Jensen and F.Baumgart, *Polyhedron*, 1991, **10**, 1037.
383. J.D.Odom, A.M.Boccanfuso, M.M.Bergama, T.S.Little, and J.R.Durig, *J. Mol. Struct.*, 1990, **238**, 159.
384. B.Krebs, F.-P.Ahlers and E.Lührs, *Z. anorg. allg. Chem.*, 1991, **597**, 115.
385. F.P.Ahlers, E.Lührs and B.J.Krebs, *Z. anorg. allg. Chem.*, 1991, **594**, 7.
386. K.Dehnicke, E.Schmock, K.F.Köhler and G.Frenking, *Angew. Chem., Int. Ed. Engl.*, 1991, **30**, 577.
387. V.Müller, K.Dehnicke, D.Fenske and G.Baum, *Z. Naturforsch., B*, 1991, **46b**, 63.
388. H.Shimizu, Y.Nakamichi and S.Sasaki, *J. Chem. Phys.*, 1991, **95**, 2036.
389. P.G.Sennikov, V.E.Shkrunin and K.G.Tokhadze, *Vysokochist. Veshchestva*, 1991, 183 (*Chem. Abs.*, 1991, **114**, 195156).
390. P.G.Sennikov, V.E.Shkrunin and K.G.Tokadze, *J. Mol. Liq.*, 1990, **46**, 29.

391. M.A.Fender, Y.M.Sayed and F.T.Procheska, *J. Phys. Chem.*, 1991, **95**, 2811.
392. A.J.Downs, A.M.Forster, G.S.McGrady and B.J.Taylor, *J. Chem. Soc., Dalton Trans.*, 1991, 81.
393. A.Ben Altabef, J.Borrajo, P.J.Aymonino and E.L.Varetti, *Can. J. Appl. Spectrosc.*, 1990, **35**, 149.
394. R.Minkwitz and V.Gerhard, *Z. anorg. allg. Chem.*, 1990, **591**, 143.
395. H.E.Askew, P.N.Gates and A.S.Muir, *J. Raman Spectrosc.*, 1991, **22**, 265.
396. R.Minkwitz, J.Nowicki, B.Jahnkow and M.Koch, *Z. anorg. allg. Chem.*, 1991, **596**, 77.
397. A.Ben Altabef, E.H.Cutin and C.O.Della Vedova, *J.Raman Spectrosc.*, 1991, **22**, 297.
398. J.Jacobs, G.S.McGrady, H.Willner, D.Christan, H.Oberhammer and P.Zylka, *J. Mol. Struct.*, 1991, **245**, 275.
399. R.Minkwitz and V.Gerhard, *Z.Naturforsch., B*, 1991, **46b**, 265.
400. A.G.Maki and Yu.Kuritsyn, *J. Mol. Spectrosc.*, 1990, **144**, 242.
401. M.H.Brooker and J.Chen, *Spectrochim. Acta, A*, 1991, **47A**, 315.
402. C.Z.Lu, J.Goldman, S.Deliwala, K.H.Che and E.Mazur, *Chem. Phys. Lett.*, 1991, **176**, 355.
403. S.Li and Y.-S.Li, *Spectrochim. Acta, A*, 1991, **47A**, 201.
404. B.S.Ault, *J. Mol. Struct.*, 1990, **238**, 111.
405. Y.P.Kuo, B.M.Cheng and Y.P.Lee, *Chem. Phys. Lett.*, 1991, **177**, 195.
406. S.M.F.Payami, A.R.Prabakharan and S.Gunesekaran, *Orient. J. Chem.*, 1990, **6**, 233.
407. L.Schriver, D.Carrere, A.Schriver and K.Jaeger, *Chem. Phys. Lett.*, 1991, **181**, 505.
408. S.Gunesekaran, K.Manoharan anf S.Mohan, *Asian J. Chem.*, 1990, **2**, 318.
409. M.Augustin and C.Herrmann, *Wiss. Z. Martin Luher Univ., Halle-Wittenberg, Math.-Naturwiss. Reihe*, 1990, **39**, 115 (*Chem. Abs.*, 1991, **115**, 49654).
410. U.A.Jayasooriya, S.F.Akettle, S.Mahasuverachai and O.Al-Jowder, *J. Chem. Phys.*, 1991, **94**, 5946.
411. B.J.Reedy, J.K.Beattie and R.T.Lowsin, *Geochim. Cosmochim. Acta*, 1991, **55**, 1609.
412. S.V.Stefanovskii, I.A.Ivanov and A.N.Gulin, *Fiz. Khim. Stekla*, 1991, **17**, 120.
413. A.K.Brisdon, R.A.Gomme and J.S.Ogden, *J. Phys. Chem.*, 1991, **95**, 2927.
414. B.Krebs, E.Lührs, R.Willmer and F.P.Ahlers, *Z. anorg. allg. Chem.*, 1991, **592**, 17.
415. R.Steudel, D.Jensen and F.Baumgart, *Polyhedron*, 1990, **9**, 1199.
416. M.J.Almond, C.A.Yates, D.A.Rice, P.J.Hendra and P.T.Brain, *J. Mol. Struct.*, 1990, **239**, 69.
417. K.K.Verma, A.Saini and O.P.Agrawal, *Synth. React. Inorg. Met.-Org. Chem.*, 1990, **20**, 559.
418. A.M.Grekhov, V.D.Khravryuchenko, N.I.Deryugina and N.I.Kashirina, *Izv. Akad. Nauk SSSR, Neorg. Mater.*, 1990, **26**, 1897.
419. A.Jayaraman and G.A.Kourouklis, *Pramana*, 1991, **36**, 133 (*Chem. Abs.*, 1991, **114**, 196820).
420. N.Santha and V.U.Nayar, *J. Raman Spectrosc.*, 1990, **21**, 765.
421. D.Philip, S.Abraham and G.Aruldhas, *J. Raman Spectrosc.*, 1990, **21**, 521.
422. B.Borgsen, F.Weller and K.Dehnicke, *Z. anorg. allg. Chem.*, 1991, **596**, 55.
423. J.B.Milne, E.J.Gabe and C.Bensimon, *Can. J. Chem.*, 1991, **69**, 648.
424. B.Krebs, S.Bonmann and K.Erpenstein, *Z. Naturforsch., B*, 1991, **46b**, 919.
425. D.Goddon, A.Groh, H.J.Hansen, M.Schneider and W.Urban, *J. Mol. Spectrosc.*, 1991, **147**, 392.
426. P.Jensen, P.J.Bunker, A.Karpfen, M.Kofranek and H.Lischka, *J. Chem. Phys.*, 1990, **93**, 6266.
427. A.McIlroy, R.Lascola, C.M.Lovejoy and D.J.Nesbitt, *J. Phys. Chem.*, 1991, **95**, 2636.
428. D.J.Nesbitt and C.M.Lovejoy, *J. Chem. Phys.*, 1991, **93**, 7716.

429. T.Zeegers-Huyskens and L.Sobczyk, *Spectrochim. Acta, A*, 1990, **46A**, 1693.
430. R.D.Hunt, L.Andrews and L.M.Toth, *J. Phys. Chem.*, 1991, **95**, 1183.
431. K.Burczyk, H.Bürger, M.Le Guennec, G.Wlodarczak and J.Demaison, *J. Mol. Spectrosc.*, 1991, **148**, 65.
432. J.Stempel and W.Kiefer, *J. Chem. Phys.*, 1991, **95**, 2391.
433. R.Minkwitz and V.Gerhard, *Z. Naturforsch., B*, 1991, **46b**, 561.
434. H.D.Stidham and M.E.Ewen, *J. Raman Spectrosc.*, 1990, **21**, 585.
435. A.Engdahl and B.Nelander, *J. Phys. Chem.*,1990, **94**, 8777.
436. M.D.Schuder, D.D.Nelson and D.J.Nesbitt, *J. Chem. Phys.*, 1991, **94**, 5796.
437. Z.Wang, A.Quinones, R.R.Lucchese and J.W.Bevan, *J. Chem. Phys.*, 1991, **95**, 3175.
438. R.D.Hunt, L.Andrews and L.M.Toth, *Inorg. Chem.*, 1991, **30**, 3829.
439. J.Ortigoso, R.Escribano, J.B.Burkholder, C.J.Howard and W.J.Lafferty, *J. Mol. Spectrosc.*, 1991, **148**, 346.
440. D.M.Adams, A.E.Heath, M.Pogson and P.W.Ruff, *Spectrochim. Acta, A*, 1991, **47A**, 1075.
441. A.I.Karelin, A.V.Dudin and V.Ya.Rosolovskii, *Russ. J. Inorg. Chem.*, 1991, **36**, 288.
442. K.O.Christe, R.Bau and D.Zhao, *Z. anorg. allg. Chem.*, 1991, **593**, 46.
443. K.Hartl, J.Novicki and R.Minkwitz, *Angew. Chem., Int. Ed. Engl.*, 1991, **30**, 328.
444. J.J.Orlando, J.B.Burkholder, A.M.R.P.Bopegedera and C.J.Howard, *J. Mol. Spectrosc.*, 1991, **145**, 278.
445. H.Bürger, G.Pawelke and E.H.Appelman, *J. Mol. Spectrosc.*, 1990, **144**, 201.
446. R.Minkwitz and V.Gerhard, *Z. Naturforsch., B*, 1991, **46b**, 884.
447. N.Santha, M.Isaac, V.U.Nayar and G.Keresztury, *J. Raman Spectrosc.*, 1991, **22**, 419.
448. K.O.Christe, J.C.P.Sanders, G.J.Schrobilgen and W.W.Wilson, *J. Chem. Soc., Chem. Comm.*, 1991, 837.
449. A.R.Mahjoub and K.Seppelt, *J. Chem. Soc., Chem. Comm.*, 1991, 840.
450. D.H.Gwo, M.Havenith, K.L.Busarow, R.C.Cohen, C.A.Schuttenmaer and R.J.Saykally, *Mol. Phys.*, 1990, **71**, 453.
451. K.O.Christe, E.C.Curtis, D.A.Dixon, H.P.Mercier, J.C.P.Sanders and G.J.Schrobilgen, *J. Am. Chem. Soc.*, 1991, **113**, 3351.

5
Vibrational Spectra of Transition Element Compounds

BY G. DAVIDSON

1. Scandium Yttrium and the Lanthanides

Raman spectra for the system ScI_3-CsI show that ScI_6^{3-} and ScI_4^- are both present, together with binuclear scandium species at high ScI_3 concentrations. For ScI_6^{3-}, ν_1 is at 119 cm^{-1}, ν_2 at 67 cm^{-1} and ν_5 near 80 cm-1. For ScI_4^-, ν_1 is at 129 cm^{-1}, ν_2 at 37 cm^{-1} and ν_4 at 54 cm^{-1}.[1]

Extra Raman lines at 210 and 248 cm^{-1} in hydrogenated $MBa_2Cu_3O_{7-\delta}$, where M = Y or Gd, were ascribed to vibrations involving the Y and Cu(I) respectively.[2]

IR and Raman spectra were assigned for $M(OH)CrO_4$, where M = La, Pr or Nd.[3] νLnO wavenumbers in Ln^{III} chloranilates, where Ln = La, Ce, Pr, Nd, Sm, Gd, Tb, Ho, Er or Yb, correlate well with the ionic radius of Ln^{3+}.[4] νLnO modes were also assigned for $Ln(PMBP)_3(phen).nH_2O$, where HPMBP = 4-benzoyl-3-methyl-1-phenyl-5-pyrazolone; Ln = La (n = 2), Pr, Nd or Sm - Lu (n = 1).[5]

There is an empirical correlation between particle size and the half-width of the t_{2g} mode of CeO_2 (465 cm^{-1}) in automotive exhaust gas analysis.[6] XLn(TPP), where X = Cl or Br; Ln = Ce, Sm, Gd, Tb, Dy, Ho, Er, Yb or Lu, and $Th(TPP)_2$ all have νMN modes in the range 421 - 456 cm^{-1}.[7]

The IR spectra of Pr and Nd complexes containing fluoro-β-diketonates show νMO bands at higher wavenumber on increasing the ring size of the chelate.[8] Adsorbed H_2 on Sm_2O_3 gives characteristic IR bands at 822 and 714 cm^{-1}.[9] EuL(phen) and $[EuL]^-$, where HL = Hacac or other β-diketones, all show νEuN at 490 - 515 cm^{-1}, and νEuO 400 - 438 cm^{-1}.[10]

νTiB for $Ti(BH_4)_3$ is at 598 cm^{-1} (Ar matrix, 15 K) or 595 cm^{-1} (vapour, *ca* 290 K). The equivalent data for $Ti(BD_4)_3$ are at 544 and 540 cm^{-1} respectively.[12] Detailed vibrational assignments have been proposed for $MeTiCl_3$, using IR data on $^{12}CH_3TiCl_3$, $^{12}CD_3TiCl_3$, $^{13}CH_3TiC_3$ and CHD_2TiCl_3.[13]

2. Titanium, Zirconium and Hafnium

Titanium or vanadium and H_2 in Ar or Kr matrices at 12 K form MH_2, where M = Ti or V. The vibrational assignments for these are summarised in Table 1.[11]

Table 1 Matrix IR data for titanium and vanadium dihydrides (/cm-1)

	MH_2		MD_2		MHD	
	Ti	V	Ti	V	Ti	V
ν_{as}	1435.5	1508.3	1041.1	1092.0	1466.6(H)	1518.2(H)
ν_s	1483.2	1532.4	1071.0	1123.6	1055.8(D)	1095.6(D)
δ	496	529	376.5	386	-	-

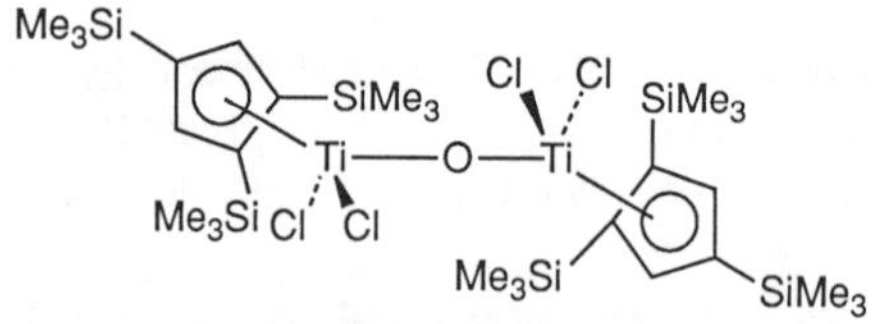

(1)

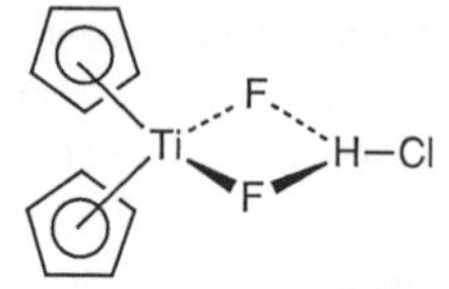

(2)

The presence of a linear TiOTi bridge in (1) was indicated by a strong IR band (ν_{as}TiOTi) at 762 cm^{-1}.[14] TiO_2 samples from a variety of sources give very different IR spectra. It was suggested that these differences were due to changing polarisation effects within the crystals.[15] The IR spectrum of Li_2TiO_3 has νTiO modes at 500 - 700 cm^{-1}.[16] $^{46}Ti/^{50}Ti$ isotopic shifts in the vibrational spectrum of $LiGaTiO_4$ show that the following bands involve motion of the Ti atom in the TiO_6 units (data refer to the ^{46}Ti form): 935, 835, 815, 684, 660, 568, 528 and 229 cm^{-1}.[17] The Raman spectra of borosilicate glasses containing Ti were consistent with the presence of TiO_4 units.[18] There appear to be TiO_4 and TiO_6 units in the structure of glasses in the system SrO-B_2O_3-Al_2O_3-TiO_2.[19-20]

The presence of νTiF modes in the matrix isolated Cp_2TiF_2/HCl complex, at 536 and 556 cm^{-1}, are consistent with the formulation (2).[21] $Cp_2Ti(F)AsF_6$ gives an IR band at 545 cm^{-1} due to νTiF_t, with a broad feature near 475 cm^{-1} derived from the Ti...F...AsF_5 stretch.[22] The low-wavenumber IR spectrum of $TiCl_2$(5-Aidtc)$_2$, where 5-AidtcH = 5-aminoindazoledithiocarbamic acid, shows that the geometry is *cis*-octahedral. The zirconium analogue, however, has *trans* geometry.[23] $Ph_3TeX.TiX_4$ and $(Ph_3TeX)_2.TiX_4$, where X = Cl or Br, have IR spectra consistent with the formation of TiX_5^-, TiX_6^{2-} respectively.[24]

Silica-bound ZrH_2, i.e. $(SiO_2)ZrH_2$ has νZrH bands at 2254 and 2190 cm^{-1}. For (SiO_2)ZrHCl, νZrH is at 2266 cm^{-1}.[25] Several skeletal mode assignments were made from the IR spectra of $M(O_2)$(SNNS), where M = Zr or Th, SNNS = Schiff base $C_{30}H_{26}N_4S_4$, as well as for $MO(O_2)$(SNNS), where M = Mo, W or U.[26]

The martensitic/austenitic phase transformation in ZrO_2 was followed as a function of temperature and pressure by Raman and IR spectroscopy. Evidence was found for a new tetragonal modification, with D_{2h} or C_{4v} symmetry, and four molecules per unit cell.[27] The

IR and Raman spectra of Li_2MO_3, where M = Zr or Hf, show νMO modes in the ranges 500 - 700 cm^{-1} (Zr) or 550 - 770 cm^{-1} (Hf).[28] $[ZrO(O_2)F_2]^{2-}$ has νZr=O at 980 cm^{-1} (IR), 1015 cm^{-1} (Raman), $\nu Zr(O_2)$ at 580 and 645 cm^{-1} (IR), 595 and 650 cm^{-1} (Raman), and νZrF at 455 cm^{-1} (IR), 480 cm^{-1} (Raman).[29]

The IR spectrum, together with earlier Raman data, for $CoZrF_6$ gave a complete vibrational assignment.[30] The IR and Raman spectra of $(N_2H_5)_3MF_7$, where M = Zr or Hf, show the presence of dimeric or polymeric anions with high coordination numbers for the metals.[31] The Raman spectra of glasses in the system ZrF_4-BaF_2-LaF_3 contain νZrF_t near 590 cm^{-1}, and νZrF_{br} near 487 cm^{-1}, due to linked ZrF_6^{2-} units.[32] Effects of the coordination environment on ν_sZrF modes in fluorozirconate glasses, crystals and melts show that there are 7- and 8-coordinate zirconium atoms in some barium fluorozirconate glasses.[33]

Detailed calculations have been carried out on the ground state of HfH, and these predict an ω_e value of 1704 cm^{-1}.[34] The IR and Raman spectra of Cp_2HfH_2 are consistent with a polymeric structure containing both terminal and bridging Hf-H bonds.[35] $Cp_2Hf(L)Cl$ and $CpHf(L)_2Cl$, where L = bidentate 3-indole carboxylate, have νHfO near 430 cm^{-1}.[36]

3. Vanadium, Niobium and Tantalum

Earlier reference has been made to vibrational studies on VH_2, VHD and VD_2.[11]

$V(arene)_2$ in N_2 matrices at 12K show resonance Raman progressions in ν_s(ring-V-ring), at 259.5 cm^{-1} (arene = C_6H_6), 273 cm^{-1} ($C_6H_5CH_3$) or 304.7 cm^{-1} (mesitylene).[37] νV-NO modes (630 - 640 cm^{-1}) were assigned for $[V(NO)_2(L\text{-}L)]^+$, where L-L = bipy or phen.[38]

Table 2 Some vibrational assignments for $[V(NX)FCl_3(CH_3CN)]^-$ (/cm^{-1})

X =	Cl	Br	I
νV≡N	1098	1098	979
νVF	394	394	396
νVCl	348	372	372
	311	289	298
	270	272	271
δVCl	124	135	136

Some skeletal mode assignments for $[V(NX)FCl_3(CH_3CN)]^-$ are summarised in Table 2, for X = Cl, Br or I.[39] $V(N_3S_2)Cl_2L$, where L = bipy or phen, have νVN of the VN_3S_2 ring at 962, 963 cm^{-1} respectively.[40]

An empirical correlation has been established between the νVO wavenumber

(Raman) and V-O bond lengths in vanadium oxide reference compounds.[41] The complex $[Cl_3PNPCl_3][VOCl_4]$ has νVO at 1023 cm^{-1}, νVCl at 420, 403 and 363 cm^{-1}, δVCl_3 at 165 cm^{-1} and δOVCl 252 cm^{-1}.[42] The skeletal modes of $VO(O_2)NH_3^-$ were assigned under C_s symmetry and a normal coordinate analysis performed.[43] The IR and Raman bands due to the $VO(O_2)$ group of $K_3[VO(O_2)CO_3]$ have been asigned.[44]

Assignments of νVO modes were reported for the following: VO(L), where H_2L = 5,14-dihydro-6,8,15,17-tetramethyldibenzo-[*b,i*][1,4,8,11]tetra-azacyclotetradecane (970 cm^{-1});[45] VOL_2, where L = dihydrobis(imidazolyl)borate and other related species (915 - 940 cm^{-1});[46] VO(Schiff base)L, where L = 2-(2'-pyridyl)benzimidazole (960 - 970 cm^{-1});[47] bis[N-(4-chlorophenyl)salicylideniminato]oxovanadium(IV) (967 cm^{-1});[48] VO-$(PenOCH_3)_2$, where $PenOCH_3$ is the anion of the carboxylate ester of penicillamine, (974 cm^{-1});[49] and a range of VO^{2+} complexes of 16- and 18-membered macrocyclic ligands.[50]

$[(salen)VOVO(salen)]^+$, where salen = N,N'-ethylenebis(salicylideneiminato), has νV=O at 870 cm^{-1}, compared to monomeric VO(salen) or $VO(salen)^+$, both of which have identical νV=O wavenumbers of 981 cm^{-1}.[51] There is an unusually low value for νV=O (943 cm^{-1}) in $[(tpa)VO(\mu\text{-}O)VO(tpa)](ClO_4)_2$, where tpa = tris(2-pyridylmethyl)amine, possibly due the effects of the *trans* tertiary nitrogen. ν_{as}VOV is at 807 cm^{-1}.[52] The new tetranuclear cation $[\{V(salen)\}_4(\mu\text{-}O)_3]^{2+}$ has νVO bands at 945 and 910 cm^{-1}, assigned to the V-O-V-O-V-O-V chain.[53] The resonance Raman spectrum of the hydrolytic dimer of aquovanadium(III) shows that it has a V-O-V, but not a V-(OH)-V bridge.[54]

The IR spectra of matrix-isolated MVO_3, where M = Li, Na, K, Rb or Cs, are consistent with the presence of C_{2v} monomers in each case.[55] The IR and Raman spectra of β-$NaVO_3$ can be interpreted in terms of a structure containing double chains of VO_5 trigonal bipyramids.[56] The high-pressure Raman spectrum of $CsVO_3$ shows phase transitions at 10, 11.5 and 13 GPa.[57]

The nature of the surface vanadia species on a high surface area silica support can be determined from the νVO bands in the IR and Raman spectra.[58] Similar Raman data were used to characterise V_2O_5/TiO_2 catalysts.[59-60] νVO modes were assigned from the IR spectra of Ag_3VO_4, $Ag_4V_2O_7$, α-, β- and γ-$AgVO_3$.[61]

IR and Raman data were given and assigned for $MBiTh(VO_4)$, where M = Mn or Cd.[62] The IR spectra of a range of species $[H_{x-1}V(V_xMo_{12-x})O_{40}]^{4-}$ all contained very similar νMO bands.[63] In $[V_{12}As_8O_{40}(HCO_2)]^{3-}$, νV=O was at 995 cm^{-1}, with νMOM' 835, 797, 685 and 606 cm^{-1}, where M and M' = As, V.[64] $[As^VV_{12}{}^VV_2{}^{IV}O_{40}]^{7-}$ has νV^{IV}=O 1010 cm^{-1}, νV^V=O 965 cm^{-1}, $\nu_{as}O=V^V=O$ 913 cm^{-1} and ν_sV-O-M 774 cm^{-1}, where M = V or As.[65]

Matrix IR spectra were reported for vanadium chlorides. VCl_2 in argon gave bands characteristic of a linear molecule, while VCl_4 in argon showed a doublet for ν_3, due to Fermi resonance with $\nu_1+\nu_4$.[66]

Ab initio calculations on NbH suggest that ω_e for the $^5\Delta$ ground state will be near 1750 cm^{-1}.[67]

Molecular structures of Nb_2O_5 supported on various oxide supports were characterised by the observed νNb=O modes.[68] νNbO modes were identified for a variety of oxalato complexes formed in solutions of niobium oxides in alkaline or acidic oxalate solutions.[69] Two isomers of $Nb(O)Cl_3(PMe_3)_3$ have been found, differing only in bond lengths. Thus one form has νNb=O at 882 cm^{-1} (835 cm^{-1} for ^{18}O), with an NbO distance of 178.1(6) pm, while the other has νNb=O at 871 cm^{-1} (824 cm^{-1} for ^{18}O), with an NbO distance of 192.9(6) pm. An analogous pair of complexes was found for the Nb=S analogues.[70]

Skeletal mode assignments were made from IR data on $NbO(OR)_3$, where R = Me, Et, Bu or Me_2CH.[71] IR and Raman spectra were reported and assigned for $Rb_5Nb_3OF_{18}$.[72] An empirical correlation was found between νNbO wavenumbers and Nb-O distances in niobium oxide minerals.[73] IR assignments for $NbSCl_4^-$, $NbSCl_5^{2-}$, $NbSBr_4^-$ and $TaSCl_5^{2-}$ are summarised in Table 3.[74]

Table 3 Some skeletal mode assignments for niobium and tantalum sulphide complexes (/cm^{-1})

		$NbSCl_4^-$	$NbSCl_5^{2-}$	$NbSBr_4^-$	$TaSCl_5^{2-}$
νM=S	(a_1)	552	496	560	480
$νMX_4$	(a_1)	346	320	274	272
$νMX_4$	(e)	339	300	255	270
$νMCl_ax$	(a_1)	-	235	-	250

Normal coordinate analyses were reported for MX_6^- and MX_6^{2-}, where M = Nb or Ta, X = F, Cl, Br or I.[75] The Raman spectra of molten LiF-NaF-KF-K_2NbF_7 at 650°C show that NbF_7^{2-} is present, with the following wavenumbers: 626 cm^{-1} (pol.), 371 cm^{-1} (depol) and 290 cm^{-1} (depol.). Addition of oxide gave evidence for $NbOF_n^{(n-3)-}$, where n was probably 5, as indicated by the separation of solid $NbOF_5^{2-}$ of C_{4v} symmetry. There was also evidence for C_{2v} $NbO_2F_4^{3-}$.[76] Quite detailed assignments were proposed for $(M_6X_{12})X'_2$, where M = Nb or Ta, X = Cl or Br, X' = Cl, Br or I.[77] Data were also presented for the mixed metal species $[(Nb_nTa_{6-n})Cl_{12}]^{2+}$, where n = 0 - 6. The number of IR active stretching modes was consistent with the expected symmetries of the intermediate species.[78]

νTa=N bands were observed in the range 1132 - 1142 cm^{-1} in $[\{Ta(\mu\text{-}X)X(NSiMe_3)L\}_2]$, where X = Cl, Br or I, L = $N(SiMe_3)_2$.[79] Bands were seen in both IR and Raman spectra for $TaBO_4$ and $NaTa_3O_8$ which suggested the presence of TaO_8 dodecahedra.[80]

4. Chromium, Molybdenum and Tungsten

Earlier reference has been made to a vibrational study of $[H_{x-1}V(V_xMo_{12-x})O_{40}]^{4-}$.[63]

The FT Raman spectra of $M(CO)_5NH_3$ gave the following assignments: νM-NH_3 490 cm^{-1} (Cr), 490 cm^{-1} (W); νM-CO 390 cm^{-1} (Cr), 431 cm^{-1} (W).[81] Fairly complete skeletal mode assignments were given for $CpCrX_2L$, where X = Cl or Br, L = CH_3CN; X = Cl, L = THF, and $CpMX_4(CH_3CN)$, where M = Mo or W, X = Cl or Br.[82]

Resonance Raman scattering at 1010 cm^{-1} in nitridochromium(V) protoporphyrin IX was assigned as $\nu Cr^{V}{\equiv}N$.[83]

IR data show that CrO_2F_2 and MnO_3F are monomeric in the solid state, with νMO 1005 cm^{-1} (Cr), 950, 905 cm^{-1} (Mn); νMF 770, 710 cm^{-1} (Cr), 705 cm^{-1} (Mn).[84] Sharp-line electronic spectroscopy of $[Cr(NH_3)_5OH](ClO_4)_2$ shows that the band at 572 cm^{-1} is not νCrO, but rather the first overtone of the OH internal rotation mode.[85] IR and Raman spectra of single crystals of $KLn(CrO_4)_2$ gave CrO_4^{2-} assignments in terms of site (C_1) and factor group (D_2) symmetry, where Ln = Eu, Gd or Tb.[86] $[Cr_3OL_6(H_2O)_3](ClO_4)_7$, where L = glycine, α-alanine or α-aminobutanoic acid, all give an IR band due to $\nu_{as}Cr_3O$ near 600 cm^{-1}, with νCrO (of H_2O) in the range 450 - 467 cm^{-1}.[87] $[Cr_4O(SO_4)_2Cl_9]^{3-}$ has νCrCl bands at 378, 368 and 353 cm^{-1}.[88]

νMoH is at 1876 cm^{-1} in MoHBr(dpepp)(PMe_2Ph), and at 1874 cm^{-1} in MoH_2Br_2-(dpepp)(PMe_2Ph), where dpepp = $PhP(CH_2CH_2PPh_2)_2$.[89] $MoN(NPh_2)_3$ has νMo≡N at 928 cm^{-1}.[90] $[\{HB(Me_2pz)_3\}MoN(L)_2]^-$, where L = Cl^- or N_3^-, have νMo≡N at 1020 cm^{-1} (Cl) or 1023 cm^{-1} (N_3). $\{HB(Me_2pz)_3MoO_2(N_3)$ has νMoO_2 at 925 and 895 cm^{-1} in the IR, i.e. the MoO_2 group is *cis*[91] A number of skeletal mode assignments for $MBr_4(CH_3CN)_2$, where M = Mo or W, and $MoBr_3(CH_3CN)_3$ are listed in Table 4.[92] Several skeletal modes were also assigned for $M(O)(O_2)C_{14}H_{10}N_2O_2$, where M = Mo, W or U, $C_{14}H_{10}N_2O_2$ is an ONNO Schiff base.[93]

Table 4 Some skeletal vibrational assignments for molybdenum and tungsten bromo complexes (/cm^{-1})

	νMoN	νMoBr		
$MoBr_4(CH_3CN)_2$	307	269	208	
$MoBr_3(CH_3CN)_3$	295	267	227	200
$WBr_4(CH_3CN)_2$	286	248	238	213

Some skeletal modes have been assigned for *trans* -$Mo(CO_2)_2(PMe_3)_4$, using ^{18}O and ^{13}C isotopic data, e.g. νMoP 397 and 362 cm^{-1}. Modes involving Mo-(CO_2) motions are very strongly coupled to each other and other low-wavenumber features.[94]

$^{16}O/^{18}O$ isotopic shifts in the resonance Raman spectra of the molybdenum centre of xanthine oxidase were used to identify bands with contributions from Mo-O motions.[95]

$[MO(bipy)(CN)_3$, where M = Mo or W, have νM=O at 973 cm^{-1} (Mo) or 971 cm^{-1} (W).[96] νMo=O bands all lie in the range 915 - 925 cm^{-1} for $MoO(S_2)(S_2CNR_2)_2$, where R = Me, Et, Pr or Bu).[97] [μ-O{Mo(NO)("S_4")}$_2$], where "S_4" = 2,3,8,9-dibenzo-1,4,7, 10-tetrathiadecane(2-), has ν_{as}MoOMo at 788 cm^{-1}, δMoOMo at 388 cm^{-1}.[98] νMo=O and νMo-S modes were assigned for $[MoO(L)_2]^{2-}$, $[Mo_2O_2S_2(L)_2]^{2-}$, and related species, where L = 1,2-carbomethoxy-1,2-ethylenedithiolate.[99] $LMo(E)Cl_2$, where E = O or S, L = hydrotris(3,5-dimethyl-1-pyrazolyl)borate, have νMo=E at 957 cm^{-1} (O) or 525cm^{-1} (S).[100]

The IR spectra of $MoO[S_2P(OR)_2]L$, where R = Et or ^{i}Bu, L = py; R = Et, Pr or ^{i}Bu, L = bipy or phen, are all consistent with six-coordinate, octahedral structures.[101] IR bands characteristic of *cis*-MoO_2 groups were seen for MoO_2(SAPT)L, where H_2SAPT = Schiff base from 2-aminothiophenol and salicylaldehyde, L = imidazole or a substituted derivative;[102] and for MoO_2(SAE)L, where H_2SAE = *N*-(hydroxyethyl)salicylidene-imine.[103]

$Mo_2O_2(NO)_2(PPh_3)_4$ has ν_s and ν_{as} of Mo-(μ-O)$_2$-Mo at 510 and 770 cm^{-1} respectively. The terminal Mo=O groups of $Mo_2O_2Cl_4(NO)(PPh_3)_4$ give IR bands at 940 and 970 cm^{-1}.[104] $Mo_2O_5(OH)_2(HL)$, where HL is one of a wide range of amino-acids, all have IR bands due to *cis* -MoO_2 near 790 cm^{-1}, with νMo=O_t 920 - 960 cm^{-1}, and νMoOMo 590 and 730 cm^{-1}.[105] Terminal νMoO bands are seen at 965 and 954 cm^{-1} for $[Br_2OMo(\mu\text{-}S)_2MoOBr_2]^{2-}$.[106] Similar skeletal assignments were proposed for $LMo^{V}O\text{-}Cl(\mu\text{-}O)Mo^{VI}\text{-}O_2L$, where L = hydrotris(3,5-dimethyl-1-pyrazolyl)borate;[107] and $[(\eta^5\text{-}MeC_5H_4)_2Mo_2O\text{-}(NPh)(\mu\text{-}O)(\mu\text{-}NPh)]$.[108]

(3)

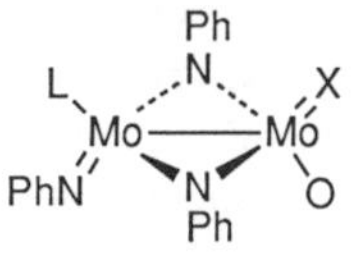

(4)

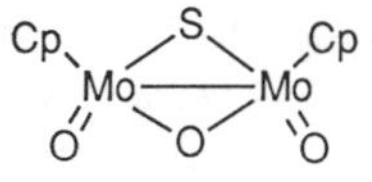

(5)

(3) has νMo=O at 927 cm^{-1}, νMoOMo at 715 cm^{-1}, while for (4), X = O, νMo=O is at 884 cm^{-1}, νMo=NPh 1327 cm^{-1}, νMo-(NPh)-Mo 1263 cm^{-1}. When X = S, νMo=S is at 474 cm^{-1}, νMo=NPh 1324 cm^{-1} and νMo-(NPh)-Mo 1264 cm^{-1}.[109] Skeletal stretching modes were assigned for dinuclear molybdenum and tungsten complexes such as (5) and its *trans* isomer.[110]

A correlation has been established between νMoO wavenumbers and Mo-O bond distances in molybdenum oxide complexes.[111] The IR spectra of oxidation products of a molybdenum(V)-silica gel at 473 - 673 K show the formation of terminal Mo=O and bridging MoOMo units.[112] The blue material formed by the reduction of acidic Li_2MoO_4 aqueous solutions was found to give νMo=O at 960 cm^{-1}, ν_{as}MoOMo at 560 cm^{-1}, and a peak at 720 cm^{-1} due to $\nu Mo(O_2)Mo$.[113]

Characteristic νMoO modes were identified for $MgMoO_4.xH_2O$, where x = 1 or 5.[114] The Raman spectrum of an unsupported Fe/Mo oxide catalyst showed the presence of $Fe_2(MoO_4)_3$ and MoO_3.[115] Vibrational spectra of several mixed alkali metal - lanthanide molybdates were analysed using $^{92}Mo/^{100}Mo$ isotopic shifts.[116-7] Other molybdate species studied were $CuLn(MoO_4)_2$, where Ln = Tm or Yb;[118] $Li_2Sn_3(MoO_4)_7$, $Li_2Sn(MoO_4)_3$, $Li_4SnMo_2O_{10}$,[119] and $GdDy(MoO_4)_3$.[120]

Terminal and bridging stretching modes were assigned for $[Mo_2O_7]^{2-}$ and $[Mo_6O_{19}]^{2-}$.[121] The IR and Raman spectra of $Rb_3(MoO_3)_{12}(PO_4).4H_2O$ show the presence of $Mo_{11}PO_{39}^{7-}$.[122] The species α-$[XM_{12}O_{40}]^{x-}$, where X = P, x = 3; X = Si, x = 4; M = Mo or W, were characterised by their IR and Raman spectra.[123] $[SMo_{12}O_{40}]^{2-}$ has νMo=O at 982 cm^{-1}, νMoOMo at 877 and 805 cm^{-1} (the last two being due to corner- and edge-sharing octahedra respectively).[124]

The resonance Raman spectrum of DMSO-reductase (DR) from *Rhodobacter sphaeroides* contains νMoS bands at 352 and 383 cm^{-1} due to the reduced form of DR, 350, 370 cm^{-1} due to the oxidised form of DR. All are characteristic of Mo(IV) dithiolene complexes.[125]

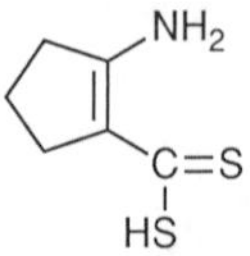

(6)

$Mo(S_2CNR_2)_2(acda)_2$ and $Mo(S_2CNR_2)(acda)$, where R = Et or Pr, Hacda = (6), all give doublet IR bands due to νMoS (350 -360; 335 - 345 cm^{-1}).[126] νM=S, νMS and νMX modes were assigned for $M_2S_4X_4^{2-}$, where M = Mo or W, X = Br; M = W, X = Cl. Data included $^{32}S/^{34}S$ isotopic shifts.[127] A normal coordinate analysis for $Mo_3S_4Se_3Cl_6^{2-}$ shows that there is very extensive mixing of all modes except νMoCl.[128]

$[Mo_2AgS_4(edt)(PPh_3)]^-$, where edt = ethane-1,2-dithiolato, gave the following IR bands: νMo-μ_2-S_t 490 cm^{-1}, νMo-μ_2-S and νMo-μ_3-S 460, 430 cm^{-1}, νMoS(edt) 345, 305 cm^{-1}, and νAgP 516, 505 cm^{-1}.[129] $[Bi(MS_4)_2]^-$ has νMS_t at 500 cm^{-1} (Mo) or 490 cm^{-1} (W); νMS_{br} 460 cm^{-1} (Mo), 440 cm^{-1} (W).[130]

Characteristic νMF and νMCl bands were assigned for $[MF_4Cl(NO)]^{2-}$, where M = Mo or W, together with νMCl for $[MCl_4(NO)_2]^{2-}$, where M = Mo or W.[131] Skeletal stretches for $MoF_5.NCMe$ and $MoF_5.NCCD_3$ are consistent with C_{4v} symmetry for the MoF_5N fragment.[132]

The IR spectra of two polymorphs of $WO(PO_4)$ show the expected structural relationships of these to α-$MO(PO_4)$ (M = Nb, Ta, V or Mo) (for the high-pressure form) and to β-$NbO(PO_4)$ (for the low-pressure form).[133] $W(O)(S)Cl_2$ has νWOW at 815 cm^{-1}, νW=S at 540 cm^{-1}, showing a preference for terminal S and bridging O.[134] νWO modes were assigned for $WO_2(ClO_4)_2$ (990 cm^{-1}), $WO_2(ClO_4)_3^-$ (990 cm^{-1}) and $WO_2(ClO_4)_4^{2-}$ (985 cm^{-1}).[135]

The IR spectra of electrochromic WO_3 films were used to monitor amounts of W=O, W-O and W-OH units present.[136] The IR and Raman spectra of amorphous WO_3 films are consistent with the presence of low-symmetry WO_6 units. Termination by W=O and W-(OH) units both occur.[137] A normal coordinate analysis of Ba_3WO_6 gave good agreement with experimental wavenumbers.[138]

νWO bands in the IR spectrum of $Na_{12}[P_4W_8O_{40}].20H_2O$ show that there is distortion of the WO_6 octahedra.[139] Characteristic skeletal stretching wavenumbers were assigned for $[SiW_{11}O_{40}(SiR)_2]^{4-}$, where R = Et, $CH{=}CH_2$, $C_{10}H_{21}$ or Ph.[140] The IR spectra of $[PW_{11}TcO_{40}]^{4-}$ and $[SiW_{11}TcO_{40}]^{5-}$ are indistinguishable from those of the corresponding $[XW_{12}O_{40}]^{n-}$.[141] The IR and Raman spectra of peroxotungstic acid were recorded and used to confirm the structure suggested by X-ray diffraction, i.e. that the polyanion $W_{12}O_{38}(O_2)_6{}^{16-}$ is present.[142]

Table 5 Vibrational assignments for WSX_5^- complexes ($/cm^{-1}$)

	νWS	νWCl	νWF
$WSCl_5^-$	515	365,320	
$WSCl_4F^-$	520	370,321	608
$WSCl_2F_3^-$	525	327	633, 615, 468
WSF_5^-	525		662, 604, 471

Vibrational assignments for WSX_5^-, where X = F or Cl, and some mixed analogues, are summarised in Table 5.[143] $WCl_4S(S_8)$ has νW=S at 551 cm^{-1}.[144] νW=S in (7) is at 513 cm^{-1}.[145] The IR and Raman spectra of $W_3S_7Br_6{}^{2-}$ gave assignments for many of the skeletal modes. The most characteristic band involving μ_3-S is at 449 cm^{-1}.[146]

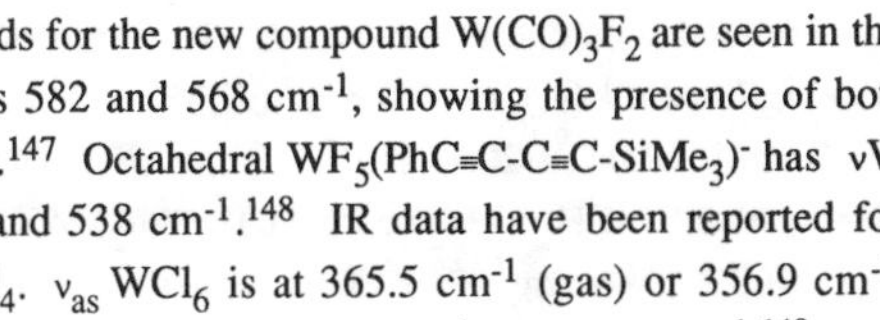

(7)

νWF bands for the new compound $W(CO)_3F_2$ are seen in the IR at 698, 668 and 642 cm^{-1}, as well as 582 and 568 cm^{-1}, showing the presence of both terminal and bridging fluoride ligands.[147] Octahedral $WF_5(PhC{\equiv}C\text{-}C{\equiv}C\text{-}SiMe_3)^-$ has νWF_{ax} at 475 cm^{-1}, νWF_{eq} 631, 595, 560 and 538 cm^{-1}.[148] IR data have been reported for WCl_6, as a gas and in solution in CCl_4. ν_{as} WCl_6 is at 365.5 cm^{-1} (gas) or 356.9 cm^{-1} (solution). The corresponding figures for $\delta_{as}WCl_6$ are 160.6 cm^{-1} and 160.0 cm^{-1}.[149]

5. Manganese, Technetium and Rhenium

Previous reference has been made to vibrational studies on $MnBiTh(VO_4)$;[62] and MnO_3F.[84]

Resonance Raman data for Mn_2 in solid xenon matrices gave $\omega_e'' = 68.1\ cm^{-1}$, $\omega_e''x_e'' = 1.05\ cm^{-1}$ for the $^1\Sigma_g^+$ ground state.[150] The IR spectrum of the MnD radical in the $^7\Sigma$ ground state shows that ω_e for ^{55}MnD is at 1104.65225(93) cm^{-1}.[151]

Resonance Raman and IR data were reported and assigned for azidomanganese(III) and azidoiron(III) porphyrins. $(N_3)Mn^{III}$(TMP) has νMnN_3 at 344 cm^{-1}.[152] Characteristic νMN wavenumbers were established for $MQ_xCl_3.nH_2O$, where Q = quinoxaline, M = Mn or Cu, x = 1, n = 1; M = Fe, x = 3, n = 0; M = Co, x = 1, n = 1.5 etc.[153]

The resonance Raman spectra of MnO_3X, where X = F or Cl, contain progressions in ν_sMnO_3 modes.[154] Oxidation of Mn(HRP), where HRP = horseradish peroxidase, gives a species whose resonance Raman spectrum contains a band at 622 cm^{-1} due to νMn=O. The low value shows that the Mn=O bond is weak, probably due to strong distal hydrogen-bonding.[155] ν_1MnO_2 and ν_2MnO_2 are near 710 and 430 cm^{-1} respectively for the side-on dioxygen adducts of Mn(P), where P = TMP, TPP or OEP. The oxomanganese(IV) porhyrins derived from these have νMn=O near 825 cm^{-1}.[156]

IR and Raman spectra were assigned for $[NO_2]_2[MnF_6]$.[157] Low-temperature IR and Raman spectra of $M(CO)_5X$, where M = Mn or Re, X = Cl or I, gave assignments for νMX (a_1) and δMX (e) modes.[158]

There have been a number of reported assignments of νTc≡N modes, e.g. in $[TcN(O_2)_2L]^{n-}$, where L = Cl, n = 1, L = bipy, n = 0;[159] $[TcNCl_4(OH_2)]^-$;[160] $[\{TcN(O_2)_2\}_2(oxalato)]^{2-}$;[161] $TcN(L)_2$, where L = (8), R = Ph, *p*-$EtOOCC_6H_4$ etc.;[162] and TcN(L), where L = *N,N'*-ethylenebis(thioacetylonylideneimine).[163] $[TcCl_4(abt)]^-$, where abt = 2-aminobenzenethiolato(2-)-*N,S*, has νTcN and νTcS at 432 and 387 cm^{-1} respectively.[164]

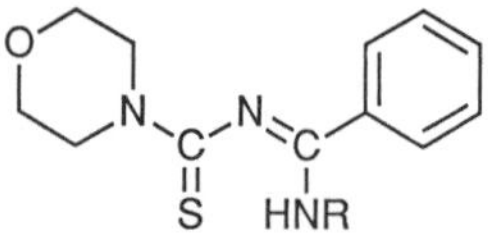

(8)

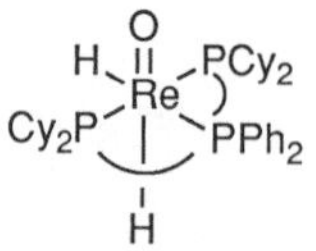

(9)

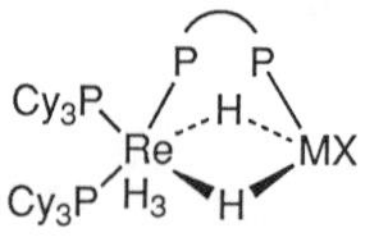

(10)

νTc=O is at 952 cm^{-1} for O=Tc(L), where L = N-(mercaptoacyl)-*N'*-[4-pentene-3-one-2]ethane-1,2-diaminato.[165] IR bands due to the $Tc(\mu\text{-}O)_2Tc$ fragment of $[(C_2O_4)_2Tc\text{-}(\mu\text{-}O)_2Tc(C_2O_4)_2]^{4-}$ are found at 730 cm^{-1} (ν_3) and 401 cm^{-1} (ν_4).[166]

(9) has νRe=O at 923 cm^{-1}, νReH at 2030 and 1716 cm^{-1}.[167] The terminal Re-H stretches in (10), where P-P = $Ph_2PC(=CH_2)PPh_2$ or $Ph_2PCH_2PPh_2$, MX = AgI, CuCl or CuCN, all lie in the range 1960 - 1986 cm^{-1}.[168] νReH modes were also assigned for ReH_5(triphos) (1949, 1914 and 1872 cm^{-1}) and $[ReH_6(triphos)]^+$ (1985, 1937 cm^{-1}), where triphos = $PhP(CH_2CH_2PPh_2)_2$.[169]

$[LReO(THF)]^+$, where L = 1,4,7-triazacyclononane, has νRe=O at 961 cm^{-1}, and νRe-OC at 654 and 615 cm^{-1}.[170] In *anti*-$[L_2Re_2Br_2(\mu\text{-}O)_2]^{2+}$, νReORe is at 706 cm^{-1}.[170] The following species have had νRe=O assignments made: Re(O)I(RC≡CR), where R = Me or Ph;[171] $[ReO(MAG_3)]^-$, where MAG_3 = mercaptoacetylglycylglycylglycine;[172] (11);[173]

$ReOCl_2[N(SPPh_2)_2]_2$ and related complexes.[174] *Cis* -$(O)_2Re(bipy)py_2^+$ has ν_{as}O=Re=O at 906 cm^{-1}.[175]

Skeletal mode assignments for CH_3ReO_3 are listed in Table 6. νReO_3 modes were also assigned for Cp^*ReO_3 (924 and 894 cm^{-1}) and $(\sigma\text{-}C_6H_2Me_3)ReO_3$ (986 and 953 cm^{-1});[176] as well as $CpReO_3$ and $(\eta^5\text{-}C_5H_4Me)ReO_3$.[177]

Table 6 Skeletal mode assignments for CH_3ReO_3 (/cm^{-1})

a_1	$\nu_s ReO_3$	998
	νReC	575
	$\delta_s ReO_3$	324
e	$\nu_{as} ReO_3$	947
	$\delta CReO$	252
	$\rho_{as} ReO_3$	238

Variations in ReO_4^- modes between crystalline and liquid $MReO_4$, where M = Li, Na or K, were followed by Raman spectroscopy.[178] The Raman spectrum of $AgReO_4$ shows that there is a phase change at 13 GPa.[179] $Fe(TPP)(OReO_3)$ has νReO at 942 and 827 cm^{-1}, δReO_3 at 336 and 321 cm^{-1}. These are consistent with unidentate attachment of perrhenate to the iron.[180]

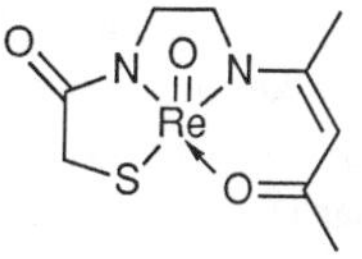

(11)

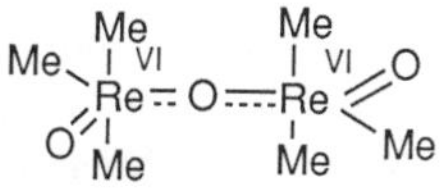

(12)

νReO modes in Sr_2MReO_6, where M = Li, Na, K, Rb or Cs, all show that there is considerable distortion of the ReO_6 octahedra.[181] (12) has νRe=O at 1005 cm^{-1}, νReORe at 803 cm^{-1}.[182]

$Re_2O_3(TSC)_2X_4$, where TSC = thiosemicarbazide, have νReX at 324 cm^{-1} (X = Cl) or 238 cm^{-1} (Br).[183] TR^3 for $Re_2X_8^{2-}$, where X = Cl or Br, in the $^1A_{2u}$ ($\delta\delta^*$) electronically excited state, show peaks due to symmetric νReRe, νReX and δReReX of the excited state, as well as a depolarised band due to δ_asReX_2, Table 7.[184]

Table 7 Vibrational assignments for the ground and δδ* excited states of $Re_2X_8^{2-}$ (/cm^{-1})

	X = Cl		X = Br	
	gd. st.	δδ*	gd. st.	δδ*
δ_sReReX	154	146	111	104
δ_{as}XReX	188	201	123	126
νReRe	274	262	275	262
ν_sReX	359	365	209	216

6. Iron, Ruthenium and Osmium

Earlier reference has been made to vibrational studies on unsupported Fe/Mo oxide catalysts;[115] FeQ_3Cl_3, where Q = quinoxaline;[153] and Fe(TPP)($OReO_3$).[180]

νMH modes were assigned for $[M(\eta^2\text{-}H_2)(H)(dppe)_2]^+$, where M = Fe, Ru or Os.[185] νFe-CO were assigned for CO adducts of a range of haem csystems.[186] The low-wavenumber FeCN vibrations of cyanoferric myeloperoxidase (MPO) and horseradish peroxidase were measured by resonance Raman spectroscopy. The order of νFeC and δFeCN differs in these : (MPO) νFeC 361 cm^{-1}; δFeCN 454 cm^{-1}; (HRP) νFeC 456 cm^{-1}, δFeCN 404 cm^{-1}.[187] The relationship between νFe-CO and the rate of CO dissociation for ferrous peroxidases and myoglobin was followed by resonance Raamn spectroscopy.[188] Resonance Raman data of CO-bound abnormal subunits of haemoglobin M Boston and haemoglobin M Saskatoon showed that νFeCO was at an unusual value (490 cm^{-1}, compared to 505 cm^{-1} in normal subunits). νFe-His in these species lay in the range 214 - 218 cm^{-1}, showing that the haem iron is bound to either E7-His or F3-His.[189] IR and resonance Raman spectra of Fe(TPP)CCl_2 show that νFe=C is at 1274 cm^{-1}.[190]

νFe-NO mode assignments were proposed for Fe(TPP)(NO) and $[Fe(TPP)(NO)]^-$, the latter having a higher value, i.e. increased back-donation.[191] Similar modes were identified in NO adducts of ferric cytochrome $P450_{cam}$ (also νFeS 350 cm^{-1}).[192] TR^3 of MbCO photodissociation showed no νFe-His band. Thus the Fe-His(F8) bond is broken within 30 psec of photodissociation.[193]

νMO modes were assigned from the IR spectra of $MF_2.4H_2O$, where M = Fe, Co, Ni or Zn. The values correlate well with M-O bond distances.[194] The IR and Raman spectra of $Ba_2H[\alpha\text{-}FeO_4W_{12}O_{36}].26H_2O$ yielded the following assignments: ν_sFeO_4 618 and 642 cm^{-1}; $\nu_{as}FeO_4$ 780 cm^{-1}; δ_sOFeO 321 cm^{-1}; δ_{as}OFeO 430 cm^{-1}.[195] Characteristic FeOFe stretching mode assignments were given for $[N6FeOFeCl_3]^+$, where N6 = N,N,N',N'-tetrakis[(2-benzimidazolyl)methyl]ethanediamine;[196] $Fe_2O(MPDP)(4,4''\text{-}Me_2bipy)_2Cl_2$, where H_2MPDP = m-phenylenedipropionic acid and related species;[197] $[Fe_2L(\mu\text{-}O)(\mu\text{-}$

$RCO_2)_2]^{2+}$, where R = Me or Ph, L = N,N-bis[benzimidazol-2-ylmethyl]-amine;[198] and $[(TPA)FeOFe(TPA)]^{4+}$, where TPA = tris-(2-pyridylmethyl)amine.[199]

Resonance Raman studies have been made of νFeO_2 and related modes in a number of O_2 adducts of porphyrins, cytochromes etc.[200-4] Addition of MeOH to Fe^{III}(OEP)(2-MeIm) gave band at 524 cm^{-1} whose isotope sensitivity revealed that it was νFe-O(Me)H.[205] Resonance Raman identification of νFe=O modes was reported for porphyrin π-cation radical forms of Fe^{III} and Fe^{IV} states of (tetramesityl-porphyrinato)iron complexes,[206] and for a variety of cytochrome oxidase reaction intermediates.[207-9]

New Raman and previous IR data were used to perform normal coordinate analyses for pyrite-type FeS_2 and RuS_2, and marcasite-type FeS_2.[210] νFeS (of L) modes were assigned (325 - 375 cm^{-1}) for $Fe(DioxH)_2L_2$, where $DioxH_2$ = dimethyl- or diphenyl-glyoxime, L = thioacetamide, thiourea or 2-aminothiazole.[211]

Resonance Raman spectra were used in characterising the [4Fe-4S] units of membrane-bound hydrogenase from *Desulphovibrio vulgaris*;[212] ferredoxins from a number of sources;[213] rubredoxins;[214] and the oxidised iron protein azoferredoxin of *Clostridium pasteurianum* nitrogenase.[215]

νMF_2 modes (ν_1 and ν_3 stretches) were assigned for $MF_2.4H_2O$, where M = Fe, Co, Ni or Zn, at ca. 296 K. These assignments are summarised in Table 7, and show that ν_3 values are higher than those for MF_2 in matrices.[216]

Table 7 M-F stretching assignments for $MF_2.4H_2O$ (/cm^{-1})

		ν_1	ν_3
M =	Fe	611	842
	Co	632	871
	Ni	633?	893
	Zn	626	871

IR spectra were reported for matrix-isolated MCl_2, where M = Fe, Co, Ni or Zn, including extensive isotopic data. Thus for $^{56}Fe^{35}Cl_2$ we have 493.2 cm^{-1}, $^{56}Fe^{35}Cl^{37}Cl$ 490.4 cm^{-1} and $^{56}Fe^{37}Cl_2$ 487.2 cm^{-1}.[217] In $Fe(NO)Cl_2(Odppe)$ νFeCl is at 351 cm^{-1}.[218] Raman and resonance Raman spectra of $FeCl_3$ (gas) are consistent with D_{3h} symmetry, with ν_4, ν_1 and ν_3 at 450, 370 and 111 cm^{-1} respectively. For Fe_2Cl_6 there are polarised bands at 422, 305, 150 and 78 cm^{-1}, and for $FeAlCl_6$ at 413 and 88 cm^{-1}.[219] All four Raman-active modes of $FeCl_4^-$ were seen for $[MgCl(THF)_5][FeCl_4]$.[220] Raman and IR spectra of $FeCl_6^{3-}$ show that the stretching wavenumbers are decreased compared to those of $FeCl_4^-$, but that the bending wavenumbers are increased.[221]

RuHCl(Cyttp) has νRuH at 2020 cm^{-1}, where Cyttp = $PhP(CH_2CH_2CH_2PCy_2)_2$.[222] νRuH modes were also assigned in $(PP_3)Ru(H)(\eta^1\text{-}BH_4)$, where $PP_3 = P(CH_2CH_2PPh_2)_3$

and related species.[223] $\nu_{as}M_2H$, ν_sM_2H and γM_2H modes were assigned in terms of D_{2d} symmetry in $H_4M_4(CO)_{12}$, where M = Ru or Os.[224] Mg_2RuH_6, Ba_2RuH_6, Ca_2OsH_6, Sr_2OsH_6 and Ba_2OsH_6 all have νMH in the range 1500 - 1850 cm^{-1}. The band positions shift to lower wavenumbers on increasing the size of the counterion. νMH increases in the series Fe to Ru to Os, showing that H^- can stabilise Os(II).[225]

$M_2^{2+}[Ru(CO)Cl_5]^{2-}$ show a steady increase in νRuC and (smaller) decrease in νRuCl on increasing the size of M^+.[226] NO adsorbed on Ru, Rh or Pt gives νM-NO bands at 310 cm^{-1} (Ru or Rh) or 325 cm^{-1} (Pt).[227] The following assignments were proposed for $Ru(NO)(NH_3)(S_4)_2^-$: νRu-NO 570 cm^{-1} (IR), 607, 527 cm^{-1} (Raman); $\nu Ru\text{-}NH_3$ 435 cm^{-1} (IR); and νRu-S 334 cm^{-1} (IR), 351, 299 cm^{-1} (Raman).[228] The complexes $\{Ru(LL)_xCl_2(DMSO)_{4-2x}\}_n$, where LL = pyrazine (pyz) or *p*phenylenediamine (ppd), x = 1 or 2, give νRuN at 274 cm^{-1} (pyz) or 280 cm^{-1} (ppd) in the resonance Raman spectra. When LL = 1,4-di-isocyanobenzene, however, *C*-donation takes place, with νRuC found at 186 cm^{-1}.[229]

Calculated vibrational wavenumbers for RuO, RuO_2, RuO_3 and RuO_4 generally gave reasonable agreement with experiment.[230] νRuO_2 and νRuX bands were reported for $[RuO_2X_3]^-$ and $[RuO_2X_4]^{2-}$, where X = Cl or Br, e.g. $RuO_2Cl_4{}^{2-}$ has ν_sRuO_2 840 cm^{-1}, ν_sRuO_2 at 824 cm^{-1}, and νRuCl at 306 cm^{-1}.[231] Solid $(PPh_4)[RuO_2Cl_2(OAc)]$ has νRuO at 872, 889 cm^{-1} (Raman), 866, 891 cm^{-1} (IR). For each pair, ν_s is the lower wavenumber feature, ν_{as} the higher.[232] a detailed study has been reported of the ν_3 (t_2) fundamental of RuO_4. Enriched isotopic species gave values of 921.6514 cm^{-1} ($^{102}RuO_4$) and 920.0674 cm^{-1} ($^{104}RuO_4$), while natural abundance samples gave the following values for $^mRuO_4$: m = 96, 926.83; 99, 924.14; 100, 923.31; 101, 922.46 cm^{-1}.[233] $[Ru(Me_2SO)_6]^{3+}$ contains both *O*- and *S*-bonded Me_2SO, with bands at 480, 430 cm^{-1} assigned as νRuO, νRuS respectively.[234]

SO_2 adsorbed on Ru-, Rh- or Pt-coated gold showed a SERS band 300 - 320 cm^{-1} due to νMS from sulphur produced by dissociative SO_2 adsorption.[235] The Raman spectra of pyrite-type RuS_2, $RuSe_2$, $OsSe_2$, PtP_2 and $PtAs_2$ show strong resonance enhancement.[236]

ν_1 Raman bands of RuF_6^- show considerable cation dependence, i.e. 663 cm^{-1} with ClO_2^+, 681 cm^{-1} with ClF_2^+.[237] νRuCl bands were assigned for $[Ru(L)(\eta^6\text{-}C_6H_6)Cl]^+$, where L = 2,3- or 2,5-bis(2-pyridyl)pyrazine (ca. 290 cm^{-1}), and related species.[238] (13), where P-O = (2-methoxyethyl)diphenylphosphine, gives a single IR band due to νRuCl (324 cm^{-1}), showing that there is a *trans*-$RuCl_2$unit.[239] νRuCl is near 340 cm^{-1} in $RuCl_2$-$(Ph_2PO_2CCMe{=}CHPh)_2$ and related complexes.[240] $[RuX_3(2mqn)(NO)]^-$, where 2mqn = 2-methyl-8-quinolinate, have νRuX at 323 cm^{-1} (Cl) or 261 cm^{-1} (Br).[241]

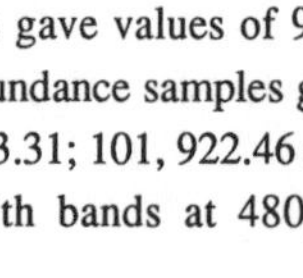

(13)

νOsH bands are at 2067 and 2005 cm^{-1} in $[OsH_3(dcpe)_2]^+$, where dcpe = 1,2-bis(dicyclohexylphosphino)ethane.[242] νOs-NO bands lie in the range 610 - 625 cm^{-1} for a large number of complexes containing the $Os(NO)^{3+}$ unit.[243] νOsN and νOsX modes were assigned for $R_2[Os^{IV}(NO)X_5]$, where R = Ph_4P^+, Ph_4As^+ or Et_4N^+, X = Cl or Br, $R[Os^{V}$-$(NO)X_5]$, where X = F, Cl or Br;[244] and *cis* -$OsCl_4(CH_3CN)_2$.[245]

νOs=N^tBu (terminal) bands are at 1208 - 1238 cm^{-1} in $Os(N^tBu)_4$, $[Os(N^tBu)2(\mu$-$N^tBu)]_2^{2+}$ and similar systems. νOs-μ-NtBu bands all lie below 1200 cm^{-1}.[246] The anionic complex $[Os(\equiv N)\{SCH_2CH_2C(O)O\}_2]^{2-}$ and related species have νOsN 1078 - 1095 cm^{-1}, consistent with such ligands being good electron-donors to the metal.[247]

Me_4Os=O has νOs=O at 1013 cm^{-1}.[248] Solid $Me_2Os(=O)_2py$ has νOsO_2 at 868 cm^{-1} suggesting octahedral coordination, i.e. polymerisation has occurred.[249] Other complexes for which assignments of OsO_2 modes have been made are: $[Os^{VI}L(O)_2]^{2+}$, where L = macrocyclic tertiary amine ligands;[250] OsO_2py_2(oxalato);[251] and OsO_2(SB), where SB = tetradentate Schiff base ligands from condensation of β-diketones with ethylene-diamine.[252] All of these contained *trans*-OsO_2units, but for $[OsO_2(S_2O_3)_2]^{2-}$ it appeared that the *cis* -OsO_2 configuration was preferred.[253]

7. Cobalt, Rhodium and Iridium

Earlier reference has been made to vibrational studies on $CoZrF_6$;[30] $CoQCl_3$, where Q = quinoxaline;[153] matrix-isolated $CoCl_2$;[217], NO adsorbed on Rh;[227] and SO_2 adsorbed on Rh-coated gold.[235]

The complex $CoH(^{14}N_2)(PMe_3)_3$ has νCoH at 1885 cm^{-1}, νCo-^{14}N at 465 cm^{-1}.[254] Terminal νCoH is at 1981 cm^{-1} in $[\{^iPr_2P(CH_2)_3PPr^i_2\}Co]_2(H)(\mu\text{-}H)_3$, together with νCoHCo at 1001 cm^{-1} and δCoH 756 cm^{-1}.[255]

The following νCo-CO modes were assigned from the IR and Raman spectra of $Co_6(CO)_8S_8$: 395 cm^{-1} (a_{1g}), 375 cm^{-1} (t_{1u}) and 360 cm^{-1} (e_g). νCoS modes were at: 315 cm^{-1} (a_{1g}), 315 cm^{-1} (t_{1u}) and 280 cm^{-1} (t_{2g}).[256] The Raman-active νCoC mode in photolabile methylcoenzyme B12 (methylcobalamin) was shown by near-IR FT Raman spectroscopy to be at 500 cm^{-1} (470 cm^{-1} in the CD_3 analogue).[257]

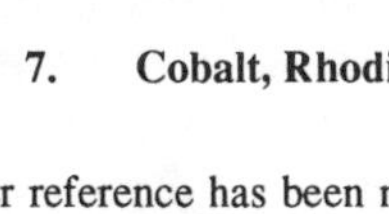

(14)

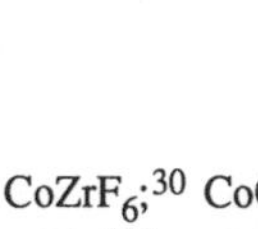

(15)

(14) has νCoGe at 255 cm^{-1} (this is the in-phase stretch of the external Ge-Co with the expanding mode of the $GeCo_2$ triangle.[258]

The IR spectra of $Cu_xCo_{2-x}(OH)_3NO_3$ show decreasing M-O force constants on replacing Cu by Co.[259] Phase transitions in Rb_2CoCl_4 and Rb_2ZnCl_4 were followed by IR spectroscopy. These were explicable in terms of deformations of the MCl_4 tetrahedron.[260]

The position and width of νRhRh in rhodium(II) *n* -alkanoates can be used to monitor solid/liquid phase transitions.[261] Theoretical studies on RhH^+ gave values for vibrational wavenumbers in many electronic levels. The $^2\Delta$ ground state ω_e value is 2243 cm^{-1}.[262] νMH modes were assigned for (15) in argon matrices, where M = Rh (2039, and 2027 cm^{-1}) or Ir (2157, 2144 and 2133 cm^{-1}).[263]

Trans -$(Ph_3P)_2RhCl_2$ has νRhP bands at 406, 423, 444 and 454 cm^{-1}, with νRhCl 310 cm^{-1}.[264] νRhCl bands (282, 262 cm^{-1}) were assigned for $\{Rh(cod)Cl\}_2(\mu\text{-dps})$, where dps = di-2-pyridyl sulphide.[265] $Cp^*MCl_2(H_2PPh)$ have νMCl_2 bands as follows: 275, 260 cm^{-1} (Rh), 285, 255 cm^{-1} (Ir).[266]

Cis -$[D_2Ir(cod)(NCPh)PPh_3]^+$ has νIrD at 1540 cm^{-1}.[267]

8. Nickel, Palladium and Platinum

Previous reference has been made to vibrational studies on $NiQCl_3.3H_2O$, where Q = quinoxaline;[153] matrix-isolated $NiCl_2$;[217] NO adsorbed on Pt;[227] SO_2 adsorbed on Pt-covered gold;[235] PtP_2 and $PtAs_2$.[236]

Trans -$NiH(Cl)(PBz_3)_2$ has νNiH at 1979 cm^{-1}.[268] The cations $[L_3MH]^+$ have been prepared, with L = Ph_3P, Ph_2MeP or Et_3P, M = Ni, Pd or Pt. νMH modes were assigned, e.g. 1965 cm^{-1} (Ni, PPh_3) or 2088 cm^{-1} (Pt, PPh_3).[269] The novel complex $Ni(CNCMe_3)_4$ has IR and Raman bands associated with νNiC_4 motions at 450, 400 and 348 cm^{-1}.[270]

$[Ni(en)_2(Ade)]^{2+}$, where Ade = adenine, has νNiN bands at 338 and 319 cm^{-1}.[271] The IR spectra of bis(L-serinato)nickel(II), with ^{58}Ni, ^{62}Ni and D isotopic data, show that the bands at 342, 297 and 222 cm-1 are due to coupled νNiO and νNiN modes.[272]

IR spectra of $NiL_2.2H_2O$, where HL = 1-phenyl-3-methyl-4-acyl-5-pyrazolone, show that νNiO increases with increasing number of carbon atoms on the 4-acyl substituent.[273]

The IR spectra of $Ni(morph)_2X_2(H_2O)_2$, where morph = 1,4-morpholine, X = Cl or Br, are consistent with octahedral nickel coordination. $ZnCl_2(morph)_2$ has skeletal modes in agreement with tetrahedral metal coordination.[274] Detailed assignments of skeletal modes were made for $[PdML_2]_n$, where M = Ni, Cu, Zn or Cd, H_2L = $H_2NCS\text{-}CSNH_2$ or MeHN-CSCSNHMe. These show that PdL_2 is acting as a tetradentate ligand with *N*-coordination to M. Extensive use was made of isotopic shifts.[275]

The matrix-IR spectrum of $NiCl_2$ in the ν_3 region is consistent with a bent molecule (angle ca. 130°) in a nitrogen matrix, but a linear molecule in argon. Thus the bonding forces are not strongly directional.[276] νNiX modes were assigned for $Ni[\,o\text{-}C_6F_4\text{-}(AsMe_2)_2]_2X_2$, where X = Cl (261 cm^{-1}) or Br (208 cm^{-1}).[277] The IR and Raman spectra of $(Et_4N)_2MCl_4$, where M = Ni or Cu, suggest that at 77 K the anions are vibronically stabilised in a static C_{2v} field.[278]

νPdN modes are assigned to 276 cm^{-1} (X = Cl) or 267 cm^{-1} (Br) for (16). The νPdX_2 modes are as expected for the *cis*-PdX_2 configuration.[279] Skeletal modes were assigned for (17), where R = phenyl or substituted phenyl (νPdN 320 - 395 cm^{-1}; νPdO 425 - 500 cm^{-1}; νPdCl 300 - 330 cm^{-1}.[280] νMP (340 - 400 cm^{-1}), νMX (280 - 310 cm^{-1} for Cl, 200 - 280

cm^{-1} for Br) modes in $MX_2(PMe_2py)_2$ show a *cis* configuration for M = Pd, X = Cl; M = Pt, X = Cl, Br or I, but a *trans* configuration for M = Pd, X = Br or I.[281]

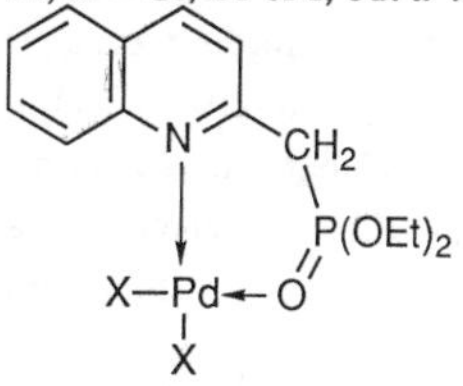

(16)

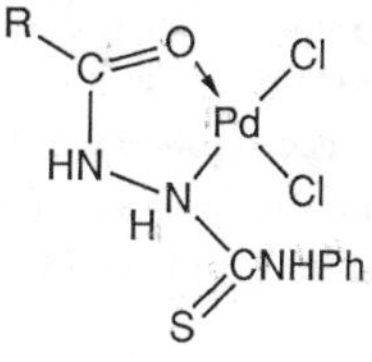

(17)

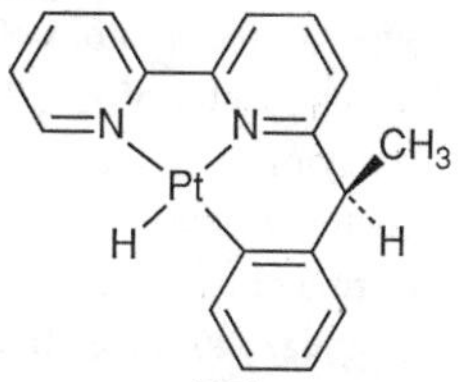

(18)

νMCl modes were assigned for $MLCl_2$, where M = Pd or Pt, L = 1'-methyl-2,4'-bipyridin-3'-ylium.[282] The Raman spectra of aqueous $PdCl_2$ solutions with a range of Cl^- concentrations gave evidence for $PdCl_x(OH_2)_{4-x}{}^{2-x}$, where x = 2, 3 or 4, as well as $Pd(OH)_2$.[283] $[Pd_2\{(CHCO_2R)_2PPh_2\}_2(\mu\text{-}Cl)]$, R = Me or Et, have νPdCl at 275 cm^{-1}.[284] The resonance Raman spectra of $[ML][MLX_2]Y_4$, where M = Pd or Pt; L = 3,7-diaza-nonane-1,9-diamine and related species, Y = BF_4, ClO_4 or PF_6, all contain long progressions in the symmetric chain stretching mode, ν_sXMX.[285] The FT Raman spectra of K_nMX_4, where M = Pd or Pt, n = 2; M = Au, n = 1, X = Cl or Br, show that in all cases the bromides have more bands than the chlorides, i.e. they have different structures.[286]

(18) has νPtH at 2153 cm^{-1}.[287] νMC (of C_6Cl_5), is near 600 cm^{-1} for a large range of complexes related to [*cis*-$Pt(C_6F_5)_2(C_6Cl_5)py]^-$.[288] The Raman spectrum of α-PtO_2 films contains sharp lines at 514 and 560 cm^{-1}. PtO has broad peaks at 438, 657 cm^{-1}.[289]

νPtCl modes were assigned in *trans*-$[Pt(NH_3)_2(L)Cl]^+$, where L = imidazole (336 cm^{-1}) or benzimidazole (342 cm^{-1}).[290] νPtX in $PtX_2(CO)(C_6H_{10})$ are at 340, 320 cm^{-1} (Cl), 231, 212 cm^{-1} (Br).[291] IR and Raman of $[Pt_2X_{10}]^{2-}$, where X = Cl or Br, show that $\nu PtCl_t$ is in the range 329 - 363 cm^{-1}; $\nu PtCl_{br}$ 272 - 299 cm^{-1}; $\nu PtBr_t$ 203 - 248 cm^{-1}, $\nu PtBr_{br}$ 174 - 187 cm^{-1}.[292] IR spectra of $Pt(dmso)(amine)Br_4$, where amine = $MeNH_2$, $EtNH_2$, py, piperidone (νPtBr) were used to follow the *cis/trans* isomerisation processes in solid and solution.[293] Resonance Raman spectra of $[Pten_2][Pten_2Br_2](ClO_4)_4$ show that the chain-axis mode is at 166 cm^{-1}.[294] The IR spectrum of *cis* -$Pt(NH_3)_2(SCN)_2I_2$ contains νPtI at 198 cm^{-1}.[295]

9. Copper, Silver and Gold

Earlier reference has been made to vibrational studies on: $MBa_2Cu_3O_{7-\delta}$, where M = Y or Gd;[2] $CuLn(MoO_4)_2$, where Ln = lanthanide;[118] $CuQCl_3.H_2O$, where Q = quinoxaline;[153] $Cu_xCo_{2-x}(OH)_3NO_3$;[259] $[PdCuL_2]_n$, where H_2L = $H_2NCSCSNH_2$ etc.;[275] $(Et_4N)_2Cu\text{-}Cl_4$;[278] $[Mo_2AgS_4(edt)_2(PPh_3)]^-$, where edt = ethane-1,2-dithiolato;[129] $AgReO_4$;[179] and $KAuX_4$, where X = Cl or Br.[286]

Ab initio calculations on the $Cu(\eta^2\text{-}H_2)Cl$ model predict νCuH 1539 and 935 cm^{-1}; νHH 3222 cm^{-1} and νCuCl 410 cm^{-1}.[296] CO adsorbed on Cu(100) and Cu(111) surfaces gives νCu-CO at 345, 346 cm^{-1}.[297] The Raman spectra of $M(cod)_2{}^+$, where M = Cu or Ag, have νM-cod in the range 350 - 420 cm^{-1}.[298]

νCuN bands are seen in the range 509 - 524 cm^{-1} (with νCuCl near 300 cm^{-1}) for $CuCl_2[C_{19}H_{18}NO_3(C_nH_{2n+1})]_2$, where n = 1, 6, 7, 9, 10, 12, showing coordination of copper to the *N*-atom of the alkoxyaniline.[299] νCu-N_3 bands were seen in the resonance Raman spectra of $Cu_2(EGTB)(N_3)(ClO_4)_3$, where EGTB = 1,1,10,10-tetrakis(2-benzimidazolylmethyl)-1,10-diaza-4,7-dioxadecane and related ligands.[300] νCuN(azide) and νCuN (L) stretches were assigned from the IR spectra of $CuL_2(N_3)_2$, $CuL(N_3)_2$, where L = a substituted pyridine.[301] νCuN (345/303 cm^{-1}) and νCuO (509 cm^{-1}) were assigned for bis(imidazole)copper(II) diacetate.[302]

νCuN modes were assigned for $[Cu\{C(CN)_3\}(Hpz)_4]^+$ (278, 242 cm^{-1}) and related complexes.[303] Skeletal modes (νCuN, νCuS and νCuX) were also assigned for $CuLCl_2$, $Cu_2L_3Cl_4$, $CuL(NO_3)$ and $CuL_2(NO_3)$, where L = 1-phenyl-4,6-dimethylpyrimidine-2-thione.[304]

Raman and IR spectra of CuO contain bands at 296 (a_g), 346 and 636 (b_g) cm^{-1} (Raman); 146, 480, 603 cm^{-1} (b_u), 164 and 542 cm^{-1} (a_u) (IR).[305] $Cu_4OCl_6(CH_3CN)_4$ has a characteristic νCu_4O band at 579 cm^{-1} in the IR, with νCuCl 291, 236 and 202 cm^{-1}.[306]

There continue to be many studies of the vibrational spectra of superconducting species and related systems: $YBa_2Cu_3O_{7-\delta}$;[307-12] $YBa_2(Cu_{1-x}M_x)_3O_{7-\delta}$, where M = Fe, Co, Ni or Zn;[313] $YBa_{2-x}Sr_xCu_3O_{7-\delta}$;[314] $Tl_mBa_2Ca_{n-1}Cu_nO_{2(n+1)+m}$;[315] $YBa_2Cu_4O_{8-\delta}$;[316] Pb-doped Bi/Sr/Ca/Cu single crystal;[317] $(Sm_{2/3}Ce_{1/3})(B_{2/3}Sm_{1/3})_2Cu_3O_9$;[318] $Nd_{2-x}Ce-CuO_{4-\delta}$.[319-20]

νCuS in $Cu(PPh_3)_2LCl$, where L is a heterocyclic thione, is seen near 300 cm^{-1}.[321] $[Cu^{II}(Et_4todit)X_2]_n$, where X = Cl or Br, Et_4todit = 4,5,6,7-tetrathiocino[1,2-*b*:3,4*b*']-diimidazolyl-1,3,8,10-tetraethyl-7,9-dithione, have νCuS near 300 and 245 cm^{-1}; νCuCl 333, 323 cm^{-1}.[322] Similar assignments were also given for $[Cu_2(Et_4todit)X_2]_n$, where X = Cl or Br, and $[Cu(Et_4todit)I]_n$.[323]

Raman and IR spectra were obtained for CuX_2^-. For X = Cl, ν_1 is at 302 cm^{-1}, ν_2 108 cm^{-1}, ν_3 407 cm^{-1}; for X = Br the corresponding figures are 194, 81 and 326 cm^{-1}. For $Cu_2I_4^{2-}$, there is a Raman band at 122 cm^{-1}, and IR features at 172 and 154 cm^{-1}.[324] $[Cu_3X_2(dppa)_3]^+$, where dppa = bis(diphenylphosphino)amine, has a ν_sXCu_3X mode at 165 cm^{-1} (Cl), 145 cm^{-1} (Br) or 125 cm^{-1} (I) in the Raman.[325] $[Cu_3(AsPh_3)_3I_4]^+$ has a far-IR band at 138 cm^{-1} due to νCuI of the Cu_3I_4 unit.[326]

The diode laser spectrum of AgD shows that ω_e is at 1759.1928(34) cm^{-1}.[327] SERS of maleimide adsorbed on colloidal silver shows a νAgN band at 248 cm^{-1}, and another feature at 158 cm^{-1}, believed to be νAg...O=C.[328]

Adsorption of O_2 on polycrystalline silver leads to dissociation and formation of AgO, with a stretching band at 351 cm^{-1}.[329] νAgS and νAgX modes were assigned for all of the following: $Ag(Me_4TMS)X$, where X = Cl, Br, I or NO_3; and νAgP for $[Ag(Me_4TMS)(PPh_3)_2]NO_3$, where Me_4TMS = tetramethylthiuram monosulphide.[330] The cubane-type complexes $[LAgX]_4$, where L = 1-phenyl-3,4-dimethylphosphole, X = Cl or Br, have νAgX at 188, 140 cm^{-1} (Cl), 152, 114 cm^{-1} (Br). For $[L_2AgX]_2$, νAgX modes are: 201, 130 cm^{-1} (Cl), 145, 129 cm^{-1} (Br), 129 cm^{-1} (I).[331]

$Au(CO)_2^+UF_6^-$ has $\nu_{as}AuC_2$ at 398 cm^{-1}.[332] For $Au_2(tmb)Cl_2$, where tmb = 2,5-dimethyl-2,5-diisocyanohexane, νAuC are at 417 (IR), 390 (Raman) cm^{-1}, νAuCl 352 (IR), 359 (Raman) cm^{-1}. The solid phase Raman spectrum gives a band at 50 cm^{-1}, assigned as an intermolecular νAu....Au mode.[333]

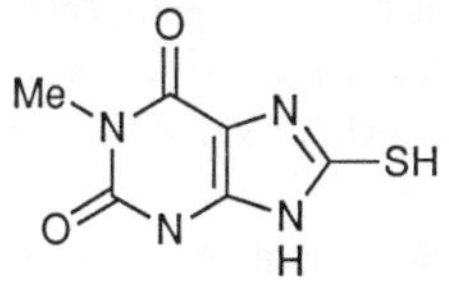

(19)

"$Au(SO_3F)_2$" has $\nu Au^{III}O_4$ at 680 cm^{-1} and $\nu Au^{I}O_2$ at 648 cm^{-1}.[334] $Au(PPh_3)(HL)$, where H_2L = (19), has νAuS at 365 cm^{-1}.[335] In $[Au(tmtu)_2Br_2]^+$, where tmtu = tetramethylthiourea, νAu^{III}-S is at 320 cm^{-1}, νAu^{III}-Br 258 cm^{-1} (doubtless subject to some mixing).[336] $[Au_2(Se_2)(Se_3)]^{2-}$ has νAuSe at 236 cm^{-1}.[337]

The Raman spectra of $M(AuF_6)_2$, where M = Mg, Ca, Sr or Ba, all contain $\nu Au^{V}F_6$ modes consistent with comparatively isolated octahedral anions.[338] The complex $Au(C_6H_4N{=}NPh\text{-}2)(R)Cl$, where R = $C_6H_4NO_2$-2 or C_6F_5, show νAuCl at 300 - 310 cm^{-1}, showing that the Cl is *trans* to an aryl group.[339] Raman and resonance Raman spectra of aqueous $AuCl_3$ solutions were used to identify the range of species present.[340] The Raman spectra of gold chloride complexes show that below 100°C $AuCl_4^-$ is dominant, while above 100°C the main species is $AuCl_2^-$.[341]

10. Zinc, Cadmium and Mercury

Reference has already been made to vibrational studies on $ZnQ_xCl_3.H_2O$, where Q = quinoxaline, x = 1 or 2;[153] matrix-isolated $ZnCl_2$;[217] Rb_2ZnCl_4;[260] $ZnCl_2(morph)_2$, where morph = 1,4-morpholine;[274] $[PdML_2]_n$, where M = Zn or Cd; H_2L = $H_2NCSCSNH_2$ and related ligands;[275] $CdBiTh(VO_4)$.[62]

$[\eta^3\text{-}HB(3\text{-}^tBupz)_3]ZnH$ has νZnH at 1770 cm^{-1} (νZnD at 1270 cm^{-1}), where 3-^{t}Bupz = 3-$C_3N_2{}^tBuN_2$.[342] The diode laser IR spectra of ZnD and CdD gave the following ω_e values: Zn 1597.2021(73) cm^{-1}; Cd 1443.9970(27) cm^{-1}.[343]

A normal coordinate analysis has been performed on $ZnCr_2O_4$. The ZnO_4 units are less ionic than CrO_6.[344] $^{64}Zn/^{70}Zn$ and $^{106}Cd/^{116}Cd$ isotopic substitution were used to identify modes due to Zn and Cd motions in $CdZnIn_2O_5$.[345] The $\nu_{as}Zn_4O$ modes in $Zn_4O(RCOO)_6$, where R = CH_3, C_2H_5, C_3H_7 etc., all are close to 520 cm^{-1}.[346]

FT Raman and IR data for $M(ER)_2$, where M = Zn or Cd, E = S or Se, R = 2,4,6-$R'_3C_6H_2$, R' = Me, ^{i}Pr or ^{t}Bu, gave values for νME modes, e.g. for $Cd[EC_6H_2{}^iPr_3\text{-}2,4,6]$, νCdS is at 285/279 cm^{-1}, νCdSe 205 cm^{-1}.[347]

Ab initio calculations of vibrational wavenumbers gave good agreement with experiment for $ZnCl_4^{2-}$ and $Zn(H_2O)_6^{2+}$.[348] Raman spectra of (Li,K)Cl - MCl_2 melts, where M = Zn or Cd, showed that MCl_4^{2-} species are present at low concentrations of MCl_2, but that $M_2Cl_6^{2-}$ ions are present at higher MCl_2 concentrations.[349] Analogous results were found for mixed (Li,K)Cl - $CdCl_2$ - $ZnCl_2$ melts.[350] Raman and far-IR spectra of aqueous solutions of ZnI_2 show the presence of $[Zn(H_2O)_6]^{2+}$, $[Zn(H_2O)_5I]^+$ (i.e. octahedral) but also $Zn(H_2O)_2I_2$ and $[Zn(H_2O)I_3]^-$ (i.e. tetrahedral).[351]

Bands are seen due to discrete CdX_4^{2-} anions in the IR and Raman spectra of $(LH_2)(CdX_4).H_2O$, where LH_2 = N-benzylpiperazinium, X = Cl, Br or I.[352] Glasses in the system $CdCl_2$ - $PbCl_2$ give Raman spectra consistent with the presence of $CdCl_6$ units.[353]

PhHgX, where X = ethylxanthate or *N,N'*-diethyldithiocarbamate, have νHgC at 455 cm^{-1}, νHgS 245 - 250 cm^{-1}.[354] ν_{as}NHgN is at 385 cm^{-1} in $Hg[Ph\text{-}C(NSiMe_3)_2]_2$.[355] Assignments to skeletal modes have been proposed for $Hg(EC_6H_2R_3\text{-}2,4,6)_2$, where E = S or Se, R = Me, ^{i}Pr or ^{t}Bu; E = Te, R = Me or ^{i}Pr.[356]

Solutions of mercury in molten HgX_2 or HgXX', where X, X' = F, Cl, Br or I, give Raman bands due to Hg_2X_2 and Hg_2XX' species. νHgHg bands were seen in all cases.[357] νHgX modes were assigned for $Hg(DMTP)_2X_2$, $[Hg(DMTP)X_2]_n$, where DMTP = 2,6-dimethyl-(4 *H*-pyran-4-thione), X = Cl, Br or I.[358]

11. Actinides

Earlier reference has been made to vibrational studies on $Th(TPP)_2$;[7] $Th(O_2)(SNNS)$, where SNNS = Schiff base $C_{30}H_{26}N_4S_4$;[26] $MBiTh(VO_4)$, where M = Mn or Cd; [62] and $U(O)(O_2)(ONNO)$, where ONNO = Schiff base $C_{14}H_{10}N_2O_2$.[93]

The 6d → 5f fluorescence spectrum of $PaCl_6^{2-}$ in a Cs_2ZrCl_6 crystal contains a progression in ν_s of $PaCl_6^{2-}$ (310 cm^{-1}).[359]

UO_2(salen)(DMF), where salen = *N,N'* -ethylenebis(salicylideneimine), has νUN (salen) 275 cm^{-1}, νUO (salen) 517 cm^{-1}, νUO (DMF) 250 cm^{-1}, $\nu_{as}UO_2$ 916/895 cm^{-1} and ν_sUO_2 840/830 cm^{-1}.[360] Raman microprobe analysis of uranium - containing minerals gave values of ν_sUO_2 which depend on the local environment, and are helpful in identifying uranium-bearing phases in high-level nuclear waste.[361] νUO_2 modes were also assigned in $(UO_2)_2(C_2O_4)X_2L_4$, where X = NCS^-, Cl^- or Br^-, L = neutral ligand,[362] $[UO_2L_2]X_2.nH_2O$, where L = 5-nitro-2-(phenylamino)benzoic acid, X =NO_3, Br, Cl or OAc,[363] and $Me_4N[UO_2(XO_4)(H_2O)_2]Z$, where X = S or Se, Z = Cl or Br. [364] IR bands were also assigned for $SrUO_4$.[365]

K_4USe_8 has νUSe modes at 167.6 and 153 cm^{-1}.[366] The $\nu_1+\nu_3$ band of UF_6 has been studied by high-resolution IR. The band centre is at 1294.329(3) cm^{-1}.[367] The IR spectrum of $CsUCl_4.4H_2O$ has been reported and assigned.[368]

Table 8 gives a summary of the vibrational assignments for matrix-isolated NpF_6 in its electronic ground state. There was also some evidence for bands associated with low-lying excited states.[369]

Table 8 Vibrational assignments for matrix-isolated NpF_6 ($/cm^{-1}$)

ν_1	646
ν_2	525
ν_3	618
ν_4	198
ν_5	208
ν_6	169

REFERENCES

1. M.M.Metallinou, L.Nalbandian, G.N.Papatheodorou, W.Voigt and H.H.Emons, *Inorg. Chem.*, 1991, **30**, 4260.
2. V.G.Hadjiev, M.V.Abrashev, M.N.Iliev and L.N.Bozukov, *Physica, C*, 1990, **171**, 257.
3. I.Bueno, C.Parada, R.Saez Puche, I.L.Botto and E.J.Baran, *J. Less-Common Metal.*, 1991, **169**, 105.
4. S.Sudoh, S.Ohta and S.Katagiri, *Sci. Rep. Hirosaki Univ.*, 1990, **37**, 23 (*Chem. Abs.*, 1991, **114**, 174027).
5. L.Yang, R.Yang and W.Dong, *Lanzhou Daxua Xuebao, Zirankexueban*, 1989, **25**, 56 (*Chem. Abs.*, 1991, **114**, 155833).
6. G.W.Graham, W.H.Heber, C.R.Peters and R.Usman, *J. Catal.*, 1991, **130**, 310.
7. T.N.Lomova, L.G.Andrianova and B.D.Berezin, *Russ. J. Inorg. Chem.*, 1991, **36**, 361.
8. S.N.Misra, W.N.Cabalfin, K.Anjaiah and G.Joseph, *Chem. Scr.*, 1989, **29**, 333.
9. Y.Skata, H.Imamura and S.Tsuchiya, *Chem. Lett.*, 1991, 349.
10. Y.Li, M.Gong, Y.Yang, M.Li and R.Chen., *Wuji Huaxue Xuebao*, 1990, **6**, 249 (*Chem. Abs.*, 1991, **114**, 177223).
11. Z.L.Xiao, R.H.Hauge and J.L.Margrave, *J. Phys. Chem.*, 1991, **95**, 2696.
12. C.J.Dain, A.J.Downs, M.J.Goode, D.G.Evans, K.T.Nicholls, D.W.H.Rankin and H.E.Robertson, *J. Chem. Soc., Dalton Trans.*, 1991, 967.
13. D.C.McKean, G.P.McQuillan, I.Torto, N.C.Bednell, A.J.Downs and J.M.Dickinson, *J. Mol. Struct.*, 1991, **247**, 73.
14. J.Okuda and E.Herdtweck, *Inorg. Chem.*, 1991, **30**, 1516.
15. M.Ocaña and C.J.Serna, *Spectrochim. Acta, A*, 1991, **47A**, 765.
16. L.V.Golubeva, N.V.Porotnikov, O.I.Kondratov and K.I.Petrov, *Russ. J. Inorg. Chem.*, 1990, **35**, 1028.
17. L.V.Golubeva, N.V.Porotnikov, O.I.Kondratov and K.I.Petrov, *Russ. J. Inorg. Chem.*, 1991, **36**, 1032.
18. N.Yao, X.Li and X.Song, *Spectrosc. Lett.*, 1990, **23**, 1153.
19. M.Nofz, T.Hübert and K.Witke, *Z. Chem.*, 1990, **30**, 335.
20. T.Hübert, U.Banach, K.Witke and P.Reich, *Phys. Chem. Glasses*, 1991, **32**, 58.
21. B.S.Ault, *Inorg. Chem.*, 1991, **30**, 2483.
22. P.K.Gowik, T.M.Klapötke and S.Cameron, *J. Chem. Soc., Dalton Trans.*, 1991, 1433.
23. N.S.S.Jalil, *Bull. Chem. Soc. Ethiop.*, 1989, **3**, 129 (*Chem. Abs.*, 1991, **114**, 93951).
24. K.K.Verma, O.P.Agrawal and A.Saini, *Synth. React. Inorg. Met.-Org. Chem.*, 1990, **20**, 851.
25. S.A.King and J.Schwartz, *Inorg. Chem.*, 1991, **30**, 3771.
26. M.T.H.Tarafder and A.R.Khan, *Polyhedron*, 1991, **10**, 973.
27. C.H.Perry, F.Lu, D.W.Liu and B.Alzyab, J. *Raman Spectrosc.*, 1990, **21**, 577.
28. L.V.Golubeva, N.V.Porotnikov, O.I.Kondratov and K.I.Petrov, *Russ. J. Inorg. Chem.*, 1991, **35**, 1480.
29. C.R.Bhattacharjee, M.Bhattacharjee, M.K.Chaudhuri and S.Choudhury, *Polyhedron*, 1990, **9**, 1653.
30. V.Rodriguez, M.Couzi and C.Sourisseau, *J. Phys. Chem. Solids*, 1991, **52**, 769.
31. A.Rahten and S.Miličev, *Thermochim. Acta*, 1990, **171**, 185.
32. L.Samek, J.Wasylak, J.M.Rincon and P.Callejas, *Bol. Soc. Esp. Ceram. Vidrio*, 1990, **29**, 229.
33. C.C.Phifer, D.J.Gosztola, J.Kieffer and C.A.Angell, *J. Chem. Phys.*, 1991, **94**, 3440.
34. K.Balasubramanian and K.K.Das, *J. Mol. Spectrosc.*, 1991, **145**, 142.
35. L.I.Strunkina, M.Kh.Minacheva, Z.S.Klemenkova and B.V.Lokshin, *Metalloorg. Khim.*,

1991, **4**, 819.
36. P.Lukose, S.C.Bhatia and A.K.Narula, *J.Organometal. Chem.*, 1991, **403**, 153.
37. A.McCamley and R.N.Perutz, *J. Phys. Chem.*, 1991, **95**, 2738.
38. R.Srivastava and S.Sarkar, *Inorg. Chim. Acta*, 1990, **176**, 27.
39. W.Massa, S.Wocadlo, S.Lotz and K.Dehnicke, *Z. anorg. allg. Chem.*, 1990, **584**, 79.
40. T.A.Kabanos and J.D.Woollins, *J. Chem. Soc., Dalton Trans.*, 1991, 1347.
41. F.D.Hardcastle and I.E.Wachs, *J. Phys. Chem.*, 1991, **95**, 5031.
42. A.Zinn, U.Patt-Siebel, U.Müller and K.Dehnicke, *Z. anorg. allg. Chem.*, 1990, **591**, 137.
43. E.M.Nour, A.N.Alnaimi and A.B.Alsada, *J. Phys. Chem. Solids*, 1990, **51**, 907.
44. P.Schwendtke and K.Volka, *Acta Fac. Rerum Nat. Univ. Comenianae, Chim.*, 1990, **38**, 41.
45. K.Sakata, F.Yamaura and M.Hashimoto, *Synth. React. Inorg. Met.-Org. Chem.*, 1990, **20**, 1043.
46. S.A.A.Zaidi, T.A.Khan and S.R.A.Zaidi, *Ind. J. Chem., Sect. A*, 1990, **29A**. 916.
47. R.N.Mohanty, V.Chakravorty and K.C.Dash, *Ind. J. Chem., Sect. A*, 1991, **30A**, 454.
48. D.E.Hamilton, *Inorg. Chem.*, 1991, **30**, 1670.
49. U.A.Bagal, C.B.Cook and T.L.Reichel, *Inorg. Chim. Acta*, 1991, **181**, 57.
50. H.S.Yadav, *Bull. Soc. Chim. France*, 1990, 641.
51. A.Hills, D.L.Hughes, G.J.Leigh and J.R.Sanders, *J. Chem. Soc., Dalton Trans.*, 1991, 61.
52. H.Toftlund, S.Larsen and K.S.Murray, *Inorg. Chem.*, 1991, **30**, 3964.
53. A.Hills, D.L.Hughes, G.J.Leigh and J.R.Sanders, *J. Chem. Soc., Chem. Comm.*, 1991, 827.
54. K.Kanamori, Y.Ookubo, K.Ino, K.Kawai and H.Michibata, *Inorg. Chem.*, 1991, **30**, 3832.
55. M.Guido and S.Nunziate Cesaro, *J. Phys. Chem.*, 1991, **95**, 2836.
56. L.V.Kristallov and L.A.Perelyaeva, *Russ. J. Inorg. Chem.*, 1991, **36**, 879.
57. G.A.Kourouklis, A.Jayaraman, G.P.Espinosa and A.S.Cooper, *J. Raman Spectrosc.*, 1991, **22**, 57.
58. M.Schraml-Marth, A.Wokaun, M.Pohl and H.L.Krauss, *J. Chem. Soc., Farad. Trans.*, 1991, **87**, 2635.
59. T.J.Dines, C.H.Rochester and A.M.Ward, *J. Chem. Soc., Farad. Trans.*, 1991, **87**, 653.
60. W.Han, W.Wang and F.Ouyang, *Shigou Xuebao, Shigou Jiagong*, 1990, **6**, 74 (*Chem. Abs.*, 1991, **115**, 100199).
61. L.V.Kristallov, V.L.Volkov and L.A.Perelyaev, *Russ. J.Inorg. Chem.*, 1990, **35**, 1031.
62. M.A.Nabar and S.S.Mangaonkar, *J. Solid State Inorg. Chem.*, 1991, **28**, 549.
63. S.Himeno, T.Osakai and A.Saito, *Bull. Chem. Soc. Japan*, 1991, **64**, 21.
64. A.Müller, J.Döring and H.Bögge, *J. Chem. Soc., Chem. Comm.*, 1991, 273.
65. A.Müller, J.Döring, M.I.Khan and V.Wittneben, *Angew. Chem., Int. Ed. Engl.*, 1991, **30**, 210.
66. I.R.Beattie, P.J.Jones, A.D.Wilson and N.A.Young, *High Temp. Sci.*, 1990, **29**, 53.
67. K.K.Das and K.Balasubramanian, *J. Mol. Spectrosc.*, 1990, **144**, 245.
68. J.M.Jehng and I.E.Wachs, *J. Phys. Chem.*, 1991, **95**, 7373.
69. J.M.Jehng and I.E.Wachs, *J. Raman Spectrosc.*, 1991, **22**, 83.
70. A.Bashall, V.C.Gibson, T.P.Kee, M.McPartlin, O.B.Robinson and A.Shaw, *Angew. Chem., Int. Ed. Engl.*, 1991, **30**, 940.
71. V.G.Kessler, S.Yu.Vasil'ev, A.I.Belokun and N.Ya.Turova, *Zh. Obshch. Khim.*, 1990, **60**, 2629.
72. A.I.Agulyanskii, V.E.Zavodnik, V.Ya.Kuznetsov, N.V.Sidorov, S.Yu.Stefanovich, D.V.Tsikaeva and V.T.Kalinnikov, *Izv. Akad. Nauk SSSR, Neorg. Mater.*, 1991, **27**, 1055.
73. F.D.Hardcastle and I.E.Wachs, *Solid State Ionics*, 1991, **45**, 201.
74. B.Siewert, G.Kollner, K.Ruhlandt-Senge, F.Schmock and U.Müller, *Z. anorg. allg. Chem.*, 1991, **593**, 160.

75. E.Hahn and R.Hebisch, *Spectrochim. Acta, A* 1991, **47A**, 1097.
76. J.H.von Barner, E.Christensen, N.J.Bjerrum and B.Gilbert, *Inorg. Chem.*, 1991, **30**, 561.
77. K.Harder and W.Preetz, *Z. anorg. allg. Chem.*, 1990, **591**, 32.
78. W.Preetz and K.Harder, *Z. anorg. allg. Chem.*, 1991, **597**, 163.
79. D.C.Bradley, M.B.Hursthouse, A.J.Howes, A.N.de M.Jelfs, J.D.Runnacles and M.Thornton-Pett, *J. Chem. Soc., Dalton Trans.*, 1991, 841.
80. A.M.Heyns, K.J.Range and M.Wildenauer, *Spectrochim. Acta, A.* 1990, **46A**, 1621.
81. T.N.Day, P.J.Hendra, A.J.Rest and A.J.Rowlands, *Spectrochim. Acta, A* 1991, **47A**, 1251.
82. M.Scheer, T.T.Nam, K.Schenzel, E.Herrmann, P.G.Jones, V.P.Fedin, V.N.Igorski and V.E.Fedorov, *Z. anorg. allg. Chem.*, 1990, **591**, 221.
83. H.Hori, M.Tsubaki, N.T.Yu and T.Yonetani, *Biochim. Biophys. Acta*, 1991, **1077**, 392.
84. W.Levason, J.S.Ogden, A.K.Saad, N.A.Young, A.K.Brisdon, P.J.Holliman, J.H.Holloway and E.G.Hope, *J. Fluorine Chem.*, 1991, **53**, 43.
85. K.W.Lee and P.E.Hoggard, *Inorg. Chem.*, 1991, **30**, 264.
86. I.Bueno, C.Parada, R.S.Puche, I.L.Botto and E.J.Baran, *J. Phys. Chem. Solids*, 1990, **51**, 1117.
87. V.R.Fisher, T.A.Nasanova, Kh.M.Yakubov, V.K.Vorunkova, L.V.Mosina and Yu.V.Yablokov, *Russ. J. Inorg. Chem.*, 1990, **35**, 1298.
88. W.Clegg, R.J.Errington, D.C.R.Hockless, A.D.Glen and D.G.Richards, *J. Chem. Soc., Chem. Comm.*, 1990, 1565.
89. T.A.George, L.Ma, S.N.Shailh, R.C.Tisdale and J.Zubieta, *Inorg. Chem.*, 1990, **29**, 4789.
90. Z.Grebeyhu, F.Weller, B.Neumüller and K.Dehnicke, *Z. anorg. allg. Chem.*, 1991, **593**, 99.
91. C.G.Young, F.Janos, M.A.Bruck, P.A.Wexler and J.H.Enemark, *Aust. J. Chem.*, 1990, **43**, 1347.
92. M.Scheer, T.T.Nam, E.Herrmann, V.P.Fedin, V.N.Ikorski and V.E.Fedorov, *Z. anorg. allg. Chem.*, 1990, **589**, 214.
93. M.T.H.Tarafder and A.R.Khan, *Polyhedron*, 1991, **10**, 819.
94. C.Jegat, M.Fouassier and J.Mascetti, *Inorg. Chem.*, 1991, **30**, 1521.
95. W.A.Oertling and R.Hille, *J. Biol. Chem.*, 1990, **265**, 17446.
96. J.Szklarzewicz, A.Samotus, N.W.Alcock and M.Moll, *J. Chem. Soc., Dalton Trans.*, 1990, 2959.
97. X.E.Yan and C.G.Young, *Aust. J. Chem.*, 1991, **44**, 361.
98. D.Sellmann, B.Seubert, F.Knoch and M.Moll, *Z. anorg. allg. Chem.*, 1991, **600**, 95.
99. D.Coucouvanis, A.Hadjikyriacou, A.Toupadakis, S.M.Koo, O.Ileperuma, M.Draganjac and A.Salifoglu, *Inorg. Chem.*, 1991, **30**, 754.
100. G.Backes, J.H.Enemark and T.M.Loehr, *Inorg. Chem.*, 1991, **30**, 1839.
101. R.Ratnani, G.Srivastava and R.C.Mehrotra, *Transition Met. Chem.*, 1991, **16**, 204.
102. R.N.Mohanty, V.Chakravortty and K.C.Dash, *Ind. J. Chem., Sect. A*, 1991, **30A**, 457.
103. R.N.Mohanty, V.Chakravortty and K.C.Dash, *Polyhedron*, 1991, **10**, 33.
104. J.O.Dziegelewski, K.Filipek and B.Jezowska-Trzebiatowska, *Inorg. Chim. Acta*, 1990, **171**, 89.
105. M.G.Felin and A.M.Levitskii, *Russ. J. Inorg. Chem.*, 1990, **35**, 971.
106. F.A.Cotton, R.L.Luck and C.S.Miertschin, *Inorg. Chem.*, 1991, **30**, 1155.
107. A.A.Eagle, M.F.Mackay and C.G.Young, *Inorg. Chem.*, 1991, **30**, 1425.
108. J.Fletcher, G.Hogarth and D.A.Tocher, *J. Organometal. Chem.*, 1991, **403**, 345.
109. G.Hogarth, P.C.Konidaris and G.C.Sanders, *J. Organometal. Chem.*, 1991, 406, 153.
110. M.Gorzelik, H.Bock, L.Gang, B.Nuber and M.L.Ziegler, *J. Organometal. Chem.*, 1991, **412**, 95.
111. F.D.Hardcastle and I.E.Wachs, *J. Raman Spectrosc.*, 1990, **21**, 683.
112. V.V.Olifirenko, A.A.Davydov and V.N.Pak, *Zh. Prikl. Khim. (Leningrad)*, 1990, **63**, 2505.

113. S.Morisaki and N.Baba, *Deuki Kagaku Oyobi Kogyo Butsuri Kagaku*, 1990, **58**, 852 (*Chem. Abs.*, 1991, **114**, 52791).
114. M.Isaac, N.Santha and V.U.Nayar, *J. Raman Spectrosc.*, 1991, **22**, 237.
115. C.G.Hill and J.H.Wilson, *J. Mol. Catal.*, 1990, **63**, 65.
116. M.V.Mokhosoev, I.I.Murzakhanova, N.M.Kozhevnikova and V.V.Fomichev, *Russ. J. Inorg. Chem.*, 1991, **36**, 724.
117. V.V.Fomichev, V.A.Efremov, O.I.Kondratov and V.B.Vyazovov, *Russ. J. Inorg. Chem.*, 1991, **36**, 883.
118. A.P.Perepelitsa, V.N.Ishchenko, Z.M.Alekseeva and V.V.Fomenko, *Russ. J. Inorg. Chem.*, 1991, **36**, 4.
119. V.V.Safonov, N.G.Chaban, N.P.Kuz'mina, A.A.Vashman and K.I.Petrov, *Russ. J. Inorg. Chem.*, 1990, **35**, 1226.
120. V.Ramakrishnan, N.Krishnamurthy and M.M.A.Jinnah, *Spectrochim. Acta, A* 1991, **47A**, 979.
121. R.G.Bhattacharyya and S.Biswas, *Inorg. Chim. Acta,* 1991, **181**, 213.
122. B.Courtney, H.G.M.Edwards and P.Robson, *J. Mol. Struct.*, 1990, **239**, 139.
123. M.Fournier, R.Thouvenot and C.Rocchiccioli-Deltcheff, *J. Chem. Soc.,* Farad. Trans., 1991, **87**, 349.
124. S.Himeno, K.Myashita, A.Saito and T.Hori, *Chem. Lett.*, 1990, 799.
125. S.Gruber, L.Kilpatrick, N.R.Bastian, K.V.Rajagopalan and T.G.Spiro, *J. Am. Chem. Soc.*, 1990, **112**, 8179.
126. S.B.Kumar and M.Chaudhury, *J. Chem. Soc., Dalton Trans.*, 1991, 1149.
127. V.P.Fedin, B.A.Kolesov, Yu.V.Mironov, O.A.Geras'ko and V.E.Fedorov, *Polyhedron,* 1991, **10**, 997.
128. V.P.Fedin, Yu.V.Mironov, M.N.Sokolov, B.A.Kolesov, V.E.Fedorov, D.S.Yufit and Yu.T.Struchkov, *Inorg. Chim. Acta,* 1990, **174**, 275.
129. N.Zhu, X.Wu and J.Lu, *J. Chem. Soc., Chem. Comm.*, 1991, 235.
130. P.K.Bharadwaj, *Inorg. Chim. Acta,* 1990, **178**, 165.
131. E.Rentscher, W.Massa, S.Vogler, K.Dehnicke, D.Fenske and G.Baum, *Z. anorg. allg. Chem.*, 1991, **592**, 59.
132. N.Bao and J.M.Winfield, *J. Fluorine Chem.*, 1990, **50**, 339.
133. E.J.Baran, I.L.Botto, N.Kinomura and N.Kumada, *J. Solid-State Chem.*, 1990, **89**, 144.
134. V.C.Gibson, T.P.Lee and A.Shaw, *Polyhedron,* 1990, **9**, 2293.
135. V.P.Babaeva and V.Ya.Rosolovskii, *Russ. J. Inorg. Chem.*, 1990, **35**, 944.
136. M.A.Habib and S.P.Maheshwari, *J. Electrochem. Soc.*, 1991, **138**, 2029.
137. T.Nanba, Y.Nishiyama and I.Yasui, *J. Mater. Res.*, 1991, **6**, 1324.
138. V.L.Balashov, A.A.Kharlanov, O.I.Kondratov and V.V.Fomichev, *Russ. J. Inorg. Chem.*, 1991, **36**, 254.
139. B.L.George and G.Aruldhas, *J. Raman Spectrosc.*, 1991, **21**, 241.
140. P.Judeinstein, C.Deprun and L.Nadjo, *J. Chem. Soc., Dalton Trans.*, 1991, 1991.
141. M.J.Abrams, C.E.Costello, S.N.Shaikh and J.Zubieta, *Inorg. Chim. Acta,* 1991, **180**, 9.
142. T.Nanba, S.Takano, I.Yasui and T.Kudo, *J. Solid-State Chem.*, 1991, **90**, 47.
143. R.Wollert, E.Rentschler, W.Massa and K.Dehnicke, *Z. anorg. allg. Chem.*, 1991, **596**, 121.
144. D.L.Hughes, J.D.Lane and R.L.Richards, *J. Chem. Soc., Dalton Trans.*, 1991, 1627.
145. X.Yang, G.K.W.Freeman, T.B.Rauchfuss and S.R.Wilson, *Inorg. Chem.*, 1991, **30**, 3034.
146. V.P.Fedin, M.N.Sokolov, O.A.Geras'ko, B.A.Kolesov, V.E.Fedorov, A.V.Mironov, D.S.Yufit, Yu.L.Slovokhotov and Yu.T.Struchkov, *Inorg. Chim. Acta,* 1990, **175**, 217.
147. S.N.Misra and S.B.Mehta, *Ind. J. Chem., A*, 1991, **30A**, 731.
148. A.Werth, K.Dehnicke, D.Fenske and G.Baum, *Z. anorg. allg. Chem.*, 1990, **591**, 125.
149. R.J.M.Konings and A.S.Booij, *J. Mol. Spectrosc.*, 1991, **148**, 513.

150. A.D.Kirkwood, K.D.Bier, J.K.Thompson, T.L.Haslett, A.S.Huber and M.Moskovits, *J. Phys. Chem.*, 1991, **95**, 2644.
151. R.D.Urban and H.Jones, *Chem. Phys. Lett.*, 1991, **178**, 295.
152. R.S.Czernuszewicz, W.D.Wagner, G.B.Ray and K.Nakamoto, *J. Mol. Struct.*, 1991, **242**, 99.
153. P.O.Lumme and S.Lindroos, *Thermochim. Acta*, 1990, **162**, 375.
154. E.L.Varetti, *J. Raman Spectrosc.*, 1991, **22**, 307.
155. R.J.Nick, G.B.Ray, K.M.Fish, T.G.Spiro and J.T.Groves, *J. Am. Chem. Soc.*, 1991, **113**, 1838.
156. A.Weselucha-Birczynska, L.M.Proniewicz, K.Bajdor and K.Nakamoto, *J. Raman Spectrosc.*, 1991, **22**, 315.
157. A.A.Artyukhov, V.D.Klimov, S.V.Krasulin, Sh.Sh.Nabiev and N.S.Tolmacheva, *Koord. Khim.*, 1990, **16**, 1348.
158. D.M.Adams, P.Ruff and D.R.Russell, *J. Chem. Soc.*, Farad. Trans., 1991, **87**, 1831.
159. J.Baldas and S.F.Colmanet, *Inorg. Chim. Acta*, 1990, **176**, 1.
160. J.Baldas, S.F.Colmanet and G.A.Williams, *Inorg. Chim. Acta*, 1991, **179**, 189.
161. J.Baldas, S.F.Colmanet and G.A.Williams, *J. Chem. Soc., Dalton Trans.*, 1991, 1631.
162. U.Abram, J.Hartung, L.Beyer, J.Stach and R.Kirmse, *Z. Chem.*, 1990, **30**, 180.
163. F.Tisato, U.Mazzi, G.Bandoli, G.Cros, M.-H.Darbieu, Y.Coulais and R.Guiraud, *J. Chem. Soc., Dalton Trans.*, 1991, 1301.
164. J.Cook, W.M.Davis, A.Davison and A.G.Jones, *Inorg. Chem.*, 1991, **30**, 1773.
165. U.Abram, S.Abram, W.Hiller, G.F.Morgan, J.R.Thornback, M.Deblaton and J.Stach, *Z. Naturforsch., B*, 1991, **46b**, 453.
166. R.Alberto, G.Anderegg and A.Albinati, *Inorg. Chim. Acta*, 1990, **178**, 125.
167. Y.Kim, J.Gallucci and A.Wojcicki, *J. Am. Chem. Soc.*, 1990, **112**, 8600.
168. S.W.Carr, X.L.R.Fontaine and B.L.Shaw, *J. Chem. Soc., Dalton Trans.*, 1991, 1025.
169. X.L.Luo and R.H.Crabtree, *J. Chem. Soc., Dalton Trans.*, 1991, 587.
170. G.Böhm, K.Wieghardt, B.Nuber and J.Weiss, *Inorg. Chem.*, 1991, **30**, 3464.
171. C.Reber and J.I.Zink, *Inorg. Chem.*, 1991, **30**, 2994.
172. T.N.Rao, D.Adhikesavalu, A.Camerman and A.R.Fritzberg, *Inorg. Chim. Acta*, 1991, **180**, 63.
173. U.Abram, S.Abram, J.Stach, R.Wollert, G.F.Morgan, J.R.Thornback and M.Deblaton, *Z. anorg. allg. Chem.*, 1991, **600**, 15.
174. R.Rossi, A.Marchi, L.Magon, U.Casellato, S.Tamburini and R.Graziano, *J. Chem. Soc., Dalton Trans.*, 1991, 263.
175. M.S.Ram, L.M.Jones, H.J.Ward, Y.H.Wong, C.S.Johnson, P.Subramanian and J.T.Hupp, *Inorg. Chem.*, 1991, **30**, 2928.
176. W.A.Herrmann, P.Kiprof, K.Rypdal, J.Tremmel, R.Blom, R.Alberto, J.Behm, R.Walbach, H.Bock, B.Solouki, J.Mink, D.Lichtenberger and N.E.Gruhn, *J. Am. Chem. Soc.*, 1991, **113**, 6527.
177. W.A.Herrmann, M.Taillefer, C.de M.de Bellefon and J.Behm, *Inorg. Chem.*, 1991, **30**, 3247.
178. M.M.Gafurov, V.D.Prisyazhnyi and A.R.Aliev, *Ukr. Khim. Zh. (Russ. Ed.)*, 1990, **56**, 1244.
179. J.K.Vassiliou, J.W.Otto, R.F.Porter and A.L.Ruoff, *Mater. Res. Symp. Proc.*, 1991, **210**, 627.
180. L.Ohlhausen, D.Cockrum, J.Register, K.Roberts, G.J.Long, G.L.Powell and B.B.Hutchinson, *Inorg. Chem.*, 1990, **29**, 4886.
181, N.G.Chernorukov, N.P.Egorov and E.V.Suleimanov, *Russ. J. Inorg. Chem.*, 1990, **35**, 1714.

182. W.A.Herrmann, C.C.Romao, P.Kiprof, J.Behm, M.R.Cook and M.Taillefer, *J. Organometal. Chem.*, 1991, **413**, 11.
183. A.A.Amindzhanov, N.M.Kurbanov and K.U.Akhmedov, *Russ. J. Inorg. Chem.*, 1990, **35**, 1454.
184. J.R.Schoonover, R.F.Dallinger, P.M.Killough, A.P.Sattelberger and W.H.Woodruff, *Inorg. Chem.*, 1991, **30**, 1093.
185. M.T.Bautista, E.P.Cappellani, S.D.Drouin, R.H.Morris, C.T.Schweitzer, A.Sella and J.Zubkowski, *J. Am. Chem. Soc.*, 1991, **113**, 4876.
186. J.He, N.Yu and S.Lin, *Wuhan Daxue Xuebao, Ziran Kexueban*, 1990, 77 (*Chem. Abs.*, 1991, **114**, 223841).
187. J.J.Lopez-Garriga, W.A.Oertling, R.T.Kean, H.Hoogland, R.Wever and G.T.Babcock, *Biochem.*, 1990, **29**, 9387.
188. M.A.Miller, J.M.Mauro, G.Smulevich, M.Coletta, J.Kraut and T.G.Traylor, *Biochem.*, 1990, **29**, 9978.
189. M.Nagai, Y.Yoneyama and T.Kitagawa, *Biochem.*, 1991, **30**, 6495.
190. D.Lu, I.R.Paeng and K.Nakamoto, *J. Coord. Chem.*, 1991, **23**, 3.
191. I.K.Choi, Y.Liu, D.W.Feng, K.J.Paeng and M.D.Ryan, *Inorg. Chem.*, 1991, **30**, 1832.
192. S.Hu and J.R.Kincaid, *J. Am. Chem. Soc.*, 1991, **113**, 2843.
193. I.E.T.Iben, B.R.Cowen, R.Sanches and J.M.Friedman, *Biophys. J.*, 1991, **59**, 908.
194. J.Swantepoel and A.M.Heyns, *Spectrochim. Acta, A* 1991, **47A**, 243.
195. H.Weiner, H.J.Lunk, J.Fuchs, B.Ziemer, R.Stösser, C.Pietzsch and P.Reich, *Z. anorg. allg. Chem.*, 1991, **594**, 191.
196. P.Gomez-Romero, E.H.Witten, W.M.Reiff and G.B.Jameson, *Inorg. Chem.*, 1990, **29**, 5211.
197. R.H.Beer, W.B.Tolman, S.G.Butt and S.J.Lippard, *Inorg. Chem.*, 1991, **30**, 2082.
198. J.D.Crane and D.E.Fenton, *J. Chem. Soc., Dalton Trans.*, 1990, 3647.
199. R.A.Leising, B.A.Brennan, L.Que, B.G.Fox and E.Münck, *J. Am. Chem. Soc.*, 1991, **113**, 3988.
200. B.A.Brennan, Q.Chen, C.Juarez-Garcia, A.E.True, C.J.O'Connor and L.Que, *Inorg. Chem.*, 1991, **30**, 1937.
201. S.Hu, A.J.Schneider and J.R.Kincaid, *J. Am. Chem. Soc.*, 1991, **113**, 4815.
202. S.Hu and J.R.Kincaid, *J. Am. Chem. Soc.*, 1991, **113**, 7189.
203. L.M.Proniewicz, I.R.Paeng and K.Nakamoto, *J. Am. Chem. Soc.*, 1991, **113**, 3294.
204. S.Han, Y.C.Chang and D.L.Rousseau, *J. Am. Chem. Soc.*, 1990, **112**, 9445.
205. V.Fidler, T.Ogura, S.Sato, K.Aoyagi and T.Kitagawa, *Bull. Chem. Soc. Japan*, 1991, **64**, 2315.
206. S.Hashimoto, Y.Mizutani, Y.Tatsuno and T.Kitagawa, *J. Am. Chem. Soc.*, 1991, **113**, 6542.
207. T.Ogura, S.Takahashi, K.Shinzawa-Itoh, S.Yoshikawa and T.Kitagawa, *J. Biol. Chem.*, 1990, **265**, 14721.
208. T.Kitagawa and T.Ogura, *Proc. SPIE - Int. Soc. Opt. Eng.*, 1991, **1403**, 563.
209. S.Han, Y.C.Ching and D.L.Rousseau, *Nature*, 1990, **348**, 89.
210. C.Sourisseau, R.Cavagnat and M.Fouassier, *J. Phys. Chem. Solids*, 1991, **52**, 537.
211. I.I.Bulgak, I.E.Rychagova, V.E.Zubareva, K.I.Turte and V.N.Shafranskii, *Russ. J. Inorg. Chem.*, 1990, **35**, 989.
212. H.Furuichi, Y.Ozaki, K.Niki and H.Akutsu, *J. Biochem. (Tokyo)*, 1990, **108**, 707.
213. G.Backes, Y.Mino, T.M.Loehr, T.E.Meyer, M.A.Cusanovich, W.V.Sweeney, A.T.Adman and J.Sanders-Loehr, *J. Am. Chem. Soc.*, 1991, **113**, 2055.
214. H.Saito, T.Imai, K.Wakita, A.Urushiyama and T.Yagi, *Bull. Chem. Soc. Japan*, 1991, **64**, 829.
215. W.Fu, T.V.Morgan, L.E.Mortenson and M.K.Johnson, *FEBS Lett.*, 1991, **284**, 165.

216. J.Swanepoel and A.M.Heyns, *Spectrochim. Acta, A*, 1990, **46A**, 1629.
217. I.R.Beattie, P.J.Jones and N.A.Young, *Chem. Phys. Lett.*, 1991, **177**, 579.
218. P.Guillaume, A.L.K.Wah and M.Postel, *Inorg. Chem.*, 1991, **30**, 1828.
219. L.Nalbandian and G.N.Papatheodorou, *High Temp. Sci.*, 1988, **28**, 49.
220. J.Shamir and P.Sobota, *J. Raman Spectrosc.*, 1991, **22**, 535.
221. J.Shamir, *J. Raman Spectrosc.*, 1991, **22**, 97.
222. G.Jia, D.W.Meek and J.C.Gallucci, *Inorg. Chem.*, 1991, **30**, 403.
223. C.Bianchini, P.J.Perez, M.Peruzzini, F.Zanobini and A.Vacca, *Inorg. Chem.*, 1991, **30**, 279.
224. C.E.Anson, U.A.Jayasooriya, S.F.A.Kettle, P.L.Stranghellini and R.Rosetti, *Inorg. Chem.*, 1991, **30**, 2282.
225. M.Kritikos and D.Noreos, *J. Solid State Chem.*, 1991, **93**, 256.
226. T.M.Buslaeva, I.V.Malyunov, N.M.Sinitsyn, M.E.Poloznikova and A.S.Solomonova, *Russ. J. Inorg. Chem.*, 1991, **36**, 538.
227. T.Wilke, X.Gao, C.G.Takoudis and M.J.Weaver, *Langmuir*, 1991, **7**, 714.
228. A.Müller, M.I.Khan, E.Krickemeyer and H.Bögge, *Inorg. Chem.*, 1991, **30**, 2040.
229. J.A.Crayston, D.C.Cupertino and T.J.Dines, *J. Chem. Soc., Dalton Trans.*, 1991, 1603.
230. H.F.Hameka, J.O.Jensen, J.G.Kay, C.M.Rosenthal and G.L.Zimmerman, *J. Mol. Spectrosc.*, 1991, **150**, 218.
231. A.C.Dengel, W.P.Griffith, A.M.El-Hendawy and J.M.Jolliffe, *Polyhedron*, 1990, **9**, 1751.
232. W.P.Griffith, J.M.Jolliffe, S.V.Ley and D.J.Williams, *J. Chem. Soc., Chem. Comm.*, 1990, 1219.
233. M.Snels, M.P.Sassi and M.Quack, *Mol. Phys.*, 1991, **72**, 145.
234. U.C.Sharma, B.C.Paul and R.K.Poddar, *Ind. J. Chem., Sect. A*, 1990, **29A**, 803.
235. T.Wilke, X.Gao, C.G.Takoudis and M.J.Weaver, *J. Catal.*, 1991, **130**, 62.
236. B.Müller and H.D.Lutz, *Phys. Chem. Miner.*, 1991, **17**, 716.
237. R.Bougon, W.V.Cicha, M.Lance, L.Meublat, M.Nierlich and J.Vigner, *Inorg. Chem.*, 1991, **30**, 102.
238. G.Di Marco, A.Bartoletta, V.Ricevuto, S.Campagna, G.Denti, L.Sabatino and G.De Rosa, *Inorg. Chem.*, 1991, **30**, 270.
239. G.M.McCann, A.Carvill, E.Lindner, B.Karle and H.A.Mayer, *J. Chem. Soc., Dalton Trans.*, 1990, 3107.
240. D.J.Irvine, S.A.Preston, D.J.Cole-Hamilton and J.C.Barnes, *J. Chem. Soc., Dalton Trans.*, 1991, 2413.
241. E.Miki, K.Harada, Y.Kamata, M.Umehara, K.Mizumachi, T.Ishimori, M.Nakahara, M.Tanaka and T.Nagai, *Polyhedron*, 1991, **10**, 583.
242. A.Mezzetti, A.Delzotto, P.Rigo and E.Farnetti, *J. Chem. Soc., Dalton Trans.*, 1991, 1525.
243. R.Bhattacharyya, A.M.Saha, P.N.Ghosh, M.Mukherjee and A.K.Mukherjee, *J. Chem. Soc., Dalton Trans.*, 1991, 501.
244. A.A.Svetlov, M.N.Sinitsyn, A.T.Fal'kenhof and Yu.V.Kokunov, *Russ. J. Inorg. Chem.*, 1990, **35**, 1007.
245. D.Fenske, G.Baum, H.W.Swidersky and K.Dehnicke, *Z. Naturforsch., B*, 1990, **45b**, 1210.
246. A.A.Danopoulos, G.Wilkinson, B.Hossain-Bates and M.B.Hursthouse, *J. Chem. Soc., Dalton Trans.*, 1991, 269.
247. J.J.Schwab, E.C.Wilkinson, S.R.Wilson and P.A.Shapley, *J. Am. Chem. Soc.*, 1991, **113**, 6124.
248. W.A.Herrmann, S.J.Eder, P.Kiprof, K.Rypdal and P.Watzlowik, *Angew. Chem., Int. Ed. Engl.*, 1990, **29**, 1445.
249. W.A.Herrmann, S.J.Eder and P.Kiprof, *J. Organometal. Chem.*, 1991, **413**, 27.
250. C.-M.Che, W.-K.Cheng and V.W.-W.Yam, *J. Chem. Soc., Dalton Trans.*, 1990, 3095.

251. C.F.Edwards and W.P.Griffith, *Polyhedron*, 1991, **10**, 61.
252. W.E.Lynch, R.L.Lintvedt and X.Q.Shui, *Inorg. Chem.*, 1991, **30**, 1014.
253. C.F.Edwards, W.P.Griffith and D.J.Williams, *J. Chem. Soc., Chem. Comm.*, 1990, 1523.
254. H.F.Klein, H.Beck, B.Hammerschmitt, U.Koch, S.Koppert, G.Cordier and H.Paulus, *Z. Naturforsch., B*, 1991, **46b**, 147.
255. M.D.Fryzuk, J.B.Ng, S.J.Rettig, J.C.Huffman and K.Jonas, *Inorg. Chem.*, 1991, **30**, 2437.
256. E.Diana, G.Gervasio, R.Rosetti, F.Valdemarin, G.Bor and P.L.Stanghellini, *Inorg. Chem.*, 1991, **30**, 294.
257. S.Nie, P.A.Marzilli, L.G.Marzilli and N.T.Yu, *J. Chem. Soc., Chem. Comm.*, 1990, 770.
258. S.G.Anema, S.K.Lee, K.M.Mackay, B.K.Nicholson and M.Service, *J. Chem. Soc., Dalton Trans.*, 1991, 1201.
257. N.Zotov, K.Petrov and M.Dimitrova-Pankova, *J. Phys. Chem. Solids*, 1990, **51**, 1199.
260. A.A.Volkov, Yu.G.Goncharov, G.V.Kozlov, V.I.Torgashev, J.Petzelt and V.Dvorak, *Izv. Akad. Nauk SSSR, Ser. Fiz.*, 1990, **54**, 1124.
261. O.Poizat, D.P.Strommen, P.Maldivi, A.-M.Giraud-Godquin and J.C.Marchon, *Inorg. Chem.*, 1990, **29**, 4851.
262. K.K.Das and K.Balasubramanian, *J. Mol. Spectrosc.*, 1991. **147**, 114.
263. T.W.Bell, D.M.Haddleton, A.McCamley, M.G.Partridge, R.N.Perutz and H.Willner, *J. Am. Chem. Soc.*, 1990, **112**, 9212.
264. C.A.Ogle, T.C.Masterman and J.L.Hubbard, *J. Chem. Soc., Chem. Comm.*, 1990, 1733.
265. G.Tresoldi, P.Piraino, E.Rotondo and F.Faraone, *J. Chem. Soc., Dalton Trans.*, 1991, 425.
266. M.Esteban, A.Pequerul, D.Carmona, F.J.Lahoz, A.Martín and L.A.Oro, *J. Organometal. Chem.*, 1991, **402**, 421.
267. C.S.Chin and B.Lee, *J. Chem. Soc., Dalton Trans.*, 1991, 1323.
268. A.L.Seligson, R.L.Cowan and W.C.Trogler, *Inorg. Chem.*,1991, **30**, 3371.
269. A.R.Siedle, R.A.Newmark and W.B.Gleason, *Inorg. Chem.*, 1991, **30**, 2005.
270. E.Bessler, B.M.Barbosa, W.Hiller and J.Weidlein, *Z. Naturforsch., B*, 1991, **46b**, 490.
271. G.Cervantes, J.J.Fiol, A.Terrón, V.Moreno, J.R.Alabart, M.Aguiló, M.Gómez and X.Solans, *Inorg. Chem.*, 1990, **29**, 5168.
272. J.Odo, M.Mifune, A.Iwado, Y.Saito, Y.Tanaka, N.Motohashi and K.Machida, *Chem. Pharm. Bull.*, 1991, **39**, 247.
273. B.A.Uzoukwu, *Synth. React., Inorg. Met.-Org. Chem.*, 1990, **20**, 1071.
274. R.B.Singh and S.Mitra, *Thermochim. Acta*, 1990, **164**, 365.
275. M.Bellaihou and H.O.Desseyn, *Spectrosc. Lett.*, 1990, **23**, 1253.
276. I.R.Beattie, P.J.Jones and N.A.Young, *Mol. Phys.*, 1991, **72**, 1309.
277. L.R.Hanton, J.Evans, W.Levason, R.J.Perry and M.Webster, *J. Chem. Soc., Dalton Trans.*, 1991, 2039.
278. N.Trendafilova, G.S.Nikolov, R.Kellner, H.Mikosch, G.Bauer and G.Bogatchev, *Chem. Phys.*, 1990, **147**, 377.
279. L.Tušek-Božić, I.Matiijašic, G.Bocelli, G.Calestani, A.Furlani, V.Scarcia and A.Papaioannou, *J. Chem. Soc., Dalton Trans.*, 1991, 195.
280. M.M.Bekheit, Y.A.Elewady, F.I.Taha and S.I.Mostafa, *Bull. Soc. Chim. France*, 1991, 178.
281. T.Suzuki, M.Kita, K.Kashiwabara and J.Fujita, *Bull. Chem. Soc Japan*, 1990, **63**, 3434.
282. P.Castan, F.Dahan, S.Wimmer and F.L.Wimmer, *J. Chem. Soc., Dalton Trans.*, 1990, 2071.
283. C.D.Tait, D.R.Janecky and P.S.Z.Rogers, *Geochim. Cosmochim. Acta*, 1991, **55**, 1253.
284. J.Vicente, M.-T.Chicote, I.Saura-Llamas, M.-J.López-Muñoz and P.G.Jones, *J. Chem. Soc., Dalton Trans.*, 1990, 3683.
285. R.J.H.Clark, D.J.Michael and M.Yamashita, *J. Chem. Soc., Dalton Trans.*, 1991, 725.
286. I.A.Degen and A.J.Rowlands, *Spectrochim. Acta, A*, 1991, **47A**, 1263.

287. G.Minghetti, M.A.Cinellu, S.Stoccoro, G.Chelucci and A.Zucca, *Inorg. Chem.*, 1990, **29**, 5137.
288. R.Usón, J.Forniés, M.Tomás, I.Ara, J.M.Casas and A.Martin, *J. Chem. Soc., Dalton Trans.*, 1991, 2253.
289. J.R.McBride, G.W.Graham, C.R.Peters and W.H.Weber, *J. Appl. Phys.*, 1991, **69**, 1596.
290. K.I.Yakovlev, N.D.Rozhkova and A.I.Stetsenko, *Russ. J.Inorg. Chem.*, 1991, **36**, 66.
291. H.Alper, Y.Huang, D.B.Dell'amico, F.Calderazzo, N.Pasqualetti and C.A.Verarini, *Organometallics*, 1991, **10**, 1665.
292. P.Hollmann and W.Preetz, *Z. anorg. allg. Chem.*, 1991, **601**, 47.
293. V.Yu.Kukushkin and A.I.Museev, *Russ. J. Inorg. Chem.*, 1990, **35**, 1136.
294. S.C.Huckett, R.J.Donohoe, L.A.Worl, A.D.F.Bulou, C.J.Burns and B.I.Swanson, *Synth. Met.*, 1991, **42**, 2773.
295. G.S.Muraveiskaya, A.A.Sidorov, G.N.Emel'yanova and E.M.Trishkina, *Russ. J. Inorg. Chem.*, 1991, **36**, 525.
296. H.S.Plitt, M.Bär, R.Ahlrichs and H.Schnöckel, *Angew. Chem., Int. Ed. Engl.*, 1991, **30**, 832.
297. C.J.Hirschmugl, G.P.Williams, F.M.Hoffmann and Y.J.Chabal, *J. Electron Spectrosc. Relat. Phenom.*, 1990, **54-5**, 109.
298. H.Masuda, M.Munakata and K.Sisumu, *J. Organometal. Chem.*, 1990, **391**, 131.
299. R.K.Gaifutdinova, R.A.Khisamutdinov, M.M.Murza, V.F.Nurgaleeva and Yu.I.Murinov, *Russ. J. Inorg. Chem.*, 1990, **35**, 1460.
300. S.M.Wang, P.J.Huang, H.Chang, C.Y.Cheng, S.L.Wang and N.C.Li, *Inorg. Chim. Acta*, 1991, **182**, 109.
301. M.A.-R.S.Goher, *Acta Chim. Hung.*, 1990, **127**, 213.
302. E.J.Baran, E.G.Ferrer and M.C.Apella, *Monatsh.*, 1991, **122**, 21.
303. J.Kožišek, M.Hvastijová, J.Kohout, J.Mroziński and H.Köhler, *J. Chem. Soc., Dalton Trans.*, 1991, 1773.
304. R.Battistuzzi and M.Borsari, *Coll. Czech. Chem. Commun.*, 1990, **55**, 2199.
305. S.Guha, D.Peebles and J.T.Wieting, *Bull. Mater. Sci.*, 1991, **14**, 539.
306. W.Hiller, A.Zinn and K.Dehnicke, *Z. Naturforsch., B*, 1990, **45b**, 1593.
307. A.Erle and G.Güntherodt, *Physica, C*, 1990, 171, 216.
308. G.Burns, F.H.Dacol, C.Feild and F.Holtzberg, *Solid State Commun.*, 1990, **75**, 893.
309. G.Burns, F.H.Dacol, C.Feild and F.Holtzberg, *Solid State Commun.*, 1991, **77**, 367.
310. E.Altendorf, J.Chrzanowski, J.C.Irwin and J.P.Franck, *Phys. Rev., B*, 1991, **43**, 2771.
311. J.C.Irwin, J.Chrzanowski, E.Altendorf, J.P.Franck and J.Jung, *J. Mater. Res.*, 1990, **5**, 2780.
312. J.Chrzanowski, E.Altendorf, J.C.Irwin, J.P.Franck and J.Jung, *Bull. Mater. Sci.*, 1991, **14**, 551.
313. B.Roughani, L.C.Sengupta, J.L.Auhel, S.Sundaram and W.C.H.Joiner, *Physica,C*, 1990, **171**, 77.
314. Y.Matsuda and S.Hikotani, *Sumitomo Kinzoku*, 1990, **42**, 15 (*Chem. Abs.*, 1991, **115**, 13899).
315. O.V.Misochko, E.I.Rashba, E.Ya.Sherman and V.B.Timofeev, *Phys. Rep.*, 1990, **194**, 387.
316. Y.Liu, Y.Lee, M.J.Sumner, R.Sooryakumar and T.R.Lemberger, *Phys. Rev., B*, 1990, **42**, 10090.
317. L.Zhong, J.Cai and H.Tang, *Proc. SPIE - Int. Soc. Opt. Eng.*, 1990, **1514**, 110.
318. M.Yoshida, S.Tajima, Y.Mizuo, T.Wada, Y.Yaegashi, A.Ichinose, M.Yamauchi, N.Koshizuka and S.Tanaka, *J. Phys. Soc., Japan*, 1991, **60**, 1204.
319. W.Sadowski, H.Hagemann, M.François, H.Bill, M.Peter, E.Walker and K.Yvon, *Physica, C*, 1990, **170**, 103.

320. Z.V.Popovic, A.Sacuto and M.Balkanski, *Solid State Commun.*, 1991, **78**, 99.
321. S.Skoulika, A.Aubry, P.Kagianidis, P.Aslandis and S.Papastefanou, *Inorg. Chim. Acta*, 1991, **183**, 207.
322. F.Bigoli, M.A.Pellinghelli, P.Deplano, E.F.Trogu, L.A.Sabatini and A.Vacca, *Inorg. Chim. Acta*, 1991, **180**, 201.
323. F.Bigoli, M.A.Pellinghelli, P.Deplano and E.F.Trogu, *Inorg. Chim. Acta*, 1991, **182**, 33.
324. I.Persson, M.Sandström, A.T.Steel, M.J.Zapetero and R.Aakesson, *Inorg. Chem.*, 1991, **30**, 4075.
325. J.Ellermann, F.A.Knoch and K.J.Meier, *Z. Naturforsch., B*, 1990, **45b**, 1657.
326. G.A.Bowmaker, A.Camus, R.D.Hart, J.D.Kildea, B.W.Skelton and A.H.White, *J. Chem. Soc., Dalton Trans.*, 1990, 3753.
327. R.D.Urban, H.Birk, P.Polomsky and H.Johns, *J. Chem. Phys.*, 1991, **94**, 2523.
328. R.Aroca, M.Scraba and J.Mink, *Spectrochim. Acta, A*, 1991, **47A**, 263.
329. X.D.Wang and R.G.Greenler, *Phys. Rev., B*, 1991, **43B**, 6808.
330. J.G.Contreras and J.A.Gnecco, *An. Quim.*, 1990, **86**, 740.
331. S.Attar, N.W.Alcock, G.A.Bowmaker, J.S.Frye, W.H.Bearden and J.H.Nelson, *Inorg. Chem.*, 1991, **30**, 4166.
332. M.Adelhelm, W.Bacher, E.G.Hohn and E.Jacob, *Chem. Ber.*, 1991, **124**, 1559.
333. D.Perreault, M.Drouin, A.Michel and P.D.Harvey, *Inorg. Chem.*, 1991, **30**, 2.
334. M.Willner, F.Mistry, G.Hwang, F.G.Herring, M.S.R.Cader and F.Aubke, *J. Fluorine Chem.*, 1991, **52**, 13.
335. E.Colacio, A.Romerosa, J.Ruiz, P.Román, J.M.Gutiérrez-Zorrilla, A.Vegas and M.Martinez-Ripoll, *Inorg. Chem.*, 1991, **30**, 3743.
336. A.C.Fabretti, A.Giusti and W.Malavasi, *J. Chem. Soc., Dalton Trans.*, 1990, 3091.
337. S.-P.Huang and M.G.Kanitzidis, *Inorg. Chem.*, 1991, **30**, 3572.
338. A.I.Popov, M.D.Val'kovskii, Yu.M.Kiselev, N.A.Chumaevskii, V.B.Sokolov and S.N.Spirin, *Russ. J. Inorg. Chem.*, 1990, **35**, 1123.
339. J.Vicente, M.D.Bermúdez, J.Escribano, M.P.Carrillo and P.G.Jones, *J. Chem. Soc., Dalton Trans.*, 1990, 3083.
340. J.A.Peck, C.D.Tait, B.I.Swanson and G.E.Brown, *Geochim. Cosmochim. Acta*, 1991, **55**, 671.
341. P.Pan and S.A.Wood, *Geochim. Cosmochim. Acta*, 1991, **55**, 2365.
342. R.Han, I.B.Gorrell, A.G.Looney and G.Parkin, *J. Chem. Soc., Chem. Comm.*, 1991, 717.
343. H.Birk, R.D.Urban, P.Polomsky and H.Jones, *J. Chem. Phys.*, 1991, **94**, 5435.
344. J.Himmrich and H.D.Lutz, *Solid State Commun.*, 1991, **79**, 447.
345. A.K.Vazhnov and K.I.Petrov, *Russ. J. Inorg. Chem.*, 1990, **35**, 1187.
346. O.Berkesi, I.Dreveni, J.A.Andor and P.L.Goggin, *Inorg. Chim. Acta*, 1991, **181**, 285.
347. M.Bochmann, K.J.Webb, M.B.Hursthouse and M.Mazid, *J. Chem. Soc., Dalton Trans.*, 1991, 2317.
348. J.A.Tossell, *J. Phys. Chem.*, 1991, **95**, 366.
349. S.V.Volkov and O.B.Babushkina, *Ukr. Khim. Zh. (Russ. Ed.)*, 1990, **56**, 678.
350. S.V.Volkov, O.B.Babushkina and N.I.Buryak, *Russ. J. Inorg. Chem.*, 1990, **35**, 1637.
351. H.Wakita, G.Johansson, M.Sandstrom, P.L.Goggin and H.Ohtaki, *J. Solution Chem.*, 1991, **20**, 643.
352. S.Bruni, F.Cariati, A.Pozzi, L.P.Battaglia and A.B.Corradi, *Inorg. Chim. Acta*, 1991, **183**, 221.
353. H.Jian, F.Gan and H.Sun, *Giusuanyan Xuebao*, 1990, **18**, 116 (*Chem. Abs.*, 1991, **114**, 48038).
354. N.Singh, N.K.Singh and C.Kaw, *Bull. Chem. Soc. Japan*, 1990, **63**, 180.
355. A.Zinn, K.Dehnicke, D.Fenske and G.Baum, *Z. anorg. allg. Chem.*, 1991, **596**, 47.

356. M.Bochmann and K.J.Webb, *J. Chem. Soc., Dalton Trans.*, 1991, 2325.
357. G.A.Voyiatzis and G.N.Papatheodorou, *Proc. Electrochem. Soc.*, 1990, **90-17**, 161.
358. G.Faraglia, Z.Guo and S.Sitran, *Polyhedron*, 1991, **10**, 351.
359. D.Piehler, W.K.Kot and N.Edelstein, *J. Chem. Phys.*, 1991, **94**, 842.
360. H.Almadfa, A.A.Said and E.M.Nour, *Bull. Soc. Chim. France*, 1991, 137.
361. B.M.Biwer, W.L.Ebert and J.K.Bates, *J. Nucl. Mater.*, 1990, **175**, 188.
362. N.A.Chumaevskii, O.U.Sharopov, N.A.Minaeva, M.T.Toshev and A.V.Sergeev, *Koord. Khim.*, 1990, **16**, 1704.
363. V.K.Rastogi, H.P.Mital and R.C.Saxena, *Ind. J. Pure Appl. Phys.*, 1989, **27**, 167.
364. L.B.Serezhkina, L.M.Bakhmet'eva and V.N.Serezhkin, *Radiokhim.*, 1990, **32**, 9.
365. M.Bickel and B.Kanellakopulos, *J. Less-Common Met.*, 1990, **163**, L19.
366. A.C.Sutorik and M.G.Kanatzidis, *J. Am. Chem. Soc.*, 1991, **113**, 7754.
367. R.N.Mulford, *J. Mol. Spectrosc.*, 1991, **147**, 260.
368. M.Karbowiak and J.Drozdzynski, *J. Less-Common Met.*, 1990, **163**, 159.
369. R.N.Mulford, H.J.Dewey and J.E.Barefield, *J. Chem. Phys.*, 1991, **94**, 4790.

6 Vibrational Spectra of Some Co-ordinated Ligands

BY G. DAVIDSON

1. Carbon and Tin Donors

C_2H_4 adsorbed on ion-exchanged mordenites gives rise to $\nu C{=}C/\delta CH_2$ bands in the ranges 1612 - 1614 cm^{-1} and 1338 - 1340 cm^{-1} for Li^+, Na^+ and Ca^{2+} mordenites. For Zn^{2+} and Cd^{2+} both bands shift to lower wavenumber, but with higher intensities.[1]

νCH and νCD modes for $MeTiCl_3$, as $^{12}CH_3$, $^{13}CH_3$, $^{12}CD_3$ and CHD_2 isotopomers, show no evidence for agostic interactions in the solid state.[2] Complete assignments have been proposed from the FTIR spectra of $Cp_2Ti(CO_2)(PMe_3)$, *trans*$Mo(CO_2)_2(PMe_3)_4$ and $Fe(CO_2)(PMe_3)_4$. Isotopic shifts and a normal coordinate analysis for the titanium complex indicate that the CO_2 ligand is η^1-*C*-coordinated.[3,4] Normal coordinate analyses for 21 compounds $(C_5H_4R)TiCl_2$, where R = H, alkyl, alkenyl, cycloalkyl etc., and IR data in the range 50 - 500 cm^{-1}, suggest that there is effective C_{2v} symmetry for the Ti environment on all cases.[5]

There is FTIR evidence that propylene is initially adsorbed as a π-complex on to vanadia-titania.[6] Several ligand modes were assigned from Raman and IR spectra of $V(arene)_2$ in N_2 complexes at 12 K, where arene = C_6H_6, $C_6H_5CH_3$ or 1,3,5-$C_6H_3Me_3$.[7]

Table 1 Some vibrational assignments for $(TCNE)M(CO)_5$ ($/cm^{-1}$)

	Free TCNE	M = Cr	M = W
$\nu C{=}C$	1567	1507	1491
$\nu C{-}C$	1282	1285	1286
$\nu C{\equiv}N$	2245	2224	2223
	2234	2196	2171

Resonance Raman spectra of $(\eta^2\text{-TCNE})M(CO)_5$, where M = Cr or W, confirm η^2-coordination for the π-acceptor TCNE. Some assignments are listed in Table 1.[8] IR and Raman intensities of bands for $(mesitylene)M(CO)_3$, where M = Cr, Mo or W, can be related to the polarities and polarisabilities of individual bonds. The results were correlated with an approximate m.o. scheme.[9]

Methylcyclopropane adsorbed on photoreduced silica/molybdena catalysts forms

Mo=CH_2 (with ν_s, $\nu_{as}CH_2$ at 2845, 3080 cm^{-1} respectively), and Mo=$CHCH_3$ (with 4 bands in the νCH region: 2850, 2890, 2910 and 2985 cm^{-1}).[10] Bands of η^2-*C,O* -coordinated CO_2 ligands in *trans*-$Mo(CO_2)_2(PMe_3)_4$ are at 1690 and 1668 cm^{-1} (νC=O). Lower wavenumber modes are highly coupled. νPC_3 modes were also assigned.[11] (1) has νC=O of the η^3-propenyl ligand at 1685 cm^{-1}.[12]

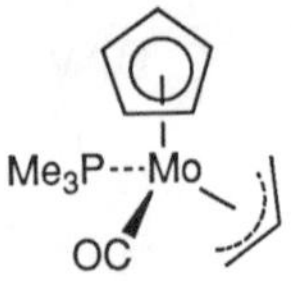

(1)

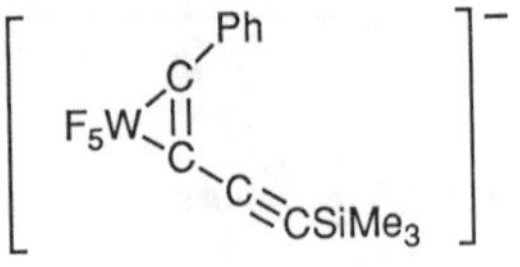

(2)

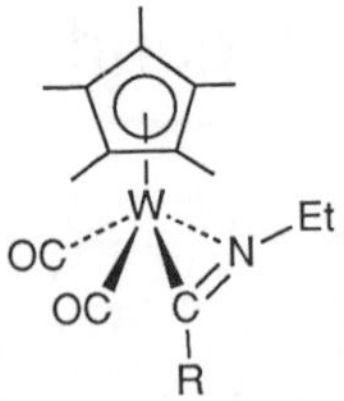

(3)

The complex (2) has νCC (of metallacycle) at 1695 cm^{-1}, νCC (exocyclic) at 2130 cm^{-1}.[13] $[W(CO)L_2(S_2CX)(\eta^2\text{-}MeC_2Me)]^+$, where X = NC_4H_8, NMe_2, NEt_2, $N(CH_2Ph)_2$ or OEt; L = PMe_3, PMe_2Ph, $PMePh_2$, PPh_3, $PEtPh_2$ etc., all have νCC near 1830 cm^{-1}, i.e. there is considerable back donation to the π* m.o. of the alkyne.[14] νC=N modes were assigned for (3), where R = Me (1663 cm^{-1}) or Et (1654 cm^{-1}).[15]

There is IR (νCO) evidence for the formation of solvent (S) coordinated acyl complexes $(OC)_4(S)MnCOR$ in the reaction of manganese carbonyl alkyls $(OC)_5MnR$ with polar solvents such as DMSO or DMF.[16] Fourier transform Raman spectra have been obtained for $CpMn(CO)_3$ and a series of (arene)$Cr(CO)_3$ complexes. The manganese complex has weak νCH bands at 3146, 3118, 3106 and 3098 cm^{-1}, as well as strong band at 3130 cm^{-1}.[17]

The IR and Raman spectra of CH_3ReO_3 gave the following assignments: ν_sCH_3, 2899 cm^{-1}; δ_sCH_3, 1205 cm^{-1}; $\nu_{as}CH_3$, 2989 cm^{-1}; $\delta_{as}CH_3$, 1363 cm^{-1}; ρCH_3, 739 cm^{-1}.[18] The bridging acyl groups in $(OC)_4Fe[\mu\text{-}C(R)O]Re(CO)_4$, where R = Ph or Me, have νCO bands at 1471 cm^{-1} (Ph) or 1523 cm^{-1} (Me).[19] $Re(O)I(RC{\equiv}CR)_2$ gave the following assignments: R = Me, νCC 1581 cm^{-1}, νCH 3085 cm^{-1}; R = Ph, νCC 1600 cm^{-1}, νCH 3080 cm^{-1}.[20] νCH in $(\eta^5\text{-}C_5H_5)ReO_3$ is at 3102 cm^{-1}.[21] The IR spectrum of Cp_2ReCl is very similar to that previously obtained for Cp_2TcCl.[22]

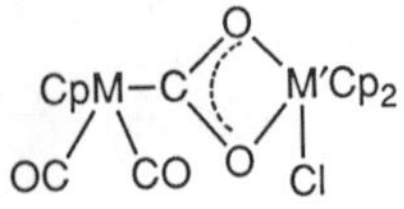

(4)

For the complexes (4), where M = Fe or Ru, M' = Ti or Zr, the νCO_2 assignments listed in Table 2 were obtained. All are derived from the μ-(η^1- *C*:η^2*O,O'* -) CO_2 ligand.[23] Conformational preferences of CpFe(CO)(PPh_3)(COMe) and related species were followed by monitoring νC=O and νC≡O modes.[24]

Table 2 νCO$_2$ vibrational assignments for complexes (4). (/cm^{-1})

M	M'	ν_{as}	ν_s
Fe	Ti	1379	1273
Fe	Zr	1363	1268
Ru	Ti	1349	1284
Ru	Zr	1349	1291

Quite detailed assignments have been proposed from IR and photoacoustic spectra of $[CpFe(CO)_2R]^+$, where R = C_2H_4, CH_2CMe_2 or 2-Me-thiophene.[25] νCC modes were assigned in CpFe(C≡CR)(dppe), where R = $SiMe_3$ (1995 cm^{-1}), CO_2Me (2054 cm^{-1}), ^{t}Bu (2081 cm^{-1}), H (1930 cm^{-1}) or Ph (2071 cm^{-1}).[26]

The electronic absorption spectrum of gaseous Cp_2Fe has fine structure related to the CH stretch and internal rotation modes.[27] (5), where R, R' and R" = H, Me, OMe etc., all have νC=O in the range 1644 - 1650 cm^{-1}.[28]

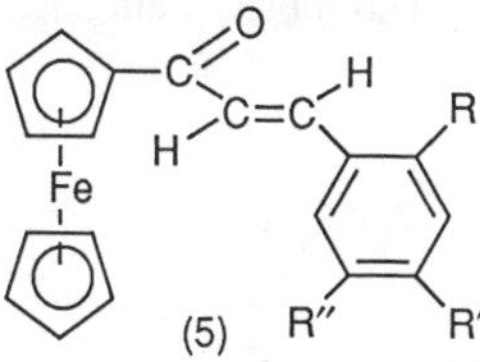

(5)

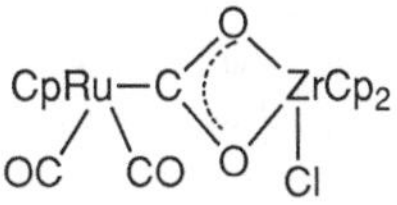

(6)

(6) has νOCO bands at 1349 and 1292 cm^{-1}, which shift to 1312, 1264 cm^{-1} with a $^{13}CO_2$ ligand.[29] νC=C is in the range 1508 - 1595 cm^{-1} in Ru(O_2CH)(CO)(HCH=CHR)-(PPh_3), where R = H, CMe_3, $SiMe_3$ or Ph. The νCO$_2$ bands of the formato ligand show bidentate chelation.[30] (7) has νC=C at 1583 cm^{-1}. The bidentate carboxylate in (8) has $\nu_{as}CO_2$ at 1529 cm^{-1}, ν_sCO_2 1468 cm^{-1}.[31]

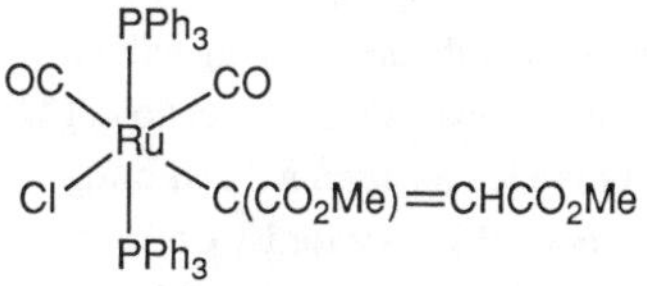

(7)

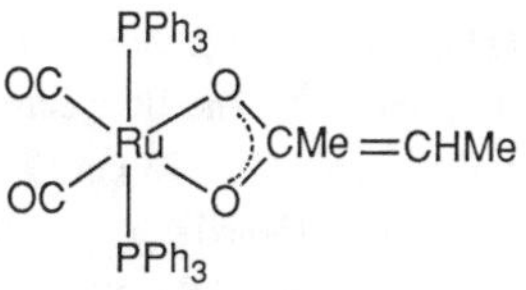

(8)

The complexes ICo(L)-C≡CR, where L = 1,3-bis(diacetylmonoximeimino)propane, R = $SiMe_3$, Ph, ^{n}Bu or ^{t}Bu, have νC≡C in the region 2110 - 2140 cm^{-1}.[32] Inelastic neutron scattering of $(C_2H_2)Co_2(CO)_6$ shows that the torsional barrier around the CC bond gives a feature at 776 cm^{-1}.[33] The IR spectra of Co(II) and Cu(II) complexes of *o*-hydroxy-benzoylmethylenetriarylphosphoranes and -arsenanes show that coordination can occur *via* the ylide carbon and the carbonyl oxygen as well as the hydroxy oxygen.[34]

νCH and νCO modes (FTIR) were used to monitor the structure and reactions of the rhodium species formed by supporting *trans*-$[Rh(C_5Me_5)(CH_3)]_2(\mu\text{-}CH_2)_2$ on a silica surface.[35] For $Rh_5(CO)_{14}CH_2R$, when R = COOMe, νC=O is at 1675 cm^{-1}, When R = CN, νCN is at 2183 cm^{-1}.[36]

C_2H_4 adsorbs on to Al_2O_3-supported Rh, Ir, Pd or Pt at room temperature to form a π-bonded species, for which νC=C and δCH_2 assignments suggest that the M-C_2H_4 π-bond strengths are in the order: Rh > Pd; Ir > Pt. At low temperatures there is some evidence for di-σ-bonded ethene.[37] νC≡C in $[Rh(C{\equiv}CR)\{E(CH_2CH_2PPh_2)_3\}]^z$, where R = Ph or CO_2Et, E = N or P, z = 0, +1, +2, shows an increase of 30 cm^{-1} on oxidation of Rh(I) (z = 0) to Rh(II) (z = +1) and another 10 cm-1 to Rh(III) (z = +2). These figures are consistent with the alkynyl ligand acting as a π-acceptor.[38]

(9), where L = P^iPr_3, gave the following assignments: R = Me, νC=C 1647 cm^{-1}; R = Ph, νC=C 1642 cm^{-1}; R = CO_2Et, νC=C 1585 cm^{-1}, νC=O 1642 cm^{-1}.[39] HREELS data on acetone adsorbed on Rh(111) showed a νCO band at 1380 cm^{-1} due to η^2-*C,O* coordination of the acetone at the surface.[40] νC=O and νC-O of the alkoxycarbonyl ligands in Cp*M(μ-pz)(μ-I)$_2$RhI(CO_2R)(CO), where M = Rh, R = Me or Et; M = Ir, R = Me, are in the ranges 1710 - 1690 cm^{-1}, 1065 - 1045 cm^{-1} respectively.[41] νC=C bands (of Cp) were assigned for (10), where M = Rh (1562/1554 cm^{-1}) or Ir (1558/1557 cm^{-1}). The δCH_2 and δCH modes of the vinyl ligand were also assigned (all in argon matrices).[42]

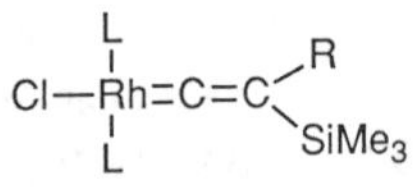

(9)

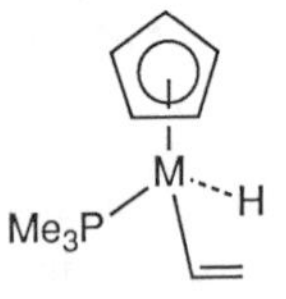

(10)

νSnCl bands of coordinated $SnCl_3^-$ ligands were assigned in $Rh(SnCl_3)(C_2H_4)_2[(p\text{-}MeOC_6H_4)_3P]_2$ (305 and 280 cm^{-1}) and related complexes.[43]

In (11), where R = 2,6-$^iPr_2C_6H_3$, νCF bands from the η^2-C_2F_4 ligand are at 1070, 1045 and 1025 cm^{-1}.[44] The IR spectrum of the triple-decker sandwich cation (12) shows significant differences (10 - 25 cm^{-1}) in equivalent modes for terminal and bridged cyclopentadienyl ligands. The Ni-Cp(t) interaction is stronger than that for Ni-Cp(br).[45]

(13) has νC≡C at 2080 cm^{-1}, νC=N at 1670 cm^{-1}.[46] $(Ph_2PCH_2CH_2PPh_2)Pd(SnMe_3)_2$ gave IR and Raman spectra from which the νSnC bands of the $SnMe_3$ ligands could be assigned, together with νCH_3 and ρCH_3.[47]

Ethane is chemisorbed on Pt/SiO_2 at room temperature to form the ethylidyne unit CH_3CPt_3, and possible di-σ-bonded $PtCH_2CH_2Pt$.[48] The IR spectra of $H_2C{=}CH_2$ and $D_2C{=}CD_2$ on Pt(111) and of HDC=CDH on Pt/SiO_2 gave evidence for isotopic scrambling between adsorbed ethylidyne species.[49]

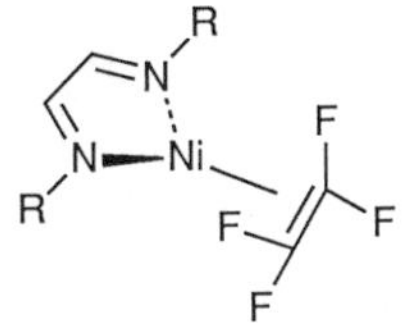

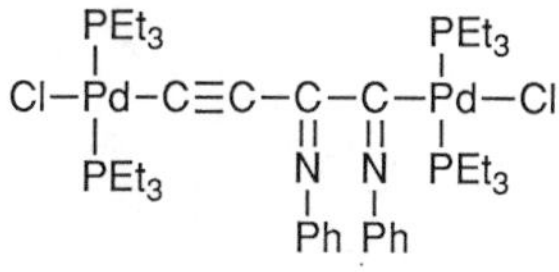

(11) (12) (13)

νC=C and νC≡O modes were assigned for cyclic olefin complexes of Pt(II), Table 3.[50] The *trans* configuration of $Pt(C{\equiv}C^tBu)_2(PPh_3)_2$ was shown by the observation of a single νC≡C band in the IR spectrum.[51]

Table 3 νC=C and νC≡O modes in some Pt(II) complexes of cyclic olefins. ($/cm^{-1}$)

	νC=C	νC≡O
$PtCl_2(CO)(C_6H_{10})$	1495 (1650)	2116
$PtBr_2(CO)(C_6H_{10})$	1493 (1650)	2108
$PtCl_2(CO)(C_5H_8)$	1470 (1615)	2118

(values of νC=C for the free ligand are shown in parentheses)

The complex $[Cu(cod)(bipy)]^+$ gives Raman bands at 1603 and 1491 cm^{-1} due to νC=C of axial and equatorial C=C respectively.[52] In $M(cod)_2{}^+$, where M = Cu or Ag, have νC=C only slightly lower than in free cod. Thus the predominant bonding mode is σ-bonding from the olefin π m.o. to a metal dσ orbital.[53] The resonance Raman spectrum of silver in C_2H_4 matrices shows analogous effects to those observed for SERS of C_2H_4 on silver surfaces.[54]

$Hg[Ph\text{-}C(NSiMe_3)_2]_2$ has νCN_2 bands at 1555 and 1172 cm^{-1}, a smaller difference than in free amidine.[55] The characteristic νCCC stretch of the allyl ligand in $(C_3H_5)_2Ln\text{-}Cl_5Mg_2(tmed)_2$, where Ln = La, Ce, Pr Nd or Sm, is always at 1545 cm^{-1}.[56]

$Te(CH_3)_6$ has IR bands due to νCH modes at 2994, 2909 and 2786 cm^{-1}.[57] Me_2X^+, where X = Cl, Br or I, gave assignments for CH_3 modes which are summarised in Table 4.[58]

Table 4 Characteristic CH_3 modes for Me_2X^+, where X = Cl, Br or I. (/cm^{-1})

X =	Cl	Br	I
$\nu_{as}CH_3$	3094	3101	3082
ν_sCH_3	2982	2985	2963
δCH_3	1430	1424	-

2. Boron Donors

νBH modes in (BP)$Ti(BH_4)_2$, where BP = a sterically hindered bisphenoxo dianion, show that both BH_4^- groups are tridentate, but in (BP)$Zr(BH_4)_2(THF)_2$ there is one bi- and one tridentate BH_4^-.[59] The IR spectra of vapour and matrix-isolated $Ti(BH_4)_3$ and $Ti(BD_4)_3$ contain bands due to tridentate BH_4^- ligands confirming electron-diffraction results.[60]

The IR spectrum of $CpZr(BH_4)_3$ clearly shows the presence of three triply-bridged borohydride ligands.[61] The IR spectra of $A^+[M(BH_4)_5]^-$, where A = NBu_4 or PPh_4, M = Zr or Hf, show the presence of both bi- and tridentate BH_4^- groups.[62]

The IR spectrum of borazine, $B_3N_3H_6$, adsorbed on a Pt(111) surface suggest that the molecular plane is oriented perpendicular to the surface.[63]

3. Carbonyl and Thiocarbonyl Complexes

A review has been made of νCO wavenumbers, especially of the less-common types of metal carbonyl bonding patterns. The following characteristic values were established: MCO (σ-CO) 2050 - 2070 cm^{-1}; M_2CO (μ_2-CO) 1870 cm^{-1}; M_3CO (μ_3-CO) 1800 cm^{-1}; MCO...M' 1730 cm^{-1}; M_3CO...M' 1380 cm^{-1}, etc.[64]

CO adsorbed on Co^{2+}/TiO_2 or Ni^{2+}/TiO_2 gives IR bands as follows: Ti^{4+}-CO, Co^{2+}-CO and Ni^{2+}-CO. For Cu^{2+}/TiO_2 there was evidence for two kinds of Cu^+-CO species.[65] The IR spectrum of CO adsorbed on TiO_2 contains νCO at 2187 and 2209 cm^{-1}. The former is believed to be due to CO adsorbed on five-coordinate Ti^{4+}, the latter to four-coordinate Ti^{4+}.[66] νCO for $R_2M(CO)_2$, where M = Ti or Zr, show that R = C_5Me_4H is a better electron-donor than C_5H_5, but poorer than C_5Me_5.[67]

The first stable derivative of the unknown $Hf(CO)_6F_2$ has been prepared, *viz.* $Hf(CO)_2(Me_2PCH_2CH_2PMe_2)_2F_2$, for which νCO bands are at 1872 and 1760 cm^{-1} (with $\nu^{13}CO$ at 1841, 1734 cm^{-1}).[68]

Metallocenes give the following reaction products (with νCO values) with CO on $AlPO_4$ or silica supports: $Cp_2V(CO)$ (1869 cm^{-1}), $[Cp_2V(CO)_2]^+$ (2046, 1992 cm^{-1}); $[CpFe(CO)_2]^+$ (2065, 2023 cm^{-1}); $CpCo(CO)_2$ (2032, 1964 cm^{-1}); and $[CpNi(CO)]^+$ (2103 cm^{-1}).[69]

CO adsorbed on $Cr/AlPO_4$ gives νCO at 2203 cm^{-1} due to a Cr(III) surface species.[70] νCO and νCN assignments were made for half-sandwich complexes of chromium (where R = η^6-mesitylene): $RCr(CO)_2(CNCF_3)$, $RCr(CO)(CNCF_3)(CNCH_3)$ and $RCr(CNCF_3)_3$. All of the results were consistent with a strong π-acceptor character for the trifluoromethyl isocyanide ligand.[71] A time-resolved IR study of the reactions of 1,3- and 1,4-pentadiene with $Cr(CO)_4$ gave kinetic data and suggested reaction pathways.[72] Resonance Raman spectra on the femtosecond time scale were used to obtain information about the reaction coordinates for CO dissociation from $M(CO)_6$, where M = Cr, Mo or W.[73] Environments of $M(CO)_6$, where M = Cr, Mo or W, in zeolites were probed by (among other techniques) IR.[74]

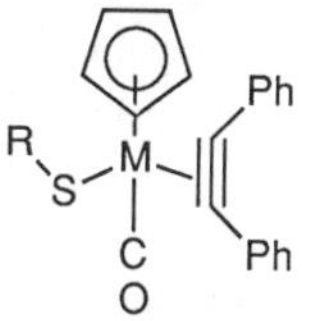

(14)

CO adsorbed on titania-supported molybdena showed νCO bands at 2198 and 2190 cm^{-1}. These were assigned to Mo^V-CO and Mo^{IV}-CO species respectively.[75] The observation of two νCO bands shows the existence of isomers due to restricted rotation about the M-SR bond in (14), where M = Mo or W; R = Me, Et, ^{n}Pr, ^{i}Pr or ^{t}Bu.[76] The photochemistry of $M(\eta^5\text{-}C_5H_5)(CO)_3Et$, where M = Mo or W, was followed by νCO IR spectra in low-temperature matrices and in room-temperature solutions.[77] IR data (νCO) for $In[Mo(CO)_3Cp]_3$ in THF solution were consistent with the formation of $[In(THF)_2\text{-}\{Mo(CO)_3Cp\}_2]^+[Mo(CO)_3Cp]^-$.[78] Similar data were used to follow the formation and decay of transient species in the flash photolysis of $M(CO)_3(PR_3)_2X_2$, where M = Mo or W, X = Cl or Br, R = phenyl or substituted phenyl.[79]

CO on W/Al_2O_3 catalysts gave features due to M^V-CO (2198 cm^{-1}), and W^{IV}-CO (2176 and 2154 cm^{-1}). NO gives W^V-NO (1843 cm^{-1}) and $W^{IV}(NO)_2$ (1780,1691 cm^{-1}).[80] The new complex $W(CO)_3F_2$ has νCO at 2117, 2006, 1980 and 1956 cm^{-1}. These values are higher than for the Cl and Br analogues, as expected.[81] Time-resolved IR spectroscopy was used to follow reactions of coordinatively-unsaturated $W(CO)_5$ with CO, CF_2Cl_2 and N_2O.[82]

Adsorption of CO on $AlPO_4$-supported metals gave the following νCO values: Mn(II) 2194 cm^{-1}; Fe(II) 2180 cm^{-1}; Co(II) 2189cm^{-1}; Ni(II) 2203, 2193 cm^{-1}.[83] Phase changes in $CH_3M(CO)_5$, where M = Mn or Re, were monitored by νCO. Transitions were detected at about 9 kbar for M = Mn, 22 kbar for M = Re.[84] The photochemical reactions of $(OC)_5MnMn(CO)_3$(α-di-imine) were followed by changes in the νCO absorptions.[85]

High-pressure Raman and IR spectra of $M_2(CO)_{10}$, where M = Mn or Re, show a phase transition at about 8 kbar (Mn) or 5 kbar (Re). This involves a change in molecular geometry from D_{4d} to D_{4h}.[86] FTIR spectra (νCO) were used to monitor the photochemical reactions of $M_2(CO)_{10}$, M = Mn or Re, with *o* and *p*-quinones in coordinating and non-coordinating solvents.[87]

High-pressure micro-Raman spectra of $CpRe(CO)_3$ show the presence of a phase transition near 9 kbar. Quite detailed vibrational assignments were given for the low- and high-pressure forms.[88] The resonance Raman spectrum of $Re[4,4'-(CO_2Et)_2\text{-bipy}](CO)_3Cl$ shows enhancement of $\nu_s(CO)$ from the CO ligands *cis* to the Cl atom.[89] The micro-Raman spectrum of $Re(CO)_5Cl$ at high pressure shows a phase transition at about 29 kbar. Up to this transition, increased pressure leads to strengthening of the Re to CO back-bonding.[90] High-pressure Raman measurements on $Re(CO)_5X$, where X = Br or I, however, gave no evidence for any phase transitions up to 43 kbar.[91]

A correlation was found between νCO and the ^{17}O n.q.r. coupling constant for a wide range of carbon monoxy myo- and haemoglobins.[92] Vibrational spectroscopy shows that the co-condensation of Fe(TPP) with CO and MeNC produces Fe(TPP)(CO)(MeNC), in which CO is a better π-acceptor than MeNC.[93] The IR spectra (νCO) of the co-condensation products of FeL, where L = phthalocyaninato (Pc) or Salen, with CO in Ar, Ar/O_2 and Ar/H_2O matrices are summarised in Table 5.[94]

Table 5 νCO wavenumbers for co-condensation products of CO and FeL (/cm^{-1})

L =	Pc	Salen
$FeL(CO)_2$	2064	2037
$FeL(CO)(O_2)$	2028	2003
$FeL(CO)(H_2O)$	2008	1986
FeL(CO)	2000	1973

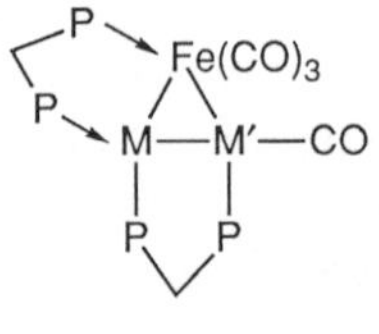

(15)

The photochemistry of (η^4-polyene)iron tricarbonyl complexes in matrices at approx. 12 K was followed by IR spectroscopy (νCO).[95] νCO for (15), where P-P = dppm, M,M' = Pd or Pt, show that M'-CO decreases by about 20 cm^{-1} on going from Pd to Pt, showing greater back-donation from Pt.[96] $[Fe(CO)_3(SPh)_3]^{4-}$, as the $[K(18\text{-crown-6})]^+$ salt gives νCO values inconsistent with either the molecular point group or factor group predictions.[97]

Cis/trans equilibria in $[Fe_2(CN)Cp_2(CO)_3]^-$ were followed by IR (νCO and νCN).[98] The ultrafast photochemistry of $[CpFe(CO)_2]_2$ was probed at the femtosecond level using a novel IR spectrometer.[99] The electrochemistry of $[CpFe(CO)_2]_2$ and $[Cp^*Fe(CO)_2]_2$ was followed by IR (νCO). Thus, $[CpFe(CO)_2]_2^+$ has νCO at 2023, 2055 and 1934 cm^{-1}, suggesting that *cis*- and *trans* -carbonyl bridged isomers are present.[100]

The photochemistry of $Fe(CO)_5$ in PVC and other films was followed by IR spectroscopy (νCO).[101] The presence of 4 terminal νCO bands in the IR spectrum of bis(η^2-acyl)hexacarbonyl di-iron complexes can be explained in terms of effective C_{3v} symmetry for these complexes and two non-equivalent $Fe(CO)_3$ fragments.[102] $[FeCo_2(CO)_9(\mu^2\text{-}PPhH)]^-$ has νCO bands at lower wavenumbers than for the neutral parent, as expected.[103]

νCO of adsorbed CO on Ru-Cu/SiO_2 is at higher wavenumber than on pure Cu/SiO_2.[104] CO on 1% Ru/SiO_2 has νCO due to linearly adsorbed CO at 2034 cm^{-1}.[105] νCO of $[Ru(CO)Cl_5]^{2-}$ are quite strongly dependent on the nature of the counter-ion, decreasing with increasing electron-donor power and size of the cation. The K^+ salt has νCO at 2075 cm^{-1}, that of $^nOct_4N^+$ at 1985 cm^{-1}.[106] The redox behaviour of (OEP)Ru(CO) and (OEP)Ru(CO)L, where L is an axial ligand has been followed by νCO. Thus (OEP)Ru(CO) in pyridine has νCO at 1931 cm^{-1}, $(OEP)Ru(CO)^-$ at 1896 cm^{-1}.[107] The reactions of $Ru(CO)_2$(triphos), where triphos = $PhP(CH_2CH_2PPh_2)_2$ and related ligands, with I_2 or MeI have been followed by νCO (IR).[108]

νCS in $Ru(Etxant)_2(CS)(PPh_3)$, where Etxant = ethyl xanthate, and related species, was seen in the range 1270 - 1290 cm^{-1}.[109]

The one-electron oxidation of $Os_2H\{(C_4H_3N)Fe(C_5H_5)\}(CO)_{10}$ gives a small shift in νCO to higher wavenumbers, suggesting that the electron is removed from the aza-ferrocenyl group, giving some change in the orbital energies of the cluster on oxidation.[110] The photochemistry of $(\mu_3\text{-}\eta^2\text{-}CH_3CH_2C{\equiv}CCH_2CH_3)(\mu\text{-}CO)Os_3(CO)_9$ and related species in argon matrices at 10 K was followed by νCO in the IR spectrum.[111] The novel cluster anion $[Os_{20}(CO)_{40}]^{2-}$ has νCO at 2045 (strong), 2002 (strong) and 1945 (weak) cm^{-1}.[112]

The π-cation radical of $[(OEP)Co(CO)]^+$ has νCO at 2137 cm^{-1}.[113] The IR spectra of CO adsorbed on reduced Co-HTsVM zeolites showed the formation of $Co^{II}CO$, $Co^I(CO)_2$, $Co^0(CO)_2$ and Co^0CO.[114]

The νCO and νCoH modes for $HCo(CO)_4$ in the presence of nitrogen or oxygen bases show no evidence of hydrogen bonding in solution or in matrices.[115] (Bipy)Cu-$Co(CO)_4$ has bridging CO in the solution as well as in the solid phase, shown by the presence of νCO at 1850 cm^{-1}.[116]

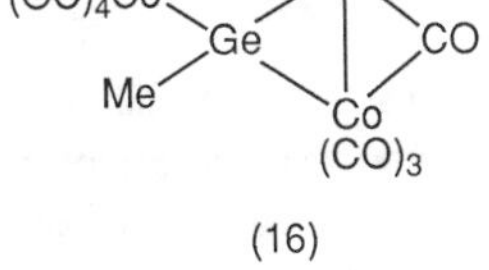

(16)

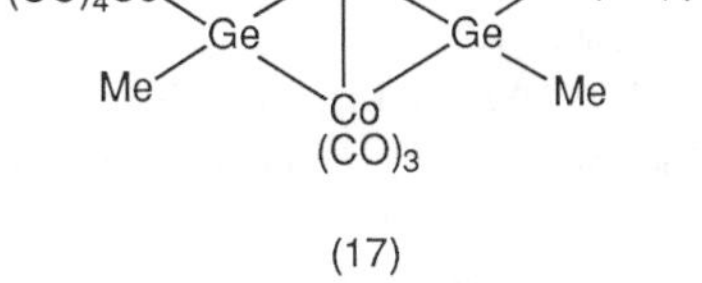

(17)

Detailed assignments of νCO modes have been made for the new cluster compounds (16) and (17).[117] $Co_6S_8(CO)_6$ has νCO bands in agreeement with O_h symmetry for the metal cluster.[118]

CO on Rh(100) at 90 K and 300 K has νCO bands due to both linear and bridging CO molecules at all coverages.[119] The IR spectra of $Rh(CO)_2L$, where HL = MeC-(NH_2)=CHCOCX_3, X = H or F, showed that increased electronegativity of the groups on L^- lead to increasing wavenumbers for νCO.[120] High-pressure reactions of the complexes $Rh_4(CO)_{12-x}\{P(OPh)_3\}_x$, where x = 1 - 4, with CO or CO/H_2 were followed by FTIR spectroscopy.[121]

Calculations of the νCO wavenumbers of CO on Pd(100) gave results in good agreement with experiment in terms of the shifts in wavenumber with increasing electrical field at the CO.[122] CO adsorbed on Pd/Cu particles has νCO at 2115 - 2120 cm^{-1} (due to Cu-CO) and a broad feature 1850 - 200 cm^{-1} (due to multiply-coordinated CO on groups of Pd atoms).[123] νCO bands of CO on thin films of Pd or Pt show that CO on Pd is both linearly (near 2060 cm^{-1}) and bridge (near 1880 cm^{-1}) bonded. On Pt, nearly all of the CO is linearly bonded.[124] The IR spectra of CO-saturated Pd sols at pH values near 6.0 gave νCO at 1950 cm^{-1} (two-fold bridge sites) and 2067 cm^{-1} (linear).[125]

CO on the (100) surface of a single-crystal Pt electrode gives on-top, asymmetric and symmetric bridged coordination. These are interconvertible by changing the applied electode potentials.[126] CO adsorbed on the (1 x 1) surface of Pt(100) gives bridging and "on-top" bonding, with νCO at 1864 and 2062 cm^{-1} respectively.[127] A νCO band at 2168 cm^{-1} is assigned to a weakly chemisorbed CO molecule on a SnO_x-supported Pt surface which is very active towards CO oxidation.[128] An FTIR study of CO_2 reduction on polycrystalline platinum in acid solutions reveals the formation of linear and bridge-bonded adsorbed CO.[129] $Pt_3(\mu\text{-}CO)_3(PCy_3)_3$ shows a strong νCO band at 1770 cm^{-1} (the e' mode of a D_{3h} cluster). Replacement of the 1,2-μ-CO by μ-SO_2 leads to a shift in νCO to higher wavenumber, showing that μ-SO_2 is a better π-acceptor ligand than is μ-CO.[130]

Adsorption studies of CO/NO mixtures on CuO/γ-Al_2O_3 catalysts (with Cu^+ and Cu^{2+} at the surface) show that CO adsorbs preferentially on Cu^+ (2120 cm^{-1}), with NO at Cu^{2+} sites (1888 cm^{-1} on CuO, 1862 cm^{-1} on $CuAl_2O_4$).[131] High coverage of CO on Cu ultrathin films on a Rh(100) surface,gives νCO at 2076 cm^{-1}, i.e. CO is adsorbed on three-dimensional copper clusters.[132] IR spectra of CO on ultra-thin films of Cu on Rh(100) and Mo(100) surfaces gives νCO corresponding to three different types of Cu site.[133] Solid Cu(CO)Cl has νCO at 2176, 2127 and 2063 cm^{-1}. The complex in solution shows only a single band at 2108 cm^{-1}.[134]

$Ag(CO)B(OTeF_5)_4$, the first isolable silver carbonyl, has νCO at 2204 cm^{-1} (shifting to 2154 cm^{-1} on ^{13}CO substitution). This shows a lack of π-acidity for CO in this complex. The value of νCO is the highest reported for any molecular metal monocarbonyl (compare BH_3CO at 2165 cm^{-1}). There was some evidence for $Ag(CO)_2B(OTeF_5)_4$, for which ν_{as}CO appears to be at 2207 cm^{-1}.[135]

νCO for CO adsorbed on Au(110) and Au(210) surfaces were found in the range

2100 - 2115 cm^{-1}.[136-7] $Au(CO)_2^+$, as the UF_6^- salt, has $\nu_{as}CO$ at 2200 cm^{-1} (2160 cm^{-1} for ^{13}CO).[138]

4. Nitrogen Donors

4.1 Molecular Nitrogen, Azido- and Related Complexes :-νNX modes were assigned for $[Na(15\text{-crown-}5)][V(NX)FCl_3(CH_3CN)]$ as follows: X = Cl 525 cm^{-1}, I 447 cm^{-1}.[139] $[Tc_2(NNPh)_2(C_6Cl_4O_2)_4]^+$, where $C_6Cl_4O_2$ = catecholate, gives characteristic bands of the hydrazide ligand at 1486 and 1521 cm^{-1}.[140]

$\nu_{as}N_3$ for $[Fe(CO)_2Cp]_2E(N_3)_2$, where E = Ge or Sn, gives two bands in each case, in regions consistent with a dominant role for the canonical form $E\text{-}N{=}N^+{=}N^-$.[141] $CoH(N_2)L_3$, (18), where L = PMe_3, has νNN at 2068 cm^{-1} ($^{14}N_2$), 1991 cm^{-1} ($^{15}N_2$). CoD-($^{14}N_2$)L_3 has νNN at 2065 cm^{-1}.[142]

Photoreactions of $Pt(N_3)_2(PPh_3)_2$ were followed by monitoring $\nu_{as}N_3$ in the IR spectrum.[143] $Pt(bipy)(NHCl)(NCl_2)Cl_2$ gives a doublet IR band at 650/672 cm^{-1} due to the coordinated NCl_2 ligand.[144]

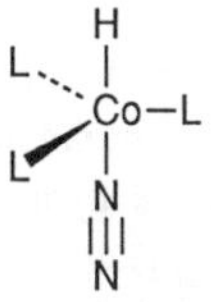

(18)

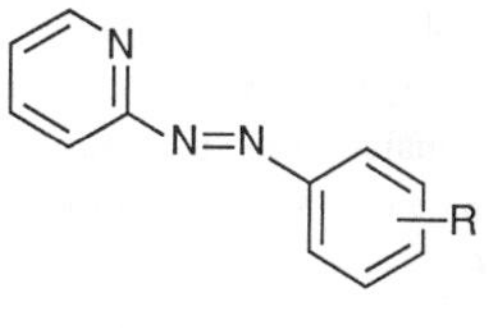

(19)

The IR spectrum of $Cu(DENA)(N_3)_2$, where DENA = *N,N* -diethylnicotinamide, shows that there are two types of N_3 ligand present, i.e. (i) μ-1 bridging (end-on), and (ii) three-coordinate, μ-1,1,3, i.e. one terminal N bridging two copper atoms and the other terminal N coordinated to a third copper.[145]

νN=N in AgL_2^+, where L = (19), R = H, 9-Me, 10-Me or 10-Cl, all show very little shift from free L. Thus there is very little dπ-π*(L) interaction in these complexes.[146]

4.2 Amines and Related Ligands :- Matrix IR spectra gave evidence for a number of $M(NH_3)_n$ complexes for M = Li or K, n = 0.5, 1, 2 and >2.[147]

The Fourier transform Raman spectra of $M(CO)_5(NH_3)$, where M = Cr or W, gave the following ν_sNH_3 mode assignments: 3309 cm^{-1} (Cr) or 3301 cm^{-1} (W).[148] $[Cr(opd)_4]^{2+}$, where opd = *o*-phenylenediamine, has $\nu C\text{-}NH_2$ at 1278 cm^{-1} (free) and 1231 cm^{-1} (coordinated), compared to 1276 cm^{-1} in the free ligand. The ligands are all unidentate in this complex.[149] Characteristic νNH bands were reported for $[CrLX_2]^+$, where X = Cl, Br, N_3 or 1/2(oxalate), L = 1,4,8,12-tetra-azacyclopentadecane, $[15]aneN_4$.[150]

The resonance Raman spectrum of *cis* -$(O)_2Re(bipy)py_2^+$ with 406.7 nm excitation is dominated by pyridine modes; with 514.5 nm excitation enhancement of bipy modes is observed.[151]

The complexes $Ru_2[1,2\text{-}(NH)_2C_6H_4](PPh_3)$(diphosphine), where diphosphine = $(Ph_2P)X$, X = C_2H_4, *cis*-C_2H_2, 1,2-C_6H_4, C_3H_6 and C_4H_8, all have νNH in the range 3355 - 3342 cm^{-1}.[152] The presence of two NH units is revealed by the observation of two νNH bonds in (20) (3348, 3332 cm^{-1}) and related complexes.[153]

Cis -$[CoF_2(NH_3)_4]^+$ has ρNH_3 modes which indicate that two NH_3 ligands are different from the other two. This is consistent with the presence of intermolecular NH...F bonds involving two of the ammonia ligands.[154] Complexes of Co(II), Ni(II), Zn(II) and Cd(II) with oxamide bis-hydrazone (OBH), i.e. $M(OBH)_3^{3+}$, all have IR spectra showing bonding to M *via* the imine nitrogen atoms of the two hydrazine fragments, i.e. (21).[155]

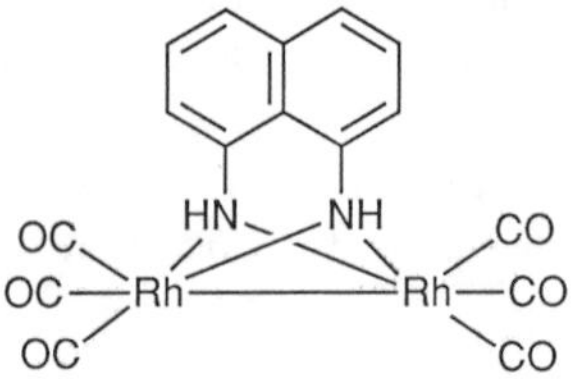

(20)

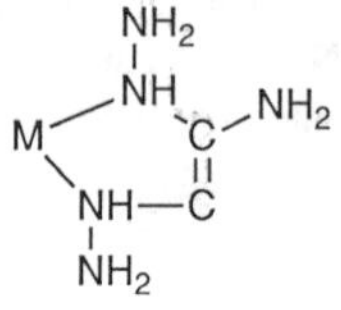

(21)

NH_3 adsorbed on Ni(111) shows characteristic δ_s and δ_{deg} bands of NH_3 at 1211 and 1645 cm^{-1} respectively.[156] A full normal coordinate analysis was performed on $Ni(NH_3)_6^{2+}$, mainly using previously published data.[157] The FTIR spectrum of "$Ni(NH_2)_2$" is consistent with the formulation $Ni_6(NH_2)_{12}$, with at most weak hydrogen-bonding between NH_2 groups.[158]

The pressure dependence of py modes in $Nipy_2Cl_2$, $Nipy_4Cl_2$ and $Znpy_2Cl_2$ was recorded by Raman spectroscopy to about 50 kbar.[159] The IR spectra of $[M(AED)_2]X_2$, where M = Ni, Cu or Cd, X = NO_3 or ClO_4, AED = 1,2-bis(β-aminoethyl)diaziridine, suggest that the AED is terdentate, *via* two N atoms of NH_2 groups, and a nitrogen of the diaziridine ring.[160] $Ni(NCS)_4(HTen)_2$, where HTen = triethylenediammonium, has IR bands due to co-ordinated HTen at 2960, 2940, 2880 cm^{-1} (νCH), 1470 cm^{-1} (δCH_2), 1060, 830 cm^{-1} (νCN) and 1000 cm^{-1} (νCC).[161]

The effect of order/disorder phase transitions on the ethylenediamine (en) modes of $[Men_2][Pten_2Cl_2](ClO_4)_4$, where M = Pd or Pt, was followed by Raman spectroscopy.[162] $M(2,4'\text{-mbipy})X_3$, where 2,4'-mbipy = 1'-methyl-2,4'bipyridinium, M = Pd or Pt, have a ligand mode at about 1595 cm^{-1}, showing coordination *via* the pyridyl nitrogen atom.[163]

The IR spectra of $CuL_2.H_2O$, where HL = 2-X-4-Cl-$C_6H_3(CH_2)_nC(O)NHNH_2$, X = Cl or Me, n = 1; X = Cl, n = 3, show bidentate coordination of hydrazide fragments.[164] IR spectra of ZnX_2L, where X = Cl, Br, I, SCN, OAc or NO_3, L = $MeC(CH_2NHMe)_3$, are all consistent with tetrahedral coordination of the zinc. The L ligands are bidentate *via* two of its nitrogen atoms.[165] *Trans* -$SiF_4(NH_3)_2$ has νNH_3 modes at 3344 (IR), 3327 (Raman) cm^{-1} (e); 3264 (IR), 3266 (Raman) cm^{-1} (a_1); δNH_3 1612 (IR), 1605 (Raman) cm^{-1} (e).[166]

4.3. Ligands containing >C=N- Groups :- The resonance Raman spectrum of bacteriochlorophyll *c* isolated from *Chloroflexus aurantiacus* has νC(9)=O at 1642 cm^{-1}, showing that this group is either hydrogen-bonded to a magnesium-coordinated OH group, or directly coordinated to the Mg itself.[167] A marked solvent effect was found for the ring-breathing mode (near 1580 cm^{-1}), the strongest resonance Raman line of bacteriochlorophyll *a*.[168] SERRS of chlorophyll *a* adsorbed on a silver electrode, together with resonance Raman spectra of chlorophyll *a* show that there is no chemical interaction between the surface and the chlorophyll.[169]

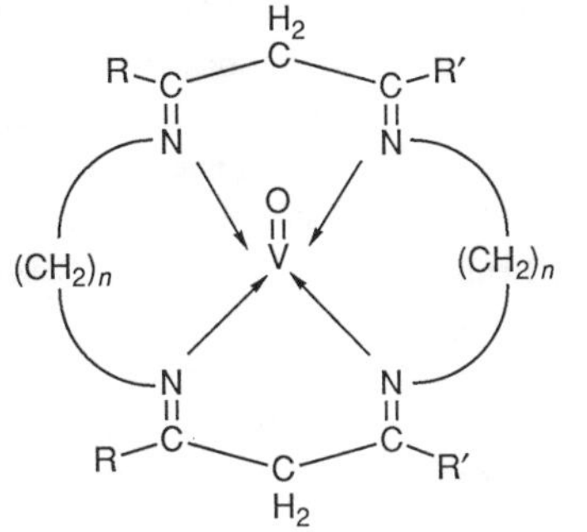

(22)

The radical cation species $[Zr(TPP)_2]^+$ and $[Zr(OEP)_2]^+$ show characteristic marker bands in their resonance Raman spectra at 1319 and 1556 cm^{-1} respectively.[170] (22), where n = 3 or 4, R = R' = Me; R = Ph, R' = Me; R = C_4H_3S, R' = CF_3, all show νC=N in the range 1610 - 1625 cm^{-1}, compared to the free ligand values of about 1660 cm^{-1}. There is therefore coordination by the azomethine nitrogens.[171]

Fe(TPP)CCl_2, Fe(TPP)$^{13}CCl_2$ and Fe(TPP)CBr_2 all have a spin-state marker band at 1569 cm^{-1} and an oxidation state marker band at 1370 cm^{-1}, showing that they are all low-spin Fe(IV) species.[172] Similar data for Fe(TSPP) and Fe(TCPP), where TSPP = tetrakis(4-sulphonatophenyl)porphinato, TCPP = tetrakis(4-carboxylatophenyl)porphinato, show that these are high-spin Fe(III) complexes.[173]

There have of course been many studies, using resonance Raman spectroscopy (often time-resolved) on biologically important iron-containing systems. These include: photoreduction of iron protoporphyrin(IX) dimethyl ester;[174] HbCO and HbO_2 photo-intermediates on the 800 - 900 fsec timescale;[175] photo-excited deoxyhaemoglobin;[176] oxidised cytochrome P450 and related complexes;[177] mutant sperm whale myoglobins;[178] horseradish peroxidase;[179] P460, the iron-containing chromophore at the active site of hydroxylamine oxidoreductase of the ammonia-oxidising bacterium *Nitrosomas europaea*;[180] catalytic intermediates of cytochrome *c* oxidase at room temperature;[181] stable form of horse heart ferricytochrome *c*;[182] bovine adrenocortical cytochrome $P\text{-}450_{11\beta}$;[183] neutrophil cytochrome *b* 558;[184] Fe(III), Fe(II) and Fe(II)(CO) forms of mutants of yeast cytochrome *c* peroxidase;[185] and single subunit oxidase of *Sulfolobus acidocaldarius*.[186]

Time-resolved IR and TR^3 were used to characterise the intermediate formed within

20 psec of photodissociation of CO from cytochrome a_3 in reduced cytochrome oxidase.[187] Changes on adsorption of cytochrome P450 enzymes on to citrate-reduced silver colloidal particles were followed by SERRS.[188] SERRS of cytochrome *c* bound to charged surfaces shows two conformational states for both oxidised and reduced cytochrome *c*.[189]

Resonance Raman and TR^3 were used to study surface interactions of substituted Ru(II) bipy complexes in the photosensitisation of colloidal TiO_2. There was evidence for the excited electron being localised on one of the ligands.[190] Resonance Raman measurements were used to probe the nature of the first electronic excited states of $[Ru(bipy)_2(bpzt)]^{3+}$, $[Ru(bipy)_2(Hbpzt)]^{2+}$, where Hbpzt = 3,5-bis(pyrazin-2-yl)-1,2,4-triazole;[191] and $[Ru(bipy)_2L]^{2+}$, where L = 3-(pyrazin-2-yl)-1,2,4-triazole and related ligands.[192]

$M(LL)_2$, where M = Co, Ni or Cu, LL = (cyclo-octane-1,5-diyl)bis(pyrazol-1-yl)-borato, have IR bands indicative of agostic M....H interactions: 2890 cm^{-1} (M = Co), 2780 cm^{-1} (Cu) or 2800 cm^{-1} (Ni).[193] Partial oxidation of Co(TPP) by I_2 was detected by observation of an IR band at 1290 cm-1 due to $Co(TPP)^{+\cdot}$.[194] Resonance Raman spectra of Co(II), Ni(II) and Zn(II) tetraphenylchlorins showed macrocyclic ligand bands characteristic of symmetry lowering and localised normal mode behaviour.[195]

νC=N in ML, where M = Ni, Cu or Zn and H_2L = Schiff base derived from 2-pyrrolecarbaldehyde and bis(2-aminophenyl)disulphide, is in the range 1575 - 1590 cm^{-1}

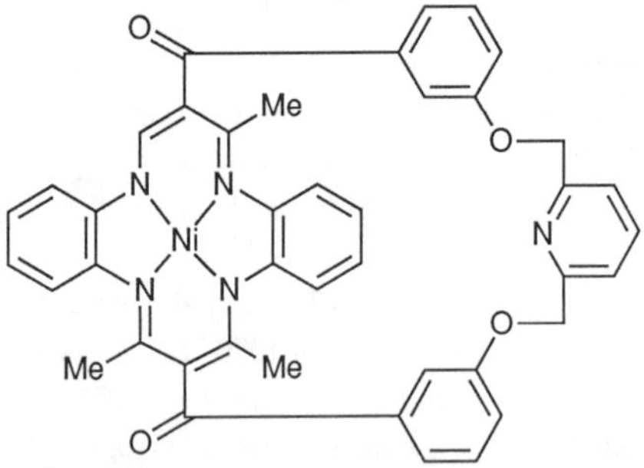

(23)

compared to 1650 cm^{-1} in the free ligand, showing coordination of the azomethine fragment.[196] The IR spectrum of (23) has νC=O at 1660 cm^{-1}, with νCOC at 1055, 1225 cm^{-1}.[197]

A normal coordinate analysis has been reported for the in-plane vibrations of nickel porphine and nickel octaethylporphyrin.[198] The resonance Raman spectrum of Ni(OEP) showed that ''ruffling' of the molecules causes perturbations which particularly effect the C_m and N_p atoms of the macrocycle.[199] A valence force-field for Ni(OEP) was transferred satisfactorily to nickel octaethylchlorin.[200] Calculations on a variety of nickel porphyrins suggest that the marker bands (ν_4, ν_3, ν_2 and ν_9) actually show a direct relationship to the angle $C_\alpha NC_\alpha$.[201] Resonance Raman studies were also reported for the Ni and Cu complexes of 2,7,12,17-tetrapropylporphycene;[202] Ni, Cu and Zn complexes of *trans*-octaethylchlorin;[203] and nickel(II) octaethylisobacteriochlorin.[204]

$[Cu_3(OH)(pz)_3py_3Cl_2.py$, where pz = pyrazolate anion, shows bands due to coordinated and free pyridine at 1607 and 1593 cm^{-1} respectively.[205] TR^3 measurements were used to probe the excited states of $[Cu(dmp)_2]^+$, where dmp = bis(2,9-dimethyl-1,10-phenanthroline)[206] $[(Ph_3P)_2Cu(dpq)]^+$ and $[(Ph_3P)_2Cu(dpq)Cu(PPh_3)_2]^{2+}$, where dpq = 2,3-di-(2-pyridyl)quinoxaline.[207]

The resonance Raman spectra of copper(II) porphyrin diones contain νC=O near 1710 cm^{-1}, together with a strong polarised band (1340 - 1350 cm^{-1}) characteristic of the dione macrocycles.[208] It has proved possible to make model compounds related to the copper-substituted haem d_1 macrocycle of *Pseudmonas aeruginosa*dissumulatory reductase, i.e. with similar resonance Raman spectra.[209]

IR and resonance Raman spectra for copper(II) complexes of octaethylisobacteriochlorin and selectively deuteriated analogues gave assignments for in-plane modes above 1000 cm^{-1}. These were significantly different from porphyrins, chlorins or bacteriochlorins.[210] TR^3 for the metalloporphyrin Cu(TMpy-P4), where TMpy-P4 = 5,10,15,20-tetrakis(4- *N*-methylpyridyl)porphyrinato, gave evidence for a short-lived (ca. 20 psec.) intermediate with characteristic bands at 1558 and 1353 cm^{-1}.[212]

SERS for 5,10,15,20-tetrakis(4-carboxyphenyl)porphyrin (TPPC4) on a silver colloid gave evidence for the formation of Ag^{II}(TPPC4).[212]

IR and Raman spectra of $M(HBT)Cl_3$, where M = Zn, Cd or Hg, HBT = 2-(α-hydroxybenzyl)thiaminium chloride, show that all are isostructural, with HBT bonded to M

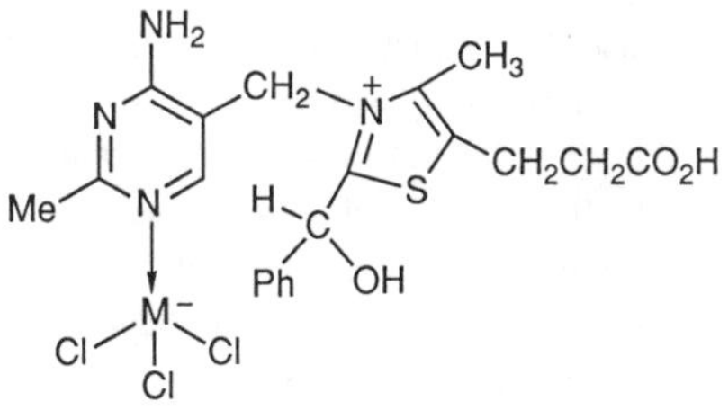

(24)

via the N(1') atom of the pyrimidine fragment (24).[213] $Cd(pyz)_2Ni(CN)_4$, where pyz = pyrazine, has IR and Raman spectra consistent with unidentate pyrazine.[214]

The resonance Raman spectrum of triple-decker sandwich species $Eu_2(OEP)_3$ reveals a doubling of bands of the macrocycles due to inequivalence of the inner and outer rings of the triple-decker sandwich.[215]

4.4 Cyanides, Isocyanides and Related Complexes :- $[Mg(MeCN)_6]_2[Bi_4Cl_{16}]$ has νCN bands at 2322 and 2290 cm^{-1} in the IR spectrum. Comparison with free MeCN shows that the ligands are coordinated *via*N atoms.[216] (25) shows νCN as a strong IR band at 1668 cm^{-1}.[217]

νCN in $CpMX_4(CH_3CN)$, where M = Mo or W, X = Cl or Br, all increase with respect to free CH_3CN, showing M-N coordination.[218] The following νCN assignments were proposed for $MoF_5.NCCH_3$, 2322 cm^{-1}; MoF_5NCCD_3 2310 cm^{-1}; $WF_5.NCCH_3$ 2319

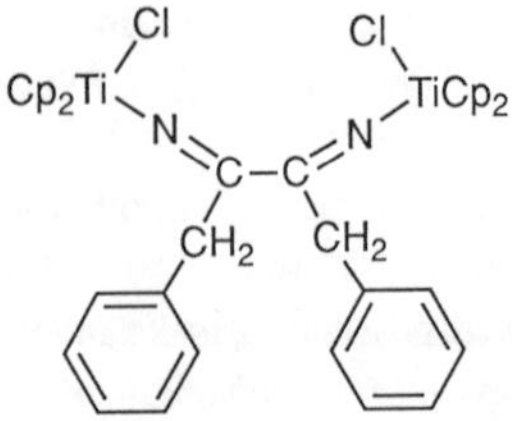

(25)

cm^{-1}.[219] νCH and νCN modes were assigned for $MBr_4(CH_3CN)_2$, where M = Mo or W, and $MoBr_3(CH_3CN)_3$. Thus, for $WBr_4(CH_3CN)_2$ we have νCH at 2975 and 2910 cm^{-1}, νCN at 2305, 2270 cm^{-1}.[220] Fast TRIR was used to probe the lower MLCT excited state of $W(CO)_5$(4-CNpy), where 4-CNpy = 4-cyanopyridine. Visible irradiation gave increased νCO values, showing oxidation of the metal.[221]

Cis - and*trans*-[(dppm){$(PhO)_3P$}(OC)Mn(μ-CN)Ru$(CO)_2(PPh_3)$(*o*-$O_2C_6Cl_4$)] show a decrease in νCN on one-electron oxidation. This is not readily explicable.[222] $Re_2(NCBH_3)_4$(μ-dppm)$_2(H_2O)_2$ has νBH at 2335 cm^{-1}, νCN at 2171 cm^{-1}, from an *N*-bonded cyanotrihydroborate ligand.[223] The low νCN wavenumbers (2140 cm^{-1}) in $[Re(NCC_6H_4Me\text{-}4)_2(Ph_2PCH_2CH_2PPh_2)]^+$ show the electron-rich nature of the $\{Re(dppe)_2\}^+$ site, as there is considerable electron-release from Re to the cyano ligand.[224] $[Re_2Cl_4(dppm)_2]_2$(μ-TCNQ) has νCN at 2190 and 2109 cm^{-1}. The former is due to non-coordinated CN of the fully-reduced TCNQ unit, the latter to coordinated CN.[225]

SERS for $[Fe(CN)_5(4\text{-picolylamine})]^{3-}$ and $[Fe_2(CN)_{10}(4\text{-picolylamine})]^{6-}$ show that both molecules are adsorbed on to a silver electrode *via* one or more cyano groups. It is possible to oxidise them to Fe(III) species on the surface, as shown by a shift in νCN to higher wavenumbers.[226] νCN values in 14 $[Fe(CN)_6]^{4-}$ salts show that there are significant interactions between the anion and cations, and also that the nature of the counterion can alter the σ- and π-contributions to the Fe-CN bonding.[227]

νCN of $Ru(bipy)_2(CN)_2$ (2060, 2072 cm^{-1}) shift to higher wavenumber on formation of $[(bipy)_2(NC)Ru\{CNRh(NH_3)_5\}]^{3+}$ (2067, 2108 cm^{-1}).[228] $[Ru(bipy)_2(CN)]_2(CN)^+$ is assigned as valence-delocalised on the basis of (i) the observation of only two νCN modes and (ii) the fact that the shift to higher energy of the terminal νCN is half that observed on oxidation of a monometallic complex.[229] The coordinated MeCN in *cis* -$OsCl_4(MeCN)_2$ has νCN at 2320 and 2290 cm^{-1}.[230]

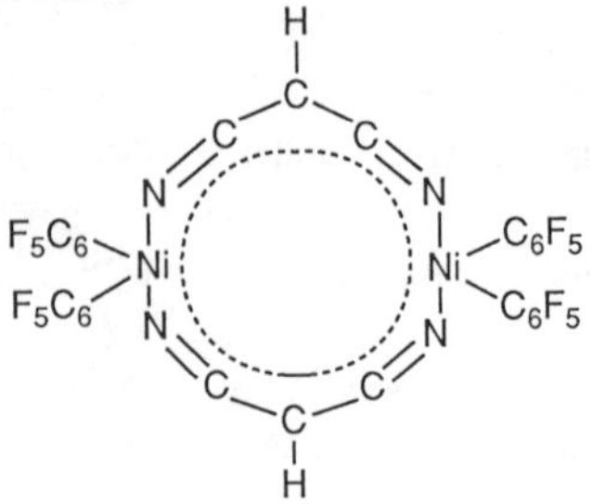

(26)

Cis - and *trans*-(Cp*Rh-μCH_2)$_2$(Me)CN have νCN near 2110 cm^{-1}, i.e. they contain terminal Rh-CN.[231] The *trans*-configurations of $[Rh(CN)_2(NH_3)_4]^+$ and $[Rh(CN)_2en_2]^+$ were shown by the observation of a single sharp νCN band in the IR spectra, at 2130, 2170 cm^{-1} respectively.[232]

(26) has νCN at 2215, 2160 cm^{-1}, compared to the free ligand value, which is 2270 cm^{-1}.[233] Characteristic bands from the ligand modes of the novel complex Ni(C≡N-CMe_3)$_4$ were reported, e.g. νCN 2020 cm^{-1} (2129 cm^{-1} in the free ligand).[234]

FTIR of CN^- adsorption on a polycrystalline surface in aqueous $NaClO_4$ electrolytes shows the formation of both linear and bridge-bonded coordinated CN^- species.[235] νCN is at 2122 cm^{-1} for $[Pt_3(2,6\text{-}Me_2C_6H_3NC)_2(\mu\text{-dppm})_3]^{2+}$, where dppm = $Ph_2PCH_2CH_2PPh_2$, showing the presence of only terminal xylylisocyanide ligands.[236]

νCN in $[Cu(PPh_3)_2L]_2$, where L = XC_6H_4NCN, X = H, 4-Cl, 3-Cl, 4-Br, 4-F, 4--Me, or 4-MeO, shows no correlation with the electron withdrawing or donating properties of X.[237] νCN bands of coordinated CH_3CN in $Cu_4OCl_6(CH_3CN)_4$ are at 2323 and 2295 cm^{-1}, with νCC at 951 and 944 cm^{-1}, δCCN 401 cm^{-1}.[238]

The formation of cyano-complexes by Zn^{2+}, Cd^{2+} and Hg^{2+} in liquid NH_3 was monitored by IR and Raman spectra in the νCN region.[239]

4.5. Nitrosyls :- The FTIR spectrum of NO on TiO_2 anatase showed that there were two different nitrosyl species present.[240] $[V(NO)_2(bipy)_2]CN$ has νNO at 1675 and 1538 cm^{-1}, showing that the $V(NO)_2$ unit is *cis* .[241]

The adsorption of nitrogen oxides with chromocene on silica catalysts produces νNO bands characteristic of dinitrosyl complexes.[242] νNO for Mo(NO)(L)I{O(CH_2)$_4$I}, where L = tris(3,5-dimethylpyrazol-1-yl)hydroborate, is at 1670 cm^{-1}.[243] The bridging NO ligands in $Mo_2O_2Cl_4(NO)(PPh_3)_4$ and $Mo_2O_2(NO)_2(PPh_3)_4$ both have νNO at 1380 cm^{-1}.[244] The following νNO assignments have been made: $MCl_4(NO)_2^{2-}$, ν_s 1769 cm^{-1} (Mo), 1734 cm^{-1} (W); ν_{as} 1639 cm^{-1} (Mo), 1612 cm^{-1} (W); $MF_4Cl(NO)^{2-}$ 1628 cm^{-1} (Mo), 1585 cm^{-1} (W).[245] The νNO bands in polymeric $\{(AlCl_2)_2(\mu\text{-OR})Mo(NO)_2(CHMe)\}_n$, where R = Me or Et, are at lower wavenumbers than for the equivalent monomers. This is due to interaction with the electron-donor chlorine atoms in the polymers.[246]

Fe(TPP)(NO) has νNO at 1681 cm^{-1} (^{14}N) or 1647 cm^{-1} (^{15}N), with marker bands characteristic of low-spin Fe(II). $Fe(TPP)(NO)^-$ has νNO at 1496 cm^{-1} (^{14}N) or 1475 cm^{-1} (^{15}N), having undergone the expected shifts on reduction of the metal.[247] NO on Ru/TiO_2 has νNO bands in the range 1750 - 1900 cm^{-1}.[248] (27) has νNO at 1635 cm^{-1}, characteristic of a Ru(0) and not a Ru(II) complex.[249]

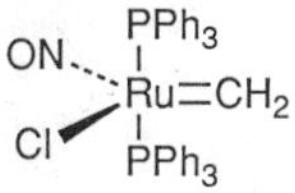

(27)

$[Co(NO)(P\text{-}P)_2]^+$, where P-P = one of a wide range of chelating diphosphines, all have νNO in the range 1785 - 1825 cm^{-1}, i.e. all contain linear Co-N-O units.[250] There is IR evidence for two types of adsorbed NO on Co(TPP)-CeO_2, with νNO 1816.7 and 1696.5 cm^{-1} respectively.[251] The unstable species $[CoH(NO)\{P(OEt)_2Ph\}_4]^+$ has been reported, with νNO at 1680 cm^{-1}.[252]

Saturation coverage of NO on Pd(111) at 100 K produces νNO bands at 1575 and 1735 cm^{-1}, assigned to bridged and on-top sites respectively.[253]

5. Phosphorus and Arsenic Donors

ν_{as} and ν_sAsMe stretching bands were assigned for $[Cp_2Ti(AsMe_3)_2]^{2+}$ and $[Cp_2Ti(As_2Me_4)]^{2+}$. In all cases a decrease was found compared to free ligand values.[254] The coordinated PMe_3 in (28) gives IR bands at 975, 960 and 745 cm^{-1}. The bands due to νCO suggest an angle OC-Ta-CO of about 100°.[255]

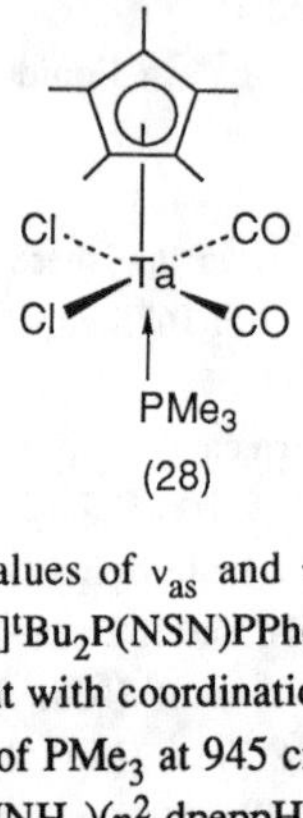

(28)

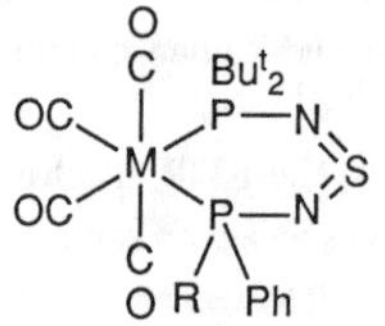

(29)

Values of ν_{as} and ν_s NSN (near 1175, 1070 cm^{-1} respectively) in the complexes $[(OC)_5M]^tBu_2P(NSN)PPh(R)[M(CO)_5]$, where M = Cr, Mo or W, R = ^{t}Bu or Ph, are consistent with coordination as in (29).[256] $Mo_4(\mu_4\text{-}CO_3)(CO)_2(O)_2(\mu_2\text{-}O)_2(\mu_2\text{-}OH)(PMe_3)_6$ has νPC of PMe_3 at 945 cm^{-1}, together with characteristic bands due to νCO and νCO_3.[257] $[MoBr(NNH_2)(\eta^2\text{-dpeppH})(PMe_2Ph)]Br_2$, where dpepp = $PhP(CH_2CH_2PPh_2)_2$, has νPH of the triphosphine ligand at 2360 cm^{-1}.[258]

(30) has an IR band due to the keto C=O stretch at 1675 cm^{-1}.[259] The IR spectrum of Rh/Al_2O_3 in the presence of CO/PH_3 shows the formation of a Rh(CO)(PH_3) unit from the original surface *gem*-dicarbonyls. The νPH_3 band is seen near 2334 cm^{-1}.[260] Cp*Rh-$Cl_2(H_2PPh)$ has νPH bands at 2395 and 2235 cm^{-1}.[261] For $RhCl(PPh_3)(Ph_2POPPh_2)$, ν_{as} POP is at 800 cm^{-1}.[262]

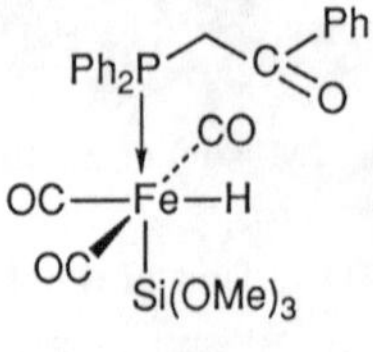

(30)

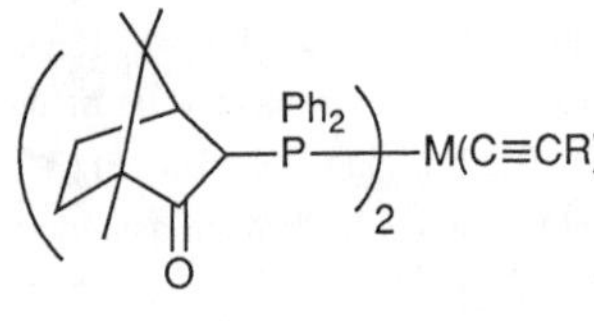

(31)

PF_3 adsorbed on Ni(111) has ν_sPF_3 at 858 cm^{-1} (at lowest coverage), increasing to 916 cm^{-1} (highest coverage). The low-coverage wavenumber is 40 cm^{-1} lower than in $Ni(PF_3)_4$.[263] The out-of-plane ring deformation mode of py in the P-coordinated complexes $MX_2(PMe_2py)_2$, where M = Pd or Pt, X = Cl, Br or I, is seen at a characteristic value of 617 - 619 cm^{-1}.[264] (31), where M = Pd or Pt, R = Ph or $CMe{=}CH_2$, have $\nu C{\equiv}C$ 2100 - 2105 cm^{-1}, $\nu C{=}O$ 1725 - 1730 cm^{-1}.[265]

A detailed assignment has been made of the IR and Raman spectra of $Me_3P{\rightarrow}GaCl_3$, using H/D and $^{69}Ga/^{71}Ga$ isotopic shifts. Some earlier assignments were revised and these revisions were supported by a normal coordinate analysis.[266]

6. Oxygen Donors

6.1 Molecular Oxygen, Peroxo-, Aquo and Related Complexes :- νOO is at 850 cm^{-1}in $[ZrO(O_2)F_2]^{2-}$, as Na^+. K^+ or NH_4^+ salts.[267] νOO is a broad IR band 800 - 900 cm^{-1} in $K_4Hf_2F_{10}(O_2).2H_2O$.[268] $VO(O_2)NH_3^-$ modes were assigned under C_s symmetry, and a normal coordinate analysis carried out.[269]

O_2 on Cr_2O_3 produces an IR band at 985 cm^{-1}, assigned to νOO of a Cr^{III}-O_2^- unit.[270] $M(O)(O_2)L$, where M = Mo, W or U, L = (32), gave νOO modes as follows: 845 cm^{-1} (Mo), 825 cm^{-1} (W) and 795 cm^{-1} (U).[271]

νOO is near 990 cm^{-1} for the side-on dioxygen adducts of Mn(P), where P = TPP, OEP or tetramesityl porphin.[272] νOO bands were assigned for $[TcN(O_2)_2Cl]^-$ and a range of related species.[273] $[\{TcN(O_2)_2\}_2(oxalato)]^{2-}$ has νOO at 900 cm^{-1}.[274]

Resonance Raman spectra of O_2 adducts of Fe(por), where por = TPP, TPP- d_8, OEP, tetramesityl- or tetrakis(pentafluorophenyl)-porphin, contained νOO bands from "end-on" isomers in the range 1223 - 1188 cm^{-1}, and "side-on" isomers 1105 - 1102 cm^{-1}.[275] νOO modes were detected in oxygen adducts of several variants of cyto-chrome P450.[276-7]

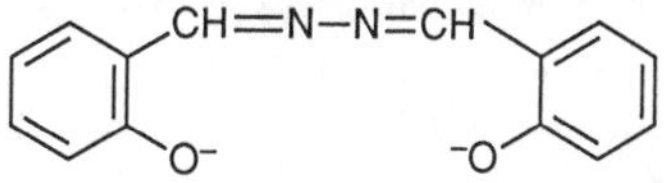

(32)

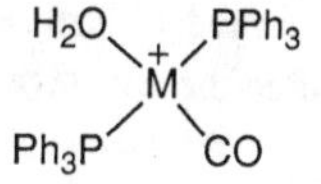

(33)

The peroxide adduct of $[Fe_2(HPTB)(\mu\text{-}OH)(NO_3)_2](NO_3)_2$, where HPTB = *N,N,N',N'* -tetrakis(2-benzimidazolylmethyl)-2-hydroxy-1,3-diaminopropane, hasνOO as a Fermi doublet centred at 895 cm^{-1} (854 cm^{-1} for $^{18}O_2$).[278]

(33) complexes have νOH of the H_2O ligand at 3373 cm^{-1} (M = Ir) or 3300 cm^{-1} (Rh).[279] The bridging OH group in (34) has νOH at 3590 cm^{-1}.[280]

νOO modes have been assigned for a range of $^{16}O/^{17}O/^{18}O$ variants of $(Ph_3P)_2Pt$-(O_2).[281] (35) and related complexes have νOO in the range 850 - 950 cm^{-1} (mixed with δCO).[282]

Reflection-absorption IR spectroscopy (RAIRS) of O_2 on silver polycrystals shows a band at 983 cm^{-1} assigned to O_2 coordinated perpendicular to the surface, and another at 622 cm^{-1} assigned to parallel-coordinated O_2.[283]

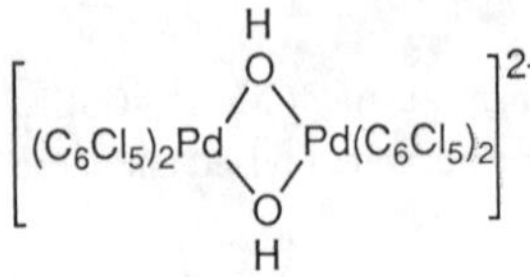

(34)

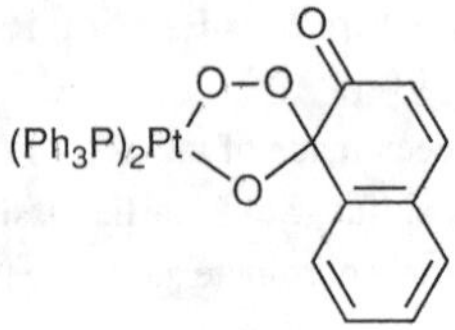

(35)

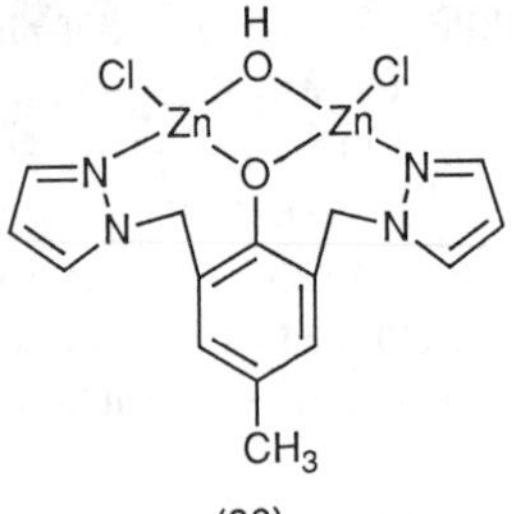

(36)

The bridging OH group in (36) has νOH as a broad band in the IR spectrum centred at 3432 cm^{-1}.[284] $B(OOH)_4^-$ has an OO stretching band at 870 cm^{-1}, together with νOH modes at 3040, 3200, 3370 and 3450 cm^{-1}.[285] νOH bands were also assigned for M_2-$[Pb(OH)_6]$, where M = K, Rb or Cs (3143 - 3366 cm^{-1}).[286]

6.2. Carboxylate and Related Complexes:- The Raman spectra of an extensive range of squarate complexes of transition metals show that D_{4h} symmetry of the oxocarbon ligand is always preserved.[287]

The Raman spectra of aqueous $Mg(OAc)_2$ solutions in the temperature range 25 - 250°C at pressures of 9MPa show the formation of $[Mg(H_2O)_5OAc]^+$ and $[Mg(H_2O)_4$-$(OAc)_2]$ complex species.[288]

Values of ν_{as} - $\nu_s(COO^-)$ for $CpLn(CF_3COO)L$, where Ln = Sm, Dy, Ho or Yb, L = (*o*-$NH_2)C_6H_4$) or analogous ligands, are all consistent with dimeric formulations and bridging carboxylato ligands.[289]

$[V_{12}As_8O_{40}(HCO_2)]^{3-}$ gave bands due to encapsulated formate at 2800 cm^{-1} (νCH), 1587 cm^{-1} ($\nu_{as}CO_2$) and 1340 cm^{-1} (ν_sCO_2).[290] The symmetric bidentate formato ligand in (37) gave IR bands at 1546, 1364 cm^{-1} due to ν_{as}, ν_sCO_2 respectively. ^{13}C substitution shifted these to 1506 and 1342 cm^{-1}, ^{18}O substitution to 1535 and 1346 cm^{-1}.[291]

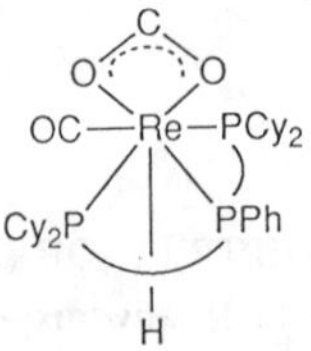

(37)

$Na_3[Fe(OCOR)_6]$, where R = Me_3C, $Me(CH_2)_5$, $Me(CH_2)_{10}$, 2,5-$Cl_2C_6H_3$, 2,5-Me_2-C_6H_3, 2-$MeOC_6H_4$ or 3-$MeOC_6H_4$, all have νCO_2 IR bands showing unidentate coordination.[292] $(PPh_4)[RuO_2Cl_2(OAc)]$ gives typical νCO_2 bands of a bidentate acetato ligand.[293] $Ru(CO)Cl(E_2CRC{=}CHR')(PPh_3)_2$, where E = O or S, R = H, Me or Ph, R' = H, CMe_3, $SiMe_3$, Ph, CO_2Me, give νCE_2 bands showing η^2- *E,E* -coordination.[294] $Ru(O)_2L_2$-$[SCH_2CHRC(O)O]$, where L = py or 1/2(bipy), R = H, NHCHO or NHCOMe, give $\nu_{as}CO_2$ showing that the carboxylate is coordinated to Ru.[295]

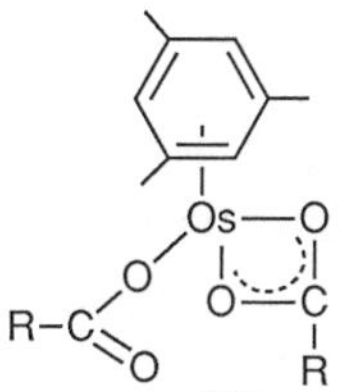

(38)

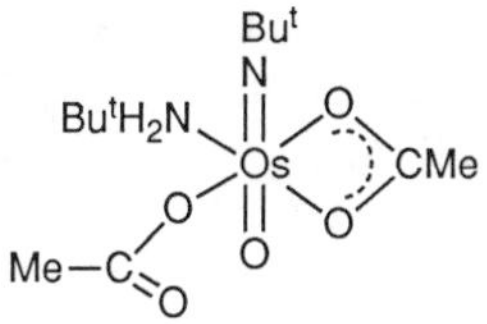

(39)

Characteristic bands due to bridging oxalato ligands were reported for $[Os(NO)(C_2O_4)_2]^-$. νNO was slightly cation-dependent.[296] IR spectra show the presence of both uni- and bidentate carboxylates in (38), where R = CH_3 or CF_3.[297] Very similar results were found in the IR spectrum of (39).[298] νCO_2 modes in $Os_3(CO)_{10}H(\mu\text{-}C_6F_5CO_2)$ and $Os_3(CO)_{11}H(C_6F_5CO_2)$ show bridging bidentate and unidentate coordination respectively.[299] $[Os(\equiv N)\{SCH_2CH_2C(O)O\}_2]^{2-}$ has νC=O at 1660 cm^{-1}, consistent with η^1-carboxylate coordination.[300]

The IR spectrum of $Rh(O_2CCF_3)_3$ shows that both bridging and unidentate trifluoroacetato groups are present.[301] Raman bands were seen for the new species $K_2Ni(C_2O_4)_2$ in aqueous solutions of nickel(II) oxalate and potassium oxalate.[302]

Characteristic IR bands in $[\{Cu(terpy)(H_2O)\}_2(ox)][\{Cu(terpy)\}_2(ox)][ClO_4]_4$ and in $Cu(terpy)(H_2O)(ox)$, where ox = oxalato, show asymmetrical bis-bidentate coordination in the former, and bidentate coordination in the latter.[303] Cu(*p*$XNC_6H_4COO)_2py$, where X = Cl or Br, give νCO_2 modes showing bidentate chelating carboxylato groups.[304]

$MCu_2(OAc)_2(Salpd)_2$, where M = Mg, Mn, Co, Ni, Cu or Zn, H_2Salpd = propane-1,3-diylbis(salicylideneimine), have the expected IR bands due to ν_{as}, ν_sCO_2 for symmetrical bridging acetato groups. The skeletal vibrations of the phenolic oxygen of the Schiff-base ligand is near 1550 cm^{-1}.[305] νCO_2 bands were assigned for $Zn_4O(OOCR)_6$, for the acetate, propionate, butyrate and valerate.[306]

The IR and Raman spectra of $(UO_2)_2(C_2O_4)X_2L_4$, where X = NCS^-, Cl^- or Br^-, L = neutral ligand, all show that the oxalato groups are planar and tetradentate.[307] IR and Raman spectra of the UO_2 derivatives of dicarboxylic acids were used to determine the lengths of the carboxylate CO bonds from ν_s and $\nu_{as}CO_2^-$.[308]

The X-ray crystal structure of (tris-lactato)aluminium(III) shows coordination *via* one carboxylate oxygen and a hydroxyl oxygen. The IR spectrum has ν_{as}, ν_sCO_2 at 1613, 1403 cm^{-1} respectively, consistent with unidentate carboxylate.[309] $Me_2In(L)$, where HL = 2-carboxybenzaldehyde, has ν_s, $\nu_{as}CO_2$ of L^- showing symmetrical bidentate coordination of the carboxylate group.[310]

$(NO_2^+)_2[Sn(OOCCF_3)_6]$ has been prepared - the νCO_2 modes do not give unequivocal evidence on the mode of coordination of the trifluoroacetate, but do suggest the presence of uni- and bidentate ligands.[311] νCO_2 modes are in agreement with bidentate chelating carboxylates in $R_2Sn(O_2CCH_2SPh)_2$, $\{[R_2Sn(O_2CCH_2SPh)]_2\}_2$, where R = Me, Et, ^nPr, ^nBu or ^nOct;[312] and $R_2Sn(O_2CC_4H_3E)$, where R = Me, Et, ^nPr or ^nBu; E = O or S.[313] Bridging carboxylates and *trans*-C_3SnO_2units, with trigonal bipyramidal geometry at the tin, were suggested by the IR spectra of $R_3Sn(O_2CCH_2S_2CNR'_2)$. where R = Me, ^nBu or Ph, $NR'_2 = NMe_2$, NEt_2, $N(CH_2)_4$, $N(CH_2CH_2)_2O$.[314]

$Sb(O_2CR)_3$, where R = H, Me, Et, $(CH_2)_2CH_3$, $CH(CH_3)_2$, $(CH_2)_3CH_3$, $CH_2CH(CH_3)_2$, $C(CH_3)_3$ or $(CH_2)_4CH_3$, all have IR bands of carboxylato ligands showing bidentate chelation and bridging functions.[315]

6.3. Keto-, Alkoxy-, Ether and Related Complexes:- IR and Raman spectra of $M(15C5)^+$, where 15C5 = 15-crown-5, complexes show that for M = Li or Na the 15C5 macrocycle contains *TGT* OCH_2CH_2O, but for M = K *TGG* conformational units.[316] Similar data were reported for $M(B15C5)^+$ species, where M = Li, Na or K, B15C5 = benzo-15-crown-5.[317]

The IR spectrum of $Ca(acac)_2$ suggest coordination of $acac^-$ to form a polymeric structure.[318] The IR and Raman spectra of polyethylene glycol complexed to $CaCl_2$ show the presence of 2 conformational sequences: (- *TGT-TG'T-GTT-TTG* -) and (- *TGT-TG'T-TGT-GTT-TTG*-).[319]

The IR spectra of Fe(II), Ni(II), Cu(II) and Zn(II) complexes of methylenebisurea (L): $M_2(OAc)_4L$, where M = Fe, Ni or Cu, $Cu(OAc)_2L_3$, and $Zn_3(OAc)_6L$ all suggest coordination of L *via* two carbonyl groups. For $M(OAc)_4L$, where M = Fe or Ni, there was evidence for coordination by N also.[320]

The Raman spectrum of Hacac adsorbed on a copper surface show the formation of a copper acetylacetonate complex.[321] Shifts in characteristic ligand bands confirm the coordination shown for (40), where X = $1/2(SO_4)$, R = PhNH.[322]

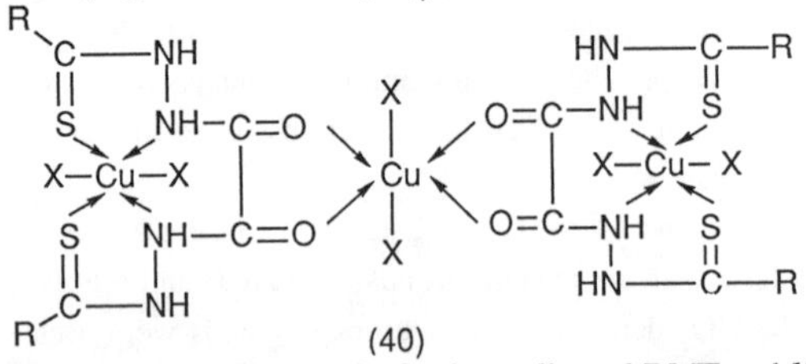

(40)

$[UO_2(DMF)_4]_3[Cr(SCN)_6]_2$ has νC=O of coordinated DMF at 1655 cm^{-1}, compared to 1680 cm^{-1} in free DMF. Coordination has therefore occurred *via* the carbonyl O atom.[323] νCOC of the crown ring in $UO_2Cl_4(DCH18C6.H_2O)_2$, where DCH18C6 = dicyclohexyl-18-crown-6, is at 1090 cm^{-1}, consistent with *O*-coordination.[324]

Bands in the νCO + νC=C region for "$Cl_2Al(acac)$" show bands due to trimers and also monomers - which of course involve unidentate $acac^-$ ligands.[325] νC=O in $AlMe(BHT)_2[O{=}C(X)R]$ and $AlMe_2(BHT)[O{=}C(X)R]$, where HBHT = 2,6-di-*tert*-butyl-4-methylphenol, X = H, OR, NR_2, R = M or Et.[326] $Tl[CH_2C(O)Me]_2(CF_3SO_3)$ has νC=O of an *O*-bonded acetonyl ligand at 1670/1650 cm^{-1}.[327]

νCO bands were assigned in $Sn(O^tBu)_4$ (937 cm^{-1}) and the perdeuteriated analogue $Sn(O^tBu\text{-}d_9)_4$(873 cm^{-1}).[328] The IR spectra of triorganostannyl esters of amic acids show that the amido carbonyl groups are bridging between R_3Sn units, (41), where R = Ph, Cy, Bu; R^1= Me or Ph; R^2, R^3 = H, Ph, Pr.[329]

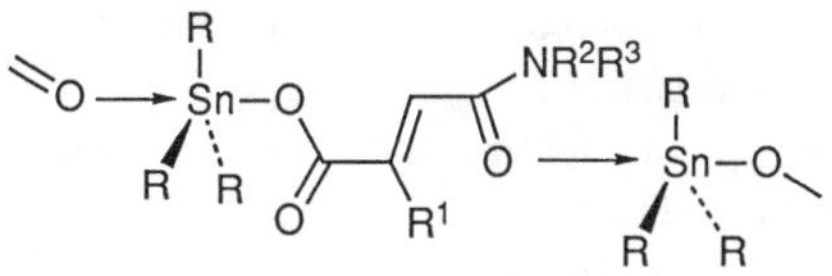

(41)

6.4. Ligands Containing O-N or O-P Bonds- The IR spectra of complexes of a range of lanthanide nitrates with urea show the presence of coordinated nitrato ligands - probably bidentate. The urea ligands are coordinated *via* oxygen.[330] A survey of νNO_3 modes of anhydrous lanthanide nitrates suggests that $Lu(NO_3)_3$ may not be isostructural with other members of the series.[331]

νPO modes in lanthanide tetraethyl imidophosphato complexes, $Ln[N\{P(O)(OEt)_2\}_2]_3$, where Ln = La, Ce, Pr, Nd or Sm, show coordination of phosphoryl oxygen to Ln.[332] $M[Ph_2P(O)NP(O)Ph_2]_3$, where M = lanthanide, have IR spectra showing that the iminodiphosphinate is coordinated *via* both oxygen atoms.[333]

NO/NO_2 on anatase (TiO_2) gives IR bands due to bridged and bidentate nitrato surface species. NO_2 also forms unidentate nitrate.[334] Bands due to nitrato ligands in $Zr(OH)_2(NO_3)_2$ and $Zr(OH)_3NO_3$ suggest that each solid complex contains nitrato groups in at least two different environments.[335]

$VCl_3(OPMe_2Ph))(bipy)$ has νP=O at 1100 cm^{-1}, showing V-O coordination.[336] NO or NO_2 on V_2O_5/TiO_2 catalysts form bridging, bidentate and unidentate nitrato species, as shown by IR spectra.[337-8]

The IR spectra of $Fe(L)NO_3$, where $H_2L = R{=}N(CH_2)_nNR'(CH_2)_nN{=}R$, R = salicylidene, n = 2 or 3, R' = H or Me, show that the nitrate is unidentate.[339] Fe(NO)(Odppe), where dppe = $Ph_2PCH_2CH_2PPh_2$, has νP=O at 1165 cm^{-1}, compared to 1180 cm^{-1} in the free ligand.[340]

The Raman spectra of aqueous solutions of aluminium, gallium or indium nitrates show that cation/anion interactions are in the order: $Al(NO_3)_3 < Ga(NO_3)_3 < In(NO_3)_3$, with the formation of definite inner-sphere nitrato complexes for Ga and In.[341]

6.5. Ligands Containing O-S or O-Se Bonds:- The novel tetranuclear chromium(III) complex $[Cr_4O(SO_4)_2Cl_9]^{3-}$ has terminal νS=O bands at 1275 and 1268 cm^{-1}, and coordinated νS-O bands at 1015 and 984 cm^{-1}.[342] A characteristic νSO_4 mode is seen at 1155 cm^{-1} in the 12-molybdosulphate(VI) complex.[343]

IR bands indicative of bridging sulphato ligands are seen for $Ir_3(\mu_3\text{-}O)(\mu\text{-}SO_4)_6(H_2O)_3$ and $[Ir_3(\mu_3\text{-}O)(\mu\text{-}SO_4)_6(HSO_4)_2(H_2O)]^{4-}$.[344]

The IR spectra of $M_2^IM^{II}(SeO_4)_2$, where M^I = K or Tl, M^{II} = Cu or Ni, show that both central atoms are six-coordinate, with bridging SeO_4 groups.[345] $Pd(LL)SO_4$, where LL = phen, bipy or *o*-phen, all give IR bands consistent with C_{2v} (bidentate) SO_4 ligands.[346]

6.6. Ligands Containing O-Cl or O-I Bonds:- IR data for $Be(IO_3)_2$ show that the Be atom is four-coordinate, by two bidentate IO_3^- ligands of C_s symmetry.[347]

There is FTIR evidence for inner-sphere coordination of perchlorate ions in CH_3CN solutions of $Ln(ClO_4)_3$, where Ln = Lu, Pr, Sm, Gd, Dy, Ho, Tm or Yb. There appeared to be a complex set of equilibria involving free, uni- and bidentate ClO_4^-.[348]

The IR bands due to ClO_4 modes in $WO_2(ClO_4)_2$ show that both of the ClO_4^- ligands are bidentate. In $[WO_2(ClO_4)_{2+n}]^{n-}$, where n = 1 or 2, data show that both uni- and bidentate perchlorato ligands are present.[349]

The coordinated ClO_4^- in the complex $[\{Cu_4(\mu_5\text{-}O)L(ClO_4)_2](ClO_4)_2$, where H_4L = (42), gives bands characteristic of unidentate coordination.[350]

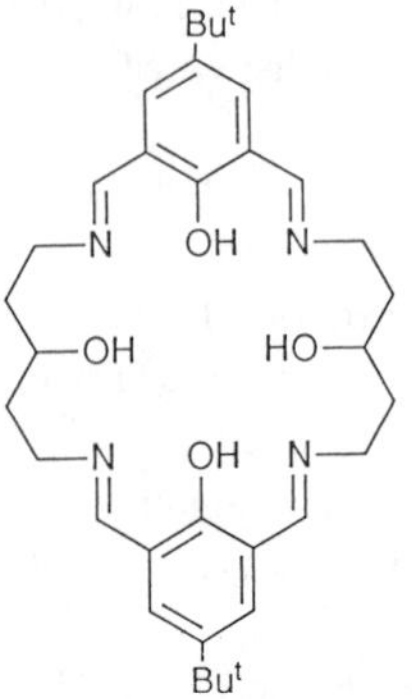

(42)

Characteristic ClO_4^- bands due to the coordination modes shown were seen for the following complexes: $[(Ph_3P)(OC)Rh(\mu_3\text{-}C_7H_4NS_2)Ag(OClO_3)]_2$ and $(cod)_2Rh_2(\mu_3\text{-}C_7H_4\text{-}NS_2)_2Ag(O_2ClO_2)$, where $C_7H_4NS_2$ = benzothiazole-2-thiolate.[351]

The IR and Raman spectra of $Sn(ClO_4)_4$ and $SnCl_2(ClO_4)_2$ show the presence of bridging bidentate ClO_4 in both, together with (respectively) unidentate ClO_4 and terminal Cl.[352] There is IR evidence for coordinated ClO_4^- in adducts of $Pb(ClO_4)_2$ with aromatic *N*-oxides.[353]

IR spectra of $Bi(ClO_4)_3$, $[Bi(ClO_4)_4]^-$, $[Bi(ClO_4)_5]^{2-}$ and $[Bi_2(ClO_4)_{10}]^{4-}$ show that the first contains only bidentate ClO_4^-, while the others all involve both uni- and bidentate coordination.[354-5]

7. Sulphur, Selenium and Tellurium Donors

A review of the coordination compounds of $R_2PS_2^-$ ligands showed that differences between νPS_2 modes were 50 - 75 cm^{-1} for isobidentate coordination, more than 100 cm^{-1} for unidentate and between 85 and 95 cm^{-1} for anisobidentate coordination.[356]

νSN in (43), where R' = Me, NR_2 = NMe_2 or $N(CH_2)_4O$; R" = Ph, NR_2 = NMe_2, $N(CH_2)_4O$, $N(CH_2Ph)_2$ or $N(C_6H_{11})_2$, were all decreased compared to those in the free ligands.[357] $^{32}S/^{34}S$ isotopic shift data for $M_2S_4X_4^{2-}$, where M = Mo, X = Br; M = W, X = Cl or Br, were used to make assignments for νSS of the $\mu_2\text{-}S_2$ bridging ligand (600 cm-1) and for the terminal S_2 ligand near 520 cm^{-1}.[358] $[M_2O_2(S_2)(S_4)]^{2-}$, where M = Mo or W have νSS of the S_2^{2-} ligand at 509 - 521 cm^{-1}, of the S_4^{2-} ligand 420 and 490 cm^{-1}.[359] (44), where R = Me, Et, Pr or Bu, all have νSS of the η_2-S_2 ligand at 550 cm^{-1}.[360]

The resonance Raman spectrum of DMSO reductase (DR) from *Rhodobacter sphaeroides* , containing Mo attached to a dithiolene residue shows νC=C at 1575 cm^1

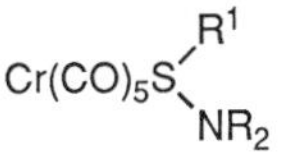

(43)

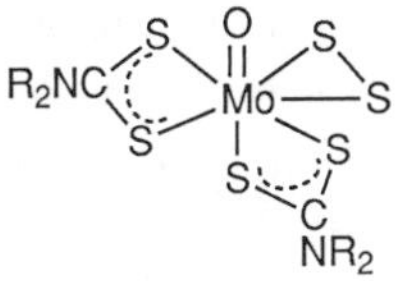

(44)

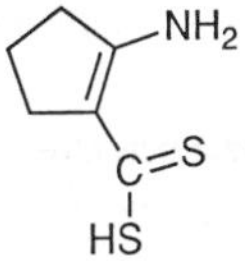

(45)

(oxidised form of DR) or 1568 cm^{-1} (reduced form of DR). This shift is consistent with reduction of Mo(VI) to Mo(IV).[361] Ligand mode assignments in the xanthate complex $[Mo(S_2COEt)(CO)_3(PPh_3)]^+$ and related species show that the xanthate is *S,S* -chelating.[362] $Mo(S_2CNR_2)_2(acda)_2$ and $Mo(S_2CNR_2)_2(acda)$, where R =Et or Pr, Hacda = (45), give bands characteristic of *S,S* -chelation for both ligands.[363]

νPF (800 - 890 cm^{-1}) and νPS (690 - 700 cm^{-1}) were assigned for $Mo_2(S_2PF_2)_2L_2$, where L = O_2CCF_3, O_2CCH_3 and S_2PF_2. νPS modes are at 720 and 730 cm^{-1} for $Mo_2(S_2P\text{-}Me_2)_4$.[364] IR and Raman spectra of $W_3S_7Br_6^{2-}$ show νSS bands of the μ_2-S_2 ligands at 558 and 552 cm^{-1}.[365]

The Raman spectra of $MnSe_2$ and MTe_2, where M = Mn, Ru or Os, contain νTeTe bands in the range 175 - 179 cm^{-1} (MTe_2) or νSeSe 267 cm^{-1} ($MnSe_2$).[366] The complexes $ReOCl_2\{N(OPPh_2)_2\}(PPh_3)$, $ReOCl_2\{N(SPPh_2)_2\}(PPh_3)$ and $ReOCl_2\{N(SPPh_2)_2\}_2$ have ν_{as}PNP of the chelating ligand at 1205, 1170 and 1210 cm^{-1} respectively, characteristic of coordination type (46).[367]

$[Ru(NO)(NH_3)(S_4)_2]^-$ has νSS bands at 458 (IR), 448 and 483 cm^{-1} (Raman).[368] $(PP_3)Rh(XH)$, where X = S, Se or Te, PP_3 = $P(CH_2CH_2PPh_2)_3$, have νXH bands as follows: X = S 2540 cm^{-1}, Se 2260 cm^{-1}, Te 1925 cm^{-1}. This last is the first definitive report of νTeH in such a complex.[369]

(47) has νC=C at 1430 cm^{-1}, and this mode shifts to 1380 cm^{-1} on one-electron oxidation of this complex.[370]

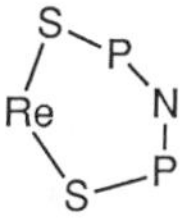

(46)

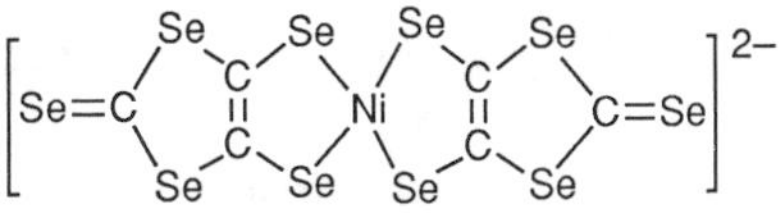

(47)

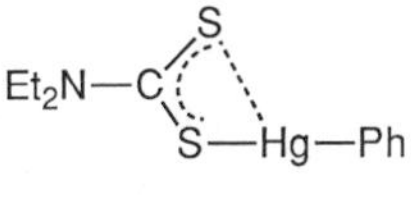

(48)

The polyselenides $[(Ph_4P)Ag(Se_4)]_n$, $[(Me_4N)Ag(Se_5)]_n$, $[(Et_4N)Ag(Se_4)]_4$,[371] $[Au_2(Se_2)(Se_3)]^{2-}$ and $[Au_2(Se_2)(Se_4)]^{2-}$ [372] all have νSeSe bands 257 - 265 cm^{-1}. νSeSe in $K_4[U(Se_2)_4]$ is at 261 cm^{-1}.[373]

$MeHgS_2PPh_2$ has $\nu_{as}PS_2$ which show unidentate coordination of the diphenyldithiophosphinate (645, 540 cm^{-1}). The wavenumbers for $PhHgS_2PPh_2$ (635, 550 cm^{-1}) suggest that here the coordination may be anisobidentate.[374] PhHgX, where X = ethylxanthate or *N,N'* -diethyldithiocarbamate, have IR spectra consistent with coordination of the type (48), with νCS 980 and 1010 cm^{-1}.[375]

νCS bands of TlSPh, TlS^tBu and $TlSC_7H_7$ are at 606, 590 and 544 cm^{-1} respectively.[376] Characteristic IR and Raman bands were identified for the unidentate xanthate derivatives $PhGe[S_2CO^iPr]_3$, $Ph_2Ge[S_2COR]_2$, where R = ^iPr or ^nBu, and Ph_3Ge-$[S_2COR]$, where R = Me, Et, ^iPr or ^nBu.[377]

8. Potentially Ambident Ligands

8.1 Cyanates, Thio- and Selenocyanates and their Iso-analogues :- The complexes $Zr(NCS)_3L(H_2O)$, where HL = 8-hydroxyquinoline, $Th(NCS)_2L'_2$ and $U(O)(NCS)_2L'_2$, where HL' = aniline-2-carboxylic acid, all have νNCS bands consistent with *N*coordination.[378] In $Zr(NCS)_4L_2$, $Th(NCS)_4L_4$, $Cr(NCS)_3(OAsPh_3)_3$, $UO_2(NCS)_4L_2$, $UO_2(NCS)_3$-$(OAsPh_3)_3(H_2O)$ and $Cr(NCS)_3py_2$, where L = pyridine- *N*oxide, $OPPh_3$, OH_2 or $OAsPh_3$, the NCS^- is *N*-coordinated to Zr(IV), Th(IV) and U(VI), but said to be*S* -coordinated to Cr(III).[379]

The IR spectra of several $M_3[Cr(NCS)_6]$ have been reported. Force constants were calculated for the N-C and C-S bonds.[380] $Cr^{II}(NCS)_6^{4-}$ has two νCN (2078, 2114 cm^{-1}), with νCS at 779 cm^{-1}, and δNCS at 470 cm^{-1}.[381] $[\{Cr(N\text{-}N)_2NCS\}_2O]^{2+}$, where N-N is bipy or phen, have a single νCN band, showing that there are two equivalent *N*-bonded NCS^- ligands.[382]

Trans -Mo(NCO)Cl(dppe) $_2$has $\nu_{as}NCO$ at 2200 cm^{-1} (2180 cm^{-1} for $N^{13}CO$), indicating probable *N*-coordination.[383] There is IR (and X-ray) evidence for one PhNCO coordinated *via*C=N and one *via* the C=O bond in *trans*-Mo(η^2-O,C-PhNCO)(η^2-C,N-PhNCO)-(*syn* -Me8[16]aneS_4).[384]

IR data (νCN) were used to characterise the S = 2 and S = 1 spin states of Fe^{II}(4,4'-dpp)$_2$(NCX)$_2$, where dpb = diphenyl-2,2'-bipyridyl, X = S or Se.[385] The IR spectra of $OsCl_5(NCSe)^{2-}$, $OsCl_5(SeCN)^{2-}$, *trans* -$[OsCl_4(NCSe)(SeCN)]^{2-}$, *trans* -$[OsCl_4I(NCSe)]^{2-}$ and *trans*$[OsCl_4I(SeCN)]^{2-}$, show that ν CN(Se) > νCN(N), νCSe(N) > νCSe(Se) and δNCSe > δSeCN.[386]

The IR spectrum of NH_3 + CO adsorbed on Rh supported on Al_2O_3 or TiO_2 shows the formation of surface Rh-NCO units.[387]

$Ni(NCS)_2(H_2O)_2(ten)_2$, where ten = triethylenediamine, has νCN (2100 cm^{-1}) and δNCS (470 cm^{-1}) of *N*-bonded NCS^- ligands.[388] $[Ni(terpy)(NCSe)_2]_2$ has νCN at 2120 and 2100 cm^{-1} due to 'end-to-end' bridges in the dimer. νCSe is at 470 cm^{-1} and δNCSe 410 cm^{-1}.[389]

Pt^{II}(2,2'-bipy)$(NCO)_2$ shows IR NCO bonds characteristic of Pt-N bonding in solid and solutions.[390] νCN bands in $Pt(NH_3)_2(SCN)_2X_2$ are at 2178 cm^{-1} (X = Br) or 2168 cm^{-1} (I), corresponding to bridging thiocyanates.[391] Characteristic NCS ligand modes confirm the structures of *cis*-[PtL(NCS)$_2$], where L = 1,2-bis(di-isopropylphosphino)ethane, and *cis* -[PtL'(NCS)(SCN)], where L' = 1,2-bis(diethylphosphino)ethane.[392]

IR spectroscopy shows that $Cu^{I}NCS$ on Al_2O_3 or SiO_2 forms bridged thiocyanate species. There was evidence for both *N*- and *S*-bonded forms of Cu(NCS) on a variety of supports.[393] *In situ* FTIR of NCS^- adsorbed on polycrystalline silver electrodes also gave evidence for both Ag-N and Ag-S coordination.[394]

νCN modes are consistent with *N*-coordination in $CH_3SCN.MF_5$ and $CH_3CN.MF_5$, where M = As or Sb.[395]

8.2. Ligands Containing N and O or P and O Atoms :- The IR and Raman spectra of solid ML_2, where M = Mg or Ca, HL = caprolactam, show that they have oligomeric chain structures containing terminal and bridging NCO groups. In solution there is evidence for association *via* NCO bridges.[396]

VOL_2, where H_2L = 1-(2-pyridylazo)2-naphthol (PAN) or 4-(2-pyridylazo)-resorcinol (PAR), have IR spectra consistent with PAN coordination *via* O and pyridyl N, and PAR coordinated *via* the pyridyl and ortho-hydroxyl groups.[397] The IR spectrum of $VO(Haafh)_2SO_4$, where Haafh = 2-aminoacetophenone-2-furoylhydrazone, shows that the Haafh is unidentate. In $M(Haafh)_2Cl_2$, however, where M = VO^{2+}, Mn, Co, Ni, Cu, Zn, Cd or Hg, the ligand is bidentate *via* the carbonyl O and azomethine N.[398]

ν_s and ν_{as} CO_2 modes of $[CrOL_6(H_2O)_3]^{7+}$, where L = glycine, α-alanine or α-amino-butanoic acid, show the presence of bidentate-bridging carboxylate groups.[399] MoO_2L-(MeOH), where H_2L = *o*-HOC_6H_4CH=NNHC(O)R, R = 4-$NO_2C_6H_4$, 3-BrC_6H_4 etc., have IR spectra consistent with tridentate hydrazones, *via* N and two O atoms.[400] Similar data support N + O coordination of SAE^{2-} in MoO_2(SAE)L, where H_2SAE = *N*-(hydroxy-ethyl)salicylidenenimine, L = imidazole or substituted derivative.[401] (49), where E = CH_2CH_2 etc., have νC=O (free) near 1700 cm^{-1} and νC=O (complexed) 1535 - 1560 cm^{-1} from the unit (50).[402]

(49)

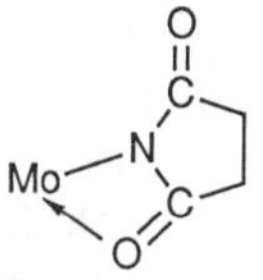

(50)

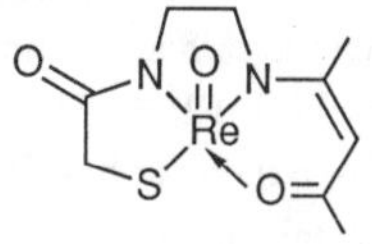

(51)

$Mo_2O_5(OH)_2$(HL), where HL is one of a wide range of amino-acids, all have IR bands due to ν_s and ν_{as} of CO_2 showing that the carboxylato groups are bidentate.[403]

The IR spectra of MnL_2, where H_2L = 2-HO-5-RC_6H_3CH=NN=C(SR')NH_2, R = H, R' = Me, Et, Pr or Bu; R = Cl, Br, Me, MeO, R' = Me, show that L^{2-} is tridentate, *via* O + 2N.[404] In $ML_2(H_2O)_2$, where M = Mn, Co, Ni or Zn, $[CuL(H_2O)_2]^+$, where HL = 5,6-bis(ethoxycarbonylmethyl)-3-seleno-1,2,4-triazine, coordination *via* an ester oxygen and a nitrogen of the selenotriazine ring is indicated.[405] IR spectra of Mn(II), Co(II), Ni(II), Cu(II) and Z(II) complexes of benzoylacetanilide show coordination *via* an NH group and an enolic O^- atom.[406] νC=O/νC=C bands in (51) are at lower wavenumber than in the free ligand, showing delocalisation of π-electron density in the chelate ring.[407]

IR spectra of FeL_2^+, CoL_2^+, $M(HL)Cl_2$ (M = Co or Ni). and $Cu(HL)X_2$ (X = Cl or Br), where HL = 2-acetylpyridine- *N*-oxide thiosemicarbazone, show that HL is bidentate (O,N-), while L- is tridentate (O,N,S-).[408]

Photochemical reactions of $CpRu(CO)_2(NO_2)$ at 13 K gave IR evidence for *exo* - and *endo* -$CpRu(CO)_2(ONO)$ (for $\lambda > 350$ nm). Photolysis at $\lambda = 314$ or 280 nm produced several species, one of which was thought to contain NO_2^- in an (*N,O*)-bridging form.[409] $RuL(PPh_3)Cl$, where $H_2L = 2\text{-}HOC_6H_4CR{=}N)_2R'$, R = H or Me, R' = $(CH_2)_2$ or $(CH_2)_3$; R = H, R' = CH_2CHMe or $CH_2CH(OH)CH_2$, give IR spectra consistent with N_2O_2 quadridentate bonding by the Schiff base ligands.[410] (52), where P-O = (2-methoxyethyl)-diphenylphosphine, has ν_{as}COC of the chelated ligand at 1067 cm^{-1}, with a corresponding feature from the *P*-bonded unidentate ligands at 1096 cm^{1}.[411]

$Co^{III}(TPP)(pip)(NO_2)$, where pip = piperidine, has ν_{as}, ν_s of NO_2 at 1420 and 1304 cm^{-1} respectively.[412] The IR spectra of $CoLpy_3(NO_3)$, where HL = MeC(=NOH)C(O)NH-$CHRCO_2H$, R = H, Me, or $CH_2CH_2SiMe_3$, and of $ZnLpy_3(NO_3)$, for R = H or Me, show that L^- is coordinated *via* the amide O and oxime N.[413] IR and Raman spectra of CoL_2-$(H_2O)_2(NCO)_2$, where L = picolinamide (PA) or isonicotinamide (INA), and of NiL_2-$(H_2O)_2(NCO)_2$, where L = INA or nicotinamide (NA), show that NA and INA are coordinated *via* heterocyclic N, while PA is coordinated *via* amino N.[414]

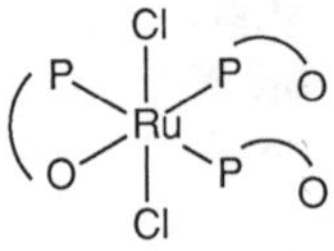

(52)

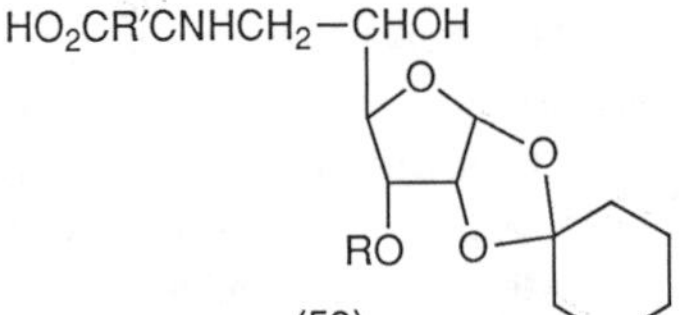

(53)

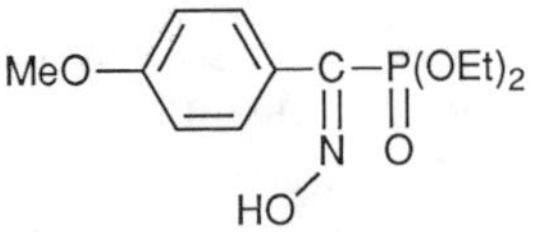

(54)

IR data on $M(4\text{-}PNH)X_2$, where 4-PNH = pyridine-4-carboxaldehyde nicotinylhydrazone, M = Co or Cu, X = Cl or I; M = Ni, X = Cl, Br or I, show that 4-PNH is bidentate *via* carbonyl O and azomethine N.[415] Similar results on $ML_2(H_2O)_2$, where M = Co, Ni or Cu, HL = (53), R' = H, Me, $CHMe_2$, CH_2OH etc., show coordination *via* OH only (Co, Ni) or carboxy O and N (Cu).[416] $M(HL)_2(H_2O)_n$, where M = Co, Ni or Zn, H_2L = pyruvylglycineoxime or pyruvyl-L-alanine oxime, n = 0 - 2, give IR spectra showing bidentate coordination of HL^- by the oxime N and amide O.[417] L (diethyl-2-pyridylmethylphosphonate) in ML_2X_2, where M = Co or Ni, X = ClO_4 or NO_3, is coordinated through N and O atoms.[418]

IR spectra of Co(II), Ni(II) or Cu(II) complexes of hydrazides of aryloxycarboxylic acids, such as 2,4-dichlorophenoxy-acetic acid, are all consistent with N + O coordination of these ligands.[419] νC=N and νP=O modes of HL (= (54)) show coordination of phosphoryl O and oxime N in $M(HL)_2X_2$, where M = Co or Ni, and in $Cu_2(HL)_2X_4$, where X = Cl or Br.[420] Coordination *via* N and O atoms has also been proposed for ML_2, where M = Co, Ni or Cu, L- = 1-aminonaphthalene-8-sulphonate;[421] and chiral complexes of Co(II), Ni(II) or Cu(II) with *N*[3-O-methyl(heptyl)-1,2-*O* -cyclohexylidene- α- *D* -glucofuranos-6-deoxy-6-yl]amino carboxylates.[422]

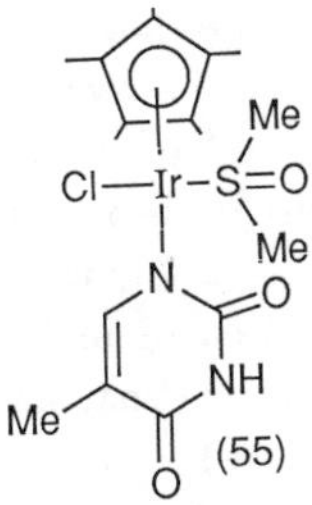

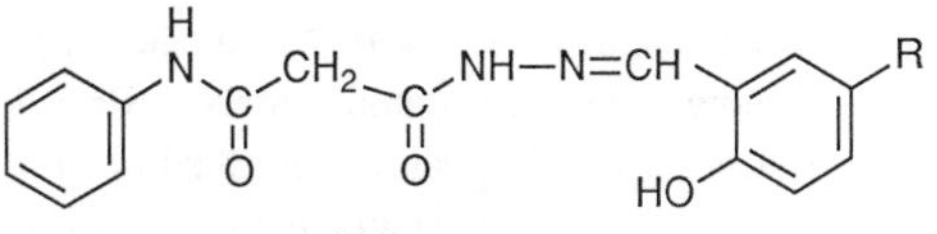

(56)

(55) has νC=O of the thyminate ligand at 1660, 1643 and 1633 cm^{-1}, showing N1 coordination. νS=O is at 1130 cm^{-1} and νNH at 3142 cm^{-1}.[423]

IR data for $[Nien_2(cyt)_2]^{2+}$, where cyt = cytosine, shows that cytosine is unidentate *via* the carbonyl oxygen only.[424] The IR spectra of Ni(II) and Cu(II) complexes of derivatives of 1-phenyl-3-methyl- *L* -naphthyl-azapyrazol-5-one (HL), i.e. ML_2, are consistent with coordination *via* azo-N and an O atom of the pyrazole ring.[425] νC=O and νC=N bands of $NiL(H_2O)_2$ and CuL, where H_2L = (56), R = H. Me, Cl or OMe, are all suggestive of coordination by oxygen atoms (phenolic, carbohydrazone and secondary amide) together with an azomethine nitrogen.[426]

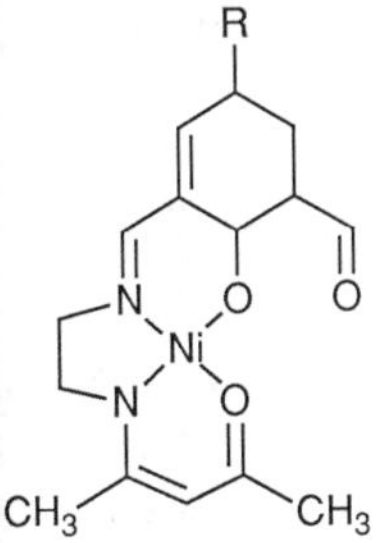

(57)

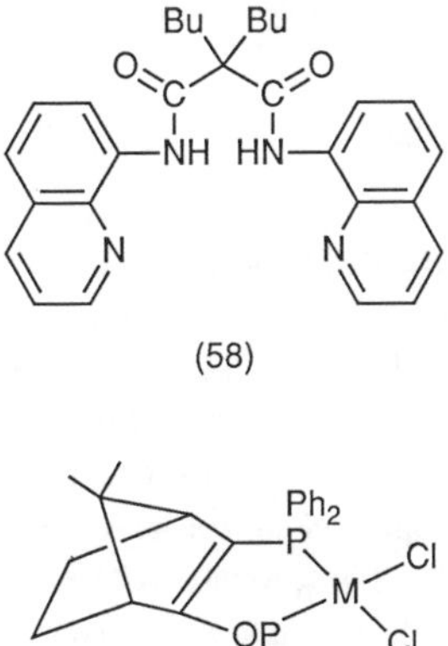

(58)

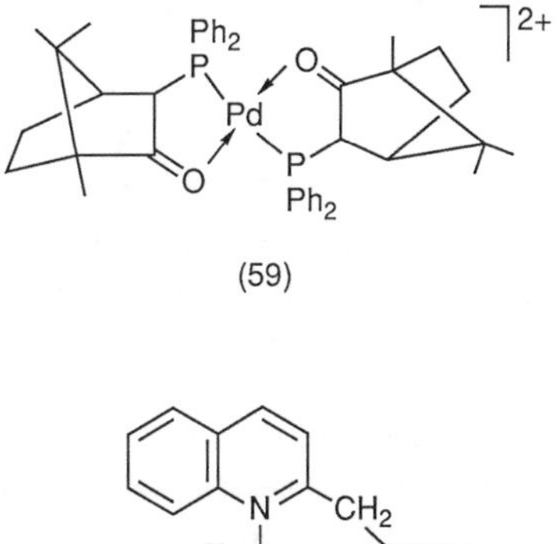

(59)

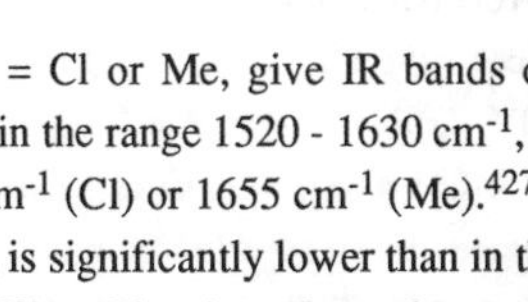

(60)

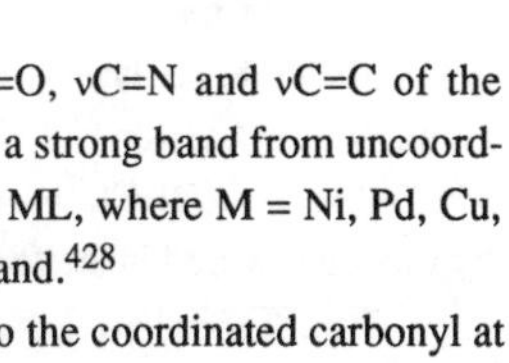

(61)

(57), where R = Cl or Me, give IR bands due to νC=O, νC=N and νC=C of the coordinated fragment in the range 1520 - 1630 cm^{-1}, as well as a strong band from uncoordinated C=O at 1670 cm^{-1} (Cl) or 1655 cm^{-1} (Me).[427] νC=O in ML, where M = Ni, Pd, Cu, Zn or Cd, H_2L = (58), is significantly lower than in the free ligand.[428]

The complex (59) and its *cis* analogue have νC=O due to the coordinated carbonyl at 1651 cm^{-1}. In (60), where M = Pd or Pt, however, no νC=O is seen, but νC=C near 1595 cm^{-1}.[429] (61), where X = Cl or Br have νP=O about 30 cm^{-1} lower than in the free ligand.[430]

$[Pt(NO_2)_4]^{2-}$ shows νNO_2 bands at 1410, 1468 and 1348 cm^{-1}, suggesting Pt-N coordination.[431] The bridging nitro groups in $Pt_4(\mu\text{-}NO_2)_4(\mu\text{-}NO)_3(NO_2)(H_2O)$ have $\nu_s NO_2$ at 1080 cm^{-1}.[432] IR bands were assigned to NO_2 and ONN vibrations in $[Pt(bipy)(ONN\text{-}C_2H_4NH_2)(NO_2)Cl]^+$, with *trans*Cl and NO_2, on the basis of $^{14}N/^{15}N$ shifts.[433] IR spectra of 2-amino-1-methyl-4-imidazolidinone ($MeNC(NH_2)NHCOCH_2$, creat), *cis* -Pt(creat)$_2$-$(NO_2)_2$ and $[Pt(creat)_4](ClO_4)_2$ show coordination *via* N, with retention of highly delocalised bonding within the ligand on complexation.[434] IR spectra of *cis* - and*trans*-$[PtL_2L'Cl]Cl$, *cis* - and *trans*$[PtL_2L'_2]Cl_2$, PtenL'Cl and $[PtenL'_2]Cl_2$, where L = NH_3 or NH_2OH; L' = guanosine, show L' coordinated *via* the N(7) imidazole atom.[435] $[Pt(Mt)_2]X_2$, where MtH = methionine, X = $1/2(PtCl_4^{2-})$ or picolinate, gave IR and Raman spectra showing tridentate coordination of Mt^-, *via*O, N and S atoms.[436]

The IR spectrum of $(Ph_3P)_2Cu(O_2CCH_2CN)_2H$ is consistent with the cyanoacetato ligand being bound through the cyano N atom, with a small amount of π-back bonding from Cu(I) to the nitrile π^* orbital.[437] (62), where R = (63) and related systems all give characteristic $\nu C{=}N$ bands for the structure shown.[438]

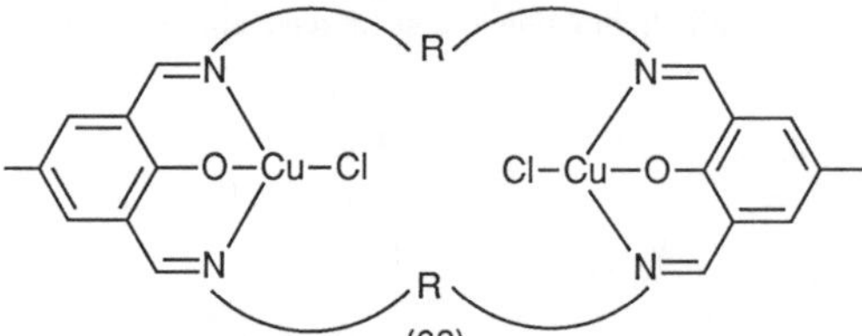

(62)

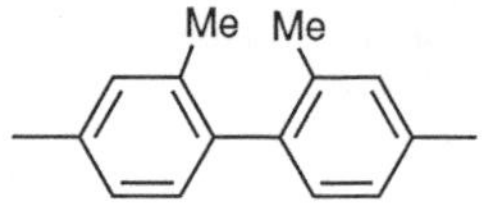

(63)

CuL, where H_2L = $[PhC(X)NHN{=}CMeCMe{=}NNHC(O)]_2(CH_2)_n$, where X = O or S, n = 0, 1, 2 or 4, have IR spectra showing N_2O_2 or N_2S_2 tetradentate coordination.[439] The IR spectra of $Cu(LH)NO_3$, where H_2L = salicylidene-ethanolamine, and Cu(LH)Q-(NO_3), where Q = 3- or 4-picoline, show tridentate coordination by HL^-, *via*N and two O atoms.[440] $ML(H_2O).nH_2O$, where M = Cu or Zn, HL = 4-hydroxysalicylidene- *L* -alanine, has IR bands from tridentate (phenolic O, imino N and carboxylate O) L^-.[441]

Raman and IR spectra of $UO_2(AIH)X_2$, where AIH = acetone isonicotinoyl-hydrazone, X = Cl, $1/2(SO_4)$ or $1/2(C_2O_4)$, show that AIH is coordinated to one uranium *via* carbonyl O and azomethine N, and to a second U through pyridine N. In $UO_2(AIH)_2$-X_2, one AIH is coordinated as above, while the other is unidentate.[442] $UO_2L(H_2O)$, where H_2L = Schiff bases of 7-hydroxy-6-formyl-methoxy-2-methylchromone with glycine, valine, leucine, serine, aspartine etc., contains L^{2-} coordinated through two O and one N atom.[443] IR data for $UO_2[XC_6H_4N(O)C(O)R]_2$, where R = $CH{=}CH_2$, $CMe{=}CH_2$, X = 4-Me, 3-Me, 4-Cl, 3-Cl, 4-Br, 3-Br, H, 4-$CH{=}CH_2$, 4-AcO or 4-MeOC(O), show that the hydroxamates are bidentate *via* two oxygen atoms.[444]

$ZnLSO_4$, where L = 1,2,4-triazole-4-aminedipropionamide or 1,3,4-thiadiazole-2-aminedipropionamide, give IR bands showing coordination of NH_2 groups to Zn, with no carbonyl interaction with the metal.[445]

Complexes of Sn(II) and Sn(IV) halides with *p*nitrosodimethylaniline and *p*nitro-sodiethylaniline have IR spectra showing coordination *via*the nitroso group.[446] IR and

Raman spectra of $SnCl_4L_2$, where L = *p*-$HOC_6H_4CONH_2$ or *o*-$BzOC_6H_4CONH_2$, show that L is unidentate *via* the amide oxygen.[447] IR spectra of $R_xSnCl_{4-x}L_y$, where R = Me or Ph, x = 0 - 3, L = R'$C_6H_4NMe_2O$ (R' = H, o-, m- or p-F, Cl,, Br, NO_2, Me, OMe or Ph), y = 1 - 3, show L coordinated *via* the O atom of the *N*-oxide.[448]

The IR spectra of $(Ph_3Pb)_2A$, where A = serinate or tyrosinate, and Me_3Pb(AcSer), where AcSer = acetylserinate, show that all contain unsymmetrical bridging carboxylate groups and coordinated NH_2.[449] IR measurements of Pb(TH-1)$_2$(NO_3)$_2$, where TH-1 = tetrazole-1-acetyl hydrazide, show that (TH-1) ligands coordinate to form 5-membered *N,O* -chelate rings. One nitrate is unidentate, the other being involved in tridentate-bridging coordination.[450]

The IR spectrum of SbF_3L_2, where L = nicotinamide, shows that L is coordinated *via* the heterocyclic nitrogen atom.[451]

8.3. Ligands Containing N and S Donor Atoms:- Schiff base modes in M(O_2)(SNNS), where M = Zr or Th, and M(O)(O_2)(SNNS), where M = Mo, W or U, and SNNS = (64), show the expected shifts on coordination. νO_2 modes were also seen, in the range 800 - 840 cm^{-1}.[452]

The absence of νSH from the IR spectrum of VO(Pen-OCH_3), where Pen-OCH_3 is the anion of the carboxylate ester of penicillamine, shows that N and S atoms are both coordinated to the vanadium.[453]

ML_2, where M = Mn, Fe, Co, Ni, Cu or Zn, HL = *N*-(fluoroanthenyl)- *N'*-(diphenylthiophosphoryl)thiourea, have IR spectra which show that L^- is coordinated *via* two sulphur atoms.[454]

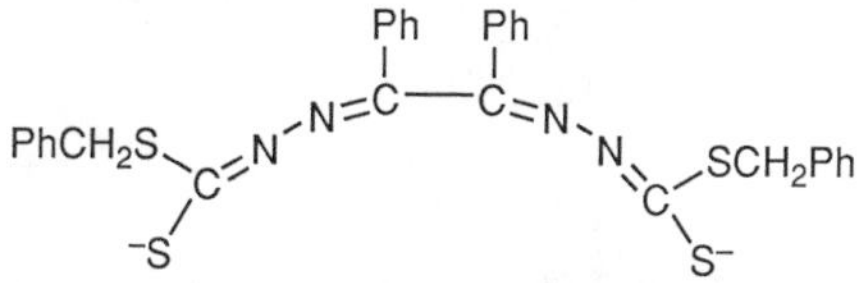

(64)

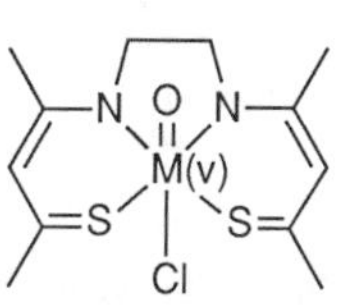

(65)

(65), where M = Tc or Re, have νC=S at slightly lower wavenumbers than for the free ligand.[455] The IR spectra of iron(II) dioximates, Fe(DioxH)$_2L_2$, where L = thioacetamide, thiourea or 2-aminothiazole, DioxH_2 = dimethyl- or diphenylglyoxime, all show that L is unidentate, *via* sulphur.[456] Fe^{II}(DBTS), where DBTS is the Schiff base from the interaction of 2,4-dihydroxybenzaldehyde and thiosemicarbazone, has νCS at 626 cm^{-1}, compared to 813 cm^{-1} in the free ligand, and so the S atom is coordinated to Fe(II).[457]

ML_2X_2, where M = Co, Ni or Cu, X = Cl, Br, NO_3 or ClO_4, L = 3-amino(pyrid-2-yl)thiourea, have IR bands suggesting *N,S*-bidentate coordination of L.[458] Similar conclusions were drawn from the IR spectra of Co(II), Ni(II) and Cu(II) complexes with *N*-benzoyl- *N'*-aryl or -heterocyclic substituted thioureas;[459] and $ML_2.2H_2O$, where M = Co, Ni, Cu, Zn, Cd or UO_2, HL = Schiff bases derived from vanillin or 2-furaldehyde and 4-phenylthiosemicarbazide.[460] Tetradentate, SNNS, coordination of L^{2-} was deduced for

$ML.nH_2O$, where M = Co, Ni or Cu, H_2L = 4,5-dimethyl-1,3-cyclohexanedione bis-S-benzyldithiocarbazoic acid.[461]

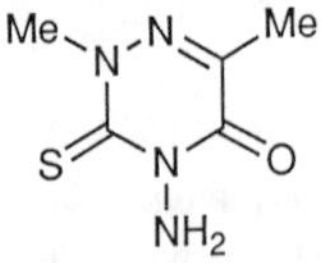

(66)

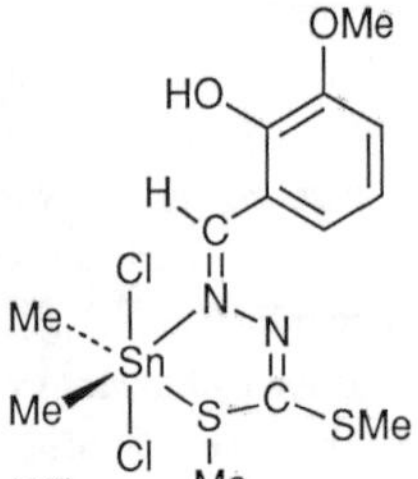

(67)

NiL_2Br_2, where L = (66), exists in two isomeric forms, one with L coordinated *via* NH_2 (νC=S 1380 cm^{-1}) and the other *via* NH_2 and the thione S (νC=S 1335 cm^{-1}).[462] IR evidence was found for uncoordinated carboxylates in $K_2[Ni(SCH_2CH(NH_2)COO)_2]$.[463] $[Ni(BBDH)(OH_2)_2]^{2+}$, where BBDH = 1,6-bis(benzimidazol-2-yl)-2,5-dithiahexane, give IR bands due to ligand modes which suggest N_2S_2 coordination by BBDH.[464]

νCN and νCS modes in $[Cu^{II}(Et_4todit)X_2]_n$, where X = Cl or Br, Et_4todit = 4,5,6,7-tetrathiocino[1,2-*b* :3,4- *b*′]-di-imidazolyl-1,3,8,10-tetraethyldithione, are consistent with *S* thioamide coordination to the metal.[465] The IR spectra of 1:1 dimeric copper(II) complexes with thiosemicarbazones of salicylaldehyde, 5-bromo-salicylaldehyde etc., are all indicative of *N,S* -chelation.[466]

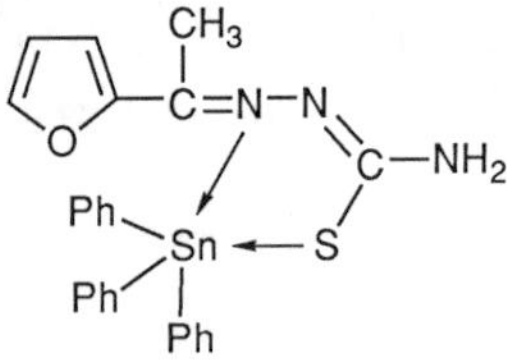

(68)

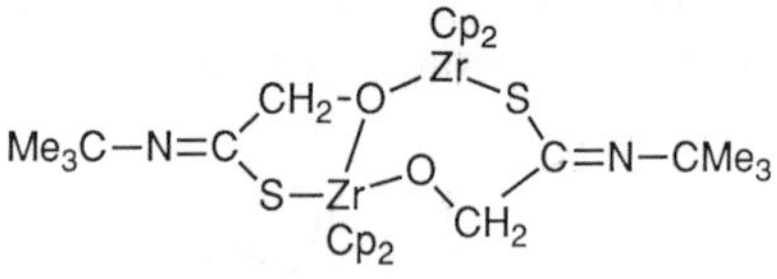

(69)

SnX_4(BT), where X = Cl, Br or I, BT = benzothiazole, have IR spectra showing that the BT is coordinated *via* N and S atoms.[467] Ligand mode assignments were consistent with the geometry (67) for this organotin(IV) Schiff base complex;[468] as they were for (68).[469]

$Bi(dapts)(N_3)$, where H_2dapts = 2,6-diacetylpyridine bis(thiosemicarbazone), show νS=O bands consistent with coordination of both sulphur atoms to the bismuth. The coordinated azide has $\nu_{as}N_3$ at 2035 cm^{-1}.[470]

8.4. Ligands Containing S and O Donor Atoms:- The Raman spectra of $M(SOCl_2)(SO_2)^{n+}(AlCl_4^-)_n$, where M = Li or Ca, were obtained in cells with M anodes and $LiAlCl_4$ as the supporting electrolyte in the presence of $SOCl_2$ and SO_2.[471]

(69) has νC=N at 1592 cm^{-1}.[472] $Mo_2L_5(H_2O)_2$, where H_2L = $HSCH_2COOH$, has an IR spectrum showing the presence of equivalent Mo(V) atoms, with *S,O* -chelation by the L^{2-}.[473]

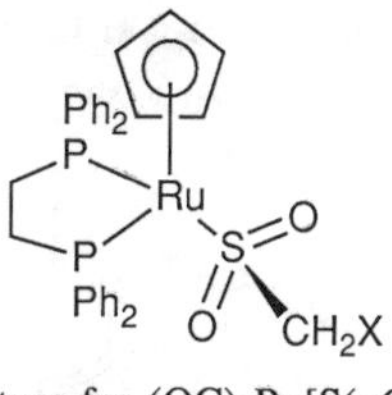

(70)

An *S*-sulphinato structure for $(OC)_5Re[S(=O)_2F]$ was shown by the characteristic νSO_2 bands at 1292 and 1138 cm^{-1}. νSF was at 650 cm^{-1}.[474]

(70), where X = Br, CN or SCN, and related species, have νSO from the η^1- *S*-bonded ligand near 1160 and 1030 cm^{-1}.[475] $[Ru(Me_2SO)_6]^{3+}$ gives νSO bands showing *O*- and *S* bonded Me_2SO for ClO_4^- and BPh_4^- salts.[476] *Cis*-$RuCl_2(DMSO)_2$ has νSO bands at 1109, 1122 cm^{-1} (*S*-bonded) and 933 cm^{-1} (*O*-bonded). Analogous features were seen in the bromo-analogue. However, complexes with TMSO (tetramethylene sulphoxide) have νSO near 1100 cm^{-1} only, i.e. no *O*-bonded form is present.[477] [*Trans*-$Ru(DMSO)_2Cl_4]^-$ also contains only *S*-bonded DMSO ligands, although $RuCl_3(DMSO)_3$ has two *S*- and one *O*-bonded DMSO ligands.[478] *Cis*-$RuCl_2$(1,5-DTCO-O)$_2$, where 1,5-DTCO-O = 1,5-dithiacyclooctane-1-oxide, has νSO of the sulphoxide at 1072 cm^{-1}, i.e. *S*-coordination.[479]

The presence of *S*-bonded thiosulphate in $(^nBu_4N)_2[OsO_2(S_2O_3)_2]$ was shown by Raman bands at 1011 and 1223 cm^{-1} (ν_s, ν_{as} SO_3 respectively).[480] $(\mu\text{-}H)_2Os_3(CO)_9(\mu,\eta^2\text{-}O_2CR)(\eta^1\text{-}O_3SCF_3)$, where R = H, CH_3 or CF_3, have νSO bands from the η_1-O_3SCF_3 near 1340, 1205 and 990 cm^{-1}.[481]

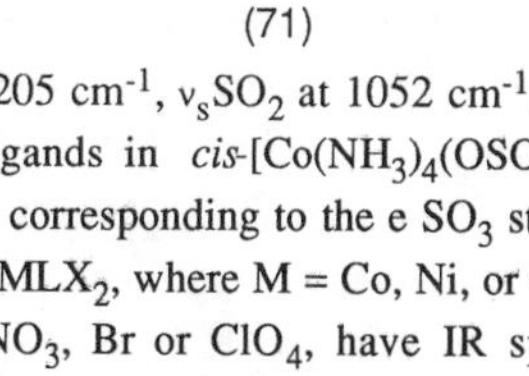

(71)

(71) has $\nu_{as}SO_2$ at 1205 cm^{-1}, ν_sSO_2 at 1052 cm^{-1}, from the μ-SO_2 bridge unit.[482] The coordinated triflate ligands in *cis*-$[Co(NH_3)_4(OSO_2CF_3)_2]^+$ and *mer*-$Co(NH_3)_3$-$(OSO_2CF_3)_3$ have IR bands corresponding to the e SO_3 stretch of the free ligand at 1400, 1332 cm^{-1} respectively.[483] MLX_2, where M = Co, Ni, or Cu, L = N-benzoyl-N'-mecaptothiocarbamide, X = Cl, NO_3, Br or ClO_4, have IR spectra showing that L is *O,S*-bidentate.[484] The IR spectrum of $[RhL(NH_3)_3]NO_3$, where H_2L = thiodiacetic acid, shows that L^{2-} is tridentate *via* S + 2O.[485]

The IR spectrum of $Ni[P_2N_3SO_2(NH_2)(OH)_3]_2.8H_2O$ suggests coordination *via* oxygen atoms of the bidentate SO_2 groups.[486] IR data for MX_2(DTPNO), where M = Ni, Cu, Zn, Cd or Hg, X = Cl, Br or I, DTPNO = 2,2'-dithiobis(pyridine-*N*-oxide), show coordination by oxygen atoms of DTPNO. The pyridine groups adopt a *cis*-configuration about the S-S bridge.[487] IR and Raman spectra of $Pt(H_2L)(HL)Cl$, where $H_2L = S(CH_2$-$CO_2H)_2$, show that H_2L is unidentate, *S*-bonded, while HL- is bidentate, i.e. *O,S*-bonded.[488]

The IR spectrum of $Cu[O_2CCH_2SC(S)NEt_2]_2.H_2O$ shows coordination *via* O and S atoms.[489] Ligand mode assignments for $Au(SO_3F)_2$ show the presence of unidentate and bidentate-bridging SO_3F groups.[490]

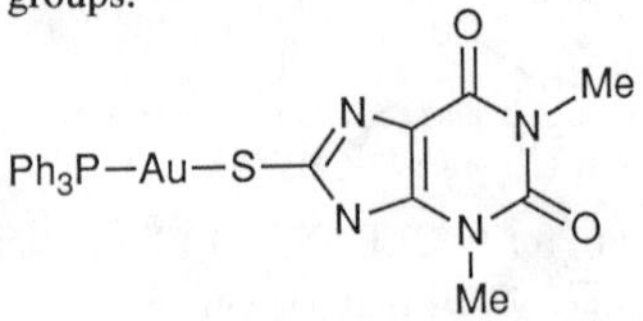

(72)

Au-S coordination in (72) was shown by the shifts in the thioamide-I and -II bands on coordination.[491] $[AuCl_2L]Cl$, where L = 4,7,13,16-tetra-oxa-1,10-dithiacyclo-octadecane and its α- and β-1,10-dioxo derivatives. The IR spectra of the last two show that they are coordinated *via* S=O oxygen atoms, the first one *via* S atoms.[492]

$[MeHg(DMSO)]^+BF_4^-$ has an IR spectrum showing solely *O*-bonded DMSO.[493] $SnCl_4(DMSO)_2$ and $[SnCl(DMSO)_3]^{3+}$ both have νSO of DMSO at 990 - 1000 cm^{-1}, and νCS 725 - 730 cm^{-1}, both indicative of Sn-O coordination.[494]

REFERENCES

1. H.Matsuzawa, H.Yamashita, M.Ito and S.Iwata, *Chem. Phys.*, 1990, **147**, 77.
2. D.C.McKean, G.P.Mcquillan, I.Torto, N.C.Bednell, A.J.Downs and J.M.Dickinson, *J. Mol. Struct.*, 1991, **247**, 73.
3. C.Jegat, M.Fouassier, M.Tranquille and J.Mascetti, *Inorg. Chem.*, 1991, **30**, 1529.
4. C.Jegat and J.Mascetti, *New J. Chem.*, 1991, **15**, 17.
5. C.Nie, C.Mao, C.Zhang, Y.Qian and S.Chen, *Guangpuxue Yu Guangpu Fenxi,* 1990, **10**, 5 (*Chem. Abs.*, 1990, **113**, 212207).
6. V.Sanchez-Exscribano, G.Buaca and V.Lorenzelli, *J. Phys. Chem.*, 1990, **94**, 8939.
7. A.McCamley and R.N.Perutz, *J. Phys. Chem.*, 1991, **95**, 2738.
8. D.J.Stufkens, T.L.Snoeck, W.Kaim, T.Roth and B.Olbrich-Deussner, *J. Organometal. Chem.*, 1991, **409**, 189.
9. R.S.Armstrong, M.J.Aroney, C.M.Barnes and K.W.Nugent, *Appl. Organometal. Chem.*, 1990, **4**, 569.
10. K.A.Vikulov, B.N.Shelimov and V.B.Kazanskii, *J. Mol. Catal.*, 1991, **65**, 393
11.. C.Jegat, M.Fouassier and J.Mascetti, *Inorg. Chem.*, 1991, **30**, 1521.
12. H.Adams, N.A.Bailey, J.T.Gauntlett, I.M.Harkin, M.J.Winter and S.Woodward, *J. Chem. Soc., Dalton Trans.*, 1991, 1117.
13. A.Werth, K.Dehnicke, D.Fenske and G.Baum, *Z. anorg. allg. Chem.*, 1990, **591**, 125.
14. P.K.Baker and K.R.Flower, *J. Organometal. Chem.*, 1991, **405**, 299.
15. A.C.Filippou, W.Grünleitner and P.Kiprof, *J. Organometal. Chem.*, 1991, **410**, 175.
16. T.L.Bent and J.D.Cotton, *Organometallics*, 1991, **10**, 3156.
17. S.M.Barnett, F.Dicaire and A.A.Ismail, *Can. J. Chem.*, 1990, **68**, 1196.
18. W.A.Herrmann, P.Kiprof, K.Rypdal, J.Tremmel, R.Blom, R.Alberto, J.Behm, R.W.Albach, M.Bock, B.Solonki, J.Mink, D.Lichtenberger and N.E.Grutin, *J. Am. Chem. Soc.*, 1991, **113**, 6527.
19. D.Xu, H.D.Kaesz and S.I.Khan, *Inorg. Chem.*, 1991, **30**, 1341.
20. D.C.Reber and J.I.Zink, *Inorg. Chem.*, 1991, **30**, 2994.
21. W.A.Herrmann, M.Taillefer, C.de M.de Bellefon and J.Behm, *Inorg. Chem.*, 1991, **30**, 3247.
22. C.Apostolidis, B.Kanellakopulos, R.Maier, J.Rebizant and M.L.Ziegler, *J. Organometal. Chem.*, 1991, **409**, 243.
23. J.C.Vites, B.D.Steffey, M.E.Guiseppetti-Dery and A.R.Cutler, *Organometallics*, 1991, **10**, 2827.
24. S.C.Mackie, Y.S.Park, H.F.Shurvell and M.C.Baird, *Organometallics*, 1991, **10**, 2993.
25. I.S.Butler, H.Li and J.P.Gao, *Appl. Spectrosc.*, 1991, **45**, 223.
26. M.P.Gamasa, J.Gimeno, E.Lastra, M.Lanfranchi and A.Tiripicchio, *J. Organometal. Chem.*, 1991, **405**, 333.
27. F.R.Blackburn, D.L.Snavely and L.Oref, *Chem. Phys. Lett.*, 1991, **178**, 538.
28. A.G.Nagy, P.Sohár and J.Márton, *J. Organometal. Chem.*, 1991, **410**, 357.
29. B.D.Steffey, J.C.Vites and A.R.Cutler, *Organometallics*, 1991, **10**, 3432.
30. H.Loumrhari, L.Matas, J.Ros, M.R.Torres and A.Perales, *J. Organometal. Chem.*, 1991, **403**, 373.
31. H.Loumrhari, J.Ros, M.R.Torres, A.Santos and A.M.Echavarren, *J. Organometal. Chem.*, 1991, **411**, 255.
32. B.Giese, M.Zeelander, M.Neuburger and F.Trach, *J. Organometal. Chem.*, 1991, **412**, 415.
33. H.Jobic, C.C.Santini and C.Coulombeau, *Inorg. Chem.*, 1991, **30**, 3088.
34. I.A.Khoruzhaya, A.A.Shvets, G.I.Kokorev, V.A.Kogan, G.P.Safaryan and S.I.Adamova, *Russ. J. Inorg. Chem.*, 1990, **35**, 1767.

35. K.Asakura, K.Itamura-Bando, Y.Iwasawa, H.Arakawa and K.Isobe, *J. Am. Chem. Soc.*, 1990, **112**, 9096.
36. F.Ragaini, F.Porta, A.Fumagalli and F.Demartin, *Organometallics*, 1991, **10**, 3785.
37. S.B.Mohsin, M.Trenary and H.J.Robota, *J. Phys. Chem.*, 1991, **95**, 6657.
38. R.L.Beddoes, C.Bitcon and M.W.Whitely, *J. Organometal. Chem.*, 1991, **402**, 85.
39. D,Schneider and H.Werner, *Angew. Chem., Int. Ed. Engl.*, 1991, **30**, 700.
40. C.Houtman and M.A.Barteau, *J. Phys. Chem.*, 1991, **95**, 3755.
41. D.Carmona, F.J.Lahoz, L.A.Oro, J.Reyes and M.Pilar, *J. Chem. Soc., Dalton Trans.*, 1990, 3551.
42. T.W.Bell, D.M.Haddleton, A.McCamley, M.G.Partridge, R.N.Perutz and H.Willner, *J. Am. Chem. Soc.*, 1991, **112**, 9212.
43. V.García and M.A.Garralda, *Inorg. Chim. Acta*, 1991, **180**, 177.
44. W.Schröder, W.Bonrath and K.R.Pörschke, *J. Organometal. Chem.*, 1991, **408**, C25.
45. I.A.Garbuzova, O.G.Garkusha, B.V.Lokshin, A.R.Kudinov and M.I.Rybinskaya, *J. Organometal. Chem.*, 1991, **408**, 247.
46. K.Onitsuka, H.Ogawa, T.Joh, S.Takahashi, Y.Yamamoto and H.Yamazaki, *J. Chem. Soc., Dalton Trans.*, 1991, 1531.
47. P.J.M.Ssebuwufu, *Inorg. Chim. Acta*, 1990, **173**, 139.
48. C. de la Cruz and N.Sheppard, *J. Catal.*, 1991, **127**, 45.
49. M.A.Chesters, C. de la Cruz, P.Gardner, E.M.McCash, J.D.Prentice and N.Sheppard, *J. Electron Spectrosc. Relat. Phenom.*, 1990, **54-5**, 739.
50. H.Alper, Y.Huang, D.Dell'Amico, F.Calderazzo, N.Pasqualetto and C.A.Veracini, *Organometallics*, 1991, **10**, 1665.
51. P.Espinet, J.Forniés, F.Martínez, M.Sotes, E.Lalinde, M.T.Moreno, A.Ruiz and A.J.Welch, *J. Organometal. Chem.*, 1991, **403**, 253.
52. M.Munakata, S.Kitagawa, H.Shimono and H.Masuda, *Inorg. Chem.*, 1991, **30**, 2610.
53. H.Masuda, M.Munakata and S.Kisumu, *J. Organometal. Chem.*, 1990, **391**, 131.
54. R.Brings, I.Mrozek and A.Otto, *J. Raman Spectrosc.*, 1991, **22**, 119.
55. A.Zinn, K.Dehnicke, D.Fenske and G.Baum, *Z. anorg. allg. Chem.*, 1991, **596**, 47.
56. W.Hu, M.Chen and P.Zhou, *Organometallics*, 1991, **10**, 98.
57. L.Ahmed and J.A.Morrison, *J. Am. Chem. Soc.*, 1990, **112**, 7411.
58. R.Minkwitz and V.Gerhard, *Z. Naturforsch., B*, 1991, **46b**, 561.
59. F.Corazza, C.Floriani, A.Chiesi-Villa and C.Guastini, *Inorg. Chem.*, 1991, **30**, 145.
60. C.J.Dain, A.J.Downs, M.J.Goode, D.G.Evans, K.T.Nicholls, D.W.H.Rankin and H.E.Robertson, *J. Chem. Soc., Dalton Trans.*, 1991, 967.
61. A.G.Császár, L.Hedberg, K.Hedberg, R.C.Burns, A.T.Wen and M.J.McGlinchey, *Inorg. Chem.*, 1991, **30**, 1371.
62. V.D.Makaev, A.P.Borisov, G.N.Boiko and B.P.Tarasov, *Izv. Akad. Nauk SSSR, Ser. Khim.*, 1990, 1207.
63. R.J.Simonson and M.Trenary, *J. Electron Spectrosc. Relat. Phenom.*, 1990, **54-5**, 717.
64. C. de la Cruz and N.Sheppard, *J. Mol. Struct.*, 1990, **224**, 141.
65. K.Hadjiivanov, D.Klissurski, M.Kantcheva and A.Davydov, *J. Chem. Soc., Farad. Trans.*, 1991, **87**, 907.
66. K.Kwan, *Bull. Korean Chem. Soc.*, 1990, **11**, 396.
67. P.Courtot, R.Pichon, J.Y.Salaun and L.Toupet, *Can. J. Chem.*, 1991, **69**, 661.
68. J.E.Ellis, S.R.Frerichs and K.M.Chi, *Organometallics*, 1990, **9**, 2858.
69. T.Lindblad and B.Rebenstorf, *J. Chem. Soc., Farad. Trans.*, 1991, **87**, 2473.
70. B.Rebenstorf and T.Lindblad, *J. Catal.*, 1991, **128**, 303.
71. D.Lentz and R.Marschall, *Z. anorg. allg. Chem.*, 1991, **593**, 181.
72. S.J.Gravelle and E.Weitz, *J. Am. Chem. Soc.*, 1990, **112**, 7839.

73. A.G.Joly and K.A.Nelson, *Chem. Phys.*, 1991, **152**, 69.
74. S.Özkar, G.A.Ozin, K.Moller and T.Bein, *J. Am. Chem. Soc.*, 1990, **112**, 9575.
75. K.Kim and S.B.Lee, *Bull. Korean Chem. Soc.*, 1991, **12**, 17.
76. J.L.Davidson and F.Sence, *J. Organometal. Chem.*, 1991, **409**, 219.
77. F.P.A.Johnson, C.M.Gordon, P.M.Hodges, M.Poliakoff and J.J.Turner, *J. Chem. Soc., Dalton Trans.*, 1991, 833.
78. L.M.Clarkson, W.Clegg, D.C.R.Hockless, N.C.Norman and T.B.Marder, *J. Chem. Soc., Dalton Trans.*, 1991, 2229.
79. R.S.Herrick, M.S.George, D.R.Duff, F.H.D'Aulnois, R.M.Jarret and J.L.Hubbard, *Inorg. Chem.*, 1991, **30**, 3711.
80. Y.Yan, Q.Xin, S.Jiang and X.Guo, *J. Catal.*, 1991, **131**, 234.
81. S.N.Misra and S.B.Mehta, *Ind. J. Chem., A*, 1991, **30A**, 731.
82. P.L.Bogdan, J.R.Wells and E.Weitz, *J. Am. Chem. Soc.*, 1991, **113**, 1294.
83. T.Lindblad and B.Rebenstorf, *Acta Chem. Scand.*, 1991, **45**, 342.
84. Y.Huang, I.S.Butler, D.F.R.Gilson and D.Lafleur, *Inorg. Chem.*, 1991, **30**, 117.
85. T. van der Graaf, D.J.Stufkens, A.Oskam and K.Goubitz, *Inorg. Chem.*, 1991, **30**, 599.
86. D.M.Adams, P.D.Hatton and A.C.Shaw, *J. Phys., Condens. Matter*, 1991, **3**, 6145.
87. T. van der Graaf, D.J.Stufkens, J.Vichova and A.M.Vlček, *J. Organometal. Chem.*, 1991, **401**, 305.
88. Y.Huang, I.S.Butler and D.F.R.Gilson, *Inorg. Chem.*, 1991, **30**, 1098.
89. L.A.Worl, R.Duesing, P.Chen, L.Della Gana and T.J.Meyer, *J. Chem. Soc., Dalton Trans.*, 1991, 849.
90. Y.Huang, I.S.Butler and D.F.R.Gilson, *Spectrochim. Acta, A*, 1991, **47A**, 909.
91. D.M.Adams, P.W.Ruff and D.R.Russell, *J. Chem. Soc., Farad. Trans.*, 1991, **87**, 1831.
92. K.D.Park, K.Guo, F.Adebodun, M.L.Chiu, S.G.Sligar and E.Oldfield, *Biochem.*, 1991, **30**, 2333.
93. T.S.Kurtikyan, G.G.Martirosyan, A.V.Gasparian, M.E.Akopyan and G.A.Zhamkochyan, *Zh. Prikl. Spektrosk.*, 1990, **53**, 421.
94. L.M.Proniewicz, T.Kuroi and K.Nakamoto, *J. Mol. Struct.*, 1990, **238**, 1.
95. S.T.Astley, M.P.V.Churton, R.B.Hitam and A.J.Rest, *J. Chem. Soc., Dalton Trans.*, 1990, 3243.
96. P.Braunstein, J.L.Richert and Y.Dusausoy, *J. Chem. Soc., Dalton Trans.*, 1990, 3801.
97. Z.Nagy-Magos, L.Markó, A.Szakács-Schmidt, G.Gervasio, E.Belluso and S.F.A.Kettle, *Bull. Soc. Chim. Belg.*, 1991, **100**, 445.
98. W.P.Fehlhammer, A.Schröder, F.Schoder, J.Fuchs, A.Vökl, B.Boyadjiev and S.Schrölkamp, *J. Organometal. Chem.*, 1991, **411**, 405.
99. P.A.Anfinrud, C.H.Han, T.Lian and R.M.Hochstrasser, *J. Phys. Chem.*, 1991, **95**, 574.
100. J.P.Bullock, M.C.Palazatto and K.R.Mann, *Inorg. Chem.*, 1991, **30**, 1284.
101. R.H.Hooker and A.J.Rest, *Appl. Organometal. Chem.*, 1990, **4**, 141.
102. G.Sundarajan, *Ind. J. Chem., Sect. A*, 1990, **29A**, 1060.
103. M.J.Don and M.G.Richmond, *Inorg. Chem.*, 1991, **30**, 1703.
104. R.Liu, B.Tesche and H.Knözinger, *J. Catal.*, 1991, **129**, 402.
105. D.Schanke, G.R.Frederiksen, E.A.Blekkan and A.Holmen, *Catal. Today*, 1991, **9**, 69.
106. T.M.Buslaeva, I.V.Malyunov, N.M.Sinitsyn, M.E.Poloznikova and A.S.Solomonova, *Russ. J. Inorg. Chem.*, 1991, **36**, 538.
107. K.M.Kadish, P.Tagliatesta, Y.Hu, Y.J.Deng, X.H.Mu and L.Y.Bao, *Inorg. Chem.*, 1991, **30**, 3737.
108. G.Jia and D.W.Meek, *Inorg. Chim. Acta*, 1990, **178**, 195.
109. A.Gutierrez-Alonso and L.Ballester-Reventos, *Polyhedron*, 1991, **10**, 1019.
110. S.P.Best, R.J.H.Clark, A.J.Deeming, R.C.S.McQueen, N.I.Powell, C.Acuna, A.J.Arce and

Y. de Sanctis, *J. Chem. Soc., Dalton Trans.*, 1991, 1111.
111. W.Moran, G.Johnson and S.Collins, *J. Mol. Struct.*, 1990, **222**, 235.
112. A.J.Amoroso, L.H.Gade, B.F.G.Johnson, J.Lewis and P.R.Raithby, *Angew. Chem., Int. Ed. Engl.*, 1991, **30**, 107.
113. Y.Hu, B.C.Han, L.Y.Bao, X.H.Mu and K.M.Kadish, *Inorg. Chem.*, 1991, **30**, 2444.
114. O.B.Ishchenko, N.V.Alekseeva, K.P.Zhdanova and F.K.Shmidt, *Zh. Fiz. Khim.*, 1991, **65**, 1200.
115. S.S.Kristiánsdóttir, J.R.Norton, A.Moroz, R.L.Sweaney and S.L.Whittenburg, *Organometallics*, 1991, **10**, 2357.
116. D.J.Darensbourg, C.-S.Chao, C.Bischoff and J.H.Reibenspies, *Inorg. Chem.*, 1990, **29**, 4637.
117. S.G.Anema, S.K.Lee, K.M.Mackay, B.K.Nicholson and M.Service, *J. Chem. Soc., Dalton Trans.*, 1991, 1201.
118. E.Diana, G.Gervasio, R.Rosetti, F.Valdermarin, G.Bor and P.L.Stanghellini, *Inorg. Chem.*, 1991, **30**, 294.
119. L.W.H.Leung, J.W.He and D.W.Goodman, *J. Chem. Phys.*, 1990, **93**, 8328.
120. A.I.Rubalo, V.P.Selina, T.G.Cherkasova and Yu.S.Varshavskii, *Koord. Khim.*, 1991, **17**, 530.
121. D.T.Brown, T.Eguchi, B.T.Heaton, J.A.Iggo and R.Whyman, *J. Chem. Soc., Dalton Trans.*, 1991, 677.
122. P.S.Bagus and G.Pacchioni, *Surf. Sci.*, 1990, **236**, 233.
123. K.I.Choi and M.A.Vannice, *J. Catal.*, 1991, **131**, 36.
124. E.Zippel, M.W.Breiter and R.Kellner, *J. Chem. Soc., Farad. Trans.*,1991, **87**, 637.
125. M.R.Mucalo and R.P.Cooney, *J. Chem. Soc., Farad. Trans.*, 1991, **87**, 1221.
126. S.Watanabe, Y.Kinomoto, F.Kitamura, M.Takahashi and M.Ito, *J. Electron Spectrosc. Relat. Phenom.*, 1990, **54-5**, 1205.
127. P.Gardner, R.Martin, M.Tueshaus and A.M.Bradshaw, *J. Electron Spectrosc. Relat. Phenom.*, 1990, **54-5**, 619.
128. P.A.Sermon, V.A.Self and E.P.S.Barrett, *J. Chem. Soc., Chem. Comm.*, 1990, 1572.
129. H.Huang, C.Fierro, D.Scherson and E.B.Yeager, *Langmuir*, 1991, **7**, 1154.
130. S.G.Bott, A.D.Burrows, O.J.Ezomo, M.F.Hallam, J.G.Jeffrey and D.M.P.Mingos, *J. Chem., Soc., Dalton Trans.*, 1990, 3335.
131. Y.Fu, Y.Tian and P.Lin, *J. Catal.*, 1991, **132**, 85.
132, J.W.He, W.K.Kuhn, L.W.H.Leung and D.W.Goodman, *J. Chem. Phys.*, 1990, **93**, 7463.
133. J.W.He, W.K.Kuhn and D.W.Goodman, *Mater. Res. Soc., Symp. Proc.*, 1991, **202**, 325.
134. M.Håkansson and S.Jagner, *Inorg. Chem.*, 1990, **29**, 6241.
135. P.K.Hurlbut, O.P.Anderson and S.H.Strauss, *J. Am. Chem. Soc.*, 1991, **113**, 8297.
136. S.C.Chang, A.Hamelin and M.J.Weaver, *J. Phys. Chem.*, 1991, **95**, 5560.
137. S.C.Chang, A.Hamelin and M.J.Weaver, *Surf. Sci.*, 1990, **239**, L543.
138. M.Adelhelm, W.Bacher, E.G.Höhn and E.Jacob, *Chem. Ber.*, 1991, **124**, 1559.
139. W.Massa, S.Wocaldo, S.Lotz and K.Dehnicke, *Z. anorg. allg. Chem.*, 1990, **589**, 79.
140. M.J.Abrams, S.K.Larsen and J.Zubieta, *Inorg. Chem.*, 1991, **30**, 2031.
141. M.J.Hampden-Smith, D.Lei and E.N.Duesler, *J. Chem. Soc., Dalton Trans.*, 1990, 2953.
142. M.F.Klein, H.Beck, B.Hammerschmitt, U.Koch, S.Koppert, G.Cordier and H.Paulus, *Z. Naturforsch., B*, 1991, **46b**, 147.
143. H.Knoll, R.Stich, H.Hennig and D.J.Stufkens, *Inorg. Chim. Acta*, 1990, **178**, 71.
144. O.N.Adrianova, T.N.Fedotova, I.F.Golovaneva, E.V.Kovaleva, A.V.Chuvaev, A.Ya.Mikhailova and I.B.Baranovskii, *Russ. J. Inorg. Chem.*, 1990, **35**, 1131.
145. M.A.S.Goher, R.Wang and T.C.W.Mak, *J. Mol. Struct.*, 1991, **243**, 179.
146. A.K.Deb, S.Choudhury and S.Goswami, *Polyhedron*, 1990, **9**, 2251.

147. A.Loutellier, L.Manceron and J.P.Perchard, *Chem. Phys.*, 1990, **146**, 179.
148. T.N.Day, P.J.Hendra, A.J.Rest and A.J.Rowlands, *Spectrochim. Acta, A*, 1991, **47A**, 1251.
149. J.Jubb, L.F.Larkworthy, L.F.Oliver, D.C.Povey and G.W.Smith, *J. Chem. Soc., Dalton Trans.*, 1991, 2045.
150. R.W.Hay and M.T.H.Tarafder, *J. Chem. Soc., Dalton Trans.*, 1991, 823.
151. M.S.Ram, L.M.Jones, H.J.Ward, Y.H.Wong, C.S.Johnson, P.Subramanian and J.T.Hupp, *Inorg. Chem.*, 1991, **30**, 2928.
152. A.Anillo, R.Obesco-Rosete, M.A.Pellinghelli and A.Tiripicchio, *J. Chem. Soc., Dalton Trans.*, 1991, 2019.
153. J.A.Cabeza, J.M.Fernández-Colinas, V.Riera, M.A.Pellinghelli and A.Tiripicchio, *J. Chem. Soc., Dalton Trans.*, 1991, 371.
154. Y.Mitsutsuka, T.Aoyama, S.Ohba, Y.Saito, F.kaneuchi and M.Tsuboi, *Bull. Chem. Soc. Japan*, 1991, **64**, 1743.
155. V.P.Sinditskii, M.P.Dutov, V.I.Sokol, A.E.Fogel'zang and V.V.Serushkin, *Russ. J. Inorg. Chem.*, 1990, **35**, 1774.
156. W.Erley, *Surf. Sci.*, 1990, **240**, 1.
157. R.Acevedo, G.Diáz, M.M.Campos-Vallette and B.Weiss, *Spectrochim. Acta, A*, 1991, **47A**, 355.
158. A.Tenten and H.Jacobs, *J. Less-Common Met.*, 1991, **170**, 145.
159. A.M.Heyns and M.W.Venter, *J. Chem. Phys.*, 1990, **93**, 7581.
160. V.P.Sinditskii, M.D.Dutov, A.E.Fogel'zang and V.V.Kuznetsov, *Russ. J. Inorg. Chem.*, 1991, **36**, 535.
161. V.N.Kokozei, Yu.A.Simonov, S.R.Petrushenko and A.A.Dvorkin, *Russ. J. Inorg. Chem.*, 1990, **35**, 1612.
162. M.Sakai, M.Hayakawa, N.Kuroda, Y.Nishina and M.Yamashima, *J. Phys. Soc. Japan*, 1991, **60**, 1619.
163. P.Castan, F.Dahan, S.Wimmer and F.L.Wimmeer, *J. Chem. Soc., Dalton Trans.*, 1990, 2971.
164. V.F.Shul'gin, O.V.Konnik and V.V.Efimova, *Izv. Vyssh. Uchebn. Zaved. Khim. Khim Tekhnol.*, 1990, **33**, 26 (*Chem. Abs.*, 1991, **114**, 54732).
165. W.Brandt, J.Wirbser, A.K.Powell and H.Vahrenkamp, *Z. Naturforsch., B*, 1991, **46b**, 440.
166. A.I.Popov, M.D.Val'kovskii, N.A.Chumaevskii, A.V.Sakharov, V.O.Gel'mbol'dt and A.A.Ennan, *Russ. J. Inorg. Chem.*, 1991, **36**, 208.
167. T.Nozawa, T.Noguchi and M.Tasumi, *J. Biochem. (Tokyo)*, 1990, **108**, 737.
168. E.Nishizawa and Y.Koyama, *Chem. Phys. Lett.*, 1991, **176**, 390.
169. L.L.Thomas, J.H.Kim and T.M.Cotton, *J. Am. Chem. Soc.*, 1990, **112**, 9378.
170. K.Kim, W.S.Lee, H.J.Kim, S.H.Cho, G.S.Girolami, P.A.Gorlin and K.S.Suslick, *Inorg. Chem.*, 1991, **30**, 2652.
171. H.S.Yadav, *Bull. Soc. Chim. France*, 1990, 641.
172. D.Lu, I.R.Paeng and K.Nakamoto, *J. Coord. Chem.*, 1991, **23**, 3.
173. S.E.J.Bell, R.E.Hester, J.N.Hill, D.R.Shawcross and J.R.L.Smith, *J. Chem. Soc., Farad. Trans.*, 1990, **86**, 4017.
174. A.L.Verma and N.K.Choudhury, *J. Raman Spectrosc.*, 1991, **22**, 427.
175. R. van den Burg and M.A.El-Sayed, *Biophys. J.*, 1990, **58**, 931.
176. R.Lingle, X.Xu, H.Zhu, S.C.Yu, J.B.Hopkins and K.D.Straub, *J. Am. Chem. Soc.*, 1991, **113**, 3992.
177. D.Morikis, P.Li, O.Bangcharoenphurpong, J.T.Sage and P.M.Champion, *J. Phys. Chem.*, 1991, **95**, 3391.
178. K.D.Egeberg, B.A.Springer, S.A.Martinis, S.G.Sligar, D.Morikis and P.M.Champion, *Biochem.*, 1990, **29**, 9783.

179. G.Smulevich, A.M.English, A.R.Mantini and M.P.Marzocchi, *Biochem.*, 1991, **30**, 772.
180. K.K.Andersson, G.T.Babcock and A.B.Hooper, *Biochem. Biophys. Res. Commun.*, 1991, **174**, 358.
181. R.W.Larsen, W.Li, R.A.Copeland, S.N.Witt, B.S.Lou, S.I.Chan and M.R.Ondrias, *Biochem.*, 1990, **29**, 10135.
182. Y.P.Myer and A.F.Saturno, *J. Protein Chem.*, 1990, **9**, 379.
183. M.Tsubaki, Y.Ichikawa, Y.Fujimoto, N.T.Yu and H.Hori, *Biochem.*, 1990, **29**, 8805.
184. J.K.Hurst, T.M.Loehr, J.T.Curnutte and H.Rosen, *J. Biol. Chem.*, 1991, **266**, 1627.
185. G.Smulevich, M.A.Miller, J.Kraut and T.G.Spiro, *Biochem.*, 1991, **30**, 9546.
186. P.Hildebrandt, G.Heibel, S.Anemüller and G.Schäfer, *FEBS Lett.*, 1991, **283**, 131.
187. W.H.Woodruff, O.Einarsdottir, R.B.Dyer, K.A.Bagley, G.Palmer, S.J.Atherton, R.A.Goldbeck, T.D.Dawes and D.S.Kliger, *Proc. Natl. Acad. Sci. USA*, 1991, **88**, 2588.
188. B.N.Rospendowski, K.Kelly, C.R.Wolf and W.E.Smith, *J. Am. Chem. Soc.*, 1991, **113**, 1217.
189. P.Hildebrandt, *J. Mol. Struct.*, 1991, **242**, 379.
190. S.Umapathy, A.M.Cartner, A.W.Parker and R.E.Hester, *J. Phys. Chem.*, 1990, **94**, 8880.
191. R.Hage, J.G.Haasnoot, J.Reedijk, R.Wang and J.G.Vos, *Inorg. Chem.*, 1991, **30**, 3263.
192. H.A.Nieuwenhuis, J.G.Haasnoot, R.Hage, J.Reedijk, T.L.Snoeck, D.J.Stufkens and J.G.Vos, *Inorg. Chem.*, 1991, **30**, 48.
193. S.Trofimenko, F.B.Hulsbergen and J.Reedijk, *Inorg. Chim. Acta*, 1991, **183**, 203.
194. M.Futamata, *Synth. Met.*, 1991, **42**, 2683.
195. O.K.Song, J.S.Ha, M.Yoon and D.Kim, *J. Raman Spectrosc.*, 1990, **21**, 645.
196. J.Castro, J.Romero, J.A.Garcia-Vázquez, M.I.Duran, A.Castiñeiras, A.Sousa and D.E.Fenton, *J. Chem. Soc., Dalton Trans.*, 1990, 3255.
197. K.Sakata and A.Ueno, *Synth. React. Inorg. Met.-Org. Chem.*, 1991, **21**, 729.
198. Y.A.Sarma, *Proc. Ind. Acad. Sci., Chem. Sci.*, 1991, **103**, 571.
199. U.Bobinger, R.Schweitzer-Stenner and W.Dreybrodt, *J. Phys. Chem.*, 1991, **95**, 7625.
200. F.Prendergast and T.G.Spiro, *J. Phys. Chem.*, 1991, **95**, 1555.
201. J.A.Shelnutt, C.J.Medforth, M.D.Berber, K.M.Barkigia and K.M.Smith, *J. Am. Chem. Soc.*, 1991, **113**, 4077.
202. W.A.Oertling, W.Wu, J.J.López-Garriga, Y.Kim and C.K.Chang, *J. Am. Chem. Soc.*, 1991, **113**, 127.
203. H.N.Fonda, W.A.Oertling, A.Salehi, C.K.Chang and G.T.Babcock, *J. Am. Chem. Soc.*, 1990, **112**, 9497.
204. D.Melamed, E.P.Sullivan, K.Prendergast, S.H.Strauss and T.G.Spiro, *Inorg. Chem.*, 1991, **30**, 1308.
205. M.Angaroni, G.A.Ardizzoia, T.Beringhelli, G.LaMonica, D.Gatteschi, N.Masciocchi and M.Moret, *J. Chem. Soc., Dalton Trans.*, 1990, 3305.
206. K.C.Gordon and J.J.McGarvey, *Inorg. Chem.*, 1991, **30**, 2986.
207. K.C.Gordon and J.J.McGarvey, *Chem. Phys. Lett.*, 1990, **173**, 443.
208. M.Mylrajan, L.A.Andersson, T.M.Loehr, W.Wu and C.K.Chang, *J. Am. Chem. Soc.*, 1991, **113**, 5000.
209. L.A.Andersson, T.M.Loehr, W.Wu, C.K.Chang and R.Timkovich, *FEBS Lett.*, 1990, **267**, 285.
210. A.D.Procyk and D.F.Bocian, *J. Am. Chem. Soc.*, 1991, **113**, 3765.
211. L.Chinsky, P.Y.Turpin, A.H.R.Al-Obaidi, S.E.J.Bell and R.E.Hester, *J. Phys. Chem.*, 1991, **95**, 5754.
212. B.Vlčková, P.Matějka, P.Panošića, V.Baumruk and K.Král, *Inorg. Chem.*, 1991, **30**, 4103.
213. N.Hadjiladis, M.Louloudi and I.S.Butler, *Spectrochim. Acta, A*, 1991, **47A**, 445.
214. T.Akyüz, S.Akyüz and J.E.D.Davies, *J. Inclusion Phenom.. Mol. Recognit. Chem.*, 1990, **9**,

349.
215. J.K.Duchowski and D.F.Bocian, *J. Am. Chem. Soc.*, 1990, **112**, 8807.
216. G.R.Willey, H.Collins and M.G.B.Drew, *J. Chem. Soc., Dalton Trans.*, 1991, 961.
217. F.Rehbaum, K.-H.Thiele and S.I.Troianov, *J. Organometal. Chem.*, 1991, **410**, 327.
218. M.Scheer, T.T.Nam, K.Schenzel, E.Herrmann, P.G.Jones, V.P.Fedin, V.N.Igorski and V.E.Fedorov, *Z. anorg. allg. Chem.*, 1990, **591**, 221.
219. N.Bao and J.M.Winfield, *J. Fluorine Chem.*, 1990, **50**, 339.
220. M.Scheer, T.T.Nam, E.Herrmann, V.P.Fedin, V.N.Ikorskii and V.E.Fedorov, *Z. anorg. allg. Chem.*, 1990, **589**, 214.
221. P.Glyn, F.P.A.Johnson, M.W.George, A.J.Lees and J.J.Turner, *Inorg. Chem.*, 1991, **30**, 3543.
222. A.Christofides, N.G.Connelly, H.J.Lawson, A.C.Loyns, A.G.Orpen, M.O.Simmonds and G.H.Worth, *J. Chem. Soc., Dalton Trans.*, 1991, 1595.
223. K.-Y.Shih, P.E.Fanwick and R.A.Walton, *Inorg. Chem.*, 1991, **30**, 3971.
224. M.F.C.G.Silva, A.J.L.Pombeiro, A.Hills, D.L.Hughes and R.L.Richards, *J. Organometal. Chem.*, 1991, **403**, C1.
225. S.L.Bartley and K.R.Dunbar, *Angew. Chem., Int. Ed. Engl.*, 1991, **30**, 448.
226. M.L.A.Temperini, J.C.Rubim, O.Sala, A.H.Jubert, M.E.Chacon-Villalba and P.J.Aymonino, *J. Raman Spectrosc.*, 1991, **22**, 301.
227. J.Fernández-Bertrán, E.Reguera-Ruíz and J.Blanco-Pascual, *Spectrochim. Acta, A*, 1990, **46A**, 1679.
228. Y.Lei, T.Buranda and J.F.Endicott, *J. Am. Chem. Soc.*, 1990, **112**, 8820.
229. J.B.Cooper, T.M.Vess, W.A.Kalsbeck and D.W.Wertz, *Inorg. Chem.*, 1991, **30**, 2286.
230. D.Fenske, G.Baum, H.W.Swidersky and K.Dehnicke, *Z. Naturforsch., B*, 1990, **45b**, 1210.
231. J.Martinez, H.Adams, N.A.Bailey and P.M.Maitlis, *J. Organometal. Chem.*, 1991, **405**, 393.
232. K.Yoshida, T.Kitada, N.Odano, M.Kashiwagi, K.Harada, A.Urushiyama and M.Nakahara, *Bull. Chem. Soc. Japan*, 1991, **64**, 895.
233. G.López, G.Sánchez, G.Garcia, J.Ruiz, J.Garcia, M.Martinez-Ripoll, A.Vegas and J.A.Hermoso, *Angew. Chem., Int. Ed. Engl.*, 1991, **30**, 716.
234. E.Bessler, B.M.Barbosa, W.Hiller and J.Weidlein, *Z. Naturforsch., B*, 1991, **46b**, 490.
235. K.Ashley, F.Weinert, M.G.Samant, H.Seki and M.R.Philpott, *J. Phys. Chem.*, 1991, **95**, 7409.
236. A.M.Bradford, N.C.Payne, R.J.Puddephatt, D.S.Yang and T.B.Marder, *J. Chem. Soc., Chem. Comm.*, 1990, 1462.
237. E.W.Ainscough, E.N.Baker, M.L.Brader, A.M.Brodie, S.L.Ingham, J.M.Waters, J.V.Hanna and P.C.Healy, *J. Chem. Soc., Dalton Trans.*, 1991, 1243.
238. W.Hiller, A.Zinn and K.Dehnicke, *Z. Naturforsch., B*, 1990, **45b**, 1593.
239. D.D.K.Chingakule, P.Gans and J.B.Gill, *J. Chem. Soc., Dalton Trans.*, 1991, 1329.
240. G.Ramis, G.Busco, V.Lorenzelli and P.Forzetti, *Appl. Catal.*, 1990, **64**, 243.
241. R.Srivastava and S.Sarkar, *Inorg. Chim. Acta*, 1990, **176**, 27.
242. S.L.Fu and J.H.Lunsford, *Langmuir*, 1991, **7**, 1172.
243. N.J.Al-Obaidi, C.J.Jones and J.A.McCleverty, *J. Chem. Soc., Dalton Trans.*, 1990, 3329.
244. J.O.Dziegielewski, K.Filipek and B.Jezowska-Trzebiatowska, *Inorg. Chim. Acta*, 1990, **171**, 89.
245. E.Rentschler, W.Massa, S.Vogler, K.Dehnicke and G.Baum, *Z. anorg. allg. Chem.*, 1991, **592**, 59.
246. A.Keller, *J. Organometal. Chem.*, 1991, **407**, 237.
247. I.K.Choi, Y.Liu, D.W.Feng, K.J.Paeng and M.D.Ryan, *Inorg. Chem.*, 1991, **30**, 1832.

248. E.Guglielminotti and F.Boccuzzi, *J. Chem. Soc., Farad. Trans.*, 1991, **87**, 337.
249. A.K.Burrell, G.R.Clark, C.E.F.Rickard, W.R.Roper and A.H.Wright, *J. Chem. Soc., Dalton Trans.*, 1991, 609.
250. A. del Zotto, A.Mezzetti and P.Rigo, *Inorg. Chim. Acta*, 1990, **171**, 61.
251. M.Zhou, D.Chen, M.Lin and S.Cao, *Xiamen Daxue Xuebao, Ziran Kexueban*, 1990, **29**, 291 (*Chem. Abs.*, 1991, **115**, 164763).
252. G.Albertin, P.Amendola, S.Antoniutti and E.Bordignon, *J. Chem. Soc., Dalton Trans.*, 1990, 2979.
253. D.T.Wickham, B.A.Banse and B.E.Koel, *Surf. Sci.*, 1991, **243**, 83.
254. P.Gowik and T.Klapőtke, *J. Organometal. Chem.*, 1991, **402**, 349.
255. V.C.Gibson, T.P.Kee and W.Clegg, *J. Chem. Soc., Dalton Trans.*, 1990, 3199.
256. M.Herberhold, S.M.Frank and B.Wrackmeyer, *J. Organometal. Chem.*, 1991, **410**, 159.
257. R.Alvarez, J.L.Atwood, E.Carmona, P.J.Pérez, M.L.Poveda and R.D.Rogers, *Inorg. Chem.*, 1991, **30**, 1493.
258. T.A.George, L.Ma, S.N.Shailh, R.C.Tisdale and J.Zubieta, *Inorg. Chem.*, 1990, **29**, 4789.
259. P.Braunstein, M.Knorr, U.Schubert, M.Lanfranchi and A.Tiripicchio, *J. Chem. Soc., Dalton Trans.*, 1991, 1507.
260. G.Lu, J.E.Darwell and J.E.Crowell, *J. Phys. Chem.*, 1990, **94**, 8326.
261. M.Esteban, A.Pequerul, D.Carmona, F.J.Lahoz, A.Martín and L.A.Oro, *J. Organometal. Chem.*, 1991, **402**, 421.
262. D.J.Irvine, C.Glidewell, D.J.Cole-Hamilton, J.C.Barnes and A.Howie, *J. Chem. Soc., Dalton Trans.*, 1991, 1765.
263. X.Guo, J.T.Yates, V.K.Agrawal and M.Trenary, *J. Chem. Phys.*, 191, **94**, 6256.
264. T.Suzuki, M.Kita, K.Kashiwabara and J.Fujita, *Bull. Chem. Soc. Japan*, 1990, **63**, 3434.
265. S.D.Perera and B.L.Shaw, *J. Organometal. Chem.*, 1991, **402**, 133.
266. D.L.W.Kwoh and R.C.Taylor, *Spectrochim. Acta, A*, 1991, **47A**, 409.
267. C.R.Bhattacharjee, M.Bhattacharjee, M.K.Chaudhuri and S.Choudhury, *Polyhedron*, 1990, **9**, 1653.
268. B.N.Chernyshov, N.A.Didenko, N.G.Bakeeva, B.K.Bukvetskii and A.V.Gerasimanko, *Russ. J. Inorg. Chem.*, 1991, **36**, 631.
269. E.M.Nour, A.N.Alnaimi and A.B.Alsada, *J. Phys. Chem. Solids*, 1990, **51**, 907.
270. A.A.Davydov, *J. Chem. Soc., Farad. Trans.*, 1991, **87**, 913.
271. M.T.H.Tarafder and A.R.Khan, *Polyhedron*, 1991, **10**, 819.
272. A.Weselucha-Birczynska, L.M.Proniewicz, K.Bajdor and K.Nakamoto, *J. Raman Spectrosc.*, 1991, **22**, 315.
273. J.Baldas and S.F.Colmanet, *Inorg. Chim. Acta*, 1990, **176**, 1.
274. J.Baldas, S.F.Colmanet and G.A.Williams, *J. Chem. Soc., Dalton Trans.*, 1991, 1631.
275. L.M.Proniewicz, I.R.Paeng and K.Nakamoto, *J. Am. Chem. Soc.*, 1991, **113**, 3294.
276. S.Hu, A.J.Schneider and J.R.Kincaid. *J. Am. Chem. Soc.*, 1991, **113**, 4815.
277. T.Egawa, T.Ogura, R.Makino, Y.Ishimura and T.Kitagawa, *J. Biol. Chem.*, 1991, **266**, 10246.
278. B.A.Brennan, Q.Chen, C.Juarez-Garcia, A.E.True, C.J.O'Connor and L.Que, *Inorg. Chem.*, 1991, **30**, 1937.
279. P.J.Stang, L.Song, Y-H.Huang and A.M.Arif, *J. Organometal. Chem.*, 1991, **405**, 403.
280. G.López, J.Ruiz, G.García, J.M.Marti, G.Sánchez and J.García, *J. Organometal. Chem.*, 1991, **412**, 435.
281. F.Porta, F.Ragaini, S.Cenini, O.Sciacovelli and M.Camporeale, *Inorg. Chim. Acta*, 1990, **173**, 229.
282. M.Pizzotti, S.Cenini, R.Ugo and F.Demartin, *J. Chem. Soc., Dalton Trans.*, 1991, 65.
283. X.D.Wang and P.Greenler, *Phys. Rev., B*, 1991, **43**, 6808.

284. C.-T.Chen, W.-K.Chang, G.-H.Lee, T.-I.Ho, Y.-C.Lin and Y.Wang, *J. Chem. Soc., Dalton Trans.*, 1991, 1569.
285. K.V.Titova and E.I.Kolmakova, *Russ. J. Inorg. Chem.*, 1991, **36**, 493.
286. B.N.Ivanov-Emin, A.M.Il'inets, B.E.Zaitsev, A.V.Kostrikin, F.M.Spiridonov and V.P.Dolganev, *Russ. J. Inorg. Chem.*, 1990, **35**, 1301.
287. P.S.Santos, J.H.Amaral and L.F.C.de Oliveira, *J. Mol. Struct.*, 1991, **243**, 223.
288. J.Semmler, D.E.Irish and T.Ozeki, *Geochim. Cosmochim. Acta*, 1990, **54**, 947.
289. Z.Wu, F.Zhao and Z.Zhou, *Polyhedron*, 1990, **9**, 2143.
290. A.Müller, J.Döring and H.Bögge, *J. Chem. Soc., Chem. Comm.*, 1991, 273.
291. Y.Kim, J.Gallucci and A.Wojcicki, *J. Am. Chem. Soc.*, 1990, **112**, 8600.
292. R.B.Lanjewar and A.N.Garg, *Ind. J. Chem, A*, 1991, **30A**, 350.
293. W.P.Griffith, J.M.Jolliffe, S.V.Ley and D.J.Williams, *J. Chem. Soc., Chem. Comm.*, 1990, 1219.
294. H.Loumrhari, J.Ros, R.Yáñez and M.R.Torres, *J. Organometal. Chem.*, 1991, **408**, 233.
295. W.S.Bigham and P.A.Shapley, *Inorg. Chem.*, 1991, **30**, 4093.
296. R.Bhattacharyya, A.M.Saha, P.N.Ghosh, M.Mukherjee and A.K.Mukherjee, *J.Chem. Soc., Dalton Trans.*, 1991, 501.
297. H.Werner, S.Stahl and W.Kohlmann, *J. Organometal. Chem.*, 1991, **409**, 285.
298. A.A.Danopoulos, G.Wilkinson, B.Hussain-Bates and M.B.Hursthouse, *J. Chem. Soc., Dalton Trans.*, 1991, 1855.
299. H.G.Ang, W.L.Kwik and E.Morrison, *J. Fluorine Chem.*, 1991, **51**, 83.
300. J.J.Schwab, E.C.Wilkinson, S.R.Wilson and P.A.Shapley, *J. Am. Chem. Soc.*, 1991, **113**, 6124.
301. N.F.Gol'dshleger, A.P.Moravskii and Yu.M.Shul'ga, *Izv. Akad. Nauk SSSR, Ser. Khim.*, 1991, 258.
302. R.I.Bickley, H.G.M.Edwards and S.J.Rose, *J. Mol. Struct.*, 1991, **243**, 341.
303. I.Castro, J.Faus, M.Julve and A.Gleizes, *J. Chem. Soc., Dalton Trans.*, 1991, 1937.
304. B.T.Usubaliev, F.N.Musaev, F.I.Guliev, E.M.Movsumov and D.M.Gambalev, *Russ. J. Inorg. Chem.*, 1991, **36**, 547.
305. C.Fukuhara, K.Tsuneyoshi, N.Matsumoto, S.Kida, M.Mikuriya and H.Mori, *J. Chem. Soc., Dalton Trans.*, 1990, 3473.
306. O.Berkesi, I.Dreveni, J.A.Andor and P.L.Goggin, *Inorg. Chim. Acta*, 1991, **181**, 285.
307. N.A.Chumaevskii, O.U.Sharopov, N.A.Minaeva, M.T.Toshev and A.V.Sergeev, *Koord. Khim.*, 1990, **16**, 1704.
308. N.A.Chumaevskii, O.U.Shaparov, N.A.Minaeva, Yu.N.Mikhailov, M.T.Toshev and A.V.Sergeev, *Koord. Khim.*, 1990, **16**, 1698.
309. G.G.Bombi, B.Coran, A.A.Sheikh-Osman and G.C.Valle, *Inorg. Chim. Acta*, 1990, **171**, 79.
310. N.W.Alcock, I.A.Degnan, S.M.Roe and M.G.H.Wallbridge, *J. Organometal. Chem.*, 1991, **414**, 285.
311. M.I.Khalil, *Inorg. Chem.*, 1990, **29**, 5131.
312. G.K.Sandhu, N.Sharma and E.R.T.Tiekink, *J. Organometal. Chem.*, 1991, **403**, 119.
313. C.Vatsa, V.K.Jain, T.Kesavadas and E.R.T.Tiekink, *J. Organometal. Chem.*, 1991, **410**, 135.
314. S.W.Ng and V.G.K.Das, *J. Organometal. Chem.*, 1991, **409**, 143.
315. A.P.Pisarevskii, L.I.Martynenko and N.G.Dzyubenko, *Russ. J. Inorg. Chem.*, 1991, **36**, 371.
316. V.A.Trofimov, A.Yu.Tsivadze, D.K.Kireeva and N.B.Generalova, *Koord. Khim.*, 1990, **16**, 1458.
317. I.K.Kireeva, N.B.Generalova, V.A.Trofimov and A.Yu.Tsivadze, *Russ. J. Inorg. Chem.*,

1991, **36**, 830.
318. N.P.Kuz'mina, M.V.Chechernikova and L.I.Martynenko, *Russ. J. Inorg. Chem.*, 1990, **35**, 1576.
319. K.Horikoshi, K.Hata, N.Kawabata, S.Ikawa and S.Koneka, *J. Mol. Struct.*, 1990, **239**, 33.
320. T.A.Azizov, Zh.U.Makhmudov, Kh.T.Sharipov, B.M.Beglov, S.Usmanov and Kh.Isakov, *Russ. J. Inorg. Chem.*, 1990, **35**, 1156.
321. G.Xue, J.Dong and Q.Sheng, *J. Chem. Soc., Dalton Trans.*, 1991, 447.
322. A.A.El-Asmy, T.Y.Al-Ansi and R.R.Amin, *Bull. Soc. Chim. France*, 1991, 39.
323. T.G.Cherkasova, E.S.Tatarinova, O.A.Kuznetsova and B.G.Tryasunov, *Russ. J. Inorg. Chem.*, 1991, **36**, 71.
324. G.Yang, Y.Fan and Y.Han, *Sci. China, Ser. B*, 1990, **33**, 1418.
325. J.Lewinski, S.Pasynkiewicz and J.Lipkowski, *Inorg. Chim. Acta*, 1990, **178**, 113.
326. M.B.Power, S.G.Bott, D.L.Clark, J.L.Atwood and A.R.Barron, *Organometallics*, 1990, **9**, 3086.
327. J.Vicente, J.-A.Abad, G.Cara and P.G.Jones, *Angew. Chem., Int. Ed. Engl.*, 1990, **29**, 1125.
328. M.J.Hampden-Smith, T.A.Wark, A.Rheingold and J.C.Huffman, *Can. J. Chem.*, 1991, **69**, 121.
329. A.Samuel-Lewis, P.J.Smith, J.H.Aupers and D.C.Povey, *J. Organometal. Chem.*, 1991, **402**, 319.
330. Yu.Ya.Kharitonov, K.S.Sulaimankulov and N.B.Khudaibergenova, *Russ. J. Inorg. Chem.*, 1990, **35**, 1207.
331. Yu.N.Medvedev, B.E.Zaitsev and B.N.Ivanov-Emin, *Russ. J. Inorg. Chem.*, 1991, **36**, 105.
332. V.L.Rudzevich, A.O.Gudima and V.A.Kalibabchuk, *Russ. J. Inorg. Chem.*, 1990, **35**, 1560.
333. A.O.Gudima, E.O.Berezhno and V.A.Kalibabchuk, *Koord. Khim.*, 1990, **16**, 1147.
334. T.J.Dines, C.H.Rochester and A.M.Ward, *J. Chem. Soc., Farad. Trans.*, 1991, **87**, 643.
335. O.A.Govorukhina, S.D.Nikitina, V.N.Brusentsova and V.A.Masloboev, *Russ. J. Inorg. Chem.*, 1990, **35**, 1804.
336. R.A.Henderson, A.Hills, D.L.Hughes and D.J.Lowe, *J. Chem. Soc., Dalton Trans.*, 1991, 1755.
337. T.J.Dines, C.H.Rochester and A.M.Ward, *J. Chem. Soc., Farad. Trans.*, 1991, **87**, 1473.
338. T.J.Dines, C.H.Rochester and A.M.Ward, *J. Chem. Soc., Farad. Trans.*, 1991, **87**, 1617.
339. X.Wang, M.E.Kotun and J.C.Fanning, *Wuji Huaxue Xuebao*, 1989, **5**, 1 (*Chem. Abs.*, 1991, **114**, 113901).
340. P.Guillaume, A.L.K.Wah and M.Postel, *Inorg. Chem.*, 1991, **30**, 1828.
341. G.V.Kuzhevnikova, K.A.Burkhov, L.S.Ilich and E.I.Fedorova, *Russ. J. Inorg. Chem.*, 1990, **35**, 1078.
342. W.Clegg, R.J.Errington, D.C.R.Hockless, A.D.Glen and D.G.Richards, *J. Chem. Soc., Chem. Comm.*, 1990, 1565.
343. S.Himeno, K.Miyashita, A.Saito and T.Hori, *Chem. Lett.*, 1990, 799.
344. A.N.Zhilyaev, T.A.Fomina, P.A.Koz'min, T.B.Larina, M.D.Surazhskaya and I.B.Baranovskii, *Russ. J. Inorg. Chem.*, 1991, **36**, 222.
345. B.Papankova and H.Langfelderova, *J. Therm. Anal.*, 1989, **35**, 2347.
346. F.T.Esmadi and M.K.Dohaidel, *Polyhedron*, 1990, **9**, 1633.
347. M.Maneva and M.Georghiev, *Spectrosc. Lett.*, 1990, **23**, 831.
348. J.C.G.Bünzli and V.Kasparek, *Inorg. Chim. Acta*, 1991, **182**, 101.
349. V.P.Babaeva and V.Ya.Rosolovskii, *Russ. J. Inorg. Chem.*, 1990, **35**, 944.
350. V.McKee and S.S.Tandon, *J. Chem. Soc., Dalton Trans.*, 1991, 221.
351. M.A.Ciriano, J.J.Pérez-Torrente, L.A.Oro, A.Tiripicchio and M.Tiripicchio-Camellini, *J. Chem. Soc., Dalton Trans.*, 1991, 255.

352. M.Fourati, M.Chaabouni, C.Belin, J.L.Pascal and J.Potier, *New. J. Chem.*, 1990, **14**, 695.
353. G.C.Junior, W.Ferraresi de Givani, L.C.Garla and V.H.Betarello, *Eclectica Quim.*, 1989, **14**, 49 (*Chem. Abs.*, 1991, **114**, 198373).
354. T.A.Ivanova, V.P.Babaeva and V.Ya.Rosolovskii, *Russ. J. Inorg. Chem.*, 1990, **35**, 1398.
355. T.A.Ivanova, V.B.Rybakov, V.E.Mistryukov and Yu.N.Mikhailov, *Russ. J. Inorg. Chem.*, 1991, **36**, 6470.
356. I.Silaghi-Dumitrescu, R.Greco, L.Silaghi-Dumitrescu and I.Haiduc, *Stud. Univ. Babes-Bolyai, Chem.*, 1989, **34**, 97.
357. C.Diaz, J.Cuevas and G.Gonzalez, *Z. anorg. allg. Chem.*, 1991, **592**, 7.
358. V.P.Fedin, B.A.Kolesov, Yu.V.Mironov, O.A.Geras'ko and V.E.Fedorov, *Polyhedron*, 1991, **10**, 997.
359. R.Bhattacharyya, P.K.Chakrabarty, P.N.Ghosh, A.K.Mukherjee, D.Podder and M.Mukherjee, *Inorg. Chem.*, 1991, **30**, 3948.
360. X.F.Yan and C.G.Young, *Aust. J. Chem.*, 1991, **44**, 361.
361. S.Gruber, L.Kilpatrick, N.R.Bastian, K.V.Rajagopalan and T.G.Spiro, *J. Am. Chem. Soc.*, 1990, **112**, 8179.
362. A.E.Sánchez-Peláez and M.F.Perpiñán, *J. Organometal. Chem.*, 1991, **405**, 101
363. S.B.Kumar and M.Chaudhury, *J.Chem. Soc., Dalton Trans.*, 1991, 1149.
364. M.Q.Islam, W.E.Hill and T.R.Webb, *J. Fluorine Chem.*, 1990, **48**, 429.
365. V.P.Fedin, M.N.Sokolov, O.A.Geras'ko, B.A.Kolesov, V.E.Fedorov, A.V.Mironov, D.S.Yufit, Yu.L.Slovokhotov and Yu.T.Struchkov, *Inorg. Chim. Acta*, 1990, **175**, 217.
366. B.Müller and H.D.Lutz, *Solid State Commun.*, 1991, **78**, 469.
367. R.Rossi, A.Marchi, L.Magon, U.Casellato, S.Tamburini and R.Graziano, *J. Chem. Soc., Dalton Trans.*, 1991, 263.
368. A.Müller, M.I.Khan, E.Krickemeyer and H.Bögge, *Inorg. Chem.*, 1991, **30**, 2040.
369. M.Di Vaira, M.Peruzzini and P.Stoppioni, *Inorg. Chem.*, 1991, **30**, 1001.
370. G.Matsubayashi and A.Yokozawa, *J. Chem. Soc., Dalton Trans.*, 1990, 3013.
371. S.-P.Huang and M.G.Kanitzidis, *Inorg. Chem.*, 1991, **30**, 1455.
372. S.-P.Huang and M.G.Kanitzidis, *Inorg. Chem.*, 1991, **30**, 3572.
373. A.C.Sutorik and M.G.Kanatzidis, *J. Am. Chem. Soc.*, 1991, **113**, 7754.
374. J.Zukerman-Schpector, E.M.Vazquez-Lopez, A.Sanchez, J.S.Casas and J.Sordo, *J. Organometal. Chem.*, 1991, **405**, 67.
375. N.Singh, N.K.Singh and C.Kaw, *Bull. Chem. Soc. Japan*, 1990, **63**, 1801.
376. B.Krebs and A.Brömmelhaus, *Z. anorg. allg. Chem.*, 1991, **595**, 164.
377. J.E.Drake, A.G.Mislankar and M.L.Y.Wong, *Inorg. Chem.*, 1991, **30**, 2178.
378. M.T.H.Tarafder, A.R.Khan and B.Nath, J. *Bangladesh Chem. Soc.*, 1988, **1**, 149.
379. B.Nath, *J. Bangladesh Chem. Soc.*, 1988, **1**, 59.
380. J.Zsako, C.Varhelyi and M.Mate, *Stud. Univ. Babes-Bolyai, Chem.*, 1989, **34**, 74 (*Chem. Abs.*, 1991, **115**, 122661).
381. T.N.Mali, R.D.Hancock, J.C.A.Boeyens and E.L.Oosthuizen, *J. Chem. Soc., Dalton Trans.*, 1991, 1161.
382. R.A.Holwerda, T.F.Tekut, B.G.Gafford, J.H.Zhang and C.J.O'Connor, *J. Chem. Soc., Dalton Trans.*, 1991, 1051.
383. A.Hills, D.L.Hughes, C.J.Macdonald, M.Y.Mohammad and C.J.Pickett, *J. Chem. Soc., Dalton Trans.*, 1991, 121.
384. T.Yoshida, T.Adachi, K.Kawazu, A.Yamamoto and N.Sasaki, *Angew. Chem., Int. Ed. Engl.*, 1991, **30**, 9482.
385. D.C.Figg, R.H.Herber and I.Felner, *Inorg. Chem.*, 1991, **30**, 2535.
386. W.Preetz and U.Sellerberg, *Z. anorg. allg. Chem.*, 1990, **589**, 158.
387. J.A.Anderson and C.H.Rochester, *J. Chem. Soc., Farad. Trans.*, 1990, **86**, 3809.

388. Yu.A.Simonov, V.V.Skopenko, V.N.Kokozei, S.R.Petrusenko and A.A.Dvorkin, *Russ. J. Inorg. Chem.*, 1991, **36**, 352.
389. T.Rojo, R.Cortés, L.Lezama, M.I.Arriortua, K.Urtiaga and G.Villeneuve, *J. Chem. Soc., Dalton Trans.*, 1991, 1779.
390. M.J.Coyer, R.H.Herber and S.Cohen, *Inorg. Chim. Acta*, 1990, **175**, 47.
391. G.S.Muraveiskaya, A.A.Sidorov, G.N.Emel'yanova and E.M.Trishkina, *Russ. J. Inorg. Chem.*, 1991, **36**, 525.
392. Q.Jiang, L.Zhang, G.Li, Z.Zhou, G.Hu and K.Yu, *Huaxue Xuebao*, 1990, **48**, 778 (*Chem. Abs.*, 1991, **114**, 74030).
393. J.H.Clark and C.W.Jones, *Inorg. Chim. Acta*, 1991, **179**, 41.
394. M.G.Samant, K.Kunimatsu, R.Viswanathan, H.Seki, G.Pacchion, P.S.Bagus and M.R.Philpott, *Langmuir*, 1991, **7**, 1261.
395. R.Minkwitz, M.Koch, J.Nowicki and H.Borrmann, *Z.anorg. allg. Chem.*, 1990, **590**, 93.
396. I.A.Garbuzova, L.A.Chekulaeva, L.B.Danilevskaya, A.V.Markov, V.A.Kotel'nikov, B.V.Lokshin, V.V.Gavrilenko, V.V.Kurashev, I.I.Tverdokhleisova and O.I.Sutkevich, *Izv. Akad. Nauk SSSR, Ser. Khim.*, 1990, 1570.
397. M.S.Sastry, U.P.Singh, R.Ghose and A.K.Ghose, *Synth. React. Inorg. Met.-Org. Chem.*, 1991, **21**, 73.
398. B.Singh and A.K.Srivastava, *Synth. React. Inorg. Met.-Org. Chem.*, 1991, **21**, 457.
399. V.R.Fisher, T.A.Nasonova, Kh.M.Yakutsov, V.K.Voronkova, L.V.Mosina and Yu.V.Yablokov, *Russ. J. Inorg. Chem.*, 1990, **35**, 1294.
400. V.L.Abramenko, A.D.Garnovskii, V.S.Sergienko, M.A.Porai-Koshits and N.A.Minaeva, *Koord Khim.*, 1990, **16**, 1500.
401. R.N.Mohanty, V.Chakravorty and K.C.Dash. *Polyhedron*, 1991, **10**, 33.
402. S.Kurishima, N.Matsuda, N.Tamura and T.Ito, *J. Chem. Soc., Dalton Trans.*, 1991, 1135.
403. M.G.Felin and A.M.Levitskii, *Russ. J. Inorg. Chem.*, 1990, **35**, 971.
404. M.A.Yampol'skaya, A.I.Shames, N.V.Gerbeleu, Yu.A.Simonov and M.S.Byrke, *Koord. Khim.*, 1991, **17**, 337.
405. J.Maslowska and J.Jaroszynska, *Zesz. Nauk-Politech. Lodz, Technol. Chem. Spozyw.*, 1988, **508**, 23 (*Chem. Abs.*, 1991, **114**, 155830).
406. S.M.Abu-El-Wafa, M.Geber, A.A.Saleh and A.A.El-Dken, *J. Chem. Soc. Pak.*, 1989, **11**, 270.
407. U.Abram, S.Abram, J.Stach, R.Wollert, G.F.Morgan, J.R.Thornback and M.Deblaton, *Z. anorg. allg. Chem.*, 1991, **600**, 15.
408. D.X.West and J.N.Albert, *Transition Met. Chem.*, 1991, **16**, 1.
409. C.M.Gordon, R.D.Feltham and J.J.Turner, *J. Phys. Chem.*, 1991, **95**, 2889.
410. L.Ruiz-Ramirez and L.Gasque-Silva, *Rev, Soc. Quim. Mex.*, 1989, **33**, 300 (*Chem. Abs.*, 1991, **114**, 239250).
411. G.M.McCann, A.Carvill, E.Lindner, B.Karle and H.A.Mayer, *J. Chem. Soc., Dalton Trans.*, 1990, 3107.
412. K.Yamamoto, *Chem. Lett.*, 1991, 1509.
413. V.V.Skopenko, I.O.Iritskii, R.D.Lampeka and O.A.Zhmurko, *Dokl. Akad. Nauk Ukr. SSR, Ser, B: Geol. Khim. Biol. Nauki*, 1990, 59.
414. G.V.Tsintsadze, M.V.Shavtvaladze and A.Yu.Tsivadze, *Koord. Khim.*, 1990, **16**, 1666.
415. N.K.Singh and R.Tripathi, *Synth. React. Inorg. Met.-Org. Chem.*, 1990, **20**, 923.
416. Yu.A.Zhdanov, V.A.Kogan, Yu.A.Alekseev, V.M.Kharkovskii, V.G.Saletov and A.G.Starikov, *Dokl. Akad. Nauk SSSR*, 1990, **312**, 115.
417. V.V.Skopenko, R.D.Lampeka and I.O.Iritskii, *Dokl. Akad. Nauk SSSR*, 1990, **312**, 123.
418. J.Ochocki, E.Zyner, B.Zurowska and J.Mrozinski, *Pr. Nauk Akad. Ekon. im Oskara Langego Wroclawiu*, 1990, **526**, 173 (*Chem. Abs.*, 1991, **114**, 198429).

419. V.F.Shul'gin, O.V.Konnik and V.L.Kokoz, *Russ. J. Inorg. Chem.*, 1990, **35**, 1066.
420. V.V.Skopenko, L.N.Morozova, R.D.Lampeka and A.A.Tolmachev, *Russ. J. Inorg. Chem.*, 1991, **36**, 42.
421. I.I.Seifullina, L.S.Skorokhod, V.V.Minin and G.N.Larin, *Russ. J. Inorg. Chem.*, 1991, **36**, 386.
422. Yu.A.Zhdonov, V.A.Kogan, Yu.E.Alekseev, V.M.Khar'kovskii, V.G.Zaletov and A.G.Starikov, *Russ. J. Inorg. Chem.*, 1991, **36**, 518.
423. R.Krämer, K.Polborn and W.Beck, *J. Organometal. Chem.*, 1991, **410**, 111.
424. G.Cervantes, J.J.Fiol, A.Terrón, V.Moreno, J.R.Alabart, M.Aguiló, M.Gómez and X.Solans, *Inorg. Chem.*, 1990, **29**, 5168.
425. B.E.Zaitsev, E.V.Nikiforov, M.A.Ryabov and P.I.Abramenko, *Russ. J. Inorg. Chem.*, 1990, **35**, 1763.
426. J.A.Jahagirdar, B.G.Patil and B.R.Havinale, *Ind. J. Chem., A*, 1990, **29A**, 924.
427. M.Kwiatkowski, E.Kwiatkowski, A.Olechnowicz, D.M.Ho and E.Deutsch, *J. Chem. Soc., Chem. Comm.*, 1990, 3063.
428. J.-C.Chambron and K.Hiratani, *J. Chem. Soc., Dalton Trans.*, 1991, 1483.
429. D.A.Knight, D.J.Cole-Hamilton and D.Cupertino, *J. Chem. Soc., Dalton Trans.*, 1990, 3051.
430. L.Tušek-Božić, I.Matijašic, G.Bocelli, G.Calestani, A.Furlani, V.Scarcia and A.Papaioannou, *J. Chem. Soc., Dalton Trans.*, 1991, 195.
431. N.M.Sinitsyn, T.M.Buslaeva, L.V.Samarova, G.G.Novitskii and N.A.Kotenova, *Russ. J. Inorg. Chem.*, 1990, **35**, 1446.
432. I.G.Fomina and V.E.Abashkin, *Russ. J. Inorg. Chem.*, 1990, **35**, 1431.
433. O.N.Adrianova, M.I.Gel'fman, E.V.Kovaleva, O.N.Evstaf'eva, I.F.Golovanova and N.V.Asadulina, *Russ. J. Inorg. Chem.*, 1990, **35**, 1754.
434. N.Trendafilova, A.P.Kurbakova, I.A.Efimenko, M.Mitewa and P.R.Bontchev, *Spectrochim. Acta, A*, 1991, **47A**, 577.
435. O.M.Adamov, A.I.Stetsenko, I.B.Pushko and V.G.Pogareva, *Koord. Khim.*, 1990, **16**, 857.
436. M.F.Mogulevskaya and N.F.Nechepurenko, *Koord. Khim.*, 1990, **16**, 128.
437. D.L.Darensbourg, E.M.Longridge, E.V.Atnip and J.H.Reibenspies, *Inorg. Chem.*, 1991, **30**, 357.
438. B.Srinivas, N.Arulsamy and P.S.Zacharias, *Polyhedron*, 1991, **10**, 731.
439. V.G.Yusupov, M.M.Karimov, B.B.Umarov, K.N.Zelenin, G.M.Larin, V.V.Minin and N.A.Parpiev, *Uzb. Khim. Zh.*, 1990, 35 (*Chem. Abs.*, 1991, **115**, 149039).
440. S.Kulemou, V.N.Shafranskii, M.S.Popov and N.M.Samus, *Izv. Akad. Nauk Mold. SSR, Ser. Biol. Khim. Nauk*, 1989, 61 (*Chem. Abs.* 1991, **114**, 155828).
441. Z.Wu, Z.Gui and Z.Yen, *Synth. React. Inorg. Met.-Org. Chem.*, 1990, **20**, 335.
442. R.A.Lal, S.S.Bhattacharjee and M.Husain, *Synth. React. Inorg. Met.-Org. Chem.*, 1990, **20**, 809.
443. A.M.El-Roudi, *Bull. Fac. Sci. Assiut Univ.*, 1989, **18**, 31 (*Chem. Abs.*, 1991, **114**, 134847).
444. A.A.Stratulat and V.I.Gorgos, *Koord. Khim.*, 1990, **16**, 1292.
445. S.I.Neikovskii, V.A.Krasovskii and F.M.Tulyupa, *Russ. J. Inorg. Chem.*, 1990, **35**, 1295.
446. P.O.Ikekwere, R.M.Fadamiro, O.A.Odunola and K.S.Patel, *Synth. React. Inorg. Met.-Org. Chem.*, 1990, **20**, 1213.
447. R.A.Slavinskaya, T.A.Kovaleva and N.A.Dolgova, *Koord. Khim.*, 1990, **16**, 1028.
448. T.A.K.Al-Allaf and M.A.M.Al-Tayy, *J. Organometal. Chem.*, 1990, **391**, 37.
449. B.Mundus-Glowacki and F.Huber, *Z. Naturforsch., B*, 1991, **46b**, 270.
450. V.P.Sinditskii, V.I.Sokol, M.D.Dutov and A.E.Fogel'zang, *Russ. J. Inorg. Chem.*, 1990, **35**, 1266.

451. R.L.Davidovich, V.B.Loginova, L.A.Zemnukhova and L.V.Teplukhina, *Koord. Khim.*, 1991, **17**, 29.
452. M.T.H.Tarafder and A.R.Khan, *Polyhedron*, 1991, **10**, 973.
453. U.A.Bagal, C.B.Cook and T.L.Riechel, *Inorg. Chim. Acta*, 1991, **180**, 57.
454. B.A.Bovykin, M.I.Shenbor, V.I.Tikhnov and N.V.Semeryazhko, *Vopr. Khim. Khim. Tekhnol.*, 1989, **90**, 43 (*Chem. Abs.*, 1991, **114**, 177108).
455. F.Tisato, U.Mazzi, G.Bandoli, G.Cros, M.-H.Darbieu, Y.Coulais and R.Guiraud, *J. Chem. Soc., Dalton Trans.*, 1991, 1301.
456. I.I.Bulgak, I.E.Rychagova, V.E.Zubareva, K.I.Turte and V.N.Shafranskii, *Russ. J. Inorg. Chem.*, 1990, **35**, 989.
457. C.Sin'de, L.Chzhifen, U.Tszen'shen and Y.Tszen'khuan, *Russ. J. Inorg. Chem.*, 1991, **36**, 705.
458. R.K.Patel and R.N.Patel, *J. Ind. Chem. Soc.*, 1990, **67**, 538.
459. G.Y.Sarkis, *J. Iraqi Chem. Soc.*, 1988, **13**, 103 (*Chem. Abs.*, 1991, **114**, 134832).
460. P.Chatterjee and B.V.Agarwala, *Bull. Chem. Soc., Ethiop.*, 1990, **4**, 39.
461. G.M.Abu El-Reash, F.I.Taha, A.M.Shallaby and O.A.El-Gamal, *Synth. React. Inorg. Met.-Org. Chem.*, 1991, **21**, 697.
462. G.López, G.Sánchez, G.García, E.Pérez, J.Casabó, E.Molins and C.Miravitlles, *Inorg. Chim. Acta*, 1990, **178**, 213.
463. N.Baidya, D.Ndreu, M.M.Olmstead and P.K.Mascharak, *Inorg. Chem.*, 1991, **30**, 2448.
464. A.Castiñeiras, W.Hiller, J.Strähle, R.Carballo, M.R.Bermejo and M.Gayoso, *Z. Naturforsch., B*, 1990, **45b**, 1267.
465. F.Bigoli, M.A.Pellinghelli, P.Deplano, E.F.Trogu, A.Sabatini and A.Vacca, *Inorg. Chim. Acta*, 1991, **180**, 201.
466. S.Laly and G.Parameswaran, *Thermochim. Acta*, 1990, **168**, 43.
467. N.S.S.Jalil, *Synth. React. Inorg. Met.-Org. Chem.*, 1990, **20**, 1285.
468. S.Gopinathan, M.P.Degaonkar and C.Gopinathan, *Ind. J. Chem., A*, 1990, **29A**, 971.
469. A.Kumari, R.V.Singh and J.P.Tandon, *Ind. J. Chem., A*, 1991, **30A**, 468.
470. L.P.Battaglia, A.B.Corradi, C.Pelizzi, G.Pelosi and P.Tarasconi, *J. Chem. Soc., Dalton Trans.*, 1990, 3857.
471. W.P.Hagen and D.G.Sargeant, *J. Power Sources*, 1991, **34**, 1.
472. G.Erker, M.Mena, C.Krüger and R.Noe, *J. Organometal. Chem.*, 1991, **402**, 67.
473. M.S.Sastry and S.K.Kulshreshtha, *J. Inorg. Biochem.*, 1991, **41**, 79.
474. W.Heilemann and R.Mews, *Inorg. Chim. Acta*, 1990, **173**, 137.
475. W.A.Schenk and P.Urban, *J. Organometal. Chem.*, 1991, **411**, C27.
476. U.C.Sharma, B.C.Paul and R.K.Poddar, *Ind. J. Chem., A*, 1990, **29A**, 803.
477. E.Alessio, B.Milani, G.Mestroni, M.Calligaris, P.Faleschini and W.M.Attia, *Inorg. Chim. Acta*, 1990, **177**, 255.
478. E.Alessio, G.Balducci, M.Calligaris, G.Costa, W.M.Attia and G.Mestroni, *Inorg. Chem.*, 1991, **30**, 609.
479. B.W.Arbuckle, P.K.Bharadwaj and W.K.Musker, *Inorg. Chem.*, 1991, **30**, 440.
480. C.F.Edwards, W.P.Griffith and D.J.Williams, *J. Chem. Soc., Chem. Comm.*, 1990, 1523.
481. G.R.Frauenhoff, S.R.Wilson and J.R.Shapley, *Inorg. Chem.*, 1991, **30**, 78.
482. T.Krümmling, T.Bartik, T.Hoffmann and E.Wenschuh, *Z. anorg. allg. Chem.*, 1991, **592**, 141.
483. D.C.Ware, B.G.Siim, K.G.Robinson, W.A.Denny, P.J.Brothers and G.R.Clark, *Inorg. Chem.*, 1991, **30**, 3750.
484. R.K.Patel and R.N.Patel, *J. Inst. Chem., (India)*, 1990, **62**, 69.
485. I.A.Efimenko, E.I.Shubochkina, A.P.Kurbakova, V.E.Mistryukov, Yu.N.Mikhailov and A.S.Kanishcheva, *Koord. Khim.*, 1991, **17**, 95.

486. I.A.Rozanov, L.Ya.Medvedeva and L.V.Goeva, *Russ. J. Inorg. Chem.*, 1991, **36**, 827.
487. J.G.Contreras and H.Cortes, *Bol. Soc. Chil. Quim.*, 1990, **35**, 169.
488. I.A.Efimenko, Yu.N.Mikhailov, A.P.Kurbakova, Kh.I.Gasanov and V.E.Mistryukov, *Koord. Khim.*, 1990, **16**, 1574.
489. P.M.Solozhenkin, A.A.Gornostar and A.V.Ivanov, *Dokl. Akad. Nauk SSSR*, 1990, **310**, 649.
490. H.Willner, F.Mistry, G.Hwang, F.G.Herring, M.-S.R.Cader and F.Aubke, *J. Fluorine Chem.*, 1991, **52**, 13.
491. E.Colacio, A.Romerosa, J.Ruiz, P.Román, J.M.Gutierrez-Zorrilla, A.Vegas and M.Martinez-Ripoll, *Inorg. Chem.*, 1991, **30**, 3743.
492. E.B.Fesenko, L.I.Budarin and E.S.Levchenko, *Teor. Eksp. Khim.*, 1990, **26**, 637.
493. H.Schmidbaur, H.J.Öller, S.Gamper and G.Müller, *J. Organometal. Chem.*, 1990, **394**, 757.
494. T.G.Cherkasova, E.S.Tatarinova, O.A.Kuznetsova and B.G.Tryasunov, *Russ. J. Inorg. Chem.*, 1990, **35**, 1440.

7
Moessbauer Spectroscopy

BY S. J. CLARK, J. D. DONALDSON, AND S. M. GRIMES

1. Introduction

The format of this year's report is the same as that used last year. This introductory section provides details of books and review articles published during the review period. It is followed by sections that deal with the theoretical aspects of Moessbauer spectroscopy and with advances in instrumentation and methodology. Sections 4, 5 and 6 contain detailed reviews in iron-57, tin-119, and other Moessbauer effects respectively, while Section 7 reviews papers published on conversion electron and back-scattering gamma-resonance spectroscopy. In view of the current interest in low T_c superconducting oxides, Moessbauer data relating to these materials are examined in detail.

The isotopes (energies in keV in parenthesis) that have been mentioned, some of them in review articles, during the review period include: Fe-57(14.412), Zn-67(93.26), Ru-99(89.36), Ag-109(88.0), Sn-119(23.875), Sb-121(37.15), Te-125(35.46), I-127(57.6), I-129 (27.72), -Cs-133(81.0), Pr-141(145.4), Nd-145(67.3), Sm-149(45.8), Eu-151(21.64), Gd-155(86.54), Tb-159(58.0), Dy-161(80.7), Er-166 (80.56), Er-167(79.32), Tm-169(8.41), Yb-170(84.26), Yb-174(66.7), W-182 (100.1), W-183(46.5 and 99.1), Ir-191(129.3), Ir-193(73.0), Au-197(77.36), and Np-237(59.54).

Books and Reviews. The fourteenth volume of the literature service "Moessbauer Effect Data and Reference Journal" was published in 1991.[1]

Berry has contributed a chapter on Moessbauer spectroscopy to Volume 5 of "Physical Methods in Chemistry".[2] The topics examined include: the theoretical principles, instrumentation, and applications. The theoretical principles behind Moessbauer spectroscopy were also reviewed by Eldridge and O'Grady, and its applications in studies of redox reactions, catalysis, electrolysis, electrodeposition and passivation were described.[3]

Recent developments in experimental studies on interface and ultra-thin film magnetism were surveyed.[4] The usefulness of the Moessbauer effect as a tool for the study of interface electronic structure was described. The magnetism of ultra-thin iron <110> films were the subject of a review by Przyblski *et al*,[5] and Kalvius and his co-workers compared the use of gamma-resonance and muon-spin rotation spectroscopies in magnetic studies.[6] A review of the uses of Moessbauer spectroscopy in structural chemistry was carried out by Morrish,[7] *in situ* studies on the structures of supported catalysts were reviewed by Lazar,[8] and Berry described the characterization of homogeneous catalysts.[9]

Chemical systems which were reviewed in the past year include: the study of colloidal soil samples,[10] the characterization of pillared clays,[11] studies on wuestite ($Fe_{1-x}O$),[12] aluminium-substituted iron oxides and hydroxides,[13] and anhydrous iron phosphates and oxyphosphates.[14]

A critical review of recent results obtained from the study of protein dynamics by Rayleigh Scattering of Moessbauer Radiation (RSMR), was published.[15] The main experimental results, approximations, and conclusions about dynamical properties of protein and interprotein water were described. The effects of the magnetic radiofrequency field induced modulation of Moessbauer γ-radiation (RF sidebands) and fast relaxation of the magnetic hyperfine field due to the RF field (RF collapse) have been reviewed.[16] Moessbauer spectroscopy allows the study of the kinetics of RF crystallization and identification of the phases formed in amorphous metals during RF-induced crystallization.

2. Theoretical

Two approximate formulae for calculating the eigenvalues of pure quadrupole interaction in Moessbauer effect studies have been proposed and used to determine eigenvalue coefficients for various excited and ground states of nuclei with different spins.[17] All the pure quadrupole interaction eigenvalues between excited and ground states of nuclei with spin I = 3/2-9/2 and the electric field gradient with different asymmetry parameters (η = 0-1.0) were calculated using these formulae.

A generalization of the pseudo-assembly method was used to describe ordering processes in binary solid solutions containing two different types of lattice sites.[18] The theory allows the description of the phenomena of long- and short-range order in silicates and other minerals of variable composition. Energy constants of solid solution and thermodynamic functions of mixing for Fe-Mg-orthopyroxenes were determined based on available Moessbauer spectral data.

A superoperator formalism has been used to calculate Moessbauer profiles in the presence of a magnetic hyperfine field and a random electric field gradient tensor.[19] The distribution of the tensor components is described by Czjzek's model, and the magnetic hyperfine fields were of a single magnitude with simple magnetic texture. Possible applications to the Mossbauer spectra of amorphous materials were discussed.

A powerful approach has been developed for obtaining arbitrary-shape static hyperfine distributions from thickness correlated Moessbauer data.[20] The distributions were taken to be sums of Gaussian components and the corresponding spectra were shown to be sums of Voigt-lines. Three cases were worked out in detail, *viz*: (1) centre shifts; (2) quadrupole splitting with linear coupling to centre shifts; and (3) hyperfine fields with linear couplings to centre shifts and quadrupole splittings. An application of hyperfine field distributions to the spectra of Fe-Ni alloys was given. Another expression for the approximation of the Voigt integral for Moessbauer spectrometry has also been presented.[21] An expression was proposed which gives better results in the range $2 < \alpha < \infty$ than previously published approximations. The full width at half maximum of the Lorentzian Γ_L and Gaussian Γ_G profiles were estimated whenever the Gaussian contribution to the Voigt profile is small (i.e. $\alpha > 2$).

Non-Lorentzian behaviour in the Moessbauer spectra of various biological and polymeric systems has been explained using the Brownian motion of harmonically bound overdamped oscillators. A quantum picture of this model was considered and the lineshape expressed in closed form for the extreme overdamped case.[22] This model also predicts a relationship between the mean square displacement and the temperature.

The wave-number-dependent Lamb-Moessbauer factors of large and small particles in a glass of neutral hard spheres have been investigated within mode-coupling theory by varying the size ratio, total packing fraction and concentration.[23] The wave-number dependence of the small particles' Lamb-Moessbauer factors shows characteristic deviations from a Gaussian in contact to the almost Gaussian behaviour of large particles. Mixed hyperfine interactions in amorphous materials have also been studied.[24] Iron-57 Moessbauer absorption profiles were calculated assuming distributions of all hyperfine parameters: hyperfine magnetic fields, isomer shifts and electric field gradients. The effect of mixed hyperfine interactions was taken into account in all orders of perturbation theory. The shapes of the magnetic hyperfine field distribution (HMFD) were reconstructed from the simulated spectra, and were found to show a double-peaked structure similar to the distributions found in experiments on amorphous alloys with low iron content.

Moessbauer lineshape distortions due to γ-ray noncollimation and the finite dimensions of source and detector have been determined using a two-dimensional angular distribution that weights a Lorentzian line-shape function.[25] This distribution function, along with a few approximations, allow the simple calculation of isomer shift, line broadening, and line height of the Moessbauer spectra. The expressions, are valid for finite dimensions of source, detector, and absorber, and generalize previous calculations. When a finite source ($R_S = 0.4$ cm) is 10 cm from a detector ($R_d = 1.4$ cm), the noncollimation of the γ-rays generates distortions of the absorption line that are still within the experimental error.

When observed through a thick resonant sample, the exponential decay of a Moessbauer γ-state appears to be greatly accelerated. Application of a phase change to the emitted γ radiation field was shown to regenerate the accelerated decay signal.[26] The amplitude of this novel echo signal can be much higher than the corresponding source intensity, and sample saturation by an intense source is not necessary. Experimental results for the ^{57}Fe Moessbauer resonance are in good agreement with the theory.

3. Methodology

Two papers have described methods for the calibration of gamma-resonance spectra. A German patent has described a laser interferometer calibration method which employs two interfering beams.[27] One beam is directed onto the mobile arm of the interferometer and the second onto a photodetector. Accuracy is improved by having the beam directed onto the mobile arm so that it undergoes a multiple passage between the mobile and immobile elements of the arm. In the other paper, the correspondence between velocity wave-form and background form of the Moessbauer spectrum was discussed, and a method presented to determine the sign of the calibration constant according to the background form.[28]

A new method has been developed to improve the resolution of gamma-resonance spectra.[29] The effective linewidth of the source and absorber are narrowed by simple integral transformation of the experimental spectrum, i.e. the Lorentzian linewidth is transformed into a Lorentzian power linewidth. Several applications of the proposed method were demonstrated: sharpening a SnO_2

spectrum to show a doublet; determination of basesline and total spectral area; determination of the spectral density of states; reconstruction of a hyperfine field distribution; fitting of spectra to a discrete set of lines; testing for the influence of relaxation.

A deconvolution method suitable for use with small computers has been developed.[30] It operates by extending an existing method so that a broadening distribution folded with a Lorentzian line is derived on the basis of a Fourier analysis, but its computation is carried out in real space. With broadened spectra, both the Lorentzian signal and the broadening distribution could be recovered under certain constraints of the method. A program for the calculation of complex hyperfine Moessbauer parameters of polycrystalline samples was described.[31] It was designed to simulate spectra in the case of hyperfine Hamiltonians nondiagonal in the standard momentum-state representation, and was tested for ^{57}Fe in iron and ^{237}Np in NpB_2.

A new experimental method for observing the Moessbauer effect in long-lived nuclear levels and nuclear coherent states has been developed.[32] The Moessbauer effect occurs for the 88 keV transition of Ag in Ag single crystals an order of magnitude smaller in the horizontal geometry than in the vertical geometry. One method used to search for the Moessbauer effect in ultra-narrow γ-ray transitions (lifetimes in the order of seconds), is based on the increased self-absorption of γ-rays emitted in a particular direction as the sample temperature is lowered.

Possible quantum effects have been observed from Moessbauer nuclei placed in a finite space between two parallel conducting plates 1 mm apart.[33] Frequency shifts of the order $10^{-8} \sim 10^{-9}$ eV for the gamma transition of ^{121}Sb and ^{133}Cs from the first excited states to the ground states. Time-reversal (*T*) violation experiments have been carried out using Moessbauer transitions.[34] Photon transmission through a system of magnetized foils that have a Moessbauer transition with $M1$ and $E2$ strength (e.g. ^{99}Ru or ^{197}Au) are considered to be true tests of *T* violation.

Thermomechanical frequency modulation of γ-radiation has been observed.[35] When a radiofrequency current was passed through a paramagnetic Moessbauer absorber containing ^{57}Fe, sidebands were generated on the 14.4 keV transition at $\omega\gamma \pm 2n\omega$. Thermomechanical oscillations induced by the radiofrequency current produced only sidebands of even orders. Nuclear-Bragg scattering from a synthetic α-$^{57}Fe_2O_3$ crystal has been observed.[36] Moessbauer absorption spectra of the scattered beam were measured by oscillating the scatterer crystal while maintaining its Bragg condition. Time and energy spectra have been studied of Moessbauer filter systems consisting of pure nuclear reflection $^{57}FeBO_3(333)$ in connection with a broad resonance absorber.[37] The nuclear-Bragg reflected synchrotron radiation pulses were studied to obtain simpler wave packets suitable for subsequent analysis of ^{57}Fe-containing materials.

Delayed coherent nuclear forward scattering of synchrotron radiation by polycrystalline iron foils has been measured.[38] Quantum beats were observed in the time spectra, which could only arise from coherent scattering. The 5 meV bandwidth provided by the silicon crystal monochromator used removed the need for a nuclear resonant filter. This direct approach offers the possibility of carrying out a wide range of Moessbauer experiments using synchrotron radiation.

4. Instrumentation

Three papers have described complete Moessbauer spectrometer systems. Two Russian patents have described spectrometers using (a) a velocity transducer having an output signal length which has a period which is a multiple of the reference signal period,[39] and (b) an additional modulator, the centreline of which is parallel to the centreline of the main modulator, an absorber being attached to the rod of the second modulator.[40] In this second case the measurement accuracy is improved by using the second modulator to compensate for external disturbances. A British, transputer-based, Moessbauer system has been developed.[41] The data collection and velocity generation electronics were supplied by a purpose-built 16-bit T212 single-board computer which plugs into a standard IBM-AT compatible PC. A second 32-bit T800 single board computer, which also plugs into the PC, is used to run a Newton-Raphson nonlinear least-squares fitting program. Channel dwell times down to 25 s were possible, which allow maximum velocities of about 300 mm s^{-1} to be achieved with a standard Doppler drive system.

A transport control module has been developed for a Moessbauer spectrometer source that achieves ~0.01% linearity of the transport rate.[42] The triangular reference signal in the module is realized by means of a bipolar digital-to-analog convertor. A low cost data acquisition system based on a microprocessor development system has been described.[43] A low dead time, about 0.16 microseconds per channel, was obtained.

A proportional counter filled with pure helium gas has been operated at liquid helium temperatures.[44] This provides a new technique for direct detection of low-energy radiation emitted by cryogenic samples. The performance of the detector in the temperature range 1-5 K was described.

A Moessbauer monochromator for synchrotron radiation has been described.[45] A 67-layer multilayer structure of ^{57}Fe-^{56}Fe of thickness 3 nm and 1.5 nm respectively was the calculated optimum with respect to the reflectivity of γ-rays in the 14.4 keV energy range. The width of the synchrotron radiation spectra was 700 meV, for a mirror receiving angle of 3 angular minutes.

5. Iron-57

General Topics.- Fundamental Studies and Metallic Iron. The reference material method has been used to determine the recoil free fraction of α-iron.[46] Values for the *f*-factor and absorber linewidth of 0.793 and 1.21 Γ were obtained. From the relation between experimental linewidth and effective thickness, a value of 0.406 was obtained for the *f*-factor of sodium nitroprusside.

Uniform, optically flat, thin films of polycrystalline α-iron were obtained by decomposing a stream of $Fe(CO)_5$ with laser pulses.[47] The Moessbauer spectra of the surface and the bulk were fitted by assuming an angle of 37 between the magnetization direction and the film normal.

The gamma-resonance spectra of ultrafine, 2 nm, iron particles dispersed in oil showed a ferromagnetic and a paramagnetic component at 4.2 K.[48] When, after keeping the sample at room temperature, the paramagnetic component disappeared, it was deduced that particles smaller than a certain diameter are paramagnetic. Nanoclusters of Fe^0 were prepared by two different methods of loading zeolites[49] The clusters were analyzed by Moessbauer and magnetic measurements.

Three papers have described investigations on iron hydrides. Two pressure regions were identified during an *in situ* study of iron hydride made in a diamond anvil cell at room temperature.[50] At 3.5 GPa, the spectra showed a sudden change from a single hyperfine pattern (attributed to normal α-iron with negligible hydrogen content) to a superposition of three patterns (residual α-Fe plus new iron hydride). Hydrides and deuterides of iron, and hydrides of cobalt have been prepared at 4-9 GPa and 350°.[51] One deuteride analogue of the dhcp ε-FeH_x ($x \approx$ 1.00) phase was found with almost identical Moessbauer parameters to the hydride, along with three other new hydride (deuteride) components, one of which was non-magnetic at 4.2 K. In the third paper describes amorphous powders were obtained by borohydride reduction.[52] Hydrogen included in globules and in the interglobular space reduced iron(III) to iron(II) and to α-iron and induced a solid state phase transition during crystallization.

Impurity Studies and Ion-exchange. Several papers have described studies using iron as an impurity probe species. Iron-impurity ions in fluorite-type CaF_2, SrF_2, and BaF_2 single crystals were studied.[53] Information was obtained on the lattice sites. Measurements on ^{57}Fe doped CuO films showed that the magnetic transition temperature is slightly lower than that of bulk CuO.[54] Gamma-resonance spectroscopy with ^{57}Fe-impurity ions was also used in structural studies on the complexes MCl_2Py_2 (where M = Co, Ni, py = pyridine).[55] While the low-temperature $CoCl_2Py_2$ structural phase was identified as the γ-phase, no structural phase transition was identified in $NiCl_2Py_2$. The suspected low temperature transition was found to be magnetic in character.

Lipid systems have been studied using Moessbauer probes.[56] Sequential "melting" of reorientational motions of various system fragments was studied in lecithin and lecithin-cholesterol multilayer membrane systems over the temperature range 80-320 K. The probes used were ^{57}Fe labelled ferrocene, ferrocencarbaldehyde, and the cholesterol ester of ferrocenacetic acid. The temperature dependences of areas of the spectral lines was determined for probes localized in various microdomains of the lecithin matrix, and the effect of hydration on lipid matrix dynamics was investigated.

Emission Moessbauer studies using a ^{57}Co-doped single crystal of La_2CuO_{4-y} were reported.[57] Measurements made in the paramagnetic region (295 K) showed that the electric field gradient lies at an angle of 39°C to the *c*-axis and has a positive sign, and that the quadrupole splitting was 1.76 mm s^{-1}. Measurements at 78 K, in the antiferromagnetic region, showed a hyperfine field of 460 kOe lying in the basal plane of the crystal structure, also with a positive sign. Magnetic and Moessbauer studies were carried out on well characterized $LaCo_{1-x}Zr_xO_3$ (x = 0.1, 0.3, 0.5) at 80-700 K.[58] Cobalt ions were found to exist predominantly in the low-spin Co(III)($t_{2g}^6e_g^0$) and Co(II)($t_{2g}^6e_g^1$) states at 80 K, which partially transform into high-spin Co^{3+}($t_{2g}^4e_g^2$) up to ≈500 K. Above 500 K, when x = 0.3 or 0.5, Co^{3+} ions captured an electron from Co(II) and transformed into Co^{2+}($t_{2g}^5e_g^2$) and Co(III) ion pairs.

Coincidence Moessbauer, using K X-ray-gated and time-resolved emission Moessbauer spectroscopies, was used to study ^{57}Co-labelled $CoSe_4$ and $CoSeO_4.H_2O$.[59] A larger area intensity of ^{57}Fe(II) was observed for the K X-ray-gated spectrum than for the non-gate one. This difference was attributed to different local radiolytic effects of the deexcitation processes between the K X-ray and Auger electron spectra. Coincidence Moessbauer spectroscopy was also used in a study on ^{57}Co labelled $CoFe_2O(OAc)_6(H_2O)_3$.[60] Measurements were made at 78 and 298 K with three time-windows of 0-50, 50-150 and 150-300 ns. The results indicate that the ^{57}Fe atoms produced by electron capture decay were incorporated into a chemical environment similar to that of the parent Co atoms, forming a trinuclear $Fe^{II}Fe_2^{III}$ structure soon after the decay. Moessbauer emission spectroscopy was also used to study spin fluctuations in the frustrated spin system $CsCoCl_3$,[61] and to study the "Co-Mo-S" phase in carbon-supported CoS catalysts.[62]

Cluster formation between iron and lithium was studied by a wide range of techniques, including Moessbauer spectroscopy.[63] Iron and lithium ions were trapped in cold, frozen pentane and warmed until agglomeration took place. At room temperature small α-iron crystallites were embedded in a matrix of nanocrystalline lithium. Gentle oxidation of the pyrophoric powder gave α-Fe particles encapsulated in a $Li_2O/LiOH/Fe_2O_3$ coating. The effects of further heat treatments up to 470°C were described and possible applications of metastable clusters of normally immiscible metals were discussed.

A ferrisilicate analogue of zeolite BETA, which is almost aluminium free ($SiO_2/Al_2O_3 > 2000$), has been studied at room temperature and 4.2 K.[64] A single resonance observed at room temperature and 4.2 K,($\delta = 0.22$ mm s^{-1}; $\Delta = 0.32$ mm s^{-1} at 4.2 K) and the sharp narrow sextet obtained at 4.2 K in an applied field, show that most of the ferric ions are well dispersed into tetrahedral sites. Ferrisilicate analogues of ZSM-11 were also characterized.[65] Again the Fe^{3+} occupied tetrahedral sites. An *in situ* gamma-resonance study of framework substituted (Fe)ZSM-5 has been described.[66] The effects of various gas treatments on the substituted iron were described. The bound diffusion of ultrafine ferric hydroxide particles embedded in Dowex 50W-x1,2 cation exchangers was studied.[67] The Moessbauer data were treated as arising from continuous diffusion in an arbitrary potential.

Frozen Solutions, Liquid Crystals and Polymers. Gamma-resonance spectra have been obtained of iron(III) hydroxy complexes on frozen 15M NaOH solutions.[68] A symmetrical iron(III) doublet with chemical shifts of 0.68 and 0.56 mm s^{-1} was observed for these red-orange materials and compared with a broad single line at 0.68 mm s^{-1} observed for dry, colourless, samples prepared by evaporation at 120-140°.

The chemical consequences of the electron capture decay of ^{57}Co and the converted isomeric transition of ^{119m}Sn in frozen H_2O and D_2O have been compared. Basic conclusions were made concerning the effect of electron capture by the matrix on the distribution of the stabilization form of the daughter atoms. Experimental data on deuterated systems confirmed the model of coulombic fragmentation and acceptance of electrons.

Structural fluctuations in glass-forming liquids were studied through the Moessbauer spectra of $^{57}Fe^{2+}$ dissolved in a glycerol-water mixture at 80-275 K.[70] In the supercooled liquid state the spectral lineshapes depended strongly on temperature, and could be fitted to a jump-diffusion model with a Cole-Davidson distribution of fluctuation times. Comparison with results from other techniques showed that the α-relaxation of viscous liquids is responsible for the line broadening of the spectra. The presence of a second, β-relaxation, process was deduced.

Absorption and emission Moessbauer spectra of frozen samples of γ-Fe_2O_3 with co-precipitated cobalt and iron ions have been recorded in order to study the early stages of the coating process.[71] Surface effects on the γ-Fe_2O_3 were found to induce the oxidation of iron(II) to iron(III), and produce magnetic ordering in this material.

Moessbauer studies have been carried out on a 1% solution of 1,1'-dionanoylferrocene (DNF) in the smectic B liquid crystal material 4-*n*-butoxybenzylidine-4'-*n'*-octylaniline (40.8).[72] The observation of two quadrupole doublets, with the same splitting but very different intensities, led to the conclusion that the DNF is located in two different sites within the liquid crystal. Moessbauer data have also been reported for liquid-crystal disubstituted derivatives of ferrocene.[73] Debye temperatures, order parameters, and intramolecular and lattice contributions to the nuclear vibrational anisotropy were obtained.

A Moessbauer and IR spectroscopic study of complexes of iron(III) with the nitrogen-containing polymers poly(2-vinylpyridine) (P2V), poly(2-methyl-5-vinylpyridine) (P2M5V), poly(4-vinylpyridine) (P4V) and vinylimidazole-vinylpyrrolidone (VI-VP) copolymer was reported.[74] The complexing ability of the polymers followed the order: P2M5V > P4V > P2V and, unlike the copolymer, did not depend on the molar ratio of the initial components. The Moessbauer parameters are typical for high-spin compounds of iron(III). Ferric chloride complexes with poly(*N*-methyl-2,5-pyrrolylene), poly(2,5-thienylene) and poly(3-methyl-2,5-thienylene) have been studied.[75] Variable temperature Moessbauer spectra indicated the formation of $FeCl_2$ and $FeCl_4^-$. Moessbauer characterization of ferricyanide oxidized polypyrrole[76] and experimental evidence for a strong interaction between the polypyrrole matrix and an $FeCl_3$ dopant[77] was also presented.

The preparation of polyferro- and polycobaltsiloxanes has been described.[78] Moessbauer, ESR, and magnetic susceptibility measurements were used to characterize $\{[RSi(OH)_xO_{(3-x)/2}]_yMO_{n/2}\}_m$, ($M = Co^{2+}$, n = 2; $M = Fe^{3+}$, n = 3; R = Ph, vinyl; y = Si/M ratio in polymer) and some mixed Co-Fe analogues. These polymers were paramagnetic and the iron-containing materials exhibited antiferromagnetic interactions.

Intercalation Compounds containing Iron. A Japanese study on iron chloride-graphite intercalation compounds synthesized from $FeCl_3$-KCl melts has been described.[79] Room temperature Moessbauer spectra consisted of two intense sets of quadrupole doublets due to two different Fe^{2+} sites, which had the same isomer shift as anhydrous $FeCl_2$ but different quadrupole splittings. Iron(II) made up > 87% of all iron ions intercalated. The $FeCl_2$ intercalation process was discussed. Progressive reduction of the stage-1 graphite intercalation compound $C_{14}FeCl_{3.04}$ with *n*-butyllithium has been studied.[80] The zero valent iron layers of stage-1 which were formed contained small amounts of Fe^{3+} and Fe^{2+}. These reduced compounds had the average formula $C_{14}FeCl_{3-x}(LiCl)_x$, where x = 0-4. When x = 2 or 3 the compounds behaved as ferro-spin glasses.

Intercalation of 2-aminoethylferrocene (L) into the layered host lattices MoO_3, 2H-TaS_2, and α-$Zr(HPO_4)_2.H_2O$ was reported.[81] The Moessbauer spectrum of $MoOL_{0.36}$ was a doublet with the parameters $\delta = 0.442$ mm s^{-1} and $\Delta = 2.344$ mm s^{-1}, indicating the presence of unoxidized ferrocene molecules within the oxide layers. The superconducting intercalation compound $Fe_x(H_2O)_yTaS_2$ has been studied using X-ray, Moessbauer spectroscopic and dilatometric methods.[82] For the fully hydrated form, the Moessbauer spectra show a quadrupole doublet with hyperfine parameters similar to those found in hydrated octahedral iron(II) salts. Unequal intensities in the components of the doublet were thought to orignate from a tilting of the 4-fold octahedral axis with respect to the crystallographic *c*-axis.

Compounds of Iron.- High-spin Iron(II) Compounds. The ^{57}Fe and ^{119}Sn Moessbauer effects have been used in an investigation of the 728 K phase transition in $FeSnF_6$.[83] From the iron spectra, recorded over the temperature range 4.2-635 K, the existence of a high temperature phase transition was confirmed. Magnetic phases in $Fe_xNi_{1-x}Cl_2$ (for x = 0.06 and 0.10) were studied.[84] When x = 0.10 the iron(II) moments align along the crystallographic *c*-axis, but when x = 0.06, there was a temperature dependent mixture of parallel and perpendicular phases. Gamma-resonance spectra have been measured for $FeSO_4{\cdot}7H_2O$ and $(NH_4)_2Fe(SO_4)_2{\cdot}6H_2O$ from 80 K to room temperature.[85] Evidence was found for the occurrence of phase transitions.

I

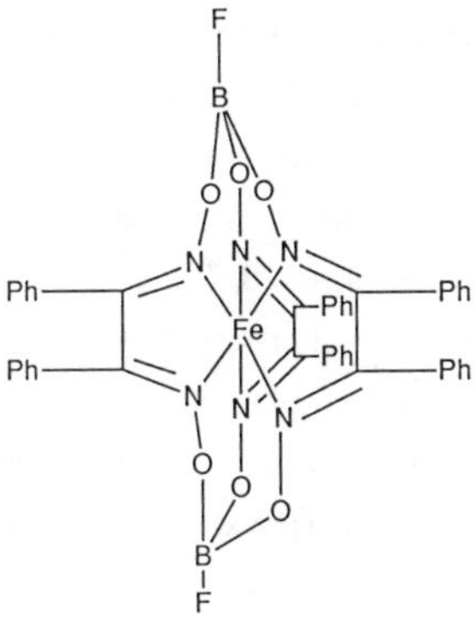

Nord and Ericsson have carried out a series of studies on mixed metal-iron phosphates.[86-88] Moessbauer spectra were presented for $(Ni,Fe)_2P_2O_7$ and α-$(Mg_{0.9}Fe_{0.1})_2P_2O_7$ and for the systems $(Fe,M)_2P_4O_{12}$ (M = Mg, Co, or Ni). For the tetrametaphosphates the preferences for site *M*1 over *M*2 followed the sequence: (Ni^{2+}, Co^{2+}) > Fe^{2+} > Mg^{2+}.[88]

A number of papers have described work carried out on phthalocyanine compounds of iron(II). Below 30 K, antiferromagnetic ordering and spin transfer effects were observed for iron(II) in iron oligophthalocyanine.[89] Magnetic dilution (reducing the Fe(II) fraction from 70% to 30%) left the hyperfine splitting unchanged, indicating that indirect interactions occur through the independent clusters of iron(III). In conjunction with EPR spectroscopy, the Cu+Fe analogue and the corresponding polyphthalocyanines were also studied.[90] The data showed that the metal was positioned either in the azaporphyrin macrocycle or outside of this ring.

Two Japanese papers have studied iron phthalocyanines in zeolite hosts. In the first, pyridine inactive FePc was found to remain with unchanged Moessbauer parameters, after formation of the $FePc(Py)_2$ adduct.[91] The ship-in-bottle synthesis was also used to prepare $FePc(CMe_3)_4$ in a NaY zeolite supercage.[92] Spectroscopic results showed the considerable distortion of the Pc planar structure. Other ferrous phthalocyanines for which Moessbauer data were reported include: $FeEt_4Pc$,[93] $Fe\{(RO)_8Pc\}$ (where $(RO)_8Pc$ = octaalkoxy-substituted phthalocyanines),[94] and monomeric and bridged phthalocyaninatoiron(II) complexes.[95,96]

Moessbauer spectra have been reported for L_2Fe, where L = hydrotris[3-(2'-thienyl)pyrazol-1-yl) borate.[97] While the hyperfine parameters are typical for a distorted high-spin octahedral iron(II) complex, the asymmetric line widths in the quadrupole doublet indicate the presence of a reduced relaxation rate for the effective paramagnetic hyperfine field. Clathrochelate tin-containing iron(II) dioxomates, $[FeQ_3(SnCl_3)_2](HL)_2$ (L = an amine), have been studied with the ^{57}Fe and ^{119}Sn Moessbauer effects.[98] According to the iron Moessbauer data, the complexes have a trigonal-antiprismatic geometry with a distortion angle of 40-55°. During the review period, γ-resonance spectral data were also reported for high-spin iron(II) complexes, including: clathrochelates with boron-containing dioximates **(I)**[99] and similar cyclooctanedione dioximates,[100] complexes with 2,9-di(*o*-alkoxyphenyl)-1,10-phenanthroline ligands,[101] and with 2,6-diacetylpyridine bis(acylhydrazones),[102] biphenyl-bridged complexes of binucleating ligands containing amide groups,[103] some Schiff-base iron(II) sulphates,[104] iron 1,2-benzenedithiolate complexes,[105] and the Fe(II) complex with pyrazine-2,3-dicarboxylic acid dihydrazide.[106]

High-Spin Iron(III) Compounds. The Moessbauer spectra of the α and β forms of $(NH_4)_2FeF_5$ have been studied in the temperature range 4.2-300 K.[107] While the α-form showed typical one-dimensional behaviour with a magnetic ordering temperature of 7.5 K, the spectra of the β-form (magnetic ordering at 13 K) revealed two different iron sites below 187 K. Moessbauer data for Na_2MgFeF_7, recorded in applied fields, have revealed an antiferromagnetic exchange between Fe^{3+} ions.[108] Below 4.2 K, however, a slow spin relaxation, consistent with a paramagnetic state, was found. Mixed fluorides in the series $Fe_{1-x}GaF_3$ (x = 0.1) have been studied.[109] The Moessbauer spectra were consistent with 6-coordinated high-spin iron(III), and high field measurements provided clear evidence for antiferromagnetic interactions. The authors also report antiferromagnetic behaviour, below T_N = 55 K, in $CsBaFe_3F_{12}$.[110]

Surface and bulk magnetic properties of Fe_3BO_6 crystals were studied with the Moessbauer effect in the critical-temperature region.[111] The surface magnetic ordering temperature was different from that in the bulk material. A new ferric arsenate, $FeAsO_4{\cdot}{}^3/_4H_2O$, has been prepared and studied.[112] Moessbauer and magnetic measurements indicated antiferromagnetic ordering below 49 K, and a paramagnetic Curie temperature of -128 K. Moessbauer data have been reported for the iron phosphates $RbFeP_2O_7$ and $CsFeP_2O_7$,[113] and $Fe_2(P_2O_7)(HPO_4)$ and $Fe_4(P_2O_7)_2$.[114]

A series of iron(III) complexes with substituted malonic acids have been prepared and characterized.[115] These complexes, with the formulae $H[FeL_2(H_2))_2]{\cdot}H_2O$ (L = malonate), $H_2[FeL_2(H_2O){\cdot}H_2O$, $Na_3FeL_3{\cdot}H_2O$, and $H_3FeL'_2{\cdot}nH_2O$ (H_2L = R-substituted malonic acid; R = Me, Et, Me_2), all have small quadrupole doublets (ΔE_Q = 0.43-0.84 mm s^{-1}), and isomer shifts (δ = 0.62-0.76 mm s^{-1}, relative to sodium nitroprusside) typical of high-spin iron(III). The same authors prepared the carboxylate complexes $Na_3[Fe(OOCR)_6$ (R = Me_3C, $Me(CH_2)_5$, $Me(CH_2)_{10}$,

2,5-$Cl_2C_6H_3$, 2,5-$Me_2C_6H_3$, 2-$MeOC_6H_4$, and 3-$MeOC_6H_4$), and Na_3FeL_3 (H_2L = pyridine-2,6-dicarboxylic acid, quinic acid (H_2L') and Na_4FeL'.[116] All gave quadrupole doublets except for the quinic acid complexes which contained a small fraction of iron(II). The effects of a bridging carboxylate ligand on (μ-oxo)bis(μ-carboxylato)diiron(III) complexes were studied,[117] and the Moessbauer parameters of bis(pyridine-2,6-dicarboxylato)iron(III) dihydrate were reported.[118]

Paramagnetic relaxation effects in mixed crystals have been studied by Moessbauer spectroscopy.[119] All the crystals studied, (Fe,Al)(acac)$_3$, (Fe,Co)(acac)$_3$, $K_3[(Fe,Al)(ox)_3]_3 \cdot H_2O$, and $NH_4(Fe,Al)(SO_4)_2 \cdot 12H_2O$, contained high-spin compounds in a diamagnetic host. The factors determining the paramagnetic relaxation time, other than the Fe-Fe distance and the temperature, were examined using the Moessbauer linewidth as an indicator. Relaxation effects were also examined in $[Fe_2L(H_2O_4)](ClO_4)_4 \cdot H_2O$, which contains pairs of Fe^{3+} ions within a binucleating seven-coordinate macrocycle.[120] Relaxation effects were present from 4.2-300 K, and application of a 3 T field at 4.2 K showed that V_{zz} is positive and $\eta \approx 0$. Fitting to a stochastic model gave relaxation times of 10^{-9}-10^{-10} s.

Other complexes containing high-spin iron(III) and for which Moessbauer data were presented during the review period include: a series of complexes with five-coordinate Schiff base ligands;[121] four Fe(III) chelates with *o*-substituted azo dyes;[122] Fe(III) complexes with dihydroxamic and aminohydroxamic acids;[123] Fe(III) complexes with imides as primary and N-aryl-N′-benzoyldithiocarbamides as secondary ligands;[124] the tetranuclear iron-oxo complexes $[Fe_4O_2(OOCR)_7(bpy)_2]ClO_4$ (R = Me, Ph);[125] tetraoxolene radicals stabilized by Fe^{3+},[126] $[FeL(HL)\text{-}py_2] \cdot 2L'$ (where H_2L = α-benzildioxime, L′ = 1-propen-2-ol);[127] tri(dibenzoylmethane)Fe^{3+};[128] mono- and binuclear iron(III) complexes with 2,6-diacetylpyridine bis(acylhydrazones);[129] the antiferromagnetically spin-coupled $[(Cu(A)Fe(salen)Cu(A)]ClO_4 \cdot 2H_2O$ (where A = an imidazolate);[130] soluble iron(III) phthalocyanines;[131] ferric complexes with D-glucopyranosyl esters of glycine;[132] and $FeCl_3$ complexed with 1,2,4-triazole-3-thiones;[133] and pyridine.[134]

Mixed Valence Compounds and Unusual Electronic States. The mixed valence diphosphate $Fe_3(P_2O_7)_2$ has been prepared and characterized.[135] This compound is antiferromagnetic with ferromagnetic couplings within the Fe_3O_{12} clusters. The Moessbauer data was characterized by an Fe^{2+} splitting of 4.32 mm s^{-1} at 300 K. An iron-phosphate glass with the nominal composition $25Fe_2O_3$-$75P_2O_5$ was found to contain both Fe(II) and Fe(III).[136] Where the surface of the glass was exposed to air, the fraction of Fe(II) was higher than in the bulk. Crystal structures and

Moessbauer spectra were reported for the mixed valence phases $Zn_2(Zn_{1-x}Fe_x)Fe^{III}(PO_4)_3{\cdot}2H_2O$,[137] and $(Me_4N)_2H_{2-x}[Fe^{II}_{2-x}Fe^{III}_x(H_2O)_6UMo_{12}O_{42}]nH_2O$.[138]

Solvated and non-solvated mixed-valence trinuclear iron stearates have been prepared and their Moessbauer spectra studied.[139] A temperature-dependent mixed-valence phase observed in the non-solvated compound was absent from the solvated compound. Magnetic and Moessbauer measurements have been made on homo- and heterovalent iron(II,III) tartrate and citrate complexes.[140] The mixed-valent complexes were characterized by electron delocalization and exchange between the iron centres.

Valence trapping in the oxo-centred complexes $[Fe_2O(OAc)_6(py)_3]{\cdot}py$ and $[Fe_3O_2(OAc)_6(py)_3]{\cdot}$-$CHCl_3$ has been studied in a diamond anvil cell at pressures up to 95 kbar at 298 K.[141] Below about 20 kbar both complexes were represented by a single quadrupole doublet, indicative of rapidly interconverting valence states. Above 80 kbar, both complexes were valence trapped with 2Fe(III):Fe(II), while intermediate pressures combined both types of spectra. A phase diagram was derived using a spin-Hamiltonian approach employing a molecular field approximation for the intermolecular interactions.

Gamma-resonance spectroscopy has also been used to provide information about the mixed-valence complex $[Fe_30(OAc)_6(TACN)_3]{\cdot}CHCl_3$ (TACN = 1,4,7-triazocyclononane), which contains bidendate acetate ligands,[142] and on the related ^{57}Co-labelled $[CoFe_2O(CH_2XCOO)_6(H_2O)_3]$ (where X = Br, I).[143]

Molecular orbital calculations and Moessbauer spectroscopy were used to study the electronic states in *trans*-$[Fe(o\text{-}C_6H_4(AsMe_2)_2)_2Cl_2]^{2+}$, which contains iron(IV).[144] The calculated hyperfine parameters for the $^3B_{2g}$ and $^3B_{3g}$ states, $\delta = 0.20$ mm s^{-1}, $\Delta = 2.83\text{-}2.84$ mm s^{-1}, were in good agreement with the experimental values, $\delta = 0.17$ mm s^{-1} and $\Delta = 3.15$ mm s^{-1}.

Spin-Crossover Systems and Unusual Spin States. Studies on the spin-crossover transition in Fe(II) triazole complexes were reported by two groups. The polymeric complexes $FeL_3X_2{\cdot}nH_2O$ (L = 1,2,4-triazole, 4-amino-1,2,4-triazole; X = Br, BF_4, ClO_4) have a distorted octahedral structure.[145] They showed a reversible transition from the low-spin $^1A_1$ (S = 0) to the $^2T_2$, (S = 2) state, which is confirmed by their thermochromism (rose → white). In $[Fe(NCS)_2(4,4'\text{-bis-}1,2,4\text{-triazole})_2]{\cdot}H_2O$, the Fe(II) ion has distorted tetrahedral symmetry, and shows a sudden high-spin ⇀ low-spin at transition at 123.5 K on cooling and 144.5 K on warming.[146] Moessbauer ligand-field spectra and magnetic behaviour of both the hydrated and non-hydrated compounds were discussed. A thermal- and light-induced spin transition from a high-spin state into a metastable low-spin state has been found in $[Fe(mtz)_6](CF_3SO_3)_2$, where mtz = 1-methyl-1*H*-tetrazole).[147]

Single crystal X-ray, EXAFS, XANES and Moessbauer methods have been used to characterize FeL_2.MeOH, where HL = 5-nitrosalicylidene-2-pyridylethylamine.[148] This Schiff base complex showed a thermally induced $^5T_{2g} \rightleftharpoons {}^1A_{1g}$ spin conversion with a several unique features, *viz*: an Fe(II) centre in an N_4O_2 ligand environment, and a discontinuous spin-conversion with two stages 30 K apart, separated by a spin equilibrium zone, in which there are equal concentrations of high- and low-spin molecules.

Moessbauer, EPR and magnetic methods were used to study an iron(II) complex with the *N*-analogue of the macrocyclic ether 18-crown-6.[149] A thermally controlled intramolecular reverse electron transfer was found for the complex of intermediate-spin iron(II) (S = 1) $\rightleftharpoons$ high-spin iron(III) (S = 5/2).

The temperature and pressure dependences of the Moessbauer spectra have been studied for the iron(III) trisdithiocarbamate complexes Fe[R_2dtc], where R = *N,N'*-disubstitution with Me, Et, *i*-Pr, benzyl, *n*-hexyl, and cyclohexyl groups.[150] Spectra recorded from 1.3 or 4.2 K to room temperature indicated that the hyperfine parameters for the high-spin and low-spin states at equilibrium in some of these complexes are similar. At lower temperatures the lineshape revealed the onset of paramagnetic relaxation. Room temperature pressure dependences were measured at up to 80 kbar in a diamond-anvil cell. These compounds are completely low-spin even at pressures of a few kilobars. The pressure dependence of the Moessbauer absorption area was modelled with by the Debye theory of lattice vibrations.

Maeda *et al* have investigated the spin-crossover behaviour of several ferric complexes. Two crystal forms exist for $[FeL]BPh_4 \cdot Me_2O$, where H_2L = bis(salicylidene)triethylenetetramine.[151] Above 200 K, the monoclinic form was in a transition spin-state between high- and low-spin, while twin crystals were high-spin over the temperature range 78 to 320 K. Both $[Fe(bzpa)_2]ClO_4$ (where Hbzpa = (1-benzoylpropen-2-yl)(2-pyridylmethyl)amine) and $[Fe(acen)(3,4\text{-}Me_2pyridine)]_2$-$BPh_4$ (H_2acen = ethylenebis(acetylacetone)) were found to be $S = {}^5/_2 \rightleftharpoons S = {}^1/_2$ complexes.[152] The bzpa complex gave time-averaged Moessbauer spectra. Crystallographic, magnetic and Moessbauer studies were made on $[Fe(acpa)_2]X$, where Hacpa = *N*-(1-acetyl-2-propylidene)(2-

pyridylmethyl)amine; X = BPh_4^- and PF_6^-.[153] The Moessbauer data implied a rapid spin conversion rate for the BPh_4^- compared with the Moessbauer lifetime and with the PF_6^- salt.

Discontinuous spin transitions between the high-and low-spin states of $(Et_4N)_2\{[(tsp)(tspH)Fe]_2$-succinate} and K[Fe(tsp)(tspH)propionate]·$3H_2O$ have been characterized.[154] Between 300 and 390 K, both compounds show a discontinuous transition from the low-spin state to the high-spin state, returning slowly to low-spin after cooling to room temperature. A significant asymmetry in the Moessbauer spectra was explained by Blume-type relaxation effects. A discontinuous $^6A_1 \rightleftharpoons {}^2T_2$ spin transition found in Fe(HPhthsa)Cl_2 (H_2Phthsa = pyridoxal 4-methylthiosemicarbazone) was also characterized by magnetic and Moessbauer measurements.[155] Two macrocyclic complexes were found to undergo spin transitions:[156] $[Fe(TAAB)(MeCN)_2][BF_4]_2$ (TAAB = tetra-benzo[*b,f,j,n*][1,5,9,13]tetraazacyclohexadecine) exist as in a spin equilibrium of S = 2 $\rightleftharpoons$ S = 0; and $[Fe(TAAB)(MeCN)(NO)][BF_4]_2$ has a $^3/_2 \rightleftharpoons {}^1/_2$ spin-crossover near 160 K.

Magnetic susceptibility measurements, and Moessbauer and FTIR spectroscopies have been used to provide evidence for the intermediate spin (S = 1) state in Fe(4,4'-dpb)$_2X_2$ (dpb = 4,4'-diphenyl-2,2′-bipyridine; X = NCS, NCSe).[157] While the spin states below 325 K could not be characterized, at higher temperatures, the FTIR data can be accounted for by the increasing population of a high-spin (S = 2) state and the Moessbauer spectra were accounted for by a rapid relaxation between two spin states. The existence of an intermediate spin (S = 1) state for Fe(II) is attributed to a major distortion from octahedral symmetry of the metal orbitals, allowing a (near) degeneracy of the d_z2 and d_{xy} levels around the iron.[158]

An iron(III) complex with the unusual S = 5 ground state has been reported twice.[159,160] The hexanuclear complex $[Fe_6O_2(OH)_2(OAc)_{10}Q_2]$ (HQ = 1,1-bis(*N*-methylimidazol-2-yl)ethanol) contains 2 μ_3-O Fe^{III}_3 complexes bridged together at two vertices. The Moessbauer parameters were fitted to two high-spin iron(III) doublets in a 2:1 area ratio with δ = 0.383 and 0.406 mm s^{-1} and ΔE_Q = 0.729 and 1.1056 mm s^{-1} respectively at 300 K. A full fit of the magnetization data established the S = 5 ground state.

Acetonitrile solutions of $[Fe^{III}Cu^{II}(BPMP)Cl_2](BPh_4)$ (BPMP = 2,6-bis[2-pyridylmethyl)amino)-methyl]-4-methylphenol) were studied with Moessbauer and EPR spectroscopies.[161] Both techniques showed that the complex is ferromagnetically coupled with an S = 3 spin ground state. Analysis of the Moessbauer spectra in applied fields up to 6 T, yielded the zero-field splitting parameters D_1 = +1.2 cm^{-1} and E_1 = 0.11 cm^{-1}, and δ = 0.48 mm s^{-1}, ΔE_Q = 0.67 mm s^{-1}, and magnetic hyperfine coupling constant A_0 = -28.8 MHz. A method was used which allows the determination of the spin of the coupled system directly from the Moessbauer spectra.

Mixed-spin relaxation processes have been applied to paramagnetic Fe^{3+} (^6S) ions have been calculated and compared with experimental Moessbauer magnetic hyperfine data obtained at low temperatures.[162] Mixed-spin behaviour was also found in $[FeL_3][FeCl_4]$, $[FeL_3][Fe(NCS)_4]$, and $[FeL_3][FeCl_4]_2$, where L = 3,3'-bipyridazine.[163] Characterization by magnetic and gamma-resonance methods showed that these were mixed low-spin cationic, high-spin anionic systems.

Low-Spin and Covalent Compounds. The Moessbauer spectroscopic quadrupole splittings for a series of six-coordinate low-spin iron(II) complexes containing bidentate phosphine ligands have been examined.[163] Partial quadrupole splittings (PQS) were used to assign *cis* and *trans* geometries to several complexes. The bidentate phosphine ligands were found to have bonding properties (both electronic and steric) that lie between those of triorganic phosphines and those of phosphites which have a similar stereochemistry around the phosphorus atoms.

Moessbauer spectra of alkyl- and arylamine coordinated pentacyanoferrate(II) complexes have been studied.[164] All exhibit a well-resolved quadrupole doublet. An unusually high quadrupole splitting was observed for the isobutylamine coordinated complex, ΔE_Q = 1.75 mm s^{-1}, compared with 0.67-0.86 mm s^{-1} for the other complexes. Partial isomer shift and δ were correlated with the partial quadrupole splitting. The Moessbauer data for a range of low-spin iron(II) complexes, $[FeXY(diphosphine)_2]^{n+}$ (n = 0 or 1; X and/or Y = Cl, Br, MeCN, MeNC, H, N_2, H_2, CO, etc) are reported.[165] The isomer shifts and quadrupole splittings were analyzed in terms of ligand partial values, assuming that addivity holds, and the partial values are discussed in terms of the bonding properties of the ligands. While the partial isomer shifts were of limited use, because of the small range of values, partial quadrupole splitting values were more informative, and best calculated relative to p.q.s.(H) = 0. Spectra for $FeL_2(H_2PO_4)_2$ and $Fe(phen)(H_2PO_4)_2$ (L = pyridine or γ-picoline, phen = 1,10-phenanthroline), were also reported.[166]

The recoilless absorption probability factor, f', and recoilless reemission, f'', have been measured for low-spin iron(III) in $Na_2[Fe(CN)_5NO]\cdot 2H_2O$ single crystals, and found to be different with $f'' > f'$.[167] The π-bonding ability of imidazole has been studied using *catena*-$[Mg(H_2O)_2$-$(1\text{-Meim})_2$-$(\mu\text{-CN})Fe(CN)_4(1\text{-Meim})]\cdot H_2O$, where 1-Meim = 1-methylimidazole.[168] This complex gave a very large quadrupole splitting of 2.62 mm s^{-1} at 291 K. A highly anisotropic low-spin iron(III) compound, $[Fe(L)_2](BF_4)_3$ (L = bis[tris(*n*-methylimidazol-2-yl)hydroxymethane), was studied in frozen solution.[169] Unusually, it shows a sharp six-line pattern, giving an overall splitting of 14.2 mm s^{-1} in an applied field at 4.2 K. Moessbauer spectra for low-spin iron(III) in [Cu(L)Fe(acen)Cu(L)]ClO_4]ClO_4·$2H_2O$, where H_2L = 4-(6-methyl-8-oxo-2,5-diazanonane-1,5,7-trienylimidazole and H_2acen = *N,N'*-bis(acetylacetonylidene)ethylenediamine, were also reported.[130]

A series of twenty *trans*-$Fe(CO)_3L_2$ complexes (L = phosphine or phosphite) have been characterized by Moessbauer, IR and ^{31}P NMR spectroscopies.[170] A linear correlation between the isomer shifts and quadrupole splittings showed that the P-to-Fe σ donation is offset by Fe-to-P π back donation. The important role of π back-donation in the Fe-P bonding was shown, and the strong interaction between *trans*-P ligands was demonstrated. Gamma-resonance spectra were also reported for monomeric $Fe(CO)_3$-SCPh-N-CH(OOCEt) and dimeric $Fe_2(CO)_6$-SCPh-N-CH-(OOCEt),[171] the mixed-metal cluster compounds $Fe_2(CO)_6(\mu\text{-}CO)(\mu\text{-}PPh_2)[\mu\text{-}M(PPh_3)]$, where M = Cu, Ag, Au;[172,173] and $[cy_2NH_2][HFe_3(CO)_{11}]\cdot OEt_2$ (where cy = cyclohexyl).[174]

Silver and his co-workers have studied correlations between the ^{13}C and ^{57}NMR data for iron(II) cyclopentadienyl and arene sandwich compounds and their ^{57}Fe Moessbauer quadrupole splittings.[175] For substituted neutral ferrocenes there is a linear relationship between the quadrupole splitting and the ^{57}Fe NMR chemical shift. For ferrocenyl carbenium ions, the relationship suggested a change in structure and bonding from species like $[Fe(\eta\text{-}C_5H_5)(\eta\text{-}C_5H_4{}^+CMe_2)]$ to the more fulvenoid structure $[Fe(\eta\text{-}C_5H_5)(\eta\text{-}C_5H_4CH_2)]^+$. Bridged ferrocenes (ferrocenophanes) showed effects which were associated with ring tilting. Weak ferrocenyl Fe-X (X = Ge, Si, P, or Pd) interactions in [1]ferrocenophanes or 1,2,3-trithia[3]ferrocenophanes resulted in quadrupole splittings about 0.4 mm s^{-1} smaller than that of ferrocene.[176] While increased splittings relative to ferrocene were found for ferrocenyl iron bonded to Hg or H^+, compounds of Pd and Pt trichalcogenoferrocenophanes showed little variation in quadrupole splitting but higher isomer shifts. Ring-tilting effects were observed in both systems.

The 17-electron iron(III) complexes $[Fe(Cp^*)(dppe)X]PF_6$, where Cp^* = η^5-pentamethylcyclopentadienyl, dppe = 1,2-bis(diphenylphosphino)ethane and X = alkyl or halide, has been studied.[177] ESR, Moessbauer and NMR experiments indicated that the Cp^* ring and the alkyl or halide ligands contribute significantly to the delocalization of the odd electron. Gamma-resonance spectroscopy was used to characterize the thermal decomposition products of $[FeCp(SPh)CO]_2$,[178] to study molecular reorientation dynamics in $(thiourea)_3$·ferrocene,[179] and to an investigation of ring-oxadiethylene-bridged bis(cyclopentadienyl) chlorides of Ti and Zr.[180]

II

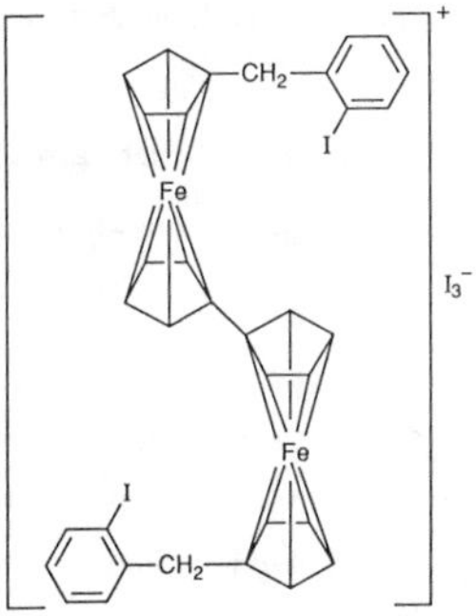

Effects of structural variations on the electron delocalization in mixed-valence biferrocenium systems have been investigated.[181-185] The distortion of the cyclopentadienyl rings from the parallel positions, resulting from ligand substitutions, was found to correlate well with the critical temperature for the delocalization-localization transition.[181] Effects of substituents on the intra-molecular electron transfer rate were explored.[184,185] This was demonstrated for 1′,1‴-dibenzylbi-ferrocenium triiodide (**II**), which crystallizes into two crystal habits. The temperature dependencies of the Moessbauer spectra of these two polymorphs were quite different. While the needle-like phase gave spectra characteristic of a valence-detrapped mixed-valence species at temperatures down to 25 K, the plate-like phase showed valence-trapped behaviour even at 300 K.

A fusion-type averaging process of the mixed-valence state has been found from the Moessbauer spectrum of 1',1'"-diethylbiferrocenium triiodide.[186] A structural change in the anion and a heat capacity anomaly above ~200 K were closely associated with the valence-detrapping of the cation. Also reported were the Moessbauer spectra of $[TCNE]^{\cdot-}$ and $[TCNQ]^{\cdot-}$ salts of $[Fe(C_5Et_5)_2]^{\cdot-}$,[187] and of $[FcCH{:}NR][TCNQ]_2$ salts (R = naphthyl, 5,6,7,8-tetrahydronaphthyl, anthracenyl, pyrenyl).[188]

Biological Systems and Related Compounds. Moessbauer spectroscopy and titration methods were used to study the binding between protoporphyrin IX and a series of pyridine ligands and imidazole.[189] With the exception of pyridine-*N*-oxide, all nitrogen bases yielded low-spin octahedral complexes. The quadrupole splitting was related to the magnitude of the overall binding constants. Frozen solution Moessbauer spectroscopy was used to study $[(PPIX)Fe^{II}$-$(THF)_2)]$.[190] This complex was high-spin (S = 2) with $\Delta E_Q = 2.49$ mm s^{-1}.

Bisaxially coordinated (tetraphenylporphinato)iron(II) complexes of the type $(TPP)FeL_2$ have been prepared and their bridged structure confirmed by Moessbauer spectroscopy.[191] A range of β-substituted iron(III) TPP derivatives were also studied.[192] The complexes $[Fe(NO_2)(L)(TPivPP)]$

(where H_2TPivPP = α,α,α,α-tetrakis(*o*-pivaloamidophenyl)porphyrin; L = pyridine, imidazole) have been shown to be low-spin ferric porphyrinates.[193] The related complex, low-spin five-coordinate [KL][Fe(NO_2)(L)(TPivPP)]·H_2O·PhCl, where L = Kryptofix-222, was also studied.[194] This complex has an isomer shift of 0.41 mm s^{-1} and an unusually large quadrupole doublet of 2.28 mm s^{-1} at 4.2 K.

Hindered porphyrin systems with a range of axial ligands with controlled orientations have been prepared.[195] Moessbauer data were used to obtain crystal field parameters for the low-spin ferric complexes [Fe(TMP)L_2]ClO_4 (L = 4-(dimeythylamino)pyridine; H_2TMP = *meso*-tetramesityl-porphyrin); [Fe(OEP)L_2]ClO_4 (H_2OEP = octaethylporphyrin); and [Fe(TMP)L'_2]ClO_4 (L' = 1-methylimidazole). Moessbauer data were also reported for both-faces hindered and highly symmetrical 5,10,15,20-tetrakis[2',6'-bis(pivaloyloxy)phenyl]porphyrinatoiron(II).[196]

Parak has used ^{57}Fe Moessbauer spectroscopy and X-ray methods as complementary techniques in the investigation of protein dynamics.[197,198] While X-ray analysis measures structural distributions without any time resolution, Moessbauer spectroscopy labels dynamics on a time scale faster than 100 ns. Even when extrapolated to 0 K, myoglobin has no well-deformed structure but rather a number of conformational substates. The dynamic properties of ^{57}Fe myoglobin crystals were found similar to those in glycerol crystals. Above 200 K, the shape and area of the spectra could be explained by a jump diffusion model. In the high-viscosity region well below 200 K, the myoglobin molecules had a similar structure, indicating a highly degenerated ground state energy. Haemoglobin dynamics in ^{57}Fe-enriched rat erythrocytes were shown to be similar to those previously found in myoglobin crystals and other protein systems.[199] The results strongly indicated the heme motion was mainly influenced by the average viscosity of the sample, suggesting that these properties can be simply extrapolated into their physiological surroundings.

Murine tetrameric oxyHb and insect monomeric erythrocruorin were studied.[200] The profiles of the oxyHb spectra did not depend on the nature of the samples (whole blood, or aqueous or water-glycerol solution), or the rate at which the solutions were frozen. These spectra were well described by two doublets with equal δ and $\Gamma_{1/2}$ values, but with different ΔE_Q and relative intensities. Preliminary results of a Moessbauer effect study of the effects of γ-irradiation on human adult oxyHb in erythrocytes have been published.[201] After dosing, five spectral components were fitted. Two components were identified as oxyHb and deoxyHb and the other three were attributed to hematin and/or μ-oxodimers, metHb hydroxide and/or hemichromes, and a high-spin iron(III) complex. Changes in the component ratios with dose, up to 600 kGy, were evaluated. The Moessbauer effect was also used to study amphiphilic heme derivatives having four steroid groups.[202]

The dinuclear iron centre of the hydroxylase component (protein A) of methane monooxygenase from *Methylococcus capsulatus* has been studied by EXAFS, Moessbauer, and EPR spectroscopy.[203] For $Fe^{II}Fe^{II}$ hydroxylase samples, Moessbauer showed that 85% was fully reduced. A diiron(II) model compound with features relevant to the active sites of the reduced forms of the polyiron-oxo proteins hemerythrin, ribonucleotide reductase, and methane monooxygenase has been studied.[204] Two overlapping quadrupole doublets showed the inequivalence of the iron atoms. A model intermediate for non-heme iron hydrogenases, $[Fe(TPA)0]^{3+}$ (TPS = tris(2-pyridylmethyl)amine) was characterized.[205] The Moessbauer data, ΔE_Q = 0.52 mm s^{-1} and δ = 0.07 mm s^{-1}, show that it is a Kramers species containing the non-Kramers ion Fe^{IV} [sic], and has electronic features similar to heme peroxidase compounds.

A Moessbauer investigation has been carried out on iron in *Pseudomonas aeruginosa*. Cells grown under different conditions were found to have different forms of iron predominating.[206] The iron core of bacterioferritin was not altered on isolation, but contained a significant proportion of its iron as small clusters during the early stage of the stationary phase of cell growth.[207] Gamma-resonance data from frozen solutions of ovotransferrin were obtained at various temperatures in applied magnetic fields.[208] The results were fitted to a simple model which required admixture of the free ion ^6S and ^4P states, indicating a weak cubic crystal field.

The photocycle and structure of iron-containing bacteriorhodopsin was studied by Moessbauer spectroscopy.[209] Two iron environments were found, *viz*: a high-spin Fe(III) bound to the acid side chains of the protein or the phosphate groups of the lipid and close-coupled at 4 K; and high-spin Fe(II) linked to the carboxy groups of the protein. The observed dynamics showed that the purple membrane becomes flexible only above 220 K.

Moessbauer and magnetic properties of iron(II,III) pyropheophytin chlorides were studied.[210] The central iron(III) was high-spin and the temperature-dependent asymmetry of the Moessbauer spectra were interpreted in terms of slow electronic spin relaxation. These parameters were similar to those of iron(III) porphyrins. These results could be applied to the electronic structure of iron-substituted chlorophylls.

Reaction centres in the photosynthetic bacterium *Rhodopseudomonas viridis* were investigated.[211] The cytochrome irons were low-spin Fe(III) and the non-heme iron of the electron acceptor was partly high-spin Fe(II) and partly low-spin Fe(II) (or high-spin Fe(III)). The redox state of iron associated with the oxygen-evolving core of the cyanobacterium *Phormidium laminosum* was identified.[212] At 77 K two quadrupole doublets were found. A high-spin ferrous site, which oxidized to low-spin ferric, was identified as the iron of the Q-Fe complex and a low-spin ferrous site was identified as the cytochrome b559. Spectra were recorded for the PS II particles from the thermophilic cyanobacterium *Synechoccus elongatus*.[213] The Moessbauer parameters of the three doublets are characteristic of proteins with an Fe-S centre (δ = 0.40, ΔE_Q = 0.85), non-heme iron of the reaction centre of higher plants (δ = 1.35, ΔE_Q = 2.35), and of the oxidized cytochrome b559 (δ = 0.25, ΔE_Q = 1.65).

Lattice effects in the Moessbauer spectra of salts of the iron-sulphur cluster $[Fe_4S_4(S\text{-}t\text{-}Bu)_4]^{2-}$.[214] While the isomer shifts were relatively constant, the quadrupole splittings were cation dependent. The coordination of alkyl esters of amino acids to $\{Fe_4S_4\}^{2+}$ clusters was studied by the same researchers.[215] These data were applied to the binding of such clusters in biological systems. Gamma-resonance data were reported for the high-potential Fe-S protein from *Chromatium vinosum*, which contains a cubane prosthetic group which oscillates between the +3 and +2 states.[216]

The tetrameric Fe-S cluster compounds $[Fe_4S_6(SEt)_4]^{4-}$ and $[Fe_4S_6(L)_2]^{4-}$ (where H_2L = *o*-durenedithiol) have been described and their Moessbauer spectra presented.[217] The electronic ground states in a series of Fe-S clusters have been studied.[218] Of these mixed-valence compounds $Fe_7S_6(PEt_3)_4Cl_3$ has the monocapped prismane structure and $Fe_6S_6(PEt_3)_4X_2$ (X = Cl, Br, I, PhS) and $Fe_6Se_6(PEt_3)_4Cl_2$ have a "basket" configuration based on the $[Fe_6(\mu_2\text{-}S)(\mu_3\text{-}S)(\mu_4\text{-}S)]^{2+,1+}$ core unit. The Moessbauer spectra were analyzed in terms of a 1:1:1 iron site population. Iron-selenium-thiolate cluster compounds have been prepared and characterized by methods including Moessbauer spectroscopy.[219] Clusters containing the $Fe_2(\mu_2\text{-}Se)_2$, $Fe_3(\mu_2\text{-}Se)_4$, and $[Fe_4(\mu_3\text{-}Se)_4]^+$ cores were described. A cuboidal mixed-metal complex, $[Mo_3FeS_4(H_2O)_{10}](p\text{-}MeC_6H_4SO_3)_4\cdot 7H_2O$, has been studied.[220] Zero-field Moessbauer spectra at 4.2 K were consistent with spin-coupled Fe(III), with an effective overall zero spin.

The action of *Azobactor vinelandii* intact cells on highly dispersed iron hydroxides has been monitored.[221] Not more than 10% of the iron was bound to the cells. The iron sulphide produced by the action of *Desulphobulbus propionicus* has been determined.[222] Moessbauer spectroscopy was also useful in determining the number of sulphate-containing bacteria. Spectral, magnetic, Moessbauer, and chemotherapeutical properties of iron(III) complexes with some isonicotinic acid hydrazide derivatives were reported.[223]

Oxide and Chalcogenide Compounds of Iron. - Hydroxides. Moessbauer phase analysis has been used in the construction of potential-pH diagrams for the iron-water system. The formation of the transient compound green rust 2 was examined[224] and the oxidation and end product of $Fe(OH)_2$ in basic sulphate solution was investigated.[225] In another study of the oxidation of $Fe(OH)_2$ at high pH the presence of an $^{IV}Fe^{3+}$ component was identified.[226] This small component increased progressively with the oxidation of the $Fe(OH)_2$, maximizing with the onset of α-FeOOH formation. The results suggested that $Fe(OH)_4^-$ ions could be the building blocks for α-FeOOH in alkaline solution. The formation of small FeOOH particles from $FeSO_4$ was also studied.[227]

Ferric hydroxides and hydroxo complexes have been studied along with their behaviour in alkaline electrolytes.[228] Under certain conditions soluble ferric hydroxo forms in strong alkali show a transformation from polynuclear forms, which are characterized by a symmetrical Moessbauer doublet, to mononuclear hydroxo complexes which are characterized by a symmetrical singlet with the same parameters.

A Moessbauer study of the thermal decomposition of lepidocrocite was described.[229] The transitions from lepidocrocite to maghemite and haematite (γ-FeOOH $\rightarrow$ γ-Fe_2O_3 $\rightarrow$ α-Fe_2O_3) were monitored. variable temperature Moessbauer spectra for maghemite with extremely small particles showed superparamagnetism over a wide temperature range. Even below 55 K, two components were resolved from the magnetically split spectra, even in an applied magnetic field the Δm_I = 0 transitions were present, indicating a strong spin canting with respect to the applied field direction. Moessbauer spectra of fine feroxyhyte (δ'-FeOOH) particles have been recorded at 4.2 K in strong magnetic fields.[230] The behaviour was consistent with ferrimagnetism.

Wuestite and Related Compounds. Gamma-resonance spectra have been reported as part of an investigation into a novel method for the passivation and thermal stabilization of unstable 3d-oxides such as FeO.[231]

Haematite, Maghemite and Related Oxides. Three natural haematites, α-Fe_2O_3, were studied in the temperature range 80-400 K.[232] At low temperatures antiferromagnetic-like and weakly-ferromagnetic-like spin states co-existed in all three samples. The spin structures were related to the crystallinity of the samples and a gradual reorientation of both spins towards the basal plane was found in the Morin-transition region. Moessbauer and X-ray absorption spectroscopies have been used to study the local environment of iron in six heavy ion-irradiated iron oxides.[233] Simulations of the Moessbauer spectra suggested the presence of fivefold coordinated iron in the amorphous irradiated compounds, with a distribution of magnetic interactions due to variations in the number of iron second-nearest neighbours.

The effects on the Morin transition of coating α-Fe_2O_3 with Ru, Ni, Mn, Co and Fe have been studied.[234] The effective anisotropy constants and hyperfine fields of fine particles coated with Ru, Ni or Mn were larger than those of particles without these metals. These results were explained in terms of an increase in the surface anisotropy on adsorption. Sodium-modified ferric oxide was manufactured by the sol-gel method.[235] The phase transition from γ-Fe_2O_3 (maghemite) to α-Fe_2O_3 was shifted to higher tempeatures by substituting a small amount of sodium.

Two papers have examined the preparation of γ-iron(III) oxide by the thermal decomposition of iron(II) carboxylates. Whether the starting material was ferrous tartrate[236] or ferrous malonate,[237] the metastable γ-Fe_2O_3 was finally converted to α-Fe_2O_3. Samples of γ-Fe_2O_3 doped with Co or Gd have been studied.[238] Moessbauer spectroscopy showed that Gd occupied A and B sites, and in Co-doped samples the effective magnetic fields were different at the two sites. Multi-domain clusters were found samples which have a multi-domain configuration.

Several papers have described work on γ-Fe_2O_3 based magnetic recording materials. The magnetization distribution in γ-Fe_2O_3 recording materials has been studied with the Moessbauer effect.[239] A spatial distribution function was proposed and the orientation ratios and distribution curve were obtained experimentally. The initial stages of coating acicular maghemite particles, coprecipitated with Co^{2+} and Fe^{2+} ions, were studied.[240] Moessbauer absorption spectra showed the rapid oxidation and magnetic ordering of Fe(II) to Fe(III). Emission spectra using $^{57}Co^{2+}$, found that these ions were mainly not magnetically ordered, but were, at first, preferentially incorporated in the surface layer of the oxide. Spin orientations in magnetic recording materials prepared with different coatings were determined using applied magnetic fields.[241] Emission Moessbauer spectroscopy was used to study Co adsorbed on the surface of acicular particles of γ-Fe_2O_3.[242] The coercivity increased up to about 2 wt.% Co, above which it levelled off. Samples with high coercivity showed a strong magnetic interaction between the adsorbed Co and the substrate. Moessbauer and XPS measurements have shown that, after annealing Co ions in the Co-ferrite layer of γ-Fe_2O_3 migrate to new equilibrium positions.[243] Also during the review period investigations were carried out on solid solutions of $(Fe_{1-x}In_x)_2O_3$[244,245] and $(Fe_{1-x}Sr_x)_2O_3$,[245] and the Morin transition in Ru-doped α-Fe_2O_3 was examined.[246]

Magnetite and Spinel-type Oxides. The formation of uniform spherical particles of magnetite has been studied.[247] Moessbauer data showed that longer aging time gave stoichiometric uniform particles. Two papers have studied the effects of surfactants on Fe_3O_4. A high concentration of surface defects was noted,[248] and the anisotropy was shown to be independent of the particle size.[249] The effects of insertion of lithium ions into magnetite were investigated.[250] Two non-interacting phases were found. High vacancy content ferrites represented by $x(MFe_2O_4)$-$.y(Fe_3O_4).z(\gamma\text{-}Fe_2O_3)$, where $x + y + z = 1$ ($z > 0.50$) and M = Zn(II), Ni(II), or Cd(II), have been prepared.[251] As the particle size increased the high-vacancy-content magnetite was transferred from superparamagnetic to ferrimagnetic particles at high Fe(II) concentrations.

The formation and cation distribution in the spinel system $FeCr_2O_4$ has been studied.[252] Intermediate Fe(II) components Fe_3O_4 and FeO were identified as intermediates and, under low partial pressures of oxygen some Fe(III) was found. Cation distributions in the mixed spinel ferrites $Mg_xMn_{1.1-x}Sn_{0.1}Fe_{1.8}O_4$ have been studied using the ^{57}Fe and ^{119}Sn Moessbauer effects.[253] Two spin-freezing transitions have been found in amorphous $CoFeO_4$ at 284 K and 86 K,[254] and two hyperfine fields, due to *A*- and *B*-site iron in $Co_{1-x}Ge_xFe_{2-2x}O_4$ ($x \leq 0.4$) were discovered.[255]

Several papers were published on the Moessbauer spectra of Ni-Zn ferrites. Thermal decomposition of mixed nickel-zinc iron oxalate gave $(Ni_{0.5}Zn_{0.5})Fe_2O_4$ which the Moessbauer results showed to contain an inhomogeneous distribution of Ni and Zn prior to annealing at 750°.[256] Solution precipitation yielded a similarly distorted structure.[257] The microscopic magnetic properties of the cubic microwave ferrite $(Ni_{0.59}Zn_{0.41})Fe_2O_4$ have been studied.[258] A broad unresolved spectrum results from the overlap of the magnetic hyperfine fields at the tetrahedral and octahedral sites. This was due to a distribution of magnetic fields at each site due to different numbers of neighbouring zinc atoms. A plasma-sprayed coating of the same composition was found to be separated into Ni-rich and Zn-rich phases.[259] The Ni-phases were distributed between the two sites, and the zinc-rich phase, similar to zinc ferrite, contained a large fraction of Fe(II). Annealing up to 650°C restored the properties to those of the unsprayed ferrite. Magnetic gamma-resonance spectra of $NiAl_xFe_{2-x}O_4$ have also been described.[260]

Copper ferrite prepared by wet-chemistry was shown to be superparamagnetic at 298 K, samples prepared by ceramic methods gave normal Zeeman-split sextets.[261] Atomic migration between the *A* and *B* sites in $CuCr_{0.25}Fe_{1.75}O_4$ was found to be stimulated by the Cr addition.[262] A similar cubic structure was found in the Fe_2O_3-$(Al_2O_3)_x$-$(CuO)_{1-x}$ system (x = 0-0.8).[263] Increasing the aluminium concentration raised the hyperfine field at the *A*-sites and decreased the field at the *B*-sites.

Other spinel-related systems investigated during the review period include: quenched (Zn,Mg)Fe_2O_4;[264] phase transitions in $Ca_{0.5}Zn_{0.5}Fe_2O_4$;[265] Mg-Ti ferrites;[266] Co-Ti substituted barium ferrite;[267,268] the ferric oxide-(Al,Ga) oxide-silicon oxide system;[269] spinel phases close to FeV_2O_4;[270] and manganese-zinc ferrites.[271]

Superconducting Ceramics Containing Iron. As in previous years a great number of papers have described the use of Moessbauer spectroscopy in the study of high critical temperature superconducting ceramics. This year a high proportion of the work described has involved the use of ^{57}Co-doping and emission Moessbauer spectroscopy.

The asymmetry ratio of the quadrupole doublet for ^{57}Fe in a non-cubic site of a diamagnetic or paramagnetic orthorhombic compound has been examined.[272] The resulting calculations were discussed in connection with experiments on oriented powders and single crystals of high-T_c superconductors. The lattice contribution to the electric field gradient (e.f.g.) tensor components have been determined at the empty Cu(1) and Cu(2) sites in $YBa_2Cu_3O_{7-x}$ using emission Moessbauer spectroscopy.[273] A $3d^{10}$ configuration was found for Cu^+ in the Cu(1) sites, and a considerable contribution found of the unfilled 3d shell to the e.f.g. at the Cu(2) sites suggested a $3d^9$ configuration for Cu^{2+}. The crystal structure and Moessbauer spectra of crystals of $YBa_2(Cu_{0.98}Fe_{0.02})_3O_{7-y}$ have been studied in their superconducting state at 68-96 K.[274] The electric field gradient tensor was determined along with the mean-square atomic displacements of the iron atoms. A $Fe^mO_n^{m-2n}$ cluster model and the X_α distorted wave method has been used to model the ^{57}Fe centre in $YBa_2(Cu_{1-x}Fe_x)_3O_{7-y}$ compounds.[275] The results were used to interpret the Moessbauer spectra. Other workers have attributed the quadrupole splitting in $YBa_2(Cu_{1-x}Fe_x)_3O_{7-y}$ compounds to the presence of holes.[276]

Several papers have described work on the site preferences of iron in $YBa_2(Cu_{1-x}Fe_x)_3O_{7-y}$ superconducting ceramics. The ^{151}Eu, ^{119}Sn, ^{57}Fe and ^{57}Co Moessbauer isotopes were all used in a Hungarian study.[277] The Co and Fe were shown to prefer the Cu(1) site in the III+ or IV+

valence state. Above 700°C migration of Fe and Co between the Cu(1) and Cu(2) sites was observed.[277,278] Swedish work on samples with x = 0.005 and a range of oxygen contents suggested that over 84% of the iron substitute for Cu(1) and the rest for Cu(2).[279] Charge states of IV+ and III+, respectively, were obtained for the two sites. While similar results were obtained by Japanese researchers for ^{57}Fe- and ^{57}Co-doped in $YBa_2(Cu_{1-x}Fe_x)_3O_{7-y}$,[280] in $YBa_2Cu_3O_6$ Fe and Co also occupied Cu(2) sites, indicating an antiferromagnetic long-range order. The preference of iron for Cu(1) sites was also reported by Russian[281] and American workers.[282]

Because ^{57}Fe substituted on the chain (in Cu(1) sites) cannot sense the magnetic field, observation of a Zeeman sextet at room temperature is widely regarded as evidence for Fe on the Cu(2) site.[283] A large proportion of magnetic species has been previously generated by heating in an inert atmosphere. Evidence was found that this magnetic species was not located in the Cu-O sheets, but occupies an interstitial site which is not equidistant from the two neighbouring sheets.

The oxygen content of these 1:2:3 ceramics has been found to play an important part in determining the Moessbauer spectra of these compounds. Absorption and emission Moessbauer spectra have identified at least three spectral components, the relative intensities of which depend on the oxygen concentration.[284] While Fe(Co) atoms mainly substitute at the Cu(1) site, some occupy the Cu(2) plane sites, indicating antiferromagnetic order in oxygen deficient compounds. A systematic study of highly-controlled samples with low iron concentrations has been described for the whole oxygen concentration range.[285] At the high oxygen limit ($y \rightarrow 7$) Fe^{4+} predominated at both crystallographic sites, while as the low limit ($y \rightarrow 6$) Fe^{3+} was dominant. Oxygen deficiency has been associated with antiferromagnetic ordering.[285,286]

A direct correlation between the superconductivity and the isomer shift of the Moessbauer probe has been found in a study of the effect of oxygen content in $YBa_2Cu_3O_{6+x}$.[287] The magnitude of the isomer shift increased continuously with decreasing oxygen content before levelling off just below x = 6.4. This corresponds to a superconductor to insulator transition. Other work on the effects of different oxygen stoichiometries were also presented.[288-292]

Samples of $YBa_2(Cu_{1-x}Fe_x)_3O_{6+y}$ have been prepared in an inert atmosphere followed by low-temperature annealing on oxygen to increase the proportion of iron on the Cu(2) sites.[293] These materials are orthorhombic and have a higher T_c than the conventionally prepared materials.

Other authors have reported an increase in T_c depending on the heat treatment procedure used,[294] and a stabilization of the orthorhombic phase after a special heat treatment.[295]

Cyclic measurements of the *in-situ* temperature dependence of emission Moessbauer spectra of $YBa_2Cu_3(^{57}Co)O_{6+y}$ were made from room temperature to 500°C.[296] Changes in the intensities of the three quadrupole doublets as a function of temperature were attributed to changes in the electronic states of iron, changes in the oxygen coordination, and to differences in the Debye temperature for different iron sites. A similar study using both ^{57}Co and ^{57}Fe to probe the Cu(1) site was also described.[297] Conversion of site *B* to site *C* was attributed to a zigzag motion of the chain, which was retarded in the ^{57}Fe-doped compound. The properties of the e.f.g. tensor at major Co(Fe) sites were discussed.

A sharp peak in the temperature dependence of the Moessbauer linewidth in $YBa_2(Cu_{1-x}Fe_x)_3O_{6.8}$ ($x = 0.017$) has been found near 110 K.[298] The Lorentzian line broadening suggested that the anomaly was due to an order-disorder phase transition. Two minima in the recoilless fraction and Debye temperature were found near 110 K and 220 K in a compound with $x = 0.12$.[299] Lattice softening was thought to be the cause. In similar materials a structural transition found at ~240 K was attributed to a disorder transition,[300] and a thermal expansion anomaly was related to Moessbauer spectra.[301]

Felner and his co-workers have used Moessbauer spectroscopy to determine the antiferromagnetic phase transition temperature, and d.c. magnetometry the superconducting transition in the ceramics $YBa_2(Cu_{1-x}M_x)_3O_z$ and $(Y_{1-x}Pr_x)Ba_2Cu_3O_z$.[302] Antiferromagnetism was only found at concentrations where the superconductivity disappeared. There was no conclusive evidence for the overlap of superconductivity and antiferromagnetism. Similar results were obtained for $(Y_{1-x}Pr_x)Ba_2Cu_{3-y}M_yO_z$ (M = Fe, Co, Zn).[303]

Gamma-resonance spectroscopy has been used in conjunction with Cu(1) site NQR, Cu(2) site zero external field nuclear antiferromagnetic resonance, and Gd EPR to study a magnetic transition in antiferromagnetic $(Y,Gd)Ba_2(Cu_{1-x}Fe_x)_3O_6$.[304] The hyperfine field distribution at Fe in the Cu(1) sites was attributed to Fe clustering. The temperature dependences of the magnetic hyperfine interactions of three iron species in $YBa_2(Cu_{0.9}Fe_{0.1})_3O_{7-y}$ have been studied.[305] The analysis was carried out using the eigenvalues and eigenvectors of the Hamiltonian of the combined electronic quadrupole moments and magnetic hyperfine interactions for determination of the line positions and intensities. Evidence was found that the magnetic splittings arise from antiferromagnetic ordering.

Other papers have investigated the origin of magnetic ordering in ^{57}Fe-doped $YBa_2CuFe_3O_{7-y}$,[306] magnetic ordering in an oxygen-rich ceramic,[307] quantitative analyses of paramagnetic and magnetic spectra of a series of 1:2:3 ceramics,[308] and compared results from Moessbauer and microwave spin resonance spectra of ceramics over the temperature range 4-300 K.[309]

A variety of modifications to the composition of $YBa_2(Cu_{1-x}Fe_x)_3O_{7-y}$ ceramics have been described. Hydrogen-charged $YBa_2Cu_3O_7H_x$:2 at.% ^{57}Fe was studied.[310] Analysis of the Moessbauer spectra at different temperatures showed that antiferromagnetism at the Cu(2) sites is induced by hydrogen in a very similar way to the removal of oxygen. Reduction with hydrogen removed oxygen and converted 5-coordinate Cu(1) sites into planar sites.[311] Defects have been induced in $YBa_2Cu_{3-x}Fe_xO_7$ by bombardment with 3.5 GeV-Xe ions.[312] As a result a fourth quadrupole doublet was induced ($\delta \approx 0.3$ mm s^{-1}; $\Delta E_Q \approx 1.1$ mm s^{-1}) with an intensity which increases with dosage.

Intercalation compounds of halogens in $YBa_2Cu_3O_{7-y}$ are superconducting.[313] Bromide intercalated compounds were insensitive to Fe doping, and had T_c at 88 K. On intercalation, initially non-superconducting samples became superconducting. The iron and tin Moessbauer isotopes were used to study $YBa_2Cu_3(O,N)_{7-y}$.[314] While four quadrupole doublets were found in $YBa_2(Cu,Fe)_3O_{7-y}$ only three were found in $YBa_2(Cu,Fe)_3N_{6.5}$. Single crystals of $YBa(Cu_{0.94}Fe_{0.04}$-$Al_{0.02})_{2.85}O_{7-y}$ have been prepared and their Moessbauer spectra described.[315] Ceramics substituted with V and Mn were studied at 300 K.[316] At low concentrations of vanadium, Fe preferentially occupied the Cu(1) sites and V the Cu(2) site. With the same concentration of Mn, however, Mn occupied the Cu(1) site and Fe the Cu(2) site.

The preparation of $EuBa_2Cu_{3-x}Fe_xO_{7-y}$ has been investigated with gamma-resonance spectroscopy.[317] In contrast to the yttrium-containing system, T_c is strongly dependent on the degree of substitution of Cu by Fe. Europium barium copper oxide has been substituted by both iron and tin, and was studied using the ^{151}Eu, ^{119}Sn, and ^{57}Fe Moessbauer effects.[318,319] An anomaly in temperature dependence of the total area of the ^{57}Fe spectrum was noted between 120 and 300 K, and anomalous changes were found in the iron parameters around T_c. Three sets of authors have studied $GdBa_2Cu_{3-x}Fe_xO_{7-y}$.[320-322] The results are similar to those obtained for the corresponding yttrium compounds.

Several ceramics with compositions close to that of $YBa_2Cu_3O_{7-y}$ have been studied to provide information about that material. The effects of Fe and Zn substitution on $YBa_2Cu_4O_8$ were described;[323,324] site assignments in $YBaCuFeO_{5+y}$ were studied;[325] and the structural implications of the simple bilayer compounds $La_2CrCu_2O_6$, $La_2CaCu_2O_6$, and $La_{1.6}Sr_{0.4}CaCu_2O_6$ doped with ^{57}Fe were investigated.[326]

A Moessbauer study on the Fe-doped 2223 phase of the $(Bi,Pb)_2Sr_2Ca_{n-1}Cu_nO_y$ superconductor has been reported.[327] Formation of even small concentrations of the 2201 ceramic phase was easily detected due to the preference of iron for the octahedral copper sites. In the system $(Bi_{1.6}Pb_{0.4})Sr_2Ca_3Cu_{4-x}Fe_xO_y$, three doublets were found and related to different Cu sites.[328] These sites correspond to iron in two square planar Cu-O sites with different local charge environments, and iron in a planar pyramidal site. Moessbauer spectra of the 2212 ceramic $Bi_2Sr_2CaCu_{2-x}Fe_xO_{8+y}$ have been fitted to two doublets.[329,330] One was identified as Fe replacing one Cu ion in the Cu-O plane, and the second, concentration dependent resonance, to clusters of two or more nearest Fe atoms. A new $Bi_2Sr_2FeO_y$ phase has been identified in this class of superconducting ceramic,[331] and the system $Bi_4Sr_3Ca_3(Cu_{1-x}Fe_x)_4O_{4y}$ was studied.[332]

The Moessbauer effect has been used to study the magnetic structures in $(Bi_{2-x}Pb_x)Sr_2Bi_{n-1}Fe_nO_y$ (with x = 0.5, 1; n = 2, 3) over the temperature range 85-560 K.[333] For both values of n the iron was high-spin Fe(III) and the transition to the paramagnetic state occurred at 520-550 K. Antiferromagnetism in non-superconducting phases related to $(Bi,Pb)_2Sr_2(Ca,Cu)O_y$ was also studied.[334,335] Moessbauer spectra were obtained for superconducting phases in the ceramics $TlBaCa_3Cu_3O_{8.5}$,[336] and $Nd_{2-x}Ce_xCuO_4$.[337]

Other Oxides. A new, electrochemical, method has been used to prepare the cubic perovskite $SrFeO_3$.[338] The Moessbauer spectrum shows that the final product contains only Fe(IV) with a zero quadrupole splitting. Defect ordering in $SrCr_{0.1}Fe_{0.9}O_{3-y}$ has been found to be strongly dependent on its preparation conditions.[339] The introduction of Cr stabilized the high-temperature state of $Sr_2Fe_2O_5$. Oxygen-deficient $Sr_2Fe_{2-x}Cr_xO_{5+y}$ was similarly studied.[340] The iron and europium Moessbauer effects were used to study the perovskites $Eu_{0.2}R_{0.8}Fe_3$ and $EuFe_{0.8}M_{0.2}O_3$ (R = rare earth; M = Sc, Cr, Mn, Co).[341] The ^{57}Fe hyperfine field parameters were related to the R ions. Moessbauer scattering and X-ray diffraction experiments on $Pb(M_{0.5}Nb_{0.5})O_3$ (M = Sc, Fe) were described.[342] Large differences were found between the Debye-Waller factors determined by the two methods.

A detailed Moessbauer study of the bipyramidal *2b* site in barium hexaferrite, $BaFe_{12}O_{19}$.[343] Of all the sites, the recoil-free fraction of iron(III) in the *2b* site decreases fastest with temperature, with θ_M = 247 K. These results indicate that Fe(III) on this site oscillates between two equivalent (4*e*) positions either side of the symmetry plane normal to the *c*-axis. Other compounds from the BaO-Fe_2O_3 system investigated were $Ba_2Fe_2O_5$, $Ba_3Fe_2O_6$, $Ba_7Fe_4O_{13}$ and $Ba_3Fe_2O_8$,[344] and $BaFe_{12}O_{19}$ and $BaFe_2O_4$.[345] The preparation and Moessbauer spectrum of lithiated barium hexaferrite was reported.[346] Substitution of Ba^{2+} for trivalent rare-earth ions in M-, W- and X-type hexaferrites was studied.[347] Significant differences were only found in the M-type hexaferrite system. Gamma-resonance data have been reported for the strontium hexaferrites $SrGa_6Fe_6O_{19}$,[348,349] and for $SrFe_{12}O_{19}$ single crystals.[350] Polycrystalline and oriented single crystals of $PbFe_{12}O_{19}$ were also investigated.[351] Hyperfine fields and isomer shifts were obtained for all five iron sites.

Moessbauer effect and dielectric constant measurements have been made on a single crystal of yttrium iron garnet, $Y_3Fe_5O_{12}$ (YIG).[352] Neither technique was able to detect a supposed phase transition at 125 K. Vibration dynamics in YIG with Fe^{3+} and Y^{3+} substituted by other cations have been studied by measuring the Moessbauer recoil free fraction.[353] For all the substituted compounds, above T_c, there was a Fe^{3+} (*a,d*) vibration anisotropy due to the elongation of oxygen polyhedra. Moessbauer spectra have been recorded for Ga-,[354] and (Gd,Al)-doped[355] YIG, submicron-sized rare-earth garnets,[356] and the substituted garnet system $(R_{3-x}Ca_x)[Sn_xFe_{2-x}]Fe_3O_{12}$, where R = a rare-earth.[357]

Moessbauer spectroscopy and neutron diffraction were used together to determine cation distributions in four synthetic $FeTi_2O_5$ oxides.[358] Transverse-field muon spin relaxation and Moessbauer experiments were used the study the spin-glass $Fe_{2-x}Ti_{1+x}O_5$ near its spin-glass temperature.[359] Kajima and his co-workers have described thin films from the ferromagnetic Fe_2O_3-Bi_2O_3-$PbTiO_3$.[360-362] Phases from the system $Ca_{1-x}R_xFe_xTi_{1-x}O_3$ (R = La, Eu) were also investigated.[363]

Four papers have reported investigations on heteropolyacids containing iron. Moessbauer data has identified distorted sites for iron, with the iron being high-spin in $H_5[FeO_4W_{12}O_{36}]\cdot6H_2O$ and in $H_3[Fe(OH)_6Mo_6O_{18}]\cdot17H_2O$ and low-spin in $FeH_2[FeO_4W_{12}O_{36}]\cdot17H_2O$.[364] A tetragonal distorted central FeO_4 tetrahedron was found in $Ba_2H[\alpha\text{-}FeO_4W_{12}O_{36}]\cdot26H_2O$.[365] The Moessbauer spectra of several ^{57}Fe-enriched heteropolymolybdo anions were also described.[366,367]

Several series of iron-containing oxides having the layered K_2NiF_4-type structure were published. In $(A,La)_2M_{0.5}Fe_{0.5}O_4$ (A = Ca, Sr, Ba; M = Li, Mg, Zn, Ga, Ni), the FeO_6 octahedra are sufficiently distorted to stabilize the high-spin Fe(IV) state.[368] Mixed valency states were found to coexist within $Sr_{1+x}Er_{1-x}FeO_{4-y}$, the proportion of Fe^{4+} increasing with x.[369] Relationships between oxygen non-stoichiometry and the magnetic and electron transport properties of $La_{1-x}Sr_xNi_{1-x}Fe_yO_{4-z}$ were investigated,[370] and K_2FeO_4 was studied.[371]

Other studies on iron oxides carried out during the review period and using the Mossbauer effect include: characterization of amorphous ferromagnetic oxides from the Sr-Fe-O and Bi-Fe-O systems;[372] investigation of the properties of the Ruddlesden-Popper phases $Y_2SrFeCuO_{6.5}$ and $LaSr_3Fe_3O_9$;[373] magnetic properties of $NbFeO_4$ and $TaFeO_4$;[374] an examination of Fe^{3+} adsorbed onto Sb_2O_5;[375] spin-relaxation in $BaSn_xTi_{2-x}Fe_4O_{11}$;[376] the system Fe_2O_3-SnO_2;[377] and the oxyhalides $FeMn_7O_{10}Cl_3$[378] and $M_3Fe_2O_5X_2$ (M = Ca, Sr; X = Cl, Br).[379]

Inorganic Oxide Glasses Containing Iron. The effects of pressure on iron were studied using silicate melts Na_2O-Fe_2O_3-SiO_2 in oxygen to pressures of 4 GPa and temperatures of 1700°C.[380] Moessbauer examination of the quenched melts at 298 K found no traces of Fe(II) or Fe(III). The doublet observed was attributed to Fe^{3+}(IV) with an isomer shift which increases with pressure, and a second component which was observed above 1.5 GPa.

The Moessbauer effect was used to study transformations in Sr-containing borosilicate glass after heat treatment at up to 900°C.[381] The quadrupole splitting of Fe^{3+} decreased at 250-300°C due to reduced distortion of the glass matrix. Nucleation and crystallization temperatures were determined, from the temperature dependence of the spectral area, to be 540 and 750°C, respectively.

Iron coordination in calcium iron borate glasses was investigated by substituting Ca with Fe.[382] A tendency was found for an increase in the formal charge and coordination number of the iron ions as the basicity of the glass increases. Glass formation has been observed in the Y_2O_3-Fe_2O_3-B_2O system.[383] Hyperfine splitting, found in the Moessbauer spectra of glasses with high iron concentrations, suggested the formation of Fe-rich microclusters.

The valence and coordination of iron in the CaO-MgO-Fe_2O_3-SiO_2-GeO_2 system has been studied by a variety of methods, including gamma-resonance spectroscopy.[384] The possibility of isomorphous substitution of Fe^{3+} for Si^{4+} in silicate gels was investigated.[385] Magnetization and Moessbauer results showed that iron could be substituted into the tetrahedral silicate structure.

Chalcogenides. Natural and synthetic samples of 4C monoclinic pyrrhotite (Fe_7S_8) was studied by Moessbauer spectroscopy down to 4.2 K.[386] A low-temperature transition was found at 30-35 K, there being five Fe^{2+} sites (six resonance lines) at liquid helium temperatures, and only four at higher temperatures. Magnetic hyperfine field distribution functions for intermediate pyrrhotite ($Fe_{1-x}S$) were used to distinguish seven non-equivalent Moessbauer sites in this material.[387]

Moessbauer spectroscopy has been used to study the intermediates formed at the FeS_2 cathode of a Li/FeS_2 battery system.[388] At 4.2 K evidence was found for $Li_3Fe_2S_4$, a phase very similar to Li_2FeS_2, and some superparamagnetic iron. The cubic spinels $FeCr_2S_4$[389] and $Ni_xFe_{1-x}Cr_2S_4$ (x = 0-0.4)[390] have been characterized. From the Moessbauer data they were shown to contain Fe(II) in tetrahedral sites. For the x = 0 material Neel and Debye temperatures of 180 K and 250 K, respectively, were reported. Below T_N, appearance of a quadrupole splitting and line broadening suggested the onset of Jahn-Teller distortion and relaxation effects. Tetrahedrites with compositions between $Cu_{12}Sb_4S_{13}$ and $Cu_{10}Fe_2Sb_4S_{13}$ have been synthesized and characterized.[391] A new mixed valence species [Fe^{II},2Fe^{III},Co^{III}] has been identified.[392]

A cubic spinel, $Cu_{0.5}Fe_{0.5}Cr_2S_{3.8}Se_{0.2}$, has been studied.[393] The temperature dependences of the magnetic hyperfine field and the magnetization were explained by the Neel theory of ferrimagnetism using three exchange integrals. Moessbauer data was obtained for a bicyclic cluster, $[Na_9Fe_{60}Se_{38}]^{9-}$ constructed by fusion of Fe_2Se_2 rhombohedra.[394]

Iron impurity centres in ZnTe have been studied with Moessbauer absorption and emission spectroscopy.[395] Isomer shift data showed that Fe^{2+}-Fe^{2+} bonds are more covalent in ZnTe than in ZnS. A transient Fe^+ state was observed after the decay of ^{57}Co below 130 K in stoichiometric ZnTe which may have a Fe^0 state as precursor. Other workers have studied the δ'-iron telluride, $Fe_{1.33}Te_2$ from room temperature to 1023 K.[396] The diffusion coefficients were calculated from the Moessbauer line broadening.

Applications of Iron-57 Moessbauer Spectroscopy. - Catalysts. Moessbauer spectroscopy and temperature programmed reduction were used to study the reaction between the catalyst and support in Fe_2O_3/γ-Al_2O_3 and Fe_2O_3-La_2O_3/γ-Al_2O_3 catalysts.[397] The La modified system showed a much stronger component-support interaction which stabilized the Fe(II)O(II) phase. The active sites for redox reactions in zeolitic catalysts have been studied.[398] Methods used to alter the sites occupied by Fe^{2+}/Fe^{3+} in zeolite Y structures were described. The characterization of Fe-ZSM-12 zeolite catalysts was also discussed.[399]

The largest number of papers published on the Moessbauer spectra of catalysts concerned carbon monoxide hydrogenation. Promoter effects of silicon on precipitated iron catalysts used for Fischer-Tropsch synthesis were examined.[400] A strong effect of silicon on the iron particle size, together with Fe-Si interactions, were noted. *In situ* Moessbauer spectroscopy has been used to study the stabilization of various CO + H_2 catalysts.[401] Three types of carbon were noted and related to the stability of the active iron component. Iron distributions in Fe-Cu-K Fischer-Tropsch catalysts were shown to be dependent on the preparation conditions and starting materials.[402] The effects of the iron phases on the catalyst system equilibrium were established. Iron/magnesia Fischer-Tropsch catalysts,[403,404] olefin selective carbon-supported Fe, Fe-Me and K-Fe-Mn hydrogenation catalysts,[405] iron catalysts supported on montmorillonites[406] and mordenite[407] were all studied. Silica-supported RhFe, PtFe and IrFe catalysts were studied by EXAFS, TEM and Moessbauer methods.[408] Fe_2Rh_4, Fe_3Pt, and Fe_4Pt species derived from carbonyl clusters were identified. The ^{57}Fe Moessbauer effect was used in conjunction with that of ^{193}Ir to monitor the prereduction of SiO_2-supported Fe-Ir in hydrogen, and following subsequent treatments.[409]

Carbon dioxide hydrogenation has been studied by Lee and his co-workers.[410,411] Moessbauer spectroscopy was used to determine the bulk phase compositions of Cr, Mn, and Mo promoted Fe catalysts at several stages. Severe oxidation of these catalysts was observed. Iron palmitate used for the liquid-phase homogeneous hydrogenation of cyclohexene was also reported.[412]

Two series of papers have been published, in Chinese, on the use of Moessbauer spectroscopy for the study of coal hydroliquefaction catalysts used for the treatment of lignites. The first series covered the transformation and action mechanism of iron sulphide in hydrogenation;[413] the transformation and hydrogenation activity of ferric oxide;[414] and a new active phase, γ-Fe, in brown coal liquefaction.[415] In the second series the transformation of iron sulphide catalysts during liquefaction[416] and the effects of added sulphur on the transformation of iron catalysts.[417] An American report has compared the use of Moessbauer and XAFS spectroscopy in the study of iron-based coal liquefaction catalysts, including $FeCl_3$-impregnated coal, highly dispersed Fe_2O_3 on carbon black, Fe cation exchanged into lignite, and a sulphated haematite.[418] The Moessbauer method was found to be more accurate but much slower.

Other catalyst systems investigated with the Moessbauer effect during the review period include: sulphidation of alumina-supported Fe and Fe-Mo-O catalysts;[419] the emission Moessbauer spectra of the sulphided Ni-containing catalysts ^{57}Co:Ni-Mo/C and ^{57}Co:Ni/C;[420] and $FeCl_3$ in benzoylation and sulphonylation systems.[421]

Minerals and Coal. Filtration, density gradient centrifugation, sequential chemical extraction, and Moessbauer spectroscopy have been compared for the speciation of particulate iron in fresh water.[422] Centrifugation and Moessbauer spectroscopy gave inconclusive results.

A combined X-ray and Moessbauer method has been developed for the determination of the geometry of local crystallographic sites of iron.[423] The resultant equations were used to describe the Fe^{3+} site geometry in several minerals including: X-ray amorphous fine-grained hydroxides and oxides of iron, minerals formed under high pressure, and glaucanites. Correlations between Moessbauer spectral parameters and crystallographic properties have also been obtained for Mg-Al oxide and silicate minerals containing iron as minor components.[424] The results were used to resolve inconsistencies found in the Moessbauer parameters and cation site occupancies of some silicate minerals.

An examination of the Moessbauer spectra of kaolinites, halloysites and the firing products of kaolinite has been described.[425] Determination of the Moessbauer parameters was complicated by the low iron content, which caused slow paramagnetic relaxation, and the presence of interfering minerals in some samples. Both minerals gave Moessbauer parameters within the same range: for Fe^{2+}-free kaolinite $\delta = 0.35$ mm s^{-1}, $\Delta E_Q = 0.51$ mm s^{-1}; and for Fe^{2+} $\delta = 1.11$ mm s^{-1}, $\Delta E_Q = 2.54$ mm s^{-1}. Spectra recorded at 4.2 K alllowed the identification of goethite in kaolins of complex mineralogy at concentrations as low as 0.1%. Because the Moessbauer spectra reflect the mineralogy of the firing products as well as the temperature and redox conditions during firing, it may be used to assess these.

The role of iron in the weathering of a climosequence of New Zealand soils derived from schist was investigated with the Moessbauer effect.[426] The spectra showed that the main changes in iron across the sequence involve the oxidation of Fe^{2+} in muskovite and its initial weathering products to Fe^{3+} in these minerals and in oxyhydroxides. Products of the reaction between iron and fulvic acids from sandy loam and a clay soil have been analyzed.[427] The effects of pH on the three spectral components, magnetically dilute Fe(III), Fe(II) and Fe(III) doublets were described. The effects of heating on an Egyptian talc ore were also studied.[428]

Iron-bearing minerals have been determined in two British and three South American coals of differing ash contents.[429] It was suggested that the paramagnetic phases in the ash may be more positively identified and may provide information about the thermal history of the ash. Moessbauer spectroscopy was used to identify the effects and products of desulphurization of coal by microwave irradiation-magnetic separation,[430] residues remaining after hydrogenation of some east German brown coals were investigated,[431] and iron species in 11 Polish bituminous coals and a Bulgarian anthracite were analyzed.[432]

Corrosion Studies. Tripathi and Dhoot have studied the corrosion of mild steel under different conditions. Treatment with NaCl and LiCl initially produced γ-FeOOH which later converted to magnetite,[433] while exposure to Na_2SO_4 solution containing ascorbic acid produced high-spin iron(II) in a distorted environment, which was associated with and Fe(II)-ascorbate complex.[434] Other Indian workers described mild steels from the bleach section of a paper mill.[435] All the samples showed paramagnetism. Corrosion products were mainly $Fe_{3-x}O_4$ and α-FeOOH (>85%) with small amounts of γ-FeOOH.

The effects of an external magnetic field on the products of the corrosion of a low-carbon steel ($Fe_{3-x}O_4$ and γ-FeOOH) in air have been studied by Moessbauer spectroscopy.[436] The magnetic field changed the dispersity of the phases and increased the proportion of the paramagnetic components in the rust. Passivation of iron and steels in sulphite solution was studied by Moessbauer spectroscopy and electrochemical measurements.[437] The main phase in the passive layer was identified as γ-FeOOH, although polarization at pH 3.37 yielded some sulphate and sulphite. Corrosion products in domestic (Indian) tap water were also studied.[438] It was observed that the water became contaminated during transport by steel pipelines with a product characterized as γ-FeOOH.

Other Applications. Gamma-resonance spectroscopy has been used to compare the ferrite phase in high-iron cement with the pure $C_2A_xF_{1-x}$.[439] Although the two materials gave the same X-ray diffraction patterns their Moessbauer spectra were quite different: for the ferrite phases δ = -0.32-0.35 mm s^{-1} and ΔE_Q= 0.80-0.95 mm s^{-1}, while the pure $C_2A_xF_{1-x}$ phases gave δ = 0.15-0.40 mm s^{-1}, ΔE_Q = 1.75-1.60 mm s^{-1}. The abnormal spectra of the ferrite phase was attributed to the substitution of Si^{4+} for Al^{3+} or Fe^{3+} on the tetrahedral lattice sites. An X-ray and Moessbauer study on two dry and hydrated portland cements and their clinker was also reported.[440]

A technique for measuring the Moessbauer spectra of metallurgical slags with a spatial resolution of ~500 microns has been developed.[441] A high activity ^{57}Co source was used with the gamma-ray beam collimated down to a diameter of 0.5 mm. A melt, $Fe_x0-Al_2O_3-SiO_2$, containing 62 mol.% FeO was studied. Crystalline Fe_2SiO_4 and an Fe^{2+}-containing glass phase were identified in proportions which varied with depth through the slag.

Chinese Lishan clay has been thoroughly characterized by Moessbauer spectroscopy, XRD, XRF, neutron diffraction and thermal and chemical analysis.[442] It was proved that this was the clay used for making the terra-cotta warriors and horses of the Qin dynasty. Transformations induced by the firing of the clay were characterized and the sintering temperature for the terra-cotta sculptures was shown to be 950-1030°C. A similar firing study on clays from the sites of four ancient Chinese kilns was also described.[443] Ancient Egyptian pottery sherds from New Kingdom Memphis were studied by Moessbauer and XRF spectra.[444] The provenance of the clay and the manufacturing technology used for the production of each type of pottery could be determined and a group of potteries were found to have been imported from a neighbouring region.

The chemical states of iron in airborne particles collected by the Anderson sampler in metropolitan Japan were investigated.[445] Fine particles (<1.1 micron), known to originate from combustion processes were found to have smaller Fe(II)/Fe(II) ratios than coarse (>7.0 micron) particles in all sampling sites. The Fe(II)/Fe(III) ratio in urban samples was always smallar than those from suburban samples. Seasonal variations of iron in the Polish atmosphere were studied by Moessbauer spectroscopy.[446] The results showed that the iron occurred as ultrafine superpara-magnetic particles of Fe_2O_3.

6. Tin-119

General and Alloys. Three isomer shift calibration studies have been reported. A theoretical investigation was carried out using the relationship between $\Delta I_\gamma/I_\gamma$ and $\Delta\rho(0)$ due to the interaction of the intensity of the γ-radiation and electron density on the nucleus for the ^{117m}Sn isotope.[447] Only tentative values of $\Delta\rho(0)$ could be obtained from the measured maximum values of $\Delta I_\gamma/I_\gamma$. Japanese workers have carried out a combined Moessbauer spectroscopy and molecular orbital calculation on the electron density in tin compounds.[448] By comparing the valence electron contact densities with isomer shifts, reported for several tin compounds in rare gas matrices, the value of $\Delta R/R = (1.57 \pm 0.03)x10^{-4}$ was obtained for the 23.87 keV M1 transition in ^{119}Sn. First-principles self-consistent local-density calculations of the electronic structures of clusters representing Sn(II) and Sn(IV) were carried out.[449] Relating the calculated electron densities with the Moessbauer isomer shifts yielded $\Delta R/R = (1.58 \pm 0.14)x10^{-4}$.

A comparison has been made between the chemical consequences of electron capture (^{57}Co) and converted isomeric transition (^{119m}Sn) in frozen H_2O and D_2O.[69] Experimental data on deuterated systems confirm the model of coulombic fragmentation and acceptance of electrons. The Moessbauer effect has been achieved in a rigid microemulsion containing submicroscopic droplets of aqueous $SnCl_4$.[450] Two types of Sn(IV) were identified, *viz.*: $SnCl_4$ in the bulk solution and on the cavity wall. Matrix isolated Me_3SnX (X = Cl, Br, I) molecules have been studied.[451] Unlike the crystalline state, where these compounds are five-coordinated, the coordination number is four in the matrix-isolated state.

Moessbauer emission spectroscopy has been used to study the fates of ^{119}Sb and ^{119m}Te after proton- and α-reactions in SnS and SnSe.[452] While proton irradiated ^{120}SnS and ^{120}SnSe gave emission spectra ascribable to ^{119}Sn at the tin site of the matrixes, the α-irradiated samples (containing ^{117}Sn) gave spectra which were assigned to Sn with a nearby defect plus Sn(IV) in a effect structure. The same authors studied ^{119m}Te and ^{119}Sb doped into S, Se, and Te.[453] Both S(^{119m}Te) and Se(^{119m}Sb) gave spectra with peaks due to Sn(IV) and Sn(II) surrounded by S or Se respectively. For the Te(^{119}Sb) samples the Moessbauer showed that most of the Sb atoms are initially well dispersed in the Te matrix, but slow crystallization and fusion results in aggregation into $^{119}Sb_2Te_3$.

The meaning of the tin oxidation number was analyzed using phases isolated in the SnS-In_2S_3-SnS_2 diagram.[454] Tight-binding calculations were performed to characterize Sn(II) and Sn(IV) in these phases. Both calculation and the Moessbauer data showed that the difference between the Sn(II) and Sn(IV) states corresponds to about 0.7 Sn 5s electrons. Gamma-resonance spectroscopy was also used to characterize $Pb_{1-x}Sn_xTe$,[455] $(Sn_xPb_{1-x})_2P_2S_6$,[456] and ternary compounds in the SnS-SnI_2 system.[457] Two electron exchange between tin impurities in $PbS_{1-x}Se_x$,[458] a two-electron donor level in $Pb_{98}Sn_xGe_{1-x}NaS_{100}$,[459] and the charge states in In_2S_3:Sn[460] were all investigated.

The ^{119}Sn Moessbauer effect has been used in several studies into tin-doped metallic and pseudometallic systems. Tin ion implanted into silicon,[461] tin-doped single crystal erbium,[462] and theoretical calculations on ^{119}Sn site Moessbauer spectra of chromium[463] were described. Bistability, local symmetries and charge states of tin-related donors in (Al,Ga)As and GaAs under pressure were studied.[464,465] Above 2.4 GPa a shallow Sn donor state and a deep Sn DX state were easily distinguished by their isomer shifts. Moessbauer and magnetic data from GaAs under pressure provide strong evidence that the Sn DX centre localizes at least two electrons in its ground state.

The half-metallic system UNiSn was found to be paramagnetic at high temperatures and undergo a second-order transition to a magnetically ordered state at 43 K.[466] The ^{119}Sn Moessbauer spectra showed that the hyperfine field seen at the tin sites does not decrease to zero at T = 43 K is approached from below, but continues to be detectable up to 55 K, suggesting that magnetic correlations persist well above the transition temperature. However, the fraction of tin nuclei that see the hyperfine field decreased dramatically at 43 K, indicating that the system orders antiferromagnetically. The same authors also studied $U(In_{3-x}Sn_x)_3$.[467] As x increased T_N was depressed and appears to go to zero between x = 0.4 and x = 0.5. The Moessbauer data, however, showed that the magnetic hyperfine field at the tin nuclei persists up to x = 0.6. A new magnetic structure was proposed to account for these data.

The hydride of Nb_3Sn has been prepared at hydrogen pressures up to 7 GPa.[468] While hydrogenation reduced the electron density around the tin, the reduction was much smaller than for the corresponding Au and Ir compounds. Aging in Cu-33Zn-5.85Sn and Cu-22Zn-10.5Sn was studied,[469] and the ternary compounds RAgSn (R = La, Ce, Pr, Nd, Sm, Gd, Tb, Ho, and Er) were investigated.[470]

Three papers have described work carried out on supported PtSn catalysts. *In-situ* Moessbauer spectroscopy was used the characterize PtSn dehydrogenation catalysts supported on activated carbon, SiO_2, MgO and Al_2O_3.[471] The state of tin, whether as tin metal, as a Pt-Sn alloy, or a bound oxide depended strongly on the support used. Alloy formation on Al_2O_3 and SiO_2 was studied in materials containing different ratios of Pt and Sn.[472] Zero-valent tin was more easily obtained on SiO_2 after reduction with hydrogen than on Al_2O_3. The third paper describes catalysts prepared by controlled surface reaction of Et_4Sn with hydrogen preadsorbed on Pt/Al_2O_3.[473] Reduction of the $Pt\text{-}SnEt_3$ surface complex yielded a Pt_xSn alloy ($3 < x < 4$).

Oxides Reactive rf sputtering has been used to prepare polycrystalline tin oxide films.[474] Moessbauer spectroscopy indicated that the most conducting films contained only SnO_2, and that a SnO phase appeared at low O_2/Ar ratios, when the conductivity was lower. X-ray diffraction, TEM and the Moessbauer effect were used to verify the nanocrystalline state of SnO_2 materials prepared by hydrothermal growth.[475] The relationship between the tin hyperfine parameters and temperature was studied.

Impurity centres in hydrogen annealed V_2O_3:Sn(0.5 at.%) have been studied.[476] Two forms of tin were identified, *viz*: Sn_V^{4+} (bulk) and Sn_S^{2+} (surface) which was rapidly oxidized to Sn_S^{4+}. At low temperatures a magnetic spectrum reflected to antiferromagnetic ordering of the host lattice. A tin-containing aluminophosphate molecular sieve was characterized.[477] About 20% of the Sn^{4+} substituted for phosphorous within the network, with the remainder forming SnO_2. The presence on SnO_2 (and thus SnS) in fresh and spent tin-containing CoMoX hydroliquefaction catalysts was demonstrated.[478]

Cation distributions in a series of $Mg_xMn_{1.1-x}Sn_{0.1}Fe_{1.8}O_4$ spinel ferrites were studied.[253] All these materials showed broad tin Moessbauer spectra due to the ferrimagnetically ordered cations at the A and B sites. Changes in the site distributions were followed through changes in the spectral shape. Orthorhombic, Sb-doped, Cd_2SnO_4 has been studied.[479] Addition of Sb induced changes in the s-electron density which suggested the presence of interstitial SnO_2 or Sn. Gamma-resonance spectroscopy was used to characterize $Zn_2Sn_{1-x}Zr_xO_4$.[480] Two publications on tin-containing glasses were reported.[481,482]

Twice as many papers were published on tin-containing high-temperature superconducting ceramics as in the previous review period. Tin-doped $YBa_2Cu_3O_{7-y}$ was studied after annealing in atmospheres of O_2 or N_2.[314,483] The Moessbauer spectra gave a broadened line which was fitted to two quadrupole doublets, corresponding to two different O-vacancy sites. The N_2 annealed samples showed different isomer shifts for each doublet, indicating different types of site, which may be due to N occupying the O vacancies.

It is generally agreed that tin substitutes for Cu as Sn^{4+}.[277,484-488] Chinese workers have reported that the relative rate at which Sn^{4+} occupies Cu(1) or Cu(2) sites in $YBa_2Cu_{3-x}Sn_xO_{7-y}$ varies with x.[485-487] They reported a preference for the Cu(2), Ba, and Y sites before the Cu(1) site in samples with high tin concentrations ($x \leq 0.4$),[486] and found that replacement of Cu by Sn increased the oxygen content of the oxides.[487] Russian workers have studied the same materials (at low tin levels) using ^{119}Sn Moessbauer and ^{63}Cu NQR and spin-lattice relaxation.[488] It was reported that Sn^{4+} substitutes for Cu at the Cu(1) site.

Gamma-resonance spectra have been recorded on $YBa_2Cu_{3-x}Sn_xO_{7-y}$ (x= 0.5, 2, 5 wt.%) over the temperature range 12-300 K.[489] Debye temperatures of 290 and 320 K were obtained for the samples with 5 and 2 wt.% Sn respectively. The temperature dependent recoil-free fraction $f(T)$ showed a small sharp drop near T_c which is dependent on the Sn concentration, and a second drop at 200-220 K. These were interpreted as being due to phonon softening at these temperatures. An independent study has examined the temperature dependences of f and isomer shift (δ) of

^{119}Sn in $YBa_2Cu_{3-x}Sn_xO_{7-y}$, by using the phonon distributions in the normal state.[490] Using a mixing of Debye and Einstein models it was shown that ln*f* and δ were quenched near T_c and that T_c cannot be interpreted by normal phonons.

Kuzmann and his co-workers have studied the system $EuBa_2Cu_3O_{7-y}$ using the ^{151}Eu, ^{119}Sn, ^{57}Fe, and ^{57}Co Moessbauer isotopes.[277,318,319] Comparison of samples with all three dopants with separately doped materials, and their yttrium analogues, was used to obtain information on the site preferences of each isotope. Below T_c,low-temperature phase transformations and phonon softening were shown from the anomalous temperature dependence of the isomer shifts and area fractions of ^{119}Sn in $EuBa_2(Cu_{1-x}Sn_x)_3O_{7-y}$ ceramics.[277] A Russian analysis of spectra obtained at 80-100 K has identified two types of Sn^{4+} in $EuBa_2(Cu_{1-x}Sn_x)_3O_{7-y}$.[491] One corresponds to Cu^+ coordinated to six oxygens and the other to Cu^{2+} coordinated to five oxygen atoms.

Some information on 2223-type superconductors has been reported. The room temperature Moessbauer spectra of $Bi(Pb)_2Sr_2Ca_2Cu_3Sn_{0.015}O_{10-y}$ showed an intense doublet with $\delta = 0.18$ mm.s^{-1} and $\Delta = 1.01$ mm.s^{-1} indicating that Sn^{4+} substituted for Cu in the triangular CuO_3 site of the CuO_4 layer.[492] The temperature dependences of the isomer shift and resonance area were consistent with the "combined Debye and Einstein model". Only the normal vibration, without softening, was observed in the CuO_4 layer. Moessbauer data were reported for $(Bi_{0.8}M_{0.15}Sb_{0.05})_2Sr_2Ca_2Cu_3O_y$ (M = Pb, Sn),[493] and the so-called superconductor $Sn_2Ba_2Sr_{0.5}Y_{0.5}Cu_3O_x$ was shown to be a mixture of $BaSnO_3$ and $YBaSrCu_3O_{7-y}$.[494]

Coordination and Organometallic Compounds. High-performance ionic conductors α-$PbSnF_4$ and $BaSnF_4$ have been studied using X-ray diffraction, Moessbauer spectroscopy and EXAFS.[495] The large Moessbauer quadrupole splitting confirmed the pseudooctahedral SnF_5E coordination around tin,where the lone pair (*E*) makes the materials highly anisotropic. Fluorides isolated from the SrF_2-SnF_2 system have been investigated.[496] The Goldanskii-Karyagin effect was observed for the fluorides $SrSnF_4$, $SrSn_2F_6$ and $SrSn_4F_{10}$. Tin oxyfluoride, Sn_2OF_2 was studied over the temperature range 4.2-383 K.[497] The effective vibrating mass of Sn(II) was deduced from the temperature dependence of the isomer shift. The range of values for the Moessbauer parameters of $SnCl_2 \cdot 2H_2O$ has been determined to be $3.55 \leq \delta \leq 3.65$ mm.s^{-1} and $1.10 \leq \Delta E \leq 1.30$ mm.s^{-1}.[498] The spread in the published data was attributed to variations in the water of crystallization and oxidation of tin(II).

Gamma-resonance spectra of tin(II) complexes with salicylideneimines,[499] oxabis(ethylene-nitrilo)tetramethylenephosphonic acid,[500] and diethylenetriaminepentaacetate and pentacin,[501] were reported. It was shown that TcO_4^- converted Sn(II) to Sn(IV).[500,501] Numerous complexes containing the $SnCl_3^-$ moiety were characterized with the tin Moessbauer effect. These complexes include: Pd^{II}-Sn complexes occurring in propan-2-ol[502] and nitromethane[503]; and in sulphuric acid solutions;[504] the Pd(I) complex $PPd(PPh_3)(SnCl_3)]_2(CO)$;[505] Pt-Sn complexes from malic acid solutions;[506] complexes formed between Ru, Os, Rh, and Ir and Sn(II) in sulphuric acid solutions;[507] and a Rh complex which contains bridging $SnCl_2$ groups and a terminal $SnCl_3$.[508]

The ^{57}Fe and ^{119}Sn Moessbauer resonances have been used to study $FeSnF_6$ at 4.2-635 K.[83] From the tin spectra it was proposed that the lattice dynamical parameters of $FeSnF_6$ and SnF_4 are similar, despite the structural differences. The series $K_2Sn(OH)_{6-m}F_m$ has been investigated.[509] Both the isomer shift and quadrupole splitting were linearly related to *m* and the average Pauling electronegativity of the ligands. The symmetry of the SnF_6^{2-} octahedra in $(N_2H_6)_2SnF_6F_2$;[510] layered Sn(IV) hydrogen phosphates;[511] $SnCl_4$(1,3-diethylthiourea);[512] and clathrochelate tin-containing iron(II) dioximates[98] were also examined.

Partial quadrupole splittings (p.q.s.) have been used to study several series of tin(IV) complexes. A series of diethyltin(IV) complexes with carbohydrate ligands was prepared.[513] Comparison of the experimental splittings with those calculated on the basis of partial quadrupole splittings revealed three classes of Sn(IV) environment, *viz*: those with pure trigonal bipyramidal, with pure octahedral, and with a 1:1 mixture of both octahedral and bipyramidal arrangements. Twenty-five organotin compounds with 4-, 5-, and 6-coordinations were studied.[514] Quadrupole splittings and asymmetry parameters calculated on the basis of the p.q.s. concept agreed with molecular structures obtained from X-ray and NMR studies. The mean Sn-Br distance in *trans*-$SnBr_4L_2$ species has been related to the p.q.s. and the results used to aid the assignment of υ(SnBr) vibrations in the IR and Raman spectra of such compounds.[515]

During the review period a large number of papers have described the use of Moessbauer spectroscopy as one of several techniques used to study the structures of organotin compounds. Compounds which were studied include: organotin hexacyanoferrates;[516] organophosphorus tin complexes;[517] complexes of rhenocene hydride with methylchlorostannates;[518] triorganotin esters of amic[519] and 3-ureidopropinoic acids;[520] compounds tested for anti-tumour activity;[521-523] cyclic[524] and bridged[525,526] diorganotin(IV) compounds; stannylimides;[527] complexes of tin with DNA and model ligands mimicking nucleic acid phosphate sites;[528] DMSO and HMPA adducts of $RSnCl_3$;[529] and the products of the thermal decomposition of organotin oxides and carboxylates.[530]

7. Other Elements

This section reviews the published data for elements other than iron and tin. In each of the three main sub-sections (main group elements, transition metal elements, and lanthanide and actinide elements) the isotopes are discussed in order of increasing atomic number.

Main Group Elements. - Antimony (Sb-121) The relationship between the Moessbauer isomer shifts and the electron contact densities were investigated for several rapidly frozen solutions of antimony compounds.[448] The isomer shifts are compared with the valence electron densities at the Sb nucleus calculated with an ab initio MO method. A value of $\Delta R/R = -(10.2 \pm 1.0)\times10^{-4}$ was obtained for the 37.15 keV transition of ^{121}Sb.

Moessbauer spectroscopy has been used in the study of several antimony phosphate compounds, in conjunction with X-ray determinations. The materials studied were: mixed-valent $Sb_2(PO_4)_3$,[531] a series of compounds containing layers of $[(Sb^{III}F)XO_4]$ (X = P, As),[532] and $MSb(PO_4)_2$ and $MSbO_xPO_4$ (M = Li, Na, K, Cs).[533] A linear correlation has been found between the chemical isomer shift and the quadrupole coupling constant for a series of antimony(III) tris-thiolates.[534] This behaviour was rationalized by considering an effect on the s/p character of the lone pair of electrons from the influence of secondary bonds.

The crown ether complexes SbX_3(15-crown-5) (X = F, Cl, Br) were characterized by infra-red and ^{121}Sb Moessbauer spectroscopy.[535] A method of orbital population analysis using Moessbauer isomer shifts and quadrupole coupling was developed for Sb(III) halides and their crown ether complexes. X-ray structures with Moessbauer data have been published for a four membered Sb-N ring system (**III**) in which one Sb is associated with other molecules via chlorine bridges;[536] and for $[PhSbCl_4OMe][1,10\text{-phenanthroline}]$.[537]

III

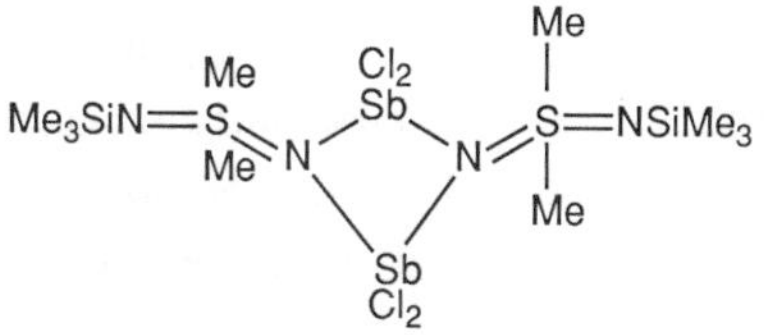

The electronic configuration of antimony ions in perovskite superconducting materials has been studied.[538] The isomer shift was that of Sb(V) and changed with the antimony content. The isomer shifts of the antimony dopant were 7.00 mm s^{-1} in $BaPb_{0.75}Sb_{0.25}O_3$, 7.25 mm s^{-1} in $BaPb_{0.75}Bi_{0.25}O_3$ and 7.70 mm s^{-1} in $(Ba_{0.6}K_{0.4})BiO_3$. In the 2223 superconductor $(Bi_{0.8}X_{0.15}Sb_{0.25})_2$-$Sr_2Ca_2Cu_3O_y$ (X = Pb or Sn) was found to be present as Sb^{5+}.[493]

Tellurium (Te-125). An isomer shift calibration for ^{125}Te has been made using the measured changes $\Delta I_\gamma/I_\gamma$ of the nuclear isomer ^{123m}Te.[447] Only approximate values of $\Delta\rho(0)$ could be obtained. Iodine impurities in silicon have been studied using the ^{125m}Te Moessbauer emission spectra following the decay of implanted ^{125}I.[539] Moessbauer experiments were performed as a function of source temperature with the γ-ray detector along the <100>, <110> or <111> direction of the crystal. The spectra were fitted with a quadrupole doublet, and the anisotropic Debye temperatures were calculated. From the sign of the electric field gradient, it was suggested that the effect was due to the presence of a nearest-neighbour vacancy.

The complex $[PPh_4]_2[Pd(Te_4)_2]\cdot$DMF contains anions with a nearly square-planar coordination of Pd with the two chelated Te_4^{2-} ligands.[540] A large quadrupole splitting in the ^{125}Te Moessbauer spectra indicates a significantly asymmetric population of the 5p orbitals similar to that observed for transition metal ditellurides. Moessbauer data for the low temperature β-phases in the ternary systems (Ag,Au)Te have also been recorded.[541]

Iodine (I-127 and I-129). It has been suggested that Mg_3TeO_6 would be a suitable material for an iodine Moessbauer source.[542] This material is simple to prepare, contains Te^{6+} in a perfectly cubic site, and has a high Debye temperature (320 K).

The ^{127}I Moessbauer effect has been used to study iodine-doped poly(tetrathiafulvalenes).[543] The iodine was intercalated in the form of I_3^- and I_5^-, with the I_3^- content decreasing with increasing doping level. Both the ^{127}I and ^{129}I isotopes were used to study polyiodide anions in oriented and non-oriented samples of N-$(CH)_n$ and S-$(CH)_n$.[544] Both I_3^- and I_5^- were present with identical hyperfine parameters. Relative abundances of the two iodine species and anisotropic binding strengths were determined from the relative intensities of the subspectra. In N-$(CHI_{0.14})_n$ there was 23% I_3^- and 75% I_5^-, a small I^- component was due to I^- found not in the acetylene matrix but in the Be windows of the absorber holder. Light-polarizing films of iodine-doped and I + Br-doped PVA were investigated with the ^{129}I Moessbauer effect.[545] Iodine was present as I^-, I_3^- and I_5^-, with abundances which change when the films stretch.

Ohmic contact formation in Au/Te/Au/GaAs structures has been investigated by ^{129}I Moessbauer spectroscopy.[546] Good, ohmic contacts were found to be correlated to the formation of a high density of defect structures of the type $Te_{As}V_{Ga}^{-}$.

The valence states of iodine in ^{129}I-doped samples of superconducting $YBa_2Cu_3O_{6.1}$ ceramics.[547] Iodine was mainly present as I^- with an isomer shift suggesting that its charge in this system lies in the range -0.87 to 0.

Caesium (Cs-133) Two Russian papers have described the use of $^{133}Ba(^{133}Cs)$ emission Moessbauer spectroscopy to study the barium sites in $YBa_2Cu_3O_{7-y}$ ceramics.[548,549] The parameters for the electric field gradient tensors in $YBa_2Cu_3O_7$ and $YBa_2Cu_3O_6$ were calculated (using a point charge model), but were not in agreement with the experimental values.

Transition Metal Elements. - Zinc (Zn-67) A series of papers have described work carried out on the emission Moessbauer spectra of $^{67}Cu(^{67}Zn)$ in the Cu(1) and Cu(2) sites in $YBa_2Cu_3O_7$ and $YBa_2Cu_3O_6$ ceramics.[549-553] Because of ambiguity in the Sternheimer antiscreening factor for Zn^{2+} and the atomic charge scaling no quantitative agreement could be obtained between the experimental and theoretical values of the principal component of the electric field gradient tensor. Comparison of these values for the copper sites makes it possible to overcome the problem and determine possible charge states of the atoms at the lattice sites. The conclusion was that the hole is localized at the O(4) site, although it is partially transferred to the O(2) and O(3) sites. Similar work was also carried out for $Bi_2Sr_2CaCu_2O_8$.[552]

Silver (Ag-109) The Moessbauer effect in the first excited nuclear level of ^{109}Ag has been observed in a ^{109}Cd-doped single silver crystal.[554] The temperature dependence of the self-absorption of the 88 keV γ-rays allowed the observation of a 0.2% Moessbauer effect at 4.9 K. This nuclear level has a 40 s half-life, and a natural linewidth of $\sim 10^{-17}$ eV, which makes it significant as far as the practical limit for detection of narrow-linewidth Moessbauer signals is concerned.

Tungsten (W-183) Precise Moessbauer line-shape analysis was carried out for tungsten metal.[555] By using an exceptionally intense source (~70 Ci for ^{183}Ta) and carefully chosen constraints between sets of Moessbauer spectra, the theory of final-state effects was quantitatively tested. The temperature dependence of the recoil-free fraction for ^{183}W in tungsten metal was determined, over the temperature range 80-1067 K, with approximately 1% accuracy, and the Debye temperature determined as 336.5 K. A value of 8.76 ± 10 was obtained for the internal conversion coefficient for the 46.5 keV transition in ^{183}W.

Iridium (Ir-191 and Ir-193) Moessbauer lineshape measurements were made for the 129 keV transition of ^{191}Ir in iridium metal using intense sources and carefully chosen constraints between sets of data.[555] The interference parameter for the 129 keV transition in ^{191}Ir in iridium metal was given as -7.7(1.0) x 10^{-3}. This value is approximately 10% greater than the theoretical calculated value.

Hydrides of intermetallic Nb_3Ir were prepared using hydrogen pressures up to 7 GPa and studied with the ^{193}Ir Moessbauer effect.[468] Hydrogenation was found to lead to a reduction in electron density at the Moessbauer nucleus. Pre-reduction of SiO_2-supported Fe-Ir catalysts in hydrogen, and the formation of iron carbides following subsequent treatment in CO and hydrogen was studied using the ^{57}Fe and ^{193}Ir Moessbauer effects.[409]

The ^{193}Ir and ^{197}Au Moessbauer spectra of *trans*-$(Ph_3P)_2Ir(CO)X$ *trans,cis*-$(Ph_3P)_2(H)_2Ir(CO)X$ (**Table 1**) have been compared.[556] The results showed that the substituents on the heterocycle influence the electron density at the iridium nucleus and that the bridging pyrazolato ligand transmits electronic effects from gold to iridium through three bonds. The Moessbauer parameters were sensitive to X but not to the presence of conformers.

Table 1 Moessbauer parameters of Ir-Au pyrazine (pz) complexes (relative to ^{193}Os and Pt/Au)

Substituent	δ (mm s^{-1})	Δ (mm s^{-1})	Γ
trans-$(Ph_3P)_2Ir(CO)X$			
Cl	0.05(2)	7.74(3)	0.75(3)
3,5-Me_2pz-*N*	-0.38(1)	8.41(2)	0.72(4)
3,5-$Me_2$4-NO_2pz-*N*	-0.37(1)	8.30(2)	0.70(4)
3,5-$(CF_3)_2$pz-*N*	-0.43(1)	7.93(2)	0.76(4)
[μ(3,5-Me_2pz-*N,N'*)]AuCl	-0.50(1)	7.22(2)	1.64(3)
idem - gold spectrum	1.02(1)	6.43(2)	1.97(3)
[μ(3,5-Me_2pz-*N,N'*)]AuBr	-0.49(2)	7.45(4)	0.77(9)
idem - gold spectrum	0.98(3)	6.58(4)	1.98(9)
trans,cis-$(Ph_3P)_2(H)_2Ir(CO)X$			
3,5-Me_2pz-*N*	-0.19(1)	3.20(1)	0.66(2)
3,5-$Me_2$4-NO_2pz-*N*	-0.17(1)	3.16(2)	0.64(3)
3,5-$(CF_3)_2$pz-*N*	-0.25(2)	3.62(1)	0.65(2)

Gold (Au-197) The ^{197}Au Moessbauer spectra have been measured for an artificial lattice of 750 layers of Au 1 nm/Ni 1 nm at 16 K.[557] An absorption of about 0.1% was observed, and the shape of the spectra was very different from that of pure gold metal. Low-frequency vibrational modes in sintered Cu and Ag powders have been studied with the ^{197}Au Moessbauer effect.[558] A decrease in resonance absorption area was found for the sinter compared with the bulk materials. One of the Cu sinters showed a broad spectral component superimposed on the normal spectrum, which was attributed to low-frequency oscillations of the particles in the sinters. The gold Moessbauer effect was also used to study Nb_3Au hydrides.[468]

Low temperature β-phases of the systems (Ag,Au)X (X = S, Se, Te) have been investigated.[541,559] The nature of the linear X-Au-X bonds in these systems was discussed on the basis of the Moessbauer data and bond distances. The quadrupole splittings of the ^{197}Au Moessbauer spectra decrease in the order $Ag_3AuS_2 < Ag_3AuSe_2 < Ag_3AuTe_2$, despite the decrease in the X-Au-X bond distances in the same order. It was found that the Au-Se and Au-Te bonds were made up from 6s6p hybrid orbitals mixed with substantial $5d_z2$ character. The same authors compared the $Ag_{1-x}Au_xX$ (X = I, S, Se) systems and obtained similar results.[560] Stable phases formed in the systems were identified, and the trend in the quadrupole splittings was S < Se < I.

A series of $(R_3P)AuQ$ compounds (R = Ph or cyclohexyl; Q^- is a substituted acetylide, a thiolate, or a heterocycle derivative) were prepared and studied.[561] The Moessbauer data show that the Au(I) is two-coordinated, but although sensitive to the type of carbon ligand, the parameters do not allow distinction between the P-Au-C, P-Au-S and P-Au-N coordination. Gamma-resonance spectra and single crystal X-ray analysis of the complex $[Me_3CN(AuPPh_3)_3]BF_4$ were also reported.[562]

The gold-doped ceramics $YBa_{2-x}Au_xCu_3O_{7-y}$ (x = 0.2, 0.4, 0.6) were investigated.[563] Only the sample with x = 0.4 was superconducting (T_c = 90 K). The gold substitutes for the Cu(1) sites and has a valency of 3+ in the superconducting state. When gold substitutes into the Cu(2) sites as Au(I), the superconductivity vanished.

Lanthanide and Actinide Elements. - During the review period papers were published on Moessbauer effect studies using isotopes of neodymium, europium, gadolinium, terbium, dysprosium, erbium, thulium, ytterbium, and neptunium.

Neodymium (Nd-145) Detailed information on the magnitude and orientation of the neodymium sublattice magnetization has been obtained from studies on a $Nd_2Fe_{14}B$ single crystal by using the 72.5 keV Moessbauer transition of ^{145}Nd.[564] At 4.2 K, the average Nd moment was tilted at and angle of 36° relative to the *c*-axis, non-collinear with the iron moment. The temperature dependence of the Nd hyperfine field showed a sharp drop at the spin reorientation temperature accounting for the discontinuity in the total magnetization curve.

Europium (Eu-151) Europium impurity nuclei in Ga_2X_3 (X = S, Se) effect crystals were studied with Moessbauer and EST techniques.[565] The results indicate that Eu interacts was a vacancy forming a double $(Eu^{2+}\text{-}\square_{Ga}{}^{2-})^{\bullet}$-type centre. Two ternary compounds, EuPtSi and EuPdSi, were characterized.[566] The Moessbauer spectra showed a hyperfine split pattern in EuPtSi at 4.2 K. For both compounds the isomer shifts were temperature independent and characteristic of Eu(II). The valence state of Eu-doped in MgS was also reported.[567]

Magnetic susceptibility, electrical resistivity, and ^{151}Eu Moessbauer measurements have been presented for RPt_2Ge_2 (R = La, Ce, Pr, Eu, and Gd).[568] Europium in $EuPt_2Ge_2$ is divalent and orders antiferromagnetically at 12 K. The intermediate-valence character of Eu in $EuNi_2Si_{0.5}Ge_{1.5}$ has been studied by Eu L_{III} X-ray absorption and ^{151}Moessbauer spectroscopy.[569] A sharp temperature-induced valence change was observed. A linear correlation between the mean valence, determined from the Eu L_{III}-edge measurements, and the Moessbauer isomer shift was only found when the variation of the recoil-free fraction at the different europium sites was taken into account.

A light green phase prepared by reduction of $Ca_9Eu(PO_4)_7$ has been characterized by magnetic susceptibility, X-ray, optical, IR, and Moessbauer methods.[570] This new phase was based on the whitlockite structure and contained Eu^{3+}. Low temperature magnetic measurements suggested a change to Eu^{2+} below 65 K. The chemistry of some europium porphyrins has been studied.[571] Under the synthetic conditions used europium was present only as Eu(III).

Mixed perovskite systems $Eu_{0.2}R_{0.8}FeO_3$ (R = rare earth) and $EuFe_{0.8}M_{0.2}O_3$ (M = Cr, Mn, Co) have been studied using the ^{151}Eu and ^{57}Fe isotopes.[341] The europium isomer shift in $Eu_{0.2}R_{0.8}FeO_3$ depend linearly on the unit cell volume $V^{1/4}$, and a relationship was found between the quadrupole splitting and the lattice parameter $1/a^3$. The state of europium ions in sodium borosilicate glass was examined.[488]

The largest number of papers published on ^{151}Eu Moessbauer effect studies describe investigations on high temperature superconducting ceramic. The anisotropy of the lattice vibration of europium in $EuBa_2Cu_3O_7$ was studied on aligned samples.[572,573] Mean square displacements of Eu along the *c*-axis were larger than those in the *ab*-plane. An anomaly was found in the isomer shift above T_C, corresponding to an anomaly in the *s*-electron density near the superconducting transition temperature. Application of a d.c. electric current has been found to change the Moessbauer parameters of $EuBa_2Cu_3O_7$ at 83 K.[574] Changes in isomer shift, quadrupole splitting and asymmetry parameter at the ^{151}Eu nucleus were though to be caused by the movement of electrons along specific directions of the layered structure.

Gamma-resonance studies were carried out on $EuBa_2Cu_3O_7$ with S, K, Fe, Pr, Eu and Ca substitutions.[575] All substitutions affected the europium quadrupole interaction. Substitution of oxygen by sulphur increased the quadrupole interaction by 15% and increased η from 0.3 to 0.8, proving that sulphur replaces oxygen in the immediate neighbourhood of the rare earth. Changes in the superconducting behaviour of $(Eu_{1-x}Pr_x)Ba_2Cu_3O_7$ with x have been noted by Polish workers.[576] The effects of zinc substitution on $Ba_2Eu(Cu_{1-x}Zn_x)_3O_{7-y}$ were described.[577] A single Eu^{3+} resonance line at 295 K found for all these compounds, was almost independent of both the zinc concentration and temperature (down to 10 K). These observations imply that the Cu-O network responsible for superconductivity is weakly coupled to the europium sublattice.

A Hungarian investigation has used the ^{151}Eu, ^{57}Fe, and ^{119}Sn Moessbauer isotopes to study $EuBa_2(Cu_{1-x-y}Sn_xFe_y)_3O_{7-\delta}$.[318,319] and Eu^{3+}-doped $YBa_2Cu_3O_{7-\delta}$.[277] Information was obtained about site preferences and structural changes. Below T_c, low-temperature phase transformation and phonon softening were identified from the anomalous temperature dependence of isomer shifts and the area fraction of the Moessbauer spectra.[277]

Cerium-doped (Eu,La)-Cu-O superconducting compounds have been studied.[578,579] For the T'-phase compounds $(Eu_{1-x}La_x)_{2-y}Ce_yO_{4-\delta}$ the isomer shift and Debye temperatures identified Eu^{3+}, and showed that the Eu-O distance is larger than those in hole-doped superconductors. The system $La_{2-x}Eu_xCuO_4$ was also prepared and studied.[580]

Gadolinium (Gd-155) Gamma-resonance studies on the ternary compounds $Gd_2Mn_{17}C_x$ and $Gd_2Fe_{17}N_x$ have been described. In the carbon compound the ^{155}Gd Moessbauer spectra showed a broadening/splitting which increases with *x*, and reflects an increasing contribution from a component with a large quadrupole interaction.[581] In the nitrogen compound (with *x* close to 3) the electric field gradient at the gadolinium nuclei was 12.6 x 10^{21} Vm^{-2}, substantially larger than in $Gd_2Fe_{14}B$.[582]

Moessbauer and magnetization measurements have been made on $GdIrSi_3$.[583] This compound becomes antiferromagnetic at 15.5 K, and the Gd moments lie in the basal plane. In $GdMn_2Ge_2$, however, the Gd moments lie along the *c*-axis.[584] Measurements in applied magnetic fields indicated that the hyperfine field is negative.

The effects of iron doping on $GdBa_2(Cu_{1-x}Fe_x)_3O_y$ were investigated by X-ray diffraction and Moessbauer spectroscopy.[320] The ^{155}Gd spectra show influence of an extra charge associated with Fe on Cu(2) sites near Gd nuclei. While the isomer shift remained constant the electric field gradient at the Gd nuclei varied smoothly with x. Antiferromagnetic order coexisted with the superconductivity. The properties of mixed $(Gd_{1-x}Pr_x)Ba_2Cu_3O_{7-y}$ superconductors have been described in two papers. The Gd-rich phases were superconducting up to x = 0.45.[585] Magnetic ordering transitions were found in the x = 0.925 phase, and it was shown that Gd substituted for Pr. Some work was also been done on Gd-doped Bi-based systems $Bi_{3.5}Gd_{0.5}Ca_3Sr_3Cu_4O_y$ and $Bi_2Sr_2Ca_{1-x}Gd_xCu_2O_y$ (x = 0.25, 0.5).[587] From the observed quadrupole split spectra it was concluded that Gd occupies Ca sites in both series. Evidence was found for paramagnetic relaxation of isolated Gd ions at low temperatures.

Terbium (Tb-159) Precise lineshape analyses have been made using the 58 keV transition in ^{159}Tb. Spectra were acquired using a Dy_2O_3 source with the Tb_4O_7 and $TbAl_2$ absorbers at 81-297 K. Recoil-free fraction measurements showed that, if the theoretical value of 11 for the internal conversion coefficient is accepted, the results cannot be applied to a Debye model. The mean lifetime of the first excited state of ^{159}Tb was determined to be 77 ps, with a value for the interference parameter of -0.0058, larger than predicted from theory.

Dysprosium (Dy-161) The ^{161}Dy Moessbauer isotope and magnetic susceptibility measurements were used to investigate $DyIrSi_3$.[583] This material orders antiferromagnetically at 7.5 K, and the data suggest that it has an incommensurate spin structure. The reduced average Dy moment of 8.7 μB was attributed to crystal field effects. Measurements on orthorhombic $DyCu_2$ found a weak magnetic anisotropy parallel to the *a*-axis, but no relaxation above T_N.[589]

Erbium (Er-166) Relaxation behaviour in $ErCu_2$ has been studied by the ^{166}Er Moessbauer spectroscopy.[589] This material shows a strong magnetic anisotropy parallel to the *b*-axis and a long relaxation curve from T_N (11.5 K) up to 45 K. A very good correspondence was found between the inelastic neutron scattering and the ^{166}Er Moessbauer results. The same authors also studied $Er_2Fe_{17}N_x$ at 4.2 K.[590] From the quadrupole splitting it was determined that the interstitial solution of nitrogen atoms strongly increases the second order crystal field coefficient ($A_{20} \approx$ -400 K a_0^{-2}).

Thulium (Tm-169) Nuclear Bragg diffraction of synchrotron radiation at the 8.41 keV resonance of ^{169}Tm was reported for the first time.[591] The dynamic theory of Moessbauer optics was applied to evaluate the time spectra. The measurements revealed information about the hyperfine fields which was not previously accessible by conventional Moessbauer spectroscopy on polycrystalline samples.

Ytterbium (Yb-170) The magnetic properties of "green phase" Yb_2BaCuO_5 were examined using ^{170}Yb Moessbauer spectroscopy down to 0.05 K and magnetic susceptibility measurements down to 1.5 K.[592] Magnetic polarization of the Yb^{3+} at each of the two sites was due to couplings with Cu^{2+} moments which magnetically ordered at 15 K. Saturated molecular fields at the two Yb sites were 12 and 22 kG.

The temperature dependence of the fluctuation rate of the Cu(2)-produced internal field acting on a dilute $^{170}Yb^{3+}$ probe substituted at the Y^{3+} sites was studied.[593] These field fluctuations were used to track the collective fluctuations of the nearest neighbour correlated Cu(2) moments. Results were presented for oxygen levels, $x \approx 6.0$ and 6.35, and for $T \leq 80$ K.

Neptunium (Np-237) Emission gamma-resonance spectra of ^{237}Np have been used to study in $^{241}AmO_2$ samples.[594] Relative contributions from Np^{4+} and Np^{5+} were determined, and the spectral lineshape was dependent on the composition of the atmosphere above the oxide. The temperature dependence of the quadrupole splitting was given by the charge state around the neptunium ion. Emission spectra were also recorded for AmO_2-ThO_2 and AmO_2-UO_2 at 77-296 K using resonant absorbers of NpO_2 and $NpAl_2$.[595] Moessbauer absorption spectra were measured in NpO_2-ThO_2 at 77-230 K. A four-fold decrease in the absorption linewidth was observed in response to a decrease in the NpO_2 content from 30% to 10%. This was explained by fast electron exchange between the Np^{4+} and Np^{5+} states. Covalent effects have been evaluated in actinide compounds.[596] The correlation between 5f and 6d Np orbital populations and the Moessbauer isomer shifts were studied.

8. Backscatter and Conversion Electron Moessbauer Spectroscopy.

As in previous years, this section presents a summary of the developments in methodology and instrumentation followed by descriptions of the recent applications of the technique in studies of chemical compounds.

A short review discussing the depth-selective conversion electron Moessbauer technique has been published.[597] The topics covered include: principles of the technique followed by a description of the ultrahigh vacuum electron spectrometer and applications.

A total of seven papers have appeared in the literature concerned with the subject of instrumentation.[598-604] Schaaf and coworkers describe an experimental set-up which enables conversion electron Moessbauer, conversion X-ray Moessbauer and transmission Moessbauer spectroscopic measurements to be made simultaneously.[598] Another paper describes the design of a universal flow-type proportional counter for complex Moessbauer studies at 100-750 K of both surface layers and the bulk of the crystal.[599] This set-up has also been used for recording conversion electron and characteristic X-rays in backscattering geometry along with gamma quanta in the γ-ray transmission mode. Spectral data for Fe_3BO_6 were obtained at 498 K during recording of all three types of radiation.

The design of a proportional counter[600] capable of recording spectra at any temperature from 1.75-300K has now been made possible by the use of different types of gases and gas mixtures - purified He, purified Ne, He+5%N and He+10%CO. The performance of a new parallel plate avalanche counter was used to measure the CEM spectra of iron electrodeposited over a steel substrate.[601] The electron counter was operated with ketone gas inside the ionisation chamber at a pressure of 34 torr and an applied voltage of 650 volts, with the purpose of showing the depth profile of the CEM spectra. A high temperature (77-300 K) proportional counter electron detector was designed for CEMS measurements[602] and showed a signal-to-noise ratio superior to that of smaller detectors due to greater ease of beam alignment to avoid collisions between incident photons and the counter walls.

Mirzababaev has reported[603] on the separation of the resonance and Rayleigh scattering lines of gamma-rays in a single Moessbauer spectrum. In order to represent both types of lines in the spectrum, quanta scattered by electrons and by nuclei must be recorded in detectors in different ways. This was made possible by using a resonance detector in which an ultrathin metallic foil is used as a converter. In the experiment gamma rays from a source of ^{57}Co in Cr were reflected from an Fe scatterer enriched with ^{57}Fe; the reflected radiation hit a detector in which a stainless steel (80% ^{57}Fe) foil is used as the converter.

The next part of this section describes those papers dealing with the use of CEMS to study iron containing materials with particular emphasis on implantation studies, films, oxides, stainless steels and alloys. This is followed by a description of those applications involving the isotope of tin.

Iron-57.- Implantation studies and films A total of seven papers have appeared in the literature[604-610] describing implantation studies involving iron. Diffraction of nuclear gamma rays in a multilayer synthetic structure of ^{57}Fe-Sc, prepared by magnetron sputtering containing 20 layers of ^{57}Fe and Sc 2.0 and 3.3 nm thick respectively, has been studied.[604] The parameters of the structure were studied using X-ray diffractometry, the Hall effect, and conversion electron Moessbauer spectroscopy. Aluminium-implanted iron was investigated[605] by CEMS and evidence of increasing disorder in the implanted region with a transition into the amorphous state was observed.[605] The effects of annealing environment on surface chemical phases of iron-implanted sintered alumina have been investigated.[606] The influence of oxygen partial pressure was studied by comparing samples annealed in air and in Ar environments. CEMS and Rutherford backscattering spectrometry clearly reveal the importance of the annealing atmosphere on the nature of the surface phases in the temperature range 600-1660°C.

McHargue has published two papers, one on the effect of annealing on the structure and distribution of charged states in iron implanted into alumina at 77 K[607] and the other on the chemical state of iron ions implanted into silicon carbide. Implantation of iron into alumina produces an amorphous state for fluences greater than about 3 x 10^{16} Fe cm^{-2}. Iron in the as-implanted specimens was distributed among two Fe^{2+}, two Fe^{4+}, and the Fe^{0} states. The conversion electron Moessbauer spectra for a sample containing 7 x 10^{16} Fe/cm^2 showed all the iron was converted to Fe^{3+} after annealing in oxygen for one hour at 700°C and above. In the second paper the results indicated a distribution of the implanted Fe ions among several sites with slightly differing local environments. The structural composition dependence on amorphous $Si_{1-x}Fe_x$ prepared by ion implantation, with iron concentrations above and below x = 0.2 was studied by CEMS.[609] Three different regions were identified: a silicon-rich one where iron has a strong covalent character in the network structure; an intermediate region where the covalent character decreased continuously at the iron sites; and finally, an iron-rich region, where covalent effects were absent, supporting a dense random packing description of the amorphous structure.

Modifications of α-iron by ion implantation of B^+ ions and ion beam mixing at the B-Fe interface were studied[610] using transmission and conversion electron Moessbauer spectroscopies. Different fluences of 30keV B^+ ions were implanted at room temperature and liquid nitrogen temperature. Ion implantation at liquid nitrogen temperature increased the amorphous fractions with crystallization of the amorphous phases occurring in the region of 373-573 K.

Eleven papers reported the use of CEMS in the study of iron-containing films.[611-621] Electric field gradients measured at the surface of Fe(110) using *in situ* monolayer probe conversion electron Moessbauer spectroscopy have been interpreted using a simple model. Ultrathin films of epitaxial *fcc* Fe(100) were grown under molecular beam conditions on Cu(100) and $Cu_3Au(001)$ substrates at 36°C and their magnetic properties studied.[612] All films were found to be ferromagnetic at 295 and 30K by *in situ* ^{57}Fe conversion electron Moessbauer spectroscopy in ultrahigh vacuum. Most of the ^{57}Fe nuclei experience a large magnetic hyperfine field indicating large atomic magnetic iron moments which are oriented within the film plane. CEMS investigations[613] at interfaces in antiferromagnetically coupled films of Fe/X (X = Cr, Gd, Y) have been reported. The Fe-Cr interaction was found to extend only up the second Fe neighbour at the Fe/Cr interface, whereas the interaction range at the Fe/Gd interface was four times larger, and the Fe/Y system behaviour was very similar to the free Fe-surface. Metallic iron layers (approximately 60 nm thick) deposited on sapphire substrates were mixed at room temperature with various ions (Ne, Ar, Kr, Xe) using wide dose and energy ranges,[614] and the microstructural and chemical characterization of the ion beam mixed iron/aluminium interfaces were investigated using a combination of CEMS and XPS techniques.

The thermal evolution of reactively sputtered iron nitride thin films was investigated by X-ray diffraction, conversion electron and transmission electron Moessbauer spectroscopy and magnetic measurements.[615] Independently of the original nitrogen content of the films, a similar composition of the end product (an α-Fe, and γ-Fe_4N) was achieved, however, the magnetic characteristics were different and depended on the nitrogen content of the as-deposited films.

Ma, Liu and Mei[616,617] have reported, in two papers, studies on compositionally modulated iron-silicon/silicon amorphous films using CEMS. In one experiment,[616] only the silicon layer thicknesses were varied, and the FeSi layer was fixed, whilst in the other[617] the FeSi layers themselves were varied. The results showed that with decreasing Si layer thickness, the hyperfine fields of the samples increased and the thickness of the interface dead layers arising from the atomic interdiffusion effect decreased. The thickness of the interface dead layers was, however, found to be independent of the thickness of the FeSi layers.

Mei *et al*[618] extended this work to include an investigation of the compositionally modulated Fe-B-Si/Si amorphous films prepared by rf sputtering. The saturation magnetisation of the modulated films decreases exponentially with an increase in the thickness of the Si layers, which arises from dead layers. Moessbauer measurements showed that the dead layers are the paramagnetic phase formed by the interdiffusion of Fe and Si atoms at the interfaces. The size effect depends only on the thickness of the Fe-B-Si layers and the dead layer effect depends on both layer thicknesses.

Three papers by Miyazaki[619-621] report studies on the structural changes and magnetic properties of iron-aluminium-silicon Sendust alloys and iron-aluminium-silicon-nickel super-Sendust alloys using CEMS. The films annealed at 500° exhibited excellent soft magnetic properties useful for recording-head materials and CEMS measurements revealed that the disordered structure of α-type in as-sputtered films transformed into the ordered structure of DO_3-type at a temperature of 500°C. With the Fe-Al-Si-Ni super-Sendust film containing 2.7at.% Ni atoms, it was found[621] through conversion electron Moessbauer spectroscopy that the disordered structure transformed into an ordered structure consisting of a mixture of B2 and DO_3 types at 500°.

Oxides. Four of the nine papers reporting on conversion electron Moessbauer studies of oxygen iron-containing compounds describe measurements on iron oxide phases.[622-625] In oxidation of polished steel in air at 300°, Fe_2O_3 forms preferentially followed by Fe_3O_4 formation.[622] An energy window selective method was used to increase the sensitivity of integral CEMS to detect the oxide on the surface by a factor of 2.5 in layers at least 10 nm thick. Powder samples of $^{57}Fe_2O_3$, $^{56}Fe_2O_3$, and haematite of natural isotope abundance implanted with ^{54}Fe, ^{56}Fe, and ^{57}Fe ions were studied.[623] The formation and thermal stability of Fe_3O_4 and FeO precipitates were investigated and possible differences in local states of implanted ions, and of those displaced from substitutional positions in the target are explained. Several (111) oriented $^{56}Fe_3O_4$ films containing a 0.5 nm thick $^{57}Fe_3O_4$ probe layer at or below the surface were grown[624] epitaxially on the α-alumina (0001) surface by a reactive vapour deposition method, and CEMS measurements were made at 6, 78, and 300 K by using a recently developed He-filled proportional counter. The well-crystallized surface was surprisingly stable even in air and were characterized by the Moessbauer parameters that are almost the same as for the bulk.

A glass containing 0.5-2.0 wt.% Fe_2O_3 revealed a metastable phase[625] containing Fe with strongly covalent chemical bonds. The nature of the hyperfine splitting of the phosphate centre indicated that Fe^{2+} and P^{3+} were far apart but interrelated, producing complementary radiation centres. A Moessbauer study of the magnetic properties of a recently prepared artificial new material 'bismuth iron garnet' was carried out.[626] To examine the temperature dependence of sublattice magnetisation, CEMS measurements were made at 6, 78, 290, and 670K. The magnetic and quadrupole hyperfine interactions as compared with the yttrium indium garnet indicated continuous effects of lattice expansion owing to the large spherical size of the Bi^{3+} ions.

Zhou and Zheng, in two papers[627,628] have reported Moessbauer studies on the oxidation of iron immersed in hydrochloric acid and the anti-oxidation effect by corrosion inhibitors. The oxide scale was found to be mainly γ-FeOOH, with a small amount of β-FeOOH, and the ability of the corrosion inhibitor to increase the oxidation resistance in air is proportional to its inhibition efficiency in HCl.

In two papers by Meisel[629] and co-workers[630] conversion electron Moessbauer spectroscopy combined with Auger electron and X-ray photon spectroscopies have been used to investigate the iron stearate on a passivation layer of steel[629] and on oxidized silicon wafers.[630] Heating of these latter samples in air up to 523 K leads to a desorption of the fatty acid chains, while the ferric ions are left on the substrate surface.

Stainless Steels and Alloys. The results of a conversion electron Moessbauer spectroscopic study of steels 90MnV8 and 210Cr46 and Fe after implantation with N, C and Al ions have been reported[631] In another paper, the CEM spectra for unimplanted and N^+ implanted 9Cr18 bearing stainless steel were investigated.[632] The ε-$(Fe,Cr)_{2+x}(C,N)$ and ζ-Fe_2N were the main phases formed after N^+ implantation, which result is essentially the same as that for steels having simple compositions. Some problems of forward and backward scattering of Moessbauer radiation in stainless steel foils with a natural content of ^{57}Fe isotopes were investigated.[633]

For the first time the scattering of Moessbauer radiation was studied in the presence of ultrasonic oscillations in the scatterer. Conversion electron Moessbauer spectroscopy was used[634] to study radiation damage by 100 keV protons in stainless steel SS302. At low doses the defect component is a quadrupole doublet with positive isomer shift, whilst for higher doses, (>5 x 10^{18} protons/cm^2), there is an additional single satellite line with a larger isomer shift.

In Fe-0.1C-1.05Mn-0.7Ni-0.4Cr-0.3Mo-0.3%Si steel foils, conversion electron Moessbauer spectroscopy showed[635] that as the (110) plane number density increased, and as the (200), (222), (211), and (332) plane number densities decreased, the amount of retained austenite (3.4-5%) on the surface decreased. Transmission Moessbauer measurements showed, however, that the bulk austenite fraction (5-5.7%) increased as the (110) texture increased and the (200), (211), (222) and (332) textures decreased. Finally conversion electron Moessbauer spectroscopy was used[636] to analyze the surface of laser-treated Fe-B-Si alloy Metglas 2605S2. The magnetic structure is changed by laser treatment from a flat spin structure within the ribbon plane to a closure domain structure.

Other Elements. - As in previous years, very few results have been reported on conversion electron Moessbauer studies using isotopes other than ^{57}Fe. This year there are reports only on the use of the ^{119}Sn isotope.

Characterization of tin oxide films as gas sensors by conversion electron Moessbauer spectroscopy is the subject of the first paper.[637] The deposition of more than ten monolayers of Pd on SnO_2(110) crystal faces in ultrahigh vacuum was observed by Auger spectroscopy. After initial layer-by-layer growth, cluster formation occurs, and the oxygen from the substrate was transferred to the clusters and the presence of metallic tin was deduced. After annealing above 650°C in ultrahigh vacuum, there is evidence of tin oxide reappearing.

In the second paper, results of hyperfine interactions of ^{111}In implanted tin oxide thin films are presented.[638] Thin tin oxide films were prepared by thermal evaporation of high-purity tin in a low pressure oxygen atmosphere, and radioactive $^{111}In^+$ was implanted into the films. Perturbed angle correlation measurements were made in air after different annealing treatments at temperatures up to 1023 K and the two phases identified and attributed to SnO and SnO_2. The temperature dependence of these phases agreed with CEMS measurements made on tin oxide films after similar annealing programmes.

References

1. "Moessbauer Effect Reference and Data Journal", ed. J.G. Stevens, V.E. Stevens, R.W. White, and J.L. Gibson, Moessbauer Effect Data Center, Univ. N. Carolina, Ashville, N.C. 28814, 1991, Vol. 14.
2. F.J. Berry, in "Phys. Methods Chem." (2nd. Ed.), Vol. 5, Ed. B.W. Rossiter and J.F. Hamilton, Wiley, New York, 1990, p.273.
3. J.I. Eldrige and W.E. O'Grady, in "Tech. Charact. Electrodes", Electrochem. Processes", ed. R. Varma, and J.L. Selman, Wiley, New York, 1991, p343.
4. S. Teruya, *Surf. Sci. Rep.*, 1991, **12**, 49.
5. M. Przybyski, J. Korecki, and U. Gradmann, *Appl. Phys. A*, 1991, **52**, 33.
6. G.M. Kalvius, L. Asch, and J. Chappert, *IOP Short Meet. Ser.*, 1989, **22**, 25.
7. A.H. Morrish, *Struct. Chem.*, 1991, **2**, 211.
8. K. Lazar, *Struct. Chem.*, 1991, **2**, 245.
9. F.J. Berry, *Stud. Surf. Sci. Catal.*, 1990, **57**, A299.
10. B.A. Goodman, *NATO ASI Ser., Ser B*, 1990, **215**, 119.
11. N.H. Gangas, D. Petridis, and A. Simopoulos, in "Pillared Layered Struct.: Curr. Trends Appl., [Proc. Workshop], Meeting Date 1989", Ed. I.V. Mitchell, Elsevier, London, 1990, p.209.
12. G.J. Long and F. Grandjean, *Adv. Solid-State Chem.*, 1991, **2**, 187.
13. S. Mitra, T. Pal, and T. Pal, *Indian J. Pure Appl. Phys.*, 1991, **29**, 313.
14. C. Gleitzer, *Eur. J. Solid State Inorg. Chem.*, 1991, **28**, 77.
15. Yu.F. Krupyanskii, V.I. Gol'danskii, I.V. Kurinov, I.P. Suzdalev, *Stud. Biophys.*, 1990, **136**, 133.
16. M. Kopcewicz, *Struct. Chem.*, 1991, **2**, 313.
17. M.Z. Jin, X.W. Liu, and Y.Q. Jia, *Nuovo Cimento Soc. Ital. Fis.*, 1991, **13D**, 157.
18. M.V. Ivanov, *Geokhimiya*, 1990, 1647.
19. I. Turek, *Nucl. Instrum. Methods Phys. Res., Sect. B.*, 1990, **B52**, 187.
20. D.G. Rancourt and J.Y. Ping, *Nucl. Instrum. Methods Phys. Res., Sect. B.*, 1991, **B58**, 85.
21. H. Flores-Llamas, A. Cabral-Prieto, H. Jiminez-Dominguez, and M. Torres-Valderrama, *Nucl. Instrum. Methods Phys. Res., Sect. A.*, 1991, **A300**, 159.
22. A. Kumar, *Phys. Lett. A*, 1991, **154**, 461.
23. J.S. Thakur and J. Bosse, *Phys. Rev. A*, 1991, **43**, 4388.
24. I. Turek, *Hyperfine Interact.*, 1990, **62**, 343.
25. H. Flores-Llamas and R. Zamorano-Ulloa, *Nucl. Instrum. Methods Phys. Res., Sect. B.*, 1991, **B58**, 272.
26. P. Helisto, I. Tittonen, M. Lippmaa, and T. Katila, *Phys. Rev. Lett.*, 1990, **66**, 2037.
27. M.E. Vakhonin, R.N. Grall, O.A. Gordeev, V.A. Gotlib, S.M. Irkaev, and V.V. Kupriyanov, German (East) Patent, DD 279593.
28. S. Zhou, *Hejishu*, 1991, **14**, 234.
29. A.M. Afanas'ev and E. Yu. Tsymbal, *Hyperfine Interact.*, 1990, **62**, 325.
30. A. Cabral-Prieto, H. Jiminez-Dominguez, and M. Torrez-Valderama, *Nucl. Instrum. Methods Phys. Res., Sect. B.*, 1990, **B54**, 532.
31. R. Chipaux, *Comput. Phys. Commun.*, 1990, **60**, 405.
32. G.R. Hoy, *Gov. Rep. Announce. Index (U.S.)*, 1990, **90**, No. 053,046.
33. I.T. Chong, *J. Phys. Soc. Jpn.*, 1991, **60**, 833.
34. S. Schaefer and E.G. Adelberger, *Z. Phys. A*, 1991, **339**, 305.
35. T.W. Sinor, O.Y. Nabas, J.D. Standifird, and C.D. Collins, *Phys. Rev. Lett.*, 1991, **66**, 1934.
36. S. Kikuta, Y. Yoda, Y. Kudo, K. Izumi, T. Ishikawa, S. Tetsuya, K. Carlos, H. Ohno, H. Takei, and K. Nakamura, *Jpn. J. Appl. Phys., Part 2*, 1991, **30**, L1686.
37. U. van Buerck, R.L. Moessbauer, E. Gerdau, W. Sturhahn, H.D. Rueter, R. Rueffer, A.I. Chumakov, M.V. Zelepukhin, and G.V. Smirnov, *Europhys. Lett.*, 1990, **13**, 371.

38. J.B. Hastings, D.P. Siddons, U. Van Buerck, R. Hollatz, and U. Bergmann, *Phys. Rev. Lett.*, 1991, **66**, 770.
39. M.E. Vakhonin, S.M. Irkaev, V.V. Kupriyanov, and V.A. Semenkin, German (East) Patent, DD 276800.
40. M.E. Vakhonin, R.N. Gall, S.M. Irkaev, and V.V. Kupriyanov, German (East) Patent, DD 280235.
41. F.W.D. Woodhams and S.M. Reader, *Meas. Sci. Technol.*, 1991, **2**, 217.
42. V.I. Sinyavskii and V.V. Leksin, *Prib. Fiz.-Tekh. Eksp.*, 1991, 55.
43. J.T.T. Kumaran and C. Bansal, *Nucl. Instrum. Methods Phys. Res., Sect. B.*, 1990, **B43**, 357.
44. R. Katano, T. Fujii, Y. Isozumi, and S. Ito, *KEK Rep.*, 1990, **90-11**, 100.
45. A.I. Chumakov and G.V. Smirnov, *Pis'ma Zh. Eksp. Teor. Fiz.*, 1991, **53**, 258.
46. J. Huang, Y. Zhang, S. Liu, and M. Jing, *Yuanzineng Kexue Jishu*, 1991, **25**, 20.
47. J.V. Armstrong, M. Enrech, C. Decrouez, J.G. Lunney, and J.M.G. Coey, *IEEE Trans. Magn.*, 1990, **26**, 1629.
48. T. Furubayashi and I. Nakatani, *IEEE Trans. Magn.*, 1990, **26**, 1855.
49. H.M. Ziethen, H. Winkler, A. Schiller, V. Schuenemann, A.X. Trautwein, A. Quazi, and F. Schmidt, *Catal. Today.*, 1991, **8**, 427.
50. I. Choe, R. Ingalls, J.M. Brown, Y. Sato-Sorensen, and R. Mills, *Phys. Rev. B*, 1991, **44**, 1
51. G. Schneider, M. Baier, R. Wordel, F.E. Wagner, V.E. Antonov, E.G. Ponyatovskii, Yu. Kopilovskii, and E. Makarov, *J. Less-Common. Met.*, 1991, **172-174**, 333.
52. S. Nikolov, I. Dragieva, and D. Buchkov, *AIP Conf. Proc.*, 1991, **231**, 294.
53. U.Yu. Yuldashev and A.G. Parsakhanov, *Dokl. Akad. Nauk UzSSR*, 1991, 26.
54. M. Sohma and K. Kawaguchi, *Solid State Commun.*, 1991, **79**, 47.
55. M. Elmassalami and R.C. Thiel, *Physica B*, 1991, **168**, 137.
56. V.I. Rochev, G.N. Kosova, E.F. Makarov, and N.K. Kivrina, *Izv. Akad. Nauk SSSR*, 1991, **55**, 1832.
57. S. Jha, M.I. Youssif, D. Suyanto, G.M. Glenn, R.A. Dunlap, and S.W. Cheong, *J. Phys.: Condens. Matter*, 1991, **3**, 3807.
58. U.C. Johri, R.M. Singru, and D. Bahadur, *Phys. Status Solidi A*, 1990, **161**, 357.
59. Y. Watanabe, M. Nakada, K. Endo, H. Nakahara, and H. Sano, *Bull. Chem. Soc. Jpn.*, 1990, **63**, 2790.
60. T. Sato, M. Nakada, K. Endo, M. Katada, and H. Sano, *J. Radioanal. Nucl. Chem.*, 1991, **154**, 95.
61. M. Mekata, *J. Magn. Magn. Mater.*, 1990, **90-91**, 247.
62. M.W.J. Craje, V.H.J. De Beer, and A.M. Van Der Kraan, *Appl. Catal.*, 1991, L7.
63. G.N. Glavee, C.F. Kernizan, K.J. Klabunde, C.M. Sorensen, and G.C. Hadjapanayis, *Chem. Mater.*, 1991, **3**, 967.
64. R. Kumar, S.K. Date, E. Bill, and A. Trautwein, *Zeolites*, 1991, **11**, 211.
65. J.S. Reddy, K.R. Reddy, R. Kumar, and P. Ratnasamy, *Zeolites*, 1991, **11**, 553.
66. K. Lazar, G. Borbely, and H. Beyer, *Zeolites*, 1991, **11**, 214.
67. A.S. Plachinda, V.E. Sedov, V.I. Khromov, L.V. Bashkeev, and I.P. Suzdalev, *Chem. Phys. Lett.*, 1990, **175**, 101.
68. N.S. Kopelev, A.A. Kamnev, Yu.D. Pefil'ev, and Yu.D. Kiselev, *Vestn. Mosk. Univ., Ser.2*, 1991, **32**, 102.
69. Yu.D. Perfil'ev, H. Alhatib, and L.A. Kulikov, *Vestn. Mosk. Univ., Ser.2*, 1991, **32**, 148.
70. G.U. Nienhaus, H. Frauenhofer, and F. Parak, *Phys. Rev. B*, 1991, **43**, 3345.
71. P.B. Fabrichnyi, L.P. Fefilat'ev, G. Demazeau, J. Etourneau, V.V. Avdeev, and M.A. Kolotyrkina, *Solid State Chem.*, 1991, **78**, 863.
72. R.P. Marande, *Hyperfine Interact.*, 1990, **62**, 225.
73. V.G. Bekeshev, V.Ya. Rochev, and E.F. Makarov, *Mol. Cryst. Liq. Cryst.*, 1990, **192**, 131.

74. E.A. Bekturov, S.E. Kudaibergenov, G.S. Kanapyanova, S.S. Saltybaeva, A.I. Skushnikova, A.L. Pavlova, and E.S. Domnina, *Polym. J. (Tokyo)*, 1991, **23**, 339.
75. T. Yamamoto, K. Sanechika, and H. Sakai, *J. Macromol. Sci., Chem.*, 1990, **A27**, 1147.
76. D.R. Rosseinsky, N.J. Morse, R.C.T. Slade, G.B. Hix, R.J. Mortimer, and D.J. Walton, *Electrochim Acta*, 1991, **36**, 733.
77. Z. Kucharski, H. Winkler, A.X. Trautwein, C. Budrowski, and J. Przyluski, *Synth. Met.*, 1991, **41**, 397.
78. A.A. Zhdanov, M.M. Levitskii, A.Yu. D'yakanov, O.I. Shchegolikhina, A.D. Kolbanovskii, R.A. Stukan, A.G. Knizhnik, and A.L. Buchachenko, *Izv. Akad. Nauk SSSR, Ser. Khim.*, 1990, 2512.
79. Z.D. Wang, M. Inagaki, and M. Takano, *Carbon*, 1991, **29**, 423.
80. C. Meyer, R. Yazami, and G. Chouteau, *J. Phys. (Paris)*, 1990, **51**, 1239.
81. K. Chataknodu, C. Formstone, M.L.H. Green, D. O'Hare, J.M. Twyman, and P.J. Wiseman, *J. Mater. Chem.*, 1991, **1**, 205.
82. K. Aggarwal, F.J. Litterst, G.M. Kalvius, W. Olberding, A. Lerf, and W. Biberacher, *Hyperfine Interact.*, 1990, **55**, 1171.
83. C. Mirambet, L. Fournes, J. Grannec, and P. Hagenmuller, *Mater. Res. Bull.*, 1991, **26**, 797.
84. R.J. Pollard, C. Bauer, J. Laban, and V.H. McCann, *J. Phys.: Condensed Mater.*, 1991, **3**, 5741.
85. E.L. Saw, C.T. Yap, and R. Chevalier, *Acta Phys. Slovaca*, 1990, **40**, 349.
86. T. Ericsson and A.G. Nord, *Acta Chem. Scand.*, 1990, **44**, 990.
87. A.G. Nord, T. Ericsson, and P. Kierkegaard, *Neues Jahrb. Mineral. Monatsh.*, 1990, 159.
88. A.G. Nord, and T. Ericsson, *Neues Jahrb. Mineral. Monatsh.*, 1991, 177.
89. Yu.B. Kopylovskii, I.P. Suzdalev, A.I. Serle, and V.R. Epshtein, *Zh. Fiz. Khim.*, 1991, **65**, 2252.
90. V.R. Epshtein, Yu.B. Kopylovskii, A.V. Kucherov, I.P. Suzdalev, and A.I. Serle, *Zh. Fiz. Khim.*, 1990, **64**, 2699.
91. M. Tanaka, Y. Minai, T. Watanabe, and T. Tominaga, *J. Radioanal. Nucl. Chem.*, 1991, **154**, 197.
92. M. Ichikawa, T. Kimura, and A. Fukuoka, *Stud. Surf. Sci. Catal.*, 1991, **60**, 335.
93. A. Beck, K.M. Mangold, and M. Hanack, *Chem. Ber.*, 1991, **124**, 2315.
94. L.R. Subramanian, A. Guel, M. Hanack, B.K. Mandal, and E. Witke, *Synth. Met.*, 1991, **42**, 2669.
95. M. Hanack, A. Hirsch, and H. Lehmann, *Angew. Chem., Int. Ed. Engl.*, 1990, **29**, 1467.
96. A. Hirsch and M. Hanack, *Chem. Ber.*, 1991, **124**, 833.
97. J.C. Calabrese, P.J. Domaille, S. Trofimenko, and G.J. Long, *Inorg. Chem.*, 1991, **30**, 2795.
98. Y.Z. Voloshin, N.A. Kostromina, A.Y. Nazarenko, and E.V. Polshin, *Inorg. Chim. Acta.*, 1991, **185**, 83.
99. S.V. Lindeman, Yu.T. Struchkov, and Ya.Z. Voloshin, *Inorg. Chim. Acta*, 1991, **184**, 107.
100. Yu.Z. Voloshin, N.A. Kostromina, A.Yu. Nazarenko, V.N. Shuman, and E.V. Pol'shin, *Ukr. Khim. Zh. (Russ. Ed.)*, 1990, **56**, 1238.
101. M.A. Masood, R. Jagannathan, and P.S. Panthappally, *J. Chem. Soc., Dalton Trans.*, 1991, 2553.
102. A. Bonardi, C. Carini, C. Merlo, C. Pelizzi, P. Tarasconi, F. Vitali, and F. Cavatorta, *J. Chem. Soc., Dalton Trans.*, 1990, 2771.
103. T. Sarojini and A. Ramachandraiah, *Indian J. Chem.*, 1990, **29A**, 1174.
104. J.F. Sheuo, H.H. Wei, and Y.D. Yao, *J. Chin. Chem. Soc. (Taipei)*, 1990, **37**, 443.
105. D. Sellmann, M. Geck, and M. Moll, *J. Am. Chem. Soc.*, 1991, **113**, 5239.
106. G.S. Sanyal and S. Garai, *Indian J. Chem.*, 1991, **30A**, 554.
107. Y. Calage, M.C. Moron, J.L. Fourquet, and F. Palacio, *J. Magn. Magn. Mater.*, 1991, **98**, 79.

108. Q.A. Pankhurst, C.E. Johnson, and B.M. Wanklyn, *J. Magn. Magn. Mater.*, 1991, **97**, 126.
109. M. Lahlou-Mimi, M. Leblanc, and J.M. Greneche, *J. Magn. Magn. Mater.*, 1991, **92**, 375.
110. J. Renaudin, G. Ferey, M. Lahlou-Mimi, J.M. Greneche, Y. Mary, and A. De Kozak, *J. Magn. Magn. Mater.*, 1991, **92**, 381.
111. A.S. Kamzin and L.A. Grigor'ev, *Fiz. Tverd. Tela*, 1990, **32**, 3278.
112. R.J.B. Jakeman, M.J. Kwiecien, W.M. Reiff, A.K. Cheetham, and C.C. Torardi, *Inorg. Chem.*, 1991, **30**, 2806.
113. J.M.M. Millet and B.F. Mentzen, *Eur. J. Solid State Inorg. Chem.*, 1991, **28**, 493.
114. W.M. Reiff and C.C. Torardi, *Hyperfine Interact.*, 1990, **53**, 403.
115. R.B. Lanjewar and A.N. Garg, *Indian J. Chem.*, 1991, **30A**, 166.
116. R.B. Lanjewar and A.N. Garg, *Indian J. Chem.*, 1991, **30A**, 350.
117. R.H. Beer, W.B. Tolman, S.G. Bott, and S.J. Lippard, *Inorg. Chem.*, 1991, **30**, 2082.
118. J.F. Hseu, J.J. Chen, C.C. Chuang, H.H. Wei, M.C. Cheng, Yu Wang, and Y.D. Yao, *Inorg. Chim. Acta.*, 1991, **184**, 1.
119. S. Yamauchi, Y. Sakai, and T. Tominaga, *J. Radioanal. Nucl. Chem.*, 1990, **146**, 185.
120. V.H. McCann, J.B. Ward, V. McKee, K. Faulalo, and D.H. Jones, *Hyperfine Interact.*, 1990, **56**, 1465.
121. X. Wang, M.E. Koutun, and J.C. Fanning, *Wuji Huaxue Xuebao*, 1989, **5**, 1.
122. S. Vatsala, *Proc. Indian Natl. Sci. Acad., Part A*, 1990, **56**, 339.
123. M.K. Das, K. Chaudhury, N. Roy, and P. Sarkar, *Transition Metal Chem.*, 1990, **15**, 468.
124. V. Mishra, *Acta Chim. Hung.*, 1990, **127**, 155.
125. J.K. McCusker, J.B. Vincent, E.A. Schmitt, M.L. Mino, K. Shin, D.K. Coggin, P.M. Hagen, J.C. Huffman, G. Christou, and D.N. Hendrickson, *J. Am. Chem. Soc.*, 1991, **113**, 3012.
126. A. Dei, D. Gatteschi, L. Pardi, and U. Russo, *Inorg. Chem.*, 1991, **30**, 3289.
127. I.I. Bulgak, K.I. Turte, and V.E. Zubareva, *Koord. Khim.*, 1991, **17**, 243.
128. J. Zhou, Z. Tai, and Q. Lu, *Huaxue Tongbao*, 1991, 42.
129. A. Bonardi, C. Merlo, C. Pelizzi, G. Pelizzi, P. Tarasconi, and F. Cavatorta, *J. Chem. Soc., Dalton Trans.*, 1991, 1063.
130. N. Matsumoto, H. Tamaki, K. Inoue, M. Koikawa, Y. Maeda, H. Okawa, and S. Kida, *Chem. Lett.*, 1991, 1393.
131. M. Hanack, A. Gul, A. Hirsch, B.K. Mandal, L.R. Subramanian, and E. Witke, *Mol. Cryst. Liq. Cryst.*, 1990, **187**, 625.
132. M. Tonkovicm, S. Horvat, S. Music, and O. Hadzija, *Polyhedron*, 1990, **9**, 2895.
133. B.V. Trzhtsinskaya, N.N. Chipanina, E.S. Domnina, E.B. Apakina, N.A. Khlopenko, K.G. Belyaeva, and S.V. Novichikhin, *Koord. Khim.*, 1991, **17**, 1083.
134. M. Januszczyk, J. Janicki, H. Wojakowska, R. Krzyminiewski, and J. Pietrzak, *Inorg. Chim. Acta*, 1991, **186**, 27.
135. M. Ijjaalo, G. Venturini, B. Malaman, and C. Gleitzer, *C.R. Acad. Sci., Ser. 2*, 1990, **310**, 1419.
136. L. Armelao, M. Bettinelli, G.A. Rizzi, and U. Russo, *J. Mater. Chem.*, 1991, **1**, 805.
137. R. Schmidt, B. Eisnmann, R. Kniep, J. Ensling, P. Guetlich, and E. Seidel, *Z. Naturforsch. B*, 1990, **45**, 1255.
138. M.A. Petrukhina, V.S. Sergienko, L.A. Kulikov, and I.V. Tat'yanina, *Koord. Khim.*, 1990, **16**, 1637.
139. T. Nakamoto, M. Katada, and H. Sano, *Chem. Lett.*, 1991, 1323.
140. A.M. Glebov, A.S. Khramov, E.V. Kirillova, Sh.T. Yusupov, Yu.I. Sal'nikov, and Z.N. Yusupov, *Zh. Neorg. Khim.*, 1990, **35**, 2290.
141. J.K. McCusker, H.G. Jang, M. Zvagulis, W. Ley, H.G. Drickamer, and D.N. Hendrickson, *Inorg. Chem.*, 1991, **30**, 1985.

142. P. Poganiuch, S. Liu, G.C. Papaefthymiou, and S.J. Lippard, *J. Am. Chem. Soc.*, 1991, **113**, 4645.
143. T. Sato, K. Ishishita, M. Katada, H.Sano, Y. Aratono, C. Sagawa, and M. Saeki, *Chem. Lett.*, 1991, 403.
144. Y. Watanabe, H. Kashiwagi, *Ibarakai Kogyo Koto Senmon Gakko Kenkyu Iho*, 1990, 53.
145. L.G. Lavrenova, V.N. Ikorskii, V.A. Varnek, I.M. Olezneva, and S.V. Larionov, *Koord. Khim.*, 1990, **16**, 654.
146. W. Vreugdenhil, J.H. van Diemen, R.A. de Graff, J.G. Haasnoot, J. Reedijk, A.M. van der Kraan, O. Kahn, and J. Zarembowitch, *Polyhedron*, 1990, **9**, 2971.
147. P. Guetlich and P. Poganiuch, *Angew. Chem., Int. Ed. Engl.*, 1991, **30**, 975.
148. A. Rakotonandrasana, D. Boinnard, J.M. Savariault, J.P. Tuchagues, V. Petrouleas, C. Cartier, and M. Verdaguer, *Inorg. Chim. Acta*, 1991, **180**, 19.
149. M. Miteva, P. Bonchev, V. Russanov, E. Zhecheva, D. Mekhandzhiev, and K. Kabassanov, *Polyhedron*, 1991, **10**, 763.
150. J.M. Fiddy, I. Hall, F. Grandjean, G.J. Long, and U. Russo, *J. Phys.*, 1990, **2**, 10091; 10109.
151. Y. Maeda, H. Oshio, Y. Tanigawa, T. Oniki, and Y. Takashima, *Bull. Chem. Soc. Jpn.*, 1991, **64**, 1522.
152. Y. Maeda, H. Oshio, K. Toriumi, and Y. Takashima, *J. Chem. Soc., Dalton Trans.*, 1991, 1227.
153. H. Oshio, K. Toriumi, Y. Maeda, and Y. Takashima, *Inorg. Chem.*, 1991, **30**, 4252.
154. C. Mickalk, T. Klink, K. Knese, A. Seidel, and P. Thomas, *Isotopenpraxis*, 1991, **27**, 76.
155. N.S. Gupta, M. Mohan, N.K. Jha, and W.E. Antholine, *Inorg. Chim. Acta*, 1991, **184**, 13.
156. K.B. Yatsimirskii, E.V. Rybak-Akimova, P.R. Bonchev, E. Zhecheva, and V. Rusanov, *Zh. Neorg. Khim.*, 1990, **35**, 3131.
157. D.C. Figg, R.H. Herber and I. Felner, *Inorg. Chem.*, 1991, **30**, 2535.
158. J.K. McCusker, C.A. Christmas, P.M. Hagen, R.K. Chadha, D.F. Harvey, and D.N. Hendrickson, *J. Am. Chem. Soc.*, 1991, **113**, 6114.
159. D.F. Harvey, C.A. Christmas, J.K. McCusker, P.M. Hagen, R.K. Chadha, and D.N. Hendrickson, *Angew. Chem., Int. Ed. Engl.*, 1991, **30**, 598.
160. C. Juarez-Garcia, M.P. Hendrich, T.R. Holman, L. Que, and E. Munck, *J. Am. Chem. Soc.*, 1991, **113**, 518.
161. H.C. Singh and J.N. Vidyarthi, *Indian J. Pure Appl. Phys.*, 1991, **29**, 774.
162. D. Onggo, A.D. Rae, and H.A. Goodwin, *Inorg. Chim. Acta*, 1990, **178**, 151.
163. J. Silver, *Inorg. Chim. Acta*, 1991, **184**, 235.
164. R.B. Lanjewar and A.N. Garg, *Bull. Chem. Soc. Jpn.*, 1991, **64**, 2502.
165. D.J. Evans, M. Jiminez-Tenorio, G. Leigh, and G. Jeffery, *J. Chem. Soc., Dalton. Trans.*, 1991, 1785.
166. M.S. Sastry and S.K. Kulsheashtha, *Indian J. Chem., Sect.A*, 1990, **29A**, 914.
167. V. Angelov, V. Rusanov, Ts. Bonchev, T. Woike, S. Haussuehl, *Z. Phys. B*, 1991, **83**, 39.
168. C.R. Johnson, C.M. Jones, S.A. Asher, and J.E. Abola, *Inorg. Chem.*, 1991, **30**, 2120.
169. K.S. Baek, P.G. Debrunner, Y.M. Xia, F.J. Wu, and D.M. Kurtz, *Hyperfine Interact.*, 1990, **54**, 447.
170. H. Inoue, T. Takei, G. Heckmann, and E. Fluck, *Z. Naturforsch., B*, 1991, **46**, 682.
171. P. Paladeau, J.P. Venien, M. Bouzid, and J.P. Pradere, *Mater. Res. Bull.*, 1990, **25**, 1129.
172. M. Ferrer, R. Reina, O. Rossell, M. Seco, and X. Solans, *J. Chem. Soc., Dalton Trans.*, 1991, 347.
173. R. Reina, O. Rossell, M. Seco, J. Ros, M. Yanez, and A. Perales, *Inorg. Chem.*, 1991, **30**, 3973.
174. R.B. King, G.S. Chorghade, N.K. Bhattacharyya, E.M. Holt, and G.J. Long, *J. Organomet. Chem.*, 1991, **411**, 419.
175. A. Houston, J.R. Miller, R.M.G. Roberts and J. Silver, *J. Chem. Soc., Dalton Trans.*, 1991, 457.

176. J. Silver, *J. Chem. Soc., Dalton Trans.*, 1990, 3513.
177. C. Roger, P. Hamon, L. Toupet, H. Rabaa, J.Y. Saillard, J.R. Hamon, and C. Lapinte, *Organometallics*, 1991, **10**, 1045.
178. Yu.V. Maksimov, V.V. Mateev, V.D. Tyurin, A.N. Muratov, A.I. Nekhaev, and I.P. Suzdalev, *Izv. Akad. Nauk SSSR, Ser. Khim.*, 1991, 771.
179. S.J. Heyes, N.J. Clayden, and C.M. Dobson, *J. Phys. Chem.*, 1991, **95**, 1547.
180. S. Chen, Y. Chen, and J. Wang, *Huaxue Xuebao*, 1990, **48**, 582.
181. S.L. Lee, F.Y. Li, and T.Y. Dong, *Chem. Phys. Lett.*, 1990, **175**, 170.
182. T.Y. Dong and C.Y. Chiao, *J. Chem. Soc., Chem. Commun.*, 1990, 1332.
183. T.Y. Dong, H.M. Lin, M.Y. Hwang, T.Y. Lee, L.H. Tseng, S.M. Peng, and G.H. Lee, *J. Organomet. Chem.*, 1991, **414**, 227.
184. T.Y. Dong, C.C. Schei, M.Y. Hwang, and T.Y. Lee, *J. Organomet. Chem.*, 1991, **410**, C39.
185. R.J. Webbm T.Y. Dong, C.G. Pierpont, S.R. boone, R.K. Chadha, and D.N. Hendrickson, *J. Am. Chem. Soc.*, 1991, **113**, 4806.
186. S. Nakashima, A. Nishimori, Y. Masuda, H. Sano, and M. Sorai, *J. Phys. Chem. Solids*, 1991, **52**, 1169.
187. K.M. Chi, J.C. Calabrese, W.F. Reiff, and J.S. Miller, *Organometallics*, 1991, **10**, 688.
188. C.J. Lee and H.H. Wei, *J. Chin. Chem. Soc. (Taipei)*, 1991, **38**, 143.
189. G. Al-Jaff, J. Silver, and M.T. Watson, *Inorg. Chim. Acta*, 1990, **176**, 307.
190. O.K. Medhi and J. Silver, *Inorg. Chim. Acta*, 1990, **176**, 247.
191. M. Mezger, M. Hanack, A. Hirsch, J. Kleinwaechter, K.M. Mangold, and L.R. Subramanian, *Chem. Ber.*, 1991, **124**, 841.
192. A. Malek, L. Latos-Graztnski, B. Lechoslaw, J. Tadeusz, amd A. Zadlo, *Inorg. Chem.*, 1991, **30**, 3222.
193. H. Nasri, Y. Wang, B.H. Huynh, F.A. Walker, and W.R. Scheidt, *Inorg. Chem.*, 1991, **30**, 1483.
194. H. Nasri, Y. Wang, B.H. Huynh, and W.R. Scheidt, *J. Am. Chem. Soc.*, 1991, **113**, 717.
195. M.K. Safo, G.P. Gupta, F.A. Walker, and W.R. Scheidt, *J. Am. Chem. Soc.*, 1991, **113**, 5497.
196. E. Tsuchida, T. Komatsu, E. Hasegawa, and H. Nishide, *J. Chem. Soc., Dalton Trans.*, 1990, 2713.
197. F. Parak, *NATO ASI Ser., Ser. A*, 1989, **178**, 197.
198. F. Parak and G.U. Nienhaus, *J. Non-Cryst. Solids*, 1991, **131-133**, 362.
199. E.N. Frolov, M. Fischer, E. Graffweg, M.A. Mirishly, V.I. Goldanskii, and F.G. Parak, *Eur. Biophys. J.*, 1991, **19**, 253.
200. V.E. Prusakov, R.A. Stukan, and V.I. Gol'danskii, *Biomed. Sci. (London)*, 1991, **2**, 127.
201. M.I. Oshtrakh and V.A. Semionkin, *Radiat. Environ. Biophys.*, 1991, **30**, 33.
202. E. Tsuchida, T. Komatsu, T. Babe, T. Nakata, H. Nishhide, and H. Inoue, *Bull. Soc. Chem. Jpn.*, 1990, **63**, 2323.
203. J.G. DeWitt, J.G. Bentsen, A.C. Rosenzweig, B. Hedman, J. Green, S. Pilkington, G.C. Papaethymiou, H. Dalton, K.O. Hodgson, and S.J. Lippard, *J. Am. Chem. Soc.*, 1991, **113**, 9219.
204. W.B. Tolman, S. Liu, J.G. Bentsen, and S.J. Lippard, *J. Am. Chem. Soc.*, 1991, **113**, 152.
205. R.A. Leising, B.A. Brennan, L. Que, B.G. Fox, and E. Munck, *J. Am. Chem. Soc.*, 1991, **113**, 3988.
206. N.M.K. Reid, D.P.E. Dickson, C. Greenwood, A. Thompson, F.H.A. Kadir, and G.R. Moore, *Biochem. J.*, 1990, **282**, 263.
207. F.H.A. Kadir, N.M.K. Read, D.P.E. Dickson, C. Greenwood, A. Thompson, and G.R. Moore, *J. Inorg. Biochem.*, 1991, **43**, 753.
208. K. Spartalian, G. Lang, and R.C. Woodworth, *Biochemistry*, 1991, **30**, 1004.

209. M. Engelhard, K.D. Kohl, K.H. Mueller, B. Hess, J. Heidemeier, M. Fischer, and F. Parak, *Eur. Biophys. J.*, 1990, **19**, 11.
210. H. Hori, K. Kadano, K. Fukuda, H. Inoue, T. Shirai, and E. Fluck, *Radiochim. Acta*, 1990, **51**, 11.
211. F. Parak, A. Birk, E. Frolov, V. Goldanskii, I. Sinning, and H. Michel, *Jerusalem Symp. Quantum Chem. Biochem.*, 1990, **22**, 413.
212. D.L. Williamson, R. Picorel, and M. Seibert, in "Curr. Res. Photosynth., Proc. Int. Conf. Photosynth., 8th, Meeting Date 1989", Vol. 1, Dordrecht, 1990, Ed. M. Baltscheffsky, p.575.
213. B.K. Semin, E.R. Lovyagina, A.Yu. Aleksandrov, Yu.N. Kaurov, A.A. Novakova, *FEBS Lett.*, 1990, **270**, 184.
214. D.J. Evans, A. Hills, D.L. Hughes, G.J. Leigh, A. Houlton, and J. Silver, *J. Chem. Soc., Dalton Trans.*, 1990, 2735.
215. D.J. Evans and G.J. Leigh, *J. Inorg. Biochem.*, 1991, **42**, 25.
216. W.R. Dunham, W.R. Hagen, J.A. Fee, R.H. Sands, J.B. Dunbar, and C. Humblet, *Biochim. Biophys. Acta*, 1991, **1079**, 253.
217. S. Al-Ahmad, J.W. Kampf, R.W. Dunham, and D. Coucouvanis, *Inorg. Chem.*, 1991, **30**, 1163.
218. B.S. Snyder, M.S. Reynolds, R.H. Holm, G.C. Papaefthymiou, and R.B. Frankel, *Polyhedron*, 1991, **10**, 203.
219. S.B. Yu, G.C. Papaefthymiou, and R.H. Holm, *Inorg. Chem.*, 1991, **30**, 3476.
220. P.W. Dimmock, D.P.E. Dickson, and A.GF. Sykes, *Inorg. Chem.*, 1990, **29**, 5120.
221. N.V. Verkhovtseva, V.P. Babanin, and A.M. Shipilin, *Biofizika*, 1991, **36**, 609.
222. N.V. Verkhovtseva, N.V. Malinina, and Yu.V. Rodionov, *Biol. Nauki (Moscow)*, 1990, 155.
223. Y. Kumar, P.D. Sethi, and C.L. Jain, *J. Indian Chem. Soc.*, 1990, **67**, 796.
224. A.A. Olowe and J.M.R. Genin, *Rapp. Tech. - Cent. Belge Etude Corros.*, 1989, **157**, 363.
225. A.A. Olowe, P. Refait, and J.M.R. Genin, *Corros. Sci.*, 1991, **32**, 1003.
226. C.M. Cardile and D.G. Lewis, *Aust. J. Soil Sci.*, 1991, **29**, 399.
227. I. Motov, T. Tabakova, D. Andreeva, and T. Tomov, *Z. Phys. D*, 1991, **19**, 275.
228. A.A. Kamnev, B.B. Ezhov, N.S. Kopelev, Yu.M. Kiselev, and Yu.D. Perfilev, *Electrochim. Acta.*, 1991, **36**, 1253.
229. P.M.A. De Bakker, E. De Grave, R.E. Vandenberghe, L.H. Bowen, R.J. Pollard, and R.M. Persoons, *Phys. Chem. Miner.*, 191, **18**, 131.
230. R.J. Pollard and Q.A. Pankhurst, *J. Magn. Magn. Mater.*, 1991, **99**, L39.
231. S.K. Date and C.E. Deshpande, *Proc. Indian natl. Acad., Part A*, 1989, **55**, 762.
232. E. De Grave and R.E. Vandenberghe, *Phys. Chem. Miner.*, 1990, **17**, 344.
233. F. Studer, C. Houpert, M. Toulemonde, and E. Dartyge, *J. Solid State Chem.*, 1991, **91**, 238.
234. K. Sun, M.X. Lin, D.L. Hou, and H.L. Lou, *Phys. Status Solidi A*, 1991, **126**, 469.
235. Y. Tamanobe, K. Yamaguchi, K. Matsumoto, and T. Fujii, *Jpn. J. App. Phys., Part 1*, 1991, **30**, 478.
236. A. Venkataraman, V.A. Mukhedkar, and A.J. Mukhedkar, *J. Therm. Anal.*, 1989, **35**, 2115.
237. A.K. Nikumbh, P.L. Sayanekar, and M.G. Chaskar, *J. Magn. Magn. Mater.*, 1991, **97**, 119
238. A.K. Nikumbh, *J. Mater. Sci.*, 1990, **25**, 3773.
239. Z. Huang and R. Lin, *Yiqi Yibao Xuebao*, 1990, **21**, 362.
240. P.B. Fabrichnyi, L.P. Fefilat'ev, G. Demazeau, J. Etourneau, V.V. Avdeev, E.F. Levina, V.V. Popov, and M.A. Kremenchugskaya, *Solid State Commun.*, 1991, **78**, 257.
241. F.T. Parker and A.E. Berkowitz, *Phys. Rev. B*, 1991, **44**, 7437.
242. T. Tsuji, K. Ando, K. Naito, and Y. Matsui, *J. Appl. Phys.*, 1991, **69**, 4472.
243. K. Sun and H.L. Luo, *Phys. Status Solidi A*, 1991, **123**, 291.
244. M. Ristic, S. Popovic, M. Tonkovic, and S. Music, *J. Mater. Sci.*, 1991, **26**, 4225.

245. R. Gerardin, A. Alebouyeh, and O. Evrard, *Mater. Res. Bull.*, 1991, **26**, 455.
246. Y. Zhuang, J. Liu, F. Zhao, and J.G. Stevens, *Phys. Lett. A*, 1991, **155**, 127.
247. L. Chen, J. Jiang, F. Lin, and K. Jiang, *Huaxue Xuebao*, 1991, **49**, 529.
248. L.M. Letyuk, M.N. Shipko, A.N. Fedorov, and V.S. Tikhonov, *Izv. Vyssh. Uchebn. Zaved., Chern. Metall.*, 1990, 72.
249. S. Mahmood, I. Abu-Aljarayesh, and N.A. Yusuf, *Magn. Gidrodin.*, 1991, 40.
250. P.H. Domingues, E. Nunez, and J.M. Neto, *J. Magn. Magn. Mater.*, 1991, **96**, 101.
251. T. Kodama, T. Itoh, M. Tabata, and Y. Tamaura, *J. Appl. Phys.*, 1991, **69**, 5915.
252. A. Eggert and E. Riedel, *Z. Naturforsch. B*, 1990, **45**, 1522; 1991, **46**, 653.
253. K. Yamada, H. Ohshita, S. Funabiki, H. Sakai, and S. Ichiba, *Bull. Chem. Soc. Jpn.*, 1990, **63**, 3193.
254. S.N. Okumo, S. Hashimoto, K. Inomata, S. Morimoto, *J. Appl. Phys.*, 1991, **69**, 5072.
255. H.H. Joshi, R.B. Jotania, R.G. Kulkarni, and R.V. Upadhyay, *Solid State Commun.*, 1991, **78**, 539.
256. S. Fischer, H. Langbein, C. Michalk, K. Knese, and U. Heinecke, *Cryst. Res. Technol.*, 1991, **26**, 563.
257. C. Michalk, K. Knese, S. Fischer, W. Toepelmann, and H. Scheler, *Isotopenpraxis*, 1991, **27**, 73.
258. T.A. Dooling and D.C. Cook, *J. Appl. Phys.*, 1991, **69**, 5352.
259. T.A. Dooling and D.C. Cook, *J. Appl. Phys.*, 1991, **69**, 5355.
260. C.S. Lee, C.Y. Hong, Y.H. Kwon, C.H. Moon, and Y.M. Kim, *Sae Mulli*, 1990, **30**, 739.
261. P.B. Pandya, H.H. Joshi, and R.G. Kulkarni, *J. Mater. Sci. Lett.*, 1991, **10**, 474.
262. K.S. Baek, S.W. Nam, H. Chou, H.L. Park, and H.N. Ok, *J. Korean Phys. Soc.*, 1991, **24**, 155.
263. S.H. Lee and K.P. Chae, *J. Korean Phys. Soc.*, 1990, **23**, 405.
264. H.H. Joshi, R.G. Kulkarni, and R.V. Upadhyay, *Indian J. Phys.*, 1991, **65A**, 310.
265. D.A. Ladds, G.A. Saunders, P.J. Ford, D.P. Almond, C. Fanggao, Z. Othaman, S.J. Bending, S. Smith, and B.F. Chapman, *Solid State Commun.*, 1991, **78**, 413.
266. S. Unnikrishnan and D.K. Chakrabarty, *Phys. Ststus Solidi A*, 1990, **121**, 265.
267. Q.A. Pankhurst, *J. Phys.: Condens. Matter*, 1991, **3**, 1323.
268. T. Fujimoto, T. Kimura, K. Ohdan, and K. Haneda, *Nippon Oyo Jiki Gakkaishi*, 1990, **14**, 81.
269. S.H. Lee and S.H. Choi, *Solid State Commun.*, 1991, **78**, 957.
270. F. Jeannot, C. Gleitzer, M. Lenglet, J. Durr, and J.B. Goodenough, *Mater. Res. Bull.*, 1990, **25**, 1377.
271. T. Pannaparayil, R. Marande, and S. Komarneni, *J. Appl. Phys.*, 1991, **69**, 5349.
272. F. Hartmann-Boutron and Y. Gros, *Physica C*, 1990, **172**, 287.
273. P.P. Seregin, F.S. Nasredinov, V.F. Masterov, and G.T. Daribaeva, *Phys. Status Solidi B*, 1990, **159**, K97.
274. I.V. Zubov, A.S. Ilyushin, V.S. Moisa, A.A. Novakova, S.V. Red'ko, L.I. Leonyuk. and A.A. Zhukov, *Sverkhoprovodimost: Fiz., Khim., Tekh.*, 1990, **3**, 1442.
275. E.I. Yur'eva, V.P. Zhukov, and V.A. Gubanov, Zhukov, *Sverkhoprovodimost: Fiz., Khim., Tekh.*, 1991, **4**, 1120.
276. S.S. Tsarevskii, O.A. Anikeenok, and A.B. Liberman, *Sverkhoprovodimost: Fiz., Khim., Tekh.*, 1990, **3**, 2347.
277. E. Kuzmann, S. Nagy, Z. Homonnay, A. Vertes, I. Halasz, M. Gal, B. Csakvari, K. Torkos, and J. Bankuti, *Struct. Chem.*, 1991, **2**, 267.
278. A. Nath, Z. Homonnay, G.W. Jang, S.I. Nagy, Y. Wei, and C.C. Chan, *NIST Spec. Publ.*, 1991, **804**, 407.
279. A. Siedel, L. Haeggstroem, P. Min, S. Eriksson, and L.G. Johansson, *Phys. Scr.*, 1991, **44**, 71.

280. S. Nasu, M. Yoshida, Y. Oda, T. Kohara, T. Shinjo, K. Asayama, F.E. Fujita, S. Katsuyama, Y. Ueda, and K. Kosonge, *J. Magn. Magn. Mater.*, 1990, **90-91**, 664.
281. O.A. Shlyakhtin, K.V. Pokholok, M.N. Oleinikov, and Yu.D. Tret'yakov, *Sverkhoprovodimost: Fiz., Khim., Tekh.*, 1989, **2**, 32.
282. M. De Marco, G. Trbovich, X.W. Wang, J. Hao, M. Naughton, and M. White, *J. Appl. Phys.*, 1991, **69**, 4886.
283. A. Nath, Z. Homonnay, S.D. Tyagi, Y. Wei, G.W. Jang, and C.C. Chan, *Physica C*, 1990, **171**, 406.
284. S. Nasu, M. Yoshida, Y. Oda, K. Asayama, F.E. Fujita, K. Ueda, T. Kohara, T. Shinjo, and S. Katsuyama, "Adv. Supercond. 2, Proc. Int. Symp. Supercond., 2nd. Meeting 1989", ed. T. Ishigura and K. Kajimura, Springer, Tokyo 1990, p.559.
285. R.A. Brand, C. Sauer, H. Luetgemeier, P.M. Meuffels, and W. Zinn, *Hyperfine Interact.*, 1990, **55**, 1229.
286. A.M. Balagurov, G.M. Mironova, I.S. Lyubutin, V.G. Terziev, and A.Yu. Shapiro, *Sverkhoprovodimost: Fiz., Khim., Tekh.*, 1990, **3**, 615.
287. Z. Homonnay, A. Nath, Y. Wei, and T. Jian, *Physica C*, 1991, **174**, 223.
288. I.V. Zubov, A.S. Ilyushin, I.A. Nikanorova, A.A. Novakova, I.E. Graboi, A.P. Kaul, V.V. Moshchalkov, and I.G. Muttik, *Sverkhoprovodimost: Fiz., Khim., Tekh.*, 1989, **2**, 35.
289. I.S. Lyubutin, V.G. Terviev, E.M. Smirnovskaya, and A. Ya. Shapiro, *Sverkhoprovodimost: Fiz., Khim., Tekh.*, 1990, **3**, 2350.
290. J.J. Bara, B.F. Bogacz, C.S. Kim, A. Szytula, and Z. Tomokowicz, *Supercond. Sci. Technol.*, 1991, **4**, 102.
291. S. Srivivasan, V. Sridharan, and T. Nagarajan, *Bull. Mater. Sci.*, 1991, **14**, 697.
292. S. Suharan, D.H. Jones, C.E. Johnson, Q.A. Pankhurst, and M.F. Thomas, *Hyperfine Interact.*, 1990, **55**, 1387.
293. M.G. Smith, R.D. Taylor, and H. Oesterreicher, and *Phys. Rev. B*, 1990, **42**, 4202.
294. S.C. Bhargava, R. Sharma, J.S. Chakrabarty, C.V. Tomy, and S.K. Malik, *Bull. Mater. Res.*, 1991, **14**, 675.
295. T. Shibata, S. Katsuyama, K. Yoshimura, and K. Kosuge, *Jpn. J. Appl. Phys., Part 2*, 1991, **30**, L175.
296. V.A. Adrianov, M.G. Kozin, I.L. Romashkina, S.I. Semenov, V.S. Rusakov, O.A. Shlyakhtin, and V.S. Shpinel, *Sverkhoprovodimost: Fiz., Khim., Tekh.*, 1991, **4**, 1128.
297. Z. Homonnay and A. Nath, *J. Supercond.*, 1990, **3**, 433.
298. A.M. Afanas'ev, E.Yu. Tsymbal, V.M. Cherepanov, M.A. Chuev, S.S. Yakimov, W. Zinn, C. Sauer, and A.A. Bush, *Solid State Commun.*, 1990, **76**, 1099.
299. S. Xia, R. Ma, G. Cao, Z. Li, and Y. Feng, *Chin. J. Met. Sci. Technol.*, 1989, **5**, 434.
300. A.N. Ozernoi, S.O. Akhmetova, M.F. Vereshchak, and A.K. Zhetbaev, *Sverkhoprovodimost: Fiz., Khim., Tekh.*, 1991, **4**, 970.
301. I.V. Zubov, A.S. Ilyushin, I.A. Nikanorova, A.A. Novakova. I.E. Graboi, A.R. Kaul, V.V. Moshchalkov, and I.G. Muttik, *Sverkhoprovodimost: Fiz., Khim., Tekh.*, 1989, **2**, 51.
302. I. Felner, E.R. Bauminger, D. Hechel, U. Yaron, and I. Nowik, *Physica A*, 1990, **168**, 229
303. I. Felner. I. Nowik, E.R. Bauminger, D. hechel, and U. Yaron, *Phys. Rev. Lett.*, 1990, **65**, 1945.
304. L. Bottyan, H. Luetgemeier, J. Dengler, S. Pekker, A. Rockenbauer, A. Janossy, and D.L. Nagy, *Springer Ser. Solid-State Sci.*, 1990, **99**, 230.
305. E. Baggio-Saitovitch and F.J. Litterst, *J. Phys.: Condens. Mater.*, 1991, **3**, 4057.
306. S. Suharan. C.E. Johnson, Q.A. Pankhurst, and M.F. Thomas, *Solid State Commun.*, 1991, **78**, 897.
307. I.S. Lyubutin, V.S. Terziev, and O.N. Morozov, *Pis'ma Zh. Eksp. Teor. Fiz.*, 1990, **52**, 1146.
308. S.C. Bhargava, J.L. Dormann, S. Sayouri, J. Jove, G. Priftis, H. Pankowska, O. Gorochov, and R. Suryanarayanan, *Bull. Mater. Sci.*, 1991, **147**, 687.

309. E. Baggio-Saitovitch, F.J. Litterst, K. Nagamine, K. Nishiyama, and E. Torikai, *Hyperfine Interact.*, 1991, **63**, 259.
310. I. Felner, B. Brosh, S.D. Goren, C. Korn, and V. Volterra, *Phys. Rev. B*, 1991, **43**, 10368.
311. M. DeMarco, M. Qi, J.H. Wang, M. Chaparala, and M.J. Naughton, *Solid State Commun.*, 1991, **78**, 385.
312. D. Bourgault, N. Nguyen, D. Groult, S. Bouffard, J. Provost, M. Hervieu, M. Toulemonde, and B. Raveau, *Radiat. Eff. Defects Solids*, 1990, **114**, 315.
313. A.G. Klimenko, V.I. Kuznetsov, Ya.Ya. Medikov, A.P. Nemudryi, Yu.T. Pavlyukhin, and N.G. Khainovskii, *Sverkhoprovodimost: Fiz., Khim., Tekh.*, 1989, **2**, 5.
314. M. DeMarco, X.W. Wang, G. Trbovich, M.J. Naughton, and M. Chaparala, *Mater. Res. Soc. Symp. Proc.*, 1990, **169**, 1025.
315. A.A. Bush, I.V. Gladyshev, I.V. Zubov, A.S. Ilyushin, R.N. Kuz'min, V.S. Moisa, A.A. Novakova, I.S. Pogosova, and S.V. Red'ko, *Sverkhoprovodimost: Fiz., Khim., Tekh.*, 1989, **2**, 70.
316. S.N. Shringi, R.V. Vadnere, and O.M. Prakash, *Bull. Mater. Sci.*, 1991, **14**, 709.
317. G.A. Chesnokov, V.A. Novichkov, A.A. Stebunov, M.Kh. Sabirov, R.A. Stukan, A.G. E. Knizhnik, and E.F. Makarov, *Zh. Neorg. Khim.*, 1990, **35**, 2723.
318. E. Kuzmann, Z. Homonnay, A. Vertes, I. Halasz, J. Bankuti, and I. Kirschner, *Hyperfine Interact.*, 1990, **55**, 1337.
319. E. Kuzmann, S. Nagy, E. Csikos, A. Vertes, and I. Halasz, *J. Radioanal. Nucl. Chem.*, 1990, **146**, 385.
320. H.J. Bornemann, *Kernforschungszent. Karlsruhe*, 1990, [Ber.] Kfk, KfK 4637.
321. A. Seidel, L. Haeggstroem, and T. Lundstroem, *Phys. Scr.*, 1991, **44**, 74.
322. H. Tang, Z.Q. Qiu, and J.C. Walker, *J. Appl. Phys.*, 1991, **69**, 5379.
323. I. Felner and B. Brosh, *Phys. Rev. B*, 1991, **43**, 10364.
324. I. Felner, I. Nowik, B. Brosh, D. Hechel, and E.R. Bauminger, *Phys. Rev. B*, 1991, **43**, 9737.
325. C. Meyer, F. Hartmann-Boutron, Y. Gros, and P. Strobel, *Solid State Commun.*, 1990, **76**, 163.
326. C. Meyer, F. Hartmann-Boutron, Y. Gros, and P. Strobel, *Physica C*, 1991, **181**, 1.
327. S.C. Bargava, J.S. Chakrabarty, R. Sharma, C.V. Tomy, and S.K. Malik, *Solid State Commun.*, 1990, **76**, 1209.
328. Y. Li, R. Ma, and G. Cao, *Physica C*, 1991, **177**, 36.
329. C. Saragovi, C. Fainstein, P. Etchegoin, and S. Duhalde, *Prog. High Temp. Supercond.*, 1990, **25**, 658.
330. C. Saragovi, C. Fainstein, P. Etchegoin, and S. Duhlade, *Physica C*, 1990, **168**, 493.
331. S. Azdad, O. Gorochov, J.L. Dormann, S. Sayouri, H. Pankowska, and R. Suryanarayanan, *J. Less-Common Met.*, 1990, **164-165**, 588.
332. B.G. Zemskov, A.N. Martynyuk, Yu.V. Permyakov, N.S. Zaugol'nikova, and A.A. Artemova, *Sverkhoprovodimost: Fiz., Khim., Tekh.*, 1990, **3**, 1083.
333. M. Pissas, S. Simopoulos, A. Kostikas, and D. Niarchos, *Physica C*, 191, **176**, 227.
334. T. Sinnemann, M. Mittag, M. Rosenberg, A. Ehmann, T. Fries, G. Mayer-von Kuerthy, and S. Kemmler-Sack, *J. Magn. Magn. Mater.*, 1991, **95**, 175.
335. T. Sinnemann, L. Ressler, M. Rosenberg, T. Fries, A. Ehmann, and S. Kemmler-Sack, *J. Magn. Magn. Mater.*, 1991, **98**, 99.
336. Y. Zeng, X. Yan, M. Lin, X. Meng, H. Cao, X. Wu, Q. Tu, and Z. Lin, *Phys. Rev. B*, 1991, **44**, 867.
337. A. Calles, E. Yepez, J.J. Castro, A. Salcido, A. Cabrera, R. Gomez, S. Aburto, V. Marquina, and M.L. Marquina, *Mater. Sci. Monogr.*, 1991, **70**, 243.
338. A. Wattiaux, L. Fournes, A. Demourgues, N. Bernaben, J.C. Grenier, and M. Pouchard, *Solid State Commun.*, 1991, **77**, 489.
339. T.C. Gibb, *J. Mater. Chem.*, 1991, **1**, 23.
340. T.C. Gibb and M. Matsuo, *J. Solid State Chem.*, 1990, **86**, 164; **88**, 485.

341. M. Jin, X. Liu, W. Zhang, W. Su, W. Xu, and M. Liu, *Solid State Commun.*, 1990, **76**, 985.
342. C.N.W. Darlington, *J. Phys.: Condens. Matter*, 1991, **3**, 4173.
343. Y.L. Chen, X.D. Li, and B.F. Xu, *Hyperfine Interact.*, 1990, **62**, 219.
344. M.N. Shipko, L.M. Letyuk, N.L. Aksel'rod, A.N. Fedorov, and E.V. Tkachenko, *Izv. Akad. Nauk SSSR, Neorg. Mater.*, 1990, **26**, 2551.
345. K. Melzer, A. Martin, and H.C. Semmelhack, *Wiss. Z. - Karl-Marx-Univ. Leipzig, Math.-Naturwiss. Reihe*, 1990, **39**, 404.
346. L.P. Bicelli, S. Maffi, F. Leccabue, A. Deriu, and G. Calestani, *J. Magn. Magn. Mater.*, 1991, **94**, 267.
347. F. Leccabue, R. Panizzieri, S. Garcia, N. Suarez, J.L. Sanchez, O. Ares, X.R. Hua, and *J. Mater. Sci.*, 1990, **25**, 2765.
348. R.K. Gubaidullin, *Izv. Akad. Nauk. SSSR, Neorg. Mater.*, 1990, **26**, 1927.
349. R.K. Gubaidullin, *Izv. Uchebn. Zaved., Fiz.*, 1991, **34**, 5.
350. J. Fontcuberta, W. Reiff, and X. Obradors, *J. Phys.: Condens Mater.*, 1991, **3**, 2131.
351. G.K. Thompson and B.J. Evans, *J. Magn. Magn. Mater.*, 1991, **95**, L142.
352. S. Hirakata, M. Tanaka, K. Kohn, E. Kita, K. Siratori, S. Kimura, and A. Tasaki, *J. Phys. Soc. Jpn.*, 1991, **60**, 294.
353. S.V. Sinitsin and M.N. Uspenskii, *Hyperfine Interact.*, 1990, **55**, 1155.
354. N.A. Kulagin and U.A. Ulmanis, *Izv. Akad. Nauk. SSSR, Neorg. Mater.*, 1990, **26**, 2438.
355. D. Barb, L. Diamandescu, R. Puflea, M. Sorescu, and D. Tarina, *Mater. Lett.*, 1991, **12**, 109.
356. V.K. Sankaranarayanan and N.S. Gajbhiye, *J. Solid State Chem.*, 1991, **93**, 134.
357. M. Vaidya and P.H. Umadikar, *J. Phys. Chem. Solids*, 1991, **52**, 827.
358. R.G. Teller, M.R. Antonio, A.E. Grau, M. Gueguin, and E. Kostiner, *J. Solid State Chem.*, 1990, **88**, 334.
359. C. Boekma, I.M. Suarez, J.C. Lam, T.J. Hoffman, E.N. La Joie, S.F. Weathersby, J.A. Flint, R.L. Lichti, C.P. Wang, and D.W. Cooke, *Hyperfine Interact.*, 1991, **64**, 467.
360. A. Kajima, *Kitakyushu Kogyo Koto Senmon Gakko Kenkyu Hokoku*, 1991, **24**, 65.
361. A. Kajima, T. Kaneda, H. Ito, T. Fujii, I. Okamoto, T. Kimura, and K. Ohdan, *J. Appl. Phys.*, 1991, **69**, 3663.
362. A. Kajima, T. Kaneda, H. Ito, T. Fujii, and I. Okamoto, *J. Appl. Phys.*, 1991, **70**, 3760.
363. A. Correia dos Santos and F.M.A. Da Costa, *Eur. J. Solid State inorg. Chem.*, 1991, **28**, 635.
364. C. Pietzsch, H. Weiner, S. Schoenherr, H.J. Lunk, and R. Stoesser, *Isotopenopraxis*, 1991, **27**, 65.
365. H. Weiner, H.J. Lunk, J. Fuchs, B. Ziemer, R. Stoesser, C. Pietzsch, and P. Reich, *Z. Anorg. Allg. Chem.*, 191, **594**, 191.
366. Z. Wang and Z. Yu, *Wuji Huaxue Xuebao*, 1990, **6**, 278.
367. Z. Yu and Z. Wang, *Huaxue Xuebao*, 1991, **49**, 473.
368. L. Fournes, G. Demazeau, L.M. Zhu, N. Chevreau, and N. Pouchard, *Hyperfine Interact.*, 1990, **53**, 335.
369. C.H. Yo, K.S. Ryu, M.S. Pyun, S.J. Lee, and J.G. Choi, *J. Korean Chem. Soc.*, 1991, **35**, 99.
370. R. Benloucif, N. Nguyen, J.M. Greneche, and B. Raveau, *J. Phys. Chem. Solids*, 1991, **52**, 381.
371. C. Feng, Z. Zhou, F. Jiang, and C. Du, *Huaxue Shijie*, 1991, **32**, 102.
372. K. Tanaka, K. Hirao, and N. Soga, *J. Appl. Phys.*, 1991, **69**, 7752.
373. J.S. Kim, J.Y. Lee, J.S. Swinnea, H. Steinfink, W.M. Reiff, P. Lightfoot, S. Pei, and J.D. Jorgensen, *NIST Spec. Publ*, 1991, **804**, 301.
374. G. Pourroy, A. Riera, I. Malats, P. Poix, and R. Poinsot, *J. Solid State Chem.*, 1990, **88**, 476.

375. E.V. Benvenutti, Y. Gushikem, A. Vasquez, S.C. De Castro, and G.A.p. Zaldivar, *J. Chem. Soc., Chem. Commun.*, 1991, 1325.
376. G.M. Irwin and E.R. Sanford, *Phys. Rev. B*, 1991, **44**, 4423.
377. S. Music, S. Popovic, M. Metikos-Hukovic, and V. Gvozdic, *J. Mater. Sci. Lett.*, 1991, **10**, 197.
378. P. Euzen, P. Palvadeau, M. Qeignec, and J. Rouxel, *C.R. l'Academie Sci., Ser.II Univers*, 1991, **312**, 367.
379. J.F. Ackerman, *J. Solid State Chem.*, 1991, **92**, 496.
380. M. Brearly, *J. Geophys. Res. [Solid Earth Planets]*, 1990, **95**, 15703.
381. S.P. Ekimov, A.G. Tutov, and I.S. Faddeev, *Fiz. Khim. Stekla*, 1990, **16**, 901.
382. V.G. Chekhovskii, J. Keiss, F. Egorov, A.P. Sizonenko, I.K. Khodoseirch, and P. Pauks, *Fiz. Khim. Stekla*, 1991, **17**, 64.
383. N.P. Padture and D.L. Pye, *Glastech. Ber.*, 1991, **64**, 128.
384. M.Yu. Yunusov, R.N. Turaev, A.B. Klyukin, and A.A. Ismatov, *Fiz. Khim. Stekla*, 1990,**16**, 867.
385. T. Pannaparayil, R. Marande, and S. Komarneni, *J. Appl. Phys.*, 1991, **69**, 6040.
386. C. Jeandey, J.L Oddou, J.L. Mattei, and G. Fillion, *Solid State Commun.*, 1991, **78**, 195.
387. E.R. Ruiz, V.I. Nikolaev, and V.S. Rusakov, *J. Radioanal. Nucl. Chem.*, 1991, **153**, 423.
388. C.H.W. Jones, P.E. Kovacs, R.D. Sharma, and R.S. McMillan, *J. Phys. Chem.*, 1990, **95**, 774.
389. C.S. Kim, K.S. Park, M.Y. Ha, and H.S. Lee, *Sae Mulli*, 190, **30**, 263.
390. C.S. Kim, D.Y. Kim, M.Y. Ha, J.Y. Park, and H.N. Ok, *J. Korean Phys. Soc.*, 1990, **23**, 31.
391. E. Makovicky, K. Forcher, W. Lottermoser, and G. Amthauer, *Mineral. Petrol.*, 1990, **43**, 73.
392. E.K.H. Roth, J.M. Greneche, and J. Jordanov, *J. Chem. Soc., Chem. Commun.*, 1991, 105.
393. H.N. Ok, K.S. Baek, H. Choi, and C.S. Kim, *J. Korean Phys. Soc.*, 1991, **24**, 255.
394. J.F. You and R.H. Holm, *Inorg. Chem.*, 1991, **30**, 1431.
395. C. Garcin, A. Gerard, and P. Imbert, *J. Phys. Chem. Solids*, 190, **51**, 1281.
396. M. Magara, T. Tsuji, and K. Naito, *Solid State Ionics*, 190, **40-41**, 284.
397. Y, Zhou and Y. Ding, *Cuihua Xuebao*, 1991, **12**, 167.
398. J.A. Dumesic and W.S. Millman, *ACS Symp. Ser.*, 1990, **437**, 66.
399. W. Wang and W. Pang, *Cuihua Xuebao*, 1991, **12**, 156.
400. Z. Huang, Y. Zeng, Z. Li, Z. Zhang, and D. Wang, *Ranliao Huaxue Xuebao*, 1991, **19**, 35
401. L. Guczi and K. Lazar, *Catal. Lett.*, 190, **7**, 53.
402. G. Filoti, V. Spanu, I. Udrea, M. Udrea, C. Capat, and I.V. Nicolescu, *Rev. Roum. Chim.*, 1990, **35**, 215.
403. M.V. Cagnoli, S.G. Marchetti, N.G. Gallegos, A.M. Alvarez, A.A. Yeramian, and R.C. Mercader, *Mater. Chem. Phys.*, 1991, **27**, 403.
404. P. Putanov, E. Kis, G. Boskovic, and K. Lazar, *Appl. Catal.*, 1991, **73**, 17.
405. J.J. Venter and M.A. Vannice, *Catal. Lett.*, 1990, **7**, 219.
406. V. Baez, E. Lujano, M.R. Goldwasser, Z.J. Perez, M.L. Cubeiro, and P.C. Franco, *Rev. Soc. Venez. Catal.*, 1990, **3**, 81.
407. H.G. Yun, S.I. Woo, and J.S. Chung, *Appl. Catal.*, 1991, **68**, 97.
408. M. Ichikawa, L. Rao, T. Kimura, and A. Fukuoka, *J. Mol. Catal.*, 1990, **62**, 15.
409. F.J. Berry, S. Jobson, T. Zhang, and J.F. Marco, *Catal. Today*, 1991, **9**, 137.
410. M.D. Lee, J.F. Lee, C.S. Chang, and T.Y. Dong, *Appl. Catal.*, 1991, **72**, 267.
411. C.K. Kuei, W.S. Chen, and M.D. Lee, *J. Chin. Chem. Soc. (Taipei)*, 1991, **38**, 127.
412. M.M. Mansurov, G.L. Semenova, N.F. Noskova, A.R. Brodskii, and A.Zh. Kazimova, *Zh. Prikl. Khim.*, 1990, **63**, 2003.
413. S. Weng, J. Gao, Y. Wu, C. Zhao, and Z. Wu, *Ranliao Huaxue Xuebao*, 1990, **18**, 97.
414. S. Weng, Y. Wu, J. Gao, C. Zhao, and Z. Wu, *Ranliao Huaxue Xuebao*, 1990, **18**, 103.

415. S. Wenr, J. Gao, L. Cheng, Z. Wang, and C. Zhao, *Ranliao Huaxue Xuebao*, 1990, **18**, 193.
416. L. Wang, Z. Cui, and S. Liu, *Ranliao Huaxue Xuebao*, 1990, **18**, 268.
417. L. Wang, Z. Cui, and S. Liu, *Ranliao Huaxue Xuebao*, 1990, **19**, 87.
418. G.P. Huffman, B. Ganguly, M. Taghiei, F.E. Huggins, and N. Shah, *Prepr. Pap. - Am. Chem. Soc., Div. Fuel Chem.*, 1991, **36**, 561.
419. W.L.T.M. Ramselaar, M.W.J. Craje, R.H. Hadders, E. Gerkema, V.H.J. De Beer, and A.M. Van der Kraan, *Appl. Catal.*, 1990, **65**, 69.
420. M.W.J. Craje, V.H.J. De Beer, and A.M. Van der Kraan, *Catal. Today*, 1991, **10**, 337.
421. Yu.A. Moskvichev, A.K. Grigorichev, N.A. Bobrov, and A.G. Solomonov, *Izv. Vyssh. Uchebn. Zaved., Khim. Khim. Tekhnol.*, 1990, **33**, 19.
422. S. Osaki, Y. Kuroki, S. Sugihara, and Y. Takashima, *Mem. Fac. Sci., Kyushi Univ., Ser. C*, 1990, **17**, 201.
423. G.N. Goncharov, *Geokhimiya*, 1991, 388.
424. R.G. Burns and T.C. Solberg, *ACS Symp Ser., Vol. Date 1988*, 1989, **415**, 262.
425. E. Murad and U. Wagner, *Neues Jahrb. Mineral., Abh.,*, 1991, **162**, 281.
426. L.P. Aldridge and G.J. Churchman, *Aust. J. Soil Res.*, 1991, **29**, 387.
427. B.A. Goodman, M.V. Cheshire, and J. Chadwick, *J. Soil. Sci.*, 1991, **42**, 25.
428. N.A. Eissam, M.A. Sallam, M.Y. Hassan, and M.S. Kany, *Ann. Geol. Survey Egypt*, 1990, **16**, 73.
429. P.J. Michael, P. Monsef-Mirzai, and W.R. McWhinnie, *Fuel*, 1991, **70**, 119.
430. S. Weng, Y. Xu, Z. Wang, C. Zhao, W. Changgen, J. Wang, and J. Yang, *Yingyong Kexue Xuebao*, 1990, **8**, 292.
431. C. Michalk, K. Knese, W. Boehlmann, and R. Meusinger, *Erdoel Kohle, Erdgas, Petrochem.*, 1990, **43**, 491.
432. J. Komraus and E.S. Popel, *Izv. Khim.*, 1990, **23**, 104.
433. R.P. Tripathi, M.L. Jangid, K. Dhoot, and S. Sharma, *Indian J. Pure Appl. Phys.*, 1991, **29**, 119.
434. R.P. Tripathi and K. Dhoot, *Indian J. Pure Appl. Phys.*, 1991, **29**, 653.
435. B. Gaur, A.K. Singh, and N.J. Rao, *Corros. Sci.*, 1991, **32**, 167.
436. T. Peev, B. Mandzhukova, I. Mandzhukov, and V. Rusanov, *Werkst. Korros.*, 19911, **42**, 90.
437. E. Kuzmann, M.L. Varsanyi, A. Vertes, and W. Meisel, *Electrochim. Acta*, 1991, **36**, 911.
438. H.V. Varma and M. Mathur, *Bull. Electrochem. Soc.*, 1990, **6**, 747.
439. Y. Zeng and N. Yang, *Cem. Concr. Res.*, 1991, **21**, 31.
440. M. Sharma, Vishwamittar, K.S. Harchand, and D. Raj, *Cem. Concr. Res.*, 1991, **21**, 484.
441. C.A. McCammon, V. Chaskar, and G.G. Richards, *Meas. Sci. Technol.*, 1991, **2**, 657.
442. G. Qin, Z. Gao, X. Pan, and Z. Yao, *Hejishu*, 1990, **13**, 530.
443. Z. Gao, S. Chen, X. Pan, and G. Li, *Chin. Sci. Bull*, 1990, **35**, 1954.
444. N.A. Eissa, H.A. Sallam, N. Salah, and N. El Enany, *Acta Phys. Hung.*, 1991, **69**, 25.
445. M. Matsuo and T. Kobayashi, *Nippon Kagaku Kaishi*, 1991, 436.
446. B. Kopcewiczz and M. Kopcewicz, *Struct. Chem.*, 1991, **2**, 303.
447. K. Makariunas, E. Makariuniene, and A. Dragunas, *Liet. Fiz. Rinkinys*, 1990, **30**, 546.
448. M. Yanaga, E. Endo, H. Nakahara, S. Ikuta, T. Miura, M. Takahashi, and M. Takeda, *Hyperfine Interact.*, 1990, **62**, 359.
449. J. Terra and D. Guenzberger, *J. Phys.: Condens. Matter*, 1991, **3**, 6763.
450. K. Burger, A. Vertes, I. Dekany, and Z. Nemes-Vetessy, *Struct. Chem.*, 1991, **2**, 277.
451. S. Bukshpan, *Chem. Phys. Lett.*, 1991, **177**, 269.
452. S. Ambe and F. Ambe, *Bull. Chem. Soc. Jpn.*, 1991, **64**, 1289.
453. S. Ambe and F. Ambe, *Bull. Chem. Soc. Jpn.*, 1990, **63**, 3260.
454. I. Lefebvre, M. Lannoo, J. Olivier-Fourcade, and J.C. Jumas, *Phys. Rev. B*, 1991, **44**, 1004.

455. V. Fano, I. Ortalli, M. Battaglioli, and G. Meletti, *Proc. Int. Conf. Thermoelectr., 9th.*, 1990, 57.
456. D. Baltrunas, R. Mikaitis, V.Yu. Slivka, and Yu.M. Vysochanskii, *Phys. Status Solidi A*, 1990, **119**, 71.
457. E.Yu. Turaev, *Izv. Akad. Nauk UzSSR, Ser. Fiz.-Mat. Nauk,* 1990, 83.
458. F.S. Nasredinov, E.Yu. Turaev, P.P. Seregin, H.B. Rakhmatullaev, and M.K. Bakhadirkhanov, *Phys. Status Solidi A*, 1990, **121**, 571.
459. E.Yu. Turaev and P.P. Seregin, *Izv. Akad. Nauk. UzSSR, Ser. Fiz.-Mat. Nauk*, 1990, 83.
460. V.E. Tezlevan, P.V. Nistiryuk, S.I. Radautsan, S.A. Ratseev, *Fiz. Tverd. Tela*, 1990, **32**, 3157.
461. P. Pringhoej, A. Nylandsted Larsen, and J.W. Petersen, *Mater. Res. Symp. Proc.*, 1990, **163**, 585.
462. S.K. Godovikov, S.A. Nikitin, and A.M. Tishin, *Phys. Lett. A*, 191, **158**, 265.
463. G. Le Caer and S.M. Dubiel, *J. Magn. Magn. Mater*, 1990, **92**, 251.
464. D.L. Williamson and P. Gibart, *Semicond. Sci. Technol.* , 1991, **6**, B70.
465. P. Gibart and D.L. Williamson, *Proc. SPIE-Int. Soc. Opt. Eng.*, 1991, **1362**, 938.
466. T. Yuen, C.L. Lin, P. Schlottmann, N. Bykovetz, P. Pernambuco-Wise, and J.E. Crow, *Physica B*, 1991, **171**, 362.
467. T. Yuen, N. Bykovetz, G.Y. Jiang, C.L. Lin, P.P. Wise, and J.E. Crow, *Physica B*, 1991, **171**, 367.
468. M. Baier, R. Wordel, F.E. Wagner, T.E. Antonova, and V.E. Antonov, *J. Less-Common. Met.*, 1991, **172-174**, 358.
469. J.E. Frackoviak, J. Morgeil, and J. Dutkiewicz, *Conf. Appl. Crystallogr., [Proc.], 13th.*, 1988, 314.
470. A. Adam, J. Sakurai, Y. Yamaguchi, H. Fujiwara, K. Mibu, and T. Shinjo, *J. Magn. Magn. Mater.*, 1990, **90-91**, 544.
471. W. Yang, L. Liwu, and F.J. Berry, *Fenzi Cuihua*, 1991, **5**, 209.
472. Y.X. Li, K.J. Klabunde, and B.H. Davis, *J. Catal.*, 1991, **128**, 1.
473. C. Vertes, E. Talas, I. Czako-Nagy, J. Ryczkowski, S. Gobolos, A. Vertes, and J. Margitfalvi, *Appl. Catal.*, 1991, **68**, 149.
474. B. Stjerna, C.G. Granquist, A. Seidel, and L. Haeggstroem, *J. Appl. Phys.*, 1990, **68**, 6241.
475. D. Zhang, D. Wang, G. Wang, Z. Wang, and Y. Wu, *Wuli Xuebao*, 1991, **40**, 844.
476. M.I. Afanasov, A.A. Shvyryaev, I.A. presnyakov, G. Demazeau, and P.B. Fabrichnyi, *Zh. Neorg. Khim.*, 1990, **35**, 2912.
477. N.J. Tapp and C.M. Cardie, *Zeolites*, 1990, **10**, 680.
478. B.K. Sharma, K.N. Bhattacharya, J. Mishra, P. Samuel, D.K. Mukherjee, and P. Bussiere, *Indian J. Chem., Sect. A*, 1990, **29A**, 1118.
479. K.J.D. MacKenzie, C.M. Cardile, and R.H.. Meinhold, *J. Phys. Chem. Solids*, 1991, **52**, 969.
480. R.A. Grigoryan, N.S. Ovanesyan, G.G. Babayan, and L.A. Grigoryan, *Arm. Khim. Zh.*, 1990, **43**, 232.
481. S. Music, Z. Bais, K. Furic, and V. Mohacek, *J. Mater. Sci. Lett.*, 1991, **10**, 889.
482. S.D. Forder, *Diss. Abstr. Int. B*, 1991, **51**, 4421.
483. M. DeMarco, G. Trbovich, X. Wang, P.G. Mattocks, and M. Naughton, in "Supercond. Appl., [Proc. Int. Conf.], 3rd.", Ed. H.S. Kwok, Y-H.Kao, and D.T. Shaw, Plenum, New York, 1989, p.419.
484. A.I. Akimov, E.A. Vasil'ev, A.N. Plevako, and T.M. Tkachenko, *Vestsi Akad. Navuk. BSSR, Ser. Fiz.-Mat. Navuk*, 1991, 116.
485. G. Wang, H. Zhang, Q. Zhang, and Y. Wu, *Jinshu Xuebao*, 1990, **26**, B366.
486. D.Y. Zhang, G.M. Wang, Y.X. Wang, Z.H. Wang, and Y.H. Zhang, *Solid State Commun.*, 1990, **75**, 629.

487. H. Zhang, Y. Zhao, X. Zhao, G. Wang, S. Liu, Z. He, S. Wang, and Q. Zhang, *Sci. China, Ser. A*, 1990, **33**, 1358.
488. A.V. Strekalovskaya, Yu.I.Zhdanov, E.A. Shabunin, K.N. Mikhlaev, N.P. Filippova, A.M. Sorkin, S.V. Verkhovskii, V.A. Tsurin, and V.L. Kozhevnikov, *Fiz. Met. Metalloved.*, 1990, 204.
489. Y.L. Chen, B.R. Li, A. Chen, and B.F. Xu, *Hyperfine Interact.*, 1990, **55**, 1249.
490. Y. Matsumoto, T. Chiba, T. Nishida, *Kinki Daigaku Kyushu Kogakubu Kenkyu Hokoku, Rikogaku-hen*, 1991, **19**, 19.
491. A.V. Dubovitskii, N.D. Kushch, M.K. Makova, A.T. Mailybaev, V.A. Merzhanov, S.I. Pesotskii, N.V. Kireev, E.F. Makarov, and R.A. Stukan, *Mater. Sci. Forum*, 1990, **62-64**, 25.
492. T. Nishida, M. Katada, N. Miura, Y. Deshimaru, T. Otani, N. Yamazoe, Y. Matsumoto, and Y. Takashima, *Jpn. J. Appl. Phys., Part 2*, 1991, **30**, L735.
493. R.A. Brand, P Sen, C. Sauer, J.P. Sanchez, A. Rojek, and W. Zinn, *Hyperfine Interact.*, 1990, **55**, 1243.
494. T.G.N. Babu, F.J. Berry, and C. Greaves, *Solid State Commun.*, 1991, **79**, 375.
495. G. Denes, Y.H. Yu, T. Tyliszczak, and A.P. Hitchcock, *J. Solid State Chem.*, 1991, **91**, 1.
496. L. Fournes, J. Grannec, C. Mirambet, B. Lestienne, and P. Hagenmuller, *J. Solid State Chem.*, 1991, **93**, 30.
497. L. Fournes, J. Grannec, C. Mirambet, B. Darriet, and P. Hagenmuller, *Z. Znorg. Allg. Chem.*, 1991, **601**, 93.
498. A.S. Khramov and A.N. Glebov, *Koord. Khim.*, 1991, **17**, 246.
499. A.M. Van den Bergen, J.D. Cashion, G.D. Fallon, and B.O. West, *Aust. J. Chem.*, 1990, **43**, 1559.
500. V.V. Sergeev, E.I. Medvedeva, R.A. Stukan, T.A. Babushkina, and G.E. Kodina, *Koord. Khim.*, 1991, **17**, 249.
501. V.V. Sergeev, T.A. Babushkina, A.V. Nikitina, R.A. Stukan, and G.E. Kodina, *Koord. Khim.*, 1991, **17**, 1215.
502. R.Kh. Karymova, L.Ya. Al't, I.A. Agapov, and V.K. Duplyakin, *Zh. Neorg. Khim.*, 1991, **36**, 469.
503. R.Kh. Karymova, L.Ya. Al't, I.A. Agapov, and V.K. Duplyakin, *Zh. Neorg. Khim.*, 1991, **36**, 464.
504. P.G. Antonov, I.A. Agapov, and M.N. Maksimov, *Zh. Obshsh. Khim.*, 1990, **60**, 2421.
505. M.E. Lamberova, A.S. Khramov, V.K. Polovnyak, and N.S. Akhmetov, *Zh. Obshch. Khim.*, 1990, **2069.**
506. P.G. Antonov, T.P. Lutsko, and I.A. Agapov, *Zh. Obshch. Khim.*, 1990, **60**, 1846.
507. P.G. Antonov, E.N. Emel'yanova, I.A. Agapov, and V.P. Kotel'nikov, *Zh.Prikl. Khim.*, 1990, **63**, 2174.
508. S.G. Bott, J.C. Machell, D.M.P. Mingos, and M.J. Watson, *J. Chem. Soc., Dalton Trans.*, 1991, 859.
509. H.S. Chen, *J. Chin. Chem. Soc. (Taipei)*, 1991, **38**, 139.
510. C. Garcia, D. Barbusse, R. Fourcade, and B. Ducourant, *J. Fluorine. Chem.*, 1991, **51**, 245.
511. M.J. Hudson and A.D. Workman, *J. Mater. Chem.* , 1991, **1**, 375.
512. D. Tudela, M.A. Khan, and J.J. Zuckerman, *J. Chem. Soc., Dalton Trans.*, 191, 999.
513. L. Nagy, L. Korecz, I. Kiricsi, L. Zsikla, and K. Burger, *Struct. Chem.*, 1991, **2**, 231.
514. K. Jurkschat, A. Tzschach, H. Weichmann, P. Rajczy, M.A. Mostafa, l. Korecz, and K. Burger, *Inorg. Chim. Acta*, 1991, **179**, 83.
515. D. Tudela and M.A. Khan, *J. Chem. Soc., Dalton Trans.*, 1991, 1003.
516. A. Bonardi, C. Clarini, C. Pelizzi, G. Pelizzi, G. Predieri, P. Tarasconi, M.A. Zoroddu, and K.C. Molloy, *J. Organomet. Chem.*, 1991, **401**, 283.
517. B.F.T. Passos, M.F. De Jesus Filho, C.A.L. Figueiras, A. Abras, and E. Galvao da Silva, *Hyperfine Interact.*, 1990, **53**, 379.

518. A.N. Protskii, B.N. Bulychev, G.L. Soloveichik, I.V. Molodnitskaya, and S.V. Kukharenko, *Metalloorg. Khim.*, 1990, **3**, 1048.
519. A. Samuel-Lewis, P.J. Smith, J.H. Aupers, and D.C. Povey, *J. Organomet. Chem.*, 191, **402**, 319.
520. K.M. Lo, V.G.K. Das, W.H. Yip, T.C.W. Mak, *J. Organomet. Chem.*, 1991, **412**, 21.
521. A. Meriem, M. Biesemans, R. Willem, B. Mahieu, D. De Vos, P. Lelieveld, and M. Gielen, *Bull. Soc. Chim. Belg.*, 1991, **100**, 367.
522. M. Boualam, R. Willem, J. Gelen, A. Sebald, P. Lelieveld, S. De Vos, and M. Gielen, *Appl. Organomet. Chem.*, 1990, **4**, 335.
523. R. Barbieri, A. Silvestri, S. Filippeschi, M. Magistrelli, and F. Huber, *Inorg. Chim. Acta*, 1990, **177**, 141.
524. T. Mancilla, N. Farfan, D. Castillo, L. Molinero, A. Meriem, R. Willem, B. Mahieu, and M. Gielen, *Main Group Met. Chem.*, 1989, **12**, 213.
525. G.H. Sadhu and N.S. Boparov, *J. Organomet. Chem.*, 1991, **411**, 89.
526. J. Lorberth, S.H. Shin, M. Otto, S. Wocaldo, W. Massa, and N.S. Yashina, *J.Organomet. Chem.*, 1991, **407**, 313.
527. S.W. Ng, A.J. Kuthubutheen, Z. Arifin, C. Wei, V.G.K. Das, B. Schulze, K.C. Molloy, W.H. Yip, and T.C.W. Mak, *J. Organomet. Chem.*, 1991, **403**, 101.
528. R. Barbieri, A. Silvestri, and V. Piro, *J. Chem. Soc., Dalton Trans.*, 1990, 3605.
529. F. Fu, J. Tian, H. Pan, R. Willem, and M. Gielen, *Bull. Soc. Chim. Belg.*, 1990, **99**, 789.
530. S.J. Blunden, R. Hill, and S.E. Sutton, *Appl. Organomet. Chem.*, 1991, **5**, 159.
531. A. Jouanneaux, A. Verbaere, D. Guyomard, Y. Piffard, S. Oyetola, and A.N. Fitch, *Eur. J. Solid State Inorg. Chem.*, 1991, **28**, 755.
532. K. Holz, F. Obst, and R, Mattes, *J. Solid State Chem.*, 1991, **90**, 353.
533. A.V. Baluev, A.L. Evdokimov, I.I. Kozhina, V.S. Miliakhina, and B.I. Rogozev, *Vestn. Leningr. Univ. Ser. 4: Fiz., Khim.*, 1990, 36.
534. G. Alonzo, M. Consiglio, F. Maggio, N. Bertazzi, *Phosphorus, Sulphur Silicon Relat. Elem.*, 1991, **56**, 287.
535. M. Schaefer, J. Pebler, B. Borgsen, F. Weller, and K. Dehnicke, *Z. Naturforsch., B*, 1990, **45**, 1243.
536. H.W. Roesky, K. Huebner, M. Noltemeyer, and M. Schaefer, *Angew. Chem.*, 1991, **103**, 856.
537. G. Alonzo, N. Bertazzi, G. Bombieri, G. Bruno, F. Nicolo, *J. Crystallogr. Spectrosc. Res.*, 1991, **21**, 635.
538. M. Eibshutz, W.M. Reiff, R.J. Cava, J.J. Krajewski, and W.F. Peck, *Appl. Phys. Lett.*, 1991, **58**, 2848.
539. L. Niesen and B. Stenekes, *J. Phys.: Condens. Matter*, 191, **3**, 3617.
540. H. Wolkers, K. Dehnicke, D. Fenske, A. Khassanov, and S.S. Hafner, *Acta Crystallogr. Sect. C*, 1991, **C47**, 1627.
541. M. Ando, H. Sakai, M. Kinoshita, and Y. Maeda, *Kyoto Daigaku Genshiro Jikkensho Gakujutsu Koenkai Hobunshu*, 1991, **30**, 65.
542. W. Wu, S. Li, and Y. Li, *Heijishu*, 1990, **13**, 513.
543. H. Gruber, M. Abdel-Hamied, G. Wortmann, H.K. Roth, E. Fanghaenel, and K. Klostermann, *Synth. Met.*, 1991, **44**, 55.
544. M. Abdel-Hamied, G. Wortmann, and H. Naarmann, *Synth. Met.*, 1991, **41**, 175.
545. M. Seto, Y. Maeda, T. Matsuyama, H. Yamaoka, and H. Sakai, *Kyoto Daigaku Genshiro Jikkensho Gakujutsu Koenkai Hobunshu*, 1991, **25**, 71.
546. K. Wuyts, J. Watte, R.E. Silverans, M. Van Hove, and M. Van Rossum, *Appl. Phys. Lett.*, 1991, **59**, 1779.
547. Yu.A. Osipyan, O.V. Zharikov, A.M. Gromov, V.K. Kulakov, R.K. Nikolaev, N.S. Sidorov, Yu.S. Grushko, Yu.V. Ganzha, and M.F. Kovalev, *Physica C*, 1990, **17**, 311.
548. G.T. Daribaeva, V.F. Masterov, F.S. Nasredinov, and P.P. Seregin, *Fiz. Tverd. Tela*, 1990, **32**, 3430.

549. F.S. Nasredinov, V.F. Masterov, N.P. Seregin, and P.P. Seregin, *Zh. Eksp. Teor. Fiz.*, 1991, **99**, 1027.
550. S.I. Bondarevskii, V.F. Masterov, and P.P. Seregin, *Fiz. Tverd. Tela*, 1990, **32**, 3150.
551. N.P. Seregin, F.S. Nasredinov, V.F. Masterov, and G.T. Daribaeva, *Supercond. Sci. Technol.*, 1991, **4**, 283.
552. G.T. Daribaeva, V.F. Masterov, F.S. Nasredinov, and P.P. Seregin, *Fiz. Tverd. Tela*, 1990, **32**, 2306.
553. P.P. Seregin, N.P. Seregin, V.F. Masterov, F.S. Nasredinov, and G.T. Daribaeva, *Sverkhprovodimost: Fiz. Khim., Tekh.*, 191, **4**, 1136.
554. S. Rezaie-Serej, G.R. Hoy, and R.D. Taylor, *Proc. Int. Conf. Lasers, Vol. Date 1989*, 1990, 52.
555. B.R. Bullard, J.G. Mullen, and G. Schupp, *Phys. Rev. B*, 1991, **43**, 7405.
556. A.L. Bandini, G. Banditelli, F. Bonati, S. Calogero, and F.E. Wagner, *J. Organomet. Chem.*, 1991, **410**, 241.
557. S. Nasu, T. Abe, T. Shibatani, Y. Maeda, Y. Kawase, S. Uehara, and H. Yoshida, *Kyoto Daigaku Genshiro Jikkensho, [Tech. Rep.]*, 1990, **KURRI-TR-341**, 25.
558. M. Hayashi, E. Gerkema, A.M. Van der Kraan, and I. Tamura, *Phys. Rev. B*, 1990, **42**, 9771.
559. H. Sakai, M. Ando, S. Ichiba, and Y. Maeda, *Chem. Lett.*, 1991, 223.
560. H. Sakai, M. Ando, M. Kinoshita, and Y.Maeda, *Kyoto Daigaku Genshiro Jikkensho, [Tech. Rep.]*, 1990, **KURRI-TR-341**, 8.
561. F. Bonati, A. Burini, B.R. Pietroni, E. Torregiani, S. Calogero, and F.E. Wagner, *J. Organomet. Chem.*, 1991, **408**, 125.
562. A. Grohmann, J. Riede, and H. Schmidbauer, *J. Chem. Soc., Dalton Trans.*, 1991, 783.
563. A.T. Mailybaev, E.F. Makarov, N.V. Kireev, and B.G. Tsinoev, *Izv. Akad. Nauk SSSR, Ser. Fiz.*, 1990, **54**, 1756.
564. I. Nowik, K. Muraleedharan, G. Wortmann, B. Perscheid, G. Kaindl, and N.C. Koon, *Solid State Commun.*, 1990, **76**, 967.
565. I.M. Askerov, G.K. Aslanov, A.O. Mekhrabov, and Kh.F. Khadzhiev, *Hyperfine Interact.*, 1990, **60**, 699.
566. D.T. Adroja, B.D. Padalia, S.K. Malik, R. Nagarajam, and R. Vijayaraghavan, *J. Magn. Magn. Mater.*, 1990, **89**, 375.
567. O. Missous, F. Loup, J. Fesquet, H. Prevost, J. Gasiot, J.P. Sanchez, and C.A. Dos Santos, *Eur, J. Solid State Inorg. Chem.*, 1991, **28**, 163.
568. I. Das, E.V. Sampathkumaran, R. Nagarajan, and R. Vijayaraghavan, *Phys. Rev. B*, 1991, **43**, 13159.
569. G. Wortmann, U. Nowik, B. Perscheid, G. Kaindl, and I. Felner, *Phys. Rev. B*, 1991, **43**, 5261.
570. B.I. Lazoryak, B.N. Viting, P.B. Babrichnyi, L. Furnes, R. Salmon, and P. Hagenmueller, *Kristallografiya*, 1990, **35**, 1403.
571. F.W. Oliver, C. Thomas, E. Hoffman, D. Hill, T.P.G. Sutter, P. Hambright, S. haye, A.N. Thomas, and N. Quoc, *Inorg. Chim. Acta*, 1991, **186**, 119.
572. T. Muraki, M. Taniwaki, and K. Shiramine, *Kogakubu Kenkyu Hokuku (Hokkaido Daigaku)*, 1990, 121.
573. T. Muraki, M. Taniwaki, and K. Shiramine, "Adv. Supercond 2, Poc. Int. Symp. Supercond., 2nd. 1989", Ed. T. Ishiguro and K. Kajimura, Pub. Springer: Tokyo, 1990, p.563.
574. Y.Q. Jia, X.W. Liu, and M.Z. Jin, *Phys. Lett. A*, 191, **155**, 214.
575. E.R. Bauminger, I. Felner, Y. Lehavi, and I. Nowik, *Hyperfine Interact.*, 1990, **55**, 1199.
576. K. Latka, A. Szytula, Z. Tomkowicz, A. Zygmunt, and R. Duraj, *Physica C*, 1990, **171**, 287.
577. C.V. Tomy, R. Nagarajan, S.K. Malik, R. Prasad, N.C. Soni, and K. Adhikary, *Bull. Mater. Sci.*, 1991, **14**, 691; *Solid State Commun.*, 1990, **75**, 59.

578. M. Yoshimoto, H. Koinuma, T. Hashimoto, J. Tanaka, S. Tanabe, and N. Soga, *Physica C*, 1991, **181**, 284.
579. H. Itoh, M. Kusuhashi, and M. Taniwaki, *Jpn. J. Appl. Phys., Part 2*, 1990, **29**, L1604.
580. F.J. Berry, C. Greaves, and R. Lobo, *Hyperfine Interact.*, 1990, **55**, 1213.
581. M.W. Dirken, R.C. Thiel, T.H. Jacobs, and K.H.J. Buschow, *J. Less-Common Met.*, 191, **168**, 269.
582. M.W. Dirken, R.C. Thiel, R. Coehorn, T.H. Jacobs, and K.H.J. Buschow, *J. Magn. Magn. Mater.*, 1991, **94**, L15.
583. J.P. Sanchez, K. Tomala, and K. Latka, *J. Magn. Magn. Mater.*, 1991, **99**, 95.
584. J.P. Sanchez, K. Tomala, and A. Szytula, *Solid State Commun.*, 1991, **78**, 419.
585. Z. Tomkowicz, K. Latka, A. Szytula, A. Bajorek, M. Balanda, R. Kmiec, R. Kruk, and A. Zygmunt, *Physica C*, 1991, **174**, 71.
586. G. Wortmann and I. Felner, *Solid State Commun.*, 1990, **75**, 981.
587. E.V. Sampathkumaran, G. Wortmann, and G. Kaindl, *Bull. Mater. Sci.*, 1991, **14**, 703.
588. B.R. Bullard and J.G. Mullen, *Phys. Rev. B*, 1991, **43**, 7416.
589. P.C.M. Gubbens, K.H.J. Buschow, M. Divis, J. Lange, and M. Loewenhaupt, *J. Magn. Magn. Mater.*, 1991, **98**, 141.
590. P.C.M. Gubbens, A.A. Moolenaar, G.J. Boender, A.M. Van der Kraan, T.H. Jacobs, and K.H.J. Buschow, *J. Magn. Magn. Mater.*, 1991, **97**, 69.
591. W. Sturhahn, E. Gerdau, R. Hollatz, R. Rueffer, H.D. Rueffer, and W. Tolksdorf, *Europhys. Lett.*, 1991, **14**, 821.
592. J.A. Hodges and J.P. Sanchez, *J. Magn. Magn. Mater.*, 1990, **92**, 201.
593. J.A. Hodges, P. Bonville, P. Imbert, G. Jehanno, *Physica B*, 1991, **171**, 332.
594. V.M. Filin, V.F. Gorbunov, and S.A. Ulanov, *Radiokhimiya*, 1991, **32**, 85.
595. V.M. Filin, V.F. Gorbunov, and S.A. Ulanov, *J. Radioanal. Nucl. Chem.*, 1990, **143**, 125.
596. J. Jove and G. Ionova, *J. Radioanal. Nucl. Chem.*, 1990, **143**, 73.
597. I.P. Jain, Y.K. Vijay, and R. Chandra, *Vacuum*, 1990, **41**, 1776.
598. P. Schaaf, A. Kraemer, L. Blaes, G. Wagner, F. Aubertin, and U. Gonser, *Nucl. Instrum. Methods Phys. Res., Sect. B*, 1991, **53**, 184.
599. A.S. Kamzin, and L.A. Grigor'ev, *Prib. Tekh. Eksp.*, 1991, **2**, 74.
600. K. Fukumura, A. Nakanishi, T. Kobayashi, R. Katano, and Y. Iozumi, *Nucl. Instrum. Methods Phys. Res. Sect. B*, 1991, **61**, 127.
601. M. Villagran, L. Sbriz, and G. Valconi, *J. Radioanal. Nucl. Chem.*, 1991, **153**, 375.
602. T. Fujii, R. Katano, S. Ito, and Y. Isozumi, *Bull. Inst. Chem. Res. Kyoto Univ.*, 1990, **68**, 110.
603. R.M. Mirzababaev, *Izv. Vyssh. Uchebn. Zaved. Fiz.*, 1990, **33**, 121.
604. A.I. Chumakov, G. V. Smirnov, S.S. Andreev, N.N. Salashchenko, and S.I. Shinkarev, *Pis'ma Zh. Eksp. Teor. Fiz.*, 1991, **54**, 220.
605. H. Reuther and E. Richter, *Phys. Res.*, 1990, **13**, 460.
606. C. Donnet, G. Marest, N. Moncoffre, and J. Tousset, *Nucl. Instrum. Methods Phys. Res. Sect. B*, 1991, **59-60**, 1177.
607. C.J. McHargue, P.S. Sklad, J.C. McCallum, C.W. White, A. Perez, G. Marest, *Mater. Res. Soc. Symp. Proc.*, 1990, **157**, 555.
608. C.J. McHargue, A. Perez, and J.C. McCallum, *Nucl. Instrum. Methods Phys. Res., Sect. B*, 1991, **59-60**, 1362.
609. F.H. Sanchez, M.B. Fernandez van Raap, and J. Desimoni, *Phys. Rev. B*, 1991, **44**, 4290.
610. M. Hans, G. Frech, G.K. Wolf, and F.E. Wagner, *Nucl. Instrum. Methods Phys. Res., Sect. B*, 1991, **53**, 161.
611. J. Korecki and W. Karas, *J. Magn. Magn. Mater.*, 1990, **92**, L11.
612. W.A.A. Macedo, W. Keune, and E.D. Ellerbrock, *J. Magn. Magn. Mater.*, 1991, **93**, 552.
613. J. Landes, *Forschungszent. Juelich: Ber., Juel 2489*, 1991.

614. E. Abonneau, A. Perez, G. Fuchs, and M. Treilleux, *Nucl. Instrum. Methods Phys. Res., Sect. B*, 1991, **59-60**, 1183.
615. D.H. Mosca, S.R. Teixeira, P.H. Dionisio, I.J.R. Baumvol, W.H. Schreiner, and W.A. Monteiro, *J. Appl. Phys.*, 1991, **69**, 261.
616. X.D. Ma, Y.H. Liu, and L.M. Mei, *J. Phys.: Condens. Matter*, 1991, **3**, 7139.
617. X.D. Ma, Y.H. Liu, and L.M. Mei, *J. Magn. Magn. Mater.*, 1991, **95**, 199.
618. L. Mei, W. Li, S. Bi, and Y. Liu, *Wuli Xuebao*, 1991, **40**, 303.
619. M. Miyazaki, M. Ichikawa, T. Komatsu, K. Matusita, and K. Nakajima, *J. Appl. Phys.*, 1991, **69**, 1556.
620. M. Miyazaki, T. Komatsu, and K. Matusita, *Yoyuen oyobi Koon Kagaku*, 1991, **34**, 193.
621. M. Miyazaki, M. Ichikawa, T. Komatsu, K. Matusita, K. Nakajima, and S. Okamoto, *J. Appl. Phys.*, 1991, **69**, 7207.
622. E. Galvao da Silva, G. P.A. Costa, A.A.G. Campos, and A.L.T. Azevedo, *Scr. Metall. Mater.*, 1991, **25**, 331.
623. H. Bincyzcka, B. Fornal, G. Marest, N. Moncoffre, and J. Stanek, *Radiat. Eff. Defects Solids*, 1991, **116**, 97.
624. T. Fujii, M. Takano, R. Katano, Y. Bando, and Y. Isozumi, *J. Appl. Phys.*, 1990, **68**, 1735
625. S.P. Ekimov, E.A. Lisitsina, N.D. Solov'eva, and D.M. Yudin, *Fiz. Khim. Stekla*, 1990, **16**, 644.
626. T. Fujii, M. Takano, Y. Bando, and T. Okuda, *Funtai oyobi Funmatsu Yakin*, 1991, **38**, 427.
627. S. Zhou and J. Zheng, *Corrosion (Houston)*, 1991, **47**, 197.
628. S. Zhou and J. Zheng, *Yingyong Kexue Xuebao*, 1991, **9**, 104.
629. W. Meisel, *Wiss. Z.-Tech. Hochsch. Ilmenau*, 1991, **37**, 127.
630. W. Meisel, P. Tippmann-Krayer, H. Moehwald, and P. Guetlich, *Fresenius' J. Anal. Chem.*, 1991, **341**, 289.
631. H. Reuther, *Zentrlinst. Kernforsch. Rossendorf Dresden, [Ber.] ZfK*, ZfK-731, 1990, 96.
632. M. Wu, Z. Tong, Q. Zhang, Y. Jin, and S. Wang, *Nucl. Instrum. Methods Phys. Res., Sect.B*, 1990, **B52**, 44.
633. L.A. Kocharyan, A. Sh. Grigoryan, R.G. Gabrielyan, and E.M. Arutyunyan, *Izv. Akad. Nauk. Arm.SSR. Fiz.*, 1990, **25**, 352.
634. G. Mukhopadhyay, D. Das, C. K. Majumdar, K.R.P.M. Rao, *Philos. Mag. Lett.*, 1991, **63**, 315.
635. S. Usinas, and G.S.A Minas, Scr. Metall. Mater., 1991, 25, 321.
636. U. Gonser, and P. Schaaf, *Fresenius' J. Anal. Chem.*, 1991, **34**, 131.
637. S.S. Sharma, K. Nomura, and Y. Ujihira, *J. Mater. Sci.*, 1991, **26**, 4104.
638. M. Renteria, A.G. Bibiloni, M.S. Moreno, J. Desimoni, R.C. Mercader, A. Bartos, M. Uhrmacher, and K.P. Lieb, *J.Phys.: Condens. Matter*, 1991, **3**, 3625.

8
Gas-phase Molecular Structures Determined by Electron Diffraction

BY D. W. H. RANKIN AND H. E. ROBERTSON

1 Introduction

In this chapter we report structural data determined by electron diffraction for gas-phase molecules, and published during 1991. There are also a few references to earlier publications which were not available to us last year. There has been an increase of about 25% in the number of publications, while the number of structures reported has increased from just over 60 to nearly 100. Studies of transition metal halides are particularly numerous, but there are still few reports of structures of transition metal organometallics.

As usual, distances and errors (quoted in parentheses after values of refined parameters) are given as reported in the original papers, and in each case we state whether the parameters are r_a, r_g, r_α etc. The compounds included in the report are listed below, showing the sections in which they can be found.

Section 2, Groups 1, 12 and 13: $(RbOH)_2$, $CdMe_2(Me_2NCH_2CH_2NMe_2)$, $Hg(1,12\text{-}C_2B_{10}H_{11})_2$, $B_2(NMe_2)_4$, $B_2(OMe)_4$, AlF_3, Ga_2H_6, Tl_2O.

Section 3, Group 14: $SiCl_2$, $SiBr_2$, $SnCl_2$, $SnBr_2$, SnI_2, CF_2I_2, $CF_2(CH_3)_2$, $CF_2(CF_3)_2$, $CH_2(COF)_2$, $CF_2(COF)_2$, C_{60}, $HC(SiH_3)_3$, SiH_2ISiH_2I, $SiHI_2SiHI_2$, $SiBu^tMe_2H$, $SiHMe_2OMe$, $SiClMe_2OMe$, cyclo-$(CH_2)_3SiMe_2$, cyclo-$(CH_2SiMe_2)_2$, cyclo-$SCH_2SiMe_2(CH_2)_2$, $Me_3SiNCNSiMe_3$, $Me_3GeNCNGeMe_3$, $GeMe_4$, cyclopropyl-GeH_3.

Section 4, Group 15: Me_2NF, Me_3NO, $MeNHNH_2$, cyclo$(CH_2NNO_2)_3$, cyclo-$CH_2(CH_2NNO_2)_2$, cyclo-$(CH_2)_4NNO_2$, $(CH_2Cl)_2NNO_2$, $PF_2S(CH_2)_3SPF_2$, PF_2SEt, PCl_2EtS, As_2Me_4.

Section 5, Group 16: (COF)OO(COF), FC(O)SCl, $CClF_2SCl$, CCl_2FSCl, cyclo-$SCF_2SC{=}O$, cyclo-$SCF_2SC{=}S$, cyclo-$(CH_2CH_2S)_3$, cyclo-$(CH_2)_5SO$, cyclo-$(NSF)_3$, cyclo-$(NSF)_2NC(CF_3)$, cyclo-$NSF(NCF)_2$, $CF_2(SF_3)_2$, $SF_5C{\equiv}CH$, $SF_5CH{=}CH_2$, SF_5CN, SF_5OCN, SF_5OF, F_5SOOSF_5, $F_5SeOOSeF_5$, $F_5TeOOTeF_5$.

Section 6, Transition metals, lanthanides and actinides: $MnCl_2$, $FeCl_2$, $CoCl_2$, $NiCl_2$, $MnBr_2$, $FeBr_2$, $CoBr_2$, $NiBr_2$, ScF_3, VF_3, CrF_3, PrF_3, GdF_3, HoF_3, FeF_3, UCl_3, UI_3, UCl_4, UBr_4, UI_4, $ThCl_4$, $MoCl_5$, WBr_4S, WBr_4Se, $CrO_2(NO_3)_2$, Me_4OsO, $MeReO_3$, $(C_5Me_5)ReO_3$, $Ti(BH_4)_3$, $C_5H_5)Zr(BH_4)_3$, $Cr(CH_2CMe_3)_4$, $U(C_8H_8)_2$.

2 Compounds of Elements in Groups 1, 12 and 13

The structures of the dimers of potassium and caesium hydroxides were reported two years ago, and the structure of $(RbOH)_2$ has now been determined.[1] In the vapour at 873 K, 44% of the rubidium hydroxide is present as dimer and the rest as monomer. Analysis of diffraction and vibrational data simultaneously gave the Rb-O distance as 2.50(2) Å in the equilibrium structure in the harmonic approximation ($r_e{}^h$) with the ORbO angles 83(2)° in a planar Rb_2O_2 heavy-atom skeleton. The Rb-O distances are almost exactly mid-way between the K-O and Cs-O distances in the other alkali hydroxides.

The adduct $Me_2Cd.Me_2NCH_2CH_2NMe_2$ has potential value as a volatile source of cadmium for the growth of II-VI semi-conductors. Its structure has now been determined.[2] The cadmium atom is four-coordinate, being bound to two nitrogen atoms of the ethylenediamine ligand and two methyl groups, with r_α(Cd-N) 2.47(5) Å and r(Cd-C) 2.11(2) Å. The long Cd-N bonds indicate the weakness of the adduct, and this is confirmed by the wide CCdC angle [132(11)°] and the correspondingly low NCdN angle [84(3)°]. The coordination of cadmium is distorted tetrahedral, and the CdNCCN ring has C_2 symmetry. The C-N and C-C bond lengths are 1.469(6) and 1.53(4) Å respectively. Angles within the ring were not determined accurately.

The structure of the compound in which two 1,12-$C_2B_{10}H_{11}$ fragments are linked by a mercury atom joining two carbon atoms was described in 1977. The equivalent compound in which the mercury links between two boron atoms (defined as being B(2) in each cluster) has now been determined, and provides a rare determination of an Hg-B bond distance.[3] The refined distance [r_a, 2.155(18) Å] is more-or-less as expected, assuming boron to be 0.05 Å larger than carbon. The structure is linear at mercury, and other important parameters are the mean B-C distance, 1.706(30), B(2)-B(7) 1.815(45), B(2)-B(3) 1.795(12), and angle HgBC 133(4)°. It was necessary to assume that the carbaborane fragments had D_{5d} symmetry.

The structures of gaseous $B_2(NMe_2)_4$ and $B_2(OMe)_4$ have been determined.[4] The amine adopts the expected fully staggered conformation with D_2 symmetry, the NBBN dihedral angle being 90.0(1.1)°. There is a planar configuration at boron and also at each nitrogen atom, but the N-dimethyl groups are twisted so that the BBNC dihedral angles are 19.7(11)°. The NBN angles are 124.0(5)°, so there is significant distortion away from regular trigonal co-ordination at boron. The bond lengths (r_a) are 1.762(11) for B-B and 1.408(3) Å for r(B-N). In the case of $B_2(OMe)_4$, the OBBO dihedral angle is only 49.5(12)°, and the conformation is thus approximately half way between fully staggered and fully eclipsed. The two methyl groups associated with one boron atom are twisted above and below the local BBO_2 planes, with BBOC dihedral angles of 12.1(27) and 21.9(19)°. In this case, there is no significant distortion of the boron atoms away from regular trigonal co-ordination, OBO angles being 119.9(4)°. The bond lengths *r*(B–B) and *r*(B–O) are 1.720(6) and 1.369(3) Å respectively.

Electron diffraction data for aluminium trifluoride at 1300 K have been analysed with vibrational frequencies to give refined force constants and an $r_e{}^h$ Al–F distance of 1.618(3) Å.[5] The molecule is undoubtedly planar, but if it is allowed to become pyramidal, the FAlF angle refined to 119.7(4)°.

A full account of the preparation, characterisation and reactions of digallane has now appeared.[6] In the r_a structure, studied at 255 K, the Ga···Ga distance is 2.580(2) Å, and the terminal and bridging Ga–H distances are 1.519(35) and 1.710(38) Å respectively. The Ga–H_b–Ga angle is 97.9(32)°. There is no evidence of any other oligomer in the gas phase, but vibrational spectrosocopy suggests that another species, probably $(GaH_3)_4$, is present in the solid phase. A similar species is probably a major constituent of toluene solutions of gallane.

A method of analysis of electron diffraction data and vibrational frequencies in terms of a potential function based on Schwinger thermodynamic perturbation theory has been described, and applied to a number of bent XY_2 molecules.[7] This analysis tends to give rather shorter bond lengths and wider angles than have been obtained previously. In the case of Tl_2O, the Tl–O bond lengths refined to 2.077(2) Å (r_e), compared with 2.090(3) Å in an earlier analysis, and the angle TlOTl refined to 145.1, instead of 141.8(9)°. Similar results for tin dihalides are reported in section 3 of this report.

3 Compounds of the Elements in Group 14

The structures of silicon dichloride and silicon dibromide have been determined by joint analysis of electron diffraction and vibrational data.[8] In the case of $SiCl_2$ the $r_e{}^h$ Si–Cl distance refined to 2.080(4), and with allowance for anharmonicity an r_e distance of 2.076(4) Å was obtained. The equilibrium ClSiCl angle was 104.2(6)°. In the case of $SiBr_2$, the $r_e{}^h$ and r_e Si–Br distances were 2.239(5) and 2.227(6) Å and the equilibrium BrSiBr angle was 103.1(4)°.

Diffraction data for the dichloride, dibromide and di-iodide of tin have been reanalysed using potential functions based on Schwinger thermodynamic perturbation theory.[7] These analyses give Sn–Cl, Sn–Br and Sn–I equilibrium distances of 2.335(1), 2.501(1) and 2.688(4) Å respectively, all very slightly shorter than values obtained earlier. The angles at tin refined to 99.1, 100.0 and 105.3° in the chloride, bromide and iodide respectively, all these angles being between 1 and 2° larger than had been found with second-order perturbation theory.

Difluorodiiodomethane, CF_2I_2, is the first compound containing two adjacent C–I bonds whose structure has been determined in the gas phase.[9] The C–I bond length, 2.148(4) Å, (the type of distance is not given) is about 0.01 Å longer than in CF_3I, and the ICI angle, 112.5(3)°, is only a little greater than the tetrahedral angle, and is almost identical to the ClCCl angle in CF_2Cl_2. The C–F bond distance is 1.336(5) Å, and the FCF angle 109.5(10)°. There is thus no evidence for any unusual structural consequences of the adjacent large iodine atoms.

Gas-phase structures of 2,2-difluoropropane and perfluoropropane have been reported.[10] On going from propane to difluoropropane the CCC angle widens from 112(1) to 115.3(4)°, and the C–C bond shortens from 1.531(3) to 1.512(3) Å. Further fluorination, to give perfluoropropane, leads to a small additional widening of the CCC angle to 115.9(7)°, but the C–C bond now lengthens, to 1.546(4) Å. The electronic effects of fluorine on the central atom and steric effects are thus apparent. Other parameters for difluoropropane (r_a, $<\alpha$), determined by joint analysis of electron diffraction and rotational data, include r(C–F) 1.370(2) Å and <FCF 106.2(4)°. In perfluoropropane, which was studied by electron diffraction only, the mean C–F bond length is 1.330(2) Å, and the FCF angles in the methylene and methyl groups are 107.0(13) and 109.3(2)° respectively.

The structures and conformations of malonyldifluoride, $CH_2(COF)_2$ and difluoromalonyldifluoride, $CF_2(COF)_2$ have been studied by electron diffraction and *ab initio* calculations.[11] The major conformer [90(10)%] of malonyldifluoride has one COF group lying in the CCC plane, with C=O *cis* to C–C. The other COF group is rotated by 112(2)° from the *cis* position, so that the C=O bond almost eclipses a C–H bond. There is probably a small amount of a second conformer in which both C=O bonds eclipse C–H bonds, giving C_2 symmetry. In difluoromalonyldifluoride, the C_2 symmetry form is more stable, comprising 70(15)% of the molecules. The COF groups are rotated 120(2)° from the position in which the C=O bonds eclipse C–C bonds, and in the higher-energy form, one COF group is twisted 120°, while the other C=O bond eclipses the opposite C–C bond. Bond lengths (r_a) in malonyldifluoride include C–C 1.502(5), C=O 1.177(3) and C–F 1.349(4) Å, and angles ($<_\alpha$) include CCC 110.2(10) and CCO 129.1(8)°. In difluoromalonyldifluoride the mean C-F bond length is 1.328(2), *r*(C-C) is 1.531(4) and *r*(C=O) is 116.8(3) Å. The CCC and CCO angles are 110.6(5) and 128.2(7)° respectively. The *ab initio* calculations reproduce the experimental geometries well, and predict the most stable conformations correctly, but the calculated relative stabilities of conformers depend on the size of the basis set used.

Buckminster-fullerene, C_{60}, has attracted a great deal of attention, although the signal-to-noise ratio of publications on the subject has been low. The race to determine the structure of the molecule in the gas phase was won by the Oregon electron diffraction group, who found that a temperature of 1000 K was necessary to obtain sufficient gaseous material.[12] The data were completely consistent with the expected icosahedral structure, and the two different bond lengths (r_g) were found to be 1.458(6) Å for bonds fusing five- and six-membered rings and 1.401(10) Å for bonds fusing two six-membered rings. The diameter of the sphere containing all the atoms is 7.110(10) Å.

The search for new precursors for the production of hydrogenated amorphous silicon carbide has led to the development of a new synthesis of trisilylmethane, which can now be prepared in sufficient quantities for structural studies.[13] The Si-C bond length (r_g) is 1.878(1) Å and the SiCSi angle 111.0(2)°, so there is no evidence of significant strain. The three silyl groups are twisted by 22(2)° from the staggered positions, reducing the molecular symmetry from C_{3v} to C_3.

Both 1,2-diiododisilane and 1,1,2,2-tetraiododisilane exist as mixtures of *anti* and *gauche* conformers in the gas phase.[14] For diiododisilane, the percentage of the *gauche* form was found to be 76(16), while it was 60(29)% for tetraiododisilane. As the expected percentage for zero energy difference is 67%, the energy differences between the two forms are very small and not significantly different from zero. The major parameters (r_g, $<\alpha$) for diiododisilane are *r*(Si-Si) 2.380(34), *r*(Si-I) 2.429(13) Å, <ISiSi 107.5(12)°. For tetraiododisilane *r*(Si-Si) was 2.389(37) Å, *r*(Si-I) 2.440(9) Å, <ISiSi 107.2(10) and <ISiI 111.4(6)°.

Small steric effects are apparent in the valence angles of *t*-butyldimethylsilane in the gas phase.[15] The three-fold axis of the *t*-butyl group is tilted 3.6(10)° away from the Si-methyl groups, so that the three SiCC angles are 114.1(9), 109.0(6) and 108.3(8)°. The angles between the Si-C bonds are also not all equal, with the C(Bu)SiC(Me) angles 111.1(4)° and <C(Me)SiC(Me) 105.5(10)°. As expected, the steric strain is minimised by the adoption of a staggered conformation about the Si-C(butyl)

bond. Other parameters (r_a) reported include r[Si–C(Bu)] 1.886(2) r[Si–C(Me)] 1.882(2), r(C–C) 1.546(1) Å and <CCC 108.5(3)°.

Structures of dimethylmethoxysilane[16] and dimethylchloromethoxysilane[17] have been reported. In the first of these compounds, the only conformer identified in the gas phase was the *gauche* form, with the HSiOC dihedral angle 78(10)°, whereas the *anti* configuration, with the ClSiOC dihedral angle of 180°, was the only one found for the chloro compound. Parameters (r_a, $<_\alpha$) reported for dimethylmethoxysilane include r(Si–O) 1.642(3), r(C–O) 1.425(6) r(Si–C) 1.866(3) Å, <COSi 125.2(15), <OSiC 111.1(9) and <CSiC 113.1(19). Molecular mechanics calculations for this compound give geometrical parameters which agree reasonably well with those found experimentally, but these calculations predict the *anti* conformer to be the more stable. Major parameters (r_g, $<_\alpha$) for dimethylchloromethoxysilane include r(Si–O) 1.620(10), r(C–O) 1.430(11), r(Si–C) 1.867(15), r(Si–Cl) 2.073(5) Å, <COSi 127.2(21), <OSiC 107.9(6), <CSiC 110.4(35) and <OSiCl 112.3(15)°.

The four-membered rings of 1,1-dimethylsilacyclobutane and 1,1,3,3-tetramethyl-1,3-disilacyclobutane are both puckered, with rings bent by 34.2(30) and 22.1(22)° respectively from the planar configurations.[18] The two compounds differ in many other structural ways as well, mainly because the disila-compound has all four ring bonds of equal length, whereas the monosila-compound has two ring bonds much longer than the other two. Major parameters (r_g, $<_\alpha$) for the disilacyclobutane include Si–C bond lengths of 1.901(4) Å in the ring, and 1.866(4) Å to the methyl groups, and ring angles of 90.1(6)° at silicon and 87.8(5)° at carbon. The angle between the two Si–C (methyl) bonds is 117.4(26)°. In dimethylsilacyclobutane the C–C bonds are 1.565(4) Å long, and the mean Si–C distance 1.880(2) Å, with the difference between ring and methyl bonds not significantly different from zero. Ring angles are CSiC 76.6(8), SiCC 88.2(4) and CCC 96.7(10)°. The MeSiMe angle is 113(2)°.

The five-membered ring containing sulphur and $SiMe_2$ separated by one CH_2 group has also been studied.[19] Conformations of saturated five-membered rings are notoriously difficult to determine, and the problems are particularly severe in this case, for which five envelope structures and five twisted conformations can be drawn. The conformation which fits the data best can be derived from a twisted conformation of C_2 symmetry, in which the sulphur and silicon are regarded as being equivalent. The mean Si–C distance is given as 1.88(2) Å, the mean S–C bond length is 1.86(3) Å, and the C–C distance is 1.53(3) Å.

The structures of *bis*(trimethylsilyl)- and *bis*(trimethylgermyl)-

carbodiimide have been determined and compared with those of many related silicon-nitrogen and germanium-nitrogen compounds.[20] In the r_a structures, both molecules are bent at nitrogen, with angles SiNC 142.0(11) and GeNC 131.4(9)°. Vibrational spectroscopy has indicated linearity at nitrogen in the silicon compound, and it is not possible to tell whether the bent structures determined by electron diffraction represent equilibrium structures, or whether there is a linear structure with a large-amplitude bending vibration and consequent large shrinkage effect. Other parameters for the silylcarbodiimide include *r*(Si-N) 1.732(3), *r*(C=N) 1.216(4), *r*(Si-C) 1.867(1) Å, <NSiC 107.7(4)°, and the dihedral angle SiNNSi, 44(2)°. Corresponding parameters for the germyl compound are *r*(Ge-N) 1.840(6), *r*(C=N) 1.219(5), *r*(Si-C) 1.947(2) Å, <NSiC 107.7(6)° and τ(GeNNGe) 59.4°.

In tetramethylgermane, the r_g(Ge-C) bond length is 1.958(4) Å, and the methyl groups are twisted 23.0(15)° from the staggered form.[21] On this basis, the torsional barrier is estimated to be 1.3 kJ mol^{-1}. The bond length is not significantly different from that in $Ge(C_6H_5)_4$ and agrees closely with the value predicted by a modified Schomaker-Stevenson rule.

The structure of cyclopropylgermane has been studied both by electron diffraction and by *ab initio* molecular orbital calculations.[22] The length of the C-C bond adajcent to the germyl group was found to be 1.521(7) Å, while the third C-C bond was 1.502(9) Å long (r_a). The Ge-C bond length was 1.924(2) Å, and made an angle of 55.5(16)° with the ring plane. The results reflect the strong π-donor character of cyclopropyl groups, and the π-acceptor character of germyl groups, which is nevertheless less strong than that of silyl groups.

4 Compounds of Elements in Group 15

A joint analysis[23] of rotation constants and electron diffraction data for fluorodimethylamine yields an r_z structure with *r*(N-F) 1.447(6), *r*(N-C) 1.462(7) Å, <CNC 112.0(10) and <FCN 103.6(5)°. The N-F bonds in this compound are thus more than 0.07 Å longer than in NF_3, and an intermediate value has been reported for $MeNF_2$. *Ab initio* calculations reproduce both the experimental N-F bond lengths and the vibrational frequencies of these three compounds very well, and show that, unusually, the longest N-F bonds are associated with the largest N-F force constants.

The nitrogen-oxygen bond of an amine oxide is normally described as N^+-O^- rather than N=O. Experimental determination of the N-O bond length in Me_3NO[24] shows that it is intermediate between normal single and double bonds. The r_a distance is 1.379(3) Å, the C-N distance is 1.496(2) Å, the

the angle ($<_\alpha$) CNO is 108.9(2)°.

Structures of two conformers of methylhydrazine have been determined by analysis of gas electron diffraction data and rotation constants for the two conformers.[25] Even with all this information, it was necessary to fix many parameters at values calculated *ab initio*. For the so-called *inner* conformer, refined parameters (r_g, $<_z$) included *r*(N-N) 1.433(12), *r*(N-C) 1.463(12) Å and <CNN 133.5(2)°. For the *outer* conformer, the bond lengths were slightly different, with *r*(N-C) 1.466(2) and *r*(N-N) 1.431 Å (the difference between these bond lengths was fixed), but the CNN angle was substantially smaller, at 109.5(2)°. All other angles, including torsion angles, were fixed at calculated values. Unfortunately, dihedral angles are not given, and the only diagrams of the conformers show two different *inner* arrangements, rather than one *inner* and one *outer*.

The structures of three cyclic nitramines have been determined.[26] In 1,3,5-trinitro-1,3,5-triazacyclohexane (more commonly known as RDX), the six-membered C_3N_3 ring adopts a chair conformation, with the nitro groups in axial positions. The overall symmetry is C_3, as the three nitro groups are twisted nearly 20° away from the positions in which the O···O vectors are parallel to the nearest C···C vectors. The ring nitrogen atoms are very nearly planar, the sum of their valence angles being 356°. In 1,3-dinitro-1,3-diazacyclopentane, the five-membered rings have a half-chair conformation of C_2 symmetry, but the sums of angles at the ring nitrogens are only 339°, with the nitro groups in equatorial positions. In contrast, the five-membered ring of *N*-nitropyrrolidine has an envelope structure of C_s symmetry, again with the nitro group in an equatorial position, and the sum of angles at the ring nitrogen atom is 342°. Other parameters for these three nitramines are listed in Table 1.

Parameters for *bis*(*N*-chloromethyl)nitramine are also given in Table 1.[27] In this case, the amine nitrogen is very nearly planar with the sum of angles 357.4°. The C_2NNO_2 skeleton is thus almost planar, and the molecule is close to having overall C_2 symmetry, with the two C-Cl bonds on opposite sides of the skeleton. This compound has the longest N-N bond and shortest C-N bond of the group of nitramines reported in this review. This can be attributed to the chlorine atoms reducing the electron-donating properties of the amino group.

The compound $(F_2P)S(CH_2)_3S(PF_2)$ is a potentially useful bidentate ligand, and its structure has been determined in the gas phase.[28] However, the determination of its conformation is difficult, because there are no less than six dihedral angles to be considered, and so $(F_2P)SEt$ was studied

Table 1 - Geometrical parameters (r_g, $<_a$) in some nitramines

Parameter	$(CH_2NNO_2)_3$	$CH_2(CH_2NNO_2)_2$	$(CH_2)_4NNO_2$	$(CH_2Cl)_2NNO_2$
r(C-N)/Å	1.464(6)	1.483(8)	1.477(8)	1.443(4)
r(N-N)/Å	1.413(5)	1.393(8)	1.363(4)	1.422(5)
r(N=O)/Å	1.213(2)	1.226(2)	1.225(2)	1.216(3)
r(C-C)/Å		1.528(6)	1.534(5)	
r(C-Cl)/Å				1.803(4)
<ONO/°	125.5(10)	128.2(18)	126.3(19)	128.0(14)
<CNN/°	116.3(5)	114.8(8)	116.0(11)	116.4(4)
<CNC/°	123.7(6)	109.7(17)	110.3(14)	124.8(9)
<NCN/°	109.4(6)			
<CCN/°		100.2(24)	99.6(4)	
<ClCN/°				111.6(5)
ø(NN)/°	19.1(23)	12.9(30)	16.8(16)	2.6(15)
ø(CN)/°				78.9(7)

as well, to provide some insight into the preferred arrangement of the $(F_2P)SCH_2$-group. In both compounds, the conformation adopted involves fairly close contact (about equal to the sum of van der Waals radii) between fluorine atoms and hydrogen atoms of adjacent CH_2 groups. If the zero angle for torsion about the P-S bond is defined as the position in which the projection of the S-C bond lies between the two P-F bonds, the refined values for this torsion angle are 9(3) and 31(4)° for the ethane and propane derivatives respectively, and the PSCC dihedral angles are 96(4) and 85(2)°. In the bidentate ligand, the SCCC dihedral angles are 143(3)°, and the molecule has overall C_2 symmetry. Other significant parameters (r_a) for $(F_2P)SEt$ are *r*(P-F) 2.085(3), *r*(P-F) 1.587(4), *r*(S-C) 1.825(6) Å, <FPF 96.2(4), <FPS 101.1(2), <PSC 100.3(6) and <SCC 108.4(9)°. In $(F_2P)S(CH_2)_3S(PF_2)$, the most significant parameters were *r*(P-S) 2.117(6), *r*(P-F) 1.577(4), *r*(S-C) 1.844(9) Å, <FPF 99.1(14) and <FPS 101.1(5)°.

In the gas phase there are two conformers of PCl_2EtS co-existing.[29] Because of the similarity of the scattering powers of sulphur and chlorine, and the similarity of the P=S and P-Cl bond lengths, it is not possible to determine the proportions of the *gauche* and *trans* conformers from the diffraction data. The proportions of the conformers were therefore fixed at 82% *gauche* and 18% *trans* as calculated *ab initio*. Differences between geometric parameters were also fixed at calculated values, and the following parameters (r_a) were obtained for the major conformer: r(P-Cl) 2.030(1), r(P=S) 1.897(2), r(P-C) 1.808(5), r(C-C) 1.505(6) Å, <PCC 114.4(8), <ClPC(mean) 103.1(5), <CPS 116.1(12) and <ClPCl, 102.0(4)°. In the *trans* conformer, the calculated PCC angle is 3.9° larger than that in the *gauche* conformer. Other geometrical differences that were calculated were much smaller than this.

The vapour of tetramethyldiarsine at room temperature consists of 60% of the *gauche* and 40% of the *trans* conformer.[30] The major parameters are r_a(As-As) 2.433(2), r(As-C) 1.973(2) Å, <CAsC 95.4(5) and <CAsC 95.3(11)°.

5 Compounds of Elements in Group 16

Almost one quarter of the references in this review are concerned with Group 16 elements, and of the fifteen papers reviewed, all but two of them originate from one electron diffraction group, and deal with fluorinated compounds. *Bis*(fluorocarbonyl) peroxide has the *syn-syn* conformation, in which the C=O bonds eclipse the central O-O bond.[31] This experimental conclusion, based on electron diffraction data, is supported by *ab initio* calculations and interpretation of vibrational spectra. The most important other parameters (r_a, $<_\alpha$) are the O-O bond length, 1.419(9) Å, and the dihedral angle COOC, which is 83.5(14)°. The small dihedral angle is unusual, comparable only to angles of 88 and 81° reported for F_2O_2 and Cl_2O_2. These results are discussed in terms of a number of factors, including the form of the lone pairs of electrons on oxygen atoms, the extent of anomeric effects, and steric demands of substituents in some peroxides.

(Fluorocarbonyl)sulphenyl chloride, FC(O)SCl, exists as a mixture of two conformers in the gas phase, with 88(5)% of the molecules having their C-F and S-Cl bonds *trans* to one another, with the remainder in the *cis* configuration.[32] This corresponds to a free-energy difference of 5.0(13) kJ mol^{-1}, compared with 5.9(4) kJ mol^{-1} based on an interpretation of matrix infra-red spectra. *Ab initio* calculations suggest that there might be a difference of between 5.2 and 6.7 kJ mol^{-1} depending on the level of sophistication of the calculation. Not surprisingly, the relative stabilities of conformations in various thioformic acid derivatives depend strongly on the substituents. For the purposes of structural refinements, the parameters (r_a, $<_\alpha$) for the *trans* conformer were refined, and differences between parameters for the *cis* and *trans* forms were fixed at calculated values. Refined parameters were r(C=O) 1.179(4), r(C-F)

1.342(4), *r*(S-C) 1.756(5), *r*(S-Cl) 1.996(3) Å, <CSCl 100.3(5), <SC=O 130.9(5) and <SCF 105.3(3)°.

Chlorodifluoromethanesulphenyl chloride and dichlorofluoromethanesulphenyl chloride have also been studied,[33] by electron diffraction and *ab initio* calculations. In the case of $CClF_2SCl$, the *trans* conformer (in which the C-Cl and S-Cl bonds are mutually *trans*) is predominant, comprising 73(5)% of the mixture, the remainder of which has the *gauche* arrangement. The *trans* form is thus more stable by 4.2(8) kJ mol^{-1}. In the case of CCl_2FSCl, the *trans* conformer (which has the C-F bond *trans* to S-Cl) is less stable than the *gauche* structure by 2.1(16) kJ mol^{-1}, corresponding to 82(10)% of the *gauche* form, for which the ClSCF dihedral angle is 62(7)°. Thus in both compounds the preferred form has a C-Cl bond *trans* to S-Cl. In both cases, differences between parameters for the more and less abundant forms were fixed at calculated values, and it was also necessary to fix the differences between the two different S-X bond lengths and two SCX angles (X = F or Cl) in the *gauche* conformers at calculated values. The refined parameters (r_a, $<_\alpha$) for $CClF_2SCl$ (*trans* conformer) were *r*(C-F) 1.333(3), *r*(C-Cl) 1.748(12), *r*(S-C) 1.813(15), *r*(S-Cl) 2.014(3) Å, <CSCl 99.3(4), <FCF 108.1(6), <SCCl 109.8(4), <FCCl 103.8(3) and <SCF 112.3(6)°. The angles at the carbon atom were very conformation-dependent, and the SCCl angle calculated for the *gauche* form was 111.8° (8° larger than in the *trans* form) and the two SCF angles calculated for the *gauche* form were 103.8 and 112.3°. For the *gauche* conformer of CCl_2FSCl, the bond lengths were *r*(C-F) 1.329(5), *r*(C-Cl) 1.750(8), *r*(S-C) 1.811(16) and *r*(S-Cl) 2.004(3) Å. The angles were CSCl 101.7(7), SCF 111.3(26) (101.7 in the *trans* conformer), SCCl 101.3(6) and 110.8(30) (110.2 in the *trans* conformer), ClCCl 111.7(19) and FCCl 110.6(8)°.

The four-membered rings in 4,4-difluoro-1,3-dithietane-2-on and 4,4-difluoro-1,3-dithietane-2-thion are both planar,[34] like that of tetrafluorodithietane, cyclo-SCF_2SCF_2. The ring angles ($<_\alpha$) in the carbonyl compound are 81.1(10)° at sulphur, 97.7(13)° at the four-coordinated carbon atom, and 100.1(13)° at the three-coordinated carbon. The corresponding angles in the thiocarbonyl compound are 83.1(5), 94.6(6) and 99.2(6)°. The ring bonds to the three-coordinated carbon atom are shorter than those to the other carbon in both compounds, at 1.791(12) Å compared with 1.821(12) Å in the carbonyl, and 1.758(6) Å compared with 1.823(6) Å in the thiocarbonyl. The difference in the latter case is unusually large, although it is as great as 0.167 Å in the crystal structure of the analogous cation in which the doubly-bonded sulphur atom has been replaced by fluorine. The C-F bond lengths in the two compounds are 1.343(4) and 1.338(4) Å, and C=O and C=S distances are 1.179(7) and 1.598(5) Å respectively.

Determining the conformation of large rings can be a very difficult and sometimes impossible task. Numerous models must be considered, and it is often the case that the experimental data are inadequate to distinguish between them. Four models were fitted to electron diffraction data for 1,4,7-trithiacyclononane.[35] The model with highest symmetry, D_3, was incompatible with the data and could be discarded. The best fit to the data was provided by a model with C_1 symmetry, but the fit for one with C_2 symmetry was only slightly worse, and one with C_3 symmetry was not sufficiently bad to be completely excluded. Molecular mechanics (MM2) calculations were therefore performed, and these indicated that the C_1 model had the lowest energy, but that the C_3 conformation was only 0.13 kJ mol^{-1} higher in energy. The other two models were between 8 - 10 kJ mol^{-1} higher in energy. On balance, it seems probable that the lowest symmetry conformer is the major constituent of the gas at 500 K, but it is possible that other conformers are also present. The results of the molecular mechanics calculations were used to supply loose constraints to the refinements, and under these conditions, the following r_a parameters were obtained, using the C_1 model: r(S-C) 1.820(1), r(C-C) 1.533(4) Å, <CSC 103.8(7) and <SCC 115.0 and 115.7(5)°.

The conformations of six-membered rings are much easier to determine, and it has been shown unequivocally that thiane-1-oxide adopts a chair conformation with the oxygen atom in an axial position and molecular C_s symmetry.[36] The bond lengths (r_g), going round the ring starting at the sulphur, are S-C 1.816(4), C-C 1.538(3) and C-C 1.539(3); the S=O distance is 1.483(3) Å. Ring angles, again starting at sulphur, are CSC 91.1(7), SCC 112.5(1), CCC 111.5(4) and CCC 114.8(7)°; the OSC angle is 108.1(3)°.

The structures of three-thiatriazines, six-membered rings in which nitrogen alternates with sulphur or carbon, have been determined. Trifluorotri-thiatriazine, $(NSF)_3$, has C_{3v} symmetry, the S_3N_3 ring adopting a chair conformation with all three fluorine atoms in axial positions.[37] Because the puckering of the ring is not very pronounced, and the NSF angles are only just over 100°, the fluorine atoms are very substantially displaced from the band containing the sulphur and nitrogen atoms. If one SF group in this compound is replaced by CCF_3, one obtains 1,3-difluoro-1,3,2,4,6-dithiatriazine.[38] This also has a chair conformation with the fluorine atoms bound to sulphur in axial positions. However, it is a reclining chair, and the dihedral angles around the ring lie between 7-18°. Of course, the carbon atom in the ring is sp^2

hybridised, with planar coordination. The compound in which two of the SF groups of $(NSF)_3$ have been replaced by CF groups contains two sp^2 carbon atoms with planar coordination, and so the whole molecule is planar with the exception of the sulphur atom and the fluorine bonded to it, which is in an axial position.[39] The plane of the NSN fragment is bent 16.4(9)° out of the plane of the rest of the ring. Despite the changes in ring shape forced by replacement of sulphur by carbon, the bond length and angles for these three thiatriazines, listed in Table 2, are remarkably consistent.

Table 2 - Geometrical parameters (r_a) for some thiatriazines

	$(NSF)_3$	$(NSF)_2NC(CF_3)$	$NSF(NCF)_2$
r(S-N)/Å	1.582(4)	1.580(4)(mean)	1.592(7)
r(S-F)/Å	1.624(7)	1.630(10)	1.633(14)
r(C-C)/Å		1.510(10)	
r(C-N)/Å		1.314(9)	1.315(6)(mean)
r(C-F)/Å		1.333(10)	1.311(5)
<SNS/°	124.3(6)	121.7(2)	
<NSN/°	112.7(12)	111.3(12)	109.8(17)
<NSF/°	100.9(9)	98.6(21)(mean)	99.9(31)
<NCN/°		131.5(16)	131.2(12)
<SNC/°		120.6(9)	115.0(9)
<CNC/°			114.6(17)
<FCF/°		107.4(4)	

In *bis*(trifluorosulphur)difluoromethane, $CF_2(SF_3)_2$, each sulphur atom has trigonal bipyramidal geometry, with the lone pair of electrons and the bond to carbon occupying two of the equatorial positions.[40] The overall molecular symmetry is C_2 and the SF_3 groups are orientated in such a way that the $SCSF_{eq}$ dihedral angles are 130.8(9)°. The two axial fluorine atoms in each SF_3 group are therefore inequivalent, and the two groups are tilted away from the CF_2 group so that the CSF_{ax} angles differ by 4.0(10)°. One axial fluorine atom in each group is in a position to interact with the sulphur atom of the other group. In fact, the non-bonded sulphur···fluorine contact is only 2.66 Å, but whether this is caused by or causes the small SCF angle of 108.2(5)° cannot be determined. Other parameters (r_g, $<_\alpha$) reported include *r*(C-F) 1.318(5), $r(S\text{-}F)_{eq}$ 1.562(6), $r(S\text{-}F)_{ax}$ 1.664(4), *r*(S-C) 1.888(7) Å, <FCF 109.8(18), $<CSF_{ax}$(mean) 87.1(3), $<CSF_{eq}$ 97.2(11) and $<F_{eq}SF_{ax}$ 88.1(3)°.

Structures of a whole series of SF_5 compounds have been reported this year, and parameters relating to the SF_5 groups are listed in Table 3. When one fluorine atom of SF_6 is replaced by an electronegative group, the axial and equatorial bonds remain approximately equal in length and the angle between them remains close to 90°. Significant distortions are only observed when less electronegative groups are substituted, when the equatorial bonds are lengthened (the '*cis*' influence) and the equatorial atoms are displaced away from the substituent. For example, study of $SF_5C{\equiv}CH$ and $SF_5CH{=}CH_2$[41] shows that the vinyl group is less electronegative than the ethynyl group. Even in these cases, the differences in bond lengths are barely significant (errors quoted are 3σ) despite the fact that the structures are based on combined analyses of electron diffraction data and rotation constants. In $SF_5C{\equiv}CH$ r_z(S–C) is 1.736(6) and r(C≡C) 1.200(7) Å; in $SF_5CH{=}CH_2$, r_z(S–C) is longer, at 1.787(9) Å, reflecting the change in hybridisation of the carbon atom, r(C=C) is 1.337(17) Å and <SCC is 124.5(15)°.

Table 3 - Structure of the SF_5 group in SF_5X compounds

Compound	$r(S\text{-}F_{ax})$/Å	$r(S\text{-}F_{eq})$/Å	$<F_{eq}SF_{ax}$/°
$SF_5C{\equiv}CH$ (r_z)	1.560(12)	1.578(3)	88.9(2)
$SF_5CH{=}CH_2$ (r_z)	1.562(13)	1.586(4)	88.4(3)
SF_5CN (r_a)	1.558(6)	1.566(6)	90.1(2)
SF_5OCN (r_a)	1.554(2)[a]	1.554(2)[a]	90.4(6)
SF_5OF (r_a)	1.555(3)[a]	1.555(3)[a]	90.1(8)
SF_5OOSF_5 (r_a)	1.561(3)[a]	1.561(3)[a]	88.8(2)

[a] Assumed to be equal

Sulphur cyanide pentafluoride, SF_5CN, is isoelectronic with $SF_5C{\equiv}CH$, but the cyanide group is significantly more electronegative than the ethynyl group and the difference between axial and equatorial S–F distances is only 0.008 Å, as determined by analysis of electron diffraction data and rotation constants.[42] The molecule has C_{4v} symmetry, and the S–C and C≡N bond lengths (r_a) are 1.765(5) and 1.152(5) Å. Other parameters are given in Table 3. The cyanate, SF_5OCN, has also been studied,[43] and its geometry has been compared with the isomeric isocyanate, SF_5NCO. The bond lengths within the OCN fragment are very different in the two compounds, as in SF_5OCN there is formally an O–C single bond [1.271(13) Å] and a C≡N triple bond [1.162(13) Å], whereas SF_5NCO can be regarded as having two double bonds, with r(N=C) 1.234(8) and *r*(C=O) 1.179(7) Å. In both cases, the NCO group is almost linear [the angles at carbon are 175.3(36) and 173.8(37)°], but the SOC and SNC angles are 120.4(13) and 124.9(12)°. In SF_5OCN this bend results in an interaction between the equatorial fluorines and the carbon atom that is sufficiently strong to cause the C_4 axis of the SF_5 group to be tilted 3.5(11)°, so that it no longer coincides with the S-O bond. The S-O bond length is 1.653(6) Å, and parameters for the SF_5 group are listed in Table 3.

The structure of SF_5OF has been determined from electron diffraction data and *ab initio* calculations at the 6-31G* level have been performed.[44] The difference between the axial and equatorial bond lengths was not sufficient to be measured, and they were assumed to be equal. Parameters for the SF_5 group are listed in Table 3, and the other parameters reported are *r*(S-O) 1.671(7), *r*(O-F) 1.408(9) Å and <SOF 108.3(11). The O-F distance is approximately the same as in OF_2, but the angle at oxygen is 5° wider in the sulphur compound, presumably because of steric repulsion. The SF_5 group is tilted slightly, by 2.1(13)°, away from the fluorine atom on oxygen.

Structures of the peroxides F_5MOOMF_5, where M = S, Se and Te, have been determined.[45] The three structures, details of which are given in Tables 3 and 4, are remarkably similar. In each case, the dihedral angle about the O-O bond is just under 130°, and the MF_5 groups are in staggered conformations with respect to the O-O bonds. The M-F bond lengths are all very close to those in the corresponding MF_6 compounds, and no distinction between axial and equatorial bond lengths could be made. The M-O distances were all much longer in these peroxides than in the corresponding oxides, the differences being 0.073, 0.085 and 0.077 Å.

Table 4 - Structures (r_a) of *bis*(pentafluorochalcogen) peroxides

Parameter	F_5SOOSF_5	$F_5SeOOSeF_5$	$FeTeOOTeF_5$
r(O-O)/Å	1.43(2)	1.42(3)	1.45(4)
r(M-F)(mean)/Å	1.561(3)	1.685(3)	1.822(4)
r(M-O)/Å	1.660(6)	1.783(10)	1.911(21)
<MOO/°	110.3(11)	110.7(13)	109.3(16)
δMOOM/°	129(2)	126(2)	127(2)
<$F_{ax}MF_{eq}$/°	88.8(2)	88.6(3)	88.1(3)

6 Compounds Containing Transition Elements, Lanthanides or Actinides

There has been much debate about whether transition dihalides have linear or bent structures. New data, analysed in a new way, for the dichlorides and dibromides of mangenese, iron, cobalt and nickel[46] show that the equilibrium structures are all linear, but that there are large-amplitude anharmonic bending vibrations. The analyses use both diffraction data and vibration frequencies, although the diffraction data alone, if taken to wide enough angles, are sufficient to refine cubic force constants and the bond Morse anharmonicity parameters. Equilibrium (r_e) and r_g bond distances are given, the latter being longer than the former by about 0.02 Å. The reported (r_g) bond lengths are Mn-Cl 2.202, Fe-Cl 2.151, Co-Cl 2.113, Ni-Cl 2.076, Mn-Br 2.344, Fe-Br 2.294, Co-Br 2.241 and Ni-Br 2.201 Å, all with uncertainties of 0.007 Å or less. The vapours of iron and cobalt dichloride and manganese, iron and cobalt dibromide all showed the presence of small amounts of dimers, between 11 and 4% abundant. The largest amount occurred for iron dibromide, enabling the structure of Fe_2Br_4 to be determined. This molecule had two bridging bromine atoms, with the bridge bond length 2.537(22) Å and the terminal bond length 2.294(7) Å. The angle subtended at iron by the two bonds to bridging bromine atoms was 92(3)°. The structure was non-planar, and the four-membered ring may be regarded as folded upwards along the axis joining the two bridging bromine atoms, by 59(6)°, with the terminal Fe-Br bonds then folded down by 40(7)°.

The trifluorides of scandium, vanadium, chromium, praseodymium, gadolinium and holmium have all been studied by electron diffraction.[47] In the first instance, r_g structures were determined from diffraction data alone, and this showed that the three lanthanide trifluorides were all pyramidal, but that the three transition metal trifluorides were close to being planar. The next stage of the analysis involved refinement of the r_e^h structures, and the diagonal harmonic force-field elements, using electron diffraction data and observed vibrational frequencies as data. This showed that all three transition-metal compounds were truly planar, with r_e^h bond lengths Sc-F 1.808(5), V-F 1.721(14) and Cr-F 1.705(8)°. The lanthanide trifluorides were confirmed as being pyramidal, with r_e^h bond lengths 2.056(5) for PrF_3, 2.016(6) for GdF_3 and 1.978(10) Å for HoF_3. The FPrF, FGdF and FHoF angles were 105.0(15), 109.9(23) and 108.2(32)° respectively. In another paper[5] similar results for AlF_3 (discussed in Section 2 of this review) and FeF_3 are reported. Iron trifluoride has a planar structure with r_e^h(Fe-F) 1.722(4) Å. In all of these compounds, the r_g bond lengths are 0.03 to 0.04 Å longer than the harmonic equilibrium distances.

In the work discussed in the last paragraph, it was shown that transition metal trifluorides have planar structures, whereas those of the lanthanides are pyramidal. New studies of uranium trichloride[48] and triiodide[49] reveal pyramidal structures with very small angles between the bonds. In the r_g structures, the ClUCl angle is 95(3)° and the IUI angle is 88.0(3)°. The U-Cl and U-I bond lengths are 2.549(8) and 2.88(1) Å respectively. These compounds were studied at 783 and 1060 K; under these conditions, uranium triiodide accounted for only 20% of the molecules in the vapour, the rest being I_2.

Studies of three uranium tetrahalides, UCl_4,[50] UBr_4[51] and UI_4[52] show that all three molecules have C_{2v} symmetry. The structures are most simply regarded as being derived from trigonal bipyramids with equatorial lone pairs of electrons. The axial bonds are distorted towards the lone pair, but the equatorial bonds are distorted away from it, so that the angles between the axial bonds are typically around 150°, while the angles between the equatorial bonds are around 90°. In the case of UCl_4, r_a(U-Cl) is given as 2.538(5) Å, $<_g(Cl_{eq}UCl_{eq})$ 84(5)° and $<_g(Cl_{eq}UCl_{ax})$ 93(1)°. The angle between the two axial bonds is not given. This is probably because the distance between the two axial atoms is apparently more than twice the U-Cl bond length, and must therefore be inconsistent with the rest of the structure. In UBr_4, r_a(U-Br) is 2.681(5) Å, and the angles ($<_g$) between the two equatorial bonds and the two axial bonds are 81 and 150° respectively. The study of UI_4 was complicated by the possible presence of molecular iodine in the vapour. If it was assumed to be absent, the U-I distance (r_a) refined to 2.918(5)°, and the angles between the U-I bonds were 82° for the equatorial bond and 138° for the axial bonds. If the mole fraction of I_2 was allowed to refine, it reached 0.66(6). The U-I distance was then 2.973(5) Å, and the two angles were 84 and 130°. For none of the uranium tetrahalides was there any evidence that the axial and equatorial bonds were significantly different in length.

In contrast to the uranium tetrahalides, thorium tetrachloride has been shown to be tetrahedral,[53] although an earlier study of its matrix infra-red spectrum suggested that it had C_{2v} symmetry, with a difference of 0.12 Å between the length of the two different types of bond. Working with the assumption of tetrahedral symmetry, the r_a thorium-chlorine bond length refined to 2.565(6) Å.

Data for molybdenum pentachloride were collected in 1976, but it has, until recently, proved impossible to find a structure that fitted the data. However, the demonstration that WMe_6 has a prismatic structure of D_{3h} symmetry led to a suggestion that $MoCl_5$ might also have a prismatic structure, with one corner removed. This structure, of C_s symmetry, has now been shown to be consistent with the experimental information.[54] A molecule of such low symmetry requires seven geometrical parameters, but only five could be refined. Two different sets of results are presented, which are consistent in all their main features. Four of the chlorine atoms lie in a rectangle which has two sides 3.22 Å, and two 3.07 Å long. The molybdenum atom lies on the normal to the rectangle passing through its centre, with two ClMoCl angles close to 85°, and two close to 90°. The bond to the remaining chlorine atoms makes an angle of about 18° to the normal to the base rectangle, with the result that this bond makesan angle of 88° to two of the other bonds, but angles of about 112° to the other two bonds. The fifth chlorine atom therefore lies not far from the position in which it would be found in a symmetric prism, but it is distorted towards a position it would have in a square pyramid of C_{4v} symmetry. If the four basal bond lengths are assumed to be equal, they refined to 2.269(1) Å, leaving the fifth bond length at 2.207(6) Å. In the second refinement reported, three of the bonds are 2.234(2) Å long, while the length of the other two is 2.291(5) Å.

Structures of $WSBr_4$ and $WSeBr_4$ have been reported,[55] thus completing a series of studies of compounds of the type WYX_4, where Y is oxygen, sulphur or selenium and X is fluorine, chlorine or bromine. All the compounds have square pyramidal structures of C_{4v} symmetry, with the tungsten atom above the plane of the halogen atoms. Indeed, there is remarkable consistency over the full range of structures, with the angle YWX only in the range 102.5 to 105.0°. This is believed to arise from balance between steric and electrostatic repulsions among the ligands. Parameters (r_a) reported for $WSBr_4$ are *r*(W=S) 2.109(11), *r*(W-Br) 2.433(3) Å, <SWBr 103.5(7), <BrWBr(*cis*) 86.9(3) and <BrWBr(*trans*) 153.0(14). For $WSeBr_4$ *r*(W=Se) is 2.220(22), *r*(W-Br) 2.427(9) Å and angles SeWBr, BrWBr(*cis*) and BrWBr(*trans*) are 102.5(9), 87.3(4) and 154.9(19)° respectively.

Chromyl nitrate, $CrO_2(NO_3)_2$, has a most unusual structure in the gas phase.[56] This structure may be regarded as derived from a tetrahedron, with each nitrate group bound through one oxygen atom, but a second atom of each nitrate is also weakly coordinated to chromium (the estimated bond order is between 0.19 and 0.29), so that the final structure may be regarded as a strongly distorted octahedron. The nitrogen-oxygen bond lengths (r_g) are 1.341(4) Å for the single bond, 1.254(4) Å for the doubly-bonded oxygen coordinated to chromium and 1.193(4) Å for the free oxygen. The chromium-oxygen bond lengths are 1.586(2) for the doubly-bonded oxo-ligands, 1.957(5) for the single bond and 2.254(20) Å for the weak interaction with the second nitrate oxygen atom. The angles O=Cr=O and O-Cr-O are 112.6(35) and 140.4(33)° respectively. The Cr-O-N angle is 97.5(5)° and the O=N=O angle 128.1(36)°. The $CrONO_2$ group is only slightly non-planar, the NO_2 group being twisted just 16(3)°, and the overall molecular symmetry is C_2. *Ab initio* calculations, extremely difficult for a molecule of this type, reproduce the most important features of the structure reasonably well.

Tetramethyloxoosmium has a square-pyramidal structure[57] with r_a(Os-C) 2.096(3), *r*(Os=O) 1.681(4) Å and <OOsC 112.2(5)°. This compound is the first methylosmium oxide to be prepared, but derivatives of other transition elements are well-known. Methyltrioxorhenium has also been studied.[58] This molecule has C_{3v} symmetry, with r_a(Re=O) 1.709(3), *r*(Re-C) 2.060(9) Å, <CReO 106.0(2) and <OReO 113.0(3)°. The rhenium-carbon distance is the shortest so far reported for a bond involving an sp^3-hybridised carbon atom. In contrast, the rhenium-carbon distance in $Re(C_5Me_5)O_3$ is the longest known, at 2.405(6) Å (r_a).[58] This is attributed to the size of the pentamethylcyclopentadienyl ligand and the *trans* influence of the oxo ligands. Other parameters reported are *r*(Re=O) 1.716(3), *r*(C-C)(ring) 1.439(7) and *r*(C-C)(methyl) 1.509(9) Å. The OReO angle is 107.0(3)°.

Titanium *tris*(tetrahydroborate), $Ti(BH_4)_3$, has been studied, both as $Ti(BH_4)_3$ and as $Ti(BD_4)_3$.[59] The deuteriated compound was used to improve resolution of the distances giving overlapping peaks in the radial distribution curve, making use of the fact that atom pairs involving deuterium have smaller amplitudes of vibration than those involving hydrogen. The tetrahydroborate groups are tridentate, and the molecule has overall C_{3h} symmetry, with a planar TiB_3 skeleton. The Ti-B distance (r_a) is 2.175(4) Å, and the Ti-D distance is 1.984(5) Å.

Vibrational spectroscopy indicates that the BH_4 groups in $Zr(C_5H_5)(BH_4)_3$ are triply bridged, but electron diffraction data fit doubly-bridged and triply-bridged models equally well.[60] Values of the parameters obtained in the refinements suggest that the triply-bridged model is the more likely. The molecule has a C_5 axis through the metal atom and the cyclopentadienyl group, coinciding with a C_3 axis through the $Zr(BH_4)_3$ fragment. The principal parameters (r_a) are *r*(Zr-B) 2.403(29) *r*(Zr-C) 2.519(18), *r*(C-C) 1.418(2) Å and <BZrB 103.4(12)°. The distance between the zirconium atom and the bridging hydrogen atoms is 2.197(28) Å.

The sterically crowded compound, *tetrakis*(neopentyl)chromium, $Cr(CH_2CMe_3)_4$, is one of a small number of compounds of chromium in oxidation state IV.[61] Structural models with S_4, D_{2d}, C_2 and C_1 symmetry were investigated, but only the S_4 model gave satisfactory agreement with the experimental data. Although with this symmetry two of the CCrC angles may be different from the other four, there was no significant distortion from regular tetrahedral co-ordination at the chromium atom. The most important r_a parameters reported are r(Cr-C) 2.038(8), r(C-C) 1.545(3) Å (the C-C bonds were all assumed to be of equal length) and <CrCC 128.0(8)°. This angle must be one of the largest known for four-coordinated carbon atoms, and is enforced by the crowding of four bulky ligands. However, there is no evidence of significant angular distortion within the neopentyl groups, all the CCC angles being within 1.3° of the tetrahedral angle.

The final compound in this report is uranocene, $U(C_8H_8)_2$, studied at 570 K.[62] The eight-membered rings are planar and parallel, giving the molecule D_{8h} symmetry, and the C-C bond lengths (r_a) of 1.385(3) Å indicate that the rings are aromatic. The uranium-carbon distance is 2.665(5) Å and the distance between the rings is 3.91(1) Å. These parameters are very close to those that have been observed in the crystalline phase.

References

1. G.V. Girichev, S.B. Lapshina and I.V. Tumanova, *Zh. Strukt. Khim.*, 1990, **31**(6), 132.

2. M.J. Almond, M.P. Beer, K. Hagen, D.A. Rice and P.J. Wright, *J. Mater. Chem.*, 1991, **1**, 1065.

3. V.S. Mastryukov, A.A. Remorova, A.V. Golubinskii, M.V. Popik, L.V. Vilkov, V.Ts. Kampel and V.I. Bregadze, *Metalloorg. Khim.*, 1991, **4**, 132.

4. P.T. Brain, A.J. Downs, P. Maccallum, D.W.H. Rankin, H.E Robertson and G.A. Forsyth, *J. Chem. Soc., Dalton Trans.*, 1991, 1195.

5. M. Hargittai, N. Yu. Subbotina and A.G. Gershikov, *J. Mol. Struct.*, 1991, **245**, 147.

6. C.R. Pulham, A.J. Downs, M.J. Goode, D.W.H. Rankin and H.E. Robertson, *J. Am. Chem. Soc.*, 1991, **113**, 5149.

7. K.V. Ermakov, B.S. Butayev and V.P. Spiridonov, *J. Mol. Struct.*, 1991, **248**, 143.

8. A.G. Gershikov, N. Yu. Subbotina and M. Hargittai, *J. Mol. Spectrosc.*, 1990, **143**, 293.

9. H.-G. Mack, H. Oberhammer, E.O. John, R.L. Kirchmeier and J.M. Shreeve, *J. Mol. Struct.*, 1991, **250**, 103.

10. H.-G. Mack, M. Dakkouri and H. Oberhammer, *J. Phys. Chem.*, 1991, **95**, 3136.

11. A. Jin, H.-G. Mack, A. Waterfeld and H. Oberhammer, *J. Am. Chem. Soc.*, 1991, **113**, 7847.

12. K. Hedberg, L. Hedberg, D.S. Bethune, C.A. Brown, M. de Vries, H.C. Dorn and R.D. Johnson, *Science*, 1991, **254**, 410.

13. H. Schmidbaur, J. Zech, D.W.H. Rankin and H.E. Robertson, *Chem. Ber.*, 1991, **124**, 1953.

14. E. Røhmen, K. Hagen, R. Stølevik, K. Hassler and M. Pöschl, *J. Mol. Struct.*, 1991, **244**, 41.

15. G.A. Forsyth, D.W.H. Rankin and H.E. Robertson, *J. Mol. Struct.*, 1991, **263**, 311.

16. O.A. Rusaeva, V.S. Mastryukov, L.V. Khristenko, L.V. Vilkov and Yu. A Pentin, *Zh. Fiz. Khim.*, 1991, **65**, 152.

17. O.A. Rusaeva, V.S. Mastryukov, L.V Khristenko and Yu. A. Pentin, *Zh. Fiz. Khim.*, 1991, **65**, 1377.

18. Q. Shen, P.G. Apen and R.L. Hilderbrandt, *J. Mol. Struct.*, 1991, **246**, 229.

19. V.S. Mastryukov, A.V. Golubinskii, L.V. Vilkov, S.V. Kirpichenko, E.N. Suslova and M.G. Voronkov, *Zh. Strukt. Khim.*, 1991, **32(2)**, 148.

20. A. Hammel, H.V. Volden, A. Haaland, J. Weidlein and R. Reischmann, *J. Organomet. Chem.*, 1991, **408**, 35.

21. E. Csákvári, B. Rozsondai and I. Hargittai, *J. Mol. Struct.*, 1991, **245**, 349.

22. M. Dakkouri, *J. Am. Chem. Soc.*, 1991, **113**, 7109.

23. D. Christen, O.D. Gupta, J. Kadel, R.L. Kirchmeier, H.G. Mack, H. Oberhammer and J.M. Shreeve, *J. Am. Chem. Soc.*, 1991, **113**, 9131.

24. A. Haaland, H. Thomassen and Y. Stenstrøm, *J. Mol. Struct.*, 1991, **263**, 299.

25. N. Murase, K. Yamanouchi, T. Egawa and K. Kuchitsu, *J. Mol. Struct.*, 1991, **242**, 409.

26. I.F. Shishkov, L.V. Vilkov, M. Kolonits and B. Rozsondai, *Struct. Chem.*, 1991, **2**, 57.

27. I.F. Shishkov, L.V. Vilkov and I. Hargittai, *J. Mol Struct.*, 1991, **248**, 125.

28. G.A. Bell, A.J. Blake, R.O. Gould and D.W.H. Rankin, *J. Fluorine Chem.*, 1991, **51**, 305.

29. D.G. Anderson, S. Cradock, G.A. Forsyth, D.W.H. Rankin, J.F. Sullivan, T.J. Hizer and J.R. Durig, *J. Mol. Struct.*, 1991, **244**, 51.

30. A.J. Downs, N.I. Hunt, G.S. McGrady, D.W.H. Rankin and H.E. Robertson, *J. Mol. Struct.*, 1991, **248**, 393.

31. H.-G. Mack, C.O. Della Védova and H. Oberhammer, *Angew. Chem. Int. Ed. Engl. 30.*, 1991, 1145.

32. H.-G. Mack, H. Oberhammer and C.O. Della Védova, *J. Phys. Chem.*, 1991, **95**, 4238.

33. C. Renschler, H.-G. Mack, C.O. Della Védova and H. Oberhammer, *J. Phys. Chem.*, 1991, **95**, 6912.

34. H.-G. Mack, H. Oberhammer and A. Waterfeld, *J. Mol. Struct.*, 1991, **249**, 297.

35. R. Blom, D.W.H. Rankin, H.E. Robertson, M. Schröder and A. Taylor, *J. Chem. Soc., Perkin Trans. 2*, 1991, 773.

36. G. Forgács, I. Hargittai, I. Jalsovszky and A. Kucsman, *J. Mol. Struct.*, 1991, **243**, 123.

37. E. Jaudas-Prezel, R. Maggiulli, R. Mews, H. Oberhammer and W.-D. Stohrer, *Chem. Ber.*, 1990, **123**, 2117.

38. E. Jaudas-Prezel, R. Maggiulli, R. Mews, H. Oberhammer, T. Paust and W.-D. Stohrer, *Chem. Ber.*, 1990, **123**, 2123.

39. E. Fischer, E. Jaudas-Prezel, R. Maggiulli, R. Mews, H. Oberhammer, R. Paape and W.-D. Stohrer, *Chem. Ber.*, 1991, **124**, 1347.

40. I. Weiss, H. Oberhammer, D. Viets, R. Mews and A. Waterfeld, *J. Mol. Struct.*, 1991, **248**, 407.

41. P. Zylka, D. Christen, H. Oberhammer, G.L. Gard and R.J. Terjeson, *J. Mol Struct.*, 1991, **249**, 285.

42. J. Jacobs, G.S. McGrady, H. Willner, D. Christen, H. Oberhammer and P. Zylka, *J. Mol. Struct.*, 1991, **245**, 275.

43. P. Zylka, H.-G. Mack, A. Schmuck, K. Seppelt and H. Oberhammer, *Inorg. Chem.*, 1991, **30**, 59.

44. E. Jaudas-Prezel, D. Christen, H. Oberhammer, S.P. Mallela and J.M. Shreeve, *J. Mol. Struct.*, 1991, **248**, 415.

45. P. Zylka, H. Oberhammer and K. Seppelt, *J. Mol. Struct.*, 1991, **243**, 411.

46. M. Hargittai, N. Yu. Subbotina, M. Kolonits and A.G. Gershikov, *J. Chem. Phys.*, 1991, **94**, 7278.

47. E.Z. Zasorin, A.A. Ivanov. L.I. Ermolaeva and V.P. Spiridonov, *Zh. Fiz. Khim.*, 1989, **63**, 669.

48. V.I. Bazhanov, Yu. S. Ezhov and S.A. Komarov, *Zh. Strukt. Khim.*, 1990, **31**(6), 152.

49. V.I. Bazhanov, S.A. Komarov, V.G. Sevast'yanov, M.V. Popik, N.T. Kuznetsov and Yu. S. Ezhov, *Vysokochistye Veshchestva*, 1990(1), 109.

50. V.I. Bazhanov, S.A. Komarov and Yu. S. Ezhov, *Zh. Fiz. Khim.*, 1989, **64**, 2247.

51. Yu. S. Ezhov, V.I. Bazhanov, S.A. Komarov, M.S. Popik, V.G. Sevast'yanov and F. Yuldashev, *Zh. Fiz. Khim.*, 1989, **63**, 3094.

52. V.I. Bazhanov, Yu. S. Ezhov, S.A. Komarov, V.G. Sevast'yanov and F. Yuldashev, *Vysokochistye Veshchestva*, 1989(5), 197.

53. V.I. Bazhanov, Yu. S. Ezhov, S.A. Komarov and V.G. Sevast'yanov, *Zh. Strukt. Khim.*, 1990, **31**(6), 153.

54. J. Brunvoll, S. Gundersen, A.A. Ischenko, V.P. Spiridonov and T.G. Strand, *Acta Chem. Scand.*, 1991, **45**, 111.

55. E.M. Page, D.A. Rice, K. Hagen, L. Hedberg and K. Hedberg, *Inorg. Chem.*, 1991, **30**, 4758.

56. C.J. Marsden, K. Hedberg, M.M. Ludwig and G.L. Gard, *Inorg. Chem.*, 1991, **30**, 4761.

57. K. Rypdal, W.A. Herrmann, S.J. Eder, R.W. Albach, P. Watzlowik, H. Bock and B. Solouki, *Organometallics*, 1991, **10**, 1331.

58. W.A. Herrmann, P. Kiprof, K. Rypdal, J. Tremmel, R. Blom, R. Alberto, J. Behm, R.W. Albach, H. Bock, B. Solouki, J. Mink, D. Lichtenberger and N.E. Gruhn, *J. Am. Chem. Soc.*, 1991, **113**, 6527.

59. C.J. Dain, A.J. Downs, M.J. Goode, D.G. Evans, K.T. Nicholls, D.W.H. Rankin and H.E. Robertson, *J. Chem. Soc., Dalton Trans.*, 1991, 967.

60. A.G. Császár, L. Hedberg, K. Hedberg, R.C. Burns, A.T. Wen and M.J. McGlinchey, *Inorg. Chem.*, 1991, **30**, 1371.

61. A. Haaland, K. Rypdal, H.V. Volden and R.A. Andersen, *Acta Chem. Scand.*, 1991, **45**, 955.

62. S.A. Komarov, V.G. Sevast'yanov, N.T. Kuznetsov and Yu. S. Ezhov, *Vysokochistye Veshchestva*, 1990(1), 106.